T0325383

SECOND EDITION

URINARY TRACT
INFECTIONS

MOLECULAR PATHOGENESIS AND CLINICAL MANAGEMENT

SECOND EDITION

URINARY TRACT
INFECTIONS

MOLECULAR PATHOGENESIS AND CLINICAL MANAGEMENT

EDITED BY

Matthew A. Mulvey
University of Utah School of Medicine
Salt Lake City, Utah

David J. Klumpp
Feinberg School of Medicine
Northwestern University
Chicago, Illinois

Ann E. Stapleton
University of Washington School of Medicine
University of Washington
Seattle, Washington

ASM
PRESS

Washington, DC

Library of Congress Cataloging-in-Publication Data
Names: Mulvey, Matthew A., editor. | Klumpp, David J., editor. |
  Stapleton, Ann E., editor.
Title: Urinary tract infections : molecular pathogenesis and clinical management /
  edited by Matthew A. Mulvey, David J. Klumpp, Ann E. Stapleton.
Description: Second edition. | Washington, DC : ASM Press, [2017] | Includes index.
Identifiers: LCCN 2016047855 (print) | LCCN 2016048642 (ebook) | ISBN
  9781555817398 (hardcover) | ISBN 9781555817404 (ebook)
Subjects: LCSH: Urinary tract infections--Molecular aspects. | Urinary tract
  infections–Pathogenesis. | Urinary tract infections–Microbiology.
Classification: LCC RC901.8 .U755 2017 (print) | LCC RC901.8 (ebook) |
  DDC 616.6/071–dc23
LC record available at https://lccn.loc.gov/2016047855

10 9 8 7 6 5 4 3 2 1

Address editorial correspondence to
ASM Press, 1752 N St., N.W.,
Washington, DC 20036-2904, USA

Send orders to ASM Press, P.O. Box 605, Herndon, VA 20172, USA
Phone: 800-546-2416; 703-661-1593
Fax: 703-661-1501
E-mail: books@asmusa.org
Online: http://www.asmscience.org

Cover: Uropathogenic *Escherichia coli* (yellow) invading a bladder umbrella cell (red), with the host plasma membrane and the extracellular milieu colorized in blue. Image was obtained using freeze-fracture deep-etch electron microscopy of an infected mouse bladder, and originally appeared in Mulvey MA, Schilling JD, Martinez JJ, Hultgren SC. 2000. *Proc Natl Acad Sci USA* **97**:8829–8835. Copyright (2000) National Academy of Sciences.

To the memories of Walter Stamm, MD and Carleen Collins, PhD and to the courage of Richard Grady, MD and Laura Hart, MD

Contents

Contributors   ix
Foreword   xvii
Preface   xix

I. CLINICAL ASPECTS OF URINARY TRACT INFECTIONS

1 **Anatomy and Physiology of the Urinary Tract: Relation to Host Defense and Microbial Infection**   3
Duane R. Hickling, Tung-Tien Sun, and Xue-Ru Wu

2 **Clinical Presentations and Epidemiology of Urinary Tract Infections**   27
Suzanne E. Geerlings

3 **Diagnosis, Treatment, and Prevention of Urinary Tract Infection**   41
Paula Pietrucha-Dilanchian and Thomas M. Hooton

4 **Urinary Tract Infections in Infants and Children**   69
Theresa A. Schlager

5 **The Vaginal Microbiota and Urinary Tract Infection**   79
Ann E. Stapleton

6 **Asymptomatic Bacteriuria and Bacterial Interference**   87
Lindsay E. Nicolle

7 **Bacterial Prostatitis: Bacterial Virulence, Clinical Outcomes, and New Directions**   121
John N. Krieger and Praveen Thumbikat

8 **Urosepsis: Overview of the Diagnostic and Treatment Challenges**   135
Florian M. E. Wagenlehner, Adrian Pilatz, Wolfgang Weidner, and Kurt G. Naber

II. ORIGINS AND VIRULENCE MECHANISMS OF UROPATHOGENIC BACTERIA

9 **Reservoirs of Extraintestinal Pathogenic *Escherichia coli***   161
Amee R. Manges and James R. Johnson

10 **Origin and Dissemination of Antimicrobial Resistance among Uropathogenic *Escherichia coli***   179
Lisa K. Nolan, Ganwu Li, and Catherine M. Logue

11 Population Phylogenomics of Extraintestinal Pathogenic
    Escherichia coli 207
    Jérôme Tourret and Erick Denamur

12 Virulence and Fitness Determinants of Uropathogenic
    Escherichia coli 235
    Sargurunathan Subashchandrabose and Harry L. T. Mobley

13 Uropathogenic *Escherichia coli*-Associated Exotoxins 263
    Rodney A. Welch

14 Structure, Function, and Assembly of Adhesive Organelles
    by Uropathogenic Bacteria 277
    Peter Chahales and David G. Thanassi

15 Pathoadaptive Mutations in Uropathogenic *Escherichia coli* 331
    Evgeni Sokurenko

16 Invasion of Host Cells and Tissues by Uropathogenic Bacteria 359
    Adam J. Lewis, Amanda C. Richards, and Matthew A. Mulvey

17 *Proteus mirabilis* and Urinary Tract Infections 383
    Jessica N. Schaffer and Melanie M. Pearson

18 Epidemiology and Virulence of *Klebsiella pneumoniae* 435
    Steven Clegg and Caitlin N. Murphy

19 Gram-Positive Uropathogens, Polymicrobial Urinary Tract Infection,
    and the Emerging Microbiota of the Urinary Tract 459
    Kimberly A. Kline and Amanda L. Lewis

20 Integrated Pathophysiology of Pyelonephritis 503
    Ferdinand X. Choong, Haris Antypas, and Agneta Richter-Dahlfors

**III. HOST RESPONSES TO URINARY TRACT INFECTIONS
        AND EMERGING THERAPEUTICS**

21 Susceptibility to Urinary Tract Infection: Benefits and Hazards of the
    Antibacterial Host Response 525
    *Ines Ambite, Karoly Nagy, Gabriela Godaly, Manoj Puthia, Björn Wullt,
    and Catharina Svanborg*

22 Innate Immune Responses to Bladder Infection 555
    Byron W. Hayes and Soman N. Abraham

23 Host Responses to Urinary Tract Infections and Emerging
    Therapeutics: Sensation and Pain within the Urinary Tract 565
    Lori A. Birder and David J. Klumpp

24 Drug and Vaccine Development for the Treatment and Prevention of
    Urinary Tract Infections 589
    *Valerie P. O'Brien, Thomas J. Hannan, Hailyn V. Nielsen,
    and Scott J. Hultgren*

Index 647

Contributors

Soman N. Abraham
Departments of Pathology, Molecular Genetics & Microbiology,
and Immunology
Duke University Medical Center
Durham, NC 27710
Program in Emerging Infectious Diseases
Duke-National University of Singapore
Singapore 169857

Ines Ambite
Department of Microbiology, Immunology and Glycobiology
Institute of Laboratory Medicine
Lund University
Lund, S-223 62
Sweden

Haris Antypas
Swedish Medical Nanoscience Center
Department of Neuroscience
Karolinska Institutet
SE-171 77, Stockholm
Sweden

Lori A. Birder
Departments of Medicine and Pharmacology and Chemical Biology
University of Pittsburgh School of Medicine
Pittsburgh, PA 15261

Peter Chahales
Center for Infectious Diseases and Department of Molecular Genetics
and Microbiology
Stony Brook University
Stony Brook, NY 11794

Ferdinand X. Choong
Swedish Medical Nanoscience Center
Department of Neuroscience
Karolinska Institutet
SE-171 77, Stockholm
Sweden

Steven Clegg
Department of Microbiology
University of Iowa College of Medicine
Iowa City, IA 52242

Erick Denamur
UMR 1137 INSERM and Université Paris Diderot, IAME,
Sorbonne Paris Cité
75018 Paris
France

Suzanne E. Geerling
Department of Internal Medicine, Division of Infectious Diseases
Center for Infection and Immunity Amsterdam (CINIMA)
Academic Medical Center
1105 AZ Amsterdam
The Netherlands

Gabriela Godaly
Department of Microbiology, Immunology and Glycobiology
Institute of Laboratory Medicine
Lund University
Lund, S-223 62
Sweden

Thomas J. Hannan
Department of Pathology & Immunology
Washington University Medical School
St. Louis, MO 63110

Byron W. Hayes
Department of Pathology
Duke University Medical Center
Durham, NC 27710

Duane R. Hickling
Division of Urology, Ottawa Hospital Research Institute
The Ottawa Hospital, University of Ottawa
Ottawa, ON K1Y 4E9
Canada

Thomas M. Hooton
Department of Medicine
University of Miami Miller School of Medicine
Miami, FL 33136

Scott J. Hultgren
Department of Molecular Microbiology
Center for Women's Infectious Disease Research
Washington University Medical School
St. Louis, MO 63110

James R. Johnson
Infectious Diseases Section
Veterans Affairs Medical Center
Minneapolis, MN 55417
Department of Medicine
University of Minnesota
Minneapolis, MN 55455

Kimberly A. Kline
Singapore Centre on Environmental Life Sciences Engineering
School of Biological Sciences
Nanyang Technological University
Singapore 637551

David J. Klumpp
Departments of Urology and Microbiology-Immunology
Feinberg School of Medicine
Northwestern University
Chicago, IL 60610

John N. Krieger
Department of Urology
University of Washington School of Medicine
Seattle, WA 98195

Adam J. Lewis
Division of Microbiology and Immunology
Pathology Department
University of Utah School of Medicine
Salt Lake City, UT 84112

Amanda L. Lewis
Department of Molecular Microbiology
Washington University School of Medicine
St. Louis, MO 63110

Ganwu Li
Department of Veterinary Microbiology and Preventive Medicine
College of Veterinary Medicine
Iowa State University
Ames, IA 50011

Catherine M. Logue
Department of Veterinary Microbiology and Preventive Medicine
College of Veterinary Medicine
Iowa State University
Ames, IA 50011

Amee R. Manges
School of Population and Public Health
University of British Columbia
Vancouver, BC V6T 1Z3
Canada

Harry L.T. Mobley
Department of Microbiology and Immunology
University of Michigan Medical School
Ann Arbor, MI 48109

Matthew A. Mulvey
Division of Microbiology and Immunology
Pathology Department
University of Utah School of Medicine
Salt Lake City, UT 84112

Caitlin N. Murphy
Department of Microbiology
University of Iowa College of Medicine
Iowa City, IA 52242

Kurt G. Naber
Technical University
80333 Munich
Germany

Karoly Nagy
Department of Urology
South-Pest Hospital
Budapest 1204
Hungary

Lindsay E. Nicolle
Department of Internal Medicine and Medical Microbiology
University of Manitoba
Winnipeg, MB R3T 2N2
Canada

Hailyn V. Nielsen
Department of Molecular Microbiology
Center for Women's Infectious Disease Research
Washington University Medical School
St. Louis, MO 63110

Lisa K. Nolan
Department of Veterinary Microbiology and Preventive Medicine
College of Veterinary Medicine
Iowa State University
Ames, IA 50011

Valerie P. O'Brien
Department of Molecular Microbiology
Center for Women's Infectious Disease Research
Washington University Medical School
St. Louis, MO 63110

Melanie M. Pearson
Department of Microbiology
New York University Langone Medical Center
New York, NY 10016

Paula Pietrucha-Dilanchian
Department of Medicine
University of Miami Miller School of Medicine
Miami, FL 33136

Adrian Pilatz
Clinic for Urology, Pediatric Urology and Andrology
Justus-Liebig-University Gießen
D-35390 Gießen
Germany

Manoj Puthia
Department of Microbiology, Immunology and Glycobiology
Institute of Laboratory Medicine
Lund University
Lund, S-223 62
Sweden

Amanda C. Richards
Division of Microbiology and Immunology
Pathology Department
University of Utah School of Medicine
Salt Lake City, UT 84112

Agneta Richter-Dahlfors
Swedish Medical Nanoscience Center
Department of Neuroscience
Karolinska Institutet
SE-171 77, Stockholm
Sweden

Jessica N. Schaffer
Department of Microbiology
New York University Langone Medical Center
New York, NY 10016

Theresa A. Schlager
Department of Emergency Medicine
University of Virginia
Charlottesville, VA 22908

Evgeni Sokurenko
University of Washington
Seattle, WA 98195

Ann E. Stapleton
Division of Allergy and Infectious Diseases
Department of Medicine
University of Washington
Seattle, WA 98195

Sargurunathan Subashchandrabose
Department of Microbiology and Immunology
University of Michigan Medical School
Ann Arbor, MI 48109

Tung-Tien Sun
Departments of Cell Biology, Biochemistry and Molecular Pharmacology
Departments of Dermatology and Urology
New York University School of Medicine
New York, NY 10016

Catharina Svanborg
Department of Microbiology, Immunology and Glycobiology
Institute of Laboratory Medicine
Lund University
Lund, S-223 62
Sweden

David G. Thanassi
Center for Infectious Diseases and Department of Molecular Genetics
and Microbiology
Stony Brook University
Stony Brook, NY 11794

Praveen Thumbikat
Department of Urology
Northwestern University School of Medicine
Chicago, IL 60611

Jérôme Tourret
Département d'Urologie, Néphrologie et Transplantation Groupe
Hospitalier Pitié-Salpêtrière
Assistance Publique-Hôpitaux de Paris
Université Pierre et Marie Curie
UMR 1137 INSERM and Université Paris Diderot, IAME, Sorbonne Paris Cité
75018 Paris
France

Florian M.E. Wagenlehner
Clinic for Urology, Pediatric Urology and Andrology
Justus-Liebig-University Gießen
D-35390 Gießen
Germany

Wolfgang Weidner
Clinic for Urology, Pediatric Urology and Andrology
Justus-Liebig-University Gießen
D-35390 Gießen
Germany

Rodney A. Welch
Department of Medical Microbiology and Immunology
University of Wisconsin School of Medicine and Public Health
Madison, WI 53706

Xue-Ru Wu
Departments of Urology and Pathology
New York University School of Medicine
Veterans Affairs, New York Harbor Healthcare Systems, Manhattan Campus
New York, NY 10016

Björn Wullt
Department of Microbiology, Immunology and Glycobiology
Institute of Laboratory Medicine
Lund University
Lund, S-223 62
Sweden

Foreword

Urinary tract infection (UTI), the second most common infection of humans after pneumonia, has likely plagued the population for as long as humans have walked the face of the earth. The inception of antibiotics provided adequate therapy but did not prevent infections from recurring. While symptoms of the infections were well documented, little was known about the primary infecting pathogen, *Escherichia coli*. In 1976, Svanborg and colleagues demonstrated that *E. coli* causing acute pyelonephritis adhered in greater numbers to uroepithelial cells (1). Further research by several groups revealed that the adherence factor P fimbria was responsible. This structure was found to be comprised of a multiprotein complex with the actual adhesin placed at the tip of the fimbria. With the advent of molecular techniques, other advances arrived quickly. For example, in 1981, Welch and colleagues demonstrated that a knock out of the hemolysin gene attenuated *E. coli* in an intraperitoneal model (2). As discoveries abounded, some 14 years later, I teamed up with infectious diseases physician John W. Warren to edit a 15-chapter book titled *Urinary Tract Infections: Molecular Pathogenesis and Clinical Management*. This treatise covered the clinical aspect of UTI (5 chapters) and the molecular mechanisms of bacterial pathogenesis of UTI (10 chapters).

Now two decades have passed, and it was essential to update this broad topic. Editors Matthew A. Mulvey, David J. Klumpp, and Ann E. Stapleton have taken on the task of a second edition. The editors assembled an all-star lineup to cover the topic of clinical aspects of UTIs in eight chapters that include the anatomical and physiological aspects of UTI, clinical presentations, diagnosis and treatment, infections in children, involvement of the vaginal microbiome, asymptomatic UTI, prostatitis, pyelonephritis, and urosepsis (the most serious complication). In the second section (12 chapters), experts deal with reservoirs of infection, antimicrobial resistance, phylogeny, virulence, and fitness factors including exotoxins, structure of adhesins, adaptive mutations, and intracellular persistence, and this section includes chapters on other important uropathogens: *Proteus mirabilis*, *Klebsiella pneumoniae*, and Gram-positive pathogens. In the final section on host responses to UTI and emerging therapeutics (4 chapters), authors summarize the host response to UTI, innate immunity, sensation and pain in the bladder, and drug and vaccine development. Overall, this volume brings us up to date on the broad topic of UTI. Those interested in these common infections, whether it be in the laboratory or the clinic, will find the second edition of *Urinary Tract Infections: Molecular Pathogenesis and Clinical Management* an indispensable book that should be on your shelf or on your computer. It is gratifying to see this critical topic brought up to date.

Harry L. T. Mobley
Frederick G. Novy Distinguished University Professor and Chair,
Department of Microbiology and Immunology
University of Michigan Medical School
Ann Arbor, Michigan

1. **Svanborg Edén C, Jodal U, Hanson LA, Lindberg U, Sohl Åkerlund A.** 1976. Variable adherence to normal human urinary-tract epithelial cells of *Escherichia coli* strains associated with various forms of urinary-tract infection. *Lancet* **1**:480–492.
2. **Welch RA, Dellinger EP, Minshew B, Falkow S.** 1981. Haemolysin contributes to virulence of extra-intestinal *E. coli* infections. *Nature* **294**:665–657.

Preface

For centuries the pain and other symptoms associated with urinary tract infections (UTIs) were erroneously ascribed to the wrath of gods, bile, phlegm, bad air, or numerous other culprits. The treatments for UTIs were at times equally off-target and included the use of bleeding and enemas, while the administration of narcotics and certain herbs provided palliative support. In the late 1800s, as evidence for the germ theory of disease mounted in the wake of Pasteur and Koch, the idea that microbes were responsible for UTIs took hold. This led to the development of more efficacious treatment options, culminating in the discovery and optimization of antibiotics that continue to this day. These achievements coincided with stunning advancements in our understanding of cellular functions and developmental processes within the urinary tract, inflammatory responses, microbiology, and the roles of both innate and adaptive host defenses. Still, despite this progress, UTIs continue to rank among the most common of infectious diseases, with most UTIs being attributable to strains of uropathogenic *Escherichia coli* (UPEC).

In 1996, for the first edition of this book, Harry L. T. Mobley and John W. Warren assembled an all-star cast of authors to highlight the multiple host and bacterial factors that impact the pathogenesis and treatment of UTIs. A lot has happened since, including remarkable progress in our ability to sequence and manipulate both bacterial and host genomes. The first *E. coli* genome, belonging to the nonpathogenic strain MG1655, was sequenced in 1997, followed a few years later by the urosepsis isolate and reference UPEC strain CFT073. Today, several thousand *E. coli* genomes have been sequenced, including many UPEC isolates. These data have revealed a huge amount of diversity among UPEC isolates, while also shedding light on the evolution and adaptability of uropathogens. These developments overlapped with the adoption of new, more facile approaches to manipulate UPEC genomes, greatly enhancing our ability to disrupt and functionally test specific pathogen-associated loci. This work is providing leads for the generation of more efficacious therapeutics for the treatment and prevention of UTIs.

Though powerful, antibiotics have not provided a cure-all for UTIs. Many individuals endure multiple recurrent UTIs despite antibiotic treatments, while circumstances such as catheterization render others prone to chronic infections. Many UPEC isolates are now resistant to multiple antibiotics, including some drugs that should be reserved as last resort choices. In terms of medical costs and loss of life, the rapid emergence and expansion of multidrug-resistant UPEC and related strains in recent years is considered by some to be more problematic than methicillin-resistant *Staphylococcus aureus* (MRSA) was over the past two decades. The rising tide of antibiotic-resistant UPEC strains is showing no signs of subsiding, being driven in part by the overuse and misuse of antibiotics in both the clinic and in agriculture. Epidemiology informed by sequencing data is showing how antibiotic resistance and other genetic elements move among UPEC strains, facilitated by human activities such as global travel and the utilization of high-throughput animal processing and food distribution networks. To better combat UTIs, and antibiotic-resistant strains in particular, scientists are working to create effective anti-UTI vaccines and new antibiotics that have fewer off-target effects. Some researchers are optimizing the use of probiotic bacterial strains that can interfere with UPEC colonization of the urinary tract, while others aim to develop antivirulence strategies that modify virulence mechanisms and host responses rather than the bacteria themselves. The realization that UPEC can act as facultative intracellular pathogens in both humans and mice is also spurring the development of new treatment approaches while at the same time challenging long-held views concerning the etiology of chronic and recurrent UTIs.

Advances in bacterial genomics have been complemented by the development of new approaches to identify UTI susceptibility factors in human populations. This work, coupled with robust UTI model systems, is beginning to explain why some individuals are more prone to UTIs, making links with innate host defense regulators, adaptive immunity, inflammatory responses, and pain perception within the urinary tract. Clinically, we are gaining a much more complete understanding of the host and bacterial factors that contribute to the onset and progression of UTIs, as well as variables that can confound treatments. These variables include patient age, sex, and catheterization, as well as the makeup of protective microbial communities within the vaginal microbiota, the gut, and potentially even the bladder itself.

In this book, leading experts have reviewed the clinical diagnostics and management of UTIs in adults and children, along with associated complications such as urosepsis and prostatitis. In other chapters we take a detailed look at the origins of UPEC and associated antibiotic-resistance factors, with consideration of bacterial population dynamics, genome architecture, and evolution. The mechanisms by which uropathogens colonize the urinary tract and cause disease are thoroughly examined, with analysis of the adhesive organelles and myriad other bacterial and host factors that affect UPEC survival and virulence within the urinary tract. This includes an assessment of innate and adaptive host responses that are triggered during the course of a UTI, and the protective effects of microbial communities within the urogenital tract. The molecular biology and clinical importance of other uropathogens, including *Klebsiella pneumoniae*, *Proteus mirabilis*, and Gram-positive opportunists such as *S. aureus*, are also discussed in detail. Finally, we turn our attention to emerging antibacterial therapeutics, including the use of probiotics and bacterial interference measures. Much of the information presented in the following pages builds on work that was just coming to light when the first edition of this book was published nearly 2 decades ago. We are eager to see where the next 20 years take the field and hope that this new book, like the first edition, serves as both a resource for the community and a stimulus for future research endeavors.

Matthew A. Mulvey
David J. Klumpp
Ann E. Stapleton

CLINICAL ASPECTS OF URINARY TRACT INFECTIONS

Anatomy and Physiology of the Urinary Tract: Relation to Host Defense and Microbial Infection

1

DUANE R. HICKLING,[1] TUNG-TIEN SUN,[2] and XUE-RU WU[3,4]

NORMAL ANATOMY AND PHYSIOLOGY OF THE URINARY TRACT

The mammalian urinary tract is a contiguous hollow-organ system whose primary function is to collect, transport, store, and expel urine periodically and in a highly coordinated fashion (1, 2). In so doing, the urinary tract ensures the elimination of metabolic products and toxic wastes generated in the kidneys. The process of constant urine flow in the upper urinary tract and intermittent elimination from the lower urinary tract also plays a crucially important part in cleansing the urinary tract, ridding it of microbes that might have already gained access (3). When not eliminating urine, the urinary tract acts effectively as a closed system, inaccessible to the microbes. Comprised, from proximal to distal, of renal papillae, renal pelvis, ureters, bladder, and urethra, each component of the urinary tract has distinct anatomic features and performs critical functions.

[1]Division of Urology, Ottawa Hospital Research Institute, The Ottawa Hospital, University of Ottawa, Ottawa, ON K1Y 4E9, Canada; [2]Departments of Cell Biology, Biochemistry and Molecular Pharmacology, Departments of Dermatology and Urology, New York University School of Medicine, New York, NY, 10016; [3]Departments of Urology and Pathology, New York University School of Medicine, New York, NY, 10016; [4]Veterans Affairs, New York Harbor Healthcare Systems, Manhattan Campus, New York, NY 10016.

Urinary Tract Infections: Molecular Pathogenesis and Clinical Management, 2nd Edition
Edited by Matthew A. Mulvey, David J. Klumpp, and Ann E. Stapleton
© 2017 American Society for Microbiology, Washington, DC
doi:10.1128/microbiolspec.UTI-0016-2012

The Upper Urinary-Collecting System

The renal papilla, into which each renal tubule-rich pyramid drains, is considered the first gross structure of the upper collecting system. In humans and other higher mammals, renal papillae are individually cupped by a minor calyx, which in turn narrows into an infundibulum. Infundibuli vary in number, length, and diameter but consistently combine to form either 2 or 3 major calyces. These branches are termed upper, middle, and lower-pole calyces depending upon which pole of the kidney they drain. The renal pelvis represents the confluence of these major calyceal branches and itself can vary greatly in size and location (intra-renal vs extra-renal) (Fig. 1). It should be noted that, in rodents, there is only one renal papilla with a corresponding calyx.

The ureters are bilateral fibromuscular tubes that drain urine from the renal pelvis to the bladder. They are generally 22–30 cm in length and course through the retroperitoneum. They originate at the ureteropelvic junction (UPJ) behind the renal artery and vein and then progress inferiorly along the anterior portion of the psoas muscle. As the ureters enter the pelvic cavity they turn medially and cross in front of the common iliac bifurcation. The ureters pierce the bladder wall obliquely (termed the ureterovesical junction or UVJ and travel in this orientation for 1.5 to 2.0 cm within the bladder wall to terminate in the bladder lumen as ureteral orifices (4). The intramural ureter is compressed by the bladder wall passively during storage and dynamically during emptying. This, in effect, prevents vesicoureteric reflux during steady state and micturition (Fig. 2). Along the length of the ureter there are three segments that physiologically narrow: the ureteropelvic junction, the ureterovesical junction, and where the ureters cross the common iliac vessels. These areas are clinically relevant as they represent the most common locations where ureteral calculi become trapped, causing obstruction.

Bladder and Urethra

The bladder is a hollow, distensible pelvic viscus that is tetrahedral when empty and ovoid when filled. It is composed primarily of smooth muscle and collagen and, to a much

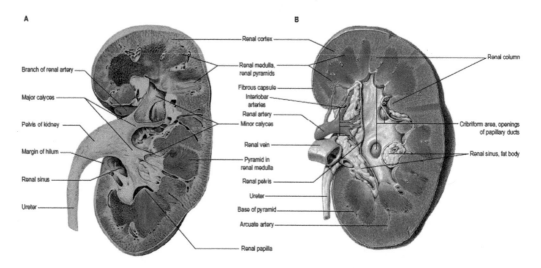

FIGURE 1 Normal anatomy of the kidney and upper urinary tract. (Reprinted from reference 163, with permission of the publisher.)

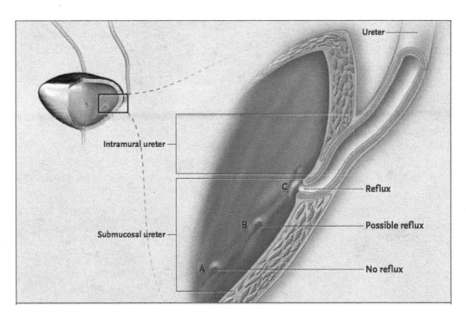

FIGURE 2 The ureterovesical junction. In this figure, A represents an orthotopic ureteral orifice. There is adequate length of ureteral tunnel in the bladder and therefore no reflux. Lateral and/or superior insertion of the ureteral orifice (B & C) can lead to inadequate submucosal ureter length and, potentially, reflux. (Reprinted from reference 162, with permission of the publisher.)

lesser degree, elastin (5). Its superior portion is defined by the urachus, a fibrous remnant of the allantois. The urachus attaches the bladder apex to the anterior abdominal wall. In males the bladder lies between the rectum and pubic symphysis and in females, between the rectum and uterus/vagina. Anterioinferiorly and laterally, the bladder is surrounded by retropubic and perivesical fat and connective tissue. This area is termed the space of Retzius. The trigone of the bladder is a triangular region of smooth muscle between the two ureteral orifices and the internal-urethral meatus. Grossly, thickened muscle between the ureteral orifices (interureteric crest) and between each ureteral orifice and the internal-urethral meatus (Bell's muscle) distinguishes the trigone from the rest of the bladder. The classic view of bladder and trigone development proposes that the trigone originates from the mesoderm-derived Wolffian ducts and the remainder of the bladder is formed from the endoderm-derived urogenital sinus (6).

Recent molecular developmental studies challenge this concept. Thus, the Wolffian ducts have been shown to undergo apoptosis during ureteral transposition and therefore do not contribute to trigone formation (7). Instead, a number of recent mouse models and tissue-transposition studies (7, 8), as well as *in vitro* studies of urothelial cells (9), suggest that the trigone is endodermal in origin. In males, the bladder base rests on the endopelvic fascia and the pelvic floor musculature, and the bladder neck is 3 to 4 cm behind the symphysis pubis and is fixed by the endopelvic fasciae and the prostate. Here, there is a layer of smooth muscle that surrounds the bladder neck and forms what is known as the involuntary internal-urethral sphincter. In females, the base of the bladder and urethra rest on the anterior wall of the vagina. The internal-urethral sphincter is not as well developed in females (10).

The urethra is contiguous with the bladder neck and begins at the distal end of the internal-urethral sphincter. In males the urethra is typically between 13 and 20 cm in length and is divided into prostatic, membranous, and penile portions. The prostatic urethra is 3–4 cm in length and runs vertically through the length of the prostate. The membranous urethra spans 2 to 2.5 cm from the apex of prostate to the perineal membrane. This portion of the urethra is completely surrounded by striated muscle known as the external-urethral sphincter. The penile portion of the urethra is contained within the corpus spongiosum. It is on average 15-cm long, it dilates slightly in the glans penis (fossa navicularis) and terminates at the external-urethral meatus. The female urethra, 3.8 to 5.1 cm long, is considerably shorter than the male one, and passes obliquely from the bladder neck to external-urethral meatus along the anterior vaginal wall. The distal two-thirds of the female urethra are invested by a slow-twitch striated muscle termed the external-urethral sphincter (10, 11).

Vagina

Although not part of the urinary tract, the vagina plays an integral role in UTI pathogenesis. It is a fibromuscular tube lined by epithelial cells. It extends from the opening of the labia minora (vestibule) to the uterus with its anterior wall approximately 7.5-cm long and its posterior wall approximately 9-cm long. The anterior wall is related to the bladder base superiorly and urethra inferiorly. Posteriorly, the vaginal wall is separated from the rectum by the recto-uterine pouch superiorly and Denonvillier's fascia and the perineal body inferiorly. The inner vagina is covered by a non-keratinized, stratified, squamous epithelium. With the onset of puberty the vaginal epithelium thickens and its superficial cells accumulate glycogen. There are no mucous glands, but transudate from the underlying lamina propria and mucus from the cervical glands lubricate the vagina. The muscular layers are composed of smooth muscle found in both longitudinal and circular orientation (12). Normal vagina in reproductive women is populated by the lactobacilli which produces lactic acid, producing a low pH condition highly unfavorable for the growth and colonization by uropathogenic microbes (13). This constitutes one of the major host defenses, as alterations in vaginal flora are considered a key predisposing factor to UTIs.

Microscopic Anatomy and Physiology of the Urinary Tract

The luminal surface of the urinary tract, from minor calyx to prostatic urethra, is lined by a specialized epithelium known as urothelium. Although the term urothelium has been used to describe the epithelia covering the mucosal surfaces of the bulk of the urinary tract, which share the expression of a group of integral membrane proteins called uroplakins, recent data indicate that the urothelia of the ureters, bladder, and possibly other areas are distinguishable with respect to the detailed morphological and biochemical features, their *in vitro* proliferative behaviors when placed under identical tissue-culture conditions, and their embryological origin. For example, it has been shown that ureteral urothelium contains lower amounts of uroplakins and fewer cytoplasmic fusiform vesicles than bladder urothelium (9, 14, 15). It is now clear that these phenotypic differences are due to intrinsic divergence instead of extrinsic modulation (9). These findings are consistent with the fact that the renal pelvic and ureteral urothelia are derived from the mesoderm, whereas bladder and urethral urothelia are derived from the endoderm. It is therefore misleading to describe urothelial cells derived from various regions of the urinary tract as if they were all equal, or to study a particular type of urothelial cells and generalize that the results must be applicable to the urothelial cells from other zones of the urinary tract. To avoid confusion, it is always necessary to be explicit about the tissue origin of the urothelial cells, e.g., bladder-urothelial cells or ureteral-urothelial cells (16).

The most extensively studied urothelium is that of the bladder. This epithelium performs important biological functions, including the formation of a physically stable apical surface and a highly effective permeability barrier, even as the surface area of the urinary tract undergoes dramatic changes during different phases of the micturition cycle. These attributes are thought to be a function of slow urothelial-cell turnover (~200 days) (16, 17) and the elaboration of the uroplakin-containing urothelial plaques and highly efficient tight junctions. The superficial urothelium consists of a single layer of large, multinucleated, and highly differentiated 'umbrella' cells. Umbrella cells accumulate a large amount of uroplakin proteins that form urothelial plaques. These plaques cover approximately 90% of the apical/luminal surface and are also present in high concentrations in association with the cytoplasmic fusiform vesicles (18, 19). Urothelial plaques are essentially two-dimensional crystals of hexagonally packed 16-nm uroplakin (UP) proteins (18, 20–23). There are four major UPs: Ia, Ib, II, and IIIa and one minor UP: IIIb (24–27). UPIa/II and UPIb/IIIa (or IIIb) heterodimer formation is necessary before these proteins can exit the endoplasmic reticulum and eventually form 16-nm particles and then plaques (19, 24, 28, 29). UP-knockout studies reveal that uroplakins are critical to the formation of urothelial plaques, formation of a normal UVJ, and a normal permeability barrier function (30–32). The cytoplasmic fusiform vesicles rich in urothelial plaques are thought to play a major role in delivering the uroplakin plaques to the apical surface (33, 34) (Fig. 3).

Uroplakins appear to play a major role during the pathogenesis of urinary tract infections.

UPIa presents a high level of terminally exposed, unmodified mannose residues and has been identified as the sole urothelial receptor to interact with the FimH lectin of the type 1-fimbriated uropathogenic *E. coli* (UPEC) (35–37). In addition to the bladder, UPIa has been found on the mucosal surfaces of the ureters, renal pelvis, and major and minor calyces (9, 14). It has been proposed that interaction of the FimH adhesin of type 1-fimbriated UPEC with UPIa at these locations help bacteria resist the flow of urine and, coupled with bacterial flagella formation, may facilitate the ascent of bacteria from the bladder into the upper urinary tract (16, 37).

Although both FimH and flagella are known to exhibit phase variation (38, 39), the temporal expression of these virulence factors in relation to the ascent of UPEC along the urinary tract has not been established. UPEC that cause pyelonephritis typically express P fimbriae in addition to type-1 fimbriae (40). Once bacteria reach the kidney, the P fimbriae interact with glycolipids in the renal tubular cells removing the need for type-1 fimbriae/UPIa interaction. UPIIIa has also been shown to be important in UTI pathogenesis. Thumbikat et al. demonstrated that the phosphorylation of UPIIIa's cytoplasmic tail is a critical step in urothelial signaling associated with bacterial invasion and host-cell apoptosis (41).

The intermediate and basal layers of the urothelium contain smaller, less well- differentiated epithelial cells. It is within the basal-cell layer that urothelial stem cells are believed to reside (16, 42, 43). The intermediate and basal layers may service as a reservoir for rapid umbrella-cell regeneration.

Different urothelial layers not only differ in morphology, proliferative potential, and degree of differentiation, they also seem have divergent abilities to support intracellular bacterial growth and propagation. For instance, intracellular bacterial communities (IBCs) of the UPEC strains are found almost exclusively in the urothelial umbrella-cell layer, but not in intermediate and basal layers (44). The cells in the latter two layers, however, can harbor so-called quiescent

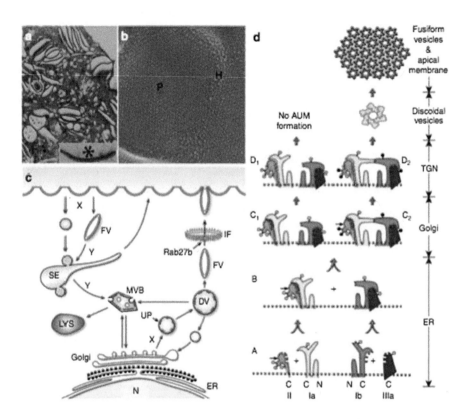

FIGURE 3 Assembly, intracelluar trafficking and structure of uroplakins. (a) Luminal portion of a superficial umbrella cell of mouse urothelium visualized by transmission electron microscopy (inset: an urothelial plaque exhibiting asymmetric unit membrane or AUM). (b) Quick-freeze deep-etch showing 16-nm uroplakin particles arranged in hexagonal arrays comprising the urothelial plaques (P) interconnected by particle-free hinges (H). (c) Vesicular trafficking in umbrella cells. Uroplakin heterodimer formation takes place in the endoplasmic reticulum (ER) and undergoes modification in the Golgi apparatus. Assembled uroplakins then amass in small vesicles and bud off the trans-Golgi network (TGN), forming discoidal vesicles (DVs). The next- stage, fusiform vesicles (FVs) pass through an intermediate-filament (IF) network and ultimately fuse with the apical membrane, a process mediated by Rab27b. Apical plaque-associated UPs are internalized via endocytic pathways and/or modified FVs that form sorting endosomes (SE) and multivesicular bodies (MVB), which merge with lysosomes (LYS) for degradation. (d) A hypothetical model of uroplakin assembly into 2-D crystals. Stages A and B: The four major uroplakins (UPIa, Ib, II, and IIIa) are modified with high-mannose glycans in the ER and heterodimerize forming UPIa/II and UPIb/IIIa and undergo major conformational changes. Symbols: the small, horizontal arrows on UPII denote the furin cleavage site at the end of the prosequence; the open and closed circles denote high-mannose and complex glycans, respectively. With *in vivo* urothelium (pathway on the right), the glycans on two of the three N-glycosylation sites on the prosequence of UPII become complex glycans in the TGN (stage C2), and the cleavage of the prosequence by furin in the TGN (stage D2) then triggers oligomerization to form a 16-nm particle. In cultured urothelial cells (pathway on the left), the differentiation-dependent glycosylation of pro-UPII is defective, preventing the formation of the uroplakin heterotetramer and the 16-nm particle, thus the lack of asymmetric-unit membrane. (Reprinted and adapted from reference 16, with permission of the publisher.)

intracellular reservoirs (QIRs), a possible source for recurrent UTIs. Whether differences in the intracellular architecture, in particular vesicular trafficking, e.g., endocytic and exocytic machineries, between various urothelial-cell layers (45–48) are responsible for the observed differences in bacterial growth remains to be seen.

NORMAL URINE TRANSPORT AND MICTURITION

Urine production is a function of both renal-glomerular filtration and tubular reabsorption and is tightly regulated by systemic hydration state and electrolyte balance. Urinary filtrate is passed through the nephron as it winds through the cortex and medulla and is concentrated via a counter-current mechanism. Urine exits the kidney at the renal papillae and is transported through the upper collecting system. The smooth muscle surrounding the calyces, renal pelvis, and ureters is of the syncytial type without discrete neuromuscular junctions. Instead, smooth-muscle excitation is spread from one muscle cell to the next. In humans, atypical smooth-muscle cells, located near the pelvicalyceal border, are thought to act as the pacemakers of urinary-tract peristalsis (49, 50). These cells initiate unidirectional peristaltic contractions which, in turn, promote the forward flow of urine. Recently, Hurato et al. demonstrated that disruption of the pelvicalyceal region from the more distal urinary-tract segments prevented downstream peristalsis. Furthermore, hyperpolarization-activated cation-3 (HCN3), an isoform of a channel family known for initiating electrical activity in the brain and heart, was isolated in the same spatial distribution as the atypical smooth-muscle cells of the pelvicalyceal junction. Inhibition of this channel protein caused a loss of electrical activity in the pelvicalyceal junction and led to randomized electrical activity and loss of coordinated peristalsis (51). Whether HCN3-positive cells are the same as the atypical smooth muscle remains to be seen. Normal ureteral contractions occur two to six times per minute and it is the advancing contraction wave that forces the urine bolus down the length of the ureters and then into the bladder (52). Some uropathogenic bacteria appear to have evolved a way to overcome the normally protective forward flow of urine that results from the peristaltic ureteral contractions. Recent studies demonstrated that most UPEC have the ability to impair ureteric contractility via a calcium-dependent mechanism and that mechanism is dependent upon FimH-urothelial interaction (53, 54).

The micturition cycle is best thought of as two distinct phases: urine storage/bladder filling and voiding/bladder emptying (55). The viscoelastic properties of the bladder allow for increases in bladder volume with little change in detrusor or intravesical pressures. Additionally, during bladder filling, spinal sympathetic reflexes (T12–L2) are activated that, through modulation of parasympathetic-ganglionic transmission, inhibit bladder contractions and increase bladder-outlet resistance via smooth-muscle activation (56). Bladder-outlet resistance also increases during filling secondary to increased external urethral-sphincter activity via a spinalsomatic reflex (guarding reflex) (57). As the bladder reaches its capacity, afferent activity from tension, volume, and nociceptive receptors are conveyed via Aδ and C fibers through the pelvic and pudenal nerves to the sacral spinal cord (56). Afferent signals ascend in the spinal cord to the pontine micturition center in the rostral brainstem. Here signals are processed under the strong influence of the cerebral cortex and other areas of the brain. If voiding is deemed appropriate, the voiding/bladder-emptying reflex is initiated. The pattern of efferent activity that follows is completely reversed, producing sacral parasympathetic outflow and inhibition of sympathetic and somatic pathways. First the external urethral-sphincter relaxes and shortly thereafter a coordinated contraction of the bladder causes the expulsion of urine (56, 58) (Fig. 4).

The forward flow of urine is imperative to the maintenance of a healthy urinary tract. Any structural or functional process that impedes the flow of urine has the potential to promote urine stasis, hence UTI pathogenesis. In the next few sections, we will elaborate upon those anatomic and physiologic abnormalities that can affect either storage or emptying of urine and, in turn, promote UTI pathogenesis.

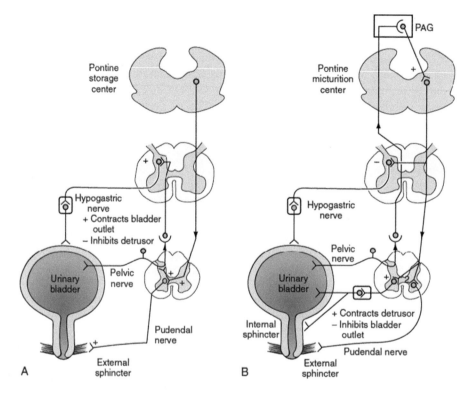

FIGURE 4 Mechanism of storage and voiding. A. Storage of urine. Low-level bladder afferent firing, secondary to bladder distension, increases sympathetic outflow to the bladder outlet and external urethral sphincter ('guarding reflex'). Sympathetic signaling also acts to inhibit detrusor-muscle contractions. B. Voiding. At bladder capacity, high-level bladder afferent activity activates the pontine-micturition center. This, in turn, inhibits the guarding reflex. The activated pontine-micturition center, under appropriate conditions, will lead to parasympathetic outflow to the bladder and internal-sphincter smooth muscle. Urinary sphincter relaxation is soon followed by a large, coordinated detrusor contraction leading to expulsion of urine from the bladder. (Reprinted and adapted from reference 58, with permission of the publisher.)

ANATOMIC ABNORMALITIES

Medullary Sponge Kidney

Medullary sponge kidney (MSK) is a renal disorder that is characterized by distal collecting-duct dilatation and multiple cysts and diverticula within the renal medullary pyramids. It is associated with a higher risk of nephrocalcinosis, urolithiasis, renal failure, and UTI (59–61). The prevalence of this disorder in the general population is unknown. However, a large series of intravenous pyelograms (IVPs), performed for any reason, revealed radiologic signs of MSK in

0.5% to 1% (62). The incidence of MSK in individuals who are known to form urolithiasis is higher, ranging from 2.6% to 12% (61, 63, 64). Clinical presentations of MSK include renal colic (51.8%), UTI (7.1%), and/or gross hematuria (16.1%) (64). MSK is diagnosed radiographically and has traditionally been accomplished via IVP. Pathognomic features on IVP include elongated ectatic papillary tubules, papillary contrast blush, and persistent medullary opacification which, taken together, give a 'bouquet of flowers' appearance. Today, IVPs have been replaced in favor of ultrasonography, computed tomography, and magnetic-resonance imaging.

MSK can be diagnosed with these imaging modalities but with much less sensitivity (65–67). MSK was once thought to be an isolated congenital abnormality. However, there is increasing data that links MSK with other malformative disorders such as hemihypertrophy, Beckwith-Wiedemann syndrome, congenital dilatation of intrahepatic bile ducts, and hepatic fibrosis, and autosomal-dominant polycystic-kidney disease (68–70). This has led some to suggest that MSK is a developmental disorder of renal embryogenesis. Gambaro et al. have hypothesized that MSK may be a consequence of disruption of the ureteral-bud/metanephric-blastema interface, which is critical to normal renal and ureteric development (71).

The increased risk of UTI in MSK patients has not been systematically studied. One might hypothesize that the increased risk of UTI could be due to urinary stasis within the ectatic collecting ducts, renal dysfunction, formation of urolithiasis, or any combination of the above. A better understanding of UTI pathogenesis in these patients is needed.

Calyceal Diverticula

A calyceal diverticulum is a congenital, urothelium-lined cavity within the renal parenchyma. They are relatively uncommon, occurring in 0.21% to 0.45% of people undergoing renal imaging (72). Diverticula development is believed to be due to failure of small ureteral bud regression (73). The majority of diverticula are unilateral, less than 1 cm in diameter, and within the posterior aspect of the upper collecting system (74). Diverticula are distributed in the upper (70%), lower (18%), and mid (12%) calyx. Urine moves passively in a retrograde manner through narrow infundibulum to fill the diverticulum. This pooling of urine within the diverticulum predisposes to calculi (9.5% to 39%) and recurrent urinary tract infections (25%) (73, 75). Additionally, obstruction at the diverticular neck can lead to rupture and hemorrhage, abscess formation, and potentially life-threatening sepsis. If symptomatic, a percutaneous approach to diverticulum ablation (+/- stone removal) is preferred.

Ureteral Obstruction

There are a number of intrinsic and extrinsic causes of ureteral obstruction (Table 1). Obstruction of the ureters can cause urinary stasis and, in severe cases, renal dysfunction; both of which are risk factors for UTI. Urinary stasis is believed to prolong the time for which bacteria can adhere to and invade the urothelium, whereas renal dysfunction prevents adequate concentration of antibiotics in the urine (76). Hematogenous infection of the kidney, a rare entity under normal conditions, is increased with ureteral obstruction (77). Transient ureteral obstruction, followed by *E.coli* infection of the lower urinary tract, has also been shown to predispose to ascending pyelonephritis in rats (78). However, the mechanisms underlying ureteral obstruction and increased ascending infection are not well understood. It has been posed that the release of ureteral obstruction alters urodynamics (i.e., secondary VUR) and

TABLE 1 **Anatomic causes of ureteral obstruction**

Intrinsic	Extrinsic
Calculi	Retroperitoneal fibrosis
Sloughed renal papillae	Neoplasm
trauma	Pregnancy
Congenital	Pelvic lipomatosis
• Stricture	Aortic aneurysm
• Ureterocele	Abscess
• Megaureter	Lymphocele
• Retrocaval ureter	Urinoma
• Prune-belly syndrome	
Neoplasm	
• Urothelial carcinoma	
• Metastatic carcinoma	
Inflammatory	
• Tuberculosis	
• Schistosomiasis	
• Amyloidosis	
• Endometriosis	
• Ureteritis cystica	
• Fungal bezoar	

may delay antegrade emptying, which acts to promote ascending infection (78). It has also been suggested that obstruction causes papillary necrosis and that sloughed papilla may act as a nidus for infection (76). Other intrinsic causes (Table 1) for ureteral obstruction may also act as a nidus for recurrent UTI. For example, urinary calculi can provide a surface for bacteria to adhere and proliferate upon.

Ureters may become obstructed secondary to ectopic insertion. In females, ectopic insertion can occur anywhere from the bladder neck to the perineum including the vagina, uterus, and rectum. This typically leads to incontinence. In males, the ectopic ureter may insert anywhere in the urogenital system above the external sphincter. Insertion can occur in the vas deferens, seminal vesicles, or ejaculatory ducts. Because insertion is above the external sphincter it does not cause incontinence, but it can be associated with infection. In a duplicated urinary tract the upper-renal moiety is associated with an ectopic ureter and this is thought to occur secondary to late ureteral budding from the mesonephric duct. The ureter subsequently inserts medial and inferior to its normal orthotopic position in the trigone. As a consequence of this abnormal insertion, the ureter must travel obliquely for greater length through the bladder and therefore can become obstructed (79).

Primary Vesicoureteric Reflux

Primary VUR is defined by the retrograde flow of urine, from bladder to the upper urinary tract, in the absence of obvious pathogenic cause. It occurs in approximately 1% of the general population and is responsible for 12% of antenatally detected hydronephrosis (80, 81). Children with normal perinatal ultrasonography who develop UTI have VUR 37.4% of the time (82). VUR is associated with recurrent UTIs, renal malformations, hypertension, and renal scarring and impairment known as reflux nephropa-

thy (83). Primary VUR represents a congenital defect in which the structure, and therefore function, of the UVJ is compromised. The length of intravesical ureter has been shown to be critical for the normal anti-reflux mechanism of UVJ (84, 85). Children with VUR were found to have a tunnel length-to-diameter ratio of 1.4:1 compared to 5:1 ratio in non-VUR children (85). VUR appears to be hereditary as VUR is present in 32% of siblings and 66% of offspring of known VUR patients (86, 87). Mouse models have confirmed that mutations in some of the genes expressed by the developing kidney and urinary tract can cause VUR (88, 89). Additionally, UP II and UP IIIa knockout mice been shown to have severe VUR (30, 32). Linkages between human VUR and genes involved in renal/urinary tract development have also been shown. Specifically, ACE, AGTR2, and RET polymorphisms have been positively associated with VUR (90–93). However, newer data do not support some of these findings (94, 95). Case-control studies of UPII and UPIIIa have failed to show significant association between single-nucleotide polymorphisms and VUR (96, 97). Jiang et al. genotyped all four UP genes in a population of 76 VUR patients. Of the 18 single-nucleotide polymorphisms identified, only 2 had a weak association with VUR. These data suggest that missense changes of UP genes cannot play a dominant role in causing VUR in humans (98). It has been speculated that major UP mutations are not compatible with human life (16). A large clinical and DNA database has recently been established from families containing sibling pairs with documented VUR (99). This will be an important resource for researchers and will hopefully bring light to the genetic components predisposing to VUR and reflux nephropathy.

Bladder-Outlet Obstruction

Bladder-outlet obstruction (BOO) in men can develop for a number of reasons including bladder calculi, medications, prostate cancer, urethral scarring, and benign prostatic hyperplasia (BPH). While most causes for BOO are relatively uncommon, BPH will develop in nearly all men by the age of 80 (100). Histologically, BPH represents a variable proliferative process of the stromal and epithelial elements of the periurethral and transition zones of the prostate (101). The pathophysiology of BOO in men with BPH is thought to be both static, due to physical blockage of the urethra and bladder outlet, and dynamically related to smooth-muscle tension. Interestingly, prostate volume does not seem to be an important determinant of BPH symptom severity (102, 103). Instead, the proportion of prostatic smooth muscle in the inner-transition zone of prostate appears to be an important determinant of clinical BPH (104). The molecular pathogenesis of BPH is not well understood. A number of growth factors have been associated with BPH, including fibroblast growth factor 2 and insulin growth factor. A number of cytokines and inflammatory mediators are also up-regulated in BPH including interleukin (IL)-1α, -2, -8, -15, and -17 and nuclear-factor κB. Madigan et al. compared molecular differences between symptomatic and asymptomatic BPH, and found 4 genes involved in the innate anti-viral immune response to be upregulated in symptomatic BPH: CFI, OAS2, APOBEC3G, and IFIT1 (105). A causal relationship between BPH/BOO and UTI risk has been difficult to establish as there is little data pertaining to this area. However, it is generally thought that urinary retention secondary to BOO increases UTI risk. Chronic urinary retention is defined as a non-painful bladder that remains palpable or percussable after urination and implies a significant residual volume of urine greater than 300 ml (106). Chronic urinary retention is a relatively uncommon finding in the general population, but incidence increases dramatically with age, lower urinary tract symptom severity, and prostate volume (107).

Abnormal Pelvic Anatomy

Although not extensively studied, differences in pelvic anatomy may predispose UTI. A single case-control study of 213 young women assessed perineal measurements and urethral length in those with and without a history of recurrent UTI (rUTI). In women not using spermicidals, and after controlling for sexual intercourse frequency, the urethra-to-anus distance and posterior fourchette-to-anus distance were found to be significantly shorter in those with a history of rUTI (4.8 vs 5.0 cm, P = .03 and 2.6 vs 2.8 cm, P = .04, respectively). This difference did not exist between groups using spermicidal products. Urethral length was measured with the aid of a urethral catheter and was not associated with a statistically significant increased risk of rUTI (3.6 vs 3.5 cm, P = .41). The authors concluded that perineal and urethral anatomy are likely important in the absence of other risk factors for rUTI (108).

Aging Female

The prevalence of UTI has been shown to increase with age (109). Anatomic and functional theories to explain this phenomenon include bladder dysfunction, pelvic-organ prolapse, urinary and fecal incontinence, and changes in estrogen status.

When healthy postmenopausal women with a history of rUTI were compared to non- recurrent controls, three strong risk factors emerged: incontinence (41% cases vs 9% controls; P < .001), pelvic-organ prolapse (cystocele) (19% vs 0%; P < .001) and postvoid residual (28% vs 2%; P < .001). Urinary incontinence was most strongly associated with rUTI in multivariate analysis (odds ratio [OR] 5.79; 95% confidence interval [CI] 2.05–16.42) (110).

There are two main theories that strive to explain normal female urinary continence. In the integral theory, proposed by Petros and Ulmsten, the urethra is closed from behind via the pelvic-floor muscles and their ability to stretch the vaginal hammock against the pubourethral ligaments (111). With this theory, weakness in the pubourethral ligaments predisposes to stress urinary incontinence (112). Delancey has suggested an alternate theory for female urinary continence called the hammock theory. This theory proposes that both urethral support and constriction are important and depend upon support from the anterior vaginal wall, the endopelvic fascia between the arcus tendineus facia pelvis, and the pelvic-floor muscles. Weakening or attenuation of these supports can therefore predispose to incontinence (113). It is unclear exactly how incontinence predisposes to rUTI. Anti-incontinence procedures may cause obstruction and urinary retention, increasing the risk of UTI. It is also conceivable that a continuously damp perineum and introitus may facilitate uropathogen colonization and ascent into the urinary tract leading to rUTI.

Anterior vaginal-wall prolapse (cystocele) is defined as pathologic descent of the anterior vaginal wall and overlying bladder base. The etiology of anterior vaginal prolapse is likely multifactorial and the result of damage or impairment of the pelvic muscles and/or connective tissues that normally provide support. Nitti et al. have shown that cystocele is an important cause of bladder-outlet obstruction and incomplete bladder emptying in females (114).

With the decline of circulating estrogen that accompanies menopause, physiologic and structural changes can occur to the vulvovaginal epithelium. In the low-estrogen state, the normally predominant lactobacilli diminish due to decreased vaginal-epithelial glycogen. Lactobacilli, via anaerobic metabolism of glycogen, normally produce lactic acid and hydrogen peroxide. These are both essential in maintaining an acidic and hostile vaginal environment to *E. coli* and other potentially uropathogenic organisms (115).

PHYSIOLOGIC ABNORMALITIES

Diabetes Mellitus

Individuals with diabetes mellitus are at a high risk for UTI and rUTIs. Epidemiologic studies show a 1.2 to 2.2-fold increase in the relative risk of UTI in diabetics when compared with non-diabetics (116–118). However, the underlying mechanisms that increase UTI susceptibility in diabetes are not completely understood. Hyperglycosuria has long been held as a causative factor in diabetic UTI pathogenesis with the theory being that higher glucose levels promote bacterial growth. However, there are no studies to date that clearly demonstrate a direct relationship between elevated urinary glucose and increased UTI risk. Clinical studies have also demonstrated that there is no dose-response relationship between serum HbA1c levels and UTI risk (119–121). One *ex vivo* study found that type-1 fimbriated *E. coli* adhered more to urothelial cells of diabetic women than non-diabetic controls; it was hypothesized that increased glycosylation of UPs might explain this finding (122). Abnormal glycosylation might also affect soluble glycoproteins in the urine, making them less effective competitive inhibitors for the type-1 fimbriae (123), which could in turn increase the adherence of UPEC to the urothelial surface. Diabetic cystopathy (DC) is a condition marked by the insidious onset of impaired bladder sensation, decreased detrusor contractility, increased post-void residual volume and, in severe cases, detrusor areflexia. Voiding dysfunction and urinary retention can lead to decreased clearance of bacteria via micturition and therefore predispose diabetic patients to UTI. Traditionally, autonomic neuropathy was felt to be the pathophysiological cause of DC (124). A more contemporary view of DC centers on a multifactorial etiology including alterations in detrusor muscle, urethra, autonomic nerves, and urothelium secondary to oxidative stress and hyperglycemia-induced polyuria (125, 126).

Neutrophil and lymphocyte dysfunction generally found in diabetics may also impair the ability to clear bacteria from the urinary tract in the later stages of infection. Using a streptozocin-induced diabetic-mouse model, Rosen et al. provided evidence to support this theory. These authors were able to show that *E. coli* titers of diabetic C3H/HeJ mouse kidneys and bladders were 10,000-fold higher than non-diabetic C3H/HeJ controls (127).

Pregnancy

The prevalence of bacteruria in pregnancy is similar to non-gravid females and ranges from 2% to 10%. However, 25% to 35% of pregnant woman with bacteruria will progress to pyelonephritis (128, 129). Screening and treatment of bacteruria in this population is recommended to prevent pyelonephritis and its associated adverse perinatal outcomes such as prematurity and/or low birth weight (128, 129). Physiologic and anatomic changes in pregnancy occur that lead to urinary stasis, secondary vesicoureteric reflux, and increased risk of UTI.

Hydronephrosis develops in the majority of pregnant women and is present in 15% during the first trimester, 20% in the second trimester, and 50% in the third trimester (130). The finding of hydronephrosis in the first trimester, before the gravid uterus reaches the pelvic brim to cause obstruction, supports a hormonal etiology. Progesterone has been hypothesized to promote ureteral dilation and therefore development of hydronephrosis (131). Hydronephrosis of pregnancy may also be mechanical. The gravid uterus after the 20th week of gestation may extrinsically compress the ureters. This idea is supported by both the increased incidence and degree of hydronephrosis after 20-weeks' gestation. Further support comes from the finding that the right collecting system is dilated two to three times more commonly then the left and that the gravid uterus is typically dextrorotated (130).

Ureteropelvic-Junction Obstruction

Ureteropelvic-junction obstruction (UPJO) is the most common cause of obstructive nephropathy in children with an estimated incidence of 1 in 1,000 1,500 (132). The widespread use of prenatal screening ultrasonography has led to the early detection of hydronephrosis and, subsequently, the majority of UPJO are diagnosed in the neonatal and infant period. However, UPJO can present at any age with symptoms including flank pain, episodic upper abdominal pain (Dietl's crisis), renal failure, hematuria, and recurrent UTI. UPJO is generally considered a functional abnormality. Accumulating evidence suggests that abnormal smooth-muscle development and differentiation in the renal pelvis and proximal ureter is responsible for creating an aperistaltic segment at the level of the UPJ. Analyses of human UPJ tissues obtained at the time of surgical correction for UPJO (pyeloplasty) have demonstrated decreased neuronal innervation, significant fibrosis, and abnormal smooth-muscle cell arrangement (133, 134).

Murine models of UPJO have also provided valuable insight into UPJO pathogenesis. Mouse lines harboring conditional knockouts of sonic hedgehog, ADAMTS-1, and calcineurin B have smooth-muscle deficiency in the renal pelvis and ureters and display phenotypes similar to human UPJO (132, 135, 136). Further studies of the genetic and cellular defects underlying UPJO are needed. In some cases of UPJO, an aberrant renal vessel is observed crossing the UPJ. Whether or not crossing vessels alone can cause extrinsic compression sufficient for UPJO remains unclear (137). Recent clinical studies demonstrate that the rate of UTI in children with high-grade UPJO is low (138, 139). Therefore, most experts do not recommend antibiotic prophylaxis in this population. However, rUTIs in the context of UPJO and urinary stasis are generally considered an indication for either antibiotic prophylaxis and/or surgical correction.

Dysfunctional Voiding

Dysfunctional voiding (DV) is characterized by an intermittent and/or fluctuating flow rate due to involuntary intermittent contractions of the peri-urethral striated or levator muscles during voiding in neurologically normal individuals (140). The prevalence of adult DV in the general population is unknown. In patients referred for urodynamic evaluation, Jorgensen et al. found DV prevalence rate of 0.5% (141). Groutz et al. retrospectively reviewed the videourodynamics of 1,015 consecutive adults and found that 2% met their criteria for DV (142). These figures likely underestimate the true prevalence of this condition. Of the 21 patients who met videourodynamic criteria for DV in the Groutz study, 14 were asked about childhood voiding. All 14 denied any form of dysfunctional voiding as a child (142). This clearly challenges the previously held notion that all adult DV stems from behavior learned and carried over from childhood. In fact, there are data that now suggest that DV can be learned in adulthood. Using urodynamic criteria of obstruction (maximum flow rate ≤12 ml/sec and detrusor pressure at maximum flow ≥25 cm H_2O), Cameron et al. demonstrated bladder-outlet obstruction in 48.1% of women with interstitial cystitis/bladder-pain syndrome. Although electromyograph (EMG) data was not reported, the authors postulated that painful voiding leads to pelvic-floor spasticity and subsequent DV (143). DV has also been shown to have an increased incidence in sexual-abuse victims and has been linked to the exposure of psychological stressors (144, 145). The goal of DV treatment is to facilitate a patient's return back to normal micturition. To achieve this any combination of behavioral, cognitive, and pharmacologic therapies can be utilized. Recurrent UTIs may occur in up to 42% of women with DV (146). Minardi et al. have recently demonstrated that in women with both DV and recurrent UTI, pelvic-floor physiotherapy can significantly reduce the rate of UTI recurrence (147).

Primary Bladder-Neck Obstruction

Primary bladder-neck obstruction (PBNO) is a condition in which the flow of urine is obstructed due to incomplete bladder-neck opening. Diagnostic features of PBNO include high-voiding pressures and low urine flow (maximum detrusor pressure >20 cm H_2O and maximum urine flow <12 mL/s), fluoroscopic absence of bladder-neck opening/funneling and minimal EMG activity during volitional voiding. Recently, Brucker et al. evaluated urodynamic differences between DV and PBNO. Patients with DV had a higher mean Qmax (12 ml/s vs 7 ml/s; P = 1.27) and a lower mean postvoiding residual (PVR) (125 ml vs 400 ml; P = .012) when compared to PBNO. In their analysis, the authors found that EMG alone would have led to misdiagnosis in 20% of DV and 14.3% of PBNO patients. This data supports the use of fluoroscopy in diagnosing and differentiating DV and PBNO (148). It has been theorized that the etiology of PBNO is either morphologic (smooth-muscle hypertrophy or fibrosis) or neurogenic in nature (149). Between 8.7% and 16% of women presenting with bladder-outlet obstruction will have PBNO as their underlying cause (114, 150). Treatment options for PBNO include observation, pharmacotherapy, and surgical intervention. Obstruction of urine flow at the bladder neck can lead to elevated PVR and presumably increase the risk of UTI. However, UTI incidence in patients with PBNO has not been systematically studied.

Neurologic Patients

Individuals with suprasacral and subpontine spinal-cord injuries develop bladder overactivity and involuntary contractions of the external urinary sphincter. The resulting loss of coordinated micturition, termed detrusor-external sphincter dyssynergia (DESD), can lead to dangerously high storage pressures and the development of serious urologic and nephrologic complications including secondary VUR, incomplete bladder emptying, bacteriuria, and recurrent UTI. Ideally, a combination of clean intermittent catheterization (CIC) and anticholinergic medications are used to reduce intravesical storage pressures and allow for bladder emptying (151). Alternative treatments may be needed for those who cannot tolerate anticholinergic agents or cannot reliably catheterize (e.g., quadriplegic or non-compliant patients).

External sphincterotomy has historically been considered the next best alternative to CIC/anticholinergic therapy for DESD. However, it is now recognized that this technique is associated with a high complication rate (151, 152). Urethral stenting, sphincter-injected botulinum toxin A and, in refractory cases, urinary diversion can also be considered for the treatment of DESD (153, 154).

Other Physiological Defects

It has long been speculated that genetic modifiers may play a role in influencing host susceptibility to recurrent UTIs. For instance, the female relatives of women with recurrent UTIs are more prone to UTIs than the general population (155). The severity and disease course in experimental animals, particularly mice, can also be dependent on different strains and genetic backgrounds under study (156). The availability of knockout mice deficient for functionally divergent genes with convergent UTI phenotype is suggestive of a polygenic nature of genetic predisposition to UTIs. Thus far, knockout-mouse strains lacking/deficient for toll-like receptors (e.g., TLR2, 4, and 11), anti-microbial peptides (e.g., defensins and cathelicidin), anti-bacterial adherence factor (e.g., Tamm-Horsfall protein), and growth factors (e.g., transforming growth factor [TGF]-β1 and vascular endothelial growth factor [VEGF]) have been found to have increased spontaneous or experimentally induced UTIs (155,

157–162). Interestingly, single-nucleotide poly morphisms (SNPs) of some these genes have also been identified in humans. As reviewed by Zaffanello and colleagues, most of these gene SNPs are actually associated with developmental anomalies of the urinary tract, including urinary-tract malformation and VUR (155). Perhaps not surprisingly, there is evidence that other genes involved in the innate and adaptive immunity can also alter host defense and hence its susceptibility to UTIs. The effects of these genes in the global host responses to microbial infections in multi-organ systems should be distinguished from those only affecting the urinary tract. Such a distinction is not only important for investigative purposes, but also for clinical decision making on whether management strategies should be global or local.

SUMMARY AND PERSPECTIVE

The ability of the urinary tract to defend against microbial infections relies on its normal anatomic architecture as well as a functional physiological state. Defects in either or both aspects, whether they be congenital or acquired, can increase the access, retention, and spread, and the decreased clearance of microbes from the urinary tract, leading to microbial infections and even recurrent episodes. A thorough understanding of anatomy and physiology of the urinary tract is therefore necessary for those who investigate and those who diagnose and treat urinary tract infections. An appreciation of the normal architecture and functions of the upper and lower collecting systems and external genitalia is also required before one can hope to identify and/or systematically study abnormalities that may predispose to UTIs. While the pathogenic processes behind anatomic abnormality-associated UTIs, such as obstruction and reflux, are well understood, those behind the physiological risk factors that predispose to uncomplicated and recurrent UTIs in anatomically normal individuals are much less clear, and this is an area that clearly requires much greater research efforts. Fundamental biological and biochemical studies, genetic analyses, gene targeting with genetically engineered animals, and gene-wide association studies in humans could be instrumental in further advancing our knowledge regarding the physiological functions of the urinary tract and its disease pathogenesis. Successful translation of such knowledge to the bedside should in turn help more effectively treat UTI and/or reduce its recurrence.

ACKNOWLEDGMENTS

Conflicts of interest: We declare no conflicts.

CITATION

Hickling DR, Sun T-T, Wu X-R. 2015. Anatomy and physiology of the urinary tract: relation to host defense and microbial infection. Microbiol Spectrum 3(4):UTI-0016-2012.

REFERENCES

1. Fowler CJ, Griffiths D, de Groat WC. 2008. The neural control of micturition. *Nat Rev Neurosci* **9:**453–466.
2. Elbadawi A. 1996. Functional anatomy of the organs of micturition. *Urol Clin North Am* **23:** 177–210.
3. O'Grady F, Cattell WR. 1966. Kinetics of urinary tract infection. II. The bladder. *Br J Urol* **38:**156–162.
4. Anderson JK, Cadeddu JA. 2012. Surgical anatomy of the retroperitoneum, adrenals, kidneys, and ureters, p 3–6. *In* Wein AJ, Kavoussi LR, Novick AC, Partin AW, Peters CA (ed), *Campbell-Walsh Urology*, 10th ed. Saunders Elsevier, Philadelphia.
5. Macarak EJ, Howard PS. 1999. The role of collagen in bladder filling. *Adv Exp Med Biol* **462:**215–223.
6. Tanagho ER, Pugh RC. 1963. The anatomy and function of the ureterovesical junction. *Br J Urol* **35:**151–165.
7. Viana R, Batourina E, Huang H, Dressler GR, Kobayashi A, Behringer RR, Shapiro E, Hensle T, Lambert S, Mendelsohn C. 2007. The development of the bladder trigone, the center of the anti-reflux mechanism. *Development* **134:**3763–3769.
8. Tanaka ST, Ishii K, Demarco RT, Pope JC IV, Brock JW III, Hayward SW. 2010. Endodermal origin of bladder trigone inferred from mesenchymal-epithelial interaction. *J Urol* **183:**386–391.
9. Liang FX, Bosland MC, Huang H, Romih R, Baptiste S, Deng F-M, Wu XT, Shapiro E, Sun TT. 2005. Cellular basis of urothelial squamous metaplasia: roles of lineage heterogeneity and cell replacement. *J Cell Biol* **171:** 835–844.
10. Chung BI, Sommer G, D BJ. 2012. Anatomy of the lower urinary tract and male genitalia, p 59–60. *In* Wein AJ, Kavoussi LR, Novick AC, Partin AW, Peters CA (ed), *Campbell-Walsh Urology*, 10th ed. Saunders Elsevier, Philadelphia.
11. Standring S. 2008. The anatomical basis of clinical practice. *In Gray's Anatomy*, 40th ed. Churchill Livingstone Elsevier, Philadelphia.
12. Standring S. 2008. Female reproductive system, p 1279–1304. *In Gray's Anatomy*. 40th ed. Churchill Livingstone Elsevier, Philadelphia.
13. Stapleton AE, Au-Yeung M, Hooton TM, Fredricks DN, Roberts PL, Czaja CA, Yarova-Yarovaya Y, Tiedler T, Cox M, Stamm WE. 2011. Randomized, placebo- controlled phase 2 trial of a *Lactobacillus crispatus* probiotic given intravaginally for prevention of recurrent urinary tract infection. *Clin Infect Dis* **52:**1212–1217.
14. Riedel I, Liang FX, Deng F-M, Tu L, Kreibich G, Wu XR, Sun TT, Hergt M, Moll R. 2005. Urothelial umbrella cells of human ureter are heterogeneous with respect to their uroplakin composition: different degrees of urothelial maturity in ureter and bladder? *Eur J Cell Biol* **84:**393–405.
15. Romih R, Korošec P, de Mello W Jr, Jezernik K. 2005. Differentiation of epithelial cells in the urinary tract. *Cell Tissue Res* **320:**259–268.
16. Wu XR, Kong XP, Pellicer A, Kreibich G, Sun T-T. 2009. Uroplakins in urothelial biology, function, and disease. *Kidney Int* **75:**1153–1165.
17. Walker B. 1960. Renewal of cell populations in the female mouse. *Am J Anat* **107:**95–105.
18. Kachar B, Liang F, Lins U, Ding M, Wu XR, Stoffler D, Aebi U, Sun T-T. 1999. Three-dimensional analysis of the 16 nm urothelial plaque particle: luminal surface exposure, preferential head-to-head interaction, and hinge formation. *J Mol Biol* **285:**595–608.
19. Hu CC, Liang FX, Zhou G, Tu L, Tang CH, Zhou J, Kreibich G, Sun TT. 2005. Assembly

of urothelial plaques: tetraspanin function in membrane protein trafficking. *Mol Biol Cell* **16:**3937–3950.

20. **Vergara J, Longley W, Robertson JD.** 1969. A hexagonal arrangement of subunits in membrane of mouse urinary bladder. *J Mol Biol* **46:**593–596.

21. **Hicks RM, Ketterer B.** 1969. Hexagonal lattice of subunits in the thick luminal membrane of the rat urinary bladder. *Nature* **224:** 1304–1305.

22. **Taylor KA, Robertson JD.** 1984. Analysis of the three-dimensional structure of the urinary bladder epithelial cell membranes. *J Ultrastruct Res* **87:**23–30.

23. **Walz T, Häner M, Wu XR, Henn C, Engel A, Sun TT, Aebi U.** 1995. Towards the molecular architecture of the asymmetric unit membrane of the mammalian urinary bladder epithelium: a closed "twisted ribbon" structure. *J Mol Biol* **248:**887–900.

24. **Wu XR, Sun TT.** 1993. Molecular cloning of a 47 kDa tissue-specific and differentiation-dependent urothelial cell surface glycoprotein. *J Cell Sci* **106:**31–43.

25. **Lin JH, Wu XR, Kreibich G, Sun TT.** 1994. Precursor sequence, processing, and urothelium-specific expression of a major 15-kDa protein subunit of asymmetric unit membrane. *J Biol Chem* **269:**1775–1784.

26. **Yu J, Lin JH, Wu XR, Sun TT.** 1994. Uroplakins Ia and Ib, two major differentiation products of bladder epithelium, belong to a family of four transmembrane domain (4TM) proteins. *J Cell Biol* **125:**171–182.

27. **Deng FM, Liang FX, Tu L, Resing KA, Hu P, Supino M, Hu CC, Zhou G, Ding M, Kreibich G, Sun TT.** 2002. Uroplakin IIIb, a urothelial differentiation marker, dimerizes with uroplakin Ib as an early step of urothelial plaque assembly. 2002. *J Cell Biol* **59:**685–694.

28. **Wu XR, Medina JJ, Sun TT.** 1995. Selective interactions of UPIa and UPIb, two members of the transmembrane 4 superfamily, with distinct single transmembrane- domained proteins in differentiated urothelial cells. *J Biol Chem* **270:**29752–29759.

29. **Tu L, Sun TT, Kreibich G.** 2002. Specific heterodimer formation is a prerequisite for uroplakins to exit from the endoplasmic reticulum. *Mol Biol Cell* **13:**4221–4230.

30. **Hu P, Deng FM, Liang FX, Hu CM, Auerbach AB, Shapiro E, Wu XR, Kachar B, Sun TT.** 2000. Ablation of uroplakin III gene results in small urothelial plaques, urothelial leakage, and vesicoureteral reflux. *J Cell Biol* **151:**961–972.

31. **Hu P, Meyers S, Liang FX, Deng FM, Kachar B, Zeidel ML, Sun TT.** 2002. Role of membrane proteins in permeability barrier function: uroplakin ablation elevates urothelial permeability. *Am J Physiol Renal Physiol* **283:** F1200–1207.

32. **Kong XT, Deng FM, Hu P, Liang FX, Zhou G, Auerbach AB, Geieser N, Nelson PK, Robbins ES, Shapiro E, Kachar B, Sun TT.** 2004. Roles of uroplakins in plaque formation, umbrella cell enlargement, and urinary tract diseases. *J Cell Biol* **167:**1195–1204.

33. **Minsky BD, Chlapowski FJ.** 1978. Morphometric analysis of the translocation of lumenal membrane between cytoplasm and cell surface of transitional epithelial cells during the expansion-contraction cycles of mammalian urinary bladder. *J Cell Biol* **77:**685–697.

34. **Lewis SA, de Moura JL.** 1982. Incorporation of cytoplasmic vesicles into apical membrane of mammalian urinary bladder epithelium. *Nature* **297:**685–688.

35. **Zhou G, Mo WJ, Sebbel P, Min G, Neubert TA, Glockshuber R, Wu XR, Sun TT, Kong XP.** 2001. Uroplakin Ia is the urothelial receptor for uropathogenic *Escherichia coli*: evidence from *in vitro* FimH binding. *J Cell Sci* **114:**4095–4103.

36. **Xie B, Zhou G, Chan SY, Shapiro E, Kong XP, Wu XR, Sun TT, Costello CE.** 2006. Distinct glycan structures of uroplakins Ia and Ib: structural basis for the selective binding of FimH adhesin to uroplakin Ia. *J Biol Chem* **281:**14644–14653.

37. **Wu XR, Sun TT, Medina JJ.** 1996. *In vitro* binding of type 1-fimbriated *Escherichia coli* to uroplakins Ia and Ib: relation to urinary tract infections. *Proc Natl Acad Sci U S A* **93:** 9630–9635.

38. **Chen SL, Hung CS, Pinkner JS, Walker JN, Cusumano CK, Li Z, Bouckaert J, Gordon JI, Hultgren SJ.** 2009. Positive selection identifies an *in vivo* role for FimH during urinary tract infection in addition to mannose binding. *Proc Natl Acad Sci U S A* **106:**22439–22444.

39. **Liu B, Hu B, Zhou Z, Guo D, Guo X, Ding P, Feng L, Wang L.** 2012. A novel non- homologous recombination-mediated mechanism for *Escherichia coli* unilateral flagellar phase variation. *Nucleic Acids Res* **40:**4530–4538.

40. **Lane MC, Mobley HL.** 2007. Role of P-fimbrial-mediated adherence in pyelonephritis and persistence of uropathogenic *Escherichia coli* (UPEC) in the mammalian kidney. *Kidney Int* **72:**19–25.

41. **Thumbikat P, Berry RE, Zhou G, Billips BK, Yaggie RE, Zaichuk T, Sun TT, Schaeffer AJ,**

Klumpp DJ. 2009. Bacteria-induced uroplakin signaling mediates bladder response to infection. *PLoS Pathog* **5**:e1000415. doi:10.1371/journal.ppat.1000415

42. Ho PL, Kurtova A, Chan KS. 2012. Normal and neoplastic urothelial stem cells: getting to the root of the problem. *Nat Rev Urol* **9**:583–594.

43. Shin K, Lee J, Guo N, Kim J, Lim A, Qu L, Mysorekar IU, Beachy PA. 2011. Hedgehog/Wnt feedback supports regenerative proliferation of epithelial stem cells in bladder. *Nature* **472**:110–114.

44. Hannan TJ, Totsika M, Mansfield KJ, Moore KH, Schembri MA, Hultgren SJ. 2012. Host-pathogen checkpoints and population bottlenecks in persistent and intracellular uropathogenic *Escherichia coli* bladder infection. *FEMS Microbiol Rev* **36**:616–648.

45. Bishop BL, Duncan MJ, Song J, Li G, Zaas D, Abraham SN. 2007. Cyclic AMP- regulated exocytosis of *Escherichia coli* from infected bladder epithelial cells. *Nat Med* **13**:625–630.

46. Guo X, Tu L, Gumper I, Plesken H, Novak EK, Chintala S, Swank RT, Pastores G, Torres P, Izumi T, Sun TT, Sabatini DD, Kreibich G. 2009. Involvement of Vps33a in the fusion of uroplakin-degrading multivesicular bodies with lysosomes. *Traffic* **10**:1350–1361.

47. Chen Y. 2003. Rab27b is associated with fusiform vesicles and may be involved in targeting uroplakins to urothelial apical membranes. *Proc Natl Acad Sci U S A* g14012–14017.

48. Zhou G, Liang FX, Romih R, Wang Z, Liao Y, Ghiso J, Luque-Garcia JL, Neubert TA, Kreibich G, Alonso MA, Schaeren-Wiemers N, Sun TT. 2012. MAL facilitates the incorporation of exocytic uroplakin-delivering vesicles into the apical membrane of urothelial umbrella cells. *Mol Biol Cell* **23**:1354–1366.

49. Lang RJ, Hashitani H, Tonta MA, Bourke JL, Parkington HC, Suzuki H. 2010. Spontaneous electrical and Ca2+ signals in the mouse renal pelvis that drive pyeloureteric peristalsis. *Clin Exp Pharmacol Physiol* **37**:509–515.

50. Iqbal J, Tonta MA, Mitsui R, Li Q, Kett M, Li J, Parkington HC, Hashitani H, Lang RJ. 2012. Potassium and ANO1/TMEM16A chloride channel profiles distinguish atypical and typical smooth muscle cells from interstitial cells in the mouse renal pelvis. *Br J Pharmacol* **165**:2389–2408.

51. Hurtado R, Bub G, Herzlinger D. 2010. The pelvis-kidney junction contains HCN3, a hyperpolarization-activated cation channel that triggers ureter peristalsis. *Kidney Int* **77**:500–508.

52. Weiss RM. 2012. Physiology and pharmacology of the renal pelvis and ureter. Ch. 59. *In* Wein AJ, Kavoussi LR, Novick AC, Partin AW, Peters CA (ed), *Campbell-Walsh Urology*, 10th ed. Saunders Elsevier, Philadelphia.

53. Floyd RV, Winstanley C, Bakran A, Wray S, Burdyga TV. 2010. Modulation of ureteric Ca signaling and contractility in humans and rats by uropathogenic *E. coli*. *Am J Physio Renal Physiol* **298**:F900–908.

54. Floyd RV, Upton M, Hultgren SJ, Wray S, Burdyga TV, Winstanley C. 2012. *Escherichia coli*-mediated impairment of ureteric contractility is uropathogenic *E. coli*-specific. *J Infect Dis* **206**:1589–1596.

55. Wein AJ. 1981. Classification of neurogenic voiding dysfunction. *J Urol* **125**:605–609.

56. de Groat WC, Yoshimura N. 2001. Pharmacology of the lower urinary tract. *Ann Rev Pharmacol Toxicol* **41**:691–721.

57. Park JM, Bloom DA, McGuire EJ. 1997. The guarding reflex revisited. *Br J Urol* **80**:940–945.

58. Yoshimura N, Chancellor MB. 2012. Physiology and pharmacology of the bladder and urethra, p 1787–1789. *In* Wein AJ, Kavoussi LR, Novick AC, Partin AW, Peters CA (ed), *Campbell-Walsh Urology*, 10th ed. Saunders Elsevier, Philadelphia.

59. Miller NL, Humphreys MR, Coe FL, Evan AP, Bledsoe SB, Handa SE, Lingeman JE. 2010. Nephrocalcinosis: re-defined in the era of endourology. *Urol Res* **38**:421–427.

60. Ginalski JM, Portmann L, Jaeger P. 1990. Does medullary sponge kidney cause nephrolithiasis? *Am J Roentgenol* **155**:299–302.

61. Laube M, Hess B, Terrier F, Vock P, Jaeger P. 1995. Prevalence of medullary sponge kidney in patients with and without nephrolithiasis. *Praxis (Bern 1994)* **84**:1224–1230.

62. Palubinskas AJ. 1963. Renal pyramidal structure opacification in excretory urography and its relation to medullary sponge kidney. *Radiology* **81**:963–970.

63. Forster JA, Taylor J, Browning AJ, Biyani CS. 2007. A review of the natural progression of medullary sponge kidney and a novel grading system based on intravenous urography findings. *Urol Int* **78**:264–269.

64. McPhail EF, Gettman MT, Patterson DE, Rangel LJ, Krambeck AE. 2012. Nephrolithiasis in medullary sponge kidney: evaluation of clinical and metabolic features. *Urology* **79**:277–281.

65. Toyoda K, Miyamoto Y, Ida M, Tada S, Utsunomiya M. 1989. Hyperechoic medulla of the kidneys. *Radiology* **173**:431–434.

66. **Ginalski JM, Schnyder P, Portmann L, Jaeger P.** 1991. Medullary sponge kidney on axial computed-tomography: comparison with excretory urography. *Eur J Radiol* **2:**104–107.

67. **Hida T, Nishie A, Asayama Y, Ishigami K, Fujita N, Inokuchi J, Naito S, Ando S, Honda H.** 2012. MR imaging of focal medullary sponge kidney: case report. *Magn Reson Med Sci* **11:**65–69.

68. **Chesney RW, Kaufman R, Stapleton FB, Rivas ML.** 1989. Association of medullary sponge kidney and medullary dysplasia in Beckwith-Wiedemann syndrome. *J Pediatr* **115:**761–764.

69. **Kerr DN, Warrick CK, Hart-Mercer J.** 1962. A lesion resembling medullary sponge kidney in patients with congenital hepatic fibrosis. *Clin Radiol* **13:**85–91.

70. **Torres VE, Erickson SB, Smith LH, Wilson DM, Hattery RR, Segura JW.** 1988. The association of nephrolithiasis and autosomal dominant polycystic kidney disease. *Am J Kidney Dis* **11:**318–325.

71. **Gambaro G, Fabris A, Citron L, Tosetto E, Anglani F, Bellan F, Conte M, Bonfante L, Lupo A, D'Angelo A.** 2005. An unusual association of contralateral congenital small kidney, reduced renal function and hyperparathyroidism in sponge kidney patients: on the track of the molecular basis. *Nephrol Dial Transplant* **20:**1042–1047.

72. **Hulbert JC, Reddy PK, Hunter DW, Castaneda-Zuniga W, Amplatz K, Lange PH.** 1986. Percutaneous techniques for the management of caliceal diverticula containing calculi. *J Urol* **135:**225–227.

73. **Monga M, Smith R, Ferral H, Thomas R.** 2000. Percutaneous ablation of caliceal diverticulum: long-term followup. *J Urol* **163:**28–32.

74. **Auge BK, Munver R, Kourambas J, Newman GE, Wu NZ, Preminger GM.** 2002. Neo-infundibulotomy for the management of symptomatic caliceal diverticula. *J Urol* **167:**1616–1620.

75. **Bellman GC, Silverstein JI, Blickensderfer S, Smith AD.** 1993. Technique and follow-up of percutaneous management of caliceal diverticula. *Urology* **42:**21–25.

76. **Heyns CF.** 2012. Urinary tract infection associated with conditions causing urinary tract obstruction and stasis, excluding urolithiasis and neuropathic bladder. *World J Urol* **30:**77–83.

77. **Beeson PB, Guze LB.** 1956. Experimental pyelonephritis. I. Effect of ureteral ligation on the course of bacterial infection in the kidney of the rat. *J Exp Med* **104:**803–815.

78. **Bitz H, Darmon D, Goldfarb M, Shina A, Block C, Rosen S, Brezis M, Heyman SN.** 2001. Transient urethral obstruction predisposes to ascending pyelonephritis and tubulo- interstitial disease: studies inrats. *Urol Res* 67–73.

79. **Fernbach SK, Feinstein KA, Spencer K, Lindstrom CA.** 1997. Ureteral duplication and its complications. *Radiographics* **17:**109–127.

80. **Chapman CJ, Bailey RR, Janus ED, Abbott GD, Lynn KL.** 1985. Vesicoureteric reflux: segregation analysis. *Am J Med Genet* **20:**577–584.

81. **Farhat W, McLorie G, Geary D, Capolicchio G, Bägli D, Merguerian P, Khoury A.** 2000. The natural history of neonatal vesicoureteral reflux associated with antenatal hydronephrosis. *J Urol* **164:**1057–1060.

82. **Hannula A, Venhola M, Renko M, Pokka T, Huttunen N-P, Uhari M.** 2010. Vesicoureteral reflux in children with suspected and proven urinary tract infection. *Pediatr Nephrol* **25:** 1463–1469.

83. **Risdon RA, Yeung CK, Ransley PG.** 1993. Reflux nephropathy in children submitted to unilateral nephrectomy: a clinicopathological study. *Clin Nephrol* **40:**308–314.

84. **Vermillion CD, Heale WF.** 1973. Position and configuration of the ureteral orifice and its relationship to renal scarring in adults. *J Urol* **109:**579–584.

85. **Paquin AJ Jr.** 1959. Ureterovesical anastomosis: the description and evaluation of a technique. *J Urol* **82:**573–583.

86. **Hollowell JG, Greenfield SP.** 2002. Screening siblings for vesicoureteral reflux. *J Urol* **168:** 2138–2141.

87. **Noe HN, Wyatt RJ, Peeden JN Jr, Rivas ML.** 1992. The transmission of vesicoureteral reflux from parent to child. *J Urol* **148:**1869–1871.

88. **Murawski IJ, Myburgh DB, Favor J, Gupta IR.** 2007. Vesico-ureteric reflux and urinary tract development in the Pax21Neu+/ mouse. *Am J Physiol Renal Physiol* **293:**F1736–1745.

89. **Yu OH, Murawski IJ, Myburgh DB, Gupta IR.** 2004. Overexpression of RET leads to vesicoureteric reflux in mice. *Am J Physiol Renal Physiol* **287:**F1123–1130.

90. **Ohtomo Y, Nagaoka R, Kaneko K, Fukuda Y, Miyano T, Yamashiro Y.** 2001. Angiotensin converting enzyme gene polymorphism in primary vesicoureteral reflux. *Pediatr Nephrol* **16:**648–652.

91. **Rigoli L, Chimenz R, di Bella C, Cavallaro E, Caruso R, Briuglia S, Fede C, Salpietro CD.**

2004. Angiotensin-converting enzyme and angiotensin type 2 receptor gene genotype distributions in Italian children with congenital uropathies. *Pediatr Res* **56:**988–993.

92. Loré F, Talidis F, Di Cairano G, Renieri A. 2001. Multiple endocrine neoplasia type 2 syndromes may be associated with renal malformations. *J Intern Med* **250:**37–42.

93. Yang Y, Houle AM, Letendre J, Richter A. 2008. RET Gly691Ser mutation is associated with primary vesicoureteral reflux in the French-Canadian population from Quebec. *Hum Mutat* **29:**695–702.

94. Yoneda A, Cascio S, Green A, Barton D, Puri P. 2002. Angiotensin II type 2 receptor gene is not responsible for familial vesicoureteral reflux. *J Urol* **168:**1138–1141.

95. Shefelbine SE, Khorana S, Schultz PN, Huang E, Thobe N, Hu ZJ, Fox GM, Jing S, Cote GJ, Gagel RF. 1998. Mutational analysis of the GDNF/RET-GDNFR alpha signaling complex in a kindred with vesicoureteral reflux. *Hum Genet* **102:**474–478.

96. Giltay JC, van de Meerakker J, van Amstel HK, de Jong TP. 2004. No pathogenic mutations in the uroplakin III gene of 25 patients with primary vesicoureteral reflux. *J Urol* **171:**931–932.

97. Jenkins D, Bitner-Glindzicz M, Malcolm S, Allison J, de Bruyn R, Flanagan S, Thomas DF, Belk RA, Feather SA, Bingham C, Southgate J, Woolf AS. 2006. Mutation analyses of Uroplakin II in children with renal tract malformations. *Nephrol Dial Transplant* **21:**3415–3421.

98. Jiang S, Gitlin J, Deng FM, Liang FX, Lee A, Atala A, Bauer SB, Ehrlich GD, Feather SA, Goldberg JD, Goodship JA, Goodship TH, Hermanns M, Hu FZ, Jones KE, Malcolm S, Mendelsohn C, Preston RA, Retik AB, Schneck FX, Wright V, Ye XY, Woolf AS, Wu XR, Ostrer H, Shapiro E, Yu J, Sun TT. 2004. Lack of major involvement of human uroplakin genes in vesicoureteral reflux: implications for disease heterogeneity. *Kidney Int* **66:**10–19.

99. Lambert HJ, Stewart A, Gullett AM, Cordell HJ, Malcolm S, Feather SA, Goodship JA, Goodship TH, Woolf AS; UK VUR Study Group. 2011. Primary, nonsyndromic vesicoureteric reflux and nephropathy in sibling pairs: a United Kingdom cohort for a DNA bank. *Clin J Am Soc Nephrol* **6:**760–766.

100. Berry SJ, Coffey DS, Walsh PC, Ewing LL. 1984. The development of human benign prostatic hyperplasia with age. *J Urol* **132:**474–479.

101. Lepor H. 2005. Pathophysiology of benign prostatic hyperplasia in the aging male population. *Rev Urol* **7**(Suppl 4):S3–S12.

102. Nitti VW, Kim Y, Combs AJ. 1994. Correlation of the AUA symptom index with urodynamics in patients with suspected benign prostatic hyperplasia. *Neurourol Urodyn* **13:**521–527.

103. Barry MJ, Cockett AT, Holtgrewe HL, McConnell JD, Sihelnik SA, Winfield HN. 1993. Relationship of symptoms of prostatism to commonly used physiological and anatomical measures of the severity of benign prostatic hyperplasia. *J Urol* **150:**351–358.

104. Shapiro E, Becich MJ, Hartanto V, Lepor H. 1992. The relative proportion of stromal and epithelial hyperplasia is related to the development of symptomatic benign prostate hyperplasia. *J Urol* **147:**1293–1297.

105. Madigan AA, Sobek KM, Cummings JL, Green WR, Bacich DJ, O'Keefe DS. 2012. Activation of innate anti-viral immune response genes in symptomatic benign prostatic hyperplasia. *Genes Immun* **13:**566–572.

106. Abrams P, Cardozo L, Fall M, Griffiths D, Rosier P, Ulmsten U, Kerrebroeck PV, Victor A, Wein A. 2003. The standardisation of terminology in lower urinary tract function: report from the standardisation subcommittee of the International Continence Society. *Urology* **61:**37–49.

107. Jacobsen SJ, Jacobson DJ, Girman CJ, Roberts RO, Rhodes T, Guess HA, Lieber MM. 1997. Natural history of prostatism: risk factors for acute urinary retention. *J Urol* **158:**481–487.

108. Hooton TM, Stapleton AE, Roberts PL, Winter C, Scholes D, Bavendam T, Stamm WE. 1999. Perineal anatomy and urine-voiding characteristics of young women with and without recurrent urinary tract infections. *Clin Infect Dis* **29:**1600–1601.

109. Nicolle LE. 2003. Asymptomatic bacteriuria: when to screen and when to treat. *Infect Dis Clin North Am* **17:**367–394.

110. Raz R, Gennesin Y, Wasser J, Stoler Z, Rosenfeld S, Rottensterich E, Stamm WE. 2000. Recurrent urinary tract infections in postmenopausal women. *Clin Infect Dis* **30:**152–156.

111. Petros PE, Ulmsten UI. 1990. An integral theory of female urinary incontinence. Experimental and clinical considerations. *Acta Obstet Gynecol Scand Suppl* **153:**7–31.

112. Petros PP, Ulmsten U. 1998. An anatomical classification–a new paradigm for management of female lower urinary tract dysfunction. *Eur J Obstet Gynecol Reprod Biol* **80:**87–94.

113. **DeLancey JO.** 1997. The pathophysiology of stress urinary incontinence in women and its implications for surgical treatment. *World J Urol* **15**:268–274.

114. **Nitti VW, TU LM, Gitlin J.** 1999. Diagnosing bladder outlet obstruction in women. *J Urol* **161**:1535–1540.

115. **Stika CS.** 2010. Atrophic vaginitis. *Dermatol Ther* **23**:514–522.

116. **Ronald A, Ludwig E.** 2001. Urinary tract infections in adults with diabetes. *Inter J Antimicrob Agents* **17**:287–292.

117. **Chen SL, Jackson SL, Boyko EJ.** 2009. Diabetes mellitus and urinary tract infection: epidemiology, pathogenesis and proposed studies in animal models. *J Urol* **182**(6 Suppl):S51–56.

118. **Hirji I, Guo Z, Andersson SW, Hammar N, Gomez-Caminero A.** 2012. Incidence of urinary tract infection among patients with type 2 diabetes in the UK General Practice Research Database (GPRD). *J Diabetes Complications* **26**:513–516.

119. **Boyko EJ, Fihn SD, Scholes D, Chen CL, Normand EH, Yarbro P.** 2002. Diabetes and the risk of acute urinary tract infection among postmenopausal women. *Diabetes Care* **25**:1778–1783.

120. **Boyko EJ, Fihn SD, Scholes D, Abraham L, Monsey B.** 2005. Risk of urinary tract infection and asymptomatic bacteriuria among diabetic and nondiabetic postmenopausal women. *Am J Epidemiol* **161**:557–564.

121. **Geerlings SE, Stolk RP, Camps MJ, Netten PM, Collet TJ, Hoepelman AI; Diabetes Women Asymptomatic Bacteriuria Utrecht Study Group.** 2000. Risk factors for symptomatic urinary tract infection in women with diabetes. *Diabetes Care* **23**:1737–1741.

122. **Geerlings SE, Meiland R, van Lith EC, Brouwer EC, Gaastra W, Hoepelman AIM.** 2002. Adherence of type 1-fimbriated *Escherichia coli* to uroepithelial cells: more in diabetic women than in control subjects. *Diabetes Care* **25**:1405–1409.

123. **Taganna J, de Boer AR, Wuhrer M, Bouckaert J.** 2011. Glycosylation changes as important factors for the susceptibility to urinary tract infection. *Biochem Soc Trans* **39**:349–354.

124. **Frimodt-Møller C.** 1980. Diabetic cystopathy: epidemiology and related disorders. *Ann Intern Med* **92**:318–321.

125. **Yoshimura N, Chancellor MB, Andersson KE, Christ GJ.** 2005. Recent advances in understanding the biology of diabetes-associated bladder complications and novel therapy. *BJU Int* **95**:733–738.

126. **Daneshgari F, Liu G, Birder L, Hanna-Mitchell AT, Chacko S.** 2009. Diabetic bladder dysfunction: current translational knowledge. *J Urol* **182**(6 Suppl):S18–26.

127. **Rosen DA, Hung CS, Kline KA, Hultgren SJ.** 2008. Streptozocin-induced diabetic mouse model of urinary tract infection. *Infect Immun* **76**:4290–4298.

128. **Romero R, Oyarzun E, Mazor M, Sirtori M, Hobbins JC, Bracken M.** 1989. Meta-analysis of the relationship between asymptomatic bacteriuria and preterm delivery/low birth weight. *Obstet Gynecol* **73**:576–582.

129. **Bolton M, Horvath DJ, Li B, Cortado H, Newsom D, White P, Paritida-Sanchez S, Justice SS.** 2012. Intrauterine growth restriction is a direct consequence of localized maternal uropathogenic *Escherichia coli* cystitis. *PLoS One* **7**:e33897. doi:10.1371/journal.pone.0033897

130. **Faúndes A, Brícola-Filho M, Pinto e Silva JL.** 1998. Dilatation of the urinary tract during pregnancy: proposal of a curve of maximal caliceal diameter by gestational age. *Am J Obstet Gynecol* **178**:1082–1086.

131. **Marchant DJ.** 1972. Effects of pregnancy and progestational agents on the urinary tract. *Am J Obstet Gynecol* **112**:487–501.

132. **Chang CP, McDill BW, Neilson JR, Joist HE, Epstein JA, Crabtree GR, Chen F.** 2004. Calcineurin is required in urinary tract mesenchyme for the development of the pyeloureteral peristaltic machinery. *J Clin Invest* **113**:1051–1058.

133. **Zhang PL, Peters CA, Rosen S.** 2000. Ureteropelvic junction obstruction: morphological and clinical studies. *Pediatr Nephrol* **14**:820–826.

134. **Murakumo M, Nonomura K, Yamashita T, Ushiki T, Abe K, Koyanagi T.** 1997. Structural changes of collagen components and diminution of nerves in congenital ureteropelvic junction obstruction. *J Urol* **157**:1963–1968.

135. **Yu J, Carroll TJ, McMahon AP.** 2002. Sonic hedgehog regulates proliferation and differentiation of mesenchymal cells in the mouse metanephric kidney. *Development* **129**:5301–5312.

136. **Kuwayama F, Miyazaki Y, Ichikawa I.** 2002. Embryogenesis of the congenital anomalies of the kidney and the urinary tract. *Nephrol Dial Transplant* **17**:45–47.

137. **Yiee JH, Johnson-Welch S, Baker LA, Wilcox DT.** 2010. Histologic differences between extrinsic and intrinsic ureteropelvic junction obstruction. *Urology* **76**:181–184.

138. **Roth CC, Hubanks JM, Bright BC, Heinlen JE, Donovan BO, Kropp BP, Frimberger D.**

2009. Occurrence of urinary tract infection in children with significant upper urinary tract obstruction. *Urology* **73**:74–78.

139. **Islek A, Güven AG, Koyun M, Akman S, Alimoglu E.** 2011. Probability of urinary tract infection in infants with ureteropelvic junction obstruction: is antibacterial prophylaxis really needed? *Pediatr Nephrol* **26**:1837–1841.

140. **Haylen BT, de Ridder D, Freeman RM, Swift SE, Berghmans B, Lee J, Monga A, Petri E, Rizk DE, Sand P, Schaer GN.** 2010. An International Urogynecological Association (IUGA)/ International Continence Society (ICS) joint report on the terminology for female pelvic floor dysfunction. *Neurourol Urodyn* **29**:4–20.

141. **Jørgensen TM, Djurhuus JC, Schrøder HD.** 1982. Idiopathic detrusor sphincter dyssynergia in neurologically normal patients with voiding abnormalities. *Eur Urol* **8**:107–110.

142. **Groutz A, Blaivas JG, Pies C, Sassone AM.** 2001. Learned voiding dysfunction (non- neurogenic, neurogenic bladder) among adults. *Neurourol Urodyn* **20**:259–268.

143. **Cameron AP, Gajewski JB.** 2009. Bladder outlet obstruction in painful bladder syndrome/interstitial cystitis. *Neurourol Urodyn* **28**:944–948.

144. **Pannek J, Einig EM, Einig W.** 2009. Clinical management of bladder dysfunction caused by sexual abuse. *Urol Int* **82**:420–425.

145. **Wood SK, Baez MA, Bhatnagar S, Valentino RJ.** 2009. Social stress-induced bladder dysfunction: potential role of corticotropin-releasing factor. *Am J Physiol Regul Integr Comp Physiol* **296**:R1671–1678.

146. **Carlson KV, Rome S, Nitti VW.** 2001. Dysfucntional voiding in women. *J Urol* **165**:143–148.

147. **Minardi D, d'Anzeo G, Parri G, Polito M Jr, Piergallina M, El Asmar Z, Marchetti M, Muzzonigro G.** 2010. The role of uroflowmetry biofeedback and biofeedback training of the pelvic floor muscles in the treatment of recurrent urinary tract infections in women with dysfunctional voiding: a randomized controlled prospective study. *Urology* **75**:1299–1304.

148. **Brucker BM, Fong E, Shah S, Kelly C, Rosenblum N, Nitti VW.** 2012. Urodynamic differences between dysfunctional voiding and primary bladder neck obstruction in women. *Urology* **80**:55–60.

149. **Leadbetter GW Jr, Leadbetter WF.** 1959. Diagnosis and treatment of congenital bladder- neck obstruction in children. *N Engl J Med* **260**:633–637.

150. **Kuo HC.** 2005. Videourodynamic characteristics and lower urinary tract symptoms of female bladder outlet obstruction. *Urology* **66**:1005–1009.

151. **Ahmed HU, Shergill IS, Arya M, Shah PJ.** 2006. Management of detrusor–external sphincter dyssynergia. *Nat Clin Pract Urol* **3**:368–380.

152. **Kim YH, Kattan MW, Boone TB.** 1998. Bladder leak point pressure: the measure for sphincterotomy success in spinal cord injured patients with external detrusor-sphincter dyssynergia. *J Urol* **159**:493–497.

153. **Abdul-Rahman A, Ismail S, Hamid R, Shah J.** 2010. A 20-year follow-up of the mesh wallstent in the treatment of detrusor external sphincter dyssynergia in patients with spinal cord injury. *BJU Int* **106**:1510–1513.

154. **Chen SL, Bih LI, Huang YH, Tsai SJ, Lin TB, Kao YL.** 2008. Effect of single botulinum toxin A injection to the external urethral sphincter for treating detrusor external sphincter dyssynergia in spinal cord injury. *J Rehabil Med* **40**:744–748.

155. **Zaffanello M, Malerba G, Cataldi L, Antoniazzi F, Franchini M, Monti E, Fanos V.** 2010. Genetic risk for recurrent urinary tract infections in humans: A systematic review. *J Biomed Biotechnol* **2010**:321082.

156. **Hung CS, Dodson KW, Hultgren SJ.** 2009. A murine model of urinary tract infection. *Nat Protoc* **4**:1230–1243.

157. **Bates JM, Raffi HM, Prasadan K, Mascarenhas R, Laszik Z, Maeda N, Hultgren SJ, Kumar S.** 2004. Tamm-Horsfall protein knockout mice are more prone to urinary tract infection: rapid communication. *Kidney Int* **65**:791–797.

158. **Mo L, Zhu XH, Huang HY, Shapiro E, Hasty DL, Wu XR.** 2004. Ablation of the Tamm-Horsfall protein gene increases susceptibility of mice to bladder colonization by type 1-fimbriated *Escherichia coli. Am J Physiol Renal Physiol* **286**:F795–802.

159. **Chromek M, Slamová Z, Bergman P, Kovács L, Podracká L, Ehrén I, Hökfelt T, Gudmundsson GH, Gallo RL, Agerberth B, Brauner A.** 2006. The antimicrobial peptide cathelicidin protects the urinary tract against invasive bacterial infection. *Nat Med* **12**:636–641.

160. **Morrison G, Kilanowski F, Davidson D, Dorin J.** 2002. Characterization of the mouse beta defensin 1, Defb1, mutant mouse model. *Infect Immun* **70**:3053–3060.

161. **Zhang D, Zhang G, Hayden MS, Greenblatt MB, Bussey C, Flavell RA, Ghosh S.** 2004. A toll-like receptor that prevents infection by uropathogenic bacteria. *Science* **303**:1522–1526.

162. **Diamond DA and Mattoo TK.** 2012. Endoscopic treatment of primary vesicoureteral reflux. *N Engl J Med* **366:**1218–1226.

163. **Standring S.** 2008. Kidneys and ureter, p 1230–1231. *Gray's Anatomy.* 40th ed. Churchill Livingstone Elsevier, Philadelphia.

Clinical Presentations and Epidemiology of Urinary Tract Infections

2

SUZANNE E. GEERLINGS[1]

CLINICAL SYNDROMES AND DEFINITIONS

Urinary tract infection (UTI) is one of the most common bacterial infections. Bacteria live around the urethra and colonize the bladder, but are washed out during micturition. The shorter distance to the bladder in women (as compared to men) makes it easier for bacterial colonizers to reach the bladder. Furthermore, the urethral opening in women is close to the rectum. Urogenital manipulations associated with daily living or medical interventions facilitate the movement of bacteria to the urethra (1).

The diagnosis of a UTI can be made by a combination of symptoms and a positive urine analysis or culture. In most patient groups, the threshold for bacteriuria is considered to be 1,000 colony-forming units (cfu)/ml, based on studies correlating midstream-urine specimens with catheterized collection to demonstrate bladder bacteriuria. However, up to 20% of women with classical urinary symptoms can have negative cultures, depending on the cut-off value used (1).

The differentiation between uncomplicated and complicated UTIs has implications for therapy because the risks of complications or treatment failure are increased for patients with a complicated UTI. In general, the following

[1]Department of Internal Medicine, Division of Infectious Diseases, Center for Infection and Immunity Amsterdam (CINIMA), Academic Medical Center, 1105 AZ Amsterdam, The Netherlands.
Urinary Tract Infections: Molecular Pathogenesis and Clinical Management, 2nd Edition
Edited by Matthew A. Mulvey, David J. Klumpp, and Ann E. Stapleton
© 2017 American Society for Microbiology, Washington, DC
doi:10.1128/microbiolspec.UTI-0002-2012

definitions are used: an uncomplicated UTI is an episode of cystitis in a woman who is not pregnant, is not immunocompromised, has no anatomical and functional abnormalities of the urogenital tract, and does not exhibit signs of tissue invasion and systemic infection. All UTIs that are not uncomplicated are considered to be complicated UTIs (2). Therefore, episodes of acute cystitis occurring in healthy nonpregnant women with no history suggestive of an abnormal urinary tract are generally classified as uncomplicated, whereas all others are classified as complicated (3). This distinction has been used to guide the choice and duration of antimicrobial treatment, with broader-spectrum agents and longer courses of treatment often recommended for persons with complicated UTIs. However, this classification scheme does not account for the diversity of complicated UTIs (3). A classification scheme that stratifies patients with UTI into multiple, homogeneous categories has been proposed but is not (yet) routinely used in practice (2, 3).

Another differentiation of UTIs is between community- and hospital-acquired UTIs. UTIs in patients acquired within the hospital or hospitalized for treatment are generally complicated UTIs. More often uropathogens other than *Escherichia coli* are the causative microorganisms. Furthermore, more-resistant pathogens are cultured compared to community-acquired UTI. Earlier antimicrobial treatment remains the strongest predictor for resistant causative microorganisms (4). Epidemics of nosocomial UTI have been described by recognizing unusual antibiotic-resistance profiles. Most are caused by transmission of outbreak strains between patients on the hands of hospital staff. An estimated 80% of the hospital-acquired UTIs are associated with catheters (1). In a recent study, patients with healthcare-associated (not hospital) UTI were older, had more co-morbidities, and had received previous antimicrobial treatment more frequently compared to patients with community-acquired UTI. Extended-spectrum beta-lactamase (ESBL) *E. coli* and *Pseudomonas aeruginosa* infections were also more frequently cultured (5). Therefore, patients from nursing homes can be considered as having hospital-acquired UTIs.

Asymptomatic Bacteriuria

Asymptomatic bacteriuria (ASB), which is defined as the presence of a positive urine culture with at least 10^5 cfu/ml collected from a patient without symptoms of a UTI, is a common, but usually benign, phenomenon, especially in women (6). Both host and bacterial factors influence the probability that ASB will resolve spontaneously or progress to symptomatic UTI.

A Swedish study among 116 schoolgirls with ASB showed that at baseline renal parenchymal reduction was found in 10.3%, while reflux was found in 20.7%, but only 30% of the 116 patients had a history referable to an earlier UTI. A 3-year follow-up of these 116 schoolgirls with ASB (treated or untreated) showed that the risk of developing renal damage as a result of ASB in a schoolgirl with a roentgenographically normal urinary tract seemed to be small (7).

Long-term follow-up studies have shown that ASB with *E. coli* is not associated with a decline in renal function or the development of end-stage renal failure in a population of generally healthy adult women (8). Although *E. coli* bacteriuria may increase the risk of future hypertension, the pathogenesis is not fully understood (8, 9).

Following the guidelines, screening and treatment is only recommended for pregnant women, or for patients prior to selected invasive genitourinary procedures (6). Clinical trials in spinal cord-injury patients, diabetic women (10), patients with indwelling urethral catheters, and elderly nursing-home residents have consistently found no benefits with treatment of ASB. Negative outcomes with antimicrobial treatment do occur, including adverse drug effects and reinfection with organisms of increasing resistance (11).

In renal-transplant patients, no differences in renal-function prognosis between patients with and without ASB following kidney transplantation could be demonstrated. However, the incidence of pyelonephritis was much higher in the group of patients with ASB. Therefore, screening protocols may be beneficial in this patient group (12).

Since nearly all studies on ASB are performed in women, it is not possible to draw conclusions about the association between ASB, the incidence of UTIs, or the development of renal-function decline in men.

Uncomplicated Urinary Tract Infection/Cystitis

Cystitis (infection of the bladder or lower UTI) has the following symptoms: dysuria with or without frequency, urgency, suprapubic pain, or hematuria. Women with a suspected uncomplicated UTI mostly present to primary care. The most-common symptom is urinary frequency. Dysuria is also common with urethritis or vaginitis, but cystitis is more likely when symptoms include frequency, urgency, or hematuria and when the onset of symptoms is sudden or severe, without the presence of vaginal irritation and discharge (13, 14). Many women also feel extremely unwell and have restricted physical activity. Patients without adequate treatment, in other words, those not given antibiotics, and those with antibiotic-resistant organisms, complain of at least one symptom that is moderately severe or worse lasting for five days (15).

The probability of cystitis is greater than 50% in women with any symptoms of UTI and greater than 90% in women who have dysuria and frequency without vaginal discharge or irritation (3, 13). Therefore, additional urine analysis is not always needed in this patient group.

Acute uncomplicated cystitis rarely progresses to severe disease, even if untreated. A trial with nitrofurantoin and placebo showed that the result for combined symptomatic improvement and cure after three days was present in 27 of the 35 women in the nitrofurantoin group, but also in 19 of the 35 patients in the placebo group. In the same study, only one case in the placebo group (1/38 = 12.6%) progressed to pyelonephritis (16). Therefore, the primary goal of treatment is to ameliorate symptoms. After start of treatment, the symptoms of a lower UTI resolved quickly; the mean duration of urinary frequency was 3.46 days, for hematuria 1.88 days, and for urgency 3.6 days (15).

In 2011, the Infectious Diseases Society of America (IDSA) updated its guidelines for antimicrobial treatment in acute uncomplicated cystitis in women. It is interesting, in view of the worldwide problem of increasing antimicrobial resistance, that these guidelines recommend that ecological adverse effects of an antimicrobial agent (selection for antimicrobial-resistant organisms) must also be considered together with efficacy in selecting the choice of the antimicrobial agent and duration of therapy (17).

Complicated Urinary Tract Infections

It is possible to differentiate between UTI with systemic symptoms and UTI in a host, which has an increased risk to develop complications (complicated host). Systemic symptoms can be noticed by signs of tissue invasion, like fever, flank pain, and delirium. These UTIs can be called febrile UTI, because it is difficult to differentiate between urosepsis, pyelonephritis, and prostatitis (18). However, all of these syndromes require a therapeutic antibiotic drug level in both tissue and urine.

A complicated host is defined as one that has an increased risk for complications of the UTI, to which the following groups belong: men, pregnant women, immunocompromised patients, or those who have an anatomical or functional abnormality of the urogenital tract (e.g., renal stones, urinary catheter, neurogenic bladder, spinal cord injury, renal transplant). Treatment of a UTI (cystitis) in a complicated host requires a therapeutic antibiotic drug level in the urine only, but generally a longer treatment duration is recommended. These different 'complicated' hosts are described below in the section Special patient groups.

Urinary Tract Infections with Systemic Symptoms or Febrile UTI

Acute pyelonephritis

Typical clinical manifestations suggestive of pyelonephritis (infections of the kidney or upper UTI) are fever (temperature >38°C) and chills, mental confusion as a sign of delirium, flank pain, costovertebral-angle tenderness, and nausea or vomiting (13, 14). The two routes by which bacteria can invade and spread within the urinary tract are the ascending route and the hematogenous route. There is no clear evidence for a lymphatic route. In practice, nearly all upper UTIs are caused by the ascending route from the bladder to the kidney. Although some patients can remember recent symptoms of cystitis or these symptoms are still present, this is often not the case. It should be recognized that symptoms may vary greatly. Flank tenderness may be more intense when an obstructive disease is present. Normal kidney function can be present, but progressive destruction of the kidney may give rise to clinical manifestations of renal insufficiency.

Prostatitis

Prostatitis ranges from a straightforward clinical entity in its acute form to a complex, debilitating condition when chronic. Diagnosis of acute and chronic bacterial prostatitis is primarily based on history, physical examination, urine culture, and urine-specimen testing. Patients with acute prostatitis complain of symptoms associated with lower UTI, such as frequency and dysuria. They may also experience lower urinary-tract obstruction due to prostatic edema. Therefore, the differential diagnosis of prostatitis includes (amongst others): acute cystitis, benign prostatic hyperplasia, urinary-tract stones, and bladder cancer. On physical examination, patients may have a high temperature and lower abdominal or suprapubic discomfort due to bladder infection. The rectal examination shows an exquisitely tender prostate on palpation (19, 20), but a normal rectal examination cannot exclude this diagnosis (21).

Urosepsis

Urosepsis is defined as sepsis caused by infection of the urinary tract. In urosepsis (as in other types of sepsis), the severity of sepsis mainly depends on the host response. The underlying UTI is almost exclusively a complicated one with involvement of parenchymatous urogenital organs (e.g., kidneys, prostate). The leading cause for developing an uroseptic shock in urological patients is urinary obstruction. It is reported that 17% of patients develop urosepsis after urological interventions (22).

SPECIAL PATIENT GROUPS

Children

UTI is one of the most common bacterial infections in children. UTI in young children and infants are often presented with nonspecific clinical signs, such as fever, irritability, and vomiting, making the diagnosis difficult. Urine collection and interpretation of urine tests in children is not easy and does not always lead to unequivocal confirmation of the diagnosis. Failure to diagnose UTI or delaying treatment of a UTI may result in a clinical deterioration with additional long-term renal damage. Renal anatomical abnormalities that are frequently associated with UTIs are vesicoureteric reflux, double systems, hydronephrosis, hydroureter, and urethral obstructions (23, 24).

In a cohort study, encopresis was found to be significantly associated with recurrent UTI (25). Therefore, dysfunctional elimination syndromes and constipation should be treated in infants and children who have had a UTI.

Concerning treatment duration, in 10 randomized controlled trials with 625 children with cystitis (aged 3 months to 18 years), no significant differences were found in persistence of clinical symptoms or bacteriuria, recurrent UTI, compliance with medication, or development of bacterial resistance between short duration (2–4 days) oral antibiotic treatment and the earlier-recommended treatment duration of 7–14 days (26).

Men

As a rule, a UTI in a man is considered to be a complicated UTI because the prostate is often involved (21). However, in general, men with a bacterial UTI can be separated into three groups, each with its own therapy:

1. Young men with a UTI without systemic symptoms, where the patient's medical history and physical examination do not suggest a causative factor. The UTIs in this group can be considered uncomplicated UTIs, but are very uncommon (27).
2. Men with a UTI and systemic symptoms or with a medical history and physical examination that suggest a causative factor. These UTIs must be considered complicated UTIs. The systemic symptoms indicate invasion of the tissue in the prostate (acute bacterial prostatitis) or the kidney (pyelonephritis) (28).
3. Men with complaints that fit a chronic bacterial prostatitis. In these cases, it is advised to wait for the results of the culture. For men with a chronic bacterial prostatitis, a fluoroquinolone is recommended as first choice because these drugs are more effective than trimethoprim/sulfamethoxazole (28, 29). Since it is not an acute illness, the results of the culture (urine, if necessary, after massage of the prostate or semen) can be awaited before therapy is initiated.

Pregnant Women

ASB occurs in 2 to 10% of pregnant women (6). ASB during pregnancy can lead to serious complications for both mother and child. The incidence of ASB is similar in both pregnant and nonpregnant women (30). However, pregnant women with ASB more often develop pyelonephritis, probably due to the anatomic and physiologic changes that occur during pregnancy, which may facilitate bacterial growth and the ascent of bacteria to the kidneys (31). If left untreated, 20 to 40% of pregnant women with ASB will develop pyelonephritis (30, 32, 33). Furthermore, during pregnancy there is an elevated risk of a more severe course of a UTI with adverse consequences for mother and child (34).

Other possible adverse effects, such as preterm delivery and delivering a low-birthweight infant, are less well established. Although preterm delivery is the main cause of neonatal mortality and morbidity worldwide, the causal mechanisms remain unknown. One of the hypotheses is that endotoxins released by bacteria cause uterine contractions leading to preterm delivery.

Antibiotics are effective again ASB during pregnancy and lower the incidence of pyelonephritis as well as prematurity and dysmaturity (35, 36).

In view of the lack of reported teratogenic effects and the resistance percentages of the causative microorganisms, the beta-lactam antibiotics are a good choice for the treatment of a UTI during pregnancy. Amoxicillin-clavulanic acid or nitrofurantoin are first-choice drugs for the treatment of cystitis during pregnancy (however, nitrofurantoin must not be used just before delivery). In view of the high resistance percentage of the uropathogens for amoxicillin, this drug is not suitable for empirical treatment. A 2nd or 3rd generation cephalosporin is the drug of first choice and amoxicillin-clavulanic acid is second choice for treatment of a pyelonephritis during pregnancy (34).

It is a sign of maternal colonization with group B streptococcus (GBS) whenever a GBS is found in the urine culture. Intravenous-antibiotic treatment of the mother during delivery reduces the number of neonatal infections with GBS (37). Based on the literature, it is recommended that pregnant and nonpregnant women with cystitis should be treated for 3–7 days (36). In general, it is recommended to hospitalize a pregnant woman with a pyelonephritis and to administer antibiotics intravenously. After a fever-free period of 24–48 hours, oral antibiotics can be given; the total duration of therapy must be at least 10 days (38).

Patients with a Urinary Catheter

Catheter-associated (CA) infection refers to infection occurring in a person whose urinary tract is currently catheterized or has been catheterized within the past 48 hours. UTI refers to significant bacteriuria in a patient with symptoms or signs attributable to the urinary tract and no alternative source. Bacteriuria is a nonspecific term that refers to UTI and ASB combined. In the urinary-catheter literature, CA-bacteriuria is mainly comprised of CA-ASB (39).

Indwelling urinary catheters are widely used in hospitalized patients for patients with urinary retention and for frequent monitoring of urine output in critically ill patients. Most patients are catheterized for 2–4 days, but many have a catheter inserted for a longer duration as, for example, spinal cord-injury patients. Unfortunately, the use of indwelling catheters is not without risks. Many catheterized patients develop bacteriuria, with an incidence of 3 to 10% per day (40). Duration of catheterization is the most important risk factor for the development of CA-bacteriuria; almost all patients with long-term catheterization (>1 month) will have bacteriuria. Although most patients with bacteriuria are asymptomatic, symptoms of UTI will develop in some patients. UTI is one of the most common hospital-acquired infec-

tions, and 80% of these are associated with the use of indwelling catheters (41). A surveillance study in the Netherlands found that 1.2% of all hospitalized patients had a catheter-associated UTI (42). In a prospective study, the incidence of UTI was 15.6% in about 1,500 patients catheterized for at least 24 hours (43). The proportion of patients with catheter-related bacteriuria in whom symptomatic UTI and bacteremia will develop was estimated through quantitative synthesis of previous reports. Of patients who had indwelling catheters for 2–10 days, bacteriuria was expected to develop in 26%, bacteriuria and symptoms of a UTI would develop in 24%, and bacteremia from a urinary-tract source would develop in 3.6%. Each episode of symptomatic UTI infection was expected to cost an additional $676 US, and catheter-related bacteriuria was at least $2,836 US. Given the clinical and economic burden of catheter-related UTI, all health care workers should try to reduce this common complication (44).

In addition to bacteriuria and CA-UTI, long-term catheterization can also lead to the following complications: bacteremia, catheter obstruction, renal and bladder stone formation, incontinence, and, with prolonged use, bladder cancer (39, 45).

The insertion of an indwelling catheter increases the susceptibility of a patient to UTIs, as it provides easier access of microorganisms to the urinary tract. Most of these uropathogens are fecal or skin bacteria from a patient's own native or transitory microflora. Bacteria can enter the bladder at the time of catheter insertion, through the catheter lumen, or along the catheter-urethral interface. Most microorganisms that cause CA-UTI enter the bladder extraluminally by ascending along the catheter-mucosa interface and are primarily endogenous. Microorganisms can also enter the bladder intraluminally, by contamination of the collecting tube or drainage bag. These organisms are often exogenous, derived from cross-contamination of organisms on the hands of healthcare personnel (39, 40, 46).

Indwelling catheters facilitate colonization of uropathogens by enhancing microbial adhesion. The catheter provides an attachment surface for bacterial adhesins that recognize host-cell receptors on the surfaces of the host cell or catheter. In addition, urinary catheters may damage the uroepithelial mucosa, which leads to exposure of new binding sites for bacterial adhesins. Once attached to the catheter surface or uroepithelium, bacteria undergo phenotypical changes, replicate, and form microcolonies that eventually mature into biofilms. These biofilms protect uropathogens from antimicrobials and the host-immune response, and migrate over the catheter surface to the bladder within 1–3 days. Bacteriuria in patients with short-term catheterization is commonly caused by a single organism, mostly *E. coli*, while infections in long-term catheterization are polymicrobial (39, 40, 46). Bacteriuria contains a large reservoir of antimicrobial-resistant organisms and can be a source of cross-infection. The most effective way to reduce CA-UTI is to avoid urinary catheterization (39).

The guidelines of the IDSA define CA-UTI as the presence of symptoms: new-onset or worsening of fever, rigors, altered mental status, malaise, or lethargy of no other identified cause; flank pain; costovertebral angle tenderness; acute hematuria; or pelvic discomfort, and more than 1,000 cfu/ml of one or more bacterial species (39).

Results of studies included in a Cochrane review on short-term urinary-catheter use in female patients with abdominal surgery and a urethral catheter for 24 hours, show weak evidence that antibiotic prophylaxis, compared to giving antibiotics when clinically indicated, reduced the rate of symptomatic UTI [relative risk (RR) 0.20 (95% confidence interval [CI] 0.06–0.66)]. There was also limited evidence that prophylactic antibiotics reduced bacteriuria in nonsurgical patients (47).

Regarding the question as to whether antibiotic prophylaxis is better than giving antibiotics when clinically indicated (i.e., having a symptomatic UTI), the available evidence is too limited to be a basis for clinical practice (39). For patients using intermittent catheterization the data were inconclusive. For patients using indwelling urethral catheterization, only a single crossover trial with 34 elderly inpatients investigated this issue and results showed fewer episodes of symptomatic UTI in the prophylaxis (norfloxacin) group (48). For patients using intermittent catheterization, the limited evidence suggested that antibiotic prophylaxis reduces the number of episodes of bacteriuria (asymptomatic and symptomatic). For patients using urethral catheterization, no data were available (47). Based on these observations, the contradictory results, and the concerns about rising antimicrobial resistance, prophylactic antimicrobials are not routinely recommended for catheter placement, removal, or replacement. This recommendation is also supported by the low rate of serious complications in the large number of patients undergoing long-term intermittent catheterization with a clean technique in the setting of chronic bacteriuria (39).

CA-UTIs are often polymicrobial and caused by multiple-drug-resistant uropathogens. Urine cultures are recommended prior to treatment in order to confirm that an empiric regimen provides appropriate coverage and to allow tailoring of the regimen based on antimicrobial susceptibility data (39).

Only a few small studies have investigated which causative uropathogens are present in patients with CA-UTIs. The numbers are too small to translate the results into a strong recommendation for empirical treatment. In patients on long-term catheterization, empirical treatment with fluoroquinolones or gentamicin may be warranted to cover less-common microorganisms such as *Serratia*, *Providencia*, and *Acinetobacter*. However, a study from the Netherlands demonstrated that patients with a urinary catheter are at increased risk to have a fluoroquinolone-resistant microorganism (4), which only leaves the toxic aminoglycosides for empirical treatment in this patient group. Earlier antimicrobial treatment remains the strongest predictor for resistant causative microorganisms (4). Therefore, in a patient with a catheter who has only local symptoms and exhibits no signs of a systemic infection, we recommend to wait for the results of the cultures.

It is desirable to limit the duration of treatment, especially for milder infections and infections that respond promptly to treatment, to reduce the selection pressure for drug-resistant flora, especially in patients on long-term catheterization. Therefore, 5–7 days is the recommended duration of antimicrobial treatment for patients with CA-UTI who have prompt resolution of symptoms, and 10–14 days is recommended in those with a delayed response, irrespective of whether or not the patient remains catheterized (39).

Diabetes mellitus

Diabetic patients have an increased risk for UTI (49, 50). A recent study in primary care patients from the Netherlands demonstrated that relapses and reinfections were reported in 7.1% and 15.9%, respectively, of women with diabetes mellitus (DM) versus 2.0% and 4.1%, respectively, of women without DM. There was a higher risk of recurrent UTI in women with DM compared to women without DM (odds ratio [OR] 2.0; 95% CI 1.4–2.9). Women who had had DM for at least 5 years (OR 2.9; 95% CI 1.9–4.4) or who had retinopathy (OR 4.1; 95% CI 1.9–9.1) were at risk of recurrent UTI (51). This increased recurrence rate was confirmed in one study (52) but not in another (53). In an American study in women with DM type 1 it was found that sexual activity, rather than measures of diabetes control and complications, was the main risk factor for UTI (54).

Diabetic patients more often develop complications, such as bacteremia (55) and a longer hospitalization (50, 56), of their UTI compared to nondiabetic patients. For this reason, cystitis in a patient with DM is considered a complicated UTI.

No prospective trial has investigated the optimal treatment (agent and duration) in these patients. Concerning the recurrence rate of UTI in diabetic compared to nondiabetic women, two studies using Dutch registration databases containing pharmacy-dispensing data from two different time periods show contradictory results (57, 58). In the largest study (58), the prescriptions of 10,366 women with diabetes and 200,258 women without diabetes were compared. Women with diabetes more often received a long treatment, but still had a higher recurrence rate of UTI, compared with those without diabetes.

It is reported that ASB in women with DM is benign and that 20% of diabetic subjects with ASB remained bacteriuric with the original infecting organism throughout the period of observation. Women infected with gram-negative organisms were more likely to have persistent bacteriuria. Many women with resolution of initial bacteriuria, with or without antibiotics, became bacteriuric again during follow-up. Furthermore, ASB in women with DM does not result in renal function decline (59). However, more women with ASB will develop a symptomatic UTI compared to those without (60). Also, in another study with male and female patients with DM type 1 and 2, the presence of ASB was associated with an increased risk of hospitalization for urosepsis (61).

In the above-mentioned prospective study (59), because no evidence was found that ASB alone can lead to a decline in renal function (in women with type 1 and type 2 DM), it is unlikely that treatment of ASB will lead to a decrease in the incidence of diabetic nephropathy. This is in accordance with a study on women with DM and with ASB, in which a comparison was made between women who received antibiotic therapy and women who received placebo. In that study, no difference was seen in serum creatinine levels after a mean follow-up of 2 years (10).

Treatment of ASB in patients with DM is not needed, because in these women ASB does not result in renal function decline, and most of these women do not develop a symptomatic UTI. Therefore, screening for ASB is not indicated in these patients. This is in accordance with the IDSA guideline for the diagnosis and treatment of ASB in adults (6).

RISK FACTORS

Risk factors for uncomplicated and recurrent cases of lower and upper UTI include sexual intercourse, use of spermicides, previous UTI, a new sex partner, and a history of UTI in a first-degree female relative (3, 62–65).

Case-control studies have shown no significant associations between recurrent UTI and precoital- or postcoital-voiding patterns, daily beverage consumption, frequency of urination, delayed-voiding habits, wiping patterns, tampon use, douching, use of hot tubs, type of underwear, or body-mass index (65); however, at least some of this absence of findings might reflect a misclassification of behaviors (3, 66).

A genetic predisposition to recurrent UTI is suggested by the strong association between a history of UTI in one or more first-degree female relatives and an increased risk of recurrent UTIs (63). Certain toll-like-receptor polymorphisms and other genetic variations, particularly those affecting the immune response, are associated with an increased risk for UTI (66, 67). Other studies have shown that acute pyelonephritis was more common in the family members of children with a history of acute pyelonephritis (15%) than in relatives of control subjects (3%) (68).

RECURRENT URINARY TRACT INFECTIONS

Recurrent UTI is a common health care problem and is defined in the literature by three episodes of UTI in the last 12 months or two episodes in the last 6 months. About 20 to 30% of women who have a UTI will have a recurrent UTI (69, 70).

Looking at the causative microorganism, it was recently demonstrated that uropathogenic *E. coli* adhere, invade, and replicate within the murine bladder urothelium to form intracellular bacterial communities. The presence of exfoliated intracellular bacterial communities and filamentous bacteria in the urine of women with acute cystitis suggests that this pathogenic pathway, characterized in the murine model, may occur in humans. The findings support the occurrence of an intracellular bacterial niche in some women with cystitis that may have important implications for UTI recurrence and treatment (71).

In general, in men and postmenopausal women, it is recommended to exclude anatomical or functional abnormalities of the urogenital tract as a cause of recurrent UTI. In premenopausal women the yield of most diagnostic procedures is low (72).

There are four patterns of response of bacteriuria to therapy: cure, bacteriologic persistence, bacteriologic relapse, or reinfection. Bacteriologic persistence is the persistence of bacteriuria with the same microorganism after 48 hours of treatment. Relapse is an infection with the same microorganism that caused initial infection and usually occurs within 1–2 weeks after the cessation of treatment. A relapse indicates that the infecting organism has persisted in the urinary tract. Reinfection is an infection after sterilization of the urine. Most of the time there is a change in bacterial species. Reinfection can be defined as a 'true' recurrence. Both persistence and relapse may be related to inadequate treatment. It is very important to determine whether recurrent UTIs are relapses or reinfections and to make a differentiation between these patterns, since this has treatment consequences. In a persistent UTI the cause must be evaluated. In a relapse of the UTI, the treatment can be given for a longer period.

The first consideration in prevention is to address modifiable behavioral practices. Other effective strategies are generally considered as antimicrobial or nonantimicrobial. Low-dose antimicrobial therapy remains an effective intervention to manage frequent, recurrent, acute, and uncomplicated UTI. The antimicrobial may be given as continuous daily or every-other-day therapy, usually at bedtime, or as postcoital prophylaxis (69). Topical vaginal estrogen is a potential intervention to decrease the number of recurrent episodes for postmenopausal women (73).

EPIDEMIOLOGY

The self-reported annual incidence of UTI in women is 12%, and by the age of 32 years, 50% of all women report having had at least one UTI (3, 74). In a study of young college women, the incidence of cystitis (lower UTI) was 0.70 episodes per person-year (62). Among young healthy women with cystitis, the infection recurs in 25% of women within 6 months after the first UTI. Although the risk of second UTI is strongly influenced by sexual behavior, women with a first UTI caused by *E. coli* are more likely than those with a non-*E. coli* first UTI to have a second UTI within 6 months (75). In a population-based study with 1,017 postmenopausal women, the incidence of cystitis was 0.07 episodes per person-year (76).

In general, about 50 to 70% of women will have a UTI sometime during their lifetime, and 20 to 30% of women who have a UTI will have a recurrent UTI (69, 70). In certain periods of life (childhood, honeymoon, pregnancy, elderly), an increased incidence of UTI has been described (Fig. 1).

Acute pyelonephritis (upper UTI) is much less common than cystitis. Incidence of pyelonephritis is highest among young women, followed by infants and the elderly population. The annual rates of outpatient pyelonephritis in women and men were 12–13 and

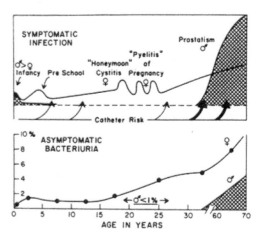

FIGURE 1 **Overview of the incidence of symptomatic UTI and the prevalence of asymptomatic bacteriuria according to age and sex (curves, females; hatched areas, males) (79).**

2–3 cases per 10,000, respectively (77). UTIs account for approximately 5 to 7% of all cases of severe sepsis (22).

The prevalence of complaints compatible with chronic prostatitis/chronic pelvic pain syndrome (CP/CPPS) is high with an overall rate of 8.2%, with prevalence ranging from 2.2 to 9.7%. Two studies suggest that about one-third of men reporting prostatitis symptoms had resolution after 1 year (19).

The prevalence of ASB depends on the patient group. Different studies report a prevalence of approximately 1 to 5% among healthy young women, increasing to over 20% in the elderly and 12 to 26% in women with DM. In a systematic review and meta-analysis, in which 22 studies were included, ASB was present in 439 of 3,579 (12.2%) patients with DM and in 121 of 2,702 (4.5%) healthy control subjects. The point prevalence of ASB was higher in both women (14.2% DM vs 5.1% controls) and men (2.3 % DM vs 0.8% controls) (78).

The high incidence of UTI results in considerable health care costs. The estimated annual direct and indirect cost of UTI in the USA in 1995 was $1.6 billion (equivalent to $2.3 billion in 2010) (1). A nosocomial

UTI necessitates one extra hospital day per patient, resulting in almost 1 million extra hospital days each year in the USA (1).

SUMMARY

UTI is one of the most common bacterial infections and the incidence in women is much higher than in men. The diagnosis of a UTI can be made based on a combination of symptoms and a positive urine analysis or culture. Most UTIs are uncomplicated UTIs, defined as cystitis in a woman who is not pregnant, is not immunocompromised, has no anatomical and functional abnormalities of the urogenital tract, and does not exhibit signs of tissue invasion and systemic infection. All UTIs that are not uncomplicated are considered to be complicated UTIs. Differentiation between uncomplicated and complicated UTIs has implications for therapy, because the risks of complications or treatment failure are increased for patients with a complicated UTI. ASB is defined as the presence of a positive urine culture collected from a patient without symptoms of a UTI. Concerning complicated UTI, it is possible to make a differentiation between UTI with systemic symptoms (febrile UTI) and UTI in a host, who has risk factors to carry resistant microorganisms and/or with difficulty in successfully treating the infection. Febrile UTIs are urosepsis, pyelonephritis, and prostatitis. A complicated host is defined as having an increased risk of complications including: men, pregnant women, immunocompromised patients, or those who have an anatomical or functional abnormality of the urogenital tract (e.g., spinal cord-injury patients, renal stones, urinary catheter).

CITATION

Geerlings SE. 2016. Clinical presentations and epidemiology of urinary tract infections. Microbiol Spectrum 4(5):UTI-0002-2012.

REFERENCES

1. **Foxman B.** 2010. The epidemiology of urinary tract infection. *Nat Rev Urol* **7:**653–660.
2. **Johansen TE, Botto H, Cek M, Grabe M, Tenke P, Wagenlehner FM, Naber KG.** 2011. Critical review of current definitions of urinary tract infections and proposal of an EAU/ESIU classification system. *Int J Antimicrob Agents* **38**(Suppl):64–70.
3. **Hooton TM.** 2012. Clinical practice. Uncomplicated urinary tract infection. *N Eng J Med* **366:**1028–1037.
4. **van der Starre WE, van Nieuwkoop C, Paltansing S, van't Wout JW, Groeneveld GH, Becker MJ, Koster T, Wattel-Louis GH, Delfos NM, Ablij HC, Leyten EM, Blom JW, van Dissel JT.** 2011. Risk factors for fluoroquinolone-resistant *Escherichia coli* in adults with community-onset febrile urinary tract infection. *J Antimicrob Chemother* **66:**650–656.
5. **Aguilar-Duran S, Horcajada JP, Sorli L, Montero M, Salvadó M, Grau S, Gómez J, Knobel H.** 2012. Community-onset healthcare-related urinary tract infections: comparison with community and hospital-acquired urinary tract infections. *J Infect* **64:**478–483.
6. **Nicolle LE, Bradley S, Colgan R, Rice JC, Schaeffer A, Hooton TM; Infectious Diseases Society of America; American Society of Nephrology; American Geriatric Society.** 2005. Infectious Diseases Society of America guidelines for the diagnosis and treatment of asymptomatic bacteriuria in adults. *Clin Infect Dis* **40:**643–654.
7. **Lindberg U, Claesson I, Hanson LA, Jodal U.** 1978. Asymptomatic bacteriuria in schoolgirls. VIII. Clinical course during a 3-year follow-up. *J Pediatr* **92:**194–199.
8. **Meiland R, Stolk RP, Geerlings SE, Peeters PH, Grobbee DE, Coenjaerts FE, Brouwer EC, Hoepelman AI.** 2007. Association between *Escherichia coli* bacteriuria and renal function in women: long-term follow-up. *Arch Intern Med* **167:**253–257.
9. **Meiland R, Geerlings SE, Stolk RP, Hoepelman AI, Peeters PH, Coenjaerts FE, Grobbee DE.** 2010. *Escherichia coli* bacteriuria in female adults is associated with the development of hypertension. *Int J Infect Dis* **14:**e304–307.
10. **Harding GK, Zhanel GG, Nicolle LE, Cheang M; Manitoba Diabetes Urinary Tract Infection Study Group.** 2002. Antimicrobial treatment in diabetic women with asymptomatic bacteriuria. *N Engl J Med* **347:**1576–1583.

11. **Nicolle LE.** 2006. Asymptomatic bacteriuria: review and discussion of the IDSA guidelines. *Int J Antimicrob Agents* **28**(Suppl 1):S42–S48.

12. **Fiorante S, López-Medrano F, Lizasoain M, Lalueza A, Juan RS, Andrés A, Otero JR, Morales JM, Aguado JM.** 2010. Systematic screening and treatment of asymptomatic bacteriuria in renal transplant recipients. *Kidney Int* **78**:774–781.

13. **Bent S, Nallamothu BK, Simel DL, Fihn SD, Saint S.** 2002. Does this woman have an acute uncomplicated urinary tract infection? *JAMA* **287**:2701–2710.

14. **Stamm WE, Counts GW, Running KR, Fihn S, Turck M, Holmes KK.** 1982. Diagnosis of coliform infection in acutely dysuric women. *N Engl J Med* **307**:463–468.

15. **Little P, Merriman R, Turner S, Rumsby K, Warner G, Lowes JA, Smith H, Hawke C, Leydon G, Mullee M, Moore MV.** 2010. Presentation, pattern, and natural course of severe symptoms, and role of antibiotics and antibiotic resistance among patients presenting with suspected uncomplicated urinary tract infection in primary care: observational study. *BMJ* **340**:b5633.

16. **Christiaens TC, De Meyere M, Verschraegen G, Peersman W, Heytens S, De Maeseneer JM.** 2002. Randomised controlled trial of nitrofurantoin versus placebo in the treatment of uncomplicated urinary tract infection in adult women. *Br J Gen Pract* **52**:729–734.

17. **Gupta K, Hooton TM, Naber KG, Wullt B, Colgan R, Miller LG, Moran GJ, Nicolle LE, Raz R, Schaeffer AJ, Soper DE; Infectious Diseases Society of America; European Society for Microbiology and Infectious Diseases.** 2011. International clinical practice guidelines for the treatment of acute uncomplicated cystitis and pyelonephritis in women: A 2010 update by the Infectious Diseases Society of America and the European Society for Microbiology and Infectious Diseases. *Clin Infect Dis* **52**:e103–120.

18. **van Nieuwkoop C, van't Wout JW, Spelt IC, Becker M, Kuijper EJ, Blom JW, Assendelft WJ, van Dissel JT.** 2010. Prospective cohort study of acute pyelonephritis in adults: safety of triage towards home based oral antimicrobial treatment. *J Infect* **60**:114–121.

19. **Krieger JN, Lee SW, Jeon J, Cheah PY, Liong ML, Riley DE.** 2008. Epidemiology of prostatitis. *Int J Antimicrob Agents* **31**(Suppl 1):S85–90.

20. **Sharp VJ, Takacs EB, Powell CR.** 2010. Prostatitis: diagnosis and treatment. *Am Fam Physician* **82**:397–406.

21. **Ulleryd P, Zackrisson B, Aus G, Bergdahl S, Hugosson J, Sandberg T.** 1999. Prostatic involvement in men with febrile urinary tract infection as measured by serum prostate-specific antigen and transrectal ultrasonography. *BJU Int* **84**:470–474.

22. **Wagenlehner FM, Pilatz A, Weidner W.** 2011. Urosepsis--from the view of the urologist. *Int J Antimicrob Agents* **38**(Suppl):51–57.

23. **Ditchfield MR, Grimwood K, Cook DJ, Powell HR, Sloane R, Gulati S, De Campo JF.** 2004. Persistent renal cortical scintigram defects in children 2 years after urinary tract infection. *Pediatr Radiol* **34**:465–471.

24. **Jodal U.** 1987. The natural history of bacteriuria in childhood. *Infect Dis Clin North Am* **1**:713–729.

25. **Shaikh N, Hoberman A, Wise B, Kurs-Lasky M, Kearney D, Naylor S, Haralam MA, Colborn DK, Docimo SG.** 2003. Dysfunctional elimination syndrome: is it related to urinary tract infection or vesicoureteral reflux diagnosed early in life? *Pediatrics* **112**:1134–1137.

26. **Michael M, Hodson EM, Craig JC, Martin S, Moyer VA.** 2003. Short versus standard duration oral antibiotic therapy for acute urinary tract infection in children. *Cochrane Database Syst Rev* **(1)**:CD003966. doi:10.1002/14651858.CD003966:CD003966.

27. **Krieger JN, Ross SO, Simonsen JM.** 1993. Urinary tract infections in healthy university men. *J Urol* **149**:1046–1048.

28. **Lipsky BA.** 1999. Prostatitis and urinary tract infection in men: what's new; what's true? *Am J Med* **106**:327–334.

29. **Sabbaj J, Hoagland VL, Cook T.** 1986. Norfloxacin versus co-trimoxazole in the treatment of recurring urinary tract infections in men. *Scand J Infect Dis Suppl* **48**:48–53.

30. **Patterson TF, Andriole VT.** 1997. Detection, significance, and therapy of bacteriuria in pregnancy. Update in the managed health care era. *Infect Dis Clin North Am* **11**:593–608.

31. **Macejko AM, Schaeffer AJ.** 2007. Asymptomatic bacteriuria and symptomatic urinary tract infections during pregnancy. *Urol Clin North Am* **34**:35–42.

32. **Kass EH.** 1960. Bacteriuria and pyelonephritis of pregnancy. *Arch Intern Med* **105**:194–198.

33. **Millar LK, Cox SM.** 1997. Urinary tract infections complicating pregnancy. *Infect Dis Clin North Am* **11**:13–26.

34. **Christensen B.** 2000. Which antibiotics are appropriate for treating bacteriuria in pregnancy? *J Antimicrob Chemother* **46**(Suppl 1):29–34; discussion 63–65.

35. **Smaill F.** 2001. Antibiotics for asymptomatic bacteriuria in pregnancy. *Cochrane Database Syst Rev* (**2**):CD000490. doi:10.1002/14651858. CD000490.

36. **Vazquez JC, Villar J.** 2000. Treatments for symptomatic urinary tract infections during pregnancy. *Cochrane Database Syst Rev* (**3**): CD002256. doi:10.1002/14651858.CD002256.

37. **Schrag SJ, Zell ER, Lynfield R, Roome A, Arnold KE, Craig AS, Harrison LH, Reingold A, Stefonek K, Smith G, Gamble M, Schuchat A; Active Bacterial Core Surveillance Team.** 2002. A population-based comparison of strategies to prevent early-onset group B streptococcal disease in neonates. *N Engl J Med* **347:** 233–239.

38. **Wing DA.** 2001. Pyelonephritis in pregnancy: treatment options for optimal outcomes. *Drugs* **61:**2087–2096.

39. **Hooton TM, Bradley SF, Cardenas DD, Colgan R, Geerlings SE, Rice JC, Saint S, Schaeffer AJ, Tambayh PA, Tenke P, Nicolle LE; Infectious Diseases Society of America.** 2010. Diagnosis, prevention, and treatment of catheter-associated urinary tract infection in adults: 2009 International Clinical Practice Guidelines from the Infectious Diseases Society of America. *Clin Infect Dis* **50:**625–663.

40. **Chenoweth CE, Saint S.** 2011. Urinary tract infections. *Infect Dis Clin North Am* **25:**103–115.

41. **Knoll BM, Wright D, Ellingson L, Kraemer L, Patire R, Kuskowski MA, Johnson JR.** 2011. Reduction of inappropriate urinary catheter use at a Veterans Affairs hospital through a multifaceted quality improvement project. *Clin Infect Dis* **52:**1283–1290.

42. **van der Kooi TI, Manniën J, Wille JC, van Benthem BH.** 2010. Prevalence of nosocomial infections in The Netherlands, 2007–2008: results of the first four national studies. *J Hosp Infect* **75:**168–172.

43. **Tambyah PA, Maki DG.** 2000. Catheter-associated urinary tract infection is rarely symptomatic: a prospective study of 1,497 catheterized patients. *Arch Intern Med* **160:**678–682.

44. **Saint S.** 2000. Clinical and economic consequences of nosocomial catheter-related bacteriuria. *Am J Infect Control* **28:**68–75.

45. **Warren JW.** 1997. Catheter-associated urinary tract infections. *Infect Dis Clin North Am* **11:**609–622.

46. **Jacobsen SM, Stickler DJ, Mobley HL, Shirtliff ME.** 2008. Complicated catheter-associated urinary tract infections due to *Escherichia coli* and *Proteus mirabilis*. *Clin Microbiol Rev* **21:**26–59.

47. **Niël-Weise BS, van den Broek PJ.** 2005. Antibiotic policies for short-term catheter bladder drainage in adults. *Cochrane Database Syst Rev* (**3**):CD005428. doi:10.1002/14651858. CD005428.

48. **Rutschmann OT, Zwahlen A.** 1995. Use of norfloxacin for prevention of symptomatic urinary tract infection in chronically catheterized patients. *Eur J Clin Microbiol Infect Dis* **14:**441–444.

49. **Boyko EJ, Fihn SD, Scholes D, Chen CL, Normand EH, Yarbro P.** 2002. Diabetes and the risk of acute urinary tract infection among postmenopausal women. *Diabetes Care* **25:** 1778–1783.

50. **Shah BR, Hux JE.** 2003. Quantifying the risk of infectious diseases for people with diabetes. *Diabetes Care* **26:**510–513.

51. **Gorter KJ, Hak E, Zuithoff NP, Hoepelman AI, Rutten GE.** 2010. Risk of recurrent acute lower urinary tract infections and prescription pattern of antibiotics in women with and without diabetes in primary care. *Fam Pract* **27:**379–385.

52. **Lawrenson RA, Logie JW.** 2001. Antibiotic failure in the treatment of urinary tract infections in young women. *J Antimicrob Chemother* **48:**895–901.

53. **Carrie AG, Metge CJ, Collins DM, Harding GK, Zhanel GG.** 2004. Use of administrative healthcare claims to examine the effectiveness of trimethoprim-sulfamethoxazole versus fluoroquinolones in the treatment of community-acquired acute pyelonephritis in women. *J Antimicrob Chemother* **53:**512–517.

54. **Czaja CA, Rutledge BN, Cleary PA, Chan K, Stapleton AE, Stamm WE; Diabetes Control and Complications Trial/Epidemiology of Diabetes Interventions and Complications Research Group.** 2009. Urinary tract infections in women with type 1 diabetes mellitus: survey of female participants in the epidemiology of diabetes interventions and complications study cohort. *J Urol* **181:**1129–1134; discussion 1134–1135.

55. **Carton JA, Maradona JA, Nuño FJ, Fernandez-Alvarez R, Pérez-Gonzalez F, Asensi V.** 1992. Diabetes mellitus and bacteraemia: a comparative study between diabetic and non-diabetic patients. *Eur J Med* **1:**281–287.

56. **Horcajada JP, Moreno I, Velasco M, Martínez JA, Moreno-Martínez A, Barranco M, Vila J, Mensa J.** 2003. Community-acquired febrile urinary tract infection in diabetics could deserve a different management: a case-control study. *J Intern Med* **254:**280–286.

57. **Goettsch WG, Janknegt R, Herings RM.** 2004. Increased treatment failure after 3-days' courses of nitrofurantoin and trimethoprim for urinary tract infections in women: a population-based retrospective cohort study using the PHARMO database. *Br J Clin Pharmacol* **58**:184–189.

58. **Schneeberger C, Stolk RP, Devries JH, Schneeberger PM, Herings RM, Geerlings SE.** 2008. Differences in the pattern of antibiotic prescription profile and recurrence rate for possible urinary tract infections in women with and without diabetes. *Diabetes Care* **31**:1380–1385.

59. **Meiland R, Geerlings SE, Stolk RP, Netten PM, Schneeberger PM, Hoepelman AI.** 2006. Asymptomatic bacteriuria in women with diabetes mellitus: effect on renal function after 6 years of follow-up. *Arch Intern Med* **166**:2222–2227.

60. **Geerlings SE, Stolk RP, Camps MJ, Netten PM, Collet JT, Schneeberger PM, Hoepelman AI.** 2001. Consequences of asymptomatic bacteriuria in women with diabetes mellitus. *Arch Intern Med* **161**:1421–1427.

61. **Karunajeewa H, McGechie D, Stuccio G, Stingemore N, Davis WA, Davis TM.** 2005. Asymptomatic bacteriuria as a predictor of subsequent hospitalisation with urinary tract infection in diabetic adults: The Fremantle Diabetes Study. *Diabetologia* **48**:1288–1291.

62. **Hooton TM, Scholes D, Hughes JP, Winter C, Roberts PL, Stapleton AE, Stergachis A, Stamm WE.** 1996. A prospective study of risk factors for symptomatic urinary tract infection in young women. *N Engl J Med* **335**:468–474.

63. **Scholes D, Hawn TR, Roberts PL, Li SS, Stapleton AE, Zhao LP, Stamm WE, Hooton TM.** 2010. Family history and risk of recurrent cystitis and pyelonephritis in women. *J Urol* **184**:564–569.

64. **Scholes D, Hooton TM, Roberts PL, Gupta K, Stapleton AE, Stamm WE.** 2005. Risk factors associated with acute pyelonephritis in healthy women. *Ann Intern Med* **142**:20–27.

65. **Scholes D, Hooton TM, Roberts PL, Stapleton AE, Gupta K, Stamm WE.** 2000. Risk factors for recurrent urinary tract infection in young women. *J Infect Dis* **182**:1177–1182.

66. **Hawn TR, Scholes D, Wang H, Li SS, Stapleton AE, Janer M, Aderem A, Stamm WE, Zhao LP, Hooton TM.** 2009. Genetic variation of the human urinary tract innate immune response and asymptomatic bacteriuria in women. *PLoS One* **4**:e8300. doi:10.1371/journal.pone.0008300.

67. **Hawn TR, Scholes D, Li SS, Wang H, Yang Y, Roberts PL, Stapleton AE, Janer M, Aderem A, Stamm WE, Zhao LP, Hooton TM.** 2009. Toll-like receptor polymorphisms and susceptibility to urinary tract infections in adult women. *PLoS One* **4**:e5990. doi:10.1371/journal.pone.005900.

68. **Lundstedt AC, Leijonhufvud I, Ragnarsdottir B, Karpman D, Andersson B, Svanborg C.** 2007. Inherited susceptibility to acute pyelonephritis: a family study of urinary tract infection. *J Infect Dis* **195**:1227–1234.

69. **Albert X, Huertas I, Pereiro II, Sanfelix J, Gosalbes V, Perrota C.** 2004. Antibiotics for preventing recurrent urinary tract infection in non-pregnant women. *Cochrane Database Syst Rev* **(3)**:CD001209. doi:10.1002/14651858.CD001209.pub2.

70. **Gupta K, Hooton TM, Roberts PL, Stamm WE.** 2001. Patient-initiated treatment of uncomplicated recurrent urinary tract infections in young women. *Am Intern Med* **135**:9–16.

71. **Rosen DA, Hooton TM, Stamm WE, Humphrey PA, Hultgren SJ.** 2007. Detection of intracellular bacterial communities in human urinary tract infection. *PLoS Med* **4**:e329. doi:10.1371/journal.pmed.0040329.

72. **van Haarst EP, van Andel G, Heldeweg EA, Schlatmann TJ, van der Horst HJ.** 2001. Evaluation of the diagnostic workup in young women referred for recurrent lower urinary tract infections. *Urology* **57**:1068–1072.

73. **Perrotta C, Aznar M, Mejia R, Albert X, Ng CW.** 2008. Oestrogens for preventing recurrent urinary tract infection in postmenopausal women. *Obstet Gynecol* **112**:689–690.

74. **Foxman B, Brown P.** 2003. Epidemiology of urinary tract infections: transmission and risk factors, incidence, and costs. *Infect Dis Clin North Am* **17**:227–241.

75. **Foxman B, Gillespie B, Koopman J, Zhang L, Palin K, Tallman P, Marsh JV, Spear S, Sobel JD, Marty MJ, Marrs CF.** 2000. Risk factors for second urinary tract infection among college women. *Am J Epidemiol* **151**:1194–1205.

76. **Jackson SL, Boyko EJ, Scholes D, Abraham L, Gupta K, Fihn SD.** 2004. Predictors of urinary tract infection after menopause: A prospective study. *Am J Med* **117**:903–911.

77. **Czaja CA, Scholes D, Hooton TM, Stamm WE.** 2007. Population-based epidemiologic analysis of acute pyelonephritis. *Clin Infect Dis* **45**:273–280.

78. **Renko M, Tapanainen P, Tossavainen P, Pokka T, Uhari M.** 2011. Meta-analysis of the significance of asymptomatic bacteriuria in diabetes. *Diabetes Care* **34**:230–235.

79. **Kunin CM.** 1987. *Detection, Prevention and Management of Urinary Tract Infections*, 4th ed. Lea & Febiger, Philadelphia, PA.

Diagnosis, Treatment, and Prevention of Urinary Tract Infection

3

PAULA PIETRUCHA-DILANCHIAN[1] and THOMAS M. HOOTON[1]

DIAGNOSIS

Urinary tract infection (UTI) may be symptomatic or asymptomatic. The diagnosis of symptomatic UTI is usually straightforward, based on symptoms and signs and support from laboratory data. However, the diagnosis of UTI may be quite difficult to make in patients who are not able to recognize symptoms, such as patients with sensory deficits such as spinal cord injury or who are catheterized, or patients with cognitive disorders. Asymptomatic bacteriuria, defined by high levels of bacteriuria in a person without symptoms attributable to the urinary tract, is discussed elsewhere.

Clinical

Acute uncomplicated cystitis in women is manifested by dysuria, urgency, and/or frequency, often with suprapubic pain and hematuria. Symptoms and signs of complicated cystitis are similar but can be subtle in the very young and very old. The differential diagnosis of acute dysuria in a sexually active young woman also includes acute urethritis due to *Chlamydia trachomatis*, *Neisseria gonorrhoeae*, or herpes simplex virus; vaginitis due to *Candida* species or

[1]Department of Medicine, University of Miami Miller School of Medicine, Miami, FL 33136.
Urinary Tract Infections: Molecular Pathogenesis and Clinical Management, 2nd Edition
Edited by Matthew A. Mulvey, David J. Klumpp, and Ann E. Stapleton
© 2017 American Society for Microbiology, Washington, DC
doi:10.1128/microbiolspec.UTI-0021-2015

Trichomonas vaginalis; and noninfectious urethritis (1, 2). These latter conditions can usually be distinguished from acute cystitis with data from the history and physical examination and simple laboratory tests. For example, a history of sexually transmitted infections or recurrent genital vesicular rash, vaginal irritation or discharge, or slow onset of symptoms warrants consideration of a sexually transmitted infection such as herpes or chlamydia. However, differentiation of urinary symptoms in sexually active adolescent women may be difficult, and testing for both UTI and sexually transmitted infection is recommended (3, 4).

Assessment of patient risk factors may also be helpful in the evaluation of women with acute uncomplicated cystitis. Thus, cystitis in healthy women is often associated with recent sexual intercourse, recent use of a spermicidal product, asymptomatic bacteriuria, or previous history of cystitis (5, 6). Women with a previous history of physician-diagnosed UTI are able to accurately self-diagnose a subsequent UTI (7). Women with recurrent UTI often have a mother, sister, or daughter with a similar history (8).

The diagnosis of UTI in healthy women can be made with reasonable certainty by evaluation of symptoms and signs (9). In a meta-analysis of studies of uncomplicated UTI in women, it was found that the probability of cystitis was greater than 90% in the setting of dysuria and frequency without vaginal discharge or irritation (9). In this meta-analysis, laboratory findings did not add substantial information to help the clinician in diagnosing UTI. The accuracy of diagnosis based on patients' symptoms supports the findings of several studies demonstrating that selected women with cystitis symptoms can be successfully managed without in-person assessment (9, 10). These findings and conclusions should not be generalized to patients with complicated UTI.

Acute uncomplicated pyelonephritis in women is suggested by fever (temperature ≥38.5°C), chills, flank pain, nausea and vomiting, and costovertebral angle tenderness. Cystitis symptoms may or may not be present. Patients with acute complicated pyelonephritis may also present with sepsis, multiple organ system dysfunction, shock, and/or acute renal failure. In some cases, complicated pyelonephritis may be associated with weeks to months of insidious, nonspecific signs and symptoms such as malaise, fatigue, nausea, or abdominal pain. In the setting of what appears to be pyelonephritis, a pelvic examination or imaging studies may be indicated, and pregnancy testing may be appropriate.

In the elderly, symptoms commonly seen in the younger population with uncomplicated cystitis, such as dysuria, frequency, urgency, or suprapubic pain may not be reported. For example, changes in mental status, or vague symptoms of malaise, may be the presenting symptoms of UTI. Moreover, there is a high prevalence of bacteriuria in this population at any given time (11), such that the urine culture is less useful in confirming the diagnosis than it is in younger populations.

In patients with neurogenic bladder and sensory deficits, UTI may present a diagnostic dilemma to the clinician. In addition to the classic signs associated with uncomplicated cystitis and pyelonephritis, such patients may also complain of nonspecific symptoms such as fatigue, irritability, nausea, headache, abdominal or back pain, or other vague symptoms. Such patients, especially those requiring catheterization for bladder management, often have chronic pyuria and bacteriuria, reducing the specificity of the urinalysis and urine culture (12).

Men with cystitis or pyelonephritis manifest the same symptoms as women, but prostatic inflammation, if present, may cause obstructive symptoms such as hesitancy, nocturia, slow stream, and dribbling in association with UTI symptoms. As with women, sexually transmitted diseases such as *N. gonorrhoeae* and *C. trachomatis* must be included in the differential diagnosis of men presenting with UTI symptoms (13).

Laboratory

Although UTI can often be accurately diagnosed based on clinical symptoms and signs, especially in healthy women, laboratory testing can be helpful. The most useful laboratory tests for UTI are dipsticks that test for leukocyte esterase (LE) and nitrites and voided urine cultures. Tests for LE and nitrites are often performed in the clinic to provide support to the clinical diagnosis, which may or may not be straightforward, or to help "rule out" UTI and thus avoid antimicrobials. Voided urine cultures provide information, usually after empiric treatment has been started, to help confirm or rule out the diagnosis of UTI to assess the antimicrobial susceptibility of infecting organisms so that treatment can be tailored, which is especially useful if the patient's therapeutic options are limited by medication intolerance. Urine cultures are more useful in men and women with complicated UTI than in women with uncomplicated cystitis because UTIs in the former group are less likely to be associated with classic UTI symptoms and more likely to be associated with antimicrobial-resistant uropathogens. In fact, in women with uncomplicated cystitis, urine cultures are generally not recommended due to the reliability of the patient's history in establishing the diagnosis, the expense, and the inconvenience to the patient and delayed availability of culture results (2, 14).

Voided urine culture

Bladder bacteriuria is the "gold standard" for the diagnosis of UTI, but direct sampling of bladder urine with and without catheterization or suprapubic bladder aspiration is usually not performed due to inconvenience and discomfort to the patient. Instead, culture of voided urine is usually performed to infer the status of bladder urine in patients who do not have neurogenic bladder and in whom a urine culture is thought to be indicated. Direct bladder urine sampling may be useful when multiple previous voided specimens appear to be contaminated (occasionally encountered in the outpatient setting) or when patients with severe UTI cannot provide a voided urine specimen in a timely manner to start empiric antimicrobials.

Voided urine collection techniques and contamination

Voided urine specimens can become contaminated from organisms present on the distal urethra, vagina, labia, or periurethral skin, and it has been recommended that the voided midstream clean catch technique be used to minimize contamination. Thus, contamination is often suspected when multiple organisms are isolated from the voided specimen or when bacteria with low pathogenicity, such as lactobacilli, are isolated. However, it has been found that coliform organisms grown in voided urine specimens, even when coisolated with other organisms, are usually reflective of bladder bacteriuria, and thus their presence should not be discounted (15). To avoid overgrowth of contaminating bacteria and accurately determine uropathogen quantity in the voided urine specimen, it should be transported in a timely fashion to the microbiology laboratory for further processing. Since uropathogenic bacteria proliferate rapidly at room temperature, samples should be processed within a couple of hours of collection or refrigerated if there will be a delay in transport.

Several studies have been performed to evaluate the optimal method for collecting urine specimens in women to reduce urethral contamination. One study in 110 young, asymptomatic, healthy college women compared several methods to reduce contamination: spreading the labia with no cleansing prior to collection, perineal cleansing with tap water-moistened cotton swabs wiped from front to back of the perineum, midstream collection with no cleansing, and a combination of techniques (16). Spreading the labia with use of no other techniques resulted in the least contamination of voided specimens.

Another study randomized 242 women with UTI symptoms into 3 groups to evaluate contamination rates of voided urine (17). The first group was instructed to urinate into a clean container without cleansing and not midstream. The second group was instructed to collect a midstream urine sample with perineal cleansing and spreading of the labia. The third group was given the same instructions as group 2, with the addition of using a vaginal tampon. Contamination rates for the three groups were nearly identical (29%, 32%, and 31%, respectively), and the authors concluded that the midstream clean-catch technique does not decrease contamination rates.

In a more recent study, 113 pregnant women were asked to collect three urine samples consecutively: morning (first void); midstream (void without further instructions); and clean-catch (void after cleaning) (18). Gram-positive rods or mixed bacteria in the Gram stain and mixed growth or skin flora in the urine culture were considered contaminants. There was no consistency in quantity of contaminants found in midstream samples compared with morning and clean-catch samples. No significant differences were found between the other end points in all three urine samples. The authors concluded that the contamination rate of midstream samples is comparable with the contamination rates of morning and clean-catch samples and thus that more complex, impractical, and time-consuming morning and clean-catch samples are not superior to the midstream (non-clean-catch) sample in pregnant women.

These studies suggest that collection of a clean-catch voided urine specimen is of little value; instructing women to spread their labia and collect midstream urine seems most reasonable when collecting a voided urine specimen for culture. In men, culture results of bladder urine showed excellent agreement with those of clean-catch, midstream-void, and uncleansed first-void specimens (19).

Interpretation of urine culture results

Early studies demonstrated the value of quantitative urine culture in discriminating between true UTI and contaminated voided urine specimens (20–28). Bacterial counts $\geq 10^5$ cfu/ml in voided urine were predictive of bladder bacteriuria in asymptomatic women and women with pyelonephritis, whereas lower counts were more likely to be associated with contamination (22–26). However, later investigations demonstrated that women with symptoms of cystitis often had lower counts (29–31). In a subsequent study of women with symptoms of uncomplicated cystitis, voided urine colony counts of coliform organisms as low as 10^2 cfu/ml were found to be indicative of bladder bacteriuria (15). In this study, the sensitivity and specificity of coliforms at $\geq 10^2$ cfu/ml in voided urine for detection of bladder bacteriuria as determined by suprapubic aspiration or urethral catheterization were 95% and 85%, respectively. In a more recent study comparing voided midstream and paired catheter urine cultures in healthy premenopausal women with symptoms of cystitis, the sensitivity and specificity of low colony counts of *Escherichia coli* in midstream urine was confirmed (32). Of note, however, enterococci and group B streptococci, both frequently isolated from midstream urines in the study, had very low positive predictive values, suggesting that they only rarely cause cystitis in this population.

Up to 95% of patients with pyelonephritis urine will have $\geq 10^4$ cfu/ml uropathogens in a voided urine culture (33). In men with urinary symptoms, a quantitative count of $\geq 10^3$ cfu/ml in a voided urine specimen best differentiates sterile from infected bladder urine (19). In urine specimens obtained by urethral catheterization from symptomatic or asymptomatic men and women, periurethral contamination is less of a problem, and lower quantitative counts of $\geq 10^2$ CFU/ml are considered to be significant (12). In those with long-term indwelling catheters, the catheter urine may be unreliable (34, 35), so a urine specimen should be obtained from a freshly placed catheter. In this regard, clinical and bacteriological outcomes are improved when long-term indwelling catheters are replaced before initiating antimicrobial therapy for symptomatic UTI (36). Catheter urine specimens should be directly aspirated with a sterile needle and syringe and not from the catheter bag since bacteria multiply in these bags (12).

Of note, most clinical laboratories do not routinely report colony counts in voided urine as low as 10^2 CFU/ml, so it is reasonable to use a quantitative count $\geq 10^3$ CFU/ml in symptomatic women or men, whether catheterized or not, as an indicator of UTI, because this threshold is a reasonable compromise between sensitivity in detecting bladder bacteriuria and feasibility for the microbiology laboratory in quantifying organisms (12). In this regard, in dysuric women (37) and catheterized individuals (38), low colony count bacteriuria that is not treated will often result in colony counts of $\geq 10^5$ cfu/ml several days later. For the diagnosis of asymptomatic bacteriuria in women and men, on the other hand, $\geq 10^5$ cfu/ml is a reasonable criterion, even though lower counts probably represent true bladder bacteriuria, since increased specificity is desirable (to avoid overtreatment). Asymptomatic bacteriuria should not be screened for except in research studies and in selected clinical situations such as pregnant women or pending urologic surgery (11).

Rapid detection strategies for bacteriuria
The availability of reliable tests for rapid detection of bacteriuria and antimicrobial resistance would likely result in a significant reduction of antimicrobial use and improved quality of care for patients with UTI. In a pilot study, real-time PCR testing provided pathogen identification 43 hours prior to standard culture results (39). Concordance of real-time PCR and urine cultures for Gram-positive and Gram-negative bacteria and fungi was 90%, 97%, and 97%, respec-

tively. However, the product being tested was not able to provide pathogen quantification, and the diagnostic utility of the test was limited with a sensitivity of 82% and specificity of 60% for detecting infection ($>10^5$ cfu/ml). In another pilot study, the real-time PCR test had limited ability to detect infection with multiple pathogens (40). Rapid whole-genome sequencing is being used to investigate outbreaks (41, 42), but it also holds promise for rapid detection of bacteriuria and antimicrobial susceptibility. Such technology may be limited in UTI, however, due to specificity issues inherent to the voided urine culture and cost of the test.

Pyuria and nitrite testing
Pyuria (≥ 10 leukocytes per high-power field of unspun urine) (43) is found in the majority of patients with UTI. Pyuria is almost always present in pyelonephritis, whereas leukocyte casts, while specific for pyelonephritis, are infrequently seen. Pyuria may not be detected in some cases of complicated UTI if the collecting system distal to the infection is blocked. Pyuria may also be present in women with urethritis caused by *C. trachomatis*, *N. gonorrhoeae*, or herpes simplex and occasionally in women with vaginitis caused by *T. vaginalis* or candidal infections. It is also usually seen in patients with indwelling or intermittent urethral catheterization and may be seen in association with renal calculi, renal tract neoplasms, and infections such as renal tuberculosis (1). In the catheterized patient, pyuria is not diagnostic of catheter-associated bacteriuria or symptomatic UTI, and the presence, absence, or degree of pyuria alone does not, by itself, differentiate catheter-associated asymptomatic bacteriuria from catheter-associated symptomatic UTI (12). The absence of pyuria in a symptomatic patient suggests a diagnosis other than UTI, although it may be absent in as many as 10% of women with acute uncomplicated cystitis (7).

Urine dipsticks are used extensively in the primary care and emergency care settings as aids to the diagnosis of UTI (44, 45). They measure LE, a leukocyte enzymatic product and indirect measure of pyuria, and nitrites, a bacterial metabolic product of nitrate breakdown. Several prospective studies, meta-analyses, and systematic analyses to assess the utility of the dipstick tests in diagnosing UTI have been published, but results have been contradictory (9, 45–50). As pointed out by St John (45), few of these studies have taken into account the wide variation in prevalence or pretest probability of UTI in different populations, an important factor in determining the effect of a positive or negative test result on the posttest probability of ruling a UTI diagnosis in or out. Furthermore, studies have often used a microbiologic definition of UTI based on different colony count thresholds. No studies have evaluated the value of urine dipstick testing of voided urine in patients in whom the presence or absence of true bladder bacteriuria has been determined by urinary catheterization or suprapubic aspiration.

St John et al. performed a systematic review of the literature to evaluate the diagnostic value of urinary dipstick tests to exclude UTI in adults (45). They performed a meta-analysis of 23 studies that used a colony count threshold of 10^5 cfu/ml to define UTI (presence or absence of urinary symptoms was not considered in the definition). The LE or nitrite test combination, with one or the other test positive, was used in 14 studies and showed the highest sensitivity and the lowest negative likelihood ratio. While there was significant heterogeneity between the studies, 7 of 14 demonstrated significant decreases in pretest to posttest probability with a pooled posttest probability of 5% for the negative result. The authors concluded that in certain circumstances there is evidence for the use of the dipstick urinalysis as a rule-out test for UTI.

Little et al. performed a validation study in 434 adult women with suspected cystitis to determine whether previously documented clinical and dipstick variables and algorithms predict laboratory diagnosis of UTI (46). They found that 66% of patients had confirmed UTI. The predictive values of nitrite, LE (+ or greater), and blood (trace or greater) were confirmed (independent multivariate odds ratios = 5.6, 3.5, and 2.1, respectively). A previously developed dipstick rule, based on the presence of nitrite, or both leukocytes and blood, was moderately sensitive (75%) but less specific (66%); the positive and negative predictive values were 81% and 57%, respectively. The authors concluded that dipstick results can modestly improve diagnostic precision but poorly rule out UTI.

A systematic review performed in adults by Hurlbut and Littenberg focused on evidence to support using dipstick tests to diagnose UTI, as defined by quantitative culture, and found that the best rule-in test was the LE or nitrite combination (47). This analysis revealed that dipstick positivity, as defined by nitrite, LE, or both being positive, is the most accurate index test. However, they found that a negative urine dipstick test cannot exclude the diagnosis of UTI in patients with high prior probabilities of UTI.

Bent et al. performed a review of studies of acute uncomplicated UTI in women to determine the accuracy and precision of history-taking and physical examination for the diagnosis of UTI in women; nine studies met inclusion criteria (9). The authors concluded that in women who present with one or more symptoms of UTI, the probability of infection is approximately 50%. Specific combinations of symptoms (e.g., dysuria and frequency without vaginal discharge or irritation) raise the probability of UTI to more than 90%, effectively ruling in the diagnosis based on history alone. They concluded that history-taking, physical examination, and dipstick urinalysis are not able to reliably lower the posttest probability of disease to a level where a UTI can be ruled out when a patient presents with one or more UTI symptoms.

In a meta-analysis by Deville et al. (48), 70 studies were evaluated to summarize the available evidence of the diagnostic accuracy of the urine dipstick test, taking into account various predefined potential sources of heterogeneity. In a variety of clinical situations, they found that combining the results of LE and nitrite increased the sensitivity from 68 to 88%, and the corresponding posttest probabilities or predictive values of the negative test ranged from 84 to 98%, with the value being greater than 96% in the majority of the settings studied. They concluded that the urine dipstick test alone seems to be useful in all populations to exclude the presence of infection if the results of both nitrites and LE are negative but that the usefulness of the dipstick test alone to rule in infection remains doubtful, even with high pretest probabilities.

Patel et al. performed a prospective study in a hospital setting to determine whether using local evidence-based guidelines for the use of urine dipstick testing at the point of care could reduce the proportion of culture-negative urines arriving in the laboratory (49). At the point of care, 1,076 urine samples were dipstick-tested using an automatic strip reader. Quantitative results for LE, nitrite, blood, and protein were compared with the results of conventional laboratory microscopy and culture. Of these samples, 175 (16.3%) were negative for all 4 markers; only 3 of these (1.7% of all true positives) were positive by culture. The absence of all four tests was found to have a greater than 98% negative predictive value. Two years after distributing the algorithm and promoting access to reagent strips and strip readers, a reduction in the urine workload was seen against an otherwise increasing laboratory specimen load.

In another systematic review of the diagnostic accuracy of symptoms and signs in women with acute uncomplicated cystitis, 16 studies with 3,711 patients were evaluated across 3 reference standards (10^2, 10^3, or 10^5 cfu/ml) (50). The presence of dysuria, frequency, hematuria, nocturia, and urgency all increased, whereas the presence of vaginal discharge decreased the probability of UTI. The presence of hematuria had the highest diagnostic utility, raising the posttest probability of UTI from 65 to 76% at $\geq 10^2$ cfu/ml, and when combined with a positive dipstick for nitrites, to 93%. The authors concluded that individual symptoms and signs have a modest ability to raise the pretest risk of UTI, but diagnostic accuracy improves considerably when combined with dipstick tests, particularly tests for nitrites.

In a prospective study of suspected cystitis in 331 women, a cystitis decision aid consisting of 3 criteria (presence of dysuria, leukocytes [greater than trace] and/or nitrites [any positive]) was used to predict UTI ($\geq 10^2$ cfu/ml [all had UTI symptoms]) and to decide whether to treat empirically or not (51). Empirical antimicrobials were given without culture if fewer than two variables were present; in those with fewer than two criteria, a urine culture was recommended before deciding about the need for antimicrobials. The rate of positive urine culture was 23% with zero criteria, 43% with one, 69% with two, and 89% with three. Following decision aid recommendations would have reduced antimicrobial prescriptions by 24%, unnecessary prescriptions by 40%, and urine cultures by 59% compared with physician care. However, many women reported that they would not have been willing to wait for urine culture results before being prescribed antimicrobials, especially those with severe symptoms.

In summary, at least in the setting of presumptive acute uncomplicated cystitis, use of the LE and nitrite tests can provide helpful data to the clinician trying to decide whether a patient has a UTI, but their use is limited in that a negative result cannot rule out the diagnosis of UTI. Given the low risks of severe complications associated with uncomplicated cystitis, it is reasonable to reserve diagnostic urine cultures (and delay use of antimicrobials) to those women who have negative LE and nitrite tests. The LE and nitrite tests may also be useful in alerting the clinician that another diagnosis other than UTI should be considered in evaluation of the patient who presents with symptoms of UTI.

TREATMENT

Symptomatic UTIs warrant treatment, whereas bacteriuria in asymptomatic individuals does not, except in very select circumstances (11). Acute uncomplicated cystitis is a benign condition in which early symptom resolution has been observed in 25 to 42% of women (52, 53) and serious complications, including pyelonephritis, are rare in untreated patients. Thus, given that cystitis is associated with significant morbidity (54), the primary goal of antimicrobial treatment is rapid resolution of symptoms rather than prevention of complications (55). Strategies that postpone or avoid antimicrobial treatment in women with presumptive acute uncomplicated cystitis should be considered.

Recently updated guidelines of The Infectious Diseases Society of America (IDSA) (55) emphasize the importance of considering collateral damage effects of antimicrobials (the selection of drug-resistant organisms and colonization or infection with multidrug-resistant organisms [56]) in selecting a treatment regimen. The guideline suggests thresholds for the prevalence of resistance in a community above which a drug is not recommended (20% for trimethoprim-sulfamethoxazole [TMP-SMX] and 10% for fluoroquinolones), but clinicians rarely have access to such information. Of note, passive laboratory-based surveillance methods tend to overestimate true resistance rates since they are skewed by urine cultures obtained from patients who may have failed initial therapy or who have specific risk factors for resistance, such as recent travel or antimicrobial use (57–60).

Antimicrobial Resistance

Empiric antimicrobial choices are influenced by geographical location and individual predictors of resistance. Surveillance studies conducted in North America and Europe found >20% resistance rates to *E. coli*, the most common uropathogen causing uncomplicated UTI, for ampicillin in all areas, and for trimethoprim and TMP-SMX in most areas (61, 62). In the United States, TMP-SMX resistance rates among *E. coli* causing uncomplicated UTI range from 15 to 42% in various regions of the country (61), with a similar range among European countries and Brazil (62). High rates of resistance to TMP-SMX have raised questions about its role as a first-line agent for uncomplicated UTI, and it is not listed as such in some European guidelines (63). Rates of resistance to nitrofurantoin, fosfomycin, and fluoroquinolones were low in these surveys (<10%). The prevalence of *E. coli* resistance to nitrofurantoin is generally less than 5%, although nitrofurantoin is inactive against *Proteus* species and some *Enterobacter* and *Klebsiella* strains. Prior use of TMP-SMX and travel outside of the United States within the previous 3 to 6 months have been shown to be predictors of resistance to TMP-SMX (55, 64–66).

Of concern, there has been a trend toward increasing resistance rates to fluoroquinolones over the past few years (55). This was emphasized in a recent antimicrobial susceptibility study of more than 12 million *E. coli* isolates from outpatients in the United States, showing that fluoroquinolone resistance had increased from 3 to 17% over a 10-year period (67). Another 4-year prospective study of urinary isolates showed that resistance to the fluoroquinolones increased by 15% for *Pseudomonas aeruginosa* and 10% for *E. coli* (68). In a recent study of 176 *E. coli* urine isolates from female college students with uncomplicated UTIs, 29.6% of the isolates were resistant to TMP-SMX, whereas none were resistant to nitrofurantoin (69). Of note, in women without a prior history of UTI, 1.8% of the isolates were resistant to ciprofloxacin compared with 11.8% in those with a prior history.

Since 2000, CTX-M extended-spectrum β-lactamase (ESBL)-producing strains of *E. coli* have emerged worldwide as important causes of community-onset urinary tract and bloodstream infections (70–74). The sudden worldwide increase of such strains is mainly due to a single clone (ST131), and foreign travel to high-risk areas such as the Indian subcontinent appears to play a role in the spread of this clone across different continents (74, 75). These strains are often resistant to the fluoroquinolones, TMP-SMX, and most cephalosporins and are a challenge to treat in the outpatient setting. However, limited data show that fosfomycin, nitrofurantoin, and to a lesser extent, amoxicillin-clavulanate have *in vitro* and clinical activity against these strains (73, 76–79). In one *in vitro* study of CTX-M ESBL-producing *Enterobacteriaceae* isolates from 46 outpatient urine samples, all were susceptible to ertapenem, 91% to fosfomycin, and 74% to nitrofurantoin (77). Moreover, 89% of the isolates were susceptible to the combination of cefdinir plus amoxicillin-clavulanate, prompting the authors to reason that the clavulanate had an inhibitory effect on the ESBL, resulting in effective cefdinir activity against most isolates. In another *in vitro* study of 100 ESBL-producing *E. coli* strains from ambulatory patients with UTI, the proportions of strains susceptible to ertapenem were 100%, fosfomycin 97%, and nitrofurantoin 94% (78). The therapeutic potential of fosfomycin and amoxicillin-clavulanate was demonstrated in one study of community-acquired cystitis caused by ESBL-producing *E. coli* in which cure rates with fosfomycin were 93% (all isolates were susceptible) and with amoxicillin-clavulanate were 93% for those with susceptible isolates (MIC ≤8 μg/ml) and 56% for those with intermediate or resistant isolates (79).

Short-course treatment regimens are as effective as longer regimens in women with acute uncomplicated cystitis (55, 80–82). Although cystitis in elderly women is often considered to be complicated and to warrant longer treatment, short-course regimens also appear to be effective in otherwise healthy ambulatory elderly women (81). Longer regimens are recommended for treatment of acute uncomplicated pyelonephritis and complicated UTIs.

Acute Uncomplicated Cystitis

Guidelines for the treatment of uncomplicated cystitis and pyelonephritis have recently been updated by the IDSA (55). Given the benign nature of cystitis in healthy women, the guideline panel suggested that the risk of collateral damage (56) of drugs be given equal weight to drug efficacy in treatment recommendations. Previous guidelines from the IDSA (82) had recommended the use of TMP-SMX for the empiric treatment of uncomplicated cystitis, but given the widespread emergence of resistance to this agent among common urinary pathogens, alternative agents that exhibit little collateral damage and are well tolerated are also currently recommended. Thus, the guideline panel listed four antimicrobials with equal weight for first-line empiric treatment of cystitis (Table 1): nitrofurantoin monohydrate/macrocrystals, 100 mg twice daily for 5 days; TMP-SMX 160/800 twice daily for 3 days (unless resistance prevalence is known or suspected to be greater than 20% or TMP-SMX was used within the last 3 months); fosfomycin trometamol, 3 g orally in one dose; or pivmecillinam, 400 mg orally twice daily for 5 days (not approved for use in the United States).

Nitrofurantoin, on the market for more than half a century, has had a resurgence in use in recent years, given its continued activity against most uropathogenic *E. coli* (UPEC), along with concomitant increasing resistance to other commonly used antimicrobial agents for urinary tract infection (83). Moreover, a recent prospective randomized, controlled study found that a 5-day course of nitrofurantoin was clinically equivalent to a 3-day course of TMP-SMX for the treatment

TABLE 1 Empiric antimicrobials for acute uncomplicated cystitis[a]

Antimicrobial category	Antimicrobial and dosing	Comments
First-line antimicrobials	Nitrofurantoin monohydrate/ macrocrystals 100 mg twice daily for 5 days (with meals)	Minimal *in vitro* resistance Minimal ecological adverse effects Avoid if pyelonephritis is suspected Rare risk of serious pulmonary, hepatic, and neurologic adverse events
	TMP-SMX[b] 160/800 mg twice daily for 3 days	Avoid if resistance is >20% or exposure in past 3 to 6 mo
	Trimethoprim[b] 100 mg twice daily for 3 days	Probably fewer ecological adverse effects than seen with fluoroquinolones
	Fosfomycin trometamol (Monurol) 3-g sachet in a single dose	Minimal *in vitro* resistance, but most labs do not test Minimal ecological adverse effects Avoid if pyelonephritis is suspected
Second-line antimicrobials	Ciprofloxacin[b] 250 mg twice daily for 3 days Levofloxacin[b] 250–500 mg once daily for 3 days	Minimal *in vitro* resistance, but prevalence is rising in United States and high in some regions of the world Propensity for ecological adverse effects When possible, reserve for uses other than cystitis
	β-lactam (e.g., amoxicillin-clavulanate, cefdinir, cefaclor, and cefpodoxime-proxetil) for 3–7 days	Less effective than TMP-SMX and fluoroquinolones *E. coli* resistance prevalence low to high Probably fewer ecological adverse effects than seen with parenteral broad-spectrum cephalosporins

[a]Table adapted from reference 14. Antimicrobial recommendations based on IDSA guidelines (55).
[b]Pregnancy category C: animal studies have shown an adverse effect on the fetus; use only if potential benefit justifies the potential risk to the fetus.

of uncomplicated cystitis in nonpregnant women (84), and it is now recommended as a first-line regimen for cystitis in treatment guidelines (55, 63). However, there are increasing questions about its safety as reflected by the number of case reports of adverse events (85–89) and by guidelines calling for restricted use of nitrofurantoin (90, 91). In this regard, nitrofurantoin can cause pulmonary, hepatic, or neurologic toxicity and should be avoided if the patient has previously had a reaction to this agent. However, such side effects appear to be rare (92).

According to the manufacturer of Macrobid, nitrofurantoin is contraindicated in patients with a creatinine clearance (CrCl) below 60 ml/min because of an increased risk of toxicity with impaired excretion of the drug. However, a recent literature review noted that data supporting this contraindication are nonexistent (93). The authors concluded that the limited data available would support considering using this drug in patients with a CrCl of 40 ml/min or higher. Nitrofurantoin

should be avoided if there is suspicion of early pyelonephritis.

Nitrofurantoin is listed as a potentially inappropriate medication for patients age 65 and older in the 2012 Beers criteria (91) due to the potential for pulmonary toxicity, the availability of safer alternatives, and the lack of efficacy in patients with CrCl <60 ml/min due to inadequate drug concentration in the urine. This panel's recommendation is to avoid its use in the elderly for long-term suppression and in those with CrCl <60 ml/min. In a recent study of 21,317 women treated with nitrofurantoin and 7,926 women treated with trimethoprim, the authors evaluated whether nitrofurantoin was less effective in patients with renal impairment (defined as the start of a second antibacterial within 1 month after the start of treatment) and whether renal impairment was associated with an increased risk of adverse events (defined as hospitalization within 90 days of drug use) (94). The authors concluded that nitrofurantoin treatment was not associated with a higher risk of ineffectiveness in women with UTI and moderate renal impairment (30 to 50 ml/min/1.73 m^2) but that there was a significant association between renal impairment (<50 ml/min/1.73 m^2) and pulmonary adverse events leading to hospitalization.

Nitrofurantoin continues to have an important role in the management of cystitis, especially because resistance to other agents has increased. As with any antimicrobial agent for urinary tract infection, however, patients should be counseled about potential toxic effects before any use and intermittently during long-term use.

Given the high prevalence of resistance to TMP-SMX in certain regions of the United States and other parts of the world, caution has been expressed about its use empirically (55, 63). In this regard, multiple studies have shown that patients infected with TMP-SMX-resistant strains who are treated with TMP-SMX are twice as likely to fail treatment (95–97). However, TMP-SMX remains very effective, with an estimated overall clinical cure rate of 85% even in regions where the prevalence of resistance is 30% (97), and it is inexpensive and well tolerated. Trimethoprim at 100 mg twice daily for 3 days is considered equivalent to TMP-SMX (82). As noted previously, however, because of concerns about the high prevalence of resistance in some areas of the world, recent European guidelines do not include TMP-SMX as a first-line empirical agent for uncomplicated cystitis (63).

Fosfomycin, which also appears to cause little collateral damage, was given equal weight to nitrofurantoin and TMP-SMX as a first-line agent for cystitis even though it may be less clinically effective than the latter two drugs, based on data from the package insert (55). However, a meta-analysis of 27 randomized controlled trials found no difference between fosfomycin and comparators in microbiological or clinical outcomes (76). Of note, fosfomycin is the only single-dose regimen listed in the recommendations, but it remains fairly expensive in the United States. Fosfomycin should not be used if pyelonephritis is suspected (55).

The choice between these recommended first-line agents should be individualized based on patient circumstances (allergy, tolerability, compliance), local community resistance prevalence (if known), availability, cost, and patient and provider threshold for failure (55). If these factors preclude the use of first-line antimicrobials, fluoroquinolones (e.g., ciprofloxacin or levofloxacin in 3-day regimens) are reasonable alternative agents and are highly effective for UTI. However, fluoroquinolones are discouraged for treatment of cystitis because of concerns about collateral damage, including worldwide increases in the prevalence of resistance, and their important role in treating many other conditions. Unfortunately, this class has become the most commonly used drug in the United States for the treatment of cystitis, likely exacerbating the growing worldwide problems of fluoroquinolone resistance (98). There are no data to suggest that one fluoroquinolone is superior to another.

Alternatively, a β-lactam such as amoxicillin-clavulanate, cefdinir, cefaclor or cefpodoxime-proxetil for 3 to 7 days may be used (55). β-lactams and fluoroquinolones were deemphasized for empiric treatment of cystitis because these classes have been associated with a higher risk of collateral damage, such as selection for ESBL-producing strains, multidrug-resistance *Staphylococcus aureus*, and *Clostridium difficile* colitis (55, 56, 99). In general, β-lactam antimicrobials have demonstrated inferior efficacy compared with TMP-SMX or fluoroquinolones in regimens of the same duration (55, 82). Although broad-spectrum oral β-lactams demonstrate *in vitro* activity against most uropathogens causing uncomplicated cystitis, clinical data are sparse (55). Amoxicillin-clavulanate (co-amoxiclav) has a high rate of gastrointestinal side effects, and a recent trial in women with acute cystitis demonstrated that a 3-day regimen of amoxicillin-clavulanate at 250 mg twice daily was significantly inferior to a 3-day regimen of ciprofloxacin at 250 mg twice daily, even in women infected with uropathogens susceptible to the drugs with which they were treated (100). Cefpodoxime at 100 mg twice daily for 3 days was also demonstrated to be inferior in this population compared with ciprofloxacin at 250 mg twice daily (101). It has been speculated that this phenomenon might be related to the shorter half-life of the oral β-lactams, compared with TMP-SMX and fluoroquinolones, and/or the poorer ability to eradicate the infecting strain from the gastrointestinal tract, the reservoir for recurrent UTI (100, 101). Moreover, there are concerns about the possibility of ecologic adverse effects with oral broad-spectrum cephalosporins, as has been observed with parenteral cephalosporins, although this has not been proven. Because of these considerations, β-lactams, if they are to be used as second-line agents for cystitis treatment, are generally recommended for durations of 7 days. Ampicillin or amoxicillin should not be used for empiric treatment, given the poor efficacy and high prevalence of resistance to these agents (61, 62, 82).

The response to antimicrobial treatment is usually rapid, and dysuria is usually diminished within a few hours after the start of treatment (102). However, for women with severe dysuria it is reasonable to use a urinary analgesic such as over-the-counter oral phenazopyridine three times daily as needed to relieve discomfort. A 2-day course is usually sufficient.

Antimicrobial stewardship considerations for the treatment of uncomplicated cystitis have prompted an exploration of treatment strategies for acute cystitis to reduce antimicrobial exposure (70). A recent randomized, controlled trial evaluated 5 management approaches in 309 nonpregnant women aged 18 to 70 years that included immediate antimicrobials at presentation, delay of antimicrobials until symptom reassessment at 48 hours, or prescription of antimicrobials based on a symptom score, a dipstick algorithm, or availability of midstream urine culture results (103). There were no significant differences in duration or severity of symptoms among the five groups ($P = 0.177$), but antimicrobial use was 97% for immediate empiric therapy, 77% for therapy delayed 48 hours, 90% for the symptom score strategy, 80% using a dipstick algorithm, and 81% when antimicrobials were withheld pending urine culture results ($P = 0.011$). Total antimicrobial use was significantly less for each of these four approaches compared with immediate empiric antimicrobials. An accompanying qualitative interview study of 21 of these women was performed to explore their views of the acceptability of the alternative approaches to management of UTI (104). The investigators found that women preferred not to take antimicrobials and were open to alternative management approaches, including antimicrobial delay, but that in doing so the clinician must address the particular worries that women might have and explain the rationale for not using antimicrobials immediately.

In a pilot study of alternative approaches to cystitis management, German investigators randomized 80 women age 18 to 65 years with acute cystitis symptoms to either 3 days of ibuprofen or 3 days of ciprofloxacin (105). Symptom resolution was similar at day 7 for both arms—75% of ibuprofen subjects and 61% of ciprofloxacin subjects. Of the ibuprofen subjects, 33% required subsequent antimicrobial therapy because of persistent or recurrent symptoms compared with 18% of the ciprofloxacin subjects (not significant). A more definitive trial is planned (106). Of interest regarding a possible role of anti-inflammatory modalities in UTI treatment, development of chronic cystitis by UPEC in a mouse model of UTI is facilitated by severe acute inflammatory responses early in infection, including severe pyuria and bladder inflammation with mucosal injury, and a distinct serum cytokine signature, which subsequently predispose to recurrent cystitis

(107). Treatment of the mice with dexamethasone prior to UPEC infection suppresses the development of chronic cystitis.

Acute Uncomplicated Pyelonephritis

The 2011 IDSA guidelines for nonpregnant, nonmenopausal women with acute pyelonephritis recommend that all patients should have a urine culture performed before empiric antimicrobials are started (Table 2) (55). The empiric treatment regimen selected for pyelonephritis should have broad-spectrum *in vitro* activity against likely uropathogens and be started quickly to minimize progression. In choosing an empiric agent, one should consider the antimicrobial susceptibility of prior UTI strains, local resistance data, history of exposure to the same class of antimicrobial in the past 3 to 6 months (choose an alternative agent), and severity of illness and comorbidities. Most episodes of acute uncomplicated pyelonephritis are now treated in the outpatient setting (108, 109); in one recent population-based study of acute pyelonephritis in adult women, only 7% were hospitalized (108). Some patients require

stabilization with parenteral fluids and antimicrobials in an office or urgent care facility before going home on oral antimicrobials under close supervision.

There are very few published studies comparing treatment regimens for acute uncomplicated pyelonephritis (55, 109–111). A randomized, double-blind study showed that a 7-day course of ciprofloxacin at 500 mg orally twice daily, compared with a 14-day course of TMP-SMX 160/800 orally twice daily for 14 days, was highly effective (110). Clinical cure rates were 96% in the ciprofloxacin group and 83% in the TMP-SMX group ($P = 0.002$), and microbiologic cure rates were 99% and 89%, respectively ($P = 0.004$). These outcome differences were mainly due to the much higher prevalence of resistance of causative uropathogens to TMP-SMX. An initial intravenous dose of ceftriaxone significantly improved cure rates in women with a TMP-SMX-resistant uropathogen. In another double-blind randomized study of men and women with complicated UTI that included men and women with acute pyelonephritis (112), a subgroup analysis of the pyelonephritis patients

TABLE 2 Empiric oral antimicrobials for outpatient management of acute uncomplicated pyelonephritis[a]

Antimicrobial and dosing[b]	Comments
Ciprofloxacin[c] 500 mg twice daily or 1 g (extended release) once daily for 7 days	Oral fluoroquinolone is empiric drug of choice
	Minimal *in vitro* resistance, but prevalence is rising in United States and high in some regions of the world
Levofloxacin[c] 750 mg once daily for 5 days	Propensity for ecological adverse effects
TMP-SMX[c] 160/800 mg twice daily for 14 days	Highly effective if strain is susceptible, but do not use empirically due to high prevalence of resistance
	Probably fewer ecological adverse effects than fluoroquinolones
	7–10 day regimens likely to be effective in women who defervesce rapidly
Oral β-lactam (e.g., amoxicillin-clavulanate, cefdinir, cefaclor, and cefpodoxime-proxetil)	Data limited, but inferior efficacy compared with TMP-SMX and fluoroquinolones
Duration 10–14 days	Oral β-lactams should be used only when other recommended agents cannot be used
	Probably fewer ecological adverse effects than parenteral broad-spectrum cephalosporins

[a]Table adapted from reference 14. Antimicrobial recommendations based on IDSA guidelines (55, 82).
[b]If there is concern about oral medication tolerance; resistance based on community resistance prevalence >10% (fluoroquinolone) or unknown (TMP-SMX) or prior exposure in past 3 to 6 months; or if using an oral β-lactam, give initial intravenous dose of 1 g of ceftriaxone or 5 to 7 mg gentamicin/kg of body weight.
[c]Pregnancy category C: animal studies have shown an adverse effect on the fetus; use only if potential benefit justifies the potential risk to the fetus.

(113) suggested that an oral 5-day regimen of once-daily 750 mg of levofloxacin was effective for mild to moderate pyelonephritis compared with a 10-day regimen of 500 mg of ciprofloxacin twice daily: clinical efficacy of levofloxacin was 86% can ciprofloxacin was 81% (some subjects in both groups received initial parenteral therapy). In another study of complicated UTI and acute pyelonephritis, once daily dosing of 1 g of extended-release ciprofloxacin was found to be effective in a 7- to 14-day regimen (114).

The 2011 IDSA UTI guidelines recommend fluoroquinolones as the only oral antimicrobials for outpatient empiric treatment of acute uncomplicated pyelonephritis (55). Patients who are not hospitalized should receive either oral ciprofloxacin at 500 mg twice daily for 7 days, 1,000 mg of ciprofloxacin extended release daily for 7 days, or levofloxacin at 750 mg once daily for 5 days as first-line empiric therapy (55). The guidelines recommend, given the increasing worldwide prevalence of resistance to fluoroquinolones, that oral treatment may be preceded by an intravenous antimicrobial such as 1 g of ceftriaxone or a consolidated 24-h dose of an aminoglycoside (55). If the prevalence of resistance to fluoroquinolones in the region is known to be >10%, ceftriaxone or a consolidated 24-h aminoglycoside should be considered for empiric treatment. Alternatively, if the organism is known to be susceptible, TMP-SMX 160/800 orally twice daily for 14 days can be given (55); if susceptibility is not known, TMP-SMX should be preceded by a parenteral agent as above. Although the duration of treatment with TMP-SMX approved by the U.S. FDA is 14 days, clinical experience suggests that 7 to 10 days is effective in women who have a rapid response to treatment (14). In contrast, β-lactam regimens shorter than 14 days have been associated with unacceptably high failure rates in some studies (82), and they should not be used alone empirically until susceptibility data are available. As with TMP-SMX, clinical experience suggests that 7 to 10 days is effective when using ceftriaxone in women who have a rapid clinical response.

Patients with pyelonephritis who require hospitalization generally should be treated empirically with intravenous antimicrobials including a fluoroquinolone, an extended-spectrum β-lactam with or without an aminoglycoside, a carbapenem, or an aminoglycoside with or without ampicillin (Table 3) (55). In one trial of men and women hospitalized with severe nonobstructive UTI (women with acute uncomplicated pyelonephritis comprised 47%), oral ciprofloxacin was as effective as intravenous ciprofloxacin in initial empiric treatment, even among those with bacteremia (115). However, given increasing concerns about the resistance of uropathogens even in community-acquired infections, it may be advisable to include a broader-spectrum agent, such as a carbapenem or piperacillin-tazobactam, in the empiric regimen of hospitalized patients until antimicrobial susceptibilities are available. In areas where ESBL bacteria are prevalent causes of UTI, carbapenems should be considered for empiric treatment of pyelonephritis until drug susceptibility data are known. This is especially true in travelers to the Indian subcontinent and Africa, which represents a major risk for rectal colonization with CTX-M-ESBL-producing *E. coli* (75).

TABLE 3 Empiric antimicrobials for inpatient management of acute uncomplicated pyelonephritis[a]

Antimicrobial and Dosing
Ciprofloxacin[b] 400 mg intravenously (i.v.) twice daily
Levofloxacin[b] 500 to 750 mg i.v. once daily
Ceftriaxone 1 to 2 g i.v. once daily
Cefepime 1 g i.v. twice daily
Piperacillin-tazobactam 3.375 g i.v. every 6 hours
Meropenem 500 mg i.v. every 8 hours
Imipenem-cilastatin[b] 500 mg i.v. every 6 to 8 hours
Doripenem 500 mg i.v. every 8 hours
Ertapenem i.v. 1 g once daily

[a]Table adapted from reference 14.
[b]Pregnancy category C: animal studies have shown an adverse effect on the fetus; use only if potential benefit justifies the potential risk to the fetus.

The recommended duration of treatment for pyelonephritis requiring inpatient treatment is 7 to 14 days, depending on the severity of illness and the rapidity of treatment response. Transition to oral medications (usually a fluoroquinolone), based on urine culture results, should be done as soon as the condition allows. In severely ill patients, imaging on admission with renal ultrasound or CT is indicated to rule out a complicating factor that may need to be corrected in conjunction with antimicrobials. Imaging is also indicated in patients with worsening illness or persistent fever 48 to 72 hours after initiation of appropriate antimicrobial treatment or if symptoms are suggestive of a complicating factor such as with renal colic or history of renal stones, diabetes, history of prior urologic surgery, immunosuppression, or repeated episodes of pyelonephritis or urosepsis. Blood cultures are frequently positive in hospitalized patients with pyelonephritis, but clinical outcomes are similar with or without bacteremia (115).

PREVENTION OF RECURRENT UTI

Antimicrobial-Sparing Approaches

Several strategies have been used to prevent recurrent cystitis. Although prevention strategies other than antimicrobial approaches have not been adequately tested in well-designed studies, they carry a low risk of adverse effects and may be helpful, and thus it is reasonable to consider some of them as a way to minimize antimicrobial exposure (14, 116). The goal of management of women with recurrent cystitis should be to improve their quality of life while minimizing antimicrobial exposure; preventive antimicrobial therapy should be considered a last resort (14).

Women who are sexually active or who use spermicides should be counseled about the possible association between these behaviors and recurrent UTI and be advised, as appropriate, to reduce sexual activity or decrease or eliminate the use of spermicide-containing products (14). Although not evaluated in controlled trials, it is also reasonable to recommend early postcoital voiding, liberal fluid intake, not routinely delaying urination, wiping from front to back after defecation, avoiding tight fitting underwear, and avoiding douching (14).

There are plausible biological mechanisms for a beneficial cranberry effect, because some studies have shown inhibition of adherence of uropathogens to uroepithelial cells (117–120) and some studies have shown clinical benefit. However, these clinical studies to date have not definitively demonstrated efficacy, and most have been limited by suboptimal study design (121–127). A recent randomized, placebo-controlled trial of 319 women presenting with acute cystitis showed that drinking 8 ounces of 27% cranberry juice twice daily did not decrease the 6-month incidence of recurrent UTI compared with placebo juice (121). Although the trial was well designed, the study was underpowered to detect a difference between cranberry juice and placebo since sample size calculations were based on an anticipated UTI recurrence rate of 30%, but the actual observed rate in the placebo arm was only 14% (128). A recently updated Cochrane meta-analysis concluded that cranberry products did not significantly reduce the occurrence of symptomatic UTI in women with recurrent UTI (Relative risk [RR] 0.74, 95% confidence interval 0.42 to 1.31) (127). Nevertheless, it seems that most women with recurrent UTI have tried cranberry (and other nonantimicrobial approaches), and some believe that it works, based on anecdotal experience. On the other hand, some believe that it does not work or they are intolerant of the juice.

Interestingly, asymptomatic bacteriuria, either naturally acquired in healthy women (129) or iatrogenic in patients with neurogenic bladders (130), may be protective against recurrent UTI.

Antimicrobial Prophylaxis

Antimicrobial prophylaxis has been demonstrated to be highly effective in reducing the risk of recurrent UTI in women (1, 14, 116, 131–136) and has been advocated for women who experience two or more symptomatic UTIs within 6 months or three or more over 12 months (10, 131–135). However, the decision as to whether to consider antimicrobial prophylaxis should be driven by the degree of discomfort experienced by the woman and concern about drug side effects and selection for antimicrobial resistance. Postcoital prophylaxis, continuous prophylaxis, and intermittent self-diagnosis/self-treatment have all been demonstrated to be highly effective in the management of recurrent uncomplicated cystitis (14, 116, 132–134). The choice of prophylaxis depends on the frequency and pattern of recurrences (e.g., intercourse-related or not) and patient preference and compliance. The choice of antimicrobial should be based on the susceptibility patterns of the strains causing the patient's recent

UTIs, any history of drug allergies, and the patient and provider's comfort level with known side effects of the agents. No antimicrobial regimen has been demonstrated to be better than another (137). Before starting any prophylaxis regimen, eradication of a previous UTI should be ensured by obtaining a negative urine culture 1 to 2 weeks after treatment (14).

Postcoital prophylaxis (Table 4)

Postcoital prophylaxis should be considered when recurrent UTI appear to be associated with sexual intercourse and, depending upon the frequency of sexual activity, usually results in receipt of smaller amounts of antimicrobials than continuous prophylaxis. The only published placebo-controlled trial of postcoital prophylaxis for recurrent uncomplicated UTI showed a decrease in cystitis recurrence rates with postcoital 40/200 mg of TMP-SMX compared with placebo (0.3 versus 3.6 per patient-year) (138). Other antimicrobials appear to be as effective

TABLE 4 Antimicrobials for recurrent acute uncomplicated cystitis[a,b,c,d]

Antimicrobial and dosing	Comments
Post-coital prophylaxis: single dose as soon as feasible after intercourse Nitrofurantoin 50–100 mg TMP-SMX[e] 40 mg/200 mg or 80 mg/400 mg TMP[e] 100 mg Cephalexin 250 mg	Use if UTIs are temporally related to intercourse Ensure no bacteriuria with a negative urine culture at least 1 week after UTI treatment Less exposure to antimicrobials compared to continuous prophylaxis Fluoroquinolones (e.g., ciprofloxacin[e] 125 mg) highly effective but not recommended
Continuous prophylaxis: daily (except fosfomycin) bedtime dose Nitrofurantoin 50–100 mg TMP-SMX[e] 40 mg/200 mg (thrice weekly also effective) TMP[e] 100 mg Cephalexin 125–250 mg Fosfomycin 3-g sachet every 10 days	Ensure no bacteriuria with a negative urine culture at least 1 week after UTI treatment 6-month trial recommended, then stop and observe; if recurrences continue, prophylaxis may be restarted Rare toxicities with long-term exposure to nitrofurantoin include pulmonary hypersensitivity, chronic hepatitis, and peripheral neuropathy; counsel and monitor patient Fluoroquinolones (e.g., ciprofloxacin[e] 125 mg) highly effective but not recommended Antimicrobial resistance in colonizing strains or breakthrough infections reported in some studies

[a]Table adapted from reference 14.
[b]The choice of antimicrobial should be based on the susceptibility pattern of the organism causing the patient's recent UTIs and history of drug allergies.
[c]Culture breakthrough UTIs to assess susceptibility of the infecting uropathogen.
[d]See text for discussion on self-diagnosis/self-treatment strategy.
[e]Pregnancy category C: animal studies have shown an adverse effect on the fetus; use only if potential benefit justifies the potential risk to the fetus.

based on uncontrolled studies of postcoital prophylaxis, including TMP-SMX, nitrofurantoin, cephalexin, and the fluoroquinolones. In a randomized trial in sexually active young women, postcoital 125 mg of ciprofloxacin was as effective in reducing recurrent cystitis as daily ciprofloxacin at 125 mg (139).

Postcoital prophylaxis is also effective in pregnant women, but the choice of drugs is more limited since TMP-SMX or fluoroquinolones should not be used. The preferred regimen in a pregnant woman is a single postcoital dose of either cephalexin or nitrofurantoin (both FDA pregnancy category B) (140). Fosfomycin, also FDA pregnancy category B, is also likely to be effective in pregnancy.

Continuous prophylaxis (Table 4)

Continuous prophylaxis should be considered, after counseling the patient about the risks of long-term use of antimicrobials, when UTIs are not clearly associated with sexual intercourse or if postcoital prophylaxis is ineffective. A Cochrane meta-analysis evaluated 10 trials involving 430 healthy nonpregnant women with 2 to 3 or more UTIs during the previous 12-month period who were treated with continuous or postcoital prophylaxis for 6 to 12 months (137). Antimicrobial agents reviewed included norfloxacin, ciprofloxacin, nitrofurantoin, TMP-SMX, cephalexin, cefaclor, and pefloxacin. During active prophylaxis, the relative risk of clinical recurrences was 0.15 (95% confidence interval: 0.08 to 0.28) in favor of antimicrobial prophylaxis. Likewise, the relative risk of having a microbiological recurrence was 0.21 (95% confidence interval: 0.13 to 0.33) in favor of antimicrobial prophylaxis. Side effects, such as mucosal candidiasis and gastrointestinal symptoms, were significantly more common in the antimicrobial groups. Nitrofurantoin has rarely been associated with severe side effects, as noted above. No conclusions could be made with regard to the optimal antimicrobial, duration, or schedule.

Based on observations that UTIs seem to cluster in some women (141, 142), it is reasonable to recommend a 6-month trial of low-dose antimicrobials administered nightly or thrice weekly (116, 132) and then stopping to observe. However, it appears that most women revert back to the previous pattern of recurrent infections once prophylaxis is stopped (10, 137). Use of TMP-SMX or other agents for as long as 5 years has been reported to be effective and well tolerated (142–144). Nitrofurantoin has also been shown to be safe and well tolerated in prophylaxis regimens up to 12 months, but it may not be as effective as some other agents in such cases for the long term (145). Moreover, as noted previously, there are concerns about toxicity with long-term use. In a double-blinded, randomized, placebo-controlled study of 317 non-pregnant women with a history of recurrent UTI, fosfomycin (3 g every 10 days for 6 months) was associated with a significant reduction in UTIs per patient-year (0.14 versus 2.97) ($P < 0.001$) (146).

Self-diagnosis and self-treatment

An alternative strategy for managing recurrent cystitis in women is self-diagnosis and self-treatment with a short-course regimen of an antimicrobial. The intent of this strategy is not to prevent recurrent UTI but instead to give the woman more autonomy in managing her recurrences by providing instructions on diagnosis and prescriptions for her to obtain antimicrobials without having to consult with her physician. Studies with TMP-SMX (147), norfloxacin (148), and ofloxacin or levofloxacin (7) have shown that recurrent UTI can be accurately self-diagnosed by women up to 95% of the time and that short-course antimicrobial therapy is highly effective in curing the infections. Use of this management strategy should be restricted to women who have clearly documented recurrent episodes of cystitis and who are motivated, compliant, and have a good relationship with their medical provider. Women should be reminded to call their provider if they have symptoms suggestive of pyelonephritis or if their symptoms are not completely resolved by 48 hours on treatment. Urine cultures should be obtained periodically before treatment to confirm the presence of UTI and drug susceptibilities.

Special Considerations about Antimicrobial Prophylaxis

The emergence of antimicrobial resistance to the agent being used for UTI prophylaxis is not frequently seen (116) but may occur in some patients. In one study of ciprofloxacin prophylaxis, 5 of 62 patients receiving daily prophylaxis for 3 months had breakthrough UTIs, 2 with organisms resistant to ciprofloxacin (139). In a study of long-term TMP-SMX prophylaxis, breakthrough infections were due to organisms resistant to TMP-SMX (143). Similarly, in a randomized trial comparing daily oral TMP-SMX to a lactobacillus probiotic for UTI prophylaxis in postmenopausal women, after 1 month of TMP-SMX prophylaxis, resistance to TMP-SMX, trimethoprim, and amoxicillin had increased from approximately 20 to 40% to approximately 80 to 95% in *E. coli* from the feces and urine of asymptomatic women and among *E. coli* causing a UTI (149). Resistance did not increase during lactobacilli prophylaxis.

Almost all antimicrobials used for UTIs have been associated with an increased risk of *C. difficile* diarrhea (150), and alarming increases in both the incidence and severity of *C. difficile* diarrhea have been reported over the past decade (151). The risk of acquiring *C. difficile* colitis should be discussed with patients in whom antimicrobials are being considered for prevention of recurrent UTI, and strong consideration should be given to nonantimicrobial preventive approaches for patients with a history of recurrent *C. difficile* diarrhea.

Nitrofurantoin is commonly used for prophylaxis of recurrent UTI, but as noted previously, it is rarely associated with pulmonary (acute and chronic), hepatitis, and neurologic disease, which can be potentially serious and even fatal (83, 85–92, 116).

Knowledge of such potential adverse effects is essential to enable early recognition and withdrawal of the drug. Patients on long-term nitrofurantoin should be reviewed and monitored regularly for pulmonary or other complications so that adverse effects can be recognized early and the drug withdrawn. Nitrofurantoin should be used with caution in the elderly (91) and is not recommended for patients with CrCl of <60 ml/ minute, although it may be effective at lower CrCl (93). The increasing baseline prevalence of resistance among UPEC strains to trimethoprim or TMP-SMX may complicate prophylaxis with these agents. Cephalosporins have also been shown to be effective in prophylaxis of uncomplicated cystitis (116). However, in a recent study of uropathogens causing uncomplicated cystitis in Europe and Brazil, only 82% of 2,315 *E. coli* strains were susceptible to cefuroxime (152).

Fluoroquinolones are highly effective in prevention of recurrent UTI, but they are not routinely recommended for prophylaxis for uncomplicated UTI given their propensity for causing collateral damage and their association with gastrointestinal, neurological, and cardiovascular side effects, including a prolonged QTc interval and, infrequently, Torsades de pointe (116). The use of fluoroquinolones has also been associated with tendonitis and, rarely, rupture, mainly of the Achilles tendon. Patients undergoing long-term prophylaxis with fluoroquinolones, especially those undergoing strenuous physical activity and those with chronic use of steroids, should be counseled about the risk of tendon rupture. Fluoroquinolones are classified by the manufacturer as pregnancy category C and thus should not be used during pregnancy.

Recurrent UTI in Postmenopausal Women

Replacement topical estrogen (153, 154), but not systemic estrogen (155), has been shown to reduce the risk of UTI in postmenopausal women. In a randomized trial of 93 post-menopausal women with a history of recurrent UTI, topically applied intravaginal estriol cream (0.5 mg estriol nightly for 2 weeks, then twice weekly for 8 months) significantly reduced the incidence of UTI compared with placebo (0.5 versus 5.9 episodes per patient-year, respectively [$P < 0.001$]) (153). Women treated with estrogen cream had an increase in the prevalence of lactobacilli and a decrease in *E. coli* vaginal colonization. Another randomized trial in postmenopausal women with recurrent UTI comparing an estrogen-releasing vaginal ring (Estring) to no estrogen treatment in 103 women over 36 weeks found a significantly higher proportion of women remaining free of UTI in the estrogen group (45% infection-free in the estrogen group versus 20% in the control group) ($P = 0.008$) (154). In contrast, a trial of estriol-containing vaginal pessaries applied twice weekly compared with nitrofurantoin once daily in 171 postmenopausal women found a significantly higher rate of symptomatic UTI in the estrogen group (67%) compared with the nitrofurantoin group (52%) (156). Interestingly, there was no change in the vaginal lactobacillus colonization and pH in the estrogen group as had been noted by these authors in their previous study of topical estriol (153).

In a recent study to better understand the mechanisms whereby estrogens exert a beneficial effect, investigators evaluated the influence of supplemental estrogen on urothelial cells from menstruating and post-menopausal women before and after a 2-week period of estrogen supplementation (157). They found that estrogen induced the expression of antimicrobial peptides, thereby enhancing the antimicrobial capacity of the urothelium and restricting bacterial multiplication. In addition, estrogen prevented excessive loss of superficial cells during infection, thus maintaining the integrity of the epithelial barrier to infection. The investigators also found in an experimental model of UTI in mice that ovariectomy rendered

mice more susceptible to UTI. These data further support the use of vaginal estrogen for prophylaxis against recurrent UTI in postmenopausal women (158).

Investigational Strategies to Prevent Acute Uncomplicated UTI

Probiotics

Probiotics may protect the vagina from colonization by uropathogens through steric hindrance or blocking potential sites of attachment, production of hydrogen peroxide, which is microbicidal to uropathogens, maintenance of a low pH, and/or induction of anti-inflammatory cytokine responses in epithelial cells. There are few randomized controlled trials of lactobacillus probiotics for prevention of UTI in women, and most of these studies did not determine whether the probiotic resulted in vaginal colonization with the probiotic strain (159, 160). However, a recent meta-analysis included 127 patients in 2 studies and found a statistically significant decrease in recurrent UTI in patients given *Lactobacillus* (risk ratio 0.51; 95% confidence interval 0.26 to 0.99, $P = 0.05$) (160). In addition, a recent placebo-controlled pilot trial in 100 premenopausal women with recurrent cystitis evaluated the effect of a strain of *Lactobacillus crispatus* (which constitutes nearly 90% of the vaginal microbial flora), administered as a vaginal capsule daily for 5 days, then once weekly for 10 weeks after treatment of acute cystitis (161). The probiotic was well tolerated, achieved high levels of vaginal colonization, and was associated with a reduced rate of recurrent UTI (15% versus 27% of women in the placebo group). Although some lactobacillus probiotic products have been commercialized, further trials that are adequately designed are needed to establish the role of probiotics in the prevention of UTI.

Vaccines

Although numerous UTI vaccines have been proposed over the past few years, a safe, effective, and easy-to-administer vaccine to reduce the risk of recurrent UTI remains elusive. Vaccine studies have focused on the lipopolysaccharide side chain (O) antigen, virulence factors including P fimbriae, alpha hemolysin, various components of type 1 fimbria (including fimH adhesin), and UPEC outer membrane protein fractions enriched for iron receptors (162). Studies have also been done with vaccines consisting of bacterial components or whole cells. Whole-cell vaccines, made from combinations of heat-killed uropathogenic strains delivered by injection or by a vaginal suppository, have to date had only partial success, and the protective effect appears to wane over several weeks (163).

Development of an effective vaccine will likely depend on a better understanding of the pathogenesis of UTI. It remains unclear why we are unable to generate an effective adaptive immune response after an initial UPEC infection, leaving us susceptible to recurrent UTI with the same UPEC strain (164). We need a better understanding of the mechanisms by which UPEC may be impeding the host's protective adaptive immune response, such as interference with Toll-like receptor signaling, the formation of intracellular bacterial communities, the suppression of cytokine secretion, impedance of antibody binding, and reduction of the secretion of secretory IgA (164). A better understanding of UPEC immune-modulating mechanisms and the mucosal immune response to infection could help guide development of more effective vaccine delivery systems and adjuvants as well as other strategies to prevent UTI (164).

Attachment inhibitors

The adhesion protein fimH, which resides on the distal tips of type 1 pili, filamentous adhesive structures encoded by virtually all UPEC isolates, binds mannose residues on host glycoprotein receptors and allows UPEC to adhere to uroepithelial cells and invade host bladder cells (165). On the basis

of murine models that mimic aspects of human disease, FimH is critical for UPEC pathogenesis (166). FimH is also under positive selection in human clinical isolates of UPEC, further supporting its role in human disease (167). Many emerging approaches to prevent UTIs, therefore, are aimed at blocking adhesion of bacteria to the urothelium and, thus, infection (162).

Small synthetic molecules known as pilicides interfere with pilus assembly and *in vitro* reduce UPEC adherence to bladder epithelial cells as well as type 1 pili-dependent biofilm formation (168), but their efficacy in animal infection models has not been reported. Soluble receptor analogues, or mannosides, bind FimH and prevent it from interacting with host receptors (169), and orally available mannoside derivatives have been developed that show promise as therapeutics (169, 170). In a murine UTI model, these agents prevented bacterial invasion into bladder tissue (166, 170).

Although D-mannose has been available in health food stores and online for years, and its use appears to be increasing for the prevention of recurrent UTI, there have been no human UTI prevention studies performed until recently. In a study of 308 women with recurrent UTI, 2 g daily of D-mannose powder dissolved in water was compared with 50 mg of nitrofurantoin nightly and no prophylaxis (171). Recurrent UTI occurred in 14.6% of women in the D-mannose group, 20.4% of those taking nitrofurantoin, and 60.8% of those with no prophylaxis. Pharmacokinetic studies were not performed. Further studies are warranted to develop safe compounds with good bioavailability, high affinity for fimH, and effective prevention of recurrent cystitis.

CITATION

Pietrucha-Dilanchian P, Hooton TM. 2016. Diagnosis, treatment, and prevention of urinary tract infection. Microbiol Spectrum 4(6):UTI-0021-2015.

REFERENCES

1. **Stamm WE, Hooton TM.** 1993. Management of urinary tract infections in adults. *N Engl J Med* **329:**1328–1334.
2. **Gupta K, Trautner B.** 2012. In the clinic. Urinary tract infection. *Ann Intern Med* **156:** ITC3-1–ITC3-15.
3. **Huppert JS, Biro F, Lan D, Mortensen JE, Reed J, Slap GB.** 2007. Urinary symptoms in adolescent females: STI or UTI? *J Adolesc Health* **40:**418–424.
4. **Prentiss KA, Newby PK, Vinci RJ.** 2011. Adolescent female with urinary symptoms: a diagnostic challenge for the pediatrician. *Pediatr Emerg Care* **27:**789–794.
5. **Hooton TM, Scholes D, Hughes JP, Winter C, Roberts PL, Stapleton AE, Stergachis A, Stamm WE.** 1996. A prospective study of risk factors for symptomatic urinary tract infection in young women. *N Engl J Med* **335:**468–474.
6. **Scholes D, Hooton TM, Roberts PL, Stapleton AE, Gupta K, Stamm WE.** 2000. Risk factors for recurrent UTI in young women. *J Infect Dis* **182:** 1177–1182.
7. **Gupta K, Hooton TM, Roberts PL, Stamm WE.** 2001. Patient-initiated treatment of uncomplicated recurrent urinary tract infections in young women. *Ann Intern Med* **135:**9–16.
8. **Scholes D, Hawn TR, Roberts PL, Li SS, Stapleton AE, Zhao LP, Stamm WE, Hooton TM.** 2010. Family history and risk of recurrent cystitis and pyelonephritis in women. *J Urol* **184:**564–569.
9. **Bent S, Nallamothu BK, Simel DL, Fihn SD, Saint S.** 2002. Does this woman have an acute uncomplicated urinary tract infection? *JAMA* **287:**2701–2710.
10. **Nicolle LE.** 2008. Uncomplicated urinary tract infection in adults including uncomplicated pyelonephritis. *Urol Clin North Am* **35:**1–12.
11. **Nicolle LE, Bradley S, Colgan R, Rice JC, Schaeffer A, Hooton TM, Infectious Diseases Society of America, American Society of Nephrology, American Geriatric Society.** 2005. Infectious Diseases Society of America guidelines for the diagnosis and treatment of asymptomatic bacteriuria in adults. *Clin Infect Dis* **40:**643–654.
12. **Hooton TM, Bradley SF, Cardenas DD, Colgan R, Geerlings SE, Rice JC, Saint S, Schaeffer AJ, Tambayh PA, Tenke P, Nicolle LE, Infectious Diseases Society of America.** 2010. Diagnosis, prevention, and treatment of catheter-associated urinary tract infection in adults: 2009 International Clinical Practice Guidelines from the Infectious Diseases Society of America. *Clin Infect Dis* **50:**625–663.
13. **Lipsky BA.** 1989. Urinary tract infections in men. Epidemiology, pathophysiology, diagnosis, and treatment. *Ann Intern Med* **110:**138–150.
14. **Hooton TM.** 2012. Clinical practice. Uncomplicated urinary tract infection. *N Engl J Med* **366:**1028–1037.
15. **Stamm WE, Counts GW, Running KR, Fihn S, Turck M, Holmes KK.** 1982. Diagnosis of coliform infection in acutely dysuric women. *N Engl J Med* **307:**463–468.
16. **Baerheim A, Digranes A, Hunskaar S.** 1992. Evaluation of urine sampling technique: bacterial contamination of samples from women students. *Br J Gen Pract* **42:**241–243.
17. **Lifshitz E, Kramer L.** 2000. Outpatient urine culture: does collection technique matter? *Arch Intern Med* **160:**2537–2540.
18. **Schneeberger C, van den Heuvel ER, Erwich JJ, Stolk RP, Visser CE, Geerlings SE.** 2013. Contamination rates of three urine-sampling methods to assess bacteriuria in pregnant women. *Obstet Gynecol* **121:**299–305.
19. **Lipsky BA, Ireton RC, Fihn SD, Hackett R, Berger RE.** 1987. Diagnosis of bacteriuria in men: specimen collection and culture interpretation. *J Infect Dis* **155:**847–854.
20. **Marple CD.** 1941. The frequency and character of urinary tract infections in an unselected group of women. *Ann Intern Med* **14:**2220–2239.
21. **Barr RH, Rantz LA.** 1948. The incidence of unsuspected urinary tract infection in a selected group of ambulatory women. *Calif Med* **68:**437–440.
22. **Kass EH.** 1956. Asymptomatic infections of the urinary tract. *Trans Assoc Am Physicians* **69:**56–64.
23. **Kass EH.** 1960. The role of asymptomatic bacteriuria in the pathogenesis of pyelonephritis, p 399–412. *In* Quinn EL, Kass EH (ed), *Biology of Pyelonephritis.* Little, Brown, Boston, MA.
24. **Kass EH.** 1960. Bacteriuria and pyelonephritis of pregnancy. *Arch Intern Med* **105:**194–198.
25. **Kass EH.** 1962. Pyelonephritis and bacteriuria. A major problem in preventive medicine. *Ann Intern Med* **56:**46–53.
26. **Norden CW, Kass EH.** 1968. Bacteriuria of pregnancy: a critical appraisal. *Annu Rev Med* **19:**431–470.
27. **Sanford JP, Favour CB, Mao FH, Harrison JH.** 1956. Evaluation of the positive urine culture; an approach to the differentiation of significant bacteria from contaminants. *Am J Med* **20:**88–93.
28. **Monzon OT, Ory EM, Dobson HL, Carter E, Yow EM.** 1958. A comparison of bacterial counts of the urine obtained by needle aspiration of the bladder, catheterization and midstream-voided methods. *N Engl J Med* **259:**764–767.

29. Gallagher DJ, Montgomerie JZ, North JD. 1965. Acute infections of the urinary tract and the urethral syndrome in general practice. *BMJ* **1**:622–626.

30. Mabeck CE. 1969. Studies in urinary tract infections. I. The diagnosis of bacteriuria in women. *Acta Med Scand* **186**:35–38.

31. Stamey TA, Timothy M, Millar M, Mihara G. 1971. Recurrent urinary infections in adult women. The role of introital enterobacteria. *Calif Med* **115**:1–19.

32. Hooton TM, Roberts PL, Cox ME, Stapleton AE. 2013. Voided midstream urine culture and acute cystitis in premenopausal women. *N Engl J Med* **369**:1883–1891.

33. Rubin RH, Shapiro ED, Andriole VT, Davis RJ, Stamm WE, Infectious Diseases Society of America and the Food and Drug Administration. 1992. Evaluation of new anti-infective drugs for the treatment of urinary tract infection. *Clin Infect Dis* **15**(Suppl 1):S216–S227.

34. Bergqvist D, Brönnestam R, Hedelin H, Ståhl A. 1980. The relevance of urinary sampling methods in patients with indwelling Foley catheters. *Br J Urol* **52**:92–95.

35. Tenney JH, Warren JW. 1988. Bacteriuria in women with long-term catheters: paired comparison of indwelling and replacement catheters. *J Infect Dis* **157**:199–202.

36. Raz R, Schiller D, Nicolle LE. 2000. Chronic indwelling catheter replacement before antimicrobial therapy for symptomatic urinary tract infection. *J Urol* **164**:1254–1258.

37. Arav-Boger R, Leibovici L, Danon YL. 1994. Urinary tract infections with low and high colony counts in young women. Spontaneous remission and single-dose vs multiple-day treatment. *Arch Intern Med* **154**:300–304.

38. Stark RP, Maki DG. 1984. Bacteriuria in the catheterized patient. What quantitative level of bacteriuria is relevant? *N Engl J Med* **311**:560–564.

39. Lehmann LE, Hauser S, Malinka T, Klaschik S, Weber SU, Schewe JC, Stüber F, Book M. 2011. Rapid qualitative urinary tract infection pathogen identification by SeptiFast real-time PCR. *PLoS One* **6**:e17146. doi:10.1371/journal.pone.0017146.

40. Lehmann LE, Hauser S, Malinka T, Klaschik S, Stüber F, Book M. 2010. Real-time polymerase chain-reaction detection of pathogens is feasible to supplement the diagnostic sequence for urinary tract infections. *BJU Int* **106**:114–120.

41. Köser CU, Holden MT, Ellington MJ, Cartwright EJ, Brown NM, Ogilvy-Stuart AL, Hsu LY, Chewapreecha C, Croucher NJ, Harris SR, Sanders M, Enright MC, Dougan G, Bentley SD, Parkhill J, Fraser LJ, Betley JR, Schulz-Trieglaff OB, Smith GP, Peacock SJ. 2012. Rapid whole-genome sequencing for investigation of a neonatal MRSA outbreak. *N Engl J Med* **366**:2267–2275.

42. Török ME, Reuter S, Bryant J, Köser CU, Stinchcombe SV, Nazareth B, Ellington MJ, Bentley SD, Smith GP, Parkhill J, Peacock SJ. 2013. Rapid whole-genome sequencing for investigation of a suspected tuberculosis outbreak. *J Clin Microbiol* **51**:611–614.

43. Stamm WE. 1983. Measurement of pyuria and its relation to bacteriuria. *Am J Med* **75**(1B):53–58.

44. Hooton TM, Stamm WE. 1997. Diagnosis and treatment of uncomplicated urinary tract infection. *Infect Dis Clin North Am* **11**:551–581.

45. St John A, Boyd JC, Lowes AJ, Price CP. 2006. The use of urinary dipstick tests to exclude urinary tract infection: a systematic review of the literature. *Am J Clin Pathol* **126**:428–436.

46. Little P, Turner S, Rumsby K, Jones R, Warner G, Moore M, Lowes JA, Smith H, Hawke C, Leydon G, Mullee M. 2010. Validating the prediction of lower urinary tract infection in primary care: sensitivity and specificity of urinary dipsticks and clinical scores in women. *Br J Gen Pract* **60**:495–500.

47. Hurlbut TA III, Littenberg B. 1991. The diagnostic accuracy of rapid dipstick tests to predict urinary tract infection. *Am J Clin Pathol* **96**:582–588.

48. Devillé WL, Yzermans JC, van Duijn NP, Bezemer PD, van der Windt DA, Bouter LM. 2004. The urine dipstick test useful to rule out infections. A meta-analysis of the accuracy. *BMC Urol* **4**:4.

49. Patel HD, Livsey SA, Swann RA, Bukhari SS. 2005. Can urine dipstick testing for urinary tract infection at point of care reduce laboratory workload? *J Clin Pathol* **58**:951–954.

50. Giesen LG, Cousins G, Dimitrov BD, van de Laar FA, Fahey T. 2010. Predicting acute uncomplicated urinary tract infection in women: a systematic review of the diagnostic accuracy of symptoms and signs. *BMC Fam Pract* **11**:78.

51. McIsaac WJ, Moineddin R, Ross S. 2007. Validation of a decision aid to assist physicians in reducing unnecessary antibiotic drug use for acute cystitis. *Arch Intern Med* **167**:2201–2206.

52. Christiaens TC, De Meyere M, Verschraegen G, Peersman W, Heytens S, De Maeseneer JM. 2002. Randomised controlled trial of nitrofurantoin versus placebo in the treatment of uncomplicated urinary tract infection in adult women. *Br J Gen Pract* **52**:729–734.

53. **Ferry SA, Holm SE, Stenlund H, Lundholm R, Monsen TJ.** 2007. Clinical and bacteriological outcome of different doses and duration of pivmecillinam compared with placebo therapy of uncomplicated lower urinary tract infection in women: the LUTIW project. *Scand J Prim Health Care* 25:49–57.

54. **Foxman B, Frerichs RR.** 1985. Epidemiology of urinary tract infection: I. Diaphragm use and sexual intercourse. *Am J Public Health* 75:1308–1313.

55. **Gupta K, Hooton TM, Naber KG, Wullt B, Colgan R, Miller LG, Moran GJ, Nicolle LE, Raz R, Schaeffer AJ, Soper DE, Infectious Diseases Society of America, European Society for Microbiology and Infectious Diseases.** 2011. International clinical practice guidelines for the treatment of acute uncomplicated cystitis and pyelonephritis in women: a 2010 update by the Infectious Diseases Society of America and the European Society for Microbiology and Infectious Diseases. *Clin Infect Dis* 52:e103–e120.

56. **Paterson DL.** 2004. "Collateral damage" from cephalosporin or quinolone antibiotic therapy. *Clin Infect Dis* 38(Suppl 4):S341–S345.

57. **Gupta K, Hooton TM, Miller L, Uncomplicated UTI IDSA Guideline Committee.** 2011. Managing uncomplicated urinary tract infection: making sense out of resistance data. *Clin Infect Dis* 53:1041–1042.

58. **Richards DA, Toop LJ, Chambers ST, Sutherland MG, Harris BH, Ikram RB, Jones MR, McGeoch GR, Peddie B.** 2002. Antibiotic resistance in uncomplicated urinary tract infection: problems with interpreting cumulative resistance rates from local community laboratories. *N Z Med J* 115:12–14.

59. **Ti TY, Kumarasinghe G, Taylor MB, Tan SL, Ee A, Chua C, Low A.** 2003. What is true community-acquired urinary tract infection? Comparison of pathogens identified in urine from routine outpatient specimens and from community clinics in a prospective study. *Eur J Clin Microbiol Infect Dis* 22:242–245.

60. **Baerheim A, Digranes A, Hunskaar S.** 1999. Are resistance patterns in uropathogens published by microbiological laboratories valid for general practice? *APMIS* 107:676–680.

61. **Zhanel GG, Hisanaga TL, Laing NM, DeCorby MR, Nichol KA, Weshnoweski B, Johnson J, Noreddin A, Low DE, Karlowsky JA, Hoban DJ, NAUTICA Group.** 2006. Antibiotic resistance in *Escherichia coli* outpatient urinary isolates: final results from the North American Urinary Tract Infection Collaborative Alliance (NAUTICA). *Int J Antimicrob Agents* 27:468–475.

62. **Naber KG, Schito G, Botto H, Palou J, Mazzei T.** 2008. Surveillance study in Europe and Brazil on clinical aspects and Antimicrobial Resistance Epidemiology in Females with Cystitis (ARESC): implications for empiric therapy. *Eur Urol* 54:1164–1175.

63. **Wagenlehner FM, Hoyme U, Kaase M, Fünfstück R, Naber KG, Schmiemann G.** 2011. Uncomplicated urinary tract infections. *Dtsch Arztebl Int* 108:415–423.

64. **Brown PD, Freeman A, Foxman B.** 2002. Prevalence and predictors of trimethoprim-sulfamethoxazole resistance among uropathogenic *Escherichia coli* isolates in Michigan. *Clin Infect Dis* 34:1061–1066.

65. **Metlay JP, Strom BL, Asch DA.** 2003. Prior antimicrobial drug exposure: a risk factor for trimethoprim-sulfamethoxazole-resistant urinary tract infections. *J Antimicrob Chemother* 51:963–970.

66. **Colgan R, Johnson JR, Kuskowski M, Gupta K.** 2008. Risk factors for trimethoprim-sulfamethoxazole resistance in patients with acute uncomplicated cystitis. *Antimicrob Agents Chemother* 52:846–851.

67. **Sanchez GV, Master RN, Karlowsky JA, Bordon JM.** 2012. *In vitro* antimicrobial resistance of urinary *Escherichia coli* isolates among U.S. outpatients from 2000 to 2010. *Antimicrob Agents Chemother* 56:2181–2183.

68. **Omigie O, Okoror L, Umolu P, Ikuuh G.** 2009. Increasing resistance to quinolones: a four-year prospective study of urinary tract infection pathogens. *Int J Gen Med* 2:171–175.

69. **Olson RP, Harrell LJ, Kaye KS.** 2009. Antibiotic resistance in urinary isolates of *Escherichia coli* from college women with urinary tract infections. *Antimicrob Agents Chemother* 53:1285–1286.

70. **Nicolle LE.** 2011. Update in adult urinary tract infection. *Curr Infect Dis Rep* 13:552–560.

71. **Schito GC, Naber KG, Botto H, Palou J, Mazzei T, Gualco L, Marchese A.** 2009. The ARESC study: an international survey on the antimicrobial resistance of pathogens involved in uncomplicated urinary tract infections. *Int J Antimicrob Agents* 34:407–413.

72. **Ho PL, Yip KS, Chow KH, Lo JY, Que TL, Yuen KY.** 2010. Antimicrobial resistance among uropathogens that cause acute uncomplicated cystitis in women in Hong Kong: a prospective multicenter study in 2006 to 2008. *Diagn Microbiol Infect Dis* 66:87–93.

73. **Meier S, Weber R, Zbinden R, Ruef C, Hasse B.** 2011. Extended-spectrum β-lactamase-producing Gram-negative pathogens in community-acquired urinary tract infections: an increasing

challenge for antimicrobial therapy. *Infection* **39:**333–340.

74. **Peirano G, Pitout JD.** 2010. Molecular epidemiology of *Escherichia coli* producing CTX-M beta-lactamases: the worldwide emergence of clone ST131 O25:H4. *Int J Antimicrob Agents* **35:**316–321.

75. **Peirano G, Laupland KB, Gregson DB, Pitout JD.** 2011. Colonization of returning travelers with CTX-M-producing *Escherichia coli. J Travel Med* **18:**299–303.

76. **Falagas ME, Vouloumanou EK, Togias AG, Karadima M, Kapaskelis AM, Rafailidis PI, Athanasiou S.** 2010. Fosfomycin versus other antibiotics for the treatment of cystitis: a meta-analysis of randomized controlled trials. *J Antimicrob Chemother* **65:**1862–1877.

77. **Prakash V, Lewis JS II, Herrera ML, Wickes BL, Jorgensen JH.** 2009. Oral and parenteral therapeutic options for outpatient urinary infections caused by enterobacteriaceae producing CTX-M extended-spectrum beta-lactamases. *Antimicrob Agents Chemother* **53:**1278–1280.

78. **Auer S, Wojna A, Hell M.** 2010. Oral treatment options for ambulatory patients with urinary tract infections caused by extended-spectrum-beta-lactamase-producing *Escherichia coli. Antimicrob Agents Chemother* **54:**4006–4008.

79. **Rodríguez-Baño J, Alcalá JC, Cisneros JM, Grill F, Oliver A, Horcajada JP, Tórtola T, Mirelis B, Navarro G, Cuenca M, Esteve M, Peña C, Llanos AC, Cantón R, Pascual A.** 2008. Community infections caused by extended-spectrum beta-lactamase-producing *Escherichia coli. Arch Intern Med* **168:**1897–1902.

80. **Milo G, Katchman EA, Paul M, Christiaens T, Baerheim A, Leibovici L.** 2005. Duration of antibacterial treatment for uncomplicated urinary tract infection in women. *Cochrane Database Syst Rev* (2):CD004682.

81. **Lutters M, Vogt-Ferrier NB.** 2008. Antibiotic duration for treating uncomplicated, symptomatic lower urinary tract infections in elderly women. *Cochrane Database Syst Rev* (3):CD001535.

82. **Warren JW, Abrutyn E, Hebel JR, Johnson JR, Schaeffer AJ, Stamm WE, Infectious Diseases Society of America (IDSA).** 1999. Guidelines for antimicrobial treatment of uncomplicated acute bacterial cystitis and acute pyelonephritis in women. *Clin Infect Dis* **29:**745–758.

83. **De Zeeuw J, Schumacher J, Gillissen AG.** 2012. *Nitrofurantoin-induced pulmonary injury.* http://www.uptodate.com/contents/nitrofurantoin-induced-pulmonary-injury.

84. **Gupta K, Hooton TM, Roberts PL, Stamm WE.** 2007. Short-course nitrofurantoin for the treatment of acute uncomplicated cystitis in women. *Arch Intern Med* **167:**2207–2212.

85. **Reynolds TD, Thomas J.** 2013. Nitrofurantoin related pulmonary disease: a clinical reminder. *BMJ Case Rep* **2013:**bcr2013009299. doi:10.1136/bcr-2013-009299.

86. **Madani Y, Mann B.** 2012. Nitrofurantoin-induced lung disease and prophylaxis of urinary tract infections. *Prim Care Respir J* **21:**337–341.

87. **Williams EM, Triller DM.** 2006. Recurrent acute nitrofurantoin-induced pulmonary toxicity. *Pharmacotherapy* **26:**713–718.

88. **Mendez JL, Nadrous HF, Hartman TE, Ryu JH.** 2005. Chronic nitrofurantoin-induced lung disease. *Mayo Clin Proc* **80:**1298–1302.

89. **Williams EM, Triller DM.** 2006. Recurrent acute nitrofurantoin-induced pulmonary toxicity. *Pharmacotherapy* **26:**713–718.

90. **Harrabi H.** 2012. Uncomplicated urinary tract infection. *N Engl J Med* **367:**185, author reply 185.

91. **Campanelli CM, American Geriatrics Society 2012 Beers Criteria Update Expert Panel.** 2012. American Geriatrics Society updated Beers Criteria for potentially inappropriate medication use in older adults. *J Am Geriatr Soc* **60:**616–631.

92. **Guay DR.** 2001. An update on the role of nitrofurans in the management of urinary tract infections. *Drugs* **61:**353–364.

93. **Oplinger M, Andrews CO.** 2013. Nitrofurantoin contraindication in patients with a creatinine clearance below 60 mL/min: looking for the evidence. *Ann Pharmacother* **47:**106–111.

94. **Geerts AF, Eppenga WL, Heerdink R, Derijks HJ, Wensing MJ, Egberts TC, De Smet PA.** 2013. Ineffectiveness and adverse events of nitrofurantoin in women with urinary tract infection and renal impairment in primary care. *Eur J Clin Pharmacol* **69:**1701–1707.

95. **Raz R, Chazan B, Kennes Y, Colodner R, Rottensterich E, Dan M, Lavi I, Stamm W, Israeli Urinary Tract Infection Group.** 2002. Empiric use of trimethoprim-sulfamethoxazole (TMP-SMX) in the treatment of women with uncomplicated urinary tract infections, in a geographical area with a high prevalence of TMP-SMX-resistant uropathogens. *Clin Infect Dis* **34:**1165–1169.

96. **Hooton TM.** 2003. The current management strategies for community-acquired urinary tract infection. *Infect Dis Clin North Am* **17:**303–332.

97. **Gupta K, Hooton TM, Stamm WE.** 2001. Increasing antimicrobial resistance and the management of uncomplicated community-acquired urinary tract infections. *Ann Intern Med* **135:**41–50.

98. **Kallen AJ, Welch HG, Sirovich BE.** 2006. Current antibiotic therapy for isolated urinary tract infections in women. *Arch Intern Med* **166:**635–639.

99. **Pépin J, Saheb N, Coulombe MA, Alary ME, Corriveau MP, Authier S, Leblanc M, Rivard G, Bettez M, Primeau V, Nguyen M, Jacob CE, Lanthier L.** 2005. Emergence of fluoroquinolones as the predominant risk factor for *Clostridium difficile*-associated diarrhea: a cohort study during an epidemic in Quebec. *Clin Infect Dis* **41:**1254–1260.

100. **Hooton TM, Scholes D, Gupta K, Stapleton AE, Roberts PL, Stamm WE.** 2005. Amoxicillin-clavulanate vs ciprofloxacin for the treatment of uncomplicated cystitis in women: a randomized trial. *JAMA* **293:**949–955.

101. **Hooton TM, Roberts PL, Stapleton AE.** 2012. Cefpodoxime vs ciprofloxacin for short-course treatment of acute uncomplicated cystitis: a randomized trial. *JAMA* **307:**583–589.

102. **Klimberg I, Shockey G, Ellison H, Fuller-Jonap F, Colgan R, Song J, Keating K, Cyrus P.** 2005. Time to symptom relief for uncomplicated urinary tract infection treated with extended-release ciprofloxacin: a prospective, open-label, uncontrolled primary care study. *Curr Med Res Opin* **21:**1241–1250.

103. **Little P, Moore MV, Turner S, Rumsby K, Warner G, Lowes JA, Smith H, Hawke C, Leydon G, Arscott A, Turner D, Mullee M.** 2010. Effectiveness of five different approaches in management of urinary tract infection: randomised controlled trial. *BMJ* **340:**c199.

104. **Leydon GM, Turner S, Smith H, Little P, UTIS Team.** 2010. Women's views about management and cause of urinary tract infection: qualitative interview study. *BMJ* **340:**c279.

105. **Bleidorn J, Gágyor I, Kochen MM, Wegscheider K, Hummers-Pradier E.** 2010. Symptomatic treatment (ibuprofen) or antibiotics (ciprofloxacin) for uncomplicated urinary tract infection? Results of a randomized controlled pilot trial. *BMC Med* **8:**30. doi:10.1186/1741-7015-8-30.

106. **Gágyor I, Hummers-Pradier E, Kochen MM, Schmiemann G, Wegscheider K, Bleidorn J.** 2012. Immediate versus conditional treatment of uncomplicated urinary tract infection: a randomized-controlled comparative effectiveness study in general practices. *BMC Infect Dis* **12:**146. doi:10.1186/1471-2334-12-146.

107. **Hannan TJ, Mysorekar IU, Hung CS, Isaacson-Schmid ML, Hultgren SJ.** 2010. Early severe inflammatory responses to uropathogenic *E. coli* predispose to chronic and recurrent urinary tract infection. *PLoS Pathog* **6:**e1001042. doi:10.1371/journal.ppat1001042.

108. **Czaja CA, Scholes D, Hooton TM, Stamm WE.** 2007. Population-based epidemiologic analysis of acute pyelonephritis. *Clin Infect Dis* **45:**273–280.

109. **van der Starre WE, van Dissel JT, van Nieuwkoop C.** 2011. Treatment duration of febrile urinary tract infections. *Curr Infect Dis Rep* **13:**571–578.

110. **Talan DA, Stamm WE, Hooton TM, Moran GJ, Burke T, Iravani A, Reuning-Scherer J, Church DA.** 2000. Comparison of ciprofloxacin (7 days) and trimethoprim-sulfamethoxazole (14 days) for acute uncomplicated pyelonephritis pyelonephritis in women: a randomized trial. *JAMA* **283:**1583–1590.

111. **Johnson JR, Lyons MF II, Pearce W, Gorman P, Roberts PL, White N, Brust P, Olsen R, Gnann JW Jr, Stamm WE.** 1991. Therapy for women hospitalized with acute pyelonephritis: a randomized trial of ampicillin versus trimethoprim-sulfamethoxazole for 14 days. *J Infect Dis* **163:**325–330.

112. **Peterson J, Kaul S, Khashab M, Fisher AC, Kahn JB.** 2008. A double-blind, randomized comparison of levofloxacin 750 mg once-daily for five days with ciprofloxacin 400/500 mg twice-daily for 10 days for the treatment of complicated urinary tract infections and acute pyelonephritis. *Urology* **71:**17–22.

113. **Klausner HA, Brown P, Peterson J, Kaul S, Khashab M, Fisher AC, Kahn JB.** 2007. A trial of levofloxacin 750 mg once daily for 5 days versus ciprofloxacin 400 mg and/or 500 mg twice daily for 10 days in the treatment of acute pyelonephritis. *Curr Med Res Opin* **23:**2637–2645.

114. **Talan DA, Klimberg IW, Nicolle LE, Song J, Kowalsky SF, Church DA.** 2004. Once daily, extended release ciprofloxacin for complicated urinary tract infections and acute uncomplicated pyelonephritis. *J Urol* **171:**734–739.

115. **Mombelli G, Pezzoli R, Pinoja-Lutz G, Monotti R, Marone C, Franciolli M.** 1999. Oral vs intravenous ciprofloxacin in the initial empirical management of severe pyelonephritis or complicated urinary tract infections: a prospective randomized clinical trial. *Arch Intern Med* **159:**53–58.

116. **Lichtenberger P, Hooton TM.** 2011. Antimicrobial prophylaxis in women with recurrent urinary tract infections. *Int J Antimicrob Agents* **38**(Suppl):36–41.

117. **Sobota AE.** 1984. Inhibition of bacterial adherence by cranberry juice: potential use for the treatment of urinary tract infections. *J Urol* **131:**1013–1016.

118. **Schmidt DR, Sobota AE.** 1988. An examination of the anti-adherence activity of cranberry juice

on urinary and nonurinary bacterial isolates. *Microbios* **55:**173–181.

119. **Zafriri D, Ofek I, Adar R, Pocino M, Sharon N.** 1989. Inhibitory activity of cranberry juice on adherence of type 1 and type P fimbriated *Escherichia coli* to eucaryotic cells. *Antimicrob Agents Chemother* **33:**92–98.

120. **Howell AB, Vorsa N, Der Marderosian A, Foo LY.** 1998. Inhibition of the adherence of P-fimbriated *Escherichia coli* to uroepithelial-cell surfaces by proanthocyanidin extracts from cranberries. *N Engl J Med* **339:**1085–1086.

121. **Barbosa-Cesnik C, Brown MB, Buxton M, Zhang L, DeBusscher J, Foxman B.** 2011. Cranberry juice fails to prevent recurrent urinary tract infection: results from a randomized placebo-controlled trial. *Clin Infect Dis* **52:**23–30.

122. **Beerepoot MA, ter Riet G, Nys S, van der Wal WM, de Borgie CA, de Reijke TM, Prins JM, Koeijers J, Verbon A, Stobberingh E, Geerlings SE.** 2011. Cranberries vs antibiotics to prevent urinary tract infections: a randomized double-blind noninferiority trial in premenopausal women. *Arch Intern Med* **171:**1270–1278.

123. **Kontiokari T, Sundqvist K, Nuutinen M, Pokka T, Koskela M, Uhari M.** 2001. Randomised trial of cranberry-lingonberry juice and *Lactobacillus* GG drink for the prevention of urinary tract infections in women. *BMJ* **322:**1571.

124. **Stothers L.** 2002. A randomized trial to evaluate effectiveness and cost effectiveness of naturopathic cranberry products as prophylaxis against urinary tract infection in women. *Can J Urol* **9:**1558–1562.

125. **Stapleton AE, Dziura J, Hooton TM, Cox ME, Yarova-Yarovaya Y, Chen S, Gupta K.** 2012. Recurrent urinary tract infection and urinary *Escherichia coli* in women ingesting cranberry juice daily: a randomized controlled trial. *Mayo Clin Proc* **87:**143–150.

126. **Wang CH, Fang CC, Chen NC, Liu SS, Yu PH, Wu TY, Chen WT, Lee CC, Chen SC.** 2012. Cranberry-containing products for prevention of urinary tract infections in susceptible populations: a systematic review and meta-analysis of randomized controlled trials. *Arch Intern Med* **172:**988–996.

127. **Jepson RG, Williams G, Craig JC.** 2012. Cranberries for preventing urinary tract infections. *Cochrane Database Syst Rev* **10:**CD001321.

128. **Eells SJ, McKinnell JA, Miller LG.** 2011. Daily cranberry prophylaxis to prevent recurrent urinary tract infections may be beneficial in some populations of women. *Clin Infect Dis* **52:**1393–1394, author reply 1394–1395.

129. **Cai T, Mazzoli S, Mondaini N, Meacci F, Nesi G, D'Elia C, Malossini G, Boddi V, Bartoletti R.** 2012. The role of asymptomatic bacteriuria in young women with recurrent urinary tract infections: to treat or not to treat? *Clin Infect Dis* **55:**771–777.

130. **Darouiche RO, Green BG, Donovan WH, Chen D, Schwartz M, Merritt J, Mendez M, Hull RA.** 2011. Multicenter randomized controlled trial of bacterial interference for prevention of urinary tract infection in patients with neurogenic bladder. *Urology* **78:**341–346.

131. **Nicolle LE, Ronald AR.** 1987. Recurrent urinary tract infection in adult women: diagnosis and treatment. *Infect Dis Clin North Am* **1:**793–806.

132. **Hooton TM.** 2001. Recurrent urinary tract infection in women. *Int J Antimicrob Agents* **17:**259–268.

133. **Nicolle LE.** 1992. Prophylaxis: recurrent cystitis in non pregnant women. *Infection* **20** (suppl 3):s164–s170.

134. **Gupta K, Trautner BW.** 2013. Diagnosis and management of recurrent urinary tract infections in non-pregnant women. *BMJ* **346:**f3140.

135. **Ronald AR, Conway B.** 1988. An approach to urinary tract infections in ambulatory women. *Curr Clin Top Infect Dis* **9:**76–125.

136. **Wagenlehner FM, Vahlensieck W, Bauer HW, Weidner W, Piechota HJ, Naber KG.** 2013. Prevention of recurrent urinary tract infections. *Minerva Urol Nefrol* **65:**9–20.

137. **Albert X, Huertas I, Pereiró II, Sanfélix J, Gosalbes V, Perrota C.** 2004. Antibiotics for preventing recurrent urinary tract infection in non-pregnant women. *Cochrane Database Syst Rev* (3):CD001209.

138. **Stapleton A, Latham RH, Johnson C, Stamm WE.** 1990. Postcoital antimicrobial prophylaxis for recurrent urinary tract infection. A randomized, double-blind, placebo-controlled trial. *JAMA* **264:**703–706.

139. **Melekos MD, Asbach HW, Gerharz E, Zarakovitis IE, Weingaertner K, Naber KG.** 1997. Post-intercourse versus daily ciprofloxacin prophylaxis for recurrent urinary tract infections in premenopausal women. *J Urol* **157:**935–939.

140. **Pfau A, Sacks TG.** 1992. Effective prophylaxis for recurrent urinary tract infections during pregnancy. *Clin Infect Dis* **14:**810–814.

141. **Kraft JK, Stamey TA.** 1977. The natural history of symptomatic recurrent bacteriuria in women. *Medicine (Baltimore)* **56:**55–60.

142. **Stamm WE, McKevitt M, Roberts PL, White NJ.** 1991. Natural history of recurrent urinary tract infections in women. *Rev Infect Dis* **13:**77–84.

143. **Nicolle LE, Harding GK, Thomson M, Kennedy J, Urias B, Ronald AR.** 1988. Efficacy of five

years of continuous, low-dose trimethoprim-sulfamethoxazole prophylaxis for urinary tract infection. *J Infect Dis* **157:**1239–1242.

144. **Harding GK, Ronald AR, Nicolle LE, Thomson MJ, Gray GJ.** 1982. Long-term antimicrobial prophylaxis for recurrent urinary tract infection in women. *Rev Infect Dis* **4:**438–443.

145. **Brumfitt W, Hamilton-Miller JM.** 1998. Efficacy and safety profile of long-term nitrofurantoin in urinary infections: 18 years' experience. *J Antimicrob Chemother* **42:**363–371.

146. **Rudenko N, Dorofeyev A.** 2005. Prevention of recurrent lower urinary tract infections by long-term administration of fosfomycin trometamol. Double blind, randomized, parallel group, placebo controlled study. *Arzneimittelforschung* **55:**420–427.

147. **Wong ES, McKevitt M, Running K, Counts GW, Turck M, Stamm WE.** 1985. Management of recurrent urinary tract infections with patient-administered single-dose therapy. *Ann Intern Med* **102:**302–307.

148. **Schaeffer AJ, Stuppy BA.** 1999. Efficacy and safety of self-start therapy in women with recurrent urinary tract infections. *J Urol* **161:**207–211.

149. **Beerepoot MA, ter Riet G, Nys S, van der Wal WM, de Borgie CA, de Reijke TM, Prins JM, Koeijers J, Verbon A, Stobberingh E, Geerlings SE.** 2012. Lactobacilli vs antibiotics to prevent urinary tract infections: a randomized, double-blind, noninferiority trial in postmenopausal women. *Arch Intern Med* **172:**704–712.

150. **Loo VG, Bourgault AM, Poirier L, Lamothe F, Michaud S, Turgeon N, Toye B, Beaudoin A, Frost EH, Gilca R, Brassard P, Dendukuri N, Béliveau C, Oughton M, Brukner I, Dascal A.** 2011. Host and pathogen factors for *Clostridium difficile* infection and colonization. *N Engl J Med* **365:**1693–1703.

151. **Blossom DB, McDonald LC.** 2007. The challenges posed by reemerging *Clostridium difficile* infection. *Clin Infect Dis* **45:**222–227.

152. **Schito GC, Naber KG, Botto H, Palou J, Mazzei T, Gualco L, Marchese A.** 2009. The ARESC study: an international survey on the antimicrobial resistance of pathogens involved in uncomplicated urinary tract infections. *Int J Antimicrob Agents* **34:**407–413.

153. **Raz R, Stamm WE.** 1993. A controlled trial of intravaginal estriol in postmenopausal women with recurrent urinary tract infections. *N Engl J Med* **329:**753–756.

154. **Eriksen B.** 1999. A randomized, open, parallel-group study on the preventive effect of an estradiol-releasing vaginal ring (Estring) on recurrent urinary tract infections in postmenopausal women. *Am J Obstet Gynecol* **180:**1072–1079.

155. **Perrotta C, Aznar M, Mejia R, Albert X, Ng CW.** 2008. Oestrogens for preventing recurrent urinary tract infection in postmenopausal women. *Cochrane Database Syst Rev* (2): CD005131.

156. **Raz R, Colodner R, Rohana Y, Battino S, Rottensterich E, Wasser I, Stamm W.** 2003. Effectiveness of estriol-containing vaginal pessaries and nitrofurantoin macrocrystal therapy in the prevention of recurrent urinary tract infection in postmenopausal women. *Clin Infect Dis* **36:**1362–1368.

157. **Lüthje P, Brauner H, Ramos NL, Ovregaard A, Gläser R, Hirschberg AL, Aspenström P, Brauner A.** 2013. Estrogen supports urothelial defense mechanisms. *Sci Transl Med* **5:**190ra80.

158. **Hannan TJ, Hooton TM, Hultgren SJ.** 2013. Estrogen and recurrent UTI: what are the facts? *Sci Transl Med* **5:**190fs23.

159. **Barrons R, Tassone D.** 2008. Use of *Lactobacillus* probiotics for bacterial genitourinary infections in women: a review. *Clin Ther* **30:**453–468.

160. **Grin PM, Kowalewska PM, Alhazzan W, Fox-Robichaud AE.** 2013. *Lactobacillus* for preventing recurrent urinary tract infections in women: meta-analysis. *Can J Urol* **20:**6607–6614.

161. **Stapleton AE, Au-Yeung M, Hooton TM, Fredricks DN, Roberts PL, Czaja CA, Yarova-Yarovaya Y, Fiedler T, Cox M, Stamm WE.** 2011. Randomized, placebo-controlled phase 2 trial of a *Lactobacillus crispatus* probiotic given intravaginally for prevention of recurrent urinary tract infection. *Clin Infect Dis* **52:**1212–1217.

162. **Sivick KE, Mobley HL.** 2010. Waging war against uropathogenic *Escherichia coli*: winning back the urinary tract. *Infect Immun* **78:**568–585.

163. **Barber AE, Norton JP, Spivak AM, Mulvey MA.** 2013. Urinary tract infections: current and emerging management strategies. *Clin Infect Dis* **57:**719–724.

164. **Brumbaugh AR, Mobley HL.** 2012. Preventing urinary tract infection: progress toward an effective *Escherichia coli* vaccine. *Expert Rev Vaccines* **11:**663–676.

165. **Martinez JJ, Mulvey MA, Schilling JD, Pinkner JS, Hultgren SJ.** 2000. Type 1 pilus-mediated bacterial invasion of bladder epithelial cells. *EMBO J* **19:**2803–2812.

166. **Totsika M, Kostakioti M, Hannan TJ, Upton M, Beatson SA, Janetka JW, Hultgren SJ, Schembri MA.** 2013. A FimH inhibitor prevents acute bladder infection and treats chronic cystitis caused by multidrug-resistant uropathogenic *Escherichia coli* ST131. *J Infect Dis* **208:**921–928.

167. **Chen SL, Hung CS, Pinkner JS, Walker JN, Cusumano CK, Li Z, Bouckaert J, Gordon JI, Hultgren SJ.** 2009. Positive selection identifies an *in vivo* role for FimH during urinary tract infection in addition to mannose binding. *Proc Natl Acad Sci USA* **106**:22439–22444.

168. **Han Z, Pinkner JS, Ford B, Obermann R, Nolan W, Wildman SA, Hobbs D, Ellenberger T, Cusumano CK, Hultgren SJ, Janetka JW.** 2010. Structure-based drug design and optimization of mannoside bacterial FimH antagonists. *J Med Chem* **53**:4779–4792.

169. **Klein T, Abgottspon D, Wittwer M, Rabbani S, Herold J, Jiang X, Kleeb S, Lüthi C, Scharenberg M, Bezençon J, Gubler E, Pang L, Smiesko M, Cutting B, Schwardt O, Ernst B.** 2010. FimH antagonists for the oral treatment of urinary tract infections: from design and synthesis to *in vitro* and *in vivo* evaluation. *J Med Chem* **53**:8627–8641.

170. **Cusumano CK, Pinkner JS, Han Z, Greene SE, Ford BA, Crowley JR, Henderson JP, Janetka JW, Hultgren SJ.** 2011. Treatment and prevention of urinary tract infection with orally active FimH inhibitors. *Sci Transl Med* **3**:109ra115.

171. **Kranjčec B, Papeš D, Altarac S.** 2013. D-mannose powder for prophylaxis of recurrent urinary tract infections in women: a randomized clinical trial. *World J Urol* **32**:79–84.

Urinary Tract Infections in Infants and Children

4

THERESA A. SCHLAGER[1]

EPIDEMIOLOGY

The most important variables influencing prevalence of urinary tract infection (UTI) are age and sex (1). In neonates, the rate for premature infants (2.9%) and very-low-birthweight infants (4 to 25%) exceeds that for full-term infants (0.7%). Male preponderance persists for the first 3 months of life, after which the prevalence rate among females exceeds that in males. The prevalence rate reported in girls 1 to 5 years of age is 1 to 3%, whereas few infections occur in boys of those ages. This is the age range in which children are most likely to experience a first symptomatic infection. Symptomatic infections occur 10 to 20 times more commonly in preschool-aged girls than in preschool-aged boys.

Febrile infants and children commonly have UTI, with rates of infection inversely proportionate to age. In four separate studies, specific rates of UTI were 7.5% in 442 febrile episodes infants younger than 8 weeks (2), 5.3% in 945 febrile episodes in those younger than 1 year of age (3), 4.1% of 501 episodes in children younger than 2 years (4); and 1.7% of 664 episodes in children younger than 5 years (5).

The recurrence rate for UTI in girls is substantial regardless of the presence or absence of a urinary tract abnormality. The greatest risk of recurrence is

[1]Department of Emergency Medicine, University of Virginia, Charlottesville, VA 22908.
Urinary Tract Infections: Molecular Pathogenesis and Clinical Management, 2nd Edition
Edited by Matthew A. Mulvey, David J. Klumpp, and Ann E. Stapleton
© 2017 American Society for Microbiology, Washington, DC
doi:10.1128/microbiolspec.UTI-0022-2016

during the first few months after an infection (6). In the United States, approximately 75 to 80% of white and 50% of African American school-aged girls have recurrence of UTI within 3 months of the first infection. Recurrences are less frequent in males, affecting approximately one-third of those with a UTI. Key risk factors for recurrent UTI include vesicoureteral reflux (VUR) and bowel and bladder dysfunction, especially in children with both conditions (7).

Enterobacteriaceae are the most common cause of uncomplicated UTI, uropathogenic *Escherichia coli* (UPEC) being responsible for 70 to 90% of first UTIs in outpatients (1, 8). Although *E. coli* is the most common cause of infection in children with underlying urinary tract abnormalities, other microorganisms, including *Klebsiella* spp., *Proteus* spp., *Enterococcus* spp., *Pseudomonas* spp., and *Enterobacter* spp. may be prevalent. *Staphylococcus saprophyticus*, a coagulase-negative staphylococcus, accounts for 15% or more of UTIs in female adolescents. Group B streptococci are unusual urinary tract pathogens but are occasionally isolated from the infected urine of neonates and adolescents. *Staphylococcus aureus*, including methicillin-resistant *S. aureus* (MRSA), rarely cause pyelonephritis or cystitis in outpatient practice. Recovery of *S. aureus* from the urine suggests an additional site of infection, such as renal abscess, osteomyelitis, and bacterial endocarditis.

Adenovirus and *E. coli* are the most common causes of acute hemorrhagic cystitis. Adenovirus cystitis presents with dysuria and gross hematuria and is a benign, self-limited disease. *Schistosomiasis haematobium* would be considered in a child returning from an endemic area who presents with microscopic or gross hematuria and dysuria with distinctive eggs on microscopy. *Mycobacterium tuberculosis* infections of the urinary tract is uncommon but should be considered in the appropriate clinical setting when sterile pyuria is present. *Lactobacillus* spp., *Corynebacterium* spp., and alpha-hemolytic streptococci are common periurethral flora; and in most cases their isolation from urine represents contamination.

Nosocomial UTI, largely due to gram-negative organisms and yeast, remains a significant cause of morbidity in hospitalized pediatric patients. Urinary catheterization of >3 days is an important risk factor for infection. MRSA outbreaks have been reported in intensive-care units with isolation of MRSA from blood, nasopharynx, and urine of infected neonates (9).

PATHOGENESIS

The sequence of events leading to an ascending infection of the urinary tract is thought to begin when *E. coli* derived from the gastrointestinal tract colonize the periurethral mucosa (10). Periurethral *E. coli* then ascend into the bladder, ureters, and kidneys by an undefined mechanism, establishing a risk for bladder urine or renal parenchymal infection. Presence of pathogens on the periurethral mucosa does not necessarily result in infection. Despite the heavy colonization of the periurethal mucosa of infants younger than 1 year of age, most of them do not experience UTI (10). In a prospective study of healthy toilet-trained girls (3 to 6 years of age) after their first UTI, weekly periurethral cultures revealed a large diversity of *E. coli* clones colonizing the periurethral region. The presence of *E. coli* on the periurethra did not predict a subsequent UTI (11).

The virulence of UPEC compared to commensal *E. coli* results from specific virulence genes often encoded on distinct pathogenicity islands on the bacterial chromosome (12). These virulence genes encode proteins that allow the expression of bacterial-adherence factors. UPEC strains express adherence factors as Afa/Dr, S/F1c, *pap* (P), M hemagglutinin, and type-1 pili adhesins. These adherence factors promote attachment of *E. coli* to bladder-epithelial cells signaling host-cell inflammatory mediators resulting in exfoliation and clearance of infected bladder cells through voiding. In children, virulence profiles of *E. coli* isolates causing asymptomatic bacteriuria (ASB) differ from those of isolates causing symptomatic UTI (13). Classic early studies in Sweden showed that ASB strains from girls had a reduced capacity for adherence *in vitro*, as compared with isolates collected from girls with symptomatic infections (13). A more recent study of UPEC virulence determinants among UTI and ASB isolates from children in Korea confirmed these findings, showing that although *fimH* was equally frequent among UTI and ASB isolates, UTI strains showed a significantly higher prevalence of *papEF* (adherence determinant) and *fyuA* (siderophore), as compared with ASB isolates (14).

As the urinary tract is devoid of mucosal surfaces, it is thought that host defense may be highly dependent on epithelial cell-derived mediators, such as Tamm-Horsfall protein (THP), soluble IgA, lactoferrin, lipocalin, and bactericidal antimicrobial peptides, including alpha and beta-defensins and cathelicidin that have evolved to combat urinary tract pathogens (12). Several of these host defenses may have multiple functions, such as inducing bacterial elimination, immune-cell recruitment, and activation.

The ability to empty the bladder completely and regularly is the most important host defense against infection. Lower urinary-tract dysfunction, a broad term to replace "voiding dysfunction" encompasses a wide range of disturbances of lower urinary-tract function in neurologically normal children (15). Lower urinary-tract dysfunction is associated with an increased risk of UTI in the school-aged child. Children with anatomic abnormalities, neurogenic bladder, and extrinsic compression from tumors or constipation are also at increased risk for UTI. It is estimated that 10 of 1,000 (1%) uncircumcised male infants will develop a UTI during the first year of life, compared with 1 of 1,000 (0.1%) circumcised male infants. The American Academy of Pediatrics (AAP) Task Force

on Circumcision recommends consideration of circumcision in newborns with significant perinatal hydronephrosis or reflux (16). However, the task force did not recommend routine circumcision and emphasized potential medical benefits as well as risks for newborn circumcision.

VESICOURETERAL REFLUX

Infected bladder urine may ascend to the upper urinary tract and renal parenchyma via VUR. VUR is the most common urologic abnormality in children (17) and is associated with an increased likelihood of pyelonephritis during a UTI. Although VUR often resolves spontaneously, some children will develop reflux nephropathy, hypertension, and adrenal insufficiency. Primary VUR is reflux caused by a congenital anatomic deformity of the ureterovesical junction without other urinary tract abnormalities. With the routine use of antenatal ultrasound, hydronephrosis has been identified as the most common fetal urinary tract abnormality.

VUR with hydronephrosis is found particularly in males and is associated with renal dysfunction, even in the absence of UTI (18). Specifically, primary VUR results from a deficiency of the longitudinal muscle of the submucosal ureter, with shortening of the intramural portion of the ureter as it traverses the bladder wall. Closure of the ureter with bladder filling and micturition is restricted, facilitating retrograde flow of urine from the bladder into the ureter and pelvicaliceal system. A decreasing incidence of VUR with age reflects spontaneous resolution of VUR. Resolution of VUR is closely associated with improvement of urodynamic parameters as voiding pressures and detrusor activity. In a prospective 5-year follow-up study of children younger than 5 years of age who had primary VUR and radiographically normal kidneys, grade I VUR resolved in 82%, grade II in 80%, and grade III in 46% of ureters. Rate of resolution of grades IV and

V over a 5-year period were approximately 30% and 13%, respectively (19).

Secondary VUR refers to reflux resulting from increased bladder pressure (posterior urethral valves), abnormal attachment of the ureter (ectopic ureter), or associated urinary tract abnormalities (prune-belly syndrome) that effect insertion of the ureter. Resolution of secondary VUR requires correction of the underlying cause.

Of children with upper urinary-tract infection or pyelonephritis diagnosed by renal-cortical scintigraphy, 38 to 57% will develop renal scarring (19). Children at greatest risk for scarring include those with grades III–V reflux, younger children (less than 1 year of age), an abnormal renal ultrasound, and recurrent UTI. A recent large study of risk factors after first UTI showed that factors placing children and adolescents at high risk for the development of renal scarring include an abnormal renal-ultrasonographic finding or the combination of high fever and UTI caused by bacteria other than *E. coli* (19).

Only a few possible candidate genes have been investigated for an association with renal scarring. Gene polymorphisms in transforming-growth factor-beta-1 (TGF beta-1) and vascular-endothelial growth factor (VEGF) gene promoters were associated with post-UTI renal-scar formation in children; however, further research with larger populations will be required to corroborate the role of these genes in renal scarring (20, 21).

CLINICAL PRESENTATION

Symptoms for UTI in the pediatric population vary from asymptomatic to systemic infection. In the first few months of life, symptoms may be nonspecific and include fever or hypothermia, irritability, poor feeding, vomiting, jaundice, failure to thrive, hematuria, or malodorous urine (1). From 3 months to 24 months of age, symptoms may refer to the urinary tract as frequency or malodorous urine. Often, however, fever without a clear source of infection is the only sign of a UTI. Respiratory or gastro-intestinal infection may coexist with UTI, nearly 5% of respiratory-syncytial virus-positive infants may also have a UTI. In the 2- to 5-year-old child, fever and abdominal pain are the most common symptoms. Bedwetting in a child who has previously exhibited nighttime bladder control is an important symptom. After 5 years of age the rates of dysuria, urgency, suprapubic tenderness, and costovertebral-angle tenderness increase with frequency. However, it is not possible to reliably differentiate cystitis (lower-tract infection) from pyelonephritis (upper-tract infection) on the basis of clinical presentation alone in children. As many as 25% of children without symptoms of pyelonephritis are found by ureteral catheterization or bladder-washout test to have renal bacteriuria (22). The presence of high fever is an imprecise indicator of upper-tract involvement. When a child presents with symptoms of UTI, other diagnoses to consider are foreign body, trauma, meatal abnormalities such as phimosis, urethritis, vaginitis, and sexual activity.

LABORATORY WORKUP AND DIAGNOSIS

The gold standard for diagnosing UTI is the urine culture. Collection of a urine sample in an infant or nontoilet-trained child should be obtained by suprapubic-bladder aspiration (SPA) or transurethral-bladder catheterization when antimicrobial therapy is warranted after collection (23). SPA or transurethral catheterization is unlikely to be contaminated and, therefore, is the preferred method to document UTI. When a urine specimen is being obtained in a child who is not to receive immediate antimicrobial therapy, a urine specimen may be collected into a bag. A urine specimen collected by the bag technique that yields no growth rules out a UTI.

Urine collected by bag that yields growth requires collection by SPA or catheterization for diagnosis of a UTI. Children who are toilet-trained may provide a midstream-urine specimen for culture.

Urine specimens should be plated immediately; if immediate processing is not possible, specimens should be refrigerated at 4°C and processed within 24 hours. Incubation of urine at room temperature for as little as 1 hour results in loss of sensitivity of the leukocyte-esterase test and loss of specificity of the nitrite test. The urinalysis does not substitute for a culture to document a UTI, but is a valuable tool for selecting antimicrobial therapy pending culture results. Positive components of a "dipstick" urinalysis that suggest a UTI pending culture include a positive leukocyte-esterase or nitrite test. In office practice, the urine "dipstick" is readily available while the urine microscopy is sent to an outside laboratory for analysis. Greater than five white blood cells per high-power field or the presence of bacteria on Gram-stained specimen performed in a technically skilled laboratory suggests a UTI pending culture. Pyuria (purulent fluid in the urine due to presence of white blood cells) alone is not adequate to diagnose a UTI and does not substitute for a urine culture. The presence of leukocytes may be related to vaginal or cervical secretions, glomerulonephritis, or tuberculosis and should be differentiated from a UTI.

Quantitative culture establishes the diagnosis of UTI and provides an isolate for susceptibility testing. Quantitative cultures are performed on both nonselective (sheep-blood agar) and selective (MacConkey agar) media inoculated with a calibrated loop that delivers an inoculum of 0.01 ml (for a catheter/SPA specimen) or 0.001 ml (for a voided specimen). The urine is streaked over the entire agar surface to permit quantification of colony-forming units (cfu). The cfu count is multiplied by 100 (when a 0.01-ml loop is used) or by 1,000 (when a 0.001-ml loop is used) to yield cfu/ml.

Kass in 1957 demonstrated that adult females with a clinical diagnosis of pyelonephritis had a greater than 10^5 bacteria per milliliter of urine, whereas asymptomatic patients had smaller counts of bacteria, which were considered contaminants (24). Although no comparable studies have been conducted in children, the presence of greater than or equal to 10^5 cfu of urinary tract pathogen per milliliter of urine is a widely accepted standard for documenting a UTI. Hellerstein refined the criteria for pediatric patients and included collection technique and gender as important variables (25). Most children with normal urinary tracts are infected with a single organism. Thus, Hellerstein adapted these criteria as follows, noting the criteria are operational rather than absolute: any number of organisms from a suprapubic aspiration or growth of greater than 10^4 cfu/ml of a urinary pathogen from a catheterized or clean catch specimen likely represents a UTI.

MANAGEMENT AND TREATMENT

The goals of management are 1) prompt diagnosis of concomitant bacteremia or meningitis, particularly in the infant less than 2 months of age, with management in hospital and broad-spectrum antimicrobials pending cultures of blood, spinal fluid, and urine; 2) prevention of progressive renal disease by eradication of the bacterial pathogen, identification of abnormalities of the urinary tract, and prevention of recurrent infections; and 3) resolution of the acute symptoms of infection.

Delay in initiation of antimicrobial therapy is associated with an increased risk of renal scarring (26), and children suspected of UTI should be started on antimicrobial therapy pending culture results. If the infant or young child is toxic-appearing or unable to tolerate oral fluids, initial antimicrobial therapy is administered parenterally and hospitalization considered (23). If the infant or young child is well-appearing and tolerating oral fluids, including antimicrobials, outpatient management may be considered if compliance with antimicrobials and follow-up is insured.

The choice of empirical antibiotics is guided by local resistance patterns for *E. coli*, the most common cause of UTI (27). *E. coli* producing extended-spectrum beta lactamases (ESBL) are becoming more common and have been associated with preexisting neurological disease, recent hospitalization (within one month), and the exposure to antimicrobials within the preceding three months (28). Standard antimicrobials for treatment of UTI include a third-generation cephalosporin or aminoglycoside for parenteral use. In a patient with compromised renal function, the use of potentially nephrotoxic antimicrobials (aminoglycoside) requires caution, and serum creatinine, along with aminoglycoside concentrations, must be monitored. For oral treatment, second- and third-generation cephalosporins are appropriate treatment for pyelonephritis. Based on local resistance patterns, amoxicillin-clavulanate, trimethoprim-sulfamethoxazole (for infants greater than 2 months of age), and first-generation cephalosporins may be used as alternative agents. Antimicrobials that are excreted in the urine but do not attain therapeutic concentrations in the serum, such as nitrofurantoin, should not be used to treat febrile infants and young children with UTI.

Studies on resolution of UTI in children support a 10–14 day course of therapy compared to short-course therapy (23). Antimicrobial-sensitivity testing may be helpful in the infant or young child experiencing their first UTI to insure bacteriologic cure is following a clinical response to antimicrobials. As many antimicrobials are excreted in the urine in high concentrations, an intermediately sensitive organism on *in vitro* testing may be fully eradicated in the urine. If *in vitro* sensitivity is not performed, if organism sensitivity is intermediate/resistant, or the child has not improved after 48 hours of treatment, a repeat urine culture may be informative (23, 29).

IMAGING

The lack of consensus and well-designed prospective clinical trials makes it difficult to recommend evidence-based guidelines for imaging children after their first UTI. The AAP Subcommittee on Urinary Tract Infection (23) and the American Urologic Association (AUA) Pediatric Vesicoureteral Reflux Guideline Panel (30) developed imaging guidelines based on review of the literature, current practice, risk assessment, and panel consensus. These guidelines include:

1. Imaging of the urinary tract after a first UTI to identify abnormalities of the urinary tract that may be amenable to medical or surgical therapy. Imaging includes a urinary tract ultrasound to identify hydronephrosis and other structural abnormalities. In addition, a study to detect VUR by voiding cystourethrography with fluoroscopy to characterize reflux, posterior-urethral valves in boys, or bladder abnormalities in girls. Postnatal ultrasound and voiding cystourethrography is also recommended in infants diagnosed with hydronephrosis on fetal ultrasound. This recommendation is made in recognition of the need for a large prospective study to determine the sensitivity of prenatal fetal hydronephrosis in predicting postnatal VUR (18).

2. Renal-cortical scintigraphy is superior over ultrasound for identifying renal-cortical abnormalities from pyelonephritis and renal scarring. Limitations of use of scintigraphy include expense, radiation exposure, possible need for patient sedation, and limited availability. As a result, scintigraphy is not routinely recommended. A recent study found that although scintigraphy was useful for demonstrating renal-parenchymal localization of infection, it did not alter antimicrobial therapy or

change outcomes (31). Future research is directed at alternative modalities, such as power Doppler ultrasonography, for detection of renal scarring since Doppler ultrasound does not require injection or radiation and is easily performed on an outpatient basis (32).

PROPHYLAXIS

VUR increases the risk of pyelonephritis when a bladder infection occurs and increases the risk of renal scarring when pyelonephritis develops. Using renal scintigraphy, there is a higher rate of renal scarring after a UTI in children with VUR than without VUR. The international reflux study in children demonstrated similar long-term outcome of medical (prophylaxis) or surgical treatment of children with nonobstructive severe VUR (33). Primary endpoints were new renal scars and renal growth. This data has led to the recommendation for low-dose antimicrobial prophylaxis in children with VUR (23, 34).

It is unclear, however, whether antimicrobial prophylaxis reduces the frequency of UTI or renal scarring in children with VUR (30, 35). A number of prospective, randomized, controlled trials showed little or no benefit of prophylaxis toward reducing the frequency of UTI or renal scarring in children with VUR (31, 36–38). However, these studies had the following flaws: children often had only 1 UTI, compliance with prophylaxis was not insured, and only one study included high-grade (IV, V) reflux. In addition, the number of children less than 1 year of age, the age of highest risk for morbidity with UTI, was small.

In recognition of limited data from previous studies, the AUA and AAP have recommended prophylaxis for children less than one year of age with VUR and history of UTI. Recommendation for prophylaxis for children over a year of age with VUR and history

of UTI has been based on VUR grade, presence of renal scarring, and parental preference. However, a recent meta-analysis that included the RIVUR study (39, 40) supported antimicrobial prophylaxis in all children with VUR, regardless of grade (41). Additional study of this issue is clearly needed.

Probiotic prophylaxis in a small number of children with VUR randomized to receive *Lactobacillus acidophilus* or trimethoprim/sulfamethoxazole resulted in a similar number of recurrent UTI in both treatment groups (42). Evidence of cranberry products for prophylaxis of recurrent UTI in the pediatric population is lacking (43). Further studies of prophylactic agents enrolling large number of patients with similar VUR grade, renal scarring, age, and gender is required to confidentially provide recommendations for children with VUR. The benefit of antimicrobial prophylaxis in children with a history of UTI and normal imaging remains to be proven.

ACKNOWLEDGMENTS

The author received funding from The Pendleton Endowment Research Fund, University of Virginia, Charlottesville, Virginia.

CITATION

Schlager TA. 2016. Urinary tract infections in infants and children. Microbiol Spectrum 4(5):UTI-0022-2016.

REFERENCES

1. **Lohr J, Downs S, Schlager T.** 2008. Genitourinary tract infections, urinary tract infections, p 343–347. *In* Long SS, Pickering LK, Prober CG (ed), *Principles and Practice of Pediatric Infectious Diseases*, 3rd ed. Churchill Livingston/Elsevier, Inc, Philadelphia, PA.
2. **Crain EF, Gershel JC.** 1990. Urinary tract infections in febrile infants younger than 8 weeks of age. *Pediatrics* **86:**363–367.
3. **Hoberman A, Chao HP, Keller DM, Hickey R, Davis HW, Ellis D.** 1993. Prevalence of urinary tract infection in febrile infants. *J Pediatr* **123:**17–23.

4. **Roberts KB, Charney E, Sweren RJ, Ahonkhai VI, Bergman DA, Coulter MP, Fendrick GM, Lachman BS, Lawless MR, Pantell RH, et al.** 1983. Urinary tract infection in infants with unexplained fever: A collaborative study. *J Pediatr* **103:**864–867.

5. **Bauchner H, Philipp B, Dashefsky B, Klein JO.** 1987. Prevalence of bacteriuria in febrile children. *Pediatr Infect Dis J* **6:**239–242.

6. **Winberg J, Andersen HJ, Bergström T, Jacobsson B, Larson H, Lincoln K.** 1974. Epidemiology of symptomatic urinary tract infection in childhood. *Acta Paediatr Scand Suppl* **252:**1–20.

7. **Keren R, Shaikh N, Pohl H, Gravens-Mueller L, Ivanova A, Zaoutis L, Patel M, deBerardinis R, Parker A, Bhatnagar S, Haralam MA, Pope M, Kearney D, Sprague B, Barrera R, Viteri B, Egigueron M, Shah N, Hoberman A.** 2015. Risk factors for recurrent urinary tract infection and renal scarring. *Pediatrics* **136:**e13–e21.

8. **Feld LG, Mattoo TK.** 2010. Urinary tract infections and vesicoureteral reflux in infants and children. *Pediatr Rev* **31:**451–463.

9. **Nambiar S, Herwaldt LA, Singh N.** 2003. Outbreak of invasive disease caused by methicillin-resistant *Staphylococcus aureus* in neonates and prevalence in the neonatal intensive care unit. *Pediatr Crit Care Med* **4:**220–226.

10. **Bollgren I, Winberg J.** 1976. The periurethral aerobic bacterial flora in healthy boys and girls. *Acta Paediatr Scand* **65:**74–80.

11. **Schlager TA, Hendley JO, Lohr JA, Whittam TS.** 1993. Effect of periurethral colonization on the risk of urinary tract infection in healthy girls after their first urinary tract infection. *Pediatr Infect Dis J* **12:**988–993.

12. **Weichhart T, Haidinger M, Hörl WH, Säemann MD.** 2008. Current concepts of molecular defence mechanisms operative during urinary tract infection. *Eur J Clin Invest* **38**(Suppl 2):29–38.

13. **Edén CS, Eriksson B, Hanson LA, Jodal U, Kaijser B, Janson GL, Lindberg U, Olling S.** 1978. Adhesion to normal human uroepithelial cells of *Escherichia coli* from children with various forms of urinary tract infection. *J Pediatr* **93:**398–403.

14. **Yun KW, Kim HY, Park HK, Kim W, Lim IS.** 2014. Virulence factors of uropathogenic *Escherichia coli* of urinary tract infections and asymptomatic bacteriuria in children. *J Microbiol Immunol Infect* **47:**455–461.

15. **Ballek NK, McKenna PH.** 2010. Lower urinary tract dysfunction in childhood. *Urol Clin North Am* **37:**215–228.

16. **Shapiro E.** 1999. American Academy of Pediatrics policy statements on circumcision and urinary tract infection. *Rev Urol* **1:**154–156.

17. **Coleman R.** 2011. Early management and long-term outcomes in primary vesico-ureteric reflux. *BJU Int* **108**(Suppl 2):3–8.

18. **Gloor JM, Ramsey PS, Ogburn PL Jr, Danilenko-Dixon DR, DiMarco CS, Ramin KD.** 2002. The association of isolated mild fetal hydronephrosis with postnatal vesicoureteral reflux. *J Matern Fetal Neonatal Med* **12:**196–200.

19. **Shaikh N, Ewing AL, Bhatnagar S, Hoberman A.** 2010. Risk of renal scarring in children with a first urinary tract infection: a systematic review. *Pediatrics* **126:**1084–1091.

20. **Hussein A, Askar E, Elsaeid M, Schaefer F.** 2010. Functional polymorphisms in transforming growth factor-beta-1 (TGFbeta-1) and vascular endothelial growth factor (VEGF) genes modify risk of renal parenchymal scarring following childhood urinary tract infection. *Nephrol Dial Transplant* **25:**779–785.

21. **Zaffanello M, Tardivo S, Cataldi L, Fanos V, Biban P, Malerba G.** 2011. Genetic susceptibility to renal scar formation after urinary tract infection: a systematic review and meta-analysis of candidate gene polymorphisms. *Pediatr Nephrol* **26:**1017–1029.

22. **Chandra M, Maddix H.** 2000. Urodynamic dysfunction in infants with vesicoureteral reflux. *J Pediatr* **136:**754–759.

23. **Anonymous.** 1999. Practice parameter: the diagnosis, treatment, and evaluation of the initial urinary tract infection in febrile infants and young children. American Academy of Pediatrics. Committee on Quality Improvement. Subcommittee on Urinary Tract Infection. *Pediatrics* **103:**843–852.

24. **Kass EH.** 1957. Bacteriuria and the diagnosis of infections of the urinary tract, with observations on the use of methionine as a urinary antiseptic. *AMA Arch Intern Med* **100:**709–714.

25. **Hellerstein S.** 1982. Recurrent urinary infections in children. *Pediatr Infect Dis* **1:**271–281.

26. **Ransley PG, Risdon RA.** 1981. Reflux nephropathy: effects of antimicrobial therapy on the evolution of the early pyelonephritic scar. *Kidney Int* **20:**733–742.

27. **Montini G, Tullus K, Hewitt I.** 2011. Febrile urinary tract infections in children. *N Engl J Med* **365:**239–250.

28. **Fan NC, Chen HH, Chen CL, Ou LS, Lin TY, Tsai MH, Chiu CH.** 2014. Rise of community-onset urinary tract infection caused by extended-spectrum β-lactamase-producing *Escherichia coli* in children. *J Microbiol Immunol Infect* **47:**399–405.

29. Saadeh SA, Mattoo TK. 2011. Managing urinary tract infections. *Pediatr Nephrol* **26:** 1967–1976.

30. Peters CA, Skoog SJ, Arant BS Jr, Copp HL, Elder JS, Hudson RG, Khoury AE, Lorenzo AJ, Pohl HG, Shapiro E, Snodgrass WT, Diaz M. 2010. Summary of the AUA Guideline on management of primary vesicoureteral reflux in children. *J Urol* 184:1134–1144.

31. Montini G, Zucchetta P, Tomasi L, Talenti E, Rigamonti W, Picco G, Ballan A, Zucchini A, Serra L, Canella V, Gheno M, Venturoli A, Ranieri M, Caddia V, Carasi C, Dall'amico R, Hewitt I. 2009. Value of imaging studies after a first febrile urinary tract infection in young children: data from Italian renal infection study 1. *Pediatrics* **123:**e239–246.

32. Kawauchi A, Yamao Y, Ukimura O, Kamoi K, Soh J, Miki T. 2001. Evaluation of reflux kidney using renal resistive index. *J Urol* **165:** 2010–2012.

33. Jodal U, Smellie JM, Lax H, Hoyer PF. 2006. Ten-year results of randomized treatment of children with severe vesicoureteral reflux. Final report of the International Reflux Study in Children. *Pediatr Nephrol* **21:**785–792.

34. Craig JC, Simpson JM, Williams GJ, Lowe A, Reynolds GJ, McTaggart SJ, Hodson EM, Carapetis JR, Cranswick NE, Smith G, Irwig LM, Caldwell PH, Hamilton S, Roy LP; Prevention of Recurrent Urinary Tract Infection in Children with Vesicoureteric Reflux and Normal Renal Tracts (PRIVENT). 2009. Antibiotic prophylaxis and recurrent urinary tract infection in children. *N Engl J Med* 361:1748–1759.

35. DeMuri GP, Wald ER. 2008. Imaging and antimicrobial prophylaxis following the diagnosis of urinary tract infection in children. *Pediatr Infect Dis J* 27:553–554.

36. Pennesi M, Travan L, Peratoner L, Bordugo A, Cattaneo A, Ronfani L, Minisini S, Ventura A, North East Italy Prophylaxis in VUR Study Group. 2008. Is antibiotic prophylaxis in children with vesicoureteral reflux effective in preventing pyelonephritis and renal scars? A randomized, controlled trial. *Pediatrics* **121:** e1489–e1494.

37. Roussey-Kesler G, Gadjos V, Idres N, Horen B, Ichay L, Leclair MD, Raymond F, Grellier A, Hazart I, de Parscau L, Salomon R, Champion G, Leroy V, Guigonis V, Siret D, Palcoux JB, Taque S, Lemoigne A, Nguyen JM, Guyot C. 2008. Antibiotic prophylaxis for the prevention of recurrent urinary tract infection in children with low grade vesicoureteral reflux: results from a prospective randomized study. *J Urol* 179:674–679; discussion 679.

38. Hari P, Hari S, Sinha A, Kumar R, Kapil A, Pandey RM, Bagga A. 2015. Antibiotic prophylaxis in the management of vesicoureteric reflux: a randomized double-blind placebo-controlled trial. *Pediatr Nephrol* 30:479–486.

39. Hoberman A, Chesney RW; RIVUR Trial Investigators. 2014. Antimicrobial prophylaxis for children with vesicoureteral reflux. *N Engl J Med* 371:1072–1073.

40. RIVUR Trial Investigators, Hoberman A, Greenfield SP, Mattoo TK, Keren R, Mathews R, Pohl HG, Kropp BP, Skoog SJ, Nelson CP, Moxey-Mims M, Chesney RW, Carpenter MA. 2014. Antimicrobial prophylaxis for children with vesicoureteral reflux. *N Engl J Med* **370:** 2367–2376.

41. de Bessa J Jr, de Carvalho Mrad FC, Mendes EF, Bessa MC, Paschoalin VP, Tiraboschi RB, Sammour ZM, Gomes CM, Braga LH, Bastos Netto JM. 2015. Antibiotic prophylaxis for prevention of febrile urinary tract infections in children with vesicoureteral reflux: a meta-analysis of randomized, controlled trials comparing dilated to nondilated vesicoureteral reflux. *J Urol* 193(Suppl 5):1772–1777.

42. Lee SJ, Shim YH, Cho SJ, Lee JW. 2007. Probiotics prophylaxis in children with persistent primary vesicoureteral reflux. *Pediatr Nephrol* 22:1315–1320.

43. Shamseer L, Vohra S, American Academy of Pediatrics Provisional Section on Complementary Holistic and Integrative Medicine. 2007. Complementary, holistic, and integrative medicine: cranberry. *Pediatr Rev* 28:e43–e45.

The Vaginal Microbiota and Urinary Tract Infection

5

ANN E. STAPLETON[1]

INTRODUCTION

Urinary tract infections (UTIs) are a common clinical problem across the lifespan of women. Although UTIs are not systematically tracked, making estimates of U.S. incidence somewhat challenging, the most recent combined National Hospital Ambulatory Medical Care Survey and National Ambulatory Medical Care Survey data from 2009–2010 suggest that approximately 10 million outpatient visits for a diagnosis of UTI occur annually in the United States among both women and men (1). Women are disproportionately affected, with an estimated lifetime risk of UTI of 60% (2). In otherwise healthy, sexually active premenopausal women, these infections occur approximately every other year (3). As women age, UTIs become more common (2). A population-based study of community-onset UTI among nearly 31,000 residents of Calgary showed that the incidence of UTI among women demonstrates an initial peak in the twenties (30 per 1,000), decreases slightly during the later reproductive years, then steadily increases with every decade of life starting in late middle age, reaching a maximum of 125/1,000 at and above age 80 (4). Although the cost of treating UTI has not been recently

[1]Division of Allergy and Infectious Diseases, Department of Medicine, University of Washington, Seattle, WA 98195.

Urinary Tract Infections: Molecular Pathogenesis and Clinical Management, 2nd Edition
Edited by Matthew A. Mulvey, David J. Klumpp, and Ann E. Stapleton
© 2017 American Society for Microbiology, Washington, DC
doi:10.1128/microbiolspec.UTI-0025-2016

estimated, the last published estimate in 2010 indicated that the annual U.S. domestic cost exceeded two billion dollars (5–7).

Despite decades of studies defining the epidemiology, risk factors, and pathogenic mechanisms in UTIs, current evidence-based prevention strategies still rely upon the use of low-dose prophylactic antimicrobials as the cornerstone of prevention of recurrent UTIs in women of any age (8). For postmenopausal women, vaginal estrogen therapy may be considered, but this is often as an adjunctive to antimicrobial-based prophylaxis (8). Given the inexorable increase in antimicrobial resistance documented worldwide for the past few decades, antimicrobial-based preventive strategies have become less attractive and potentially contributory to worsening resistance rates (9). As discussed in detail below, the vaginal microbiota (VMB) plays a key role in the pathogenesis of UTI: alterations in the VMB are associated with risks for UTI, and effects on the VMB associated with treatment of UTIs can affect the success of this therapy. Thus, a better understanding of the role of the VMB in UTI may lead to improved interventions to prevent and treat these infections.

THE ROLE OF THE VAGINA IN THE PATHOGENESIS OF UTI

In women, the vagina plays a key role in the pathogenesis of UTIs. The intestinal microbiota is the ultimate source of bacterial strains causing cystitis and pyelonephritis in the majority of cases (10). The initial step in the pathogenesis of UTI is colonization of the vaginal introitus and periurethra with the infecting uropathogens, followed by ascension of uropathogens via the urethra to the bladder and sometimes the kidneys to cause infection (10–12). Thus, understanding factors that affect the microbiota of the vagina is key to understanding the pathogenesis of UTI and to designing interventions to prevent UTIs.

PROTECTIVE ROLE OF LACTOBACILLI IN THE VAGINA

Studies of the VMB have long indicated that certain bacterial species and/or microbial characteristics are associated with disease-free conditions, or "health," of the genitourinary tract (13). Studies specifically related to the pathogenesis of UTI largely date from decades preceding the availability of culture-independent methodologies for characterizing microbial communities, larger-scale microbial sequencing methods, and widespread interest in the microbiome. In culture-based studies of vaginal samples from women without urogenital disease conditions, *Lactobacillus* species comprise 90% of the organisms present (14, 15), and 80 to 90% of these lactobacilli produced H_2O_2, mostly attributable to *Lactobacillus crispatus* and *Lactobacillus jensenii* (14, 15). Beginning with early studies utilizing DNA hybridization and quantitative real-time PCR (RT-PCR) of lactobacilli cultured from vaginal samples, numerous studies have identified *L. crispatus* as the predominant species present in healthy premenopausal women (14). Many experts view the presence of *L. crispatus* as an overall marker of a healthy VMB (16).

Several subsequent studies have characterized the VMB in states of health and disease using culture-independent methods, largely based on 16S ribosomal RNA bacterial gene sequencing (13, 17). One of the first large-scale, non-culture-dependent studies of vaginal microbial communities assessed approximately 400 asymptomatic North American women, using pyrosequencing of bar-coded 16S rRNA genes (18). This study confirmed that most vaginal microbial communities (73%) were dominated by one or more species of *Lactobacillus* and that this genus constituted over half of all sequences obtained (18). As of 2014, a systematic review of VMB studies performed using molecular characterization methodology identified 63 studies meeting criteria for inclusion and noted that these studies have definitively demonstrated that a *Lactobacillus*-dominated VMB is correlated with what was termed a "healthy vaginal micro-environment" (17).

Conversely, the absence of vaginal lactobacilli has been associated with several disease states. In earlier clinical studies, vaginal samples collected from women with and without various urogenital disease conditions were cultured using standard microbiological methods for isolating lactobacilli and were compared for characteristics such as the presence or absence of *Lactobacillus* spp., especially H_2O_2-producing lactobacilli. Women lacking vaginal lactobacilli per vaginal cultures, particularly those with vaginal samples with decreased relative amounts of or total absence of peroxide-producing lactobacilli, were found to be at increased risk for a variety of urogenital disease conditions, including bacterial vaginosis, HIV infection, and *Neisseria gonorrhoeae*, as well as vaginal colonization with *Escherichia coli*, the most common cause of UTI in women (19–24).

Proposed mechanisms through which lactobacilli may prevent vaginal colonization by uropathogens include competitive exclusion of uropathogens by adherence of *Lactobacillus* species to uroepithelial cells; lowering of vaginal pH by production of lactic acid; production of bacteriocins, surfactants, and other antimicrobial products; and finally, production of H_2O_2 (25–32). *L. crispatus* is highly adherent to vaginal epithelial cells and produces high quantities of H_2O_2 (29, 33, 34). Lactic acid concentrations are high in a lactobacillus-dominated VMB, and these conditions produce a potently antimicrobial environment *in vitro* (35). Hydrogen peroxide alone is microbicidal for many bacterial species, and this microbicidal activity is 10- to 100-fold greater when it is combined with chloride anion and myeloperoxidase, both of which are found in the vagina (36, 37). This vaginal antimicrobial defense system (H_2O_2, chloride anion, and myeloperoxidase) has potent *in vitro* activity against *E. coli* as well as other microorganisms (38). Lactobacilli also produce surfactants that inhibit growth of *E. coli* and other uropathogens (27, 28).

ALTERATIONS IN MICROBIOTA ASSOCIATED WITH UTI

The VMB has been demonstrated to be altered at the time of UTI, in women with recurrent UTI, and following treatment of the infection, even in women without a history of recurrent UTI. As noted above, the critical event preceding UTI is colonization of the vaginal introitus with intestinal microbiota, most commonly *E. coli* (36, 39, 40). Multiple culture-based studies showed that women with recurrent UTI often have increased rates of colonization with *E. coli* and depletion of the normally predominant H_2O_2-producing lactobacilli (39, 41–43), suggesting that vaginal colonization with H_2O_2-producing lactobacilli may prevent *E. coli* vaginal introital colonization and UTI. In a case-control study, we found that women with recurrent UTI who lacked vaginal H_2O_2-producing lactobacilli had a 5-fold increased risk of *E. coli* vaginal colonization compared to women with H_2O_2-producing lactobacilli (41). In another study of reproductive-age women, 15% of 301 women who had vaginal colonization with *L. crispatus* or *L. jensenii*, both H_2O_2-producers, were colonized with *E. coli*, compared with 27% of women who did not have these lactobacilli species present ($P = 0.01$) (14). A Canadian study showed that the VMB of women with recurrent UTI demonstrated a diminished lactobacillus morphotype composition resembling bacterial vaginosis pathophysiology (44). Finally, data from studies of premenopausal women in Seattle demonstrated that only 50% of women have H_2O_2-producing lactobacilli in the vagina at the time of presentation with recurrent UTI (45).

ALTERATIONS OF THE VMB ASSOCIATED WITH LOSS OF ESTROGEN

Among the manifold effects of the loss of estrogen at the time of menopause are changes in the vaginal microbiome of most women, decreasing the relative amounts of *Lactobacillus* present (46–49). As noted above, rates of UTI begin to rise at the climacteric, and recurrent UTIs are considered among the features of the genitourinary syndrome of menopause, which is characterized by thinning of the vaginal epithelium, various symptoms associated with vulvovaginal atrophy, and relative loss of lactobacilli in the VMB (50). In a study of 463 community-dwelling postmenopausal women, colonization with *E. coli* was more frequent in women without estrogen replacement and inversely associated with the presence of *Lactobacillus* (51). Further, these alterations of the VMB were associated with having a history of recurrent UTI (51).

Conversely, in the same study of community-dwelling postmenopausal women, retention of lactobacilli in the VMB at menopause was associated with having received systemic or topical vaginal hormone replacement therapy in the preceding year (51). Studies of treatment of menopause-associated estrogen deficiency with topical estrogen preparations such as estrogen cream or an estradiol-releasing vaginal ring (Estring) has been demonstrated to reduce the rate of recurrent UTI and restore vaginal lactobacilli in most women (46–49, 52). A recent study of microbial communities in postmenopausal women with atrophic vaginitis showed depletion of *Lactobacillus* spp. prior to therapy, with low-dose estrogen therapy increasing the relative amounts of *Lactobacillus* spp. and decreasing vaginal pH (53). As a result, topical vaginal estrogen therapy is recommended in the management of recurrent UTI in older women, in addition to low-dose antimicrobial therapy (54).

ALTERATIONS OF THE VMB ASSOCIATED WITH ANTIMICROBIAL THERAPY

Exposure to antimicrobials has been demonstrated in multiple studies to alter the VMB, such as during pregnancy (55, 56) or after therapy for bacterial vaginosis (56, 57). In one study, use of antimicrobials in the preceding weeks was associated with increased risk of UTI (58). The choice of antimicrobial for treatment of UTI may also affect the VMB. For example, beta lactam antibiotics are associated with less efficacy in eradicating vaginal colonization with *E. coli*, and clinically, these agents are more associated with rapid recurrence of UTI after therapy compared with other agents (45, 59). Of note, the Infectious Diseases Society of America considered adverse ecological effects of antimicrobials, termed collateral damage, as key factors in ranking recommended therapies for acute uncomplicated UTI (60).

EFFECT OF CONTRACEPTIVE METHOD ON THE VMB AND RISK OF UTI

Numerous studies have demonstrated that spermicidal products containing compounds such as nonoxynol-9 disrupt the VMB by depleting lactobacilli and increasing *E. coli* colonization and increasing the risk of UTI (8, 61–64). This adverse effect appears to be mediated by a direct toxic effect on vaginal *Lactobacillus* spp. (65). Data regarding other methods of contraception are more limited. Oral contraceptives appear to reduce the rate of bacterial vaginosis (66) but are neutral with respect to risk of UTI (61, 63, 64). Data regarding the intrauterine device and effects on the VMB overall are conflicting, with no published studies considering the effect of this method on the risk of UTI at the time of this review (66).

CLINICAL IMPLICATIONS

The interactions between the VMB and the risks of UTI, treatment choices, and possible means of UTI prevention lead to several fairly evident clinical implications. First, lifestyle considerations may be discussed with patients. Though spermicidal contraceptives are less commonly used at present, these may be substituted with methods having a neutral effect on UTI, such as hormonal contraceptives (8). Patients and their providers may be counseled regarding current knowledge of adverse effects of antimicrobial agents and their effect upon risk of UTI. Postmenopausal women without contraindications to the use of topical hormone therapies may be offered such medications for primary or secondary prevention of UTI (8). In clinical practice, UTI episodes and subsequent clusters of recurrent UTI that are the hallmark of the natural history of recurrent UTI (67) may arise apparently spontaneously, for no known reason. However, given the evidence that alterations in the VMB are likely associated with these clinical events, the clinician may still attempt to identify and ameliorate known causes of VMB alterations.

Finally, the use of oral or intravaginal probiotics to attempt to restore protective vaginal lactobacilli is an attractive option that would, in theory, likely help reduce the use of antimicrobials in preventive strategies. Unfortunately, data on this approach are limited overall, and studies regarding reduction of UTI by means of oral probiotics are conflicting (8). An alternative approach of a vaginal *L. crispatus* probiotic directly administered to the vagina has shown promise in a small randomized, double-blind, placebo-controlled phase 2 study, in which receiving the lactobacillus probiotic ($n = 50$) was associated with decreased rates of recurrent UTI compared with placebo ($n = 50$) (68). Unfortunately, in a recent Cochrane review of all probiotic approaches to UTI prevention, few studies were adequate for inclusion, and the above-described study of a vaginal probiotic was the only one of its kind (69). Additional randomized, double-blind, placebo-controlled studies of adequate sample size using a carefully selected probiotic strain are needed to ascertain whether this approach is effective.

SUMMARY

The vagina is a key anatomical site in the pathogenesis of UTI, serving as a potential reservoir for infecting bacteria ascending from the intestinal source of uropathogenic bacteria. The vagina is also an important site at which interventions that positively affect the VMB can be instituted to attempt to decrease risk of UTI. The VMB is a dynamic and often critical factor in this pathogenic interplay, because changes in the characteristics of the VMB resulting in the loss of normally protective *Lactobacillus* spp. increase the risk of UTI. These alterations may result from estrogen deficiency, antimicrobial therapy, contraceptives, and episodes of UTI itself. Interventions designed to maintain homeostasis of the VMB and/or to ameliorate adverse effects of exposures such as antimicrobials or menopause hold promise for future preventive or therapeutic options, potentially avoiding additional use of antibiotics.

CITATION

Stapleton AE. 2016. The vaginal microbiota and urinary tract infection. Microbiol Spectrum 4(6):UTI-0025-2016.

REFERENCES

1. **CDC.** 2010. *NAMCS and NHAMCS Web Tables. Table 1. Annual number and percent distribution of ambulatory care visits by setting type according to diagnosis group: United States, 2009-2010.* http://www.cdc.gov/nchs/ahcd/web_tables.htm.
2. **Foxman B.** 2014. Urinary tract infection syndromes: occurrence, recurrence, bacteriology, risk factors, and disease burden. *Infect Dis Clin North Am* **28:**1–13.
3. **Hooton TM, Scholes D, Hughes JP, Winter C, Roberts PL, Stapleton AE, Stergachis A, Stamm WE.** 1996. A prospective study of risk factors for symptomatic urinary tract infection in young women. *N Engl J Med* **335:**468–474.
4. **Laupland KB, Ross T, Pitout JD, Church DL, Gregson DB.** 2007. Community-onset urinary tract infections: a population-based assessment. *Infection* **35:**150–153.
5. **Engel JD, Schaeffer AJ.** 1998. Evaluation of and antimicrobial therapy for recurrent urinary tract infections in women. *Urol Clin North Am* **25:**685–701, x.

6. **Foxman B, Barlow R, D'Arcy H, Gillespie B, Sobel JD.** 2000. Urinary tract infection: self-reported incidence and associated costs. *Ann Epidemiol* **10:**509–515.

7. **Hooton TM, Stamm WE.** 1997. Diagnosis and treatment of uncomplicated urinary tract infection. *Infect Dis Clin North Am* **11:** 551–581.

8. **Hooton TM, Gupta K.** 2016. Recurrent urinary tract infection in women. *UpToDate,* Waltham, MA. (accessed 5 June 2016.)

9. **Gupta K, Bhadelia N.** 2014. Management of urinary tract infections from multidrug-resistant organisms. *Infect Dis Clin North Am* **28:**49–59.

10. **Czaja CA, Stamm WE, Stapleton AE, Roberts PL, Hawn TR, Scholes D, Samadpour M, Hultgren SJ, Hooton TM.** 2009. Prospective cohort study of microbial and inflammatory events immediately preceding *Escherichia coli* recurrent urinary tract infection in women. *J Infect Dis* **200:**528–536.

11. **Russo TA, Stapleton A, Wenderoth S, Hooton TM, Stamm WE.** 1995. Chromosomal restriction fragment length polymorphism analysis of *Escherichia coli* strains causing recurrent urinary tract infections in young women. *J Infect Dis* **172:**440–445.

12. **Beerepoot M, Geerlings S.** 2016. Non-antibiotic prophylaxis for urinary tract infections. *Pathogens* **5:**5.

13. **Ravel J, Brotman RM.** 2016. Translating the vaginal microbiome: gaps and challenges. *Genome Med* **8:**35.

14. **Antonio MA, Hawes SE, Hillier SL.** 1999. The identification of vaginal *Lactobacillus* species and the demographic and microbiologic characteristics of women colonized by these species. *J Infect Dis* **180:**1950–1956.

15. **Eschenbach DA, Davick PR, Williams BL, Klebanoff SJ, Young-Smith K, Critchlow CM, Holmes KK.** 1989. Prevalence of hydrogen peroxide-producing *Lactobacillus* species in normal women and women with bacterial vaginosis. *J Clin Microbiol* **27:**251–256.

16. **Lepargneur JP.** 2016. *Lactobacillus crispatus* as biomarker of the healthy vaginal tract. *Ann Biol Clin (Paris)* **74:**421–427.

17. **van de Wijgert JH, Borgdorff H, Verhelst R, Crucitti T, Francis S, Verstraelen H, Jespers V.** 2014. The vaginal microbiota: what have we learned after a decade of molecular characterization? *PLoS One* **9:**e105998. doi:10.1371/journal.pone.0105998.

18. **Ravel J, Gajer P, Abdo Z, Schneider GM, Koenig SS, McCulle SL, Karlebach S, Gorle R, Russell J, Tacket CO, Brotman RM, Davis CC, Ault K, Peralta L, Forney LJ.** 2011. Vaginal microbiome of reproductive-age women. *Proc Natl Acad Sci USA* **108**(Suppl 1):4680–4687.

19. **Hawes SE, Hillier SL, Benedetti J, Stevens CE, Koutsky LA, Wolner-Hanssen P, Holmes KK.** 1996. Hydrogen peroxide-producing lactobacilli and acquisition of vaginal infections. *J Infect Dis* **174:**1058–1063.

20. **Martin HL, Richardson BA, Nyange PM, Lavreys L, Hillier SL, Chohan B, Mandaliya K, Ndinya-Achola JO, Bwayo J, Kreiss J.** 1999. Vaginal lactobacilli, microbial flora, and risk of human immunodeficiency virus type 1 and sexually transmitted disease acquisition. *J Infect Dis* **180:**1863–1868.

21. **Schwebke JR.** 2001. Role of vaginal flora as a barrier to HIV acquisition. *Curr Infect Dis Rep* **3:**152–155.

22. **Taha TE, Hoover DR, Dallabetta GA, Kumwenda NI, Mtimavalye LA, Yang LP, Liomba GN, Broadhead RL, Chiphangwi JD, Miotti PG.** 1998. Bacterial vaginosis and disturbances of vaginal flora: association with increased acquisition of HIV. *AIDS* **12:**1699–1706.

23. **van De Wijgert JH, Mason PR, Gwanzura L, Mbizvo MT, Chirenje ZM, Iliff V, Shiboski S, Padian NS.** 2000. Intravaginal practices, vaginal flora disturbances, and acquisition of sexually transmitted diseases in Zimbabwean women. *J Infect Dis* **181:**587–594.

24. **Zheng HY, Alcorn TM, Cohen MS.** 1994. Effects of H2O2-producing lactobacilli on *Neisseria gonorrhoeae* growth and catalase activity. *J Infect Dis* **170:**1209–1215.

25. **Boris S, Barbés C.** 2000. Role played by lactobacilli in controlling the population of vaginal pathogens. *Microbes Infect* **2:**543–546.

26. **Herthelius M, Gorbach SL, Möllby R, Nord CE, Pettersson L, Winberg J.** 1989. Elimination of vaginal colonization with *Escherichia coli* by administration of indigenous flora. *Infect Immun* **57:**2447–2451.

27. **McGroarty JA, Reid G.** 1988. Detection of a *Lactobacillus* substance that inhibits *Escherichia coli*. *Can J Microbiol* **34:**974–978.

28. **McGroarty JA, Tomeczek L, Pond DG, Reid G, Bruce AW.** 1992. Hydrogen peroxide production by *Lactobacillus* species: correlation with susceptibility to the spermicidal compound nonoxynol-9. *J Infect Dis* **165:**1142–1144.

29. **Osset J, Bartolomé RM, García E, Andreu A.** 2001. Assessment of the capacity of *Lactobacillus* to inhibit the growth of uropathogens and block their adhesion to vaginal epithelial cells. *J Infect Dis* **183:**485–491.

30. **Reid G, Cook RL, Bruce AW.** 1987. Examination of strains of lactobacilli for properties that

may influence bacterial interference in the urinary tract. *J Urol* **138**:330–335.

31. **Reid G, Heinemann C, Velraeds M, van der Mei HC, Busscher HJ.** 1999. Biosurfactants produced by *Lactobacillus. Methods Enzymol* **310**:426–433.

32. **Stamey TA, Kaufman MF.** 1975. Studies of introital colonization in women with recurrent urinary infections. II. A comparison of growth in normal vaginal fluid of common versus uncommon serogroups of *Escherichia coli. J Urol* **114**:264–267.

33. **Andreu A, Stapleton AE, Fennell CL, Hillier SL, Stamm WE.** 1995. Hemagglutination, adherence, and surface properties of vaginal *Lactobacillus* species. *J Infect Dis* **171**:1237–1243.

34. **Butler D, Silvestroni A, Stapleton A.** 2016. Cytoprotective effect of *Lactobacillus crispatus* CTV-05 against uropathogenic *E. coli. Pathogens* **5**:27.

35. **O'Hanlon DE, Moench TR, Cone RA.** 2013. Vaginal pH and microbicidal lactic acid when lactobacilli dominate the microbiota. *PLoS One* **8**:e80074. doi:10.1371/journal.pone.0080074.

36. **Hooton TM, Stamm WE.** 1996. The vaginal flora and urinary tract infections. *In* Warren JW, Mobley HL (ed), *Urinary Tract Infections: Molecular Pathogenesis and Clinical Management.* ASM Press, Washington, DC.

37. **Schaeffer AJ, Jones JM, Amundsen SK.** 1980. Bacterial effect of hydrogen peroxide on urinary tract pathogens. *Appl Environ Microbiol* **40**:337–340.

38. **Klebanoff SJ, Hillier SL, Eschenbach DA, Waltersdorph AM.** 1991. Control of the microbial flora of the vagina by H_2O_2-generating lactobacilli. *J Infect Dis* **164**:94–100.

39. **Pfau A, Sacks T.** 1981. The bacterial flora of the vaginal vestibule, urethra and vagina in premenopausal women with recurrent urinary tract infections. *J Urol* **126**:630–634.

40. **Stamey TA, Sexton CC.** 1975. The role of vaginal colonization with enterobacteriaceae in recurrent urinary infections. *J Urol* **113**:214–217.

41. **Gupta K, Stapleton AE, Hooton TM, Roberts PL, Fennell CL, Stamm WE.** 1998. Inverse association of H2O2-producing lactobacilli and vaginal *Escherichia coli* colonization in women with recurrent urinary tract infections. *J Infect Dis* **178**:446–450.

42. **Hooton TM, Fihn SD, Johnson C, Roberts PL, Stamm WE.** 1989. Association between bacterial vaginosis and acute cystitis in women using diaphragms. *Arch Intern Med* **149**:1932–1936.

43. **Hooton TM, Roberts PL, Stamm WE.** 1994. Effects of recent sexual activity and use of a diaphragm on the vaginal microflora. *Clin Infect Dis* **19**:274–278.

44. **Kirjavainen PV, Pautler S, Baroja ML, Anukam K, Crowley K, Carter K, Reid G.** 2009. Abnormal immunological profile and vaginal microbiota in women prone to urinary tract infections. *Clin Vaccine Immunol* **16**:29–36.

45. **Hooton TM, Scholes D, Gupta K, Stapleton AE, Roberts PL, Stamm WE.** 2005. Amoxicillin-clavulanate vs ciprofloxacin for the treatment of uncomplicated cystitis in women: a randomized trial. *JAMA* **293**:949–955.

46. **Raz R.** 2011. Urinary tract infection in postmenopausal women. *Korean J Urol* **52**:801–808.

47. **Raz R.** 2001. Hormone replacement therapy or prophylaxis in postmenopausal women with recurrent urinary tract infection. *J Infect Dis* **183**(Suppl 1):S74–S76.

48. **Raz R, Stamm WE.** 1993. A controlled trial of intravaginal estriol in postmenopausal women with recurrent urinary tract infections. *N Engl J Med* **329**:753–756.

49. **Cauci S, Driussi S, De Santo D, Penacchioni P, Iannicelli T, Lanzafame P, De Seta F, Quadrifoglio F, de Aloysio D, Guaschino S.** 2002. Prevalence of bacterial vaginosis and vaginal flora changes in peri- and postmenopausal women. *J Clin Microbiol* **40**:2147–2152.

50. **Muhleisen AL, Herbst-Kralovetz MM.** 2016. Menopause and the vaginal microbiome. *Maturitas* **91**:42–50.

51. **Pabich WL, Fihn SD, Stamm WE, Scholes D, Boyko EJ, Gupta K.** 2003. Prevalence and determinants of vaginal flora alterations in postmenopausal women. *J Infect Dis* **188**:1054–1058.

52. **Eriksen B.** 1999. A randomized, open, parallel-group study on the preventive effect of an estradiol-releasing vaginal ring (Estring) on recurrent urinary tract infections in postmenopausal women. *Am J Obstet Gynecol* **180**:1072–1079.

53. **Shen J, Song N, Williams CJ, Brown CJ, Yan Z, Xu C, Forney LJ.** 2016. Effects of low dose estrogen therapy on the vaginal microbiomes of women with atrophic vaginitis. *Sci Rep* **6**:24380. doi:10.1038/srep24380.

54. **Mody L, Juthani-Mehta M.** 2014. Urinary tract infections in older women: a clinical review. *JAMA* **311**:844–854.

55. **Stokholm J, Schjørring S, Eskildsen CE, Pedersen L, Bischoff AL, Følsgaard N, Carson CG, Chawes BL, Bønnelykke K, Mølgaard A, Jacobsson B, Krogfelt KA, Bisgaard H.** 2014. Antibiotic use during pregnancy alters the commensal vaginal microbiota. *Clin Microbiol Infect* **20**:629–635.

56. **Mayer BT, Srinivasan S, Fiedler TL, Marrazzo JM, Fredricks DN, Schiffer JT.** 2015. Rapid

and profound shifts in the vaginal microbiota following antibiotic treatment for bacterial vaginosis. *J Infect Dis* **212:**793–802.

57. **Macklaim JM, Clemente JC, Knight R, Gloor GB, Reid G.** 2015. Changes in vaginal microbiota following antimicrobial and probiotic therapy. *Microb Ecol Health Dis* **26:**27799.

58. **Smith HS, Hughes JP, Hooton TM, Roberts P, Scholes D, Stergachis A, Stapleton A, Stamm WE.** 1997. Antecedent antimicrobial use increases the risk of uncomplicated cystitis in young women. *Clin Infect Dis* **25:**63–68.

59. **Hooton TM, Roberts PL, Stapleton AE.** 2012. Cefpodoxime vs ciprofloxacin for short-course treatment of acute uncomplicated cystitis: a randomized trial. *JAMA* **307:**583–589.

60. **Gupta K, Hooton TM, Naber KG, Wullt B, Colgan R, Miller LG, Moran GJ, Nicolle LE, Raz R, Schaeffer AJ, Soper DE, Infectious Diseases Society of America, European Society for Microbiology and Infectious Diseases.** 2011. International clinical practice guidelines for the treatment of acute uncomplicated cystitis and pyelonephritis in women: a 2010 update by the Infectious Diseases Society of America and the European Society for Microbiology and Infectious Diseases. *Clin Infect Dis* **52:**e103–e120.

61. **Eschenbach DA, Patton DL, Meier A, Thwin SS, Aura J, Stapleton A, Hooton TM.** 2000. Effects of oral contraceptive pill use on vaginal flora and vaginal epithelium. *Contraception* **62:**107–112.

62. **Gupta K, Hillier SL, Hooton TM, Roberts PL, Stamm WE.** 2000. Effects of contraceptive method on the vaginal microbial flora: a prospective evaluation. *J Infect Dis* **181:**595–601.

63. **Hooton TM, Hillier S, Johnson C, Roberts PL, Stamm WE.** 1991. *Escherichia coli* bacteriuria and contraceptive method. *JAMA* **265:**64–69.

64. **Hooton TM, Scholes D, Roberts PL, Stapleton A, Stergachis A, Stamm WE.** 1994. A prospective cohort study of the association between UTI and contraceptive method. *Abstr Intersci Conf Antimicrob Agents Chemother* **34:**134.

65. **Hooton TM, Fennell CL, Clark AM, Stamm WE.** 1991. Nonoxynol-9: differential antibacterial activity and enhancement of bacterial adherence to vaginal epithelial cells. *J Infect Dis* **164:**1216–1219.

66. **Achilles SL, Hillier SL.** 2013. The complexity of contraceptives: understanding their impact on genital immune cells and vaginal microbiota. *AIDS* **27**(Suppl 1)**:**S5–S15.

67. **Stamm WE, McKevitt M, Roberts PL, White NJ.** 1991. Natural history of recurrent urinary tract infections in women. *Rev Infect Dis* **13:**77–84.

68. **Stapleton AE, Au-Yeung M, Hooton TM, Fredricks DN, Roberts PL, Czaja CA, Yarova-Yarovaya Y, Fiedler T, Cox M, Stamm WE.** 2011. Randomized, placebo-controlled phase 2 trial of a *Lactobacillus crispatus* probiotic given intravaginally for prevention of recurrent urinary tract infection. *Clin Infect Dis* **52:**1212–1217.

69. **Schwenger EM, Tejani AM, Loewen PS.** 2015. Probiotics for preventing urinary tract infections in adults and children. *Cochrane Database Syst Rev* **12:**CD008772.

Asymptomatic Bacteriuria and Bacterial Interference

6

LINDSAY E. NICOLLE[1]

INTRODUCTION

Asymptomatic bacteriuria is the presence of bacteria in the normally sterile urine of the bladder or kidneys, together with the absence of clinical signs or symptoms attributable to urinary tract infection (1). Asymptomatic bacteriuria is also referred to as asymptomatic urinary tract infection and, occasionally, bladder colonization. In this chapter, the terms asymptomatic urinary infection, asymptomatic bacteriuria, and bacteriuria are used interchangeably, and the discussion is also relevant to asymptomatic candiduria. Bacteriuria, when present, may be restricted to the bladder, or may involve one or both kidneys. Bacteria in the prostate in men will not be discussed.

Asymptomatic bacteriuria is very common, and is usually benign. An individual may have transient bacteriuria of any duration, or bacteriuria may persist for days to years with the same or differing organisms. Resolution of bacteriuria may occur spontaneously or as a consequence of antimicrobial therapy given for any indication. Recurrent bacteriuria is frequent. Bacteriuria has been associated with harmful outcomes in a few well-characterized populations, for whom screening and treatment of bacteriuria prevents adverse outcomes. On the other hand, antimicrobial therapy prescribed inappropriately

[1]Department of Internal Medicine and Medical Microbiology, University of Manitoba, Winnipeg, MB, R3T 2N2 Canada.

Urinary Tract Infections: Molecular Pathogenesis and Clinical Management, 2nd Edition
Edited by Matthew A. Mulvey, David J. Klumpp, and Ann E. Stapleton
© 2017 American Society for Microbiology, Washington, DC
doi:10.1128/microbiolspec.UTI-0001-2012

to the many individuals with asymptomatic bacteriuria who are not at risk for adverse outcomes contributes to antimicrobial pressure, which promotes the development of antimicrobial resistance. In addition, asymptomatic bacteriuria appears to prevent development of symptomatic urinary tract infection for some populations. This observation has stimulated exploration of the potential therapeutic benefit of establishing asymptomatic bacteruria with an avirulent strain to prevent recurrent symptomatic infection, referred to as bacterial interference.

DIAGNOSIS

Quantitative Urine Culture

Asymptomatic bacteriuria is diagnosed by isolation of one or more organisms meeting appropriate quantitative counts from a urine specimen collected in a manner that minimizes contamination (Table 1). For most patients, a voided urine specimen is obtained and the relevant quantitative count is $\geq 10^5$ colony-forming units (cfu)/ml. The groundbreaking studies of Kass (2) in the 1950s, confirmed the validity of a quantitative count of $\geq 10^5$ cfu/ml as the threshold to differentiate urinary infection from contamination and facilitated clinical and epidemiologic studies addressing asymptomatic bacteriuria. Two consecutive urine specimens collected

TABLE 1 Quantitative urine culture criteria for diagnosis of asymptomatic bacteriuria

Population	Quantitative count	References
Voided urine specimens		
Healthy women[a]	$\geq 10^5$ cfu/ml	2, 3
Ambulatory men	$\geq 10^5$ cfu/ml	2, 3, 204
Catheter specimens		
In and out catheter	$\geq 10^2$ cfu/ml	7, 29
Intermittent catheter	$\geq 10^3$ cfu/ml	30
Indwelling catheter	$\geq 10^5$ cfu/ml	32, 33, 36
Condom: elderly men	$\geq 10^5$ cfu/ml	24, 25
Suprapubic aspirate	any number	5, 6

[a]Two consecutive specimens preferred.

by in and out catheter from asymptomatic outpatients distinguished specimens growing either $<10^5$ cfu/ml (usually $<10^4$ cfu/ml) of bacteria, which seldom persisted in a second specimen, and $\geq 10^5$ cfu/ml, where *E. coli* was more likely to be isolated and tended to persist. Only 1% of patients had counts between 10^4 cfu/ml and 10^5 cfu/ml. Subsequent studies confirmed that growth of $\geq 10^5$ cfu/ml from a voided urine specimen correlated with a similar quantitative count in a catheter specimen (3, 4) or suprapubic aspirate (5, 6).

Voided Urine Specimens

Women

Voided urine specimens collected from women are invariably contaminated with quantitative counts $\geq 10^2$ cfu/ml of one or more organisms which normally colonize the periurethral area or vagina (7, 8). Rigorous collection methods using repeated periurethral cleaning or midstream collection do not decrease the frequency of contaminated specimens (3, 9–11). In fact, use of the antiseptic chlorhexidine for vulvar cleansing prior to specimen collection resulted in falsely low quantitative counts of the infecting organism in the urine (11). It is now accepted that collection of a clean catch voided urine specimen without perineal cleaning is appropriate for most women.

A second specimen is recommended for women to confirm bacteriuria when $\geq 10^5$ cfu/ml of a potential uropathogen is isolated from an initial specimen. This recommendation was based on the observation of a 20% error rate in a single voided specimen compared with a catheter urine, but 96% accuracy with two consecutive voided specimens compared with the catheter specimen (4, 12). For schoolchildren, an initial specimen with a gram-negative organism isolated at $\geq 10^5$ cfu/ml was confirmed by a second specimen obtained within 2 weeks in only 61% (13). When three voided urine specimens were collected from pregnant women to confirm bacteriuria with the same organism at $\geq 10^5$ cfu/ml, the second urine specimen remained positive for 91% following an initial positive specimen, while 96% of third specimens remained positive following two consecutive positive specimens (9). However, in another study the prevalence of bacteriuria with a gram-negative organism in pregnant women decreased from 7.0% on a first specimen to 4.4% with a second specimen (14), while a Swedish study reported 15% of pregnant women had a negative second culture (15). Only 42% of healthy, sexually active nonpregnant women aged 18–40 years had *E. coli* $\geq 10^5$ cfu/ml confirmed on a second urine specimen obtained one week or one month after the first (16). For these women, isolation of <10^5 cfu/ml *E. coli* on a first specimen was followed by $\geq 10^5$ cfu/ml isolated on the next culture in only 3%. In a cohort of 40- to 64-year–old women in Finland, a second specimen obtained within 2 weeks confirmed bacteriuria in 90% (17). Diabetic women of mean age 56 years, had persistence of an organism in 69% of repeat specimens obtained within 2 weeks (18), while 56% of 18- to 75-year–old diabetic women had persistent *E. coli* bacteriuria at 2–4 months (19). Swedish women resident in the community with a mean age of 83 years, had bacteriuria with a single gram-negative organism $\geq 10^5$ cfu/ml confirmed on a second urine specimen obtained within 2 weeks in 85% (20). Female residents of a long-term care facility with a mean age of 83.4 years had bacteriuria confirmed on 90% of second specimens repeated within 2 weeks (21). However, a second specimen at 2 weeks was positive for only 53% of women of mean age 85 years in Swedish nursing homes, although the impact of intercurrent antimicrobials was not described (22).

The variability in persistence of bacteriuria on a second urine specimen obtained from women following an initial positive specimen is likely attributable to differences in patient characteristics, the spectrum of species reported as bacteriuria, time elapsed between specimen collection, and any concurrent antimicrobial therapy. It seems likely, for most women, that a single appropriately collected voided urine specimen with *E. coli* or other gram-negative organism isolated at >10^5 cfu/ml represents true bacteriuria, rather than contamination. A second positive specimen then identifies persistent bacteriuria. Many episodes of bacteriuria are likely transient, especially in sexually active young women.

Men

Contamination of voided specimens collected from men is less frequent, with only 12.3% of men reported to have <10^4 cfu/ml of bacteria isolated from voided specimens (23). Circumcision and povidone-iodine meatal cleaning did not decrease the frequency of contaminated cultures, but contamination was less frequent with midstream urines, where 8.4% of specimens had low counts isolated compared with 13% of initial void urine specimens. Isolation of bacteria at $\geq 10^5$ cfu/ml was, however, similar for initial voided or midstream urine specimens (7.1% vs 7.8%, respectively). When urine specimens are collected from elderly men using a freshly applied clean condom catheter, isolation of $\geq 10^5$ cfu/ml of one or two organisms has a positive predictive value of 85% to 100% and negative predictive value of 86% to 94% for isolation of the same organism in a concurrent catheter specimen (24, 25). In men with spinal cord injury, most of whom were receiving prophylactic antimicrobials or antiseptics and using external urine-collecting devices, 81% of organisms isolated from a suprapubic aspirate were present in counts $\geq 10^5$ cfu/ml from the voided specimen (26).

Older, ambulatory asymptomatic men with an isolation of single gram-negative organism $\geq 10^5$ cfu/ml from a voided urine specimen had bacteriuria confirmed in 95% when a second voided specimen was obtained within 1 month (27). However, of 22 Japanese men, only 46% had a persistent positive culture when a urine specimen was repeated within 2 months, including 3 of 10 men who had a gram-positive organism initially isolated and 7 of 12 men who had a gram-negative organism (28). Co-operative institutionalized elderly men in Sweden with an initial positive screening culture had bacteriuria confirmed on a repeat culture at 2 weeks in 53%, but interval antimicrobial therapy was not described (22). A single voided urine specimen is generally considered adequate to identify bacteriuria in men, although a repeat specimen may be advisable when a gram-positive organism is isolated.

Other Urine Specimens

Growth of any organism in any quantitative count from a urine specimen collected by direct puncture of the urinary tract, including suprapubic or renal pelvis aspiration, is diagnostic of bacteriuria (5, 6). When urethral catheterization is used for specimen collection a small number of periurethral organisms may be introduced into the bladder. Evaluation of paired specimens collected by in and out catheter and suprapubic aspiration identified $\geq 10^2$ cfu/ml as the most reliable quantitative count for identification of bacteriuria in specimens obtained by in and out catheter, including intermittent catheterization (7, 29, 30).

Indwelling urinary devices, including urethral catheters, nephrostomy tubes, and ureteric stents uniformly acquire biofilm on the device surface (31). Some organisms present in the biofilm but not in urine may contaminate a specimen collected through the device. When low counts of bacteria ($<10^2$ cfu/ml) are isolated from urine specimens collected through short term indwelling catheters, the quantitative count progresses to $\geq 10^5$ cfu/ml by 72 hours in 96% of patients who remain catheterized and do not receive antibiotics (32). Presumably, the low counts reflect initial biofilm formation on the catheter followed by bladder bacteriuria as the biofilm ascends to the bladder. Compared with bladder urine, a higher number of organisms are isolated and at higher quantitative counts when urine specimens are collected through a biofilm-laden chronic indwelling device (33–35). Replacing an *in situ* indwelling catheter allows urine specimen collection from the freshly inserted catheter where biofilm has not yet formed, so bladder urine rather than urine contaminated by biofilm is sampled. A specimen obtained from a chronic indwelling catheter was 90% sensitive but only 43% specific to identify organisms isolated from urine collected through a freshly placed catheter (35). When multiple organisms are isolated from urine collected through the replacement catheter, organisms present in quantitative counts $<10^5$ cfu/ml tend not to persist (36). Thus, only organisms isolated in counts $\geq 10^5$ cfu/ml should be interpreted as bacteriuria for patients with an indwelling urethral catheter (29).

Pyuria and Other Inflammatory Markers

Pyuria is a non-specific marker for inflammation within the genitourinary tract. Asymptomatic bacteriuria is usually accompanied by pyuria, but the prevalence of pyuria varies among different bacteriuric populations (Table 2). Pyuria accompanies bacteriuria in 20% to 50% of young women and girls, 50% of pregnant women (1), and 70% of diabetic women of any age (37). From 60% to 100% of bacteriuric institutionalized elderly men or women have pyuria (38–40), 30% to 75% of bacteriuric patients with a short-term indwelling catheter (41), and virtually all patients with a chronic indwelling catheter (42).

Other inflammatory markers are present in the urine of patients with asymptomatic bacteriuria but usually vary substantially among patients and in the same patient over time. Urine interleukin-6 (IL-6) levels >20 units/ml were present in 86% of young women initially identified with bacteriuria in childhood and with persistent bacteriuria for 2 to 16 years' follow-up (43). A similar proportion of women with pyelonephritis had these levels, but none of the bacteriuric patients had elevated serum levels of IL-6.

TABLE 2 Prevalence of pyuria in populations with asymptomatic bacteriuria

Population	Pyuria	Diagnostic method	References
Community populations:			
Swedish schoolgirls	26%	hemocytometer >50 cells/mm	94
Canadian pre-school and schoolgirls	37%	≥5 leukocytes/hpf[a]	205
Female college students	24%	hemocytometer ≥8 cells/mm^3	16
Pregnant women	47%	leukocyte esterase/reagent strip/nitrite	157
	81%	urinalysis >10 leukocytes/hpf plus reagent strip positive	
Diabetic women	70%	hemocytometer ≥10 cells/mm^5	37
Elderly male/female, Sweden	47%	leukocyte esterase dipstick	46
Institutionalized elderly:			
Male nursing home U.S.	27–67%	≥5 leukocytes/hpf	74
Condom catheter	50%		
Incontinent/no catheter			
Canadian nursing home – male/female	93%	≥10 leukocytes/hpf	206
Mexican women/nursing homes	77%	≥10 leukocytes/hpf	207
U.S. nursing home, women	93%	>10 leukocytes/mm^3, uncentrifuged urine	38
U.S. nursing home, men/women	59%	leukocyte esterase dipstick	39
U.S. nursing home, men/women	100%	leukocyte esterase and/or nitrite reagent strip	208
U.S. nursing homes, incontinent women	65%	>10 leukocytes/hpf	168
Other:			
Spinal cord injured/intermittent catheterization	86%	hemocytometer >10 cells/mm^3	209
Indwelling catheter, short-term	30–75%	hemocytometer >20 cells/mm^3	41
Indwelling catheter, chronic	66%	≥5 leukocytes/hpf	42
Ileal neobladder	99%	cytoflowmetry >10/μL	58

[a]hpf: high-powered field.

IL-6 was present in the urine of 16% of healthy young women with asymptomatic bacteriuria; significantly increased urine interleukin-8 (IL-8) levels in these women correlated with neutrophil count (44). Chemokine ligand 5 (CXCL-5), chemokine ligand 6 (CXCL-6), and intracellular adhesion molecule-1 (ICAM-1) were not increased. When persistent asymptomatic bacteriuria was established with the avirulent *E. coli* 83972 strain, 21 of 23 male and female patients had pyuria and elevated urine IL-8, while IL-6, interleukin 1 receptor antagonist (IL-1RA), monocyte chemotactic protein-1 (MCP-1), and interleukin 1 (IL-1) ∝ varied over time in the same patient (45). Bacteriuric patients older than 80 years of age living in the community had significantly increased urine interleukin-12 (IL-12), chemokine ligand-1 (CXCL-1), IL-8, IL-6, and IL-10 compared with elderly individuals without bacteriuria, but there was a wide variation among patients (46). Urine levels of CXCL-1, IL-8, and IL-6 in this population were generally lower in urine from asymptomatic patients compared with patients who had symptomatic infection. For institutionalized elderly residents, 43% of bacteriuric patients had urine IL-6 present, whether asymptomatic or symptomatic (47).

EPIDEMIOLOGY

Prevalence

The prevalence of asymptomatic bacteriuria in healthy patients varies with age and gender (Table 3). Reported surveys may not be comparable because of differences in

TABLE 3 Prevalence of asymptomatic bacteriuria in patients with a normal genitourinary tract

Population	Bacteriuria (%) Female	Male	References
Infants	0–1.8	0.5–2.7	48, 210
Preschool	0.8–2.0	0	48, 66
School age	1.1–1.8	0–0.03	13, 48
Adolescent	0.8–1.8	0	48
Premenopausal women	1.8–5.2	–	12, 16, 48
Nuns	0.7	–	12
Pregnant women	1.8–6.7	–	15, 48
Postmenopausal women 55 to 75 years	4.5	–	211
Elderly, community:			
50 to 59 years	4.4–8.6	0.6	48
60 to 69 years	6.6–8.6	1.5	48
≥70 years	10.8–5.0	6.0–15	48–50
Elderly institutionalized	25–53	19–37	48, 49

diagnostic criteria (i.e., number of positive specimens required) or laboratory methods. In the first year of life, bacteriuria is more common in boys than girls. Subsequently, asymptomatic bacteriuria is absent in boys with a normal urinary tract but the prevalence increases with increasing age in girls. For sexually-active young women the prevalence is about 1.5% at age 20 years, increasing to 3% to 5% by age 50 years (48). Married

American women between 25 and 44 years of age had a prevalence of bacteriuria of 4.6%, while nuns of the same age had a prevalence of only 0.7% (12). At least one episode of asymptomatic bacteriuria was identified in 22% of 348 young, healthy, sexually active American women monitored with weekly urine cultures for 4 weeks, then monthly to 6 months (16). The prevalence in pregnant women ranges from 2% to 7% and is similar to age-matched nonpregnant women. Bacteriuria continues to increase as women age, and becomes common in men older than 50 years. By age 80 years, between 5% and 10% of men and 15% and 20% of women living in the community have bacteriuria (48–50). The prevalence of bacteriuria in residents of long-term–care facilities without indwelling urinary catheters is exceptionally high — 25% to 50% of women and 15% to 40% of men (49). In one institution, male residents had a monthly prevalence of 26% to 47%, while 24% of new residents were bacteriuric at the time of admission to the unit (51).

Patients with genitourinary abnormalities who experience recurrent symptomatic urinary infection also have an increased likelihood of asymptomatic bacteriuria, irrespective of gender or age (Table 4). Diabetic women, but not men, have a higher prevalence of bacteriuria than non-diabetic individuals (37).

TABLE 4 Prevalence of asymptomatic bacteriuria in populations with abnormalities of the genitourinary tract

Population	Bacteriuria (%) Female	Male	Both	References
Diabetes mellitus	7.9–17.7	1.5–0.2		37, 48
Multiple sclerosis	11–23			191
Spinal cord injury:			55–73	48, 212, 213
spontaneous voiding			22–26	
intermittent catheter	65–89	23–74		
sphincterotomy/condom		98		
Indwelling catheter:				
short-term			9–16	41, 48
chronic			98–100	42, 92, 177
Urologic stents:				
urethral		45–100		54
ureteral		17–24		55, 56
Ileal neobladder	94	50	49–57	58, 59

Men with spinal cord injuries have a prevalence of bacteriuria of 25% to 50% whether voiding is managed by intermittent catheter or by sphincterotomy with condom drainage (52, 53). When a chronic indwelling catheter is used, the prevalence of bacteriuria is 100%. Between 50% and 70% of patients with urethral stents are bacteriuric (54). At removal of ureteral stents, 17% to 24% of patients have bacteriuria, while the stents themselves are culture positive in 34% to 42% of patients (55, 56). Following urinary diversion using an abdominal stoma conduit, virtually all patients have bacteriuria, usually with mixed gram-positive skin flora (57). Between 30% and 80% of patients with orthoptic bladder substitution are bacteriuric (58, 59), and bacteruria is also common following augmentation cystoplasty, particularly for patients using clean intermittent catheterization (57).

Incidence

The incidence of asymptomatic bacteriuria is less well described (Table 5). It is reported to be 100-fold higher in sexually active women compared with school-aged girls. The incidence of bacteriuria in pregnant Swedish women was 1.3% between 8 and 16 weeks of pregnancy and 0.2% from 16 weeks to term (15). Another study reported 2% of pregnant women without bacteriuria at initial screening developed bacteriuria during the remainder of the pregnancy (60). For elderly bacteriuric women resident in a nursing home, monthly urine cultures identified an incidence of new infection of 0.87 to 1.67/patient-year (21). The incidence of asymptomatic bacteriuria was 45/100 patient-years in elderly institutionalized men, and 10% of nonbacteriuric men acquired bacteriuria every 3 months (49). Acquisition of bacteriuria while an indwelling catheter remains *in situ* is between 3% and 7% per day. The acquisition of new strains remains 4% to 7% per day, or 3.2 new organisms per month, when there is a chronic indwelling catheter (61). In patients with spinal cord injuries, the incidence of bacteriuria was 5/100 person-days for patients using indwelling catheters,

TABLE 5 Incidence of asymptomatic bacteriuria in selected populations

Population	Female	Male	Both	References
Schoolgirls	0.35/100 person-years	–		13
Premenopausal women, sexually active	39–53/100 person-years	–		16
Pregnant women	1–1.3% after 12 weeks	–		15
Postmenopausal women				214
non-diabetic	3.0/100 person-years	–		
diabetic	6.7/100 person-years			
Bacteriuric institutionalized elderly	87–67/100 patient (pt)–years	45/100 pt-years		21, 51
Intermittent catheter			1/146 catheterizations; 2.95/100 pt-years	48, 62
Indwelling catheter:				
Short-term		11.1/100 days	2% to 7%/day	48, 215
Long-term		2.72/100 pt-days	0.56/pt-week; 5/100 pt-days	62, 72
Suprapubic	0.96/100 pt-days			62
Spinal cord injured, normal voiding			0.33/100 pt-days	62
Condom drainage:				
Elderly		6.1/100 pt-days		216
Spinal cord		2.41/100 pt-days		62

2.95/100 person-days for patients with an intermittent catheter, and 2.41/100 person-days for men using condom catheter drainage (62).

The incidence of urinary infection in elderly institutionalized populations has also been described as a turnover of bacteriuria. Female residents in U.S. life-care communities, self-care, and nursing homes from whom urine cultures were obtained every 6 months had conversion from negative to positive cultures in 5%, 11%, and 8%, respectively, at 6 months, and from positive to negative in 33%, 34%, and 31%, respectively (63). Female residents in U.S. community housing and long-term care from whom monthly urine cultures were obtained had a probability of transition from positive to negative urine culture on two consecutive monthly specimens of 0.3, and from negative to positive of 0.12 (64). In a Greek nursing home population, 23% of initially negative women and 11% of men had bacteriuria at one year, while 27% of positive women and 22% of positive men became negative (65). Residents with bacteriuria present on the initial culture but a negative culture at 6 months had recurrent bacteriuria at one year in 77% of men and 44% of women.

PATHOGENESIS

Host Factors

Genetic and behavioral factors associated with acute uncomplicated urinary infection in healthy girls and women are also major contributing factors for asymptomatic bacteriuria. Women and girls with asymptomatic bacteriuria are more likely to have recurrent symptomatic urinary infection (13, 16, 66, 67) and recurrent bacteriuria (68). A population-based study enrolled Swedish women initially screened between 38 and 60 years of age. When evaluated 6 years after enrollment, 18% of women with bacteriuria at the initial screening had bacteriuria again, compared with only 3.2% without initial bacteriuria (risk ratio 6.92; 95% CI 3.53–13.53) (87). At 12 years' follow-up, there was still significantly increased bacteriuria in women with bacteriuria on the initial screening (RR 3.13; 95% CI 1.45–6.73). Another long-term Swedish study reported that 20% of women bacteriuric at enrollment had bacteriuria at 15 years' follow-up, but only 3 (7.5%) women without bacteriuria at enrollment (69).

Nonsecretors of the blood group substance have an increased risk for symptomatic infection, but this association has not been evaluated for asymptomatic bacteriuria. Genetic variation in components of the host innate immune response appear to influence the risk of acquisition of bacteriuria as well as the clinical presentation of infection (45, 70). Selected genetic polymorphisms of TLR-2 and CXCR1 have been reported to be associated with asymptomatic bacteriuria in healthy young women (44).

The most important behavioral risk factors for asymptomatic bacteriuria in premenopausal women are sexual intercourse and spermicide use, with or without a diaphragm (16). For young university students, the relative risk for bacteriuria was 1.6 (95% CI 1.4–1.9) with recent diaphragm use with spermicide, 1.4 (95% CI 1.1–1.7) for spermicide use with cervical cap, and 1.3 (95% CI 1.1–1.4) for recent sexual intercourse. For older premenopausal women risks were 1.5 (95% CI 1.2–1.8) for diaphragm with spermicide, 2.4 (95% CI 1.4–4.4) for spermicide use alone, and 1.2 (95% CI 1.1–1.5) for recent sexual intercourse. Spermicide kills or inhibits many organisms of the normal vaginal flora, facilitating replacement with potential uropathogens, while sexual intercourse allows organisms colonizing the periurethral and vaginal mucosa to ascend into the bladder.

Following menopause, the age-related increase in bacteriuria prevalence continues, but is not accelerated (71). Loss of the estrogen effect on the genitourinary mucosa correlates with changes in colonizing flora of the vagina and periurethral area, but these changes have not been shown to promote bacteriuria (72). The presence of bacteriuria in elderly women living in the community correlates with urinary incontinence, reduced mobility, and systemic estrogen treatment and, for men, is associated with prostate disease, prior stroke, and living in a service flat (i.e., requiring more care) (50). Elevated post-void residual urine volume was not associated with bacteriuria in residents of a Swedish long-term care facility (73). Bacteriuria does not correlate with age in the institutionalized elderly population. The highest prevalence of bacteriuria in institutionalized elderly individuals is among the most functionally impaired residents (49). Male nursing home residents who use a condom catheter for voiding management have an increased prevalence and incidence of bacteriuria compared with those without condom drainage (74). Bacteria was present in 29% of men using continuous condom drainage, 15% with a night condom only, 10% of incontinent men not using a condom, and 3% who were continent.

Bacteriuria in women with diabetes correlates with duration of diabetes and the presence of long-term complications such as retinopathy and neuropathy, but not with parameters of glucose control such as hemoglobin A1C (37, 75). The increased prevalence of bacteriuria observed in diabetic women is likely secondary to impaired voiding attributed to diabetic neuropathy rather than metabolic or immune changes accompanying diabetes. Obstruction and urine stasis with abnormalities such as cystoceles, bladder diverticuli and, in men, prostate hypertrophy, is associated with asymptomatic bacteriuria. A review of 342 urodynamic studies undertaken in a British center, however, reported no significant association of residual urine volume, stratified as <100 cc or >100 cc and bacteriuria for men (9.9% and 17.6%, respectively) or women (24.6% vs 20.0%) (76). A retrospective review of 176 renal transplant patients reported that independent associa-

tions with bacteriuria in addition to age and female sex were days of bladder catheterization following transplant surgery, presence of genitourinary anatomic abnormalities, and urinary infection within one month of transplant (77). Routine stenting at the time of transplant surgery also increases the likelihood of bacteriuria.

Intermittent catheterization promotes bacteriuria by repeated introduction of organisms into the bladder (62). Acquisition of bacteriuria in patients with urethral catheters or other indwelling urologic devices is a consequence of biofilm formation on the device. The duration the device remains *in situ* predicts bacteriuria. There may also be a genetic influence on the clinical presentation of infection in patients with complicated urinary tract infection. Patients with long-term *E. coli* 83972 bacteriuria associated with a low urinary neutrophil count, IL-6 and MCP-1 were more likely to have TLR4 polymorphisms and IRF3 genotypes previously reported to be associated with asymptomatic bacteriuria (45).

Organism Factors

E. coli is the most common organism isolated from healthy women with asymptomatic bacteriuria. *E. coli* occurred in 72% of bacteriuric school-aged girls (13), 77% of young, sexually active American women (16), 65% to 84% of pregnant women (9, 58), 75% of postmenopausal women (78), and 60% of diabetic women (18). Other organisms commonly isolated are *Klebsiella* spp. and *P. mirabilis*; *Streptococcus agalactiae* occurs more frequently in diabetic and pregnant women.

E. coli strains isolated from asymptomatic healthy women may have a lower frequency of virulence factors, such as specific lipopolysaccharide types, adhesins, motility determinants, toxins, and other proteins compared with strains isolated from symptomatic urinary tract infection (45, 78). Decreased virulence characteristics are also

TABLE 6 Studies of *E. coli* virulence factors in strains isolated from asymptomatic bacteriuria[a]

Population	*E. coli* virulence factors	References
Nosocomial; male/female; age 54 year; bacteriuria (ABU) vs cystitis (c), pyelonephritis (p)	ABU strains decreased genes for pap G II, sfaS, afaBC, fewer fimbrial markers, hlyA, iha, fyuA. Expression of adhesins, hemolysin, and siderophores lower for ABU than c/p.	78
11 ABU isolates, 3 virulent UPEC, one fecal strain	ABU strains genetically related to UPEC had altered virulence genes – point mutations, DNA rearrangements, deletions; ABU strains unrelated to UPEC lacked most virulence genes.	203
ABU, pyelonephritis (p) cystitis (c) strains; complicated and uncomplicated	Comprehensive genotyping could not differentiate between strains isolated from ABU or c/p.	84
Pregnant women: asymptomatic and pyelonephritis strains	ABU strains had decreased adherence and MRHA, with different serotypes compared with pyelonephritis; hemolysin and MSHA similar.	79
Nosocomial, catheterized vs noncatheterized (male/female, 56 year)	Catheter-acquired and not catheter-ABU strains had similar virulence profiles, including biofilm formation.	87
Diabetic women	Virulence factors for ABU *E. coli* similar for diabetic and non-diabetic with ABU.	80, 82
Diabetic women	Virulence characteristics of ABU strains similar to fecal strains from healthy women.	81
Premenopausal women, asymptomatic and cystitis (c) strains	Proportion of strains with pap G (39% ABU vs 41% c) and class I, II, III papG alleles similar.	16
Neobladder	Neobladder strains distinct from UPEC; more similar to GI strains.	88
Spinal cord injured (indwelling, condom, and intermittent)	76% MSHA, 31% MRHA, 17% P, 27% hemolysin, 27% aerobactin, 54% virulent serotypes.	83

[a]sfaS: sialosyl-specific fimbriae; pap GII: P fimbria adhesin; afaBC: afa fimbriae; hly A: hemolysin; iha: siderophore receptor; fyuA: yersiniabactin receptor; ABU: asymptomatic bacteriuria; MRHA: mannose-resistant hemagglutination; MSHA: mannose-sensitive hemagglutination.

described for strains isolated from pregnant women (79), diabetic women (80–82), and women with spinal cord injury (83) (Table 6). There is, however, overlap in virulence expression among *E. coli* strains isolated from asymptomatic or symptomatic infection, and some reports describe similar virulence factors for asymptomatic and symptomatic strains (16, 84). *E. coli* strains isolated from asymptomatic bacteriuria may originate either from a clonal lineage lacking the virulence genes generally associated with uropathogenic *E. coli* (UPEC), or by attenuation of a UPEC strain so urovirulence characteristics are no longer present or expressed (45, 85). The prototypic, well-characterized, avirulent *E. coli* strain 83972, originally isolated from a girl with persistent asymptomatic bacteriuria, is a UPEC strain with attenuation of the adhesin virulence determinants fimH, P, and F1C fimbriae (85, 86).

E. coli is also a common organism isolated from bacteriuria in patients with a complicated genitourinary tract. However, the proportion of isolates accounted for by *E. coli* is lower and a much wider variety of organisms occurs (Table 7). *E. coli* appears to be less frequent in males in some populations. For instance, in patients using intermittent catheterization, *E. coli* accounted for 19% and 53% of isolates (61) and 42% and 68% (52) from males and females, respectively. *E. coli* strains isolated from hospital-acquired asymptomatic bacteriuria possessed similar virulence profiles whether an indwelling catheter was present or not (87). Two *E. coli* isolates causing long-term ileal neobladder colonization genetically resembled commensal/gastrointestinal adapted *E. coli* rather than UPEC, potentially explaining the persistence of these strains in the gut-derived neobladder (88).

TABLE 7 **Microbiology of asymptomatic bacteriuria in selected populations with complicated urinary tract infection**

Population (reference)	% of urine specimens					
	Spinal cord/ male (217)	Spinal cord male/ female (218)	Chronic catheter (92)	Urethral stent (55)	Long-term care; women (168)	Long-term care; men (51)
E. coli	18	14.5	34	33.3	56	11
K. pneumoniae	15.1	17.1	22		11	5.9
P. mirabilis	4.9	9.4	26		14	30
P. aeruginosa	16	5.7	48	23.8	6	19
Providencia spp.	4.6	3.5	16		5	16
Other Enterobacteriaceae	15.1	7.4	10.5		8	6.7
Enterococcus spp.	13.5	17.8	35	33.4	12	5.0
Group B streptococcus						
S. aureus	6.0	3.5			2	2.5
CNS[a]		8.7	75	100		1.7
Other gram-positive	3.3	2.3	76			1.7
Candida spp.		0.8		23.8		
Acinetobacter spp.		7.9				

[a]CNS: coagulase-negative staphylococcus.

Attributes of other organisms isolated from patients with complicated asymptomatic bacteriuria are not well described. *Klebsiella pneumoniae, Citrobacter* species, *P. mirabilis*, other Enterobacteriaceae, *Pseudomonas* spp, Enterococcus, and coagulase-negative staphylococci are all common (Table 7). Relatively avirulent organisms, including coagulase negative staphylococci (other than *Staphylococcus saprophyticus*), *Enterococcus* species, and *Candida* spp. are more common with asymptomatic bacteriuria, particularly when devices with biofilm are present. Strains of *K. pneumoniae* isolated from bacteriuric patients with and without catheters were characterized by less serum resistance, but had a similar frequency of Type 1 fimbriae and capsular K2 serotype compared to cystitis strains (89). For *Providencia stuartii*, the MR/K hemagglutinin correlates with catheter adherence and persistence of infection (90). *P. mirabilis* is common and frequently associated with catheter obstruction resulting from crystalline biofilm formation. One report isolated *P. mirabilis* from 86% of obstructed catheters (91). Patients with complex urologic abnormalities or indwelling devices frequently have polymicrobial bacteriuria. When a chronic indwelling catheter is present, 77% or more of urine specimens have more than one organism isolated (42, 92). Bacteria isolated from patients with complicated genitourinary tracts are more likely to be antimicrobial resistant. This higher prevalence of resistance is attributed to repeated antimicrobial exposure as well as health care acquisition for many patients.

NATURAL HISTORY OF BACTERIURIA

Microbiology

Potential microbiologic outcomes of asymptomatic bacteriuria include spontaneous resolution, resolution with concomitant antimicrobial therapy given for any indication, persistence of bacteriuria with the same or different organisms, or progression to symptomatic urinary tract infection. Of these, progression to symptomatic infection is the least common. In the absence of indwelling devices, *E. coli* is the organism most often associated with persistent bacteriuria. The likelihood of resolution or persistence is dependent on the duration of follow-up as well as patient characteristics. Patients frequently have recurrent bacteriuria following resolution.

Infants and children

A Swedish study of infants with asymptomatic bacteriuria identified on screening and confirmed by suprapubic aspiration monitored 50 bacteriuric patients with six urine cultures in the first year, four in the second, two in the third and fourth, and yearly thereafter (93). There were 12 girls and 25 boys followed for the full 6 years; 45 infants with normal urography at enrollment were not treated with antimicrobial therapy. Bacteriuria cleared spontaneously in 36 infants and following antibiotics given for concomitant respiratory tract infections in 8. Ten of 50 children who cleared bacteriuria had a recurrence during follow-up.

Of 116 Swedish school-aged girls with bacteriuria on screening confirmed by a second culture, 11% who did not receive antimicrobials had spontaneous resolution by 1 year (94). Another Swedish study followed 54 girls aged 3 years to 15 years referred to a specialty outpatient clinic with asymptomatic *E. coli* bacteriuria, for a median of 2.5 years (95). All patients with evidence of renal scarring were included in the follow-up cohort, and a random sample of girls with normal kidneys. Spontaneous strain change identified by epidemiologic typing of *E. coli* was uncommon–only 1/11.6 patient-years of follow-up. Eleven (46%) of 24 strain changes occurred following antimicrobials given for other infections. A US cohort of 156 school-aged girls with persistent bacteriuria defined by 3 consecutive cultures were all treated with antimicrobials and followed for 10 years; 50% of white and 20% of black girls had recurrent bacteriuria by 1 year, and 70% and 45% by 3 years, respectively; 80% of recurrences were with a different organism (96). Thus, asymptomatic bacteriuria may persist for a prolonged time in some children, and recurrence is common following resolution with or without antibiotics.

Healthy women

Ten of 65 bacteriuric *E. coli* episodes identified on a single urine culture in 348 sexually active American women aged 18 years to 40 years were followed by symptomatic cystitis (6 with *E. coli* isolated) (16). A second culture obtained 1 week or 1 month later showed bacteriuria had resolved spontaneously in 34, 17 had persistent bacteriuria with the same strain, and 4 had *E. coli* bacteruria with a different strain. However, *E. coli* bacteriuria persisted for 2 months or longer in only 5 women. Of 45 untreated healthy women aged 20 years to 65 years with bacteriuria on two specimens, 11% cleared spontaneously by 14 days, 20% by 4 weeks, 33% at 6 months, and 36% at 1 year (97, 98). The variation in persistence of bacteriuria appears largely to reflect the number of urine specimens obtained to define the bacteriuric population and the frequency of interval antimicrobial therapy.

Pregnant women

Bacteriuria in pregnant women confirmed by two or three consecutive cultures generally persists with the same organism throughout pregnancy or until antimicrobial therapy is given. Of 145 untreated pregnant women less than 32 weeks gestation in whom bacteriuria was identified by three positive cultures, only 20 (14%) cleared spontaneously during the remainder of the pregnancy, 98 remained bacteriuric, and 27 developed pyelonephritis and were treated (60). In 106 women bacteriuric on three consecutive urine specimens, 72 remained bacteriuric and untreated throughout the pregnancy, and only 8 (7.5%) cleared spontaneously, while 26 developed symptomatic infection (9). Women with bacteriuria during pregnancy are likely to have recurrent or persistent bacteriuria after delivery. At 2 to 4 years after delivery 27% of 285 bacteriuric women who had received antimicrobial treatment and had sterile urine at discharge were bacteriuric (99).

Diabetic women

Diabetic women with untreated bacteriuria and urine cultures repeated every 3 months remained bacteriuric in 50% at 12 months, 40% at 24 months, and 50% at 36 months (100). Persistent bacteriuria with the same *E. coli* strain occurred in 30% at 12 months, 20% at 24 months and 20% at 36 months (82). In another study, 53 diabetic women with *E. coli* bacteriuria confirmed in a second specimen at 2 months to 4 months had the same strain in 9 (63%) of 16 pairs available for typing (19). Thus, *E. coli* bacteriuria frequently persists for extended periods in women with diabetes.

Older women and men: community
A population-based study in Sweden screened persons aged 72 years to 79 years; bacteriuric patients were subsequently screened intermittently and nonsystematically. In patients who underwent repeated screening, bacteriuria persisted for at least 12 months only in women and only when the organism isolated was *E. coli* (101). Another Swedish study screened men and women older than 80 years of age in the community with urine cultures repeated at 6 months and 18 months (50). When bacteriuria was present at the initial screening, 60% of patients remained bacteriuric with any organism at both 6 and 18 months and, for *E. coli* bacteriuria, 76% had the same strain at 6 months and 40% at 18 months. Male outpatients at a Veteran's hospital followed every 3 months for 1 to 4.5 years had at least one spontaneous resolution in 76%, while 38% had persistence of bacteriuria with the same species for 2 to 21 months; 21% had intermittent bacteriuria (102).

Institutionalized elderly
A cohort of elderly American women and men who were residents in assisted-living facilities or nursing homes had urine cultures persistently positive with a single species for 18 months in 6% of women and 2.6% of men (63). Ten of 12 women with recurrent *E. coli* bacteriuria following antimicrobial therapy had the pre-therapy strain isolated while 11 of 14 untreated women had a persistent *E. coli*

strain for up to 18 months (103). Five of 7 initially bacteriuric women resident in an American nursing home had persistence of the initial strain during 6 months of monthly urine cultures (64). A Canadian long-term care facility reported persistent bacteriuria with the same species in 71% of untreated women monitored with monthly urine cultures for 12 months, but epidemiologic typing was not performed; 25% of the women had spontaneous resolution but experienced re-infection at 1 month to 1 year (21). In another Canadian study, bacteriuria persisted during 1-year follow-up in 77% of 1,387 urine specimens obtained monthly from 83 initially bacteriuric men and women (40). While 20 (24%) patients had persistent bacteriuria with the same organism only 13 (16%) had periods of at least 4 months free of bacteriuria. For 11 of these 13 patients, the non-bacteriuric period followed an antimicrobial course given for another indication. Nineteen bacteriuric elderly institutionalized men who did not receive antimicrobial therapy remained bacteriuric at 24 months, and 55% of all monthly urine cultures were positive during follow-up (49). At 24 months, 84% had bacteriuria with the same species isolated at the first culture, and 3 had replacement with a new organism at 5, 7, and 13 months. Thus, spontaneous resolution of bacteriuria is uncommon in elderly institutionalized patients, while the same strain frequently persists for months or years.

Other populations
Bacteriuria persists in patients with a chronic indwelling catheter but there is continuous replacement of strains (29, 92, 104). Different species remain for different durations. Enterococcus species persisted for a mean of 2.9 weeks, *E. coli*, *P. mirabilis*, *K. pneumoniae*, and *P. aeruginosa* for 4 to 6 weeks, and *Providencia stuartii* for 10.4 weeks in a prospective study of patients with chronic catheters when weekly urine cultures were obtained (92). In contrast, 75% of non-enterococcal gram positive cocci persisted less than 1 week.

Patients maintained on intermittent catheterization experience repeated episodes of bacteriuria with new organisms, but persistence of a single organism for extended periods may also occur. For 50 spinal cord injury patients using indwelling catheters followed for 24 to 270 days, relapsing bacteriuria following antibiotics was 35.6/1,000 days and reinfection 55.8/1,000 days (104). Patients with relapsing bacteruria were more likely to acquire symptomatic infection. In a prospective study, 18 (67%) patients with a continent ileal reservoir had persistent bacteriuria, 5 of whom had *E. coli* persisting for 5 or more months (62).

Morbidity and Mortality

Potential adverse outcomes following acquisition of asymptomatic bacteriuria include short-term events, such as symptomatic urinary tract infection, and long-term outcomes, including chronic renal failure, hypertension, or increased mortality.

Infants and Children

Early screening studies of infants and preschool children reported an association of bacteriuria with obstructive uropathy and vesicoureteral junction deformities (105). Only 2 of 50 infants with bacteriuria identified on population based screening and followed prospectively developed pyelonephritis, both within 2 weeks of the initial identification of bacteriuria (95). Infants with normal urography were not given antimicrobial therapy and, with a median follow-up of 32 months, none of 36 patients developed renal damage. Renal concentrating ability at the end of follow-up was comparable to a reference non-bacteriuric population. An American study that enrolled 1,817 infants and preschool girls reported evidence of upper tract damage in 17% of bacteriuric infants and 13% of bacteriuric preschool children (105). Vesicoureteral reflux was present in 46% of infants and 9% of preschool girls. At 3 to 5 years follow-up, infants with

bacteriuria and a normal urinary tract, with or without reflux, were more likely to experience recurrent bacteriuria, but the kidneys remained anatomically normal. Infants with high-risk lesions, such as obstructive uropathy and vesicoureteral junction ectopy and deformity, were more likely to experience both bacteriuria and recurrent symptomatic infection.

School-aged girls with bacteriuria on screening were more likely to have renal parenchymal reduction, pyelonephritic scarring, vesicoureteral reflux, and past history of symptomatic urinary infection (13, 67, 106). One of 28 Swedish school-aged girls with untreated *E. coli* bacteriuria developed pyelonephritis during 9 months' follow-up, but the strain isolated from the symptomatic episode was different from that isolated at screening (94). In this cohort, at 3 years, renal scarring or arrested renal growth only occurred in patients who experienced symptomatic pyelonephritis (107). In fact, treatment of asymptomatic bacteriuria was associated in the short-term with an increased incidence of symptomatic infection (107, 108). During 9 to –18 years' follow-up of 60 American school-aged girls identified with asymptomatic bacteriuria on screening, five had reflux repair, two had nephrectomy, and one, with atrophic pyelonephritis, had reduced inulin clearance (109). Renal scars or caliectasis were present in 16 bacteriuric girls and no controls, but blood pressure remained similar for both groups during the extended follow-up. Girls with asymptomatic bacteriuria were more likely to have bacteriuria identified at subsequent follow-up, including during pregnancy. The authors concluded that bacteriuria in school-aged girls identifies patients at risk of recurrent symptomatic infection and renal scars, but at low risk of reduced renal function.

Thus, infants and school-aged girls with asymptomatic bacteriuria identified at routine screening are more likely to have genitourinary abnormalities, including renal scars and vesicoureteral reflux, and at greater risk of experiencing symptomatic urinary infection. Persistent untreated asymptomatic bacteriuria, however, is not associated with an increased risk for new or progressive renal scars, symptomatic infection with the same organism, hypertension, or renal failure. The few children at risk for progression of renal abnormalities are identified by the occurrence of symptomatic infection.

Healthy non-pregnant women

Young, sexually active American women with *E. coli* $\geq 10^5$ cfu/ml isolated in one urine specimen developed symptomatic infection with the same strain in 8% of bacteriuric episodes, invariably within 1 week of initial identification of bacteriuria (16). British women with untreated asymptomatic bacteriuria had symptomatic infection develop in 36% by 1 year compared with 7% of control women without bacteriuria (97). A Swedish study re-evaluated 40 women 15 years after identification of bacteriuria, and compared these with 40 age-matched controls without bacteriuria (69). The median age was 58 years (35–72 years) at the long-term assessment. In the 15-year interval, 22 (55%) women with bacteriuria and 4 (10%) controls were treated for symptomatic urinary infection, including 3 episodes of pyelonephritis. Diagnosis of hypertension and renal function at 15 years were similar for patients with and without bacteriuria. There was a significant reduction in renal concentrating ability compared to the initial assessment, but the rate of decline was similar for both bacteriuric and non-bacteriuric patients. Renal imaging of selected women identified no evidence for progression or development of new renal abnormalities. A 24-year Swedish follow-up study also reported no differences in renal disease between women initially identified as bacteriuric or nonbacteriuric (68). American (12), Japanese (110), and Jamaican (111) studies have reported no association of hypertension with bacteriuria at long-term follow-up.

Women enrolled in prospective population based surveys in Wales and Jamaica were combined for an analysis of mortality stratified by bacteriuria or not at initial screening (112). There was significantly increased mortality at 10 to 13 years for women who bacteriuric at the first survey [RR 2.5 (1.73, 2.68)], or at two consecutive surveys [RR 4.2 (2.23, 8.04)]. However, when adjusted for age and weight, the association was lower [RR 1.5 (0.96, 2.32) and 2.0 (1.05, 3.92)], respectively. Other potential confounding factors were not addressed. The analysis was not stratified by geographic origin and pooling of these different populations for the analysis was likely not appropriate. The 24-year Swedish study did not report any association of mortality with asymptomatic bacteriuria (68).

Thus, bacteriuric women are at increased risk of symptomatic urinary tract infection, but the strain isolated from asymptomatic infection seldom progresses to symptomatic infection. When this does occur, the symptomatic episode usually develops shortly following acquisition of the strain. Women with bacteriuria are not at risk for long-term negative renal outcomes, hypertension, or increased mortality. Several studies report an increased risk of symptomatic infection following antimicrobial therapy in women (113, 114). This observation has been attributed to disruption of normal vaginal flora by the antimicrobial, thus promoting colonization with potential uropathogens (114). However, an alternate explanation is that eradication of asymptomatic bacteriuria in a woman with a biologic predisposition to both bacteriuria and recurrent symptomatic infection is followed by reinfection with a more virulent organism, resulting in the symptomatic infection.

Diabetic women

Diabetic women with asymptomatic bacteriuria are also at increased risk for symptomatic urinary infection (83, 115). In a Dutch cohort study of 589 diabetic women followed for 18 months, 20% developed symptomatic urinary infection. For women with type 2 diabetes 19% without and 34% with bacteriuria at enrollment developed symptomatic infection, and for type 1 diabetes 12% and 15%, respectively (115). Bacteriuric women with type II diabetes had poorer renal function at enrollment compared with those without bacteriuria, but the decline in renal function at 6 years was similar for patients with and without bacteriuria at enrollment (116). Thus, bacteriuric diabetic women are more likely to experience symptomatic urinary infection, but are not at increased risk for long-term negative renal outcomes.

Pregnant women

From 19% to 36% of untreated pregnant bacteriuric women will develop acute pyelonephritis later in pregnancy, usually with the same organism isolated early in the pregnancy (1, 9, 14, 60) (Table 8). Pyelonephritis occurring at the end of the second or early third trimester precipitates preterm delivery and increases neonatal death rates (60, 117). Asymptomatic bacteriuria in pregnant women is also associated with intra-uterine growth retardation and low birth weight (118).

Progression of asymptomatic bacteriuria to pyelonephritis in these women is attributed to the physiologic hormonal changes of pregnancy, as well as ureteric obstruction (119). Progesterone induces relaxation of the genitourinary autonomic musculature, leading to impaired voiding and urine stasis. Decreased muscle tone also increases the likelihood of reflux of urine from the bladder

into the kidneys. Pressure of the fetal head on the pelvic brim, usually on the right side, obstructs drainage from the kidney. These changes are maximal at the end of the second trimester and early third trimester, when pyelonephritis carries the greatest risk for preterm delivery and poor fetal outcomes.

Elderly patients

The 15- (69) and 24-year (68) Swedish follow-up studies of women in the community enrolled both pre and post-menopausal women. Outcomes did not differ for older compared with younger women. A prospective, longitudinal study in ambulatory outpatient men (102) in the United States reported 5 of 34 bacteriuric men were treated for symptomatic infection during 1- to 4.5-years follow-up, but symptoms were not described.

Ambulatory women in a supervised living facility experienced chronic genitourinary or non-specific symptoms with similar frequency and intensity whether bacteriuria was present or not (120). Bacteriuria in elderly institutionalized women was not associated with alterations in mental status or increased rates of hospitalization for urinary infection (121). There is evidence of increased systemic immune activation in some frail elderly patients with asymptomatic bacteriuria (122). Elevated circulating levels of soluable tumor necrosis factor receptors (sTNFR-1) and a higher number of blood neutrophils were reported in frail bacteriuric men and women, compared to individuals without bacteriuria. Bacteriuric elderly patients with increased systemic antibody to *Enterobacteriaceae* have decreased survival compared with bacteriuric patients without elevated antibody (40, 123). However, the mortality of patients with increased antibody is similar to elderly residents without bacteriuria (123). The clinical relevance of these observations is not clear, and they may represent epiphenomena accompanying aging-associated alterations of immune function.

The clinical diagnosis of symptomatic urinary infection is often problematic in

TABLE 8 Asymptomatic bacteriuria in pregnancy and the frequency of pyelonephritis

Number with bacteriuria (prevalence)	Pyelonephritis in pregnancy (%)		
	Antimicrobial treated	Untreated	References
265 (5.3%)	3.2	24.8	218
145 (4.0%)	2.8	28.7	219
144 (4.7%)	2.8	28	220
110 (6.5%)	4.3	19.5	161
179 (NS)	–	31	117
173 (4.9%)	–	36	221
203 (6.0%)	1	26	9

NS: not stated

elderly institutionalized populations with impaired communication, a high frequency of chronic genitourinary symptoms, and underlying urologic abnormalities (124). One adverse outcome is the promotion of antimicrobial resistance in this population when unnecessary antimicrobial treatment is given for nonspecific clinical deterioration in the bacteriuric resident (121, 125). Given the very high prevalence of bacteriuria in these populations, clinical deterioration without localizing signs or symptoms is frequently diagnosed as urinary infection. However, for residents without indwelling catheters, localizing genitourinary signs or symptoms should be present to support a diagnosis of symptomatic infection (49, 120, 126, 127).

Early studies in Finland (128) and Greece (129) reported increased 5-year mortality for bacteriuric compared with non-bacteriuric institutionalized elderly women and men. Subsequent reports with follow-up of 5 and 9 years of elderly residents in Finland (130), Sweden (131), the United States (132), and Canada (133) reported no differences in survival between bacteriuric and nonbacteriuric residents. It is, in fact, somewhat surprising that asymptomatic bacteriuria is not associated with decreased survival, as functional impairment is an independent predictor of mortality in elderly institutionalized populations and the more functionally impaired residents are more likely to be bacteriuric (134).

Indwelling urinary catheters

Catheter-acquired bacteriuria in patients with a short-term indwelling catheter is infrequently associated with symptomatic infection. In 1,497 patients with a newly inserted indwelling catheter, 235 of whom developed bacteriuria, potential symptoms of urinary infection occurred with similar frequency in patients with or without bacteriuria. Only one episode of bacteremia was attributed to catheter-acquired bacteriuria (135). Another prospective study reported presumed symptomatic urinary tract infection in 1.43 to

1.61/100 catheter-days, with concordant blood and urine isolates for 0.53% of all catheters and 4.8% of catheters with bacteriuria (136). In critical care units, less than 3% of bacteremic episodes in patients are attributed to catheter-acquired urinary infection (29). Increased mortality was reported in hospitalized patients with indwelling catheters in one study, but receiving antimicrobials either before or during catheterization did not predict lower mortality in the multivariate analysis (137). Other studies report no association of bacteriuria with mortality. Confounding attributed to the substantial clinical differences between catheterized and noncatheterized patients seems the likely explanation for any mortality differences (29).

Elderly institutionalized residents with chronic indwelling catheters have increased morbidity and mortality compared with bacteriuric residents without an indwelling catheter, presumably at least partially attributable to bacteriuria (138). The usual presentation of symptomatic infection in these residents is fever alone (127). Urinary infection is usually a diagnosis of exclusion, as localizing symptoms are seldom present. Febrile episodes of presumed urinary source occur at a rate of 0.69 to 1.1/100 catheter-days (127, 139) and bacteremia is 3- to 39-times more frequent in bacteriuric residents with indwelling catheters (140, 141). Local complications, such as purulent urethritis, paraurethral abscesses, and, in men, epididymoorchitis and prostatitis, may also develop. Transient bacteremia has been described in 4% of residents when a chronic indwelling catheter is replaced, but is of low quantitative count and not associated with morbidity (142–144).

Intermittent catheterization

A prospective observational study of 14 children with neurogenic bladder managed with intermittent catheterization reported 70% of urine specimens had $\geq 10^4$ cfu/ml of organisms isolated. Persistent carriage for 4 weeks or longer was common, but only 5 symptomatic infections were observed during

323 patient weeks, less than 1 per patient-year. There was no deterioration of renal function during follow-up (145). In a small number of men with spinal cord injury, symptomatic urinary infection was uniformly attributed to organisms previously isolated from weekly urine cultures, with a median time of 72 days from first isolation to symptomatic infection for patients not receiving antibiotics (146). In this study, neither the intensity of pyuria nor trend over time predicted progression to symptomatic infection.

Invasive genitourinary procedures

Patients who are bacteriuric and undergo a traumatic genitourinary procedure have a high frequency of post-procedure bacteremia, with sepsis in 4% to −10% (147, 148). Transurethral resection of the prostate is the procedure best described, but any urologic procedure with a high likelihood of mucosal bleeding likely carries a similar risk. This may include extracorporeal shock wave lithotripsy, bladder tumor resection, ureteroscopy, percutaneous stone extraction, and some open procedures (148).

Other populations

Retrospective cohort studies do not report increased morbidity or impaired graft survival with bacteriuria alone in renal transplant patients (149). When asymptomatic bacteriuria is present and there is decline in renal graft function, alternate explanations for deterioration are invariably present, including symptomatic pyelonephritis or urologic complications. A study completed before antimicrobial prophylaxis became the standard of care for hematologic patients with profound neutropenia reported bacteremia developed in a high proportion of patients who acquired bacteriuria (150). However, urosepsis is uncommon with current approaches of antimicrobial prophylaxis in these patients. Morbidity has not been attributed to bacteriuria in patients with continent ileal reservoirs or other neobladders (151).

Some clinicians recommend treatment of asymptomatic bacteriuria prior to elective surgical procedures outside the genitourinary tract to prevent post-operative wound infections. There are conflicting observations from cohort studies, including report of an increased occurrence of post-operative surgical infections when there is pre-operative bacteriuria (152) or of no increase (153, 154). The organisms isolated from surgical wound infections following bone and joint surgery differ from those isolated from the urine of bacteriuric patients prior to surgery (154). An association of bacteriuria with post-operative wound infection may be attributable to confounding by host risk factors associated with surgical site infections, such as age and functional impairment, which are also risk factors for bacteriuria.

SCREENING FOR AND TREATMENT OF ASYMPTOMATIC BACTERIURIA

Early clinical trials evaluating antimicrobial treatment or nontreatment of asymptomatic bacteriuria in pregnant women consistently reported compelling benefits with antimicrobial therapy (Table 8). However, prospective, randomized clinical trials in all other non-surgical populations have reported no benefits with treatment of bacteriuria (Table 9). In addition, treatment of asymptomatic bacteriuria in school-aged girls (94, 95, 108) adult women, (113, 114), and diabetic women (18) is reported, in the short term, to be followed by an increased frequency of symptomatic urinary tract infection. In the absence of evidence for improved outcomes with treatment of bacteriuria, screening of populations to identify asymptomatic bacteriuria is also not indicated.

Children

A controlled trial randomized bacteriuric American school-aged girls to antimicrobial treatment or no treatment (155). At 2 years,

TABLE 9 Symptomatic infection in patients enrolled into prospective, randomized clinical trials of treatment or non-treatment of asymptomatic bacteriuria in populations other than pregnant women[a]

Population	Trial characteristics	Symptomatic urinary infection		References
		Treated	Not treated	
Schoolgirls	N = 63; ampicillin, NF, or T/S; 24 mo	6.9%	6.3%	155
Women	N = 94, 12 months, cultures, q 6 months	37%	36%	97, 98
Elderly ambulatory women	N = 61; 6 months	7.9%	16.4%	169
Elderly institutionalized men	N = 36; 24 months; monthly cultures	12.5%	10%	51
Elderly institutionalized women, incontinent	N = 50; 12 months; monthly cultures	42%	35%	21
Diabetic women	N = 105; double-blind first 6 weeks; 36 months, cultures every 3 months	0.93/1,000 days	1.1/1,000 days	18
Spinal cord patients	N = 46; mean 993, 1,180 days; cultures 4 and 14 days post-therapy	1.86/100 pt days	1.11/100 pt6 days	182
Chronic indwelling catheter	N = 35; weekly cultures; 15–43 wk and 12–44 wk	0.18 febrile days/ patient-week	0.22 febrile days/pt-wk	177

[a]NF: nitrofurantoin; T/S: trimethoprim/sulfamethoxazole.

84% of controls and 74% of treated school-aged girls had persistent or recurrent bacteriuria, 63% and 41% at 3 years, and 59% and 46% at 4 years, respectively. Similar numbers of children in both groups developed clinical pyelonephritis or radiologic evidence of new scars by 2 years following enrollment. The radiologic changes were all minor and did not affect renal growth. Swedish school-aged girls without vesicoureteral reflux or renal parenchymal reduction who remained bacteriuric for 6 months were treated with 10 days of nitrofurantoin or received no antibiotics (94). For the treated girls, 93% had a negative urine culture at 1 year compared with 11% of those not treated. Only one girl in each group developed pyelonephritis. The authors concluded that therapy for bacteriuria is not necessary for school-aged girls and screening for covert bacteriuria of childhood could not be recommended. An analysis of the costs and benefits of screening for bacteriuria in toilet trained asymptomatic children concluded there was no evidence that detection and treatment of asymptomatic bacteriuria prevents subsequent pyelonephritis, renal scarring or renal failure, and that screening for bacteriuria in asymptomatic children was costly and ineffective (156).

Premenopausal, Nonpregnant Women

A prospective, 12-month trial randomized 94 women aged 20 years to 65 years to treatment with nitrofurantoin, followed by ampicillin if nitrofurantoin was not effective, or no antimicrobial treatment (97, 98). Two weeks following treatment 76% of treated and 20% of untreated patients had sterile urine, and, at 12 months, 45% of treated and 36% untreated. Symptomatic urinary tract infection occurred with equal frequency in 37% of treated and in 36% of untreated patients, –during 12 months' follow-up. However, 12 of 15 patients with reinfection developed symptomatic infection compared with only 20 of 44 with persistent or relapsing infection, a significant difference (97). Thus, treatment of asymptomatic bacteriuria had no clinical benefits and may have been harmful.

Pregnant Women

Screening for and treatment of asymptomatic bacteriuria in early pregnancy decreases the risk of pyelonephritis during the pregnancy by 90% (Table 8). Treatment is also associated with decreased preterm labor and improved fetal survival (118). Thus, all pregnant women should be screened for bacteriuria in early pregnancy and treated if bacteriuria is present (1, 119). A culture method should be used for screening. Tests for pyuria, such as dipstick or urinalysis, are insensitive and identify only 50% of pregnant women with bacteriuria (157–160). A second urine culture should be obtained to confirm bacteriuria. From 20% to 50% of pregnant women may not have bacteriuria persist on a second specimen (9, 60, 161) and antimicrobial therapy can be avoided in these women. A strategy of screening and treatment was concluded to be cost effective in the American context if the prevalence of bacteriuria exceeded 2% and the risk of pyelonephritis exceeded 13% (162).

The antimicrobial regimen chosen for treatment of bacteriuria should be safe for the fetus, effective for the organism isolated, and given for as short a duration as possible (Table 10). Recommended first-line agents include amoxicillin, amoxicillin/clavulanic acid, cephalexin, or nitrofurantoin. Pivmecillinam and fosfomycin trometamol are also safe in pregnancy. Trimethoprim/sulfamethoxazole

TABLE 10 Recommended antimicrobial regimens for the treatment of asymptomatic bacteriuria in pregnancy

Agent	Dose	Duration
Preferred:		
Amoxicillin	500 mg tid	7 days
Cephalexin	500 mg qid	7 days
Nitrofurantoin	50–100 mg qid or 100 mg bid	5 days
Pivmecillinam[a]	400 mg bid	3–7 days
Fosfomycin trometamol	3 g	single dose
Other options[b]		
TMP/SMZ[c]	80/400 mg	3 days
Amoxicillin/ clavulanate	500 mg tid or 875 mg bid	7 days
Cefuroxime	500 mg bid	5 days
Cefixime	400 mg od	5–7 days

[a]Not licensed in North America;
[b]For women with intolerance or resistance to preferred regimens;
[c]Avoid in first trimester; TMP/SMZ: trimethoprim/sulfamethoxazole; tid: three times daily; qid: four times daily; bid: twice daily; od: once daily.

(TMP/SMX) is effective but associated with a small increased risk for fetal abnormalities when given in the first trimester, so this agent is not recommended as first-line therapy. The fluoroquinolones are contraindicated in pregnancy because of potential abnormalities of fetal cartilage development. A Cochrane review concluded that pregnant women should be treated with standard durations of antimicrobial regimens until more data becomes available (163). Although single-dose antimicrobial treatment was not significantly inferior at 4 to 7 days, there was a trend to lower efficacy and the studies were judged to be of poor quality. A subsequent prospective, randomized comparative trial enrolling pregnant women in Thailand, the Philippines, Vietnam, and Argentina randomized bacteriuric pregnant women to nitrofurantoin monohydrate/macrocrystals 100 mg twice a day for 1 day or for 7 days (164). The microbiologic cure rates at 14 days post-treatment were 76% and 86%, respectively, a significant difference. Thus, short courses of antimicrobial therapy for treatment of asymptomatic bacteriuria in pregnancy are not currently recommended.

Following one episode of either asymptomatic bacteriuria or symptomatic urinary infection in a pregnant woman, urine cultures should be monitored monthly throughout the remainder of the pregnancy (158, 165). If a second episode of asymptomatic or symptomatic infection occurs, prophylactic antimicrobial therapy should be initiated and continued for the duration of the pregnancy and, optimally, 6 weeks post-partum. Cephalexin or nitrofurantoin are preferred, but nitrofurantoin should be discontinued at 32 to 34 weeks gestation.

The management of pregnant women with genitourinary abnormalities which promote an increased frequency of asymptomatic bacteriuria, is not well studied. Some of these women (e.g., women with spinal cord injuries) managed with intermittent catheterization, always have a high prevalence of bacteruria. Current guidelines recommend screening for bacteriuria and treatment of all positive urine cultures in such women (166). However, the effectiveness of this approach in preventing pyelonephritis and adverse fetal outcomes balanced against the risk of acquisition of resistant organisms has not been rigorously evaluated.

Women with Diabetes

A prospective, randomized 36-month trial of treatment or non-treatment of bacteriuria in diabetic women, the majority of whom had type 2 diabetes, reported no benefits with antimicrobial treatment (18). Continued screening for and treatment of bacteriuria, including prophylactic antimicrobial therapy following repeated recurrences, did not alter the frequency of symptomatic infection, including pyelonephritis, compared with non-treatment. Hospitalization for urinary infection or any other cause was also similar with or without treatment, and there was no evidence for renal functional decline in either group. Women randomized to treatment, however, had substantially greater antimicrobial exposure and, in the short term, an increased incidence of pyelonephritis. A prospective, cohort study followed 53 Polish diabetic patients with bacteriuria and 54 diabetic patients without bacteriuria for 14 years, with clinical and bacteriological assessment every 3–6 months. Acute pyelonephritis occurred with similar frequency in both groups (11% and 9.3%, respectively), and there were no differences in renal functional decline or development of hypertension (167). Thus, routine screening for or treatment of asymptomatic bacteriuria for diabetic women is not beneficial.

Older Women in the Community

There are no prospective, randomized trials of treatment or nontreatment of bacteriuria specifically evaluating older community residents. Some post-menopausal women were enrolled into a prospective, randomized study of healthy women who reported no benefits with treatment (97, 98). For diabetic women, there were no differences in outcomes with treatment of bacteriuria for older compared with younger women (18). The absence of adverse outcomes in long term prospective cohort studies also supports a lack of benefit for screening and treatment (68, 69, 102). Thus, current evidence does not support screening for or treatment of asymptomatic bacteriuria in elderly women or men resident in the community, but clinical trial evidence is limited.

Elderly Institutionalized Populations

Prospective, randomized, clinical trials enrolling men and women resident in long-term care facilities have reported no benefits with treatment of asymptomatic bacteriuria (21, 168, 169). Studies have enrolled residents of differing functional levels and are relevant across the spectrum of institutionalized elderly patients. Treatment of bacteriuria does not decrease the likelihood of subsequent symptomatic urinary tract infection or alter mortality. The prevalence of asymptomatic bacteriuria is decreased only modestly in the treated arm, and for only a short duration. There was also no improvement of chronic incontinence in women with treatment of bacteriuria (168). An increased likelihood of reinfection with a more resistant organism follows antimicrobial treatment, and there are increased antimicrobial adverse drug effects and costs (21). Thus, screening for and treatment of asymptomatic bacteriuria in elderly long-term care residents is not recommended. As most elderly institutionalized individuals with bacteriuria have pyuria, the presence of pyuria does not identify patients with a poorer outcome or for whom antimicrobial treatment is indicated (123).

Indwelling Urethral Catheter

Most patients with short-term indwelling catheters receive concomitant antimicrobial therapy (32, 170). Antimicrobial therapy given for any indication delays the onset of bacteriuria and decreases the prevalence of bacteriuria at catheter removal (170). A randomized, controlled trial of catheter irrigation with either neomycin-polymyxin or saline reported no decrease in bacteriuria with the antimicrobial solution, perhaps because irrigation required disconnecting the catheter tubing (171). Bacteria isolated from patients receiving the antimicrobial irrigation solution had increased antimicrobial resistance. A clinical trial enrolling intensive care unit patients reported no differences in mortality, duration of intensive care unit stay or duration of hospitalization in patients randomized to catheter replacement with treatment of bacteriuria compared with the catheter remaining *in situ* without antimicrobial treatment (172).

Clinical trials evaluating treatment of asymptomatic bacteriuria when an indwelling catheter is discontinued report conflicting results. A non-blinded Swiss study of patients not otherwise receiving antimicrobials reported a significantly decreased frequency of symptomatic infection (4.9% vs 21.6%) and of bacteriuria when three doses of TMP/SMX were given at catheter removal (173). However, a blinded trial in male surgical patients comparing a single dose of ciprofloxacin, TMP/SMX, or no antibiotics at catheter removal reported no differences in bacteriuria or symptomatic infections for 14 days following removal, but increased isolation of ciprofloxacin resistant organisms from infections (174). Another blinded pilot study enrolling patients not otherwise receiving antimicrobials reported a 48 hour course of ciprofloxacin or placebo at catheter removal was followed by a similar frequency of symptomatic infection in the two groups, but patients who received ciprofloxacin were more likely to have resistant organisms isolated (175). In women requiring a short-term indwelling catheter, with a negative urine culture at catheter insertion, and with bacteriuria persisting 48 hours following

catheter removal, TMP/SMX treatment at 48 hours significantly decreased the frequency of symptomatic urinary infection during the subsequent 14 days, from 17% of untreated patients to none (176). This was a highly selected patient population and the generalizability of the effect needs confirmation in further clinical trials.

A prospective, randomized trial in nursing home residents with chronic indwelling catheters reported that cephalexin treatment of bacteriuria did not decrease subsequent symptomatic episodes compared with no antibiotics, but in the treated population cephalexin resistant organisms were isolated from 30% of re-infections (177). A prospective, non-comparative study of consecutive courses of antimicrobial treatment given to eradicate bacteriuria in elderly patients with chronic catheters also reported no decrease in febrile episodes compared with the historical pretreatment period and immediate recurrence of bacteriuria following therapy, often with an organism of increased resistance (178). Patients with chronic indwelling catheters given trimethoprim/sulfamethoxazole or no antimicrobials had a similar incidence of bacteriuria (59). A randomized, double-blind study enrolled 89 patients with neurologic conditions and indwelling catheters to one of 3 bladder irrigations–sterile saline, acetic acid, or neomycin/polymyxin (179). During the 8-week study, none of the three irrigating solutions altered the frequency of bacteriuria or pyuria in the 52 patients who completed the study.

Thus, studies consistently report no benefits with treatment of asymptomatic bacteriuria while the catheter remains *in situ*. The risks or benefits of treatment of asymptomatic bacteriuria acquired during catheterization and persisting following catheter removal remain controversial. Currently, treatment of bacteriuric patients at the time of catheter removal cannot be recommended, but further studies to address this question are necessary.

Patients with Spinal Cord Injury

A randomized, double-blind, placebo-controlled, 16-week trial enrolled 112 men and 17 women with acute spinal cord injury managed with intermittent catheterization (180). Treated patients received TMP/SMX one-half tablet daily, with additional treatment for breakthrough bacteriuria. The 50% weekly prevalence of bacteriuria decreased to 30% in antimicrobial treated patients compared with 56% of placebo patients. At least one episode of symptomatic infection was experienced by 7% of men receiving TMP/SMX and 35% with placebo, but there was no decrease for women. However, 95% of recurrent bacteriuria episodes in the treatment group and 50% in placebo were with TMP/SMX resistant organisms. Men with chronic spinal cord injury for at least 6 months and using any type of bladder management were randomized to continue or discontinue daily TMP/SMX prophylaxis and followed with weekly urine cultures for a minimum of 3 months (109). Episodes of asymptomatic bacteriuria were 0.241 per week in controls and 0.243 with continued TMP/SMX, while 76% of control urine cultures and 65% from TMP/SMX patients were positive. Symptomatic urinary tract infection occurred with similar frequency in the two groups. Organisms isolated were resistant to TMP/SMX in 78% of controls and 94% receiving TMP/SMX.

In a non-comparative study with uniform treatment of asymptomatic bacteriuria in a cohort of primarily men with spinal cord injury, patients managed with intermittent catheterization or condom drainage reported prompt recurrence of bacteriuria after antimicrobial treatment (181). Following a course of antibiotics of 7 or 14 days, 93% of patients were again bacteriuric by 1 month. When the duration of antimicrobial therapy was increased to 4 weeks, 85% were bacteriuric within 1 month. A prospective, randomized controlled trial of antimicrobial treatment or no treatment of bacteriuria in a small number of patients with spinal cord injury and bacteriuria reported a similar frequency of symptomatic urinary tract infection in the two groups during an average of 50 days follow-up (182).

Thus, clinical trials report no benefits with treatment of asymptomatic bacteriuria in patients with spinal cord injury irrespective of bladder management. The National Institute on Disability and Rehabilitation Research Consensus Statement recommends that asymptomatic bacteriuria in patients with spinal cord injury should not be treated (183).

Invasive Genitourinary Procedures

Antimicrobial therapy initiated for bacteriuric patients prior to an invasive urologic procedure decreases the risk of post-procedure bacteremia and sepsis (1). This is a strategy of peri-operative surgical prophylaxis rather than treatment of asymptomatic bacteriuria, per se (148). Antimicrobial therapy is indicated for any bacteriuric patient undergoing an invasive genitourinary procedure associated with mucosal trauma and bleeding (184–188). Clinical trials do not identify a preferred regimen, but a single dose given 1 hour prior to the procedure is usually effective, similar to other surgical prophylaxis regimens (189).

Other Populations

Renal transplant patients receive antimicrobial prophylaxis, usually with TMP/SMX, for 6 months following transplantation, and this reduces the frequency of asymptomatic bacteriuria as well as of symptomatic infections. A non-randomized retrospective case-control study from Switzerland described 334 episodes of *E. coli* or Enterococcal bacteriuria in 77 renal transplant recipients (190); 101 (30%) episodes were treated with antimicrobials. Subsequent symptomatic urinary tract infection was similar for treated or untreated patients, but resistant organisms were isolated from 78% of the treated group with recurrent infection. Current renal transplant guidelines do not recommend routine screening for bacteriuria in renal transplant patients.

A prospective cohort study stratified women with multiple sclerosis into three groups–those using intermittent catheterization, those with post-void residual (PVR) >100 ml but not using intermittent catheterization, and women with no evidence of bladder dysfunction (191). Bacteriuria present at study enrollment was treated and urine cultures repeated monthly for 6 months, with further appropriate antimicrobial therapy given for eradication of recurrent bacteriuria. After a first episode of recurrent bacteriuria, long-term prophylaxis with nitrofurantoin 100 mg daily was initiated for all women, and if there was further recurrent bacteriuria on nitrofurantoin, norfloxacin prophylaxis was instituted. The prevalence of any episode of bacteriuria during 6 study months was 90% of women using intermittent catheterization, 34% with increased PVR but not using a catheter, and 24% in the normal-voiding group. Bacteriuria developed on prophylactic antimicrobial therapy in 31% of the intermittent-catheter group and 22% of women with elevated PVR, but in none of the patients with normal bladder function; 14% of patients in the intermittent-catheter group developed symptomatic infection but none in the other groups. Thus, a strategy of treatment of bacteriuria and antimicrobial prophylaxis was not effective for patients using intermittent catheterization.

A double-blind, placebo-controlled study of patients with asymptomatic candiduria, about one-half of whom had indwelling catheters, reported a significantly lower prevalence of candiduria at the end of 14 days fluconazole treatment, but no difference 2 weeks following the end of treatment (192).

BACTERIAL INTERFERENCE

Bacterial interference describes the therapeutic strategy of establishing persistent asymptomatic bacteriuria using an avirulent bacterial strain with the goal of preventing symptomatic infection. This approach was suggested following observations from clinical and epidemiologic studies that repeatedly describe an increased frequency of acute symptomatic urinary tract infection shortly after antimicrobial therapy in girls (108), adult women (97), diabetic women (18) with bacteriuria, and in the general adult female population (113, 114). Asymptomatic bacteriuria is hypothesized to prevent symptomatic infection by interfering with persistence of virulent uropathogens in the bladder. Potential mechanisms include inhibiting adherence by blocking uroepithelial cell receptors, competition for nutrients, or production of toxins that inhibit the virulent strain (193).

An initial study described eight women with chronic symptomatic urinary infection and catheter instillation into the bladder of either the avirulent *E. coli* 83972 strain (85, 86, 194), the same strain transformed with pap and pil adhesins, or both strains together (195). The avirulent *E. coli* 83972 strain persisted beyond 30 days in more than 50% of patients, while the virulent transformant elicited a vigorous local inflammatory response leading to elimination within 2 days. These observations were the basis for further clinical studies of the *E. coli* 83972 strain, and an *E. coli* HU2117 strain derived from *E. coli* 83972 by deletion mutation of the papG gene (196, 197). Preliminary studies confirmed that extended bacteriuria was achieved in selected patients and appeared safe (195, 198–200), and that incomplete bladder emptying was necessary to establish bacteriuria (198). Coating indwelling catheters with the strain was one effective means of introducing the organism into the bladder (198).

Three prospective, randomized clinical trials have evaluated bacterial interference using the *E. coli* 83972 strain in selected patients experiencing frequent recurrent symptomatic infection. A blinded, prospective, randomized trial in the United States enrolled male patients with spinal cord injury, most with indwelling catheters (201); there were 21 experimental patients and 6 controls. Following a course of antibiotic therapy to resolve pre-existing bacteriuria, the therapeutic strain was instilled by catheter twice daily for 3 consecutive days. Colonization was successful in 13 (62%) experimental patients for an average duration of 3.5 months; 4 (31%) patients remained colonized for more than 12 months. For the clinical efficacy analysis, the 8 patients who failed colonization were combined with the 6 control patients and compared with successfully colonized patients while they remained colonized. The frequency of at least one symptomatic infection was significantly lower in the *E. coli* 83972 colonized patients. There were no serious toxicities, but 1 of 13 colonized patients developed symptomatic infection with a mixed bacterial culture, including the therapeutic strain isolated.

These same investigators proceeded to a multicenter randomized trial of patients with chronic spinal cord injury managed with indwelling or intermittent catheters and who had experienced two or more symptomatic infections in the previous year (196). Colonization with *E. coli* 83972 was achieved for only 30% of the experimental group. There was a significant difference in time to first symptomatic episode in colonized or saline inoculation placebo patients, but the analysis used a one-sided confidence interval and 66% of the experimental group and 37% of the placebo group were not included in the analysis. The occurrence of symptomatic infection in the two study arms did not diverge until more than 100 days following inoculation.

A randomized, blinded Swedish study using the *E. coli* 83972 strain enrolled only patients without an indwelling catheter (202). Twenty of 26 patients enrolled completed the full trial and were included in the analysis. Of the 20 patients who completed the trial, 12 patients were women, 12 patients used an intermittent catheter, and 8 patients used no catheter. Antibiotics were given to clear pre-existing bacteriuria and the thera-peutic strain instilled by catheter once daily for 3 days. In the first phase of the study, 13 of 18 (72%) patients were successfully colonized; the 5 not colonized were combined with 6 patients randomized to saline control for the outcomes analysis. At 1 year, colonized patients had significantly fewer symptomatic episodes. In a second phase of the study, patients were crossed over after 12 months or following a symptomatic episode. There were also significantly fewer symptomatic episodes in colonized patients after the crossover.

Bacterial interference to prevent symptomatic urinary infection is an attractive concept, which could limit morbidity and antimicrobial exposure for some patients with frequent symptomatic infections. Evidence to date suggests this approach may be effective for a limited number of selected individuals for a limited duration. However, the randomized clinical trials supporting efficacy are compromised by problems with study design and analysis. Further clinical studies are needed to characterize patient populations most likely to benefit from this approach, standardize effective instillation techniques, and explore strategies to increase the duration of colonization. The impact of host genetic variation on the likelihood of establishing colonization should be evaluated. Given the propensity for genetic evolution of *E. coli* strains isolated from persistent bacteriuria, deliberate instillation of a potentially pathogenic organism remains conceptually problematic (85, 203). Other safety issues include the need for antimicrobial therapy prior to instillation and repeated bladder catheterization to establish bacteriuria.

ACKNOWLEDGMENTS

Conflicts of interest: I declare no conflicts.

CITATION

Nicolle LE. 2015. Asymptomatic bacteriuria and bacterial interference, Microbiol Spectrum 3(5):UTI-0001-2012.

REFERENCES

1. **Nicolle LE, Bradley S, Colgan R, Rice JC, Schaeffer A, Hooton TM.** 2005. IDSA Guideline for the diagnosis and treatment of asymptomatic bacteriuria in adults. *Clin Infect Dis* **40:**643–654.

2. **Kass EH.** 1956. Asymptomatic infections of the urinary tract. *Trans Assoc Amer Phys* **69:** 56–64.

3. **Kass EH.** 1957. Bacteriuria and the diagnosis of infections of the urinary tract. *Arch Intern Med* **100:**709–714.

4. **Kass EH.** 1962. Pyelonephritis and bacteriuria: A major problem in preventive medicine. *Ann Intern Med* **56:**46–53.

5. **Beard RW, McCoy DR, Newton JR, Clayton SG.** 1965. Diagnosis of urinary infection by suprapubic bladder puncture. *Lancet* **ii:**610–611.

6. **Stamey TA, Govan DE, Palmer JM.** 1965. The localization and treatment of urinary tract infections: The role of bactericidal urine levels as opposed to serum levels. *Medicine (Baltimore)* **44:**1–36.

7. **Platt R.** 1983. Quantitative definition of bacteriuria. *Am J Med* **75**(1B)**:**44–52.

8. **Kunin CM, White LV, Hua TH.** 1993. A reassessment of importance of "low count" bacteriuria in young women with acute urinary symptoms. *Ann Intern Med* **119:**454–460.

9. **Savage WE, Hajj SN, Kass EH.** 1967. Demographic and prognostic characteristics of bacteriuria in pregnancy. *Medicine* **46:**385–407.

10. **Leisure MK, Dudley SM, Donowitz LG.** 1993. Does a clean-catch urine sample reduce bacterial contamination? *N Engl J Med* **328:**289–290.

11. **Roberts AP, Robinson RE, Beard RW.** 1967. Some factors affecting bacterial colony counts in urinary infection. *BMJ* **1:**400–403.

12. **Kunin CM, McCormack RC.** 1968. An epidemiologic study of bacteriuria and blood pressure among nuns and working women. *N Engl J Med* **278:**635–642.

13. **Kunin CM, Deutscher R, Paquin A.** 1964. Urinary tract infection in school children: An epidemiologic, clinical and laboratory study. *Medicine (Baltimore)* **43:**91–130.

14. **Kaitz AL, Hodder EW.** 1961. Bacteriuria and pyelonephritis of pregnancy. *N Engl J Med* **265:** 667–672.

15. **Stenqvist K, Kahlen-Nilsson I, Lidin-Janson G, Lincoln K, Oden A, Bignell S, Svanborg-Eden C.** 1989. Bacteriuria in pregnancy. Frequency and risk of acquisition. *Amer J Epidemiol* **129:** 372–379.

16. **Hooton TM, Scholes D, Stapleton AE, Roberts PL, Winter C, Gupta K, Samadpour M, Stamm WE.** 2000. A prospective study of asymptomatic bacteriuria in sexually active young women. *N Engl J Med* **343:**992–997.

17. **Takala J, Jausimes H, Sievers K.** 1977. Screening for and treatment of bacteriuria in a middle aged female population. *Acta Med Scand* **202:** 69–73.

18. **Harding GKM, Zhanel GG, Nicolle LE, Cheang M.** 2002. Antimicrobial treatment in diabetic women with asymptomatic bacteriuria. *N Engl J Med* **347:**1576–1583.

19. **Geerlings SE, Brouwer EC, Gaastra W, Hopelman AIM.** 2000. Is a second urine specimen necessary for the diagnosis of asymptomatic bacteriuria? *Clin Infect Dis* **31:**e3–4.

20. **Rodhe N, Lofgren S, Matussek A, Andre M, Englund L, Kuhn I, Molstad S.** 2008. Asymptomatic bacteriuria in the elderly. High prevalence and high turnover of strains. *Scand J Infect Dis* **26:**1–7.

21. **Nicolle LE, Mayhew WJ, Bryan L.** 1987. Prospective, randomized comparison of therapy and no therapy for asymptomatic bacteriuria in institutionalized women. *Amer J Med* **8:**27–33.

22. **Hedin K, Petersson C, Wideback K, Kahlmeter G, Molstad S.** 2002. Asymptomatic bacteriuria in a population of elderly in municipal institutional care. *Scand J Prim Health Care* **20:**166–168.

23. **Lipsky BA, Inui TS, Plorde JJ, Berger RE.** 1984. Is the clean-catch midstream void procedure necessary for obtaining urine culture specimens from men. *Amer J Med* **76:**257–262.

24. **Ouslander JG, Greengold BA, Silverblatt FJ, Garcia JP.** 1987. An accurate method to obtain urine for culture in men with external catheters. *Arch Intern Med* **147:**286–288.

25. **Nicolle LE, Harding GKM, Kennedy J, McIntyre M, Aoki F, Murray D.** 1988. Urine specimen collection with external devices for diagnosis of bacteriuria in elderly incontinent men. *J Clin Microbiol* **26:**1115–1119.

26. **Deresinski SC, Perkash I.** 1985. Urinary tract infections in male spinal cord injured patients. Part one: Bacteriologic diagnosis. *J Amer Paraplegia Soc* **8:**4–6.

27. **Gleckman R, Esposito A, Crowly M, Natsios G.** 1979. Rehability of a single urine culture in establishing diagnosis of asymptomatic bacteriuria in adult males. *J Clin Microbiol* **9:**596–597.

28. **Freedman LR, Phair JP, Seki M, Hamilton HB, Nefzger MD.** 1965. The epidemiology of urinary tract infections in Hiroshema. *Yale J Biol Med* **37:**262–282.

29. Hooton TM, Bradley SF, Cardenas DD, Colgan R, Geerlings SE, Rice JC, Saint S, Schaeffer AJ, Tambyah PA, Tenke P, Nicolle LE. 2010. Diagnosis, prevention and treatment of catheter associated urinary tract infection in adults. *2009 International Clinical Practice Guidelines. Clin Infect Dis* **50:**625–683.

30. Gribble MJ, McCallum NM, Schechter MT. 1988. Evaluation of diagnostic criteria for bacteriuria in acutely spinal cord injured patients undergoing intermittent catheterization. *Diagn Microbiol Infect Dis* **9:**197–206.

31. Saint S, Chemoweth CE. 2003. Biofilms and catheter-associated urinary tract infections. *Infect Dis Clin N Amer* **17:**411–432.

32. Stark RP, Marki DG. 1984. Bacteriuria in the catheterized patient: What quantitative level of bacteriuria is relevant. *N Engl J Med* **311:** 560–564.

33. Tenney JH, Warren JW. 1988. Bacteriuria in women with long term catheters: Paired comparison of indwelling and replacement catheters. *J Infect Dis* **157:**199–202.

34. Bergquist D, Bronnestam R, Hedelin H, Stahl A. 1980. The relevance of urinary sampling methods in patients with indwelling Foley catheters. *Brit J Urol* **52:**91–95.

35. Grahn D, Norman DC, White ML, Cantrell M, Yoshikawa TT. 1985. Validity of urinary catheter specimen for diagnosis of urinary tract infection in the elderly. *Arch Intern Med* **145:** 1858–1860.

36. Tenney JH, Warren JW. 1987. Long-term catheter-associated bacteriuria: species at low concentration. *Urology* **30:**444–446.

37. Zhanel GG, Nicolle LE, Harding GKM. 1995. Prevalence of asymptomatic bacteriuria in women with diabetes mellitus. *Clin Infect Dis* **21:**316–322.

38. Boscia JA, Abrutyn E, Levison ME, Pitsakis PG, Kaye D. 1989. Pyuria and asymptomatic bacteriuria in elderly ambulatory women. *Ann Intern Med* **110:**404–405.

39. Ouslander JG, Schapira M, Schnelle J, Fingold S. 1996. Pyuria among chronically incontinent but otherwise asymptomatic nursing home residents. *J Am Geriatr Soc* **44:**420–423.

40. Nicolle LE, Duckworth H, Brunka J, Urias B, Kennedy J, Murray D, Harding GKM. 1998. Urinary antibody level and survival in bacteriuric institutionalized elderly subjects. *J Am Geriatr Soc* **46:**947–953.

41. Tambyah PA, Maki DG. 2000. The relationship between pyuria and infection in patients with indwelling urinary catheters. *Arch Intern Med* **160:**673–682.

42. Steward DK, Wood GL, Cohen RL, Smith JW, Mackowiak PA. 1985. Failure of the urinalysis and quantitative urine culture in diagnosing symptomatic urinary tract infections in patients with long-term urinary catheters. *Am J Infect Control* **13:**154–160.

43. Hedges S, Stenqvist K, Lidin-Janson G, Martinell J, Sandberg T, Svanbork C. 1992. Comparison of urine and serum concentrations of interleukin-6 in women with acute pyelonephritis or asymptomatic bacteriuria. *J Infect Dis* **166:**653–656.

44. Hawn TR, Scholes D, Wang H, Li SS, Stapleton AE, Janer M, Aderem A, Stamm WE, Zhao LP, Hooton TM. 2009. Genetic variation of the human urinary tract innate immune response and asymptomatic bacteriuria in women. *PLoS ONE* **4:**e300.

45. Hernandez JG, Sunden F, Connolly J, Svanborg C, Wullt B. 2011. Genetic control of the variable innate immune response to asymptomatic bacteriuria. *PLoS ONE* **6:**e28289.

46. Rodhe N, Lofgren S, Strindhall J, Matussek A, Molstad S. 2009. Cytokines in urine in elderly subjects with acute cystitis and asymptomatic bacteriuria. *Scand J Primary Health Care* **27:**74–79.

47. Nicolle LE, Brunka J, Orr P, Wilkins J, Harding GKM. 1993. Urinary immunoreactive interleukin–1 ∝ and interleukin–6 in bacteriuric institutionalized elderly subjects. *J Urol* **149:**1049–1053.

48. Nicolle LE. 2003. Asymptomatic bacteriuria: When to screen and when to treat. *Infect Dis Clin North Am* **17:**367–394.

49. Nicolle LE. 2009. Urinary tract infections in the elderly. *Clin Geriatr Med* **25:**423–436.

50. Rodhe N, Molstad S, Englund L, Svardsudd K. 2006. Asymptomatic bacteriuria in a population of elderly residents living in a community setting: prevalence, characteristics and associated factors. *Family Practice* **23:** 303–307.

51. Nicolle LE, Bjornson J, Harding GKM, MacDonell JA. 1983. Bacteriuria in elderly institutionalized men. *N Engl J Med* **309:**1420–1426.

52. Bakke A, Digranes A. 1991. Bacteriuria in patients treated with clean intermittent catheterization. *Scand J Infect Dis* **23:**577–582.

53. Waites KB, Canupp KC, De Vivo MJ. 1993. Epidemiology and risk factors for urinary tract infection following spinal cord injury. *Arch Phys Med Rehabil* **74:**691–695.

54. Riedl CR, Plas E, Hubner WA, Zimmer H, Ulrish W, Pfluger H. 1999. Bacterial colonization of urethral stents. *Eur Urol* **36:**53–59.

55. Kehinde EO, Rotimi VO, Al-Hunayan A, Abdul-Halim H, Boland F, Al-Awadi KA. 2004. Bacteriology of urinary tract infection associated with indwelling J ureteral stents. *J Endourol* **18:**891–896.

56. Akay AF, Aflay U, Gedik A, Sahin H, Birean MK. 2007. Risk factors for lower urinary tract infection and bacterial stent colonization in patients with a double J ureteral stent. *Int Urol Nephrol* **39:**95–90.

57. Wullt B, Agace W, Mansson W. 2004. Bladder bowel and bugs–bacteriuria in patients with intestinal urinary diversion. *World J Urol* **22:**186–195.

58. Suriano F, Gallucci M, Flammia P, Musco S, Alcini A, Imbalzanot G, Dicuonzo G. 2007. Bacteriuria in patients with an orthoptic ileal neobladder: urinary tract infection or asymptomatic bacteriuria? *Brit J Urol Internat* **101:**1576–1579.

59. Wood DP, Bianco FJ Jr, Pontes JE, Heath MA, daJusta D. 2003. Incidence and significance of positive urine cultures in patients with an orthoptic neobladder. *J Urol* **169:**2196–2199.

60. Elder HA, Santamarina BAG, Smith S, Kass EH. 1971. The natural history of asymptomatic bacteriuria during pregnancy. *Am J Obstet Gynecol* **111:**441–462.

61. Breitenbucher RB. 1984. Bacterial changes in the urine samples of patients with long term indwelling catheters. *Arch Intern Med* **144:**1585–1588.

62. Esclarin de Rus A, Leoni EG, Cabrera RH. 2000. Epidemiology and risk factors for urinary tract infection in patients with spinal cord injury. *J Urol* **164:**1285–1289.

63. Abrutyn E, Mossey J, Levison M, Boscia J, Pitsakis P, Kaye D. 1991. Epidemiology of asymptomatic bacteriuria in elderly women. *J Am Ger Soc* **39:**388–393.

64. Monane M, Gurwitz JH, Lipsitz LA, Glynn RJ, Choodnovskiy I, Avorn J. 1995. Epidemiology and diagnostic aspects of bacteriuria: a longitudinal study in older women. *J Am Ger Soc* **43:**618–622.

65. Kasviki-Charvati P, Drolette-Kefakis B, Papanayiotou PC, Dontas AS. 1982. Turnover of bacteriuria in old age. *Age Aging* **11:**169–174.

66. Lincoln K, Winberg J. 1964. Studies of urinary tract infection in infancy and childhood II. Quantitative estimation of bacteriuria in unselected neonates with special reference to the occurrence of asymptomatic infections. *Acta Paediat* **53:**307–316.

67. Silverberg DS, Allard MJ, Ulan RA, Beamish WE, Lentle BC, McPhee MS, Grace MG. 1973. City-wide screening for urinary abnormalities in schoolgirls. *Can Med Assoc J* **109:**981–985.

68. Bengtsson C, Bengtsson U, Bjorkelund C, Lincoln K, Sigurdsson JA. 1998. Bacteriuria in a population sample of women: 24 year follow-up study. *Scand J Urol Nephrol* **32:**284–289.

69. Tencer J. 1988. Asymptomatic bacteriuria–a long-term study. *Scand J Urol Nephrol* **22:**31–34.

70. Wullt B, Bergsten G, Fischer H, Godaly G, Karpman D, Leijonhufuud I, Lundstedt A-C, Samuelsson P, Samuelsson M, Svensson M-L, Svanborg C. 2003. The host response to urinary tract infection. *Infect Dis Clin N Am* **17:**279–301.

71. Hextall A, Hooper R, Cardozo L, Stringer C, Workman R. 2004. Dose the menopause influence the risk of bacteriuria? *Int Urogynecol J* **12:**332–336.

72. Pabich WL, Fihn SD, Stamm WE, Scholes D, Boyko EJ, Gupta K. 2003. Prevalence and determinants of vaginal flora alterations in postmenopausal women. *J Infect Dis* **188:**1054–1058.

73. Barabas G, Molstad S. 2005. No association between elevated post-void residual volume and bacteriuria in residents of nursing homes. *Scand J Primary Health Care* **23:**52–56.

74. Ouslander JG, Greengold B, Chen S. 1987. External catheter use and urinary tract infections among incontinent male nursing-home patients. *J Am Ger Soc* **35:**1063–1070.

75. Geerlings SE. 2008. Urinary tract infections in patients with diabetes mellitus: epidemiology, pathogenesis and treatment. *Internat J Antimicrob Agents* **31:**S54–S57.

76. Hampson SJ, Nobel JG, Rickards D, Milroy EJG. 1992. Does residual urine predispose to urinary tract infection? *Brit J Urol* **70:**506–508.

77. Sorto R, Irizar SS, Delgadillo G, Alberu J, Correa-Rotter R, Morales-Buenrostro LE. 2010. Risk factors for urinary tract infections during the first year after kidney transplantation. *Transplant Proceedings* **42:**280–281.

78. Mabbett AN, Ulett GC, Watts RE, Tree JJ, Totsika M, Ong CLY, Wood JM, Monaghan W, Looke DF, Nimmo GR, Svanborg C, Schembri MA. 2009. Virulence properties of asymptomatic bacteriuric *Escherichia coli*. *Int J Med Microbiol* **299:**53–63.

79. Stenqvist K, Sandberg T, Lidin-Janson G, Orskov F, Orskov I, Svanborg-Eden C. 1987. Virulence factors of *Escherichia coli* in urinary isolates from pregnant women. *J Infect Dis* **156:**870–877.

80. Geerlings SE, Brouwer EC, Gaastra W, Stolk R, Diepersloot RJ, Hoepelman AI. 2001.

Virulence factors of *Escherichia coli* isolated from urine of diabetic women with asymptomatic bacteriuria: correlation with clinical characteristics. *Antonie Van Leeuwenhoek* **80:** 119–127.

81. **Dalal S, Nicolle L, Marrs CF, Zhang L, Harding G, Foxman B.** 2009. Long-term *Escherichia coli* asymptomatic bacteriuria among women with diabetes mellitus. *Clin Infect Dis* **49:**491–497.

82. **Brauner A, Flodin U, Hylander B, Ostenson C-G.** 1993. Bacteriuria, bacterial virulence and host factors in diabetic patients. *Diabetic Med* **10:**550–554.

83. **Benton J, Chawa J, Parry S, Stickler D.** 1992. Virulence factors in *Escherichia coli* from urinary tract infections in patients with spinal injuries. *J Hosp Infect* **22:**117–127.

84. **Abraham S, Chapman TA, Zhang R, Chin J, Mabbett AN, Totsika M, Schembri MA.** 2012. Molecular characterization of *Escherichia coli* strains that cause symptomatic and asymptomatic urinary tract infections. *J Clin Microbiol* **50:**1027–1030.

85. **Klemm P, Roos V, Ulett GC, Svanborg C, Schembri MA.** 2006. Molecular characterization of the *Escherichia coli* asymptomatic bacteriuria strain 83972: the taming of a pathogen. *Infect Immun* **74:**781–785.

86. **Roos V, Schembri MA, Ulett GC, Klemm P.** 2006. Asymptomatic bacteriuria *Escherichia coli* strain 83972 carries mutations in the <u>foc</u> locus and is unable to express F1C fimbrae. *Microbiology* **152:**1799–1806.

87. **Watts RE, Hancock V, Ong C-LY, Vejborg RM, Mabbett AN, Totsika M, Looke DF, Nimmo GR, Klemm P, Schembri MA.** 2010. *Escherichia coli* isolates causing asymptomatic bacteriuria in catheterized and non-catheterized individuals possess similar virulence properties. *J Clin Microbiol* **48:**2449–2458.

88. **Sahl JW, Lloyd AL, Redman JC, Cebula TA, Wood DP, Mobley HLT, Rasko DA.** 2011. Genomic characterization of asymptomatic *E. coli* isolated from the neobladder. *Microbiology* **157:**1088–1102.

89. **Podsehun R, Sievers D, Fischer A, Ullmann U.** 1993. Serotypes, hemagglutinins, siderophore synthesis and serum resistance of Klebsiella isolates causing human urinary tract infections. *J Infect Dis* **168:**1415–1421.

90. **Mobley HL, Chippendale GR, Tenney JH, Mayrer AR, Crips LJ, Penner JL, Warren JW.** 1988. MR/K hemagglutination of *Providencia stuartii* correlates with adherence to catheters and with persistence in catheter-associated bacteriuria. *J Infect Dis* **157:**264–271.

91. **Mobley HL, Warren JW.** 1987. Urease-positive bacteriuria and obstruction of long term urinary catheters. *J Clin Microbiol* **25:**2216–2217.

92. **Warren JW, Tenney JH, Hoopes JM, Muncie HL, Anthony WC.** 1982. A prospective microbiologic study of bacteriuria in patients with chronic indwelling urethral catheters. *J Infect Dis* **146:**719–723.

93. **Wettergren B, Hellstrom M, Stokland E, Jodal U.** 1990. Six year follow-up of infants with bacteriuria on screening. *BMJ* **301:**845–848.

94. **Linderg U.** 1975. Asymptomatic bacteriuria in school girls. The clinical course and response to treatment. *Acta Peadiatr Scand* **64:**718–724.

95. **Hansson S, Caugant D, Jodal U, Svanborg-Eden C.** 1989. Untreated asymptomatic bacteriuria in girls. I–Stability of urinary isolates. *BMJ* **298:**853–855.

96. **Kunin CM.** 1970. The natural history of recurrent bacteriuria in school-aged girls. *N Engl J Med* **282:**1443–1448.

97. **Asscher AW, Sussman M, Waters WE, Evans JAS, Campbell H, Evans KT, Williams JE.** 1969. Asymptomatic significant bacteriuria in the non-pregnant women II. Response to treatment and follow-up. *BMJ* **1:**804–806.

98. **Asscher AW, Sussman M, Waters WE, Evans JAS, Campbell H, Evans KT, Williams JE.** 1969. The clinical significant of asymptomatic bacteriuria in the nonpregnant woman. *J Infect Dis* **120:**17–26.

99. **Raz R, Sakran W, Chazan B, Kunin C.** 2003. Long-term follow-up of women hospitalized for acute pyelonephritis. *Clin Infect Dis* **37:** 1014–1020.

100. **Nicolle LE, Zhanel GG, Harding GKM.** 2006. Microbiology outcomes in women with diabetes and untreated asymptomatic bacteriuria. *World J Urol* **24:**61–65.

101. **Nordenstam G, Sundh V, Lincoln K, Svanborg A, Svanborg Eden C.** 1989. Bacteriuria in representative population samples of persons aged 72–79 years. *Amer J Epidemiol* **130:**1176–1186.

102. **Mims AD, Norman DC, Yamamura RH, Yoshikawa TT.** 1990. Clinically inapparent (asymptomatic) bacteriuria in ambulatory elderly men: epidemiological, clinical, and microbiological findings. *J Am Ger Soc* **38:** 1209–1214.

103. **LiPuma JJ, Stull TL, Dason SE, Pidcock KA, Kaye D, Korzeniowski OM.** 1989. DNA polymorphisms among *Escherichia coli* isolated from bacteriuric women. *J Infect Dis* **159:**526–532.

104. **Elden H, Hizmetli S, Nacitarhan V, Kunt B, Goker I.** 1997. Relapsing significant bacteriuria: effect on urinary tract infection in patients with spinal cord injury. *Arch Phys Med Rehabil* **78:**468–470.

105. **Siegel SR, Siegel B, Sokoloff BZ, Kanter MH.** 1980. Urinary infection in infants and preschool children. Five year follow-up. *Am J Dis Child* **134:**369–372.

106. **Lindberg U, Claesson I, Hanson LA, Jodal U.** 1976. Asymptomatic bacteriuria in school-aged girls. Clinical and laboratory findings. *Acta Paediatr Scand* **64:**425–431.

107. **Lindberg U, Claesson I, Hanson LA, Jodal U.** 1978. Asymptomatic bacteriuria in school-aged girls VIII: Clinical course during a 3-year follow-up. *J Pediatr* **92:**194–199.

108. **Hansson S, Jodal U, Lincoln K, Svanborg Eden C.** 1989. Untreated asymptomatic bacteriuria in girls: II Effect of phenoxymethylpenicillin and erythromycin given for intercurrent infections. *BMJ* **298:**856–859.

109. **Gillenwater JY, Harrison RB, Kunin CM.** 1979. Natural history of bacteriuria in school-aged girls: A long-term case-control study. *N Engl J Med* **301:**396–399.

110. **Switzer S.** 1961. Bacteriuria in a healthy population and its relation to hypertension and pyelonephritis. *N Engl J Med* **264:**7–10.

111. **Miall WE, Kass EH, Ling J, Stuart KL.** 1962. Factors influencing arterial pressure in the general population in Jamaica. *BMJ* **ii:**497–506.

112. **Evans DA, Kass EH, Hennekens CH, Rosner B, Miao L, Kendrick MI, Miall WE, Stuart KL.** 1982. Bacteriuria and subsequent mortality in women. *Lancet* **i:**156–158.

113. **Foxman B, Somsel P, Tallman P, Gillespie B, Raz R, Colodner R, Kandula D, Sobel JD.** 2001. Urinary tract infection among women aged 40 to 65: behavioral and sexual risk factors. *J Clin Epidemiol* **54:**710–718.

114. **Smith HS, Hughes JP, Hooton TM, Roberts P, Scholes D, Stergachis A, Stapleton A, Stamm WE.** 1997. Antecedent antimicrobial use increases the risk of uncomplicated cystitis in young women. *Clin Infect Dis* **25:**63–68.

115. **Geerlings SE, Stolk RP, Camps MJ, Netten PM, Collet JT, Schneeberger PM, Hoepelman AI.** 2001. Consequences of asymptomatic bacteriuria in women with diabetes mellitus. *Arch Int Med* **161:**1421–1427.

116. **Melland R, Geerlings SE, Stolk RP, Netten PM, Schneeberger PM, Hoepelman AI.** 2006. Asymptomatic bacteriuria in women with diabetes mellitus: effect on renal function after 6 years of follow-up. *Arch Intern Med* **166:**2222–2227.

117. **Brumfitt W.** 1976. The effects of bacteriuria in pregnancy on maternal and fetal health. *Kidney Int* **Suppl 8:**S113–S119.

118. **Romero R, Oyarzun E, Mazor M, Sirtori M, Hobbins JC, Bracken M.** 1989. Meta-analysis of the relationship between asymptomatic bacteriuria and pre-term delivery/low birth weight. *Obstet Gynecol* **73:**576–582.

119. **Patterson TF, Andriole VT.** 1997. Detection, significance, and therapy of bacteriuria in pregnancy. *Infect Dis Clin N Amer* **11:**593–608.

120. **Boscia JA, Kobasa WD, Abrutyn E, Levison ME, Kaplan AM, Kaye D.** 1986. Lack of association between bacteriuria and symptoms in the elderly. *Am J Med* **81:**979–982.

121. **Das R, Towle V, Van Ness PH, Juthani-Mehta M.** 2011. Adverse outcomes in nursing home residents with increased episodes of observed bacteriuria. *Infect Control Hosp Epidemiol* **32:**84–86.

122. **Prio TK, Bruunsgaard H, Roge B, Pederson BK.** 2002. Asymptomatic bacteriuria in elderly humans is associated with increased levels of circulatory TNF receptors and elevated numbers of neutrophils. *Exper Geront* **37:**693–699.

123. **Nicolle LE, Brunka J, McIntyre M, Murray P, Harding GKM.** 1992. Asymptomatic bacteriuria, urinary antibody and survival in the institutionalized elderly. *J Am Ger Soc* **40:**607–613.

124. **Loeb M, Bentley DW, Bradley S, Crosley K, Garibaldi R, Gantz N, McGeer A, Muder RR, Mylotte J, Nicolle LE, Nurse B, Paton S, Simor AE, Smith P, Strausbaugh L.** 2001. Development of minimum criteria for the initiation of antibiotics in residents of long term care facilities: results of a consensus conference. *Infect Control Hosp Epidemiol* **22:**120–124.

125. **Nicolle LE.** 2000. Urinary tract infection in long-term care facility residents. *Clin Infect Dis* **31:**757–761.

126. **Loeb M, Brazil K, Lohfeld L, McGeer A, Simor A, Stevenson K, Zoutman D, Smith S, Liu X, Walter SD.** 2005. Effect of a multifaceted intervention on number of antimicrobial prescriptions for suspected urinary tract infections in residents of nursing homes: cluster randomized controlled trial. *BMJ* **33:**669.

127. **Orr P, Nicolle LE, Duckworth H, Brunka J, Kennedy J, Murray D, Harding GKM.** 1996. Febrile urinary infection in the institutionalized elderly. *Amer J Med* **100:**71–77.

128. **Sourander LB, Kasanen A.** 1972. A 5 year follow-up of bacteriuria in the aged. *Geront Clin* **14:**274–281.

129. **Dontas AS, Kasvik-Charvati P, Papanayiotou PC, Marketos SG.** 1981. Bacteriuria and survival in old age. *N Engl J Med* **304:**939–948.

130. **Heinamaki P, Haavesto M, Hakulinem T, Mattila K, Rajola S.** 1986. Mortality in relation to urinary characteristics in the very aged. *Gerontology* **32:**167–171.

131. **Nordenstam GR, Bradber CA, Oden AS, Svanborg-Eden CM, Svanborg A.** 1986. Bacteriuria and mortality in an elderly population. *N Eng J Med* **314:**1152–1156.

132. **Abrutyn E, Mossey J, Berlin JA, Boscia J, Levison M, Pitsakis P, Kaye D.** 1994. Does asymptomatic bacteriuria predict mortality and does antimicrobial treatment reduce mortality in elderly and ambulatory women? *Ann Intern Med* **120:**827–833.

133. **Nicolle LE, Henderosn E, Bjornson J, McIntyre M, Harding GK, MacDonell JA.** 1987. The association of bacteriuria with resident characteristics and survival in elderly institutionalized men. *Ann Intern Med* **106:** 682–686.

134. **High K, Bradley S, Loeb M, Palmer R, Quagliarello V, Yoshikawa T.** 2005. A new paradigm for clinical investigation of infectious syndromes in older adults: assessing functional status as a risk factor and outcome measure. *Clin Infect Dis* **40:**114–122.

135. **Tambyah PA, Maki DG.** 2000. Catheter-associated urinary tract infection is rarely symptomatic: a prospective study of 1,497 catheterized patients. *Arch Intern Med* **160:** 678–682.

136. **Srinivasan A, Karchmer T, Richards A, Song X, Perl TM.** 2006. A prospective trial of a novel, silicone-coated, silver-coated Foley catheter for the prevention of nosocomial urinary tract infection. *Infect Control Hosp Epidemiol* **27:**38–43.

137. **Platt R, Polk BF, Murdock B, Rosner B.** 1982. Mortality associated with nosocomial urinary tract infection. *N Engl J Med* **307:**637–642.

138. **Nicolle LE.** 2012. Urinary catheter-associated infections. *Infect Dis Clin N Am* **26:**13–27.

139. **Warren JW, Damron D, Tenney JH, Hoopes JM, Deforge B, Muncie HL Jr.** 1987. Fever, bacteremia, and death as complications of bacteriuria in women with long term urethral catheters. *J Infect Dis* **155:**1151–1158.

140. **Muder RR, Brennen C, Wagener MM, Goetz AM.** 1992. Bacteremia in a long term care facility: a five year prospective study of 163 consecutive episodes. *Clin Infect Dis* **14:**647–654.

141. **Rudman D, Hontanosas A, Cohen Z, Mattson DE.** 1998. Clinical correlates of bacteremia in a veteran's administration extended care facility. *J Am Geriatr Soc* **36:**726–732.

142. **Bregenzer T, Frei R, Widmer AF, Seiler W, Probst W, Mattarelli G, Zimmerli W.** 1997. Low risk of bacteremia during catheter replacement in patients with long-term urinary catheters. *Arch Intern Med* **157:**521–525.

143. **Jewes LA, Gillespie WA, Leadbetter A, Myers B, Simpson RA, Stower MJ, Viant AC.** 1988. Bacteriuria and bacteremia in patients with long-term indwelling catheters-a domiciliary study. *J Med Microbiol* **26:**61–65.

144. **Polastri F, Auckenthaler R, Loew F, Michel JP, Lew DP.** 1990. Absence of significant bacteremia during urinary catheter manipulation in patients with chronic indwelling catheters. *J Am Geriatr Soc* **38:**1203–1208.

145. **Schlager T, Dilks S, Trudell J, Whittam TS, Hendley O.** 1995. Bacteriuria in children with neurogenic bladder treated with intermittent catheterization: natural history. *J Pediatr* **126:** 490–496.

146. **Garouiche RD, Cadle RM, Zenon GJ, Markowski J, Rodriguez M, Musher DM.** 1993. Progression from asymptomatic to symptomatic urinary tract infection in patients with SCI: A preliminary study. *J Am Paraplegia Soc* **16:**219–224.

147. **Grabe M.** 1987. Antimicrobial agents in transurethral prostatic resection. *J Urol* **138:**245–252.

148. **Grabe M.** 2011. Antibiotic prophylaxis in urological surgery; a European viewpoint. *Int J Antimicrob Agents* **385:**58–63.

149. **Alangaden GJ.** 2007. Urinary tract infections in renal transplant recipients. *Curr Infect Dis Rep* **9:**475–479.

150. **Gurwith MJ, Brunton JL, Lank BA, Harding GK, Ronald AR.** 1979. A prospective controlled investigation of prophylactic trimethoprim/sulfamethoxazole in hospitalized granulocytopenic patients. *Am J Med* **66:**248–256.

151. **Akerlund S, Campanello M, Kaijser B, Jonsson O.** 1994. Bacteriuria in patients with a continent ileal reservoir for urinary diversion does not regularly require antibiotic treatment. *Brit J Urol* **74:**177–181.

152. **Lee S-F, Kim K-T, Park Y-S, Kim Y-B.** 2010. Association between asymptomatic urinary tract infection and postoperative spine infection in elderly women: A retrospective analysis study. *J Korean Neurosurg Soc* **47:**265–270.

153. **Glynn MK, Sheehan JM.** 1984. The significance of asymptomatic bacteriuria in patients undergoing hip/knee arthroplasty. *Clin Ortho Relat Res* **185:**151–154.

154. **Ritter MA, Fechtman RW.** 1987. Urinary tract sequelae: Possible influence on joint infections following total joint replacement. *Orthopedics* **10**:467–469.

155. **Savage CDL, Howie G, Adler K, Wilson MI.** 1976. Controlled trial of therapy in covert bacteriuria of childhood. *Lancet* **I**:358–361.

156. **Kemper KJ, Avner ED.** 1992. The case against screening urinalysis for asymptomatic bacteriuria in children. *Am J Dis Chil* **146**:343–346.

157. **McNair RD, MacDonald SR, Dooley SL, Peterson LR.** 2000. Evaluation of the centrifuged and Gram-stained smear, urinalysis, and reagent strip testing to detect asymptomatic bacteriuria in obstetric patients. *Am J Obstet Gynecol* **182**:1076–1079.

158. **The US Preventive Services Task Force.** 1990. Screening for asymptomatic bacteriuria, hematuria and proteinuria. *Am Fam Physician* **42**:389–395.

159. **Pels RJ, Bor DH, Woolhandler S, Himmelstein DU, Lawrence RS.** 1989. Dipstick urinalysis screening of asymptomatic adults for urinary tract disorders II. Bacteriuria. *J Am Med Assoc* **262**:1221–1224.

160. **Loretzon S, Hovelius B, Miorner H, Tendler M, Aberg A.** 1990. The diagnosis of bacteriuria during pregnancy. *Scand J Prim Health Care* **8**:81–83.

161. **LeBlanc AL, McGanity WJ.** 1964. The impact of bacteriuria in pregnancy–a survey of 1300 pregnant patients. *Biol Med (Paris)* **22**:336–347.

162. **Uncu Y, Uncu G, Esmer A, Bilgel N.** 2002. Should asymptomatic bacteriuria be screened in pregnancy? *Clin Exp Obstet Gynecol* **29**:281–285.

163. **Widmer M, Gulmezoglu AM, Mignini L, Roganti A.** 2011. Duration of treatment for asymptomatic bacteriuria during pregnancy. *Cochrane Database Syst Rev* Issue 12, Art No CD000491.

164. **Lumbiganon P, Villar J, Laopaiboon M, Widmer M, Thinkhamrop J, Carroli G, Vy ND, Mignini L, Feston M, Prasertcharoensak W, Limpongsanurak T, Liabsuetrakul T, Sirivatanapa P.** 2009. One day compared with 7-day nitrofurantoin for asymptomatic bacteriuria in pregnancy. *Obstet Gynecol* **113**:339–345.

165. **Whalley PJ, Cunningham FG.** 1977. Short-term versus continuous antimicrobial therapy for asymptomatic bacteriuria in pregnancy. *Obstet Gynecol* **49**:262–265.

166. **ACOG Committee on Obstetric Practice.** 2002. Obstetric management of patients with spinal cord injuries. *Obstet Gynecol* **100**:625–627.

167. **Semetkowska-Jurkiewicz E, Horoszek-Maziarz S, Galinski J, Manitius A, Krupa-Wojciechowska B.** 1995. The clinical course of untreated asymptomatic bacteriuria in diabetic patients–14 year follow-up. *Mater Med Pol* **27**:91–95.

168. **Ouslander JG, Schapira M, Schnelle JF, Uman G, Finegold S, Tuico E, Nigom JG.** 1995. Does eradicating bacteriuria affect the severity of chronic urinary incontinence in nursing home residents? *Ann Intern Med* **122**:749–754.

169. **Boscia JA, Kobasa WD, Knight RA, Abrutyn E, Levison ME, Kaye D.** 1987. Therapy vs no therapy for bacteriuria in elderly, ambulatory, non-hospitalized women. *JAMA* **257**:1067–1071.

170. **Hustinx WNM, Mintjes-de Groot AJ, Verkooyen RP, Verbrugh HA.** 1991. Impact of concurrent antimicrobial therapy on catheter-associated urinary tract infection. *J Hosp Infect* **18**:45–56.

171. **Warren JW, Platt R, Thomas RJ, Rosner B, Kass EH.** 1979. Antibiotic irrigation and catheter-associated urinary tract infections. *N Engl J Med* **299**:570–573.

172. **Leone M, Perrin A-S, Granier I, Visintini P, Blasco V, Antonini F, Albanese J, Martin C.** 2007. A randomized trial of catheter change and short course antibiotics for asymptomatic bacteriuria in catheterized ICU patients. *Intensive Care Med* **33**(4):726–729.

173. **Pfefferkorn U, Lea S, Moldenhauer J, Peterli R, von Flue M, Ackermann C.** 2009. Antibiotic prophylaxis at urinary catheter removal prevents urinary tract infections: a prospective randomized trial. *Ann Surg* **249**:573–575.

174. **van Hees BG, Vijverberg PL, Hoorntie LE, Wiltink EH, Go PM, Tersmette M.** 2011. Single dose oral antibiotic prophylaxis for urinary catheter removal does not reduce the risk of urinary tract infection in surgical patients: a randomized double blind placebo-controlled trial. *Clin Microbiol Infect* **17**:1091–1094.

175. **Wazait HD, Patel HR, van der Meulen JH, Ghei M, Al-Buheissi S, Kelsey M, Miller RA, Emberton M.** 2004. A pilot randomized double-blind placebo-controlled trial on the use of antibiotics on urinary catheter removal to reduce the rate of urinary tract infection: the pitfalls of ciprofloxacin. *Brit J Urol* **94**:1048–1050.

176. **Harding GK, Nicolle LE, Ronald AR, Preiksaitis JK, Forward KR, Cheang M.**

1991. Management of catheter-acquired urinary tract infection in women: Therapy following catheter removal. *Ann Intern Med* **114**: 713–719.

177. **Warren JW, Anthony WC, Hoopes JM, Muncie HL Jr.** 1982. Cephelexin for susceptible bacteriuria in afebrile long-term catheterized patients. *JAMA* **248**:454–458.

178. **Alling B, Brandberg A, Secberg S, Svanborg A.** 1975. Effect of consecutive antibacterial therapy on bacteriuria in hospitalized geriatric patients. *Scand J Infect Dis* **7**:201–209.

179. **Waites KB, Canapp KC, Roper JF, Camp SM, Chen Y.** 2006. Evaluation of 3 methods of bladder irrigation to treat bacteriuria in persons with neurogenic bladder. *J Spinal Cord Med* **29**:217–226.

180. **Gribble MJ, Puterman ML.** 1993. Prophylaxis of urinary tract infection in persons with recent spinal cord injury: A prospective, randomized double-blind, placebo controlled study of trimethoprim/sulfamethoxazole. *Am J Med* **95**:141–152.

181. **Waites KB, Canupp KC, DeVivo MJ.** 1993. Eradication of urinary tract infection following spinal cord injury. *Paraplegia* **31**:654–652.

182. **Mohler JL, Cowen DL, Flanigan RC.** 1987. Suppression and treatment of urinary tract infection in patients with an intermittently catheterized neurogenic bladder. *J Urol* **138**: 336–340.

183. **National Institute on Disability and Rehabilitation Research Consensus Statement.** 1992. The prevention and management of urinary tract infections among people with spinal cord injuries. *J Am Paraplegia Soc* **15**:194–204.

184. **Grabe M, Forsgren A, Bjork T, Hellsten S.** 1987. Controlled trial of a short and a prolonged course with ciprofloxacin in transurethral prostatic surgery. *Eur J Clin Microbiol* **6**:11–17.

185. **Grabe M, Forsgren A, Hellsten S.** 1984. The effect of a short antibiotic course in transurethral prostatic resection. *Scand J Urol Nephrol* **18**:37–42.

186. **Allen WR, Kumar A.** 1985. Prophylactic mezlocillin for transurethral prostatectomy. *Brit J Urol* **57**:46–49.

187. **Olsen JH, Friis-Moller A, Jensen SK, Korner B, Hvidt V.** 1983. Cefotaxime for prevention of infection complications in bacteriuric men undergoing transurethral prostatic resection: a controlled comparison with methenamine. *Scand J Urol Nephrol* **17**:299–301.

188. **Cafferkey MT, Falkiner FR, Gillespie WA, Murphy DM.** 1982. Antibiotics for the prevention of septicemia in urology. *J Antimicrob Chemother* **9**:471–477.

189. **Naber KG, Bergman B, Bishop MC, Bjerklund-Johnsen TE, Botto H, Lobel B, Cruz FJ, Selvaggi FP.** 2001. EAU guidelines for the management of urinary and male genital tract infections. *Eur Urol* **40**:576–588.

190. **El Amari EB, Hadaya K, Buhler L, Berney T, Rohner P, Martin PY, Mentha G, van Delden C.** 2011. Outcome of treated and untreated asymptomatic bacteriuria in renal transplant recipients. *Nephrol Dial Transplant* **26**:4109–4114.

191. **Fakas N, Souli M, Koratzanis G, Karageorgiou C, Giamarellou H, Kanellakopoulou K.** 2010. Effects of antimicrobial prophylaxis on asymptomatic bacteriuria and predictors of failure in patients with multiple sclerosis. *J Chemother* **22**:36–42.

192. **Sobel JD, Kauffman CA, McKinsey D, Zervos M, Vazquez JA, Karchmer AW, Lee J, Thomas C, Panzer H, Dismukes WE.** 2000. Candiduria a randomized double-blind study of treatment with fluconazole and placebo. *Clin Infect Dis* **30**:19–24.

193. **Reid G, Howard J, Gan BS.** 2001. Can bacterial interference prevent infection? *Trends Microbiol* **9**:424–428.

194. **Hull RA, Rudy DC, Donovan WH, Wieser IE, Stewart C, Darouiche RO.** 1999. Virulence properties of *Escherichia coli* 83972, a prototype strain associated with asymptomatic bacteriuria. *Infect Immun* **67**:429–432.

195. **Anderson P, Engberg I, Lidin-Janson G, Lincoln K, Hull R, Hull S, Svanborg C.** 1991. Persistence of *Escherichia coli* bacteriuria is not determined by bacterial adherence. *Infect Immun* **29**15–2921.

196. **Darouiche RO, Green BG, Donovan WH, Chen D, Schwartz M, Merritt J, Mendez M, Hull RA.** 2011. Multicentre randomized controlled trial of bacterial interference for prevention of urinary tract infection in patients with neurogenic bladder. *Urology* **78**:341–347.

197. **Trautner BW, Hull RA, Thornby JI, Darouiche RO.** 2007. Coating urinary catheters with an avirulent strain of *Escherichia coli* as a mean to establish asymptomatic colonization. *Infect Control Hosp Epidemiol* **28**:92–94.

198. **Wullt B, Connell H, Rollano P, Mansson W, Colleen S, Svanborg C.** 1998. Urodynamic factors influence the duration of *Escherichia coli* in deliberately colonized cases. *J Urol* **159**:2057–2062.

199. **Hull R, Rudy D, Donovan W, Svanborg C, Wieser I, Stewart C, Darouiche R.** 2000. Urinary tract infection prophylaxis using *Escherichia coli* 83972 in spinal cord injured patients. *J Urol* **163**:872–877.

200. **Sunden F, Hakansson L, Ljunggren E, Wullt B.** 2006. Bacterial interference–is deliberate colonization with *Escherichia coli* 83972 an alternative treatment for patients with recurrent urinary tract infection? *Int J Antimicrob Agents* **28**(Suppl 1):S26–S29.

201. **Darouiche RO, Thorby JI, Cerra-Stewart C, Donovan WH, Hull RA.** 2005. Bacterial interference for prevention of urinary tract infection: A prospective, randomized, placebo-controlled, double-blind pilot trial. *Clin Infect Dis* **41**:1531–1534.

202. **Sunden F, Hakansson L, Ljunggren E, Wullt B.** 2010. *Escherichia coli* 83972 bacteriuria protects against recurrent lower urinary tract infections in patients with incomplete bladder emptying. *J Urol* **184**:179–185.

203. **Zdziarski J, Svanborg C, Wullt B, Hacker J, Dobrindt U.** 2008. Molecular bases of commensalism in the urinary tract; low virulence or virulence attenuation? *Infect Immun* **76**:695–703.

204. **Lipsky B.** 1989. Urinary tract infections in men: Epidemiology, pathophysiology, diagnosis, and treatment. *Ann Intern Med* **110**:138–150.

205. **Silverberg DS, Jackson FL, Bryan LE.** 1976. Antibody-coated bacteria in the urine of pre-school and school-aged girls with asymptomatic bacteriuria. *Can Med Assoc J* **15**:1091–1093.

206. **Rogers K, Nicolle LE, McIntyre M, Harding GKM, Hoban D, Murray D.** 1991. Pyuria in institutionalized elderly subjects. *Can J Infect Dis* **2**:142–146.

207. **Aguirre-Avlos G, Zavala-Silva ML, Diaz-Nava A, Amaya-Tapia G, Aguilar-Benavides S.** 1999. Asymptomatic bacteriuria and inflammatory response to urinary tract infection in elderly ambulatory women in nursing homes. *Arch Med Res* **30**:29–32.

208. **Juthani-Mehta M, Tinetti M, Perrelli E, Towle V, Quagliarello V.** 2007. Role of dipstick testing in the evaluation of urinary tract infection in nursing home residents. *Infect Control Hosp Epidemiol* **28**:889–891.

209. **Gribble MJ, Puterman ML, McCallum MN.** 1989. Pyuria: Its relationship to bacteriuria in spinal cord injured patients on intermittent catheterization. *Arch Phys Med Rehab* **70**:376–379.

210. **Randolph MF, Greenfield M.** 1964. The incidence of asymptomatic bacteriuria and pyuria in infancy: A study of 400 infants in private practice. *J Pediatr* **65**:57–66.

211. **Jackson SL, Boyko EJ, Scholes D, Abraham L, Gupta K, Fihn SD.** 2004. Predictors of urinary tract infection after menopause: A prospective study. *Am J Med* **117**:903–911.

212. **Lee BB, Haran MJ, Hunt LM, Simpson JM, Marial OK, Rutkowski SB, Middleton JW, Kotsiou G, Tudehope M, Cameron ID.** 2007. Spinal-injured neuropathic bladder antisepsis trial. *Spinal Cord* **45**:542–550.

213. **Sanderson PJ, Weissler S.** 1990. A comparison of the effect of chlorhexidine antisepsis, soap and antibiotics on bacteriuria, perineal colonization and environmental contamination in spinally injured patients. *J Hosp Infect* **15**:235–243.

214. **Boyko EJ, Fihn SD, Scholes D, Abraham L, Monsey B.** 2005. Risk of urinary tract infection and asymptomatic bacteriuria among diabetic and nondiabetic postmenopausal women. *Am J Epidemiol* **161**:557–564.

215. **Saint S, Kaufman SR, Rogers MAM, Baker PD, Ossenkop K, Lipsky BA.** 2006. Condom versus indwelling urinary catheters: A randomized trial. *J Am Ger Soc* **54**:1055–1061.

216. **Sandock DS, Gothe BG, Bodner DR.** 1995. Trimethoprim/sulfamethoxazole prophylaxis against urinary tract infection in the chronic spinal cord injury patient. *Paraplegia* **32**:156–160.

217. **Bennett CJ, Young MN, Darrington H.** 1995. Differences in urinary tract infections in male and female spinal cord injury patients on intermittent catheterization. *Paraplegia* **33**:69–72.

218. **Little PJ.** 1966. The incidence of urinary infection in 5000 pregnant women. *Lancet* **2**:925–928.

219. **Kincaid-Smith P, Bullen M.** 1965. Bacteriuria in pregnancy. *Lancet* **1**:395–399.

220. **Gratacos E, Torres PJ, Vila J, Alonso PL, Cararach V.** 1994. Screening and treatment of asymptomatic bacteriuria in pregnancy prevent pyelonephritis. *J Infect Dis* **169**:1390–1392.

221. **Wren BG.** 1969. Subclinical renal infection and prematurity. *Med J Aust* **2**:596–600.

Bacterial Prostatitis: Bacterial Virulence, Clinical Outcomes, and New Directions

7

JOHN N. KRIEGER[1] and PRAVEEN THUMBIKAT[2]

PROSTATITIS SYNDROMES AND BACTERIAL VIRULENCE

According to the internationally accepted National Institutes of Health (NIH) classification, there are four prostatitis syndromes (1). Category I is acute bacterial prostatitis that presents as an acute urinary tract infection, most commonly caused by *E. coli*. Category II is chronic bacterial prostatitis that presents as recurrent bacterial infections caused by the same bacterial species, most commonly *E. coli*. Category III is chronic prostatitis/chronic pelvic pain syndrome, the most common clinical syndrome, characterized by pelvic pain complaints in the absence of urinary tract infection recognized by standard testing. Category IV is asymptomatic prostatitis in which patients are diagnosed with prostatic inflammation in the absence of symptoms, for example, prostatic inflammation (pathologist's definition of "prostatitis") in an asymptomatic patient undergoing biopsy to evaluate possible prostate cancer.

E. coli in Acute Bacterial Prostatitis

Young men rarely develop urinary tract infections. In male children, urinary infections occur in patients with complicating factors, such as abnormal

[1]Department of Urology, University of Washington School of Medicine, Seattle, WA 98195; [2]Department of Urology, Northwestern University School of Medicine, Chicago, IL 60611.

Urinary Tract Infections: Molecular Pathogenesis and Clinical Management, 2nd Edition
Edited by Matthew A. Mulvey, David J. Klumpp, and Ann E. Stapleton
© 2017 American Society for Microbiology, Washington, DC
doi:10.1128/microbiolspec.UTI-0004-2012

anatomy, voiding disorders, or urinary tract instrumentation. However, clinical series suggest that healthy young men presenting with acute urinary tract infections seldom have anatomical or functional urinary tract abnormalities (2, 3). Many men presenting with acute infections are diagnosed with acute bacterial prostatitis according to the NIH classification, "Patients with acute bacterial prostatitis present with acute symptoms of a urinary tract infection, characteristically including urinary frequency and dysuria. Some patients have symptoms suggestive of systemic infection, such as malaise, fever, and myalgias. Bacteriuria and pyuria are related to infection of the prostate and bladder caused by well-recognized uropathogenic bacteria, especially *E. coli*" (1, 4), the most common cause of urinary tract infection and bacterial prostatitis.

Acute bacterial prostatitis is well recognized clinically, but we have very little understanding of the virulence-associated characteristics of the etiological bacteria. Limited data available reflect cases from referral centers, isolates from laboratory-based surveillance with minimal clinical data, or mixed populations including women and patients with acute, recurrent, and chronic infections (4–8). To better understand the critical issues, we conducted a structured literature review of *E. coli* prostatitis, then used these insights to extend our understanding by evaluating phylogenetic characteristics, virulence-associated traits, and antimicrobial sensitivity in a well-characterized population of previously healthy young men without predisposing urological disorders and no behavioral factors that increase infection risk.

We searched the National Library of Medicine (Medline) database using the search terms: "prostatitis," and "urinary tract infection" in binary combinations with the terms: "male," "bacterial," "treatment," "antibiotic," "antimicrobial," and "*E. coli*." Titles and abstracts were reviewed to identify articles that included evaluation of potential virulence-associated characteristics of *E. coli* in clini-

cally defined populations. All papers identified were English-language, full-text papers. We also searched the reference lists of identified articles for further papers. Relevant articles were then reviewed in detail using a standard data collection sheet. These searches identified a total of 2,926 articles published from January 1, 1963 to February 28, 2010. Review of the titles and abstracts identified 104 articles that merited detailed analysis. Of these 104 articles, only ten (4–6, 8–14) evaluated potential factors associated with *E. coli* virulence in strains from men with prostatitis. Most articles included strains collected more than 15 years ago prior to dissemination of widespread antimicrobial resistance among clinical *E. coli* strains. Table 1 summarizes the clinical characteristics of the patient populations. The relevant articles were published from 1987–2007, including a series of five articles evaluating one strain collection for different virulence-associated factors (4, 6, 8, 11, 13) and another series of three articles (5, 12, 14) evaluating another strain collection. These ten studies evaluated a total of 426 isolates including 333 isolates from patients with acute prostatitis and 86 isolates from patients with chronic prostatitis, plus seven isolates from patients that could not be classified as either acute or chronic. Only one article included information on antimicrobial sensitivity. Most strain collections were from laboratory-based surveillance and included limited clinical information on patient presentation. No study included information on therapy or clinical outcomes.

Critical questions in understanding bacterial virulence

Part of the problem is that most patients with acute prostatitis receive therapy in primary care settings. Patients often respond well to treatment (15) but this has limited etiological studies because bacteria from such acute patients are unlikely to be collected for detailed investigation. Patients in referral practices are often older, refractory to therapy, or have complications such as prostatic abscess

TABLE 1 Studies of *E. coli* virulence factors in bacterial prostatitis

Study, year	City, country	No. strains	Prostatitis category, no. strains	Antimicrobial susceptibility	Clinical outcomes	References
Dowling, 1987	New Orleans and Covington, LA, USA	17	Acute, 12 Chronic, 5	None	None	9
Dalet, 1991	Barcelona, Spain	247	Acute, 173 Chronic, 74	None	None	10
Terai, 1997	Kyoto, Japan	107[a]	Acute, 107	None	None	5
Andreu, 1997	Barcelona, Spain	30[b]	Acute, 7 Chronic, 23	Yes	None	11
Mitsumori, 1999	Kyoto, Japan	107[a]	Acute, 107	None	None	12
Ruiz, 2002	Barcelona, Spain[a]	37[b]	NA[d]	None	None	13
Ishitoya, 2003	Kyoto, Japan	107[a]	Acute, 107	None	None	14
Johnson, 2005[c]	Barcelona, Spain	23[b]	NA	None	None	8
Parham, 2005	Barcelona, Spain	37[b]	NA	None	None	6
Soto, 2007	Barcelona, Spain	32[b]	NA	None	None	4
Current study	Seattle, WA, USA	18	Acute	Yes	Yes	

[a]Data from reference 17. Copyright © Elsevier Ltd. *Urology* **77:**1420–1425, 2011.
[b]The same strain collection was evaluated for different virulence factors in three reports.
[c]The same strain collection was evaluated for different virulence factors in five reports. The paper from Johnson and associates (8) evaluated only one clinical isolate per patient.
[d]NA, Information not available.

or urinary retention (16). Many studies were completed prior to availability of important new insights into multiple mechanisms necessary for virulence of extraintestinal pathogenic *E. coli* (ExPEC) strains. To better understand the microbial virulence factors associated with acute bacterial prostatitis, we completed a comprehensive microbiological and molecular investigation of a unique strain collection isolated from healthy young men with no factors compromising urinary tract anatomy or function.

Bacterial prostatitis in previously healthy young men

We examined previously healthy men, aged 17-to-40 years old with acute bacterial prostatitis, who had positive urine cultures for *E. coli* (17). This age range was selected to exclude children who have a high prevalence of congenital and functional abnormalities and also to exclude older men who are likely to have lower urinary tract symptoms related to prostatic hypertrophy and other urological conditions. Men with congenital genitourinary disorders, spinal cord injuries, or history of genitourinary tract manipulation were excluded. We also excluded patients with findings suggesting anatomical or functional urinary tract disorders, previous genitourinary trauma, previous urinary tract infections, neurological diseases, or other conditions that might affect urinary tract function.

Following a standardized history, including a detailed sexual history that assessed behavioral risk factors for infection (such as a history of insertive anal intercourse (18)) and physical examination, midstream urine was obtained for urinalysis and culture using methods to isolate pathogens present at ≥50 colony-forming units (cfu)/ml. Isolation of a pathogen ≥50 cfu/ml was considered significant. Isolates were obtained from the clinical laboratory and maintained in the research laboratory prior to batch evaluation for bacterial characteristics and potential virulence factors. Empirical therapy was prescribed and modified based on antimicrobial sensitivity and clinical response. Recommended follow-up studies included urine culture after treatment, lower urinary tract localization cultures, uroflow study, and post-void residual by ultrasound. This study was approved by the University of Washington Human Subjects Committee.

Characterization of the *E. coli* isolates

Isolates were typed based on the API20E Enteric Identification System (bioMerieux, Inc.). When necessary, a 16S rRNA gene fragment was sequenced using primers 27f (5′-GAGTTTGATCCTGGCTCA-3′), 1492r (5′-TACGGYTACCTTGTTACGACTT-3′) and 341f (5′-CCTACGGGAGGCAGCAG-3′) to confirm API20E results. Genomic characterization included the detection of fitness- and virulence-associated genes of extraintestinal pathogenic *E. coli* (*afa/draBC, bmaE, cdtB, cnfl, clbA-Q, cvaC, fimH, fyuA, hlyA, ibeA, iroN-B, iutA, kpsMT*(I), *kpsMT*(II), *kpsMT* (II) K1, *kpsMT*(II) K5, *malX, papAH, papG, sfa/focDE, sfaS, focG, rfc, traT*) or, when necessary, of enteroaggregative *E. coli* (EAEC) (*pAA, aggR, astA, pic*) by polymerase chain reaction (PCR) as recently described (19–21). Strains were also tested phenotypically for expression of α-hemolysin, microcins, colicins, and the siderophore aerobactin (22). Expression of the polyketide Colibactin, CDT, and CNF was analyzed using HeLa cells as described (23, 24). Allocation of isolates to the major *E. coli* phylogenetic groups was based on results of a triplex PCR (25). Multilocus sequence typing (MLST) was performed as described (http://mlst.ucc.ie). Sequence types (STs) were assigned using the *E. coli* MLST database hosted at the University College Cork, Ireland (http://mlst.ucc.ie). Information on all *E. coli* isolates was deposited at the MLST database.

Clinical presentation

Of 19 patients who had their initial isolates available, 18 (94.7%) had *E. coli* confirmed by API analysis or 16S rRNA gene sequencing. The remaining patient had an *Enterobacter aerogenes* strain that was excluded from this analysis. During this period, the University of Washington Prostatitis Clinic evaluated more than 700 patients with chronic prostatitis. Patients averaged 26.9 ± 5.9 years old (1 standard deviation, median 26 years old) and had an average symptom duration of 6 ± 7 days (median 3 days, range: 2 to 21 days) at

presentation. All presented with urinary symptoms including: dysuria in 12 men (67%), urinary frequency in 6 men (33%), cloudy or malodorous urine in 3 men (17%), urinary urgency in 3 men (17%), suprapubic pain in 3 men (17%), urethral discharge in 2 men (11%), flank pain in 2 men (11%), and scrotal pain in 1 man (6%). In addition, 4 men (22%) had systemic symptoms including: fever and chills in all 4 (22%), night sweats in 2 (11%), and malaise in 2 (11%). One man had a history of depression, but none had a history of other significant medical problems. Of the 18 men, 12 (67%) were circumcised. One man had scrotal tenderness but no other abnormalities were noted on physical examination. No patient had an anatomical or functional urological abnormality.

Urinalysis, colony count, and antimicrobial sensitivity

Pyuria was characteristic with a median of 50 leukocytes/400x microscopic field of centrifuged urinary sediment (range 10 to >100 per 400x field). Hematuria was noted frequently, with a median of 15 red blood cells per 400x microscopic field (range 0 to >100 red blood cells per 400x field). For the initial bacteriuria episodes, the median was 10^5 cfu/ml (range 2 x 10^2 to >10^5 cfu/ml). Of the 18 initial *E. coli* isolates, 13 (72%) were sensitive to all antimicrobials tested. Resistance exhibited by 5 isolates included: ampicillin (4 isolates), sulfonamides (4), trimethoprim-sulfamethoxazole (4), tetracyclines (3), and cephalothin (3). The two men with symptoms of urethral discharge both had negative tests for *Neisseria gonorrhoeae* and *Chlamydia trachomatis*. Each isolate was sensitive *in vitro* to the initial antimicrobial therapy.

Clinical outcomes

Initial treatment included: levofloxacin (11 patients), trimethoprim-sulfamethoxazole (5 patients), doxycycline (1 patient), and the combination of doxycycline plus cefixime (1 patient). The patients treated with doxycycline had presumed urethritis initially, but were changed to levofloxacin or trimethoprim-sulfamethoxazole (one patient each) after laboratory results were available. The mean duration of treatment was 18 ± 9 days (median 14, range: 7 to 30 days). Follow-up averaged 19 + 8 months (median 18 months, range 6 to 30 months). Each patient experienced a prompt initial response, including long-term resolution in 16 patients who each had at least one negative culture or urinalysis following therapy.

Two patients developed chronic bacterial prostatitis. These included a 21-year old man who presented with urethral discharge, urinary frequency, and orchalgia. Initially he was treated with cefixime plus doxycycline, but was changed to trimethoprim-sulfamethoxazole for a 10-day course after a pan-sensitive *E. coli* was cultured. He did well for 11 months, then developed recurrent urethral discharge and dysuria with culture of 5×10^4 cfu/ml of pan-sensitive *E. coli*. He received 21 days of trimethoprim-sulfamethoxazole with slow symptomatic resolution. Computerized tomography with contrast, post-void residual urine, and uroflow studies were normal. After completing therapy, he had localization cultures compatible with chronic bacterial prostatitis: 90 cfu/ml of Gram-positive flora in first-void urine, *E. coli* 30 cfu/ml in mid-stream urine, *E. coli* $>10^5$ cfu/ml in expressed prostatic secretions, and *E. coli* 60 cfu/ml in post prostate-massage urine. The *E. coli* was sensitive to all tested antimicrobials. He was then treated with levofloxacin, 500 mg twice daily, for 60 days. He did well for 6 months, then presented with recurrence of low-grade symptoms. Cultures again showed chronic bacterial prostatitis with a pan-sensitive *E. coli* in all specimens: 1,700 cfu/ml in first-void urine, 30 cfu/ml in mid-stream urine, $>10^5$ cfu/ml in expressed prostatic secretions, and 60 cfu/ml in post prostate-massage urine. He was treated with levofloxacin for 3 weeks, then changed to trimethoprim-sulfamethoxazole (160/800 mg) twice daily for 3 months because of quinolone side effects. His symptoms resolved, with no recurrence over the next 14 months.

The remaining patient was a 25-year-old man who presented with dysuria and gross hematuria. Urine culture grew $>10^5$ cfu/ml of sensitive *E. coli*. He received levofloxacin, 500 mg twice daily, for 7 days. After 6 days, his symptoms had resolved and a urine culture had no growth. Symptoms recurred 2 weeks later and *E. coli* (1,800 cfu/ml) was cultured from his mid-stream urine. He received two additional months of levofloxacin, 500 mg twice daily, with complete resolution and has had no further problems.

Characteristics of the acute prostatitis *E. coli* isolates

The relatedness of acute prostatitis with the initial *E. coli* isolates and uropathogenic *E. coli* (UPEC) strains was examined by multilocus-sequence typing (MLST) (Table 2). Thirteen strains belonged to phylogenetic group B2 and five to phylogenetic group D. These lineages typically included UPEC, other ExPEC, or commensal variants.

Of the 18 strains, five (28%) belonged to sequence type (ST) 127; and two each were allocated to STs 80, 95, and 394, respectively. The seven remaining strains belonged to different STs, including two previously unknown types (ST992 and ST998, Table 2). Generally, the predominant *E. coli* pathotypes represented by the acute prostatitis STs can be described as ExPEC and mainly UPEC. ST394 represents an exception that includes enteroaggregative *E. coli* (EAEC) strains from patients with diarrhea, with few exceptions in the *E. coli* MLST database (as of March 31, 2009). Interestingly, characteristic DNA regions of EAEC, i.e., the pAA virulence plasmid and the regulator gene *aggR*, were detected in one of the two ST394 prostatitis isolates.

The prevalence of ExPEC virulence-related genes was studied by PCR including traits related to adherence, toxins, siderophores, bacterial capsules, and miscellaneous traits (Table 3). Adhesins proved very common. The *fimH* gene coding for the type 1 fimbrial adhesin was present in all strains tested. Genes of the P- and S-/F1C fimbriae-

TABLE 2 Sequence types (ST), clonal complexes, phylogenetic groupings, and predominant pathotypes of 18 *E. coli* isolates from healthy young men with acute bacterial prostatitis[a]

ST	Clonal complex	Phylogenetic (ECOR) group	Predominant pathotype within the ST, general characteristics[b]	Number of isolates
127	–	B2	ExPEC, mainly UPEC	5
80	568	B2	ExPEC, mainly UPEC, often K1 capsule	2
95	95	B2	ExPEC, often K1 capsule	2
12	12	B2	ExPEC, mainly UPEC	1
73	73	B2	ExPEC, mainly UPEC	1
992	–	B2	ExPEC	1
998	–	B2	ExPEC	1
394	394	D	EAEC	2
62	–	D	ExPEC, often K1 encapsulated UPEC	1
69	69	D	ExPEC, mainly UPEC	1
405	405	D	ExPEC, mainly UPEC	1

[a]Data from reference 17. Copyright © Elsevier Ltd. *Urology* **77:**1420–1425, 2011.
[b]Based on comparison with the MLST database as materials and methods, characteristics of the acute prostatitis *E. coli* isolates section.

encoding gene clusters were present in 14 (78%) and 11 (61%) isolates, respectively. Most isolates carried gene clusters encoding toxins: α-hemolysin (72%), cytotoxic necrotizing factor-1 (61%), and the polyketide Colibactin (72%); whereas only one strain (6%) carried the gene cluster coding for the cytolethal-distending toxin 1. The isolates differed in the presence of determinants coding for characteristic ExPEC siderophores, e.g., yersiniabactin (94%), salmochelin (61%), and aerobactin (33%). Certain capsule-associated genes proved common, with 83% and 72% of isolates carrying group II capsule-encoding genes and the *traT* gene, respectively, both of which are involved in survival of host defense mechanisms. Only phylogenetic group B2 isolates carried determinants coding for the polyketide Colibactin, the cytotoxic necrotizing factor 1, S-/F1C fimbriae, and salmochelin. Phenotypic analysis of α-hemolysin, aerobactin, CDT, CNF, Colibactin, and colicin expression confirmed that most of these virulence- or fitness-associated genes functioned and were efficiently expressed, at least under laboratory conditions (Fig. 1). Detection of Colibactin was masked by the hemolytic activity in most of the strains encoding the α-hemolysin. These strains had a median of 12.5 virulence-associated traits (range 2 to 16, mean 12.4 ± 3.5).

New findings on virulence of acute prostatitis *E. coli* isolates

Studying bacterial isolates from healthy young men with acute prostatitis without compromising urinary tract abnormalities provided an unusual opportunity to define the importance of bacterial characteristics in the pathogenesis of acute urinary tract infection. We found that these uropathogens had an impressive repertoire of virulence traits. Our findings amplify and extend previous studies to suggest that that these virulence factors contribute to development of acute prostatitis in healthy young men.

This study adds substantially to our understanding of acute prostatitis and urinary tract infections. In contrast to earlier series (Table 1), clinical information was available to describe this unusual cohort. The 18 men (average age 27 years old) represented 2.5% of the patients who completed evaluation in our prostatitis clinic. Patients presented with urinary tract symptoms, typically dysuria and increased frequency. Four men (22%) also had systemic symptoms such as fever, chills, and night sweats. Pyuria was characteristic and cultures had a median of 10^5 cfu/ml *E. coli*. Clinical evaluation assured that no patient had urological risk factors, such as abnormal anatomy or function, or history of urinary tract instrumentation, that increase urinary tract infection risk.

TABLE 3 Virulence-associated genes in 18 *E. coli* isolates from healthy young men with acute prostatitis[a]

Virulence factor category	Virulence gene/trait	Prevalence of gene/trait [%]
Adhesins	*fimH*	100
	papA/C/E/F/H	77.8
	papG allele II	38.9
	papG allele III	38.9
	papG allele I	0
	sfa/focDE	61.1
	focG	22.2
	sfaS	33.3
	afa/draBC	0
	gafD	0
	bmaE	0
Toxins	*clbA-N*	72.2
	hlyA	72.2
	*cnf*1	61.1
	cdtB	5.6
Siderophores	*fyuA*	94.4
	iroN	61.1
	iutA	33.3
Capsule	*kpsMT* (II)	83.3
	kpsMT K5	50
	kpsMT K1	33.3
	kpsMT (III)	5.6
Miscellaneous	*malX*	77.8
	traT	72.2
	ibeA	22.2
	cvaC	0
Expression of	α-hemolysin	66.7
	Aerobactin	33.3
	CNF 1	61.1
	CDT-1	5.6
	Colicins	5.6

[a]Data from reference 17. Copyright © Elsevier Ltd. *Urology* **77**:1420–1425, 2011.

Follow-up averaging 19 months assured that 16 (89%) of 18 patients resolved their infections following initial treatment. The two remaining patients required prolonged therapy to resolve chronic bacterial prostatitis. Although it is likely that the pan-sensitive *E. coli* recovered multiple times from the two patients with chronic prostatitis were the same strain (within a patient), patients presented at different times and were not always cultured at our clinical laboratory. We did not have sequential isolates banked for DNA fingerprinting or other approaches to prove that these patients experienced recurrence rather than reinfection. This clinical experience supports previous data that urinary tract infections are uncommon in healthy young men and that patients with acute prostatitis who have no urological risk factors are usually managed successfully (2, 3).

The phylogenetic background, virulence- and fitness-associated gene content of the acute prostatitis isolates matches that of UPEC and other ExPEC. Even compared to ExPEC strains from patients with other syndromes described in the literature (8, 10, 14, 19, 26–30), the acute prostatitis isolates were highly virulent. These isolates had accumulated a median of 12.5 virulence associated genetic traits, characteristic of ExPEC gene clusters. Given the variable proportion of virulence genes detected (other than the fim gene present in all strains and fyuA in 94%), we could not determine the precise contributions of individual virulence factors in acute bacterial prostatitis. Overall, the findings are consistent with the literature and extend the range of virulence factors detected by its comprehensive approach. Our data support the recent finding that certain EAEC share similarity with UPEC (31) and demonstrate that such isolates can cause urogenital infections.

Synthesis with the ExPEC literature

Our findings are consistent with the emerging literature on *E. coli* infections. This species represents an important extraintestinal pathogen in humans, mammals, and birds; commonly causing urinary tract infections, sepsis, neonatal meningitis, and other infections (32). *E. coli* strains causing most extraintestinal infections, so-called ExPEC, tend to exhibit multiple accessory traits, including adhesins, toxins, extracellular polysaccharides, invasins, and siderophores that are absent or uncommon among non-pathogenic and commensal strains (28, 29, 32). Besides well-known virulence or antagonistic factors (33, 34), specialized metabolic capabilities may also contribute to urovirulence (30, 35, 36). Such traits contribute to ExPEC fitness

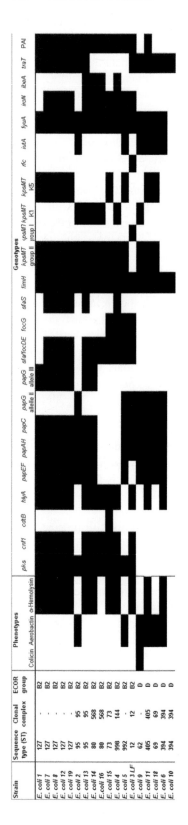

FIGURE 1 Geno- and phenotypic diversity of acute prostatitis *E. coli* strains. White and black denote the presence and absence, respectively, of virulence-associated genes or traits, as detected by PCR or phenotypic tests. CPE refers to cytopathic effect.

by enhancing their ability to resist host defenses to successfully colonize and cause injury.

E. coli can be divided into at least four major phylogenetic groups (A, B1, B2, and D) (26, 37). Our isolates belonged to phylogenetic groups B2 and, to a lesser extent, group D as classically reported for human ExPEC (19, 38). MLST recently indicated a strong association between clonal complexes and virulence-factor content. Many ExPEC virulence genes, such as *pap*, *sfa/foc*, and *hly*, were clustered in clonal complexes with high virulence potential, particularly those of phylogenetic group B2 (27).

The polyketide Colibactin was initially described as a genotoxin restricted to ExPEC and commensal *E. coli* strains of phylogenetic group B2 (23). The biological role of this polyketide for virulence or commensalism is not yet clear. Johnson and coworkers reported recently that the Colibactin determinant was associated with bacteremia isolates and with multiple other virulence genes (39). The Colibactin biosynthesis genes were only found in a highly virulent subset of phylogenetic group B2 isolates. Presence of the Colibactin island was also associated with gene clusters encoding adhesins, siderophore systems, and lipopolysaccharide biosynthesis among phylogenetic group B2 ExPEC isolates from diverse clinical sources and with high lethality in an experimental mouse model (40).

Our observations extend previous studies of *E. coli* isolates from patients with acute prostatitis by substantially increasing the range of virulence-associated traits evaluated. The acute prostatitis *E. coli* strains in this study exhibited a marked clustering of virulence determinants: the genes coding for α-hemolysin (t*hlyA*), Colibactin (*clb*), cytotoxic necrotizing factor 1 (*cnf1*), P-fimbriae (*pap*), S-/F1C-fimbriae (*sfa/foc*), yersiniabactin (*fyuA*), salmochelin (*iroN*); group II capsules as well as a factor involved in serum resistance (*traT*) were frequently present. These observations include a more comprehensive evaluation of virulence-associated traits than previous reports, and are consistent with earlier results that *E. coli* isolates from prostatitis cases in elderly patients are likely to express α-hemolysin P-fimbriae, yersiniabactin (*fyuA*) and aerobactin (*iutA*) compared to fecal isolates (4). In addition, our findings are consistent with reports that prostatitis isolates often have virulence-associated traits including: adhesins such as P-fimbriae and afimbrial adhesins, (8, 9, 11–14), toxins such as α-hemolysin and cytotoxic necrotizing factor-1 (4, 8, 11–13), a limited range of phylogenetic groups, and presence of pathogenicity islands (6).

Most *E. coli* causing acute bacterial prostatitis produced toxins with carcinogenic potential. These observations may help explain epidemiological data suggesting that a history of prostatitis is associated with significantly increased risk for prostate cancer (41, 42). Exposure to potential genotoxins from these highly virulent *E. coli*, such as CDT, Colibactin, or as yet unidentified toxins, could enhance subsequent carcinogenesis. Similarly, CNF1, through constitutive activation of the Rho family of small GTPases, could result in aberrant cell signaling, in turn enhancing risk of host-cell transformation.

In contrast to the marked accumulation of virulence-associated traits, antibiotic resistance was less important for the *E. coli* isolates causing acute bacterial prostatitis in these young men. This finding is consistent with the low rate of antimicrobial usage in our healthy outpatient population and with the observation that these patients had not had previous urinary tract procedures, urological risk factors, or extensive contact with the healthcare system.

Conclusions on *E. coli* virulence in acute bacterial prostatitis

Acute bacterial prostatitis is an important clinical syndrome. However, a structured literature review identified only ten articles that evaluated *E. coli* virulence factors in strains from prostatitis patients. Most articles

evaluated strains collected more than 15 years ago, prior to wide dissemination of antimicrobial resistance. Only one article included information on antimicrobial sensitivity and no study included information on therapy or clinical outcomes. Many studies were completed prior to availability of important new insights into multiple mechanisms necessary for virulence of extraintestinal pathogenic *E. coli* strains. To better understand the microbial-virulence factors associated with acute bacterial prostatitis, we completed a comprehensive microbiological and molecular investigation of a unique strain collection isolated from healthy young men with no factors compromising urinary tract anatomy or function. The phylogenetic background and accumulation of multiple extraintestinal pathogenic *E. coli* virulence-associated genes in the acute prostatitis isolates indicate that these strains belong to a highly virulent subset of uropathogenic variants able to circumvent and withstand the normal host response in young healthy individuals with no urological risk factors for urinary tract infection. An exceptional repertoire of virulence is needed to overcome the formidable host defenses in this low-risk population. In contrast, antimicrobial resistance traits appear to confer little added advantage for the *E. coli* causing acute prostatitis in these patients.

Uropathogenic *E. coli* in chronic pelvic pain syndrome

The third prostatitis disease category, CP/CPPS, accounts for approximately 90% of all chronic prostatitis and is considered to be predominantly non-bacterial. Its defining clinical manifestations include chronic pain in the perineum, rectum, prostate, penis, testicles, and abdomen (43). Despite the stated non-bacterial nature of CP/CPPS, up to 8% of patients with CP/CPPS harbor uropathogens that have traditionally been deemed to be of no significance (44). Numerous studies have also identified bacterial DNA in prostate samples from CP/CPPS patients (44–48).

CP/CPPS accompanied by uropathogens is differentiated from chronic bacterial prostatitis by the requirement for clinical symptoms of pelvic pain and the lack of recurrent urinary tract infections. While a number of bacterial species have been described as being associated with CP/CPPS (47, 49–52) conclusive evidence of a role for these bacteria has remained elusive. Recent studies have suggested that uropathogenic *E. coli*, the major uropathogen involved in acute and chronic bacterial prostatitis may also play a role in CP/CPPS (53).

Influence of virulence factors

It has been suggested that the virulence of major uropathogens such as UPEC is dependent on the expression of multiple virulence factors (54, 55). Phylogenetic analysis suggests that prostatitis-causing UPEC strains largely belong to the B2 phylogenetic group and exhibit a wide variety of virulence traits, including nonhemagglutinin adhesin-siderophore receptor (*ihA*), type-1 fimbriae (*fimH*), the salmochelin-siderophore receptor (*iroN*) and outer-membrane protease T (*ompT*) (8, 11, 56, 57). Although CP/CPPS is associated with bacteria only in about 8% of cases (44), infection has often been postulated as an initiating factor (58). CP/CPPS-associated *E. coli* strains have not been comprehensively catalogued or analyzed with regard to their virulence characteristics. In a recent study from our laboratory (53), we demonstrated that a bacterial isolate from a patient with CP/CPPS can initiate and sustain the development of chronic pelvic pain, a distinguishing characteristic of CP/CPPS. The *E. coli* strain (strain CP1) was isolated at Northwestern University urology clinics using the Meares–Stamey 4-glass collection technique (59) from the expressed prostatic secretion (EPS) and post massage-voided urine (VB3 fraction) of a patient with CP/CPPS. The patient presented with chronic pelvic pain with no concurrent UTI. CP1 was one of six separate *E. coli* isolates collected over a two-year time span that exhibited

an identical biotype and antimicrobial-susceptibility pattern. Molecular and phenotypic characterization of CP1 demonstrated that it lacked a number of the virulence-associated traits that typify acute prostatitis isolates, and UPEC strains generally. Major *E. coli* phylogenetic group was determined by a 3-locus PCR-based method (25) and multiplex PCR-based assays were used to ascertain *E. coli*-associated virulence-factor (VF) genes and variants (19). The *E. coli* species is divided into four main phylogenetic groups, commonly designated A, B1, B2, and D (60). Most *E. coli* strains responsible for UTIs or other extraintestinal infections belong to group B2 or, less frequently, group D (38). Our studies place CP1 within group B1, whose members normally lack extra-intestinal virulence factors (61), but when selected for in clinical disease states can exhibit high levels of virulence (8). In addition to its phylogenetic background, CP1's virulence-gene repertoire also is atypical in comparison to acute prostatitis isolates, which characteristically are enriched for *hly, cnf1, sfa,* and *iroN* (56), all of which are absent in CP1 (Table 4). The atypical nature of CP1 may reflect specific adaptations for chronic prostate colonization that differentiate it from the strains obtained from acute prostatitis. Characteristics that still place CP1 in the larger UPEC family include its possession of virulence genes such as *fimH, usp, iut,* and *iha*, which are common among UPEC strains.

E. coli pathogenesis

It has been hypothesized that chronic bacterial prostatitis is characterized by the presence of biofilms (62). *In vitro* studies on prostatitis bacteria have demonstrated a greater tendency for the development of biofilm-like structures that are assumed to adhere to the epithelium of the ductal system (4). In addition, epithelial invasion and proliferation have been described as virulence strategies utilized by UPEC and are well characterized in the bladder *in vitro* and *in vivo* (63, 64). The possession of a similar

TABLE 4 Characteristics of *E. coli* study strains[a]

Characteristic	Strain NU14	Strain CP1
Clinical source	cystitis	prostatitis
Phylogenetic group	B2	B1
Virulence genes[b]		
papACEFG	+	−
sfa/focDC	+	−
iha	−	+
fimH	+	+
hlyD	+	−
cnf1	+	−
fyuA	+	+
IroN	+	−
iutA	−	+
IIp	+	−
IIb	+	−
ibeA	+	−
traT	+	−
ompT	+	−
usp	+	+
malX (PAI)	+	+

[a]Copyright © American Society for Microbiology. *Infect Immun* **79:**628–635, 2011.
[b]Virulence genes pap, P fimbriae structural and adhesive subunits; sfa/foc, S and F1C fimbriae; iha, adhesin-siderophore; fimH, type 1 fimbriae; hly, hemolysin; cnf, cytotoxic necrotizing factor; fyu, yersiniabactin receptor; iroN, siderophore receptor; iut, aerobactin; IIp and IIb, kpsM group 2 capsule; ibe, invasion of brain endothelium; traT, serum resistance associated; ompT, outer membrane protease T; malx, pathogenicity island marker; usp, uropathogenic specific protein. Both strains were negative for focG, F1C fimbriae; afa/dra, Dr adhesions; cdt, cytolethal distending toxin; K1, kpsM group 2 capsule; III, kpsM group 3 capsule; rfc, O4 lipopolysaccharide and cva, colicin V.

strategy, in addition to classical mechanisms of biofilm formation, would equip prostatitis strains with the ability to persist within the prostate and establish a unique ecological niche. Our laboratory has recently demonstrated that strain CP1 appears to attach to epithelial cells utilizing adhesins other than type-1 pili or P and S fimbriae and was fully capable of persistence within the cytoplasm of prostate epithelial cells. Interestingly, the strain, while highly invasive and demonstrating persistence *in vitro*, was not unique in these abilities and performed as well as a prototypical cystitis UPEC strain in our *in vitro* studies. *In vivo* studies in mice identified the ability of the CP1 strain to persist in the prostate for almost a month, in contrast to the cystitis UPEC strain that was

eliminated in two weeks. The tropism to the prostate and persistence of the CPPS-derived UPEC strain is suggestive of as yet undefined pathoadaptive mechanisms that contribute to virulence. The limitation of these studies is that they are presently restricted to a single CPPS UPEC isolate and need to be demonstrated in multiple isolates before intracellular persistence as a survival strategy for UPEC becomes a consideration for clinical therapy in prostatitis. However, these studies suggest that even small numbers of bacteria in the prostate can mediate pathology, given the right virulence signature and pathogenic strategy.

Pathogenesis of chronic pelvic pain

Bacterial infections have been hypothesized to have a role in the pathogenesis of CPPS but there have been limited studies showing a cause-effect relationship that supports the hypothesis. We recently demonstrated that a UPEC strain isolated from a CPPS patient was capable of eliciting pelvic pain behavior in a host-specific manner (53). The UPEC isolate induced sustained chronic pain in a strain of mice (NOD) genetically prone to develop different organ-specific autoimmune diseases (65), but was incapable of inducing pain behavior in wild-type mice (C57BL/6J), despite equivalent levels of bacterial colonization and histological inflammation. Interestingly, the differential pain responses are not simply a function of the host but also a function of the bacterial strain. That is, infection of the NOD male by a cystitis UPEC strain is insufficient to activate mechanisms leading to chronic pain. These results are intriguing and suggest that the pathogenesis of pelvic pain in CPPS is complex and contrary to a focus primarily on the pathogen, the answer perhaps lies in identifying scenarios where aberrant host-pathogen interactions mediate the pathogenesis of chronic pelvic pain. Of note in these studies is the observation that despite the propensity of the UPEC strain to persist, it is eventually cleared from the prostate, leaving behind a host that still exhibits persistent and chronic pain behavior. Thus, the inability to demonstrate a strong cause-effect relationship between bacterial species and CPPS may be less due to a lack of such a relationship and more due to elimination of the offending pathogen from the prostate at the time of clinical sampling. The role of uropathogenic *E. coli* in chronic pelvic pain is complex and is not nearly as well understood as in acute prostatitis or urinary tract infections. Preliminary evidence seems to point towards a critical role for host-pathogen interactions in mediating the process of chronic pelvic pain. These results implicate characteristics of both the host and the pathogen as key features in determining the development of pain symptoms in CP/CPPS.

ACKNOWLEDGMENTS

Conflicts of interest: We declare no conflicts.

CITATION

Krieger JN, Thumbikat P. 2016. Bacterial prostatitis: bacterial virulence, clinical outcomes, and new directions. Microbiol Spectrum 4(1):UTI-0004-2012.

REFERENCES

1. Krieger JN, Nyberg L Jr, Nickel JC. 1999. NIH consensus definition and classification of prostatitis. *JAMA* **282:**236–237.

2. Krieger JN, Ross SO, Simonsen JM. 1993. Urinary tract infections in healthy university men. *J Urol* **149:**1046–1048.

3. Abarbanel J, Engelstein D, Lask D, Livne PM. 2003. Urinary tract infection in men younger than 45 years of age: is there a need for urologic investigation? *Urology* **62:**27–29.

4. Soto SM, Smithson A, Martinez JA, Horcajada JP, Mensa J, Vila J. 2007. Biofilm formation in uropathogenic *Escherichia coli* strains: relationship with prostatitis, urovirulence factors and antimicrobial resistance. *J Urol* **177:**365–368.

5. Terai A, Yamamoto S, Mitsumori K, Okada Y, Kurazono H, Takeda Y, Yoshida O. 1997. *Escherichia coli* virulence factors and serotypes in acute bacterial prostatitis. *Int J Urol* **4:**289–294.

6. Parham NJ, Pollard SJ, Chaudhuri RR, Beatson SA, Desvaux M, Russell MA, Ruiz J, Fivian A, Vila J, Henderson IR. 2005. Prevalence of pathogenicity island IICFT073 genes among extraintestinal clinical isolates of *Escherichia coli*. *J Clin Microbiol* **43:**2425–2434.

7. Johnson JR, Scheutz F, Ulleryd P, Kuskowski MA, O'Bryan TT, Sandberg T. 2005. Host-pathogen relationships among *Escherichia coli* isolates recovered from men with febrile urinary tract infection. *Clin Infect Dis* **40:**813–822.

8. Johnson JR, Kuskowski MA, Gajewski A, Soto S, Horcajada JP, Jimenez de Anta MT, Vila J. 2005. Extended virulence genotypes and phylogenetic background of *Escherichia coli* isolates from patients with cystitis, pyelonephritis, or prostatitis. *J Infect Dis* **191:**46–50.

9. Dowling KJ, Roberts JA, Kaack MB. 1987. P-fimbriated *Escherichia coli* urinary tract infection: A clinical correlation. *South Med J* **80:**1533–1536.

10. Dalet F, Segovia T, Del Río G. 1991. Frequency and distribution of uropathogenic *Escherichia coli* adhesins: A clinical correlation over 2,000 cases. *Eur Urol* **19:**295–303.

11. Andreu A, Stapleton AE, Fennell C, Lockman HA, Xercavins M, Fernandez F, Stamm WE. 1997. Urovirulence determinants in *Escherichia coli* strains causing prostatitis. *J Infect Dis* **176:**464–469.

12. Mitsumori K, Terai A, Yamamoto S, Ishitoya S, Yoshida O. 1999. Virulence characteristics of *Escherichia coli* in acute bacterial prostatitis. *J Infect Dis* **180:**1378–1381.

13. Ruiz J, Simon K, Horcajada JP, Velasco M, Barranco M, Roig G, Moreno-Martínez A, Martínez JA, Jiménez de Anta T, Mensa J, Vila J. 2002. Differences in virulence factors among clinical isolates of *Escherichia coli* causing cystitis and pyelonephritis in women and prostatitis in men. *J Clin Microbiol* **40:**4445–4449.

14. Ishitoya S, Yamamoto S, Kanamaru S, Kurazono H, Habuchi T, Ogawa O, Terai A. 2003. Distribution of afaE adhesins in *Escherichia coli* isolated from Japanese patients with urinary tract infection. *J Urol* **169:**1758–1761.

15. Turner JA, Ciol MA, Von Korff M, Berger R. 2004. Prognosis of patients with new prostatitis/pelvic pain syndrome episodes. *J Urol* **172:**538–541.

16. Kravchick S, Cytron S, Agulansky L, Ben-Dor D. 2004. Acute prostatitis in middle-aged men: A prospective study. *BJU Int* **93:**93–96.

17. **Krieger JN, Dobrindt U, Riley DE, Oswald E.** 2011. Acute *Escherichia coli* prostatitis in previously health young men: bacterial virulence factors, antimicrobial resistance, and clinical outcomes. *Urology* **77:**1420–1425.

18. **Wong ES, Stamm WE.** 1983. Sexual acquisition of urinary tract infection in a man. *JAMA* **250:**3087–3088.

19. **Johnson JR, Stell AL.** 2000. Extended virulence genotypes of *Escherichia coli* strains from patients with urosepsis in relation to phylogeny and host compromise. *J Infect Dis* **181:**261–272.

20. **Schmidt H, Knop C, Franke S, Aleksic S, Heesemann J, Karch H.** 1995. Development of PCR for screening of enteroaggregative *Escherichia coli*. *J Clin Microbiol* **33:**701–705.

21. **Müller D, Greune L, Heusipp G, Karch H, Fruth A, Tschäpe H, Schmidt MA.** 2007. Identification of unconventional intestinal pathogenic *Escherichia coli* isolates expressing intermediate virulence factor profiles by using a novel single-step multiplex PCR. *Appl Environ Microbiol* **73:**3380–3390.

22. **Zdziarski J, Svanborg C, Wullt B, Hacker J, Dobrindt U.** 2008. Molecular basis of commensalism in the urinary tract: low virulence or virulence attenuation? *Infect Immun* **76:** 695–703.

23. **Nougayrède JP, Homburg S, Taieb F, Boury M, Brzuszkiewicz E, Gottschalk G, Buchrieser C, Hacker J, Dobrindt U, Oswald E.** 2006. *Escherichia coli* induces DNA double-strand breaks in eukaryotic cells. *Science* **313:**848–851.

24. **Oswald E, Sugai M, Labigne A, Wu HC, Fiorentini C, Boquet P, O'Brien AD.** 1994. Cytotoxic necrotizing factor type 2 produced by virulent *Escherichia coli* modifies the small GTP-binding proteins Rho involved in assembly of actin stress fibers. *Proc Natl Acad Sci U S A* **91:**3814–3818.

25. **Clermont O, Bonacorsi S, Bingen E.** 2000. Rapid and simple determination of the *Escherichia coli* phylogenetic group. *Appl Environ Microbiol* **66:**4555–4558.

26. **Boyd EF, Hartl DL.** 1998. Chromosomal regions specific to pathogenic isolates of *Escherichia coli* have a phylogenetically clustered distribution. *J Bacteriol* **180:**1159–1165.

27. **Jaureguy F, Landreau L, Passet V, Diancourt L, Frapy E, Guigon G, Carbonnelle E, Lortholary O, Clermont O, Denamur E, Picard B, Nassif X, Brisse S.** 2008. Phylogenetic and genomic diversity of human bacteremic *Escherichia coli* strains. *BMC Genomics* **9:**560.

28. **Johnson JR, Russo TA.** 2005. Molecular epidemiology of extraintestinal pathogenic (uropathogenic) *Escherichia coli*. *Int J Med Microbiol* **295:**383–404.

29. **Oelschlaeger TA, Dobrindt U, Hacker J.** 2002. Virulence factors of uropathogens. *Curr Opin Urol* **12:**33–38.

30. **Roos V, Klemm P.** 2006. Global gene expression profiling of the asymptomatic bacteriuria *Escherichia coli* strain 83972 in the human urinary tract. *Infect Immun* **74:**3565–3575.

31. **Wallace-Gadsden F, Johnson JR, Wain J, Okeke IN.** 2007. Enteroaggregative *Escherichia coli* related to uropathogenic clonal group A. *Emerg Infect Dis* **13:**757–760.

32. **Russo TA, Johnson JR.** 2003. Medical and economic impact of extraintestinal infections due to *Escherichia coli*: focus on an increasingly important endemic problem. *Microbes Infect* **5:**449–456.

33. **Aoki SK, Pamma R, Hernday AD, Bickham JE, Braaten BA, Low DA.** 2005. Contact-dependent inhibition of growth in *Escherichia coli*. *Science* **309:**1245–1248.

34. **Gillor O, Kirkup BC, Riley MA.** 2004. Colicins and microcins: the next generation antimicrobials. *Adv Appl Microbiol* **54:**129–146.

35. **Roesch PL, Redford P, Batchelet S, Moritz RL, Pellett S, Haugen BJ, Blattner FR, Welch RA.** 2003. Uropathogenic *Escherichia coli* use d-serine deaminase to modulate infection of the murine urinary tract. *Mol Microbiol* **49:**55–67.

36. **Snyder JA, Haugen BJ, Buckles EL, Lockatell CV, Johnson DE, Donnenberg MS, Welch RA, Mobley HL.** 2004. Transcriptome of uropathogenic *Escherichia coli* during urinary tract infection. *Infect Immun* **72:**6373–6381.

37. **Ochman H, Selander RK.** 1984. Standard reference strains of *Escherichia coli* from natural populations. *J Bacteriol* **157:**690–693.

38. **Picard B, Garcia JS, Gouriou S, Duriez P, Brahimi N, Bingen E, Elion J, Denamur E.** 1999. The link between phylogeny and virulence in *Escherichia coli* extraintestinal infection. *Infect Immun* **67:**546–553.

39. **Johnson JR, Johnston B, Kuskowski MA, Nougayrède JP, Oswald E.** 2008. Molecular epidemiology and phylogenetic distribution of the *Escherichia coli pks* genomic island. *J Clin Microbiol* **46:**3906–3911.

40. **Le Gall T, Clermont O, Gouriou S, Picard B, Nassif X, Denamur E, Tenaillon O.** 2007. Extraintestinal virulence is a coincidental by-product of commensalism in B2 phylogenetic group *Escherichia coli* strains. *Mol Biol Evol* **24:**2373–2384.

41. **Krieger JN, Lee SW, Jeon J, Cheah PY, Liong ML, Riley DE.** 2008. Epidemiology of

prostatitis. *Int J Antimicrob Agents* **31**(Suppl 1): S85–90.

42. **De Marzo AM, Nakai Y, Nelson WG.** 2007. Inflammation, atrophy, and prostate carcinogenesis. *Urol Oncol* **25**:398–400.

43. **Collins MM, Stafford RS, O'Leary MP, Barry MJ.** 1998. How common is prostatitis? A national survey of physician visits. *J Urol* **159**:1224–1228.

44. **Nickel JC, Alexander RB, Schaeffer AJ, Landis JR, Knauss JS, Propert KJ; Chronic Prostatitis Collaborative Research Network Study Group.** 2003. Leukocytes and bacteria in men with chronic prostatitis/chronic pelvic pain syndrome compared to asymptomatic controls. *J Urol* **170**:818–822.

45. **Leskinen MJ, Rantakokko-Jalava K, Manninen R, Leppilahti M, Marttila T, Kylmälä T, Tammela TL.** 2003. Negative bacterial polymerase chain reaction (PCR) findings in prostate tissue from patients with symptoms of chronic pelvic pain syndrome (CPPS) and localized prostate cancer. *Prostate* **55**:105–110.

46. **Krieger JN, Riley DE, Roberts MC, Berger RE.** 1996. Prokaryotic DNA sequences in patients with chronic idiopathic prostatitis. *J Clin Microbiol* **34**:3120–3128.

47. **Krieger JN, Riley DE.** 2002. Bacteria in the chronic prostatitis-chronic pelvic pain syndrome: molecular approaches to critical research questions. *J Urol* **167**:2574–2583.

48. **Hochreiter WW, Duncan JL, Schaeffer AJ.** 2000. Evaluation of the bacterial flora of the prostate using a 16S rRNA gene based polymerase chain reaction. *J Urol* **163**:127–130.

49. **Karatas OF, Turkay C, Bayrak O, Cimentepe E, Unal D.** 2010. *Helicobacter pylori* seroprevalence in patients with chronic prostatitis: A pilot study. *Scand J Urol Nephrol* **44**:91–94.

50. **Motrich RD, Cuffini C, Oberti JP, Maccioni M, Rivero VE.** 2006. *Chlamydia trachomatis* occurrence and its impact on sperm quality in chronic prostatitis patients. *J Infect* **53**:175–183.

51. **Mändar R, Raukas E, Türk S, Korrovits P, Punab M.** 2005. Mycoplasmas in semen of chronic prostatitis patients. *Scand J Urol Nephrol* **39**:479–482.

52. **Potts JM, Sharma R, Pasqualotto F, Nelson D, Hall G, Agarwal A.** 2000. Association of *Ureaplasma urealyticum* with abnormal reactive oxygen species levels and absence of leukocytospermia. *J Urol* **163**:1775–1778.

53. **Rudick CN, Berry RE, Johnson JR, Johnston B, Klumpp DJ, Schaeffer AJ, Thumbikat P.** 2011. Uropathogenic *Escherichia coli* induces chronic pelvic pain. *Infect Immun* **79**:628–635.

54. **Kanamaru S, Kurazono H, Ishitoya S, Terai A, Habuchi T, Nakano M, Ogawa O, Yamamoto S.** 2003. Distribution and genetic association of putative uropathogenic virulence factors iroN, iha, kpsMT, ompT and usp in *Escherichia coli* isolated from urinary tract infections in Japan. *J Urol* **170**:2490–2493.

55. **Johnson JR.** 2003. Microbial virulence determinants and the pathogenesis of urinary tract infection. *Infect Dis Clin North Am* **17**:261–278, viii.

56. **Yamamoto S.** 2007. Molecular epidemiology of uropathogenic *Escherichia coli*. *J Infect Chemother* **13**:68–73.

57. **Kanamaru S, Kurazono H, Nakano M, Terai A, Ogawa O, Yamamoto S.** 2006. Subtyping of uropathogenic *Escherichia coli* according to the pathogenicity island encoding uropathogenic-specific protein: comparison with phylogenetic groups. *Int J Urol* **13**:754–760.

58. **Pontari MA, Ruggieri MR.** 2004. Mechanisms in prostatitis/chronic pelvic pain syndrome. *J Urol* **172**:839–845.

59. **Meares EM, Stamey TA.** 1968. Bacteriologic localization patterns in bacterial prostatitis and urethritis. *Invest Urol* **5**:492–518.

60. **Herzer PJ, Inouye S, Inouye M, Whittam TS.** 1990. Phylogenetic distribution of branched RNA-linked multicopy single-stranded DNA among natural isolates of *Escherichia coli*. *J Bacteriol* **172**:6175–6181.

61. **Johnson JR, Delavari P, Kuskowski M, Stell AL.** 2001. Phylogenetic distribution of extraintestinal virulence-associated traits in *Escherichia coli*. *J Infect Dis* **183**:78–88.

62. **Costerton JW, Stewart PS, Greenberg EP.** 1999. Bacterial biofilms: A common cause of persistent infections. *Science* **284**:1318–1322.

63. **Berry RE, Klumpp DJ, Schaeffer AJ.** 2009. Urothelial cultures support intracellular bacterial community formation by uropathogenic *Escherichia coli*. *Infect Immun* **77**:2762–2772.

64. **Martinez JJ, Mulvey MA, Schilling JD, Pinkner JS, Hultgren SJ.** 2000. Type 1 pilus-mediated bacterial invasion of bladder epithelial cells. *EMBO J* **19**:2803–2812.

65. **Kikutani H, Makino S.** 1992. The murine autoimmune diabetes model: NOD and related strains. *Adv Immunol* **51**:285–322.

Urosepsis: Overview of the Diagnostic and Treatment Challenges

8

FLORIAN M. E. WAGENLEHNER,[1] ADRIAN PILATZ,[1] WOLFGANG WEIDNER,[1] and KURT G. NABER[2]

DEFINITIONS OF UROSEPSIS

Urosepsis is defined as sepsis caused by an infection in the urogenital tract (Table 1). In urosepsis, as in other types of sepsis, the severity of sepsis depends mostly upon the host response. Patients who are more likely to develop urosepsis include elderly patients; diabetics; immunosuppressed patients, such as transplant recipients; cancer patients receiving chemotherapy or corticosteroids; and patients with acquired immunodeficiency syndromes. However, all patients can be affected by bacterial species capable of inducing inflammation within the urinary tract.

Sepsis is a systemic response to infection. The signs and symptoms of systemic inflammatory response syndrome (SIRS), which were initially considered to be 'mandatory' for the diagnosis of sepsis (1, 2), are now considered to be alerting symptoms (3). Many other clinical or biological symptoms must be considered (Table 2). The classifications of sepsis differentiate severity levels:

a. severe sepsis is defined as sepsis associated with organ dysfunction;
b. septic shock is persistence of hypoperfusion or hypotension despite fluid resuscitation, and

[1]Clinic for Urology, Pediatric Urology and Andrology, Justus-Liebig-University Gießen, D-35390 Gießen, Germany; [2]Technical University, 80333 Munich, Germany.
Urinary Tract Infections: Molecular Pathogenesis and Clinical Management, 2nd Edition
Edited by Matthew A. Mulvey, David J. Klumpp, and Ann E. Stapleton
© 2017 American Society for Microbiology, Washington, DC
doi:10.1128/microbiolspec.UTI-0003-2012

TABLE 1 Definitions

Infection: Presence of organisms in a normally sterile site that is usually, but not necessarily, accompanied by an inflammatory host response

Bacteremia: Bacteria present in blood as confirmed by culture. May be transient

Systemic inflammatory response syndrome (SIRS): Response to a wide variety of clinical insults, which can be infectious, as in sepsis, but may be non-infectious in etiology (e.g., burns, pancreatitis). This systemic response is manifested by two or more of the following conditions:

Temperature >38°C or <36°C

Heart rate >90 beats/min

Respiratory rate >20 breaths/min or $PaCO_2$ <32 mm Hg (<4.3k Pa)

WBC >12,000 cells/mm^3 or <4,000 cells/mm^3 or ≥10% immature (band) forms

Sepsis: Activation of the inflammatory process due to infection

Hypotension: A systolic blood pressure of <90 mmHg or a reduction of >40 mmHg from baseline in the absence of other causes of hypotension

Severe sepsis: Sepsis associated with organ dysfunction, hypoperfusion, or hypotension. Hypoperfusion and perfusion abnormalities may include, but are not limited to, lactic acidosis, oliguria, or an acute alteration of mental status

Septic shock: Sepsis with hypotension despite adequate fluid resuscitation along with the presence of perfusion abnormalities that may include, but are not limited to, lactic acidosis, oliguria, or an acute alteration in mental status. Patients who are on inotropic or vasopressor agents may not be hypotensive at the time that perfusion abnormalities are measured

Refractory septic shock: Septic shock that lasts for more than 1 hour and does not respond to fluid administration or pharmacological intervention

c. refractory septic shock is defined by an absence of response to therapy.

For therapeutic purposes, the diagnostic criteria of sepsis should identify patients at an early stage of the syndrome, prompting urologists and intensive-care specialists to search for and treat infection, apply appropriate therapy, and monitor for organ failure and other complications (4).

EPIDEMIOLOGY OF UROSEPSIS

Urinary tract infections (UTIs) can manifest in a wide clinical range from bacteriuria with no or limited clinical symptoms, to sepsis, severe sepsis, or septic shock. In approximately 30% of all septic patients, the infectious focus is localized in the urogenital tract and arises from infections of the parenchymatous urogenital organs, e.g., kidneys, prostate, or testicles. This may comprise obstructive diseases of the urinary tract, such as ureteral stones, stenosis of the collecting system, tumor formations, or anomalies of the urinary system. Urosepsis may also occur after operations in the urogenital tract. In patients with nosocomial UTI treated in urology, the prevalence of urosepsis was, on average, about 12% in a multinational surveillance study (5).

Severe sepsis is a critical situation with a reported mortality rate ranging from 20% to 50% (6–8). Severe sepsis and septic shock are also the major causes of admission and death in intensive-care units (ICUs) (7). In the U.S., there is a steady increase of 71% in the number of cases of severe sepsis, from 415,280 cases in 2003 to 711,736 in 2007. The total hospital costs for all patients with severe sepsis also increased by 57%, from $15.4 billion in 2003 to $24.3 billion in 2007 (9). Sepsis is more common in men than in women (6). In recent years, the incidence of sepsis has increased (6, 10), but the associated mortality has decreased, suggesting improved management of patients (6, 10). Most severe sepsis cases reported in the literature are related to pulmonary (50%) or abdominal infections (24%), with the urogenital tract accounting for only 5% (11) to 7% (12). Uro-

TABLE 2 Clinical diagnostic criteria of sepsis and septic shock

Criterion I:	Presence of bacteremia (positive blood culture) or clinical suspicion of sepsis. Bacteremia can be of low inoculum (<10 bacteria/ml) or of short duration. Multiple blood cultures are recommended.	
Criterion II:	Systemic inflammatory response syndrome [SIRS]	
	Body temperature	$\geq 38°C$ or $\leq 36°C$
	Tachycardia	≥ 90 beats/min
	Tachypnea	≥ 20 breaths/min
	Respiratory alkalosis	$PaCO_2 \leq 32$ mm Hg
	Leukocytes	$\geq 12,000/\mu l$ or $\leq 4,000/\mu l$
	Segmented neutrophils	>10%
Criterion III:	Multiple organ dysfunction syndrome [MODS]	
	Circulation:	Arterial systolic blood pressure ≤ 90 mm Hg or mean arterial blood pressure ≤ 70 mm Hg during ≥ 1 hour despite adequate fluid resuscitation and adequate intravascular volume or use of vasopressors, in order to maintain systolic blood pressure ≥ 90 mm Hg.
	Kidney:	Urine production <0.5 ml/kg body weight/hour during 1 hour despite adequate fluid resuscitation.
	Lung:	$PaO_2 \leq 75$ mm Hg (at ambient air) or $PaO_2/FiO_2 \leq 300$ (acute lung injury), or $PaO_2/FiO_2 \leq 200$ (acute respiratory distress syndrome) at assisted ventilation [PaO_2, arterial O_2-partial pressure; FiO_2, inspiratory O_2 concentration].
	Thrombocytopenia:	Platelets <80,000/μl or decrease of platelets $\geq 50\%$ within 3 days.
	Metabolic acidosis:	Blood pH $\leq 7,30$ or base excess ≥ 5 mmol/l; plasma lactate ≥ 1.5 fold of normal.
	Encephalopathy:	Somnolence, agitation, coma, confusion.

Following these criteria sepsis can be classified in three grades:

Simple sepsis	Criterion I + ≥ 2 Criterion II	Lethality
	-2 Criterion II	7%
	-3 Criterion II	10%
	-4 Criterion II	17%
Severe sepsis	Criterion I + ≥ 2 Criteria II + ≥ 1 Criterion III per affected organ (kidney, lung, liver) lethality increases + 15% to + 20%	
Septic shock	Criterion I + ≥ 2 Criteria II + therapy-refractory arterial hypotension ≤ 90 mmHg Lethality 50% to 80%	

sepsis, however, may still show high mortality rates of 25% to 60% in special patient groups (13). Hofmann (14) analyzed 59 patients (54% females) treated for uroseptic shock and hospitalized over a 10-year period: 78% of patients showed urinary obstruction as predisposing factors, mainly due to nephrolithiasis, and the remaining 22% patients also had uropathies with significant impact on urodynamics. Seventeen percent of patients developed urosepsis after urological interventions. Ninety-two percent of patients needed operative intervention and at that time 24% underwent nephrectomy. Of the 12% patients who died due to the critical illness, no intervention was performed. A consistent finding, however, is that the mortality associated with sepsis from a urinary source is substantially lower than all other sources.

PATHOPHYSIOLOGY OF UROSEPSIS

Microorganisms reach the urinary tract mostly by way of the intraluminal-ascending route, more rarely by hematogenous or lymphatic routes. Inflammation is the physiologic response of the body to infection and is mediated by the release of soluble substances by cells of the immune system. For urosepsis to be established from the urinary tract, the pathogens or pathogenic factors have to reach the bloodstream (15). The risk of bacteremia is increased in severe urogenital infections, such as pyelonephritis and acute bacterial prostatitis (ABP), and is facilitated by obstruction of the urinary tract. The systemic inflammatory response syndrome (SIRS) is than triggered: an initially overwhelming proinflammatory reaction, activated by mediators such as bacterial toxins, is accompanied by a counter-regulatory anti-inflammatory response syndrome (CARS) (16).

Complete bacteria and components of the bacterial cell wall act as exogenous pyrogens on eukaryotic target cells of patients. These include lipopolysaccharides (LPS), especially the lipid A component (endotoxin) of the outer membrane of Gram-negative bacteria; the peptidoglycan, teichon- or lipoteichoic acids of Gram-positive bacteria; and toxins like toxic-shock syndrome toxin 1 and *Staphylococcus aureus* toxin A (17). Many of these factors bind to cellular receptors and co-receptors of the innate immune system (e.g., CD 14, "Toll-like receptors" TLR2 and TLR4, CD 18, and selectin) on the surface of macrophages, neutrophils, endothelial cells, and others. Intracellular messenger molecules, such as nuclear factor kappa-light-chain-enhancer of activated B cells (NF-κB) or protein-kinase C, are activated, which induce transcription of mediator genes and thus the synthesis of numerous endogenous mediators, such as cytokines.

Cytokines as Markers of the Septic Response

Cytokines are produced with different kinetics and are classified into pro-inflammatory and anti-inflammatory cytokines. Tumor necrosis factor (TNF)-α and interleukin (IL)-1 are the most important pro-inflammatory cytokines and exhibit similar biologic properties. They influence the temperature-regulatory centers in the hypothalamus, which results in fever. They also have an effect on the *formatio reticularis* in the brain stem that renders the patient somnolent and comatose.

Release of adrenocorticotrophic hormone (ACTH) in the pituitary gland is increased, which stimulates the adrenal gland. These factors also stimulate hematopoietic growth factors leading to the formation of new neutrophils and the release of stored ones. The neutrophils are additionally activated and produce bactericidal substances, such as proteases and oxygen radicals. B and T lymphocytes are stimulated for synthesis of antibodies and cellular immune reaction.

In the continuing septic process, however, apoptosis of B cells, CD4-helper cells, and follicular-dendritic cells cause an anti-inflammatory immune suppression, called transient-immune paralysis (18). In the liver, the production of acute-phase proteins (C-reactive protein, α1-antitrypsin, and complement factors) is triggered. The muscular protein is degraded and the released amino acids are used for antibody synthesis. In endothelial cells, the production of platelet-activating factor (PAF) and nitric oxide (NO) is triggered, leading to a decreased vessel tone. The endothelial cells are damaged and increased permeability results. Surface receptors of endothelial cells and neutrophils are upregulated, increasing the mutual adhesiveness. Additionally, the endothelial procoagulatory activity and the synthesis of a plasminogen activator-inhibitor substance is increased, which activates the blood-coagulation system (11, 17, 19–22). The cytokines IL-4 and IL-10, among others, act as anti-inflammatories and inhibit the formation of IL-1 and TNF (11, 21).

All of these factors may act on target organs and are responsible for the local and systemic effects within the target organs (11, 19, 21, 22). Sepsis may indicate an immune system that is severely compromised and unable to eradicate pathogens or a nonregulated and excessive activation of inflammation, or both. A genetic predisposition likely explains different outcomes in septic patients. Mechanisms of organ failure and death in patients with sepsis remain only partially understood (11).

The Importance of the Neuro-Endocrine Axis

The immune system and central nervous system communicate and regulate each other via the autonomic nervous system and the hypothalamo-pituitary-adrenocortical axis (23). The sympathetic system innervates all lymphoid organs and thereby modulates the immune system. Under pathological circumstances noradrenalin can stimulate α_2-receptors on the surface of macrophages and thus stimulate release of TNF-α (24). Stimulation of β-receptors inhibits release of pro-inflammatory cytokines.

Release of pro-inflammatory cytokines in response to infection stimulates hypothalamic centers activating the sympathetic-nerve system and the hypothalamo-pituitary-adrenocortical axis, inducing the expression of corticotropin-releasing hormone (CRH) or arginine-vasopressin in the hypothalamus and ACTH in the pituitary gland. The consecutive release of cortisol from the adrenal gland leads to an anti-inflammatory response by suppressing NF-κB and increasing anti-inflammatories IL-4 and IL-10 (25). The parasympathetic system acts via the vagal-nerve system sensing inflammation and informing specific centers in the brain. Release of acetylcholine reduces pro-inflammatory cytokines as well as releasing high-mobility-group-box 1 (HMGB1), a critical pro-inflammatory mediator (26, 27).

Thus, the neuro-endocrine axis appears to be a critical regulator of systemic inflammation important for fine tuning of the inflammatory response, which might also become an important target for pharmacological intervention.

PATHOPHYSIOLOGICAL CONDITIONS ALTERING RENAL FUNCTION

Pyelonephritis is the most frequent cause for urosepsis. The renal function is therefore most important in the consideration of urosepsis (28). Impairment of renal function can be acute or chronic and unilateral or bilateral. Post-renal obstruction is one of the most frequent causes of urosepsis in urological cases. The obstruction causes sepsis on the one side and severely influences the pharmacokinetics of drugs, such as antibiotics, in the urinary tract on the other side. In this context the pharmacokinetics of drugs at the affected site is also significantly influenced by the total renal function and thus by the function of the contralateral kidney. Furthermore, pharmacokinetics and the concentration of an antibiotic in the kidney is influenced by its arterial and venous plasma concentration, the various renal (proximal and distal) tubular concentrations, the renal concentration in tissue, and the final urine concentration. The renal concentration of an antibiotic in tissue is complex and is a function of renal blood flow, glomerular filtration, tubular secretion and reabsorption, pyelovenous- and lymphous backflow, and the number of intact nephrons. The renal-tissue concentration of a drug is therefore difficult to assess; representative concentrations have been investigated and one of these could be the concentrations in the renal lymph that might resemble interstitial concentrations (29). Renal-lymph concentrations in unobstructed, normal, and unobstructed, but infected kidneys, have been determined for a variety of ß-lactam antibiotics, aminoglycosides, and nitrofurantoin. Concentrations in the renal lymph were generally lower than the corresponding arterial-plasma concentrations, which suggests that there is no concentration effect in the renal-interstitial space.

In acute, complete, unilateral ureteral obstruction there is still some residual glomerular filtration and tubular secretion present. An experimental study in dogs (29) showed that within the first hours of obstruction, glomerular-filtration rate, as well as the effective renal-plasma flow, decreased significantly to an average of 14% in the obstructed kidney, compared to the unobstructed kidney. The persisting turnover of urine is primarily due to pyelovenous backflow and also, but less importantly, due to drainage into the renal lymphatics. Then the concentration of a substance, such as certain antibiotics, in the renal lymph are higher than the corresponding arterial-plasma concentrations in the acute phase of an unilateral obstruction (first hours to first week) and become equal to the arterial-plasma concentrations in chronic-obstructive disease (longer than 1 week) (30).

The glomerular filtration rate is influenced by the balance of inward and outward pressures at the glomerular arterioles and Bowman's capsule. Inward forces are significantly increased with postrenal obstruction and are the cause of reduced filtration rate (29, 31). An increasing number of nephrons will cease filtering if the ureteral pressure exceeds one-third of the mean blood pressure. The acute unilateral occlusion of a ureter results in a characteristic triphasic relationship between renal blood flow and ureteral pressure. The first phase, lasting approximately 1.5 hours, shows a rise in both ureteral pressure and renal blood flow, followed by a second phase with decline in renal blood flow and a continued increase in ureteral pressure, lasting from approximately 1.5 to 5 hours, followed by a third phase resulting in a further decline of renal blood flow accompanied by a progressive decrease in ureteral pressure. Phase I is characterized by an initial afferent-arteriole vasodilatation followed by an efferent-arteriole vasoconstriction in phase II and afferent-arteriole vasoconstriction in phase III. This third phase is not seen in bilateral ureteral obstruction, which leads to a progressive rise in ureteral pressure despite a decrease in renal blood flow. Single-nephron glomerular-filtration rate declines in unilateral and bilateral ureteral occlusion, which is secondary to an increase in afferent-arteriolar resistance in unilateral occlusion, and secondary to a rise in intratubular pressure in bilateral occlusion (32). These differences

between unilateral and bilateral obstruction are most probably due to substances that accumulate in bilateral obstruction or unilateral obstruction of a solitary kidney, but do not accumulate in unilateral obstruction with a functioning contralateral kidney. One important factor of these changes is the atrial-natriuretic peptide (32).

Using different antibiotic pharmacological models, mainly with β-lactam agents, the following findings in the different settings of renal functions and renal obstructions can be summarized (33):

1. In the case of a severe unilateral renal insufficiency (glomerular filtration rate 1 ml/min) but normal contralateral renal function (glomerular filtration rate 60 ml/min) the urine antibiotic concentrations of both kidneys are high, whereby the impaired kidney achieves half the urine concentration of the intact kidney.

2. In the case of a bilateral renal insufficiency, the urinary antibiotic concentrations significantly decrease down to plasma levels in the case of severe impairment (glomerular filtration rate 2 ml/min). This difference in unilateral and bilateral renal insufficiency can be explained by the intact-nephron theory: the single nephrons, e.g., the concentration ability, remain intact in the case of physiologic prerenal conditions. In bilateral renal insufficiency there is an increased offer of solutes per nephron that results in diuresis with impaired ability for concentration (34).

3. In acute obstruction, urinary concentrations of filtered and secreted substances will at first reach a high plateau. If the ureteral pressure rises and exceeds one-third of the mean blood pressure, an increasing number of nephrons will cease filtering, resulting in decrease of glomerular filtration and also in decrease of urinary concentrations. This process is very much enhanced by

infection of an obstructed kidney with the result that, in urosepsis due to obstruction, high doses of antimicrobials mainly excreted by the kidneys are necessary.

4. In acute unilateral obstruction the urinary antibiotic concentrations of the obstructed kidney are almost as high as those of the normal unobstructed kidney, which is also due to the maximal urinary concentration in acute obstruction.

5. Even in chronic unilateral obstruction rather high urinary antibiotic concentrations are achieved, depending on the function of the contralateral kidney.

Although antibiotic concentration is an important pharmacological parameter, it does not necessarily reflect antibacterial activity of an antimicrobial substance at the site of infection. (see below)

ANTIMICROBIAL THERAPY

Antimicrobials are among the most important drugs in the management of patients with severe infections (35). Inappropriate use of antimicrobials may cause therapeutic failure in the individual patient and, additionally, may contribute towards promoting the emergence of resistant pathogens that might also readily spread in the hospital setting (36). An adequate initial (i.e., in the first hour) antibiotic therapy has been shown to correlate with improved outcome in septic shock (37–39) and is therefore critical also in severe UTI. Inappropriate antimicrobial therapy in severe UTI is linked to a higher mortality rate (40), as it has been shown with other infections as well (41, 42). Empirical antibiotic therapy therefore needs to follow certain rules (43), which might be based upon the expected bacterial spectrum, the institutional-specific resistance rates and the individual patient's requirements. Empirical initial treatment should provide broad antimicrobial coverage and should later be adapted on the basis of culture results.

Parameters for Antimicrobial Treatment in Urosepsis

If the infection involves renal or other urogenital parenchymal tissues or the patient has urosepsis, adequate serum concentrations are necessary to produce sufficiently high tissue concentrations. Therefore, administration of high-dose intravenous antimicrobials is necessary. Even in uncomplicated pyelonephritis and uncomplicated cystitis there is interstitial and intracellular invasion by the uropathogens (44–47). The antibiotic concentrations in tissue are dependent on the plasma concentrations, the specific tissue architecture, the pharmacokinetic parameters of the antibiotic drug (charge and size of molecule, protein binding, pH in the infectious focus), and the distribution of the infection in the tissue (stroma, epithelium).

Alteration of Pathophysiological Conditions During Urosepsis

Sepsis, and the treatment thereof, increases renal preload and leads to third-spacing. Both alterations result in higher clearances of antibacterial drugs (48). The increased volume of distribution as a result of edema in sepsis will lead to underexposure, especially of hydrophilic antimicrobials such as β-lactams and aminoglycosides, which exhibit a volume of distribution mainly restricted to the extracellular space (49). On the other hand, sepsis may cause multiple-organ dysfunction, such as hepatic or renal dysfunction, resulting in decreased clearance of antibacterial drugs. Individualized dosing is therefore necessary. As β-lactams are time-dependent antibacterials, an increase in the volume of distribution or clearance will require increased dosing or administration by continuous infusion. Fluoroquinolones, on the other hand, display largely concentration-dependent activity. The volume of distribution in sepsis is not much altered by fluid shifts in the case of fluoroquinolones and therefore no alterations of standard doses are necessary, unless renal dysfunction occurs (48). In order to optimize the antibacterial activities in septic patients, the pathophysiological effects of the sepsis syndrome and the pharmacokinetic/ pharmacodynamic (PK/PD) properties of the antibacterial substances need careful consideration. Therapeutic drug monitoring would be a beneficial method to optimize the individual dosing regimens.

The co-administration of various drugs may also frequently alter the pharmacokinetic behavior of some antibacterial agents. Although some interaction with co-administered drugs involve absorption or distribution, the most clinically relevant interactions during anti-infective treatment involve the metabolic and the renal elimination phase. Hydrophilic anti-infective agents are often eliminated unchanged by renal glomerular filtration and tubular secretion, and are therefore involved in competition for excretion with co-administered drugs. Therapeutic failure with these compounds may be due to hemodynamically active co-administered drugs, such as dopamine, dobutamine, and furosemide, which increase their renal clearance by means of enhanced cardiac output and/or renal blood flow (50).

PK/PD Parameters for Treatment of Severe UTI

In patients with early-onset of ventilator-associated pneumonia, Pea et al. (51) found a reduced area under the concentration-time curve (AUC) over the 12-hour dosage interval after administration of parenteral levofloxacin 500 mg twice daily. The authors explained the reduced exposure by a greater clearance of levofloxacin leading to a shorter elimination half-life. Cumulative urinary excretion confirmed the greater excretion of unchanged drug compared with healthy volunteers. Therefore, from a pharmacokinetic point of view, critically ill and septic

patients should be considered a particular subpopulation (50), in whom usually recommended dosages have to be reconsidered.

Beta-lactams

While pharmacodynamic studies in UTI are relatively scarce, at least one study has documented that therapeutic success following beta-lactam therapy depends on the time the antimicrobial concentration remains above the minimum inhibitory concentration (MIC; T>MIC). There was a significant correlation between the cumulative T>MIC in serum and bacteriological cure, wherein a cumulative T>MIC of 30 hours provided a maximal cure rate of 80% to 90% (52).

Fluoroquinolones and Aminoglycosides

For drugs with concentration-dependent time-kill activity, such as the aminoglycosides and the fluoroquinolones, a positive outcome appears to be more dependent on the $C_{max}/$ MIC or AUC/MIC ratio. While it remains unclear which ratio is a better predictor of outcome, in either case a high ratio is desirable. In pharmacodynamic UTI studies of ciprofloxacin in mice infected with *Escherichia coli* (52), there was an obvious correlation between reduced bacterial counts and the AUC/MIC ratio. In severely ill patients (mainly patients with pneumonia) treated with ciprofloxacin, a significant breakpoint for probabilities of both clinical and microbiologic cures was an AUC/MIC ratio of 125 and higher, leading to a significantly higher cure rate (53). In any infection of the urogenital tract, however, a significant level of the bacteria (sometimes more than 10^6/ml) is also freely floating in the urine. Therefore, high urinary concentrations of the antibiotic are needed as well (52, 54, 55). Antimicrobials primarily eliminated via renal excretion achieve high urinary concentrations, sometimes 100 to 1,000 times of the concomitant serum concentrations. Theoretically, these antimicrobials represent optimal choices for the treatment of UTIs. Besides favorable pharmacokinetics, however, an agent suitable for the treatment of severe UTI and urosepsis should also provide optimal pharmacodynamic properties at the site of infection, i.e., in the urine. Therefore, even agents modestly eliminated by renal mechanisms but with high intrinsic potency (low MIC) against the causative uropathogens, are also important considerations in antimicrobial selection (56). Correlation of those considerations to human data in complicated lower UTI showed that pharmacodynamic targets representing 90%-probability thresholds for bacterial eradication were far below the 125 AUC/MIC breakpoint, suggesting that, in addition to its plasma concentration, the high concentration of fluoroquinolones in the urine might have played a significant role in eradicating bacteria (57).

The antimicrobial activity of many antibiotics is, however, reduced in urine as compared to standard growth medium, depending on pH and urinary contents (58). Therefore, the urinary concentrations cannot be correlated directly with standard MIC. By determining the urinary-bactericidal titer (UBT), i.e., the highest urinary dilution still bactericidal, pharmacokinetic and pharmacodynamic parameters of an antibiotic in urine are linked together. The reciprocal value of an UBT corresponds to the ratio of urinary concentration/minimal bactericidal concentration$_{urine}$.

In a study of complicated UTI and pyelonephritis, the UBTs were measured for levofloxacin and doripenem. The results showed that microbiological failures in patients treated with levofloxacin correlated well with low urinary-bactericidal activity, whereas there was no such correlation for doripenem (59). Such data evaluating antibiotic activity in patients with urosepsis are missing.

Biofilm Infection

Biofilm infection plays a considerable role in urosepsis, not only in association with urinary catheters, but also in scar tissue, stones, prostatitis, and in any obstructed urinary tract (60–63). For most uropathogens, the ability to form biofilms has been shown (64–70). In a study investigating virulence factors in *E. coli* strains causing bacteremia, 53% of the strains were able to produce biofilm (71). In an *in vitro* biofilm-catheter infection model, kill-curves for *Pseudomonas aeruginosa* treated with different antibiotic substances were investigated (72, 73). Beta-lactam antibiotics (piperacillin, ceftazidime) were not able to eradicate the biofilm-cells, even when concentrations up to 128-fold the minimal-bactericidal concentrations were administered. With fluoroquinolones (ciprofloxacin, levofloxacin), eradication was possible within 24 hours; however, only in concentrations that reached 32-fold the minimal-bactericidal concentrations (72, 73). Therefore, generally high dosages of antimicrobials need to be applied in conjunction with the attempt to eliminate the biofilm. If there is a chance to remove the biofilm, this should be done, e.g., by removing infected stones or catheters.

Bacterial Spectrum in Urosepsis

There are not many publications on the specific bacterial spectrum in urosepsis. Mainly the bacterial spectrum of complicated and nosocomially acquired UTI are taken as representative for urosepsis as well, which in general might be correct (74). The German septicemia study published in 2002 (75) discriminated the bacterial spectrum of blood-culture isolates according to their origin and showed that if the urinary tract was the source for the septicemia, the bacterial spectrum consisted of about 61% *E. coli*, 16% other enterobacteria, 8% S. *aureus*, and 6% enterococci, underlining the predominant role of *E. coli* (Fig. 1) (75). If host defense is impaired, less-virulent organisms such as enterococci, coagulase-negative staphylococci, or *P. aeruginosa* may also cause urosepsis.

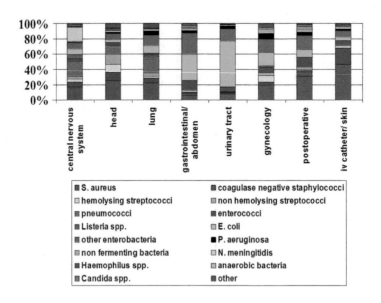

FIGURE 1 **Bacterial spectrum of pathogens detected in patients with septicaemia (n = 6,128).**

Selection of Antimicrobials for Empiric Therapy

Since effective antimicrobial therapy is best initiated during the first hour when sepsis is diagnosed, the empiric intravenous therapy should be initiated immediately after microbiological sampling. For the selection of appropriate antimicrobials it is important to know the site-of-origin underlying diseases, and whether the sepsis is primary or secondary and community or nosocomially acquired. In addition, the preceding antimicrobial therapies must be recorded as precisely as possible.

Resistance surveillance should always be performed locally to determine the best suitable empiric treatment. Surveillance studies have been performed using blood cultures as the data source. For example, the resistance to ciprofloxacin from the German blood-culture studies of 1983–1985, 1991–1992, 2000–2001, and 2006–2007 and the Paul Ehrlich resistance-surveillance study of 2007 are shown in Fig. 2, and to cefotaxime in Fig. 3 (75–79). *E. coli* resistance to ciprofloxacin is currently about 30% and to cefotaxime approximately 10%.

The European Antibiotic Resistance Surveillance Study comprises a network of over 900 microbiological laboratories serving some 1,500 hospitals in 33 European countries that collects routinely generated antimicrobial-susceptibility testing data on invasive infections caused by seven important bacterial pathogens. The 2011 results showed for *E. coli* 24% ciprofloxacin resistance, 9% 3rd-generation cephalosporin resistance, 12% aminoglycoside resistance, 0% carbapenem resistance, and 13% multidrug resistance. *Klebsiella pneumoniae* showed 13% ciprofloxacin resistance, 8% 3rd-generation cephalosporin resistance, 8% aminoglycoside resistance, 2% carbapenem resistance, and 8% multidrug resistance. *P. aeruginosa* had 13% ciprofloxacin resistance, 8% ceftazidime resistance, 3% piperacillin/tazobactam resistance, 7% gentamicin resistance, 8% carbapenem resistance, and 4% multidrug resistance. *S. aureus* showed 24% methicillin resistance (MRSA) and 0% vancomycin resistance (VRSA). *Enterococcus faecalis* had 1% ampicillin resistance, 5% vancomycin resistance, and 29% high-level gentamicin resistance, while *Enterococcus faecium* showed

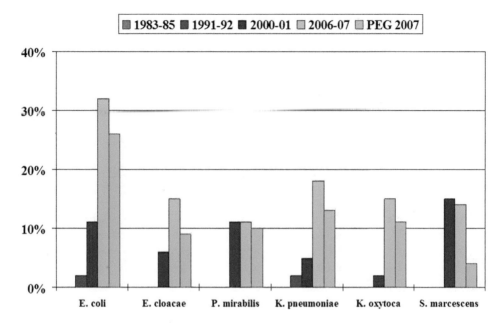

FIGURE 2 **Ciprofloxacin resistance from the German blood culture studies 1983–1985, 1991–1992, 2000–2001, 2006–2007, and the Paul-Ehrlich resistance surveillance study 2007.**

96% ampicillin resistance, 37% vancomycin resistance, and 37% high-level gentamicin resistance (80). Thus, comparing the 2011 data to earlier reports, the numbers of MRSA have decreased by 52% since 2004, the numbers of vancomycin-resistant *E. faecalis* have increased, the number of vancomycin-resistant *E. faecium* have remained constant

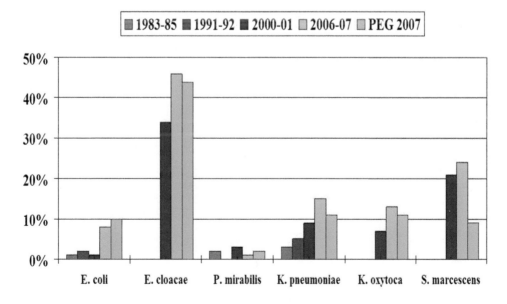

FIGURE 3 **Cefotaxime resistance from the German blood culture studies 1983–1985, 1991–1992, 2000–2001, 2006–2007, and the Paul-Ehrlich resistance surveillance study 2007.**

at a high level, and the proportion of *E. coli* isolates that are resistant to 3rd-generation cephalosporins, ciprofloxacin, and aminoglycosides and multidrug-resistant isolates are the highest annual proportions reported to date (80). Taking into account that *E. coli* is the most frequent pathogen causing severe UTI, it is concerning that there is a continuous increase of antibiotic resistance by *E. coli*.

There is currently no consensus towards a certain resistance threshold for an antibiotic to be recommended for empiric treatment. Given the paramount importance of administering a susceptible antibiotic initially in sepsis, the resistance threshold should therefore be below 10%. Depending on the local susceptibility patterns, a third- or fourth-generation cephalosporin, piperacillin in combination with a β-lactamase inhibitor (BLI), or a carbapenem may be appropriate. In case of no or partial response in secondary urosepsis, i.e., after nosocomial UTI (especially after urological interventions or in patients with long-term indwelling urinary catheters), an antipseudomonal, 3rd-generation cephalosporin or piperacillin and BLI in combination with an aminoglycoside or a carbapenem may be necessary to cover a broader bacterial spectrum, including multi-restistant pathogens (Tables 3 and 4). If the pretreatment history is known, the same group of antimicrobials should be avoided. All alternatives have to be selected in consideration of the local susceptibility patterns.

TABLE 3 Antibiotics recommended for the treatment of urinary tract infections

Antibiotic group	Substance	Dosage	
		Oral	IV[a]
Aminopenicillin + BLI[b]	Ampicillin/sulbactam	0.750 g twice daily	0.75–3 g 3 times daily
	Amoxicillin/clavulanic acid	1 g twice daily *or* 0.625 g 3 times daily	1.2–2.2 g 3 times daily
Acylureidopenicillin + BLI	Piperacillin/tazobactam	—	2.5–4.5 g 3 times daily
	Piperacillin/Combactam	—	5 g 3 times daily
Cephalosporin Gr. 1	Cephalexin	Prophylaxis only	—
Cephalosporin Gr. 2	Cefuroxime axetil	500 mg twice daily	—
	Cefuroxime	—	0.75–1.5 g 3 times daily
	Cefotiam	—	1–2 g 2–3 times daily
Cephalosporin Gr. 3	Cefpodoxime proxetil	200 mg twice daily	—
	Ceftibuten	200–400 mg daily	—
Cephalosporin Gr. 3a	Cefotaxime	—	1–2 g 2–3 times daily
	Ceftriaxone	—	1–2 g daily
Cephalosporin Gr. 3b	Ceftazidime	—	1–2 g 2–3 times daily
Cephalosporin Gr. 4	Cefepime	—	2 g twice daily
Carbapenem Gr. 1	Imipenem	—	0.5–1 g q 6–8 h
	Meropenem	—	0.5–1 g 3 times daily
	Doripenem	—	0.5 g 3 times daily
Carbapenem Gr. 2	Ertapenem	—	1 g daily
Fluoroquinolone Gr. 2	Ciprofloxacin	500–750 mg twice daily	400 mg twice daily
	Ciprofloxacin XR	1000 mg daily	—
Fluoroquinolone Gr. 3	Levofloxacin	500–750 mg daily	500 mg daily
Antimycotic group			
Azole derivatives	Fluconazole	400–800 mg daily	400–800 mg daily
	Voriconazole	4–6 mg/kg BW[c] daily	4–6 mg/kg BW daily
Pyrimidine analog	Flucytosine		100–150 mg/kg BW 4 times daily
Echinocandin	Caspofungin		50–70 mg daily

[a]*IV*, intravenous.
[b]*BLI*, β-lactamase inhibitor.
[c]*BW*, body weight.

TABLE 4 Antibiotics recommended for the treatment of urosepsis

Most frequent pathogens/species	Initial, empirical antimicrobial therapy	Therapy duration
E. coli Other enterobacteria After urological interventions – multi-resistant pathogens	Cephalosporin (group 3a/b) Fluoroquinolone[a]	3–5 days after defervescence or control/ elimination of complicating factor
Pseudomonas spp. Proteus spp. Serratia spp. Enterobacter spp.	Anti-pseudomonas active acylaminopenicillin/BLI Carbapenem ± Aminoglycoside	

[a]Only in regions where fluoroquinolone resistance is below 10%.

Correct dosing in respect to the altered systemic and, especially, renal pathophysiology in patients with urosepsis and length of therapy, are equally important.

Evidence from *in vitro* experiments and animal and human studies indicate that antibiotic therapy may induce the release of endotoxin. Antibiotics that bind to penicillin-binding protein (PBP)-2, e.g., imipenem, are associated with little release of endotoxin, whereas antibiotics that bind to PBP-3, e.g., ceftazidime, are associated with far greater release. Whether these differences are clinically relevant could, however, not be demonstrated in a clinical study in patients with Gram-negative urosepsis (81).

SURVIVING SEPSIS CAMPAIGN GUIDELINES

In 2004, the Surviving Sepsis Campaign (SSC) first introduced guidelines for the management of severe sepsis and septic shock, as well as strategies for bedside implementation (20, 82, 83); these were updated in 2008 (84) and 2012 (85, 86). The treatment recommendations were organized in so-called sepsis bundles, such as a resuscitation bundle (tasks to begin immediately and to be accomplished within 6 hours) and a management bundle (tasks to be completed within 24 hours).

Key recommendations were listed by category and comprise the following recommendations:

Early goal-directed resuscitation of the septic patient during the first 6 hours after recognition. Early goal-directed therapy (8) is simply a protocol derived from components that have long been recommended as standard care for the septic patient to optimize hemodynamics and oxygen supply (Table 5). Recently, the early goal-directed therapy approach has been challenged by two multicenter studies performed in the U.S. and Australia/New Zealand showing that protocol-based resuscitation of patients in whom septic shock was diagnosed in the emergency department did not improve outcomes (87, 88).

Blood cultures before antibiotic therapy.

Imaging studies performed promptly to confirm potential source of infection.

Administration of broad-spectrum antibiotic therapy within 1 hour of diagnosis of septic shock and severe sepsis without septic shock.

Reassessment of antibiotic therapy with microbiology and clinical data to narrow coverage, when appropriate.

TABLE 5 Target parameters of early goal-directed therapy (8)

Parameter	Target
Central venous pressure (CVP)	8–12 mm Hg
Mean arterial pressure (MAP)	65–90 mm Hg
Central venous oxygen (CVO$_2$)	≥70%
Hematocrit (HCT)	>30%
Urine output	>40 ml/h

A usual 7 to 10 days duration of antibiotic therapy guided by clinical response.

Source control with attention to the balance of risks and benefits of the chosen method.

Administration of crystalloid-fluid resuscitation and consideration of adding albumin in certain patient groups.

Fluid challenge to restore mean circulating-filling pressure.

Vasopressor preference for norepinephrine to maintain an initial target of mean arterial pressure ≥65 mm Hg and epinephrine when an additional vasopressor is needed.

Avoiding use of steroid therapy.

In the absence of tissue hypoperfusion, coronary artery disease, or acute hemorrhage, the hemoglobin target is 7 to 9 g/dL.

A low tidal volume and limitation of inspiratory-plateau pressure strategy for acute lung injury (ALI)/acute respiratory distress syndrome (ARDS).

Application of at least a minimal amount of positive end-expiratory pressure in acute lung injury.

Head of bed elevation in mechanically ventilated patients unless contraindicated.

Protocols for weaning and sedation/analgesia.

Minimizing use of either intermittent-bolus sedation or continuous-infusion sedation.

Avoidance of neuromuscular blockers, if at all possible.

A protocoled approach to blood-glucose management.

Continuous veno-veno hemofiltration or intermittent hemodialysis is equivalent.

Prophylaxis for deep-vein thrombosis.

Use of stress-ulcer prophylaxis to prevent upper gastrointestinal bleeding in patients at risk.

Oral or enteral feedings.

Consideration of limitation of support where appropriate.

ALGORITHM MANAGEMENT OF UROSEPSIS

A rapid diagnosis of urosepsis is critical (8). Effective treatment eliminates the infectious focus, and improves organ perfusion. Treatment of urosepsis comprises four basic strategies:

a. supportive therapy (stabilization and maintaining blood pressure),
b. antimicrobial therapy in the first hour,
c. control or elimination of the complicating factor, and
d. specific sepsis therapy (89).

All four strategies need to be started as early as possible. A diagnosis and management algorithm is therefore helpful (Fig. 4):

The initial patient aspect is often directive. The clinical picture of a septic patient frequently, but not always, involves warm skin, bounding pulses, and hyperdynamic circulation. If the patient is hypovolemic, has pre-existing myocardial dysfunction, or is at late stage of the septic process, hypotension, vasoconstriction, and peripheral cyanosis may be present. The internationally accepted criteria for diagnosis of SIRS and sepsis (Table 1) should rapidly be checked in order to initiate further investigations.

If sepsis is suspected, early (i.e., first hour) supportive therapy with stabilization of the blood pressure and sufficient tissue oxygenation is mandatory. The management of fluid and electrolyte balance is a crucial aspect of patient care in sepsis syndrome, particularly when the clinical course is complicated by shock. An early (immediate) goal-directed therapy has been shown to reduce mortality in one study (8), but has been questioned in two other studies (87, 88). Volemic expansion and vasopressor therapy have considerable impact on the outcome. Early

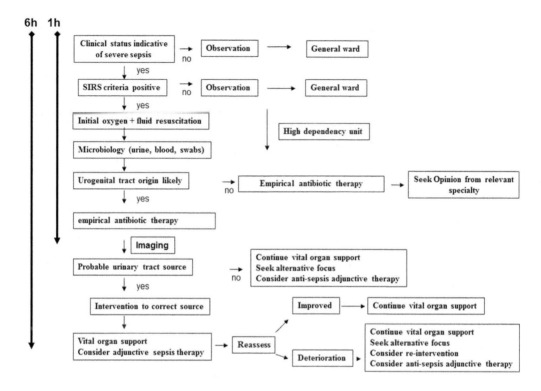

FIGURE 4 Algorithm for management of urosepsis.

(immediate) intervention to maintain adequate tissue perfusion and oxygen delivery by prompt institution of fluid therapy, stabilization of arterial pressure, and provision of sufficient oxygen-transport capacity are highly effective (90) (Table 5).

Additional symptoms pointing to the urogenital tract should be examined: flank pain, costovertebral tenderness, renal colic, pain at micturition, urinary retention, and prostatic or scrotal pain. A digital-rectal examination of the prostate is therefore mandatory to rule out acute prostatitis. Urinary analysis as well as urine and blood cultures must be included as part of the first routine laboratory tests. In the case of urosepsis, the clinical evidence of UTI is based on symptoms, physical examination, sonographic and radiological features, and laboratory data, such as bacteriuria and leukocyturia.

Immediately after microbiological sampling of urine, blood, and suspicious secretion, tissue, and abscess fluids, empirical broad-spectrum antibiotic therapy should be instigated parenterally.

If urosepsis is the putative diagnosis, sonographic examination of the urogenital organs should be followed, including sonographic examination of the prostate to rule out prostatic abscess.

Further radiographic investigations (CT scan; urography) of the urinary tract are generally applied to specify the complicating factor.

If a complicating factor warranting treatment is identified, control and/or removal of the complicating factor should follow immediately. Drainage of any obstruction in the urinary tract and removal of foreign bodies, such as urinary catheters or stones, may themselves cause resolution of symptoms and lead to recovery. These are key components of the strategy. This condition is an absolute emergency. The way by which the complicating factor is controlled should be as least invasive as possible. The definitive elimination of the complicating factor can be performed after days or weeks, until the general condition of the patient has improved.

In parallel with the urological control of the septic focus, further intensive medical treatment encompassing specific sepsis therapy should be instigated.

SPECIAL CLINICAL PICTURES OF SEVERE URO-GENITAL INFECTIONS

Cystitis

Cystitis is frequently limited to the bladder mucosa and hence shows no systemic signs or symptoms. An ascending infection can, however, clinically result. Cystitis in the intensive-care setting is almost exclusively catheter associated and can cause hematuria, ascending infection, and acute prostatitis. Spontaneous elimination is frequently found after removal of the indwelling catheter, but occurs less frequently in elderly patients (91).

Pyelonephritis

The pathophysiological early alterations in pyelonephritis have recently been addressed in a uropathogenic *E. coli*-induced pyelonephritis animal model, where it was shown that epithelial signaling produced an increase in cellular oxygen consumption and affected microvascular flow by clotting, causing localized ischemia (92). The subsequent ischemic damage led to actin re-arrangement and epithelial sloughing, leading to paracellular bacterial movement. A denuded tubular-basement membrane was able to hinder immediate dissemination of bacteria, giving the host time to isolate the infection by clotting. Interestingly, suppression of clotting by heparin treatment caused fatal urosepsis (92). Clinically, these findings may be relevant in antibiotic delivery in pyelonephritis patients and to the use of anticoagulants in sepsis and should therefore be followed up in clinical studies.

Clinical symptoms of pyelonephritis are unilateral or bilateral flank pain, and systemic symptoms such as fever (>38°C), chills, or malaise. Focal nephritis is limited to one or more renal lobules, comparable to lobular pneumonia. Ultrasonographic findings are of a circumscribed lesion with interrupted echoes, which break through the normal cortex-medulla organization. The CT scan shows typical wedge-shaped, poorly limited areas of diminished sonographic density. As differential diagnoses, renal abscess, tumor, and renal infarction must be taken into account. Emphysematous pyelonephritis characteristically shows gas formation in the renal parenchyma and perirenal space. Diabetes mellitus or obstructive renal disease are predisposing factors. The most frequently isolated organisms are *E. coli*, *Klebsiella pneumoniae*, and *Enterobacter cloacae*. Fermentation of glucose in Enterobacteriaceae occurs via two different metabolic pathways: mixed-acid fermentation and the butylene-glycol pathway. Organisms of the *Klebsiella-Enterobacter-Hafnia-Serratia* group, and to a lesser extent *E. coli*, use the butylene-glycol pathway and produce copious amounts of CO_2, which appears clinically as gas formation (93). Aggravated by diminished tissue perfusion, the contralateral side is often affected as well.

Renal and Perirenal Abscess

Clinical symptoms are rigors, fever, back or abdominal pain, flank tenderness, mass lesion and redness of the flank, and protection of upper lumbar and paraspinal muscles. Respiratory insufficiency, hemodynamic instability, or reflectory paralytic ileus can occur. Frequent signs of renal-abscess formation are fever and increasing inflammatory parameters such as leukocytosis for more than 72 hours, despite antibiotic therapy. In this case, further imaging has to be done for diagnosis of renal abscess. Urinary culture may be negative in 14% to 20% of cases (94). Frequently isolated organisms are *E. coli, K. pneumoniae, Proteus* species, and *S. aureus* from hematogenous spread. The fascial limitations are open distally and the perirenal fat is in close contact with the pelvic fat tissue. A perinephritic abscess may therefore point to groin or perivesical tissue, or to the contralateral side, thus penetrating the peritoneum. Inflammation of flank, thigh, back, buttocks, and lower abdomen may occur. Blood cultures are positive in 10% to 40% of cases, and urinary cultures are positive in 50% to 80% (95). Prompt surgical treatment must be performed, such as drainage or, ultimately, nephrectomy.

Acute Prostatitis and Prostatic Abscess

Acute prostatitis and prostatic abscess are bacterial infections of the prostate gland. The bacterial spectrum consists of 53% to 80% *E. coli* and other enterobacteria, 19% Gram-positive bacteria, and 17% anaerobic bacteria (96). Acute bacterial prostatitis can ensue after transrectal prostate biopsy for diagnosis of prostate cancer. Recently, various studies have reported an apparent increase in the incidence of infective complications after transrectal prostate biopsies, reaching up to 5% of men suffering from symptomatic UTI and resulting in bacterial prostatitis in about 3% of patients undergoing prostate biopsies (67, 97–100). The most important risk factor seems to be the increase of fecal fluoroquinolone-resistant *E. coli* (100).

Symptoms are high fever, rigors, dysuria, urinary retention, and perineal pain. Rectal palpation reveals an enlarged, tender pros-

tate. Prostate massage is contraindicated. In acute prostatitis the pathogens are usually detected in urine. However, the urine may be sterile in prostatic-abscess formation. Therapy consists of a combination of antibiotic therapy with broad-spectrum antibiotics, as well as the insertion of a suprapubic catheter, if there is urinary retention. If the patient had a history of fluoroquinolone treatment in the past months, fluoroquinolones should not be administered empirically. In the case of a prostatic abscess, urologic drainage is necessary (96).

Epididymitis/Orchitis

Epididymitis is usually an ascending infection and can also involve the testis as well. Possible causes are subvesical obstruction, transurethral resection of the prostate, or an indwelling, transurethral urinary catheter, in which case the pathogens are identical with the pathogens in the urine. Of note, epididymitis is frequently involved in urogenital tuberculosis. Orchitis with the formation of a sterile hydrocele can appear in the course of polyserositis or heart insufficiency and may point to a generalized systemic disease.

Cavernitis

Cavernitis of the penis is a rare phlegmonous infection of the cavernous bodies. Possible causes are indwelling transurethral urinary catheters, penile operations, auto-injection for erectile dysfunction, pelvic operations, or trauma. Pathogens may represent skin flora or uropathogens. Treatment consists of suprapubic catheterization, broad-spectrum antibiotic therapy, and, if needed, operative debridement.

Fournier's gangrene

Fournier's gangrene is a necrotizing fasciitis of dartos and Colles' fascias. It is mainly seen in men in the fourth to seventh decade but also occurs in women or the newborn. Causes are operations or trauma in the genital or perineal region, including microlesions, or infectious processes from the rectal or urethral areas. Important predisposing factors are diabetes mellitus, liver insufficiency, chronic alcoholism, hematologic diseases, or malnutrition. Patient-related predictors of mortality are increasing age, increased comorbidity, preexisting conditions, such as congestive heart failure, renal failure, and coagulopathy, and hospital admission via transfer (101). Fatality rates were 7.5% in one large North American study (102).

The use of a Fournier's gangrene-severity index has been shown to correlate well with the course of the disease. A Fournier's gangrene-severity index-threshold value of nine was significantly associated with outcome (103): A score greater than nine showed a 75% probability of death, while a score of nine or less was associated with a 78% probability of survival (103).

The infectious process follows anatomically preformed spaces. The superficial perineal fascia is fixed dorsally at the transverse deep-perineal muscle and laterally at the iliac bone and merges ventrally in the superficial abdominal fascia. Hence, a ventrally open and craniodorsally and laterally closed space is formed (Colles' space) that facilitates the spread of infection. In contrast to gas gangrene, the fascial borders are respected in Fournier's gangrene. A mixed bacterial flora is seen, consisting of Gram-positive cocci, enterobacteria, and anaerobic bacteria. The released toxins facilitate platelet aggregation and entrapment of complement, which, in conjunction with the release of heparinase by anaerobic bacteria, leads to small-vessel thrombosis and tissue necrosis. The destruction of tissue enhances the potential of acute renal failure. Fournier's gangrene is a rapidly progressing infection leading to septic shock, if not treated in time.

Therapy consists of immediate, operative debridement, followed by subsequent operations, until the infectious process has been controlled. A suprapubic catheter is advisable and a colostomy needs to be performed in some cases in which continuous fecal contamination of the wound is inevitable. A combination of antibiotic therapy with broad-spectrum β-lactam antibiotics, fluoroquinolones, and clindamycin is recommended.

PREVENTION OF UROSEPSIS

Septic shock is the most frequent cause of death for patients hospitalized for both community- and nosocomial-acquired infection (20% to 40%). Sepsis initiates the cascade that progresses to severe sepsis and then septic shock in a clinical continuum. Urosepsis treatment calls for the early combination of treatment of the cause (obstruction), adequate life-supporting care, and appropriate antibiotic therapy. In such a situation, it is recommended that urologists collaborate early with intensive-care and infectious-disease specialists for the best management of the patient (104).

The most effective methods to prevent nosocomial urosepsis are the same as those used to prevent other nosocomial infections. As most urosepsis cases are due to obstruction of the urinary tract at some level, the development of the full picture of septic shock can frequently be prevented by performing an early de-obstruction procedure. A patient with a so-called "infected hydronephrosis" is an absolute emergency. Before starting the de-obstruction procedure, an empirical antibiotic treatment needs to be administered.

There are no randomized data available. However, in a historical cohort study it was shown that, despite antimicrobial therapy, appropriate urological intervention was very important. Of 49 patients with urosepsis due to pyonephrosis, 22% died despite intensive care, but no patient died if pyonephrosis was treated by nephrectomy or, in a few cases, by nephrostomy drainage before urosepsis developed (105).

CONCLUSION

Sepsis syndrome in urology remains a severe situation with a mortality rate as high as 20% to 40%. A recent campaign, 'Surviving Sepsis Guidelines', aimed at reducing mortality by 25% in the next few years (20, 84). Early recognition of the symptoms may decrease the mortality by timely treatment of urinary tract disorders, e.g., obstruction by urolithiasis. Adequate life-support measures and appropriate antibiotic treatment, including optimized dosing, provide the best conditions for improving patients' survival. The prevention of sepsis syndrome is dependent on good practice to avoid nosocomial infections and using antibiotic prophylaxis and therapy in a prudent and well-accepted manner.

ACKNOWLEDGMENTS

Conflicts of interest: F. Wagenlehner is a consultant at the following companies: Achaogen, Astellas, Astra-Zeneca, Bionorica, Cubist/MSD, Galenus, Leo Pharma, Medpace, MerLion, OM-Pharma/Vifor, Rempex Pharm, Rosen Pharma, Shionogi.

K. Naber is a consultant at the following companies: Basilea, Bayer, Bionorica, Boehringer Ingelheim, Cubist/MSD, Daiichi Sankyo, Galenus, Leo Pharma, Melinta, MerLion, OM-Pharma/Vifor, Paratek, Pierre Fabre, Rempex Pharm, Rosen Pharma, Shionogi, Zambon.

A. Pilatz and W. Weidner declare no conflicts of interest.

CITATION

Wagenlehner FME, Pilatz A, Weidner W, Naber KG. 2015. Urosepsis: Overview of the diagnostic and treatment challenges. Microbiol Spectrum 3(5):UTI-0003-2012.

REFERENCES

1. **Bone RC, Balk RA, Cerra FB, Dellinger RP, Fein AM, Knaus WA, Schein RM, Sibbald WJ.** 1992. Definitions for sepsis and organ failure and guidelines for the use of innovative therapies in sepsis. The ACCP/SCCM Consensus Conference Committee. American College of Chest Physicians/Society of Critical Care Medicine. *Chest* **101:**1644–1655.

2. **Bone RC, Sprung CL, Sibbald WJ.** 1992. Definitions for sepsis and organ failure. *Crit Care Med* **20:**724–726.

3. **Levy MM, Fink MP, Marshall JC, Abraham E, Angus D, Cook D, Cohen J, Opal SM, Vincent JL, Ramsay G.** 2003. 2001 SCCM/ESICM/ACCP/ATS/SIS International Sepsis Definitions Conference. *Crit Care Med* **31:**1250–1256.

4. **Wagenlehner FM, Pilatz A, Naber KG, Weidner W.** 2008. Therapeutic challenges of urosepsis. *Eur J Clin Invest* **38**(Suppl 2):45–49.

5. **Bjerklund Johansen TE, Cek M, Naber K, Stratchounski L, Svendsen MV, Tenke P; PEP and PEAP Study Investigators; European Society of Infections in Urology.** 2007. Prevalence of hospital-acquired urinary tract infections in urology departments. *Eur Urol* **51:**1100–1112.

6. **Martin GS, Mannino DM, Eaton S, Moss M.** 2003. The epidemiology of sepsis in the United States from 1979 through 2000. *N Engl J Med* **348:**1546–1554.

7. **Angus DC, Linde-Zwirble WT, Lidicker J, Clermont G, Carcillo J, Pinsky MR.** 2001. Epidemiology of severe sepsis in the United States: analysis of incidence, outcome, and associated costs of care. *Crit Care Med* **29:**1303–1310.

8. **Rivers E, Nguyen B, Havstad S, Ressler J, Muzzin A, Knoblich B, Peterson E, Tomlanovich M; Early Goal-Directed Therapy Collaborative Group.** 2001. Early goal-directed therapy in the treatment of severe sepsis and septic shock. *N Engl J Med* **345:**1368–1377.

9. **Lagu T, Rothberg MB, Shieh MS, Pekow PS, Steingrub JS, Lindenauer PK.** 2012. Hospitalizations, costs, and outcomes of severe sepsis in the United States 2003 to 2007. *Crit Care Med* **40:**754–761.

10. **Brun-Buisson C, Meshaka P, Pinton P, Vallet B; EPISEPSIS Study Group.** 2004. EPISEPSIS: a reappraisal of the epidemiology and outcome of severe sepsis in French intensive care units. *Intensive Care Med* **30:**580–588.

11. **Hotchkiss RS, Karl IE.** 2003. The pathophysiology and treatment of sepsis. *N Engl J Med* **348:**138–150.

12. **Brunkhorst FM.** 2006. Epidemiology, economy and practice – results of the German study

on prevalence by the competence network sepsis (SepNet) (In German). *Anasthesiol Intensivmed Notfallmed Schmerzther* **41:**43–44.

13. **Rosser CJ, Bare RL, Meredith JW.** 1999. Urinary tract infections in the critically ill patient with a urinary catheter. *Am J Surg* **177:** 287–290.

14. **Hofmann W.** 1990. Urosepsis and uroseptic shock. *Z Urol Nephrol* **83:**317–324.

15. **Wagenlehner FM, Pilatz A, Weidner W.** 2011. Urosepsis–from the view of the urologist. *Int J Antimicrob Agents* **38**(Suppl)**:**51–57.

16. **Astiz ME, Rackow EC.** 1998. Septic shock. *Lancet* **351:**1501–1505.

17. **Dinarello CA.** 1989. The endogenous pyrogens in host-defense interactions. *Hosp Pract (Off Ed)* **24:**111–115, 118, 121 passim.

18. **Matute-Bello G, Liles WC, Radella F II, Steinberg KP, Ruzinski JT, Jonas M, Chi EY, Hudson LD, Martin TR.** 1997. Neutrophil apoptosis in the acute respiratory distress syndrome. *Am J Respir Crit Care Med* **156:**1969–1977.

19. **Van Amersfoort ES, Van Berkel TJ, Kuiper J.** 2003. Receptors, mediators, and mechanisms involved in bacterial sepsis and septic shock. *Clin Microbiol Rev* **16:**379–414.

20. **Dellinger RP, Carlet JM, Masur H, Gerlach H, Calandra T, Cohen J, Gea-Banacloche J, Keh D, Marshall JC, Parker MM, Ramsay G, Zimmerman JL, Vincent JL, Levy MM; Surviving Sepsis Campaign Management Guidelines Committee.** 2004. Surviving Sepsis Campaign guidelines for management of severe sepsis and septic shock. *Crit Care Med* **32:**858–873.

21. **Gogos CA, Drosou E, Bassaris HP, Skoutelis A.** 2000. Pro- versus anti-inflammatory cytokine profile in patients with severe sepsis: a marker for prognosis and future therapeutic options. *J Infect Dis* **181:**176–180.

22. **Russell JA.** 2006. Management of sepsis. *N Engl J Med* **355:**1699–1713.

23. **Weismüller K, Bauer M, Hofer S, Weigand MA.** 2010. [The neuroendocrine axis and the pathophysiology of sepsis]. *Anasthesiol Intensivmed Notfallmed Schmerzther* **45:**574–578; quiz 579.

24. **Spengler RN, Allen RM, Remick DG, Strieter RM, Kunkel SL.** 1990. Stimulation of alpha-adrenergic receptor augments the production of macrophage-derived tumor necrosis factor. *J Immunol* **145:**1430–1434.

25. **John CD, Buckingham JC.** 2003. Cytokines: regulation of the hypothalamo-pituitary-adrenocortical axis. *Curr Opin Pharmacol* **3:**78–84.

26. **Borovikova LV, Ivanova S, Zhang M, Yang H, Botchkina GI, Watkins LR, Wang H, Abumrad N, Eaton JW, Tracey KJ.** 2000. Vagus nerve stimulation attenuates the systemic inflammatory response to endotoxin. *Nature* **405:**458–462.

27. **Huston JM, Gallowitsch-Puerta M, Ochani M, Ochani K, Yuan R, Rosas-Ballina M, Ashok M, Goldstein RS, Chavan S, Pavlov VA, Metz CN, Yang H, Czura CJ, Wang H, Tracey KJ.** 2007. Transcutaneous vagus nerve stimulation reduces serum high mobility group box 1 levels and improves survival in murine sepsis. *Crit Care Med* **35:**2762–2768.

28. **Wagenlehner FM, Weidner W, Naber KG.** 2007. Pharmacokinetic characteristics of antimicrobials and optimal treatment of urosepsis. *Clin Pharmacokinet* **46:**291–305.

29. **Naber KG, Madsen PO.** 1973. Renal function during acute total ureteral occlusion and the role of the lymphatics: an experimental study in dogs. *J Urol* **109:**330–338.

30. **Naber KM, Madsen PO, Bichler KH, Sauerwein D.** 1973. Determination of renal tissue levels of antibiotics. *Infection* **1:**208–213.

31. **Jaenike JR.** 1970. The renal response to ureteral obstruction: a model for the study of factors which influence glomerular filtration pressure. *J Lab Clin Med* **76:**373–382.

32. **Gulmi FA, Falsen FD, Vaughan ED.** 2002. Pathophysiology of urinary tract obstruction, p 411–462. *In* Walsh PC, Retik RA, Vaughan ED, Wein AJ (ed), *Campbell's Urology*, 8th ed, **vol 1.** W B Saunders Co, Philadelphia.

33. **Naber KG, H.** 1973. Renale Pharmakokinetik. *Symposium Pyelonephritis:*271–285.

34. **Bricker NS, Morrin PA, Kime SW Jr.** 1960. The pathologic physiology of chronic Bright's disease. An exposition of the "intact nephron hypothesis". *Am J Med* **28:**77–98.

35. **Paterson DL.** 2003. Restrictive antibiotic policies are appropriate in intensive care units. *Crit Care Med* **31:**S25–28.

36. **Eggimann P, Pittet D.** 2001. Infection control in the ICU. *Chest* **120:**2059–2093.

37. **Kreger BE, Craven DE, McCabe WR.** 1980. Gram-negative bacteremia. IV. Re-evaluation of clinical features and treatment in 612 patients. *Am J Med* **68:**344–355.

38. **Kreger BE, Craven DE, Carling PC, McCabe WR.** 1980. Gram-negative bacteremia. III. Reassessment of etiology, epidemiology and ecology in 612 patients. *Am J Med* **68:**332–343.

39. **Kumar A, Roberts D, Wood KE, Light B, Parrillo JE, Sharma S, Suppes R, Feinstein D, Zanotti S, Taiberg L, Gurka D, Kumar A, Cheang M.** 2006. Duration of hypotension before initiation of effective antimicrobial therapy is the critical determinant of survival

in human septic shock. *Crit Care Med* **34:**1589–1596.

40. **Elhanan G, Sarhat M, Raz R.** 1997. Empiric antibiotic treatment and the misuse of culture results and antibiotic sensitivities in patients with community-acquired bacteraemia due to urinary tract infection. *J Infect* **35:**283–288.

41. **Kollef MH, Ward S.** 1998. The influence of mini-BAL cultures on patient outcomes: implications for the antibiotic management of ventilator-associated pneumonia. *Chest* **113:**412–420.

42. **Luna CM, Vujacich P, Niederman MS, Vay C, Gherardi C, Matera J, Jolly EC.** 1997. Impact of BAL data on the therapy and outcome of ventilator-associated pneumonia. *Chest* **111:**676–685.

43. **Singh N, Yu VL.** 2000. Rational empiric antibiotic prescription in the ICU. *Chest* **117:**1496–1499.

44. **Mulvey MA, Schilling JD, Hultgren SJ.** 2001. Establishment of a persistent *Escherichia coli* reservoir during the acute phase of a bladder infection. *Infect Immun* **69:**4572–4579.

45. **Ivanyi B, Rumpelt HJ, Thoenes W.** 1988. Acute human pyelonephritis: leukocytic infiltration of tubules and localization of bacteria. *Virchows Arch A Pathol Anat Histopathol* **414:**29–37.

46. **Deguchi T, Kuriyama M, Maeda S, Sakai S, Ban Y, Kawada Y, Nishiura T.** 1990. Electron microscopic study of acute retrograde pyelonephritis in mice. *Urology* **35:**423–427.

47. **Chippendale GR, Warren JW, Trifillis AL, Mobley HL.** 1994. Internalization of *Proteus mirabilis* by human renal epithelial cells. *Infect Immun* **62:**3115–3121.

48. **Roberts JA, Lipman J.** 2006. Antibacterial dosing in intensive care: pharmacokinetics, degree of disease and pharmacodynamics of sepsis. *Clin Pharmacokinet* **45:**755–773.

49. **Pea F, Viale P, Furlanut M.** 2005. Antimicrobial therapy in critically ill patients: a review of pathophysiological conditions responsible for altered disposition and pharmacokinetic variability. *Clin Pharmacokinet* **44:**1009–1034.

50. **Pea F, Furlanut M.** 2001. Pharmacokinetic aspects of treating infections in the intensive care unit: focus on drug interactions. *Clin Pharmacokinet* **40:**833–868.

51. **Pea F, Pavan F, Di Qual E, Brollo L, Nascimben E, Baldassarre M, Furlanut M.** 2003. Urinary pharmacokinetics and theoretical pharmacodynamics of intravenous levofloxacin in intensive care unit patients treated with 500 mg b.i.d. for ventilator-associated pneumonia. *J Chemother* **15:**563–567.

52. **Frimodt-Møller N.** 2002. Correlation between pharmacokinetic/pharmacodynamic parameters and efficacy for antibiotics in the treatment of urinary tract infection. *Int J Antimicrob Agents* **19:**546–553.

53. **Forrest A, Nix DE, Ballow CH, Goss TF, Birmingham MC, Schentag JJ.** 1993. Pharmacodynamics of intravenous ciprofloxacin in seriously ill patients. *Antimicrob Agents Chemother* **37:**1073–1081.

54. **Frimodt-Møller N.** 2002. How predictive is PK/PD for antibacterial agents? *Int J Antimicrob Agents* **19:**333–339.

55. **Hvidberg H, Struve C, Krogfelt KA, Christensen N, Rasmussen SN, Frimodt-Møller N.** 2000. Development of a long-term ascending urinary tract infection mouse model for antibiotic treatment studies. *Antimicrob Agents Chemother* **44:**156–163.

56. **Naber KG.** 2001. Which fluoroquinolones are suitable for the treatment of urinary tract infections? *Int J Antimicrob Agents* **17:**331–341.

57. **Deguchi T, Nakane K, Yasuda M, Shimizu T, Monden K, Arakawa S, Matsumoto T.** 2010. Microbiological outcome of complicated urinary tract infections treated with levofloxacin: a pharmacokinetic/pharmacodynamic analysis. *Int J Antimicrob Agents* **35:**573–577.

58. **Naber KG.** 1997. Antibacterial activity of antibacterial agents in urine: an overview of applied methods, p 74–83. *In* Bergan T (ed), *Urinary Tract Infections*, vol 1. Karger Publishers, Basel, Switzerland.

59. **Wagenlehner FM, Wagenlehner C, Redman R, Weidner W, Naber KG.** 2009. Urinary bactericidal activity of doripenem versus that of levofloxacin in patients with complicated urinary tract infections or pyelonephritis. *Antimicrob Agents Chemother* **53:**1567–1573.

60. **Anderson GG, Martin SM, Hultgren SJ.** 2004. Host subversion by formation of intracellular bacterial communities in the urinary tract. *Microbes Infect* **6:**1094–1101.

61. **Justice SS, Hung C, Theriot JA, Fletcher DA, Anderson GG, Footer MJ, Hultgren SJ.** 2004. Differentiation and developmental pathways of uropathogenic *Escherichia coli* in urinary tract pathogenesis. *Proc Natl Acad Sci U S A* **101:**1333–1338.

62. **Kumon H.** 2000. Management of biofilm infections in the urinary tract. *World J Surg* **24:**1193–1196.

63. **Nickel JC, Olson ME, Costerton JW.** 1991. Rat model of experimental bacterial prostatitis. *Infection* **19**(Suppl 3)**:**S126–130.

64. **Zogaj X, Bokranz W, Nimtz M, Römling U.** 2003. Production of cellulose and curli fimbriae

by members of the family Enterobacteriaceae isolated from the human gastrointestinal tract. *Infect Immun* **71**:4151–4158.

65. **Sabbuba NA, Mahenthiralingam E, Stickler DJ.** 2003. Molecular epidemiology of *Proteus mirabilis* infections of the catheterized urinary tract. *J Clin Microbiol* **41**:4961–4965.

66. **Jansen AM, Lockatell V, Johnson DE, Mobley HL.** 2004. Mannose-resistant *Proteus*-like fimbriae are produced by most *Proteus mirabilis* strains infecting the urinary tract, dictate the *in vivo* localization of bacteria, and contribute to biofilm formation. *Infect Immun* **72**:7294–7305.

67. **Wagenlehner FM, van Oostrum E, Tenke P, Tandogdu Z, Çek M, Grabe M, Wuult B, Pickard R, Naber KG, Pilatz A, Weidner W, Bjerklund-Johansen TE; GPIU.** 2013. Infective complications after prostate biopsy: outcome of the Global Prevalence Study of Infections in Urology (GPIU) 2010 and 2011, a prospective multinational multicentre prostate biopsy study. *Eur Urol* **63**:521–527.

68. **Debbia EA, Dolcino M, Marchese A, Piazzi A, Berio A.** 2004. Enhanced biofilm-production in pathogens isolated from patients with rare metabolic disorders. *New Microbiol* **27**:361–367.

69. **Seno Y, Kariyama R, Mitsuhata R, Monden K, Kumon H.** 2005. Clinical implications of biofilm formation by *Enterococcus faecalis* in the urinary tract. *Acta Med Okayama* **59**:79–87.

70. **Bokranz W, Wang X, Tschäpe H, Römling U.** 2005. Expression of cellulose and curli fimbriae by *Escherichia coli* isolated from the gastrointestinal tract. *J Med Microbiol* **54**:1171–1182.

71. **Rijavec M, Müller-Premru M, Zakotnik B, Zqur-Bertok D.** 2008. Virulence factors and biofilm production among *Escherichia coli* strains causing bacteraemia of urinary tract origin. *J Med Microbiol* **57**:1329–1334.

72. **Goto T, Nakame Y, Nishida M, Ohi Y.** 1999. *In vitro* bactericidal activities of beta-lactamases, amikacin, and fluoroquinolones against *Pseudomonas aeruginosa* biofilm in artificial urine. *Urology* **53**:1058–1062.

73. **Goto T, Nakame Y, Nishida M, Ohi Y.** 1999. Bacterial biofilms and catheters in experimental urinary tract infection. *Int J Antimicrob Agents* **11**:227–231; discussion 237–239.

74. **Foz A.** 1976. Sepsis of urological origin: microbiological aspects. *Antibiot Chemother* **21**:69–72.

75. **Rosenthal EJ.** 2002. Epidemiology of septicaemia pathogens. *Dtsch Med Wochenschr* **127**:2435–2440.

76. **Rosenthal EJ.** 1986. Septicemia causative organisms 1983–1985. The results of a multicenter study. *Dtsch Med Wochenschr* **111**:1874–1880.

77. **Rosenthal EJ.** 1993. The epidemiology of septicemia causative agents. A blood culture study of the Paul Ehrlich Society for Chemotherapy e. V. *Dtsch Med Wochenschr* **118**:1269–1275.

78. **Rosenthal EJ.** 1995. Antibiotika-Empfindlichkeit von Septikämieerregern 1991 bis 1992. Blutkulturstudie der Paul-Ehrlich-Gesellschaft für Chemotherapie. *Chemotherapie Journal* **4**:67–71.

79. **Kresken M, Hafner D, Schmitz FJ, Wichelhaus TA; für die Studiengruppe.** 2009. *Resistenzsituation bei klinisch wichtigen Infektionserregern gegenüber Antibiotika in Deutschland und immitteleuropäischen Raum. Bericht über die Ergebnisse einer multizentrischen Studie der ArbeitsgemeinschaftEmpfi ndlichkeit-sprüfungen & Resistenz der Paul-Ehrlich-Gesellschaft für Chemotherapie e.V. aus dem Jahre 2007.* Antiinfectives Intelligence, Rheinbach, Germany.

80. **Group IE-NS.** 2012. EARS-Net Report 2011, on *Health Services Executive (HSE) –Health Protection Surveillance Centre (HPSC).* Accessed 17 May 2012.

81. **Luchi M, Morrison DC, Opal S, Yoneda K, Slotman G, Chambers H, Wiesenfeld H, Lemke J, Ryan JL, Horn D.** 2000. A comparative trial of imipenem versus ceftazidime in the release of endotoxin and cytokine generation in patients with gram-negative urosepsis. Urosepsis Study Group. *J Endotoxin Res* **6**:25–31.

82. **Dellinger RP, Carlet JM, Gerlach H, Ramsey G, Levy M.** 2004. The surviving sepsis guidelines: not another "groundhog day". *Crit Care Med* **32**:1601–1602.

83. **Dellinger RP, Carlet JM, Masur H, Gerlach H, Calandra T, Cohen J, Gea-Banacloche J, Keh D, Marshall JC, Parker MM, Ramsay G, Zimmerman JL, Vincent JL, Levy MM.** 2004. Surviving Sepsis Campaign guidelines for management of severe sepsis and septic shock. *Intensive Care Med* **30**:536–555.

84. **Dellinger RP, Levy MM, Carlet JM, Bion J, Parker MM, Jaeschke R, Reinhart K, Angus DC, Brun-Buisson C, Beale R, Calandra T, Dhainaut JF, Gerlach H, Harvey M, Marini JJ, Marshall J, Ranieri M, Ramsay G, Sevransky J, Thompson BT, Townsend S, Vender JS, Zimmerman JL, Vincent JL.** 2008. Surviving Sepsis Campaign: international guidelines for management of severe sepsis and septic shock: 2008. *Crit Care Med* **36**:296–327.

85. Dellinger RP, Levy MM, Rhodes A, Annane D, Gerlach H, Opal SM, Sevransky JE, Sprung CL, Douglas IS, Jaeschke R, Osborn TM, Nunnally ME, Townsend SR, Reinhart K, Kleinpell RM, Angus DC, Deutschman CS, Machado FR, Rubenfeld GD, Webb SA, Beale RJ, Vincent JL, Moreno R, Surviving Sepsis Campaign Guidelines Committee including The Pediatric Subgroup. 2013. Surviving Sepsis Campaign: international guidelines for management of severe sepsis and septic shock, 2012. *Intensive Care Med* **39**:165–228.

86. Dellinger RP, Levy MM, Rhodes A, Annane D, Gerlach H, Opal SM, Sevransky JE, Sprung CL, Douglas IS, Jaeschke R, Osborn TM, Nunnally ME, Townsend SR, Reinhart K, Kleinpell RM, Angus DC, Deutschman CS, Machado FR, Rubenfeld GD, Webb SA, Beale RJ, Vincent JL, Moreno R. 2013. Surviving sepsis campaign: international guidelines for management of severe sepsis and septic shock: 2012. *Crit Care Med* **41**:580–637.

87. ProCESS Investigators, Yealy DM, Kellum JA, Huang DT, Barnato AE, Weissfeld LA, Pike F, Terndrup T, Wang HE, Hou PC, LoVecchio F, Filbin MR, Shapiro NI, Angus DC. 2014. A randomized trial of protocol-based care for early septic shock. *N Engl J Med* **370**:1683–1693.

88. AriseInvestigators, ANZICS Clinical Trials Group ACT, Peake SL, Delaney A, Bailey M, Bellomo R, Cameron PA, Cooper DJ, Higgins AM, Holdgate A, Howe BD, Webb SA, Williams P. 2014. Goal-directed resuscitation for patients with early septic shock. *N Engl J Med* **371**:1496–1506.

89. Grabe M, Bjerklund-Johansen TE, Botto H, Wullt B, Çek M, Naber KG, Pickard RS, Tenke P, Wagenlehner FM. 2012. Guidelines on urological infections. *European Association of Urology(EAU) Guidelines*. Arnhem, The Netherlands.

90. Glück T, Opal SM. 2004. Advances in sepsis therapy. *Drugs* **64**:837–859.

91. Grabe M, Bishop MC, Bjerklund-Johansen TE, Botto H, Çek M, Lobel B, Naber KG, Palou J, Tenke P, Wagenlehner F. 2009. Guidelines on urological infections. *European Association of Urology (EAU) Guidelines*. Arnhem, The Netherlands.

92. Melican K, Boekel J, Månsson LE, Sandoval RM, Tanner GA, Källskog O, Palm F, Molitoris BA, Richter-Dahlfors A. 2008. Bacterial infection-mediated mucosal signalling induces local renal ischaemia as a defence against sepsis. *Cell Microbiol* **10**:1987–1998.

93. Koneman EW, Allen SD, Janda WM, Schreckenberger PC, Winn WC. 1997. Enterobacteriaceae: Carbohydrate utilization, p 172–176. *In* Koneman EW, Allen SD, Janda WM, Schreckenberger PC, Winn WC Jr, (ed), *Color Atlas and Textbook of Diagnostic Microbiology*, 5th ed. Lippincott, Philadelphia.

94. Elkin M. 1975. Renal cystic disease–an overview. *Semin Roentgenol* **10**:99–102.

95. Sheinfeld J, Erturk E, Spataro RF, Cockett AT. 1987. Perinephric abscess: current concepts. *J Urol* **137**:191–194.

96. Naber KG, Wagenlehner FM, Weidner W. 2008. Acute bacterial prostatitis, p 17–30. *In* Shoskes DA (ed), *Current Clinical Urology Series, Chronic Prostatitis/Chronic Pelvic Pain Syndrome*. Humana Press, Totowa, NJ.

97. Loeb S, Carter HB, Berndt SI, Ricker W, Schaeffer EM. 2011. Complications after prostate biopsy: data from SEER-Medicare. *J Urol* **186**:1830–1834.

98. Loeb S, van den Heuvel S, Zhu X, Bangma CH, Schröder FH, Roobol MJ. 2012. Infectious complications and hospital admissions after prostate biopsy in a European randomized trial. *Eur Urol* **61**:1110–1114.

99. Nam RK, Saskin R, Lee Y, Liu Y, Law C, Klotz LH, Loblaw DA, Trachtenberg J, Stanimirovic A, Simor AE, Seth A, Urbach DR, Narod SA. 2010. Increasing hospital admission rates for urological complications after transrectal ultrasound guided prostate biopsy. *J Urol* **183**:963–968.

100. Steensels D, Slabbaert K, De Wever L, Vermeersch P, Van Poppel H, Verhaegen J. 2011. Fluoroquinolone-resistant *E. coli* in intestinal flora of patients undergoing transrectal ultrasound-guided prostate biopsy-should we reassess our practices for antibiotic prophylaxis? *Clin Microbiol Infect* **18**:575–581.

101. Sorensen MD, Krieger JN, Rivara FP, Klein MB, Wessells H. 2009. Fournier's gangrene: management and mortality predictors in a population based study. *J Urol* **182**:2742–2747.

102. Sorensen MD, Krieger JN, Rivara FP, Broghammer JA, Klein MB, Mack CD, Wessells H. 2009. Fournier's gangrene: population based epidemiology and outcomes. *J Urol* **181**:2120–2126.

103. Laor E, Palmer LS, Tolia BM, Reid RE, Winter HI. 1995. Outcome prediction in patients with Fournier's gangrene. *J Urol* **154**:89–92.

104. Persky L, Liesen D, Yangco B. 1992. Reduced urosepsis in a veterans' hospital. *Urology* **39**:443–445.

105. Schilling A, Marx FJ, Hofstetter A, Jesch F. 1977. Septic shock in the urologic patient. IV. monitoring and therapy (author's transl). *Urologe A* **16**:351–355.

ORIGINS AND VIRULENCE MECHANISMS OF UROPATHOGENIC BACTERIA

Reservoirs of Extraintestinal Pathogenic *Escherichia coli*

9

AMEE R. MANGES[1] and JAMES R. JOHNSON[2,3]

INTRODUCTION

The existence of important infection-causing ExPEC lineages (i.e., genetically closely related *E. coli* clones or clonal groups) has been recognized for the past 40 years. However, identification and detailed characterization of specific lineages has only recently been advanced, owing to the development of new methods for bacterial genotyping. Since ExPEC can colonize and persist in the human intestinal tract without detriment to the host, the operational definition of ExPEC is critical for defining the chain of ExPEC transmission from non-human reservoir to human intestinal colonization to active infection.

ExPEC clonal groups have been defined differently over time largely in relation to the available phenotyping and genotyping technologies. Early studies identified 10 to 15 O-antigen–based serogroups (of the approximately 180 that occur in *E. coli*) as being associated with human extraintestinal infections (1, 2). The addition of K and H antigen typing provided finer resolution. DNA-based genotyping methods, such as multi-locus sequence typing (MLST) have further advanced our understanding of these lineages. For the purposes of this review, ExPEC is defined as isolates (i) recovered from

[1]School of Population and Public Health, University of British Columbia, Vancouver, BC V6T 1Z3, Canada; [2]Infectious Diseases Section, Veterans Affairs Medical Center, Minneapolis, MN 55417; [3]Department of Medicine, University of Minnesota, Minneapolis, MN 55455.
Urinary Tract Infections: Molecular Pathogenesis and Clinical Management, 2nd Edition
Edited by Matthew A. Mulvey, David J. Klumpp, and Ann E. Stapleton
© 2017 American Society for Microbiology, Washington, DC
doi:10.1128/microbiolspec.UTI-0006-2012

extraintestinal infections, (ii) possessing known ExPEC virulence factors, (iii) exhibiting MLST sequence types associated with extraintestinal infections, and/or (iv) exhibiting classic ExPEC-associated serogroups/serotypes in conjunction with typical ExPEC virulence factor profiles or phylogenetic group membership (e.g., groups B2 and D).

Certain major *E. coli* lineages have been implicated repeatedly in epidemiologic studies as the cause of a large proportion of human extraintestinal *E. coli* infections (Fig. 1). Many such groups are multidrug-resistant and likely contribute significantly to the ongoing population-level increase in antimicrobial-resistant human *E. coli* infections. Here, for consistency, we identify specific ExPEC lineages by O:K:H serotype (if known), phylogenetic group of origin (A, B1, B2, or D), and sequence type (ST), as determined by MLST; for example, O25:H4-B2-ST131. Several important ExPEC lineages are highlighted based

on their large contribution to overall disease burden, including: O25b:H4-B2-ST131, O25a:H4(a)-D-ST648, O11/O17/O77:K52:H18-D-ST69, O15:K52:H1-D-ST393, (serotype: various)-D-ST117, O1/O2/O18:K1:H7-B2-ST95, O6:K2:H1-B2-ST73, (serotype: various)-A-ST10, (serotype: various)-D-ST405, and O75:K+:H5-B2-ST14. Since many of these groups have been identified in surveillance studies focused on extended spectrum β-lactamase (ESBL)-producing *E. coli*, ESBL-producing lineages tend to be overrepresented.

IMPORTANT ExPEC LINEAGES

E. coli O25:H4-B2-ST131, reported first in 2008, is an important new globally emerging pathogen (3, 4). This clonal group is of major public health concern because of its typically extensive antimicrobial resistance profile, which often includes ESBL production, specifically of CTX-M-15, plus fluoroquinolone resistance, and its widespread distribution. *E. coli* O25:H4-B2-ST131 has been identified primarily from human infections and has been associated with travel, which may explain its international spread over the past decade. In surveys of human *E. coli* infections this group accounts for a large fraction of cases overall (5), and up to 88% of antimicrobial-resistant infections, depending on the specific resistance phenotype (6).

E. coli O11/O17/O77:K52:H18-D-ST69 (also termed CgA, for "clonal group A") was identified initially in an apparent outbreak of extraintestinal infections in Berkeley, California, during which it accounted for 11% of all urinary tract infections (UTIs) and 52% of antimicrobial-resistant UTIs (7). It has since been identified around the world (8). This group often exhibits multidrug resistance and has been responsible for urinary tract and more severe human extraintestinal infections (8–11). Several studies found this group to cause approximately 10% to 20% of all human *E. coli* infections (8, 12, 13). A recent global survey suggested that the

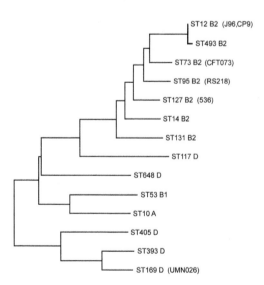

FIGURE 1 The tree, which is drawn to scale, was inferred using the Neighbor-Joining method, based on concatenated DNA sequences from the seven MLST loci. *E. fergusonii* was used to root the tree, but for simplicity has been excluded. All positions containing gaps and missing data were eliminated. There were a total of 3423 positions in the final dataset.

O11/O17/O77:K52:H18-D-ST69 group is concentrated in North America and may have emerged in the 1990s (13).

E. coli O15:K52:H1-D-ST393 was first recognized during an outbreak of extraintestinal infections from 1986 to 1987 in London, UK (14), and has subsequently been identified across Europe (15, 16) and globally (17). In the initial epidemic, O15:K52:H1 caused 26% of all extraintestinal infections during a 1-year interval (5, 14). This group also is typically multidrug-resistant and has caused community-acquired extraintestinal infections of all types (14, 16, 18, 19). It is closely related to O11/O17/O77:K52:H18-D-ST69; both possess similar antimicrobial resistance and virulence factor patterns and appear to share a common ancestor (20).

E. coli O6:H1-B2-ST73 was recently reported as a leading cause (17% overall) of human extraintestinal infections in the UK (21). This clonal group, like many others, is associated occasionally with ESBL production (21–23). It may represent a long-standing, human-adapted ExPEC group, since it has caused UTIs in women in many geographic areas and over many years (24).

E. coli (serotype: various)-D-ST405 is another globally disseminated, extensively antimicrobial-resistant group (3, 25–27). ST405 may be the next most important contributor after ST131 to the global dissemination of CTX-M-15 (3). Furthermore, an ST405 *E. coli* isolate was recently encountered that produced multiple ESBLs, including QepA1, CTX-M-15, and RmtB (28). ST405 also has been associated with New Delhi metallo (NDM) β-lactamases (29). ST405 has been associated with person-to-person transmission (26), but evidence for non-human reservoirs is still lacking (24).

E. coli (serotype: various)-A-ST10 or *E. coli* ST10 clonal complex (i.e., ST10 and closely related STs), although typically encountered as an antimicrobial-susceptible, low-virulence human intestinal colonizer, also has been associated with human infections and ESBL production. This clonal group,

members of which have caused hospital and community-acquired infections, was found to account for 7% of all human clinical *E. coli* isolates in Calgary, Alberta, Canada (30).

OTHER NOTABLE ExPEC CLONAL GROUPS

E. coli O1/O2/O18:K1:H7-B2-ST95 is both a recognized avian pathogenic *Escherichia coli* (APEC) clonal group (31) and a prominent human urinary tract and meningitis pathogen, accounting for 6% of human extraintestinal clinical isolates in one study (21). *E. coli* (serotype: various)-D-ST117 is another recognized APEC lineage that has also been identified as a cause of human UTI (32, 33). *E. coli* O75:K+:H5-B2-ST14, in the past an important ampicillin-resistant clonal group, has recently been associated with fluoroquinolone resistance in Australia (34). *E. coli* (serotype: various)-D-ST648 has been associated with human disease in China (35), Canada, and the Netherlands (36), and with NDM β-lactamases (37). Additionally, *E. coli* ST167, ST410, and ST38 have also been identified in epidemiologic studies of human extraintestinal infections; however, less information is available concerning the dissemination and reservoirs for these *E. coli* clonal groups (23, 36, 38–41). Given the wide distribution of several of these *E. coli* clonal groups, efforts to identify their reservoirs are needed.

THE HUMAN RESERVOIR OF ExPEC

An appropriate starting point for reviewing the sources of human ExPEC is the human intestinal tract. The human intestinal tract has long been recognized as a reservoir of *E. coli* (42). ExPEC also are members of the intestinal microbiota in a large fraction of healthy individuals (42). However, in contrast to non-ExPEC intestinal colonizers, once ExPEC gain access to other body sites they can persist and cause disease. ExPEC that cause UTIs typically follow a fecal-to-perineal/vaginal-to-urethral transmission pathway within the individual host (43, 44). ExPEC also sometimes cause sepsis by entering the bloodstream directly, via translocation across the intestinal barrier (45). Importantly, the intestinal microbiota also can serve as a reservoir of antimicrobial resistance determinants for ExPEC (46, 47). Exposure to an antimicrobial agent can temporarily increase the prevalence of antimicrobial-resistant *E. coli* in the intestine, although this increase may be short-lived for most antimicrobials (48).

Several studies have documented direct person-to-person transmission of ExPEC strains between household members and sexual contacts, in many instances with linkage to subsequent extraintestinal infections (49–53). The exchange of ExPEC between sexual partners in particular could be one explanation for the phenomenon of recurrent UTIs in some women, since the male partner (who is not treated) may remain a reservoir for re-infection of the woman. Although strain sharing among household and sexual contacts could increase the risk of future infection in a given individual, spread of ExPEC or antimicrobial-resistant *E. coli* from non-human or food reservoirs through the entire population represents a larger public health threat. Understanding how these organisms reach the human intestinal tract could lead to strategies to minimize transmission, thereby reducing the pool of antimicrobial-resistant and pathogenic *E. coli* in the gut available to cause disease.

ENVIRONMENTAL RESERVOIRS OF ExPEC

Surface water, rainwater, wild animals, sewage, and wastewater effluents have all been investigated as possible environmental sources of ExPEC. Sewage may contribute significantly to the environmental dissemination or circulation of ExPEC due to the presence of highly concentrated human fecal waste containing ExPEC. To study environmental reservoirs is challenging because these sources tend to contain *E. coli* from multiple sources (e.g., sewage may contain human, animal, and industrial waste). Microbial source tracking is complicated by the difficulty of assigning the *E. coli* recovered in any given environment to a particular source. Furthermore, the methods used to discriminate and describe ExPEC and other types of *E. coli* can be highly variable. In this section, water and other potential environmental sources of ExPEC are reviewed.

Survival of ExPEC during sewage treatment, leading to possible environmental contamination, has been investigated in several studies, which identified ExPEC-associated phylogenetic groups, virulence genes, and lineages among *E. coli* from sewage samples. For example, in a study of four sewage treatment plants 60% of *E. coli* sewage isolates belonged to ExPEC phylogenetic groups B2 and D and a large fraction possessed *hly*A (74%) and *iro*N (82%), while smaller proportions possessed P fimbrial genes *pap*AH, *pap*EF, *pap*C, and the cytotoxic necrotizing factor 1 (*cnf*1) gene (54). Another study found a high prevalence of group D *E. coli* (37%) and quinolone resistance (56%) among *E. coli* sewage isolates (55). Both studies identified putative ExPEC isolates in the final effluent, suggesting that treated water released into the environment may contain viable ExPEC (54, 55).

Notably, two important ExPEC lineages were identified in sewage sources, including *E. coli* O25:H4-B2-ST131 (treated wastewater and river water) (56, 57) and *E. coli* O11/O17/O77:K52:H18-D-ST69 (sewage effluent) (58). It is clear that, despite sewage treatment, ExPEC persist throughout the treatment process, with their presence risking contamination of surface water (54). Investigation of the movement of ExPEC in treated sewage and other wastewater sources is needed, since these sources may participate in the environmental cycling of ExPEC clonal groups that cause disease in humans and animals.

Surveillance studies have consistently identified antimicrobial-resistant and pathogenic *E. coli* in natural bodies of water and recreational waterways (59–61). In one study, 40% of *E. coli* isolates from coastal marine sediments represented phylogenetic groups B2 and D, and most of these contained at least 1 of 11 ExPEC virulence genes (61). Hamelin et al. used a DNA microarray containing probes to 348 antimicrobial resistance genes and virulence genes from all *E. coli* pathotypes to characterize *E. coli* from recreational and river waterways. Isolates were defined as uropathogenic if positive for at least one of the following: *pap* genes, *hly*A, S fimbriae-encoding genes, *chu*A, *fep*C, *cnf*1, *irp*1, *irp*2, *fyu*A, *iro*N, or *usp* (60, 62). From Lake Ontario recreational water, 73% of *E. coli* isolates were uropathogenic, with most belonging to phylotypes B2 and D, whereas from six different river, estuary, and offshore lake locations, 26% of *E. coli* were uropathogenic (60, 62). Similarly, Mühldorfer et al. found that 41% of *E. coli* isolates from surface water samples from the Elbe River in Germany contained ExPEC-associated genes (63). Finally, rainwater contamination with ExPEC, as defined based on ExPEC virulence genes, was documented for 68% of rain barrels in Australia, which was concerning since many of the barrels were used as a potable water source (64).

Wild animals, particularly wild birds, also have been investigated as an environmental source of antimicrobial-resistant and pathogenic *E. coli* (65–69). A study of ESBL-positive *E. coli* from yellow-legged gulls in France identified CTX-M-1-positive *E. coli* representing the ST23 complex and the ST533 clonal group, which previous studies associated with UTIs (65, 70–73). Additionally, *E. coli* ST648 was identified in wild birds in Germany (74). A review of ESBL-producing *E. coli* recovered from wildlife sources documented the diversity of *E. coli* colonizing different animal hosts, but highlighted the occurrence primarily in wild birds of ExPEC-associated STs, including ST69, ST131, ST405, ST10, and ST648 (75).

Thus, compelling evidence exists that ExPEC occurs in sewage and other environmental sources, including wild animals and water. However, to date such ExPEC have not been tested in established animal models of human extraintestinal disease for their potential human extraintestinal pathogenicity. This step would provide important confirmation that wild animals, sewage, and water could act as reservoirs for ExPEC capable of causing infections in humans.

COMPANION ANIMAL RESERVOIRS OF ExPEC

Companion animals, which develop similar types of extraintestinal infections as their human guardians, have also been identified as a reservoir of human ExPEC (76–78). Surveys of the *E. coli* causing infections in pets, and surveillance studies at veterinary hospitals, have documented the circulation of genetically related and frequently antimicrobial-resistant ExPEC in dogs, cats, and other animals that live in close contact with humans (79, 80).

Intestinal colonization by ExPEC has been demonstrated in companion animals. A study of *E. coli* from canine fecal samples identified approximately 50% of isolates as ExPEC by virulence factor genotyping and direct comparison with reference human ExPEC isolates (81). Environmental contamination with feces, and consequent transmission of fecal-source *E. coli* between dogs, in particular, likely occurs frequently. Multidrug-resistant *E. coli* isolates from rectal swabs from a population of dogs were similar (according to phylogenetic group and pulsed field gel electrophoresis [PFGE]) to the *E. coli* causing extraintestinal infections in animals (82). The ExPEC isolates were closely related, persistent, and present in multiple species, including cats, dogs, and horses (82).

Closely related or indistinguishable ExPEC strains have been recovered from humans and their companion animals, suggesting transmission in either direction. For example, in one study, an *E. coli* strain (O1:K1:H7-B2-ST95) was shared extensively (present in 45% of samples) among household members and the family dog, and caused a UTI in the mother (83). Similarly, an *E. coli* O6:K2:H1-B2-ST73 strain was shared between family members and the family dog, which developed UTI due to this strain (51, 84). The human ExPEC lineages O6:K15:H31-B2-ST127 and O4:K54:H5-B2-ST493 also have been identified in dogs, cats, and humans (85). These results were confirmed using a set of *E. coli* recovered from dogs alone, where the canine isolates exhibited ExPEC virulence factors related to those of human reference ExPEC isolates (85, 86). Additionally, *E. coli* O75:K+:H5-B2-ST14 was recently identified in pet dogs (34).

ESBL-producing *E. coli*, some from ExPEC-associated clonal groups, also have been recovered from companion animals in diverse locales. ESBL (specifically, CTX-M)-positive *E. coli* recovered from stray dogs in Korea included members of the ST10, ST38, ST93, and ST95 clonal groups, whereas CMY-2-positive *E. coli* included ST405 and ST648 (87). CTX-M-positive *E. coli* ST648 was recovered from companion animals in Japan (88). Closely related ESBL (CTX-M-15)-positive *E. coli* O25:H4-B2-ST131 isolates were identified in humans, companion animals (primarily dogs), and other animals from several European countries (89). Likewise, in Australia, 42% of fluoroquinolone-resistant human ExPEC isolates belonged to three ExPEC lineages (O25:H4-B2-ST131, O11/O17/O77:K52:H18-D-ST69 and O15:K52:H1-D-ST393), representatives of which also were identified among contemporaneous clinical isolates from companion animals (80). These findings demonstrate that the extensive clonal spread of *E. coli* O25:H4-B2-ST131 is not limited to humans. Another study provided strong evidence for direct transmission of an *E. coli* O25:H4-B2-ST131 strain among multiple companion animals within a single household (52). This strain's PFGE profile closely resembled the PFGE profiles of *E. coli* O25:H4-B2-ST131 isolates from human infections from other studies, implying cross-species transmission at some point (90). Additional ExPEC strains were also shared between dogs and cats in this study, again suggesting cross-species transmission within a household (51).

These results support the hypothesis that ExPEC from companion animals and humans represent overlapping populations, with certain ExPEC lineages capable of causing infections in and/or colonizing either host group. Related to this, antimicrobial resistance among ExPEC may be further selected by antimicrobial therapy given to companion animals, leading to additional exposure of humans to antimicrobial-resistant ExPEC through routine contact with colonized companion animals.

AVIAN PATHOGENIC *E. COLI* AND HUMAN EXTRAINTESTINAL *E. COLI*

Evidence is increasing that some human-associated ExPEC are related to APEC, the cause of colibacillosis, an extraintestinal infection of poultry. Considerable genetic similarity has been demonstrated between APEC isolates from infected poultry and ExPEC isolates from infected humans (32, 91–95). For example, a human sepsis-associated O111:H4-D-ST117 ExPEC strain was closely related to known APEC strains by PFGE, virulence factor profiling and phylotyping (32). A population-level association between human ExPEC and APEC (or *E. coli* derived from poultry) is suggested by these groups' overlapping antimicrobial resistance patterns, resistance genes, and virulence factors (32, 70, 72, 83, 91, 93, 95–100). As for experimental evidence of cross-species pathogenicity, several studies have demonstrated that APEC can cause disease in mammalian models of human ExPEC infections (e.g., the mouse model of ascending UTI) (101, 102), and human-source ExPEC can cause disease in certain avian disease models (72, 103).

FOODBORNE RESERVOIRS OF HUMAN ExPEC

Foodborne transmission of food animal or retail meat-associated ExPEC has been the topic of many recent investigations, with poultry and poultry products receiving the most attention (83, 97, 98, 100, 104–106). A review of outbreaks of ExPEC-related infections identified a sizable number of community-wide clusters (5). Although the sources of the outbreaks were not identified, these *E. coli* infection clusters typically result from point-source dissemination and food or waterborne transmission. The proposed chain of foodborne ExPEC transmission involves ExPEC from food animals or meat products subclinically colonizing the human consumer's intestinal tract after ingestion of undercooked meat or cross-contamination in the kitchen during food handling. Once an incoming ExPEC strain has established residence in the human intestinal tract, it persists there until circumstances favor an extraintestinal infection, e.g., sexual intercourse or urinary catheter insertion (53). The often-lengthy interval between ExPEC acquisition and disease development, if disease ever occurs, makes transmission of ExPEC and antimicrobial-resistant *E. coli* from external reservoirs (e.g., foods) to humans very difficult to detect. However, although direct transmission has not been demonstrated, abundant circumstantial evidence links *E. coli* recovered from food animal sources to human ExPEC isolates. The most troubling aspect of these observations is the extensive antimicrobial resistance of *E. coli* isolates from food animals, particularly chickens (107, 108).

One likely contributing factor to the emergence of antimicrobial-resistant *E. coli* in food animals, especially poultry, is the large increase in human consumption of retail chicken over the past 30 years (109). This has led to changes in the scale and methods of food animal production, including increased use of antimicrobial agents for growth promotion, infection prevention, and treatment (110), which likely has selected for and amplified multidrug-resistant ExPEC in the food animal reservoir (111). Resistant *E. coli*, once established in food animals, can spread among animals and the local environment, maintaining the circulation of antimicrobial-resistant ExPEC within the herd, flock, or production facility (111).

The plausibility of foodborne transmission of antimicrobial-resistant *E. coli* to humans is strengthened by the finding that antimicrobial-resistant *E. coli* from chicken carcasses widely contaminate the kitchen during meal preparation and can appear in the intestinal tract of individuals who prepare food dishes from the carcasses (112, 113). Furthermore, volunteers fed a diet of irradiated food exhibited markedly reduced levels of antimicrobial-resistant intestinal *E. coli*, with reversion to baseline post-intervention (114), implying that a conventional diet sustains a certain prevalence of intestinal resistant *E. coli* through a steady input of resistant strains. Thus, the available evidence, albeit indirect, strongly supports a foodborne component to the antimicrobial-resistant *E. coli* problem in humans.

Poultry Sources

Based on existing evidence, poultry is the food animal source most closely linked to human ExPEC (83, 97–100, 104–106). In women, UTI caused by antimicrobial-resistant ExPEC has been epidemiologically linked with high levels of self-reported chicken consumption (106). Among different meat types, poultry generally exhibits the highest overall levels of *E. coli* contamination, and poultry-associated *E. coli* tend to be more extensively antimicrobial-resistant than those from other meats (107, 108). Poultry-associated *E. coli* also often possess virulence genes characteristic of human ExPEC, suggesting a potential to cause human disease.

A 2006 study by the U.S. National Antimicrobial Resistance Monitoring Systems (NARMS) identified 200 ExPEC isolates among 1,275 *E. coli* isolates from retail meat samples. The proportion of isolates representing ExPEC varied by meat type, being highest in ground turkey (23.5%), followed by chicken breast meat (20.2%), pork chops (8.3%), and ground beef (3.4%) (115). More than half of these isolates belonged to serogroups associated with human extraintestinal infections. Serogroup O25 (17.9%) was the most common among chicken isolates, while serogroup O2 (36.2%) was most common among turkey isolates. Furthermore, 42% of retail meat ExPEC isolates represented ExPEC-associated phylogenetic groups B2 (25%) and D (23%) (115). These findings confirmed those of an earlier study from 2002 to 2004 involving NARMS isolates, which additionally showed that isolates from different meat types were genetically distinct, suggesting that they originated from the respective animal species (116). Importantly, however, susceptible and resistant isolates from a given meat type did not differ genetically, suggesting that the resistant isolates emerged from susceptible isolates within the different food animal hosts, possibly from on-farm antimicrobial use (116).

Specific ExPEC lineages have been found in poultry and poultry meat sources. Indistinguishable *E. coli* O25:H4-B2-ST131 strains were identified in a human UTI case and retail chicken meat sample in Canada (33). Mora et al. also demonstrated similarity by PFGE and virulence gene content between one chicken and one human CTX-M-9-positive *E. coli* O25:H4-B2-ST131 isolate (117). ESBL-positive *E. coli* O25:H4-B2-ST131 was also recovered from poultry farms in Spain, where human and poultry-source O25:H4-B2-ST131 isolates exhibited moderate (75%) PFGE similarity (118). Several other studies identified antimicrobial susceptible ST131 isolates in chickens and/or chicken meat (100, 105, 119, 120).

O11/O17/O77:K52:H18-D-ST69 likewise has been linked to non-human reservoirs, primarily chicken (33, 101). Moreover, in an experimental study, CgA *E. coli* isolates

recovered in Denmark from human infections and retail chicken meat were equally able to cause UTI in a mouse model, suggesting that poultry-source CgA *E. coli* are just as pathogenic for mammals as are human-derived CgA strains (97).

A recent study from the Netherlands identified *E. coli* (serotype: various)-A-ST10 producing CTX-M-1 ESBL in human blood cultures and poultry, whereas TEM-52-producing ST10 isolates were recovered from human urine samples and poultry (105). A similar study from the Netherlands also identified ESBL-producing *E. coli* ST10 in chicken meat, other meat types, rectal swabs from healthy humans, and human blood cultures (100). A study from Canada identified multidrug-resistant ST10 isolates (albeit of limited PFGE similarity) in human clinical samples, chicken feces, and retail chicken meat (121).

E. coli O1/O2/O18:K1:H7-B2-ST95 isolates with related PFGE profiles have been identified in humans and poultry (33, 93) and, separately, in honeydew melon and multiple human infections (33). In another study, 58% of 108 APEC and human ExPEC isolates from serogroups O1, O2, or O18, representing diverse host species, belonged to ST95 (70). When assessed in animal models, O18-B2-ST95 *E. coli* isolates from human neonates with meningitis could cause colisepticemia in poultry (72); conversely, O18-B2-ST95 *E. coli* isolates from cases of avian colibacillosis (APEC) could cause neonatal meningitis in a rat model (102). This indicates that the APEC and NMEC-associated ST95 group may have zoonotic potential.

ESBL-producing ST117 isolates were also identified in human and poultry reservoirs (100, 105), and genetically closely related O114:H4-ST117 strains, as assessed by PFGE, were identified in a human UTI case and retail chicken meat in Canada (33). CTX-M-32-positive O25a:H4-D-ST648 isolates, one from a human infection and one from a poultry source, exhibited closely related PFGE profiles (118). Additionally, human clinical and turkey meat samples containing *E. coli* ST410 were identified in Spain (122).

Pork and Pig Sources

Contamination of pork products with ExPEC has also been reported; however, fewer studies have been conducted to date (33, 98, 123–126). One epidemiologic study found that women experiencing a drug-resistant UTI were more likely to report frequent pork consumption (106). In another study, *E. coli* recovered from pigs tended to be from phylogenetic groups A and B1 and exhibited lower overall numbers of virulence genes compared with poultry isolates (118).

E. coli ST131 was detected among pork and UTI isolates in one study from Denmark and Norway (127). Additionally, O11/O17/O77:K52:H18-D-ST69 has been linked to pork (33, 123). Multidrug-resistant and possibly related (by PFGE) ST10 isolates have been identified in human clinical samples, pig feces, and retail pork meat (121), whereas in another study O2:H(non-motile)-A-ST10 isolates were identified in pigs (118).

Beef and Cattle Sources

The evidence for a beef cattle reservoir for human ExPEC is fairly weak. In general, the prevalence of *E. coli* resembling ExPEC is low in beef cattle sources (115, 128). One study identified a single cow isolate that was similar by PFGE to a human O11/O17/O77:K52:H18-D-ST69 isolate (129). The reason for limited colonization of healthy beef cattle with ExPEC, compared with their extensive colonization with *E. coli* such as shiga-toxin producing *E. coli* (130), is unknown.

PUBLIC HEALTH PERSPECTIVES

Investigators of the human versus extra-human origins of ExPEC have used both population ecology surveys and experimental pathogenesis studies to examine host species distribution and specificity of different ExPEC types. What has emerged is a picture of a large and complex group of *E. coli* variants, within which a small number of highly successful lineages (in terms of virulence, prevalence, and geographic range) have become established and account for the bulk of extraintestinal *E. coli* infections in humans and some domestic animals. The importance of these lineages is highlighted in one recent study in which just three clonal groups (O25:H4-B2-ST131, O15:K52:H1-D-ST393, and O11/O17/O77:K52:H18-D-ST69) accounted for 19% of 500 consecutive extraintestinal *E. coli* isolates in 5 hospitals in Spain in February 2009, and for 30% of multidrug-resistant isolates (9). ST131 *E. coli* alone was estimated to cause approximately 17% of extraintestinal infections in the source populations of another surveillance program (131). More importantly, these extraintestinal disease-causing lineages are becoming increasingly multidrug-resistant. Identification of the reservoirs and transmission pathways for these important human-associated ExPEC groups conceivably could help to curb their transmission and to reduce their extent of antimicrobial resistance, thereby leading to fewer and easier to treat human ExPEC infections.

It is increasingly clear that certain prominent ExPEC groups can occur in non-human sources, including food animals and meat products, although the extent of this phenomenon, its importance to human health, and the strength of the supporting evidence vary by source and *E. coli* group. The strongest evidence for a food animal reservoir and foodborne transmission of human ExPEC exists for poultry and pigs, where genetically similar ExPEC have been recovered from retail chicken, turkey, and pork products and human infections. Although transmission of ExPEC from food animals or retail meats to humans has yet to be directly demonstrated, transmission of ExPEC has been documented between cohabiting humans, and between companion animals and humans (49, 51–53, 85).

Development and amplification of multi-drug resistance in animal-associated *E. coli* is likely facilitated by the frequent or continuous selection pressure provided by antimicrobial use during food animal production and veterinary medicine (132). Antimicrobial use in human clinical medicine provides additional selection pressure once such organisms enter the human population. Thus, the global increase of antimicrobial resistance among human-associated ExPEC likely has multiple ecological origins, with important inputs from the transmission of antimicrobial-resistant ExPEC from food animals, companion animals, and environmental sources to humans. Each of these ExPEC reservoirs provides an environment for amplification, followed by further amplification in humans, then possible transmission back to these reservoirs, in a continuous feedback loop (133).

Certainly, antimicrobial stewardship is important for reducing the selection pressure for antimicrobial resistance in food animals and other environments sources. Antimicrobial usage in food animals varies widely throughout Europe, despite the 2006 European Union antimicrobial growth promoter ban (134). However, given the linkage of antimicrobial resistance genes and the unavoidable need for antimicrobial use for therapy and prophylaxis in some cases, in humans and animals alike, stewardship is at best a harm reduction strategy, slowing rather than preventing further emergence and spread of resistant *E. coli*. Recent efforts have focused on eliminating or reducing the use in food animals of antimicrobials classified as very important to human medicine (e.g., extended-spectrum cephalosporins) (135), to reduce antimicrobial resistance; however, this is not always been readily accomplished (e.g., the U.S. Food and Drug Administration's delayed ban on fluoroquinolone use in poultry) (136). Reducing *E. coli* contamination levels on meat products is another possible intervention.

Another preventive approach deserving consideration is vaccines. Food animal-associated ExPEC lineages possess specific traits that allow them to cause extraintestinal disease, to survive within specific reservoirs (e.g., in avian hosts), and to disseminate widely among human hosts. Elucidation of such factors is essential for identifying good vaccine targets. Vaccines could be additionally effective as a public health intervention if directed toward the animal hosts that act as reservoirs (137).

Ecologic studies comparing ExPEC from multiple sources have inherent limitations, including the difficulty of determining direction of transmission (from animals to human or vice versa) and pinpointing actual transmission events. Nonetheless, the public health community should heed the growing body of evidence supporting foodborne transmission of ExPEC with human pathogenic potential and should move toward implementing policies and practices that would limit the evolution, selection/amplification, and dissemination of antimicrobial-resistant ExPEC from food animals and other reservoirs to humans.

ACKNOWLEDGMENTS

Conflicts of interest: We declare no conflicts.

CITATION

Manges AR, Johnson JR. 2015. Reservoirs of extraintestinal pathogenic *Escherichia coli*. Microbiol Spectrum 3(5):UTI-0006-2012.

REFERENCES

1. **Bettelheim KA, Ismail N, Shinbaum R, Shooter RA, Moorhouse E, Farrell W.** 1976. The distribution of serotypes of *Escherichia coli* in cow-pats and other animal material compared with serotypes of *E. coli* isolated from human sources. *J Hyg (Lond)* **76:**403–406.
2. **Bettelheim KA.** 1997. *E. coli in the normal flora of humans and animals*, p 85–109. Cambridge University Press, Cambridge.
3. **Coque TM, Novais A, Carattoli A, Poirel L, Pitout JDD, Peixe L, Baquero F, Canton R, Nordmann P.** 2008. Dissemination of clonally related *Escherichia coli* strains expressing

extended-spectrum beta-lactamase CTX-M-15. *Emerg Infect Dis* **14:**195–200.

4. **Nicolas-Chanoine MH, Blanco J, Leflon-Guibout V, Demarty R, Alonso MP, Canica MM, Park YJ, Lavigne JP, Pitout J, Johnson JR.** 2008. Intercontinental emergence of *Escherichia coli* clone O25:H4-ST131 producing CTX-M-15. *J Antimicrob Chemother* **61:**273–281.

5. **George DB, Manges AR.** 2010. A systematic review of outbreak and non-outbreak studies of extraintestinal pathogenic *Escherichia coli* causing community-acquired infections. *Epidemiol Infect* **138:**1679–1690.

6. **Rogers Ba, Sidjabat HE, Paterson DL.** 2011. *Escherichia coli* O25b-ST131: a pandemic, multiresistant, community-associated strain. *J Antimicrob Chemother* **66:**1–14.

7. **Manges AR, Johnson JR, Foxman B, O'Bryan TT, Fullerton KE, Riley LW.** 2001. Widespread distribution of urinary tract infections caused by a multidrug-resistant *Escherichia coli* clonal group. *N Engl J Med* **345:**1007–1013.

8. **Johnson JR, Murray AC, Kuskowski MA, Schubert S, Prere M, Picard B, Colodner R, Raz R.** 2005. Distribution and characteristics of *Escherichia coli* clonal group A. *Emerg Infect Dis* **11:**141–145.

9. **Blanco J, Mora A, Mamani R, López C, Blanco M, Dahbi G, Herrera A, Blanco JE, Alonso MP, García-Garrote F, Chaves F, Orellana MA, Martínez-Martínez L, Calvo J, Prats G, Larrosa MN, González-Lopez JJ, López-Cerero L, Rodríguez-Baño J, Pascual A.** 2011. National survey of *Escherichia coli* causing extraintestinal infections reveals the spread of drug-resistant clonal groups O25b:H4-B2-ST131, O15:H1-D-ST393 and CGA-D-ST69 with high virulence gene content in Spain. *J Antimicrob Chemother* **66:**2011–2021.

10. **Johnson JR, Manges AR, O'Bryan TT, Riley LW.** 2002. A disseminated multidrug-resistant clonal group of uropathogenic *Escherichia coli* in pyelonephritis. *Lancet* **359:**2249–2251.

11. **Manges AR, Perdreau-Remington F, Solberg O, Riley LW.** 2006. Multidrug-resistant *Escherichia coli* clonal groups causing community-acquired bloodstream infections. *J Infect* **53:**25–29.

12. **Dias RCS, Marangoni DV, Smith SP, Alves EM, Pellegrino FLPC, Riley LW, Moreira BM.** 2009. Clonal composition of *Escherichia coli* causing community-acquired urinary tract infections in the State of Rio de Janeiro, Brazil. *Microbial Drug Resistance* **15:**303–308.

13. **Johnson JR, Menard ME, Lauderdale TL, Kosmidis C, Gordon D, Collignon P, Maslow JN, Andrasevic AT, Kuskowski MA.** 2011. Global distribution and epidemiologic associations of *Escherichia coli* clonal group A, 1998–2007. *Emerg Infect Dis* **17:**2001–2009.

14. **Phillips I, King A, Rowe B, Eykyn S, Gransden WR, Frost JA, Gross RJ.** 1988. Epidemic multiresistant *Escherichia coli* infection in West Lambeth Health District. *Lancet* **8593:**1038–1041.

15. **Olesen B, Kolmos HJ, Orskov F, Orskov I.** 1995. A comparative study of nosocomial and community-acquired strains of *Escherichia coli* causing bacteraemia in a Danish University Hospital. *J Hospital Infect* **31:**295–304.

16. **Prats G, Navarro F, Mirelis B, Dalmau D, Margall N, Coll P, Stell AL, Johnson JR.** 2000. *Escherichia coli* serotype O15:K52:H1 as a uropathogenic clone. *J Clin Microbiol* **38:**201–209.

17. **Johnson JR, Stell AL, Bryan TTO, Kuskowski M, Nowicki B, Johnson C, Maslow JN, Kaul A, Kavle J, Prats G.** 2002. Global molecular epidemiology of the O15:K52:H1 extraintestinal pathogenic *Escherichia coli* clonal group: evidence of distribution beyond Europe. *J Clin Microbiol* **40:**1913–1923.

18. **Neil PMO, Talboys CA, Roberts AP, Azadian BS.** 1990. The rise and fall of *Escherichia coli* O15 in a London Teaching Hospital. *J Med Microbiol* **33:**23–27.

19. **Olesen B, Scheutz F, Menard M, Skov MN, Kolmos HJr, Kuskowski MA, Johnson JR.** 2009. Three-decade epidemiological analysis of *Escherichia coli* O15:K52:H1. *J Clin Microbiol* **47:**1857–1862.

20. **Johnson JR, Owens KL, Clabots CR, Weissman SJ, Cannon SB.** 2006. Phylogenetic relationships among clonal groups of extraintestinal pathogenic *Escherichia coli* as assessed by multi-locus sequence analysis. *Microbes Infect* **8:**1702–1713.

21. **Gibreel TM, Dodgson AR, Cheesbrough J, Fox AJ, Bolton FJ, Upton M.** 2012. Population structure, virulence potential and antibiotic susceptibility of uropathogenic *Escherichia coli* from Northwest England. *J Antimicrob Chemother* **67:**346–356.

22. **Fam N, Leflon-Guibout V, Fouad S, Aboul-Fadl L, Marcon E, Desouky D, El-Defrawy I, Abou-Atta A, Klena J, Nicolas-Chanoine MH.** 2011. CTX-M-15-producing *Escherichia coli* clinical isolates in Cairo (Egypt), including isolates of clonal complex ST10 and clones ST131, ST73, and ST405 in both community and hospital settings. *Microb Drug Resist* **17:**67–73.

23. **Suzuki S, Shibata N, Yamane K, Wachino Ji, Ito K, Arakawa Y.** 2009. Change in the prevalence of extended-spectrum-β-lactamase-

producing *Escherichia coli* in Japan by clonal spread. *J Antimicrob Chemother* **63**:72–79.

24. **Manges AR, Tabor H, Tellis P, Vincent C, Tellier PP.** 2008. Endemic and epidemic lineages of *Escherichia coli* that cause urinary tract infections. *Emerg Infect Dis* **14**:1575–1583.

25. **Jones GL, Warren RE, Skidmore SJ, Davies VA, Gibreel T, Upton M.** 2008. Prevalence and distribution of plasmid-mediated quinolone resistance genes in clinical isolates of *Escherichia coli* lacking extended-spectrum β-lactamases. *J Antimicrob Chemother* **62**:1245–1251.

26. **Mihaila L, Wyplosz B, Clermont O, Garry L, Hipeaux MC, Vittecoq D, Dussaix E, Denamur E, Branger C.** 2010. Probable intrafamily transmission of a highly virulent CTX-M-3-producing *Escherichia coli* belonging to the emerging phylogenetic subgroup D2 O102-ST405 clone. *J Antimicrob Chemother* **65**:1537–1539.

27. **Smet A MA, Persoons D, Dewulf J, Heyndrickx M, Claeys G, Lontie M, Van Meensel B, Herman L, Haesebrouck F, Butaye P.** 2010. Characterization of extended-spectrum beta-lactamases produced by *Escherichia coli* isolated from hospitalized and nonhospitalized patients: emergence of CTX-M-15-producing strains causing urinary tract infections. *Microb Drug Resist* **16**:6.

28. **Tian GB, Rivera JI, Park YS, Johnson LE, Hingwe A, Adams-Haduch JM, Doi Y.** 2011. Sequence type ST405 *Escherichia coli* isolate producing QepA1, CTX-M-15, and RmtB from Detroit, Michigan. *Antimicrob Agents Chemother* **55**:3966–3967.

29. **Mushtaq S, Irfan S, Sarma JB, Doumith M, Pike R, Pitout J, Livermore DM, Woodford N.** 2011. Phylogenetic diversity of *Escherichia coli* strains producing NDM-type carbapenemases. *J Antimicrob Chemother* **66**:2002–2005.

30. **Peirano G, Schreckenberger PC, Pitout JDD.** 2011. Characteristics of NDM-1-producing *Escherichia coli* isolates that belong to the successful and virulent clone ST131. *Antimicrob Agents Chemother* **55**:2986–2988.

31. **Antao EM, Ewers C, Gürlebeck D, Preisinger R, Homeier T, Li G, Wieler LH.** 2009. Signature-tagged mutagenesis in a chicken infection model leads to the identification of a novel avian pathogenic *Escherichia coli* fimbrial adhesin. *PloS One* **4**:e7796.

32. **Mora A, Lopez C, Herrera A, Viso S, Mamani R, Dhabi G, Alonso MP, Blanco M, Blanco JE, Blanco J.** 2011. Emerging avian pathogenic *Escherichia coli* strains belonging to clonal groups O111:H4-D-ST2085 and O111:H4-D-ST117 with high virulence-gene content and zoonotic potential. *Vet Microbiol* **156**:347–352.

33. **Vincent C, Boerlin P, Daignault D, Dozois CM, Dutil L, Galanakis C, Reid-Smith RJ, Tellier PP, Tellis Pa, Ziebell K, Manges AR.** 2010. Food reservoir for *Escherichia coli* causing urinary tract infections. *Emerg Infect Dis* **16**:88–95.

34. **Platell JL, Trott DJ, Johnson JR, Heisig P, Heisig A, Clabots CR, Johnston B, Cobbold RN.** 2012. Prominence of an O75 clonal group (clonal complex 14) among non-ST131 fluoroquinolone-resistant *Escherichia coli* causing extraintestinal infections in humans and dogs in Australia. *Antimicrob Agents Chemother* doi:101128/AAC06120–11.

35. **Zong Z, Yu R.** 2010. *Escherichia coli* carrying the blaCTX-M-15 gene of ST648. *J Med Microbiol* **59**:1536–1537.

36. **van der Bij AK PG, Goessens WH, van der Vorm ER, van Westreenen M, Pitout JD.** 2011. Clinical and molecular characteristics of extended-spectrum-beta-lactamase-producing *Escherichia coli* causing bacteremia in the Rotterdam Area, Netherlands. *Antimicrob Agents Chemother* **55**:3.

37. **Hornsey M PL, Wareham DW.** 2011. A novel variant, NDM-5, of the New Delhi metallo-β-Lactamase in a multidrug-resistant *Escherichia coli* ST648 isolate recovered from a patient in the United Kingdom. *Antimicrob Agents Chemother* **55**:3.

38. **Coelho A, Mora A, Mamani R, López C, González-López JJ, Larrosa MN, Quintero-Zarate JN, Dahbi G, Herrera A, Blanco JE, Blanco M, Alonso MP, Prats G, Blanco J.** 2011. Spread of *Escherichia coli* O25b:H4-B2-ST131 producing CTX-M-15 and SHV-12 with high virulence gene content in Barcelona (Spain). *J Antimicrob Chemother* **66**:517–526.

39. **Kim J, Bae IK, Jeong SH, Chang CL, Lee CH, Lee K.** 2011. Characterization of IncF plasmids carrying the blaCTX-M-14 gene in clinical isolates of *Escherichia coli* from Korea. *J Antimicrob Chemother* **66**:1263–1268.

40. **Mshana SE, Imirzalioglu C, Hain T, Domann E, Lyamuya EF, Chakraborty T.** 2011. Multiple ST clonal complexes, with a predominance of ST131, of *Escherichia coli* harbouring blaCTX-M-15 in a tertiary hospital in Tanzania. *Clin Microbiol Infect* **17**:1279–1282.

41. **Oteo J, Diestra K, Juan C, Bautista V, Novais A, Pérez-Vázquez M, Moyá B, Miró E, Coque TM, Oliver A, Cantón R, Navarro F, Campos J.** 2009. Extended-spectrum β-lactamase-producing *Escherichia coli* in Spain belong to a large variety of multilocus sequence typing types, including ST10 complex/A, ST23 complex/A and ST131/B2. *International J Antimicrob Agents* **34**:173–176.

42. Caugant DA, Levin BR, Lidin-Janson G, Whittam TS, Svanborg Eden C, Selander RK. 1983. Genetic diversity and relationships among strains of *Escherichia coli* in the intestine and those causing urinary tract infections. *Prog Allergy* **33**:203–227.

43. Obata-Yasuoka M, Ba-Thein W, Tsukamoto T, Yoshikawa H, Hayashi H. 2002. Vaginal *Escherichia coli* share common virulence factor profiles, serotypes and phylogeny with other extraintestinal E. coli. *Microbiology* **148**:2745–2752.

44. Yamamoto S, Tsukamoto T, Terai A, Kurazono H, Takeda Y, Yoshida O. 1997. Genetic evidence supporting the fecal-perineal-urethral hypothesis in cystitis caused by *Escherichia coli*. *J Urol* **157**:1127–1129.

45. MacFie J, O'Boyle C, Mitchell CJ, Buckley PM, Johnstone D, Sudworth P. 1999. Gut origin of sepsis: a prospective study investigating associations between bacterial translocation, gastric microflora, and septic morbidity. *Gut* **45**:223–228.

46. Bailey JK, Pinyon JL, Anantham S, Hall RM. 2010. Commensal *Escherichia coli* of healthy humans: a reservoir for antibiotic-resistance determinants. *J Med Microbiol* **59**:1331–1339.

47. Balis E, Vatopoulos AC, Kanelopoulou M, Mainas E, Kontogianni V, Balis E, Vatopoulos AC, Kanelopoulou M, Mainas E, Hatzoudis G, Kontogianni V. 1996. Indications of in vivo transfer of an epidemic R plasmid from *Salmonella enteritidis* to *Escherichia coli* of the normal human gut flora. *Microbiology* 4–7.

48. Raum E, Lietzau S, von Baum H, Marre R, Brenner H. 2008. Changes in *Escherichia coli* resistance patterns during and after antibiotic therapy: a longitudinal study among outpatients in Germany. *Clin Microbiol Infect* **14**:41–48.

49. Foxman B, Zhang L, Tallman P, Andree BC, Geiger AM, Koopman JS, Gillespie BW, Palin KA, Sobel JD, Rode CK, Bloch CA, Marrs CF. 1997. Transmission of uropathogens between sex partners. *J Infect Dis* **175**:989–992.

50. Johnson JR, Brown JJ, Carlino UB, Russo TA. 1998. Colonization with and acquisition of uropathogenic *Escherichia coli* as revealed by polymerase chain reaction-based detection. *J Infect Dis* **177**:1120–1124.

51. Johnson JR, Clabots C. 2006. Sharing of virulent *Escherichia coli* clones among household members of a woman with acute cystitis. *Clin Infect Dis* **43**:101–108.

52. Johnson JR, Clabots C, Kuskowski MA. 2008. Multiple-host sharing, long-term persistence, and virulence of *Escherichia coli* clones from human and animal household members. *J Clin Microbiol* **46**:4078–4082.

53. Manges AR, Johnson JR, Riley LW. 2004. Intestinal population dynamics of UTI-causing *Escherichia coli* within heterosexual couples. *Curr Issues Intest Microbiol* **5**:49–57.

54. Anastasi EM, Matthews B, Gundogdu A, Vollmerhausen TL, Ramos NL, Stratton H, Ahmed W, Katouli M. 2010. Prevalence and persistence of *Escherichia coli* strains with uropathogenic virulence characteristics in sewage treatment plants. *Appl Environ Microbiol* **76**:5882–5886.

55. Mokracka J, Koczura R, Jabłońska L, Kaznowski A. 2011. Phylogenetic groups, virulence genes and quinolone resistance of integron-bearing *Escherichia coli* strains isolated from a wastewater treatment plant. *Antonie van Leeuwenhoek* **99**:817–824.

56. Dhanji H, Murphy NM, Akhigbe C, Doumith M, Hope R, Livermore DM, Woodford N. 2011. Isolation of fluoroquinolone-resistant O25b:H4-ST131 *Escherichia coli* with CTX-M-14 extended-spectrum beta-lactamase from UK river water. *J Antimicrob Chemother* **66**:512–516.

57. Dolejska M, Frolkova P, Florek M, Jamborova I, Purgertova M, Kutilova I, Cizek A, Guenther S, Literak I. 2011. CTX-M-15-producing *Escherichia coli* clone B2-O25b-ST131 and Klebsiella spp. isolates in municipal wastewater treatment plant effluents. *J Antimicrob Chemother* **66**:2784–2790.

58. Boczek La, Rice EW, Johnston B, Johnson JR. 2007. Occurrence of antibiotic-resistant uropathogenic *Escherichia coli* clonal group A in wastewater effluents. *Appl Environ Microbiol* **73**:4180–4184.

59. Abhirosh C, Sherin V, Thomas AP, Hatha AA, Mazumder A. 2011. Potential public health significance of faecal contamination and multidrug-resistant *Escherichia coli* and Salmonella serotypes in a lake in India. *Public health* **125**:377–379.

60. Hamelin K, Bruant G, El-Shaarawi A, Hill S, Edge Ta, Fairbrother J, Harel J, Maynard C, Masson L, Brousseau R. 2007. Occurrence of virulence and antimicrobial resistance genes in *Escherichia coli* isolates from different aquatic ecosystems within the St. Clair River and Detroit River areas. *Appl Environ Microbiol* **73**:477–484.

61. Luna GM, Vignaroli C, Rinaldi C, Pusceddu A, Nicoletti L, Gabellini M, Danovaro R, Biavasco F. 2010. Extraintestinal *Escherichia coli* carrying virulence genes in coastal marine sediments. *Appl Environ Microbiol* **76**:5659–5668.

62. Hamelin K, Bruant G, El-Shaarawi A, Hill S, Edge TA, Bekal S, Fairbrother JM, Harel J, Maynard C, Masson L, and Brousseau R.

2006. A virulence and antimicrobial resistance DNA microarray detects a high frequency of virulence genes in *Escherichia coli* from Great Lakes recreational waters. *Appl Environ Microbiol* **6**:4200–4206.

63. **Mühldorfer I, Blum G, Donohue-Rolfe A, Heier H, Olschläger T, Tschäpe H, Wallner U, Hacker J.** 1996. Characterization of *Escherichia coli* strains isolated from environmental water habitats and from stool samples of healthy volunteers. *Res Microbiol* **147**:625–635.

64. **Ahmed W, Hodgers L, Masters N, Sidhu JPS, Katouli M, Toze S.** 2011. Occurrence of intestinal and extraintestinal virulence genes in *Escherichia coli* isolates from rainwater tanks in Southeast Queensland, Australia. *Appl Environ Microbiol* **77**:7394–7400.

65. **Bonnedahl J, Drobni M, Gauthier-Clerc M, Hernandez J, Granholm S, Kayser Y, Melhus A, Kahlmeter G, Waldenström J, Johansson A, Olsen B.** 2009. Dissemination of *Escherichia coli* with CTX-M type ESBL between humans and yellow-legged gulls in the south of France. *PloS one* **4**:e5958.

66. **Bonnedahl J, Drobni P, Johansson A, Hernandez J, Melhus A, Stedt J, Olsen B, Drobni M.** 2010. Characterization, and comparison, of human clinical and black-headed gull (Larus ridibundus) extended-spectrum beta-lactamase-producing bacterial isolates from Kalmar, on the southeast coast of Sweden. *J Antimicrob Chemother* **65**:1939–1944.

67. **Edge T, Hill S.** 2007. Multiple lines of evidence to identify the sources of fecal pollution at a freshwater beach in Hamilton Harbour, Lake Ontario. *Water research* **41**:3585–3594.

68. **Fogarty LR, Haack SK, Wolcott MJ, Whitman RL.** 2003. Abundance and characteristics of the recreational water quality indicator bacteria *Escherichia coli* and enterococci in gull faeces. *J Appl Microbiol* **94**:865–878.

69. **Yan T, Hamilton MJ, Sadowsky MJ.** 2007. High-throughput and quantitative procedure for determining sources of *Escherichia coli* in waterways by using host-specific DNA marker genes. *Appl Environ Microbiol* **73**:890–896.

70. **Johnson TJ, Wannemuehler Y, Johnson SJ, Stell AL, Doetkott C, Johnson JR, Kim KS, Spanjaard L, Nolan LK.** 2008. Comparison of extraintestinal pathogenic *Escherichia coli* strains from human and avian sources reveals a mixed subset representing potential zoonotic pathogens. *Appl Environ Microbiol* **74**:7043–7050.

71. **Lau SH, Reddy S, Cheesbrough J, Bolton FJ, Willshaw G, Cheasty T, Fox AJ, Upton M.** 2008. Major uropathogenic *Escherichia coli* strain isolated in the Northwest of England

identified by multilocus sequence typing. *J Clin Microbiol* **46**:1076–1080.

72. **Moulin-Schouleur M, Répérant M, Laurent S, Brée A, Mignon-Grasteau S, Germon P, Rasschaert D, Schouler C.** 2007. Extraintestinal pathogenic *Escherichia coli* strains of avian and human origin: link between phylogenetic relationships and common virulence patterns. *J Clin Microbiol* **45**:3366–3376.

73. **Tartof SY, Solberg OD, Manges AR, Riley W, Riley LW.** 2005. Analysis of a uropathogenic *Escherichia coli* clonal group by multilocus sequence typing analysis. *J Clin Microbiol* **43**:5860–5864.

74. **Guenther S, Grobbel M, Lübke-Becker A, Goedecke A, Friedrich ND, Wieler LH, Ewers C.** 2010. Antimicrobial resistance profiles of *Escherichia coli* from common European wild bird species. *Vet Microbiol* **144**:219–225.

75. **Guenther S, Ewers C, Wieler LH.** 2011. Extended-spectrum beta-lactamases producing *E. coli* in wildlife, yet another form of environmental pollution? *Front Microbiol* **2**:246.

76. **Blanco J, Blanco M, Wong I, Blanco JE.** 1993. Haemolytic *Escherichia coli* strains isolated from stools of healthy cats produce cytotoxic necrotizing factor type 1 (CNF1). *Vet Microbiol* **38**:157–165.

77. **Chérifi A, Contrepois M, Picard B, Goullet P, Orskov I, Orskov F.** 1994. Clonal relationships among *Escherichia coli* serogroup O78 isolates from human and animal infections. *J Clin Microbiol* **32**:1197–1202.

78. **Low D, Braaten B, Ling G, Johnson D, Ruby A.** 1988. Isolation and comparison of *Escherichia coli* strains from canine and human patients with urinary tract infections. *Infect Immun* **56**:2601–2609.

79. **Platell JL, Cobbold RN, Johnson JR, Heisig A, Heisig P, Clabots C, Kuskowski Ma, Trott DJ.** 2011. Commonality among fluoroquinolone-resistant sequence type ST131 extraintestinal *Escherichia coli* isolates from humans and companion animals in Australia. *Antimicrob Agents Chemother* **55**:3782–3787.

80. **Platell JL, Cobbold RN, Johnson JR, Trott DJ.** 2010. Clonal group distribution of fluoroquinolone-resistant *Escherichia coli* among humans and companion animals in Australia. *J Antimicrob Chemother* **65**:1936–1938.

81. **Johnson JR, Stell AL, Delavari P.** 2001. Canine feces as a reservoir of extraintestinal pathogenic *Escherichia coli*. *Infect Immun* **69**:1306–1314.

82. **Gibson JS, Cobbold RN, Trott DJ.** 2010. Characterization of multidrug-resistant *Escherichia coli* isolated from extraintestinal clinical infections in animals. *J Med Microbiol* **59**:592–598.

83. Johnson JR, Kuskowski Ma, Menard M, Gajewski A, Xercavins M, Garau J. 2006. Similarity between human and chicken *Escherichia coli* isolates in relation to ciprofloxacin resistance status. *J Infect Dis* **194**:71–78.

84. Murray AC, Kuskowski MA, Johnson JR. 2004. Virulence factors predict *Escherichia coli* colonization patterns among human and animal household members. *Ann Intern Med* **140**:848–849.

85. Johnson J, Delavari P, Stell A, Whittam T, Carlino U, Russo T. 2001. Molecular comparison of extraintestinal *Escherichia coli* isolates of the same electrophoretic lineages from humans and domestic animals. *J Infect Dis* **183**: 154–159.

86. Johnson J, Stell A, Delavari P, Murray A, Kuskowski M, Gaastra W. 2001. Phylogenetic and pathotypic similarities between *Escherichia coli* isolates from urinary tract infections in dogs and extraintestinal infections in humans. *J Infect Dis* **183**:897–906.

87. Tamang MD, Nam H-M, Jang G-C, Kim S-R, Chae MH, Jung S-C, Byun J-W, Park YH, Lim S-K. 2012. Molecular characterization of extended-spectrum-β-lactamase-producing and plasmid-mediated AmpC β-lactamase-producing *Escherichia coli* isolated from stray dogs in South Korea. *Antimicrob Agents Chemother* **56**:2705–2712.

88. Harada K NY, Kataoka Y. 2012. Mechanisms of resistance to cephalosporin and emergence of O25b-ST131 clone harboring CTX-M-27 β-lactamase in extraintestinal pathogenic *Escherichia coli* from dogs and cats in Japan. *Microbiol Immunol* **56**:5.

89. Ewers C, Grobbel M, Stamm I, Kopp Pa, Diehl I, Semmler T, Fruth A, Beutlich J, Guerra B, Wieler LH, Guenther S. 2010. Emergence of human pandemic O25:H4-ST131 CTX-M-15 extended-spectrum-beta-lactamase-producing *Escherichia coli* among companion animals. *J Antimicrob Chemother* **65**:651–660.

90. Johnson JR, Miller S, Johnston B, Clabots C, Debroy C. 2009. Sharing of *Escherichia coli* sequence type ST131 and other multidrug-resistant and urovirulent *E. coli* strains among dogs and cats within a household. *J Clin Microbiol* **47**:3721–3725.

91. Johnson TJ, Kariyawasam S, Wannemuehler Y, Mangiamele P, Johnson SJ, Doetkott C, Skyberg Ja, Lynne AM, Johnson JR, Nolan LK. 2007. The genome sequence of avian pathogenic *Escherichia coli* strain O1:K1:H7 shares strong similarities with human extraintestinal pathogenic *E. coli* genomes. *J Bacteriol* **189**:3228–3236.

92. Johnson TJ, Logue CM, Johnson JR, Kuskowski MA, Sherwood JS, Barnes HJ, DebRoy C, Wannemuehler YM, Obata-Yasuoka M, Spanjaard L, Nolan LK. 2012. Associations between multidrug resistance, plasmid content, and virulence potential among extraintestinal pathogenic and commensal *Escherichia coli* from humans and poultry. *Foodborne Pathog Dis* **9**:37–46.

93. Mora A, López C, Dabhi G, Blanco M, Blanco JE, Alonso MP, Herrera A, Mamani R, Bonacorsi S, Moulin-Schouleur M, Blanco J. 2009. Extraintestinal pathogenic *Escherichia coli* O1:K1:H7/NM from human and avian origin: detection of clonal groups B2 ST95 and D ST59 with different host distribution. *BMC microbiology* **9**:132.

94. Moulin-Schouleur M, Schouler C, Tailliez P, Kao M-R, Brée A, Germon P, Oswald E, Mainil J, Blanco M, Blanco J. 2006. Common virulence factors and genetic relationships between O18:K1:H7 *Escherichia coli* isolates of human and avian origin. *J Clin Microbiol* **44**: 3484–3492.

95. Rodriguez-Siek KE, Giddings CW, Doetkott C, Johnson TJ, Fakhr MK, Nolan LK. 2005. Comparison of *Escherichia coli* isolates implicated in human urinary tract infection and avian colibacillosis. *Microbiology* **151**:2097–2110.

96. Brinas L, Moreno MA, Zarazaga M, Porrero C, Saenz Y, Garcia M, Dominguez L, Torres C. 2003. Detection of CMY-2, CTX-M-14, and SHV-12 β-Lactamases in *Escherichia coli* fecal-sample isolates from healthy chickens. *Antimicrob Agents Chemother* **47**:2056–2058.

97. Jakobsen L, Hammerum AM, Frimodt-Møller N. 2010. Detection of clonal group A *Escherichia coli* isolates from broiler chickens, broiler chicken meat, community-dwelling humans, and urinary tract infection (UTI) patients and their virulence in a mouse UTI model. *Appl Environ Microbiol* **76**:8281–8284.

98. Jakobsen L, Spangholm DJ, Pedersen K, Jensen LB, Emborg HD, Agersø Y, Aarestrup FM, Hammerum AM, Frimodt-Møller N. 2010. Broiler chickens, broiler chicken meat, pigs and pork as sources of ExPEC related virulence genes and resistance in *Escherichia coli* isolates from community-dwelling humans and UTI patients. *Int J Food Microbiol* **142**:264–272.

99. Johnson TJ, Jordan D, Kariyawasam S, Stell AL, Bell NP, Wannemuehler YM, Alarcón CF, Li G, Tivendale Ka, Logue CM, Nolan LK. 2010. Sequence analysis and characterization of a transferable hybrid plasmid encoding multidrug resistance and enabling zoonotic potential for extraintestinal *Escherichia coli*. *Infect Immun* **78**:1931–1942.

100. **Overdevest I, Willemsen I, Rijnsburger M, Eustace A, Xu L, Hawkey P, Heck M, Savelkoul P, Vandenbroucke-Grauls C, van der Zwaluw K, Huijsdens X, Kluytmans J.** 2011. Extended-spectrum β-lactamase genes of *Escherichia coli* in chicken meat and humans, The Netherlands. *Emerg Infect Dis* **17:**1216–1222.

101. **Jakobsen L, Garneau P, Bruant G, Harel J, Olsen S, Porsbo L, Hammerum A, Frimodt-Møller N.** 2011. Is *Escherichia coli* urinary tract infection a zoonosis? Proof of direct link with production animals and meat. *Eur J Clin Microbiol Infect Dis* **31:**1–9.

102. **Tivendale KA, Logue CM, Kariyawasam S, Jordan D, Hussein A, Li G, Wannemuehler Y, Nolan LK.** 2010. Avian-pathogenic *Escherichia coli* strains are similar to neonatal meningitis *E. coli* strains and are able to cause meningitis in the rat model of human disease. *Infect Immun* **78:**3412–3419.

103. **Skyberg JA, Johnson TJ, Johnson JR, Clabots C, Logue CM, Nolan LK.** 2006. Acquisition of avian pathogenic *Escherichia coli* plasmids by a commensal *E. coli* isolate enhances its abilities to kill chicken embryos, grow in human urine, and colonize the murine kidney. *Infect Immun* **74:**6287–6292.

104. **Johnson JR, Delavari P, O'Bryan TT, Smith KE, Tatini S.** 2005. Contamination of retail foods, particularly turkey, from community markets (Minnesota, 1999–2000) with antimicrobial-resistant and extraintestinal pathogenic *Escherichia coli*. *Foodborne Path Dis* **2:**38–49.

105. **Leverstein-van Hall Ma, Dierikx CM, Cohen Stuart J, Voets GM, van den Munckhof MP, van Essen-Zandbergen A, Platteel T, Fluit AC, van de Sande-Bruinsma N, Scharinga J, Bonten MJM, Mevius DJ.** 2011. Dutch patients, retail chicken meat and poultry share the same ESBL genes, plasmids and strains. *Clin Microbiol Infect* **17:**873–880.

106. **Manges AR, Smith SP, Lau BJ, Nuval CJ, Eisenberg JNS, Dietrich PS, Riley LW.** 2007. Retail meat consumption and the acquisition of antimicrobial resistant *Escherichia coli* causing urinary tract infections: a case-control study. *Foodborne Path Dis* **4:**419–431.

107. **O'May GA, Jacobsen SM, Longwell M, Stoodley P, Mobley HL, Shirtliff ME.** 2009. The high-affinity phosphate transporter Pst in *Proteus mirabilis* HI4320 and its importance in biofilm formation. *Microbiology* **155:**1523–1535.

108. **Department of Health and Human Services.** 2010. *NARMS Retail Meat Annual Report.* Food and Drug Administration. http://www.fda.gov/downloads/AnimalVeterinary/SafetyHealth/ AntimicrobialResistance/NationalAntimicrobial ResistanceMonitoringSystem/UCM293581.pdf

109. **Lawrence J.** 2006. *Long Term Meat Production and Consumption Trends.* http://www.thepoultrysite.com/articles/527/long-term-meat-production-and-consumption-trends/

110. **Coffman JR, Beran GW.** 1999. Issues specific to antibiotics, p 142–178. *In* National Research Council: Committee on Drug Use in Food APoAH, Food Safety and Public Health. *The use of drugs in food animals: Benefits and risks.* The National Academy Press, Washington, DC

111. **Aarestrup FM, Wegener HC, Collignon P.** 2008. Resistance in bacteria of the food chain: epidemiology and control strategies. *Expert Rev Anti Infect Ther* **6:**733–750.

112. **Cogan TA, Bloomfield SF, Humphrey TJ.** 1999. The effectiveness of hygiene procedures for prevention of cross-contamination from chicken carcasses in the domestic kitchen. *Letters in Applied Microbiology* **29:**354–358.

113. **Linton AH, Howe K, Bennett PM, Richmond MH, Whiteside EJ.** 1977. The colonization of the human gut by antibiotic resistant *Escherichia coli* from chickens. *J Appl Bacteriol* **43:**465–469.

114. **Corpet DE.** 1988. Antibiotic resistance from food. *N Engl J Med* **318:**1206–1207.

115. **Xia X, Meng J, Zhao S, Bodeis-Jones S, Gaines SA, Ayers SL, McDermott PF.** 2011. Identification and antimicrobial resistance of extraintestinal pathogenic *Escherichia coli* from retail meats. *J Food Protect* **74:**38–44.

116. **Johnson JR, McCabe JS, White DG, Johnston B, Kuskowski Ma, McDermott P.** 2009. Molecular analysis of *Escherichia coli* from retail meats (2002–2004) from the United States National Antimicrobial Resistance Monitoring System. *Clin Infect Dis* **49:**195–201.

117. **Mora A, Herrera A, Mamani R, López C, Alonso MP, Blanco JE, Blanco M, Dahbi G, García-Garrote F, Pita JM, Coira A, Bernárdez MI, Blanco J.** 2010. Recent emergence of clonal group O25b:K1:H4-B2-ST131 ibeA strains among *Escherichia coli* poultry isolates, including CTX-M-9-producing strains, and comparison with clinical human isolates. *Appl Environ Microbiol* **76:**6991–6997.

118. **Cortés P, Blanc V, Mora A, Dahbi G, Blanco JE, Blanco M, López C, Andreu A, Navarro F, Alonso MP, Bou G, Blanco J, Llagostera M.** 2010. Isolation and characterization of potentially pathogenic antimicrobial-resistant *Escherichia coli* strains from chicken and pig farms in Spain. *Appl Environ Microbiol* **76:**2799–2805.

119. **Dhanji H, Murphy NM, Doumith M, Durmus S, Lee SS, Hope R, Woodford N, Livermore**

DM. 2010. Cephalosporin resistance mechanisms in *Escherichia coli* isolated from raw chicken imported into the UK. *J Antimicrob Chemother* **65**:2534–2537.

120. Warren RE, Ensor VM, O'Neill P, Butler V, Taylor J, Nye K, Harvey M, Livermore DM, Woodford N, Hawkey PM. 2008. Imported chicken meat as a potential source of quinolone-resistant *Escherichia coli* producing extended-spectrum β-lactamases in the UK. *J Antimicrob Chemother* **61**:504–508.

121. Racicot Bergeron C, Prussing C, Boerlin P, Daignault D, Dutil L, Reid-Smith RJ, Zhanel GG, Manges AR. 2012. Chicken as reservoir for human extraintestinal pathogenic *Escherichia coli*, Canada. *Emerg Infect Dis* **18**:415–421.

122. López-Cerero L, Egea P, Serrano L, Navarro D, Mora A, Blanco J, Doi Y, Paterson DL, Rodríguez-Baño J, Pascual A. 2011. Characterisation of clinical and food animal *Escherichia coli* isolates producing CTX-M-15 extended-spectrum β-lactamase belonging to ST410 phylogroup A. *Int J Antimicrob Agents* **37**:365–367.

123. Jakobsen L, Garneau P, Kurbasic A, Bruant G, Stegger M, Harel J, Jensen KS, Brousseau R, Hammerum AM, Frimodt-Møller N. 2011. Microarray-based detection of extended virulence and antimicrobial resistance gene profiles in phylogroup B2 *Escherichia coli* of human, meat and animal origin. *J Med Microbiol* **60**:1502–1511.

124. Johnson JR, Kuskowski Ma, Smith K, O'Bryan TT, Tatini S. 2005. Antimicrobial-resistant and extraintestinal pathogenic *Escherichia coli* in retail foods. *J Infect Dis* **191**:1040–1049.

125. Manges AR, Smith SP, Lau BJ, Nuval CJ, Eisenberg JN, Dietrich PS, Riley LW. 2007. Retail meat consumption and the acquisition of antimicrobial resistant *Escherichia coli* causing urinary tract infections: A case-control study. *Foodborne Pathog Dis* **4**:419–431.

126. Johnson JR, Sannes MR, Croy C, Johnston B, Clabots C, Kuskowski MA, Bender J, Smith KE, Winokur PL, Belongia EA. 2007. Antimicrobial drug-resistant *Escherichia coli* from humans and poultry products, Minnesota and Wisconsin, 2002–2004. *Emerg Infect Dis* **13**:838–846.

127. Trobos M, Christensen H, Sunde M, Nordentoft S, Agersø Y, Simonsen GS, Hammerum AM, Olsen JE. 2009. Characterization of sulphonamide-resistant *Escherichia coli* using comparison of sul2 gene sequences and multilocus sequence typing. *Microbiology* **155**:831–836.

128. Johnson JR, Menard M, Johnston B, Kuskowski MA, Nichol K, Zhanel GG. 2009. Epidemic clonal groups of *Escherichia coli* as a cause of antimicrobial-resistant urinary tract infections in Canada (2002–2004). *Antimicrob Agents Chemother* **53**: 2733–2739.

129. Ramchandani M, Manges AR, DebRoy C, Smith SP, Johnson JR, Riley LW. 2005. Possible animal origin of human-associated, multidrug-resistant, uropathogenic *Escherichia coli*. *Clin Infect Dis* **40**:251–257.

130. Shere JA, Bartlett KJ, Kaspar CW. 1998. Longitudinal study of *Escherichia coli* O157:H7 dissemination on four dairy farms in Wisconsin. *Appl Environ Microbiol* **64**:1390–1399.

131. Johnson JR, Johnston B, Clabots C, Kuskowski MA, Castanheira M. 2010. *Escherichia coli* sequence type ST131 as the major cause of serious multidrug-resistant *E. coli* infections in the United States. *Clin Infect Dis* **51**:286–294.

132. Threlfall EJ, Ward LR, Frost JA, Willshaw GA. 2000. Spread of resistance from food animals to man—the UK experience. *Acta Vet Scand Suppl* **93**:63–68.

133. Levy SB, FitzGerald GB, Macone AB. 1976. Changes in intestinal flora of farm personnel after introduction of a tetracycline-supplemented feed on a farm. *N Engl J Med* **295**:583–588.

134. Grave K, Torren-Edo J, Mackay D. 2010. Comparison of the sales of veterinary antibacterial agents between 10 European countries. *J Antimicrob Chemother* **65**:2037–2040.

135. Gilbert N. 2012. Rules tighten on use of antibiotics on farms. *Nature* **481**:125.

136. Price LB, Graham JP, Lackey LG, Roess A, Vailes R, Silbergeld E. 2007. Elevated risk of carrying gentamicin-resistant *Escherichia coli* among U.S. poultry workers. *Environ Health Perspect* **115**:1738–1742.

137. Moriel DG, Bertoldi I, Spagnuolo A, Marchi S, Rosini R, Nesta B, Pastorello I, Corea VAM, Torricelli G, Cartocci E, Savino S, Scarselli M, Dobrindt U, Hacker J, Tettelin H, Tallon LJ, Sullivan S, Wieler LH, Ewers C, Pickard D, Dougan G, Fontana MR, Rappuoli R, Pizza M, Serino L. 2010. Identification of protective and broadly conserved vaccine antigens from the genome of extraintestinal pathogenic *Escherichia coli*. *Proc Natl Acad Sci USA* **107**:9072–9077.

Origin and Dissemination of Antimicrobial Resistance among Uropathogenic *Escherichia coli*

10

LISA K. NOLAN,[1] GANWU LI,[1] and CATHERINE M. LOGUE[1]

INTRODUCTION

Without doubt, antimicrobial agents, in particular antibiotics, have saved countless lives and revolutionized medicine in many respects, leaving few to question their importance to modern society. However, early optimism that antibiotic usage would conquer bacterial infections was soon eclipsed by reports of emerging resistance. Certainly, this has been the case with uropathogens, such as uropathogenic *Escherichia coli* (UPEC), where multidrug-resistant strains are emerging and causing outbreaks worldwide (1). Here, we will review some of the common mechanisms of antibiotic resistance and the prevalence and dissemination of antibiotic-resistance determinants among UPEC.

OVERVIEW TO THE ANTIMICROBIAL AGENTS USED TO CONTROL UPEC-CAUSED UTIS

Antimicrobial agents may originate from a living organism or a laboratory, exerting their 'inhibitory' or 'killing' effects via various mechanisms after

[1]Department of Veterinary Microbiology and Preventive Medicine, College of Veterinary Medicine, Iowa State University, Ames, IA 50011.

Urinary Tract Infections: Molecular Pathogenesis and Clinical Management, 2nd Edition
Edited by Matthew A. Mulvey, David J. Klumpp, and Ann E. Stapleton
© 2017 American Society for Microbiology, Washington, DC
doi:10.1128/microbiolspec.UTI-0007-2012

appropriate delivery into or on living hosts or abiotic surfaces. Though antimicrobial agents encompass a wide range of chemical and physical agents, the primary foci of this chapter are antibiotics and resistance to the antibiotics used to treat UPEC-caused urinary tract infections (UTIs). In addition, some attention is given to UPEC's resistance to silver-containing antiseptics, which may be incorporated into catheters to prevent foreign-body associated UTIs. Antibiotics, as defined by Waksman (2), are chemical substances produced by microorganisms that are antagonistic to the growth of other microorganisms. An example here would be aminoglycosides. By contrast, synthetic antibacterial agents, like sulfonamides and quinolones, do not have a biologic origin, and semisynthetic drugs are produced by modification of natural antibiotics. Examples of semisynthetic agents are carbapenems, which are produced by chemical modification of naturally produced β-lactam antibiotics. However, more recently, usage of these terms has blurred with some regarding natural, synthetic, and semisynthetic antibacterial agents as antibiotics (3).

Antibiotics are designed to exhibit selective toxicity, meaning that their toxic activity is directed towards vulnerable targets in the pathogen that do not exist in the host. Fortunately, several differences in the cell machinery and functions of prokaryotes and mammalian cells exist that can be exploited to selectively inhibit the growth or reproduction of the pathogen without harming its host. Typical targets of antibiotics include bacterial cell wall synthesis, membrane structure, protein synthesis, DNA or RNA synthesis, and certain metabolic pathways (4–6).

Origins of Antibiotic Resistance

Though there is evidence that antibiotic-resistant bacteria existed prior to clinical use of antibiotics, clinical use of these agents has certainly been accompanied by an acceleration in the emergence of antibiotic-resistant

pathogens, forcing changes in the 'drugs of choice' used to treat UTIs and other infections. Today, just a little less than three generations after the discovery of penicillin, a return to the 'pre-antibiotic era' due to the emergence of multidrug-resistant (MDR) pathogens is a disturbing possibility. For instance, some strains of *Mycobacterium tuberculosis, Acinetobacter baumannii, Staphylococcus aureus,* and *Clostridium difficile* have undergone multiple mutations, resulting in high levels of resistance to the antibiotic classes specifically recommended for their treatment, thus, leading to enhanced morbidity and mortality due to infections with these organisms.

The origins of antibiotic resistance are murky at best with the recent emergence of certain resistances being interpreted by some to mean that resistance is a modern phenomenon. This view is supported by the finding that bacteria collected between 1914 and 1950 (Murray collection) were completely sensitive to antibiotics (7), although they did contain a range of conjugative plasmids capable of carrying resistance genes. However, identification of a bacterial penicillinase before the clinical use of penicillin (3) raised an interesting question: did resistance elements originate before or as a result of our use of antibiotics? Recent metagenomic studies provide insight into this issue. Though the cold-seep sediments of the deep-sea Edison seamount are estimated to be 10,000 years old, metagenomic analysis revealed that sediment bacteria contained genes encoding antibiotic-resistance factors like TEM-type β-lactamases (8). Similarly, a study of a metagenomic library from 30,000-year-old Beringian permafrost sediments identified a highly diverse collection of genes encoding resistance to β-lactam, tetracycline, and glycopeptide antibiotics (9). In addition, resistant microorganisms have been identified from historic culture collections prepared before the advent of modern day antimicrobial usage. For example, *E. coli* that were resistant to sulfadiazine, spectinomycin, and tetracycline were identified in isolates collected prior to 1950, providing added evidence that resistance existed in nature prior to the clinical use of antibiotics and even into ancient times.

Further, antimicrobial resistance may be broadly distributed across environmental locales regardless of the existence of antimicrobial pressure. The soil microorganisms, actinomycetes, were found to be resistant to an average of seven or eight antibiotics with some strains resistant to as many as 20 antibiotics (10). Novel resistance mechanisms, including those not traditionally found in pathogens, were identified in a collection of these organisms. The identification of antibiotic-producing microorganisms in soil has led to the idea that a primary ecological role of antibiotics is to inhibit the growth of competitors. Following this line of reasoning, antibiotic-resistance determinants may have arisen to protect microorganisms from antibiotic activity. This 'fight-for-survival' view is supported by the discovery in aminoglycoside-producing organisms of aminoglycoside-modifying enzymes that display marked homology to the modifying enzymes found in aminoglycoside-resistant bacteria (11). Interestingly, recent work indicates that some bacteria may utilize low concentrations of antibiotics for intercellular signaling rather than warfare, with antibacterial resistance genes in turn providing a means to modulate, silence, or otherwise disrupt interbacterial communication (12). According to this view, resistance genes may serve purposes other than providing an edge in dealing with competing microbes.

Some antimicrobial resistance genes may have originated from bacterial "genetic juggling". Resistance to fluoroquinolones (FQs) is a case in point. Mutations in gyrase, an efflux pump, and aminoglycoside N-acetyltransferases all resulted in emergence of FQ-resistant bacteria. A comprehensive survey of the Keio *E. coli* mutant library to determine the extent of the intrinsic resistome identified a total of 140 different mutants showing enhanced susceptibility to at least one of seven different antibiotics (13). Later work identified 283 mutants within the Keio library that harbored intrinsic resistance to at least one of 22 antibiotics (14). These observations demonstrated that in the *E. coli* K12 genome, and likely in UPEC, genes are present that confer a resistance phenotype. These 'so-called' proto- or quasi-resistances are referred to as intrinsic resistance. In addition, susceptible UPEC become resistant through horizontal gene transfer (HGT) by acquisition of mobile genetic elements such as transposons, integrons, plasmids, and phages, all of which can harbor resistance genes.

In general, bacterial sampling from the mobilome (the pool of mobile genes) likely plays a crucial role in microbial evolution, providing bacteria a means to compensate for their lack of sexual reproduction, the major mechanism of genetic innovation in higher organisms (15). Acquisition of "ready-made" genes on plasmids and other mobile genetic elements enables bacteria to respond quickly to change, such as the introduction of disinfectants and antibiotics into their environment. This would not be the case if bacterial fitness were solely reliant on *de novo* evolution (15). For example, resistance (R) plasmids, occurring in some extraintestinal pathogenic *E. coli* (ExPEC), the *E. coli* pathotype that includes UPEC, have been shown to encode resistance to up to eight or more antimicrobial agents, including antibiotics, synthetic agents, disinfectants, and heavy-metal compounds (16). The acquisition of resistance transposons, which are mobile genetic elements consisting of insertion sequences (IS elements) flanking a DNA sequence that can include a resistance gene(s) that can move both intra- and intermolecularly, is another means by which bacteria can quickly respond to environmental perturbations. IS elements are usually very similar in sequence and occur in inverted orientation on either end of the transposon. Transposons contribute to the dissemination of antibiotic-resistance elements by their insertion in conjugative plasmids or by acting as conjugative transposons. One well-known example of a transposon is Tn*10*, which encodes tetracycline resistance and is found among certain ExPEC strains of avian origin (17). However, despite their possible importance to UPEC resistance, transposons have been given only scant attention for their role in the dissemination of antibiotic resistance among uropathogens.

More attention has been given to the role of integrons in the rapid dissemination of antibiotic resistance in UPEC. Integrons, which were first identified and characterized by Stokes and Hall in 1989 (18), are 'assembly platforms' into which exogenous genes are incorporated by site-specific recombination. All characterized integrons contain three key elements necessary for the capture of exogenous antimicrobial-resistance genes: a tyrosine-recombinase family integrase gene (*intI*); a primary recombinant site (*attI*); and an outward-oriented promoter. Integron-encoded integrases can recombine an antimicrobial resistance "gene cassette" in a RecA-independent manner (18). Gene cassettes are free circular-DNA structures that are not expressed on their own due to the lack of promoters. Integration occurs downstream of the resident promoter at the *attI* site, allowing the expression of the exogenously acquired resistance gene in the cassette. In general, a gene cassette contains only a single gene and an imperfect inverted repeat at the 3' end of the gene called the *attC* site, which is recognized by the site-specific integrase. Gene cassettes can be arranged in tandem, and more than 60 distinct cassettes have been identified. In UPEC, cassette-associated genes have been shown to confer resistance to trimethoprim, sulfonamides, streptomycin/spectinomycin, beta-lactams, chloramphenicol, lincosamide, gentamicin, and aminoglycosides (19).

A strong association in UPEC has been established between the presence of integrons and resistance to both multiple and single antimicrobial agents. Based on sequence analysis of encoded integrases, at least five classes of integrons have been identified with class I and II found most frequently in UPEC strains (20). The reported prevalence of integrons in UPEC strains range from 16.6% to 64% (19, 21). Class I integrons, whose prevalence range from 26.9% to 64% among UPEC isolates (22), have a functional integrase gene allowing integration of new cassette genes into the bacterial genome. Dihydrofolate reductase (*dfr*) genes that confer resistance to trimethoprim are commonly inserted into the UPEC genome in this manner (23). To date, about 20 variant sequences of the *dfr* gene have been described with *dfrA1, dfrA5, dfrA7, dfrA12,* and *dfrA17* often associated with gene cassettes found in UPEC's class I integrons (20). In Europe, *dfrA1* is the most common variant found in UPEC that contain class I integrons (24–26), while in Korea (27, 28) and Australia (29), *dfrA17* and *dfrA12* are more common. These regional differences may be the result of clonal expansion of isolates containing different *dfr* alleles and limited contact between continents.

In UPEC, sulfamethoxazole-resistance genes are also associated with class I integrons but can also occur independent of the integron (30). Currently there are three types (*sul1, sul2,* and *sul3*) of sulfamethoxazole-resistance genes that have been characterized, with *sul1* frequently observed as conserved or semi-conserved regions within class I integrons carried by UPEC isolates (31). Interestingly, combined use of trimethoprim and sulfamethoxazole has been the standard treatment for acute uncomplicated UTI (32). The fact that resistance to these drugs is encoded by class I integrons in UPEC underscores the importance of these mobile elements in the dissemination of antimicrobial resistance among UPEC isolates.

Class II integrons occur less often in UPEC than class I integrons, occurring in 4% to 20% of UPEC strains (28, 33). However, in UPEC from children in Iran, class II integrons (10.41%) were more prevalent than class I integrons (6.25%) (21). Interestingly, class II integrons cannot acquire new gene cassettes, because their integrase is inactive due to the presence of a premature in-frame stop codon (34). They do, however, contain a static array of resistance genes: *dfrA1* (trimethoprim), *sat* (streptothricin), and *aadA1* (spectinomycin/streptomycin) (30).

Exceptionally, a functional class II integrase was found in a UPEC strain (35) in which the in-frame stop codon of the integrase had been replaced by a glutamine codon. This finding suggests that such class II integrons could acquire new antibiotic-resistance genes through integration of gene cassettes. Subsequently, a polymerase chain reaction (PCR) protocol was developed whereby the functional class II integrase could be distinguished from the non-functional variant. However, use of PCR to distinguish the two has not proved reliable. To confirm these PCR results, it is now recommended that the amplicon be sequenced.

Bacterial plasmids are extra-chromosomal, double-stranded, circular DNA molecules that replicate independently of the bacterial chromosome (36). In general, plasmids do not encode genes essential for survival but rather carry accessory genes and genes enabling bacteria to adapt to particular situations, such as environments in which an antibiotic has been introduced. Some plasmids encode genes enabling bacteria to resist bactericidal or bacteriostatic antimicrobial agents, including antibiotics and heavy-metal compounds. These are called R (resistance) plasmids or R factors. R plasmids may harbor these resistance genes within transposons and integrons, and these elements plus other resistance genes may be clustered together in R plasmids forming multidrug-resistance (MDR)-encoding islands (16, 37, 38). R plasmids are widely distributed among UPEC (39–42), and UPEC's acquisition of MDR-encoding R plasmids have been major factors in the emergence of resistant UPEC clones in recent years (1, 43–47).

In avian pathogenic *E. coli* (APEC), a subpathotype of ExPEC that bears certain similarities to UPEC in disease-causing abil-

ities and other traits, R plasmids frequently co-transfer with so-called virulence plasmids (16, 37, 48–51). These virulence plasmids can harbor genes involved in host cell adherence, invasion, iron acquisition, and other pathogenic processes. Sequence analysis of these plasmids revealed that the virulence genes that they harbor are often clustered into pathogenicity islands (PAIs) (48, 49). For instance, pAPEC-O2-ColV (Fig. 1), a 180-kilobase pair (kb) virulence plasmid isolated from APEC O2, harbors a 93-kb PAI encoding several virulence traits, including the serum-survival protein Iss, the autotransporter toxin Tsh, and several iron acquisition and transport systems. During conjugation, this virulence plasmid co-transfers with pAPEC-O2-R (Fig. 2), a 101-kb INcF R plasmid, which possesses an MDR-encoding island (37). This 33-kb pair island contains 15 genes responsible for resistance to at least eight antimicrobial agents, including silver and other heavy-metal compounds, quaternary ammonium disinfectants, tetracyclines, sulfonamides, trimethoprim, aminoglycosides, and beta-lactam antimicrobial agents. Also, pAPEC-O2-R harbors a Tn*21*–like region and a class I integron. In this case, the class I integron encodes resistance to chloramphenicol (*catB3*), aminoglycoside (*aad5*), and trimethoprim (*folA*).

Interestingly, acquisition of these co-transferring R and virulence plasmids by an avirulent recipient *E. coli* strain enhanced its abilities to grow in human urine and cause UTI in the murine model of human disease (52). Thus, the occurrence of similar plasmids among UPEC and related strains could have important implications for management of UTIs. These plasmids were also transferable to a human UPEC isolate (37), and *E. coli* contaminants of retail poultry meat appear to harbor similar plasmids (53). Still other APEC-associated plasmids are hybrids of resistance and virulence plasmids, harboring both an MDR-encoding island and a PAI (38). For instance, plasmid pAPEC-O103-ColBM (Fig. 3) harbors a 35-kb MDR-encoding island encoding ten resistance elements and a 55-kb

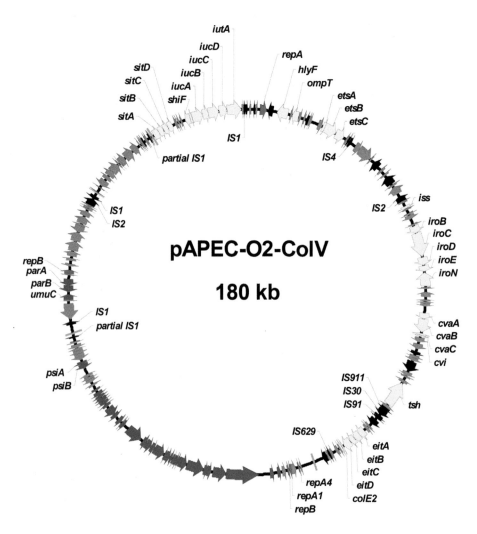

FIGURE 1 Circular genetic map of pAPEC-O2-ColV, drawn to scale. Arrows indicate predicted genes and their directions of transcription. Yellow arrows indicate virulence-associated genes. Blue arrows indicate genes involved in plasmid transfer and maintenance. Red arrows indicate genes involved in plasmid replication. Gray arrows indicate genes of unknown function. Black arrows indicate mobile genetic elements. Orange slashes indicate gaps in contiguous sequence that were unable to be resolved due to IS1 elements. Reprinted from the *Journal of Bacteriology* (48) with permission of the publisher.

conserved portion of the PAI found among APEC-virulence plasmids (38) (Fig. 3). Localized at the 5' end of the MDR-encoding region is a portion of Tn1721, including tetracycline-resistance genes (tetAB) and transposase gene (tnpA) with a downstream region flanked by exact 25-base pair (bp) inverted repeats. This 22-kb region contains sul2, encoding sulfonamide resistance;

strAB, encoding streptomycin resistance; and a class I integron. In the gene-cassette region of this class I integron is dfrA1, encoding trimethoprim resistance and the core class I integron genes intI1, aadA1, qacEΔ1, and sul1. Other antimicrobial resistance genes found in the MDR-encoding island were mphB, encoding macrolide resistance, and merEDACPTR genes, encoding mercury resistance. Among

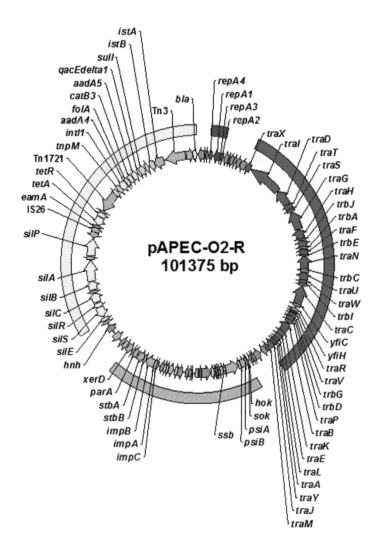

FIGURE 2 Circular genetic map of pAPEC-O2-R. Coding regions are indicated by arrows pointing in the direction of transcription. Yellow arrows indicate coding regions involved in antimicrobial resistance, blue arrows indicate coding regions involved in replication, red and pink arrows indicate coding regions involved in plasmid transfer, brown arrows indicate coding regions involved in plasmid maintenance, green arrows indicate mobile elements, blue-gray arrows indicate conserved hypothetical proteins, and gray arrows indicate unknown hypothetical proteins. Reprinted from *Antimicrobial Agents and Chemotherapy* (37) with permission of the publisher.

the antimicrobials to which pAPEC-O103-ColBM confers resistance were several agents that are frequently used for treatment of community-acquired UTI (32).

In addition to the potential impact of such a plasmid on successful treatment of UTIs, it also could impact uropathogenesis, as its PAI

encodes three iron-uptake systems (aerobactin, SitABCD, and salmochelin), the putative hemolysin HlyF, and the serum resistance-determinant Iss. Not surprisingly, this plasmid contributes to the pathogenesis of ExPEC in several *in vitro* and multiple animal models of human and avian disease (38). Thus, plas-

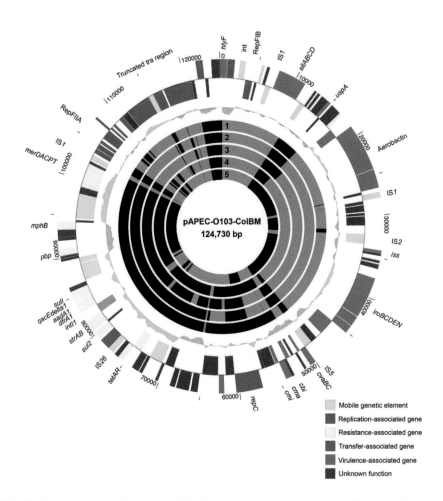

FIGURE 3 Circular map of pAPEC-O103-ColBM. The outer two circles show predicted coding regions in forward and reverse orientations, and different colors indicate different predicted functions. The next circle shows the G_C content in a 1,000-bp window with 10-bp steps. The next five circles show levels of nucleotide homology with other sequenced ColV-type plasmids. Blue indicates >90% homology with pAPEC-O103-ColBM, while black indicates ≤90% homology. The numbers on these circles indicate comparisons with pAPEC-O2-ColV (circle 1), pAPEC-O1-ColBM (circle 2), pVM01 (circle 3), pCVM29188_146 (circle 4), and pSMS-3-5_130 (circle 5). Reprinted from *Infection and Immunity* (38) with permission of the publisher.

mids harboring MDR-encoding islands and/ or PAIs are common among APEC, found in the *E. coli* contaminating retail poultry, can be transferred to human ExPEC isolates, and can promote resistance to multiple antimicrobial agents as well as uropathogenesis in models of human UTI. Collectively, these results demonstrate the 1) potential of plasmids to enhance both the resistance and virulence of UPEC, 2) the presence of plasmid-containing

E. coli strains in food destined for human consumption, and 3) the potential of contaminated food to be a reservoir of mobile resistance and virulence genes capable of contributing to the pathogenesis of human UTIs and the outcomes of these infections.

Manges and Johnson (54) summarized additional evidence of a possible link between human-associated ExPEC and APEC-contaminated poultry. This evidence includes overlapping virulence genes and antimicrobial-resistance profiles in APEC, UPEC, and other human ExPEC strains. If confirmed, these findings would lend further support to the hypothesis that contaminated poultry is a source of drug-resistant ExPEC and large R and virulence plasmids important in human disease. Collectively, such observations suggest that continued vigilance and a proactive approach to meat safety are warranted.

Most, if not all, classes of antibiotics used in the clinical treatment of UTIs can be countered by plasmid-borne antibiotic resistance harbored in one or more UPEC isolates. Notable among these are such commonly used agents as trimethoprim, sulfonamides, streptomycin/spectinomycin, beta-lactams, chloramphenicol, lincosamide, gentamicin, and aminoglycosides. Most R plasmids described among UPEC are small, ranging in size from 2–10 kb (55), and often mobilizable, meaning that they are transferable only when using the transfer apparatus encoded by a co-resident conjugative plasmid. By contrast, conjugative plasmids are relatively large, ranging in size from 30 kb to more than 200 kb in size, and harboring a transfer region of about 20–30 kb in size that encodes the machinery for cell-to-cell transfer (56). Suhartono showed that 22.2% (8/36) of UPEC isolates harbored large plasmids (55), and Rijavec et al. demonstrated that at least 17% (19/110) of UPEC strains carried conjugative plasmids encoding antibiotic resistance (57).

Plasmids with the same replication machinery cannot propagate in the same bacterial cell. Based on this property, plasmids have been categorized into incompatibility (Inc) groups or replicon types (58). To date, 26 Inc groups have been identified among the *Enterobacteriaceae*. Although classification of plasmids into Inc groups is desirable, early methods of Inc testing were laborious and time-consuming. Recently, a polymerase chain reaction (PCR)-based typing method developed by Carattoli et al. (59) allowed for the rapid identification of 18 common types

of plasmid replicons in the *Enterobacteriaceae*. Johnson et al. (60) streamlined this method and applied it to the replicon typing of commensal and pathogenic *E. coli* isolates. IncFIB and IncB/O plasmid replicons were most often identified among UPEC (60). However, IncFIB and IncP1-α replicons are associated with the greatest numbers of resistance genes. Specifically, IncFIB plasmids can provide resistance to amoxicillin, ceftriaxone, gentamicin, kanamycin, streptomycin, and ceftiofur, while IncP1-α replicons are associated with resistance to gentamicin, kanamycin, nalidixic acid, streptomycin, and tetracycline (61). Of note, many ExPEC-associated plasmid-replicon types, including IncA/C, IncP1-α, IncFIB, and IncI1, carry class I integrons (61).

Mechanisms of UPEC Resistance

Though bacterial antibiotic resistance existed prior to the clinical usage of antibiotics, the mechanisms of resistance have continued to evolve over time, at least in part due to selective pressure associated with antimicrobial use. Mechanisms of antimicrobial resistance can be categorized into five broad areas based on how a pathogen copes with the stress incurred by an antimicrobial agent (4) (Table 1). Here, some of the more common antimicrobial resistance mechanisms of UPEC will be examined. However, this list can be expected to expand as new drug-resistance mechanisms and mobility units evolve and disseminate.

Antimicrobial agents are designed to exhibit selective toxicity, meaning that their toxic activity is directed towards vulnerable targets in the pathogen that do not exist in the host. Typical targets of antimicrobial agents include bacterial cell wall synthesis, protein synthesis, DNA or RNA synthesis, certain metabolic pathways, and bacterial membrane structures (4–6).

Bacteria use a range of resistance strategies in order to protect themselves from the host's defenses as well as the activity of

TABLE 1 Mechanisms of action of antibacterial agents

- Interference with cell wall synthesis
  - β-lactams: penicillins, cephalosporins, carbapenems, monobactans
  - Glycopeptides: vancomycin, teicoplanin
- Protein-synthesis inhibition
  - Bind to 50S ribosomal subunit: macrolides, chloramphenicol, clindamycin, quinupristin-dalfopristin, linezolid
  - Bind to 30S ribosomal subunit: aminoglycosides, tetracyclines
  - Bind to bacterial isoleucyl-tRNA synthetase: mupirocin
- Interference with nucleic-acid synthesis
  - Inhibit DNA synthesis: fluoroquinolones
  - Inhibit RNA synthesis: rifampin
- Inhibition of metabolic pathway: sulfonamides, folic acid analogs
- Disruption of bacterial membrane structure: polumixins, daptomycin

antimicrobial agents. Andersson and Hughes (5) suggest that most resistance mechanisms incur a fitness cost for the bacterium, which is manifested as a decreased bacterial growth rate. Logic would suggest that in the presence of a selective agent, such as an antimicrobial agent, only the fittest cells (i.e., resistant cells) are capable of survival but that they do so at a high fitness cost. This line of thought also leads to the expectation that removal of the selective pressure will result in the susceptible strains outcompeting the resistant ones. However, data to date suggest that this is not usually the case (5).

Van Hoek et al. (62) noted that some of the more common mechanisms of drug resistance in bacteria include changes in cell wall permeability that restrict access of the drug to its target site; active efflux of the drug out of the cell; enzymatic modification of the drug; acquisition of alternative metabolic pathways to those against which the drug is designed; modification of antimicrobial targets; and overproduction of a target enzyme. Resistances via these processes is acquired in a number of ways, including the mutation of antibiotic targets. For example, resistance to macrolide antibiotics, which bind the 23S ribosomal RNA (rRNA) within the 50S ribosomal subunit of the bacterial ribosome, can be achieved by nucleotide-base substitutions in the 23S rRNA. Single-nucleotide polymorphisms (SNPs) can also result in resistance to quinolone, sulfonamide, and trimethoprim drugs. Mutations in the *rpsL* gene (encoding

a 30S ribosomal protein subunit) have been associated with streptomycin resistance, and frame-shift mutations in the chromosomal gene *ddl* (encoding a cytoplasm enzyme) resulted in glycopeptide resistance in *Enterococcus*.

Acquisition of mobile genetic elements that carry genes or cassettes encoding antimicrobial resistance is another common means by which bacteria become resistant to antimicrobial agents (5). In addition, in members of the *Enterobacteriaceae*, inherent resistance can be associated with efflux pumps, chromosomal mutations, and enzymatic inactivation of drugs (63).

Specific resistance mechanisms

In this section, we focus on the mechanisms of resistance to some of the more commonly used antimicrobial agents that are recommended for treatment of UTIs under the current guidelines from the Infectious Diseases Society of America. These guidelines were recently updated for the treatment of patients with catheter-associated infection (64) and for acute uncomplicated cystitis and pyelonephritis in women (65). Among the drugs considered below are the quinolones, which have superseded trimethoprim-sulfamethoxazole (TMP-SMX) as the first choice of therapy for UTIs in some regions where the prevalence of resistance to TMP-SMX is high (66). Mechanisms of resistance to other classes of drugs used to treat UTIs are also examined, including the β-lactams, fluoroquinolones, tetracyclines, aminoglyco-sides, nitrofurantoin, and fosfomycin (66, 67). Aside from these drugs, the resistance of UPEC to silver compounds is discussed, as these antimicrobial agents are sometimes used to coat urinary catheters to prevent nosocomial UTIs.

Trimethoprim-sulfamethoxazole (TMP-SMX)

Sulfonamides were first implemented as treatments for human infections in the mid-1930s (68). Currently, long-acting sulfonamides, such as SMX and sulfadiazine, are the most commonly used members of this drug class (69). TMP was first used in treating humans in the 1960s (68).

Mode of action of TMP-SMX. TMP and sulfonamides are antimetabolite antimicrobial agents. Each drug blocks distinct steps in folic acid synthesis (70), and their combined use is synergistic (71) and effective against a wide spectrum of bacteria (68). TMP-SMX are selectively toxic for bacteria because, unlike mammalian cells, bacteria cannot use pre-formed folic acid and so must synthesize their own (69). Sulfonamide drugs competitively inhibit dihydropteroate synthase (DHPS), the enzyme that is involved in the catalysis of the condensation reaction of p-aminobenzoic acid (PABA) and dihydro-6-hydroxymethylpterin-pyrophosphate (DHPPP) to form dihydropteroic acid, an important intermediate in the formation of dihydrofolic acid (69). Sulfonamides are PABA analogs and, as such, are competitive inhibitors of the synthase in this reaction.

TMP works by inhibiting the enzyme dihydrofolate reductase (DHFR), which catalyzes the formation of tetrahydrofolate from dihydrofolate (70). It is thought that TMP has fewer side effects than sulfonamides (70). Typically, SMX and TMP are used in combination as broad-spectrum antimicrobial agents, effective against *E. coli* and other members of the *Enterobacteriaceae* (68). It was initially thought that the use of these combination drugs would prevent emergence of drug-resistant strains; however, the worldwide prevalence of resistance to these drugs does not support this claim (70).

Resistance to TMP-SMX. Resistance to TMP-SMX or TMP alone is widespread and has been linked to UTI treatment failures (70). TMP and SMX resistances are often encoded by genes linked to mobile genetic units such as transposons, integrons, and plasmids (68). TMP resistance is mediated by several mechanisms including 1) changes in the bacterial permeability barrier or efflux of the drug and 2) altered regulation or sensitivity of the target enzyme (70).

Chromosomally encoded TMP resistance may be due to insertion of the transposon *Tn7* in the bacterial chromosome. A second, low-level resistance may be linked to a mutation that results in the failure of bacteria to methylate deoxyuridylic acid, resulting in dependence on an exogenous source of thiamine. A third mechanism of TMP resistance may be related to a mutation requiring an increase in the binding affinity of the drug to exert its toxic effect. Most often, TMP resistance has been associated with overproduction of the enzyme DHFR to combat the effect of TMP, alteration of the enzyme, or a combination of these two mechanisms. The number and type of transferable *dhfrI* genes linked with TMP resistance continues to increase with at least 20 such genes currently recognized (72).

Chromosomal sulfonamide resistance is associated with the ability of the SMX to act as a competitive inhibitor of DHPS. Sulfonamides, which are analogs of the DHPS substrate PABA, competitively inhibit DHPS activity, thereby blocking the formation of folate in the bacterial cell. Chromosomal mutations in the *dhps* gene may alter the susceptibility of DHPS to inhibition by sulfonamides. However, resistance to sulfonamides may also be transferable and linked to mobile genetic units. Transferable resistance is mediated by two drug-resistant DHPS enzymes, encoded by *sulI* and *sulII* (68). These genes are largely plasmid-borne. *sulI* is usually linked with other resistance genes on *Tn21* transposons, while *sulII* is usually linked to small IncQ and pBP1 plasmids. Indeed, almost all sulfonamide resistance appears to be due to *sulI* and *sulII* genes with integrons typically carrying *sulI* and small multi-copy plasmids carrying *sulII* (68).

Prevalence of TMP-SMX resistance. Although TMP-SMX is the primary choice of antimicrobial agent to treat uncomplicated UTI in adults and children (66), resistance is increasingly common resulting in substitution of FQs for TMP-SMX in certain countries, where the effectiveness of TMP-SMX is limited. Prevalence rates of resistance to TMP-SMX among UPEC range from 51% to 58% in pediatric and adult cases in Brazil (73, 74), 27.3% in Greece (75), 18% in uncomplicated cases of UTIs in Canada (76), 64% in Nicaragua (77), and 34% to 43% in Turkey (78, 79). Data for the U.S. shows a prevalence of 24.2% for 2010 with this number representing an overall 6.3% increase for the period of 2000–2010 (80).

Fluoroquinolones

Quinolones are a family of chemosynthetic broad-spectrum bactericidal agents, which include the FQs. Nalidixic acid is the predecessor of all the quinolones, including the FQs, which include ciprofloxacin, norfloxacin, levofloxacin, and many others (81). FQs are commonly used as a second drug choice for the treatment of UTIs and are recommended where the prevalence of resistance to TMP-SMX among uropathogens is high (66).

Mode of action of FQs. The FQs are the only antimicrobial agents that act by direct inhibition of DNA synthesis (6). They target type II topoisomerases: DNA gyrase (topoisomerase II) and DNA topoisomerase IV, both of which are essential in bacterial DNA replication (82). DNA gyrase is a tetramer composed of two types of subunits, GyrA and

GyrB. Topoisomerase IV has a similar structure with two types of subunits, ParC and ParE. DNA gyrase introduces negative superhelical twists into DNA, which are important for initiation of DNA replication. Topoisomerase IV is involved in decatenation or separation of the daughter chromosomes in the latter part of DNA replication so that segregation of the daughter cells can occur (83). Both enzymes are critical to bacterial growth, as the topology of chromosomal DNA affects DNA replication, transcription, recombination, and repair (82, 83). FQs inhibit DNA gyrase or topoisomerase IV by stabilizing the drug-DNA-enzyme complexes. As a result, replication-fork complexes accumulate leading to inhibition of DNA synthesis, strand breaks, and ultimately, cell death (83).

Antimicrobial resistance associated with FQs. FQ resistance in the *Enterobacteriaceae* is most often the result of an accumulation of mutations in DNA gyrase (GyrA) primarily and topoisomerase IV (ParC) secondarily (83). Mutations in GyrA that result in FQ resistance mainly occur in the quinolone resistance-determining region (QRDR) (83). Mutations that alter these two enzymes or that limit FQ's access to them are responsible for classically recognized chromosomally encoded FQ resistance (84). Restriction of FQ's access to these enzymes can result from decreased bacterial permeability or enhanced expression of energy-dependent efflux pumps (62).

More recently identified mechanisms of FQ resistance are plasmid-encoded and involve FQ-resistant proteins (Qnr proteins) such as the efflux pumps QepA and OqxAB (83) and the aminoglycoside acetyltransferase AAC(6′)-Ib-cr (where 'cr' stands for ciprofloxacin resistance) (85, 86), an enzyme able to inactivate both aminoglycosides and FQs (84).

Qnr was the first plasmid-linked FQ determinant identified. More recently, it has been referred to as QnrA1 (87). All known *qnr* genes reside on plasmids ranging in size from

7–320 kb, in association with *sul1*-containing class 1 integrons (83, 85, 88). These R plasmids often encode resistance to multiple drugs including β-lactams, aminoglycosides, chloramphenicol, tetracycline, sulfonamides, trimethoprim, and rifampin. Qnr proteins contribute to low-level FQ resistance via alterations in target enzymes, enhanced expression of drug efflux pumps, or changes in outer membrane porins (83). Qnr-containing isolates appear to 'supplement' FQ resistance and facilitate selection of strains bearing chromosomal mutations conferring FQ resistance when under FQ-selective pressure (83).

Another plasmid-encoded FQ-resistance mechanism is the aminoglycoside acetyltransferase AAC(6')-Ib-cr. This variant of the AAC(6')-Ib aminoglycoside acetyltransferase acetylates kanamycin, tobramycin, and amikacin, but also confers a low-level resistance to FQs (83). This enzyme is encoded by a widely distributed cassette localized within *sul1*-type class I integrons on plasmids (86, 88). Still other plasmid-encoded FQ resistance mechanisms involve efflux pumps associated with some human *E. coli* isolates. The first, known as the quinolone efflux pump (Qep), is encoded on large mobilizable plasmids (83). Another known as OqxAB, which confers resistance to multiple agents including FQs (83), was found in one study to be associated with IS*26* on a 43- to 115-kb IncF transferable plasmid (89). Though most frequently associated with bacterial isolates from animals, farm environments, and farm workers, OqxAB was also encoded, though infrequently, among a group of 261 *E. coli* isolates recovered from human blood between 1998 and 2006 (90).

Prevalence of FQ resistance. The overall prevalence of FQ resistance among *E. coli* implicated in UTIs has increased significantly, with FQ-resistance rates varying worldwide. In Brazil, FQ-resistance rates of 17.1% to 21.6% have been reported (91), while Karaca et al. (92) noted the emergence of

resistance to the FQs ofloxacin and ciprofloxacin had increased from 4.1% and 5.2% in 1995 and 1996 to 25.3% and 27.6% in 2002, respectively. A similar study in Italy reported resistance rates of approximately 17% (93). Arslan et al. (94) noted ciprofloxacin resistance varying from 17% to 38% in uncomplicated versus complicated UTIs in Turkey, while resistance to nalidixic acid, norfloxacin, and ciprofloxacin in Senegal were 23.9%, 16.4%, and 15.5%, respectively (95). In Norway, a FQ-resistance rate of 1.2% was observed among 7,302 UPEC cases (96), and a 10-year study in the U.S. reported an overall increase (14.1%) in ciprofloxacin resistance from 3% in 2000 to 17.1% in 2010 (80). Recently, Khawcharoenporn and colleagues (97) examined levofloxacin resistance among *E. coli* from catheter-associated UTIs and found that resistance was significantly higher in UPEC from nosocomial UTIs, compared to community-associated UTIs (38% vs 10%). Study of the distribution of FQ resistance suggests that its emergence occurs where its use is common and that its failure as a secondary treatment for UTIs appears to be increasing.

β-lactam antimicrobials

β-lactam antibiotics are important in healthcare and include a range of agents such as the penicillins, cephalosporins, carbapenems, monobactams, penicillin-cephalosporin hybrids, and penems (98).

Mode of action of β-lactams. All β-lactams function by targeting penicillin-binding proteins (PBPs), the bacterial enzymes responsible for the terminal stages of peptidoglycan biosynthesis. Specifically, β-lactams inhibit the transpeptidation, transglycosylation, endopeptidase, and carboxypeptidase functions of these enzymes (99). In the presence of β-lactams, PBPs form a lethal drug-PBP covalent complex that blocks normal cell wall synthesis, resulting in weakly cross-linked peptidoglycan and leading eventually to cell lysis and death of growing bacteria (100).

Antimicrobial resistance associated with β-lactams. Resistance to the β-lactam agents has become a significant challenge in the treatment of infections with both Gram-positive and -negative organisms. Bacterial resistance to these agents is achieved in three ways. These involve production of β-lactamase enzymes that hydrolyze β-lactams; target-site alteration in PBPs to diminish their sensitivity to β-lactams; and active expulsion of the drug via efflux pumps (100).

Production of β-lactamases is a major resistance mechanism used by Gram-negative organisms. These hydrolytic enzymes target the amide bond of the four-membered β-lactam ring of the drug, causing its inactivation. β-lactamases are classified into various classes, groups, and subgroups based on their structures and functions (101). They can be metal-dependent or independent (98), and there are many different types, a number of which are carried by multidrug-resistance-encoding plasmids (102).

Among the β-lactamases, new types have emerged capable of resisting the effect of newer extended-spectrum β-lactam drugs that have been developed over the past few decades. Several of the extended-spectrum β-lactamases, or ESBLs as they are known (103, 104), are derived by mutation of the active sites of so-called TEM-1, TEM-2, or SHV-1 β-lactamases (103), but others, like CTX-M β-lactamases appear to arise by alternate means (103, 105). The names of these ESBLs have varied origins. The term SHV is derived from 'sulfhydryl variable' as it relates, or was thought to relate, to inhibition of SHV activity by p-chloromercuribenzoate (105). The name of the TEM-type ESBLs was derived from the name of the patient from which the first known TEM-1-containing E. coli was isolated (103). CTX reflects the strong hydrolytic activity of CTX ESBLs against cefotaxime (103), and OXA- type ESBLs are named for their oxacillin-hydrolyzing abilities (103). Many of these ESBLs have broad activity and are able to hydrolyze many of the new extended-spectrum third-generation cephalosporins. Genes encoding the ESBLs are typically plasmid-linked within integrons (98). Most ESBL-producing organisms are resistant to penicillins, first- and second-generation cephalosporins, third-generation oxyimino-cephalosporins, and monobactams. However, most ESBL-producing bacteria are still susceptible to fourth-generation cephalosporins (98).

There are currently more than 200 TEM- and 160 SHV-derived ESBLs recognized. An accurate list of the TEM-, SHV- and OXA-derived β-lactamases is housed at http://www.lahey.org/Studies/ and is updated as new β-lactamases are identified. These enzymes have been found in a range of Enterobacteriaceae including E. coli. Excellent reviews by Poole (98) and Paterson and Bonomo (103) provide detailed overviews of β-lactam resistance and ESBLs.

β-lactam resistance associated with efflux pumps is also recognized. There are five main families of efflux mechanisms for bacterial strains; these include the major facilitator superfamily (MFS), ATP-binding cassette (ABC), resistance-nodulation division (RND), multidrug and toxic compound extrusion (MATE), and small multidrug resistance (SMR) families. The RND family has been reported to accommodate β-lactams including the third- and fourth-generation cephalosporins and carbapenems (98).

Prevalence of β-lactam resistance. Resistance to β-lactams among UPEC is prevalent worldwide. A study in Switzerland by Meier et al. (106) reported a prevalence rate of 69.6% to amoxicillin/clavulanic acid. In Japan, Shigemura et al. (107) reported decreasing susceptibility to a range of β-lactam drugs in complicated versus uncomplicated UTIs ranging from 0% to as high as 21.5% and 47.2%, respectively. In India, Taneja et al. (108) reported resistance rates in UPEC against β-lactams ranging from 70% to 93.9% with the lowest resistance observed to imipenem. In the U.S., Sanchez et al. (80) reported resistance prevalence increasing from 38.2% in 2000 to 43.4% in 2010.

Of special concern with regards to the β-lactams is the emergence of the new metallo-β-lactamase 1 (denoted New Deli NDM-1). This metallo-β-lactamase was first identified in *Klebsiella pneumoniae* and *E. coli* from an Indian patient in Sweden in 2008 (109). Strains possessing NDM-1 can hydrolyze all penicillins, cephalosporins, and carbapenems and are typically broadly resistant to other drug classes in addition to the β-lactams (110). Data suggest that these NDM-type *E. coli* strains may also be linked with UTIs and have been recovered from UTI patients in Thailand (111) as well as from the urine, blood, and feces of patients in New Zealand (112), Lebanon (113), and Denmark (114). The emergence of such a strain type in human illness including UTIs could have devastating consequences. Active monitoring for the emergence of NDM-1 strain types in UPEC warrants careful attention and proactive design of methods for its control should be undertaken.

Alternative antimicrobials for the control of β-lactam-resistant strains include fosfomycin (115) and tigecycline (116).

Nitrofurantoin

Nitrofurantoin has been used as an option for treating UTIs for more than 50 years. Its use has declined in favor of other antibiotics such as TMP-SMX and the FQs (76). However, the emergence of uropathogens that are resistant to many front-line drugs, has led some researchers to re-examine the empirical therapeutic value of nitrofurantoin (117). Upon administration, the drug becomes concentrated in urine, where it can be effective against bladder infections (cystitis) (76). However, nitrofurantoin has a lower cure rate than other drugs, but is nonetheless considered appropriate for use during pregnancy, in pediatric medicine, and as prophylaxis against recurrent UTI.

Mode of action of nitrofurantoin. Nitrofurantoin is a nitroheterocyclic compound.

The active site of the drug is where the nitro group is coupled to the heterocyclic furan. The drug is activated within bacterial cells by nitrofuran reductases that rapidly reduce nitrofurantoin to various active intermediates that interfere with protein and DNA synthesis, energy metabolism, and cell wall and carbohydrate synthesis (118). Intracellularly produced intermediate products attack the chromosomal DNA of the cell causing a range of mutations, and it is believed that hydroxylamine (one of the intermediate products produced during the reduction of the nitrofurantion) activity results in bacterial DNA and protein damage (118, 119).

Antimicrobial resistance associated with nitrofurantoin. Antimicrobial resistance to nitrofurantoin can be chromosomal or plasmid-mediated and is associated with inhibition of nitrofuran reductase (120, 121). McCalla et al. (122) showed that *E. coli* resistance to nitrofuran compounds is associated with loss of enzyme activity. However, the full mechanism of nitrofuran resistance is still not fully understood. McOsker and Fitzpatrick (123) showed that nitrofurantoin itself inhibits the synthesis of bacterial nitrofuran reductase by interacting with bacterial ribosomes. Sandegren et al. (119) also noted that resistance may be linked to the nitroreductase activity, which is of two types – type I that is insensitive to oxygen and type II that is inhibited by oxygen.

Prevalence of resistance to nitrofuran compounds. A study from Crete found that the overall prevalence of nitrofuran resistance over a five-year period ranged from a low of 3.7% to a high of 13% (124). In Brazil, the prevalence of nitrofuran resistance among UPEC was also low (2.9%) (125), while the rates of resistance in Portugal, Spain, Finland, and Sweden were 5.8%, 4.2%, 0.5%, and 0%, respectively (126). In the U.S., Sanchez et al. (80) reported that the prevalence of nitrofurantoin resistance in UPEC has ranged from 0.8% in 2000 to

1.6% in 2010. These data contrast significantly with Sire et al. (95) who reported a resistance rate of 10.1% among UPEC isolates in Senegal. Given the low resistance rates observed, recommendations from the Infectious Diseases Society of America list nitrofurantoin as an option for treating acute uncomplicated cystitis (65).

Fosfomycin
Fosfomycin is a phosphoric acid antibacterial agent, which is indicated in the treatment of uncomplicated cystitis (76). Fosfomycin is a broad-spectrum drug used in the treatment of uncomplicated UTIs, bacteriuria during pregnancy, and pediatric UTIs, and as a prophylactic agent before urological procedures (127).

Mode of action of fosfomycin. Fosfomycin was first described in 1969 (128). It has activity against both Gram-positive and -negative organisms, and most organisms seem to be sensitive to the agent or show low resistance rates. Activity of fosfomycin is related to its ability to inhibit cell wall biosynthesis by inactivating the enzyme UDP-N-acetylglucosamine enolpyruvyl transferase (MurA), which is involved in peptidoglycan biosynthesis (129).

Mechanisms of fosfomycin resistance. Bacterial resistance to fosfomycin is primarily chromosomal, however, plasmid-mediated resistance to fosfomycin has also been observed in rare instances (127). Chromosomally mediated fosfomycin resistance is associated with mutations that interfere with the L-α-glycophosphate and hexose phosphate uptake transport systems (127, 129). Plasmid-mediated resistance has been linked to carriage of *fosA* gene, which encodes a protein (FosA) involved in the conjugation of glutathione to fosfomycin, thus rendering it useless (130).

Prevalence of fosfomycin resistance. The occurrence of fosfomycin resistance among UPEC is low, if present at all. Liu et al. (131) reported a rate of 4.5% resistance among *E. coli*

in Taiwan, while earlier work by Schito (132) reported that the worldwide prevalence was 1% to 3%. Sire et al. (95) reported a prevalence of 0.7% resistance in Senegal. An extensive countrywide study in Europe reported an overall prevalence of <2% (126), and a rate of 6% was noted in Pakistan by Noor et al. (133). These findings suggest that this 'old world' drug may find 'new world' use, as resistance to the more commonly used antimicrobial agents for UTI treatment continues to emerge.

Tetracyclines
The tetracyclines were first discovered in the early 1940s. They are broad-spectrum antibiotics produced by members of the *Streptomyces* (134). Tetracyclines in general have been used to treat UTIs and other infections caused by organisms such as *Mycoplasma* and *Chlamydia*.

Mode of action of tetracyclines. The results of several studies suggest that tetracyclines function by binding the 30S subunit of microbial ribosomes. In doing so, they inhibit protein synthesis by blocking attachment of charged aminoacyl-transfer RNA (tRNA) to the A site of the ribosome, resulting in inhibition of bacterial growth (135, 136). However, there is evidence that their mode of action may be more complex than once appreciated (137).

Tetracycline has a broad spectrum of activity against a range of Gram-positive and -negative organisms, mycoplasmas, rickettsias, and protozoan parasites; however, *Pseudomonas aeruginosa* and *Proteus* spp. appear to have intrinsic resistance to tetracycline (134). Alternative tetracyclines, such as tigecycline, a derivative of minocycline, have good activity against bacteria that are resistant to tetracyclines and ESBL-producing *E. coli*. Tigecycline is, however, considered to be an alternative when aminoglycosides and carbapenems are unavailable (138).

Resistance associated with tetracyclines. Antimicrobial resistance to tetracycline occurs by ribosomal protection, energy-dependent efflux of the drug, or enzymatic drug inactivation (62). Ribosomal protection relies on alteration of drug target sites. Here, bacterial cells produce ribosomal protection proteins that bind to the ribosome and alter its conformation, thus altering the active site where tetracycline would normally act (134). Though resistance to tetracyclines is primarily due to active efflux or ribosomal protection, oxygen-dependent, TetX-catalyzed destruction of tetracycline has been described in certain organisms (139). To our knowledge, enzymatic inactivation of tetracycline is not a mechanism of tetracycline resistance that has been found among *E. coli*. There are at least 46 different tetracycline-resistance determinants recognized (134) with novel determinants identified as unique if they share less than 79% amino acid-sequence identity with all of the known genes (62).

As the list of tetracycline-resistance genes has grown over the years, the nomenclature describing these determinants has had to evolve. Where once the letters of the English alphabet provided enough coverage, modifications of the naming schemes had to be designed to accommodate new and hybrid tetracycline-resistance genes (140–143). Determinants include several *tet* (tetracycline resistance), a few *otr* (oxytetracycline resistance), and a few others (62). The designations *tet* and *otr* do not reflect inherent differences in these determinants, but instead refer to the source organism from which these determinants were originally identified (136). Typically, resistance to tetracyclines is due to acquisition of new genes on plasmids or transposons although, in a few cases, resistance may be the result of mutations (136).

Prevalence of resistance to tetracyclines. Antimicrobial resistance to tetracyclines appears to be widespread among UPEC. Thus, tetracyclines may not be the best first choice for treatment of a UTI. For instance, the prevalence of tetracycline resistance in Nigerian UPEC ranges from 88% to 100% (144, 145). Although the prevalence is less in Brazil, it is still high, ranging from >30% to 73% (73, 125), and while it is still lower in the U.S., the rate of tetracycline resistance has increased from 22.6% in 2000 to 24.9% in 2010 (a 2.3% increase over a 10-year period) (80). The high and increasing prevalence of tetracycline resistance likely reflects the multiple mechanisms by which a strain can become resistant and the linkage of these mechanisms to mobile genetic elements.

Aminoglycosides

The aminoglycosides are broad-spectrum antibiotics with concentration-dependent killing action that have been in clinical use for over 60 years (62, 146). Aminoglycosides include several natural and semisynthetic antimicrobial compounds, which contain one of several aminated sugars joined to a dibasic cyclitol via glycosidic linkages (147). The first aminoglycoside discovered was streptomycin, which was produced by *Streptomyces griseus*. Not long after, additional aminoglycosides from *Streptomyces* spp. were described including neomycin and kanamycin (62). Gentamicin was recovered from the actinomycete, *Micromonospora purpurea* (62) in the 1960s, and in the 1970s the first semisynthetic aminoglycosides were developed. These included netlimicin, dibekacin, and amikacin (62, 147). Aminoglycosides derived from *Streptomyces* spp. can be identified by their '-mycin' suffix, while *Micromonospora*-derived aminoglycosides have the '-micin' suffix (62).

These drugs are indicated in the case of serious bacterial infection or a complicated UTI (148). The drug is usually given by injection and is excreted in the urine unchanged. Often, aminoglycosides are given in combination with other drugs such as β-lactams and are valued for their ability to act in synergy with other agents (62). However, their use is complicated by their toxicity to the host and the emergence of aminoglycoside resistance among certain bacterial pathogens (62, 147).

Mechanism of action of aminoglycosides.
Aminoglycosides work by inhibiting prokaryotic protein synthesis by binding to 16S rRNA and also by disrupting the integrity of the bacterial cell membrane (149). At high concentrations, they may also affect protein synthesis in mammalian cells, resulting in their toxic effects (147).

Resistance associated with aminoglycosides. Resistance to most aminoglycosides, other than streptomycin, which is greatly impacted by alteration of ribosomal binding sites, is owed to one of two overall strategies: 1) decreased accumulation of the drug in the bacterium or 2) bacterial production of drug-modifying enzymes, (147). The specific mechanisms by which aminoglycoside resistance occurs include, 1) deactivation of the drug by N-acetylation, adenylation, or O-phosphorylation; 2) decreased outer membrane permeability to the drug; 3) decreased inner-membrane transport, active efflux, and drug trapping; 4) alteration of the 30S ribosomal subunit target by mutation; and 5) methylation of the aminoglycoside-binding site (149).

Active efflux, membrane impermeablization, and ribosomal alteration may be intrinsic properties of the resistant bacterium or the result of chromosomal mutations (62). However, the most common aminoglycoside-resistance mechanism is enzymatic inactivation of the drug. Over 50 such modifying enzymes have been identified (150) and include acetyltransferases (AAC), adenyltransferases (ANT), phosphotransferases (APH), and bifunctional enzymes (62). The genes encoding these modifying enzymes are generally located on mobile genetic elements, including plasmids, transposons, and integrons (149). This location has no doubt facilitated the emergence of aminoglycoside resistance. Indeed, the first plasmid-linked aminoglycoside-resistance gene, *armA* (the *a*minoglycoside *r*esistance *m*ethyltransferase) was found in *Klebsiella* in 2003 (151), but by 2005, Galimand et al. (152) had found

it in clinical isolates of *Citrobacter freundii*, *Enterobacter cloacae*, *E. coli*, *Klebsiella pneumoniae*, *Salmonella enterica* serotype Enteritidis, and *Shigella flexneri* from across Europe. Several other such genes have since been described (153, 154).

Efflux pump-associated resistance. Aminoglycosides are substrates for a number of efflux pumps, including chromosomally encoded members of the RND family of efflux pumps (155).

16S rRNA methylation. Another mechanism of resistance to aminoglycosides is 16S rRNA methylation. Bacteria producing rRNA methylases can alter the nucleotides normally bound aminoglycosides, preventing the drugs from disrupting ribosomal function.

Prevalence of aminoglycoside resistance. Because of the relative toxicity of these drugs, their use is often limited to the treatment of complicated infections. In a study in Iran, Mohammad-Jafari et al. (156) reported the prevalence of gentamicin resistance among *E. coli* recovered from UTIs to be 23%, while a study in Tunisia reported <14% of isolates were resistant to amikacin and <5% resistant to gentamicin (157). In contrast, Gad et al. (158) examined the aminoglycoside resistance of Gram-negative bacteria from patients with UTIs, ear, skin, and GI infections in Egypt. Overall, over 80% of these isolates were resistant to streptomycin, but only ~18% were resistant to amikacin. Intermediate degrees of resistance were found to neomycin, kanamycin, gentamicin, and tobramycin. Of the 50 *E. coli* isolates they examined, almost 40% were from UTIs. Among these UTI-associated strains, 78% to 82% were resistant to streptomycin, 48% to 54% to neomycin, 44% to 46% to kanamycin, 36% to 40% to gentamicin, 30% to tobramycin, and 4% to 16% to amikacin. Such widely varying results demonstrate that resistance rates can differ greatly depending on regional use.

Silver-containing antimicrobial agents

Heavy metals such as silver, copper, cadmium, lead, and mercury are considered toxic to bacteria and, as such, have often been used in their salt forms as antimicrobial agents in hospital settings and other environments (159). Of particular interest here is the use of silver compounds since they are used to coat urinary catheters in order to prevent nosocomial UTIs. These agents also find frequent use in burn wards, where they may be applied to wounds in creams and dressings. Silver salts can also be applied to the eyes of newborns to prevent neonatal eye infection and are commonly used in dental amalgams, water-filtration systems, impregnated plastics (cutting boards), and fabrics (160). The application of silver as a coating agent on urinary-tract catheters and heart valves has become a common practice in an effort to prevent the growth of bacterial biofilms (159, 160). The introduction of silver-coated catheters in the U.S. occurred more than a decade ago (161).

Johnson et al. (162) reported that catheter-associated UTIs account for 40% of hospital-acquired infections. A number of systematic reviews suggest that use of silver-coated catheters results in fewer UTIs (161, 163, 164). Indeed, Rupp and colleagues (164) reported a 57% decrease in the prevalence of catheter-associated UTIs when silver-coated catheters were used. However, Hooton et al. (64) concluded that there are insufficient data on which to make a recommendation for use of coated catheters for short- or long-term indwelling use.

Considering the potential of silver coatings to prevent catheter-associated UTIs, the prevalence of silver resistance among UPEC and other uropathogens warrants monitoring. Silver (160) reported on the presence of silver-resistant strains in environments where silver use as an antiseptic was high (e.g., burn wards). A silver-resistance cluster consisting of nine genes has been identified that encodes a periplasmic protein, two efflux pumps, and a chemiosmotic RND-exchange system (160). Evidence for the presence of the silver-resistance genes in IncH plasmids has been described by Gupta and colleagues (159). Plasmids of this replicon type were found among UPEC isolates in Denmark (165) and elsewhere, but in UPEC and human fecal commensal *E. coli* they are observed infrequently (60).

A recent report that may have negative implications for the future use of silver compounds in the prevention of UTIs is found in the description of the sequence of plasmid pAPEC-O2-R (37). This IncF plasmid was recovered from an APEC strain isolated from a bird with colibacillosis. pAPEC-O2-R, which contains a functional silver-resistance gene cluster, was easily transferred by conjugation into other *E. coli* strains, including a human UPEC isolate. In addition to silver resistance, this plasmid encodes resistance to quaternary ammonium compounds, tetracycline, sulfonamides, aminoglycosides, trimethoprim, and beta-lactam antimicrobial agents. IncF plasmids occur at greater prevalence than IncH plasmids in a variety of *E. coli*, including in human UPEC and fecal strains (60). Results like these suggest that the promiscuous nature of ExPEC R plasmids may have a significant bearing on the future of UTI control. Another R plasmid (the IncHI2 type plasmid pAPEC-O1-R (16)), containing a functional silver resistance operon, was found in APEC O1, an APEC strain with many similarities to human UPEC isolates (166).

Such findings, coupled with recent reports linking plasmid-containing, APEC-contaminated retail poultry to the occurrence of human disease (53), serve as important reminders that many microbes containing a variety of resistance determinants are circulating in the environment. Many times, all that may be needed for these strains to emerge is the use of some key selective agent. Thus, silver compounds to prevent UTIs must always be used thoughtfully, recognizing that their use may select for bacteria containing plasmids encoding resistance to silver and many other drugs, and perhaps even virulence genes. For this reason, continued and close monitoring for the development of silver resistance among UPEC is justifiably warranted.

CONCLUSIONS

Though the authors would be remiss if we failed to encourage prudent use of antimicrobial agents when treating UTIs and other infections, we must recognize that any use, including prudent use, may select for the emergence of resistant strains of bacteria. This line of reasoning is especially sobering in terms of organisms containing R plasmids like pAPEC-O1-R, pAPEC-O2-R, or pAPEC-O103-ColBM, as use of even a single antimicrobial agent (including antibiotics, silver, mercury, or copper-containing antiseptics and disinfectants, or even benzalkonium chloride antiseptics and disinfectants) can select for MDR pathogens with enhanced abilities to resist therapy and disinfection. This scenario is even more sobering when considering pathogenic strains harboring hybrid R-virulence plasmids - in which case use of even a single antimicrobial agent could conceivably select for pathogens *with both enhanced drug resistance and increased virulence potential.*

As noted by many researchers and experienced first-hand all too often by physicians, veterinarians, and their patients, the emergence of antimicrobial resistance among uropathogens and other pathogens is a growing problem, one that has surpassed the pace of drug discovery and development (105, 167). Continuing the successful fight against bacterial pathogens will likely require a multipronged approach involving enhanced awareness, increased vigilance, intensified control measures to limit the dissemination of resistance, strict implementation of infection control procedures, and investment in drug discovery (105). For the latter to be successful, new drugs or drug combina-

tions might incorporate an anti-plasmid approach, where small molecules target plasmid-replication control as a means to eliminate MDR-encoding plasmids and the transmission of drug-resistance genes. Such adjuncts to routine therapies could 'rejuvenate' many antimicrobial agents whose effectiveness has waned due to plasmid-mediated multidrug resistance (36). Kunin (168) also reminds us that it is better to avoid a UTI than to treat one and finds that there is no substitute for thoughtful care. Avoiding unnecessary or prolonged use of urinary catheters, avoidance of invasive procedures that put patients at greater risk of infection, and the development of improved catheter and drainage bags are all common-sense approaches that can decrease the occurrence of UTIs in hospitals. Maki and Tambyah (169) also advocate for 'engineering out' the risk of infection with improved urinary catheters and suggest that our greatest hope for reducing catheter-associated UTIs is vaccine development. Gilbert and McBain (170) find hope in good hygienic practice in appropriate settings and the wise application of biocides. Thus, there are multiple approaches that deserve study with preventive strategies and judicious application of antimicrobial agents, all playing important roles in the future control of UPEC-caused UTIs.

ACKNOWLEDGMENTS

Conflicts of interest: We declare no conflicts.

CITATION

Nolan LK, Li G, Logue CM. 2015. Origin and dissemination of antimicrobial resistance among uropathogenic *Escherichia coli.* Microbiol Spectrum 3(5):UTI-0007-2012.

REFERENCES

1. **Totsika M, Beatson SA, Sarkar S, Phan MD, Petty NK, Bachmann N, Szubert M, Sidjabat HE, Paterson DL, Upton M, Schembri MA.**

2011. Insights into a multidrug resistant *Escherichia coli* pathogen of the globally disseminated ST131 lineage: genome analysis and virulence mechanism. *PLoS One* **6**:e26578. doi:10.1371/journal.pone.0026578

2. **Kresge N, Simoni RD, Hill RL.** 2004. Selman Waksman: the Father of Antibiotics. *J Biolog Chem* **279**:101–102.

3. **Davies J, Davies D.** 2010. Origins and evolution of antibiotic resistance. *Microbiol Mol Biol Rev* **74**:417–433.

4. **Tenover FC.** 2006. Mechanisms of antimicrobial resistance in bacteria. *Am J Infect Control* **34**:S3–10.

5. **Andersson DI, Hughes D.** 2010. Antibiotic resistance and its cost: is it possible to reverse resistance? *Nat Rev Microbiol* **8**:260–271.

6. **Hooper DC.** 2001. Mechanisms of action of antimicrobials: focus on fluoroquinolones. *Clin Infect Dis* **32**:S9–15.

7. **Hughes VM, Datta N.** 1983. Conjugative plasmids in bacteria of the 'pre-antibiotic' era. *Nature* **302**:725–726.

8. **Song JS, Jeon JH, Lee JH, Jeong SH, Jeong BC, Kim SJ, Lee JH, Lee SH.** 2005. Molecular characterization of TEM-type β-lactamases identified in cold-seep sediments of Edison Seamount (south of Lihir Island, Papua New Guinea). *J Microbiol* **43**:172–178.

9. **D'Costa VM, King CE, Kalan L, Morar M, Sung WW, Schwarz C, Froese D, Zazula G, Calmels F, Debruyne R, Golding GB, Poinar HN, Wright GD.** 2011. Antibiotic resistance is ancient. *Nature* **477**:457–461.

10. **D'Costa VM, McGrann KM, Hughes DW, Wright GD.** 2006. Sampling the antibiotic resistome. *Science* **311**:374–377.

11. **Davies JE.** 1997. Origins, acquisition and dissemination of antibiotic resistance determinants. *Ciba Found Symp* **207**:15–27; discussion 27–35.

12. **Martínez JL.** 2008. Antibiotics and antibiotic resistance genes in natural environments. *Science* **321**:365–367.

13. **Tamae C, Liu A, Kim K, Sitz D, Hong J, Becket E, Bui A, Solaimani P, Tran KP, Yang H, Miller JH.** 2008. Determination of antibiotic hypersensitivity among 4,000 single-gene-knockout mutants of *Escherichia coli*. *J Bacteriol* **190**:5981–5988.

14. **Liu A, Tran L, Becket E, Lee K, Chinn L, Park E, Tran K, Miller JH.** 2010. Antibiotic sensitivity profiles determined with an *Escherichia coli* gene knockout collection: generating an antibiotic bar code. *Antimicrob Agents Chemother* **54**:1393–1403.

15. **Smets BF, Barkay T.** 2005. Horizontal gene transfer: perspectives at a crossroads of scientific disciplines. *Nature Rev Microbiol* **3**:675–678.

16. **Johnson TJ, Wannemeuhler YM, Scaccianoce JA, Johnson SJ, Nolan LK.** 2006. Complete DNA sequence, comparative genomics, and prevalence of an IncHI2 plasmid occurring among extraintestinal pathogenic *Escherichia coli* isolates. *Antimicrob Agents Chemother* **50**:3929–3933.

17. **Kariyawasam S, Nolan LK.** 2011. papA gene of avian pathogenic *Escherichia coli*. *Avian Dis* **55**:532–538.

18. **Mazel D.** 2006. Integrons: agents of bacterial evolution. *Nature Rev Microbiol* **4**:608–620.

19. **Chang LL, Chang TM, Chang CY.** 2007. Variable gene cassette patterns of class 1 integron-associated drug-resistant *Escherichia coli* in Taiwan. *Kaohsiung J Med Sci* **23**:273–280.

20. **Blahna MT, Zalewski CA, Reuer J, Kahlmeter G, Foxman B, Marrs CF.** 2006. The role of horizontal gene transfer in the spread of trimethoprim-sulfamethoxazole resistance among uropathogenic *Escherichia coli* in Europe and Canada. *J Antimicrob Chemother* **57**:666–672.

21. **Farshad S, Japoni A, Hosseini M.** 2008. Low distribution of integrons among multidrug resistant *E. coli* strains isolated from children with community-acquired urinary tract infections in Shiraz, Iran. *Pol J Microbiol* **57**:193–198.

22. **Rijavec M, Starcic Erjavec M, Ambrozic Avgustin J, Reissbrodt R, Fruth A, Krizan-Hergouth V, Zgur-Bertok D.** 2006. High prevalence of multidrug resistance and random distribution of mobile genetic elements among uropathogenic *Escherichia coli* (UPEC) of the four major phylogenetic groups. *Curr Microbiol* **53**:158–162.

23. **White PA, Rawlinson WD.** 2001. Current status of the *aada* and *dfr* gene cassette families. *J Antimicrob Chemother* **47**:495–496.

24. **Kerrn MB, Klemmensen T, Frimodt-Møller N, Espersen F.** 2002. Susceptibility of Danish *Escherichia coli* strains isolated from urinary tract infections and bacteraemia, and distribution of *sul* genes conferring sulphonamide resistance. *J Antimicrob Chemother* **50**:513–516.

25. **Towner KJ, Brennan A, Zhang Y, Holtham CA, Brough JL, Carter GI.** 1994. Genetic structures associated with spread of the type Ia trimethoprim-resistant dihydrofolate reductase gene amongst *Escherichia coli* strains isolated in the Nottingham area of the United Kingdom. *J Antimicrob Chemother* **33**:25–32.

26. **Heikkilä E, Sundström L, Skurnik M, Huovinen P.** 1991. Analysis of genetic localization of the type I trimethoprim resistance gene from *Escherichia coli* isolated in Finland. *Antimicrob Agents Chemother* **35**:1562–1569.

27. **Yu HS, Lee JC, Kang HY, Jeong YS, Lee EY, Choi CH, Tae SH, Lee YC, Seol SY, Cho DT.** 2004. Prevalence of *dfr* genes associated with integrons and dissemination of *dfrA17* among urinary isolates of *Escherichia coli* in Korea. *J Antimicrob Chemother* **53**:445–450.

28. **Yu HS, Lee JC, Kang HY, Ro DW, Chung JY, Jeong YS, Tae SH, Choi CH, Lee EY, Seol SY, Lee YC, Cho DT.** 2003. Changes in gene cassettes of class 1 integrons among *Escherichia coli* isolates from urine specimens collected in Korea during the last two decades. *J Clin Microbiol* **41**:5429–5433.

29. **White PA, McIver CJ, Rawlinson WD.** 2001. Integrons and gene cassettes in the *Enterobacteriaceae*. *Antimicrob Agents Chemother* **45**:2658–2661.

30. **Peirano G, Agersø Y, Aarestrup FM, dos Prazeres Rodrigues D.** 2005. Occurrence of integrons and resistance genes among sulphonamide-resistant *Shigella* spp. from Brazil. *J Antimicrob Chemother* **55**:301–305.

31. **Grape M, Farra A, Kronvall G, Sundström L.** 2005. Integrons and gene cassettes in clinical isolates of co-trimoxazole-resistant Gram-negative bacteria. *Clin Microbiol Infect* **11**:185–192.

32. **Karlowsky JA, Kelly LJ, Thornsberry C, Jones ME, Sahm DF.** 2002. Trends in antimicrobial resistance among urinary tract infection isolates of *Escherichia coli* from female outpatients in the United States. *Antimicrob Agents Chemother* **46**:2540–2545.

33. **Solberg OD, Ajiboye RM, Riley LW.** 2006. Origin of class 1 and 2 integrons and gene cassettes in a population-based sample of uropathogenic *Escherichia coli*. *J Clin Microbiol* **44**:1347–1351.

34. **Bennett PM.** 1999. Integrons and gene cassettes: a genetic construction kit for bacteria. *J Antimicrob Chemother* **43**:1–4.

35. **Márquez C, Labbate M, Ingold AJ, Roy Chowdhury P, Ramírez MS, Centrón D, Borthagaray G, Stokes HW.** 2008. Recovery of a functional class 2 integron from an *Escherichia coli* strain mediating a urinary tract infection. *Antimicrob Agents Chemother* **52**:4153–4154.

36. **Carattoli A.** 2009. Resistance plasmid families in *Enterobacteriaceae*. *Antimicrob Agents Chemother* **53**:2227–2238.

37. **Johnson TJ, Siek KE, Johnson SJ, Nolan LK.** 2005. DNA sequence and comparative genomics of pAPEC-O2-R, an avian pathogenic *Escherichia coli* transmissible R plasmid. *Antimicrob Agents Chemother* **49**:4681–4688.

38. **Johnson TJ, Jordan D, Kariyawasam S, Stell AL, Bell NP, Wannemuehler YM, Alcarón CF, Li G, Tivendale KA, Logue CM, Nolan LK.** 2010. Sequence analysis and characterization of a transferrable hybrid plasmid encoding multidrug resistance and enabling zoonotic potential for extraintestinal *Escherichia coli*. *Infect Immun* **78**:1931–1942.

39. **Ojo KK, Kehrenberg C, Schwarz S, Odelola HA.** 2002. Identification of a complete *dfrA14* gene cassette integrated at a secondary site in a resistance plasmid of uropathogenic *Escherichia coli* from Nigeria. *Antimicrob Agents Chemother* **46**:2054–2055.

40. **Ojo KK, Kehrenberg C, Odelola HA, Schwarz S.** 2003. Structural analysis of the tetracycline resistance gene region of a small multiresistance plasmid from uropathogenic *Escherichia coli* in Nigeria. *J Antimicrob Chemother* **52**:1043–1044.

41. **Adeniyi BA, Amajoyi CC, Smith SI.** 2006. Plasmid profiles of multidrug resistant local uropathogenic *Escherichia coli*, *Klebsiella* spp., *Proteus* spp., and *Pseudomonas* spp. isolates. *J Biol Sci* **6**:527–531.

42. **Lina TT, Rahman SR, Gomes DJ.** 2007. Multiple-antibiotic resistance mediated by plasmids and integrons in uropathogenic *Escherichia coli* and *Klebsiella pneumoniae*. *Bangladesh J Microbiol* **24**:19–23.

43. **Coque TM, Novais A, Carattoli A, Poirel L, Pitout J, Peixe L, Baquero F, Cantón R, Nordmann P.** 2008. Dissemination of clonally related *Escherichia coli* strains expressing extended-spectrum β-lactamase CTX-M-15. *Emerg Infect Dis* **14**:195–200.

44. **Deschamps C, Clermont O, Hipeaux MC, Arlet G, Denamur E, Branger C.** 2009. Multiple acquisitions of CTX-M plasmids in the rare D$_2$ genotype of *Escherichia coli* provide evidence for convergent evolution. *Microbiology* **155**:1656–1668.

45. **Zhao WH, Hu ZQ.** 2012. Epidemiology and genetics of CTX-M extended-spectrum β-lactamases in Gram-negative bacteria. *Crit Rev Microbiol* **39**:79–101.

46. **Ho PL, Lo WU, Yeung MK, Li Z, Chan J, Chow KH, Yam WC, Tong AH, Bao JY, Lin CH, Lok S, Chiu SS.** 2012. Dissemination of pHK01-like incompatibility group IncFII plasmids encoding CTX-M-14 in *Escherichia coli* from human and animal sources. *Vet Microbiol* **158**:172–179.

47. **Baral P, Neupane S, Marasini BP, Ghimire KR, Lekhak B, Shrestha B.** 2012. High prevalence of multidrug resistance in bacterial uropathogens from Kathmandu, Nepal. *BMC Res Notes* **5:**38.

48. **Johnson TJ, Siek KE, Johnson SJ, Nolan LK.** 2006. DNA sequence of a ColV plasmid and prevalence of selected plasmid-encoded virulence genes among avian *Escherichia coli* strains. *J Bacteriol* **188:**745–758.

49. **Johnson TJ, Johnson SJ, Nolan LK.** 2006. Complete DNA sequence of a ColBM plasmid from avian pathogenic *Escherichia coli* suggests that it evolved from closely related *ColV* virulence plasmids. *J Bacteriol* **188:**5975–5983.

50. **Rodriguez-Siek KE, Giddings CW, Doetkott C, Johnson TJ, Fakhr MK, Nolan LK.** 2005. Comparison of *Escherichia coli* isolates implicated in human urinary tract infection and avian colibacillosis. *Microbiology* **151:**2097–2110.

51. **Johnson TJ, Wannemuehler Y, Johnson SJ, Stell AL, Doetkott C, Johnson JR, Kim KS, Spanjaard L, Nolan LK.** 2008. Comparison of extraintestinal pathogenic *Escherichia coli* strains from human and avian sources reveals a mixed subset representing potential zoonotic pathogens. *Appl Environ Microbiol* **74:**7043–7050.

52. **Skyberg JA, Johnson TJ, Johnson JR, Clabots C, Logue CM, Nolan LK.** 2006. Acquisition of avian pathogenic *Escherichia coli* plasmids by a commensal *E. coli* isolate enhances its abilities to kill chicken embryos, grow in human urine, and colonize the murine kidney. *Infect Immun* **74:**6287–6292.

53. **Johnson TJ, Logue CM, Wannemuehler Y, Kariyawasam S, Doetkott C, DebRoy C, White DG, Nolan LK.** 2009. Examination of the source and extended virulence genotypes of *Escherichia coli* contaminating retail poultry meat. *Foodborne Pathog Dis* **6:**657–667.

54. **Manges AR, Johnson JR.** 2012. Food-borne origins of *Escherichia coli* causing extraintestinal infections. *Clin Infect Dis* **55:**712–719.

55. **Suhartono.** 2010. Examination of uropathogenic *Escherichia coli* strains conferring large plasmids. *Biodiversitas* **11:**59–64.

56. **Frost LS, Ippen-Ihler K, Skurray RA.** 1994. Analysis of the sequence and gene products of the transfer region of the F sex factor. *Microbiol Rev* **58:**162–210.

57. **Rijavec M, Starcic Erjavec M, Ambrozic Avgustin J, Reissbrodt R, Fruth A, Krizan-Hergouth V, Zgur-Bertok D.** 2006. High prevalence of multidrug resistance and random distribution of mobile genetic elements among uropathogenic *Escherichia coli* (UPEC) of the four major phylogenetic groups. *Curr Microbiol* **53:**158–162.

58. **Couturier M, Bex F, Bergquist PL, Maas WK.** 1988. Identification and Classification of bacterial plasmids. *Microbiol Rev* **52:**375–395.

59. **Carattoli A, Bertini A, Villa L, Falbo V, Hopkins KL, Threlfall EJ.** 2005. Identification of plasmids by PCR-based replicon typing. *J Microbiol Methods* **63:**219–228.

60. **Johnson TJ, Wannemuehler YM, Johnson SJ, Logue CM, White DG, Doetkott C, Nolan LK.** 2007. Plasmid replicon typing of commensal and pathogenic *Escherichia coli* isolates. *Appl Environ Microbiol* **73:**1976–1983.

61. **Johnson TJ, Logue CM, Johnson JR, Kuskowski MA, Sherwood JS, Barnes HJ, DebRoy C, Wannemuehler YM, Obata-Yasuoka M, Spanjaard L, Nolan LK.** 2012. Associations between multidrug resistance, plasmid content, and virulence potential among extraintestinal pathogenic and commensal *Escherichia coli* from humans and poultry. *Foodborne Pathog Dis* **9:**37–46.

62. **van Hoek AH, Mevius D, Guerra B, Mullany P, Roberts AP, Aarts HJ.** 2011. Acquired antibiotic resistance genes: an overview. *Front Microbiol* **2:**203.

63. **Allen HK, Donato J, Wang HH, Cloud-Hansen KA, Davies J, Handelsman J.** 2010. Call of the wild: antibiotic resistance genes in natural environments. *Nature Rev Microbiol* **8:**251–259.

64. **Hooton TM, Bradley SF, Cardenas DD, Colgan R, Geerlings SE, Rice JC, Saint S, Schaeffer AJ, Tambayh PA, Tenke P, Nicolle LE, Infectious Diseases Society of America.** 2010. Diagnosis, prevention, and treatment of catheter-associated urinary tract infection inadults: 2009 International Clinical Practice Guidelines from the Infectious Diseases Society of America. *Clin Infect Dis* **50:**625–663.

65. **Gupta A, Hooton TM, Naber KG, Wullt B, Colgan R, Miller LG, Nicolle LE, Raz R, Schaeffer AJ, Soper DE.** 2011. International clinical practice guidelines for the treatment of acute uncomplicated cystitis and pyelonephritis in women: A 2010 update by the Infectious Diseases Society of America and the European Society for Microbiology and Infectious Diseases. *Clin Infect Dis* **52:**e103–e120.

66. **Warren JW, Abrutyn E, Hebel JR, Johnson JR, Schaeffer AJ, Stamm WE.** 1999. Guidelines for antimicrobial treatment of uncomplicated acute bacterial cystitis and acute pyelonephritis in women. Infectious Diseases Society of America (IDSA). *Clin Infect Dis* **29:**745–758.

67. **Anon.** 2011. *Urinary tract infections - medications.* http://www.umm.edu/patiented/articles/how_antibiotics_used_treating_urinary_tract_infections_000036_8.htm. Accessed 04/01/2015.

68. **Huovinen P, Sundström L, Swedberg G, Sköld O.** 1995. Trimethoprim and sulfonamide resistance. *Antimicrob Agents Chemother* **39:** 279–289.

69. **Sköld O.** 2000. Sulfonamide resistance: mechanisms and trends. *Drug Resist Updat* **3:**155–160.

70. **Huovinen P.** 2001. Resistance to trimethoprim-sulfamethoxazole. *Clin Infect Dis* **32:** 1608–1614.

71. **Bushby SR, Hitchings GH.** 1968. Trimethoprim, a sulphonamide potentiator. *Br J Pharmacol Chemother* **33:**72–90.

72. **Sköld O.** 2001. Resistance to trimethoprim and sulfonamides. *Vet Res* **32:**261–273.

73. **Santo E, Savador MM, Marin JM.** 2007. Multidrug-resistant urinary tract isolates of *Escherichia coli* from Ribeirão Preto, São Paulo, Brazil. *Braz J Infect Dis* **11:**575–578.

74. **Guidoni EB, Berezin EN, Nigro S, Santiago NA, Benimi V, Toporovski J.** 2008. Antibiotic resistance patterns of pediatric community-acquired urinary infections. *Braz J Infect Dis* **12:**321–323.

75. **Anatoliotaki M, Galanakis E, Schinaki A, Stefanaki S, Mavrokosta M, Tsilimigaki A.** 2007. Antimicrobial resistance of urinary tract pathogens in children in Crete, Greece. *Scand J Infect Dis* **39:**671–675.

76. **Nicolle LE.** 2003. Urinary tract infection: traditional pharmacologic therapies. *Dis Mon* **49:**111–128.

77. **Matute AJ, Hak E, Schurink CA, McArthur A, Alonso E, Paniagua M, Van Asbeck E, Roskett AM, Froeling F, Rozenberg-Arsaka M, Hopelman IM.** 2004. Resistance of uropathogens in symptomatic urinary tract infections in León, Nicaragua. *Int J Antimicrob Agents* **23:**506–509.

78. **Guneysel O, Onur O, Erdede M, Denizbasi A.** 2009. Trimethoprim/sulfamethoxazole resistance in urinary tract infections. *J Emerg Med* **36:**338–341.

79. **Kurtaran B, Candevir A, Tasova Y, Kibar F, Inal AS, Komur S, Aksu HS.** 2010. Antibiotic resistance in community-acquired urinary tract infections: prevalence and risk factors. *Med Sci Monit* **16:**CR246–251.

80. **Sanchez GV, Master RN, Karlowsky JA, Bordon JM.** 2012. *In vitro* antimicrobial resistance of urinary *Escherichia coli* isolates among U.S. outpatients from 2000 to 2010. *Antimicrob Agents Chemother* **56:**2181–2183.

81. **Hooper DC, Wolfson JS.** 1991. Fluoroquinolone antimicrobial agents. *New Engl J Med* **324:**384–394.

82. **Drlica K, Zhao X.** 1997. DNA gyrase, topoisomerase IV, and the 4-quinolones. *Microb Mol Biol Rev* **61:**377–392.

83. **Poirel L, Cattoir V, Nordmann P.** 2012. Plasmid-mediated quinolone resistance; interactions between human, animal and environmental ecologies. *Front Microbiol* **3:**24.

84. **Luzzaro F.** 2008. Fluoroquinolones and Gram-negative bacteria: antimicrobial activity and mechanisms of resistance. *Infez Med* **16** (Suppl 2):5–11.

85. **Strahilevitz J, Jacoby GA, Hooper DC, Robicsek A.** 2009. Plasmid-mediated quinolone resistance: A multifaceted Threat. *Clin Microbiol Rev* **22:**664–689.

86. **Robicsek A, Jacoby GA, Hooper DC.** 2006. The worldwide emergence of plasmid-mediated quinolone resistance. *Lancet Infect Dis* **6:** 629–640.

87. **Martínez-Martínez L, Pascual A, García I, Tran J, Jacoby GA.** 2003. Interaction of plasmid and host quinolone resistance. *J Antimicrob Chemother* **51:**1037–1039.

88. **Cattoir V, Nordmann P.** 2009. Plasmid-mediated quinolone resistance in Gram-negative bacterial species: An update. *Curr Med Chem* **16:**1028–1046.

89. **Zhao J, Chen Z, Chen SL, Deng Y, Liu Y, Tian W, Huang X, Wu C, Sun Y, Sun Y, Zeng Z, Liu JH.** 2010. Prevalence and dissemination of *oqxAB* in *Escherichia coli* isolates from animals, farmworkers, and the environment. *Antimicrob Agents Chemother* **54:**4219–4224.

90. **Kim HB, Wang M, Park CH, Kim EC, Jacoby GA, Hooper DC.** 2009. *oqxAB* encoding a multidrug efflux pump in human clinical isolates of *Enterobacteriaceae*. *Antimicrob Agents Chemother* **53:**3582–3584.

91. **Andrade JM, Cairrão F, Arraiano CM.** 2006. RNase R affects gene expression in stationary phase: regulation of *ompA*. *Mol Microbiol* **60:** 219–228.

92. **Karaca Y, Coplu N, Gozalan A, Oncul O, Citil BE, Esen B.** 2005. Co-trimoxazole and quinolone resistance in *Escherichia coli* isolated from urinary tract infections over the last 10 years. *Int J Antimicrob Agents* **26:**75–77.

93. **Fadda G, Nicoletti G, Schito GC, Tempera G.** 2005. Antimicrobial susceptibility patterns of contemporary pathogens from uncomplicated urinary tract infections isolated in a multi-center italian survey: possible impact on guidelines. *J Chemother* **17:**251–257.

94. **Arslan H, Azap OK, Ergönül O, Timurkaynak F; Urinary Tract Infection Study Group.** 2005. Risk factors for ciprofloxacin resistance among *Escherichia coli* strains isolated from community-acquired urinary tract infections in Turkey. *J Antimicrob Chemother* **56:**914–918.

95. **Sire JM, Nabeth P, Perrier-Gros-Claude JD, Bahsoun I, Siby T, Macondo EA, Gaye-Diallo A, Guyomard S, Seck A, Breurec S, Garin B.** 2007. Antimicrobial resistance in outpatient *Escherichia coli* urinary isolates in Dakar, Senegal. *J Infect Dev Ctries* **1:**263–268.

96. **Grude N, Strand L, Mykland H, Nowrouzian FL, Nyhus J, Jenkins A, Kristiansen BE.** 2008. Fluoroquinolone-resistant uropathogenic *Escherichia coli* in Norway: evidence of clonal spread. *Clin Microbiol Infect* **14:**498–500.

97. **Khawcharoenporn T, Vasoo S, Ward E, Singh K.** 2012. High rates of quinolone resistance among urinary tract infections in the ED. *Am J Emerg Med* **30:**68–74.

98. **Poole K.** 2004. Resistance to β-lactam antibiotics. *Cell Mol Life Sci* **61:**2200–2223.

99. **Dougherty TJ, Kennedy K, Kessler RE, Pucci MJ.** 1996. Direct quantitation of the number of individual penicillin-binding proteins per cell in *Escherichia coli*. *J Bacteriol* **178:**6110–6115.

100. **Wilke MS, Lovering AL, Strynadka NC.** 2005. β-lactam antibiotic resistance: A current structural perspective. *Curr Opin Microbiol* **8:** 525–533.

101. **Bush K, Jacoby GA, Medeiros AA.** 1995. A functional classification scheme for β-lactamases and its correlation with molecular structure. *Antimicrob Agents Chemother* **39:**1211–1233.

102. **Bush K.** 2010. Alarming β-lactamase-mediated resistance in multidrug-resistant *Enterobacteriaceae*. *Curr Opin Microbiol* **13:**558–564.

103. **Paterson DL, Bonomo RA.** 2005. Extended-spectrum β-lactamases: A clinical update. *Clin Microbiol Rev* **18:**657–686.

104. **Ambler RP, Coulson AF, Frére JM, Ghuysen JM, Joris B, Forsman M, Levesque RC, Tiraby G, Waley SG.** 1991. A standard numbering scheme for the class A β-lactamases. *Biochem J* **276:**269–270.

105. **Paterson DL.** 2006. Resistance in Gram-negative bacteria: *Enterobacteriaceae*. *Am J Med* **119:**S20–S28.

106. **Meier S, Weber R, Zbinden R, Ruef C, Hasse B.** 2011. Extended-spectrum β-lactamase-producing Gram-negative pathogens in community-acquired urinary tract infections: An increasing challenge for antimicrobial therapy. *Infection* **39:**333–340.

107. **Shigemura K, Tanaka K, Adachi M, Yamashita M, Arakawa S, Fujisawa M.** 2011. Chronological change of antibiotic use and antibiotic resistance in *Escherichia coli* causing urinary tract infections. *J Infect Chemother* **17:**646–651.

108. **Taneja N, Rao P, Arora J, Dogra A.** 2008. Occurrence of ESBL & Amp-C-beta-lactamases & susceptibility to newer antimicrobial agents in complicated UTI. *Indian J Med Res* **127:**85–88.

109. **Yong D, Toleman MA, Giske CG, Cho HS, Sundman K, Lee K, Walsh TR.** 2009. Characterization of a new metallo-β-lactamase gene, bla$_{NDM-1}$, and a novel erythromycin esterase gene carried on a unique genetic structure in *Klebsiella pneumoniae* sequence type 14 from India. *Antimicrob Agents Chemother* **53:**5046–5054.

110. **Nordmann P, Poirel L, Walsh TR, Livermore DM.** 2011. The emerging NDM carbapenemases. *Trends Microbiol* **19:**588–595.

111. **Rimrang B, Chanawong A, Lulitanond A, Wilailuckana C, Charoensri N, Sribenjalux P, Phumsrikaew W, Wonglakorn L, Kerdsin A, Chetchotisakd P.** 2012. Emergence of NDM-1 and IMP-14a-producing Enterobacteriaceae in Thailand. *J Antimicrob Chemother* **67:**2626–2630.

112. **Williamson DA, Sidjabat HE, Freeman JT, Roberts SA, Silvey A, Woodhouse R, Mowat E, Dyet K, Paterson DL, Blackmore T, Burns A, Heffernan H.** 2012. Identification and molecular characterization of New Delhi metallo-β-lactamase-1 (NDM-1) and NDM-6-producing *Enterobacteriaceae* from New Zealand hospitals. *Int J Antimicrob Agents* **39:**529–533.

113. **El-Herte RI, Araj GF, Matar GM, Baroud M, Kanafani ZA, Kanj SS.** 2012. Detection of carbapanem-resistant *Escherichia coli* and *Klebsiella pneumoniae* producing NDM-1 in Lebanon. *J Infect Dev Ctries* **6:**457–461.

114. **Nielsen JB, Hansen F, Littauer P, Schønning K, Hammerum AM.** 2012. An NDM-1-producing *Escherichia coli* obtained in Denmark has a genetic profile similar to an NDM-1-producing *E. coli* isolate from the UK. *J Antimicrob Chemother* **67:**2049–2051.

115. **Falagas ME, Kastoris AC, Kapaskelis AM, Karageorgopoulos DE.** 2010. Fosfomycin for the treatment of multidrug-resistant, including extended-spectrum β-lactamase producing, *Enterobacteriaceae* infections: A systematic review. *Lancet Infect Dis* **10:**43–50.

116. **Garau J.** 2008. Other antimicrobials of interest in the era of extended-spectrum beta-lactamases: fosfomycin, nitrofurantoin, and tigecycline. *Clin Microbiol Infect* **14:**S198–202.

117. **Hames L, Rice CE.** 2007. Antimicrobial resistance of urinary tract isolates in acute

uncomplicated cystitis among college-aged women: choosing a first-line therapy. *J Am Coll Health* **56:**153–156.

118. **Hof H.** 1988. Antimicrobial therapy with nitro-heterocyclic compounds, for example, metronidazole and nitrofurantoin. *Immun Infekt* **16:** 220–225.

119. **Sandegren L, Lindqvist A, Kahlmeter G, Andersson DI.** 2008. Nitrofuranation resistance mechanism and fitness cost in *Escherichia coli.* *J Antimicrob Chemother* **62:**495–503.

120. **Breeze AS, Obaseiki-Ebor EE.** 1983. Nitro-furan reducatase activity in nitrofurantion-resistant strains of *Escherichia coli* K12: some with chromosomally determined resistance and others carrying R-plasmids. *J Antimicrob Chemother* **12:**543–547.

121. **Breeze AS, Obaseiki-Ebor EE.** 1983. Trans-ferable nitrofuran resistance conferred by R-plasmids in clinical isolates of *Escherihcia coli.* *J Antimicrob Chemother* **12:**459–467.

122. **McCalla DR, Kaiser C, Green MH.** 1978. Genetics of nitrofurazone resistance in *Escherichia coli.* *J Bacteriol* **133:**10–16.

123. **McOsker CC, Fitzpatrick PM.** 1994. Nitro-furantoin: mechanism of action and implications for resistance development in common uropathogens. *J Antimicrob Chemother* **33:**23–30.

124. **Maraki S, Mantadakis E, Michailidis L, Samonis G.** 2012. Changing antibiotic suscep-tibiliites of community-acquired uropathogens in Greece, 2005–2010. *J Microbiol Immunol Infect* **46:**202–209.

125. **Kiffer CR, Mendes C, Oplustil CP, Sampaio JL.** 2007. Antibiotic resistance and trend of urinary pathogens in general outpatients from a major urban city. *Inter Braz J Urol* **33:**42–49.

126. **Kahlmeter G.** 2000. The ECO.SENS Project: A prospective, multinational, multicentre epi-demiological survey of the prevalence and antimicrobial susceptibility of urinary tract pathogens - interim report. *J Antimicrob Chemo-ther* **46:**S15–22.

127. **Shrestha NK, Tomford JW.** 2001. Fosfo-mycin: A review. *Infect Dis Clin Pract* **10:**255–260.

128. **Hendlin D, Stapley EO, Jackson M, Wallick H, Miller AK, Wolf FJ, Miller TW, Chaiet L, Kahan FM, Foltz EL, Woodruff HB, Mata JM, Hernandez S, Mochales S.** 1969. Phosphonomycin, a new antibiotic pro-duced by strains of Streptomyces. *Science* **166:**122–123.

129. **Kahan FM, Kahan JS, Cassidy PJ, Kropp H.** 1974. The mechanism of action of fosfomycin (phosphonomycin). *Ann N Y Acad Sci* **235:** 364–386.

130. **Arca P, Rico M, Braña AF, Villar CJ, Hardisson C, Suárez JE.** 1988. Formation of an adduct between fosfomycin and glutathione: A new mechanism of antibiotic resistance in bacteria. *Antimicrob Agents Chemother* **32:**1552–1556.

131. **Liu HY, Lin HC, Lin YC, Yu SH, Wu WH, Lee YJ.** 2011. Antimicrobial susceptibilities of urinary extended-spectrum beta-lactamase-producing *Escherichai coli* and *Klebsiella pneu-moniae* to fosfomycin and nitrofurantoin in a teaching hospital in Taiwan. *J Microbiol Immunol Infect* **44:**364–368.

132. **Schito GC.** 2003. Why fosfomycin trometamol as a first line therapy for uncomplicated UTI? *Int J Antimicrob Agents* **22:**S79–S83.

133. **Noor N, Ajaz M, Rasool SA, Pirzada ZA.** 2004. Urinary tract infections associated with multidrug resistant enteric bacilli: characteri-zaton and genetical studies. *Pak J Pharm Sci* **17:**115–123.

134. **Nelson ML, Levy SB.** 2011. The history of the tetracyclines. *Ann N Y Acad Sci* **1241:**17–32.

135. **Speer BS, Shoemaker NB, Salyers AA.** 1992. Bacterial resistance to tetracycline: mecha-nisms, transfer, and clinical significance. *Clin Microbiol Rev* **5:**387–399.

136. **Chopra I, Roberts M.** 2001. Tetracycline antibiotics: mode of action, applications, mo-lecular biology, and epidemiology of bacterial resistance. *Microbiol Mol Biol Rev* **65:**232–260.

137. **Schnappinger D, Hillen W.** 1996. Tetracy-clines: antibiotic action, uptake and resistance mechanisms. *Arch Microbiol* **165:**359–369.

138. **Nix DE, Matthias KR.** 2010. Should tige-cycline be considered for urinary tract in-fections? A pharmacokinetic re-evaluation. *J Antimicrob Chemother* **65:**1311–1312.

139. **Yang W, Moore IF, Koteva KP, Bareich DC, Hiughes DW, Wright GD.** 2004. *TetX* is a flavin-dependent monooxygenase conferring resistance to tetracycline antibiotics. *J Biol Chem* **279:**52346–52352.

140. **Levy SB, McMurray LM, Barbosa TM, Burdett V, Courvalin P, Hillen W, Roberts MC, Rood JI, Taylor DE.** 1999. Nomenclature for new tetracycline resistance determinants. *Antimicrob Agents Chemother* **43:**1523–1524.

141. **Levy SB, McMurray LM, Roberts MC.** 2005. Tet protein hybrids. *Antimicrob Agents Chemo-ther* **49:**3099.

142. **Stanton TB, Humphrey SB, Scott KP, Flint HJ.** 2005. Hybrid *tet* genes and *tet* gene nomenclature: request for opinion. *Antimicrob Agents Chemother* **49:**1265–1266.

143. **van Hoek AH, Mayrohfer S, Domig KJ, Flórez AB, Ammor MS, Mayo B, Aarts HJM.** 2008. Mosaic tetracycline resistance genes and their

flanking regions in *Bifidobacterium thermophilium* and *Lactobacillus johnsonii*. *Antimicrob Agents Chemother* **52:**248–252.

144. **Aboderin OA, Abdu AR, Odetoyin BW, Lamikanra A.** 2009. Antimicrobial resistance in *Escherichia coli* strains from urinary tract infections. *J Natl Med Assoc* **101:**1268–1273.

145. **Okesola AO, Aroundegbe TI.** 2011. Antibiotic resistance pattern of uropathogenic *Escherichia coli* in South West Nigeria. *Afr J Med Med Sci* **40:**235–238.

146. **Jana S, Deb JK.** 2006. Molecular understanding of aminoglycoside action and resistance. *Appl Microbiol Biotechnol* **70:**140–150.

147. **Mingeot-Leclercq MP, Glupczynski Y, Tulkens PM.** 1999. Aminoglycosides: activity and resistance. *Antimicrob Agents Chemother* **43:**727–737.

148. **Bader MS, Hawboldt J, Brooks A.** 2010. Management of complicated urinary tract infections in the era of antimicrobial resistance. *Postgrad Med* **122:**7–15.

149. **Shakil S, Khan R, Zarrilli R, Khan AU.** 2008. Aminoglycosides versus bacteria - a description of the action, resistance mechanism, and nosocomial battleground. *J Biomed Sci* **15:**5–14.

150. **Davies J, Wright GD.** 1997. Bacterial resistance to aminoglycoside antibiotics. *Trends Microbiol* **5:**234–240.

151. **Galimand M, Courvalin P, Lambert T.** 2003. Plasmid-mediated high-level resistance to aminoglycosides in *Enterobacteriaceae* due to 16S rRNA methylation. *Antimicrob Agents Chemother* **47:**2565–2571.

152. **Galimand M, Sabtcheva S, Courvalin P, Lambert T.** 2005. Worldwide disseminated *armA* aminoglycoside resistance methylase gene is borne by composite transposon Tn*1548*. *Antimicrob Agents Chemother* **49:**2949–2953.

153. **Courvalin P.** 2008. New plasmid mediated resistances to antimicrobial agents. *Ach Microbiol* **189:**289–291.

154. **Doi Y, Wachino JI, Arakawa Y.** 2008. Nomenclature of plasmid-mediated 16S rRNA methylases responsible for panaminoglycoside resistance. *Antimicrob Agents Chemother* **52:**2287–2288.

155. **Poole K.** 2005. Efflux-mediated antimicrobial resistance. *J Antimicrob Chemother* **56:**20–51.

156. **Mohammad-Jafari H, Saffar MJ, Nemate I, Saffar H, Khalilian AR.** 2012. Increasing antibiotic resistance among uropathogens isolated during years 2006-2009: impact on the empirical management. *Inter Braz J Urol* **38:**25–32.

157. **Thabet L, Messadi AA, Meddeb B, Mbarek M, Turki A, Ben Redjeb S.** 2010. Bacteriological profile of urinary tract infections in women in Aziza Othmana Hospital: 495 cases. *Tunis Med* **88:**898–901.

158. **Gad GF, Mohamed HA, Ashour HM.** 2011. Aminoglycoside resistance rates, phenotypes, and mechanisms of Gram-negative bacteria from infected patients in Upper Egypt. *PLoS One* **6:**e17224. doi:10.1371/journal.pone.0017224

159. **Gupta A, Phung LT, Taylor DE, Silver S.** 2001. Diversity of silver resistance genes in IncH incompatability group plasmids. *Microbiology* **147:**3393–3402.

160. **Silver S.** 2003. Bacterial silver resistance: molecular biology and uses and misuses of silver compounds. *FEMS Microbiol Rev* **27:**341–353.

161. **Beattie M, Taylor J.** 2011. Silver alloy vs. uncoated urinary catheters: A systematic review of the literature. *J Clin Nurs* **20:**2098–2108.

162. **Johnson JR, Kuskowski MA, Wilt TJ.** 2006. Systematic review: antimicrobial urinary catheters to prevent catheter-associated urinary tract infection in hospitalized patients. *Ann Intern Med* **144:**116–126.

163. **Johnson JR, Johnson BD, Kuskowski MA, Pitout J.** 2010. *In vitro* activity of available antimicrobial coated Foley catheters against *Escherichia coli*, including strains resistant to extended spectrum cephalosporins. *J Urol* **184:**2572–2577.

164. **Rupp ME, Fitzgerald T, Marion N, Helget V, Puumala S, Anderson JR, Fey PD.** 2004. Effect of silver-coated urinary catheters: efficacy, cost-effectiveness, and antimicrobial resistance. *Am J Infect Control* **32:**445–450.

165. **Ejrnæs K.** 2011. Bacterial characteristics of importance for recurrent urinary tract infections caused by *Escherichia coli*. *Dan Med Bull* **58:**B4187.

166. **Johnson TJ, Kariyawasam S, Wannemuehler Y, Mangiamele P, Johnson SJ, Doetkott C, Skyberg JA, Lynne AM, Johnson JR, Nolan LK.** 2007. The genome sequence of avian pathogenic *Escherichia coli* strain O1:K1:H7 shares strong similarities with human extraintestinal pathogenic *E. coli* genomes. *J Bacteriol* **189:**3228–3236.

167. **Jayaraman R.** 2009. Antibiotic resistance: An overview of mechanisms and a paradigm shift. *Curr Sci* **96:**1475–1484.

168. **Kunin CM.** 2001. Nosocomial urinary tract infections and the indwelling catheter: what is new and what is true? *Chest* **120:**10–12.

169. **Maki DG, Tambyah PA.** 2001. Engineering out the risk for infection with urinary catheters. *Emerg Infect Dis* **7:**342–347.

170. **Gilbert P, McBain AJ.** 2003. Potential impact of increased use of biocides in consumer products on prevalence of antibiotic resistance. *Clin Microbiol Rev* **16:**189–208.

Population Phylogenomics of Extraintestinal Pathogenic *Escherichia coli*

11

JÉRÔME TOURRET[1,2] and ERICK DENAMUR[2]

INTRODUCTION

Escherichia coli can act as both a model organism that has been a workhorse for molecular biology, genetics, biochemistry, and biotechnology, and a major agent of urinary tract infection (UTI). In 1997, at the beginning of the genomics era, the first *E. coli* genome to be sequenced (1) was of course a representative of the laboratory-derived strain K-12, originally isolated in 1922 from the stools of a convalescent diphtheria patient. Five years later, the first genome of an UTI strain was published (2), being the third *E. coli* to be fully sequenced after an O157:H7 strain in 2001 (3). Among the 61 available complete *E. coli* chromosomal sequences in July 2012 (Genome sequencing projects National Center for Biotechnology Information (NCBI) http://www.ncbi.nlm.nih.gov/genome/genomes/167), only 9 correspond to UTI strains and one to an asymptomatic bacteriuria (ABU) strain, all originating from humans. Four additional *E. coli* strains causing extraintestinal diseases are represented, with 3 human-newborn meningitis-causing strains and 1 avian pathogenic *E. coli*. The other sequenced strains have very diverse origins. This reflects the fact that members of the *E. coli* species are versatile, with various life styles.

[1]Département d'Urologie, Néphrologie et Transplantation Groupe Hospitalier Pitié-Salpêtrière, Assistance Publique-Hôpitaux de Paris, Université Pierre et Marie Curie; [2]UMR 1137 INSERM and Université Paris Diderot, IAME, Sorbonne Paris Cité, 75018 Paris, France.

Urinary Tract Infections: Molecular Pathogenesis and Clinical Management, 2nd Edition
Edited by Matthew A. Mulvey, David J. Klumpp, and Ann E. Stapleton
© 2017 American Society for Microbiology, Washington, DC
doi:10.1128/microbiolspec.UTI-0010-2012

Indeed, among the 10^{20} *E. coli* estimated to be present on the surface of the earth (4), the vast majority alternate between their primary habitat (the gut of vertebrates where they live as commensals) (5) and their secondary habitat, the soil and sediments (6). But *E. coli* is also a devastating pathogen both in domestic animals and humans, where it can cause various intestinal diseases, such as severe diarrhea in children under the age of five in low- and middle-income nations, traveler's diarrhea, dysentery, hemorrhagic colitis, and hemolytic uremic syndrome. It can also cause extraintestinal diseases, mainly UTI, sepsis, newborn meningitis, and abdominal suppuration (7, 8). Although the delineation between commensal strains and strains causing intestinal infections is relatively clear, the categorical separation between commensal and extraintestinal infection-causing strains is more tenuous. The acronym ExPEC for "extraintestinal pathogenic *E. coli*" has been proposed to parallel acronyms such as EPEC (enteropathogenic *E. coli*) or EHEC (enterohemorrhagic *E. coli*) used for intestinal pathogenic strains (9). According to this definition, an ExPEC must possess recognized extraintestinal-virulence factors (VFs) or demonstrate significant virulence in an appropriate animal model of extraintestinal infection. Isolating an *E. coli* strain from a patient with an extraintestinal infection is not sufficient to define an ExPEC, as some commensal strains can cause extraintestinal infections in immunocompromised patients, and ExPEC can be isolated in commensal situations (9). ExPEC encompass the more narrow acronyms UPEC (uropathogenic *E. coli*), NMEC (neonatal meningitis *E. coli*), and APEC (avian pathogenic *E. coli*).

What can be learned from genomic and related data about the pathophysiology of infections due to the ExPEC and, more specifically, those due to UPEC? One of the goals of genomic analysis is to link some genome features to specific phenotypes, such as the ability to cause UTI. However, this is possible only when considering the species as a whole and knowing the evolutionary forces that drive the genetic diversity, i.e., mutation and recombination rates, selective pressure, and population sizes. Population genetics and phylogenetic approaches should be combined when analyzing whole-genome sequences.

POPULATION-GENETICS STRUCTURE OF *E. COLI* SPECIES

The genetic structure of a bacterial species can range from clonal to panmictic, according to the relative proportion of mutation and recombination in the generation of diversity (10): a panmictic population exhibits a high recombination rate (horizontal diversity) whereas a clonal population generates diversity almost exclusively by mutation (vertical diversity). The first attempts to determine the *E. coli* genetic population structure were based on the serotypes (O, K, and H antigens) (11) and the metabolic-enzyme polymorphisms studied by multilocus enzyme electrophoresis (MLEE) (12). The authors observed strong linkage disequilibrium among the different antigenic or protein loci studied, with non-random associations. Only some combinations of O:K:H antigens or enzyme alleles were observed, and strains exhibiting these combinations were found in distinct hosts at diverse locations and periods. It was postulated that the *E. coli* species consists of an array of stable lineages, called clones, among which little recombination of chromosomal genes occurred (5). These data were corroborated at the DNA level by random-amplified polymorphic DNA (RAPD) and ribosomal DNA (rDNA) restriction fragment-length polymorphism analyses (13). However, doubts appeared as the first DNA sequences showed clustered base substitutions within genes, which were interpreted as recombination events (14). Furthermore, the phylogenetic trees reconstructed from individual genes were sometimes incongruent with each other, i.e., strains were not grouped in the same way in the different trees (15), indicating recombination at the gene level. The reconciliation between these apparently discordant data came from the whole-genome-sequence analysis (16). Approximate Bayesian computations on more than 1,500 conserved genes between 20 *E. coli* genomes showed that the rate of genetic exchange was twice as high as the mutation rate for short fragments of 50 base-pairs (bp), confirming a strong role of recombination in genome evolution. However, because recombined fragments are short, they do not blur the phylogenetic signal due to vertical evolution if the sequences used for phylogenetic reconstruction are long enough. It can be concluded that despite an important rate of recombination, an apparent clonal structure is observed within the *E. coli* species with a clear phylogenetic signal reflecting the relationship between the strains (5). Interestingly, the level of recombination was not the same all along the chromosome and two "bastions of polymorphism" (17) with high-recombination frequency were identified at the *rfb* locus encoding the O antigen, and a region containing the *hsd* restriction and modification system and mannose-sensitive type 1 pilus (*fim*) operons (16).

Excluding regions with a high level of recombination, the phylogenetic history of the strains can be reconstructed from the concatenated sequences of a set of genes (18, 19), which is the basis of the multilocus-sequence typing (MLST). Using such an approach, Walk et al. were surprised to identify strains with an *E. coli* biochemical phenotype that were genetically very distant from the other *E. coli* strains (20). These strains were classified as cryptic *Escherichia* clades and divided in 5 groups named clades I to V, clade I being closely related to classical *E. coli* and *E. fergusonii*, and the clade V being the most divergent (21). The classical *E. coli* can then be divided in 7 main groups, A, B1, B2, C, D, E, and F. The C and F groups are closely related to the B1 and B2 groups, respectively (5, 22). The B2 and F groups are located at the base of the tree, whereas the A and B1 groups are the most recently emerged and appear as sister groups. The MLST also allows the delineation of subgroups or clonal complexes. This is particularly true within the B2 group that can be delineated in at least 9 main subgroups (I to IX) corresponding to the sequence-type complexes (STc) 131, 73, 127, 141, 144, 12, 14, 452, and 95, defined by Achtman's MLST scheme (http://mlst.warwick.ac.uk/mlst/dbs/Ecoli) (19, 19bis), respectively. Strain E2348/69 represents an additional B2 STc, as it is a representative of the EPEC-1 group (Fig. 1).

WHICH *E. COLI* STRAINS ARE CAUSING UTI AND WHAT ARE THE NECESSARY GENES?

Data from Complete Genomes

There are now more than 400 *E. coli* genomes in the NCBI database as chromosomes or scaffolds and contigs (www.http://www.ncbi.nlm.nih.gov/genome/genomes/167). It can be summarized from this database that an *E. coli* chromosome has an average guanine-cytosine content (GC%) of 50.5% (49.2% to 51.2%), but can vary extensively in length, ranging from 4.3 to 6.2 megabases (Mb) (mean 5.2 Mb) and encoding 4,084 to 6,453 proteins (mean 5,249 proteins7). This huge diversity is structured as a set of less than 2,000 genes (the core genome) that are conserved between all the strains and a set of more variable genes (the flexible gene pool). The functional pattern of these groups of genes greatly varies as genes of known function are strongly over-represented in the core genome, whereas genes of unknown function, and especially selfish DNA, such as transposable and prophage elements, are over-represented among the flexible gene pool (Fig. 2) (16). The sum of the core-genome genes and the variable genes is called the pangenome and has been estimated to be around 10 times the size of the core genome (16, 23, 24). As more strains are sequenced,

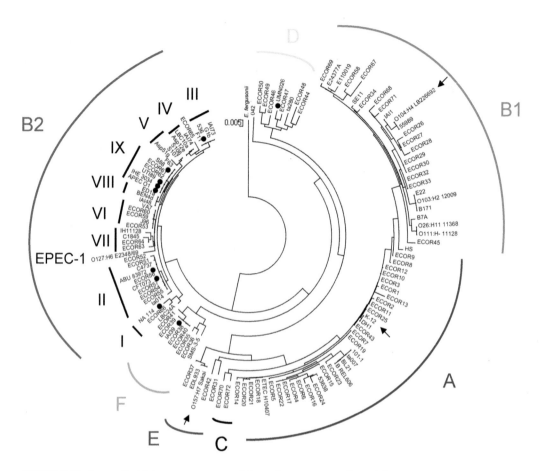

FIGURE 1 Phylogenetic history, reconstructed from 8 concatenated partial-gene sequences using the Pasteur Institut MLST schema (67), of 128 *E. coli* strains rooted on *E. fergusonii*. The *E. coli* strains have been chosen to be representative of the species' genetic diversity and life-styles. They originate from the ECOR collection (170), our laboratory collection (77), and from complete genomes available in GenBank. No *Escherichia* clade strain is represented, see (21) for their phylogeny. The strains with a black dot correspond to the strains discussed in the text for which a complete-genome sequence is available. The phylogenetic groups and subgroups (ST complexes) are indicated [see the main text for the correspondence with the ST complexes of (19)]. The EPEC strain E2348/69 belongs to the EPEC-1 group. The arrows indicate 3 famous archetypal strains: the O157:H7 EHEC strain, the laboratory-derived K-12 strain, and the O104:H4 Shiga toxin-producing strain from the 2011 German outbreak, belonging to the E, A, and B1 phylogenetic groups, respectively. This phylogeny is similar to the one obtained from core genomes (data not shown).

the core genome tends to stabilize, whereas each new sequenced strain brings new genes to the species' pangenome (25). Interestingly, it can be considered that the number of genes in the *E. coli* species as a whole equals the number of genes in the human species.

Thus, *E. coli* genome is a highly dynamic structure with a constant flux of insertions and deletions. How can such a gene flow be compatible with the chromosome organization and constancy of basic functions required for life? The genome keeps a strong structure as very few genome rearrangements have been observed. First, half of the genes gained or lost are organized "en bloc" in contiguous stretches of DNA referred to as

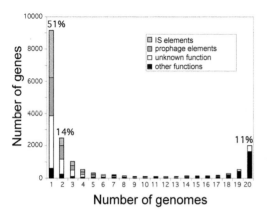

FIGURE 2 **Analysis of the presence of genes in 20 genomes of *E. coli* (16). The number of genes present in 1 to 20 (all) genomes is presented. The genes that are present in the 20 genomes represent the core genome (11% of the pan-genome), whereas the genes present in only one strain are strain-specific (51% of the pan-genome). It can be seen that very few genes are between these two extremes. When the genes are categorized according to their origin and functions, it appears that strain-specific genes are mostly from mobile elements and of unknown functions, whereas the core-genome genes are almost exclusively composed of non-mobile genes of known functions. Although some of the strain-specific genes confer adaptive functions as discussed in the text, most of these genes are non-adaptive and thus purged over time (16).**

genomic islands (26) and second, most gene acquisitions and losses take place at particular hot spots located in permissive regions where large insertions can occur without causing significant loss of fitness, respecting genes, operons, and supra-operonic structures. Some of these hot spots correspond to tRNA or phage integration hot spots, but most have no specific modular signature to date (16). This allows the combination of order and disorder during *E. coli* divergence (27).

According to this partition between the core and the variable genomes, genomic analyses can be performed on ExPEC and/or UPEC strains with two main complementary approaches: (i) the search for presence or absence of specific genes and (ii) the

identification of traces of positive selection within common genes. A prerequisite to these genomic analyses is the identification of the evolutionary history (phylogeny) of these strains. The phylogeny is indeed required to interpret the link between virulence and the presence/absence of characters (gene or allele). If one character is associated to virulence, two scenarios for their origin can be envisioned depending on the phylogeny of the strains (Fig. 3). In (A), a unique event (acquisition or loss) is necessary to explain the pattern, whereas in (B), numerous events are necessary. In (A), the character, acquired by chance, is probably a phylogenetic marker. In (B), convergence, which is a hallmark of adaptive evolution (28), is observed and thus the character has been selected and is probably directly involved in virulence. This kind of analysis is possible as *E. coli* species, as discussed above, has a clonal structure. Most of the studies have focused on the first approach whereas only a few authors have looked at the second approach.

The search for UPEC-specific genes

The latter approach described above for genomic analysis is most often used as a way to identify genes associated with virulence. However it has also been performed to identify antibiotic-resistance genes in some resistant clones that are spreading worldwide (29).

The first paper that really applied comparative genomics to UPEC strains was published in 2006, reporting the complete genome of strain 536 (O6:K15:H31) (30) and comparing it to another UPEC strain, CFT073 (O6:K2:H1), an EHEC O157 strain, and the laboratory-derived commensal strain K-12. The two UPEC strains belong to the B2 phylogroup (subgroup III/ST127 and subgroup II/ST73, respectively) whereas the O157 and the K-12 strains belong to the E and A phylogroup, respectively. A major strength of this paper was that it also presented data from animal-infection models using both

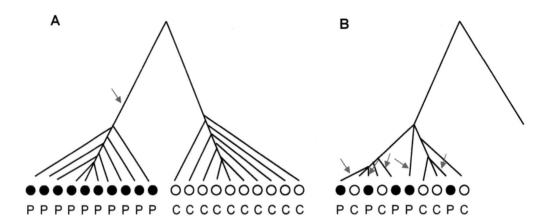

FIGURE 3 Schematic representation of two distinct evolutionary scenarios leading to association of a character with virulence. P is for pathogenic (black circle) whereas C stands for commensal (white circle). The character can be the presence of a gene or an allele within a gene. In A, the character has been acquired by chance once in the ancestor of the black strains (red arrow) and is a phylogenetic marker. In B, several independent acquisitions of the character are observed (red arrows), representing a convergence and indicating that this character has been selected and is involved in virulence. The same reasoning can be applied for the loss of a character; in this case the ancestral status is the presence of the character.

wild-type and mutant UPEC strains. Genomic differences between the strains were observed in large genomic islands (pathogenicity islands (PAIs)), but also in small gene clusters or even single genes, and corresponded to repeated insertions and deletions in certain parts of the genome. These regions were coding for all major classes of virulence-associated factors (toxins, adhesins, siderophore systems, proteases, capsules, lipopolysaccharide), as well as for specialized metabolic activities, such as D-serine catabolism and sugar-utilization systems. The cumulative impact of PAIs on virulence was tested using single and multiple PAI-deletion mutants in UTI and septicemia murine models (31). The results indicate that there is no single set of virulence factors or PAIs that define ExPEC.

In 2007, an APEC strain O1:K1:H7, belonging to the B2 subgroup IX/ST95, was totally sequenced (32). Its chromosome and 4 plasmids total a little bit more than 5.5 million bp. APEC O1 chromosome showed very high similarity with the genomes of human UPEC reference strains UTI89, which belong to the same ST, CFT073, and *E. coli* strain 536. Less than 5% of the APEC O1 open-reading frames (ORFs) were not found in any of the 3 reference UPEC strains. A total of 5 PAIs were described in APEC O1, four of which were chromosomal. Interestingly, one PAI was found on a virulence plasmid in APEC O1, but some of the VFs that it contained (e.g., salmochelin and aerobactin) have also been described on the chromosome of other UPEC. MLST confirmed that APEC O1:H1:K7 was very close to NMEC and UPEC strains showing the same serotype. The ExPEC phenotype seems to be more important than the sub-pathotypes (NMEC, UPEC or APEC), and more important than the host origin (human or avian).

Another important paper was published in 2009 with the analysis of 20 *E. coli* genomes and the genome of the closest *E. coli* related species, *E. fergusonii* (16). In this work, two additional UPEC were sequenced: UMN026 (O17:K52:H18), a representative of clonal group A (CGA) (33) belonging to the D phylogroup, and IAI39 (O7:K1), a F phylogroup strain. Other strains included

S88 (O45:K1:H7), a B2 phylogroup; subgroup IX/ST95, newborn meningitis-causing strain; one enteroaggregative O104:H4 strain; and two commensal strains. Only one and 16 genes were specifically absent or present, respectively, and associated with the ExPEC phenotype. ExPEC-specific genes corresponded to two clusters: the *pap* operon, a well-known adhesin determinant (34), and genes encoding enzymes with aldo-reductase activity. In addition, when considering intrinsic extraintestinal virulence as assessed in a mouse model of septicemia that avoids host variability (35), no gene specific to the virulent phenotype in mouse was identified. These data extended earlier studies showing that extraintestinal virulence is a multigenic process resulting from numerous gene combinations with complex epistatic interactions and multiple redundancies that take individual isolates down distinct evolutionary paths.

These three works, in addition to a more recent one in which another newborn meningitis-causing strain – IHE3034, O18:K1: H7, from the B2 subgroup IX/ST95 – was sequenced (36), clearly revealed a strong convergence of ExPEC and UPEC strains. However, this convergence is more functional than genetic. This is in agreement with what has been recently reported in numerous lineages of *E. coli* evolved experimentally at 42.2°C for 2,000 generations. The sequencing of one genome from these independent lineages identified more than 1,000 mutations, with few shared mutations among replicates, but with a strong pattern of functional convergence and a pervasive presence of epistasis shaping several adaptive trajectories (37).

While *E. coli* strains were not considered a therapeutic problem for many years, the emergence and spread of multidrug-resistant clones over the last decade has greatly complicated the situation. The more infamous multidrug-resistant strains that have been analyzed are the CGA clone mentioned above (which is resistant to ampicillin, chloramphenicol, streptomycin, sulfonamides, tetracycline,

and trimethoprim (i.e., ACSSuTTp phenotype) (38)), and the B2 subgroup I/ST131 O25b:H4 clone (harboring an extended-spectrum beta-lactamase most frequently of the CTX-M type (39), or more recently, the New Delhi metalloprotease (NDM)-1 carbapenemase (40)).

In 1999 and 2000, a single *E. coli* clonal group was found to be responsible for half of the trimethoprim-sulfamethoxazole-resistant UTIs in a community in California (33). This group was defined by a specific 4-band electrophoretic pattern after enterobacterial-repetitive intergenic-consensus (ERIC) polymerase chain reaction (PCR), and was named CGA for "clonal group A", even though it belongs to the D phylogroup. This clonal group was also isolated from the feces of healthy volunteers from the same community in California, and in other very distant locations within the U.S.A. (38, 41). Complete-genome sequencing of a representative of this CGA clone, UMN026, showed that most of the genes responsible for the resistance phenotype were clustered in an unique 23-kilo-base pair (kbp) chromosomal region, which occurred within a 105-kbp genomic island situated at the *leuX* transfer RNA (tRNA) (42). This was the first documentation of chromosomally encoded TEM (TEM is for the name of the patient from whom the gene was first isolated on a plasmid: Temoniera) beta-lactamase and *dhfrA17* genes in *E. coli*. This 23-kbp chromosomal region is characterized by numerous remnants of mobilization and rearrangement events, suggesting multiple horizontal transfers. The chromosomal region where the integration of this resistance island took place is a hot spot of integration. It is characterized by a mosaic pattern of genetic loci found in other *E. coli* strains and is localized in a region of the chromosome near the *fim* operon where recombination occurs with a high frequency (16).

Two groups have sequenced a CTX-M-producing ST131 O25b:H4 strain isolated from UTI patients in India (NA114) (43) and the U.K. (EC958) (44). As expected, they found the CTX-M gene on a plasmid in addition to numerous chromosomal-virulence genes. In strain EC958, a transposon insertion was found in the *fimB* gene encoding the activator of type 1 fimbriae resulting in a slower off-to-on switching phenotype for type 1 fimbriae that needs further characterization (44).

These two cases clearly demonstrate that a given strain can be both highly resistant and virulent, as UMN026 and TN03 (another representative of the ST131 CTX-M-producing clone) are virulent in a mouse model of bacteremia (16, 45).

Complete-genome sequencing has also been used to decipher the resistome of a NDM-1-producing strain of phylogenetic B1 group/ST101 isolated from the urine of a man transferred from Bangladesh (46). It showed that the multidrug-resistance pattern of the strain was the result of combined chromosome- and plasmid-encoded mechanisms (47).

An important question is how these genomic islands are spreading within the *E. coli* species. As mentioned above for the *fim* region, some hotspots of integration are also hotspots of recombination. This is also true for the *rfb* region, which harbors the high-pathogenicity island (HPI) involved in iron capture (48). The HPI has a role in extraintestinal virulence as demonstrated in a mouse model of septicemia (49). Using phylogenetic analyses and *in vitro* experiments, it has been shown that, once introduced in a member of the species, the HPI spread with flanking DNA regions of the *E. coli*-genomic backbone by conjugative transfer followed by *recA*-dependent homologous recombination (50). According to the low level of polymorphisms observed within the HPI sequence, the spread of the HPI must have occurred dramatically fast. This led to the current situation of an extremely high (>80%) distribution of the HPI among all ExPEC strains (48).

Allelic variation under positive selection

The role of genes that are common to all *E. coli*, i.e., the core-genome genes, has received much less attention than the flexible-

gene pool one. The first work that identified genes subject to positive selection was published in 2006. The authors reported the complete sequence of the B2 subgroup IX/ST95 UTI89 strain originating from a patient with cystitis (51). They used the PAML program (52) – which assigns likelihood scores to different hypotheses for selection based on alignments of orthologous genes and their corresponding phylogenetic trees – to identify genes under positive selection only in UPEC strains. Using this approach, they found 29 genes under positive selection, with enrichment for genes coding for cell wall/membrane biogenesis including *ompC* and *ompF*, iron acquisition systems, and recombination and repair genes. The implication of cell-surface proteins can easily be explained by immune pressure, which acts on surface/membrane structures. This point has been refined in a work using the same method but looking for genes under positive selection not restricted to UPEC strains, where the authors showed that the residues under selection occur almost exclusively in the extracellular regions of the cell-surface proteins (53). Iron-acquisition genes are important contributors to UPEC virulence (54), as shown above in the discussion of the flexible-gene pool.

The identification of genes involved in the mutation rate is less trivial and merits further attention. It has been reported that UPEC strains have higher mutation rates than other *E. coli* strains (55), and that elevated mutation rate generated by the inactivation of mismatch-repair system gene *mutS* confers a fitness advantage both *ex vivo* in human urine and *in vivo* in a mouse model of UTI (56). Furthermore, Chattopadhyay and collaborators have developed a molecular-evolution tool – zonal-phylogeny analysis – specifically designed to identify footprints of positive selection and hotspot mutations, i.e., repeated, phylogenetically unlinked mutations in the same amino-acid position (57). Using this tool, they showed that ExPEC strains accumulate hotspot mutations at a higher rate than other *E. coli* strains (58). All these data could indicate important roles for genes that regulate the mutation rates in ExPEC strains. Positive selection on the antimutator genes is, in this case, a second-order selection, with the mutators being selected because they generate mutations that increase adaptation of UPEC in their specific environments at a higher rate than nonmutators (59).

Within clone diversity

Next-generation sequencing technologies (60) now allow more isolates to be rapidly sequenced, at a relatively low cost. Scientists are beginning to look at variations at the clone level with these techniques. Recent works report DNA-sequence variations between isolates belonging to a single ST, or between even more closely related isolates, which evolve during colonization and/or infection processes, thus defining *in vivo* evolution.

The genomes of 10 English UTI ST131 isolates, some of which carry CTX-M resistance gene, have been sequenced and found to be genetically homogeneous with only 1,324 single nucleotide polymorphisms (SNPs), of which 371 are non-synonymous (61). One strain was very divergent, showing 460 strain-specific SNPs in comparison with the other strains that have only 10–60 strain-specific SNPs. Some of the reported SNPs may modulate the intrinsic extraintestinal virulence that has been observed with ST131 isolates (61). A phylogenetic tree based on SNP data grouped CTX-M-producing strains together (61), but these results did not hold up with the Indian ST131 isolate NA114 and therefore need to be confirmed using a larger worldwide panel of strains (48).

In vivo evolution of *E. coli* strains has been reported during ABU, during feces carriage, and during UTI. The prototypical ABU, *E. coli* strain 83972, was originally isolated from the urinary tract of a school girl (62), and belongs to the B2 subgroup II/ST73. It has been used for therapeutic urinary-bladder colonization in patients suffering from chronic UTI. After intravesical inocu-

lation, the strain establishes an ABU and this approach has proven to be safe and to protect the patient from super-infection with more virulent strains (63). ABU strain 83972 has recently been sequenced together with 3 re-isolates from 3 patients that received intravesical inoculation of the strain (64). Thirty-four loci were found to be polymorphic in the three re-isolates, both in the core genome and in the flexible-gene pool. These were non-synonymous substitutions within genes encoding regulatory proteins involved in metabolic pathways, iron uptake, and stress protection systems. These mutations can be considered as adaptive in the human hosts as a certain level of convergence is observed in the patient isolates and no mutations were observed in a control experiment in which the strain was cultured *in vitro* in human urine for 2,000 generations.

Another interesting study reported on the *in vivo* evolution of 14 isolates of a B2 subgroup II/ST73 strains that persisted for 3 years in the feces of a six-member household and caused a UTI in the family dog (65). Twenty mutations were observed in the 14 isolates during the course of the study, indicating a mutation rate of about 1.1 per genome per year. However, no clear evidence of adaptive changes were found in strains from either the human inhabitants or the dog, in contrast to results obtained from a comparable study with the ABU 83972 strain (64). The powerful approaches used in these studies highlight the existence of complex patterns of inter-host transfer events.

Epidemiologic Data Based on Proxy Markers

Genomics is not yet in the era where hundreds or thousands of strains can be sequenced and analyzed rapidly. However, due to the clonal structure of the population and the frequent clustering of virulence genes in genomic islands, good proxies can be used. First, the phylogeny of the strains (corresponding to the phylogenetic history of the core genome) can be approximated using MLST with only a few genes (19, 66, 67) or, more easily, using a triplex-PCR method developed by Clermont and collaborators (68) that allows the delineation of the strains in the four main phylogenetic groups; A, B1, B2, and D, in good correlation with MLST (69). More recently, this PCR-based method has been improved (69bis) to allow the detection of the seven man phylogenetic groups (Fig. 1). Second, the O:K:H serotype is also a good reflection of the clonal groups (48). Lastly, the flexible-gene pool can be studied by PCR assays or by DNA microarrays. Of course, the limit of this last approach is that the search is performed only on genes with known function. Using all these proxies, epidemiologic data have been extensively generated, allowing for the comparison of thousands of strains implicated in extraintestinal infections and in the various forms of UTI.

UPEC, NMEC, and APEC strains: close relatives with multiple phenotypic potential

Genomic surrogates were used to explore genetic relatedness within EXPECs. While NMEC and UPEC mainly belong to phylogroups B2, D, and F (35, 67, 70, 71), APECs belong more commonly to phylogenetic group A, C, and then B2 (72–77). The recently described *Escherichia* clades are very rarely (<1%) isolated from extraintestinal sites (21). Despite a very high heterogeneity of VF-association patterns in ExPECs (particularly in UPECs), some NMEC and APEC strains have been shown to be very similar, sharing the same B2 phylogroup, O1:K1:H7 serotype, and VF pattern (75). Many genomic islands of APEC strain B2 subgroup IX/ST95 O1:H1:K7 were found with a high prevalence in a collection of O1 and O18 NMEC strains (78). Close relatedness within EXPECs has also been demonstrated by MLST, with APEC and human ExPEC strains closely intertwined in any subcluster (73). A single clonal group – B2/ST95 O1:K1:H7 – harbors UPEC, APEC,

and NMEC members (79). Maybe the most convincing argument that ExPEC are a specific, cohesive group of *E. coli* strains is the fact that most of the strains do not show any host specificity. UPEC strain U17 and APEC strain E058 were found to be comparably pathogenic in a chicken-infection model and a mouse urinary-tract-infection model (76). Furthermore, the expression of a set of genes in the two models was compared to their expression during exponential phase in a rich medium. In each model (UTI or chicken infection), the two strains had a tendency to regulate some genes in the same way. Particularly, iron-acquisition genes seemed to play a critical role in both pathologic conditions.

Some APEC strains are highly virulent in a rat meningitis model, and some NMEC strains are lethal to chicken embryos (80), further illustrating the multiple and interchangeable pathogenic potential of some ExPEC strains. Interestingly, this view was recently challenged. The sequenced isolate of APEC strain ST95 O1:K1:H7 was not able to cause meningitis in a rat model, and was less virulent than UPEC strain CFT073 in a mouse-septicemia model (78). This could be explained by the fact that *in vivo* just a few mutations can drastically affect the intrinsic virulence of an isolate (81). Therefore, even if some ExPEC strains show some multi-host, multi-pathogenic potential, it is possible that some host-specific adaptations have occurred in other ExPEC strains.

It clearly appears that APEC can be a reservoir of human pathogens (zoonosis) or of horizontally transmissible VFs within the ExPEC group (82). Whether some host-specific genetic factors have arisen in this general background of interchangeability is still debated (78, 83).

Where do UPEC come from?

It is generally assumed that *E. coli* strains that are responsible for UTIs in humans usually originate from the patients' own intestinal microbiota. Even though this pathophysiological scheme is very widely accepted, only a few studies provide evidence to support it (84, 85). This scheme suggests that the very first steps of a UTI consist of acquiring a UPEC strain, and integrating it as part of the digestive-tract community.

Many genomic surrogate markers have been used to assess the presence of potential UPEC strains in humans, animals, and the environment. UPEC-associated VFs have been identified in *E. coli* isolates coming from the feces of healthy individuals, farm animals, and local or imported meat products (86). CGA-specific ERIC-PCR and RAPD-electrophoretic patterns were identified in strains isolated from water supplies and animals (87). Comparison of outer membrane-protein profile on sodium dodecyl sulfate-polyacrylamide gel electrophoresis (SDS-PAGE), pilin molecular weight, plasmid electrophoresis, and RAPD pattern evidenced a close relatedness between human and dog UPEC isolates (88, 89). Several studies showed that the VFs, and serotypes frequently found in strains recovered from human UTIs, are also highly prevalent in ExPEC strains isolated from cats, dogs, other animals, and even meat products (77, 90–95). As a specific example, strains isolated from dog UTIs belong to the B2 subgroups III (O6) and VI (O4) (96), exactly matching the human archetypal strains 536 and J96, respectively (77). Because of these similarities between human and non-human UPEC strains, and because within-household zoonotic fecal *E. coli* strain sharing has been demonstrated (97, 98), it is possible that humans and animals serve as reservoirs of ExPEC strains for each other (99), further corroborating the zoonosis hypothesis.

Interestingly, CGA was isolated both from the urine of symptomatic patients and from the feces of healthy volunteers (38). Hence, healthy carriers represent a reservoir of ExPEC. This implies that the expression "ExPEC" can be very ambiguous. When a VF-containing *E. coli* strain is isolated from a non-digestive tract pathological sample, there is no doubt that it is an ExPEC. However, the CGA strains isolated from the feces of healthy volunteers are ExPEC too, even though they were leading a perfectly commensal life. It is probable that some healthy carriers will develop a UTI due to the CGA strain they are hosting in their own intestinal microbiota. To our knowledge, no prospective follow up of a cohort of humans hosting a known pathological clonal group in their digestive tract is available. A case-control study comparing the patients developing an extraintestinal infection with the subjects that maintain the potentially harmful bacteria under control in their digestive tract would be utterly interesting. Because potential ExPEC strains can be found in a commensal setting, we believe it is very important to consider intrinsic virulence. It can be defined as the presence of known virulence factors and/or as a pathogenic potential when tested in a controlled animal model (see below).

The prototypical UPEC is a highly virulent, B2 phylogroup strain

In an attempt to identify genes that are specific to a syndrome, the presence of VFs has extensively been examined in various collections of UPEC and fecal isolates. From these studies, it clearly appears that there is a strong correlation between phylogeny and VF content. B2 phylogroup strains contain more VFs, and PAI-associated loci than phylogroup A and B1 strains (35, 100–105). It has been postulated that there was a fine-tuning between the chromosomal background and the VFs, allowing the acquisition and the expression of the VFs (106). With the exception of some emerging clones (such as ST131 and CGA), strains from phylogroup B2 that contain a high number of VFs usually exhibit low antibiotic resistance (101, 107).

There is also some degree of correlation between phylogeny and syndrome severity. UPEC strains that are responsible for pyelonephritis, prostatitis, and urosepsis belong to phylogroup B2 in the vast majority of the cases (100, 102, 103, 105, 108). In contrast, UPEC strains isolated from patients with less severe syndromes such as asymptomatic bacteriuria or cystitis belong to phylogroups A/B1 or B2 in almost the same proportion. Fecal strains more frequently belong to phylogenetic group A/B1 (5). However, B2 phylogenetic-group strains isolated from cystitis harbor a VF content that is indistinguishable from that of B2 phylogenetic-group strains isolated from pyelonephritis or urosepsis (102). This indicates that phylogeny prevails over the clinical syndrome. No single genetic trait associated with virulence (VF, PAI, or phylogenetic group) is specifically associated with extraintestinal infections, including UTIs. Even with increasingly detailed information concerning the VF composition of UPEC strains, no gene or set of genes has ever been specifically associated to UTI type. Many genes contribute to extraintestinal virulence, but overall there does not seem to be any genetic determinism for specific clinical syndromes. Exceptions are B2-phylogroup strains that are recovered from patients with prostatitis. These strains consistently exhibit a slightly higher number of VFs and greater levels of biofilm formation than other UPEC strains (107, 109, 110).

Debilitated patients are prone to be infected by less virulent A/B1 phylogroup *E. coli* strains

Many host backgrounds and medical conditions are known to increase the risk for UTI. *E. coli* strains that are isolated from immunocompromised individuals frequently belong to phylogenetic group A or B1, express few VFs, and display some antibiotic resistance (101). Such strains often infect diabetic patients (111), liver-transplant recipients (112), patients undergoing bladder reconstitution with transposition of an intestinal segment (113), patients with neurogenic bladder, and patients with indwelling catheters (114). Bacteriuria is a very common complication in these last two patient groups. *E. coli* strains isolated from symptomatic and non-symptomatic bacteriuria in neurogenic-bladder patients express the same adhesion molecules (115–117). Patients with neurogenic

bladder most frequently need intermittent or long-term bladder catheterization. The risk for developing bacteriuria has been estimated as 5% to 10% per day of catheterization (118), and most patients with indwelling catheter for more than 30 days develop bacteriuria (119). Interestingly, catheter-associated long-term colonizing strains more frequently belonged to phylogenetic group B2 and harbored more VFs than transient strains (120). However, there was no association between long-term colonization with a B2 strain and symptomatic UTI. This is an indication that phylogenetic-group B2 strains can be both dreadful pathogens and excellent colonizers. We believe that this is because the same genetic elements can be virulence factors in some clinical settings and fitness factors in others (see below). An exception to this rule might be the case of kidney-transplant recipients, as several works have reported the predominance of highly virulent clone B2 ST131 O25b:H4 in this population (121, 122). However, other studies report little, if any, differences between *E. coli* strains isolated from immunocompromised and immunocompetent hosts (123, 124). These divergent results can probably be explained by different definitions of immunosuppression used in studies.

UPEC and NMEC in infants: the same, but not exactly?

UTI is the most frequent bacterial infection in young children of less than 90 days of age. As the same population can be affected by neonatal meningitis-causing *E. coli*, UPEC and NMEC strains isolates from infants have been compared. Phylogenetic group B2 and D strains are highly predominant in both pathovars (125). Even though NMEC and UPEC strains are very closely related, the combination of the O serotype and the MLST-sequence type (the "sequence-O-type") show that some bacterial complexes are exclusively associated with one type of infection (meningitis or urosepsis), while other complexes are frequently found in both (125). This indicates that while NMEC and UPEC strains are phylogenetically closely related, a particular subset of VFs and genetic backgrounds can result in niche specialization.

When considering UTIs in infants, septicemic (non-meningitis) and non-septicemic UPEC strains showed an indistinguishable phylogenetic repartition and VF content (126). In contrast, host factors seem to influence the type UPEC of strains that affect infants. UPEC strains isolated from male uncircumcised infants are generally more closely related to UPEC strains from adults than to UPEC strains from female infants (127). This probably indicates a different mode of acquisition of UPEC strains between male and female infants.

Toward specialized UPEC strains that are both resistant and virulent

Many reports seem to indicate that there is a trade-off between virulence and resistance in the *E. coli* species (101, 128, 129). Two clonal groups that do not follow this general rule are emerging worldwide.

First, *E. coli* CGA, identified as ST69, can exhibit various O types (O11, O17 and O77) and has a high number of conserved VFs (130). While it was initially identified in UTIs in North America, it was also recently shown to be an important cause of septicemia in Europe (131). In this study, CGA strains showed a higher virulence score than the other strains. Yet, non-CGA strains were more frequently sensitive to all antibiotics, while CGA strains were more frequently resistant to trimethoprim-sulfamethoxazole.

Second, the O25b:H4 ST131-B2 clonal group encompasses strains that can be highly lethal in the mouse-septicemia model (45, 132). This clonal group was identified in community settings as well as in the hospital, in samples coming from a wide range of extraintestinal infections as well as in feces of healthy individuals (39, 133). These strains have been identified in every region of the world in association with humans, animals, and even food products (29). O25b:H4 ST131-B2 strains seem to be particularly well-adapted to the urinary tract. Indeed, these strains are responsible for a substantial number of community-acquired cystitis cases in immunocompetent hosts (134). Recently, multidrug-resistant O25b:H4-ST131 isolates were found to be responsible for about 40% of septicemia cases after transrectal prostate biopsies (135). These strains also account for up to 35% of all UTIs, and 60% of quinolone-resistant UTIs in kidney-transplant recipients (121, 122). Members of clonal group O25b:H4 ST131 have a significantly higher antibiotic-resistance score than other phylogenetic group B2 UPEC strains (122). ST-131 can express ESBLs of various types, as well as carbapenemases (e.g., NDM-1), but can also be sensitive to third-generation cephalosporin (29). Overall, CGA and ST-131 account for approximately 5% and 10%, respectively, of all UPEC strains in two European studies (136, 137).

Clinical Relevance of Intrinsic Virulence

How to measure extraintestinal virulence

The occurrence of an extraintestinal infection is the result of a complex network of interacting parameters. On the pathogen side, the phylogenetic group, the specific set of virulence factors, the initial inoculum, and the portal of entry play a role. On the host side, species, gender, age, underlying pathology, and innate and adaptive immunity status can affect the course of infection. On top of these variables, many external interventions, such as surgery, presence of an indwelling catheter or of other implanted material, and antibiotic treatment must be considered. Because most extraintestinal infections arise from the host's own fecal bacteria, the composition of the gut microbiota is also probably very important to take into account, even though this factor has not yet been well studied. None of these parameters, taken individually, can account for virulence. B2-phylogenetic group *E. coli* strains that harbor an important number of VFs can be found leading a commensal life within healthy individuals (5). Conversely, *E. coli* K-12 strains are devoid of any notable VF, and yet can invade mesenteric lymph nodes and spleens of severely immunodepressed mice (138). In order to understand the myriad factors that affect extraintestinal virulence, well-standardized animal-infection models have been developed that allow measurement of a strain's "intrinsic virulence" (71, 139–141). One very widely used model is the mouse-septicemia model, in which a fixed *E. coli* inoculum is injected subcutaneously into female mice (35, 142). Mortality is then recorded. Intrinsic extraintestinal virulence of a given bacterial strain appears as a clear-cut parameter as most of the strains tested killed either more than 8 mice out of 10, or less than 3 out of ten. A marked link between phylogeny and intrinsic virulence was evidenced. Commensal strains that belong predominantly to phylogenetic group A and B1 lack VFs and do not kill mice in the septicemia model. In contrast, phylogenetic group B2 strains possessed the highest content in VFs and were also highly lethal to mice.

Overall intrinsic extraintestinal virulence results from additive effects of VFs

Seven isogenic single-PAI-deletion mutants and several multiple-PAI-deletion mutants were created in the UPEC reference strain 536 (30, 143). The intrinsic extraintestinal virulence of these mutants in the mouse-septicemia model has been tested (31). The majority of single-PAI-deletion mutants showed the same virulence as wild-type strain 536, i.e., 100% lethality within 30 h after subcutaneous challenge. Even the deletion of the so-called "high-pathogenicity island" (i.e., PAI IV536 that harbors the yersiniabactin gene) did not result in any decrease of virulence. Only two single-PAI-deletion mutants showed reduced virulence. This indicates that 1) all PAIs don't participate in virulence with the same importance, and 2) most PAIs are individually dispensable in this animal model. Interestingly, in a similar septicemia model, Brzuszkiewicz et al. did not find any reduction in virulence in these two single-PAI-deletion mutants (30), probably because they used a higher inoculum. The study of Tourret et al. also showed that, overall, PAIs contribute to virulence in an additive manner. For example, the mutant of strain 536 that lacked all 7 of the PAIs was significantly less virulent than a mutant that lacked only 1 or 2 PAIs. One of the mutants that lacked 5 PAIs was less virulent than one of the single-PAI-deletion mutants. Interestingly, the mutant that lacked all 7 of the PAIs still retained a low but significant level of virulence. This implies that not all VFs are harbored on PAIs. Comparisons of the effects of PAI deletion in septicemia versus UTI models showed that some PAIs might be more relevant to sepsis, while others are more pertinent during UTI (30).

In total, these results reveal the genetic complexity of virulence. The interpretation of results from animal models is further complicated because the lack of VFs in a strain can be compensated by using higher inoculation doses. In addition, the relative participation of each individual VF can vary depending on the animal model used.

E. coli intrinsic virulence is not associated with clinical outcome

The analysis of intrinsic virulence of *E. coli* strains isolated from human septicemia confirmed a strong correlation between phylogeny and VF content (144). In contrast, no link was found between intrinsic virulence or VF content and clinical outcome (101, 144, 145). Indeed, strains that killed more than 80% of the mice in the septicemia model were not associated with a worse clinical outcome in humans than strains that killed no mice (144). In a large, multicentric, prospective study of more than 1,000 septicemia patients, multivariate analysis showed that risk factors for death in *E. coli* septicemias are patient-related: age, existence of a cirrhosis, hospitalization before septicemia, immunosuppression, and cutaneous origin of the septicemia (145). A urinary-tract origin was protective, even though strains isolated in septicemia with a urinary-tract origin often belonged to phylogenetic group B2 and had a higher median virulence score than those of digestive-tract origin (101, 145). Therefore, septicemia with the urinary tract as the portal of entry seems to be associated with highly virulent strains that frequently belong to phylogenetic group B2, but nonetheless have a better prognosis than septicemic patients who are infected with less-virulent phylogenetic group A or B1 strains that originated from the digestive tract. Simply put, an immunocompetent host has a stronger chance to survive to urosepsis with a highly pathogenic bacterium than an immunocompromised host with a less-virulent isolate from the intestinal tract.

Interestingly, factors influencing the occurrence of a septicemia in UTI patients are also mainly host-related. Only one bacterial factor – a capsule determinant – is known to favor blood-bacterial translocation from the urinary tract into the bloodstream (146).

Hierarchical organization of factors involved in extraintestinal infections

Extraintestinal virulence is the result of the complex interactions among bacterial, host, and environmental factors. Different combinations of these factors can result in the same clinical syndrome. However, based on studies covered in previous paragraphs, a hierarchical organization of infectious determinants can be proposed (Fig. 4). Firstly, host-related factors influence the phylogenetic group of the strains that are responsible for the extraintestinal infection. Typically, A/B1 phylogroup strains, possibly from abdominal origin, are more frequently found in debilitated patients. In contrast, patients without any predisposing condition who present with pyelonephritis or urosepsis are more frequently affected by a B2 phylogenetic-group strain. Host-related factors also affect outcome of the infection, with debilitated patients showing a poorer prognosis

than patients with no preexisting medical condition. Secondly, phylogenetic group is associated with VF content. B2-group strains have a high VF content and are often sensitive to antibiotics (with the exception of the recently emerged ST131 and CGA clones). A/B1 strains show a lower VF content, and are more frequently resistant to antibiotics. Lastly, VF content is correlated with the intrinsic virulence of the strains. Those with a high VF content kill more than 80% of the mice in the murine-septicemia model, while strains that are devoid of VF usually kill less than 10% of the mice.

E. COLI POLYMICROBIAL EXTRAINTESTINAL INFECTIONS

Until recently, it was generally assumed that infections were initiated by a single organism, followed by its proliferation. However, with the availability of molecular tools, including whole-genome analysis, it became evident that distinct *E. coli* isolates can be simultaneously present at an extraintestinal-infection site, including the urinary tract (81, 147, 148). Two kinds of diversity can be observed: polyclonal diversity caused by phylogenetically divergent clones, and monoclonal diversity involving isolates of a single clone that exhibit micro-heterogeneity. The exact proportion of *E. coli* polymicrobial infections is unknown as very few studies have been reported, but is not an unusual event, with frequencies ranging from 20% to more than 50% (81, 147, 148).

What can be learned from genomics in these infections? In polyclonal infections, phylogrouping/MLST and virulence-factor content determination are sufficient to demonstrate that, in some cases, there is an association of a highly virulent *E. coli* strain with a non-virulent *E. coli* strain (81, 149). To try to understand the effects of such polymicrobial infections, the interaction of two divergent *E. coli* strains, isolated from a patient abscess, one highly virulent and the

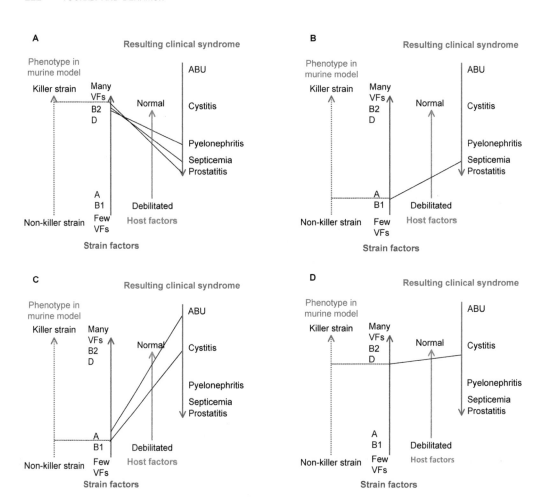

FIGURE 4 Schematic representation of interactions between bacterial-associated genotypic factors, host-related conditions, and the resulting clinical syndrome. A. Highly virulent phylogroup B2 strains can be responsible for severe clinical syndromes in patients with no medical conditions, such as pyelonephritis, urosepsis, or prostatitis. They are highly lethal to mice. NB: These strains can also be found as fecal commensals, a situation that can be explained by the "virulence by-product of commensalism" hypothesis. B. A/B1 phylogenetic-group strains can be responsible for a severe clinical syndrome in debilitated patients. However, they show little lethality in a mouse model measuring intrinsic virulence. C. In patients with no medical condition, phylogroup A/B1 strains with little virulence potential are usually found in less-severe conditions such as cystitis, asymptomatic bacteriuria, or even in non-pathogenic fecal samples. They do not show any virulence in a mouse model measuring intrinsic virulence. NB: Some B2 strains with reductive evolution inactivating numerous virulence determinants can also cause ABU. These strains are not lethal in the mouse model of septicemia (E. Denamur, personal data). D. Depending on the virulence-factors/host-condition combination, highly virulent B2 phylogroup strains can also be responsible for a non-severe clinical syndrome, such as cystitis. Such strains show high intrinsic virulence in a mouse model of septicemia. VFs: virulence factors. ABU: asymptomatic bacteriuria. A, B1, B2, D: phylogenetic groups.

other not, has been studied in mouse septicemia and UTI models (150). In these models, co-infection with the two strains (a 1-to-100 mix of the virulent and nonvirulent strains, respectively) resulted in a synergistically more severe disease than produced by either microbe on its own. This is probably due to a lure effect, in which the avirulent strain overwhelms the host bacterial-clearing systems, allowing for the low numbers of the pathogenic strain to initiate a symptomatic infection. Considering that most mammals host commensal avirulent *E. coli* strains in their digestive tract, these results raise questions about the potential role of this reservoir of "virulence enhancers" in extraintestinal infections (150).

The presence of monoclonal diversity within a site of infection is even more intriguing. In this case, the isolates from a single patient belong to the same sequence type and serotype, but differ by minor DNA-sequence variations as determined by pulsed-field gel electrophoresis. The isolates also present a wide continuum of phenotypic variability with respect to their antibiotic resistance, outer-membrane permeability, growth rate, stress resistance, and intrinsic virulence in a mouse-septicemia model (81). Many of these phenotypic differences within clones are considered to be the result of trade-offs between self-preservation and nutritional competence (SPANC) (151). Self-preservation involves stress protection and is an important, but costly, contributor to bacterial survival. Nutritional competence is the ability to metabolize and grow on an extensive set of substrates. Different *E. coli* strains can differ in how they balance SPANC, in part due to differences in expression of the stress-resistance sigma factor RpoS (152). Strains with higher RpoS levels are more resistant to external stress but metabolize fewer substrates, whereas strains with lower RpoS levels have broader nutritional capabilities but lower resistance to external stress (151). The observed phenotypic variability of within-patient isolates was indeed associated with distinct levels of RpoS in the co-existing isolates (81). Mutational analysis of multiple isolates from a single patient showed a star-like relationship of changes, indicating rapid diversification, with one to three mutational events occur-

ring per isolate. These include plasmid loss, insertion sequence (IS) insertions, small-deletion and point mutations, mainly in metabolic and membrane-related genes, but not in *rpoS*. There are strong arguments for the fact that these mutational events have been selected for as (i) all but one point mutations are non-synonymous or occur in a regulatory region, (ii) a convergence in the gene *ompA* encoding an outer-membrane protein was observed with two different molecular defects leading to the absence of protein, (iii) the identified mutation in the *rbs* operon coding for ribose metabolism has been reported to be under selection in an experimental evolution system (153) and (iv) mutator strains were present among the different isolates (81). These data parallel the observations made in experimental-evolution systems where no immune-selective pressure is present (154) and were explained by a simple mathematical model demonstrating that multiple genotypes with distinct levels of RpoS can co-exist as a results of SPANC trade-offs (81). Several lines of evidence suggest that diversification occurs during the infectious process, and does not come from infection by multiple isolates exhibiting a micro-heterogeneity.

Whatever the type of diversity observed, the fact that a non-negligible number of UTI can be caused by polymicrobial infections has strong medical implications. In addition to the fact that such polymicrobial infections can enhance virulence, it has relevant impacts for clinical laboratories and their antibiogram strategies. The classical antimicrobial-susceptibility testing is performed on few (2 to 5) colonies obtained from the patient sample to establish antibiotic treatment (155, 156). This may fail to detect the presence of other, more-resistant isolates, and may lead to therapeutic failure (157). Such diagnostic problems may be remedied by screening numerous isolates from each patient, perhaps using more streamlined antibiograms. (81).

EVOLUTIONARY FORCES SELECTING EXPEC VIRULENCE DETERMINANTS: THE "BY-PRODUCT OF COMMENSALISM" HYPOTHESIS

It is clear that multiple bacterial factors such as adhesins, iron-capture systems, toxins, and protectins, which are often localized within genomic islands, promote intrinsic extraintestinal virulence (30, 31, 158), allowing colonization of extraintestinal sites. Without treatment, bacterial colonization that is facilitated by multiple VFs will lead to severe disease and potentially to rapid host death that could consequently limit transmission of the infecting strain. In this context, how can we explain that these virulence determinants have been selected for during the more than 50-million years of *E. coli* history (159)? This apparent contradiction between the presence and maintenance of many VFs and the presumably poor selective advantage of highly virulent strains during extraintestinal infection has led to the idea that many VFs might be required for transmission to new hosts or for the colonization of niches where the bacteria reside as commensals. This view leads to the hypothesis that infections caused by facultative pathogens such as *E. coli* occur by accident (160). A combination of epidemiological, phylogenetic, genomic, and clinical data argues for an ancestral emergence of extraintestinal virulence that appears as a coincidental by-product of commensalism (161). This has been experimentally demonstrated in a streptomycin-treated mouse-colonization model using the UPEC wild-type 536 strain and isogenic-PAI mutants (30, 143). First, the wild-type strain 536 outcompetes the non-pathogenic K-12 strain in this model (30). Second, deletion of all seven PAIs drastically reduces the fitness of 536 during persistent intestinal colonization in co-challenge experiments with the wild-type strain. This mutant defect seems to be linked to the hypermotility observed in mutants devoid of all PAIs (143). In addition, this study shows that PAIs diminish bacterial fitness during growth in urine, suggesting that urinary-tract infections are unlikely to provide selective pressure for the maintenance of ExPEC PAIs (143).

All these data are in line with epidemiological observations of the commensal *E. coli* microbiota in healthy humans reporting that B2-group strains that encode numerous virulent determinants on PAIs are both dominant (the most abundant clone in each fecal sample) (162) and persistent (long-term colonizer) (163). This has clear implications for the pathophysiology of UTI. Moreno et al. compared fecal and urine clones in a cohort of women affected with acute uncomplicated cystitis. They showed that belonging to the B2 phylogroup, harboring multiple virulence factors, and being dominant and pauciclonal in the feces were factors associated with being a urinary clone (164).

Furthermore, ABU strains, which establish long-term bacteriuria without causing symptoms of UTI, are related to UPEC, but exhibit reductive evolution by point mutations, DNA rearrangements, and deletions in essential virulence genes (165). So, the long-term survival of these clones in urine is seemingly associated with an attenuation of virulence (165, 166).

CONCLUSIONS

The arrival of whole-genome sequencing has clearly helped break new ground in understanding the evolution and virulence of ExPEC, including UTI strains. However, genomic sequencing has also pointed out an unsuspected complexity in ExPEC and related *E. coli* clones. There is not one (or even a few) gene(s) that specifically define UPEC. Instead, numerous combinations of genes, associated to specific alleles, can result in an *E. coli* strain acquiring a UPEC phenotype. The barrier between commensal and UPEC strains is difficult to draw. This is probably due to genes that enable *E. coli* to thrive within the intestinal tract overlap with those that are responsible for the UPEC phenotype. In other words, UPEC virulence is potentially a by-product of commensalism. Lastly, infections are not the result of a single, unique, and stable isolate. Rather, several isolates with variable levels of diversity are often involved, dynamically changing over time.

The number of completed sequences will continue to grow exponentially and, due to the improvement of the high-throughput-sequencing technologies associated with reduced costs, we can expect an explosion of data in the next few years. High-throughput RNA-sequencing techniques will also continue to improve, allowing for more accurate study of transcriptional analysis of UPEC strains and providing detailed insight into the gene-regulatory processes that govern bacterial adaptation within extraintestinal sites. Indeed, gene regulation is a key process in adaptation (167, 168). A priority should be given now to the development of powerful bioinformatics and statistical tools. One of the key steps to further progress in the understanding of extraintestinal virulence will be the knowledge of the strains that will be analyzed and the pertinence of their sampling.

As we have seen a very important role of the host status in the gravity of the infections caused by UPEC, one can imagine that the progress of genomics could also be applied to the sequencing of the host genome, moving towards more personalized medicine (169) based on the identification of infection-susceptibility genes, in combination with microbial genomics.

ACKNOWLEDGMENTS

We acknowledge all the members of our lab that have collaborated with us during the last 10 years, especially Bertrand Picard and Olivier Tenaillon, for their valuable contribution to the work presented. We are grateful to Florence Reibel and Mathilde Lescat for providing Figure 1. Our work is partly supported by the "Fondation pour la Recherche Médicale" and the "Société Française de Néphrologie".

Conflicts of interest: We declare no conflicts.

CITATION

Tourret J, Denamur E. 2016. Population phylogenomics of extraintestinal pathogenic *Escherichia coli*. Microbiol Spectrum 4(1): UTI-0010-2012.

REFERENCES

1. **Blattner FR, Plunkett G III, Bloch CA, Perna NT, Burland V, Riley M, Collado-Vides J, Glasner JD, Rode CK, Mayhew GF, Gregor J, Davis NW, Kirkpatrick HA, Goeden MA, Rose DJ, Mau B, Shao Y.** 1997. The complete genome sequence of *Escherichia coli* K-12. *Science* **277:**1453–1462.

2. **Welch RA, Burland V, Plunkett G III, Redford P, Roesch P, Rasko D, Buckles EL, Liou SR, Boutin A, Hackett J, Stroud D, Mayhew GF, Rose DJ, Zhou S, Schwartz DC, Perna NT, Mobley HL, Donnenberg MS, Blattner FR.** 2002. Extensive mosaic structure revealed by the complete genome sequence of uropathogenic *Escherichia coli*. *Proc Natl Acad Sci U S A* **99:**17020–17024.

3. **Perna NT, Plunkett G III, Burland V, Mau B, Glasner JD, Rose DJ, Mayhew GF, Evans PS, Gregor J, Kirkpatrick HA, Pósfai G, Hackett J, Klink S, Boutin A, Shao Y, Miller L, Grotbeck EJ, Davis NW, Lim A, Dimalanta ET, Potamousis KD, Apodaca J, Anantharaman TS, Lin J, Yen G, Schwartz DC, Welch RA, Blattner FR.** 2001. Genome sequence of enterohaemorrhagic *Escherichia coli* O157:H7. *Nature* **409:**529–533.

4. **Whitman WB, Coleman DC, Wiebe WJ.** 1998. Prokaryotes: The unseen majority. *Proc Natl Acad Sci U S A* **95:**6578–6583.

5. **Tenaillon O, Skurnik D, Picard B, Denamur E.** 2010. The population genetics of commensal *Escherichia coli*. *Nat Rev Microbiol* **8:**207–217.

6. **Savageau MA.** 1983. *Escherichia coli* habitats, cell types, and molecular mechanisms of gene control. *Am Nat* **122:**732–744.

7. **Russo TA, Johnson JR.** 2003. Medical and economic impact of extraintestinal infections due to *Escherichia coli*: focus on an increasingly important endemic problem. *Microbes Infect* **5:**449–456.

8. **Kaper JB, Nataro JP, Mobley HL.** 2004. Pathogenic *Escherichia coli*. *Nat Rev Microbiol* **2**:123–140.

9. **Russo TA, Johnson JR.** 2000. Proposal for a new inclusive designation for extraintestinal pathogenic isolates of *Escherichia coli*: ExPEC. *J Infect Dis* **181**:1753–1754.

10. **Smith JM, Smith NH, O'Rourke M, Spratt BG.** 1993. How clonal are bacteria? *Proc Natl Acad Sci U S A* **90**:4384–4388.

11. **Orskov F, Orskov I, Evans DJ Jr, Sack RB, Sack DA, Wadström T.** 1976. Special *Escherichia coli* serotypes among enterotoxigenic strains from diarrhoea in adults and children. *Med Microbiol Immunol* **162**:73–80.

12. **Selander RK, Levin BR.** 1980. Genetic diversity and structure in *Escherichia coli* populations. *Science* **210**:545–547.

13. **Desjardins P, Picard B, Kaltenböck B, Elion J, Denamur E.** 1995. Sex in *Escherichia coli* does not disrupt the clonal structure of the population: evidence from random amplified polymorphic DNA and restriction-fragment-length polymorphism. *J Mol Evol* **41**:440–448.

14. **Milkman R, Crawford IP.** 1983. Clustered third-base substitutions among wild strains of *Escherichia coli*. *Science* **221**:378–380.

15. **Dykhuizen DE, Green L.** 1991. Recombination in *Escherichia coli* and the definition of biological species. *J Bacteriol* **173**:7257–7268.

16. **Touchon M, Hoede C, Tenaillon O, Barbe V, Baeriswyl S, Bidet P, Bingen E, Bonacorsi S, Bouchier C, Bouvet O, Calteau A, Chiapello H, Clermont O, Cruveiller S, Danchin A, Diard M, Dossat C, Karoui ME, Frapy E, Garry L, Ghigo JM, Gilles AM, Johnson J, Le Bouguénec C, Lescat M, Mangenot S, Martinez-Jéhanne V, Matic I, Nassif X, Oztas S, Petit MA, Pichon C, Rouy Z, Ruf CS, Schneider D, Tourret J, Vacherie B, Vallenet D, Médigue C, Rocha EP, Denamur E.** 2009. Organised genome dynamics in the *Escherichia coli* species results in highly diverse adaptive paths. *PLoS Genet* **5**:e1000344. doi:10.1371/journal.pgen.1000344

17. **Milkman R, Jaeger E, McBride RD.** 2003. Molecular evolution of the *Escherichia coli* chromosome. VI. Two regions of high effective recombination. *Genetics* **163**:475–483.

18. **Escobar-Páramo P, Sabbagh A, Darlu P, Pradillon O, Vaury C, Denamur E, Lecointre G.** 2004. Decreasing the effects of horizontal gene transfer on bacterial phylogeny: the *Escherichia coli* case study. *Mol Phylogenet Evol* **30**:243–250.

19. **Wirth T, Falush D, Lan R, Colles F, Mensa P, Wieler LH, Karch H, Reeves PR, Maiden MC, Ochman H, Achtman M.** 2006. Sex and virulence in *Escherichia coli*: An evolutionary perspective. *Mol Microbiol* **60**:1136–1151.

20. **Walk ST, Alm EW, Gordon DM, Ram JL, Toranzos GA, Tiedje JM, Whittam TS.** 2009. Cryptic lineages of the genus *Escherichia*. *Appl Environ Microbiol* **75**:6534–6544.

21. **Clermont O, Gordon DM, Brisse S, Walk ST, Denamur E.** 2011. Characterization of the cryptic *Escherichia* lineages: rapid identification and prevalence. *Environ Microbiol* **13**:2468–2477.

22. **Moissenet D, Salauze B, Clermont O, Bingen E, Arlet G, Denamur E, Mérens A, Mitanchez D, Vu-Thien H.** 2010. Meningitis caused by *Escherichia coli* producing TEM-52 extended-spectrum beta-lactamase within an extensive outbreak in a neonatal ward: epidemiological investigation and characterization of the strain. *J Clin Microbiol* **48**:2459–2463.

23. **Rasko DA, Rosovitz MJ, Myers GS, Mongodin EF, Fricke WF, Gajer P, Crabtree J, Sebaihia M, Thomson NR, Chaudhuri R, Henderson IR, Sperandio V, Ravel J.** 2008. The pangenome structure of *Escherichia coli*: comparative genomic analysis of *E. coli* commensal and pathogenic isolates. *J Bacteriol* **190**:6881–6893.

24. **Vieira G, Sabarly V, Bourguignon PY, Durot M, Le Fèvre F, Mornico D, Vallenet D, Bouvet O, Denamur E, Schachter V, Médigue C.** 2011. Core and panmetabolism in *Escherichia coli*. *J Bacteriol* **193**:1461–1472.

25. **Lukjancenko O, Wassenaar TM, Ussery DW.** 2010. Comparison of 61 sequenced *Escherichia coli* genomes. *Microb Ecol* **60**:708–720.

26. **Hacker J, Kaper JB.** 2000. Pathogenicity islands and the evolution of microbes. *Annu Rev Microbiol* **54**:641–679.

27. **Hendrickson H.** 2009. Order and disorder during *Escherichia coli* divergence. *PLoS Genet* **5**:e1000335. doi:10.1371/journal.pgen.1000335

28. **Harvey PH, Pagel MD.** 1991. *The Comparative Method in Evolutionary Biology*, Oxford University Press, New York.

29. **Rogers BA, Sidjabat HE, Paterson DL.** 2011. *Escherichia coli* O25b-ST131: A pandemic, multiresistant, community-associated strain. *J Antimicrob Chemother* **66**:1–14.

30. **Brzuszkiewicz E, Brüggemann H, Liesegang H, Emmerth M, Olschläger T, Nagy G, Albermann K, Wagner C, Buchrieser C, Emody L, Gottschalk G, Hacker J, Dobrindt U.** 2006. How to become a uropathogen: comparative genomic analysis of extraintestinal pathogenic *Escherichia coli* strains. *Proc Natl Acad Sci U S A* **103**:12879–12884.

31. **Tourret J, Diard M, Garry L, Matic I, Denamur E.** 2010. Effects of single and multiple pathogenicity island deletions on uropathogenic *Escherichia coli* strain 536 intrinsic extra-intestinal virulence. *Int J Med Microbiol* **300**:435–439.

32. **Johnson TJ, Kariyawasam S, Wannemuehler Y, Mangiamele P, Johnson SJ, Doetkott C, Skyberg JA, Lynne AM, Johnson JR, Nolan LK.** 2007. The genome sequence of avian pathogenic *Escherichia coli* strain O1:K1:H7 shares strong similarities with human extraintestinal pathogenic *E. coli* genomes. *J Bacteriol* **189**:3228–3236.

33. **Manges AR, Johnson JR, Foxman B, O'Bryan TT, Fullerton KE, Riley LW.** 2001. Widespread distribution of urinary tract infections caused by a multidrug-resistant *Escherichia coli* clonal group. *N Engl J Med* **345**:1007–1013.

34. **Lane MC, Mobley HL.** 2007. Role of P-fimbrial-mediated adherence in pyelonephritis and persistence of uropathogenic *Escherichia coli* (UPEC) in the mammalian kidney. *Kidney Int* **72**:19–25.

35. **Picard B, Garcia JS, Gouriou S, Duriez P, Brahimi N, Bingen E, Elion J, Denamur E.** 1999. The link between phylogeny and virulence in *Escherichia coli* extraintestinal infection. *Infect Immun* **67**:546–553.

36. **Moriel DG, Bertoldi I, Spagnuolo A, Marchi S, Rosini R, Nesta B, Pastorello I, Corea VA, Torricelli G, Cartocci E, Savino S, Scarselli M, Dobrindt U, Hacker J, Tettelin H, Tallon LJ, Sullivan S, Wieler LH, Ewers C, Pickard D, Dougan G, Fontana MR, Rappuoli R, Pizza M, Serino L.** 2010. Identification of protective and broadly conserved vaccine antigens from the genome of extraintestinal pathogenic *Escherichia coli*. *Proc Natl Acad Sci U S A* **107**:9072–9077.

37. **Tenaillon O, Rodríguez-Verdugo A, Gaut RL, McDonald P, Bennett AF, Long AD, Gaut BS.** 2012. The molecular diversity of adaptive convergence. *Science* **335**:457–461.

38. **Johnson JR, Manges AR, O'Bryan TT, Riley LW.** 2002. A disseminated multidrug-resistant clonal group of uropathogenic *Escherichia coli* in pyelonephritis. *Lancet* **359**:2249–2251.

39. **Nicolas-Chanoine MH, Blanco J, Leflon-Guibout V, Demarty R, Alonso MP, Caniça MM, Park YJ, Lavigne JP, Pitout J, Johnson JR.** 2008. Intercontinental emergence of *Escherichia coli* clone O25:H4-ST131 producing CTX-M-15. *J Antimicrob Chemother* **61**:273–281.

40. **Peirano G, Schreckenberger PC, Pitout JD.** 2011. Characteristics of NDM-1-producing

Escherichia coli isolates that belong to the successful and virulent clone ST131. *Antimicrob Agents Chemother* **55**:2986–2988.

41. **Johnson JR, Menard ME, Lauderdale TL, Kosmidis C, Gordon D, Collignon P, Maslow JN, Andrasević AT, Kuskowski MA; Trans-Global Initiative for Antimicrobial Resistance Analysis Investigators.** 2011. Global distribution and epidemiologic associations of *Escherichia coli* clonal group A, 1998-2007. *Emerg Infect Dis* **17**:2001–2009.

42. **Lescat M, Calteau A, Hoede C, Barbe V, Touchon M, Rocha E, Tenaillon O, Médigue C, Johnson JR, Denamur E.** 2009. A module located at a chromosomal integration hot spot is responsible for the multidrug resistance of a reference strain from *Escherichia coli* clonal group A. *Antimicrob Agents Chemother* **53**:2283–2288.

43. **Avasthi TS, Kumar N, Baddam R, Hussain A, Nandanwar N, Jadhav S, Ahmed N.** 2011. Genome of multidrug-resistant uropathogenic *Escherichia coli* strain NA114 from India. *J Bacteriol* **193**:4272–4273.

44. **Totsika M, Beatson SA, Sarkar S, Phan MD, Petty NK, Bachmann N, Szubert M, Sidjabat HE, Paterson DL, Upton M, Schembri MA.** 2011. Insights into a multidrug resistant *Escherichia coli* pathogen of the globally disseminated ST131 lineage: genome analysis and virulence mechanisms. *PLoS One* **6**:e26578. doi:10.1371/journal.pone.0026578

45. **Clermont O, Lavollay M, Vimont S, Deschamps C, Forestier C, Branger C, Denamur E, Arlet G.** 2008. The CTX-M-15-producing *Escherichia coli* diffusing clone belongs to a highly virulent B2 phylogenetic subgroup. *J Antimicrob Chemother* **61**:1024–1028.

46. **Poirel L, Lagrutta E, Taylor P, Pham J, Nordmann P.** 2010. Emergence of metallo-beta-lactamase NDM-1-producing multidrug-resistant *Escherichia coli* in Australia. *Antimicrob Agents Chemother* **54**:4914–4916.

47. **Poirel L, Bonnin RA, Nordmann P.** 2011. Analysis of the resistome of a multidrug-resistant NDM-1-producing *Escherichia coli* strain by high-throughput genome sequencing. *Antimicrob Agents Chemother* **55**:4224–4229.

48. **Schubert S, Cuenca S, Fischer D, Heesemann J.** 2000. High-pathogenicity island of *Yersinia pestis* in enterobacteriaceae isolated from blood cultures and urine samples: prevalence and functional expression. *J Infect Dis* **182**:1268–1271.

49. **Schubert S, Picard B, Gouriou S, Heesemann J, Denamur E.** 2002. Yersinia high-pathogenicity island contributes to virulence in *Escherichia coli*

causing extraintestinal infections. *Infect Immun* **70:**5335–5337.

50. **Schubert S, Darlu P, Clermont O, Wieser A, Magistro G, Hoffmann C, Weinert K, Tenaillon O, Matic I, Denamur E.** 2009. Role of intraspecies recombination in the spread of pathogenicity islands within the *Escherichia coli* species. *PLoS Pathog* **5:**e1000257. doi:10.1371/journal.ppat.1000257

51. **Chen SL, Hung CS, Xu J, Reigstad CS, Magrini V, Sabo A, Blasiar D, Bieri T, Meyer RR, Ozersky P, Armstrong JR, Fulton RS, Latreille JP, Spieth J, Hooton TM, Mardis ER, Hultgren SJ, Gordon JI.** 2006. Identification of genes subject to positive selection in uropathogenic strains of *Escherichia coli*: A comparative genomics approach. *Proc Natl Acad Sci U S A* **103:**5977–5982.

52. **Yang Z.** 2007. PAML 4: phylogenetic analysis by maximum likelihood. *Mol Biol Evol* **24:** 1586–1591.

53. **Petersen L, Bollback JP, Dimmic M, Hubisz M, Nielsen R.** 2007. Genes under positive selection in *Escherichia coli*. *Genome Res* **17:**1336–1343.

54. **Wiles TJ, Kulesus RR, Mulvey MA.** 2008. Origins and virulence mechanisms of uropathogenic *Escherichia coli*. *Exp Mol Pathol* **85:**11–19.

55. **Denamur E, Bonacorsi S, Giraud A, Duriez P, Hilali F, Amorin C, Bingen E, Andremont A, Picard B, Taddei F, Matic I.** 2002. High frequency of mutator strains among human uropathogenic *Escherichia coli* isolates. *J Bacteriol* **184:**605–609.

56. **Labat F, Pradillon O, Garry L, Peuchmaur M, Fantin B, Denamur E.** 2005. Mutator phenotype confers advantage in *Escherichia coli* chronic urinary tract infection pathogenesis. *FEMS Immunol Med Microbiol* **44:**317–321.

57. **Chattopadhyay S, Dykhuizen DE, Sokurenko EV.** 2007. ZPS: visualization of recent adaptive evolution of proteins. *BMC Bioinformatics* **8:**187.

58. **Chattopadhyay S, Weissman SJ, Minin VN, Russo TA, Dykhuizen DE, Sokurenko EV.** 2009. High frequency of hotspot mutations in core genes of *Escherichia coli* due to short-term positive selection. *Proc Natl Acad Sci U S A* **106:**12412–12417.

59. **Denamur E, Matic I.** 2006. Evolution of mutation rates in bacteria. *Mol Microbiol* **60:** 820–827.

60. **Ansorge WJ.** 2009. Next-generation DNA sequencing techniques. *N Biotechnol* **25:**195–203.

61. **Clark G, Paszkiewicz K, Hale J, Weston V, Constantinidou C, Penn C, Achtman M, McNally A.** 2012. Genomic analysis uncovers

a phenotypically diverse but genetically homogeneous *Escherichia coli* ST131 clone circulating in unrelated urinary tract infections. *J Antimicrob Chemother* **67:**868–877.

62. **Lindberg U, Claesson I, Hanson LA, Jodal U.** 1978. Asymptomatic bacteriuria in schoolgirls. VIII. Clinical course during a 3-year follow-up. *J Pediatr* **92:**194–199.

63. **Sundén F, Håkansson L, Ljunggren E, Wullt B.** 2010. *Escherichia coli* 83972 bacteriuria protects against recurrent lower urinary tract infections in patients with incomplete bladder emptying. *J Urol* **184:**179–185.

64. **Zdziarski J, Brzuszkiewicz E, Wullt B, Liesegang H, Biran D, Voigt B, Gronberg-Hernandez J, Ragnarsdottir B, Hecker M, Ron EZ, Daniel R, Gottschalk G, Hacker J, Svanborg C, Dobrindt U.** 2010. Host imprints on bacterial genomes–rapid, divergent evolution in individual patients. *PLoS Pathog* **6:**e1001078. doi:10.1371/journal.ppat.1001078

65. **Reeves PR, Liu B, Zhou Z, Li D, Guo D, Ren Y, Clabots C, Lan R, Johnson JR, Wang L.** 2011. Rates of mutation and host transmission for an *Escherichia coli* clone over 3 years. *PLoS One* **6:**e26907. doi:10.1371/journal.pone.0026907

66. **Reid SD, Herbelin CJ, Bumbaugh AC, Selander RK, Whittam TS.** 2000. Parallel evolution of virulence in pathogenic *Escherichia coli*. *Nature* **406:**64–67.

67. **Jaureguy F, Landraud L, Passet V, Diancourt L, Frapy E, Guigon G, Carbonnelle E, Lortholary O, Clermont O, Denamur E, Picard B, Nassif X, Brisse S.** 2008. Phylogenetic and genomic diversity of human bacteremic *Escherichia coli* strains. *BMC Genomics* **9:**560. doi:10.1186/1471-2164-9-560

68. **Clermont O, Bonacorsi S, Bingen E.** 2000. Rapid and simple determination of the *Escherichia coli* phylogenetic group. *Appl Environ Microbiol* **66:**4555–4558.

69. **Gordon DM, Clermont O, Tolley H, Denamur E.** 2008. Assigning *Escherichia coli* strains to phylogenetic groups: multi-locus sequence typing versus the PCR triplex method. *Environ Microbiol* **10:**2484–2496.

70. **Bingen E, Picard B, Brahimi N, Mathy S, Desjardins P, Elion J, Denamur E.** 1998. Phylogenetic analysis of *Escherichia coli* strains causing neonatal meningitis suggests horizontal gene transfer from a predominant pool of highly virulent B2 group strains. *J Infect Dis* **177:**642–650.

71. **Bonacorsi S, Clermont O, Houdouin V, Cordevant C, Brahimi N, Marecat A, Tinsley C, Nassif X, Lange M, Bingen E.** 2003. Molecular analysis and experimental virulence of French

and North American *Escherichia coli* neonatal meningitis isolates: identification of a new virulent clone. *J Infect Dis* **187:**1895–1906.

72. **Rodriguez-Siek KE, Giddings CW, Doetkott C, Johnson TJ, Fakhr MK, Nolan LK.** 2005. Comparison of *Escherichia coli* isolates implicated in human urinary tract infection and avian colibacillosis. *Microbiology* **151:**2097–2110.

73. **Moulin-Schouleur M, Répérant M, Laurent S, Brée A, Mignon-Grasteau S, Germon P, Rasschaert D, Schouler C.** 2007. Extraintestinal pathogenic *Escherichia coli* strains of avian and human origin: link between phylogenetic relationships and common virulence patterns. *J Clin Microbiol* **45:**3366–3376.

74. **Johnson TJ, Wannemuehler Y, Johnson SJ, Stell AL, Doetkott C, Johnson JR, Kim KS, Spanjaard L, Nolan LK.** 2008. Comparison of extraintestinal pathogenic *Escherichia coli* strains from human and avian sources reveals a mixed subset representing potential zoonotic pathogens. *Appl Environ Microbiol* **74:**7043–7050.

75. **Ewers C, Li G, Wilking H, Kiessling S, Alt K, Antáo EM, Laturnus C, Diehl I, Glodde S, Homeier T, Böhnke U, Steinrück H, Philipp HC, Wieler LH.** 2007. Avian pathogenic, uropathogenic, and newborn meningitis-causing *Escherichia coli*: how closely related are they? *Int J Med Microbiol* **297:**163–176.

76. **Zhao L, Gao S, Huan H, Xu X, Zhu X, Yang W, Gao Q, Liu X.** 2009. Comparison of virulence factors and expression of specific genes between uropathogenic *Escherichia coli* and avian pathogenic *E. coli* in a murine urinary tract infection model and a chicken challenge model. *Microbiology* **155:**1634–1644.

77. **Clermont O, Olier M, Hoede C, Diancourt L, Brisse S, Keroudean M, Glodt J, Picard B, Oswald E, Denamur E.** 2011. Animal and human pathogenic *Escherichia coli* strains share common genetic backgrounds. *Infect Genet Evol* **11:**654–662.

78. **Johnson TJ, Wannemuehler Y, Kariyawasam S, Johnson JR, Logue CM, Nolan LK.** 2012. Prevalence of avian-pathogenic *Escherichia coli* strain O1 genomic islands among extraintestinal and commensal *E. coli* isolates. *J Bacteriol* **194:**2846–2853.

79. **Mora A, López C, Dabhi G, Blanco M, Blanco JE, Alonso MP, Herrera A, Mamani R, Bonacorsi S, Moulin-Schouleur M, Blanco J.** 2009. Extraintestinal pathogenic *Escherichia coli* O1:K1:H7/NM from human and avian origin: detection of clonal groups B2 ST95 and D ST59 with different host distribution. *BMC Microbiol* **9:**132. doi:10.1186/1471-2180-9-132

80. **Tivendale KA, Logue CM, Kariyawasam S, Jordan D, Hussein A, Li G, Wannemuehler Y, Nolan LK.** 2010. Avian-pathogenic *Escherichia coli* strains are similar to neonatal meningitis *E. coli* strains and are able to cause meningitis in the rat model of human disease. *Infect Immun* **78:**3412–3419.

81. **Levert M, Zamfir O, Clermont O, Bouvet O, Lespinats S, Hipeaux MC, Branger C, Picard B, Saint-Ruf C, Norel F, Balliau T, Zivy M, Le Nagard H, Cruvellier S, Chane-Woon-Ming B, Nilsson S, Gudelj I, Phan K, Ferenci T, Tenaillon O, Denamur E.** 2010. Molecular and evolutionary bases of within-patient genotypic and phenotypic diversity in *Escherichia coli* extraintestinal infections. *PLoS Pathog* **6:** e1001125. doi:10.1371/journal.ppat.1001125

82. **Bauchart P, Germon P, Brée A, Oswald E, Hacker J, Dobrindt U.** 2010. Pathogenomic comparison of human extraintestinal and avian pathogenic *Escherichia coli*–search for factors involved in host specificity or zoonotic potential. *Microb Pathog* **49:**105–115.

83. **Dai J, Wang S, Guerlebeck D, Laturnus C, Guenther S, Shi Z, Lu C, Ewers C.** 2010. Suppression subtractive hybridization identifies an autotransporter adhesin gene of *E. coli* IMT5155 specifically associated with avian pathogenic *Escherichia coli* (APEC). *BMC Microbiol* **10:**236. doi:10.1136/1471-2180-10-236

84. **Yamamoto S, Tsukamoto T, Terai A, Kurazono H, Takeda Y, Yoshida O.** 1997. Genetic evidence supporting the fecal-perineal-urethral hypothesis in cystitis caused by *Escherichia coli*. *J Urol* **157:**1127–1129.

85. **Moreno E, Andreu A, Pérez T, Sabaté M, Johnson JR, Prats G.** 2006. Relationship between *Escherichia coli* strains causing urinary tract infection in women and the dominant faecal flora of the same hosts. *Epidemiol Infect* **134:**1015–1023.

86. **Jakobsen L, Garneau P, Kurbasic A, Bruant G, Stegger M, Harel J, Jensen KS, Brousseau R, Hammerum AM, Frimodt-Møller N.** 2011. Microarray-based detection of extended virulence and antimicrobial resistance gene profiles in phylogroup B2 *Escherichia coli* of human, meat and animal origin. *J Med Microbiol* **60:**1502–1511.

87. **Ramchandani M, Manges AR, DebRoy C, Smith SP, Johnson JR, Riley LW.** 2005. Possible animal origin of human-associated, multidrug-resistant, uropathogenic *Escherichia coli*. *Clin Infect Dis* **40:**251–257.

88. **Low DA, Braaten BA, Ling GV, Johnson DL, Ruby AL.** 1988. Isolation and comparison of *Escherichia coli* strains from canine and

human patients with urinary tract infections. *Infect Immun* **56**:2601–2609.

89. **Johnson JR, Stell AL, Delavari P, Murray AC, Kuskowski M, Gaastra W.** 2001. Phylogenetic and pathotypic similarities between *Escherichia coli* isolates from urinary tract infections in dogs and extraintestinal infections in humans. *J Infect Dis* **183**:897–906.

90. **Yuri K, Nakata K, Katae H, Yamamoto S, Hasegawa A.** 1998. Distribution of uropathogenic virulence factors among *Escherichia coli* strains isolated from dogs and cats. *J Vet Med Sci* **60**:287–290.

91. **Yuri K, Nakata K, Katae H, Tsukamoto T, Hasegawa A.** 1999. Serotypes and virulence factors of *Escherichia coli* strains isolated from dogs and cats. *J Vet Med Sci* **61**:37–40.

92. **Johnson JR, Kaster N, Kuskowski MA, Ling GV.** 2003. Identification of urovirulence traits in *Escherichia coli* by comparison of urinary and rectal *E. coli* isolates from dogs with urinary tract infection. *J Clin Microbiol* **41**:337–345.

93. **Kurazono H, Nakano M, Yamamoto S, Ogawa O, Yuri K, Nakata K, Kimura M, Makino S, Nair GB.** 2003. Distribution of the *usp* gene in uropathogenic *Escherichia coli* isolated from companion animals and correlation with serotypes and size-variations of the pathogenicity island. *Microbiol Immunol* **47**:797–802.

94. **Xia X, Meng J, Zhao S, Bodeis-Jones S, Gaines SA, Ayers SL, McDermott PF.** 2011. Identification and antimicrobial resistance of extraintestinal pathogenic *Escherichia coli* from retail meats. *J Food Prot* **74**:38–44.

95. **Tan C, Tang X, Zhang X, Ding Y, Zhao Z, Wu B, Cai X, Liu Z, He Q, Chen H.** 2012. Serotypes and virulence genes of extraintestinal pathogenic *Escherichia coli* isolates from diseased pigs in China. *Vet J* **192**:483–488.

96. **Johnson JR, Delavari P, Stell AL, Whittam TS, Carlino U, Russo TA.** 2001. Molecular comparison of extraintestinal *Escherichia coli* isolates of the same electrophoretic lineages from humans and domestic animals. *J Infect Dis* **183**:154–159.

97. **Johnson JR, Owens K, Gajewski A, Clabots C.** 2008. *Escherichia coli* colonization patterns among human household members and pets, with attention to acute urinary tract infection. *J Infect Dis* **197**:218–224.

98. **Johnson JR, Clabots C.** 2006. Sharing of virulent *Escherichia coli* clones among household members of a woman with acute cystitis. *Clin Infect Dis* **43**:e101–108.

99. **Johnson JR, Kuskowski MA, Owens K, Clabots C, Singer RS.** 2009. Virulence genotypes and phylogenetic background of fluoroquinolone-resistant and susceptible *Escherichia coli* urine isolates from dogs with urinary tract infection. *Vet Microbiol* **136**:108–114.

100. **Johnson JR, Stell AL.** 2000. Extended virulence genotypes of *Escherichia coli* strains from patients with urosepsis in relation to phylogeny and host compromise. *J Infect Dis* **181**:261–272.

101. **Jauréguy F, Carbonnelle E, Bonacorsi S, Clec'h C, Casassus P, Bingen E, Picard B, Nassif X, Lortholary O.** 2007. Host and bacterial determinants of initial severity and outcome of *Escherichia coli* sepsis. *Clin Microbiol Infect* **13**:854–862.

102. **Vejborg RM, Hancock V, Schembri MA, Klemm P.** 2011. Comparative genomics of *Escherichia coli* strains causing urinary tract infections. *Appl Environ Microbiol* **77**:3268–3278.

103. **Guyer DM, Kao JS, Mobley HL.** 1998. Genomic analysis of a pathogenicity island in uropathogenic *Escherichia coli* CFT073: distribution of homologous sequences among isolates from patients with pyelonephritis, cystitis, and catheter-associated bacteriuria and from fecal samples. *Infect Immun* **66**:4411–4417.

104. **Parham NJ, Pollard SJ, Chaudhuri RR, Beatson SA, Desvaux M, Russell MA, Ruiz J, Fivian A, Vila J, Henderson IR.** 2005. Prevalence of pathogenicity island IICFT073 genes among extraintestinal clinical isolates of *Escherichia coli*. *J Clin Microbiol* **43**:2425–2434.

105. **Johnson JR, O'Bryan TT, Delavari P, Kuskowski M, Stapleton A, Carlino U, Russo TA.** 2001. Clonal relationships and extended virulence genotypes among *Escherichia coli* isolates from women with a first or recurrent episode of cystitis. *J Infect Dis* **183**:1508–1517.

106. **Escobar-Páramo P, Clermont O, Blanc-Potard AB, Bui H, Le Bouguénec C, Denamur E.** 2004. A specific genetic background is required for acquisition and expression of virulence factors in *Escherichia coli*. *Mol Biol Evol* **21**:1085–1094.

107. **Krieger JN, Dobrindt U, Riley DE, Oswald E.** 2011. Acute *Escherichia coli* prostatitis in previously health young men: bacterial virulence factors, antimicrobial resistance, and clinical outcomes. *Urology* **77**:1420–1425.

108. **Kanamaru S, Kurazono H, Nakano M, Terai A, Ogawa O, Yamamoto S.** 2006. Subtyping of uropathogenic *Escherichia coli* according to the pathogenicity island encoding uropathogenic-specific protein: comparison with phylogenetic groups. *Int J Urol* **13**:754–760.

109. **Soto SM, Smithson A, Martinez JA, Horcajada JP, Mensa J, Vila J.** 2007. Biofilm

formation in uropathogenic *Escherichia coli* strains: relationship with prostatitis, urovirulence factors and antimicrobial resistance. *J Urol* **177**:365–368.

110. Ruiz J, Simon K, Horcajada JP, Velasco M, Barranco M, Roig G, Moreno-Martínez A, Martínez JA, Jiménez de Anta T, Mensa J, Vila J. 2002. Differences in virulence factors among clinical isolates of *Escherichia coli* causing cystitis and pyelonephritis in women and prostatitis in men. *J Clin Microbiol* **40:** 4445–4449.

111. Ghenghesh KS, Elkateb E, Berbash N, Abdel Nada R, Ahmed SF, Rahouma A, Seif-Enasser N, Elkhabroun MA, Belresh T, Klena JD. 2009. Uropathogens from diabetic patients in Libya: virulence factors and phylogenetic groups of *Escherichia coli* isolates. *J Med Microbiol* **58:**1006–1014.

112. Bert F, Huynh B, Dondero F, Johnson JR, Paugam-Burtz C, Durand F, Belghiti J, Valla D, Moreau R, Nicolas-Chanoine MH. 2011. Molecular epidemiology of *Escherichia coli* bacteremia in liver transplant recipients. *Transpl Infect Dis* **13:**359–365.

113. Keegan SJ, Graham C, Neal DE, Blum-Oehler G, N'Dow J, Pearson JP, Gally DL. 2003. Characterization of *Escherichia coli* strains causing urinary tract infections in patients with transposed intestinal segments. *J Urol* **169:**2382–2387.

114. Benton J, Chawla J, Parry S, Stickler D. 1992. Virulence factors in *Escherichia coli* from urinary tract infections in patients with spinal injuries. *J Hosp Infect* **22:**117–127.

115. Guidoni EB, Dalpra VA, Figueiredo PM, da Silva Leite D, Mimica LM, Yano T, Blanco JE, Toporovski J. 2006. *E. coli* virulence factors in children with neurogenic bladder associated with bacteriuria. *Pediatr Nephrol* **21:**376–381.

116. Hull RA, Rudy DC, Wieser IE, Donovan WH. 1998. Virulence factors of *Escherichia coli* isolates from patients with symptomatic and asymptomatic bacteriuria and neuropathic bladders due to spinal cord and brain injuries. *J Clin Microbiol* **36:**115–117.

117. Schlager TA, Whittam TS, Hendley JO, Wilson RA, Bhang J, Grady R, Stapleton A. 2000. Expression of virulence factors among *Escherichia coli* isolated from the periurethra and urine of children with neurogenic bladder on intermittent catheterization. *Pediatr Infect Dis J* **19:**37–41.

118. Stamm WE. 1991. Catheter-associated urinary tract infections: epidemiology, pathogenesis, and prevention. *Am J Med* **91:**65S–71S.

119. Warren JW, Tenney JH, Hoopes JM, Muncie HL, Anthony WC. 1982. A prospective microbiologic study of bacteriuria in patients with chronic indwelling urethral catheters. *J Infect Dis* **146:**719–723.

120. Schlager TA, Johnson JR, Ouellette LM, Whittam TS. 2008. *Escherichia coli* colonizing the neurogenic bladder are similar to widespread clones causing disease in patients with normal bladder function. *Spinal Cord* **46:**633–638.

121. Rice JC, Peng T, Kuo YF, Pendyala S, Simmons L, Boughton J, Ishihara K, Nowicki S, Nowicki BJ. 2006. Renal allograft injury is associated with urinary tract infection caused by *Escherichia coli* bearing adherence factors. *Am J Transplant* **6:**2375–2383.

122. Johnson JR, Johnston B, Clabots C, Kuskowski MA, Pendyala S, Debroy C, Nowicki B, Rice J. 2010. *Escherichia coli* sequence type ST131 as an emerging fluoroquinolone-resistant uropathogen among renal transplant recipients. *Antimicrob Agents Chemother* **54:**546–550.

123. Wang MC, Tseng CC, Wu AB, Lin WH, Teng CH, Yan JJ, Wu JJ. 2012. Bacterial characteristics and glycemic control in diabetic patients with *Escherichia coli* urinary tract infection. *J Microbiol Immunol Infect* **46:**24–29.

124. Merçon M, Regua-Mangia AH, Teixeira LM, Irino K, Tuboi SH, Goncalves RT, Santoro-Lopes G. 2010. Urinary tract infections in renal transplant recipients: virulence traits of uropathogenic *Escherichia coli*. *Transplant Proc* **42:**483–485.

125. Bidet P, Mahjoub-Messai F, Blanco J, Dehem M, Aujard Y, Bingen E, Bonacorsi S. 2007. Combined multilocus sequence typing and O serogrouping distinguishes *Escherichia coli* subtypes associated with infant urosepsis and/or meningitis. *J Infect Dis* **196:**297–303.

126. Bonacorsi S, Houdouin V, Mariani-Kurkdjian P, Mahjoub-Messai F, Bingen E. 2006. Comparative prevalence of virulence factors in *Escherichia coli* causing urinary tract infection in male infants with and without bacteremia. *J Clin Microbiol* **44:**1156–1158.

127. Bonacorsi S, Lefèvre S, Clermont O, Houdouin V, Bourrillon A, Loirat C, Aujard Y, Bingen E. 2005. *Escherichia coli* strains causing urinary tract infection in uncircumcised infants resemble urosepsis-like adult strains. *J Urol* **173:**195–197; discussion 197.

128. Johnson JR, Goullet P, Picard B, Moseley SL, Roberts PL, Stamm WE. 1991. Association of carboxylesterase B electrophoretic pattern with presence and expression of urovirulence factor determinants and antimicrobial resistance

among strains of *Escherichia coli* that cause urosepsis. *Infect Immun* **59:**2311–2315.

129. Cereto F, Herranz X, Moreno E, Andreu A, Vergara M, Fontanals D, Roget M, Simó M, González A, Prats G, Genescà J. 2008. Role of host and bacterial virulence factors in *Escherichia coli* spontaneous bacterial peritonitis. *Eur J Gastroenterol Hepatol* **20:**924–929.

130. Johnson JR, Murray AC, Kuskowski MA, Schubert S, Prère MF, Picard B, Colodner R, Raz R; Trans-Global Initiative for Antimicrobial Resistance Initiative (TIARA) Investigators. 2005. Distribution and characteristics of *Escherichia coli* clonal group A. *Emerg Infect Dis* **11:**141–145.

131. Skjøt-Rasmussen L, Olsen SS, Jakobsen L, Ejrnaes K, Scheutz F, Lundgren B, Frimodt-Møller N, Hammerum AM. 2012. *Escherichia coli* clonal group A causing bacteraemia of urinary tract origin. *Clin Microbiol Infect* **19:**656–661.

132. Johnson JR, Porter SB, Zhanel G, Kuskowski MA, Denamur E. 2012. Virulence of *Escherichia coli* clinical isolates in a murine sepsis model in relation to sequence type ST131 status, fluoroquinolone resistance, and virulence genotype. *Infect Immun* **80:**1554–1562.

133. Johnson JR, Johnston B, Clabots C, Kuskowski MA, Castanheira M. 2010. *Escherichia coli* sequence type ST131 as the major cause of serious multidrug-resistant E. coli infections in the United States. *Clin Infect Dis* **51:**286–294.

134. Cagnacci S, Gualco L, Debbia E, Schito GC, Marchese A. 2008. European emergence of ciprofloxacin-resistant *Escherichia coli* clonal groups O25:H4-ST 131 and O15:K52:H1 causing community-acquired uncomplicated cystitis. *J Clin Microbiol* **46:**2605–2612.

135. Williamson DA, Roberts SA, Paterson DL, Sidjabat H, Silvey A, Masters J, Rice M, Freeman JT. 2012. *Escherichia coli* bloodstream infection after transrectal ultrasound-guided prostate biopsy: implications of fluoroquinolone-resistant sequence type 131 as a major causative pathogen. *Clin Infect Dis* **54:**1406–1412.

136. Gibreel TM, Dodgson AR, Cheesbrough J, Fox AJ, Bolton FJ, Upton M. 2012. Population structure, virulence potential and antibiotic susceptibility of uropathogenic *Escherichia coli* from Northwest England. *J Antimicrob Chemother* **67:**346–356.

137. Blanco J, Mora A, Mamani R, López C, Blanco M, Dahbi G, Herrera A, Blanco JE, Alonso MP, García-Garrote F, Chaves F, Orellana MÁ, Martínez-Martínez L, Calvo J, Prats G, Larrosa MN, González-López JJ, López-Cerero L, Rodríguez-Baño J, Pascual A. 2011. National survey of *Escherichia coli* causing extraintestinal infections reveals the spread of drug-resistant clonal groups O25b:H4-B2-ST131, O15:H1-D-ST393 and CGA-D-ST69 with high virulence gene content in Spain. *J Antimicrob Chemother* **66:**2011–2021.

138. Slack E, Hapfelmeier S, Stecher B, Velykoredko Y, Stoel M, Lawson MA, Geuking MB, Beutler B, Tedder TF, Hardt WD, Bercik P, Verdu EF, McCoy KD, Macpherson AJ. 2009. Innate and adaptive immunity cooperate flexibly to maintain host-microbiota mutualism. *Science* **325:**617–620.

139. Nolan LK, Wooley RE, Brown J, Spears KR, Dickerson HW, Dekich M. 1992. Comparison of a complement resistance test, a chicken embryo lethality test, and the chicken lethality test for determining virulence of avian *Escherichia coli*. *Avian Dis* **36:**395–397.

140. Hung CS, Dodson KW, Hultgren SJ. 2009. A murine model of urinary tract infection. *Nat Protoc* **4:**1230–1243.

141. Smith SN, Hagan EC, Lane MC, Mobley HL. 2010. Dissemination and systemic colonization of uropathogenic *Escherichia coli* in a murine model of bacteremia. *MBio* **1.**

142. Johnson JR, Clermont O, Menard M, Kuskowski MA, Picard B, Denamur E. 2006. Experimental mouse lethality of *Escherichia coli* isolates, in relation to accessory traits, phylogenetic group, and ecological source. *J Infect Dis* **194:**1141–1150.

143. Diard M, Garry L, Selva M, Mosser T, Denamur E, Matic I. 2010. Pathogenicity-associated islands in extraintestinal pathogenic *Escherichia coli* are fitness elements involved in intestinal colonization. *J Bacteriol* **192:**4885–4893.

144. Landraud L, Jauréguy F, Frapy E, Guigon G, Gouriou S, Carbonnelle E, Clermont O, Denamur E, Picard B, Lemichez E, Brisse S, Nassif X. 2012. Severity of *Escherichia coli* bacteraemia is independent of the intrinsic virulence of the strains assessed in a mouse model. *Clin Microbiol Infect* **19:**85–90.

145. Lefort A, Panhard X, Clermont O, Woerther PL, Branger C, Mentré F, Fantin B, Wolff M, Denamur E; COLIBAFI Group. 2011. Host factors and portal of entry outweigh bacterial determinants to predict the severity of *Escherichia coli* bacteremia. *J Clin Microbiol* **49:**777–783.

146. Marschall J, Zhang L, Foxman B, Warren DK, Henderson JP; CDC Prevention Epicenters Program. 2012. Both host and pathogen factors predispose to *Escherichia coli* urinary-source bacteremia in hospitalized patients. *Clin Infect Dis* **54:**1692–1698.

147. **Wendt C, Messer SA, Hollis RJ, Pfaller MA, Herwaldt LA.** 1998. Epidemiology of polyclonal gram-negative bacteremia. *Diagn Microbiol Infect Dis* **32:**9–13.

148. **Johnson JR, Gajewski A, Lesse AJ, Russo TA.** 2003. Extraintestinal pathogenic *Escherichia coli* as a cause of invasive nonurinary infections. *J Clin Microbiol* **41:**5798–5802.

149. **Leflon-Guibout V, Bonacorsi S, Clermont O, Ternat G, Heym B, Nicolas-Chanoine MH.** 2002. Pyelonephritis caused by multiple clones of *Escherichia coli*, susceptible and resistant to co-amoxiclav, after a 45 day course of co-amoxiclav. *J Antimicrob Chemother* **49:** 373–377.

150. **Tourret J, Aloulou M, Garry L, Tenaillon O, Dion S, Ryffel B, Monteiro RC, Denamur E.** 2011. The interaction between a non-pathogenic and a pathogenic strain synergistically enhances extra-intestinal virulence in *Escherichia coli*. *Microbiology* **157:**774–785.

151. **Ferenci T.** 2005. Maintaining a healthy SPANC balance through regulatory and mutational adaptation. *Mol Microbiol* **57:**1–8.

152. **Hengge-Aronis R.** 2002. Signal transduction and regulatory mechanisms involved in control of the sigma(S) (RpoS) subunit of RNA polymerase. *Microbiol Mol Biol Rev* **66:**373–395.

153. **Cooper VS, Schneider D, Blot M, Lenski RE.** 2001. Mechanisms causing rapid and parallel losses of ribose catabolism in evolving populations of *Escherichia coli* B. *J Bacteriol* **183:** 2834–2841.

154. **Maharjan R, Seeto S, Notley-McRobb L, Ferenci T.** 2006. Clonal adaptive radiation in a constant environment. *Science* **313:**514–517.

155. **European Committee for Antimicrobial Susceptibility Testing (EUCAST) of the European Society of Clinical Microbiology and Infectious Diseases (ESCMID).** 2003. Determination of minimum inhibitory concentrations (MICs) of antibacterial agents by broth dilution. *Clin Microbiol Infect* **9:**ix–xv.

156. **Turnidge JD, Bell JM.** 2005. Antimicrobial susceptibility on solid media, p 8–60. *In* Lorian V (ed), *Antibiotics in Laboratory Medicine*, 5th ed. Lippincott Williams & Wilkins, Philadelphia.

157. **Proctor RA.** 2000. Editorial response: coagulase-negative staphylococcal infections: A diagnostic and therapeutic challenge. *Clin Infect Dis* **31:**31–33.

158. **Lloyd AL, Henderson TA, Vigil PD, Mobley HL.** 2009. Genomic islands of uropathogenic *Escherichia coli* contribute to virulence. *J Bacteriol* **191:**3469–3481.

159. **Lecointre G, Rachdi L, Darlu P, Denamur E.** 1998. *Escherichia coli* molecular phylogeny using the incongruence length difference test. *Mol Biol Evol* **15:**1685–1695.

160. **Levin BR.** 1996. The evolution and maintenance of virulence in microparasites. *Emerg Infect Dis* **2:**93–102.

161. **Le Gall T, Clermont O, Gouriou S, Picard B, Nassif X, Denamur E, Tenaillon O.** 2007. Extraintestinal virulence is a coincidental byproduct of commensalism in B2 phylogenetic group *Escherichia coli* strains. *Mol Biol Evol* **24:**2373–2384.

162. **Moreno E, Johnson JR, Pérez T, Prats G, Kuskowski MA, Andreu A.** 2009. Structure and urovirulence characteristics of the fecal *Escherichia coli* population among healthy women. *Microbes Infect* **11:**274–280.

163. **Ostblom A, Adlerberth I, Wold AE, Nowrouzian FL.** 2011. Pathogenicity island markers, virulence determinants malX and usp, and the capacity of *Escherichia coli* to persist in infants' commensal microbiotas. *Appl Environ Microbiol* **77:**2303–2308.

164. **Moreno E, Andreu A, Pigrau C, Kuskowski MA, Johnson JR, Prats G.** 2008. Relationship between *Escherichia coli* strains causing acute cystitis in women and the fecal E. coli population of the host. *J Clin Microbiol* **46:**2529–2534.

165. **Zdziarski J, Svanborg C, Wullt B, Hacker J, Dobrindt U.** 2008. Molecular basis of commensalism in the urinary tract: low virulence or virulence attenuation? *Infect Immun* **76:** 695–703.

166. **Klemm P, Hancock V, Schembri MA.** 2007. Mellowing out: adaptation to commensalism by *Escherichia coli* asymptomatic bacteriuria strain 83972. *Infect Immun* **75:**3688–3695.

167. **Le Gall T, Darlu P, Escobar-Páramo P, Picard B, Denamur E.** 2005. Selection-driven transcriptome polymorphism in *Escherichia coli/Shigella* species. *Genome Res* **15:**260–268.

168. **Toledo-Arana A, Dussurget O, Nikitas G, Sesto N, Guet-Revillet H, Balestrino D, Loh E, Gripenland J, Tiensuu T, Vaitkevicius K, Barthelemy M, Vergassola M, Nahori MA, Soubigou G, Régnault B, Coppée JY, Lecuit M, Johansson J, Cossart P.** 2009. The *Listeria* transcriptional landscape from saprophytism to virulence. *Nature* **459:**950–956.

169. **Hamburg MA, Collins FS.** 2010. The path to personalized medicine. *N Engl J Med* **363:**301–304.

170. **Ochman H, Selander RK.** 1984. Standard reference strains of *Escherichia coli* from natural populations. *J Bacteriol* **157:**690–693.

Virulence and Fitness Determinants of Uropathogenic *Escherichia coli*

12

SARGURUNATHAN SUBASHCHANDRABOSE[1] and HARRY L. T. MOBLEY[1]

INTRODUCTION

Urinary tract infections (UTIs) are one of the most common bacterial infections affecting humans, and uropathogenic *Escherichia coli* (UPEC) is the etiological agent in 75% to 95% of UTIs in otherwise healthy individuals (1–4). Based on the presence or absence of anatomic abnormalities and recent history of instrumentation in the urinary tract, UTIs are divided into either complicated or uncomplicated cases, respectively. UPEC is the most common cause of uncomplicated UTIs in humans. In clinical settings, UTIs are described with specific reference to the site of inflammation; cystitis indicates inflammation of the urinary bladder and pyelonephritis indicates inflammation of the renal pelvis and the kidneys. Presence of bacteria (bacteriuria) and neutrophils in the urine are hallmarks of UTIs caused by UPEC. Some patients, however, are bacteriuric unaccompanied with symptoms of UTI for long periods. This condition is referred to as asymptomatic bacteriuria (ABU) and is the most benign form of *E. coli* colonization in the human urinary tract. In patients suffering from pyelonephritis, UPEC can gain access to renal capillaries, leading to bacteremia (presence of bacteria in blood) and sepsis with the latter being the most dangerous and potentially fatal complication of

[1]Department of Microbiology and Immunology, University of Michigan Medical School, Ann Arbor, MI 48109.
Urinary Tract Infections: Molecular Pathogenesis and Clinical Management, 2nd Edition
Edited by Matthew A. Mulvey, David J. Klumpp, and Ann E. Stapleton
© 2017 American Society for Microbiology, Washington, DC
doi:10.1128/microbiolspec.UTI-0015-2012

UTIs caused by UPEC. Numerous virulence and fitness factors confer advantages to UPEC within host urinary tract and are discussed in this chapter.

Incidence of UTIs is around four-times higher in women compared to men and this difference is mainly attributed to the shorter length of urethra and a shorter distance between anus and urethral opening in women (3). Sexual activity is the primary risk factor reported in patients with uncomplicated UTIs. Almost one in every two adult women (40%) will be affected by cystitis during their lifetime and there is a 25% risk for the development of recurrent UTI within the next year (4). In the United States alone, UTIs led to an estimated 11 million physician visits, 1.7 million emergency room visits and 470,000 hospitalizations in 2006, at an estimated cost of $3.5 billion (5, 6). Better understanding of clinically relevant virulence factors is required for developing novel therapeutic and prophylactic strategies that will take us one step closer towards reducing the morbidity, loss of productivity, and healthcare costs associated with UPEC UTIs.

Most cases of uncomplicated UTIs are community-acquired and gut colonization precedes access to the urinary tract. UPEC colonizes the perineal region, followed by ascent along the urethra and subsequent colonization of the bladder resulting in cystitis (7). UPEC is adept at orchestrating two completely opposite phenomena, adhesion and motility, for successful colonization and ascending infection. Both extracellular and intracellular lifestyles are exhibited by UPEC within the urinary bladder (8–10) and intracellular growth phase is proposed to promote persistent infection of the urinary bladder. Within superficial facet cells of the urinary bladder, some strains of UPEC form intracellular bacterial communities (IBCs).

UPEC utilizes numerous virulence and fitness factors to gain entry, adhere, acquire essential nutrients, multiply in a hostile environment, cause tissue damage, and disseminate within the urinary tract. Fitness and virulence factors are related but distinct entities: factors that contribute to UPEC survival in a given niche are described as fitness factors and genes that satisfy molecular Koch's postulates are defined as virulence genes. Tenets of molecular Koch's postulates include the presence of a gene in the pathogenic members of a species, mutants lacking that gene should display a measurable loss of pathogenicity or virulence, and complementation with the wild-type copy of the gene should restore wild-type levels of pathogenicity or virulence (11). Here, we review the role of fitness and virulence factors in the pathobiology of urinary tract infections caused by UPEC. This book also contains chapters dedicated to specific virulence factors and readers will be directed to those chapters for additional information.

NUTRIENT REQUIREMENTS

Growth in Urine

Urine is a high osmolar, dilute mixture of amino acids and peptides. Continuous inflow of urine from the kidneys to the bladder via ureters results in relatively steady levels of nutrients in urine. Several groups use urine as a culture medium for UPEC in an attempt to recapitulate conditions encountered within a human host, at least in part. Although urine is rich in amino acids and peptides, amino acid biosynthetic capacity appears to be important for growth in urine. Specifically, UPEC mutants auxotrophic for arginine, glutamine, or guanine exhibit a severe growth defect in human urine and leucine, methionine, serine, phenylalanine, or proline auxotrophs display a reduced growth rate (12). Auxotrophs are mutants that lack the biosynthetic capacity for a particular nutrient; in this study, chemical mutagenesis was used for generating the mutants defective in amino acid biosynthetic pathways. A recent study with defined mutants generated by allelic exchange revealed that mutants auxotrophic for arginine and serine did not exhibit growth defects in human urine *in vitro* or in a mouse model of UTI (13). However, whether this is true for mutants auxotrophic for other amino acids remains to be tested. Transcriptional profiling experiments revealed that genes involved in amino acid and carbohydrate metabolism are among the genes significantly up-regulated during growth in human urine (14). Strain-specific differences in generation time during growth in human urine have been observed, indicating that in addition to growth in urine, there are other factors that are collectively required for urofitness. ABU 83972, an asymptomatic bacteriuria strain, exhibits rapid growth in urine and outcompetes UPEC strains such as 536, CFT073, NU14, and UTI89 in human urine (15). Additionally, ABU 83972 outcompetes NU14 in mouse bladder (15). It is possible that rapid growth can compensate for the lack of specific adherence organelles and ensure persistence within host urinary tract for at least some strains.

Competition for Metals

Metals such as iron, zinc, and manganese are essential nutrients for most forms of life, including bacteria. Mammalian immune system exploits the essentiality of metal acquisition to retard the growth of bacteria during infection. This concept is known as nutritional immunity and is part of innate immune response activated during bacterial infection (16). Here we review the mechanisms used by UPEC to surmount sequestration of iron and zinc within host urinary tract.

Iron Acquisition

Although iron is one of the most abundant metals in the earth's crust, pathogens encounter iron limitation within hosts. Indeed, iron limitation is one of the innate defenses against survival of bacteria within hosts

(17), where elemental iron is bound by host glycoproteins such as transferrin and lactoferrin or incorporated into the heme moieties of proteins such as hemoglobin and myoglobin. To ensure adequate levels of intracellular iron, UPEC up-regulates the expression of genes involved in iron acquisition in response to iron limitation encountered within the mammalian urinary tract (14, 18, 19).

UPEC can acquire ferric iron using siderophores, ferrous iron through iron transporters, and heme through outer-membrane heme receptors. A common player involved in iron uptake through all these mechanisms is the inner membrane-associated TonB-ExbB-ExbD complex. ExbB and ExbD proteins energize TonB with the proton-motive force generated at the inner membrane and TonB transduces this energy to the outer-membrane iron-receptor complexes leading to their translocation to the periplasm (Fig. 1). Iron-containing complexes are bound to binding proteins within the periplasm and are transported across the inner membrane via ABC transporters using adenosine triphosphate (ATP) as an energy source (Fig. 1). A UPEC mutant lacking the *tonB* gene is attenuated in a mouse model of UTI, demonstrating the importance of TonB-mediated iron uptake within the urinary tract (20). A high-throughput screen devised to identify inhibitors of iron uptake systems in UPEC CFT073 has revealed novel small-molecule inhibitors of TonB (21).

Regulation of Iron Uptake

Excess levels of intracellular iron are detrimental to a bacterial cell and therefore iron homeostasis is subject to rigorous regulation. The ferric-uptake regulator (Fur) regulates the expression of genes involved in iron uptake and metabolism at transcriptional level (Fig. 1). Iron-containing Fur homodimer binds to 19-base pair inverted repeats, known as the "Fur boxes", in the promoter region of Fur-regulated genes, such as siderophore

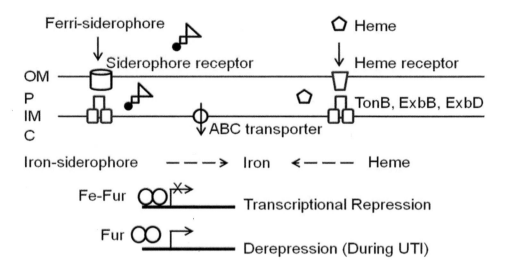

FIGURE 1 **A simple model of iron uptake in UPEC. UPEC produces iron-scavenging molecules known as siderophores. Cognate outer-membrane receptors bind ferri-siderophore complexes or heme that are then imported into the periplasm. Within the periplasm, they associate with periplasmic-binding proteins. Translocation across the inner membrane involves ABC transporters. Iron is extracted from iron-siderophore complexes and heme via multiple reactions, denoted by broken arrows. In the presence of iron, Fur represses (X) transcription of the genes involved in iron uptake. Iron limitation within the host urinary tract results in the derepression of Fur-regulated genes, including siderophore-biosynthetic genes. UPEC, uropathogenic *E. coli*; OM, outer membrane; P, periplasm; IM, inner membrane; C, cytoplasm; Fur, ferric-uptake regulator; Fe, ferric iron; and TonB, ExbB, ExbD, energy-transduction complex for transport of iron-containing complexes to the periplasm.**

biosynthesis and uptake systems and heme-uptake systems, and represses their transcription (22). When the intracellular levels of iron are low, Fur-mediated repression is relieved due to low affinity of free Fur for Fur boxes and derepression of iron-uptake systems facilitates iron acquisition (Fig. 1). Fur, however, is also known to positively regulate the expression of some genes including *acnA, ftnA, fumA,* and *sdhCDAB* (23). This paradoxical observation was reconciled after the identification of a Fur-regulated small RNA, RyhB; RyhB acts as a negative post-transcriptional regulator for some genes whose protein products either require iron as a cofactor or are involved in iron storage (23). The expression of *ryhB* is subject to negative regulation by Fur. Small-RNA-mediated regulation adds an additional layer of complexity to the regulatory scheme controlling iron homeostasis.

Siderophores

Siderophores are low-molecular-mass, high-affinity, iron-chelating molecules released into the environment by several bacteria, including UPEC. Siderophores bind ferric iron and iron-siderophore complexes are recognized by cognate outer-membrane receptors (Fig. 1). UPEC strains encode the proteins required for the biosynthesis and uptake of the following siderophores: aerobactin, enterobactin, salmochelin, and yersiniabactin. Some UPEC strains such as 536 and UTI89 produce and utilize yersiniabactin whereas another prototypical strain, CFT073, lacks the biosynthetic capacity to produce yersiniabactin. Enterobactin and salmochelin belong to the catecholate family whereas aerobactin and yersiniabactin are members of hydroxamate and five-member heterocyclic ring-containing siderophores, respectively. Independent of siderophore activity, yersiniabactin also binds copper and mitigates copper toxicity (24).

Intense competition between bacteria and host for iron is exemplified by the fact that mammalian hosts produce specific proteins that bind siderophores and prevent the reuptake of iron-siderophore complexes by bacteria. As part of innate immune response, neutrophils produce lipocalin-2 (25), a molecule that binds enterobactin and prevents enterobactin-mediated iron uptake by bacteria (26). Bacterial pathogens, including UPEC and *Salmonella enterica* subspecies Enterica, overcome this barrier by using salmochelin, a glycosylated derivative of enterobactin, which is not recognized by lipocalin-2 (27).

Production of siderophores and their correlation with the presence of genes encoding siderophore biosynthetic pathways in prototypical UPEC strains have been characterized using a quantitative metabolomics approach (28). UPEC isolates from urine and *E. coli* from corresponding rectal samples were tested for siderophore production. Enterobactin production was identical in strains isolated from both urinary tract and rectum indicating that enterobactin system does not respond to specific cues sensed by UPEC within the human urinary tract. However, salmochelin and yersiniabactin were both produced at significantly higher levels in UPEC isolated from urine compared to rectal isolates and co-expression of salmochelin and yersiniabactin siderophores were observed, indicating that these two siderophores might play a greater role compared to enterobactin during iron acquisition within the human urinary tract (28).

Receptor-Mediated Heme Uptake

Heme is an abundant source of iron within mammalian hosts. Receptor-mediated uptake of heme is another iron-uptake mechanism utilized by UPEC. Two outer-membrane heme receptors, ChuA and Hma, have been characterized (Fig. 1). Functional homologues of ChuA are found in enterohemorrhagic

E. coli and *Shigella dysenteriae* whereas Hma appears to be a UPEC-specific heme receptor. Genes encoding both ChuA and Hma are expressed during growth under iron-limited conditions, including in human urine and during experimental mouse UTI (29, 30). Mutants lacking ChuA are outcompeted by mutants lacking Hma, indicating that ChuA is the favored heme receptor *in vivo*. Nevertheless, both ChuA and Hma are required for optimal fitness in a mouse model of UTI (30). ChuA also appears to be important during IBC formation by UPEC; expression of *chuA* was upregulated during IBC formation within bladder epithelial cells and a *chuA* mutant forms significantly smaller IBCs compared to parent strain (31). In combination, these studies imply that the ability to acquire iron from heme is an important *in vivo* fitness factor.

Redundant Iron Acquisition Systems

Contribution of the apparently redundant iron-uptake systems to fitness was tested in a mouse model of ascending UTI (32). Mutants lacking *fepA, iha,* and *iroN* genes (encoding cognate receptors for catecholate siderophores) displayed similar fitness indices within the murine urinary tract indicating that they are a functionally redundant group. Aerobactin-receptor mutant and yersiniabactin-receptor mutant were both out-competed by mutants lacking either hydroxamate-siderophore receptors or catecholate-siderophore receptors, respectively. Aerobactin and yersiniabactin receptors contribute significantly to fitness *in vivo* indicating that these two systems are critical for uropathogenesis. Additionally, the *hma chuA* double mutant, defective in heme/hemin uptake, is out-competed by mutants lacking catecholate or hydroxamate receptors only within the kidneys (32). Taken together, these data demonstrate that despite being redundant, individual iron-uptake systems appear to play a key role within specific regions of the urinary tract.

Ferrous Iron Uptake

Manganese and ferrous iron are transported into the cytoplasm by the Sit system, comprised of SitA, SitB, SitC, and SitD proteins. In *Salmonella enterica* serovar Typhimurium, the *sit* genes are regulated by intracellular levels of both iron and manganese via corresponding metalloregulators, Fur and MntR, respectively (33). The *sit* genes are up-regulated during murine UTI and are expressed during human UTIs (14, 19). Several UPEC strains, including CFT073, contain the Sit system. Cystitis isolate F11 utilizes FetMP, a bipartite system to acquire ferrous iron (34). The contribution of ferrous iron-uptake systems to urofitness remains to be evaluated.

Iron Receptors as Vaccine Candidates

Several independent lines of evidence indicate iron limitation within urinary tract: genes involved in iron uptake are highly expressed during murine and human UTIs; iron receptors are found at higher levels during growth in human urine; iron-uptake receptors can be detected with antisera from mice with chronic UTI; and a subset of iron receptors are more prevalent in UPEC strains compared to commensal *E. coli* strains (14, 18, 19, 29, 35). Since iron uptake is clearly required for successful infection, the potential of iron-uptake receptors as vaccine candidates was evaluated. Outer-membrane iron receptors FyuA, Hma, Iha, IreA, IroN, and IutA, were tested as vaccine candidates against UPEC infection in a murine model (35, 36). Purified proteins were cross-linked to cholera toxin, an adjuvant, and administered intranasally. Mice immunized with Hma, IreA, and IutA displayed significantly higher protection upon challenge infection in kidneys, bladder, and kidneys and bladder, respectively, compared to adjuvant-only controls. Vaccination with FyuA protected mice against UPEC only in the kidneys. Furthermore, vaccination with Hma and IreA protected mice during challenge with a heterologous strain of UPEC. Protective antigens Hma, IreA, and IutA demonstrated significantly higher serum immunoglobulin (Ig)G levels post challenge, compared to ChuA and Iha, antigens that did not confer protection. Antigen-specific IgA was found only in the urine of mice vaccinated with IreA and IutA, both protective antigens. Taken together, specific iron-uptake receptors appear to confer protective immunity in a mouse model of UTI and are certainly among the top candidates for developing a subunit vaccine against UTIs caused by UPEC.

Zinc Acquisition

Zinc is an essential nutrient involved in a gamut of cellular functions including iron homeostasis and resistance to oxidative stress. Neutrophil-derived protein calprotectin is involved in sequestration of zinc and manganese and stifles the growth of bacterial pathogens within mammalian hosts (16). UPEC strain CFT073 encodes at least two different zinc-transport systems encoded by *znuACB* and *zupT* (37). The ZnuACB system is an ABC transporter similar to the SitABCD system that is involved in the acquisition of ferrous iron and manganese. Zinc uptake through the ZupT system is conserved among bacteria and eukaryotes and, in addition to zinc, this system can also transport cobalt, iron, and manganese.

ZnuACB system is required for wild-type levels of growth in minimal medium, whereas ZupT system is dispensable during growth under zinc-limiting conditions; this observation suggests that there are redundant zinc-transport systems albeit, with varying levels of expression and/or affinity to zinc. Furthermore, mutants devoid of either ZnuACB or both ZnuACB and ZupT revealed fitness defect in a murine model of UTI indicating that the ability to acquire zinc is critical for survival within urinary tract (37). Human urine does not appear to be a zinc-limited milieu because growth of mutants defective in zinc uptake is indistinguishable from wild-type strain during growth in urine. This observation, however, does not exclude the possibility that zinc levels in urine could be drastically altered during massive influx of neutrophils observed during UTIs. Zinc is required for defenses against oxidative stress both directly and indirectly. Indirectly, reduced levels of zinc results in the derepression of Fur-regulated genes resulting in increased intracellular levels of iron, which exacerbates the damage inflicted during oxidative stress. Not surprisingly, mutants lacking ZnuACB exhibit higher sensitivity to oxidative stress. It would be interesting to determine the protection, if any, offered by zinc-uptake receptors alone or in combination with iron-uptake receptors in a composite subunit vaccine against UPEC.

METABOLISM

Since many metabolic functions are not pathogen-specific, the role of metabolic potential of pathogens in fitness during infections remains underappreciated. Thorough understanding of the metabolic state of a pathogen is, however, of paramount importance to define the microenvironment at the site of infection and thereby facilitate the identification of novel antimicrobial agents that target specific metabolic pathways.

Proteins involved in the transport of short peptides, transport and catabolism of sialic acid, gluconate, xylose, and arabinose, and biosynthesis of arginine and serine are highly expressed in UPEC cultured in human urine (Fig. 2) (13). Despite the fact that arginine and serine biosynthetic proteins are selectively expressed during growth in urine, arginine and serine auxotrophs did not exhibit any fitness defect in a mouse model of UTI. However, mutants defective in the uptake of oligopeptides and dipeptides were outcompeted by the wild-type strain, indicating that peptide uptake is critical for fitness of UPEC during infection. Mutants lacking enzymes in the citric-acid cycle and gluco-

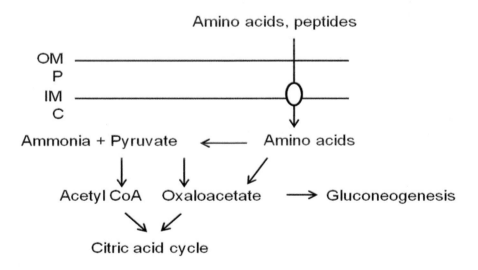

FIGURE 2 Metabolism in UPEC during infection. Amino acids and di/oligo peptides are the primary sources of carbon and nitrogen for the UPEC within urinary tract. Mutants defective in peptide transport exhibit a fitness defect *in vivo*. Pyruvate is generated from amino acids and feeds into the citric-acid cycle and gluconeogenesis. Disruption of either citric-acid cycle or gluconeogenesis is detrimental for the survival of UPEC within the urinary tract. UPEC, uropathogenic *E. coli*; OM, outer membrane; P, periplasm; IM, inner membrane; and C, cytoplasm.

neogenesis exhibit fitness defect in both bladder and kidneys (13). Defects in Entner-Doudoroff pathway, glycolysis, or pentose-phosphate pathway, however, did not affect *in vivo* fitness. In combination, these findings assert the idea that amino acids and peptides are the primary carbon sources for UPEC within the urinary tract (Fig. 2).

D-serine is one of the most abundant amino acids found in human urine. A UPEC CFT073 mutant lacking the D-serine deaminase gene (*dsdA*) required for D-serine catabolism, revealed a long lag phase when cultured in human urine. Surprisingly, the mutant exhibits a hypercolonization phenotype, manifested by increased fitness during co-infection with parental strain, in a mouse model of UTI (38). Additionally, the *dsdA* mutant is hypermotile and hyperflagellated compared to parental strain (Fig. 3). A *dsdA fliC* double mutant lacking flagella loses the fitness advantage, indicating that hyper-

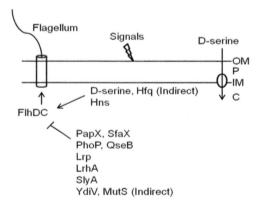

FIGURE 3 Regulation of flagellar motility in UPEC. *flhDC*, master regulator of the flagellar-biosynthesis cascade, is subject to both positive and negative regulation. Note that not all regulators affect the transcription of *flhDC* directly. Regulators such as PhoP and QseB are activated in response to specific cues encountered by UPEC. Various signals are integrated into the flagellar gene expression cascade via two-component regulatory systems and other signal-transduction cascades. UPEC, uropathogenic *E. coli*; OM, outer membrane; P, periplasm; IM, inner membrane; and C, cytoplasm.

virulence phenotype of *dsdA* mutant is mediated by excessive production of flagella. In addition to flagella, hemolysin A, and both P fimbriae (CFT073 strain contains two P fimbrial-gene clusters) contribute to the hypervirulence phenotype of *dsdA* mutant, albeit in a non-redundant manner (39).

Serine is deaminated to produce ammonia and pyruvate by enantiomer-specific (D/L) serine deaminases. Catabolism of pyruvate leads to the production of acetyl coenzyme A (CoA); acetyl CoA is converted to acetyl phosphate by phosphotransacetylase (Pta) and acetate kinase (AckA) catalyzes the conversion of acetyl phosphate to acetate (40). Due to its phosphate-donor role, acetyl phzosphate acts as an intracellular signal that causes genome-wide changes in gene expression (41). Acetate is excreted into the medium and reuptake of acetate occurs only after depletion of easily assimilable carbon sources. The hypervirulent phenotype of D-serine deaminase (DsdA)-defective mutants was abrogated when L-serine deaminases (SdaA and SdaB) were also inactivated in CFT073 (40). Indeed, a mutant defective in catabolizing both D and L forms of serine displayed a fitness defect in the urinary tract. Genes up-regulated in UPEC upon exposure to D-serine during growth in urine includes *pta* and *ackA*. The mutant defective in re-uptake of acetate, however, does not exhibit a fitness defect (40). Taken together, these data demonstrate that UPEC produces acetate during growth in urine and acetogenic potential contributes to *in vivo* fitness, but acetate reuptake does not affect *in vivo* fitness.

SURFACE STRUCTURES

Fimbriae

Fimbriae or pili are short filamentous organelles used by bacteria to adhere to various surfaces. In general, UPEC strains carry significantly higher number of fimbrial gene clusters compared to fecal/commensal strains

(42). The genome of a prototypical pyelone-phritis strain, CFT073, contains 12 fimbrial gene clusters: 10 belong to the chaperone-usher family, including type 1, two complete P, F1C, F9, and Auf fimbriae and 2 putative-type IV pili (*c2394-c2395* and *ppdD-hofBC*) (43). The chaperone-usher fimbrial-biogenesis pathway utilizes a periplasmic chaperone that binds and assists folding of the fimbrial subunits. Chaperone-fimbrial subunit complexes are targeted to the outer membrane, where the usher forms a pore through which pilus subunits are exported and assembled. Type IV pilus biogenesis requires a functional type II-secretion system and both type IV pilus gene clusters are involved in fitness during UTI (44, 45).

Type 1 and P fimbria, both assembled by the chaperone-usher pathway, are among the best characterized bacterial fimbriae. While type 1 fimbrial genes are found in the fecal/commensal backbone of the *E. coli* genome, P fimbrial genes are exclusively found in pathogenicity islands distributed among UPEC strains. Additionally, there is evidence for regulatory cross-talk between these two fimbrial types; constitutive expression of type 1 fimbria leads to down-regulation of P fimbrial gene expression (46). A brief review of fimbriae found in UPEC is included here (reviewed in 124).

Type 1 Fimbria

Type 1 fimbria mediates mannose-sensitive agglutination of guinea pig erythrocytes. Uroplakin, a highly mannosylated protein found in the uroepithelial cells is the receptor for type 1 fimbria (47). Specifically, FimH (type 1 fimbrial-tip adhesin) mediates interaction with D-mannose on glycoproteins. Type 1 fimbria is required to cause cystitis and development of IBCs in a mouse model of UTI (48–51). Within renal tubules of live rats, type 1 fimbriae promote interbacterial binding resulting in biofilm-like communities within renal tubules that cause obstruction of tubular flow (52).

Expression of type 1 fimbria is regulated by an invertible element carrying the promoter and is flanked by inverted repeats. FimB and FimE are site-specific recombinases that change the orientation of the promoter from ON-to-OFF/OFF-to-ON and ON-to-OFF, respectively. Several global regulatory proteins, such as lipoprotein receptor-related proteins (Lrp), histone-like nucleoid-structuring (H-NS), integration host factor (IHF), and cyclic adenosine monophosphate (cAMP) receptor protein (CRP)/cAMP, modulate the expression of type 1 fimbrial genes by controlling the orientation of the promoter region (53). For instance, CRP acts as a repressor of type 1 fimbrial gene expression by negatively regulating *lrp* resulting in reduced FimB-mediated recombination that flips the promoter from OFF-to-ON orientation. Additionally, CRP enhances the activity of DNA gyrase leading to OFF orientation of the promoter. Thus, expression of type 1 fimbrial genes is coordinately regulated by multiple regulatory inputs converging on recombinases that control the orientation of the promoter.

P Fimbria

UPEC are endowed with a unique type of pili encoded by the *pap* genes that are known as pyelonephritis-associated pili or P fimbriae. P fimbriae are required for mannose-resistant agglutination of human erythrocytes and PapG is the fimbrial-tip adhesin that renders specificity to P fimbriae-mediated adherence. PapG binds to glycosphingolipids containing digalactoside moieties found in the renal epithelium, specifically the P blood-group antigen (54). Although P fimbriae are associated with adherence to renal cells, their contribution to the virulence of UPEC remains equivocal (55–57). P fimbriae are required for attachment of UPEC to Bowman's capsule in a live animal. P fimbriae-mediated adhesion minimizes the impact of shear stress induced by the flow of primary filtrate within Bowman's space (52). In addition to its role in adherence, P fimbriae are also involved in regulating motility. PapX, a nonstructural gene found in the P fimbrial gene cluster, is a critical player in the cross-talk between fimbria-mediated adherence and flagella-mediated motility in UPEC (58).

Expression of P fimbrial genes is regulated by a methylation-dependent epigenetic switch (59). Competition between Lrp and deoxyadenosine methylase (Dam) methylase for binding sites at the *papBA* promoter determines the status (ON/OFF) of the switch. Methylation of the guanine, adenine, thymine, cytosine (GATC) site closer to the *papB* gene leads to the expression of P fimbrial genes; in contrast, Lrp binding to this site precludes methylation resulting in transcriptional repression. Once the switch is turned ON, there is an innate mechanism that strives to sustain the ON phase. PapB activates the transcription of *papI* and PapI promotes Lrp binding to distal GATC site facilitating the expression of P fimbrial genes. In summary, expression of both type 1 fimbria and P fimbria are tightly regulated in UPEC.

Other fimbrial gene clusters found in UPEC strains include Auf, Dr, F1C, F9, Pix, S, Yad, Yeh, Yfc, and Ygi. Dr fimbria agglutinates human erythrocytes and promotes renal trop hism by binding to the basement membrane of renal tubular epithelium and Bowman's capsule (60). F1C fimbria binds to glycosphingolipids found in kidney cells and induces the production of IL-8, a proinflammatory cytokine (61). IL-8 production is independent of LPS-mediated TLR4 signaling because that signaling cascade is absent in renal cells. F1C fimbria was highly expressed in a strain lacking both type 1 and P fimbriae suggesting cross-talk between regulatory proteins that affect expression of various fimbriae (46). S fimbria mediates UPEC binding to primary human renal proximal-tubular cells by binding to α-sialyl-2-3-β-galactoside and S fimbrial gene clusters are more frequently found in *E. coli* isolated from patients with neonatal sepsis or meningitis (62, 63). Pix pilus is a filamentous adhesin capable of agglutinating human erythrocytes (64). Recently, Ygi fimbria was demonstrated to be involved in adherence to human kidney cells and a mutant lacking this fimbria exhibits a fitness defect in a mouse model of UTI (42). Yad fimbria mediates adherence to human bladder cells but does not appear to contribute to fitness, at least independent of other fimbriae. The ability to produce multiple fimbria appears to confer selective advantage within a particular niche and active cross-talk is apparent between regulators of various fimbrial types.

Pilicides

Agents that specifically target virulence factors without affecting the survival of bacterial pathogens have been proposed as a promising alternative for conventional antimicrobial compounds (65). Since both type 1 and P pili are implicated in the pathogenesis of UTIs, chemical inhibitors of pili biogenesis (pilicides) were developed. Pilicides in the 2-bicyclic-pyrimidone family inhibit the production of both type 1 and P pili in UPEC and are potential candidates for development as therapeutic agents (66).

Motility

The ascending nature of most cases of UTI led to a speculation that flagella might be a critical virulence factor for UPEC. Two groups independently determined that possession of flagella confers a significant fitness advantage to UPEC. Lane et al. (67) constructed aflagellate, flagellate but hypomotile, flagellate but nonmotile, and flagellate, motile but nonchemotactic, mutant strains of UPEC CFT073 and tested the contribution of motility and chemotaxis to virulence in a mouse model of ascending UTI. Flagellum-mediated motility was not required for successful colonization during individual infections indicating that this is not an urovirulence trait. Nevertheless, flagellum-mediated motility increased the fitness of mutant strains during mixed infections demonstrating that this is an important fitness factor. A UPEC mutant lacking flagella did not exhibit a defect in forming IBCs within mouse bladder-facet cells compared to the wild-type strain (68). Additionally, AL511 (pyelonephritis strain) invades mouse renal collecting-duct epithelial cells utilizing a flagella and motility-driven mechanism (69). This finding demonstrates that flagella-dependent invasion might be cell type-specific and might contribute to breaching the renal epithelial barrier to gain access to the bloodstream.

An *in vivo* imaging approach was utilized to determine temporo-spatial pattern of flagellar gene expression within the murine urinary tract (70). A luciferase reporter system was used, where expression of *fliC* gene was detected and measured by a corresponding increase in light production. Maximal expression of flagellar genes is coincident with UPEC ascension into ureters and kidneys, suggesting a role for flagella in the development of pyelonephritis. During the early stages of infection, the *fliC* mutant was defective in kidney colonization despite successful colonization of the bladder, indicating an advantage conferred by flagella on the progression of ascending UTI.

Chemotaxis

The ability to sense and respond to chemical signals by altering motility is known as chemotaxis and bacteria use flagellum-mediated motility to respond to chemotactic cues. Since most UTIs arise from colonization of periurethral region with UPEC strains originating from the gastrointestinal tract, it is possible that UPEC strains could exhibit enhanced chemotactic behavior towards urine compared to fecal-commensal strains. *E. coli* strains express chemotactic genes *tap, tar, tsr*, and *trg* that respond to dipeptides, aspartate and maltose, serine, and ribose and galactose, respectively. The UPEC strains CFT073 and F11, however, lack *tap* and *trg* genes whereas the fecal/commensal strains harbor all 4 genes. Interestingly, both UPEC

and fecal-commensal strains displayed a similar degree of chemotaxis towards urine, indicating that this is not an uropathogen-specific trait (71).

Motility at Population Level

Pyelonephritogenic strains ascend to the kidneys via ureters whereas cystitis strains remain within the bladder. Understanding the population dynamics involved in this phenomenon is important to understand how UPEC causes a spectrum of conditions, including cystitis, pyelonephritis, bacteremia, and sepsis. Walters et al. (72) used a collection of CFT073 strains that harbor unique traceable sequence tags to infect mice transurethrally and assessed the distribution of individual UPEC strain as a proportion of the population in bladders, kidneys, and spleen. Ascension to kidneys and entry into the bloodstream appear to result from multiple rounds of dissemination; progression from pyelonephritis to bacteremia represents a major bottleneck in the progression of pyelonephritis to sepsis.

Interplay Between Adherence and Motility

Motility, mediated primarily by flagella, and adherence, mediated primarily by fimbriae, represent two opposing phenomenon, yet both have been shown to contribute to uropathogenesis. Understanding how UPEC regulates these two phenomena is therefore crucial for better understanding of UTIs. Lane and coworkers demonstrated that constitutive expression of type 1 fimbria negatively affects motility and the levels of flagellin (Fig. 3), the major structural subunit of bacterial flagella (73). Conversely, overexpression of flagellar proteins or hypermotility phenotype did not alter the expression of type 1 fimbria. These results indicate that the expression of type 1 fimbria exerts a negative, unidirectional effect on flagellum-mediated motility.

Inhibitor of Motility, PapX

To understand the complex relationship between type 1 fimbrial gene expression and flagella, a transposon mutagenesis screen was conducted in a strain constitutively expressing type 1 fimbria to identify mutants that regained motility phenotype (74). This screen led to the identification of *papX* gene as a negative regulator of motility in UPEC (Fig. 3). Although the UPEC strain CFT073 has two gene clusters encoding the production of P fimbria, only the *pap*1 gene cluster located on the PheV-associated pathogenicity island contains *papX*. PapX was previously shown to inhibit bacterial motility and is a functional homolog of MrpJ, another bacterial motility inhibitor, found in *Proteus mirabilis* (75). PapX binds directly to the *flhDC* promoter, thereby inhibiting the transcription of master regulator of flagellar biosynthesis, motility and chemotaxis (58, 76). Overexpression of PapX led to reduced transcription of the genes involved in flagellar biosynthesis and chemotaxis; predictably, lack of PapX led to an increase in transcription of the same set of genes. Recently, the PapX binding site at the *flhD* promoter region was mapped to a 29-base-pair site upstream of the *flhDC* translational start site (76, 77).

TOXINS

UPEC strains are known to elaborate α-hemolysin, cytotoxic necrotizing factor-1 (CNF-1), secreted autotransporter toxin (Sat), and protease involved in colonization (Pic) (78–87). Hemolysin A (HlyA) is one of the best characterized pore-forming toxins and is the founding member of the repeats in toxin (RTX) family of toxins. HlyA contains the characteristic glycine-rich nonapeptide repeats in the C-terminal domain. The *hlyCABD* operon encodes the proteins involved in the production, activation and export of HlyA. The toxin is encoded by

hlyA gene and *hlyC* gene encodes an acyl transferase that is required for activation of the toxin. HlyA is secreted via type 1-secretion system. HlyB and HlyD act in concert with TolC, an outer membrane protein, for energy-dependent secretion of HlyA. HlyA is not only a hemolysin but also exhibits toxicity towards other cell types including leukocytes.

Functional genomic analysis of IBCs revealed that *hlyA* is among the genes highly expressed during UPEC growth within bladder cells (31). In addition to its cytolytic activity, HlyA is also involved in the communication between host and pathogen. HlyA degrades paxillin, a cytoskeletal scaffold protein, and other proteins involved in nuclear factor kappa-light-chain-enhancer of activated B cells (NF-κB)-signaling cascade (81). This study elucidated the role of HlyA in the exfoliation of bladder epithelial cells during UPEC UTIs. By interfering with NF-κB-mediated proinflammatory-signaling pathway, HlyA dampens host immune response to infection. While most UPEC strains carry one copy of the hemolysin operon, pyelonephritogenic strains 536 and J96 harbor two copies of the *hly* operon and both loci are required for the full virulence (82). However, there is preferential expression of one locus over the other and selective expression could provide distinct advantages within different niches.

CNF-1 is elaborated by some UPEC isolates and it deamidates small Rho guanosine triphosphatases (GTPases) including RhoA and Rac1, resulting in constitutive activation of these proteins. Rho GTPases are critical for several cellular processes including phagocytosis and oxidative burst in neutrophils. A mutant lacking CNF-1 is outcompeted by the wild-type strain in a murine model of UTI, clearly indicating that CNF-1 enhances the fitness of UPEC during UTI (79). Individual infection with the CNF-1-deficient mutant induces less severe inflammation but the mutant colonizes at levels comparable to wild-type strain. CNF-1-mediated disruption of Rho GTPase signaling pathways leads to immune dysregulation and production of CNF-1 confers an advantage during survival in the presence of neutrophils (83).

Autotransporter Toxins

Pic, secreted autotransporter toxin (Sat), and temperature-sensitive hemagglutinin (Tsh) are members of the SPATE (serine-protease autotransporter toxins of *Enterobacteriaceae*) family of proteins elaborated by UPEC. Sat induces vacuolation in proximal renal-tubule cells and serine-protease activity is required for cytopathic effects of Sat (84, 85). However, Sat does not contribute to fitness of UPEC in a mouse model of ascending UTI. Pic is expressed during UPEC infection in a murine model and is a potential mucus secretagogue (86, 87). By targeting major human leukocyte-adhesion molecules CD43, CD44, CD45, and CD93, Pic deregulates leukocyte migration and inflammation (80). Tsh does not display serine-protease activity (86) and appears to be more closely related to Vat, a vacuolating autotransporter toxin found in avian pathogenic *E. coli*. In summary, toxins produced by UPEC not only inflict tissue damage but are also involved in the communication between pathogen and host.

BIOFILM

Biofilms are sessile, complex, differentiated, multicellular, bacterial communities that may be populated by a single or multiple species. A prominent feature that distinguishes biofilm from planktonic cells is the presence of extracellular matrix that could be composed of proteins, polysaccharides, and/or DNA. The composition of biofilm matrix differs between various species of bacteria; proteins (curli and fimbriae) and polysaccharides (cellulose, colonic acid, and poly-N-acetyl glucosamine) are found in the matrix of biofilms formed by *E. coli*. Cells in a biofilm exhibit higher degree of antibiotic resistance compared to planktonic cells, and biofilms formed on in-dwelling devices such as urinary catheters pose a huge problem in healthcare settings. Therefore, there is a considerable interest in understanding the biology of biofilm formation in UPEC. By screening a transposon-mutant library, multiple genes involved in biofilm formation were identified in the cystitis strain UTI89 (88). A subset of the genes involved in biofilm formation is also implicated in colonization of mouse urinary bladders.

Curli, amyloid-like bacterial proteins, are produced by several members of the family Enterobacteriaceae including UPEC. These complex insoluble proteins form the extracellular matrix that acts as a scaffold for cells during biofilm development. Curli are found at detectable levels in human UTI urine samples and promotes adherence of UPEC to epithelial cells derived from both human bladder and kidneys (89). Curliated strains induce production of proinflammatory cytokine IL-8 and exhibit higher levels of resistance to an antimicrobial peptide, cathelicidin (LL-37). LL-37 inhibits polymerization of curli-specific gene (Csg)A, structural subunit of curli fibers, and prevents curli-mediated adherence and biofilm formation *in vitro*. Chemical inhibitors of curli biogenesis, curlicides, were sought as a novel treatment option for UPEC (90). Molecules belonging to the ring-fused 2-pyridone class were detected to display bifunctional curlicide and pilicide activity. These findings shed light on the potential role of curli in the pathogenesis of UTIs.

UPEC CFT073 encodes a gene cluster (*c1931–c1936*) capable of producing detectable fimbria, known as F9 fimbria (91). *E. coli* K-12 strains overexpressing F9 fimbria produce denser biofilm on polystyrene surfaces than control strain. However, F9 fimbria does not seem to be involved in hemagglutination or adherence to epithelial cells. It must be noted that several studies on biofilm formation are conducted under conditions vastly different from the microenvironment within host urinary tract and therefore results must be interpreted with due diligence.

Gene Expression Within UPEC Biofilms

Genes involved in the biogenesis of type 1, P, and F1C fimbriae are down-regulated in CFT073 biofilms grown in human urine, indicating that these fimbriae are not critical for biofilm formation, at least during growth in urine. In contrast, Auf fimbrial genes and Yad fimbrial genes are up-regulated, suggesting that these fimbriae could act as determinants for biofilm mode of growth in human urine (92). Transcriptional profiles of asymptomatic bacteriuria strain 83972 in biofilm, planktonic cells, and human volunteer colonization studies were compared to delineate biofilm-specific changes in gene expression (93). These studies elucidate the differences in expression of surface structures that might be involved in biofilm formation of various UPEC and ABU strains under specific conditions, such as growth in urine *in vitro* or during colonization of human bladder.

Regulation of Biofilm Formation

RfaH is an *E. coli* transcriptional antiterminator protein that affects the expression of genes involved in lipopolysaccharide biosynthesis, capsule production, hemolysin A, and CNF-I. Loss of *rfaH* leads to an increased biofilm phenotype in UPEC strain 536 and *E. coli* K-12. Upregulation of the *flu* gene that encodes antigen 43, a protein involved in autoaggregation and biofilm formation was observed in K-12 *rfaH* mutant. However, a precise mechanism for the regulation of biofilm formation by RfaH in UPEC 536 is yet to emerge (94). Nitric oxide-sensitive transcriptional repressor, NsrR, is yet another regulator of biofilm formation in UPEC. A mutant lacking NsrR displayed significantly reduced surface-attached growth and this effect might be mediated by derepression of

NsrR-repressed genes involved in flagellar biogenesis and motility (95).

Inhibition Biofilm Formation

Prototypical UPEC strain CFT073 and all *E. coli* strains producing a group II type of capsule release a soluble high-molecular-weight polysaccharide into spent medium that reduces biofilm formation by a broad range of pathogens including *Enterococcus faecalis*, *Klebsiella pneumoniae*, *Pseudomonas aeruginosa*, *Staphylococcus aureus*, and *S. epidermidis*. Although it is not clear how capsular polysaccharides that are covalently attached to the membrane are released into the medium, this phenomenon could be harnessed to develop a coating material for catheters to minimize biofilm development (96).

OUTER MEMBRANE PROTEINS

Autotransporter Proteins

UPEC encodes several autotransporter (AT) proteins that are typically 70 to 100 amino acids in length, with a C-terminal membrane anchor region that forms the β-barrel through which the passenger domain is translocated to the cell surface; the latter phenomenon lends the name for this class of secretion system as AT proteins. AT proteins are also referred to as type V-secretion system in Gram-negative bacteria and some AT proteins affect adherence and biofilm formation.

Urokinase-type plasminogen activator (Upa)B and UpaC are among several AT proteins found in UPEC (97). However, they are not uropathogen-specific, as their homologs are widely distributed among various strains of *E. coli*. UpaC does not affect the fitness of UPEC in a mouse model of UTI. However, UpaC promotes adherence to abiotic surfaces and biofilm formation. H-NS was identified as a transcriptional repressor of the *upaC* gene. UpaB promotes adherence to biotic surfaces such as components of extracellular matrix, including fibronectin, fibrinogen, and laminin. An isogenic mutant lacking UpaB displayed reduced fitness in murine bladders suggesting that UpaB-mediated adherence might be favorable for fitness *in vivo*. UpaG is a UPEC trimeric-autotransporter protein that promotes biofilm formation on abiotic surfaces and facilitates binding to extracellular matrix proteins, fibronectin and laminin (98).

Antigen 43 is an outer-membrane protein encoded by the *flu* gene that is widely distributed among *E. coli* strains, including UPEC and is subject to phase variation. Prototypical UPEC strain CFT073 contains two variant copies of this gene that were designated as *ag43a* and *ag43b*. Ag43a increased autoaggregation and biofilm formation on abiotic surfaces, when expressed in K-12 genetic background. A CFT073 strain overexpressing *ag43a* was more successful at persistent colonization for up to 5 days post inoculation, in a mouse model of UTI. Furthermore, it was demonstrated that in the presence of type 1 fimbriae, which prevents intimate bacterial cell-cell contact, the effect of Ag43 on autoaggregation and biofilm formation is diminished (99). Immunofluorescence analyses revealed that Ag43 is expressed within IBCs and is therefore proposed to play a role in the intracellular growth phase of UPEC (100). Taken together, these observations support a model in which Ag43-mediated autoaggregation and biofilm formation are important under conditions where close contact between bacterial cells can be achieved.

CAPSULE

The majority of the extraintestinal pathogenic *E. coli* strains, including UPEC, produce group II type of capsules (101). These capsular types are highly heterogeneous and resemble various glycoconjugates found within vertebrate hosts; this molecular mimicry is part of the bacterium's immune-evasion strategy. For instance, K1 and K4 capsular types mimic polysialic acid and substituted-chondroitin backbone, respectively. The assembly and biosynthesis of capsules in UPEC are closely related to the capsule-biogenesis systems found in the genera *Haemophilus* and *Neisseria*. Genes involved in the biosynthesis, serotype specificity, and export of UPEC capsules are located in 3 distinct gene clusters, designated as regions 1, 2, and 3, respectively.

Genes involved in the biosynthesis of capsule were up-regulated during murine infection (14). A signature-tagged mutagenesis screen of UPEC strain CFT073 revealed that a mutation in the *kpsC* gene that impairs capsular-polysaccharide transport was attenuated in a murine model of UTI (102). The role of K2 capsule in the pathogenesis of UTI was assessed using a mutant deficient in the biosynthesis of capsular polysaccharides in a murine model. The mutant displayed a significant fitness defect in urine and kidneys that was reversed upon trans-complementation of the capsule-biosynthesis genes, clearly ascertaining the role of K2 capsule during UTI. Furthermore, this group also demonstrated that the K2 capsule was essential for protection from complement-mediated killing (103). The K5 capsule, also belonging to the group II family, was shown to prevent the association of neutrophils with an UPEC strain (104). In combination, these results establish that capsule is a bona-fide virulence factor in UPEC.

UTI89, another prototypical UPEC strain, produces a K1-type capsule and this capsule was shown to be required for successful development of IBCs within bladders from both wild-type and TLR4-deficient mice (105). Since defects in capsule synthesis and assembly increases the intracellular levels of sialic acid, the effect of N-acetyl-neuraminic acid regulator (NanR), a sialic acid-sensitive bacterial-transcriptional regulator, was tested in wild-type and capsule-mutant strains. Loss of NanR led to a partial increase in IBC formation in the capsule-mutant strain, indicating that perturbation in sialic acid levels has a profound effect on IBC formation.

GLOBAL REGULATORS OF GENE EXPRESSION

PhoP

Bacteria use two-component systems comprised of a sensor kinase and a cognate-response regulator to respond to external stimuli. PhoPQ is a two-component regulatory system that initiates a transcriptional program in response to environmental cues including low concentrations of calcium and magnesium. PhoQ is the inner membrane-associated sensor kinase that phosphorylates PhoP and phosphorylated PhoP acts as a transcriptional regulator of the PhoP regulon (genes regulated by PhoP). PhoP is a virulence factor in UPEC and controls the expression of around 11% of the genome (106). In CFT073, PhoP regulon includes the genes required for adaptation to acid stress and resistance to cationic antimicrobial peptides. UPEC is exposed to acidic pH in the stomach prior to gut colonization and in the urine. Cationic antimicrobial peptides are part of innate immune response to bacterial infections and play a critical role in controlling UTIs. Furthermore, PhoP represses the expression of flagellar genes and a mutant strain lacking PhoP is hypermotile (Fig. 3). In summary, PhoP affects acid resistance and virulence in UPEC by modulating polarization state of inner membrane.

Histone-Like Nucleoid Structuring Protein (H-NS)

H-NS, a DNA-binding protein, binds double-stranded DNA regions with intrinsic curvature. In UPEC, H-NS acts as a global transcriptional regulator; genes involved in motility (Fig. 3), hemolysin A production, type 1 (*fimB, fimE*) and S fimbrial expression, and iron uptake are repressed by H-NS. Not surprisingly, an *hns* mutant displayed increased virulence in mouse models of UTI and sepsis (107). StpA, another DNA-binding protein, however, partially compensates for the loss of H-NS function.

Sigma E

Sigma E facilitates transcription of genes required to combat envelope stress. Sigma E itself is regulated at several levels including sequestration by an inner membrane associated anti-sigma factor, RseA. During envelope stress, RseA is degraded by a protease (DegS) resulting in the expression of sigmaE-regulated genes. Mutants lacking DegS, therefore, constitutively express sigma E-regulated genes and are attenuated in a mouse model of UTI (108). Sigma E-regulated periplasmic chaperone (DegP) also contributes to urovirulence (109). These results demonstrate the critical need to maintain envelope homeostasis during colonization of urinary tract.

Host Factor Q-beta (Hfq)

RNA chaperone Hfq is a key player in ribo-regulation, small RNA-mediated regulation of gene expression, in bacteria. Hfq and small RNA-mediated circuits have been shown to affect global changes in transcript abundance and/or translation. Kulesus et al. (110) reported that an *hfq* mutant is highly attenuated in both bladder and kidneys concomitant with a reduced number of IBCs within the bladders of mice infected with the mutant strain. UPEC grows under acidic conditions in the presence of reactive oxygen and nitrogen species in the urinary tract. Hfq was required for wild-type levels of growth under acidic conditions as well as during oxidative or nitrosative stress. The *hfq* mutant exhibits impaired motility (Fig. 3) and resistance to cationic polypeptide, polymixin B. These phenotypes and transcriptome analyses revealed that the *hfq* mutant is compromised in membrane-stress response, similar to an *rpoE* mutant. In summary, Hfq-mediated riboregulation is critical for orchestrating gene expression during infection.

Cell-Cell Communication

Population-level changes in gene expression that occur in a bacterial cell density-dependent manner are known as quorum sensing. Bacteria use small molecules or peptides, referred to as autoinducers, to determine cell density or quorum. UPEC strains, similar to several Gram-negative bacteria, are capable of sensing autoinducer 2. Recently, a novel quorum-sensing (Qse) system that uses catecholamines (epinephrine and norepinephrine) as signals has been discovered in enterohemorrhagic *E. coli* and has been proposed as an interkingdom-signaling system. A functional QseBC-signaling system is also found in UPEC. QseC is an inner membrane-associated sensor kinase that phosphorylates QseB, the cognate-response regulator. QseB is a transcriptional regulator that affects the expression of target genes including curli and flagella. In UPEC UTI89, a mutant strain lacking QseC is attenuated and is defective in differentiating into IBCs in a murine infection model (111). Interestingly, an isogenic mutant lacking *qseB* retained wild-type levels of virulence and the *qseBC* operon was constitutively expressed in the absence of QseC due to the loss of QseC-associated phosphatase activity. In summary, QseBC system is required for coordinated gene expression in UPEC during infection.

GENOME SEQUENCES

Complete, annotated genome sequences are available for the following strains: cystitis strains F11, IAI39, UMN026, and UTI89; pyelonephritis strain 536; pyelonephritis and sepsis strain CFT073; and asymptomatic bacteriuria strains ABU 83972 and VR50.

Sequences of UPEC strains not only provide genome-wide insights into the presence of fitness/virulence factors but have also ushered in a host of genome-enabled studies such as comparative genomics (comparing genome contents), functional genomics (transcriptional profiling, high-density mutational analysis) and proteomics (translational profiling). Due to rapid advances in sequencing technology, draft-genome sequences of several UPEC isolates are also available from GenBank (NCBI; National Center for Biotechnology Information) and the Broad Institute.

Comparison of genomes of pyelonephritis isolate CFT073, fecal/commensal strain K-12 MG1655, and enterohemorrhagic *E. coli* strain EDL933 revealed that genomes of pathogenic *E. coli* are mosaic in nature with large regions of horizontally transferred genes interspersed in between the core fecal/commensal *E. coli* genome (43). Only 39% of the protein-coding regions was conserved among these three strains of *E. coli,* demonstrating that plastic nature of the genomes could play a vital role in adaptation of these strains to the milieu encountered within their respective niches. Absence of a type III-secretion system and phage-encoded toxin genes in UPEC compared to enterohemorrhagic *E. coli* (EHEC) are among the remarkable differences between these two pathotypes. Recently, the presence of putative type III-secretion system genes in the genomes of UPEC isolated from patients with cystitis was reported (112). Its role in fitness during UTI, if any, remains to be determined.

Gene clusters encoding characterized or predicted fimbriae were overrepresented in the CFT073 genome (43). A total of 12 fimbrial gene clusters were identified; 10 belong to the chaperone-usher family of fimbriae that includes the type 1 fimbria and the P fimbria and two are reminiscent of the members of the type IV pilus family. A preponderance of FimBE-like recombinases (a total of five) was observed in the CFT073 genome and it was proposed that recombination-mediated regulation of gene expression is involved in orchestrating gene expression during various phases of UTI. In general, the genomes of pathogenic *E. coli* are larger than fecal/commensal strains. Specifically, the genomes of UPEC strains 536, CFT073, and UTI89 are 4.94 million base pairs (Mb), 5.23 Mb, and 5.07 Mb in length, respectively. The genome of the fecal/commensal strain K-12 MG1655 is 0.59 Mb shorter than the CFT073 genome (43) and these additional, UPEC-specific genes are thought to confer fitness advantage during survival within the urinary tract.

Pathogenicity Islands

Genomes of several pathogenic bacteria, including UPEC, harbor regions of DNA that are acquired by horizontal gene transfer (HGT). Such regions are referred to as genomic islands and are marked by signatures including the presence of direct repeats and evidence of mobile elements, such as transposons or phages. HGT is one of the fastest means of dissemination of fitness or virulence factors among bacteria. Sequenced genomes of UPEC strains reveal a mosaic structure with multiple horizontally transferred islands (43, 113). More often than not, such islands contain fitness or virulence genes and are known as pathogenicity islands (PAIs). First insights into the presence of PAIs in bacterial pathogens were garnered from studies using UPEC strain 536.

Genomes of 3 pyelonephritis strains, 4 cystitis strains, and 3 fecal/commensal strains were compared by comparative genomic hybridization and around 52% of genome was found to be conserved between these strains, indicating that there is high level of divergence among *E. coli* strains (114). Only 131 genes, however, were found in all the 11 UPEC strains, including 536, CFT073, F11, and UTI89, that were absent in fecal/commensal strains and were designated as uropathogen-specific genes. A subset of the uropathogen-specific genes (38 genes) is up-regulated during mouse UTI suggesting that

they are involved in adaptation to growth within the host urinary tract (14). In agreement with the general trend of close proximity of PAIs to transfer RNA (tRNA) genes, 80% of the PAIs in CFT073 are associated with tRNA loci.

Pyelonephritis strain CFT073 harbors 13 genomic islands comprising 7 PAIs, 3 genomic islands (GI), and 3 phage-rich islands (PI) (115). Of these islands, PAI-*aspV*, PAI-*asnT*, and PAI-*metV* are required for *in vivo* fitness. In contrast, PAI-*serX*, GI-*asnW*, GI-*cobU*, GI-*selC*, PI-*b0847*, and PI-*potB* are dispensable without a measurable loss of fitness (115). This study led to the identification of a member of the RTX family of toxins, TosA. In an independent study, TosA was identified as an *in vivo*-expressed antigen that contributes to fitness both in the urinary tract and during bacteremia in a mouse model of UTI and bacteremia, respectively (116, 117). However, no toxic functions have been ascribed to TosA and TosA could very well belong to a growing subset of RTX family members that do not exhibit toxicity (118).

Studies on PAIs in UPEC CFT073 led to the identification of two proteins, homologs of *Shigella flexneri shiA*, that suppress host inflammatory response during UTI. The genes *sisA* and *sisB* are located in the PAI-pheV and PAI-selC, respectively (119). Strains lacking both genes incited a robust inflammatory response within mouse urinary tract compared to wild-type and this *in vivo* phenotype was reversed upon complementation with either of the two genes. Not surprisingly, the double mutant exhibited a colonization defect during the early stages of UTI that can be attributed to a severe inflammatory response. Interestingly, these anti-inflammatory proteins are localized within the cytoplasm and are predicted to be released only upon lysis of UPEC. Keeping in line with the genetic diversity observed among UPEC strains, *sisA* and *sisB* are not found in other model UPEC strains, including 536, F11, and UTI89.

The sequence of UPEC 536 is 292 Kb smaller than CFT073 genome and this difference is primarily due to PAIs that are unique to each strain. One unique feature of UPEC 536 genome is the presence of two hemolysin A operons, *hlyA1* and *hlyA2*, within PAI I and PAI II, respectively; both operons contribute to the virulence independently (113). A functional yersiniabactin biosynthesis and uptake system is found in a PAI in UPEC 536. In summary, availability of genome sequences has facilitated genome-wide comparisons that have delineated uropathogen-specific genes from the pangenome of *E. coli*.

UPEC GENE EXPRESSION *IN VIVO*

Dedicated research has led to the identification of numerous virulence factors in UPEC; however, we are just beginning to understand the response of UPEC to cues encountered during infection. The availability of genome sequences for several UPEC strains and tools to determine genome-wide changes has shed new light on the molecular pathogenesis of UTIs. Genes that are either up or down-regulated under certain conditions are classified as differentially expressed genes. Approximately 9% of the UPEC CFT073 genes are differentially expressed during murine UTI (14). Genes involved in ribosome biosynthesis are among the most highly expressed genes, indicating that UPEC multiply rapidly, even during growth in the face of inflammation. Genes encoding the type 1 fimbria, siderophores, other factors involved in iron acquisition, and transporters of capsular polysaccharides are known virulence or fitness genes that are highly up-regulated within the murine urinary tract. Genes involved in glutamine biosynthesis and import, *glnA* and *glnPQ*, are up-regulated *in vivo* clearly indicating that UPEC experiences nitrogen limitation within host urinary tract. Although urine is replete with urea (~0.5 M), nitrogen derived from urea is unavailable for UPEC because of the absence of urease activity. Transporters of osmoprotectants, proline, and glycine betaine encoded by *proP* and *proVWX*, are up-regulated *in vivo*.

Conversely, many of the genes involved in flagellar biosynthesis and chemotaxis are among the most down-regulated genes within the murine urinary tract. Genes regulated by fumarate and nitrate reductase (Fnr), a global transcriptional regulator facilitating the adaptation to anaerobic conditions, such as *frdABCD*-encoding subunits of the fumarate reductase and *aspA* aspartase, are down-regulated, suggesting that UPEC is experiencing aerobic conditions *in vivo*. Taken together, the transcriptome paints the picture of the murine urinary tract as an iron and nitrogen-limited milieu, with high osmolarity and adequate oxygenation; UPEC utilizes amino acids and peptides along with carbohydrates such as altronate, fructose, glucitol, and mannonate, as sources of carbon and nitrogen to grow rapidly within the murine urinary tract (14).

While the *in vivo* transcriptome analysis elucidated the conditions encountered by UPEC during UTI, it also raised the question of parallels between a murine model and human infection. To precisely address that question, transcriptome analysis on UPEC samples directly derived from women with clinical UTIs was performed (19). General conditions appear to be conserved between murine and human UTI. Genes involved in iron uptake, peptide and amino acid transporters, ribosome biogenesis, transcription, and translation machinery were highly expressed in the human UTI-derived samples. This transcriptional profile reflects the rapid growth of UPEC under aerobic, nitrogen, and iron-limited environment. Despite the high levels of expression of genes involved in biogenesis of type 1 fimbria in the murine model, the *fim* genes were poorly expressed in human samples. The gene *fimH*, encoding the fimbrial-tip adhesin, was expressed only in 2 out of a total of 8 samples. These results suggest that type 1 fimbria might play significantly different roles in murine and human UTIs. Also, genes involved in the biogenesis of P, F1C, and Auf fimbriae were not expressed at detectable levels *in vivo,* at least in this group of patients. These findings emphasize the need to exercise caution when extrapolating information obtained from laboratory-animal models to human health.

OVERVIEW OF SELECT TECHNIQUES USED IN MOLECULAR PATHOGENESIS RESEARCH

Transcriptional profiling and transposon mutagenesis are widely used techniques that elucidate gene expression and genes required for growth and/or survival under specific conditions, respectively. Here we introduce these techniques and project how studies utilizing these techniques could impact research on the molecular pathogenesis of UPEC.

DNA microarrays, a well-established technique, and RNA-seq, an emerging technique, are utilized to determine transcriptional profile in bacteria. Hybridization based-detection of complementary DNA (cDNA) is the basis of DNA-microarray technology. Relative abundance of a particular transcript is used to determine the genes that are up or down-regulated under a specific condition. It is also possible to determine the absolute quantity of transcripts using specialized arrays such as gene chip (Affymetrix). DNA microarrays can also be used for comparative genomic-hybridization studies that analyze gene content. While DNA-microarray technology revolutionized the ability to gain a global perspective on UPEC gene expression during specific conditions (14, 19, 31, 93), it is slowly losing ground to RNA-seq, a sequencing-based technique.

Recent advances in high-throughput (HT) DNA-sequencing technology have been applied to transcriptome analyses. RNA-seq involves massively parallel sequencing of cDNA libraries utilizing HT-sequencing platforms such as Illumina (120). Both absolute and differential expression under various conditions can be evaluated using RNA-seq. Advantages conferred by RNA-seq over DNA microarrays include global analysis in an unbiased manner at a hitherto unprecedented resolution and dynamic range. Untranslated regions of the transcripts, promoter regions, novel transcripts, including small regulatory RNAs, and operon structure can be characterized using RNA-seq. Additional information offered by RNA-seq combined with decreasing cost associated with HT sequencing is poised to make RNA-seq the top choice for transcriptional analyses. Efforts are underway in the Mobley laboratory to define the transcriptome of UPEC during human infection using RNA-seq (unpublished results).

Transposon mutagenesis, specifically signature-tagged mutagenesis (STM), has been utilized to identify genes necessary for growth and survival under conditions of interest. An STM screen in a mouse model of UPEC UTI has unraveled several novel virulence factors along with validation of known virulence factors (102). To provide complete genome saturation, a major bottleneck in STM, several modifications that integrate state-of-the-art technologies in genome analysis with transposon mutagenesis such as transposon-site hybridization and transposon-insertion-site sequencing have been developed (121). Transposon-insertion libraries, comprised of genome-saturating numbers of transposon mutants, are subjected to selection under a condition of interest. The genomic DNA from the library prior to and post selection are sequenced to detect and determine the frequency of transposon insertion at a given locus. Since these approaches are similar to STM, a negative-selection strategy, the mutants that are less represented in the output pool, compared to the input pool, are delineated as putative virulence or fitness genes. Such genes warrant further characterization to determine their exact role in fitness or virulence. Genetic determinants of fitness in UPEC during bacteremia in a murine model have been identified using transposon-directed insertion-site sequencing (TraDIS) (44). Recently, UPEC genes involved in serum resistance (122) and in colonization of a zebra-fish model of infection were also identified using similar approaches (123). Conducting such forward genetic screens using UPEC in a mouse model of UTI has the potential to unravel hitherto unrecognized fitness and virulence mechanisms involved in the molecular pathogenesis of UPEC.

CONCLUDING REMARKS

The UPEC research community has made remarkable progress since the publication of the previous edition of this book. Access to advanced technology in the post-genome era has the potential to generate an enormous wealth of data that can be utilized to better understand the infection biology of UPEC and to identify pathways that can be targeted for developing novel therapeutics. Moving forward, a prime challenge is translating the knowledge gained in virulence and fitness mechanisms into developing therapeutics and vaccines that can alleviate the humongous burden imposed on human health by UPEC.

ACKNOWLEDGMENTS

S.S. is supported by a Research Scholars Program post-doctoral fellowship from the North Central Section of the American Urological Association that is administered by the Urology Care Foundation. Research in the Mobley laboratory is supported by Public Health Service grants AI059722, AI043363, and DK094777 (H.L.T.M) from the National Institutes of Health. We apologize to our colleagues engaged in UPEC research whose work could not be mentioned here due to space limitation.

Conflicts of interest: We declare no conflicts.

CITATION

Subashchandrabose S, Mobley HLT. 2015. Virulence and fitness determinants of uropathogenic *Escherichia coli*. Microbiol Spectrum 3(4):UTI-0015-2012.

REFERENCES

1. **Sivick KE, Mobley HL.** 2010. Waging war against uropathogenic *Escherichia coli*: winning back the urinary tract. *Infect Immun* **78:**568–585.

2. **Gupta K, Hooton TM, Naber KG, Wullt B, Colgan R, Miller LG, Moran GJ, Nicolle LE, Raz R, Schaeffer AJ, Soper DE; Infectious Diseases Society of America; European Society for Microbiology and Infectious Diseases.** 2011. International clinical practice guidelines for the treatment of acute uncomplicated cystitis and pyelonephritis in women: A 2010 update by the Infectious Diseases Society of America and the European Society for Microbiology and Infectious Diseases. *Clin Infect Dis* **52:**e103–120.

3. **Foxman B.** 2003. Epidemiology of urinary tract infections: incidence, morbidity, and economic costs. *Dis Mon* **49:**53–70.

4. **Foxman B.** 2010. The epidemiology of urinary tract infection. *Nat Rev Urol* **7:**653–660.

5. **Litwin MS, Saigal CS, Yano EM, Avila C, Geschwind SA, Hanley JM, Joyce GF, Madison R, Pace J, Polich SM, Wang M; Urologic Diseases in America Project.** 2005. Urologic diseases in America Project: analytical methods and principal findings. *J Urol* **173:**933–937.

6. **DeFrances CJ, Lucas CA, Buie VC, Golosinskiy A.** 2008. 2006 National Hospital Discharge Survey. *Natl Health Stat Report* **5:**1–20.

7. **Yamamoto S, Tsukamoto T, Terai A, Kurazono H, Takeda Y, Yoshida O.** 1997. Genetic evidence supporting the fecal-perineal-urethral hypothesis in cystitis caused by *Escherichia coli*. *J Urol* **157:**1127–1129.

8. **Mulvey MA, Schilling JD, Hultgren SJ.** 2001. Establishment of a persistent *Escherichia coli* reservoir during the acute phase of a bladder infection. *Infect Immun* **69:**4572–4579.

9. **Rosen DA, Hooton TM, Stamm WE, Humphrey PA, Hultgren SJ.** 2007. Detection of intracellular bacterial communities in human urinary tract infection. *PLoS Med* **4:**e329. doi:10.1371/journal.pmed.0040329.

10. **Gunther NW IV, Lockatell V, Johnson DE, Mobley HL.** 2001. *In vivo* dynamics of type 1 fimbria regulation in uropathogenic *Escherichia coli* during experimental urinary tract infection. *Infect Immun* **69:**2838–2846.

11. **Falkow S.** 1988. Molecular Koch's postulates applied to microbial pathogenicity. *Rev Infect Dis.* **10**(Suppl 2)**:**S274–276.

12. **Hull RA, Hull SI.** 1997. Nutritional requirements for growth of uropathogenic *Escherichia coli* in human urine. *Infect Immun* **65:**1960–1961.

13. **Alteri CJ, Smith SN, Mobley HL.** 2009. Fitness of *Escherichia coli* during urinary tract infection requires gluconeogenesis and the TCA cycle. *PLoS Pathog* **5:**e1000448. doi:10.1371/journal.ppat.1000448

14. **Snyder JA, Haugen BJ, Buckles EL, Lockatell CV, Johnson DE, Donnenberg MS, Welch RA, Mobley HL.** 2004. Transcriptome of uropathogenic *Escherichia coli* during urinary tract infection. *Infect Immun* **72:**6373–6381.

15. **Roos V, Ulett GC, Schembri MA, Klemm P.** 2006. The asymptomatic bacteriuria *Escherichia coli* strain 83972 outcompetes uropathogenic *E. coli* strains in human urine. *Infect Immun* **74:**615–624.

16. **Corbin BD, Seeley EH, Raab A, Feldmann J, Miller MR, Torres VJ, Anderson KL, Dattilo BM, Dunman PM, Gerads R, Caprioli RM, Nacken W, Chazin WJ, Skaar EP.** 2008. Metal chelation and inhibition of bacterial growth in tissue abscesses. *Science* **319:**962–965.

17. **Haley KP, Skaar EP.** 2012. A battle for iron: host sequestration and *Staphylococcus aureus* acquisition. *Microbes Infect* **14:**217–227.

18. **Alteri CJ, Mobley HL.** 2007. Quantitative profile of the uropathogenic *Escherichia coli* outer membrane proteome during growth in human urine. *Infect Immun* **75:**2679–2688.

19. **Hagan EC, Lloyd AL, Rasko DA, Faerber GJ, Mobley HL.** 2010. *Escherichia coli* global gene expression in urine from women with urinary tract infection. *PLoS Pathog* **6:**e1001187. doi:10.1371/journal.ppat.1001187

20. **Torres AG, Redford P, Welch RA, Payne SM.** 2001. TonB-dependent systems ofuropathogenic *Escherichia coli*: aerobactin and heme transport and TonB are required forvirulence in the mouse. *Infect Immun* **69:**6179–6185.

21. **Yep A, McQuade T, Kirchhoff P, Larsen M, Mobley HL.** 2014. Inhibitors of TonB function identified by a high-throughput screen for inhibitors of iron acquisition in uropathogenic *Escherichia coli* CFT073. *mBio* **5:**e01089-13. doi:10.1128/mBio.01089-13

22. **Lee JW, Helmann JD.** 2007. Functional specialization within the Fur family of metalloregulators. *Biometals* **20:**485–499.

23. **Massé E, Gottesman S.** 2002. A small RNA regulates the expression of genes involved in iron metabolism in *Escherichia coli*. *Proc Natl Acad Sci U S A* **99:**4620–4625.

24. **Chaturvedi KS, Hung CS, Crowley JR, Stapleton AE, Henderson JP.** 2012. The siderophore yersiniabactin binds copper to protect pathogens during infection. *Nat Chem Biol* **8:**731–736.

25. **Flo TH, Smith KD, Sato S, Rodriguez DJ, Holmes MA, Strong RK, Akira S, Aderem A.** 2004. Lipocalin 2 mediates an innate immune response to bacterial infection by sequestrating iron. *Nature* **432:**917–921.

26. **Goetz DH, Holmes MA, Borregaard N, Bluhm ME, Raymond KN, Strong RK.** 2002. The neutrophil lipocalin NGAL is a bacteriostatic agent that interferes with siderophore-mediated iron acquisition. *Mol Cell* **10:**1033–1043.

27. **Hantke K, Nicholson G, Rabsch W, Winkelmann G.** 2003. Salmochelins, siderophores of *Salmonella enterica* and uropathogenic *Escherichia coli* strains, are recognized by the outer membrane receptor IroN. *Proc Natl Acad Sci U S A* **100:**3677–3682.

28. **Henderson JP, Crowley JR, Pinkner JS, Walker JN, Tsukayama P, Stamm WE, Hooton TM, Hultgren SJ.** 2009. Quantitative metabolomics reveals an epigenetic blueprint for iron acquisition in uropathogenic *Escherichia coli*. *PLoS Pathog* **5:**e1000305. doi:10.1371/journal.ppat.1000305

29. **Hagan EC, Mobley HL.** 2007. Uropathogenic *Escherichia coli* outer membrane antigens expressed during urinary tract infection. *Infect Immun* **75:**3941–3949.

30. **Hagan EC, Mobley HL.** 2009. Haem acquisition is facilitated by a novel receptor Hma and required by uropathogenic *Escherichia coli* for kidney infection. *Mol Microbiol* **71:**79–91.

31. **Reigstad CS, Hultgren SJ, Gordon JI.** 2007. Functional genomic studies of uropathogenic *Escherichia coli* and host urothelial cells when intracellular bacterial communities are assembled. *J Biol Chem* **282:**21259–21267.

32. **Garcia EC, Brumbaugh AR, Mobley HL.** 2011. Redundancy and specificity of *Escherichia coli* iron acquisition systems during urinary tract infection. *Infect Immun* **79:**1225–1235.

33. **Ikeda JS, Janakiraman A, Kehres DG, Maguire ME, Slauch JM.** 2005. Transcriptional regulation of *sitABCD* of *Salmonella enterica* serovar Typhimurium by MntR and Fur. *J Bacteriol* **187:**912–922.

34. **Koch D, Chan AC, Murphy ME, Lilie H, Grass G, Nies DH.** 2011. Characterization of a dipartite iron uptake system from uropathogenic *Escherichia coli* strain F11. *J Biol Chem* **286:**25317–25330.

35. **Alteri CJ, Hagan EC, Sivick KE, Smith SN, Mobley HL.** 2009. Mucosal immunization with iron receptor antigens protects against urinary tract infection. *PLoS Pathog* **5:** e1000586. doi:10.1371/journal.ppat.1000586

36. **Brumbaugh AR, Smith SN, Mobley HL.** 2013. Immunization with the yersiniabactin receptor, FyuA, protects against pyelonephritis in a murine model of urinary tract infection. *Infect Immun* **81:**3309–3316.

37. **Sabri M, Houle S, Dozois CM.** 2009. Roles of the extraintestinal pathogenic *Escherichia coli* ZnuACB and ZupT zinc transporters during urinary tract infection. *Infect Immun* **77:**1155–1164.

38. **Roesch PL, Redford P, Batchelet S, Moritz RL, Pellett S, Haugen BJ, Blattner FR, Welch RA.** 2003. Uropathogenic *Escherichia coli* use d-serine deaminase to modulate infection of the murine urinary tract. *Mol Microbiol* **49:**55–67.

39. **Haugen BJ, Pellett S, Redford P, Hamilton HL, Roesch PL, Welch RA.** 2007. *In vivo* gene expression analysis identifies genes required for enhanced colonization of the mouse urinary tract by uropathogenic *Escherichia coli* strain CFT073 *dsdA*. *Infect Immun* **75:**278–289.

40. **Anfora AT, Halladin DK, Haugen BJ, Welch RA.** 2008. Uropathogenic *Escherichia coli* CFT073 is adapted to acetatogenic growth but does not require acetate during murine urinary tract infection. *Infect Immun* **76:**5760–5767.

41. **Wolfe AJ.** 2005. The acetate switch. *Microbiol Mol Biol Rev* **69:**12–50.

42. **Spurbeck RR, Stapleton AE, Johnson JR, Walk ST, Hooton TM, Mobley HL.** 2011. Fimbrial profiles predict virulence of uropathogenic *Escherichia coli* strains: contribution of *ygi* and *yad* fimbriae. *Infect Immun* **79:**4753–4763.

43. **Welch RA, Burland V, Plunkett G III, Redford P, Roesch P, Rasko D, Buckles EL, Liou SR, Boutin A, Hackett J, Stroud D, Mayhew GF, Rose DJ, Zhou S, Schwartz DC, Perna NT, Mobley HL, Donnenberg MS, Blattner FR.** 2002. Extensive mosaic structure revealed by the complete genome sequence of uropathogenic *Escherichia coli*. *Proc Natl Acad Sci U S A* **99:**17020–17024.

44. **Subashchandrabose S, Smith SN, Spurbeck RR, Kole MM, Mobley HL.** 2013. Genome-wide detection of fitness genes in uropathogenic *Escherichia coli* during systemic infection. *PLoS Pathogens* **9:**e1003788. doi:10.1371./journal.ppat.10003788

45. **Kulkarni R, Dhakal BK, Slechta ES, Kurtz Z, Mulvey MA, Thanassi DG.** 2009. Roles of putative type II secretion and type IV pilus systems in the virulence of uropathogenic *Escherichia coli*. *PLoS One* **4:**e4752. doi:10.1371 ./journal.pone.0004752

46. **Snyder JA, Haugen BJ, Lockatell CV, Maroncle N, Hagan EC, Johnson DE, Welch RA, Mobley HL.** 2005. Coordinate expression of fimbriae in uropathogenic *Escherichia coli*. *Infect Immun* **73:**7588–7596.

47. **Wu XR, Sun TT, Medina JJ.** 1996. *In vitro* binding of type 1-fimbriated *Escherichia coli* to uroplakins Ia and Ib: relation to urinary tract infections. *Proc Natl Acad Sci U S A* **93:**9630–9635.

48. **Gunther NW IV, Snyder JA, Lockatell V, Blomfield I, Johnson DE, Mobley HL.** 2002. Assessment of virulence of uropathogenic *Escherichia coli* type 1 fimbrial mutants in which the invertible element is phase-locked on or off. *Infect Immun* **70:**3344–3354.

49. **Mysorekar IU, Hultgren SJ.** 2006. Mechanisms of uropathogenic *Escherichia coli* persistence and eradication from the urinary tract. *Proc Natl Acad Sci U S A* **103:**14170–14175.

50. **Martinez JJ, Mulvey MA, Schilling JD, Pinkner JS, Hultgren SJ.** 2000. Type 1 pilus-mediated bacterial invasion of bladder epithelial cells. *EMBO J* **19:**2803–2812.

51. **Connell I, Agace W, Klemm P, Schembri M, Mårild S, Svanborg C.** 1996. Type 1 fimbrial expression enhances *Escherichia coli* virulence for the urinary tract. *Proc Natl Acad Sci U S A* **93:**9827–9832.

52. **Melican K, Sandoval RM, Kader A, Josefsson L, Tanner GA, Molitoris BA, Richter-Dahlfors A.** 2011. Uropathogenic *Escherichia coli* P and type 1 fimbriae act in synergy in a living host to facilitate renal colonization leading to nephron obstruction. *PLoS Pathog* **7:**e1001298. doi:10.1371/journal.ppat.1001298

53. **Müller CM, Aberg A, Straseviçiene J, Emody L, Uhlin BE, Balsalobre C.** 2009. Type 1 fimbriae, a colonization factor of uropathogenic *Escherichia coli*, are controlled by the metabolic sensor CRP-cAMP. *PLoS Pathog* **5:** e1000303. doi:10.1371/journal.ppat.1000303

54. **Lund B, Lindberg F, Marklund BI, Normark S.** 1987. The PapG protein is the alpha-D-galactopyranosyl-(1-4)-beta-D-galactopyranose-binding adhesin of uropathogenic *Escherichia coli*. *Proc Natl Acad Sci U S A* **84:**5898–5902.

55. **Mobley HL, Jarvis KG, Elwood JP, Whittle DI, Lockatell CV, Russell RG, Johnson DE, Donnenberg MS, Warren JW.** 1993. Isogenic P-fimbrial deletion mutants of pyelonephritogenic *Escherichia coli*: the role of alpha Gal(1-4) beta Gal binding in virulence of a wild-type strain. *Mol Microbiol* **10:**143–155.

56. **Roberts JA, Marklund BI, Ilver D, Haslam D, Kaack MB, Baskin G, Louis M, Möllby R, Winberg J, Normark S.** 1994. The Gal(alpha 1-4)Gal-specific tip adhesin of *Escherichia coli* P-fimbriae is needed for pyelonephritis to occur in the normal urinary tract. *Proc Natl Acad Sci U S A* **91:**11889–11893.

57. **Lane MC, Mobley HL.** 2007. Role of P-fimbrial-mediated adherence in pyelonephritis and persistence of uropathogenic *Escherichia coli* (UPEC) in the mammalian kidney. *Kidney Int* **72:**19–25.

58. **Simms AN, Mobley HL.** 2008. PapX, a P fimbrial operon-encoded inhibitor of motility in uropathogenic *Escherichia coli*. *Infect Immun* **76:**4833–4841.

59. **Hernday A, Krabbe M, Braaten B, Low D.** 2002. Self-perpetuating epigenetic piliswitches in bacteria. *Proc Natl Acad Sci U S A* **99**(Suppl 4):16470–16476.

60. **Nowicki B, Moulds J, Hull R, Hull S.** 1988. A hemagglutinin of uropathogenic *Escherichia coli* recognizes the Dr blood group antigen. *Infect Immun* **56:**1057–1060.

61. **Bäckhed F, Alsén B, Roche N, Angström J, von Euler A, Breimer ME, Westerlund-Wikström B, Teneberg S, Richter-Dahlfors A.** 2002. Identification of target tissue glycosphingolipid receptors for uropathogenic, F1C-fimbriated *Escherichia coli* and its role in mucosal inflammation. *J Biol Chem* **277:**18198–18205.

62. **Kreft B, Placzek M, Doehn C, Hacker J, Schmidt G, Wasenauer G, Daha MR, van der Woude FJ, Sack K.** 1995. S fimbriae of uropathogenic *Escherichia coli* bind to primary human renal proximal tubular epithelial cells but do not induce expression of intercellular adhesion molecule 1. *Infect Immun* **63:**3235–3238.

63. **Korhonen TK, Valtonen MV, Parkkinen J, Väisänen-Rhen V, Finne J, Orskov F, Orskov I, Svenson SB, Mäkelä PH.** 1985. Serotypes, hemolysin production, and receptor recognition of *Escherichia coli* strains associated with neonatal sepsis and meningitis. *Infect Immun* **48:**486–491.

64. **Lügering A, Benz I, Knochenhauer S, Ruffing M, Schmidt MA.** 2003. The Pix pilus adhesin of the uropathogenic *Escherichia coli* strain X2194 (O2: K(-): H6) is related to Pap pili but exhibits a truncated regulatory region. *Microbiology* **149:**1387–1397.

65. **Rasko DA, Sperandio V.** 2010. Anti-virulence strategies to combat bacteria-mediated disease. *Nat Rev Drug Discov* **9:**117–128.

66. **Pinkner JS, Remaut H, Buelens F, Miller E, Aberg V, Pemberton N, Hedenstrom M, Larsson A, Seed P, Waksman G, Hultgren SJ, Almqvist F.** 2006. Rationally designed small compounds inhibit pilus biogenesis in uropathogenic bacteria. *Proc Natl Acad Sci U S A* **103:**17897–17902.

67. **Lane MC, Lockatell V, Monterosso G, Lamphier D, Weinert J, Hebel JR, Johnson DE, Mobley HL.** 2005. Role of motility in the colonization of uropathogenic *Escherichia coli* in the urinary tract. *Infect Immun* **73:**7644–7656.

68. **Wright KJ, Seed PC, Hultgren SJ.** 2005. Uropathogenic *Escherichia coli* flagella aid in efficient urinary tract colonization. *Infect Immun* **73:**7657–7668.

69. **Pichon C, Héchard C, du Merle L, Chaudray C, Bonne I, Guadagnini S, Vandewalle A, Le Bouguénec C.** 2009. Uropathogenic *Escherichia coli* AL511 requires flagellum to enter renal collecting duct cells. *Cell Microbiol* **11:**616–628.

70. **Lane MC, Alteri CJ, Smith SN, Mobley HL.** 2007. Expression of flagella is coincident with uropathogenic *Escherichia coli* ascension to the upper urinary tract. *Proc Natl Acad Sci U S A* **104:**16669–16674.

71. **Lane MC, Lloyd AL, Markyvech TA, Hagan EC, Mobley HL.** 2006. Uropathogenic *Escherichia coli* strains generally lack functional Trg and Tap chemoreceptors found in the majority of *E. coli* strains strictly residing in the gut. *J Bacteriol* **188:**5618–5625.

72. **Walters MS, Lane MC, Vigil PD, Smith SN, Walk ST, Mobley HL.** 2012. Kinetics of uropathogenic *Escherichia coli* metapopulation movement during urinary tract infection. *mBio* **3:**e00303-11. doi:10.1128/mBio.00303-11

73. **Lane MC, Simms AN, Mobley HL.** 2007. Complex interplay between type 1 fimbrial expression and flagellum-mediated motility of uropathogenic *Escherichia coli*. *J Bacteriol* **189:**5523–5533.

74. **Simms AN, Mobley HL.** 2008. Multiple genes repress motility in uropathogenic *Escherichia coli* constitutively expressing type 1 fimbriae. *J Bacteriol* **190:**3747–3756.

75. **Li X, Rasko DA, Lockatell CV, Johnson DE, Mobley HL.** 2001. Repression of bacterial motility by a novel fimbrial gene product. *EMBO J* **20:**4854–4862.

76. **Reiss DJ, Mobley HL.** 2011. Determination of target sequence bound by PapX, repressor of bacterial motility, in *flhD* promoter using systematic evolution of ligands by exponential enrichment (SELEX) and high throughput sequencing. *J Biol Chem* **286:**44726–44738.

77. **Reiss DJ, Howard FM, Mobley HLT.** 2012. A novel approach for transcription factor analysis using SELEX with high-throughput sequencing (TFAST). *PLoS One* **7:**e42761. doi:1371/journal.pone.0042761

78. **Guyer DM, Henderson IR, Nataro JP, Mobley HL.** 2000. Identification of Sat, an autotransporter toxin produced by uropathogenic *Escherichia coli*. *Mol Microbiol* **38:**53–66.

79. **Rippere-Lampe KE, O'Brien AD, Conran R, Lockman HA.** 2001. Mutation of the gene encoding cytotoxic necrotizing factor type 1 (cnf-1) attenuates the virulence of uropathogenic *Escherichia coli*. *Infect Immun* **69:**3954–3964.

80. **Ruiz-Perez F, Wahid R, Faherty CS, Kolappaswamy K, Rodriguez L, Santiago A, Murphy E, Cross A, Sztein MB, Nataro JP.** 2011. Serine protease autotransporters from *Shigella flexneri* and pathogenic *Escherichia coli* target a broad range of leukocyte glycoproteins. *Proc Natl Acad Sci U S A* **108:**12881–12886.

81. **Dhakal BK, Mulvey MA.** 2012. The UPEC pore-forming toxin α-hemolysin triggers proteolysis of host proteins to disrupt cell adhesion, inflammatory, and survival pathways. *Cell Host Microbe* **11:**58–69.

82. **Nagy G, Altenhoefer A, Knapp O, Maier E, Dobrindt U, Blum-Oehler G, Benz R, Emody L, Hacker J.** 2006. Both alpha-haemolysin determinants contribute to full virulence of uropathogenic *Escherichia coli* strain 536. *Microbes Infect* **8:**2006–2012.

83. **Davis JM, Rasmussen SB, O'Brien AD.** 2005. Cytotoxic necrotizing factor type 1production by uropathogenic *Escherichia coli* modulates polymorphonuclear leukocyte function. *Infect Immun* **73:**5301–5310.

84. **Guyer DM, Radulovic S, Jones FE, Mobley HL.** 2002. Sat, the secreted autotransporter toxin of uropathogenic *Escherichia coli*, is a vacuolating cytotoxin for bladder and kidney epithelial cells. *Infect Immun* **70:**4539–4546.

85. **Maroncle NM, Sivick KE, Brady R, Stokes FE, Mobley HL.** 2006. Protease activity, secretion, cell entry, cytotoxicity, and cellular targets of secreted autotransporter toxin of uropathogenic *Escherichia coli*. *Infect Immun* **74:**6124–6134.

86. **Heimer SR, Rasko DA, Lockatell CV, Johnson DE, Mobley HL.** 2004. Autotransporter genes *pic* and *tsh* are associated with *Escherichia coli* strains that cause acute pyelonephritis and are expressed during urinary tract infection. *Infect Immun* **72:**593–597.

87. **Navarro-Garcia F, Gutierrez-Jimenez J, Garcia-Tovar C, Castro LA, Salazar-Gonzalez H, Cordova V.** 2010. Pic, an autotransporter protein secreted by different pathogens in the *Enterobacteriaceae* family, is a potent mucus secretagogue. *Infect Immun* **78:**4101–4109.

88. **Hadjifrangiskou M, Gu AP, Pinkner JS, Kostakioti M, Zhang EW, Greene SE, Hultgren SJ.** 2012. Transposon mutagenesis identifies uropathogenic *Escherichia coli* biofilm factors. *J Bacteriol* **194:**6195–6205.

89. **Kai-Larsen Y, Luthje P, Chromek M, Peters V, Wang X, Holm A, Kádas L, Hedlund KO, Johansson J, Chapman MR, Jacobson SH, Römling U, Agerberth B, Brauner A.** 2010. Uropathogenic *Escherichia coli* modulates immune responses and its curli fimbriae interact with the antimicrobial peptide LL-37. *PLoS Pathog* **6:**e1001010. doi:10.1371/journal.ppat.1001010

90. **Cegelski L, Pinkner JS, Hammer ND, Cusumano CK, Hung CS, Chorell E, Aberg V, Walker JN, Seed PC, Almqvist F, Chapman MR, Hultgren SJ.** 2009. Small-molecule inhibitors target *Escherichia coli* amyloid biogenesis and biofilm formation. *Nat Chem Biol* **5:**913–919.

91. **Ulett GC, Mabbett AN, Fung KC, Webb RI, Schembri MA.** 2007. The role of F9 fimbriae of uropathogenic *Escherichia coli* in biofilm formation. *Microbiology* **153:**2321–2331.

92. **Hancock V, Vejborg RM, Klemm P.** 2010. Functional genomics of probiotic *Escherichia coli* Nissle 1917 and 83972, and UPEC strain CFT073: comparison of transcriptomes, growth and biofilm formation. *Mol Genet Genomics* **284:**437–454.

93. **Hancock V, Seshasayee AS, Ussery DW, Luscombe NM, Klemm P.** 2008. Transcriptomics and adaptive genomics of the asymptomatic bacteriuria *Escherichia coli* strain 83972. *Mol Gen Genom* **279:**523–534.

94. **Beloin C, Michaelis K, Lindner K, Landini P, Hacker J, Ghigo JM, Dobrindt U.** 2006. The transcriptional antiterminator RfaH represses biofilm formation in *Escherichia coli*. *J Bacteriol* **188:**1316–1331.

95. **Partridge JD, Bodenmiller DM, Humphrys MS, Spiro S.** 2009. NsrR targets in the *Escherichia coli* genome: new insights into DNA sequence requirements for binding and a role for NsrR in the regulation of motility. *Mol Microbiol* **73:**680–694.

96. **Valle J, Da Re S, Henry N, Fontaine T, Balestrino D, Latour-Lambert P, Ghigo JM.** 2006. Broad-spectrum biofilm inhibition by a secreted bacterial polysaccharide. *Proc Natl Acad Sci U S A* **103:**12558–12563.

97. **Allsopp LP, Beloin C, Ulett GC, Valle J, Totsika M, Sherlock O, Ghigo JM, Schembri MA.** 2012. Molecular characterization of UpaB and UpaC, two new autotransporter proteins of uropathogenic *Escherichia coli* CFT073. *Infect Immun* **80:**321–332.

98. **Valle J, Mabbett AN, Ulett GC, Toledo-Arana A, Wecker K, Totsika M, Schembri MA, Ghigo JM, Beloin C.** 2008. UpaG, a new member of the trimeric autotransporter family of adhesins in uropathogenic *Escherichia coli*. *J Bacteriol* 190:4147–4161.

99. **Ulett GC, Valle J, Beloin C, Sherlock O, Ghigo JM, Schembri MA.** 2007. Functional analysis of antigen 43 in uropathogenic *Escherichia coli* reveals a role in long-term persistence in the urinary tract. *Infect Immun* 75:3233–3244.

100. **Anderson GG, Palermo JJ, Schilling JD, Roth R, Heuser J, Hultgren SJ.** 2003. Intracellular bacterial biofilm-like pods in urinary tract infections. *Science* 301:105–107.

101. **Whitfield C.** 2006. Biosynthesis and assembly of capsular polysaccharides in *Escherichia coli*. *Ann Rev Biochem* 75:39–68.

102. **Bahrani-Mougeot FK, Buckles EL, Lockatell CV, Hebel JR, Johnson DE, Tang CM, Donnenberg MS.** 2002. Type 1 fimbriae and extracellular polysaccharides are preeminent uropathogenic *Escherichia coli* virulence determinants in the murine urinary tract. *Mol Microbiol* 45:1079–1093.

103. **Buckles EL, Wang X, Lane MC, Lockatell CV, Johnson DE, Rasko DA, Mobley HL, Donnenberg MS.** 2009. Role of the K2 capsule in *Escherichia coli* urinary tract infection and serum resistance. *J Infect Dis* 199:1689–1697.

104. **Burns SM, Hull SI.** 1999. Loss of resistance to ingestion and phagocytic killing by O(-) and K(-) mutants of a uropathogenic *Escherichia coli* O75:K5 strain. *Infect Immun* 67:3757–3762.

105. **Anderson GG, Goller CC, Justice S, Hultgren SJ, Seed PC.** 2010. Polysaccharide capsule and sialic acid-mediated regulation promote biofilm-like intracellular bacterial communities during cystitis. *Infect Immun* 78:963–975.

106. **Alteri CJ, Lindner JR, Reiss DJ, Smith SN, Mobley HL.** 2011. The broadly conserved regulator PhoP links pathogen virulence and membrane potential in *Escherichia coli*. *Mol Microbiol* 82:145–163.

107. **Müller CM, Dobrindt U, Nagy G, Emödy L, Uhlin BE, Hacker J.** 2006. Role of histone-like proteins H-NS and StpA in expression of virulence determinants of uropathogenic *Escherichia coli*. *J Bacteriol* 188:5428–5438.

108. **Redford P, Roesch PL, Welch RA.** 2003. DegS is necessary for virulence and is among extraintestinal *Escherichia coli* genes induced in murine peritonitis. *Infect Immun* 71:3088–3096.

109. **Redford P, Welch RA.** 2006 Role of sigma E-regulated genes in *Escherichia coli* uropathogenesis. *Infect Immun* 74:4030–4038.

110. **Kulesus RR, Diaz-Perez K, Slechta ES, Eto DS, Mulvey MA.** 2008. Impact of the RNA chaperone Hfq on the fitness and virulence potential of uropathogenic *Escherichia coli*. *Infect Immun* 76:3019–3026.

111. **Kostakioti M, Hadjifrangiskou M, Pinkner JS, Hultgren SJ.** 2009. QseC-mediated dephosphorylation of QseB is required for expression of genes associated with virulence in uropathogenic *Escherichia coli*. *Mol Microbiol* 73:1020–1031.

112. **Subashchandrabose S, Hazen TH, Rasko DA, Mobley HL.** 2013. Draft genome sequences of five recent human uropathogenic *Escherichia coli* isolates. *Pathog Dis* 69:66–70.

113. **Brzuszkiewicz E, Brüggemann H, Liesegang H, Emmerth M, Olschläger T, Nagy G, Albermann K, Wagner C, Buchrieser C, Emody L, Gottschalk G, Hacker J, Dobrindt U.** 2006. How to become a uropathogen: comparative genomic analysis of extraintestinal pathogenic *Escherichia coli* strains. *Proc Natl Acad Sci U S A* 103:12879–12884.

114. **Lloyd AL, Rasko DA, Mobley HL.** 2007. Defining genomic islands and uropathogen-specific genes in uropathogenic *Escherichia coli*. *J Bacteriol* 189:3532–3546.

115. **Lloyd AL, Henderson TA, Vigil PD, Mobley HL.** 2009. Genomic islands ofuropathogenic *Escherichia coli* contribute to virulence. *J Bacteriol* 191:3469–3481.

116. **Vigil PD, Alteri CJ, Mobley HL.** 2011. Identification of in vivo-induced antigensincluding an RTX family exoprotein required for uropathogenic *Escherichia coli* virulence. *Infect Immun* 79:2335–2344.

117. **Vigil PD, Wiles TJ, Engstrom MD, Prasov L, Mulvey MA, Mobley HL.** 2012. The repeat-in-toxin family member TosA mediates adherence of uropathogenic *Escherichia coli* and survival during bacteremia. *Infect Immun* 80:493–505.

118. **Satchell KJ.** 2011. Structure and function of MARTX toxins and other large repetitive RTX proteins. *Annu Rev Microbiol* 65:71–90.

119. **Lloyd AL, Smith SN, Eaton KA, Mobley HL.** 2009. Uropathogenic *Escherichia coli* suppresses the host inflammatory response via pathogenicity island genes *sisA* and *sisB*. *Infect Immun* 77:5322–5333.

120. **Güell M, Yus E, Lluch-Senar M, Serrano L.** 2011. Bacterial transcriptomics: what is beyond the RNA horiz-ome? *Nat Rev Microbiol* 9:658–669.

121. **van Opijnen T, Camilli A.** 2013. Transposon insertion sequencing: a new tool for systems-level analysis of microorganisms. *Nat Rev Microbiol* 11:435–442.

122. **Phan MD, Peters KM, Sarkar S, Lukowski SW, Allsopp LP, Gomes Moriel D, Achard ME, Totsika M, Marshall VM, Upton M, Beatson SA, Schembri MA.** 2013. The serum resistome of a globally disseminated multidrug resistant uropathogenic *Escherichia coli* clone. *PLoS Genet* **9**:e1003834. doi:10.1371/journal .pgen.1003834

123. **Wiles TJ, Norton JP, Russell CW, Dalley BK, Fischer KF, Mulvey MA.** 2013. Combining quantitative genetic footprinting and trait enrichment analysis to identify fitness determinants of a bacterial pathogen. *PLoS Genet* **9**: e1003716. doi:10.1371/journal.pgen.1003716

124. **Chahales P, Thanassi DG.** 2017. Structure, function and assembly of adhesive organelles by uropathogenic bacteria, p 277–329. *In* Mulvey MA, Stapleton AE, Klumpp DJ (ed), *Urinary Tract Infections*, 2nd ed. ASM Press, Washington, DC.

Uropathogenic *Escherichia coli*-Associated Exotoxins

13

RODNEY A. WELCH[1]

INTRODUCTION

From as long ago as 1929, a study by MW Lyon indicated that the *Escherichia coli* strains isolated from the urine of infected, symptomatic individuals were different than the fecal isolates from healthy individuals (1). The *E. coli* urine isolates, when grown on blood agar plate medium, are commonly lytic to erythrocytes, whereas the fecal isolates are infrequently hemolytic. This was the first suggestion that the *E. coli* involved in extraintestinal diseases (ExPEC) are phenotypically different than the commensal strains common to the gastrointestinal tract. We now appreciate that the presence of no single virulence gene, including those encoding toxins, is the *sine qua non* of ExPEC disease potential. This fact puts these *E. coli* in a much different context for disease potential than other classic human pathogens, such as *Corynebacterium diphtheria*, *Vibrio cholerae*, and *Bordetella pertussis*, where specific toxin genes essentially define these bacteria as pathogens. I bring this issue up in the introduction to this chapter to make the point that, among the variety of toxins that uropathogenic *E. coli* (UPEC) and more broadly, ExPEC possess, there are no examples of "must-have" toxin genes. Why this is the case I will leave for

[1]Department of Medical Microbiology and Immunology, University of Wisconsin School of Medicine and Public Health, Madison, WI 53706.
Urinary Tract Infections: Molecular Pathogenesis and Clinical Management, 2nd Edition
Edited by Matthew A. Mulvey, David J. Klumpp, and Ann E. Stapleton
© 2017 American Society for Microbiology, Washington, DC
doi:10.1128/microbiolspec.UTI-0011-2012

the reader to ponder while reading this review. In my conclusion, I will provide some opinions about the matter. What will now follow are summaries of the research on several families of protein toxins associated with the pathogenesis of *E. coli* that cause urinary-tract infections. This will include the repeats-in-toxin (RTX) family, cytotoxic-necrotizing factor (CNF), and the auto-transporter (AT) family/type V-secretion family.

THE *E. COLI* HEMOLYSIN, THE PROTOTYPICAL MEMBER OF THE RTX FAMILY

Before DNA-DNA hybridization technology made genotypic-based epidemiological studies of virulence genes in ExPEC possible, investigators were limited to a handful of phenotypic tests. As mentioned above, complete lysis of erythrocytes for bacterial colonies growing on red blood-agar plates is a simple test. This phenotype ultimately proved to be a strong indicator of ExPEC strains that are more commonly associated with severe upper urinary-tract infections, pyelonephritis, and bloodstream invasion (urosepsis) (2–5). Initially, investigators separated these hemolytic *E. coli* into two, somewhat confusing, categories. The alpha-hemolysin-producing *E. coli* expressed a true extracellular, filterable hemolytic activity, whereas the *E. coli* with a beta-hemolysin produced a hemolytic activity that was cell-bound and not found free in filtered culture supernatants (6). Until an example of each determinant was cloned and tested by DNA-DNA hybridization, it was unclear if the two hemolysins were different or similar. As it turns out, they are encoded by a set of homologous genes organized in a four-gene operon, *hlyCABD* (7, 8). The difference in extracellular activities reflects several factors. The *E. coli* hemolysin activity is relatively labile, where aggregation leads to rapid loss of activity in even freshly prepared supernatants kept on ice. In addition, there

is a curious variability in the promoter sequences among hemolysin determinants found in *E. coli* strains (8, 9). Among the different hemolysin determinants, there are non-homologous sequences immediately upstream of the start codon for the *hlyC* gene. In only a handful of instances have the 5′ ends of the *hlyCABD* transcript been identified (8). What is clear is that this evolutionary event has led to significant differences in the levels of *hlyCABD* transcription and subsequent expression of hemolysin activity. The classic alpha-hemolysin producers possess stronger promoters, whereas strains characterized as beta-hemolysin producers have weaker promoters. So, when investigators tried to assess extracellular hemolytic activity in the beta-hemolysin isolates, the weak expression and lability of hemolytic activity lead them to the erroneous conclusion that the hemolytic activity was cell-bound. Therefore, I have been a proponent of dropping the antiquated alpha- versus beta-hemolysin nomenclature and simply refer to all *hlyCABD*-encoded activities as the *E. coli* hemolysin.

EXPERIMENTAL EVIDENCE THAT THE *E. COLI* HEMOLYSIN IS A VIRULENCE FACTOR FOR UPEC

One of the first applications of molecular Koch's postulates was with the demonstration that the *E. coli* hemolysin is a virulence factor in a rat model of intra-abdominal peritonitis (10, 11). A recombinant hemolysin plasmid, pSF4000, was transformed into an avirulent, non-hemolytic, normal human-fecal *E. coli* strain, J198. When this construct was injected intraperitoneally into rats, the result was a 1,000-fold decrease in the lethal dose 50% ($LD_{50\%}$) compared to J198 alone. When a nonhemolytic transposon-insertion mutant plasmid, pSF4000::Tn1, was transformed into J198 and used to challenge rats, there was an increase in the $LD_{50\%}$ back to the level of J198 alone. These results have essentially stood as the primary evidence that the *E. coli* hemolysin acts as a significant virulence factor for ExPEC. A subsequent publication that lent support to this showed that the differences in hemolysin expression that result from different promoters is directly related to the relative levels of virulence in the peritonitis model (9).

Despite the compelling epidemiological evidence that implicates the *E. coli* hemolysin is a significant virulence factor for pyelonephritis and urosepsis isolates and our *in vivo* peritonitis-model results, the application of molecular Koch's postulates to demonstrate that the hemolysin is a virulence factor for UPEC is not entirely persuasive. *In vitro*, the *E. coli* hemolysin is clearly cytotoxic to primary cultured renal cells (12, 13). O'Hanley et al. showed that the hemolysin contributed to renal parenchymal damage in a murine model of urinary-tract infection (UTI), but it did not influence colonization levels in the kidney (14). Interestingly, that group also showed, using the mouse UTI model, that immunization with denatured hemolysin provided significant protection against the hemolysin-mediated damage in the mouse kidney (14). There are some technical issues with this study that should be pointed out. The UTI model involved dehydration of the

mice, with the animals denied water for 24 hours prior to bacterial inoculation. Second, the immunization protocol involved complete and incomplete Freund's adjuvant and repeated intramuscular injections. Lastly, there was elevated hemolysin expression in the *E. coli* challenge strains compared to natural hemolytic UPEC strains. The investigators took *E. coli* pyelonephritis isolates that were non-hemolytic and transformed them with a medium-copy number, recombinant plasmid, pSF4000. As was the case with my earlier peritonitis studies, exaggerated expression of hemolysin clearly can lead to *in vivo* pathology and morbidity, but what role does hemolysin play in uropathogenesis when expressed at its native levels? A recent paper from Alison O'Brien's laboratory demonstrated that mice immunized with a hemolysin toxoid had 10-fold reduced bacterial numbers in the urine when compared to sham-immunized mice in the UTI model (75). The hemolysin-immunized mice also had reduced evidence of bladder inflammation compared to the control mice (93). This report represents some of the best evidence to date that hemolysin plays a significant role in causing inflammation and a selective advantage to the UPEC strains that express this toxin (Table 1).

TABLE 1 Prominent protein toxins of uropathogenic *Escherichia coli*

Toxin	Exoprotein family	Action	General epidemiology
Hemolysin (HlyA)	Type I	Membrane pore-formation	Common to UPEC strains that cause pyelonephritis and sepsis, less common to cystitis isolates, infrequent in fecal isolates
Cytotoxic necrotizing factor-1 (CNF-1)	Genetically linked to Type I hemolysin operon	Deamidation of GTPases, RhoA, Cdc42 and Rac	~30-40% of pyelonephritis UPEC isolates, rare in fecal isolates
Secreted autotransporter toxin (Sat)	Type V	Serine protease that targets cytoskeletal proteins	Common to UPEC strains that cause pyelonephritis and sepsis, less common to cystitis isolates, infrequent in fecal isolates
Protein involved in intestinal colonization (PicU)	Type V	Serine protease that targets O-linked glycoproteins	~50% in cystitis and pyelonephritis UPEC isolates
Vacuolating autotransporter toxin (Vat)	Type V	Serine protease that causes vacuole formation in cultured bladder cells	~50% in cystitis and pyelonephritis UPEC isolates

A report by Nagy and coworkers described results with UPEC-model strain 536 in several animal models where deletion mutations of the two chromosomally encoded *hlyA* genes resulted in 536-virulence attenuation (15). The authors concluded that their experiments demonstrate that the hemolysin is a virulence factor in an ascending model of urinary-tract infection. The model they employed involved intravesical inoculations of 3–4 day-old suckling mice and the mice were not separated as to sex. Intravesical inoculation of male adult mice is difficult to do without causing urethral bleeding. Second, bacterial-colonization levels were not the end-point of their analysis, rather, death of the pups was used. Thus, this model did not measure bladder or kidney colonization or histopathology, which are the end points commonly used in the adult murine-UTI model.

In the late 1990s, Shai Pellett, Rachel Quinn, and I constructed a site-directed hemolysin knock-out mutant in the model UPEC strain J96. We tested *E. coli* J96 and J96*hlyA* in single infections in the broadly employed adult female mouse-UTI model. The results went unpublished because we observed no statistical difference in the number of colony-forming units for the two strains in either the bladder or kidneys at 48 hours post-inoculation. However, we did publish later that a hyper-colonization mutant of UPEC-model strain CFT073, CFT073*dsdA*, lost its hyper-colonization ability with mutational loss of *hlyA* in bladders and kidneys at 48 hours post-inoculation (16). Although this result suggests that the hemolysin can aid urinary-tract colonization, the hyper-colonization mutant is certainly not a natural UPEC isolate. Alison O'Brien's laboratory tested an isogenic hemolysin-mutant pair with UPEC-model strain CP9 in the murine UTI model (17). These authors performed both single and competitive infections of mice, with neither inoculation strategy revealing an impact on colonization levels in either the bladder or kidneys at three differ-ent time points; 1, 2, and 5 days post-inoculation. This makes it quite clear that hemolysin expression does not influence UPEC-colonization numbers in the mouse urinary tract. However, they were able to make an important histopathological observation. They saw that the bladders taken from CP9-infected mice 24 hours post-inoculation are more grossly inflamed and hemorrhagic than the bladders infected with the CP9*hlyA*-mutant strain. This inflammation effect wanes at later time points in the infection. They also observed that the hemolysin was responsible for extensive sloughing of the urothelium in the bladder. The sloughing of the bladder-urothelial cells during UTI is a common observation in this model and it is fundamentally similar to what is observed during human infections (18). A study that compares the hemolysin status of *E. coli* causing a clinical infection and the amount of urothelial cell sloughing and damage is needed to support this role in human infection.

E. COLI HEMOLYSIN: STRUCTURE, FUNCTION, AND REGULATION

Previously, my colleagues and I have written reviews on the structure and mechanisms of cytotoxic activity of the *E. coli* hemolysin and I suggest consulting those publications for more specific details (19–21). For the purpose of this chapter, I will provide a brief synopsis and a few new facts that have been published in the past several years. As mentioned above, the hemolysin is produced through the direct action of four genes encoded on an operon, *hlyCABD*. The actual secreted-protein product, HlyA, is 110 kilodaltons in size and is encoded by *hlyA* (7). HlyA is post-transcriptionally modified heterogeneously at two lysine residues with 14-, 15-, and 17-carbon fatty acids through the enzymatic activity of *hlyC*. The acylation is required for hemolytic and cytotoxic activities, but the modification is not required for extracellular secretion. The immature, unacylated, and mature forms of the hemolysin are respectively referred to as proHlyA and HlyA. The *hlyB* and *hlyD* genes encode proteins that are involved in the secretion of the cytoplasmically modified HlyA polypeptide across the inner and outer membranes without amino-terminal cleavage of the polypeptide (22). These proteins form a type I secretion complex together with an unlinked-gene product, TolC (23). The *E. coli* hemolysin is generally considered to be the prototypical type I secretion protein in gram-negative bacteria. The family members share a motif of nine amino-acid repeats near their C-terminal end that are involved in Ca2+-binding, and it is this structural motif that led to the name of the repeats-in-toxin, RTX family (24). These proteins are secreted without aid of the general leader-peptide-dependent secretory pathway and the most C-terminal amino acids are involved in targeting these proteins for export by the HlyB-HlyD-TolC envelope complex (25, 26).

The RTX family of toxins produced by different gram-negative pathogens has members with narrow host and target-cell specificities. For example, the nonhemolytic *Mannheimia haemolytica* leukotoxin is cytotoxic to bovine and ovine leukocytes, but not epithelial, and endothelial cells (27). Unlike the homologous RTX leukotoxins, the *E. coli* hemolysin appears to be cytotoxic to many different host-cell types and towards many species of hosts (see review 21). The differences in target and host-cell specificity appear to be due to primary-sequence and N-linked oligosaccharide-modification differences in the primary receptors for the toxins, β2 integrins, on immune cells (27–29).

The mechanism for *E. coli* hemolysin cytotoxicity is commonly believed to involve the formation of membrane pores. When erythrocytes are used as a target cell, loss of intracellular K+ ions to the environment and subsequent influx of cations and water leads to osmotic lysis (30, 31). The use of extracellular osmotic protectants of defined size suggests a hemolysin-mediated membrane pore that is 2 nanometers in diameter (32). Mahtab Maoyeri and I published observations that suggested that, although initial incubations of toxin with osmotic protectants and erythrocytes indicates a 2 nm pore, there is a hemolysin concentration and time-dependent change in the apparent size of the membrane pores (33). Helle Praetorius's laboratory recently published some elegant work that indicates there is a bi-phasic hemolytic process by the *E. coli* hemolysin (30, 34). She and her colleagues found that the initial pore formation leads to partial lysis of erythrocytes, but that the pore formation triggers purinergic-receptor activation and creation of a host-mediated pannexin pore. These events then lead to full lysis of the hemolysin-treated cells. The initial pore does not lead to osmotic swelling of the erythrocyte, rather there is a delay until the secondary-activation events occur when the erythrocytes then swell and lyse.

There are numerous studies on the *in vitro* effects of the *E. coli* hemolysin on host cells and it is difficult to assess which of these are relevant to the pathogenesis of UPEC. It seems unlikely that lysis of erythrocytes is significant, although the canonical hypothesis is that release of iron-containing heme molecules is an important role for bacterial hemolysins in disease pathogenesis. As for the *E. coli* hemolysin activity against other host-cell types, there are several variables that confound interpretations and conclusions about its role in uropathogenesis. The most significant variable is the presence of lipopolysaccharide (LPS) in *E. coli* hemolysin preparations. In general, at low, sub-lytic, or apoptotic-inducing concentrations, the hemolysin triggers pro-inflammatory events (35–37). The critical issue to be resolved is to what degree are these proinflammatory events dependent on LPS present in hemolysin preparations? Mansson et al. showed that the classic Ca2+ signaling seen in many different cell types upon exposure to hemolysin is dependent on the presence of LPS co-purified with hemolysin protein HlyA

(38). In addition, the LPS present mediates the delivery of hemolysin via the classic LPS-CD14 pathway and not the pattern-recognition molecule, TLR4, commonly stimulated by LPS (38). Although it may be entirely coincidental, it is intriguing that the hemolytic activity of the hemolysin is profoundly affected by the structure of LPS (39, 40). There is decreasing activity and loss of expression of the extracellular HlyA in strain backgrounds where there are successive truncations of the LPS to make progressively rougher strains. Associated with the loss of activity is an increase in the apparent size of the hemolysin. The conclusion drawn is that, during or after secretion of HlyA, the polypeptide becomes associated with LPS. The rough forms of LPS become more hydrophobic with the loss of the O-antigen and outer-core constituents. These species then more readily aggregate with HlyA into hemolytic- or cytotoxic-inactive forms. Shai Pellett and I find that, in order to prepare endotoxin-free hemolysin as measured by the limulus assay, HlyA has to be denatured by boiling with 1% sodium dodecyl sulfate (SDS), and separated from LPS by SDS-gel electrophoresis. HlyA is then eluted from excised-gel fragments and dialyzed, before then being dissolved in 0.9% NaCl containing 1 mM CaCl2 (38). In this purified form, HlyA hemolytic and cytotoxic activity is stable for months at 4°C, in contrast with HlyA-containing material present in filtered bacterial-culture supernatants. Extracellular HlyA in its native, *in vivo* state is probably never free of LPS and experiments to assess HlyA events independent of LPS are unrealistic and likely do not reflect the *in vivo* situation. The caveat to this supposition is the possibility that, *in vivo*, HlyA effects on the host are not the result of extracellular HlyA that comes in contact with host cells. Rather, it is easy to envision that HlyA is delivered upon close cell-to-cell contact between hemolytic *E. coli* and host cells. Aside from the flagellar apparatus common to nearly all UPEC strains, genomic-sequence analysis indicates that there are no type III secretion systems present in UPEC strains. However, common to the pathogenicity islands encoding the *hlyCABD* genes, are the pyelonephritis-associated pili (pap), and this close linkage remains suspicious. The co-expression of hemolysin and pap fimbriae that mediate host-cell adhesion could provide a means to deliver the HlyA to host cells with little or no extracellular exposure.

A second experimental variable when considering the significance of the hemolysin in uropathogenesis is the hemolysin concentrations employed in the different studies. As mentioned above, at sub-lytic or sub-necrotic concentrations, the hemolysin can influence different host-signal pathways and *in vitro* appears to be pro-inflammatory in nature. It is also clear, based on the *in vivo* studies mentioned above, that the hemolysin can cause cell destruction and inflammation in the bladder or kidney. In contrast to those reports, Wiles et al. recently showed that the hemolysin inhibits activation of host-cell Akt (protein kinase B) and may inhibit inflammation (41). Hemolysin causes uncontrolled activation of host-protein phosphatases and proteases (41, 42). The negative affect on Akt activation was proposed to lead to a reduction in NFκB. These investigators compared Akt-activation states in 5,637 human bladder-epithelial cells that were challenged with either UPEC-model strain UTI89 or an isogenic UTI89*hlyA* mutant. The inhibition of Akt activation by the *E. coli* hemolysin was not dependent on fluxes in intracellular Ca2+ nor extracellular K+ leakage, events commonly attributed as the very initial events mediated by HlyA upon membrane insertion. In this study, two other pore-former toxins, the *Staphylococcus aureus* α-toxin and the *Aeromonas hydrophilia* aerolysin, also caused inhibition of Akt activation. The authors speculate that the osmotic-stress created by three different pore-former toxins leads to the identical event (41). Dhakal and Mulvey recently showed with cultured bladder-epithelial cells or mouse peritoneal macrophages that the *E. coli* hemolysin activates host-cell serine proteases such as mesotrypsin. This results in degradation of the cytoskeletal-scaffolding protein paxillin, components of the pro-inflammatory NFκB-signaling cascade and caspase 3/7 (42). There is a significant reduction in the production of the inflammatory cytokine, interleukin (IL)-6, attributed to hemolysin intoxication. The induction of the proteolytic cascade does not occur with treatment of the cells by two other pore-forming toxins, the *S. aureus* α-toxin or the *A. hydrophilia* aerolysin. Dhakal and Mulvey propose that the hemolysin-inhibition of cellular inflammation helps explain that UPEC cause UTIs because they suppress inflammation when in the urinary tract. This is an interesting hypothesis supported with results in several laboratories (43–45).

Lastly, the significance of the hemolysin to uropathogenesis is indirectly supported by the observation that expression of critical virulence determinants for UPEC, O-antigens, type II capsules, and the ChuA hemin-utilization factor are positively co-regulated with the hemolysin operon by the specialized virulence NusG-like transcriptional anti-terminator, RfaH (46–50). As expected, a mutation in *rfaH* renders UPEC highly attenuated in rodent models of disease (51). Interestingly, it has been recently shown that RfaH represses biofilm formation (52), a trait often touted to be important for uropathogenesis (53, 54).

A SECOND CHROMOSOMAL RTX FAMILY MEMBER, UPXA IN UPEC

A second RTX-like gene, *upxA*, and linked type I transport genes are apparent in the chromosome of classic UPEC-model strain CFT073 (55). This gene was originally annotated as *upxA*, with this designation standing for uropathogen RTXA gene, which is consistent with the general pattern of including X in the names of RTX genes. This nomenclature was disregarded by Parham et al. in 2005 when they renamed *upxA*, *tosA*, which they designated for type one secretion gene A (56). Recently, Vigil et al. have published interesting results that indicate *upxA* (*tosA*) provides a critical adherence function for CFT073 in the kidney in the murine model of urinary-tract infection (57, 58). In addition, UpxA (TosA) promoted CFT073 survival during experimental murine sepsis and could serve as a protective antigen during murine urosepsis (58). Unfortunately, these authors utilized the *tosA* nomenclature (58).

CYTOTOXIC NECROTIZING FACTOR TYPE 1 (CNF1)

CNF1 was originally described as a toxin that caused dermonecrosis when CNF-producing *E. coli* strains were injected intradermally in rabbits. Its first described cytotoxic activity was the formation of multi-nucleated cells in culture (59). Caparioli et al. found that 40% of UTI *E. coli* isolates produce CNF1, whereas only 1% of the normal fecal isolates produced the toxin (60). Epidemiologically, like the *E. coli* hemolysin, CNF1 production is more often associated with UPEC strains responsible for more severe UTIs; Andreu et al. showed that 48% of the pyelonephritis isolate were *cnf1* positive (61). The curious and enigmatic observation is that when *cnf1* occurs in UPEC strains, it is always linked to the *hlyCABD* operon (62). It occurs 3′ to that operon and is co-transcribed with the *hlyCABD* genes and positively regulated by RfaH (62). The converse is not true, however, where UPEC strains can be commonly found with chromosomally encoded *hlyCABD* operons without *cnf1* being present (55).

CNF1 is a 115-kDa protein that catalyzes the deamidation of a conserved-glutamine residue in three members of the Rho family of GTP-binding proteins. This leads to the activation of three specific GTPases; RhoA, Cdc42, and Rac (63, 64). The CNF1 protein is divided into three functional domains (65).

The enzymatic C-terminal half of the CNF1 polypeptide is present in the host cytosol after cleavage and release from late endosomes (66). The very N-terminal CNF1 domain appears to be responsible for binding to a laminin-receptor-precursor protein (67, 68). An internal CNF1 domain is responsible for translocation across the endosomal membrane (69). The activation of these regulators leads to cytoskeleton rearrangements, cell-cycle disruption, and interruption of host-cell signaling pathways (70). Hofman and coworkers found that CNF1 treatment of neutrophils led to increased production of reactive-oxygen molecules, but a decrease in the ability of those cells to phagocytize bacteria (71). Later, Davis et al. showed that CNF1 synthesis leads to increased survival of UPEC in association with isolated human neutrophils (72). CNF1 treatment of cultured epithelial cells leads to uptake of noninvasive bacteria and latex particles (73). There is also a significant increase in *in vitro* Rac1-dependent invasion of epithelial cells by UPEC strains expressing CNF1 (74, 75).

As is the case for hemolysin, the role of CNF1 in uropathogenesis is not entirely clear, despite the evidence that it induces neutrophil dysfunction and increased epithelial-cell invasion. Johnson et al. reported in 2000 that an *E. coli* cystitis isolate, F11, was not attenuated in the mouse UTI model in terms of bladder and kidney colonization at either 2 or 7 days post-inoculation upon the mutational loss of *cnf1* (76). There also were no changes in the relative amount of inflammation of the infected bladders or kidneys between single challenges of F11 and F11*cnf1*. These results are in contrast to a report a year later where, in a competitive murine bladder-colonization model, the UPEC-model pyelonephritis strain CP9 outcompeted a CP*cnf1* mutant (77). However, Alison O'Brien's laboratory did show, with the same strains in single-strain, murine-bladder infections, that there is no statistical difference in the mean number of colony-forming units (CFUs) of the wild-type strain and the *cnf1* mutant at one

day post-inoculation (17). A CP9*hlyA cnf1* mutant was also tested in the single-challenge murine-UTI model and again there were no statistically significant differences in the mean of CFUs from bladders or kidneys for this mutant and the wild-type. In this same study, it was demonstrated that CNF1 does cause bladder inflammation and edema at 3 and 5 days post-inoculation (17). However, the decrease in bladder inflammation attributable to CNF1 does not result in greater colonization ability of a CP9 *cnf1* mutant. Real et al. described a study where they monitored the levels of inflammatory mediators, IL-8, MCP-1, and MIP3a, in the urine of patients with UTIs (78). They found that the patients with the strongest inflammatory responses and with high red blood cell counts in their urine were disproportionately infected with UPEC strains possessing the genes for *hlyA* and *cnf1*. A recent publication by Boyer et al. suggests that, in some contexts, CNF1 acts as an avirulence factor (79). They show, using a *Drosophila melanogaster* traumatic-wound model, that the presence of CNF1 in UPEC strain J96 induces a protective immune response and that there is an increased infectious burden in the flies when challenged with a J96*cnf1* mutant. They demonstrate that CNF1 activates Rac2, this activates innate-immune adaptors, IMD and Rip1-Rip2, and ultimately leads to increased production of protective antimicrobial peptides.

A recent, more detailed review of the enzymatic mechanism and *in vitro* and *in vivo* effects of CNF1 is available (80). Interestingly, in this review, there is discussion of how CNF1 is being extensively studied as a pharmacologic tool to control pain. CNF1 apparently holds some promise as another bacterial toxin that may have practical clinical benefit in some settings.

TOXINS IN TYPE V SECRETION FAMILY

Based on genomic DNA-sequence analysis of model UPEC strains, there are 10 or more different genes of the Type V Secretion family (T5S) present in the chromosome (81). This family is also often referred to as the autotransporter (AT) family. In UPEC model strain CFT073, there are 10 such genes (82). The function of many of these remain to be elucidated, but those that have been characterized fall into two broad functional categories, serine proteases (the SPATE AT subfamily) and adhesins such as antigen 43 (Ag43) (83) and UpaB (82). The SPATE family members are further divided into two classes, 1 and 2, which respectively represent toxic vs nontoxic proteases (84). These extracellular serine proteases are produced by a variety of *E. coli* pathovars besides UPEC.

The UPEC-toxin SPATE-family member that has received the greatest attention is the secreted autotransporter toxin (Sat). The *sat* gene is present, along with the hemolysin and pap pili determinants, within the large *pheV –* associated pathogenicity island of UPEC-model strain CFT073. The *sat* gene encodes a 142 kDa protein that possesses the three characteristic domains of SPATE proteins (85). ATs possess a long N-terminal signal sequence, a secreted-passenger domain, and canonical C-terminal autotransporter domain that provides the cell-envelope-export channel. In the case of Sat, the mature extracellular polypeptide is 107 kDa in size (86). The *sat* gene is present in 68% of *E. coli* strains isolated from pyelonephritis, but occurs in only 14% of normal fecal *E. coli* isolates (86). Guyer et al. demonstrated that Sat possessed potent cytopathic effects to cultured cells and, in the murine model of UTI, histopathological lesions in the kidney can be attributed to Sat (87). However, in single-strain, challenge experiments, a CFT073*sat* mutant did not have a statistically significant reduction in bacterial load in either the bladder or kidney compared to the parent strain CFT073. Mononcle et al. demonstrated, through construction of active mutants, that Sat proteolytic activity was responsible for cytotoxic activity, cytoskeleton rearrangements, and proteolysis of host proteins (85).

Recently, Moal et al. showed in cultured HeLa cells that the Sat proteolytic activity initiates disorganization of F-actin that results in detachment of the cell monolayers and loosening of cell-to-cell junctions (88). These events then induce host-cell autophagy. These authors speculate that Sat may contribute to the exfoliation of the urothelial cells, an event commonly seen during model UTIs.

There are two other SPATE family members commonly found in UPEC, Pic and Vat, which were first described respectively in *Shigella flexneri* and avian pathogenic *E. coli* (89, 90). In UPEC model strain CFT073, the genes for these two ATs are located respectively at the *aspV-* and *thrW*-associated pathogenicity islands. Pic possesses mucinase activity towards O-linked glycoproteins, such as CD43, CD45, and fractalkine commonly found on neutrophil surfaces (91). Pic treatment of neutrophils *in vitro* results in chemotaxis and transmigration defects. The Pic-treated neutrophils are also stimulated to produce an oxidative burst (91). Lloyd et al. demonstrated that a CFT073*pic* mutant colonizes the mouse bladder ∼ 50-fold less than wild-type CFT073 (92). Vat is vacuolating AT toxin that occurs in more than half of *E. coli* cystitis and pyelonephritis isolates (81). There are no reports of the significance of Vat in UPEC *in vitro* or *in vivo* model systems. It has been demonstrated to be a significant virulence factor in avian-pathogenic *E. coli* using respiratory- and cellulitis-infection models of disease in broiler chickens (90).

SUMMARY

A wealth of epidemiological evidence supports that the UPEC toxins reviewed in this chapter are virulence factors. It is clear that UPEC exotoxins cause dramatic *in vitro* effects to primary and cultured host cells relevant to human UTIs. It does remain a conundrum why, in the murine UTI model, do we see that UPEC *hlyA*, *cnf1*, or *sat* mutants cause less inflammation without a subsequent increase in the number of mutant bacteria recovered in the infected tissues? The pervasive rationale for this observation is that UPEC strains evolved redundant in exotoxin activities that cumulatively impede host-immune responses. There is certainly precedence for redundancy of virulence factors in UPEC as seen adhesins and systems for iron acquisition. I contend that another rationale should be considered, the murine model of UTI simply lacks sufficient functional similarity to human urinary-tract infection. The mouse model has aided in our understanding of UPEC colonization and growth in the urinary tract but, in terms of the interplay of toxins, host cells, and inflammation, the mouse model fails us in its present form.

ACKNOWLEDGMENTS

The author would like to thank Eric Battaglioli, Andrew Hryckowian, and Jay Lemke for their critical reading of this manuscript. UPEC research in my laboratory was formerly supported by NIH grant RO1 DK063250 and currently by an endowment for the Robert Turell Professorship in Infectious Diseases awarded by the University of Wisconsin School of Medicine and Public Health.

Conflicts of interest: I declare no conflict of interest.

CITATION

Welch RA. 2016. Uropathogenic *Escherichia coli*-associated exotoxins. Microbiol Spectrum 4(3):UTI-0011-2012.

REFERENCES

1. **Lyon MW.** 1917. A case of cystitis caused by *Bacillus coli-hemolyticus*. *JAMA* **69**:353–358.
2. **Brooks HJ, O'Grady F, McSherry MA, Cattell WR.** 1980. Uropathogenic properties of *Escherichia coli* in recurrent urinary-tract infection. *J Med Microbiol* **13**:57–68.

3. **Arthur M, Johnson CE, Rubin RH, Arbeit RD, Campanelli C, Kim C, Steinbach S, Agarwal M, Wilkinson R, Goldstein R.** 1989. Molecular epidemiology of adhesin and hemolysin virulence factors among uropathogenic *Escherichia coli*. *Infect Immun* **57**:303–313.
4. **O'Hanley P, Low D, Romero I, Lark D, Vosti K, Falkow S, Schoolnik G.** 1985. Gal-Gal binding and hemolysin phenotypes and genotypes associated with uropathogenic *Escherichia coli*. *N Engl J Med* **313**:414–420.
5. **Ulleryd P, Lincoln K, Scheutz F, Sandberg T.** 1994. Virulence characteristics of *Escherichia coli* in relation to host response in men with symptomatic urinary tract infection. *Clin Infect Dis* **18**:579–584.
6. **Cavalieri SJ, Bohach GA, Snyder IS.** 1984. *Escherichia coli* alpha-hemolysin: characteristics and probable role in pathogenicity. *Microbiol Rev* **48**:326–343.
7. **Felmlee T, Pellett S, Welch RA.** 1985. Nucleotide sequence of an *Escherichia coli* chromosomal hemolysin. *J Bacteriol* **163**:94–105.
8. **Welch RA, Pellett S.** 1988. Transcriptional organization of the *Escherichia coli* hemolysin genes. *J Bacteriol* **170**:1622–1630.
9. **Welch RA, Falkow S.** 1984. Characterization of *Escherichia coli* hemolysins conferring quantitative differences in virulence. *Infect Immun* **43**:156–160.
10. **Welch RA, Dellinger EP, Minshew B, Falkow S.** 1981. Haemolysin contributes to virulence of extra-intestinal *E. coli* infections. *Nature* **294**:665–667.
11. **Falkow S.** 1988. Molecular Koch's postulates applied to microbial pathogenicity. *Rev Infect Dis* **10**(Suppl 2):S274–276.
12. **Keane WF, Welch R, Gekker G, Peterson PK.** 1987. Mechanism of *Escherichia coli* alpha-hemolysin-induced injury to isolated renal tubular cells. *Am J Pathol* **126**:350–357.
13. **Mobley HL, Green DM, Trifillis AL, Johnson DE, Chippendale GR, Lockatell CV, Jones BD, Warren JW.** 1990. Pyelonephritogenic *Escherichia coli* and killing of cultured human renal proximal tubular epithelial cells: role of hemolysin in some strains. *Infect Immun* **58**:1281–1289.
14. **O'Hanley P, Lalonde G, Ji G.** 1991. Alpha-hemolysin contributes to the pathogenicity of piliated digalactoside-binding *Escherichia coli* in the kidney: efficacy of an alpha-hemolysin vaccine in preventing renal injury in the BALB/c mouse model of pyelonephritis. *Infect Immun* **59**:1153–1161.
15. **Nagy G, Altenhoefer A, Knapp O, Maier E, Dobrindt U, Blum-Oehler G, Benz R, Emody**

L, Hacker J. 2006. Both alpha-haemolysin determinants contribute to full virulence of uropathogenic *Escherichia coli* strain 536. *Microbes Infect* **8:**2006–2012.

16. Haugen BJ, Pellett S, Redford P, Hamilton HL, Roesch PL, Welch RA. 2007. *In vivo* gene expression analysis identifies genes required for enhanced colonization of the mouse urinary tract by uropathogenic *Escherichia coli* strain CFT073 dsdA. *Infect Immun* **75:**278–289.

17. Smith YC, Rasmussen SB, Grande KK, Conran RM, O'Brien AD. 2008. Hemolysin of uropathogenic *Escherichia coli* evokes extensive shedding of the uroepithelium and hemorrhage in bladder tissue within the first 24 hours after intraurethral inoculation of mice. *Infect Immun* **76:**2978–2990.

18. Elliott TS, Reed L, Slack RC, Bishop MC. 1985. Bacteriology and ultrastructure of the bladder in patients with urinary tract infections. *J Infect* **11:**191–199.

19. Welch RA, Forestier C, Lobo A, Pellett S, Thomas W Jr, Rowe G. 1992. The synthesis and function of the *Escherichia coli* hemolysin and related RTX exotoxins. *FEMS Microbiol Immunol* **5:**29–36.

20. Welch RA, Bauer ME, Kent AD, Leeds JA, Moayeri M, Regassa LB, Swenson DL. 1995. Battling against host phagocytes: The wherefore of the RTX family of toxins? *Infect Agents Dis* **4:**254–272.

21. Welch RA. 2001. RTX toxin structure and function: A story of numerous anomalies and few analogies in toxin biology. *Curr Top Microbiol Immunol* **257:**85–111.

22. Felmlee T, Pellett S, Lee EY, Welch RA. 1985. *Escherichia coli* hemolysin is released extracellularly without cleavage of a signal peptide. *J Bacteriol* **163:**88–93.

23. Wandersman C, Delepelaire P. 1990. TolC, an *Escherichia coli* outer membrane protein required for hemolysin secretion. *Proc Natl Acad Sci U S A* **87:**4776–4780.

24. Welch RA. 1991. Pore-forming cytolysins of gram-negative bacteria. *Mol Microbiol* **5:**521–528.

25. Gray L, Baker K, Kenny B, Mackman N, Haigh R, Holland IB. 1989. A novel C-terminal signal sequence targets *Escherichia coli* haemolysin directly to the medium. *J Cell Sci Suppl* **11:**45–57.

26. Koronakis V, Koronakis E, Hughes C. 1989. Isolation and analysis of the C-terminal signal directing export of *Escherichia coli* hemolysin protein across both bacterial membranes. *EMBO J* **8:**595–605.

27. Shanthalingam S, Srikumaran S. 2009. Intact signal peptide of CD18, the beta-subunit of

beta2-integrins, renders ruminants susceptible to *Mannheimia haemolytica* leukotoxin. *Proc Natl Acad Sci U S A* **106:**15448–15453.

28. Morova J, Osicka R, Masin J, Sebo P. 2008. RTX cytotoxins recognize beta2 integrin receptors through N-linked oligosaccharides. *Proc Natl Acad Sci U S A* **105:**5355–5360.

29. Lally ET, Kieba IR, Sato A, Green CL, Rosenbloom J, Korostoff J, Wang JF, Shenker BJ, Ortlepp S, Robinson MK, Billings PC. 1997. RTX toxins recognize a beta2 integrin on the surface of human target cells. *J Biol Chem* **272:**30463–30469.

30. Skals M, Jensen UB, Ousingsawat J, Kunzelmann K, Leipziger J, Praetorius HA. 2010. *Escherichia coli* alpha-hemolysin triggers shrinkage of erythrocytes via K(Ca)3.1 and TMEM16A channels with subsequent phosphatidylserine exposure. *J Biol Chem* **285:**15557–15565.

31. Jorgensen SE, Mulcahy PF, Wu GK, Louis CF. 1983. Calcium accumulation in human and sheep erythrocytes that is induced by *Escherichia coli* hemolysin. *Toxicon* **21:**717–727.

32. Bhakdi S, Mackman N, Nicaud JM, Holland IB. 1986. *Escherichia coli* hemolysin may damage target cell membranes by generating transmembrane pores. *Infect Immun* **52:**63–69.

33. Moayeri M, Welch RA. 1994. Effects of temperature, time, and toxin concentration on lesion formation by the *Escherichia coli* hemolysin. *Infect Immun* **62:**4124–4134.

34. Skals M, Jorgensen NR, Leipziger J, Praetorius HA. 2009. Alpha-hemolysin from *Escherichia coli* uses endogenous amplification through P2X receptor activation to induce hemolysis. *Proc Natl Acad Sci U S A* **106:**4030–4035.

35. Grimminger F, Scholz C, Bhakdi S, Seeger W. 1991. Subhemolytic doses of *Escherichia coli* hemolysin evoke large quantities of lipoxygenase products in human neutrophils. *J Biol Chem* **266:**14262–14269.

36. Valeva A, Walev I, Kemmer H, Weis S, Siegel I, Boukhallouk F, Wassenaar TM, Chavakis T, Bhakdi S. 2005. Binding of *Escherichia coli* hemolysin and activation of the target cells is not receptor-dependent. *J Biol Chem* **280:**36657–36663.

37. Grimminger F, Rose F, Sibelius U, Meinhardt M, Pötzsch B, Spriestersbach R, Bhakdi S, Suttorp N, Seeger W. 1997. Human endothelial cell activation and mediator release in response to the bacterial exotoxins *Escherichia coli* hemolysin and staphylococcal alpha-toxin. *J Immunol* **159:**1909–1916.

38. **Månsson LE, Kjäll P, Pellett S, Nagy G, Welch RA, Bäckhed F, Frisan T, Richter-Dahlfors A.** 2007. Role of the lipopolysaccharide-CD14 complex for the activity of hemolysin from uropathogenic *Escherichia coli*. *Infect Immun* **75**:997–1004.

39. **Wandersman C, Létoffé S.** 1993. Involvement of lipopolysaccharide in the secretion of *Escherichia coli* alpha-haemolysin and *Erwinia chrysanthemi* proteases. *Mol Microbiol* **7**:141–150.

40. **Bauer ME, Welch RA.** 1997. Pleiotropic effects of a mutation in rfaC on *Escherichia coli* hemolysin. *Infect Immun* **65**:2218–2224.

41. **Wiles TJ, Dhakal BK, Eto DS, Mulvey MA.** 2008. Inactivation of host Akt/protein kinase B signaling by bacterial pore-forming toxins. *Mol Biol Cell* **19**:1427–1438.

42. **Dhakal BK, Mulvey MA.** 2012. The UPEC pore-forming toxin α-hemolysin triggers proteolysis of host proteins to disrupt cell adhesion, inflammatory, and survival pathways. *Cell Host Microbe* **11**:58–69.

43. **Billips BK, Forrestal SG, Rycyk MT, Johnson JR, Klumpp DJ, Schaeffer AJ.** 2007. Modulation of host innate immune response in the bladder by uropathogenic *Escherichia coli*. *Infect Immun* **75**:5353–5360.

44. **Hunstad DA, Justice SS, Hung CS, Lauer SR, Hultgren SJ.** 2005. Suppression of bladder epithelial cytokine responses by uropathogenic *Escherichia coli*. *Infect Immun* **73**:3999–4006.

45. **Loughman JA, Hunstad DA.** 2011. Attenuation of human neutrophil migration and function by uropathogenic bacteria. *Microbes Infect* **13**:555–565.

46. **Leeds JA, Welch RA.** 1997. Enhancing transcription through the *Escherichia coli* hemolysin operon, hlyCABD: RfaH and upstream JUMPStart DNA sequences function together via a postinitiation mechanism. *J Bacteriol* **179**:3519–3527.

47. **Rahn A, Whitfield C.** 2003. Transcriptional organization and regulation of the *Escherichia coli* K30 group 1 capsule biosynthesis (cps) gene cluster. *Mol Microbiol* **47**:1045–1060.

48. **Nagy G, Dobrindt U, Kupfer M, Emody L, Karch H, Hacker J.** 2001. Expression of hemin receptor molecule ChuA is influenced by RfaH in uropathogenic *Escherichia coli* strain 536. *Infect Immun* **69**:1924–1928.

49. **Wang L, Jensen S, Hallman R, Reeves PR.** 1998. Expression of the O antigen gene cluster is regulated by RfaH through the JUMPstart sequence. *FEMS Microbiol Lett* **165**:201–206.

50. **Marolda CL, Valvano MA.** 1998. Promoter region of the *Escherichia coli* O7-specific lipopolysaccharide gene cluster: structural and functional characterization of an upstream untranslated mRNA sequence. *J Bacteriol* **180**:3070–3079.

51. **Nagy G, Dobrindt U, Schneider G, Khan AS, Hacker J, Emody L.** 2002. Loss of regulatory protein RfaH attenuates virulence of uropathogenic *Escherichia coli*. *Infect Immun* **70**:4406–4413.

52. **Beloin C, Michaelis K, Lindner K, Landini P, Hacker J, Ghigo JM, Dobrindt U.** 2006. The transcriptional antiterminator RfaH represses biofilm formation in *Escherichia coli*. *J Bacteriol* **188**:1316–1331.

53. **Anderson GG, Goller CC, Justice S, Hultgren SJ, Seed PC.** 2010. Polysaccharide capsule and sialic acid-mediated regulation promote biofilm-like intracellular bacterial communities during cystitis. *Infect Immun* **78**:963–975.

54. **Wright KJ, Seed PC, Hultgren SJ.** 2007. Development of intracellular bacterial communities of uropathogenic *Escherichia coli* depends on type 1 pili. *Cell Microbiol* **9**:2230–2241.

55. **Welch RA, Burland V, Plunkett G III, Redford P, Roesch P, Rasko D, Buckles EL, Liou SR, Boutin A, Hackett J, Stroud D, Mayhew GF, Rose DJ, Zhou S, Schwartz DC, Perna NT, Mobley HL, Donnenberg MS, Blattner FR.** 2002. Extensive mosaic structure revealed by the complete genome sequence of uropathogenic *Escherichia coli*. *Proc Natl Acad Sci U S A* **99**:17020–17024.

56. **Parham NJ, Pollard SJ, Chaudhuri RR, Beatson SA, Desvaux M, Russell MA, Ruiz J, Fivian A, Vila J, Henderson IR.** 2005. Prevalence of pathogenicity island IICFT073 genes among extraintestinal clinical isolates of *Escherichia coli*. *J Clin Microbiol* **43**:2425–2434.

57. **Vigil PD, Stapleton AE, Johnson JR, Hooton TM, Hodges AP, He Y, Mobley HL.** 2011. Presence of putative repeat-in-toxin gene tosA in *Escherichia coli* predicts successful colonization of the urinary tract. *MBio* **2**:e00066-11. doi:10.1128/mBio.00066-11.

58. **Vigil PD, Wiles TJ, Engstrom MD, Prasov L, Mulvey MA, Mobley HL.** 2012. The repeat-in-toxin family member TosA mediates adherence of uropathogenic *Escherichia coli* and survival during bacteremia. *Infect Immun* **80**:493–505.

59. **Caprioli A, Falbo V, Roda LG, Ruggeri FM, Zona C.** 1983. Partial purification and characterization of an *Escherichia coli* toxic factor that induces morphological cell alterations. *Infect Immun* **39**:1300–1306.

60. **Caprioli A, Falbo V, Ruggeri FM, Baldassarri L, Bisicchia R, Ippolito G, Romoli E, Donelli G.** 1987. Cytotoxic necrotizing factor production by hemolytic strains of *Escherichia coli* causing extraintestinal infections. *J Clin Microbiol* **25:**146–149.

61. **Andreu A, Stapleton AE, Fennell C, Lockman HA, Xercavins M, Fernandez F, Stamm WE.** 1997. Urovirulence determinants in *Escherichia coli* strains causing prostatitis. *J Infect Dis* **176:**464–469.

62. **Landraud L, Gibert M, Popoff MR, Boquet P, Gauthier M.** 2003. Expression of cnf1 by *Escherichia coli* J96 involves a large upstream DNA region including the hlyCABD operon, and is regulated by the RfaH protein. *Mol Microbiol* **47:**1653–1667.

63. **Flatau G, Lemichez E, Gauthier M, Chardin P, Paris S, Fiorentini C, Boquet P.** 1997. Toxin-induced activation of the G protein p21 Rho by deamidation of glutamine. *Nature* **387:**729–733.

64. **Schmidt G, Sehr P, Wilm M, Selzer J, Mann M, Aktories K.** 1997. Gln 63 of Rho is deamidated by *Escherichia coli* cytotoxic necrotizing factor-1. *Nature* **387:**725–729.

65. **Lemichez E, Flatau G, Bruzzone M, Boquet P, Gauthier M.** 1997. Molecular localization of the *Escherichia coli* cytotoxic necrotizing factor CNF1 cell-binding and catalytic domains. *Mol Microbiol* **24:**1061–1070.

66. **Knust Z, Blumenthal B, Aktories K, Schmidt G.** 2009. Cleavage of *Escherichia coli* cytotoxic necrotizing factor 1 is required for full biologic activity. *Infect Immun* **77:**1835–1841.

67. **Kim KJ, Chung JW, Kim KS.** 2005. 67-kDa laminin receptor promotes internalization of cytotoxic necrotizing factor 1-expressing *Escherichia coli* K1 into human brain microvascular endothelial cells. *J Biol Chem* **280:**1360–1368.

68. **McNichol BA, Rasmussen SB, Carvalho HM, Meysick KC, O'Brien AD.** 2007. Two domains of cytotoxic necrotizing factor type 1 bind the cellular receptor, laminin receptor precursor protein. *Infect Immun* **75:**5095–5104.

69. **Pei S, Doye A, Boquet P.** 2001. Mutation of specific acidic residues of the CNF1 T domain into lysine alters cell membrane translocation of the toxin. *Mol Microbiol* **41:**1237–1247.

70. **Munro P, Flatau G, Doye A, Boyer L, Oregioni O, Mege JL, Landraud L, Lemichez E.** 2004. Activation and proteasomal degradation of rho GTPases by cytotoxic necrotizing factor-1 elicit a controlled inflammatory response. *J Biol Chem* **279:**35849–35857.

71. **Hofman P, Le Negrate G, Mograbi B, Hofman V, Brest P, Alliana-Schmid A, Flatau G, Boquet P, Rossi B.** 2000. *Escherichia coli* cytotoxic necrotizing factor-1 (CNF-1) increases the adherence to epithelia and the oxidative burst of human polymorphonuclear leukocytes but decreases bacteria phagocytosis. *J Leukoc Biol* **68:**522–528.

72. **Davis JM, Rasmussen SB, O'Brien AD.** 2005. Cytotoxic necrotizing factor type 1 production by uropathogenic *Escherichia coli* modulates polymorphonuclear leukocyte function. *Infect Immun* **73:**5301–5310.

73. **Falzano L, Fiorentini C, Donelli G, Michel E, Kocks C, Cossart P, Cabanié L, Oswald E, Boquet P.** 1993. Induction of phagocytic behaviour in human epithelial cells by *Escherichia coli* cytotoxic necrotizing factor type 1. *Mol Microbiol* **9:**1247–1254.

74. **Visvikis O, Boyer L, Torrino S, Doye A, Lemonnier M, Lorès P, Rolando M, Flatau G, Mettouchi A, Bouvard D, Veiga E, Gacon G, Cossart P, Lemichez E.** 2011. *Escherichia coli* producing CNF1 toxin hijacks Tollip to trigger Rac1-dependent cell invasion. *Traffic* **12:**579–590.

75. **Doye A, Mettouchi A, Bossis G, Clément R, Buisson-Touati C, Flatau G, Gagnoux L, Piechaczyk M, Boquet P, Lemichez E.** 2002. CNF1 exploits the ubiquitin-proteasome machinery to restrict Rho GTPase activation for bacterial host cell invasion. *Cell* **111:**553–564.

76. **Johnson DE, Drachenberg C, Lockatell CV, Island MD, Warren JW, Donnenberg MS.** 2000. The role of cytotoxic necrotizing factor-1 in colonization and tissue injury in a murine model of urinary tract infection. *FEMS Immunol Med Microbiol* **28:**37–41.

77. **Rippere-Lampe KE, O'Brien AD, Conran R, Lockman HA.** 2001. Mutation of the gene encoding cytotoxic necrotizing factor type 1 (cnf(1)) attenuates the virulence of uropathogenic *Escherichia coli*. *Infect Immun* **69:**3954–3964.

78. **Real JM, Munro P, Buisson-Touati C, Lemichez E, Boquet P, Landraud L.** 2007. Specificity of immunomodulator secretion in urinary samples in response to infection by alpha-hemolysin and CNF1 bearing uropathogenic *Escherichia coli*. *Cytokine* **37:**22–25.

79. **Boyer L, Magoc L, Dejardin S, Cappillino M, Paquette N, Hinault C, Charriere GM, Ip WK, Fracchia S, Hennessy E, Erturk-Hasdemir D, Reichhart JM, Silverman N, Lacy-Hulbert A, Stuart LM.** 2011. Pathogen-derived effectors trigger protective immunity via activation of the Rac2 enzyme and the IMD or Rip kinase signaling pathway. *Immunity* **35:**536–549.

80. **Fabbri A, Travaglione S, Fiorentini C.** 2010. *Escherichia coli* cytotoxic necrotizing factor 1 (CNF1): toxin biology, *in vivo* applications and therapeutic potential. *Toxins (Basel)* **2:**283–296.

81. **Parham NJ, Pollard SJ, Desvaux M, Scott-Tucker A, Liu C, Fivian A, Henderson IR.** 2005. Distribution of the serine protease autotransporters of the Enterobacteriaceae among extraintestinal clinical isolates of *Escherichia coli. J Clin Microbiol* **43:**4076–4082.

82. **Allsopp LP, Beloin C, Ulett GC, Valle J, Totsika M, Sherlock O, Ghigo JM, Schembri MA.** 2012. Molecular characterization of UpaB and UpaC, two new autotransporter proteins of uropathogenic *Escherichia coli* CFT073. *Infect Immun* **80:**321–332.

83. **Henderson IR, Navarro-Garcia F, Nataro JP.** 1998. The great escape: structure and function of the autotransporter proteins. *Trends Microbiol* **6:**370–378.

84. **Dutta PR, Cappello R, Navarro-Garcia F, Nataro JP.** 2002. Functional comparison of serine protease autotransporters of enterobacteriaceae. *Infect Immun* **70:**7105–7113.

85. **Maroncle NM, Sivick KE, Brady R, Stokes FE, Mobley HL.** 2006. Protease activity, secretion, cell entry, cytotoxicity, and cellular targets of secreted autotransporter toxin of uropathogenic *Escherichia coli. Infect Immun* **74:**6124–6134.

86. **Guyer DM, Henderson IR, Nataro JP, Mobley HL.** 2000. Identification of sat, an autotransporter toxin produced by uropathogenic *Escherichia coli. Mol Microbiol* **38:**53–66.

87. **Guyer DM, Radulovic S, Jones FE, Mobley HL.** 2002. Sat, the secreted autotransporter toxin of uropathogenic *Escherichia coli*, is a vacuolating cytotoxin for bladder and kidney epithelial cells. *Infect Immun* **70:**4539–4546.

88. **Liévin-Le Moal V, Comenge Y, Ruby V, Amsellem R, Nicolas V, Servin AL.** 2011. Secreted autotransporter toxin (Sat) triggers autophagy in epithelial cells that relies on cell detachment. *Cell Microbiol* **13:**992–1013.

89. **Henderson IR, Czeczulin J, Eslava C, Noriega F, Nataro JP.** 1999. Characterization of pic, a secreted protease of *Shigella flexneri* and enteroaggregative *Escherichia coli. Infect Immun* **67:**5587–5596.

90. **Parreira VR, Gyles CL.** 2003. A novel pathogenicity island integrated adjacent to the thrW tRNA gene of avian pathogenic *Escherichia coli* encodes a vacuolating autotransporter toxin. *Infect Immun* **71:**5087–5096.

91. **Ruiz-Perez F, Wahid R, Faherty CS, Kolappaswamy K, Rodriguez L, Santiago A, Murphy E, Cross A, Sztein MB, Nataro JP.** 2011. Serine protease autotransporters from Shigella flexneri and pathogenic *Escherichia coli* target a broad range of leukocyte glycoproteins. *Proc Natl Acad Sci U S A* **108:**12881–12886.

92. **Lloyd AL, Henderson TA, Vigil PD, Mobley HL.** 2009. Genomic islands of uropathogenic *Escherichia coli* contribute to virulence. *J Bacteriol* **191:**3469–3481.

93. **Smith MA, Weingarten RA, Russo LM, Ventura CL, O'Brien AD.** 2015. Antibodies against hemolysin and cytotoxic necrotizing factor type 1 (CNF1) reduce bladder inflammation in a mouse model of urinary tract infection with toxigenic uropathogenic *Escherichia coli. Infect Immun* **83:**1661–1673.

Structure, Function, and Assembly of Adhesive Organelles by Uropathogenic Bacteria

14

PETER CHAHALES[1] and DAVID G. THANASSI[1]

INTRODUCTION

Bacteria assemble a variety of adhesive proteins (adhesins) on their surface to mediate binding to receptors and colonization of surfaces. For pathogenic bacteria, adhesins are critical for early stages of infection, allowing the bacteria to initiate contact with host cells, colonize different tissues, and establish a foothold within the host. Adhesins recognize specific receptors expressed by specific subsets of host cells. Therefore, the repertoire of adhesins expressed by a pathogen play a major role in dictating the tropism of the pathogen toward specific host tissues and organs. Moreover, binding of bacterial adhesins to host cell receptors influences subsequent events by triggering signaling pathways in both the host and bacterial cells. These signaling pathways may determine whether the bacteria remain extracellular or become internalized, and influence the intracellular trafficking of invaded bacteria and their ability to survive and replicate (1, 2). The adhesins expressed by a pathogen are also critical for bacterial-bacterial interactions and the formation of bacterial communities, including biofilms. The ability to adhere to host tissues is particularly important for bacteria that colonize sites that include

[1]Center for Infectious Diseases and Department of Molecular Genetics and Microbiology, Stony Brook University, Stony Brook, NY 11794.

Urinary Tract Infections: Molecular Pathogenesis and Clinical Management, 2nd Edition
Edited by Matthew A. Mulvey, David J. Klumpp, and Ann E. Stapleton
© 2017 American Society for Microbiology, Washington, DC
doi:10.1128/microbiolspec.UTI-0018-2013

the urinary tract, where the flow of urine functions to maintain sterility by washing away non-adherent pathogens.

Adhesins vary from monomeric proteins that are directly anchored to the bacterial surface to polymeric, hairlike fibers that extend out from the cell surface. These latter fibers are termed pili or fimbriae, and were among the first identified virulence factors of uropathogenic *Escherichia coli* (UPEC) (3). Pili were first described in the late 1940s and early 1950s as bacterial surface structures distinct from flagella (4). Duguid et al. used the term *fimbriae*, Latin for thread or fiber, to describe surface appendages that allowed *E. coli* to bind to and agglutinate erythrocytes (5). Brinton later used the term *pili*, Latin for hair, to describe the non-flagellar surface structures expressed by *E. coli* (6). Ottow subsequently proposed that the term pili be reserved for the F or conjugative pili involved in bacterial mating, and that the term fimbriae should be used to describe surface fibers involved in adhesion (4). However, today the terms pili and fimbriae are generally used interchangeably. We will refer to these structures collectively as pili.

Various schemes have been proposed to classify the different types of pili (4, 7–10). Although most of these classification schemes are no longer in common use, parts have entered the standard nomenclature. Pili were originally classified as mannose resistant (MR) or mannose sensitive (MS), based on their ability to agglutinate erythrocytes in the presence or absence of mannosides (11, 12). This classification led to the term type 1 pili, which is still in current use, to refer to MS surface fibers. The MR pili were initially divided into the P and unknown (X) pili, with the unknown pili now defined to include the S, Dr, and additional pilus adhesins (3). Uropathogenic bacteria have been closely associated with the discovery and characterization of pili. The chromosomal gene clusters responsible for expression of both type 1 and P pili were first cloned from the J96 UPEC strain (13), and the genes coding

for S pili were isolated from UPEC strain 536 (14). As discussed in detail in the following section, much of our current understanding of the structure, assembly, and functions of bacterial pili stems from studies of the type 1 and P pili originally isolated from UPEC.

Bacteria are now known to express a number of different types of pilus structures and other non-flagellar surface appendages (15). One additional structure, curli, is expressed by UPEC and imparts unique characteristics to the bacteria that influence colonization within the urinary tract, including promoting biofilm formation (16). Curli are assembled by a completely different mechanism from pili such as the type 1 and P pili, and appear as aggregated masses on the bacterial surface rather than hair-like fibers. Pilus assembly is not restricted to Gram-negative bacteria. Pili were observed on the Gram-positive bacterium *Corynebacterium renale* in the 1960s (17, 18), but this observation was largely forgotten until studies dating from 2003 by Ton-That and Scheewind to characterize pilus biogenesis in *Corynebacterium diphtheriae* (19, 20). A number of different Gram-positive bacteria are now known to assemble adhesive pili associated with virulence and this is an active area of research. The Gram-positive pili have unique structural features and assembly mechanisms compared to Gram-negative pili (21, 22).

Pili and other extended surface fibers increase the functional reach of adhesins, enabling the bacteria to act at a distance. Pili place adhesins outside capsular or other protective surface structures, allowing contact with receptors while maintaining the protective integrity of the bacterial envelope. The ability to initiate contact at a distance also provides a means for pathogenic bacteria to avoid detection or uptake by host cells. Despite these advantages of pilus adhesins, bacteria also express a range of non-pilus adhesins, which are anchored directly on the bacterial surface. Non-pilus adhesins confer intimate binding to surfaces and are often associated with formation of bacterial colonies and biofilms. Gram-negative uropathogens display several adhesins important for pathogenesis on their outer membrane, with the majority of these adhesins assembled by the autotransporter (type V) secretion pathway. Gram-positive uropathogens also display adhesins on their surface important for colonizing the urinary tract. These Gram-positive adhesins typically are covalently linked to the peptidoglycan cell wall and are termed MSCRAMMs (microbial surface components recognizing adhesive matrix molecules) (22, 23).

Table 1 lists adhesins that contribute to infection of the urinary tract by Gram-negative and Gram-positive uropathogens. In this chapter, we will describe the different types of adhesins, what is known about their structures, and how they are assembled on the bacterial surface. We will also describe the functions of specific adhesins in the pathogenesis of urinary tract infections (UTIs). For the Gram-negative adhesins, we will focus our description on UPEC, which serves as a model system and for which extensive studies have been done.

ADHESINS EXPRESSED BY GRAM-NEGATIVE UROPATHOGENS

Pili Assembled by the Chaperone/Usher Pathway

A wide range of Gram-negative bacteria use the chaperone/usher (CU) pathway to assemble a superfamily of virulence-associated adhesive surface fibers (24–27). The CU pathway takes its name from the components of its secretion machinery, which consist of a dedicated periplasmic chaperone and an integral outer membrane protein termed the usher. The CU pathway builds a diverse array of peritrichous surface fibers, ranging from thin, flexible filaments to rigid, rod-like organelles. For uropathogenic bacteria, pili assembled by the CU pathway mediate adhesion to receptors in the urinary tract,

initiating infection and promoting bacterial colonization. Pili are critical virulence factors of uropathogenic bacteria and have been the subject of intense study (Table 1). The CU pili expressed by uropathogenic bacteria are exquisitely adapted to colonization within the urinary tract, engineered to withstand and take advantage of forces encountered during colonization such as the flow of urine (28–31). In addition to binding to host molecules, CU pili are important for bacterial-bacterial interactions, biofilm formation, and adhesion to abiotic surfaces. Moreover, binding of bacteria to host cells via CU pili modulates host–signaling pathways and promotes subsequent stages of pathogenesis, including invasion inside host cells (32–38).

Genes coding for CU pili are found on both the bacterial chromosome and on plasmids, and are clustered together with a similar organization: a 5′ regulatory region that is followed by a single downstream operon encoding the required pilus structural proteins and assembly components (Fig. 1). CU gene clusters are often associated together with other virulence determinants in pathogenicity islands, which have characteristics indicating acquisition by horizontal gene transfer (39). A single bacterial genome often contains multiple CU pathways, which presumably provides the ability to adhere to a variety of different receptors and surfaces (40–42). A recent genomic analysis found that *E. coli* strains encode as many as 17 CU gene clusters and that UPEC strains encode from 9 to 12 intact CU gene clusters (43). Many of the CU gene clusters present in a bacterial genome are not expressed under laboratory growth conditions and their functions remain unknown (41). The expression of CU gene clusters is typically highly regulated, subject to phase variation, and responsive to environmental cues (44, 45). Regulatory cross talk may occur among different CU gene clusters (46–48). This cross talk likely ensures that a given bacterium only expresses a single pilus at a given time, thus controlling adhesive specificity.

TABLE 1 Adhesins of uropathogenic bacteria

Organism	Assembly pathway	Adhesin	Associated UTI disease or function	Receptor(s)	Reference(s)
GRAM-NEGATIVE BACTERIA					
Escherichia coli					
	Chaperone/usher pathway				
		P pili	Pyelonephritis	Digalactose (galabiose)	165, 365, 366
		Type 1 pili	Cystitis	Mannosylated proteins, uroplakin, β_1 and α_3 integrin	34, 141, 367
		S pili	Ascending UTI	α-sialic acid	116, 368, 369
		F1C pili	Ascending UTI	galactosylceramide on bladder epithelium and globotriaosylceramide on kidney epithelium	128, 370
		Afa/Dr pili	Recurrent/chronic UTI, cystitis, pyelonephritis	Dr[a], DAF, type IV collagen, $\alpha_5\beta_1$ integrin, CEACAM family proteins	51, 114, 123
		Yad pili	Ascending UTI	Bladder epithelial cells	43, 118
		Ygi pili	Ascending UTI	Human embryonic kidney cells	118
		F9 pili	Biofilm formation	?	120
		Auf pili	?	?	119
		Type 3 pili	CAUTI	?	371
	Autotransporter				
		Ag43	UTI persistence, biofilm formation	Collagen, laminin	263–265
		UpaB	?	Fibronectin, fibrinogen, laminin	270
		UpaC	Biofilm formation	?	270
		UpaH	Biofilm formation, bladder colonization	Collagen V, laminin, fibronectin	217, 272
		UpaG	Biofilm formation	Fibronectin, laminin	232
		FdeC	Bladder and kidney colonization	?	273, 274
	Outer membrane protein				
		Iha	Fitness in urinary tract	?	278, 279
	Type I secretion				
		TosA	Fitness in urinary tract	Kidney epithelium	284, 287
	Extracellular nucleation/precipitation				
		Curli	Biofilm formation, ?	Fibronectin, laminin, H-kininogen, fibrinogen, factor XII	174, 178, 196
Klebsiella pneumoniae					
	Chaperone/usher pathway				
		Type 1 pili	Ascending UTI, CAUTI	Mannosylated proteins	372–374
		Type 3 (MR/K) pili	CAUTI	Type V collagen	373, 375, 376
Citrobacter freundii					
	Chaperone/usher pathway				
		Type 3 pili	CAUTI	?	377
Proteus mirabilis					
	Chaperone/usher pathway				
		MR/P	Pyelonephritis, ascending UTI	Mannose-resistant	378–380
		PMF (MR/K)	Ascending UTI	?	381, 382
		UCA (NAF)	Colonization of urinary tract, complicated UTI	GalNAcβ(1-4)Gal	383–385
		ATF	?	?	386, 387

(Continued on next page)

TABLE 1 Adhesins of uropathogenic bacteria *(Continued)*

Organism	Assembly pathway	Adhesin	Associated UTI disease or function	Receptor(s)	Reference(s)
	Autotransporter				
		AipA	Bladder and kidney colonization	Collagen I, collagen IV, laminin	220
		TaaP	Bladder colonization	Collagen I, collagen IV, laminin	220
GRAM-POSITIVE BACTERIA					
Staphylococcus saprophyticus					
	Sortase-assembled MSCRAMM				
		UafA	Ascending UTI	?	307, 327
		UafB	Ascending UTI	Fibronectin, fibrinogen	308
		SdrI	UTI persistence	Fibronectin	296, 309, 329
Enterococcus faecalis					
	Sortase-assembled MSCRAMM				
		Ace	CAUTI, ascending UTI	Collagen I and IV	297, 306, 312
	Sortase-assembled pili				
		Ebp	CAUTI, ascending UTI	?	340, 345, 352
	Unknown				
		EfbA	Ascending UTI	Fibronectin	331
		Esp	Urinary tract colonization and persistence, biofilm formation	?	332, 333
Enterococcus faecium					
	Sortase-assembled pili				
		Ebp$_{fm}$	CAUTI, ascending UTI	?	339, 347

Furthermore, expression of adhesive pili has been shown to be inversely correlated with the expression of flagella for motility (49).

Much of what we know about the biogenesis and functions of CU pili comes from work on the prototypical type 1 and P pili expressed by UPEC, which bind to receptors in the bladder and kidney, respectively. We will focus on these pili as models, but will also discuss additional CU pili identified as important for pathogenesis in the urinary tract. Although we will limit this discussion to pili expressed by UPEC, CU pili have been identified as virulence factors in other uropathogens, particularly for *Proteus mirabilis* (Table 1) (50).

Structure of CU pili

The pilus fiber

Pili assembled by the CU pathway range from 2 nanometer (nm) to 10 nm in diameter and generally 1 micrometer (µm) to 3 µm in length. The pili are linear fibers built from thousands of copies of non-covalently interacting subunit proteins, termed pilins. Some pili adopt a final helical quaternary structure, resulting in the formation of rigid, rod-like organelles. Alternatively, the pili may remain as linear, flexible fibers, which, in some cases, form amorphous or 'afimbrial' structures. Many pili assembled by the CU pathway are composite structures containing both a rigid, helical rod, which extends out from the bacterial surface, as well as a thin, flexible tip structure, which is located at the distal end of the rod and contains the adhesive activity. Type 1 and P pili expressed by UPEC are prototypical composite organelles with distinct rod and tip structures (Fig. 1). The Afa/Dr family of pili expressed by UPEC and other pathogenic *E. coli* are well-studied examples of thin, flexible fibers that often have an amorphous appearance by electron microscopy (51).

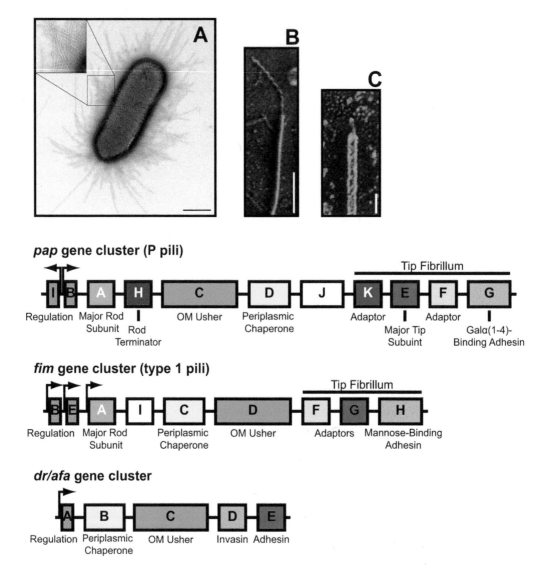

FIGURE 1 Representative CU gene clusters and pili. Gene clusters coding for P (*pap*), type 1 (*fim*) and Dr/Afa pili are depicted, with the functions of the genes indicated. Electron micrographs are shown for (A) an *E. coli* bacterium expressing type 1 pili, (B) a P pilus fiber, and (C) a type 1 pilus fiber. Scale bars equal 700 nm (A), 100 nm (B), and 20 nm (C). The images in panels A-C are reprinted from references 138, 157, and 137, respectively, with permission of the publishers.

The structures of pilins and many aspects of pilus assembly by the CU pathway are understood in atomic detail (26, 52–57). All pilins contain an immunoglobulin (Ig)-like fold termed the pilin domain (Fig. 2). Canonical Ig folds comprise seven β-strands arranged into two sheets as a β-sandwich (58).

However, pilins lack the seventh, C-terminal β-strand (the G strand) and thus are unable to complete their own fold (52–55). This missing strand results in a deep groove on the surface of the subunit, exposing its hydrophobic core. To complete their folds, pilins rely on structural information provided by

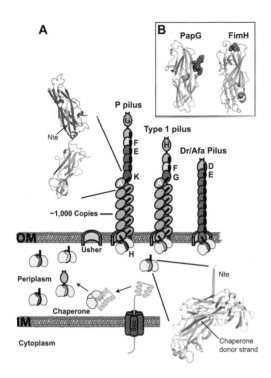

FIGURE 2 (A) Model for pilus biogenesis by the CU pathway. Pilus subunits enter the periplasm as unfolded polypeptides via the Sec system. Subunits fold upon forming binary complexes with the periplasmic chaperone (yellow). The crystal structure in the lower right depicts the chaperone-subunit donor strand exchange reaction (PapD-PapA; PDB ID: 2UY6), with the chaperone donor strand indicated in red. Pilus assembly takes place at the outer membrane usher, which catalyzes the exchange of chaperone-subunit for subunit-subunit interactions. Models for assembled P, type 1 and Afa/Dr pilus fibers are shown. The crystal structure in the upper left depicts the subunit-subunit donor strand exchange reaction that occurs in the pilus fiber (PapA-PapA; PDB ID: 2UY6), with the Nte donor strand indicated. (B) Crystal structures of the PapG (P pili; PDB ID: 1J8R) and FimH (type 1 pili; PDB ID: 1KLF) adhesin domains with bound globoside and mannose, respectively. The sugars are depicted as dark gray spheres.

interaction with the periplasmic chaperone or with neighboring subunits in the pilus fiber.

Subunit-subunit interactions in the pilus fiber are mediated by a mechanism termed donor strand exchange (54, 55). Pilins contain a conserved N-terminal extension (Nte) in addition to the pilin domain. In the pilus fiber, the Nte of one pilus subunit is 'donated' to the preceding subunit, completing the Ig fold of the preceding subunit (Fig. 2). Therefore, the pilus fiber consists of an array of Ig folds, with each subunit noncovalently bound to the preceding subunit by donor strand exchange. This arrangement provides great mechanical strength and stability to the pili, which is reflected by the property that subunit–subunit interactions in the pilus are resistant to dissociation by heat and denaturants (59, 60). A high level of mechanical strength is essential for the pili to maintain adhesion in the face of shear forces encountered from the flow of urine. The helical pilus rod provides an additional mechanism to withstand hydrodynamic forces in the urinary tract; the helical rod is able to uncoil under stress to an extended linear fiber, thereby acting as a spring or shock absorber to prevent breakage of the pilus and extend the lifetime of pilus-receptor interactions (28, 29, 31, 61).

The pilus adhesin

The receptor-binding activity of pili is conferred by the pilus adhesin. For composite pili such as type 1 and P pili, the adhesin is located in single copy at the distal end of the tip fiber (Fig. 2). Such pili have been termed monoadhesive pili (25). In contrast, for pili lacking a district tip structure, the main structural subunit that builds the pilus fiber may also contain receptor-binding sites along exposed surfaces and thus the entire pilus may function in adhesion (57, 62–64). Such fibers are termed polyadhesive pili. Afa/Dr pili are polyadhesive fibers with a single major structural subunit/adhesin; however, these pili also have a separate subunit, termed the invasin, with distinct binding activity and which promotes uptake inside host cells (38, 64, 65). The invasin subunit is present in single copy at the distal end of the pilus fiber (Fig. 2).

Crystal structures have been solved for several adhesins from monoadhesive pili (52, 66–71). In contrast to other pilus subunits, the adhesins are two domain proteins, containing an N-terminal receptor-binding or adhesin domain (in place of the Nte) and a C-terminal pilin domain. The pilin domain mediates incorporation of the adhesin into the pilus fiber and is an incomplete Ig-like fold as found for all CU pilins. Adhesin domains also have Ig-like folds, but the folds are complete (not lacking the terminal β-strand) and structurally distinct from the pilin domain. Despite their common architecture, adhesins vary greatly in sequence and employ distinct receptor binding mechanisms, reflecting their specific functions within the host (72). The FimH adhesin from type 1 pili folds as an elongated 11-stranded β-barrel with a jelly roll-like topology (52, 67). The binding site for the mannose ligand is located at the tip of the adhesin domain and is formed by a deep, negatively-charged pocket surrounded by a hydrophobic ridge (Fig. 2B). In comparison, the adhesin domain of the P pilus adhesion, PapG, adopts a structure with two sub-regions; one region having a β-barrel fold similar to FimH and the other region having a unique, largely β-sheet structure that contains the binding site for the globoside receptor (66, 68). In contrast to FimH, the receptor-binding site of PapG is located in a shallow groove along the side of the adhesin (Fig. 2B). For polyadhesive fibers such as Afa/Dr pili, both the main structural subunit and the tip-located invasin function as adhesins and both are single domain proteins with Ig-like pilin domains (57, 62, 65, 73). However, the invasin subunit lacks an Nte donor strand, thus restricting its position to the tip of the fiber. For the AfaD and DraD major subunits, distinct receptor-binding sites for CD55/decay accelerating factor (DAF) and for members of the carcinoembryonic antigen family (CEACAM) have been located along opposite sides of the pilin domain (25, 57, 62). These binding sites would be repetitively presented along the length of the assembled fiber.

Pilus adhesins such as FimH exhibit the property of shear-enhanced binding, which enables tighter binding under conditions of shear stress, including the shear stress encountered during bacterial colonization of the urinary tract (30). The application of shear stress causes FimH to switch from a low-affinity to a high-affinity binding state. This greater affinity presumably allows the bacteria to avoid being washed away by the flow of urine, and may also provide a mechanism for the bacteria to discriminate between surface-located and soluble receptors, as binding to the latter will not result in force generation on the pilus and FimH will stay in the low-affinity state. The shear-enhanced binding of FimH is mediated by a catch-bond mechanism that involves allosteric activation of the adhesin domain (74). When incorporated into the type 1 pilus tip fiber, the pilin domain of FimH interacts with the adhesin domain, causing structural alterations of the adhesin that weaken its mannose binding pocket. However, the application of force to the pilus fiber causes the FimH pilin and adhesin domains to separate, allowing the binding pocket to clamp tightly around its mannose ligand (75). Moreover, the physical properties of both the type 1 pilus tip fiber and the helical pilus rod appear to be designed to optimize the shear-enhanced behavior of FimH, and the flexibility of the pilus tip likely provides FimH maximum opportunity to find its target receptors (76, 77).

Pilus Assembly by the Chaperone/Usher Pathway

Formation of chaperone-subunit complexes in the periplasm

Pilus subunits are synthesized with an N-terminal signal sequence that directs them to the Sec general secretory pathway for translocation to the periplasm (78). The signal sequence is cleaved in the periplasm, and the subunits form stable, binary complexes with the periplasmic chaperone (Fig. 2A). The chaperone enables proper folding of the pilus subunits, prevents premature subunit-subunit interactions, and maintains the subunits in an assembly-competent state (52–55). In the absence of the chaperone, pilus subunits misfold and form aggregates that are degraded by the DegP periplasmic protease (79, 80).

The structure of the PapD chaperone and subsequent structures of chaperone-subunit complexes revealed the molecular basis for chaperone function in pilus biogenesis (52–54, 81–83). As described above, pilins have an incomplete Ig fold, lacking the C-terminal G β-strand. The chaperone contains two Ig-like domains oriented in an L or boomerang shape. The binding site for subunits resides in the cleft between the two domains and extends out along the chaperone's N-terminal domain (domain 1). The chaperone functions by a mechanism termed donor strand complementation, in which the chaperone inserts its G1 β-strand and a portion of its F1-G1 loop into the groove caused by the missing G strand of the subunit, completing the Ig fold of the pilin domain (Fig. 2A) (52, 53, 83, 84). Conserved sequence differences in the F1–G1 loop region of chaperones defines two subfamilies of CU pathways: chaperones with a short F1–G1 loop belong to the FGS (F1–G1 short) subfamily and chaperones with a long F1–G1 loop belong to the FGL subfamily (64, 85). Interestingly, these differences in the chaperones correlate with differences in the types of surface fibers assembled. FGL chaperones assemble only thin or amorphous pili comprising only one or two types of pilins (e.g., Afa/Dr pili), whereas FGS chaperones assemble both rod-like and thin pilus fibers that generally comprise multiple different pilins and may have composite architectures (e.g., type 1 and P pili).

The groove in the pilin domain caused by the missing β-strand contains a series of binding pockets, termed P1–5 (54). The G1 β-strand donated by the chaperone contains a conserved motif of alternating hydrophobic residues, and during donor strand exchange these residues insert into the P1–4 pockets of the subunit, forming a β-zipper interaction (55, 86). In FGL chaperones, the longer G1 donor strand fills the P5 binding pocket as well, but this interaction is weaker than at the other pockets (86, 87). The chaperone G1 β-strand is inserted parallel to the F strand of the subunit, forming a non-canonical Ig fold. This, together with the large size of the residues inserted by the chaperone, maintains pilins in an open, "activated" state, which enables subsequent assembly into the pilus fiber (54, 55, 87). The groove of the pilin domain is also the site of subunit-subunit interactions, which are mediated by the donor strand exchange reaction as described above for the pilus fiber (54, 55). Thus, donor strand complementation by the chaperone couples the folding of pilins with the simultaneous capping of their interactive surfaces, preventing premature fiber assembly in the periplasm. Recent studies have shown that chaperones also perform a quality control function during the initial binding of pilus subunits and that formation of chaperone-subunit complexes results in an allosteric change in the chaperone that permits binding to the outer membrane usher assembly platform (88, 89).

Assembly of the pilus fiber at the outer membrane

Chaperone-subunit complexes must interact with the outer membrane usher for release of the chaperone, assembly of subunits into the pilus fiber, and secretion of the fiber to the cell surface through the usher channel (56, 90). The usher acts as a pilus assembly catalyst, accelerating the rate of subunit incorporation into the pilus fiber (91). Subunit-subunit interactions form at the periplasmic face of the usher via the donor strand exchange mechanism (54, 55). The donated subunit Nte contains a conserved motif of alternating hydrophobic residues, similar to the chaperone G1 β-strand (92, 93). At the usher, the hydrophobic residues of the Nte from an incoming chaperone-subunit complex insert into the subunit groove of the preceding chaperone-subunit complex bound at the usher, displacing the donated G1 β-strand of the chaperone from the preceding subunit by a concerted strand displacement mechanism that initiates at the P5 pocket (54, 55, 86, 94, 95). In contrast to the donated chaperone β-strand, the Nte is inserted anti-parallel to the F strand of the preceding subunit and inserts smaller-sized residues into the subunit groove, thus completing the Ig fold of the pilin domain in a canonical fashion and allowing the subunit to adopt a highly stable final state (54, 55, 60, 87). ATP is not available in the periplasm and pilus biogenesis at the outer membrane usher does not require input from other energy sources (96, 97). The canonical Ig fold formed by donor strand exchange represents a more compact, lower energy state compared to the non-canonical Ig fold formed by donor strand complementation with the chaperone (54, 55, 87). This topological transition from the higher-energy chaperone-subunit complex to the lower-energy subunit-subunit interaction provides the driving force for fiber formation and secretion at the usher (98).

Pili are assembled in a top-down order, with the adhesin incorporated first, followed by the rest of the pilus tip and finally the rod. Each subunit specifically interacts with its appropriate neighbor subunit in the pilus, with the specificity of binding determined by the donor strand exchange reaction (99–101). In addition, the usher ensures ordered and complete pilus assembly by differentially recognizing chaperone-subunit complexes according to their final position in the pilus; i.e., chaperone-adhesin complexes have highest affinity for the usher, whereas chaperone-rod subunit complexes have low affinity (102–104). The usher channel is only wide enough to allow secretion of a linear fiber of folded pilus subunits (56, 90). Therefore, the pilus rod is constrained to a linear fiber as it passes through the usher and only converts to its final helical form upon reaching the bacterial surface.

The pilus usher

Ushers are large, integral outer-membrane proteins containing five domains: a central transmembrane β-barrel domain that forms the secretion channel, a middle domain located within the β-barrel region that forms a channel gate (the plug domain), a periplasmic N-terminal domain (NTD), and two periplasmic C-terminal domains (CTD1 and CTD2) (Fig. 3) (56, 90, 105–109). The NTD provides the initial binding site for chaperone-subunit complexes and functions in the recruitment of periplasmic complexes to the usher (105, 110). The CTDs provide a second binding site for chaperone-subunit complexes and anchor the growing pilus fiber (56). The usher is present as a dimeric complex in the OM, but only one channel is used for secretion of the pilus fiber and the function of the usher dimer remains to be determined, particularly since the usher monomer appears to be sufficient for pilus assembly (56, 90, 106, 111, 112).

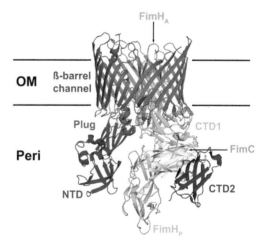

FIGURE 3 Crystal structure of the FimD-FimC-FimH type 1 pilus assembly intermediate (PDB ID: 3RFZ). The Usher NTD, plug, β-barrel channel, and CTD domains are indicated. The FimH adhesin domain (FimH$_A$) is inserted inside the usher channel, and the FimH pilin domain (FimH$_P$) and bound FimC chaperone are located at the usher CTDs.

The structure of the type 1 pilus usher FimD bound to the FimC-FimH chaperone-adhesin complex was recently solved, revealing the usher pilus assembly machine in action (Fig. 3) (56). The usher channel is formed by a 24-stranded β-barrel that is occluded by an internal plug domain (56, 90). The binding of the FimH adhesin to FimD activates the usher for pilus biogenesis (91, 103, 113), resulting in displacement of the plug to the periplasm and insertion of the FimH-adhesin domain inside the usher channel. The FimH-pilin domain remains in complex with the FimC chaperone and bound to the usher CTDs (Fig. 3). CU pili extend by step-wise addition of new chaperone-subunit complexes to the base of the fiber. New chaperone-subunit complexes are recruited by binding to the usher NTD, which is unoccupied in the FimD–FimC–FimH structure (56, 105, 110). Modelling studies suggest that binding of a chaperone-subunit complex to the usher NTD would perfectly position the Nte of the newly recruited subunit to initiate donor strand exchange with the P5 pocket of the subunit bound at the usher CTDs, providing a molecular explanation for the catalytic activity of the usher in pilus assembly (56). Following donor strand exchange, the chaperone is displaced from the subunit bound at the CTDs and released into the periplasm. To reset the usher for another round of subunit incorporation, the newly incorporated chaperone-subunit complex must transfer from the NTD to the CTDs, concomitant with translocation of the pilus fiber through the usher channel toward the cell surface. Repeated iterations of this cycle would then result in assembly and secretion of a complete pilus fiber.

Functions of CU Pili Expressed by UPEC

A number of different CU pili have been identified that contribute to colonization of the urinary tract by UPEC (Table 1). In addition to type 1 and P pili, which are described in detail in the following paragraphs, CU systems with demonstrated or putative roles in UTIs include Afa/Dr, S, F1C, F9, type 3, Auf, Yad and Ygi pili (51, 114–120). Further characterization is needed for many of these systems, which may have roles in direct adherence to host receptors or may facilitate bacterial-bacterial interactions and biofilm formation. The best characterized of these additional CU pili are the Afa/Dr family, which includes Dr, F1845, Afa, Nfa, and Aaf pili. Afa/Dr pili are thin, polyadhesive fibers that are expressed by diffusely adhering strains of diarrheagenic *E. coil* (DAEC) in addition to being prominent virulence factors of UPEC (51, 114, 121). Afa/Dr adhesins bind to the Dr[a] blood group antigen and have affinity for DAF, members of the CEACAM family, type IV collagen, and $\alpha_5\beta_1$ integrin (122–124). In contrast to the Afa/Dr polyadhesins, S and F1C pili are structurally similar to type 1 and P pili. S pili bind to sialyl-galactoside moieties on extracellular matrix proteins, such as fibronectin and laminin (125, 126). S pili are expressed by clinical UPEC isolates and expression of S pili confers binding to bladder and kidney epithelial cells, indicating potential roles in ascending UTIs (115, 127). F1C pili have affinity for globotriaosylceramide, which is present on the kidney epithelium, and for galactosylceramide, found in the bladder, kidney, and ureters (128).

Type 1 pili

Type 1 pili are expressed by most strains of *E. coli* and mediate binding to a variety of surfaces and host tissues in a mannose-sensitive manner. Type 1 pili are a major virulence factor of UPEC and antibodies to the type 1 pilus adhesin FimH provide protection against urinary tract infection by *E. coli* in both murine and primate models (129, 130). However, a definitive requirement for type 1 pili in human UTIs has remained elusive (131), likely due to the large repertoire of adhesins expressed by uropathogenic strains. UPEC use type 1 pili to bind to α-D-mannosylated proteins present in the bladder, leading to bacterial colonization, bladder epithelial cell invasion, and the development of cystitis (33, 132). In addition to urothelial cells, type 1 pili have been reported to bind to Tamm-Horsfall protein, surface glycoproteins of immune cells, extracellular matrix proteins, and abiotic surfaces (133–136).

Type 1 pili are encoded by the *fim* gene cluster (Fig. 1), which is present on the chromosome of pathogenic as well as non-pathogenic and laboratory strains of *E. coli*. Type 1 pili are built from 4 different types of pilins arranged into a rigid helical rod measuring 6.9 nm in diameter, and a short tip fiber measuring 2 nm in diameter and generally 10 nm to 19 nm in length (Fig. 1) (137, 138). The type 1 pilus rod is built from more than 1,000 copies of the FimA major pilin arranged into a right-handed helix (138). Type 1 pilus tips contain a single copy of the FimH adhesin at the distal end, followed by the FimG and FimF adaptor subunits, which are generally present in single copy (Fig. 2A) (75, 137, 138). The mannose binding site of the FimH adhesin is located in a deep pocket at the tip of the adhesin domain (Fig. 2B) (67). This places the receptor-binding site at the most distal end of the type 1 pilus organelle, which presumably facilitates access of the pilus to its receptor.

Studies using the murine urinary tract infection model have revealed many aspects of type 1 pilus function during UPEC pathogenesis. On entering the urinary tract, UPEC use their type 1 pili to bind to uroplakins, mannosylated proteins that coat the luminal surface of the bladder, allowing the bacteria to colonize the bladder and avoid being washed out by the flow of urine (33, 139). Type 1 pili not only mediate binding of UPEC to the bladder surface, but also trigger host cell signaling pathways that lead to actin rearrangement in the urothelial cells and invasion of the bacteria inside the cells by a zipper-like mechanism (32, 33, 140). Additionally, bacterial uptake is facilitated by binding of type 1 pili to β1 and α3 integrins

(34). Binding of *E. coli* to the urothelium leads to induction of innate host cell responses, including upregulation of proinflammatory cytokines and cell death pathways (33, 140, 141). The FimH adhesin acts as a pathogen-associated molecular pattern that is recognized by Toll-like receptor 4 (TLR4), present on bladder epithelial cells as well as macrophages, and stimulates immune signaling pathways through a mechanism independent of LPS (36).

Following uptake inside bladder epithelial cells, UPEC are initially contained within vesicles, which may be routed for exocytosis in a TLR4- and cyclic AMP-dependent mechanism that may be used by the host cells to expel the invading bacteria (142). Bacteria that evade expulsion enter the cytoplasm where they rapidly replicate to form aggregates termed intracellular biofilm-like communities or pods (143, 144). Bacteria within these intracellular communities are protected from innate host immune responses and shielded from antibiotics (145). Type 1 pili, which are known to contribute to the formation of extracellular biofilms (136), are also expressed by the intracellular bacteria and required for formation of the pods, separate from their function in host cell binding and invasion (35, 146). Urothelial cells respond to UPEC invasion by undergoing programmed cell death and exfoliating into the bladder lumen, a host defense mechanism to wash out the colonizing bacteria (33). However, UPEC counter this by fluxing out of the host cells and undergoing additional rounds of attachment to and invasion of neighboring cells, presumably mediated by type 1 pili as in the initial round of infection (144, 147). During this process, the *E. coli* may gain access to the underlying bladder epithelium, leading to the formation of quiescent bacterial reservoirs from which recurrent infections can be seeded to begin the infection process anew (147 , 148). Thus, type 1 pili function at multiple different points during UPEC pathogenesis in the urinary tract and have both extracellular and intracellular roles.

P pili

P pili are expressed by UPEC and are strongly associated with the ability of the bacteria to colonize the kidney and cause pyelonephritis (66, 149, 150). P pili bind to Gal(α1-4)Gal moieties present in the globo-series of glycolipids found in kidney epithelial cells. The glycolipid receptor is also part of the P blood group antigen, thus allowing P pilus-mediated agglutination of human erythrocytes (151). P pili are encoded by the chromosomal *pap* (pyelonephritis-associated pili) gene cluster (Fig. 1), which is present on pathogenicity islands of UPEC strains, and also found in *E. coli* causing neonatal meningitis and avian pathogenic strains (152). Individual *E. coli* strains may carry more than one *pap* gene cluster, located in different pathogenicity islands (153, 154). There are three predominant alleles of the P pilus adhesin PapG – class I, II, and III – which have specificities for receptor isotypes that differ in carbohydrate residues distal from the Gal(α1-4)Gal core (155, 156). Class II PapG is correlated with human kidney infections, whereas class III PapG is associated with colonization of the human bladder.

P pili are built from six different structural subunits that form a right-handed helical rod and distal tip fiber, similar to type 1 pili (Fig. 1). The P pilus tip fiber is longer and more flexible compared to type 1 pilus tips, measuring approximately 40 nm in length. The P pilus tip is composed mainly of PapE, which is present at approximately 5 to 10 copies per pilus. The PapG adhesin is present in single copy at the distal end of the tip and is joined to PapE via the PapF adaptor subunit (157, 158) (Fig. 2A). Another adaptor subunit, PapK, links the tip fiber to the pilus rod (158). The helical P pilus rod measures 8.2 nm in diameter and is built from a linear homopolymer of over 1000 copies of the PapA major pilin (159). The P pilus rod is terminated by the PapH minor pilin, which also plays a role in anchoring the pilus fiber in the OM (160, 161).

The glycolipid binding site on the PapG adhesin is formed by a shallow pocket on one side of the adhesin domain (Fig. 2B) (66, 68). This is in contrast to the tip-located mannose-binding site of FimH on type 1 pili. P pili have a longer, more flexible tip fiber compared to type 1 pili. The flexibility of the P pilus tip and side-on orientation of the PapG binding site likely function in tandem to facilitate docking of the adhesin onto the globoside moiety of the glycolipid receptor, which is oriented parallel to the membrane surface (66).

Expression of P pili promotes ascending urinary tract infection and facilitates colonization of the kidneys by *E. coli* (150, 162). Consistent with a role in pathogenesis, vaccination with P pili was shown to provide protection against pyelonephritis in both murine and primate models (163, 164). However, studies using P pilus mutants have had variable results in establishing an essential requirement for the pili in kidney infections, likely due to the many different adhesins expressed by UPEC strains (165). As for type 1 pili, P pilus-mediated adhesion of UPEC to the urothelium stimulates cytokine production and resultant inflammatory responses in the urinary tract, which likely exacerbates kidney damage during acute pyelonephritis (37, 166, 167). Binding of P pili to its glycolipid receptor in kidney epithelial cells causes release of the second messenger ceramide, which forms the membrane anchor portion of the receptor. Ceramide is as an agonist for TLR4, and thus provides a potential link between bacterial adhesion and induction of innate immune pathways (168). PapG-mediated binding also activates signal transduction pathways within the bacteria (169). These pathways result in upregulation of iron acquisition systems and may prepare UPEC for colonization of the urinary tract.

Curli

Curli fibers, also called thin aggregative fimbriae, are produced by Gram-negative enteric bacteria such as *E. coli* and *Salmonella* and form part of a complex extracellular matrix that contributes to adhesion, biofilm formation, host colonization, and invasion (16, 170–172). The expression of curli imparts special properties to biofilm structures, allowing attachment to normally resistant surfaces such as Teflon and stainless steel (170). Curli were first characterized by Normark and colleagues as novel bacterial surface structures that conferred binding to fibronectin (173). Curli bind to range of host molecules in addition to fibronectin, including laminin, human contact phase proteins, and MHC class I (174–176). Most bacteria optimally express curli at temperatures of 30°C or lower, consistent with a central role in biofilm formation and colonization of environmental surfaces. However, many clinical *E. coli* strains, including UPEC isolates, express curli at host temperature (37°C) (177, 178). In addition to conferring adhesive and aggregative properties to bacteria, curli expression is sensed by the host and modulates host immune responses (177, 179).

Curli share many properties with eukaryotic amyloid fibers. Amyloid fibers are typically associated with human neurodegenerative illnesses such as Alzheimer's, Parkinson's, and prion-mediated diseases (180, 181). In contrast to these diseases, which are thought to be due to uncontrolled protein folding, curli belong to a growing class of fibers termed 'functional amyloid', whose expression is controlled and directed for the benefit of the expressing cell (182, 183). Curli assemble as thin, tangled fibers that are extraordinarily stable and impart important physiological properties to bacteria, some of which play significant roles during host-pathogen interactions. The pathway for assembly of curli on the bacterial surface is distinct from the CU pathway and other pilus assembly systems, and instead utilizes an extracellular nucleation-precipitation mechanism, in which curli subunit proteins are first secreted to the cell surface before being incorporated into the growing fiber.

Curli structure

Curli form densely aggregated masses on the bacterial surface (Fig. 4B). Individual curli fibers measure 3 nm to 4 nm in diameter and are of varying lengths (173, 184, 185). Similar to eukaryotic amyloid, curli fibers are non-branching, rich in β-sheet structure, and highly resistant to the action of proteases and denaturants (182, 184, 186). Curli and other amyloid fibers also share the property of binding to specific dyes such as Congo red and thioflavin T (187).

Curli fibers expressed by *E. coli* are composed of repeating copies of the major subunit protein, CsgA (Fig. 4A). Each CsgA molecule contains five conserved repeating units (R1 through R5), which are predicted to form two parallel, stacked β-sheets containing five β-strands each (Fig. 4C) (16, 188). Curli fibers also contain a minor, nucleating subunit; in *E. coli* this is CsgB. CsgB shares 30% sequence identity with CsgA, both proteins are of identical predicted size, and both are built from similar repeat motifs (189). The R1 and R5 repeat units, which flank the N and C termini of CsgA, mediate intersubunit contacts and are important for CsgB-mediated nucleation of the CsgA fiber as well as for CsgA-CsgA polymerization (190). In contrast, the R2-to-R4 internal repeats govern the kinetics of fiber polymerization, slowing the rate of polymerization to limit toxicity during curli production (190). Structural analysis indicates that the individual CsgA subunits in the curli fiber stack on top of each other, forming an extended β-helix-like structure (185). Therefore, the final fiber consists of an expanse of β-sheets oriented parallel to the fiber axis, but with the individual β-strands oriented perpendicular to the fiber axis. This cross β-strand structure is a hallmark of amyloid fibers (181, 185).

The exact binding sites present on curli fibers and the mechanism by which curli bind to a wide range of receptors is unknown. A study examining synthetic peptides corresponding to overlapping regions of the CsgA sequence identified N- and C-terminal

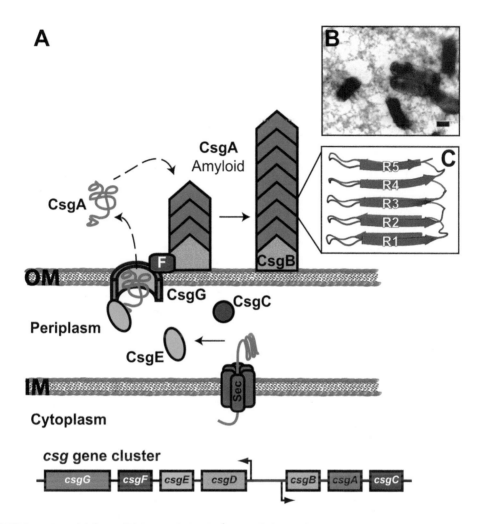

FIGURE 4 (A) Model for curli biogenesis by the extracellular nucleation/precipitation pathway. The *csg* gene cluster coding for curli biogenesis is shown at the bottom. The curli subunit proteins enter the periplasm via the Sec system and are secreted to the bacterial surface via the CsgG outer membrane channel. CsgE may act as a chaperone for the curli subunits in the periplasm, whereas CsgF assists assembly of CsgB on the cell surface. Polymerization of CsgA occurs on the cell surface and is nucleated by interaction with CsgB. (B) Electron micrograph of *E. coli* expressing curli. Scale bar equals 1 μm; reprinted from reference (205) with permission of the publisher. (C) Structure of a CsgA subunit, with the R1-R5 repeats indicated.

regions of 24 and 26 residues, respectively, that recapitulated binding to several different human proteins (176). In addition, CsgB may have a direct role in adhesion separate from its role in nucleating polymerization of CsgA. This is suggested by studies in *S. enterica*, in which a deletion of *csgB*, but not *csgA*, decreased adherence to alfalfa sprouts (191).

Curli assembly on the bacterial surface

The genes required for curli biogenesis in *E. coli* are encoded by the divergently transcribed *csgBAC* and *csgDEFG* operons (Fig. 4A) (192). The *csgBAC* operon encodes the major structural subunit CsgA and the nucleator protein CsgB (189, 192). CsgC is a periplasmic protein with structural similarity to oxido-reductases (193). The function of CsgC is not understood, but it may be important for proper function of the CsgG outer membrane channel (193). Expression of the *csgBAC* operon is dependent on the positive regulator CsgD, which is part of the *csgDEFG* operon (192, 194).

Extracellular nucleation-precipitation
Assembly of curli on the bacterial surface occurs by an extracellular nucleation-precipitation pathway (16). In the absence of the CsgB nucleator protein, curli are not assembled; instead, CsgA is released from the bacteria in an unpolymerized, soluble form (184, 189). This released CsgA can be assembled into curli fibers on recipient cells expressing only CsgB (189). This process, termed interbacterial complementation, demonstrates that curli assembly may take place entirely on the bacterial surface, and that assembly likely involves a conformational change in CsgA that is triggered by CsgB. Thus, CsgA is secreted outside the bacteria as a soluble, unstructured monomer, which is then nucleated into a fiber on the cell surface by interaction with CsgB or with the structurally altered CsgA in the growing curli fiber (Fig. 4A) (182, 195).

Curli assembly machinery
The secretion and polymerization of CsgA is dependent on the CsgE, CsgF and CsgG proteins, the functions of which are not fully understood (192). CsgG is an outer membrane lipoprotein that is thought to form the channel for secretion of CsgA and CsgB to the cell surface (196). In the absence of CsgG, curli fibers are not assembled and the CsgA and CsgB subunits become unstable (184). Consistent with a channel protein, CsgG forms oligomeric, ring-shaped complexes and overexpression of CsgG correlates with increased pore-formation in the outer membrane (196). Structural analysis of CsgG predicts that it belongs to the recently characterized class of transporters that assemble in the outer membrane as α-helical rather than β-barrel channels

(193). The CsgG-mediated secretion of CsgA is dependent on the N-terminal 22 amino acids of the mature CsgA protein. These residues are not predicted to be an integral part of the curli fiber, suggesting that they act as a secretion signal (196, 197).

CsgE is a periplasmic protein and its expression is important for stability of the CsgA and CsgB subunits (184). Consistent with a role in proper folding of the curli subunits, *csgE* mutant bacteria do not act as donors or acceptors for interbacterial complementation and the few curli fibers produced by *csgE* mutants are morphologically distinct from curli expressed by wild-type cells (184). CsgE physically interacts with CsgG at the outer membrane, and CsgE may chaperone periplasmic CsgA subunits to the CsgG secretion channel by interacting with the N-terminal CsgA signal sequence (198). CsgE may also function to prevent premature fiber assembly in the periplasm (198). The CsgF protein also interacts with CsgG at the outer membrane; however, CsgF localizes to the cell surface rather than the periplasm (199). *csgF* mutants have a distinct phenotype, producing reduced levels of curli fibers and secreting soluble, unpolymerized CsgA (184, 199). Similar to *csgB* mutants, *csgF* mutants act as donors but not acceptors for interbacterial complementation. In agreement with this behavior, CsgF influences the folding of CsgB and localization of CsgB to the bacterial surface, suggesting that CsgF functions as an extracellular chaperone for CsgB (199).

Functions of curli in UPEC
Curli are multifunctional surface fibers, conferring adhesion to specific host molecules, promoting bacterial community behaviors such as aggregation and biofilm formation, and modulating interactions with the host immune system (16, 170–172, 178, 200). Curli bind to the extracellular matrix proteins fibronectin and laminin (173, 174). Curli also bind to human contact phase proteins including H-kininogen, fibrinogen, and factor

XII (175, 176). By binding to the contact phase proteins, curliated bacteria slow clotting, which could facilitate bacterial dissemination throughout the host (177, 201). In addition, curli interact directly with molecules of the immune system. MHC class I molecules, which present antigens to T cells, bind to curli and curliated bacteria adhere better to tissue culture cells that over-produce MHC class I (202).

UPEC and other *E. coli* isolates produce curli at the host temperature of 37°C, including UPEC strains freshly isolated from the urine of infected patients (178, 201). This supports a functional role for curli in colonization of the urinary tract. Curli expression enhances adhesion to urothelial cells in cell culture, and the ability to express curli correlates with increased colonization of the urinary tract during early stages of the murine infection model (178, 203, 204). Curli-mediated binding to host molecules may also facilitate uptake inside host cells (16, 172, 173). Expression of curli genes promoted invasion of human epithelial cells by a non-pathogenic K12 strain of *E. coli*, and invasion was inhibited by addition of peptides that blocked curli formation (172, 205).

In addition to binding to specific host molecules, a major functional role of curli in UPEC is likely promoting bacterial aggregation and biofilm formation. Indeed, curli were shown to contribute to biofilm formation by UPEC distinct from the action of type 1 pili (203). Finally, curli expression by UPEC appears to be an important modulator of host immune responses during infection. Curli fibers are recognized by host cells as a PAMP (pathogen-associated molecular pattern) (200). Curli recognition is mediated by TLR2, resulting in the activation of proinflammatory molecules such as IL-6, IL-8, and TNF-α (177, 200). A recent study demonstrated multiple functions for curli during infection of the murine urinary tract: facilitating colonization, protecting the bacteria from the action of host anti-microbial peptides, and provoking an increased pro-inflammatory response (178). Taken together, these results demonstrate that curli play important and varied roles during both initial colonization and subsequent stages of the infectious process.

AUTOTRANSPORTERS AND OTHER NON-PILUS ADHESINS

In addition to assembling adhesins in the form of extended pili or curli fibers, Gram-negative uropathogens also display adhesins directly on their cell surface. The majority of these non-pilus adhesins are assembled on the outer membrane by the autotransporter (type V) secretion pathway (206, 207). Autotransporters are a widespread family of secreted proteins with activities ranging from proteases and toxins to adhesins and invasins. The term autotransporter was first used by Meyer and colleagues to describe the IgA1 protease of *Neisseria meningitidis*, and refers to the idea that a single polypeptide encodes both functional and secretion activities (208). The range of autotransporter functions is reflected by additional well-studied autotransporters, including the NalP protease of *N. meningitidis*, the VacA cytotoxin of *Helicobacter pylori*, and the Pertactin and AIDA-I adhesins of *Bordetella pertussis* and *E. coli*, respectively (209–213). Autotransporters are characterized by the presence of a conserved C-terminal translocator or β-domain that inserts into the outer membrane and directs the secretion of an N-terminal passenger or α-domain, which carries the functional activity, to the cell surface (206, 207). Following secretion, the passenger domain may remain tethered to the outer membrane by the translocator domain or may undergo proteolytic cleavage to be released into the extracellular environment (Fig. 5) (207, 209). In some cases, such as with AIDA-I and related autotransporters, the passenger domain remains associated with the cell surface even after proteolytic cleavage, through noncovalent interactions with the translocator domain (214).

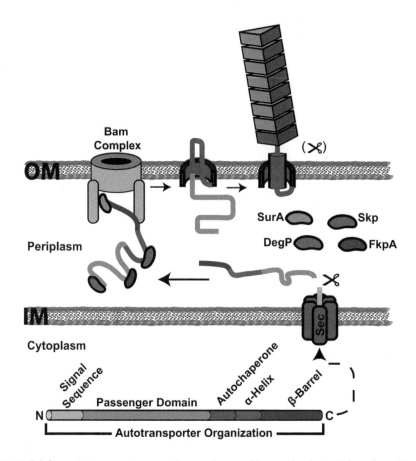

FIGURE 5 Model for autotransporter secretion and assembly on the bacterial surface. The domain organization of an autotransporter protein is shown at the bottom. Autotransporter polypeptides have an N-terminal signal sequence for translocation to the periplasm via the Sec system. The protein is maintained in an extended, largely unfolded state during transit across the periplasm, assisted by periplasmic folding factors (SurA, Skp, DegP and FkpA). The C-terminal translocator domain inserts into the outer membrane as a β-barrel channel, with the assistance of the Bam complex. The Bam complex may also assist in secretion of the passenger domain to the cell surface. In the hairpin model of secretion, the C-terminal region of the passenger domain forms a hairpin structure in the translocator channel, exposing part of the passenger to the cell surface. Folding initiates at the autochaperone region, which then nucleates folding and secretion of the rest of the passenger domain. Following secretion, the linker region adopts an α-helical structure to plug the translocator domain channel. The passenger domain may remain linked to the translocator domain or may be proteolytically cleaved.

The contributions of autotransporters to bacterial pathogenesis in the urinary tract are still being defined, and their identification has largely proceeded from genomics studies and efforts to characterize UPEC-specific virulence factors (Table 1). The relative paucity of information on surface-located adhesins compared to pili is due to the expression of pili or other extended surface structures may obscure or sterically hinder the functions of proteins present at the bacterial surface (215, 216). Thus, autotransporter adhesins are likely to be important under conditions in which pili expression is turned off. Eleven autotransporters have been identified in the genome of UPEC strain CFT073, a prototypical pyelonephritis isolate (153, 217, 218). Seven of these belong to the

AIDA-I family of autotransporter adhesins, and at least four of the UPEC autotransporters function in adhesion to host cells and contribute to fitness in the urinary tract, as discussed below. At least one of the other UPEC autotransporters, Sat, is not an adhesin but is an important protease and toxin of UPEC (219). In addition to UPEC, autotransporter adhesins that contribute to colonization of the urinary tract have also been identified in *P. mirabilis* (220).

This section will provide an overview of the structure and assembly of autotransporters, and will describe the functions of autotransporter adhesins that have been characterized in UPEC. We will also describe two additional non-pilus adhesins expressed by UPEC that are assembled by distinct mechanisms.

Autotransporter Structure

The translocator domain

Autotransporters contain a C-terminal translocator domain and N-terminal passenger domain. The translocator domain is the most conserved feature of autotransporters, whereas passenger domains exhibit a high level of sequence variation (221). Translocator domains belonging to the classical (type Va) autotransporter family are typically 250 to 300 residues in length and insert into the outer membrane to form a β-barrel channel. Crystal structures for several translocator domains have been solved, revealing a typical outer membrane β-barrel structure comprising 12 antiparallel transmembrane β-strands, enclosing a channel of approximately 10 to 13 angstrom (Å) diameter (Fig. 6) (210, 222, 223). An α-helical linker region important for secretion of the passenger domain precedes the β-barrel, and the helix and barrel together have been termed the translocation unit (224–226). In the autotransporter structures, the α-helical linker occupies the lumen of the β-barrel channel (Fig. 6) (210, 222, 223). The N terminus of the α-helix is oriented toward the bacterial

surface, suggesting that translocation of the passenger domain would occur through the β-barrel channel. However, as discussed below, there is debate about the exact mechanism of passenger domain secretion.

Some autotransporters have smaller translocation domains, consisting of only approximately 70 to 80 amino acids. These proteins trimerize to form a single 12-stranded β-barrel, with each monomer contributing 4 strands (Fig. 6) (227–229). The Hia and YadA adhesins of *H. influenzae* and *Y. pestis* are prototypical members of this subfamily (type Vc), termed trimeric autotransporter adhesins (206, 230, 231). The β-barrel formed by the Hia translocator domain has a central pore of 18 Å diameter, which is sufficient to accommodate passage of the three α-helical linker segments that connect to the extracellular passenger domains (Fig. 6) (229). Of the 11 autotransporters identified in UPEC, only UpaG belongs to the trimeric subfamily (232).

The passenger domain

The N-terminal passenger domains of autotransporters are structurally and functionally diverse, but share important conserved features. Most passenger domains are large in size and contain repetitive sequence motifs that assemble into repetitive structural elements (206, 207). The sequence motifs typically form β-sheets that are arranged into an extended β-helix, with each rung of the helix comprising three β-strands in a triangular arrangement (Fig. 6) (212, 233–235). In different autotransporters, this β-helix core structure may be modified with interspersed extended loops, globular domains, or other elements, which contain specific functions such as receptor binding sites or protease activity. However, not all passenger domains have a β-helix architecture. For example, the complete structure of the EstA autotransporter from *Pseudomonas aeruginosa* revealed a typical 12-stranded β-barrel translocator domain, but a globular passenger domain primarily composed of α-helices and loop sequences (Fig. 6) (222).

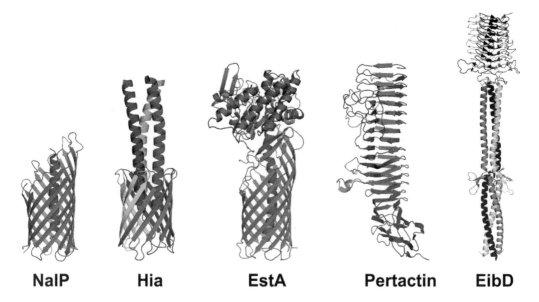

NaIP   **Hia**   **EstA**   **Pertactin**   **EibD**

FIGURE 6 Crystal structures of representative autotransporter proteins. Translocator domains from the monomeric NaIP and trimeric Hia autotransporters are shown (PDB IDs: 1UYN and 2GR7, respectively), with the β-barrel channels in blue and the α-helical linker regions in red. Passenger domains from the monomeric Pertactin and trimeric EibD autotransporters are shown (PDB IDs: 1DAB and 2XQH, respectively), with the approximate location of the Pertactin autochaperone region indicated in purple. The complete structure of the EstA autotransporter is shown (PDB ID: 3KVN), with the translocator domain in blue, the α-helical linker in red, and the globular passenger domain in gray.

Passenger domains from the trimeric autotransporter subfamily exhibit a distinct architecture compared to the classical autotransporters, but also assemble into extended structures built from repetitive sequence elements. Members of this family typically function as adhesins and the passenger domains remain attached to the outer membrane translocator rather than undergoing proteolysis. The passenger domains trimerize, matching the trimeric structure of the translocator domain, and form extended rod-shaped structures. A common architecture is shared by the trimeric passenger domains, comprising a globular N-terminal head region with extensive β-sheet structure, followed by an extended coiled-coil stalk region that connects to the translocator anchor domain in the outer membrane (Fig. 6) (230, 231, 236–238). Some trimeric autotransporters have more complex architectures, with modular arrangements of interspersed head,

neck, and stalk regions (236, 237). The head region typically contains the receptor binding site (Fig. 6), which may be present in each monomer and thus displayed in triplicate around the surface of the trimer, such as for the Hia adhesin (230). The stalk region may also have binding activity for host molecules or mediate bacterial-bacterial interactions (238, 239).

The Autotransporter Secretion Pathway

Transit across the periplasm and insertion into the outer membrane

Nascent autotransporter polypeptides are synthesized with an N-terminal signal peptide that is cleaved following translocation of the protein from the cytoplasm to the periplasm via the Sec general secretory pathway (78). The passenger domain must be kept in a largely unfolded state to remain competent for secretion. Resident periplasmic chaperones and folding factors (DegP, FkpA, SurA, and Skp) interact with the extended autotransporter polypeptide in the periplasm (Fig. 5) to maintain the passenger domain in a secretion-competent state and prevent non-productive interactions, as well as assist in proper folding and insertion of the translocator domain in the outer membrane (206, 207, 240, 241). Some autotransporters have an extended N-terminal Sec signal sequence that slows the rate of translocation across the cytoplasmic membrane through the Sec system (207, 242, 243). This slowing of translocation may facilitate proper transit of the autotransporter polypeptide across the periplasm.

Secretion of the N-terminal passenger domain to the cell surface requires the C-terminal translocator domain, which inserts into the outer membrane to form a β-barrel channel. Recent studies have defined a set of proteins in the bacterial outer membrane, termed the β-barrel assembly machine (Bam) complex, which is responsible for proper insertion of most β-barrel proteins into the outer membrane (244). In keeping with this, correct assembly of the translocator domain in the outer membrane and its proper functioning in autotransporter secretion requires the Bam complex (245–247).

Secretion of the passenger domain to the cell surface

The exact mechanism by which the passenger domain is secreted from the periplasm to the bacterial surface remains a topic of active investigation. Two main models have been proposed for this process: a 'classical' hairpin model and a newer model that invokes a central role for the Bam complex (206, 207).

In the hairpin model, secretion of the passenger domain occurs through the lumen of the β-barrel channel formed by the translocator domain. Secretion of the passenger domain initiates when a C-terminal region of the passenger domain, likely including the α-helical linker region, forms a temporary hairpin structure within the pore (Fig. 5).

Formation of the hairpin exposes part of the C terminus of the passenger domain to the cell surface, where it may begin folding. Folding of the passenger domain would then proceed vectorially from the C to the N terminus, progressively pulling the polypeptide through the channel (248, 249). The hairpin model requires the presence of two strands of the passenger domain polypeptide within the lumen of the translocator domain pore. Given the narrow dimensions of the pore, these strands would need to be in a largely unfolded and extended conformation, which is consistent with studies showing a general lack of tolerance for structured elements in passenger domains (241, 250, 251).

Autotransporter secretion across outer membrane does not require the input of energy from the cytoplasmic membrane, and folding of the passenger domain at the cell surface likely provides the energy to drive secretion through the translocator channel (248, 249). Folding at the cell surface could also act as a ratchet to prevent diffusion of the passenger back into the periplasm. Most passenger domains contain a conserved junction sequence adjacent to the α-helical linker, termed the autochaperone domain. This region, which is critical for the folding of β-helical passenger domains, would be exposed to the cell surface upon formation of the hairpin loop and may act as an intra-molecular chaperone to nucleate folding of the rest of the passenger domain on the cell surface (Figs. 5 and 6) (252–254). Progressive folding of rest of the passenger domain would then occur through a self-templating mechanism, driven by the repeated β-helix structure (255). Once secretion of the passenger domain is complete, the linker region would assume its final α-helical conformation to plug the pore, as observed in the autotransporter crystal structures (Figs. 5 and 6) (210, 222, 223).

The classical hairpin model proposes that the β-barrel channel formed by the translocator domain is sufficient for passenger secretion and that no other accessory factors are required (which is the basis for the term autotransporter). However, conflicts with this model, particularly in studies showing tolerance for secretion of some folded elements and the presence of post-translationally modified passenger domains (256–258), have led to revised models that invoke a central role for the Bam complex (210, 245, 259). In these models, the β-barrel translocator domain serves to target the autotransporter polypeptide to the Bam complex. Interaction with the Bam complex then allows folding and insertion of the translocator domain in the outer membrane, coupled with secretion of the passenger domain to the cell surface. Rather occurring through the translocator domain channel, secretion of the passenger domain would occur, at least in part, through the BamA channel or through some interface between the Bam complex and the translocator domain (207, 210). As the BamA channel is unlikely to be able to gate laterally to allow release of the passenger domain, secretion of the passenger domain would need to occur in concert with insertion of the translocator β-barrel into the outer membrane, and is likely to involve some aspects of the hairpin model (206, 207, 245, 259, 260). Thus, the Bam complex may facilitate coupled formation of the hairpin structure and insertion of the translocator domain into the outer membrane (Fig. 5), and possibly assist in secretion of structured regions. This would then establish an initiating point from which secretion of the remainder of the passenger domain would proceed through the translocator channel as proposed in the classical hairpin model.

Functions of Autotransporter Adhesins in UPEC

Ag43

Antigen 43 (Ag43) is an autotransporter adhesin encoded by the *flu* gene (also termed *agn43*). Ag43 functions in adhesion to host cells and self-associates to promote bacterial aggregation (autoaggregation), leading to flocculation in static liquid cultures and biofilm formation on surfaces (261–264). Ag43 is present in approximately 80% of UPEC strains, and many strains encode more than one copy (265). UPEC strain CFT073 expresses two Ag43 variants, Ag43a and Ag43b. The Ag43a variant appears to be the functionally relevant form in UPEC, promoting high levels of aggregation, biofilm formation, and colonization of the urinary tract (265). Expression of Ag43 is phase variable and opposite from expression of type 1 pili; expression of the longer pilus fibers on the bacterial surface sterically blocks adhesion mediated by the shorter Ag43 molecules (216, 266).

Ag43 belongs to the AIDA family of autotransporters (267). Similar to AIDA-I, the passenger domain of Ag43 is proteolytically processed following transport to the cell surface, but remains associated with the translocator domain via non-covalent interactions. Also similar to AIDA-I, the Ag43 passenger domain is glycosylated in some *E. coli* strains, including UPEC isolates (258). The importance of glycosylation for Ag43 function remains to be determined, as different studies have found variable effects of glycosylation on autoaggregation, biofilm formation, and adhesion to host cells (258, 263). Ag43 promotes adhesion to various cell lines, including human kidney cells, and binds to the extracellular matrix components collagen and laminin (263). The Ag43 passenger domain contains multiple repeats of approximately 19 residues each and folds with an extended, L-shaped β-helical structure (268). The region of Ag43 responsible for autoaggregation is located in the first N-terminal third of the mature passenger domain (261). Recent structural analysis of the Ag43 passenger domain suggests that self-association is mediated by a "Velcro-like" mechanism (268).

Several lines of evidence point to a role for Ag43 in UTIs. Anderson and colleagues found that Ag43 is expressed during bladder infection by UPEC strain UTI89 (143). Specifically, Ag43 was present on the surface of bacteria engaged in formation of intracellular biofilm-like communities following invasion of bladder epithelial cells. Consistent with this observation, the Ag43a variant was found to promote long-term persistence of UPEC strain CFT073 in the murine UTI model; a *fluA* but not *fluB* mutant of CFT073 is present at lower numbers in the bladder compared to the parental strain at day 5 post infection (265). In addition, expression of Ag43 may promote formation of linked bacterial chains that are formed by asymptomatic bacteriuria *E. coli* isolates when grown in human urine (269). The autoaggregation properties of Ag43 are likely to enhance bacterial colonization of the urinary tract as well as formation of intracellular and extracellular biofilms.

UpaB

UpaB is an autotransporter adhesin identified in UPEC strain CFT073, and is widely distributed among both uropathogenic and non-uropathogenic *E. coli* strains (270). UpaB belongs to the AIDA-I family of autotransporters and contains a predicted pertactin-like passenger domain (270). UpaB confers binding to extracellular matrix proteins, including fibronectin, fibrinogen, and laminin. A *upaB* deletion mutant of CFT073 is outcompeted by the wild-type strain for colonization of the bladder, and the mutant strain is specifically defective for an early stage of bladder colonization (270). However, a direct role for UpaB in adhesion to the urinary tract has not been demonstrated. UPEC encode an additional autotransporter related to UpaB, termed UpaC; however, UpaC is not expressed by CFT073 and a UpaC mutant had no phenotype in the murine UTI model (270).

UpaG

UpaG is a trimeric autotransporter adhesin prevalent among extraintestinal pathogenic *E. coli* (ExPEC) strains belonging to the B2 and D phylogenetic groups, including UPEC strain CFT073 (232). The structure of UpaG has been reconstructed from crystal structures of fragments of the homologous *Salmonella enterica* protein SadA (237). UpaG assembles as an extended coiled-coil fiber, ~115 nm in length, containing four YadA-like head repeats and adaptor neck regions typical of trimeric autotransporters (237). Expression of UpaG in CFT073 promotes adhesion to the T24 human bladder epithelial cell line, with specificity for fibronectin and laminin, and promotes autoaggregation and biofilm formation (232). UpaG was identified as a potential protective antigen of ExPEC, suggesting that it is expressed during infection (271). However, native expression of UpaG was not detected in CFT073 grown under *in vitro* conditions, and no role was found for UpaG in colonization of either the bladder or kidneys using the murine infection model (232).

UpaH

The UpaH autotransporter adhesin is expressed by the CFT073 UPEC strain, where it provides a competitive advantage for colonization of the bladder and contributes to biofilm formation (217). UpaH is a large-sized (~280 kDa) member of the AIDA-I family of autotransporters (272). UpaH binds to the extracellular matrix proteins collagen V, fibronectin, and laminin (272). However, a direct role for UpaH in adhesion to the urinary tract has not been demonstrated. The *upaH* gene is present in the chromosomes of many UPEC isolates, as well as in non-uropathogenic *E. coli* strains (217). Bioinformatics analysis predicts a typical 12-stranded β-barrel translocator domain and a large passenger domain with 50 imperfect sequence repeats predicted to encode an extended β-helix structure (217). Sequence variation is present in the UpaH passenger domain from different *E. coli* isolates (272). These variations were found to impact function in biofilm formation but not binding to extracellular matrix proteins.

FdeC

FdeC was identified in a screen for ExPEC vaccine antigens that provided protection in a murine sepsis model (273). The *fdeC* gene is widely distributed among ExPEC and also intestinal *E. coli* strains (274). FdeC shares a low level of sequence homology with the invasin and intimin proteins of *Yersinia pseudotuberculosis* and enteropathogenic *E. coli*, respectively, which function in adhesion to host cells (275, 276). Similar to these proteins, FdeC is anchored in the outer membrane via a presumed N-terminal β-barrel domain, with the extracellular portion of the protein forming an elongated structure comprising nine repeated Ig-like domains (274). A model was recently proposed that proteins such as intimin and invasin form a new subfamily of autotransporters (type Ve) (206, 277). In this model, the proteins are secreted in an analogous mechanism to autotransporters, but with a reverse topology (i.e., the outer membrane translocator domain is located at the N terminus instead of the C terminus as for typical autotransporters). In contrast to intimin and invasin, no obvious lectin domain is present in FdeC (274). Recombinant FdeC binds to human urothelial cell lines, as well as other types of epithelial cells, with specificity for collagen (274). FdeC is expressed during interactions with host cells and during infection of the urinary tract, and a *fdeC* mutant of UPEC strain 536 was defective for colonization of the bladder and kidneys during co-infection with the wild-type strain (274).

Other Outer Membrane-Associated Adhesins of UPEC

At least two additional non-pilus adhesins that contribute to pathogenesis in the urinary tract have been identified in UPEC. These adhesins are assembled on the bacterial outer membrane by mechanisms that are distinct from the autotransporter pathway. The Iha adhesin is secreted by the type I secretion pathway and TosA, a multifunctional

siderophore receptor and adhesin, is an integral outer membrane protein.

Iha

The IrgA homologue adhesin (Iha) protein, encoded by the *iha* gene, is prevalent among UPEC strains and has the novel phenotype of functioning in both adhesion and iron uptake (278, 279). Iha was originally identified in the diarrheagenic *E. coli* strain O157:H7 as an adhesin with homology to the *Vibrio cholerae* iron-regulated virulence factor IrgA (280). Iha is an outer membrane protein with homology to β-barrel siderophore receptor proteins such as FepA (281). *Iha* is a virulence factor of UPEC, as demonstrated by reduced fitness of a CFT073 *iha* deletion mutant for colonization of both bladder and kidneys in the murine UTI model (279). Similarly, an *iha* mutant of the UPEC clonal group A outbreak strain UCB34 (282) was attenuated for infection of the urinary tract in competition with the parental wild-type strain (278). There is also evidence supporting a role for *iha* in the pathogenesis of UTIs caused by *Proteus mirabilis* (283). *Iha* is expressed *in vivo* during infection of the murine urinary tract (278). Expression of recombinant Iha promoted adhesion to human epithelial cells, including the T24 bladder cell line, whereas overexpression of other siderophore receptors did not promote adhesion (278, 279). In addition to its function as an adhesin, Iha functions in iron uptake as an iron-regulated catecholate siderophore receptor (278). Given the likely topology of Iha as in integral outer membrane β-barrel protein, adhesive activity is presumably conferred by surface-exposed loops of the protein. Whether the adhesin function, iron uptake, or both are important for pathogenesis remains to be determined.

TosA

TosA belongs to the repeats-in-toxin (RTX) family of secreted bacterial proteins that have diverse functions, including acting as toxins, proteases, and adhesins (284). RTX proteins share the characteristic features of repetitive glycine- and aspartate-rich sequences, located in the C-terminal region of the protein, and use of the type I secretion system for export out of bacteria. Type I secretion systems function in the secretion of a variety of toxins and other virulence factors directly from the cytoplasm to the extracellular milieu in a single energized step (285, 286). The type I system comprises three components: an outer membrane channel-forming protein (TolC in *E. coli*), a periplasmic adaptor or membrane fusion protein, and an inner-membrane pump that typically belongs to the ATP-binding cassette family. In contrast to most RTX proteins, TosA remains associated with the bacterial surface following secretion, rather than being released into the external environment (287). The *tosA* gene is present in a pathogenicity island in UPEC strains, particularly those of the B2 phylogenetic group, in an operon together with genes encoding a type I secretion system (287, 288). *tosA* is expressed *in vivo* during infection of the urinary tract (287) and a CFT073 Δ*tosA* mutant was defective for colonization of both the bladder and kidneys in the mouse UTI model (289). Evidence suggests that *tosA* also enhances fitness during disseminated infections (287). Expression of TosA promotes adherence to both murine and human kidney epithelial cells, but does not appear to be important for colonization of the lower urinary tract (287).

ADHESINS EXPRESSED BY GRAM-POSITIVE UROPATHOGENS

While Gram-negative bacteria are responsible for the majority of UTIs, Gram-positive bacteria are also significant uropathogens. *Staphylococcus saprophyticus* is the second leading cause of community acquired UTIs in sexually active women, accounting for approximately 15% of all incidences (290). *Enterococcus faecalis*, a normal member of the gut flora, is also a causative agent of UTIs,

particularly in nosocomial infections (291). The adhesive organelles of Gram-positive bacteria differ significantly from those of Gram-negative bacteria in structure and assembly mechanism, but fall into the same two general classes of pilus or surface-located adhesins. Gram-positive pili, like their Gram-negative counterparts, consist of multiple pilin subunits linked together to form an extended, hair-like fiber (21, 22, 292). The majority of non-pilus adhesins associated with UTIs are multi-domain proteins known as MSCRAMMs (microbial surface components recognizing adhesive matrix molecules), which are anchored directly to the bacterial cell wall (22, 23).

MSCRAMMs

The extracellular matrix is a complex protein network that serves as the major scaffolding component of eukaryotic cells and tissues, and mediates numerous essential cellular processes including morphogenesis and differentiation (293, 294). Gram-positive bacteria utilize components of the extracellular matrix such as collagen, fibronectin, and laminin as binding ligands to promote adherence, colonization, and biofilm formation (295–297). To achieve this, Gram-positive bacteria express the MSCRAMM family of surface-associated adhesins. MSCRAMMs are multi-domain proteins that are covalently linked to the peptidoglycan cell wall, exposing both conserved and non-conserved regions to the extracellular milieu. Different MSCRAMM domains confer specific functionality, including ligand binding, cell wall anchoring, and structural integrity (22). Crystal structures have been solved for domains from several different MSCRAMMs, including SdrG from *Staphylococcus epidermidis*, Ace from *E. faecalis*, and Cna and ClfA from *Staphylococcus aureus*, providing a structural context for understanding the assembly and functions of MSCRAMMs (298–305). MSCRAMMs that have been associated with uropathogenic bacteria are Ace, expressed by *E. faecalis*, and UafA, UafB and SdrI, expressed by *S. saprophyticus* (306–309).

Structure of MSCRAMMs

All MSCRAMMs share a common domain organization. The N terminus contains a signal peptide that directs the proteins for translocation across the cytoplasmic membrane by the Sec pathway. The N-terminal signal peptide is followed by a multi-domain central region, where the major functional and structural diversity resides (22). The C terminus contains a conserved cell wall sorting signal (CWSS), which consists of the amino acid sequence LPXTG (X represents any amino acid), followed by a hydrophobic transmembrane domain, followed by a positively charged cytoplasmic tail (Fig. 6A). This CWSS is required for the covalent linkage of MSCRAMMs to the cell wall (310).

The MSCRAMM functional region is typically divided into A and B regions. The N-terminal A region is responsible for ligand binding and specificity; the C-terminal B region can have both binding and structural properties. Variations on this organization occur, including the presence of an R structural region instead of or in addition to a B region (22). The C-terminal structural domains function to project the binding domain away from the bacterial surface. Similar to the CU pilins of Gram-negative bacteria, MSCRAMMs, as well as Gram-positive pilins, have exploited the Ig fold as a common building block (58). Two Ig variants present in both MSCRAMMs and Gram-positive pilins are the DEv-IgG and IgG-rev folds (Figs. 7B and 8) (22). These folds were first observed in the A and B regions of the *S. aureus* Cna and ClfA MSCRAMMs (305, 311). A typical IgG constant domain contains two β-sheets (sheets 1 and 2) of four and three β-strands, arranged into a barrel configuration. DEv-IgG folds contain the same overall structure, but with the addition of at least two β-strands between strands D and E of sheet 1. The IgG-rev fold has a typical two sheet, seven-stranded barrel, but the strands

are arranged in a reverse order compared to typical Ig folds.

Ace and the collagen hug model

The *E. faecalis* Ace (<u>a</u>dhesin to <u>c</u>ollagen of *E. faecalis*) protein was the first MSCRAMM to be associated with UTIs (295, 297, 306, 312). Much of the structural organization of Ace was delineated based on its homology to Cna (295, 302, 303). The functional portion of Ace is made up of A and B regions. The A region is divided into two DEv-IgG subdomains, N_1 (146 amino acids) and N_2 (135 amino acids). These domains are essential for binding extracellular matrix proteins such as laminin and collagen I and IV (297). Crystal structures have been solved for the Ace N_1 and N_2 subdomains (Fig. 7B) (300, 301). Both subdomains are composed of ten β-strands forming two β-sheets in a sandwich like configuration. A missing G strand in sheet 1 of the N_1 subdomain is complemented by a C-terminal extension of the N_2 subdomain, forming an interface between the two subdomains (Fig. 7B). This interface, in addition to a collagen-binding site on the N_2 subdomain and a linker region connecting the two subdomains, forms a deep, tunnel-like, collagen-binding trench. Together, the N_1 and N_2 subdomains form a dynamic cooperative structure utilizing a "collagen hug" model of ligand binding (301).

Structural and functional studies of both Cna and Ace have led to a mechanistic understanding of ligand binding by the A region. The structures of the N_1 and N_2 subdomains of the A region show a closed form of the molecule, in which the N_1 and N_2 subdomains and the inter-domain linker interact with each other, creating a tunnel to accommodate and secure collagen (Fig. 7B) (302). This closed form of the molecule is unable to initiate binding to collagen, and evidence indicates that the N_1 and N_2 subdomains exist in equilibrium between open and closed conformations (301). In the open conformation, a shallow groove in the N_2 subdomain binds the repeating glycine-proline-hydroxylproline

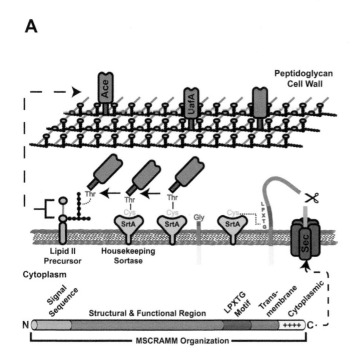

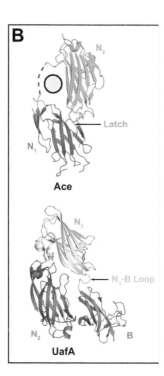

FIGURE 7 (A) Model for MSCRAMM secretion and incorporation into the cell wall. The domain organization of a typical MSCRAMM is shown at the bottom. MSCRAMMs have an N-terminal Sec signal sequence for translocation across the cytoplasmic membrane. The protein remains anchored in the cytoplasmic membrane by the CWSS. The positively charged C terminus remains in the cytoplasm, orienting the LPXTG motif to the extracellular side of the membrane. The SrtA sortase cleaves between the Thr and Gly of the MSCRAMM LPXTG motif, forming a covalent thioacyl intermediate. The MSCRAMM is then transferred to a lipid II peptidoglycan precursor and finally integrated into the cell wall at an amino acid cross-bridge. (B) Crystal structures of the Ace and UafA MSCRAMMs (PDB IDs: 2Z1P and 3IRP, respectively). The upper structure shows the N_1 and N_2 subdomains of Ace in blue and green, respectively; the yellow circle represents bound collagen. Both domains have DEv-Ig folds. The C terminus of the N_2 subdomain inserts into the N_1 subdomain, forming a latch. The lower structure depicts the N_2, N_3, and B subdomains of UafA. The N_2 and N_3 subdomains adopt DEv-Ig folds and the B subdomain adopts a variant of the IgG-rev fold. The loop connecting the N_3 and B domains (cyan) is thought to insert into the N_2 subdomain upon ligand binding to form a latch.

$(GPO)_n$, triple helical peptide of collagen with low affinity. Following binding, the C-terminal extension of the N_2 subdomain orients and inserts into a trench in the N_1 subdomain, complementing a missing N_1 β-strand and allowing the N_1 and N_2 subdomains to come into close proximity of each other, shrinking the hole and "hugging" the collagen molecule in place (301, 302). The N_2 C-terminal extension also acts as a latch, securing the complex. Truncations of the latch cause a decreased affinity for collagen, likely due to the N_1 and N_2

subdomains insufficiently securing the collagen in place (301).

The B region of Ace has not been resolved, but a high degree of sequence homology between this region and the B region of Cna provides insight into its structural organization and function. The Ace B region comprises five repeating domains, B_{1-5} (301). The crystal structure of the Cna B domain indicates that it functions as a structural element rather than in ligand binding (303). The Cna B_1 domain is divided into two IgG-rev

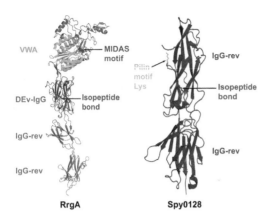

FIGURE 8 Crystal structures of representative tip and major pilins. The RrgA tip pilin of *S. pneumoniae* (PDB ID: 2WW8) is shown on the left, with the fold adopted by each of the four subdomains indicated. The VWA domain is depicted in green, with the residues forming the MIDAS motif and bound magnesium ion shown in purple. The residues involved in intramolecular isopeptide bond formation are shown in red. The Spy0128 major pilin of *S. pyogenes* (PDB ID: 3B2M) is depicted on the right in purple and the two subdomains are labeled as for RrgA. The lysine side chain of Syp0128 thought to be involved in intermolecular isopeptide bond formation is shown in cyan.

fold subdomains, D_1 and D_2. Domains B_{2-5} share the same structure, and the repeating B domains are arranged in an accordion-like fashion, which acts as a stalk supporting the A region and extending it distally from the surface of the cell (303). C terminal to the B region is the characteristic CWSS, which allows anchoring to the peptidoglycan cell wall by the housekeeping sortase enzyme.

UafA and the dock, lock, and latch model
Another MSCRAMM shown to impact bacterial adherence in the urinary tract is UafA (uro-adherence factor A), expressed by *S. saprophyticus* (307). Information obtained from MSCRAMMs such as the *S. aureus* Cna and ClfA, as well as the recent crystal structure of the functional region of UafA, have provided insight into the structural organization and function of UafA (Fig. 7B) (313).

UafA contains an N-terminal Sec signal sequence, a 72 kDa A region, a 13 kDa B region, a 148 kDa R region, and a C-terminal CWSS. Unlike Ace however, the UafA A region comprises three subdomains, N_{1-3}, and the B region contains a single, non-repeating domain. The additional R region is rich in serine and glutamine residues and of low complexity (313).

The A region of UafA is required for ligand binding and shares the same variant DEv-IgG folds in its N_2 and N_3 subdomains as seen in Ace, Cna, and ClfA (305, 313). The single B domain does not act as a stalk to extend the A region from the surface, but instead is a required component in ligand binding (313). A short loop links the B domain to the N_3 subdomain, and in the three-dimensional structure the B domain also resides adjacent to the N_2 subdomain (Fig. 7B). The low-complexity R region is thought to form the UafA stalk, supporting both the upstream A and B regions (313).

A "dock, lock, and latch" model has been proposed for UafA, based on similarities with ClfA and the *S. epidermidis* fibrinogen-binding protein SdrG (298, 299, 304). This model is mechanistically related to that of the collagen hug model of Ace and Cna, but has distinct features. The N_3 subdomain is composed of two β-sheets with a total of nine β-strands in the configuration of A, B, D, E, and C, D1, D2, F, G (313). The N_2 subdomain contains two similar β-sheets, but sheet 2 lacks a D β-strand and in its place exists a D loop, which does not appear to hydrogen bond with the adjacent E strand. The dock, lock and latch model proposes that in the apo form of the complex, the ligand is first captured by binding between the N_2 and N_3 subdomains. This docking event then triggers a conformational change to engage the loop connecting the N_3 subdomain and the B domain (Fig. 7B), causing the loop to insert into the pocket between the D and E β-strands of the N_2 domain, completing the missing β-strand and forming the "latch" (298).

The ligand for UafA has not yet been defined. UafA functions in hemagglutination and this activity is dependent on the N_2 and N_3 subdomains, as well as the B domain (313). Hemagglutination studies with proteinase K-treated erythrocytes suggest that the ligand is not a protein, but may instead be a carbohydrate or lipid molecule (313). In addition, the B domain may also have secondary ligand binding properties of its own, independent of the N_2 and N_3 subdomains (313).

Assembly of MSCRAMMs

The majority of surface exposed proteins in Gram-positive bacteria utilize a highly conserved assembly mechanism that facilitates translocation of these proteins across the cytoplasmic membrane and their covalent linkage to the peptidoglycan cell wall. As with most Gram-negative adhesins, MSCRAMMs are synthesized with an N-terminal signal sequence that directs the proteins for translocation across the cytoplasmic membrane via the Sec general secretory pathway. The signal sequence is then cleaved following translocation by the signal peptidase (22, 314). Since Gram-positive bacteria lack an outer membrane, proteins translocated across the cytoplasmic membrane will be lost to the extracellular milieu unless they are anchored to the peptidoglycan cell wall (315). Proteins destined to be anchored to the cell wall, including MSCRAMMs, contain the highly conserved CWSS, comprising the LPXTG motif, a stretch of hydrophobic residues, and a positively charged C-terminal tail. Mutations to this motif abrogate cell wall anchoring (316). Following passage across the cytoplasmic membrane via the Sec pathway, the hydrophobic region of the CWSS forms a transmembrane domain, with the positively charged tail orienting the domain such that the C terminus remains in the cytoplasm and the LPXTG motif is exposed on the extracellular surface of the membrane (Fig. 6A). Anchoring of the secreted protein to the call wall is then achieved by processing of the LPXTG motif by membrane-associated proteins known as sortases (317, 318).

Sortases are cysteine transpeptidases responsible for sorting covalently linked cell wall proteins to the bacterial surface, including many virulence factors (319). While different classes of sortases exist, most Gram-positive bacteria express a general housekeeping sortase required for displaying a broad spectrum of proteins with distinct functions (320). Sortase A (SrtA) is the best characterized of these housekeeping sortases (318). SrtA is responsible for processing a newly secreted protein's CWSS by catalyzing a transpeptidation reaction between the LPXTG motif and an amino acid cross-bridge of the peptidoglycan (Fig. 7A). SrtA cleaves the carbonyl carbon between the threonine and glycine residues of the LPXTG motif via nucleophilic attack by the conserved active site cysteine of the sortase (318). This cleavage facilitates the creation of a thioacyl bond between the SrtA cysteine and the threonine of the surface protein, resulting in the two proteins being covalently linked together (Fig. 7A). SrtA then transfers the covalently linked protein to lipid II, a membrane bound peptidoglycan precursor. An amino group of lipid II nucleophilically cleaves the sortase-surface protein thioacyl linkage, forming an isopeptide bond with the surface protein threonine and creating a lipid II-surface protein complex (Fig. 7A). The lipid II-surface protein complex is then modified by transpeptidases and transglycosylases during peptidoglycan synthesis. The lipid moiety is processed and the resulting protein and peptidoglycan fragment is integrated into to the cell wall, covalently anchoring the protein to the peptidoglycan amino acid cross-bridge and exposing it to the surface (Fig. 7A) (321, 322).

Functions of MSCRAMMs in uropathogenic bacteria

The initial stage of adherence and colonization is absolutely necessary for pathogenic

Gram-positive bacteria to establish successful infection (323). MSCRAMMs mediate this process by facilitating the recognition and binding of surface exposed host ligands to promote colonization of specific tissues. In the context of UTIs, Gram-positive uropathogens express and assemble specialized MSCRAMM molecules with a distinct tropism for urogenital epithelia. This tropism is dependent both on the ligand expressed by the host tissue, in most cases a component of the extracellular matrix, as well as the affinity for that ligand by the bacterial adhesin. However, compared to well-characterized Gram-negative adhesins such as type 1 and P pili, much less is known about the mechanisms of MCRAMMs during bacterial infection of the urinary tract.

Ace

E. faecalis, a major contributor of endocarditis and infections of the blood, wounds and abdomen, is also associated with high incidences of nosocomial catheter associated urinary tract infections (CAUTIs) (324, 325). The Ace A domain is required for recognition and binding of extracellular matrix molecules. *In vitro* mutational analysis as well as competitive inhibition studies of the A domain show that the protein has binding affinity for laminin and types I and IV collagen, each of which is a major component of the extracellular matrix (297). Ace has been identified as an important virulence determinant of *E. faecalis* in UTIs, using the murine infection model. Lebreton and colleagues showed that significantly higher doses of an *E. faecalis ace* deletion mutant are required to establish infection in mice when compared to wild-type *E. faecalis* (306). Furthermore, organ burden analysis revealed that an *ace* deletion mutant is attenuated in its ability to colonize renal tissue (306, 312). While current research implicates collagen as the major Ace binding ligand, further work needs to be done to determine the role of Ace and its binding partners in the establishment of UTIs.

UafA

S. saprophyticus holds significant prevalence as a causative agent of uncomplicated UTIs (326). Comparative genomic analysis between *S. aureus*, *S. epidermidis*, and *S. saprophyticus* revealed that *S. saprophyticus* contains the unique MSCRAMM UafA (307). Expression of *uafA* by *S. saprophyticus* promotes adherence to human bladder carcinoma cells, while deletion of the *uafA* gene causes decreased adherence (307). Furthermore, strains of *S. saprophyticus* that express UafA, in addition to other surface proteins such as SdrI and Ssp, are internalized into human bladder carcinoma cells (327). Although the precise ligand or tissue tropism of UafA remains unknown, preliminary data suggest that the ligand may be a carbohydrate or lipid molecule rather than a protein (313).

UafB

UafB is a recently discovered, plasmid-encoded MSCRAMM of *S. saprophyticus* that affects bacterial adherence to uroepithelial cell lines (308). In addition to a predicted N-terminal Sec signal sequence and C-terminal CWSS, the majority of the 2279 residue UafB protein comprises three serine-rich tandem repeats (repeats 1–3) and a single non-repeating region. The non-repeating region lies between repeats 1 and 2, and is a putative binding domain. The third repeat region is the longest of the three and is located just upstream of the CWSS. *S. saprophyticus* UafB is predicted to be glycosylated on the surface of *S. saprophyticus* (308). Although less prevalent than UafA among *S. saprophyticus* isolates, strains expressing UafB exhibit increased adhesion to human bladder carcinoma cells when compared to a *uafB* mutant (308). Analysis of the putative non-repeat binding domain indicates that UafB binds both fibronectin and fibrinogen, but not collagen types I, II, or IV; laminin; or vitronectin (308). This information may suggest that UafB has a tropism for bladder epithelium since human bladder carcinoma cells abundantly express fibronectin (328).

However, a role for UafB during infection remains to be established, as both wild-type and *uafB* knockout strains equally colonize mouse bladders in a murine ascending UTI model (308).

SdrI

SdrI is a recently discovered MSCRAMM in *S. saprophyticus* classified by its serine-aspartate repeat (SD or Sdr) region, which is indicative of the Sdr family of MSCRAMMs. Sequence analysis of SdrI shows that it contains a C-terminal CWSS, an N-terminal A region, a B region containing two repeat domains, and the SD region (329). Initial binding experiments showed that when SdrI was deleted, *S. saprophyticus* exhibited a significant deficiency in adhesion to collagen compared to wild-type bacteria (329). Further work indicated that SdrI also has affinity for fibronectin, which is dependent on its A domain (296). This is unique since SdrI does not contain any known fibronectin binding motifs. *In vivo* data using a murine UTI model revealed that *sdrI* mutant *S. saprophyticus* bacteria do not have defects in initial colonization of the bladder or kidneys when compared to wild-type bacteria; however, the mutant strain is cleared faster from these organs (309). This suggests a role for SdrI in the persistence of *S. saprophyticus* in the urinary tract, rather than in initial adhesion or colonization.

Additional surface-located adhesins

A unique class of Gram-positive surface exposed adhesins has been discovered that lack both a Sec secretion signal as well as a cell wall localization LPXTG motif (330). These adhesins are not classified as MSCRAMMs or pili. While little has been elucidated as to their structure, one such adhesin, the enterococcal fibronectin-binding protein A (EfbA), expressed by *E. faecalis*, has been implicated in mouse models of ascending urinary tract infections. An *efbA* deletion strain was significantly attenuated in its ability to bind immobilized fibronectin *in vitro* as well as mouse kidneys and bladders *in vivo* (331).

An additional protein, Esp, has been implicated in colonization and persistence of *E. faecalis* in the urinary tract (332). Esp has not been classified as an MSCRAMM, but it contains an N-terminal Sec signal sequence, a central region with a non-repeat domain followed by a number of repeating sequences, and a CWSS-like region at the C terminus with a variation of the LPXTG motif. The N-terminal region of Esp promotes biofilm formation *in vitro*, but may do this indirectly (333).

GRAM-POSITIVE PILI

Although first described in 1968, the assembly of pili by Gram-positive bacteria has only been widely recognized and characterized since the work of Ton-That and Scheewind beginning in 2003 (17–20). Multi-subunit, peptidoglycan-linked adhesive pili have now been described on the surface of a number of Gram-positive bacteria (21, 22). Like MSCRAMMs, pili play important roles in binding to and recognizing extracellular matrix molecules, colonization of host tissues, and biofilm formation. Gram-positive pili are expressed from gene clusters encoding one or more minor pilins, a major pilin, and pilus assembly machinery in the form of pilus-specific sortases. Gram-positive pili expressed by *C. diphtheria* and Streptococcal pathogens have been well studied and serve as prototypes (334–338). Although similar to Gram-negative pili in their general ultrastructure (hair-like polymeric fibers) and functions (adhesion to and colonization of surfaces), Gram-positive pili have unique features, including the presence of intramolecular and intermolecular covalent bonds that stabilize the fibers against factors encountered in the extracellular environment. Currently, *E. faecalis and E. faecium* are the only Gram-positive bacteria known to express pili associated with UTIs. These pili were first identified as biofilm determinants during endocarditis infections and are named

Ebp for endocarditis and biofilm associated pilus (339, 340).

Structure of Gram-positive pili

The pilus fiber

Gram-positive pili form thin, extended fibers one protein subunit wide (approximately 3 nm diameter) and up to several micrometers in length. The pilus fiber is built from repeating copies of covalently linked major pilus subunits, with minor pilins covalently incorporated at different points in the fiber. Minor pilins that function to recognize and bind target host ligands are known as tip pilins, while pilins that anchor the entire pilus fiber to the cell wall are referred to as base pilins. The major pilin forms the pilus shaft, which functions to extend the adhesive tip pilin distally from the bacterial surface.

A number of crystal structures have been solved for Gram-positive pilins, including from *C. diphtheria*, Streptococcal *species*, and *Bacillus cereus* (Fig. 8) (22, 334–338, 341). Pilins share a common domain organization with MSCRAMMs, comprising an N-terminal Sec secretion signal, a central structural region, and a C-terminal CWSS (Fig. 9). The central structural region typically contains multiple N subdomains that adopt DEv-IgG or IgG-rev folds (Fig. 8) (335, 342). An important characteristic of Gram-positive pilins is their ability to form covalent, intramolecular isopeptide bonds between lysine and asparagine residues of the DEv-IgG and IgG-rev folds (Fig. 8) (343). These isopeptide bonds promote resistance to proteases and increase pilin structural stability (334). DEv-IgG folds typically possess a D-type isopeptide bond, in which a lysine residue in the A β-strand of sheet 1 forms a bond with an asparagine on the antiparallel F β-strand of sheet 2 (Fig. 8). A C β-strand aspartic acid residue on sheet 2 catalyzes this reaction. IgG-rev folds exhibit an E-type bond where a lysine on the A β-strand of sheet 1 is covalently linked to the asparagine of the parallel G β-strand, also of sheet 1

(Fig. 8). A glutamine residue on the sheet 2 E β-strand catalyzes this reaction (22).

In addition to their LPXTG motif, major pilins also have a conserved pilin motif, typically of the sequence VYPK, housing an essential lysine residue (22). An intermolecular isopeptide bond is formed between the lysine of the pilin motif and the threonine of the LPXTG motif of a neighboring subunit in the pilus fiber, catalyzed by a pilus-specific sortase (Fig. 9). Tip pilins have an LPXTG motif but typically lack a pilin motif, and therefore can only be incorporated at the beginning of the pilus fiber. Base pilins, which contain both LPXTG and pilin motifs, are linked to the peptidoglycan cell wall via their LPXTG motif by the housekeeping sortase, thus anchoring the pilus fiber to the bacterial surface (Fig. 9) (336). In some instances, base pilins can also be incorporated along the length of the pilus fiber (19, 344). Some pilins, particularly major pilins, may also possess a third motif called an E-box, which contains an invariant glutamic acid residue. Little work has been done to elucidate the role of the E-box in pilus assembly, but it may facilitate integration of minor pilins throughout the pilus fiber (19).

Ebp pili

The Ebp operon, found in both *E. faecalis* and *E. faecium*, consists of four genes, *ebpA*, *ebpB*, *ebpC*, and *srtC* on what is termed a pilus island on the bacterial chromosome (Fig. 9) (340). Little is known about the structure of Ebp pili, but studies have shown that EbpA and EbpB are minor pilins, with EbpA serving as the tip pilin and EbpB as the base pilin (340). EbpC is the major structural pilin. SrtC is a class C pilus specific sortase required for polymerization of the pilus fiber (340, 345, 346). Sequence analysis of these pilins suggests that they share structural characteristics with known MSCRAMMs, having potential Dev-IgG and IgG-rev domains (347). EbpA and EbpC also possess predicted E-box motifs (340, 347).

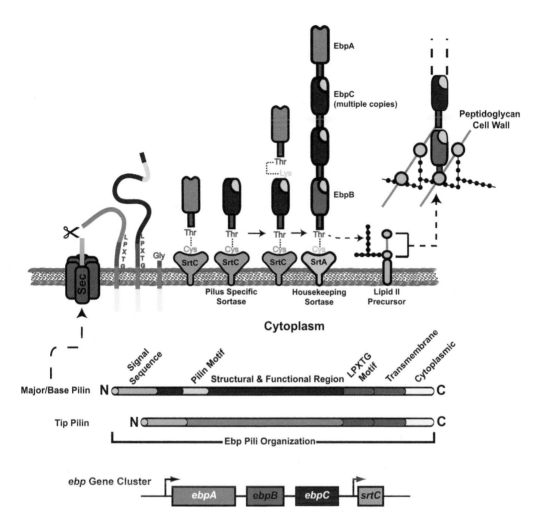

FIGURE 9 Model for Gram-positive pilus polymerization and incorporation into the cell wall. The domain organization of typical major and minor pilins is shown at the bottom, along with the *ebp* gene cluster coding for Ebp pili of *E. faecalis*. The steps of secretion across the cytoplasmic membrane and covalent linkage to a sortase are the same as for MSCRAMMs, except the pilins are processed by the SrtC pilus-specific sortase. Pilus subunits are polymerized by formation of intermolecular isopeptide bonds between the Lys of a pilin motif of one subunit and the Thr of the LPXTG motif of a preceding subunit in the fiber. Linkage to the cell wall occurs when a growing pilus fiber is transferred to a base pilin bound to the SrtA housekeeping sortase. Integration of the pilus into the cell wall follows the mechanism as described for MSCRAMMs.

Sequence and mutational analysis of EbpA predicts that the tip pilin contains a von Willebrand factor A (VWA) domain with a metal ion-dependent adhesion site (MIDAS) (Fig. 8), which facilitates adhesion to host molecules (345, 348). VWA domains are found in proteins from different kingdoms and contain a β-sheet surrounded by a number of α-helices. Many VWA domain proteins also contain a MIDAS motif (348). Functional analysis of the EbpA MIDAS motif indicates that it is essential for EbpA function (348, 349). While the ligand specificity of EbpA remains unknown, other VWA domain- and

MIDAS motif-containing proteins bind extracellular matrix proteins such as collagen (350).

Assembly of Gram-positive pili

The assembly mechanisms of Gram-positive pili in bacteria such as *C. diphtheriae* have been well studied and share many commonalities with the anchoring of MSCRAMMs to the cell wall (19, 20, 292). Additional pilus-specific features, including the pilin motif and class C sortases, are required for biogenesis of the multi-subunit pilus fiber. The initial stages of pilus assembly are essentially the same as for all LPXTG cell wall anchored proteins. Like MSCRAMMs, Gram-positive pilins contain a cleavable N-terminal signal sequence for targeted translocation across the cytoplasmic membrane via the general Sec pathway. Following translocation, the transmembrane domain and the positively charged cytoplasmic tail of the CWSS retains and orients the pilin as so to make the LPXTG motif accessible for cleavage by a sortase (Fig. 9). Gram-positive pilus gene clusters typically encode a class C pilus specific sortase (srtC), which is essential for pilus polymerization (340). Differences between the pilus specific and housekeeping (SrtA) sortases determine how the LPXTG motif of individual pilins are processed, either for covalent linkage to the lipid II peptidoglycan precursor via SrtA, or for the formation of structural pilin-pilin isopeptide bonds via SrtC.

For both major and minor pilins, SrtC cleaves between the threonine and glycine residues of the pilin LPXTG motif, forming an acyl-enzyme intermediates between the active site cysteine of the sortase and the threonine of the pilin (Fig. 9). Major pilins contain a pilin motif (VYPK) in addition to the LPXTG motif, whereas tip pilins typically only contain the LPXTG motif. To incorporate the tip pilin into the pilus fiber, SrtC catalyzes formation of an intramolecular isopeptide bond between the lysine of the pilin motif of the major pilin with the threonine of the LPXTG motif of the tip pilin (Fig. 9). This reaction releases the tip pilin from its SrtC molecule, forming a tip pilin-major pilin-SrtC complex (351). This process is then repeated to add additional major pilins to the complex, elongating the pilus structure and forming the pilus shaft to project the tip pilin distally from the bacterial surface. Pilus elongation is terminated via incorporation of a minor base pilin into the pilus structure by the same mechanism (21, 292). Following elongation, the pilus fiber must be securely anchored to the peptidoglycan cell wall. This is achieved by the SrtA housekeeping sortase, which catalyzes isopeptide bond formation between the LPXTG motif of the base pilin and a lipid II amino group (Fig. 9) (345). As for MSCRAMMs, the lipid II-pilus fiber complex is then processed by transpeptidases and transglycosylases, covalently securing the pilus to the peptidoglycan cross-bridge.

Function of Ebp pili in UTIs

The Ebp pili expressed by *E. faecalis* and *E. faecium*, two major causative agents of hospital acquired CAUTI, are the only Gram-positive pili that have been associated with UTIs (345, 352). The connection between Ebp pili-mediated biofilm formation and infection was first investigated in a rat endovascular model (340, 353). Data from this work showed that strains with deletions of each individual pilin, or a deletion of *srtC* alone, had significant defects in biofilm formation *in vivo*. Additionally, a deletion strain of just the tip pilin *ebpA* was significantly attenuated for colonization of aortic vegetations and kidneys, when compared to wild-type bacteria, suggesting a role for Ebp pili in UTIs as well as endocarditis (340). Later studies confirmed the role of Ebp pili in colonization of the urogenital tract, using *ebpA* and *ebpC* deletion strains to show a colonization defect in both the kidneys and bladders of mice (346, 352). Similar results were observed in experiments using *ebp* deletion mutants in *E. faecium* (339).

Experiments have also been done to identify the role of Ebp pili in CAUTI. Using a murine CAUTI model, an *E. faecalis* strain lacking the *ebpABC* pilus genes displayed significantly reduced adherence compared to the wild-type strain to bladders and silicone implants placed in the bladder to mimic catheters (345). Interestingly, bacteria lacking only the major pilin EpbC colonized mouse bladders and implants similar to wild-type bacteria, suggesting that expression of just the minor pilins on the bacterial surface is sufficient to mediate colonization. Although the ligand(s) for Ebp pili remain unknown, mutations introduced into the MIDAS motif of the EbpA tip pilin drastically inhibited bacterial adhesion in the CAUTI model, indicating a role for this motif in EbpA function and pilus-mediated adhesion (345).

CONCLUSION

The diverse array of adhesins expressed by uropathogenic bacteria reflects the importance of adhesion to colonization. Pathogens invading into the urinary tract must have strong adherence properties to overcome the washing action of the flow of urine. In addition, adhesins expressed by bacteria must be able to withstand mechanical forces exerted by urine flow, to avoid being sheared off once they have bound to their receptors on the urothelial surface. These factors underlie the prominence of pilus adhesins such as type 1 and P pili, which are adapted to function in the urinary tract, among the virulence factors of uropathogens. Uropathogenic bacteria encode multiple different types of adhesins, providing specificity for different niches within the urinary tract, as well as redundancy in function to ensure maintenance of adhesion under varying conditions. Bacterial colonization is also enhanced by formation of bacterial-bacterial interactions and biofilm structures, explaining the association of adhesins, including curli and Ag43 with uropathogens.

Adhesion is crucial at early stages of infection, and thus represents an attractive target for therapeutic intervention. Advances in understanding the structure and assembly of bacterial adhesins will be critical for the development of effective vaccines and antimicrobial agents that target adhesion. Pilus-based vaccines have shown promise in preventing UTIs, although none has reached clinical use. Studies using purified P pili or recombinant PapG adhesin purified in complex with its PapD chaperone demonstrated protection against pyelonephritis in primates (163, 354). Similarly, vaccination with the type 1 pilus adhesin FimH, purified in complex with its FimC chaperone, provided protection in primates against challenge with a UPEC cystitis isolate (129). Interestingly, a recent study found that rather than blocking type 1 pilus-mediated adhesion, the host antibody response may actually enhance binding of FimH to its ligands by stabilizing the adhesin's high-affinity binding state (355). An improved understanding of pilus adhesion mechanisms under conditions in the urinary tract may allow tailoring of the antigen used for vaccination to provoke a more effective immune response. In addition to pili, surface-located adhesins such as the MSCRAMMs expressed by Gram-positive bacteria also represent viable targets for vaccine development (356, 357).

An alternative approach to vaccination is to disrupt bacterial adhesion to host cells through the use of small molecule competitive inhibitors of adhesin-receptor interactions. Examples include the use of galabiose- and mannose-based inhibitors to interfere with adhesion mediated by P and type 1 pili, respectively, with the goal of preventing UPEC colonization of the urinary tract (358–361). Another alternative approach is to develop small molecule inhibitors that disrupt the machinery used for adhesin biogenesis, thereby preventing assembly of the adhesins on the bacterial surface. Once such class of small molecules developed by Almqvist and colleagues, termed pilicides, interferes with the CU assembly pathway and blocks expression of both P and type 1 pili by *E. coli* (362). Pilicides target the periplasmic chaperone and appear to disrupt pilus assembly by interfering with binding of chaperone-subunit complexes to the outer membrane usher. Modified pilicides were recently developed that also had activity against curli (203). Treatment of UPEC with one of these inhibitors blocked assembly of both curli and type 1 pili, reduced biofilm formation, and attenuated bacterial colonization in the murine urinary tract infection model (203). These compounds highlight the potential for a new class of anti-infective agents that target virulence factor secretion systems and the assembly of virulence-associated surface structures, rather than disrupting essential biological processes as for conventional antibiotics (363, 364). Such anti-virulence molecules should place less pressure on the bacteria and thus may be less prone to the development of resistance mechanisms. This strategy may also allow the selective targeting of pathogenic bacteria, avoiding detrimental side effects of broad-spectrum antibiotics on the normal bacterial flora.

ACKNOWLEDGMENTS

We thank Matthew Chapman (University of Michigan, Ann Arbor) for providing structural coordinates for CsgA. This work was supported by award number R01GM62987 from the National Institute of General Medical Sciences. P.C. is supported by award number T32AI007539 from the National Institute of Allergy and Infectious Diseases.

Conflicts of interest: We declare no conflicts.

CITATION

Chahales P, Thanassi DG. 2015. Structure, function, and assembly of adhesive organelles by uropathogenic bacteria. Microbiol Spectrum 3(5):UTI-0018-2013.

REFERENCES

1. **Pizarro-Cerda J, Cossart P.** 2006. Bacterial adhesion and entry into host cells. *Cell* **124:** 715–727.

2. **Kline KA, Falker S, Dahlberg S, Normark S, Henriques-Normark B.** 2009. Bacterial adhesins in host-microbe interactions. *Cell Host Microbe* **5:**580–592.

3. **Johnson JR.** 1991. Virulence factors in *Escherichia coli* urinary tract infection. *Clin Microbiol Rev* **4:**80–128.

4. **Ottow JCG.** 1975. Ecology, physiology and genetics of fimbriae and pili. *Annu Rev Microbiol* **29:**79–108.

5. **Duguid JP, Smith IW, Dempster G, Edmunds PN.** 1955. Non-flagellar filamentous appendages ("fimbriae") and hemagglutinating activity in *bacterium coli*. *J Pathol Bacteriol* **70:** 335–348.

6. **Brinton CC.** 1959. Non-flagellar appendages of bacteria. *Nature* **183:**782–786.

7. **Duguid JP, Anderson ES, Campbell I.** 1966. Fimbriae and adhesive properties in *Salmonellae*. *J Pathol Bacteriol* **92:**107–138.

8. **Orskov I, Orskov F.** 1990. Serologic classification of fimbriae. *Curr Top Microbiol Immunol* **151:**71–90.

9. **Low D, Braaten B, van der Woude M.** 1996. Fimbriae, p 146–157. *In* Neidhardt FC (ed), *Escherichia Coli and Salmonella; Cellular and Molecular Biology*, 2nd ed. ASM Press, Washington DC.

10. **Thanassi DG, Nuccio S-P, Shu Kin So S, Bäumler AJ.** 2007. Fimbriae: Classification and Biochemistry. *In* Bôck A, Curtiss R III, Kaper JB, Neidhardt FC, Nyström T, Rudd KE, Squires CL (ed), *EcoSal–Escherichia coli and Salmonella: cellular and molecular biology*, vol. [Online] http://www.ecosal.org. ASM Press, Washington, DC.

11. **Duguid JP, Clegg S, Wilson MI.** 1979. The fimbrial and non-fimbrial haemagglutinins of *Escherichia coli*. *J Med Microbiol* **12:**213–227.

12. **Old DC.** 1972. Inhibition of the interaction between fimbrial hemagglutinatinins and erythrocytes by D-mannose and other carbohydrates. *J Gen Microbiol* **71:**149–157.

13. **Hull RA, Gill RE, Hsu P, Minshaw BH, Falkow S.** 1981. Construction and expression of recombinant plasmids encoding type 1 and D-mannose-resistant pili from a urinary tract infection *Escherichia coli* isolate. *Infect Immun* **33:**933–938.

14. **Hacker J, Schmidt G, Hughes C, Knapp S, Marget M, Goebel W.** 1985. Cloning and char-

acterization of genes involved in production of mannose-resistant neuraminidase-susceptible (X) fimbriae from a uropathogenic O6:K15:H31 *Escherichia coli* strain. *Infect Immun* **47**:434–440.

15. **Fronzes R, Remaut H, Waksman G.** 2008. Architectures and biogenesis of non-flagellar protein appendages in Gram-negative bacteria. *EMBO J* **27**:2271–2280.

16. **Barnhart M, Chapman M.** 2006. Curli biogenesis and function. *Annu Rev Microbiol* **60**:131–147.

17. **Yanagawa R, Otsuki K.** 1970. Some properties of the pili of *Corynebacterium renale*. *J Bacteriol* **101**:1063–1069.

18. **Yanagawa R, Otsuki K, Tokui T.** 1968. Electron microscopy of fine structure of *Corynebacterium renale* with special reference to pili. *Jpn J Vet Res* **16**:31–37.

19. **Ton-That H, Marraffini LA, Schneewind O.** 2004. Sortases and pilin elements involved in pilus assembly of *Corynebacterium diphtheriae*. *Mol Microbiol* **53**:251–261.

20. **Ton-That H, Schneewind O.** 2003. Assembly of pili on the surface of *Corynebacterium diphtheriae*. *Mol Microbiol* **50**:1429–1438.

21. **Danne C, Dramsi S.** 2012. Pili of gram-positive bacteria: roles in host colonization. *Res Microbiol* **163**:645–658.

22. **Vengadesan K, Narayana SV.** 2011. Structural biology of Gram-positive bacterial adhesins. *Protein Sci* **20**:759–772.

23. **Hendrickx AP, Willems RJ, Bonten MJ, van Schaik W.** 2009. LPxTG surface proteins of enterococci. *Trends Microbiol* **17**:423–430.

24. **Nuccio SP, Baumler AJ.** 2007. Evolution of the chaperone/usher assembly pathway: fimbrial classification goes Greek. *Microbiol Mol Biol Rev* **71**:551–575.

25. **Zav'yalov V, Zavialov A, Zav'yalova G, Korpela T.** 2010. Adhesive organelles of Gram-negative pathogens assembled with the classical chaperone/usher machinery: structure and function from a clinical standpoint. *FEMS Microbiol Rev* **34**:317–378.

26. **Waksman G, Hultgren SJ.** 2009. Structural biology of the chaperone-usher pathway of pilus biogenesis. *Nat Rev Microbiol* **7**:765–774.

27. **Thanassi DG, Bliska JB, Christie PJ.** 2012. Surface organelles assembled by secretion systems of Gram-negative bacteria: diversity in structure and function. *FEMS Microbiol Rev* **36**:1046–1082.

28. **Miller E, Garcia T, Hultgren S, Oberhauser AF.** 2006. The mechanical properties of *E. coli* type 1 pili measured by atomic force microscopy techniques. *Biophys J* **91**:3848–3856.

29. **Fallman E, Schedin S, Jass J, Uhlin BE, Axner O.** 2005. The unfolding of the P pili quaternary structure by stretching is reversible, not plastic. *EMBO Rep* **6**:52–56.

30. **Thomas WE, Trintchina E, Forero M, Vogel V, Sokurenko EV.** 2002. Bacterial adhesion to target cells enhanced by shear force. *Cell* **109**:913–923.

31. **Castelain M, Ehlers S, Klinth J, Lindberg S, Andersson M, Uhlin BE, Axner O.** 2011. Fast uncoiling kinetics of F1C pili expressed by uropathogenic *Escherichia coli* are revealed on a single pilus level using force-measuring optical tweezers. *Eur Biophys J* **40**:305–316.

32. **Martinez JJ, Mulvey MA, Schilling JD, Pinkner JS, Hultgren SJ.** 2000. Type 1 pilus-mediated bacterial invasion of bladder epithelial cells. *EMBO J* **19**:2803–2812.

33. **Mulvey MA, Lopez-Boado YS, Wilson CL, Roth R, Parks WC, Heuser J, Hultgren SJ.** 1998. Induction and evasion of host defenses by type 1-piliated uropathogenic *Escherichia coli*. *Science* **282**:1494–1497.

34. **Eto DS, Jones TA, Sundsbak JL, Mulvey MA.** 2007. Integrin-mediated host cell invasion by type 1-piliated uropathogenic *Escherichia coli*. *PLoS Pathog* **3**:e100.

35. **Wright KJ, Seed PC, Hultgren SJ.** 2007. Development of intracellular bacterial communities of uropathogenic *Escherichia coli* depends on type 1 pili. *Cell Microbiol* **9**:2230–2241.

36. **Mossman KL, Mian MF, Lauzon NM, Gyles CL, Lichty B, Mackenzie R, Gill N, Ashkar AA.** 2008. Cutting edge: FimH adhesin of type 1 fimbriae is a novel TLR4 ligand. *J Immunol* **181**:6702–6706.

37. **Bergsten G, Samuelsson M, Wullt B, Leijonhufvud I, Fischer H, Svanborg C.** 2004. PapG-dependent adherence breaks mucosal inertia and triggers the innate host response. *J Infect Dis* **189**:1734–1742.

38. **Plancon L, Du Merle L, Le Friec S, Gounon P, Jouve M, Guignot J, Servin A, Le Bouguenec C.** 2003. Recognition of the cellular beta1-chain integrin by the bacterial AfaD invasin is implicated in the internalization of *afa*-expressing pathogenic *Escherichia coli* strains. *Cell Microbiol* **5**:681–693.

39. **Oelschlaeger TA, Dobrindt U, Hacker J.** 2002. Pathogenicity islands of uropathogenic *E. coli* and the evolution of virulence. *Int J Antimicrob Agents* **19**:517–521.

40. **van der Velden AW, Baumler AJ, Tsolis RM, Heffron F.** 1998. Multiple fimbrial adhesins are required for full virulence of *Salmonella typhimurium* in mice. *Infect Immun* **66**:2803–2808.

41. Korea CG, Badouraly R, Prevost MC, Ghigo JM, Beloin C. 2010. *Escherichia coli* K-12 possesses multiple cryptic but functional chaperone-usher fimbriae with distinct surface specificities. *Environ Microbiol* **12**:1957–1977.

42. Hatkoff M, Runco LM, Pujol C, Jayatilaka I, Furie MB, Bliska JB, Thanassi DG. 2012. Roles of chaperone/usher pathways of *Yersinia pestis* in a murine model of plague and adhesion to host cells. *Infect Immun* **80**:3490–3500.

43. Wurpel DJ, Beatson SA, Totsika M, Petty NK, Schembri MA. 2013. Chaperone-Usher Fimbriae of *Escherichia coli*. *PLoS ONE* **8**:e52835.

44. Blomfield IC. 2001. The regulation of pap and type 1 fimbriation in *Escherichia coli*. *Adv Microb Physiol* **45**:1–49.

45. van der Woude M, Braaten B, Low D. 1996. Epigenetic phase variation of the pap operon in *Escherichia coli*. *Trends Microbiol* **4**:5–9.

46. Xia Y, Gally D, Forsman-Semb K, Uhlin BE. 2000. Regulatory cross-talk between adhesin operons in *Escherichia coli*: inhibition of type 1 fimbriae expression by the PapB operon. *EMBO J* **19**:1450–1457.

47. Totsika M, Beatson SA, Holden N, Gally DL. 2008. Regulatory interplay between *pap* operons in uropathogenic *Escherichia coli*. *Mol Microbiol* **67**:996–1011.

48. Snyder JA, Haugen BJ, Lockatell CV, Maroncle N, Hagan EC, Johnson DE, Welch RA, Mobley HL. 2005. Coordinate expression of fimbriae in uropathogenic *Escherichia coli*. *Infect Immun* **73**:7588–7596.

49. Lane MC, Simms AN, Mobley HL. 2007. complex interplay between type 1 fimbrial expression and flagellum-mediated motility of uropathogenic *Escherichia coli*. *J Bacteriol* **189**:5523–5533.

50. Armbruster CE, Mobley HL. 2012. Merging mythology and morphology: the multifaceted lifestyle of *Proteus mirabilis*. *Nat Rev Microbiol* **10**:743–754.

51. Servin AL. 2005. Pathogenesis of Afa/Dr diffusely adhering *Escherichia coli*. *Clin Microbiol Rev* **18**:264–292.

52. Choudhury D, Thompson A, Stojanoff V, Langermann S, Pinkner J, Hultgren SJ, Knight SD. 1999. X-ray structure of the FimC-FimH chaperone-adhesin complex from uropathogenic *Escherichia coli*. *Science* **285**:1061–1066.

53. Sauer FG, Fütterer K, Pinkner JS, Dodson KW, Hultgren SJ, Waksman G. 1999. Structural basis of chaperone function and pilus biogenesis. *Science* **285**:1058–1061.

54. Sauer FG, Pinkner JS, Waksman G, Hultgren SJ. 2002. Chaperone priming of pilus subunits facilitates a topological transition that drives fiber formation. *Cell* **111**:543–551.

55. Zavialov AV, Berglund J, Pudney AF, Fooks LJ, Ibrahim TM, MacIntyre S, Knight SD. 2003. Structure and biogenesis of the capsular F1 antigen from *Yersinia pestis*: preserved folding energy drives fiber formation. *Cell* **113**:587–596.

56. Phan G, Remaut H, Wang T, Allan WJ, Pirker KF, Lebedev A, Henderson NS, Geibel S, Volkan E, Yan J, Kunze MBA, Pinkner JS, Ford B, Kay CWM, Li H, Hultgren S, Thanassi DG, Waksman G. 2011. Crystal structure of the FimD usher bound to its cognate FimC-FimH substrate. *Nature* **474**:49–53.

57. Anderson KL, Billington J, Pettigrew D, Cota E, Simpson P, Roversi P, Chen HA, Urvil P, du Merle L, Barlow PN, Medof ME, Smith RA, Nowicki B, Le Bouguenec C, Lea SM, Matthews S. 2004. An atomic resolution model for assembly, architecture, and function of the Dr adhesins. *Mol Cell* **15**:647–657.

58. Bork P, Holm L, Sander C. 1994. The immunoglobulin fold. Structural classification, sequence patterns and common core. *J Mol Biol* **242**:309–320.

59. Henderson NS, Ng TW, Talukder I, Thanassi DG. 2011. Function of the usher N-terminus in catalysing pilus assembly. *Mol Microbiol* **79**:954–967.

60. Puorger C, Eidam O, Capitani G, Erilov D, Grutter MG, Glockshuber R. 2008. Infinite kinetic stability against dissociation of supramolecular protein complexes through donor strand complementation. *Structure* **16**:631–642.

61. Bullitt E, Makowski L. 1995. Structural polymorphism of bacterial adhesion pili. *Nature* **373**:164–167.

62. Pettigrew D, Anderson KL, Billington J, Cota E, Simpson P, Urvil P, Rabuzin F, Roversi P, Nowicki B, du Merle L, Le Bouguenec C, Matthews S, Lea SM. 2004. High resolution studies of the Afa/Dr adhesin DraE and its interaction with chloramphenicol. *J Biol Chem* **279**:46851–46857.

63. Korotkova N, Le Trong I, Samudrala R, Korotkov K, Van Loy CP, Bui AL, Moseley SL, Stenkamp RE. 2006. Crystal structure and mutational analysis of the DaaE adhesin of *Escherichia coli*. *J Biol Chem* **281**:22367–22377.

64. Zavialov A, Zav'yalova G, Korpela T, Zav'yalov V. 2007. FGL chaperone-assembled fimbrial polyadhesins: anti-immune armament of Gram-negative bacterial pathogens. *FEMS Microbiol Rev* **31**:478–514.

65. Cota E, Jones C, Simpson P, Altroff H, Anderson KL, du Merle L, Guignot J, Servin A, Le Bouguenec C, Mardon H, Matthews S. 2006. The solution structure of the invasive tip complex from Afa/Dr fibrils. *Mol Microbiol* **62**: 356–366.

66. Dodson KW, Pinkner JS, Rose T, Magnusson G, Hultgren SJ, Waksman G. 2001. Structural basis of the interaction of the pyelonephritic *E. coli* adhesin to its human kidney receptor. *Cell* **105**:733–743.

67. Hung CS, Bouckaert J, Hung D, Pinkner J, Widberg C, DeFusco A, Auguste CG, Strouse R, Langermann S, Waksman G, Hultgren SJ. 2002. Structural basis of tropism of *Escherichia coli* to the bladder during urinary tract infection. *Mol Microbiol* **44**:903–915.

68. Sung MA, Fleming K, Chen HA, Matthews S. 2001. The solution structure of PapGII from uropathogenic *Escherichia coli* and its recognition of glycolipid receptors. *EMBO Rep* **2**: 621–627.

69. Buts L, Bouckaert J, De Genst E, Loris R, Oscarson S, Lahmann M, Messens J, Brosens E, Wyns L, De Greve H. 2003. The fimbrial adhesin F17-G of enterotoxigenic *Escherichia coli* has an immunoglobulin-like lectin domain that binds N-acetylglucosamine. *Mol Microbiol* **49**:705–715.

70. Merckel MC, Tanskanen J, Edelman S, Westerlund-Wikstrom B, Korhonen TK, Goldman A. 2003. The structural basis of receptor-binding by *Escherichia coli* associated with diarrhea and septicemia. *J Mol Biol* **331**:897–905.

71. Li YF, Poole S, Rasulova F, McVeigh AL, Savarino SJ, Xia D. 2007. A Receptor-binding Site as Revealed by the Crystal Structure of CfaE, the Colonization Factor Antigen I Fimbrial Adhesin of Enterotoxigenic *Escherichia coli*. *J Biol Chem* **282**:23970–23980.

72. Westerlund-Wikstrom B, Korhonen TK. 2005. Molecular structure of adhesin domains in *Escherichia coli* fimbriae. *Int J Med Microbiol* **295**:479–486.

73. Jedrzejczak R, Dauter Z, Dauter M, Piatek R, Zalewska B, Mroz M, Bury K, Nowicki B, Kur J. 2006. Structure of DraD invasin from uropathogenic *Escherichia coli*: a dimer with swapped beta-tails. *Acta Crystallogr D Biol Crystallogr* **62**:157–164.

74. Yakovenko O, Sharma S, Forero M, Tchesnokova V, Aprikian P, Kidd B, Mach A, Vogel V, Sokurenko E, Thomas WE. 2008. FimH forms catch bonds that are enhanced by mechanical force due to allosteric regulation. *J Biol Chem* **283**:11596–11605.

75. Le Trong I, Aprikian P, Kidd BA, Forero-Shelton M, Tchesnokova V, Rajagopal P, Rodriguez V, Interlandi G, Klevit R, Vogel V, Stenkamp RE, Sokurenko EV, Thomas WE. 2010. Structural basis for mechanical force regulation of the adhesin FimH via finger trap-like beta sheet twisting. *Cell* **141**:645–655.

76. Forero M, Yakovenko O, Sokurenko EV, Thomas WE, Vogel V. 2006. Uncoiling mechanics of *Escherichia coli* type I fimbriae are optimized for catch bonds. *PLoS Biol* **4**: e298.

77. Aprikian P, Interlandi G, Kidd BA, Le Trong I, Tchesnokova V, Yakovenko O, Whitfield MJ, Bullitt E, Stenkamp RE, Thomas WE, Sokurenko EV. 2011. The bacterial fimbrial tip acts as a mechanical force sensor. *PLoS Biol* **9**:e1000617.

78. Lycklama ANJA, Driessen AJ. 2012. The bacterial Sec-translocase: structure and mechanism. *Philos Trans R Soc Lond B Biol Sci* **367**: 1016–1028.

79. Jones CH, Dexter P, Evans AK, Liu C, Hultgren SJ, Hruby DE. 2002. *Escherichia coli* DegP protease cleaves between paired hydrophobic residues in a natural substrate: the PapA pilin. *J Bacteriol* **184**:5762–5771.

80. Slonim LN, Pinkner JS, Branden CI, Hultgren SJ. 1992. Interactive surface in the PapD chaperone cleft is conserved in pilus chaperone superfamily and essential in subunit recognition and assembly. *EMBO J* **11**:4747–4756.

81. Holmgren A, Brändén C. 1989. Crystal structure of chaperone protein PapD reveals an immunoglobulin fold. *Nature* **342**:248–251.

82. Kuehn MJ, Ogg DJ, Kihlberg J, Slonim LN, Flemmer K, Bergfors T, Hultgren SJ. 1993. Structural basis of pilus subunit recognition by the PapD chaperone. *Science* **262**:1234–1241.

83. Zavialov AV, Kersley J, Korpela T, Zav'yalov VP, MacIntyre S, Knight SD. 2002. Donor strand complementation mechanism in the biogenesis of non-pilus systems. *Mol Microbiol* **45**:983–995.

84. Barnhart MM, Pinkner JS, Soto GE, Sauer FG, Langermann S, Waksman G, Frieden C, Hultgren SJ. 2000. PapD-like chaperones provide the missing information for folding of pilin proteins. *Proc Natl Acad Sci USA* **97**:7709–7714.

85. Hung DL, Knight SD, Woods RM, Pinkner JS, Hultgren SJ. 1996. Molecular basis of two subfamilies of immunoglobulin-like chaperones. *EMBO J* **15**:3792–3805.

86. Remaut H, Rose RJ, Hannan TJ, Hultgren SJ, Radford SE, Ashcroft AE, Waksman G.

2006. Donor-strand exchange in chaperone-assisted pilus assembly proceeds through a concerted beta strand displacement mechanism. *Mol Cell* **22**:831–842.

87. Yu XD, Fooks LJ, Moslehi-Mohebi E, Tischenko VM, Askarieh G, Knight SD, Macintyre S, Zavialov AV. 2012. Large is fast, small is tight: determinants of speed and affinity in subunit capture by a periplasmic chaperone. *J Mol Biol* **417**:294–308.

88. Crespo MD, Puorger C, Scharer MA, Eidam O, Grutter MG, Capitani G, Glockshuber R. 2012. Quality control of disulfide bond formation in pilus subunits by the chaperone FimC. *Nat Chem Biol* **8**:707–713.

89. Di Yu X, Dubnovitsky A, Pudney AF, Macintyre S, Knight SD, Zavialov AV. 2012. Allosteric mechanism controls traffic in the chaperone/usher pathway. *Structure* **20**:1861–1871.

90. Remaut H, Tang C, Henderson NS, Pinkner JS, Wang T, Hultgren SJ, Thanassi DG, Waksman G, Li H. 2008. Fiber Formation across the Bacterial Outer Membrane by the Chaperone/Usher Pathway. *Cell* **133**:640–652.

91. Nishiyama M, Ishikawa T, Rechsteiner H, Glockshuber R. 2008. Reconstitution of Pilus Assembly Reveals a Bacterial Outer Membrane Catalyst. *Science* **320**:376–379.

92. Soto GE, Dodson KW, Ogg D, Liu C, Heuser J, Knight S, Kihlberg J, Jones CH, Hultgren SJ. 1998. Periplasmic chaperone recognition motif of subunits mediates quaternary interactions in the pilus. *EMBO J* **17**:6155–6167.

93. Sauer FG, Remaut H, Hultgren SJ, Waksman G. 2004. Fiber assembly by the chaperone-usher pathway. *Biochim Biophys Acta* **1694**:259–267.

94. Vetsch M, Erilov D, Moliere N, Nishiyama M, Ignatov O, Glockshuber R. 2006. Mechanism of fibre assembly through the chaperone-usher pathway. *EMBO Rep* **7**:734–738.

95. Yu J, Kape JB. 1992. Cloning and characterization of the *eae* gene of enterohaemorrhic *Escherichia coli* O157:H7. *Mol Microbiol* **6**:411–417.

96. Jacob-Dubuisson F, Striker R, Hultgren SJ. 1994. Chaperone-assisted self-assembly of pili independent of cellular energy. *J Biol Chem* **269**:12447–12455.

97. Thanassi DG, Stathopoulos C, Karkal A, Li H. 2005. Protein secretion in the absence of ATP: the autotransporter, two-partner secretion, and chaperone/usher pathways of Gram-negative bacteria. *Mol Membr Biol* **22**:63–72.

98. Zavialov AV, Tischenko VM, Fooks LJ, Brandsdal BO, Aqvist J, Zav'yalov VP,

Macintyre S, Knight SD. 2005. Resolving the energy paradox of chaperone/usher-mediated fibre assembly. *Biochem J* **389**:685–694.

99. Lee YM, Dodson KW, Hultgren SJ. 2007. Adaptor function of PapF depends on donor strand exchange in P-pilus biogenesis of *Escherichia coli. J Bacteriol* **189**:5276–5283.

100. Rose RJ, Verger D, Daviter T, Remaut H, Paci E, Waksman G, Ashcroft AE, Radford SE. 2008. Unraveling the molecular basis of subunit specificity in P pilus assembly by mass spectrometry. *Proc Natl Acad Sci USA* **105**:12873–12878.

101. Nishiyama M, Glockshuber R. 2010. The outer membrane usher guarantees the formation of functional pili by selectively catalyzing donor-strand exchange between subunits that are adjacent in the mature pilus. *J Mol Biol* **396**:1–8.

102. Dodson KW, Jacob-Dubuisson F, Striker RT, Hultgren SJ. 1993. Outer membrane PapC usher discriminately recognizes periplasmic chaperone-pilus subunit complexes. *Proc Natl Acad Sci USA* **90**:3670–3674.

103. Saulino ET, Thanassi DG, Pinkner JS, Hultgren SJ. 1998. Ramifications of kinetic partitioning on usher-mediated pilus biogenesis. *EMBO J* **17**:2177–2185.

104. Li Q, Ng TW, Dodson KW, So SS, Bayle KM, Pinkner JS, Scarlata S, Hultgren SJ, Thanassi DG. 2010. The differential affinity of the usher for chaperone-subunit complexes is required for assembly of complete pili. *Mol Microbiol* **76**:159–172.

105. Nishiyama M, Horst R, Eidam O, Herrmann T, Ignatov O, Vetsch M, Bettendorff P, Jelesarov I, Grutter MG, Wuthrich K, Glockshuber R, Capitani G. 2005. Structural basis of chaperone-subunit complex recognition by the type 1 pilus assembly platform FimD. *EMBO J* **24**:2075–2086.

106. Shu Kin So S, Thanassi DG. 2006. Analysis of the requirements for pilus biogenesis at the outer membrane usher and the function of the usher C-terminus. *Mol Microbiol* **60**:364–375.

107. Mapingire OS, Henderson NS, Duret G, Thanassi DG, Delcour AH. 2009. Modulating effects of the plug, helix and N- and C-terminal domains on channel properties of the PapC usher. *J Biol Chem* **284**:36324–36333.

108. Ford B, Rego AT, Ragan TJ, Pinkner J, Dodson K, Driscoll PC, Hultgren S, Waksman G. 2010. Structural homology between the C-terminal domain of the PapC usher and its plug. *J Bacteriol* **192**:1824–1831.

109. Huang Y, Smith BS, Chen LX, Baxter RH, Deisenhofer J. 2009. Insights into pilus

assembly and secretion from the structure and functional characterization of usher PapC. *Proc Natl Acad Sci USA* **106**:7403–7407.

110. **Ng TW, Akman L, Osisami M, Thanassi DG.** 2004. The usher N terminus is the initial targeting site for chaperone-subunit complexes and participates in subsequent pilus biogenesis events. *J Bacteriol* **186**:5321–5331.

111. **Li H, Qian L, Chen Z, Thahbot D, Liu G, Liu T, Thanassi DG.** 2004. The outer membrane usher forms a twin-pore secretion complex. *J Mol Biol* **344**:1397–1407.

112. **Allen WJ, Phan G, Hultgren SJ, Waksman G.** 2013. Dissection of Pilus Tip Assembly by the FimD Usher Monomer. *J Mol Biol* **425**:958–967.

113. **Munera D, Hultgren S, Fernandez LA.** 2007. Recognition of the N-terminal lectin domain of FimH adhesin by the usher FimD is required for type 1 pilus biogenesis. *Mol Microbiol* **64**:333–346.

114. **Nowicki B, Selvarangan R, Nowicki S.** 2001. Family of *Escherichia coli* Dr adhesins: decay-accelerating factor receptor recognition and invasiveness. *J Infect Dis* **183**(Suppl 1):S24–27.

115. **Marre R, Kreft B, Hacker J.** 1990. Genetically engineered S and F1C fimbriae differ in their contribution to adherence of *Escherichia coli* to cultured renal tubular cells. *Infect Immun* **58**:3434–3437.

116. **Schmoll T, Morschhauser J, Ott M, Ludwig B, van Die I, Hacker J.** 1990. Complete genetic organization and functional aspects of the *Escherichia coli* S fimbrial adhesion determinant: nucleotide sequence of the genes sfa B, C, D, E, F. *Microb Pathog* **9**:331–343.

117. **Riegman N, Kusters R, H VV, Bergmans H, Van Bergen En Henegouwen P, Hacker J, Van Die I.** 1990. F1C fimbriae of a uropathogenic *Escherichia coli* strain: genetic and functional organization of the foc gene cluster andidentification of minor subunits. *J Bacteriol* **172**:1114–1120.

118. **Spurbeck RR, Stapleton AE, Johnson JR, Walk ST, Hooton TM, Mobley HL.** 2011. Fimbrial profiles predict virulence of uropathogenic *Escherichia coli* strains: contribution of Ygi and Yad fimbriae. *Infect Immun* **79**:4753–4763.

119. **Buckles EL, Bahrani-Mougeot FK, Molina A, Lockatell CV, Johnson DE, Drachenberg CB, Burland V, Blattner FR, Donnenberg MS.** 2004. Identification and characterization of a novel uropathogenic *Escherichia coli*-associated fimbrial gene cluster. *Infect Immun* **72**:3890–3901.

120. **Ulett GC, Mabbett AN, Fung KC, Webb RI, Schembri MA.** 2007. The role of F9 fimbriae of uropathogenic *Escherichia coli* in biofilm formation. *Microbiology* **153**:2321–2331.

121. **Labigne-Roussel A, Falkow S.** 1988. Distribution and degree of heterogeneity of the afimbrial-adhesin-encoding operon (afa) among uropathogenic *Escherichia coli* isolates. *Infect Immun* **56**:640–648.

122. **Van Loy CP, Sokurenko EV, Moseley SL.** 2002. The major structural subunits of Dr and F1845 fimbriae are adhesins. *Infect Immun* **70**:1694–1702.

123. **Berger CN, Billker O, Meyer TF, Servin AL, Kansau I.** 2004. Differential recognition of members of the carcinoembryonic antigen family by Afa/Dr adhesins of diffusely adhering *Escherichia coli* (Afa/Dr DAEC). *Mol Microbiol* **52**:963983.

124. **Carnoy C, Moseley SL.** 1997. Mutational analysis of receptor binding mediated by the Dr family of *Escherichia coli* adhesins. *Mol Microbiol* **23**:365–379.

125. **Schmoll T, Hoschutzky H, Morschhauser J, Lottspeich F, Jann K, Hacker J.** 1989. Analysis of genes coding for the sialic acid-binding adhesin and two other minor fimbrial subunits of the S-fimbrial adhesin determinant of *Escherichia coli*. *Mol Microbiol* **3**:1735–1744.

126. **Virkola R, Parkkinen J, Hacker J, Korhonen TK.** 1993. Sialyloligosaccharide chains of laminin as an extracellular matrix target for S fimbriae of *Escherichia coli*. *Infect Immun* **61**:4480–4484.

127. **Korhonen TK, Parkkinen J, Hacker J, Finne J, Pere A, Rhen M, Holthofer H.** 1986. Binding of *Escherichia coli* S fimbriae to human kidney epithelium. *Infect Immun* **54**:322–327.

128. **Backhed F, Alsen B, Roche N, Angstrom J, von Euler A, Breimer ME, Westerlund-Wikstrom B, Teneberg S, Richter-Dahlfors A.** 2002. Identification of target tissue glycosphingolipid receptors for uropathogenic, F1C-fimbriated *Escherichia coli* and its role in mucosal inflammation. *J Biol Chem* **277**:18198–18205.

129. **Langermann S, Mollby R, Burlein JE, Palaszynski SR, Auguste CG, DeFusco A, Strouse R, Schenerman MA, Hultgren SJ, Pinkner JS, Winberg J, Guldevall L, Soderhall M, Ishikawa K, Normark S, Koenig S.** 2000. Vaccination with FimH adhesin protects cynomolgus monkeys from colonization and infection by uropathogenic *Escherichia coli*. *J Infect Dis* **181**:774–778.

130. **Langermann S, Palaszynski S, Barnhart M, Auguste G, Pinkner JS, Burlein J, Barren P,**

Koenig S, Leath S, Jones CH, Hultgren SJ. 1997. Prevention of mucosal *Escherichia coli* infection by FimH-adhesin-based systemic vaccination. *Science* **276**:607–611.

131. Hannan TJ, Totsika M, Mansfield KJ, Moore KH, Schembri MA, Hultgren SJ. 2012. Host-pathogen checkpoints and population bottlenecks in persistent and intracellular uropathogenic *Escherichia coli* bladder infection. *FEMS Microbiol Rev* **36**:616–648.

132. Abraham SN, Sun D, Dale JB, Beachey EH. 1988. Conservation of the D-mannose-adhesion protein among type 1 fimbriated members of the family *Enterobacteriaceae*. *Nature* **336**:682–684.

133. Pak J, Pu Y, Zhang ZT, Hasty DL, Wu XR. 2001. Tamm-Horsfall protein binds to type 1 fimbriated *Escherichia coli* and prevents *E. coli* from binding to uroplakin Ia and Ib receptors. *J Biol Chem* **276**:9924–9930.

134. Baorto DM, Gao Z, Malaviya R, Dustin ML, van der Merwe A, Lublin DM, Abraham SN. 1997. Survival of FimH-expressing enterobacteria in macrophages relies on glycolipid traffic. *Nature* **389**:636–639.

135. Kukkonen M, Raunio T, Virkola R, Lahteenmaki K, Makela PH, Klemm P, Clegg S, Korhonen TK. 1993. Basement membrane carbohydrate as a target for bacterial adhesion: binding of type 1 fimbriae of *Salmonella enterica* and *Escherichia coli* to laminin. *Mol Microbiol* **7**:229–227.

136. Pratt LA, Kolter R. 1998. Genetic analysis of *Escherichia coli* biofilm formation: roles of flagella, motility, chemotaxis and type I pili. *Mol Microbiol* **30**:285–293.

137. Jones CH, Pinkner JS, Roth R, Heuser J, Nicholoes AV, Abraham SN, Hultgren SJ. 1995. FimH adhesin of type 1 pili is assembled into a fibrillar tip structure in the *Enterobacteriaceae*. *Proc Natl Acad Sci USA* **92**:2081–2085.

138. Hahn E, Wild P, Hermanns U, Sebbel P, Glockshuber R, Haner M, Taschner N, Burkhard P, Aebi U, Muller SA. 2002. Exploring the 3D molecular architecture of *Escherichia coli* type 1 pili. *J Mol Biol* **323**:845–857.

139. Zhou G, Mo WJ, Sebbel P, Min G, Neubert TA, Glockshuber R, Wu XR, Sun TT, Kong XP. 2001. Uroplakin Ia is the urothelial receptor for uropathogenic *Escherichia coli*: evidence from in vitro FimH binding. *J Cell Sci* **114**:4095–4103.

140. Thumbikat P, Berry RE, Zhou G, Billips BK, Yaggie RE, Zaichuk T, Sun TT, Schaeffer AJ, Klumpp DJ. 2009. Bacteria-induced uroplakin signaling mediates bladder response to infection. *PLoS Pathog* **5**:e1000415.

141. Connell H, Agace W, Klemm P, Schembri M, Marild S, Svanborg C. 1996. Type 1 fimbrial expression enhances *Escherichia coli* virulence for the urinary tract. *Proc Natl Acad Sci U S A* **93**:9827–9832.

142. Song J, Bishop BL, Li G, Grady R, Stapleton A, Abraham SN. 2009. TLR4-mediated expulsion of bacteria from infected bladder epithelial cells. *Proc Natl Acad Sci USA* **106**:14966–14971.

143. Anderson GG, Palermo JJ, Schilling JD, Roth R, Heuser J, Hultgren SJ. 2003. Intracellular bacterial biofilm-like pods in urinary tract infections. *Science* **301**:105–107.

144. Justice SS, Hung C, Theriot JA, Fletcher DA, Anderson GG, Footer MJ, Hultgren SJ. 2004. Differentiation and developmental pathways of uropathogenic *Escherichia coli* in urinary tract pathogenesis. *Proc Natl Acad Sci USA* **101**:1333–1338.

145. Blango MG, Mulvey MA. 2010. Persistence of uropathogenic *Escherichia coli* in the face of multiple antibiotics. *Antimicrob Agents Chemother* **54**:1855–1863.

146. Chen SL, Hung CS, Pinkner JS, Walker JN, Cusumano CK, Li Z, Bouckaert J, Gordon JI, Hultgren SJ. 2009. Positive selection identifies an in vivo role for FimH during urinary tract infection in addition to mannose binding. *Proc Natl Acad Sci USA* **106**:22439–22444.

147. Mulvey MA, Schilling JD, Hultgren SJ. 2001. Establishment of a persistent *Escherichia coli* reservoir during the acute phase of a bladder infection. *Infect Immun* **69**:4572–4579.

148. Mysorekar IU, Hultgren SJ. 2006. Mechanisms of uropathogenic *Escherichia coli* persistence and eradication from the urinary tract. *Proc Natl Acad Sci USA* **103**:14170–14175.

149. Bock K, Breimer ME, Brignole A, Hansson GC, Karlsson K-A, Larson G, Leffler H, Samuelsson BE, Strömberg N, Svanborg-Edén C, Thurin J. 1985. Specificity of binding of a strain of uropathogenic *Escherichia coli* to Galα(1-4)Gal-containing glycosphingolipids. *J Biol Chem* **260**:8545–8551.

150. Roberts JA, Marklund B-I, Ilver D, Haslam D, Kaack MB, Baskin G, Louis M, Mollby R, Winberg J, Normark S. 1994. The Gal(alpha1-4)Gal-specific tip adhesin of *Escherichia coli* P-fimbriae is needed for pyelonephritis to occur in the normal urinary tract. *Proc Natl Acad Sci USA* **91**:11889–11893.

151. Kallenius G, Mollby R, Svenson SB, Windberg J, Lundblud A, Svenson S, Cedergen B. 1980. The P^k antigen as receptor for the haemagglutinin of pyelonephritogenic *Escherichia coli*. *FEMS Microbiol Lett* **7**:297–302.

152. **Ewers C, Li G, Wilking H, Kiessling S, Alt K, Antao EM, Laturnus C, Diehl I, Glodde S, Homeier T, Bohnke U, Steinruck H, Philipp HC, Wieler LH.** 2007. Avian pathogenic, uropathogenic, and newborn meningitis-causing *Escherichia coli*: how closely related are they? *Int J Med Microbiol* **297:**163–76.

153. **Welch RA, Burland V, Plunkett G 3rd, Redford P, Roesch P, Rasko D, Buckles EL, Liou SR, Boutin A, Hackett J, Stroud D, Mayhew GF, Rose DJ, Zhou S, Schwartz DC, Perna NT, Mobley HL, Donnenberg MS, Blattner FR.** 2002. Extensive mosaic structure revealed by the complete genome sequence of uropathogenic *Escherichia coli*. *Proc Natl Acad Sci USA* **99:**17020–17024.

154. **Lund B, Marklund BI, Stromberg N, Lindberg F, Karlsson KA, Normark S.** 1988. Uropathogenic *Escherichia coli* Can Express Serologically Identical Pili of Different Receptor-Binding Specificities. *Mol Microbiol* **2:**255–263.

155. **Stromberg N, Marklund BI, Lund B, Ilver D, Hamers A, Gaastra W, Karlsson KA, Normark S.** 1990. Host-specificity of uropathogenic *Escherichia coli* depends on differences in binding specificity to Galalpha(1-4) Gal-containing isoreceptors. *EMBO J* **9:**2001–2010.

156. **Johnson JR, Russo TA, Brown JJ, Stapleton A.** 1998. *papG* alleles of *Escherichia coli* strains causing first-episode or recurrent acute cystitis in adult women. *J Infect Dis* **177:**97–101.

157. **Kuehn MJ, Heuser J, Normark S, Hultgren SJ.** 1992. P pili in uropathogenic *E. coli* are composite fibres with distinct fibrillar adhesive tips. *Nature* **356:**252–255.

158. **Jacob-Dubuisson F, Heuser J, Dodson K, Normark S, Hultgren SJ.** 1993. Initiation of assembly and association of the structural elements of a bacterial pilus depend on two specialized tip proteins. *EMBO J* **12:**837–847.

159. **Mu XQ, Bullitt E.** 2006. Structure and assembly of P-pili: a protruding hinge region used for assembly of a bacterial adhesion filament. *Proc Natl Acad Sci USA* **103:**9861–9866.

160. **Baga M, Norgren M, Normark S.** 1987. Biogenesis of *E. coli* Pap pili: PapH, a minor pilin subunit involved in cell anchoring and length modulation. *Cell* **49:**241–251.

161. **Verger D, Miller E, Remaut H, Waksman G, Hultgren S.** 2006. Molecular mechanism of P pilus termination in uropathogenic *Escherichia coli*. *EMBO Rep* **7:**1228–1232.

162. **Hagberg L, Hull R, Hull S, Falkow S, Freter R, Svanborg Eden C.** 1983. Contribution of adhesion to bacterial persistence in the mouse urinary tract. *Infect Immun* **40:**265–272.

163. **Roberts JA, Hardaway K, Kaack B, Fussell EN, Baskin G.** 1984. Prevention of pyelonephritis by immunization with P-fimbriae. *J Urol* **131:**602–607.

164. **O'Hanley P, Lark D, Falkow S, Schoolnik G.** 1985. Molecular basis of *Escherichia coli* colonizatin of the upper urinary tract in BALB/c mice: Gal-Gal pili immunization prevents *Escherichia coli* pyelonephritis. *J Clin Invest* **83:**2102–2108.

165. **Lane MC, Mobley HL.** 2007. Role of P-fimbrial-mediated adherence in pyelonephritis and persistence of uropathogenic *Escherichia coli* (UPEC) in the mammalian kidney. *Kidney Int* **72:**19–25.

166. **Hedlund M, Svensson M, Nilsson Å, Duan R-D, Svanborg C.** 1996. Role of the ceramide-signaling pathway in cytokine responses to P-fimbriated *Escherichia coli*. *J Exp Med* **183:**1037–1044.

167. **Hedlund M, Wachtler C, Johansson E, Hang L, Somerville JE, Darveau RP, Svanborg C.** 1999. P fimbriae-dependent, lipopolysaccharide-independent activation of epithelial cytokine responses. *Mol Microbiol* **33:**693–703.

168. **Fischer H, Ellstrom P, Ekstrom K, Gustafsson L, Gustafsson M, Svanborg C.** 2007. Ceramide as a TLR4 agonist; a putative signalling intermediate between sphingolipid receptors for microbial ligands and TLR4. *Cell Microbiol* **9:**1239–1251.

169. **Zhang JP, Normark S.** 1996. Induction of gene expression in *Escherichia coli* after pilus-mediated adherence. *Science* **273:**1234–1236.

170. **Austin JW, Sanders G, Kay WW, Collinson SK.** 1998. Thin aggregative fimbriae enhance *Salmonella enteritidis* biofilm formation. *FEMS Microbiol Lett* **162:**295–301.

171. **Zogaj X, Bokranz W, Nimtz M, Romling U.** 2003. Production of cellulose and curli fimbriae by members of the family *Enterobacteriaceae* isolated from the human gastrointestinal tract. *Infect Immun* **71:**4151–4158.

172. **Gophna U, Barlev M, Seijffers R, Oelschlager T, Hacker J, Ron E.** 2001. Curli fibers mediate internalization of *Escherichia coli* by eukaryotic cells. *Infect Immun* **69:**2659–2665.

173. **Olsen A, Jonsson A, Normark S.** 1989. Fibronectin binding mediated by a novel class of surface organelles on *Escherichia coli*. *Nature* **338:**652–655.

174. **Olsen A, Arnqvist A, Hammar M, Sukupolvi S, Normark S.** 1993. The RpoS sigma factor relieves H-NS-mediated transcriptional repression of *csgA*, the subunit gene of fibronectin-binding curli in *Escherichia coli*. *Mol Microbiol* **7:**523–536.

175. Ben Nasr A, Olsen A, Sjobring U, Muller-Esterl W, Bjorck L. 1996. Assembly of human contact phase proteins and release of bradykinin at the surface of curli-expressing *Escherichia coli*. *Mol Microbiol* **20**:927–935.

176. Olsen A, Herwald H, Wikstrom M, Persson K, Mattsson E, Bjorck L. 2002. Identification of two protein-binding and functional regions of curli, a surface organelle and virulence determinant of *Escherichia coli*. *J Biol Chem* **277**:34568–34572.

177. Bian Z, Brauner A, Li Y, Normark S. 2000. Expression of and cytokine activation by *Escherichia coli* curli fibers in human sepsis. *J Infect Dis* **181**:602–612.

178. Kai-Larsen Y, Luthje P, Chromek M, Peters V, Wang X, Holm A, Kadas L, Hedlund KO, Johansson J, Chapman MR, Jacobson SH, Romling U, Agerberth B, Brauner A. 2010. Uropathogenic *Escherichia coli* modulates immune responses and its curli fimbriae interact with the antimicrobial peptide LL-37. *PLoS Pathog* **6**:e1001010.

179. Tukel C, Nishimori JH, Wilson RP, Winter MG, Keestra AM, van Putten JP, Baumler AJ. 2010. Toll-like receptors 1 and 2 cooperatively mediate immune responses to curli, a common amyloid from enterobacterial biofilms. *Cell Microbiol* **12**:1495–1505.

180. Cohen FE, Kelly JW. 2003. Therapeutic approaches to protein-misfolding diseases. *Nature* **426**:905–909.

181. Nelson R, Sawaya MR, Balbirnie M, Madsen AO, Riekel C, Grothe R, Eisenberg D. 2005. Structure of the cross-beta spine of amyloid-like fibrils. *Nature* **435**:773–778.

182. Blanco LP, Evans ML, Smith DR, Badtke MP, Chapman MR. 2012. Diversity, biogenesis and function of microbial amyloids. *Trends Microbiol* **20**:66–73.

183. Fowler DM, Koulov AV, Balch WE, Kelly JW. 2007. Functional amyloid–from bacteria to humans. *Trends Biochem Sci* **32**:217–224.

184. Chapman MR, Robinson LS, Pinkner JS, Roth R, Heuser J, Hammar M, Normark S, Hultgren SJ. 2002. Role of *Escherichia coli* curli operons in directing amyloid fiber formation. *Science* **295**:851–855.

185. Shewmaker F, McGlinchey RP, Thurber KR, McPhie P, Dyda F, Tycko R, Wickner RB. 2009. The functional curli amyloid is not based on in-register parallel beta-sheet structure. *J Biol Chem* **284**:25065–25076.

186. Collinson SK, Emody L, Muller KH, Trust TJ, Kay WW. 1991. Purification and characterization of thin, aggregative fimbriae from *Salmonella enteritidis*. *J Bacteriol* **173**:4773–4781.

187. Nilsson MR. 2004. Techniques to study amyloid fibril formation in vitro. *Methods* **34**:151–160.

188. Wang X, Chapman MR. 2008. Sequence determinants of bacterial amyloid formation. *J Mol Biol* **380**:570–580.

189. Hammar M, Bian Z, Normark S. 1996. Nucleator-dependent intercellular assembly of adhesive curli organelles in *Escherichia coli*. *Proc Natl Acad Sci USA* **93**:6562–6566.

190. Wang X, Zhou Y, Ren JJ, Hammer ND, Chapman MR. 2010. Gatekeeper residues in the major curlin subunit modulate bacterial amyloid fiber biogenesis. *Proc Natl Acad Sci USA* **107**:163–168.

191. Barak J, Gorski L, Naraghi-Arani P, Charkowski A. 2005. *Salmonella enterica* virulence genes are required for bacterial attachment to plant tissue. *Appl Environ Microbiol* **71**:5685–5691.

192. Hammar M, Arnqvist A, Bian Z, Olsen A, Normark S. 1995. Expression of two csg operons is required for production of fibronectin- and congo red-binding curli polymers in *Escherichia coli* K–12. *Mol Microbiol* **18**:661–670.

193. Taylor JD, Zhou Y, Salgado PS, Patwardhan A, McGuffie M, Pape T, Grabe G, Ashman E, Constable SC, Simpson PJ, Lee WC, Cota E, Chapman MR, Matthews SJ. 2011. Atomic resolution insights into curli fiber biogenesis. *Structure* **19**:1307–1316.

194. Romling U, Rohde M, Olsen A, Normark S, Reinkoster J. 2000. AgfD, the checkpoint of multicellular and aggregative behaviour in *Salmonella typhimurium* regulates at least two independent pathways. *Mol Microbiol* **36**:10–23.

195. Hammer N, Schmidt J, Chapman M. 2007. The curli nucleator protein, CsgB, contains an amyloidogenic domain that directs CsgA polymerization. *Proc Natl Acad Sci USA* **104**:12494–12499.

196. Robinson LS, Ashman EM, Hultgren SJ, Chapman MR. 2006. Secretion of curli fibre subunits is mediated by the outer membrane-localized CsgG protein. *Mol Microbiol* **59**:870–881.

197. Collinson SK, Parker JM, Hodges RS, Kay WW. 1999. Structural predictions of AgfA, the insoluble fimbrial subunit of *Salmonella* thin aggregative fimbriae. *J Mol Biol* **290**:741–756.

198. Nenninger AA, Robinson LS, Hammer ND, Epstein EA, Badtke MP, Hultgren SJ, Chapman MR. 2011. CsgE is a curli secretion specificity factor that prevents amyloid fibre aggregation. *Mol Microbiol* **81**:486–499.

199. Nenninger AA, Robinson LS, Hultgren SJ. 2009. Localized and efficient curli nucleation

requires the chaperone-like amyloid assembly protein CsgF. *Proc Natl Acad Sci USA* **106:** 900–905.

200. **Tukel C, Raffatellu M, Humphries A, Wilson R, Andrews-Polymenis H, Gull T, Figueiredo J, Wong M, Michelsen K, Akcelik M, Adams L, Baumler A.** 2005. CsgA is a pathogen-associated molecular pattern of *Salmonella enterica* serotype Typhimurium that is recognized by Toll-like receptor 2. *Mol Microbiol* **58:**289–304.

201. **Hung C, Marschall J, Burnham CA, Byun AS, Henderson JP.** 2014. The bacterial amyloid curli is associated with urinary source bloodstream infection. *PLoS One* **9:**e86009.

202. **Olsen A, Wick M, Morgelin M, Bjorck L.** 1998. Curli, fibrous surface proteins of *Escherichia coli*, interact with major histocompatibility complex class I molecules. *Infect Immun* **66:**944–949.

203. **Cegelski L, Pinkner JS, Hammer ND, Cusumano CK, Hung CS, Chorell E, Aberg V, Walker JN, Seed PC, Almqvist F, Chapman MR, Hultgren SJ.** 2009. Small-molecule inhibitors target *Escherichia coli* amyloid biogenesis and biofilm formation. *Nat Chem Biol* **5:**913–919.

204. **Kikuchi T, Mizunoe Y, Takade A, Naito S, Yoshida S.** 2005. Curli fibers are required for development of biofilm architecture in *Escherichia coli* K-12 and enhance bacterial adherence to human uroepithelial cells. *Microbiol Immunol* **49:**875–884.

205. **Cherny I, Rockah L, Levy-Nissenbaum O, Gophna U, Ron EZ, Gazit E.** 2005. The formation of *Escherichia coli* curli amyloid fibrils is mediated by prion-like peptide repeats. *J Mol Biol* **352:**245–252.

206. **Leo JC, Grin I, Linke D.** 2012. Type V secretion: mechanism(s) of autotransport through the bacterial outer membrane. *Philos Trans R Soc Lond B Biol Sci* **367:**1088–1101.

207. **Leyton DL, Rossiter AE, Henderson IR.** 2012. From self sufficiency to dependence: mechanisms and factors important for autotransporter biogenesis. *Nat Rev Microbiol* **10:**213–225.

208. **Jose J, Jahnig F, Meyer TF.** 1995. Common structural features of IgA1 protease-like outer membrane protein autotransporters. *Mol Microbiol* **18:**378–380.

209. **Pohlner J, Halter R, Beyreuther K, Meyer TF.** 1987. Gene structure and extracellular secretion of *Neisseria gonorrhoeae* IgA protease. *Nature* **325:**458–462.

210. **Oomen CJ, Van Ulsen P, Van Gelder P, Feijen M, Tommassen J, Gros P.** 2004. Structure of the translocator domain of a bacterial autotransporter. *EMBO J* **23:**1257–1266.

211. **Phadnis SH, Ilver D, Janzon L, Normark S, Westblom TU.** 1994. Pathological significance and molecular characterization of the vacuolating toxin gene of *Helicobacter pylori*. *Infect Immun* **62:**1557–1565.

212. **Emsley P, Charles IG, Fairweather NF, Isaacs NW.** 1996. Structure of *Bordetella pertussis* virulence factor P.69 pertactin. *Nature* **381:**90–92.

213. **Benz I, Schmidt MA.** 1992. AIDA-I, the adhesin involved in diffuse adherence of the diarrhoeagenic *Escherichia coli* strain 2787 (O126:H27), is synthesized via a precursor molecule. *Mol Microbiol* **6:**1539–1546.

214. **Charbonneau ME, Janvore J, Mourez M.** 2009. Autoprocessing of the *Escherichia coli* AIDA-I autotransporter: a new mechanism involving acidic residues in the junction region. *J Biol Chem* **284:**17340–17351.

215. **Schembri MA, Dalsgaard D, Klemm P.** 2004. Capsule shields the function of short bacterial adhesins. *J Bacteriol* **186:**1249–1257.

216. **Hasman H, Chakraborty T, Klemm P.** 1999. Antigen-43-mediated autoaggregation of *Escherichia coli* is blocked by fimbriation. *J Bacteriol* **181:**4834–4841.

217. **Allsopp LP, Totsika M, Tree JJ, Ulett GC, Mabbett AN, Wells TJ, Kobe B, Beatson SA, Schembri MA.** 2010. UpaH is a newly identified autotransporter protein that contributes to biofilm formation and bladder colonization by uropathogenic *Escherichia coli* CFT073. *Infect Immun* **78:**1659–1669.

218. **Parham NJ, Srinivasan U, Desvaux M, Foxman B, Marrs CF, Henderson IR.** 2004. PicU, a second serine protease autotransporter of uropathogenic *Escherichia coli*. *FEMS Microbiol Lett* **230:**73–83.

219. **Guyer DM, Henderson IR, Nataro JP, Mobley HL.** 2000. Identification of sat, an autotransporter toxin produced by uropathogenic *Escherichia coli*. *Mol Microbiol* **38:**53–66.

220. **Alamuri P, Lower M, Hiss JA, Himpsl SD, Schneider G, Mobley HL.** 2010. Adhesion, invasion, and agglutination mediated by two trimeric autotransporters in the human uropathogen *Proteus mirabilis*. *Infect Immun* **78:**4882–4894.

221. **Yen MR, Peabody CR, Partovi SM, Zhai Y, Tseng YH, Saier MH.** 2002. Protein-translocating outer membrane porins of Gram-negative bacteria. *Biochim Biophys Acta* **1562:**6–31.

222. **van den Berg B.** 2010. Crystal structure of a full-length auto-transporter. *J Mol Biol* **396:**627–633.

223. **Barnard TJ, Dautin N, Lukacik P, Bernstein HD, Buchanan SK.** 2007. Autotransporter structure reveals intra-barrel cleavage followed by conformational changes. *Nature Struct Mol Biol* **14:**1214–1220.

224. **Klauser T, Kramer J, Otzelberger K, Pohlner J, Meyer TF.** 1993. Characterization of the *Neisseria* Iga beta-core. The essential unit for outer membrane targeting and extracellular protein secretion. *J Mol Biol* **234:**579–593.

225. **Oliver DC, Huang G, Fernandez RC.** 2003. Identification of secretion determinants of the *Bordetella pertussis* BrkA autotransporter. *J Bacteriol* **185:**489–495.

226. **Maurer J, Jose J, Meyer TF.** 1999. Characterization of the essential transport function of the AIDA-I autotransporter and evidence supporting structural predictions. *J Bacteriol* **181:**7014–7020.

227. **Roggenkamp A, Ackermann N, Jacobi CA, Truelzsch K, Hoffmann H, Heesemann J.** 2003. Molecular analysis of transport and oligomerization of the *Yersinia enterocolitica* adhesin YadA. *J Bacteriol* **185:**3735–3744.

228. **Surana NK, Cutter D, Barenkamp SJ, St Geme JW 3rd.** 2004. The *Haemophilus influenzae* Hia autotransporter contains an unusually short trimeric translocator domain. *J Biol Chem* **279:**14679–14685.

229. **Meng G, Surana NK, St Geme JW 3rd, Waksman G.** 2006. Structure of the outer membrane translocator domain of the *Haemophilus influenzae* Hia trimeric autotransporter. *EMBO J* **25:**2297–2304.

230. **Yeo HJ, Cotter SE, Laarmann S, Juehne T, St Geme JW, Waksman G.** 2004. Structural basis for host recognition by the *Haemophilus influenzae* Hia autotransporter. *EMBO J* **23:**1245–1256.

231. **Nummelin H, Merckel MC, Leo JC, Lankinen H, Skurnik M, Goldman A.** 2004. The *Yersinia* adhesin YadA collagen-binding domain structure is a novel left-handed parallel beta-roll. *EMBO J* **23:**701–711.

232. **Valle J, Mabbett AN, Ulett GC, Toledo-Arana A, Wecker K, Totsika M, Schembri MA, Ghigo JM, Beloin C.** 2008. UpaG, a new member of the trimeric autotransporter family of adhesins in uropathogenic *Escherichia coli*. *J Bacteriol* **190:**4147–4161.

233. **Meng G, Spahich N, Kenjale R, Waksman G, St Geme JW 3rd.** 2011. Crystal structure of the *Haemophilus influenzae* Hap adhesin reveals an intercellular oligomerization mechanism for bacterial aggregation. *EMBO J* **30:**3864–3874.

234. **Khan S, Mian HS, Sandercock LE, Chirgadze NY, Pai EF.** 2011. Crystal structure of the passenger domain of the *Escherichia coli* autotransporter EspP. *J Mol Biol* **413:**985–1000.

235. **Gangwer KA, Mushrush DJ, Stauff DL, Spiller B, McClain MS, Cover TL, Lacy DB.** 2007. Crystal structure of the *Helicobacter pylori* vacuolating toxin p55 domain. *Proc Natl Acad Sci USA* **104:**16293–16298.

236. **Szczesny P, Lupas A.** 2008. Domain annotation of trimeric autotransporter adhesins–daTAA. *Bioinformatics* **24:**1251–1256.

237. **Hartmann MD, Grin I, Dunin-Horkawicz S, Deiss S, Linke D, Lupas AN, Hernandez Alvarez B.** 2012. Complete fiber structures of complex trimeric autotransporter adhesins conserved in enterobacteria. *Proc Natl Acad Sci USA* **109:**20907–20912.

238. **Leo JC, Lyskowski A, Hattula K, Hartmann MD, Schwarz H, Butcher SJ, Linke D, Lupas AN, Goldman A.** 2011. The structure of *E. coli* IgG-binding protein D suggests a general model for bending and binding in trimeric autotransporter adhesins. *Structure* **19:**1021–1030.

239. **Biedzka-Sarek M, Salmenlinna S, Gruber M, Lupas AN, Meri S, Skurnik M.** 2008. Functional mapping of YadA- and Ail-mediated binding of human factor H to *Yersinia enterocolitica* serotype O:3. *Infect Immun* **76:**5016–5027.

240. **Ruiz-Perez F, Henderson IR, Leyton DL, Rossiter AE, Zhang Y, Nataro JP.** 2009. Roles of periplasmic chaperone proteins in the biogenesis of serine protease autotransporters of Enterobacteriaceae. *J Bacteriol* **191:**6571–6583.

241. **Leyton DL, Sevastsyanovich YR, Browning DF, Rossiter AE, Wells TJ, Fitzpatrick RE, Overduin M, Cunningham AF, Henderson IR.** 2011. Size and conformation limits to secretion of disulfide-bonded loops in autotransporter proteins. *J Biol Chem* **286:**42283–42291.

242. **Desvaux M, Scott-Tucker A, Turner SM, Cooper LM, Huber D, Nataro JP, Henderson IR.** 2007. A conserved extended signal peptide region directs posttranslational protein translocation via a novel mechanism. *Microbiology* **153:**59–70.

243. **Szabady RL, Peterson JH, Skillman KM, Bernstein HD.** 2005. An unusual signal peptide facilitates late steps in the biogenesis of a bacterial autotransporter. *Proc Natl Acad Sci USA* **102:**221–226.

244. **Hagan CL, Silhavy TJ, Kahne D.** 2011. beta-Barrel membrane protein assembly by the Bam complex. *Ann Rev Biochem* **80:**189–210.

245. **Ieva R, Tian P, Peterson JH, Bernstein HD.** 2011. Sequential and spatially restricted interactions of assembly factors with an autotransporter beta domain. *Proc Natl Acad Sci USA* **108:**E383–391.

246. **Jain S, Goldberg MB.** 2007. Requirement for YaeT in the outer membrane assembly of autotransporter proteins. *J Bacteriol* **189:** 5393–5398.

247. **Rossiter AE, Leyton DL, Tveen-Jensen K, Browning DF, Sevastsyanovich Y, Knowles TJ, Nichols KB, Cunningham AF, Overduin M, Schembri MA, Henderson IR.** 2011. The essential beta-barrel assembly machinery complex components BamD and BamA are required for autotransporter biogenesis. *J Bacteriol* **193:**4250–4253.

248. **Junker M, Besingi RN, Clark PL.** 2009. Vectorial transport and folding of an autotransporter virulence protein during outer membrane secretion. *Mol Microbiol* **71:**1323–1332.

249. **Peterson JH, Tian P, Ieva R, Dautin N, Bernstein HD.** 2010. Secretion of a bacterial virulence factor is driven by the folding of a C-terminal segment. *Proc Natl Acad Sci USA* **107:**17739–17744.

250. **Jose J, Kramer J, Klauser T, Pohlner J, Meyer TF.** 1996. Absence of periplasmic DsbA oxidoreductase facilitates export of cysteine-containing passenger proteins to the *Escherichia coli* cell surface via the Iga beta autotransporter pathway. *Gene* **178:**107–110.

251. **Jong WS, ten Hagen-Jongman CM, den Blaauwen T, Slotboom DJ, Tame JR, Wickstrom D, de Gier JW, Otto BR, Luirink J.** 2007. Limited tolerance towards folded elements during secretion of the autotransporter Hbp. *Mol Microbiol* **63:**1524–1536.

252. **Ohnishi Y, Nishiyama M, Horinouchi S, Beppu T.** 1994. Involvement of the COOH-terminal pro-sequence of *Serratia marcescens* serine protease in the folding of the mature enzyme. *J Biol Chem* **269:**32800–32806.

253. **Oliver DC, Huang G, Nodel E, Pleasance S, Fernandez RC.** 2003. A conserved region within the *Bordetella pertussis* autotransporter BrkA is necessary for folding of its passenger domain. *Mol Microbiol* **47:**1367–1383.

254. **Soprova Z, Sauri A, van Ulsen P, Tame JR, den Blaauwen T, Jong WS, Luirink J.** 2010. A conserved aromatic residue in the autochaperone domain of the autotransporter Hbp is critical for initiation of outer membrane translocation. *J Biol Chem* **285:**38224–38233.

255. **Junker M, Schuster CC, McDonnell AV, Sorg KA, Finn MC, Berger B, Clark PL.** 2006. Pertactin beta-helix folding mechanism suggests common themes for the secretion and folding of autotransporter proteins. *Proc Natl Acad Sci USA* **103:**4918–4923.

256. **Veiga E, de Lorenzo V, Fernandez LA.** 2004. Structural tolerance of bacterial autotransporters for folded passenger protein domains. *Mol Microbiol* **52:**1069–1080.

257. **Skillman KM, Barnard TJ, Peterson JH, Ghirlando R, Bernstein HD.** 2005. Efficient secretion of a folded protein domain by a monomeric bacterial autotransporter. *Mol Microbiol* **58:**945–958.

258. **Sherlock O, Dobrindt U, Jensen JB, Munk Vejborg R, Klemm P.** 2006. Glycosylation of the self-recognizing *Escherichia coli* Ag43 autotransporter protein. *J Bacteriol* **188:**1798–1807.

259. **Sauri A, Soprova Z, Wickstrom D, de Gier JW, Van der Schors RC, Smit AB, Jong WS, Luirink J.** 2009. The Bam (Omp85) complex is involved in secretion of the autotransporter haemoglobin protease. *Microbiology* **155:**3982–3991.

260. **Ieva R, Skillman KM, Bernstein HD.** 2008. Incorporation of a polypeptide segment into the beta-domain pore during the assembly of a bacterial autotransporter. *Mol Microbiol* **67:**188–201.

261. **Klemm P, Hjerrild L, Gjermansen M, Schembri MA.** 2004. Structure-function analysis of the self-recognizing Antigen 43 autotransporter protein from *Escherichia coli*. *Mol Microbiol* **51:**283–296.

262. **Henderson IR, Meehan M, Owen P.** 1997. Antigen 43, a phase-variable bipartite outer membrane protein, determines colony morphology and autoaggregation in *Escherichia coli* K–12. *FEMS Microbiol Lett* **149:**115–120.

263. **Reidl S, Lehmann A, Schiller R, Salam Khan A, Dobrindt U.** 2009. Impact of O-glycosylation on the molecular and cellular adhesion properties of the *Escherichia coli* autotransporter protein Ag43. *Int J Med Microbiol* **299:**389–401.

264. **van der Woude MW, Henderson IR.** 2008. Regulation and function of Ag43 (flu). *Annu Rev Microbiol* **62:**153–169.

265. **Ulett GC, Valle J, Beloin C, Sherlock O, Ghigo JM, Schembri MA.** 2007. Functional analysis of antigen 43 in uropathogenic *Escherichia coli* reveals a role in long-term persistence in the urinary tract. *Infect Immun* **75:**3233–3244.

266. **Schembri MA, Klemm P.** 2001. Coordinate gene regulation by fimbriae-induced signal transduction. *EMBO J* **20:**3074–3081.

267. **Henderson IR, Navarro-Garcia F, Desvaux M, Fernandez RC, Ala'Aldeen D.** 2004. Type V protein secretion pathway: the autotransporter story. *Microbiol Mol Biol Rev* **68:**692–744.

268. **Heras B, Totsika M, Peters KM, Paxman JJ, Gee CL, Jarrott RJ, Perugini MA, Whitten AE,**

Schembri MA. 2014. The antigen 43 structure reveals a molecular Velcro-like mechanism of autotransporter-mediated bacterial clumping. *Proc Natl Acad Sci USA* **111**:457–462.

269. Vejborg RM, Klemm P. 2009. Cellular chain formation in *Escherichia coli* biofilms. *Microbiology* **155**:1407–1417.

270. Allsopp LP, Beloin C, Ulett GC, Valle J, Totsika M, Sherlock O, Ghigo JM, Schembri MA. 2012. Molecular characterization of UpaB and UpaC, two new autotransporter proteins of uropathogenic *Escherichia coli* CFT073. *Infect Immun* **80**:321–332.

271. Durant L, Metais A, Soulama-Mouze C, Genevard JM, Nassif X, Escaich S. 2007. Identification of candidates for a subunit vaccine against extraintestinal pathogenic *Escherichia coli*. *Infect Immun* **75**:1916–1925.

272. Allsopp LP, Beloin C, Moriel DG, Totsika M, Ghigo JM, Schembri MA. 2012. Functional heterogeneity of the UpaH autotransporter protein from uropathogenic *Escherichia coli*. *J Bacteriol* **194**:5769–5782.

273. Moriel DG, Bertoldi I, Spagnuolo A, Marchi S, Rosini R, Nesta B, Pastorello I, Corea VA, Torricelli G, Cartocci E, Savino S, Scarselli M, Dobrindt U, Hacker J, Tettelin H, Tallon LJ, Sullivan S, Wieler LH, Ewers C, Pickard D, Dougan G, Fontana MR, Rappuoli R, Pizza M, Serino L. 2010. Identification of protective and broadly conserved vaccine antigens from the genome of extraintestinal pathogenic *Escherichia coli*. *Proc Natl Acad Sci USA* **107**: 9072–9077.

274. Nesta B, Spraggon G, Alteri C, Moriel DG, Rosini R, Veggi D, Smith S, Bertoldi I, Pastorello I, Ferlenghi I, Fontana MR, Frankel G, Mobley HL, Rappuoli R, Pizza M, Serino L, Soriani M. 2012. FdeC, a novel broadly conserved *Escherichia coli* adhesin eliciting protection against urinary tract infections. *MBio* **3**: e00010–e00012.

275. Isberg RR, Voorhis DL, Falkow S. 1987. Identification of invasin: a protein that allows enteric bacteria to penetrate cultured mammalian cells. *Cell* **50**:769–778.

276. Luo Y, Frey EA, Pfuetzner RA, Creagh AL, Knoechel DG, Haynes CA, Finlay BB, Strynadka NC. 2000. Crystal structure of enteropathogenic *Escherichia coli* intimin-receptor complex. *Nature* **405**:1073–1077.

277. Oberhettinger P, Schutz M, Leo JC, Heinz N, Berger J, Autenrieth IB, Linke D. 2012. Intimin and invasin export their C-terminus to the bacterial cell surface using an inverse mechanism compared to classical autotransport. *PLoS One* **7**:e47069.

278. Leveille S, Caza M, Johnson JR, Clabots C, Sabri M, Dozois CM. 2006. Iha from an *Escherichia coli* urinary tract infection outbreak clonal group A strain is expressed in vivo in the mouse urinary tract and functions as a catecholate siderophore receptor. *Infect Immun* **74**:3427–3436.

279. Johnson JR, Jelacic S, Schoening LM, Clabots C, Shaikh N, Mobley HL, Tarr PI. 2005. The IrgA homologue adhesin Iha is an *Escherichia coli* virulence factor in murine urinary tract infection. *Infect Immun* **73**:965–971.

280. Tarr PI, Bilge SS, Vary JC Jr, Jelacic S, Habeeb RL, Ward TR, Baylor MR, Besser TE. 2000. Iha: a novel *Escherichia coli* O157:H7 adherence-conferring molecule encoded on a recently acquired chromosomal island of conserved structure. *Infect Immun* **68**:1400–1407.

281. Buchanan SK, Smith BS, Venkatramani L, Xia D, Esser L, Palnitkar M, Chakraborty R, van der Helm D, Deisenhofer J. 1999. Crystal structure of the outer membrane active transporter FepA from *Escherichia coli*. *Nature Struct Biol* **6**:56–63.

282. Manges AR, Johnson JR, Foxman B, O'Bryan TT, Fullerton KE, Riley LW. 2001. Widespread distribution of urinary tract infections caused by a multidrug-resistant *Escherichia coli* clonal group. *N Engl J Med* **345**: 1007–1013.

283. Burall LS, Harro JM, Li X, Lockatell CV, Himpsl SD, Hebel JR, Johnson DE, Mobley HL. 2004. *Proteus mirabilis* genes that contribute to pathogenesis of urinary tract infection: identification of 25 signature-tagged mutants attenuated at least 100-fold. *Infect Immun* **72**:2922–2938.

284. Linhartova I, Bumba L, Masin J, Basler M, Osicka R, Kamanova J, Prochazkova K, Adkins I, Hejnova-Holubova J, Sadilkova L, Morova J, Sebo P. 2010. RTX proteins: a highly diverse family secreted by a common mechanism. *FEMS Microbiol Rev* **34**:1076–1112.

285. Koronakis V, Eswaran J, Hughes C. 2004. Structure and function of TolC: the bacterial exit duct for proteins and drugs. *Annu Rev Biochem* **73**:467–489.

286. Delepelaire P. 2004. Type I secretion in gram-negative bacteria. *Biochim Biophys Acta* **1694**:149–161.

287. Vigil PD, Wiles TJ, Engstrom MD, Prasov L, Mulvey MA, Mobley HL. 2012. The repeat-intoxin family member TosA mediates adherence of uropathogenic *Escherichia coli* and survival during bacteremia. *Infect Immun* **80**:493–505.

288. Parham NJ, Pollard SJ, Chaudhuri RR, Beatson SA, Desvaux M, Russell MA, Ruiz J, Fivian A, Vila J, Henderson IR. 2005. Prevalence of pathogenicity island II$_{CFT073}$ genes among extraintestinal clinical isolates of *Escherichia coli*. *J Clin Microbiol* **43**:2425–2434.

289. Vigil PD, Alteri CJ, Mobley HL. 2011. Identification of in vivo-induced antigens including an RTX family exoprotein required for uropathogenic *Escherichia coli* virulence. *Infect Immun* **79**:2335–2344.

290. Ronald A. 2003. The etiology of urinary tract infection: traditional and emerging pathogens. *Dis Mon* **49**:71–82.

291. Koch S, Hufnagel M, Theilacker C, Huebner J. 2004. Enterococcal infections: host response, therapeutic, and prophylactic possibilities. *Vaccine* **22**:822–830.

292. Mandlik A, Swierczynski A, Das A, Ton-That H. 2008. Pili in Gram-positive bacteria: assembly, involvement in colonization and biofilm development. *Trends Microbiol* **16**:33–40.

293. Chagnot C, Listrat A, Astruc T, Desvaux M. 2012. Bacterial adhesion to animal tissues: protein determinants for recognition of extracellular matrix components. *Cell Microbiol* **14**:1687–1696.

294. Patti JM, Hook M. 1994. Microbial adhesins recognizing extracellular matrix macromolecules. *Curr Opin Cell Biol* **6**:752–758.

295. Rich RL, Kreikemeyer B, Owens RT, LaBrenz S, Narayana SV, Weinstock GM, Murray BE, Hook M. 1999. Ace is a collagen-binding MSCRAMM from *Enterococcus faecalis*. *J Biol Chem* **274**:26939–26945.

296. Sakinc T, Kleine B, Michalski N, Kaase M, Gatermann SG. 2009. SdrI of *Staphylococcus saprophyticus* is a multifunctional protein: localization of the fibronectin-binding site. *FEMS Microbiol Lett* **301**:28–34.

297. Nallapareddy SR, Qin X, Weinstock GM, Hook M, Murray BE. 2000. *Enterococcus faecalis* adhesin, ace, mediates attachment to extracellular matrix proteins collagen type IV and laminin as well as collagen type I. *Infect Immun* **68**:5218–5224.

298. Bowden MG, Heuck AP, Ponnuraj K, Kolosova E, Choe D, Gurusiddappa S, Narayana SV, Johnson AE, Hook M. 2008. Evidence for the "dock, lock, and latch" ligand binding mechanism of the staphylococcal microbial surface component recognizing adhesive matrix molecules (MSCRAMM) SdrG. *J Biol Chem* **283**:638–647.

299. Ponnuraj K, Bowden MG, Davis S, Gurusiddappa S, Moore D, Choe D, Xu Y, Hook M, Narayana SV. 2003. A "dock, lock, and latch" structural model for a staphylococcal adhesin binding to fibrinogen. *Cell* **115**:217–228.

300. Ponnuraj K, Narayana SV. 2007. Crystal structure of ACE19, the collagen binding subdomain of *Enterococus faecalis* surface protein ACE. *Proteins* **69**:199–203.

301. Liu Q, Ponnuraj K, Xu Y, Ganesh VK, Sillanpaa J, Murray BE, Narayana SV, Hook M. 2007. The *Enterococcus faecalis* MSCRAMM ACE binds its ligand by the Collagen Hug model. *J Biol Chem* **282**:19629–19637.

302. Zong Y, Xu Y, Liang X, Keene DR, Hook A, Gurusiddappa S, Hook M, Narayana SV. 2005. A 'Collagen Hug' model for *Staphylococcus aureus* CNA binding to collagen. *EMBO J* **24**:4224–4236.

303. Deivanayagam CC, Rich RL, Carson M, Owens RT, Danthuluri S, Bice T, Hook M, Narayana SV. 2000. Novel fold and assembly of the repetitive B region of the *Staphylococcus aureus* collagen-binding surface protein. *Structure* **8**:67–78.

304. Ganesh VK, Rivera JJ, Smeds E, Ko YP, Bowden MG, Wann ER, Gurusiddappa S, Fitzgerald JR, Hook M. 2008. A structural model of the *Staphylococcus aureus* ClfA-fibrinogen interaction opens new avenues for the design of anti-staphylococcal therapeutics. *PLoS Pathog* **4**:e1000226.

305. Deivanayagam CC, Wann ER, Chen W, Carson M, Rajashankar KR, Hook M, Narayana SV. 2002. A novel variant of the immunoglobulin fold in surface adhesins of *Staphylococcus aureus*: crystal structure of the fibrinogen-binding MSCRAMM, clumping factor A. *EMBO J* **21**:6660–6672.

306. Lebreton F, Riboulet-Bisson E, Serror P, Sanguinetti M, Posteraro B, Torelli R, Hartke A, Auffray Y, Giard JC. 2009. *ace*, Which encodes an adhesin in *Enterococcus faecalis*, is regulated by Ers and is involved in virulence. *Infect Immun* **77**:2832–2839.

307. Kuroda M, Yamashita A, Hirakawa H, Kumano M, Morikawa K, Higashide M, Maruyama A, Inose Y, Matoba K, Toh H, Kuhara S, Hattori M, Ohta T. 2005. Whole genome sequence of *Staphylococcus saprophyticus* reveals the pathogenesis of uncomplicated urinary tract infection. *Proc Natl Acad Sci U S A* **102**:13272–13277.

308. King NP, Beatson SA, Totsika M, Ulett GC, Alm RA, Manning PA, Schembri MA. 2011. UafB is a serine-rich repeat adhesin of *Staphylococcus saprophyticus* that mediates binding to fibronectin, fibrinogen and human uroepithelial cells. *Microbiology* **157**:1161–1175.

309. Kline KA, Ingersoll MA, Nielsen HV, Sakinc T, Henriques-Normark B, Gatermann S,

Caparon MG, Hultgren SJ. 2010. Characterization of a novel murine model of *Staphylococcus saprophyticus* urinary tract infection reveals roles for Ssp and SdrI in virulence. *Infect Immun* **78**:1943–1951.

310. Schneewind O, Mihaylova-Petkov D, Model P. 1993. Cell wall sorting signals in surface proteins of gram-positive bacteria. *EMBO J* **12**:4803–4811.

311. Symersky J, Patti JM, Carson M, House-Pompeo K, Teale M, Moore D, Jin L, Schneider A, DeLucas LJ, Hook M, Narayana SV. 1997. Structure of the collagen-binding domain from a *Staphylococcus aureus* adhesin. *Nat Struct Biol* **4**:833–838.

312. Nallapareddy SR, Singh KV, Sillanpaa J, Zhao M, Murray BE. 2011. Relative contributions of Ebp Pili and the collagen adhesin ace to host extracellular matrix protein adherence and experimental urinary tract infection by *Enterococcus faecalis* OG1RF. *Infect Immun* **79**:2901–2910.

313. Matsuoka E, Tanaka Y, Kuroda M, Shouji Y, Ohta T, Tanaka I, Yao M. 2011. Crystal structure of the functional region of Uro-adherence factor A from *Staphylococcus saprophyticus* reveals participation of the B domain in ligand binding. *Protein Sci* **20**:406–416.

314. Schneewind O, Missiakas DM. 2012. Protein secretion and surface display in Gram-positive bacteria. *Philos Trans R Soc Lond B Biol Sci* **367**:1123–1139.

315. Model P, Russel M. 1990. Prokaryotic secretion. *Cell* **61**:739–741.

316. Schneewind O, Model P, Fischetti VA. 1992. Sorting of protein A to the staphylococcal cell wall. *Cell* **70**:267–281.

317. Spirig T, Weiner EM, Clubb RT. 2011. Sortase enzymes in Gram-positive bacteria. *Mol Microbiol* **82**:1044–1059.

318. Mazmanian SK, Liu G, Ton-That H, Schneewind O. 1999. *Staphylococcus aureus* sortase, an enzyme that anchors surface proteins to the cell wall. *Science* **285**:760–763.

319. Maresso AW, Schneewind O. 2008. Sortase as a target of anti-infective therapy. *Pharmacol Rev* **60**:128–141.

320. Marraffini LA, Dedent AC, Schneewind O. 2006. Sortases and the art of anchoring proteins to the envelopes of gram-positive bacteria. *Microbiol Mol Biol Rev* **70**:192–221.

321. Perry AM, Ton-That H, Mazmanian SK, Schneewind O. 2002. Anchoring of surface proteins to the cell wall of *Staphylococcus aureus*. III. Lipid II is an in vivo peptidoglycan substrate for sortase-catalyzed surface protein anchoring. *J Biol Chem* **277**:16241–16248.

322. Ruzin A, Severin A, Ritacco F, Tabei K, Singh G, Bradford PA, Siegel MM, Projan SJ, Shlaes DM. 2002. Further evidence that a cell wall precursor [C(55)-MurNAc-(peptide)-GlcNAc] serves as an acceptor in a sorting reaction. *J Bacteriol* **184**:2141–2147.

323. Foster TJ, Hook M. 1998. Surface protein adhesins of *Staphylococcus aureus*. *Trends Microbiol* **6**:484–488.

324. Murray BE. 1990. The life and times of the Enterococcus. *Clin Microbiol Rev* **3**:46–65.

325. Arias CA, Murray BE. 2012. The rise of the Enterococcus: beyond vancomycin resistance. *Nat Rev Microbiol* **10**:266–278.

326. Raz R, Colodner R, Kunin CM. 2005. Who are you—*Staphylococcus saprophyticus*? *Clin Infect Dis* **40**:896–898.

327. Szabados F, Kleine B, Anders A, Kaase M, Sakinc T, Schmitz I, Gatermann S. 2008. *Staphylococcus saprophyticus* ATCC 15305 is internalized into human urinary bladder carcinoma cell line 5637. *FEMS Microbiol Lett* **285**:163–169.

328. Valle J, Mabbett AN, Ulett GC, Toledo-Arana A, Wecker K, Totsika M, Schembri MA, Ghigo JM, Beloin C. 2008. UpaG, a new member of the trimeric autotransporter family of adhesins in uropathogenic Escherichia coli. *J Bacteriol* **190**:4147–4161.

329. Sakinc T, Kleine B, Gatermann SG. 2006. SdrI, a serine-aspartate repeat protein identified in *Staphylococcus saprophyticus* strain 7108, is a collagen-binding protein. *Infect Immun* **74**:4615–4623.

330. Holmes AR, McNab R, Millsap KW, Rohde M, Hammerschmidt S, Mawdsley JL, Jenkinson HF. 2001. The pavA gene of *Streptococcus pneumoniae* encodes a fibronectin-binding protein that is essential for virulence. *Mol Microbiol* **41**:1395–1408.

331. Torelli R, Serror P, Bugli F, Paroni Sterbini F, Florio AR, Stringaro A, Colone M, De Carolis E, Martini C, Giard JC, Sanguinetti M, Posteraro B. 2012. The PavA-like Fibronectin-Binding Protein of *Enterococcus faecalis*, EfbA, Is Important for Virulence in a Mouse Model of Ascending Urinary Tract Infection. *J Infect Dis* **206**(6):952–960.

332. Shankar N, Lockatell CV, Baghdayan AS, Drachenberg C, Gilmore MS, Johnson DE. 2001. Role of *Enterococcus faecalis* surface protein Esp in the pathogenesis of ascending urinary tract infection. *Infect Immun* **69**:4366–4372.

333. Tendolkar PM, Baghdayan AS, Shankar N. 2005. The N-terminal domain of enterococcal surface protein, Esp, is sufficient for Esp-

mediated biofilm enhancement in *Enterococcus faecalis*. *J Bacteriol* **187:**6213–6222.

334. **Kang HJ, Paterson NG, Gaspar AH, Ton-That H, Baker EN.** 2009. The *Corynebacterium diphtheriae* shaft pilin SpaA is built of tandem Ig-like modules with stabilizing isopeptide and disulfide bonds. *Proc Natl Acad Sci USA* **106:**16967–16971.

335. **Spraggon G, Koesema E, Scarselli M, Malito E, Biagini M, Norais N, Emolo C, Barocchi MA, Giusti F, Hilleringmann M, Rappuoli R, Lesley S, Covacci A, Masignani V, Ferlenghi I.** 2010. Supramolecular organization of the repetitive backbone unit of the *Streptococcus pneumoniae* pilus. *PLoS One* **5:**e10919.

336. **Linke C, Young PG, Kang HJ, Bunker RD, Middleditch MJ, Caradoc-Davies TT, Proft T, Baker EN.** 2010. Crystal structure of the minor pilin FctB reveals determinants of Group A streptococcal pilus anchoring. *J Biol Chem* **285:**20381–20389.

337. **Kang HJ, Baker EN.** 2009. Intramolecular isopeptide bonds give thermodynamic and proteolytic stability to the major pilin protein of *Streptococcus pyogenes*. *J Biol Chem* **284:** 20729–20737.

338. **Krishnan V, Gaspar AH, Ye N, Mandlik A, Ton-That H, Narayana SV.** 2007. An IgG-like domain in the minor pilin GBS52 of *Streptococcus agalactiae* mediates lung epithelial cell adhesion. *Structure* **15:**893–903.

339. **Sillanpaa J, Nallapareddy SR, Singh KV, Prakash VP, Fothergill T, Ton-That H, Murray BE.** 2010. Characterization of the *ebp(fm)* pilus-encoding operon of *Enterococcus faecium* and its role in biofilm formation and virulence in a murine model of urinary tract infection. *Virulence* **1:**236–246.

340. **Nallapareddy SR, Singh KV, Sillanpaa J, Garsin DA, Hook M, Erlandsen SL, Murray BE.** 2006. Endocarditis and biofilm-associated pili of *Enterococcus faecalis*. *J Clin Invest* **116:**2799–2807.

341. **Budzik JM, Poor CB, Faull KF, Whitelegge JP, He C, Schneewind O.** 2009. Intramolecular amide bonds stabilize pili on the surface of bacilli. *Proc Natl Acad Sci USA* **106:**19992–19997.

342. **Izore T, Contreras-Martel C, El Mortaji L, Manzano C, Terrasse R, Vernet T, Di Guilmi AM, Dessen A.** 2010. Structural basis of host cell recognition by the pilus adhesin from *Streptococcus pneumoniae*. *Structure* **18:**106–115.

343. **Kang HJ, Coulibaly F, Clow F, Proft T, Baker EN.** 2007. Stabilizing isopeptide bonds revealed in gram-positive bacterial pilus structure. *Science* **318:**1625–1628.

344. **Mandlik A, Das A, Ton-That H.** 2008. The molecular switch that activates the cell wall anchoring step of pilus assembly in gram-positive bacteria. *Proc Natl Acad Sci USA* **105:**14147–14152.

345. **Nielsen HV, Guiton PS, Kline KA, Port GC, Pinkner JS, Neiers F, Normark S, Henriques-Normark B, Caparon MG, Hultgren SJ.** 2012. The metal ion-dependent adhesion site motif of the *Enterococcus faecalis* EbpA pilin mediates pilus function in catheter-associated urinary tract infection. *MBio* **3:**e00177–e00112.

346. **Sillanpaa J, Chang C, Singh KV, Montealegre MC, Nallapareddy SR, Harvey BR, Ton-That H, Murray BE.** 2013. Contribution of Individual Ebp Pilus Subunits of *Enterococcus faecalis* OG1RF to Pilus Biogenesis, Biofilm Formation and Urinary Tract Infection. *PLoS One* **8:** e68813.

347. **Sillanpaa J, Nallapareddy SR, Prakash VP, Qin X, Hook M, Weinstock GM, Murray BE.** 2008. Identification and phenotypic characterization of a second collagen adhesin, Scm, and genome-based identification and analysis of 13 other predicted MSCRAMMs, including four distinct pilus loci, in *Enterococcus faecium*. *Microbiology* **154:**3199–3211.

348. **Whittaker CA, Hynes RO.** 2002. Distribution and evolution of von Willebrand/integrin A domains: widely dispersed domains with roles in cell adhesion and elsewhere. *Mol Biol Cell* **13:**3369–3387.

349. **Lee JO, Rieu P, Arnaout MA, Liddington R.** 1995. Crystal structure of the A domain from the alpha subunit of integrin CR3 (CD11b/CD18). *Cell* **80:**631–638.

350. **Hilleringmann M, Giusti F, Baudner BC, Masignani V, Covacci A, Rappuoli R, Barocchi MA, Ferlenghi I.** 2008. Pneumococcal pili are composed of protofilaments exposing adhesive clusters of Rrg A. *PLoS Pathog* **4:**e1000026.

351. **Budzik JM, Marraffini LA, Souda P, Whitelegge JP, Faull KF, Schneewind O.** 2008. Amide bonds assemble pili on the surface of bacilli. *Proc Natl Acad Sci USA* **105:**10215–10220.

352. **Singh KV, Nallapareddy SR, Murray BE.** 2007. Importance of the *ebp* (endocarditis- and biofilm-associated pilus) locus in the pathogenesis of *Enterococcus faecalis* ascending urinary tract infection. *J Infect Dis* **195:**1671–1677.

353. **Sillanpaa J, Xu Y, Nallapareddy SR, Murray BE, Hook M.** 2004. A family of putative MSCRAMMs from *Enterococcus faecalis*. *Microbiology* **150:**2069–2078.

354. **Roberts JA, Kaack MB, Baskin G, Chapman MR, Hunstad DA, Pinkner JS, Hultgren SJ.** 2004. Antibody responses and protection from

pyelonephritis following vaccination with purified *Escherichia coli* PapDG protein. *J Urol* **171:**1682–1685.

355. **Tchesnokova V, Aprikian P, Kisiela D, Gowey S, Korotkova N, Thomas W, Sokurenko E.** 2011. Type 1 fimbrial adhesin FimH elicits an immune response that enhances cell adhesion of *Escherichia coli. Infect Immun* **79:**3895–3904.

356. **Otto M.** 2008. Targeted immunotherapy for staphylococcal infections : focus on anti-MSCRAMM antibodies. *BioDrugs* **22:**27–36.

357. **Patti JM.** 2004. A humanized monoclonal antibody targeting *Staphylococcus aureus. Vaccine* **22**(Suppl 1)**:**S39–43.

358. **Salminen A, Loimaranta V, Joosten JA, Khan AS, Hacker J, Pieters RJ, Finne J.** 2007. Inhibition of P-fimbriated *Escherichia coli* adhesion by multivalent galabiose derivatives studied by a live-bacteria application of surface plasmon resonance. *J Antimicrob Chemother* **60:**495–501.

359. **Wellens A, Garofalo C, Nguyen H, Van Gerven N, Slattegard R, Hernalsteens JP, Wyns L, Oscarson S, De Greve H, Hultgren S, Bouckaert J.** 2008. Intervening with urinary tract infections using anti-adhesives based on the crystal structure of the FimH-oligomannose-3 complex. *PLoS One* **3:**e2040.

360. **Han Z, Pinkner JS, Ford B, Obermann R, Nolan W, Wildman SA, Hobbs D, Ellenberger T, Cusumano CK, Hultgren SJ, Janetka JW.** 2010. Structure-based drug design and optimization of mannoside bacterial FimH antagonists. *J Med Chem* **53:**4779–4792.

361. **Schwardt O, Rabbani S, Hartmann M, Abgottspon D, Wittwer M, Kleeb S, Zalewski A, Smiesko M, Cutting B, Ernst B.** 2011. Design, synthesis and biological evaluation of mannosyl triazoles as FimH antagonists. *Bioorg Med Chem* **19:**6454–6473.

362. **Pinkner JS, Remaut H, Buelens F, Miller E, Aberg V, Pemberton N, Hedenstrom M, Larsson A, Seed P, Waksman G, Hultgren SJ, Almqvist F.** 2006. Rationally designed small compounds inhibit pilus biogenesis in uropathogenic bacteria. *Proc Natl Acad Sci USA* **103:**17897–17902.

363. **Cegelski L, Marshall GR, Eldridge GR, Hultgren SJ.** 2008. The biology and future prospects of antivirulence therapies. *Nat Rev Microbiol* **6:**17–27.

364. **Dobrindt U, Hacker J.** 2008. Targeting virulence traits: potential strategies to combat extraintestinal pathogenic *E. coli* infections. *Curr Opin Microbiol* **11:**409–413.

365. **Jones CH, Dodson K, Hultgren SJ.** 1996. Structure, function, and assembly of adhesive P pili, p 175–219. *In* Mobley HLT, Warren JW (ed), *Urinary tract infections: molecular pathogenesis and clinical management*, 1 ed. ASM Press, Washington, DC.

366. **Leffler H, Svanborg-Eden C.** 1981. Glycolipid receptors for uropathogenic *Escherichia coli* on human erythrocytes and uroepithelial cells. *Infect Immun* **34:**920–929.

367. **Wu XR, Sun TT, Medina JJ.** 1996. In vitro binding of type 1-fimbriated *Escherichia coli* to uroplakins Ia and Ib: relation to urinary tract infections. *Proc Natl Acad Sci USA* **93:**9630–9635.

368. **Korhonen TK, Valtonen MV, Parkkinen J, Vaisanen-Rhen V, Finne J, Orskov F, Orskov I, Svenson SB, Makela PH.** 1985. Serotypes, hemolysin production, and receptor recognition of *Escherichia coli* strains associated with neonatal sepsis and meningitis. *Infect Immun* **48:**486–491.

369. **Castelain M, Sjostrom AE, Fallman E, Uhlin BE, Andersson M.** 2010. Unfolding and refolding properties of S pili on extraintestinal pathogenic *Escherichia coli. Eur Biophys J* **39:**1105–1115.

370. **Khan AS, Kniep B, Oelschlaeger TA, Van Die I, Korhonen T, Hacker J.** 2000. Receptor structure for F1C fimbriae of uropathogenic *Escherichia coli. Infect Immun* **68:**3541–3547.

371. **Ong CL, Ulett GC, Mabbett AN, Beatson SA, Webb RI, Monaghan W, Nimmo GR, Looke DF, McEwan AG, Schembri MA.** 2008. Identification of type 3 fimbriae in uropathogenic *Escherichia coli* reveals a role in biofilm formation. *J Bacteriol* **190:**1054–1063.

372. **Struve C, Bojer M, Krogfelt KA.** 2008. Characterization of *Klebsiella pneumoniae* type 1 fimbriae by detection of phase variation during colonization and infection and impact on virulence. *Infect Immun* **76:**4055–4065.

373. **Gerlach GF, Clegg S, Allen BL.** 1989. Identification and characterization of the genes encoding the type 3 and type 1 fimbrial adhesins of *Klebsiella pneumoniae. J Bacteriol* **171:**1262–1270.

374. **Stahlhut SG, Chattopadhyay S, Struve C, Weissman SJ, Aprikian P, Libby SJ, Fang FC, Krogfelt KA, Sokurenko EV.** 2009. Population variability of the FimH type 1 fimbrial adhesin in *Klebsiella pneumoniae. J Bacteriol* **191:**1941–1950.

375. **Tarkkanen AM, Allen BL, Westerlund B, Holthofer H, Kuusela P, Risteli L, Clegg S, Korhonen TK.** 1990. Type V collagen as the target for type-3 fimbriae, enterobacterial adherence organelles. *Mol Microbiol* **4:**1353–1361.

376. **Stahlhut SG, Struve C, Krogfelt KA.** 2012. *Klebsiella pneumoniae* type 3 fimbriae aggluti-

nate yeast in a mannose-resistant manner. *J Med Microbiol* **61:**317–322.

377. **Ong CL, Beatson SA, Totsika M, Forestier C, McEwan AG, Schembri MA.** 2010. Molecular analysis of type 3 fimbrial genes from *Escherichia coli*, *Klebsiella* and *Citrobacter* species. *BMC Microbiol* **10:**183.

378. **Zunino P, Geymonat L, Allen AG, Preston A, Sosa V, Maskell DJ.** 2001. New aspects of the role of MR/P fimbriae in *Proteus mirabilis* urinary tract infection. *FEMS Immunol Med Microbiol* **31:**113–120.

379. **Li X, Johnson DE, Mobley HL.** 1999. Requirement of MrpH for mannose-resistant *Proteus*-like fimbria-mediated hemagglutination by *Proteus mirabilis*. *Infect Immun* **67:**2822–2833.

380. **Bahrani FK, Mobley HL.** 1994. *Proteus mirabilis* MR/P fimbrial operon: genetic organization, nucleotide sequence, and conditions for expression. *J Bacteriol* **176:**3412–3419.

381. **Zunino P, Sosa V, Allen AG, Preston A, Schlapp G, Maskell DJ.** 2003. *Proteus mirabilis* fimbriae (PMF) are important for both bladder and kidney colonization in mice. *Microbiology* **149:**3231–3237.

382. **Massad G, Lockatell CV, Johnson DE, Mobley HLT.** 1994. *Proteus mirabilis* fimbriae: construction of an isogenic *pmfA* mutant and analysis of virulence in a CBA mouse model of ascending urinary tract infection. *Infect Immmun* **62:**536–542.

383. **Pellegrino R, Scavone P, Umpierrez A, Maskell DJ, Zunino P.** 2013. *Proteus mirabilis* uroepithelial cell adhesin (UCA) fimbria plays a role in the colonization of the urinary tract. *Pathog Dis* **67:**104–107.

384. **Cook SW, Mody N, Valle J, Hull R.** 1995. Molecular cloning of *Proteus mirabilis* uroepithelial cell adherence (*uca*) genes. *Infect Immun* **63:**2082–2086.

385. **Lee KK, Harrison BA, Latta R, Altman E.** 2000. The binding of *Proteus mirabilis* non-agglutinating fimbriae to ganglio-series asialoglycolipids and lactosyl ceramide. *Can J Microbiol* **46:**961–966.

386. **Massad G, Fulkerson JF Jr, Watson DC, Mobley HL.** 1996. *Proteus mirabilis* ambient-temperature fimbriae: cloning and nucleotide sequence of the *aft* gene cluster. *Infect Immun* **64:**4390–4395.

387. **Zunino P, Geymonat L, Allen AG, Legnani-Fajardo C, Mask ell DJ.** 2000. Virulence of a *Proteus mirabilis* ATF isogenic mutant is not impaired in a mouse model of ascending urinary tract infection. *FEMS Immunol Med Microbiol* **29:**137–143.

Pathoadaptive Mutations in Uropathogenic *Escherichia coli*

15

EVGENI SOKURENKO[1]

ECO-EVO OVERVIEW OF BACTERIAL PATHOGENS

Adaptive evolution of bacterial pathogens happens through the action of positive selection on random genetic changes, affecting their frequency within the compartment they infect (e.g., the urinary tract) or during the transmission to new hosts. The evolution of virulence, in general, results in the ability of the microorganism to cause clinically manifested damage of the host, i.e., diseases like cystitis or pyelonephritis. This, in turn, reflects the microbial fitness during the infection – the ability to survive and reproduce during the infection itself, at the level and in a mode to induce the tissue damage. Driving forces and mechanisms behind the evolution of microbial virulence can be understood from the perspective of bacterial ecology and by comparative analysis of the microorganisms that are able or unable to cause the infection.

From ecological perspectives, the driving force behind the emergence of pathogenic lineages can be examined from the perspectives of how the ability of *Escherichia coli* to cause disease contributes to its circulation in nature. In order to circulate continuously in a given habitat, commensal bacteria, whether associated with a living host or an environment, should have three general properties. They should be able to a) *enter* their habitat by crossing

[1]University of Washington, Seattle, WA 98195.
Urinary Tract Infections: Molecular Pathogenesis and Clinical Management, 2nd Edition
Edited by Matthew A. Mulvey, David J. Klumpp, and Ann E. Stapleton
© 2017 American Society for Microbiology, Washington, DC
doi:10.1128/microbiolspec.UTI-0020-2015

existing barriers; b) *sustain* itself there by sufficient growth rate and resistance to clearance, and c) *transmit* from the occupied habitat into new ones to repeat the cycle (Fig. 1).

By definition, all pathogens are capable of entering, one way or another, the infection habitat and sustain themselves long enough to cause clinical infection. However, depending on how pathogens, including *E. coli*, transmit into and from the infection site they can be subdivided into three general categories – *opportunistic, accidental,* and *professional* pathogens (Fig. 1). Opportunistic pathogens represent the most diverse, but ambiguous from ecological and evolutionary perspectives, group of pathogens. Opportunistic pathogens appear to be able to continuously circulate as human commensals and cause clinical infections only occasionally, primarily when the defense barriers or systems are somewhat compromised. An important feature of the opportunistic pathogens is that they are able to return to the original (commensal) habitat while being shed during the infection, sometimes in high numbers. However, they generally are not able to transmit from the patient to cause another

infection directly, i.e., without returning to non-pathogenic circulation. As described below, uropathogenic *E. coli* (UPEC) represent an example of the opportunistic pathogen.

In accidental human pathogens, clinical infection is an ecological accident, where pathogens do not circulate among humans (at least not continuously) and they get into contact with individuals from a completely different host species or environment. Unlike with opportunistic pathogens, the infection in humans does not promote their transmission into the original habitat or, in any significant numbers, to another person. Among *E. coli*, enterohemorrhagic strains are a good example of accidental pathogen by being a zoonotic pathogen. Also, the ability of *E. coli* to cause infection in enclosed, dead-end compartments, like cerebrospinal fluid or bloodstream, also makes these *E. coli* accidental meningitis or sepsis pathogens. For professional pathogens, the ability to cause disease is essential for continuous circulation in nature, with *E. coli* pathotypes like *Shigella*, enteroinvasive strains, and, possibly, enterotoxigenic *E. coli* being examples. The professional pathogens can effectively transmit during the active infection to form new

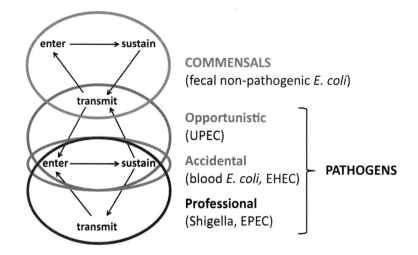

FIGURE 1 Eco-Evo categories of *E. coli*. UPEC = uropathogenic *E. coli*, EHEC = enterohemorrhagic *E. coli*, EPEC = enteropathogenic *E. coli*.

infections, i.e., from patient to patient, while upon the resolution of clinical infection, an asymptomatic state usually follows.

UPEC AS OPPORTUNISTIC HUMAN PATHOGENS

UPEC represent one of the best examples of opportunistic human pathogens, with a relatively well-understood ecological cycle (Fig. 2). The main habitat of *E. coli* is the large intestine of humans and other warm-blooded animals, from which they spread to new host organisms by fecal-oral transmission. To enter the intestinal tract of the new individual, *E. coli* has to pass the acidic environment of the stomach, which could represent a significant barrier for the transmission. It is believed that transient colonization of the oropharynx provides the means to pass through the stomach successfully (1).

The urinary bladder, ureters, and kidneys are normally well-protected, sterile compartments of the human body, but are also infection habitats for UPEC. *E. coli* is the main urinary pathogen in humans (2) and affects over 7 million women each year in the USA (3). The infecting *E. coli* usually come from the intestinal microbiota, which is introduced into the bladder by frequent sexual intercourse, diaphragm use, or bladder catheter-

ization. This leads to the breach of the urethra sphincter or an increase in the periurethral colonization (4). Bladder infection commonly lasts 5–7 days, during which time bacteria are shed with urine, often more than 10^5 colony-forming units (cfu) per ml. In one out of a few dozen cases, bacteria from the bladder ascend into the kidney, resulting in a more serious clinical infection – pyelonephritis – that is also not self-enclosed, allowing bacteria to be shed and then spread. Urinary tract infections are not considered to be contagious, i.e., resulting in direct spread from patient to patient as uropathogen, without entering the fecal-oral transmission cycle. However, it has been documented that *E. coli* can be transmitted sexually between partners, but to what extent this transmission mode is prevalent is unknown (5).

Obviously, little prevents urinary bacteria from returning back into the intestinal microflora of the same individual or into fecal-oral circulation in the population. Back in circulation, the pathogen is likely again to cause an urinary tract infection, sometimes causing infection in the same patient. However, to what extent repeated infection is due to re-infection from the intestinal niche (or transient vaginal colonization) and/or reactivation of the original infection in the bladder, is still debated (6, 7). It is certainly possible that after causing infection in one individual and returning to the fecal-oral circulation, the same strain can cause infection in another individual, but the rate is not known.

GENETIC MECHANISMS OF VIRULENCE EVOLUTION

There is a huge range in the ability of different bacterial strains to cause disease, even when they belong to the same species. In *E. coli*, for example, to stay healthy, the recommended regular dose of probiotic MutaFlor contains billions of live bacteria of strain Nissle-1917, while the ingestion of

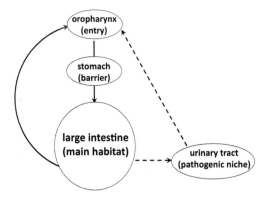

FIGURE 2 UPEC ecology.

100 or less of live cells of O157 strain or *Shigella flexneri* can result in fatal disease (8, 9). Moreover, same-species strains with a drastically different pathogenic potential could be closely related (10). For example, the probiotic strain Nissle-1917 that was isolated from feces of a healthy individual belongs to the same clonal group (based on multi-locus sequence typing) as one of the model UPEC strains, CFT073, isolated from a patient with urosepsis – a rare but highly lethal complication of urinary tract infections (11). Interestingly, this same clonal group also includes strain 83972 that was isolated from a patient with the mildest (but probably most common) form of urinary tract infection – asymptomatic bacteriuria. As described below, this strain is being examined in clinical studies for being used as a urinary tract probiotic – to be installed into the bladders of catheterized patients for asymptomatic colonization – to prevent development of symptomatic infection.

The difference in pathogenic ability reflects the diversity of phenotypic traits across strains from the same species that differ in the composition of their surface components, adhesion organelles, secreted factors, metabolic properties, etc., both quantitatively or qualitatively (12). The bacterial traits that contribute to the ability to cause disease are defined as virulence factors, which could be a specific gene or specific form of the gene, including a functionally inactive, over-expressed, or otherwise functionally distinct gene variant. Thus, the evolution of virulence is the evolution of more-pathogenic bacteria from less-pathogenic ones by acquisition of virulence factors. There are two major genetic mechanisms that increase the fitness of bacteria as pathogens: acquisition of novel genes by horizontal transfer and alteration of the existing genes by pathoadaptive mutations that result in gene loss, over-expression, or variation (Fig. 3).

Bacterial adhesion as a virulence trait can be used to illustrate the mechanisms of virulence evolution. The ability to adhere to the uroepithelial cells is considered to be one of the key virulence factors for UPEC as it allows bacteria either to resist wash off from the urinary tract by voiding or to invade the urothelium to sustain itself intracellularly (12, 13). Theoretically, because urinary epithelium and gastrointestinal mucosal surface are different tissues, *E. coli* adhesins that are adapted to colonize the intestine could be unable to ensure sufficient level of urothelial

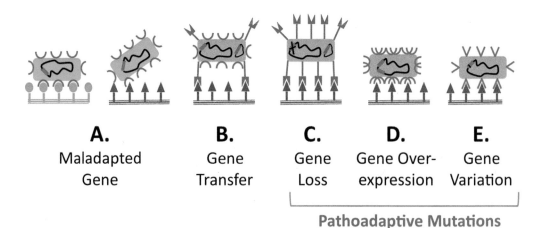

A.
Maladapted
Gene

B.
Gene
Transfer

C.
Gene
Loss

D.
Gene Over-
expression

E.
Gene
Variation

Pathoadaptive Mutations

FIGURE 3 Different genetic mechanisms of evolution of virulence exemplified by an adaptive increase in bacterial adhesiveness. Green surface = gastrointestinal mucosa; Gray surface = urothelium.

adhesion (Fig. 3A). This would render *E. coli* a relatively low fitness as a urinary pathogen, opening room for the selection for more pathogenic *E. coli* clones with increased ability to adhere to the uroepithelial cells than can be achieved by gene transfer and mutation, as described below.

Horizontal Gene Transfer

Horizontal or lateral gene transfer involves the transfer of genetic material from one strain to another, either between organisms from different species or between different strains of the same species (14). Horizontal gene transfer implies addition to the genome of novel genes that encode separate adaptive traits for the pathogen. Genetic elements that are typically of horizontal gene origin are plasmids, prophages, and chromosomal (pathogenicity) islands. For example, to increase tropism to urothelium, UPEC could use a novel set of horizontally transferred adhesive organelles (Fig. 3B). Horizontal gene transfer should be distinguished from gene exchange by homologous recombination involving different allelic variants of the same gene, even though the mechanism of the gene movement between different organisms could be the same – conjugation, transduction, and transformation.

In *E. coli* in general, plasmids are either associated with genes conferring resistance to antimicrobial agents (e.g., cephalosporins, aminoglycosides) or with genes coding for toxins secretion and colonization factors. Among virulence factors found in uropathogenic strains, serum-resistance-associated protein TraT in UPEC is coded by a plasmid (15). Prophages are temperate phages that did not enter into the lytic cycle but instead integrated into the bacterial chromosome. In fact, most of the horizontally transferred genes in *E. coli* genome are of a prophage nature (16). The majority of the prophage genes are silent during the bacterial growth/reproduction, but some carry genes that are expressed and add new phenotypic traits to

their hosts (17). In *E. coli*, most prophage-encoded virulence factors are found in diarrheal pathotypes, e.g., Shiga toxins (18), type III secretion-system-effector proteins (19, 20), and heat-labile enterotoxin (21). Interestingly, besides some association of cytolethal-distending toxin-coding prophages with UPEC (22), until recently there was no known major virulence determinants encoded by prophages in UPEC (23).

Chromosomal islands characterize a highly diverse group of DNA elements, with a broad range in size and abundance across the bacterial chromosomes (24). Pathogenicity islands are found across all *E. coli* pathotypes (25–28). Originally, however, chromosomal islands were described as pathogenicity-associated islands, coined to define large unstable regions (10–200 kb) harboring virulence determinants on UPEC chromosomes (29). Examples of the islands-coded virulence factors in UPEC is quite abundant and include hemolysin (30), siderophore synthesis (31), and several fimbrial (e.g., type P, S, or Dr fimbriae) and non-fimbrial (e.g., NFA-1) adhesins (32). For example, one trait associated with the ability to cause recurrent cystitis is the presence of Dr-family adhesins, a family that includes members that specifically recognize decay-accelerating factor (DAF) as a receptor (33). In another study, multiple same-strain recurrences were associated with a di-galactose-specific P-fimbrial adhesin, *papG* (class II), and also with the presence of *iha* (nonhemagglutinating adhesin) and *iutA* (ferric sidero-phore receptor) genes (34).

Experimental evidence for the importance of horizontally transferred genes in urovirulence is abundant, showing that deletion of certain genes (or gene blocks) significantly reduces urinary tract colonization in murine model of the infection (35). Despite evidence for the importance of specific urovirulence factors, there is no single trait or set of traits identified that can be said to be necessary and/or sufficient for the pathogenesis of *E. coli* infections in the urinary tract. This led to the concept that there are many different ways to become an uropathogen (36). Furthermore, deletion of some genes that appear to be associated with UPEC have no significant effect in the model UTI (37). Also, most of the horizontally transferred urovirulence factors are either present only in a minority of the uropathogens or can be found in a sizeable portion of the commensal strains as well, thus being adaptive for non-pathogenic environments (38). Finally, some of the genes that were shown to be important for the urovirulence belong to the genomic core, i.e., shared by all or the majority of *E. coli* and are not positioned on horizontally transferred genetic elements (39).

Thus, while the importance of horizontally transferred genes in urovirulence of *E. coli* is undisputable, it is also obvious that other evolutionary mechanisms could be involved in emergence of successful UPEC strains.

Pathoadaptive Mutations

Just the mere presence of specific virulence factor genes in the genome is not sufficient to optimize the bacterial fitness during the infection. First, the genes need to be expressed at the right time and level. For example, it was shown that pre-adapting ('conditioning') bacteria to nitrosative stress conditions (by growing in the presence of acidified sodium nitrite) significantly increased their level of bladder colonization in mice (40). Second, functional properties of allelic variants of a gene might differ significantly from each other. A striking example of this are allelic variants of DNA gyrase gene, *gyrA*, that define quinolone resistance in *E. coli* and other bacteria (41). Finally, though horizontally transferred genes come in functional blocks (operons, gene clusters), they have to function within the original regulatory, transport, and biogenesis networks of the host cell (42). As mobile genetic elements are commonly in a genetic conflict with the rest of the genome, such conflicts have to be resolved to accommodate the novel gene's function (43).

Thus, another mechanism for the increase in pathogenic potential of bacterial strains is through mutation of the existing bacterial genes, i.e., without acquisition of novel genetic material (44). Such pathoadaptive mutations are the primary focus of this review. Unlike horizontal gene transfer, mutations can affect genes that are shared by both pathogenic and commensal isolates. Such molecular adaptive evolution can help pathogens, for example, to optimize host-tissue tropism or growth rate, avoid immune recognition, or acquire antibiotic resistance. In contrast to the horizontal transfer of novel genes, gene mutations were shown to be selected specifically in the course of a single infection (45, 46). Therefore, their occurrence represents a truly virulence-specific adaptation mechanism. The likelihood of adaptive evolution of bacterial pathogens in the course of a single UTI is favored by a) the relatively large size of the bacterial population infecting the patient; b) a relatively long-term nature of the asymptomatic or recurrent forms of the infection, and c) frequent fluctuation of the population size due to voiding that provides an easy opportunity for population expansion of newly emerged adaptive clones.

THE MAJOR GENETIC MECHANISMS OF THE PATHOADAPTATION ARE GENE INACTIVATION, AMPLIFICATION, AND VARIATION

Inactivation (loss) of genes could have a highly adaptive effect for the pathogens by a variety of mechanisms. For example, inactivating a negative regulator of the trait that is protective for a pathogen will lead to over-expression for the trait, which might have an adaptive effect, similar to the induction of a mucoid phenotype in *Pseudomonas aeruginosa* by knock-out of *mucA* (47). Also, if surface expression of a sub-optimal adhesin competes with the function of adhesin with a better tissue tropism (as is known for fimbrial adhesins (48)), deletion of the former will be adaptive for the pathogen (Fig. 3C). Besides inactivation of genes suppressing expression of virulence factors, pathoadaptive-gene inactivation could also target gene expression that represents liability during the infection (e.g., those recognized by immune defense). Such adaptive 'loss-of-function' mutations could cause protein inactivation via point substitution affecting the active site of the protein, frameshift mutation, or premature stop-codon, resulting in the protein truncation as well as deletion of the entire gene or even block of genes (44). Genes' functions that are detrimental for the pathogens and selected for the inactivation are sometimes called 'anti-virulence factors' (49). However, it remains a challenge sometimes to distinguish adaptive gene loss, which is driven by positive selection, from the reductive gene loss due to lack of the gene use, which is driven by random genetic drift (50). Some examples of the adaptive gene loss in UPEC are described in the next section.

Amplification of gene copies could lead to increased production of the corresponding protein that is similar or in addition to the inactivation of the negative regular of the corresponding gene. For example, expressing higher number of the adhesin with even somewhat suboptimal receptor-specificity would potentially increase the tropism strength (Fig. 3D). Indeed, some strains of UPEC have multiple copies of operons coding for P-fimbriae – di-galactose-binding organelles associated with pyelonephritis (51). Recently, a concept of transient copy amplification of large genetic regions was introduced as a general mechanism of *E. coli* adaptation to a novel environment, providing the means to overproduce critical traits and to increase rates of adaptive evolution of these traits via mutation (52). The 'gain-of-function' effect of gene amplification could also be reached via increased rate of the gene transcription or translation rate, by mutations in regulatory regions (e.g., promoters, operators, etc.), ribosome-binding sites, or even silent (synonymous) positions of the gene that turn rare codons into common ones (53–56).

Finally, other types of pathoadaptive mutations are genetic variations that represent single-nucleotide polymorphisms (SNPs) that do not result in protein truncation or deletion, but modify its primary structure by single amino-acid changes. In this way, specific functional properties of the protein could be modified and fine-tuned to the pathogen's needs. For example, the receptor specificity of the adhesin could be changed to make it fit better to the receptor's structure in the pathogenic habitat (Fig. 3E). This type of genetic change is the most common in nature and represents the most important mechanism in the evolution overall (57). In fact, genetic variation is the most studied-to-date mechanism of pathoadaptive evolution and, in particular, in UPEC, as described extensively below.

Evidence for Occurrence of Pathoadaptive Mutations in UPEC

Observations of phenotypic variations

Several types of phenotypic variation have been reported earlier to occur in urinary *E. coli* isolates from the same patient. It has been shown that urines from patients with UTI commonly yield mixed growths of different colony types of *E. coli*, with the latter found to be variants of single infecting strains (58). Thus, whether the colony heterogeneity is caused by mutation or by phase variation (which was not then determined), care should be exercised in the interpretation of apparently mixed growths from urine. It has been observed that 30% to 75% of successive isolates from recurrent cystitis or asymptomatic bacteriuria tend to lose the O-antigen and convert from clones with an O-typable, lipopolysaccharide (LPS), smooth-colony morphology to ones with O-nontypable rough or, more often, semi-rough-colony morphology (59–62). The LPS semi-rough-colony morphology is characterized by formation of smooth colonies on agar and, therefore, these variants are indistinguishable from the true LPS-smooth variants on the agar. These strains, however, are not typable by various O-antisera and tend to autoagglutinate in normal saline after boiling. The change from smooth to semi-rough or rough clones involves a progressive loss of long O-specific side-chains, as was shown by gel-column filtration and gas-liquid chromatography (62). Interestingly, the LPS-smooth/-rough phenotypic variation is common among lung *P. aeruginosa* isolates derived from later stages of infection in patients with cystic fibrosis (63, 64). The exact genetic nature of the LPS variation is unknown, but is possibly due to different types of mutation in the multi-kb *rfb* locus (64, 65). The adaptive advantage of O-antigen loss might be in the ability of these bacteria to escape host defenses guided by anti-LPS antibodies or innate immunity, but its true value remains to be determined. Interestingly, a naturally occurring shift in the O-serotype can also be induced by a point mutation in *wbdA* genes encoding mannosyltransferase or *wzz* controlling chain length of O-antigen (66, 67).

Besides the loss of O-antigen, *E. coli* strains from patients with recurrent cystitis may lose the ability to express flagella or shift capsular antigenicity (59). Variations in colony morphology could reflect various types of adaptive differentiation within a spatially organized bacterial population, e.g., grown in parallel on a surface and in a liquid phase (68). Another type of variation described among *E. coli* strains causing long-term and, usually, mild symptomatic UTIs is the occurrence of auxotrophic clones, primarily cysteine, thymidine, glutamine, and thiamine-requiring strains (69, 70). Such auxotrophs grow on the surface of nutrient agar as very small translucent colonies. Interestingly, the emergence of auxotrophs and small colony-forming variants is typical for other pathogens causing chronic infections, like lung *P. aeruginosa* and *Staphylococcus aureus* isolates (71, 72). It was suggested

that selection for such variants is induced by certain types of antibiotic therapy, by high amino acid concentration in the patient's fluids infected by the bacteria, or by prolonged intracellular persistence of such variants.

Mutator phenotype and genome instability of UPEC

The importance of mutational changes in the adaptation to urinary tract colonization is supported by the prevalence of mutator phenotype among UPEC. In bacterial clones with the mutator phenotype, which is related to defects in DNA-mismatch repair genes, mutation rates can increase 10–1000-fold. Though mutators are relatively unstable from an evolutionary perspective, the change in mutation rate can lead to dramatically faster acquisition of adaptive mutations and, therefore, provides a selective advantage during colonization of novel niches (73).

In one study, among >600 commensal and pathogenic *E. coli* isolates tested, UPEC strains had the highest frequency of mutators (74). It was the only group of isolates that differed significantly from the commensal strains (12% vs 3.5% of mutators). Another study also corroborated that mutators were more frequent in urinary tract isolates (25%) than in feces of healthy individuals (11%) (75). No correlation was found between the mutation rates and antibiotic resistance among the strains and it was hypothesized that UTI mutators are selected because they generate mutations that increase adaptation to the urinary tract at a higher rate than do nonmutators. However, another study has shown that a high mutation rate correlated strongly with a *clinical* resistance to fluoroquinolones, suggesting its role in the emergence of antibiotic-resistant clones during the infection treatment (76).

It was shown experimentally that a strong mutator phenotype (mutation frequency >5 x 10^{-7}) increases the virulence of *E. coli* strains in intraperitoneal infection experiments in mice (77). Moreover, in a mouse UTI model, mutator strains persist longer than wild-type strains, and serial passage of a mutator strain increases its virulence over that of wild-type (78). The strains tested were *mutS* mutants that are defective in recognition of mispaired DNA. Moreover, defects in specific components of a mismatch-repair system (*mutS* and *mutH*) have been linked to at least one specific phenotype: type 1 fimbriae, which were shown experimentally to suppress reciprocal control of motility by the major adherence trait of UPEC (79). Even when the fimbrial expression was locked-ON, the *mutS* and *mutH* mutants showed an increase of flagella expression and motility – one of the urovirulence traits in UPEC.

Besides mutator phenotypes linked to genetic defects in the mismatch-repair system, it appears that the urinary tract environment itself has an intrinsic capability to induce genetic alterations in the bacteria. The phenomenon of transient (i.e., conditional) stress-induced mutator phenotype was demonstrated two decades ago by showing that *E. coli* growing in high-salt concentrations (but not a regular osmolarity) exhibit an increased rate of mutability (80). Thus, a transient mutator phenotype might be commonly induced during the infection due to the hypertonic nature of urine (81). Indeed, it was recently shown that the mutation rate is increased 100-fold during experimental UTI compared to *in vitro* conditions (82). It was demonstrated that TLR4-signaling defect significantly reduced the UTI-induced mutation frequency. This suggests that besides the host response, like a neutrophil oxidative burst as a source of reactive oxygen species (ROS), there could also be a driver of UPEC-genome instability. Interestingly, it is tempting to hypothesize that such transient genome instability during the infection could be responsible for the occasional loss of pathogenicity islands (and thus some of the putative urovirulence-associated factors) in UPEC strains that is observed under certain environmental stresses, including urinary infection (83, 84).

Adaptive gene inactivation in UPEC

As indicated above, gene loss is a common mechanism of pathoadaptive mutations in general. Usually, positive selection favors loss-of-traits expression, which represents liability for colonization of the virulence habitat or interferes with expression or function of traits adaptive during the infection (44). Some of the genes inactivation was shown experimentally to promote urinary tract colonization. For example, the deletion of a D-serine deaminase (DsdA) results in hypercolonization phenotype in a murine model of UTI due to increased virulence-gene expression (85). While the exact mechanism of up-regulation is unclear, the loss of DsdA leads to an increased intracellular accumulation of D-serine that likely acts as a signal for the virulence-gene expression. Another example is the periplasmic oxidase CueO that is involved in copper homeostasis and protection against oxidative stress, the deletion of which increases *E. coli*-colonization level in the mouse model of UTI (86). The increased fitness in the urinary tract is possibly due to increased efficiency of iron accumulated by the mutant – a trait proposed to bring advantage in the iron-limited environment of the urinary tract (87).

Direct evidence for the occurrence of adaptive gene loss in *E. coli* during the urinary tract colonization in patients comes from trials that involved human volunteers colonized with asymptomatic bacteriuria (ABU) strain 83972 (88). ABU is more common than symptomatic UTI, with *E. coli* being isolated from urine in up to 5% women at a given time, in as high of counts as 10^5 and for as long as 10 months (89). ABU episodes often precede acute or recurrent cystitis (90, 91). However, it is generally recommended to treat lasting ABU only in pregnant women and before invasive urological procedures (92). Moreover, in patients with catheterized bladders, there is a negative correlation between ABU and symptomatic UTI, suggesting protective effect of the former (93). In any case, predictive of or protective against UTI, colonization of the urinary bladder by normally intestinal commensal *E. coli* is a pathological phenomenon that occurs only transiently in healthy individuals.

Though ABU is not a classical clinical infection, because ABU bacteria have to adapt for survival to the same environment as do UPEC, they are likely to utilize the same mechanisms to increase the fitness as those bacteria causing symptomatic UTI. As in other chronic bacterial infections, the relatively long-lasting ABU provides ample opportunities to accumulate changes promoting successful colonization of the urinary bladder. This provides us with a great model to understand the genetic mechanisms likely to be involved in the UPEC adaptation to surviving in the urinary tract compartments.

In the clinical trials, ABU *E. coli* strain 83972 was installed into the bladders of several catheterized patients to induce long-term colonization (88). During a 2- to 6-month period, in patients with continuous asymptomatic colonization by the strain, the genome and transcription profiles of the established bacterial clones were compared to those of the originally installed strain. Across different strains, 34 different genetic mutations were identified, with a third being genetic knock-outs by either premature-stop codons, frameshifts, or deletions (Table 1). The loci affected were commonly involved in gene regulation and,

TABLE 1 Pathoadaptive genes inactivation in long-term bladder colonization trial (premature stop-codons; frame-shift mutations, deletions)

Gene	Function
glpB	Glycerol-3-phosphate dehydrogenase
barA	Hybrid sensory histidine kinase of two-component regulatory system BarA/UvrY
ufl	FtsI suppressor
yhdP	Predicted transporter
yifB	Putative ATP-dependent protease
ECABU_c32770 - ECABU_c32780	Aerobactin uptake system
bcsA	Cellulose synthase
frmR	frmRAB repressor

interestingly, were sometimes the same in different patients (e.g., hybrid sensory histidine kinase BarA or FrmR repressor). As a result of the gene-alterations interplay, a significant change in expression of multiple traits has been observed. One of the most notable and consistent across all patients was up-regulation of loci involved in the protection against oxidative/nitrosative stress. Interestingly, conditioning of bacteria to nitrosative stress was shown experimentally to significantly improve the urinary tract colonization properties of UPEC (40). Another consistent type of changes involved bacterial-motility (flagella) regulation, which is a highly immunogenic trait that is considered to be a major urovirulence factor (94). Thus, pathoadaptive changes are likely to target genes involved in promoting the continuous survival in the protected compartment, irrespectively to what extent the bacterial persistence results in a clinically manifested damage to the host. Not surprising, other up-regulated genes were involved in iron-uptake and nutrient transport and degradation, i.e., promotion of the bacterial growth. However, sometimes the same genes had an opposite transcriptional profile in different patients, indicating that pathoadaptive changes could be tuned to the specific host conditions.

Adaptive gene variation in UPEC

Gene-inactivation mutations are, by definition, functional in nature. While their (patho)adaptive effect might not be always straightforward to determine, their identification is relatively easy as they involve obvious genetic alterations like deletion, frameshift, and nonsense mutations. On the other hand, a more numerous type of pathoadaptive changes could involve the most common type of DNA changes – point-nucleotide substitutions (variations). In the coding regions, point mutations are either synonymous (silent) or nonsynonymous (replacement) in nature that do not or do change the coding-protein structure, respectively. Synonymous changes are generally considered functionally neutral, though in some cases they could significantly affect the rate of gene transcription or translation. While nonsynonymous mutations could also have no or marginal effect on the protein function, they are more likely to do so than silent changes. In most cases, structural alterations are detrimental to the protein function and the organism fitness and, thus, are generally under negative (purifying) selection. This is likely to be the reason why the rate of nonsynonymous changes (usually abbreviated as D_n) seen in the vast majority of *E. coli* genes is drastically lower than the rate of synonymous changes (D_s) (95). However, in at least some cases, the nonsynonymous mutations result in functional changes that increase bacterial fitness in the given place at a given time, which would be of pathoadaptive nature in pathogens.

Some of the most direct evidence for the functional and adaptive impact of the genetic variations in UPEC come from the studies of the adhesive organelles. Surface adhesins are considered to be an important virulence factor of UPEC that mediates the bacterial tropism to the urothelium surface and/or is critical for the biofilm formation. Most of the genes coding UPEC adhesins are of patchy (i.e., non-core) distribution among the strains and commonly carried by horizontally transferred chromosomal (pathogenicity) islands.

Mutations in the type 1 fimbrial adhesin, FimH

Type 1 fimbriae are hair-like surface appendages (Fig. 4A) that are comprised of the major, non-adhesive subunit FimA (pilin) and the fimbrial tip that, in turn, contains minor subunits, FimF, FimG, and, at the very end, a mannose-binding adhesin FimH that has a lectin and pilin domains (Fig. 4B). Type 1 fimbriae are among the most ubiquitous adhesive traits in *E. coli*, expressed by >90% of the strains, both commensal and pathogenic (96). It is proposed that, in commensal

E. coli, type 1 fimbriae are involved in fecal-oral transmission by allowing bacteria to colonize the oropharyngeal epithelial surface – a transient step that might increase the probability of successful passage of the *E. coli* through the stomach acid to become resident in the large intestine (1). At the same time, expression of type 1 fimbriae has been shown to have a direct role in a mouse model of urovirulence by mediating bacterial binding to uroepithelial cells (97, 98).

As type 1 fimbriae function in distinctly different habitats, the urinary tract and the oropharynx, one might expect selection to favor specific adaptation of the adhesin under these different conditions. In fact, it has been shown that type 1 fimbriae from urinary isolates of *E. coli* tend to bind mannose-containing glycoproteins on average 3 times and up to 20 times stronger than fimbriae from fecal isolates (Fig. 4C). The increased mannose-binding capability is a result of point mutations in the FimH adhesin that lead to an increased tropism of bacteria for uroepithelial cells (Fig. 4D) and increased bladder colonization by *E. coli*, indicating that this property is adaptive for the urovirulence of *E. coli* (99, 100).

The pathoadaptive significance of the FimH mutations is supported by the molecular epidemiology, microevolutionary, and population-genetics studies. Mutant forms of FimH that bind strongly to mannose are predominant among uropathogenic strains, while evolutionary original variants with a relatively low-binding capability to mannose are dominant among intestinal *E. coli* strains (99). Also, emergence of strongly binding FimH mutants was observed in the course of a single UTI (101). The phylogenetic analysis of FimH mutations strongly indicated that they emerged under a strong positive selection, i.e., are functionally adaptive in nature (102). In particular, there is a predominance of convergent (hot-spot) mutations in the same protein positions and nonsynonymous (amino acid) changes on the FimH tree branches (102, 103). This pattern

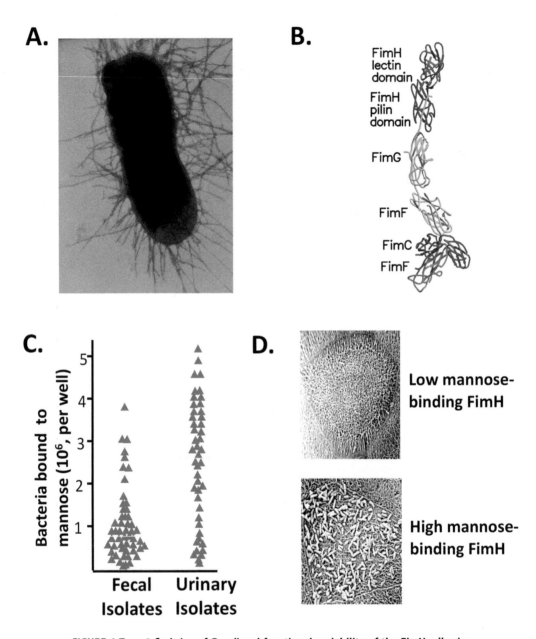

FIGURE 4 Type 1 fimbriae of *E. coli* and functional variability of the FimH adhesin.

is not seen with housekeeping genes or other genes within the type 1 fimbrial operon and is a strong indication of the action of positive selection (103).

Mutations in the Dr-family adhesins

Occurrence of point mutations leading to functional modification that are pathoadap-

tive in nature has been shown also in Dr-family adhesins (104). The Dr family of adhesins of *E. coli* combines a number of fimbrial and afimbrial adhesins that recognize as an epithelial cell a complement-regulatory protein-receptor decay-accelerating factor (DAF) (105, 106). Some Dr-family adhesins also bind to type IV collagen and carcinoembryonic antigen-related cell-adhesion molecules (CEACAMs) (107–110). Presence of these adhesins is associated with *E. coli* causing UTI and, in particular, recurrent infections (111).

Several studies over the last two decades have shown that naturally occurring point mutations in Dr-family adhesins can affect their adhesive properties (112–115). Notably, it has been demonstrated that mutations that increase the binding to collagen IV increase the *E. coli* ability to colonize kidneys long-term in a murine model of UTI (116).

In a more recent detailed study, analysis of eight types of Dr adhesins from 100 *E. coli* strains (isolated from urine as well as feces of healthy infants, diarrheal stools, and blood) found 63 nonsynonymous and only 3 synonymous mutations that were acquired by microevolution of individual adhesin types (110). Such overwhelming predominance of amino acid over silent changes is indicating adaptive significance of the former. Functional analysis of protein variants comprising one of the most common Dr adhesins, DraE hemagglutinin, revealed that the point mutations resulted in distinctly different binding phenotypes, with a tendency of increased (up to 4-fold) affinity to DAF.

Mutations in the P-fimbrial adhesin

Another fimbrial adhesin of UPEC that has shown a footprint of positive selection for potentially pathoadaptive variations is P-fimbrial adhesin, PapG. P-fimbriae mediate di-galactose-specific adhesion of *E. coli* and are associated with UPEC isolates, especially those causing pyelonephritis (from where their 'P' nomenclature actually derives) (117). These fimbriae were also shown to be important for intestinal persistence of *E. coli* in humans (118), but also were shown to provide advantage in murine model of UTI (119).

Accumulation of multiple amino-acid substitutions in the PapG adhesin was shown to be driven by positive selection, at the level similar to the FimH adhesin (120). In particular, seven SNPs – six nonsynonymous and

only one synonymous, in *papG* alleles from strains of one of the notorious UPEC clones of O1/2/18:K1 serotypes, in the class II allele group, and one nonsynonymous in the class III group. The rate of amino-acid mutations was more than two-fold higher than the rate of the silent changes, strongly suggesting the action of positive selection. Adaptive nature of the PapG variations (specifically, PapGII isotype) was corroborated by other bioinformatics tests as well (103). Unlike the FimH and DraE adhesin, however, the functional impact of the mutations in PapG and their pathoadaptive significance has not been elucidated yet. In addition to the adhesin variability, there is evidence that the gene coding for one of the regulators of the P-fimbriae, *papI*, is under positive selection, with PapI allelic variants displaying differences in their capacity to activate the P-fimbrial operons (121).

Genome-wide screens for prevalence of pathoadaptive variations in UPEC

Availability of multiple genomes of bacterial pathogens from the same species enabled us to screen for potential pathoadaptive changes across a large number of genes. With any fitness-increasing changes, accumulation of pathoadaptive variations should leave a footprint in the gene sequences indicative of the action of positive selection. Multiple population-genetics tests have been developed to identify such a footprint that can be applied across specific genes in bacterial genomes.

One such attempt has been undertaken by comparisons between seven fully sequenced *E. coli* genomes that include two UPEC strains – CFT073 and UT89 (122). As a method for detection of positive selection for point amino-acid mutations in individual genes, the authors used the Codeml program [from the Pathology Associates Medical Laboratory (PAML) package]. This test estimates the prevalence of amino-acid vs silent changes along different branches of the phylogenetic tree of each gene (123). By using this method, a total of 29 genes out of 3,470 core genes (i.e., that are shared by all strains) were identified to acquire point mutations under positive selection, specifically along the branches leading to the UPEC gene variants. The results were validated for 3 of the candidate genes by sequencing them in a panel of 39 urinary isolates from patients with UTI. The genes that mutated under positive selection in UPEC were enriched for three functional categories: cell wall/membrane biogenesis; secondary metabolites biosynthesis, transport, and catabolism; and DNA replication, recombination, and repair (genes with defined function are listed in Table 2).

Among the UPEC-selected genes was one coding FepE that regulates O-antigen chain length (124), confirming that LPS modification is likely to be one of the targets of the pathoadaptive mutations in the course of UTI. Two other genes – xseA coding the large subunit of exodeoxyribonuclease VII and recC coding a subunit of the RecBCD helicase/nuclease – play roles in DNA repair and, thus, affect the rate of mutation and recombination in UPEC. Genes under selection for mutations also included ompF and ompC that encode general porins, some of the most abundant proteins in the E. coli outer membrane (125), with the latter up-regulated when UPEC are grown in urine (126). Other likely targets for pathoadaptive mutations were amiA, cedA, and topB, which play a role in regulating cell shape or division – the phenotypes altered during the formation of communities by UPEC bacteria inside urothelial cells in the mouse model of UTI (127). Finally, loci that could be affected by pathoadaptive mutations included three genes important in iron scavenging during the UTI – fhuA coding a ferrichrome-iron receptor, and entD and entF that play a role in the synthesis of enterobactin.

More recently, another study had undertaken a genome-wide search of candidate genes affected by pathoadaptive mutations by using more E. coli genomes and an alternative

TABLE 2 Pathoadaptive gene variations found by genome-wide screening

Gene	Gene function	Reference
cedA	Cell division regulatory protein	123
fdnG	α subunit of formate dehydrogenase N	123
cycA	Serine/alanine/glycine APC transporter	123
argI	Ornithine carbamoyltransferase chain I	123
yjjN	Hypothetical zinc-type alcohol dehydrogenase	123
fruA	PTS system, fructose-specific IIBC component	123
agaI	Putative galactosamine-6-phosphate isomerase	123
topB	DNA topoisomerase III	123
xseA	Exodeoxyribonuclease VII large subunit	123
recC	Exodeoxyribonuclease V γ chain	123
fepE	Ferric enterobactin transport protein	123
cutE	Apolipoprotein N-acyltransferase	123
ompF	Outer membrane protein F	123
ompC	Outer membrane porin protein C	123, 88
amiA	Probable N-acetylmuramoyl-L-alanine amidase	123
fhuA	Ferrichrome-iron receptor	123
entD	4'-phosphopantetheinyl transferase	123
entF	Enterobactin synthetase component F	123
yoji	Hypothetical ABC transporter ATP-binding protein	123
ydfG	Probable oxidoreductase	128
mutT	Mutator MutT protein	128
alkB	Alkylated DNA repair protein	128
helD	Helicase IV	128
hrpA	ATP-dependent helicase	128
ddlB	D-alanine–D-alanine ligase B	128
araB	L-ribulokinase	128
c1323	Cytochrome b561 homolog 2	128
deoC	Deoxyribose-phosphate aldolase	128
hybE	Hydrogenase-2 operon protein	128
yjeS	Putative electron transport protein	128
ydhF	Hypothetical oxidoreductase	128
frmR	frmRAB repressor	88
mdoH	Synthesis of osmoregulated periplasmic glucans	88
marR	Multiple antibiotic resistance	88
marA	Multiple antibiotic resistance	88
mreB	Rod shape-determining protein	88
ompR	Regulator of two-component regulatory system EnvZ/OmpR	88
rfaJ	Lipopolysaccharide 1,2-glucosyltransferase	88

(Continued on next page)

TABLE 2 Pathoadaptive gene variations found by genome-wide screening *(Continued)*

Gene	Gene function	Reference
rmuC	DNA recombination protein	88
oxyR	Transcriptional dual regulator	88
yejM	Putative sulfatase	88
yghJ	Putative lipoprotein	88
cytR	Transcriptional repressor of nucleosides transport	88
rpiR	Repressor of ribose catabolism	88
gyrA	DNA gyrase	88
fusA	Translation elongation factor G	88
fsaB	Fructose-6-phosphate aldolase 2	88
rpoC	RNA polymerase	88

analytical approach (128). To evaluate frequency of pathoadaptive mutation, nucleotide variations were analyzed across 22 fully assembled *E. coli*, with 5 genomes being of UPEC isolates. Unlike the previous study, this study concentrated on a highly conserved portion of the core genes, with relative little (<5%) sequence variability, and on the analysis of repeated changes in the same amino acid positions – so-called hotspot mutations. Occurrence of hotspot mutations is strong evidence of positive selection, indicating that replacement of a specific amino acid provides an adaptive advantage under particular conditions and, thus, is repeatedly selected in different allelic backgrounds of clones competing for survival in the same environments (95, 102). In other words, occurrence of hotspot mutations represents molecular-convergent evolution of traits, where different bacterial strains are adapting to a same environment via same or similar evolutionary trajectory.

A total of 1,483 core genes were among >95% identical, with average identity level of 98.4%. Because of the highly conserved nature of this subset of the *E. coli* core genome (and thus the lack of long phylogenetic tree branches in these genes), only 2 of them were found to be under positive selection according to the previous Codeml analysis (122). However, 172 of the genes were targeted by hotspot mutations in UPEC strains, among which 47 mutated genes were specific to UPEC (genes with defined function are listed in the Table 2). The genes identified so far as adaptively evolving in UPEC represent genes involved in metabolism, respiration, and replication, i.e., gene categories that have been identified by other studies. This indicates the critical importance of the increased bacterial-growth rate in the infection establishment, where survival is the outcome of the bacterial replication and elimination by the defense systems.

It has been shown that the number of genes targeted by pathoadaptive variations is increasing with the increase in the number of genomes to be analyzed and, in the core genes, is predicted to increase more than two-fold from what has been identified so far (95). Considering that backbone genes that are not universally present in all *E. coli* and horizontally transferred mosaic genes could also accumulate pathoadaptive variations, the total overall number of genes adaptively mutating in UPEC is likely to be in the hundreds.

Interestingly, when the frequency and rate of hotspot mutations was compared across core genes from different *E. coli* eco-/pathotypes, the frequency of mutated genes was the lowest among commensal isolates, while the rate of mutations accumulations was the highest among UPEC strains (up to 4 times higher than in *Shigella*) (95, 128). Thus, pathoadaptive evolution of UPEC occurs very fast in evolutionary time and, potentially, transpired relatively recently in human history. If so, it is tempting to speculate what environmental, social, or other factors could contribute to the burst of UPEC evolution. Use of the vaginal diaphragm and spermicides, as well as the growing number of house pets (who could be a reservoir of uropathogens for humans) could be among the possible reasons.

Good leads on the potential nature of pathoadaptive nonsynonymous mutations across the genome come from the human trial of adapting the ABU strains described above (88). A total of 22 SNPs were identified to be acquired in the course of a long-term adaptation of the ABU strain yielding 83,972 mutations in 3 different patients. All but one out of them were nonsynonymous in nature (genes with defined function are listed in the Table 2). Such prevalence of nonsynonymous changes over the silent ones is a strong indication that the former have emerged and there is a strong positive selection, i.e. the genes are adaptive to the bacteria colonizing the urinary bladder. Out of 20 genes carrying nonsynonymous mutations, some were the same loci that were affected in other patients by knock-out mutations (e.g., *frmR* repressor), suggesting that, at least in certain cases, amino-acid changes might result in a loss-of-function effect. Though the functional effects of nonsynonymous mutations in the affected genes remain to be determined, they likely contributed to the shifted transcriptional profile of the strains (along with the inactivated genes). Indeed, many of the genes with amino-acid replacements are known to be part of the regulatory systems in *E. coli* – *oxyR* (up-regulator of H_2O_2-inducible genes), *mdoH* (suppressor of expression of colanic-acid capsule), *ompR* (involved in osmoregulation of outer membrane-proteins expression), *cytR* (repressor of nucleotide transport and utilization), and *rpiR* (ribose catabolism repressor). Among other genes affected by nonsynonymous changes were those involved in the peptidoglycan synthesis (*sufI* and *mreB*), LPS synthesis (*rfaJ*), membrane permeability (*ompC*, *marA*, *marR*), and DNA transcription (*gyrA*, *rpoC*).

Thus, taken together, the genome-wide analyses identified dozens of genes from multiple functional categories that either are suspected or shown directly to be involved in adaptation of *E. coli* as a successful invader of the urinary tract.

Pathoadaptive mutations and functional analysis of bacterial traits

By definition, pathoadaptive mutations are functional in nature. On one hand, pathoadaptive gene inactivation via deletion or truncation mutations has a straight-forward functional effect on the coded protein. From the perspective of understanding the gene function, this mutation type provides similar information as a standard gene knock-out approach used in laboratory research. On the other hand, pathoadaptive gene variations can provide some unique insights into the protein structure and function that would be difficult to obtain by laboratory experimentation. Indeed, pathoadaptive amino-acid replacements inactivate or otherwise modify functional properties of the protein. The inactivation amino-acid changes, for example, could guide us to the location of an active site or other critical functional region in the corresponding enzyme or regulatory or adhesive protein. Even more informative could be the modifying amino-acid variations that result in a somewhat novel functional property of the protein. A good illustration of such unique insights is provided by the structure-functional studies of the pathoadaptive variations in the type 1 fimbrial adhesin of UPEC, FimH.

As described above, UPEC isolates tend to accumulate mutations in FimH that result in a stronger mannose-specific adhesion to uroepithelial cells. Detailed experimental studies that followed the initial observations led to the discovery of shear-dependent binding properties of FimH (129). It was found that the relatively low mannose-binding of the FimH variants without mutations is significantly increased if the adhesion assay is performed under flow-induced shear instead of the static (i.e., low-shear) conditions (Fig. 5A). In contrast, adhesion of the pathoadaptive, high mannose-binding FimH mutants are relatively shear-independent (Fig. 5A).

To uncover the molecular mechanisms underlying the shear-dependent properties of FimH, further studies involved functional testing of different structural constructs of FimH and its domains, molecular-dynamic simulations of the adhesin structure, and obtaining novel crystal structure of FimH in the fimbrial-tip complex (129–132).

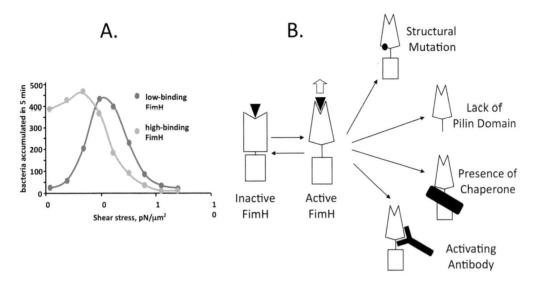

FIGURE 5 Shear-dependent (A) and conformational (B) properties of FimH adhesin.

Ultimately, it was found that the shear-enhanced binding is based on allosteric conformational properties of the mannose-binding lectin domain of FimH (Fig. 5B). It was found that the lectin domain can interact with the fimbrial tip-anchoring pilin domain of FimH and this interaction is associated with a low mannose-binding (*inactive*) conformation of the lectin domain. In contrast, separation of the domains is associated with a shift of the lectin domain to high mannose-binding (*active*) conformation. As a result of the conformational shift, the mannose-binding pocket on the top of the lectin domain converts from a wide-open to a closed-tight configuration, with up to a 200-fold increase in the binding affinity to mannose (from dissociation constant K_D ~300 μM to ~1.5 μM) (17). The domain separation and, thus, the low- to high-affinity shift, occurs with mechanical stress under shear, providing the molecular basis for the shear-dependent binding of FimH. With the shear decrease, the lectin domain docks to the pilin domain and the affinity to mannose decreases.

The pathoadaptive mutations that enhance the FimH binding to mannose under static conditions commonly alter residues in the inter-domain interface, in either of the domains. Their presence is interfering with the domains interaction, thus stabilizing the lectin-domain conformation in the high-affinity state (Fig. 5B). Thus, the binding becomes less dependent on shear and stronger under static or low-shear conditions. Interestingly, the lectin domain in the purified form (i.e., in the absence of the pilin domain) or in FimH complexed with its molecular chaperon FimC (that is wedged between the domain) (Fig. 5B) is also stabilized in the high-affinity conformation (130–133). Moreover, antibodies raised against the purified lectin domain of FimH are specific against the high affinity conformation of FimH and do not recognize the low-affinity one (134). Instead of inhibiting FimH-mediated bacterial binding, these antibodies increase by stabilizing lectin domain in the high-affinity state (Fig. 5B). Interestingly, the FimH/FimC complex has been considered as a vaccine candidate against UPEC and, indeed, showed protective effect against UTI in an animal model (135). However, to what extent the protection was due to the ability, if any, of antibodies to block FimH activity, opsonize, or aggregate bacteria was not determined. Thus, it is unclear what the optimal ways are that will enhance the putative protective properties of anti-FimH vaccine.

Taken together, these studies have shown how detailed analysis of naturally occurring mutations that emerged under natural positive selection could provide insights on how proteins work. Indeed, comparative studies of the FimH variants of UPEC, with and without the pathoadaptive mutations, had guided the efforts to unravel the force-activated, allosteric catch-bond properties of FimH protein. Several subsequent studies have shown that the phenomenon of catch-bond-mediated shear-dependent binding is widely spread among bacterial adhesins and also shared by various eukaryotic adhesive molecules, including selectins and integrins (136).

PATHOADAPTIVE MUTATIONS AND UNDERSTANDING THE PHYSIOLOGICAL CONDITIONS IN THE HOST COMPARTMENTS

Besides providing potential insights into the functional mechanisms of bacterial traits, pathoadaptive mutations also can give us details on the host conditions encountered by the pathogen during the infection. Unlike horizontally transferred virulence factors, pathoadaptive mutations are commonly acquired in the course of infection itself. Therefore, their functional effect is directly reflecting the physiological conditions to which the pathogen has to adapt in order to sustain itself in the compartment.

For example, the FimH mutations that reduce dependence of the bacterial adhesion on shear indicate a low-shear environment in the lumen of urinary tract. Indeed, urine flow is very slow along the urinary tract and only enhanced in the urethra during the voiding (137). Combined with relatively low viscosity of urine, this results in a relatively low shear-stress environment in the urinary tract that, paradoxically, inhibits rather than promotes

adhesion mediated by shear-dependent FimH variants. However, shear-independent mutant forms of FimH provide a higher level of urothelial adhesion *in vivo* and, thus, selective advantage. Similarly, the increased adhesion of DraE mutants in high-salt concentration also corresponds to the osmotic characteristics of urine. Indeed, high-salt concentrations (400–800 mM NaCl) are conditions typical of the urinary tract (138). Because salt dramatically reduces the DraE–DAF binding affinity and DAF-mediated binding to bladder-epithelial cells (110), *E. coli* adherence that is mediated by DraE is likely to be inhibited in the urinary tract unless the adhesin carries the mutations. Along the same lines, mutations that increase protection against nitrosative stress or decrease/eliminate LPS production give more insights on the defense mechanisms in the urinary tract.

Thus, on one hand, deciphering the genes affected by the pathoadaptive mutations guides us to the bacterial traits important for virulence. On the other hand, understanding their functional impact can provide us insights on the host conditions encountered by the pathogen. In turn, this information could help to improve the experimental models conditions used in the laboratory research, i.e., to make them match the physiological conditions more closely. Indeed, under conditions that are relatively close to the physiological, the pathoadapted mutant strain should have advantage relatively to the strain that has original, wild-type-gene version (Fig. 6). Therefore, finding the model conditions when a naturally occurring mutant outcompetes a wild-type makes the model more physiologically relevant.

Detection of Pathoadaptive Mutations

Because pathoadaptive mutations can occur in the course of a single infection, they could be detected by comparing different clonal variants of the pathogen from the same patient. This can be readily accomplished if

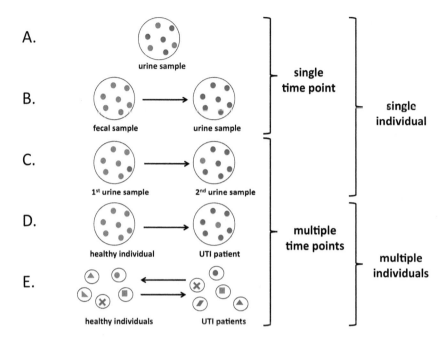

FIGURE 6 Strategies of detection of patho-adapted gene variants (red).

the mutation results in a distinct phenotypic heterogeneity of the infecting-bacterial population that is as obvious as, for example, colony morphologies, or requires specific functional screens (Fig. 6A). However, if the pathoadaptive mutation has resulted in a (nearly) complete population sweep, the mutated clone will be difficult to identify in a single clinical specimen (e.g., urine or blood). However, if a matching strain could be available from a different host compartment from where the infecting strains originated and was not under selection for pathoadaptive mutations (e.g., from intestinal microflora) pathoadaptive mutations could be potentially detected by comparison of the matching isolates (Fig. 6B). Similarly, sequential strains isolated from recurrent episodes of the infection or at different stages of a chronic colonization would be expected to have a different clone dominate at different time points (Fig. 6C). Distinct phenotypic effect of the pathoadaptive mutation could direct attention to the genetic locus affected. Alternatively, the recent advances of bacte-

rial genome sequencing makes it feasible to detect pathoadaptive changes by comparing whole genomes of the infecting clones. This approach has been used for example, for comparing the genomes of original probiotic *E. coli* strain 83972 before installation into the patient bladder and recovered after months of asymptomatic colonization.

Besides isolating strains from the same patient, pathoadaptive changes might be detected by tracing the same strain spread between different individuals (Fig. 6D). Same-strain circulation was documented between single-household members, with some individuals being asymptomatic carriers and some developing acute infection (139). In one reported case, an *E. coli* strain caused a chronic, relatively mild bladder infection in a catheterized paraplegic female patient and then an otherwise healthy caregiver developed a lethal sepsis, with the same strain being recovered from her bloodstream (140).

Even if a genetic change has been acquired during the infection, it is necessary to prove that it has been acquired under positive selection rather than being random and acquired by genetic drift. Besides an experimental approach, one can use population-genetics analysis to evaluate the adaptive significance of the accumulated mutations. Population genetics methods are designed for detecting adaptive genetic changes by comparing allelic variants of a gene of interest across different organisms of the species, especially those isolated from different habitats, like from intestinal microflora and urinary tract (Fig. 6E).

A footprint for positive selection in the allelic variants of a gene can be determined by three categories of methods that measure 1) divergence pattern of all kinds of nucleotide changes; 2) prevalence of nonsynonymous or synonymous changes; or 3) convergent evolution of the same changes. The first category includes, for example, the Tajima D, Fu, and Li statistics, as well as Hudson-Kreitman-Aguadé (HKA), which compares the number of polymorphic positions (sites) and polymorphisms expected under neutral or positive selection. The second category includes McDonald–Kreitman, D_n/D_s, and Codeml analyses that compare the total number or rate of non-synonymous vs synonymous mutations at the level of a whole gene, sliding windows, individual codons, or along phylogenetic tree branches. Prevalence of non-synonymous mutations indicates positive selection for amino-acid changes. Finally, the third category includes Zonal-Phylogeny analysis that detects repeated, phylogenetically unlinked changes in the same nucleotide or amino-acid positions (141). Such molecular-convergent evolution is a strong indication for the adaptive significance of the changes, especially when they are prevalent in isolates from specific habitats.

One of the characteristics of pathoadaptive mutations is that they are relatively unstable in nature, i.e., they circulate relatively short-term from evolutionary perspectives. This is due to their detrimental, trade-off effect of the mutations on the evolutionary original gene function. For example, along with the increased urothelial tropism, the increased mannose-affinity of mutant FimH adhesin leads to an increased sensitivity of the mutant variants to inhibition by soluble mannosylated compounds (95, 101). Mannose-rich glycoproteins are abundantly present in human saliva and this inhibition hampers the ability of the type 1 fimbriae to mediate adhesion to the oropharyngeal epithelia, an important step on the fecal-oral transmission of *E. coli*. Thus, while uropathogenic clones carrying mutant FimH return to the fecal-oral circulation, they are gradually eliminated from the circulation in the reservoir. This makes the pathoadaptive mutations exhibit a phylogenetic footprint of a relatively short-term (recent) evolutionary origin, part of which is occurrence of the mutations only in small numbers per gene (142). All this makes it relatively difficult to distinguish positive selection from neutral- or relaxed-purifying selection by using most of the tests mentioned above. However, certain features of the Zonal-Phylogeny analysis are specifically designed for distinguishing short- and long-term mutations that increases both sensitivity and specificity of the test for the detection of pathoadaptive mutations.

To make the population-genetics analyses of pathoadaptive mutations easier, different molecular-evolutionary tests were combined in a single user-friendly TimeZone software package (143). Analysis of adaptive footprints can be done in either a set of alleles of individual genes or across bacterial genomes. This package is available free from https://sourceforge.net/projects/timezone1/. Furthermore, to ease the search for genes affected by pathoadaptive mutation in uropathogenic and other *E. coli*, a prototype "*E. coli* variome" database has been established based on a set of fully assembled genomes of *E. coli* strains of different origin. The database has open anonymous user access at http://depts.washington.edu/sokurel/ecoli. It is comprised of ready-to-use information on gene presence/absence, nucleotide polymorphisms, and associated footprints of selection. The database is dynamic and allows for single- and multiple-field sorting, filtering, and searching. The *E. coli* Variome database is distinct from the database of the PathoSystems Resource Integration Center (PATRIC; see http://patricbrc.org) that was created by National Institute of Allergy and Infectious Diseases (NIAID) funding to serve as a comprehensive web-based resource for pathogen genomes. While PATRIC presents an overview of the sequenced genomes of a species through various quantitative details, the *E. coli* Variome Project focuses on diverse qualitative details of positive selection linked to pathoadaptation.

CONCLUSION

Taken together, the observational, experimental, and computational studies presented above demonstrate that pathoadaptive mutations significantly contribute into the evolution of *E. coli* urovirulence. While the presence of a specific repertoire of both core and mosaic genes is critical for the unique capabilities of UPEC, fine-tuning of their function via pathoadaptive mutations represents a highly specific mechanism for increasing the UPEC fitness in the urinary tract. From genetic perspectives, pathoadaptive mutations of minor changes explain in part why studies are lagging behind those focused on the mosaic-genes presence and absence. However, the potential functional impact of even the smallest changes in a gene could be enormous. Studying pathoadaptive mutations in bacteria pathogens is increasing and already making some critical contributions in our understanding of the molecular mechanisms of the bacterial pathogenesis. In this respect, studies of UPEC promise to be on the very front edge!

ACKNOWLEDGMENTS

Conflicts of interest: We declare no conflicts.

CITATION

Sokurenko E. 2016. Pathoadaptive mutations in uropathogenic *Escherichia coli*. Microbiol Spectrum 4(2):UTI-0020-2015.

REFERENCES

1. **Bloch CA, Stocker BA, Orndorff PE.** 1992. A key role for type 1 pili in enterobacterial communicability. *Mol Microbiol* **6**:697–701.
2. **Johnson JR, Russo TA.** 2002. Extraintestinal pathogenic *Escherichia coli*: "the other bad *E. coli*". *J Lab Clin Med* **139**:155–162.
3. **Foxman B.** 2002. Epidemiology of urinary tract infections: incidence, morbidity, and economic costs. *Am J Med* **113**(Suppl 1A):5S–13S.
4. **Brown PD, Foxman B.** 2000. Pathogenesis of urinary tract infection: the role of sexual behavior and sexual transmission. *Curr Infect Dis Rep* **2**:513–517.
5. **Foxman B, Zhang L, Tallman P, Andree BC, Geiger AM, Koopman JS, Gillespie BW, Palin KA, Sobel JD, Rode CK, Bloch CA, Marrs CF.** 1997. Transmission of uropathogens between sex partners. *J Infect Dis* **175**:989–992.
6. **Rosen DA, Hooton TM, Stamm WE, Humphrey PA, Hultgren SJ.** 2007. Detection of intracellular bacterial communities in human urinary tract infection. *PLoS Med* **4**:e329. doi:10.1371/journal.pmed.0040329
7. **Stapleton AE, Au-Yeung M, Hooton TM, Fredricks DN, Roberts PL, Czaja CA, Yarova-Yarovaya Y, Fiedler T, Cox M, Stamm WE.** 2011. Randomized, placebo-controlled phase 2 trial of a *Lactobacillus crispatus* probiotic given intravaginally for prevention of recurrent urinary tract infection. *Clin Infect Dis* **52**:1212–1217.
8. **Allen SJ, Martinez EG, Gregorio GV, Dans LF.** 2010. Probiotics for treating acute infectious diarrhoea. *Cochrane Database Syst Rev* **11**:CD003048.
9. **Todd EC, Greig JD, Bartleson CA, Michaels BS.** 2008. Outbreaks where food workers have been implicated in the spread of foodborne disease. Part 4. Infective doses and pathogen carriage. *J Food Prot* **71**:2339–2373.
10. **Weissman SJ, Johnson JR, Tchesnokova V, Billig M, Dykhuizen D, Riddell K, Rogers P, Qin X, Butler-Wu S, Cookson BT, Fang FC, Scholes D, Chattopadhyay S, Sokurenko E.** 2012. High-resolution two-locus clonal typing of extraintestinal pathogenic *Escherichia coli*. *Appl Environ Microbiol* **78**:1353–1360.
11. **Vejborg RM, Friis C, Hancock V, Schembri MA, Klemm P.** 2010. A virulent parent with

probiotic progeny: comparative genomics of *Escherichia coli* strains CFT073, Nissle 1917 and ABU 83972. *Mol Genet Genomics* **283**:469–484.

12. **Nielubowicz GR, Mobley HL.** 2010. Host-pathogen interactions in urinary tract infection. *Nat Rev Urol* **7**:430–441.

13. **Mulvey MA, Schilling JD, Hultgren SJ.** 2001. Establishment of a persistent Escherichia coli reservoir during the acute phase of a bladder infection. *Infect Immun* **69**:4572–4579.

14. **Koonin EV, Makarova KS, Aravind L.** 2001. Horizontal gene transfer in prokaryotes: quantification and classification. *Annu Rev Microbiol* **55**:709–742.

15. **Timmis KN, Boulnois GJ, Bitter-Suermann D, Cabello FC.** 1985. Surface components of *Escherichia coli* that mediate resistance to the bactericidal activities of serum and phagocytes. *Curr Top Microbiol Immunol* **118**:197–218.

16. **Canchaya C, Proux C, Fournous G, Bruttin A, Brüssow H.** 2003. Prophage genomics. *Microbiol Mol Biol Rev* **67**:238–276.

17. **Brüssow H, Canchaya C, Hardt WD.** 2004. Phages and the evolution of bacterial pathogens: from genomic rearrangements to lysogenic conversion. *Microbiol Mol Biol Rev* **68**:560–602.

18. **Dobrindt U.** 2005. (Patho-)Genomics of *Escherichia coli*. *Int J Med Microbiol* **295**:357–371.

19. **Marchès O, Ledger TN, Boury M, Ohara M, Tu X, Goffaux F, Mainil J, Rosenshine I, Sugai M, De Rycke J, Oswald E.** 2003. Enteropathogenic and enterohaemorrhagic *Escherichia coli* deliver a novel effector called Cif, which blocks cell cycle G2/M transition. *Mol Microbiol* **50**:1553–1567.

20. **Campellone KG, Robbins D, Leong JM.** 2004. EspFU is a translocated EHEC effector that interacts with Tir and N-WASP and promotes Nck-independent actin assembly. *Dev Cell* **7**:217–228.

21. **Jobling MG, Holmes RK.** 2012. Type II heat-labile enterotoxins from 50 diverse *Escherichia coli* isolates belong almost exclusively to the LT-IIc family and may be prophage encoded. *PLoS One* **7**:e29898. doi:10.1371/journal.pone.0029898

22. **Tóth I, Hérault F, Beutin L, Oswald E.** September 2003. Production of cytolethal distending toxins by pathogenic *Escherichia coli* strains isolated from human and animal sources: establishment of the existence of a new cdt variant (Type IV). *J Clin Microbiol* **41**:4285–4291.

23. **Lavigne JP, Blanc-Potard AB.** 2008. Molecular evolution of *Salmonella enterica* serovar

Typhimurium and pathogenic *Escherichia coli*: from pathogenesis to therapeutics. *Infect Genet Evol* **8**:217–226.

24. **Dobrindt U, Hochhut B, Hentschel U, Hacker J.** 2004. Genomic islands in pathogenic and environmental microorganisms. *Nat Rev Microbiol* **2**:414–424.

25. **McDaniel TK, Jarvis KG, Donnenberg MS, Kaper JB.** 1995. A genetic locus of enterocyte effacement conserved among diverse enterobacterial pathogens. *Proc Natl Acad Sci U S A* **92**:1664–1668.

26. **Fleckenstein JM, Lindler LE, Elsinghorst EA, Dale JB.** 2000. Identification of a gene within a pathogenicity island of enterotoxigenic *Escherichia coli* H10407 required for maximal secretion of the heat-labile enterotoxin. *Infect Immun* **68**:2766–2774.

27. **Henderson IR, Czeczulin J, Eslava C, Noriega F, Nataro JP.** 1999. Characterization of pic, a secreted protease of *Shigella flexneri* and enteroaggregative *Escherichia coli*. *Infect Immun* **67**:5587–5596.

28. **Al-Hasani K, Henderson IR, Sakellaris H, Rajakumar K, Grant T, Nataro JP, Robins-Browne R, Adler B.** 2000. The sigA gene which is borne on the she pathogenicity island of *Shigella flexneri* 2a encodes an exported cytopathic protease involved in intestinal fluid accumulation. *Infect Immun* **68**:2457–2463.

29. **Hacker J, Bender L, Ott M, Wingender J, Lund B, Marre R, Goebel W.** 1990. Deletions of chromosomal regions coding for fimbriae and hemolysins occur *in vitro* and *in vivo* in various extraintestinal *Escherichia coli* isolates. *Microb Pathog* **8**:213–225.

30. **O'Hanley P, Lalonde G, Ji G.** 1991. Alpha-hemolysin contributes to the pathogenicity of piliated digalactoside-binding *Escherichia coli* in the kidney: efficacy of an alpha-hemolysin vaccine in preventing renal injury in the BALB/c mouse model of pyelonephritis. *Infect Immun* **59**:1153–1161.

31. **Larbig KD, Christmann A, Johann A, Klockgether J, Hartsch T, Merkl R, Wiehlmann L, Fritz HJ, Tümmler B.** 2002. Gene islands integrated into tRNA(Gly) genes confer genome diversity on a *Pseudomonas aeruginosa* clone. *J Bacteriol* **184**:6665–6680.

32. **Johnson JR.** 1991. Virulence factors in *Escherichia coli* urinary tract infection. *Clin Microbiol Rev* **4**:80–128.

33. **Foxman B, Zhang L, Tallman P, Palin K, Rode C, Bloch C, Gillespie B, Marrs CF.** 1995. Virulence characteristics of *Escherichia coli* causing first urinary tract infection predict risk of second infection. *J Infect Dis* **172**:1536–1541.

34. Johnson JR, O'Bryan TT, Delavari P, Kuskowski M, Stapleton A, Carlino U, Russo TA. 2001. Clonal relationships and extended virulence genotypes among *Escherichia coli* isolates from women with a first or recurrent episode of cystitis. *J Infect Dis* **183**:1508–1517.

35. Lloyd AL, Henderson TA, Vigil PD, Mobley HL. 2009. Genomic islands of uropathogenic *Escherichia coli* contribute to virulence. *J Bacteriol* **191**:3469–3481.

36. Welch RA, Burland V, Plunkett G III, Redford P, Roesch P, Rasko D, Buckles EL, Liou SR, Boutin A, Hackett J, Stroud D, Mayhew GF, Rose DJ, Zhou S, Schwartz DC, Perna NT, Mobley HL, Donnenberg MS, Blattner FR. 2002. Extensive mosaic structure revealed by the complete genome sequence of uropathogenic *Escherichia coli*. *Proc Natl Acad Sci U S A* **99**:17020–17024.

37. Lane MC, Lockatell V, Monterosso G, Lamphier D, Weinert J, Hebel JR, Johnson DE, Mobley HL. 2005. Role of motility in the colonization of uropathogenic *Escherichia coli* in the urinary tract. *Infect Immun* **73**:7644–7656.

38. Diard M, Garry L, Selva M, Mosser T, Denamur E, Matic I. 2010. Pathogenicity-associated islands in extraintestinal pathogenic *Escherichia coli* are fitness elements involved in intestinal colonization. *J Bacteriol* **192**:4885–4893.

39. Haugen BJ, Pellett S, Redford P, Hamilton HL, Roesch PL, Welch RA. 2007. *In vivo* gene expression analysis identifies genes required for enhanced colonization of the mouse urinary tract by uropathogenic *Escherichia coli* strain CFT073 dsdA. *Infect Immun* **75**:278–289.

40. Bower JM, Gordon-Raagas HB, Mulvey MA. 2009. Conditioning of uropathogenic *Escherichia coli* for enhanced colonization of host. *Infect Immun* **77**:2104–2112.

41. Johnson JR, Tchesnokova V, Johnston B, Clabots C, Roberts PL, Billig M, Riddell K, Rogers P, Qin X, Butler-Wu S, Price LB, Aziz M, Nicolas-Chanoine MH, Debroy C, Robicsek A, Hansen G, Urban C, Platell J, Trott DJ, Zhanel G, Weissman SJ, Cookson BT, Fang FC, Limaye AP, Scholes D, Chattopadhyay S, Hooper DC, Sokurenko EV. 2013. Abrupt emergence of a single dominant multidrug-resistant strain of *Escherichia coli*. *J Infect Dis* **207**:919–928.

42. Navarre WW, Porwollik S, Wang Y, McClelland M, Rosen H, Libby SJ, Fang FC. 2006. Selective silencing of foreign DNA with low GC content by the H-NS protein in *Salmonella*. *Science* **313**:236–238.

43. Mc Ginty SÉ, Rankin DJ. 2012. The evolution of conflict resolution between plasmids and their bacterial hosts. *Evolution* **66**:1662–1370.

44. Sokurenko EV, Hasty DL, Dykhuizen DE. 1999. Pathoadaptive mutations: gene loss and variation in bacterial pathogens. *Trends Microbiol* **7**:191–195.

45. Hoboth C, Hoffmann R, Eichner A, Henke C, Schmoldt S, Imhof A, Heesemann J, Hogardt M. 2009. Dynamics of adaptive microevolution of hypermutable *Pseudomonas aeruginosa* during chronic pulmonary infection in patients with cystic fibrosis. *J Infect Dis* **200**:118–130.

46. Hayden HS, Lim R, Brittnacher MJ, Sims EH, Ramage ER, Fong C, Wu Z, Crist E, Chang J, Zhou Y, Radey M, Rohmer L, Haugen E, Gillett W, Wuthiekanun V, Peacock SJ, Kaul R, Miller SI, Manoil C, Jacobs MA. 2012. Evolution of *Burkholderia pseudomallei* in recurrent melioidosis. *PLoS One* **7**:e36507. doi:10.1371/journal.pone.0036507

47. Martin DW, Schurr MJ, Mudd MH, Govan JR, Holloway BW, Deretic V. 1993. Mechanism of conversion to mucoidy in *Pseudomonas aeruginosa* infecting cystic fibrosis patients. *Proc Natl Acad Sci U S A* **90**:8377–8381.

48. Holden NJ, Totsika M, Mahler E, Roe AJ, Catherwood K, Lindner K, Dobrindt U, Gally DL. 2006. Demonstration of regulatory crosstalk between P fimbriae and type 1 fimbriae in uropathogenic *Escherichia coli*. *Microbiology* **152**(Pt 4):1143–1153.

49. Bliven KA, Maurelli AT. 2012. Antivirulence genes: insights into pathogen evolution through gene loss. *Infect Immun* **80**:4061–4070.

50. Moran NA. 2002. Microbial minimalism: genome reduction in bacterial pathogens. *Cel* **108**:583–586.

51. Kao JS, Stucker DM, Warren JW, Mobley HL. 1997. Pathogenicity island sequences of pyelonephritogenic *Escherichia coli* CFT073 are associated with virulent uropathogenic strains. *Infect Immun* **65**:2812–2820.

52. Kugelberg E, Kofoid E, Reams AB, Andersson DI, Roth JR. 2006. Multiple pathways of selected gene amplification during adaptive mutation. *Proc Natl Acad Sci U S A* **103**:17319–17324.

53. Kudla G, Murray AW, Tollervey D, Plotkin JB. 2009. Coding-sequence determinants of gene expression in *Escherichia coli*. *Science* **324**:255–258.

54. Corvec S, Prodhomme A, Giraudeau C, Dauvergne S, Reynaud A, Caroff N. 2007. Most *Escherichia coli* strains overproducing chromosomal AmpC beta-lactamase belong to phylogenetic group A. *J Antimicrob Chemother* **60**:872–876.

55. Mammeri H, Eb F, Berkani A, Nordmann P. 2008. Molecular characterization of AmpC-producing *Escherichia coli* clinical isolates

recovered in a French hospital. *J Antimicrob Chemother* **61**:498–503.

56. **Smet A, Martel A, Persoons D, Dewulf J, Heyndrickx M, Catry B, Herman L, Haesebrouck F, Butaye P.** 2008. Diversity of extended-spectrum beta-lactamases and class C beta-lactamases among cloacal *Escherichia coli* isolates in Belgian broiler farms. *Antimicrob Agents Chemother* **52**:1238–1243.

57. **Mira A, Martín-Cuadrado AB, D'Auria G, Rodríguez-Valera F.** 2010. The bacterial pan-genome: A new paradigm in microbiology. *Int Microbiol* **13**:45–57.

58. **Nichols GL.** 1975. Variants of *Escherichia coli* giving the appearance of mixed growths in urine. *J Clin Pathol* **28**:728–730.

59. **Bettelheim KA, Taylor J.** 1969. A study of *Escherichia coli* isolated from chronic urinary infection. *J Med Microbiol* **2**:225–236.

60. **Lindberg U, Hanson LA, Jodal U, Lidin-Janson G, Lincoln K, Olling S.** 1975. Asymptomatic bacteriuria in schoolgirls. II. Differences in *Escherichia coli* causing asymptomatic bacteriuria. *Acta Paediatr Scand* **64**:432–436.

61. **Lidin-Janson G, Hanson LA, Kaijser B, Lincoln K, Lindberg U, Olling S, Wedel H.** 1977. Comparison of *Escherichia coli* from bacteriuric patients with those from feces of healthy school-children. *J Infect Dis* **136**:346–353.

62. **Webb L, Goodwin CS, Green J.** 1982. O antigen loss by semi-rough *E. coli* causing recurrent urinary infections, analysed by gel column filtration and gas-liquid chromatography. *Pathology* **14**:17–24.

63. **Pier GB, Desjardins D, Aguilar T, Barnard M, Speert DP.** 1986. Polysaccharide surface antigens expressed by nonmucoid isolates of *Pseudomonas aeruginosa* from cystic fibrosis patients. *J Clin Microbiol* **24**:189–196.

64. **Evans DJ, Pier GB, Coyne MJ Jr, Goldberg JB.** 1994. The rfb locus from *Pseudomonas aeruginosa* strain PA103 promotes the expression of O antigen by both LPS-rough and LPS-smooth isolates from cystic fibrosis patients. *Mol Microbiol* **13**:427–434.

65. **Liu D, Reeves PR.** 1994. *Escherichia coli* K12 regains its O antigen. *Microbiology* **140**(Pt 1):49–57.

66. **Kido N, Kobayashi H.** 2000. A single amino acid substitution in a mannosyltransferase, WbdA, converts the *Escherichia coli* O9 polysaccharide into O9a: generation of a new O-serotype group. *J Bacteriol* **182**:2567–2573.

67. **Franco AV, Liu D, Reeves PR.** 1998. The wzz (cld) protein in *Escherichia coli*: amino acid sequence variation determines O-antigen chain length specificity. *J Bacteriol* **180**:2670–2675.

68. **Buckling A, Kassen R, Bell G, Rainey PB.** 2000. Disturbance and diversity in experimental microcosms. *Nature* **408**:961–964.

69. **McIver CJ, Tapsall JW.** 1993. Further studies of clinical isolates of cysteine-requiring *Escherichia coli* and *Klebsiella* and possible mechanisms for their selection *in vivo*. *J Med Microbiol* **39**:382–387.

70. **Borderon E, Horodniceanu T.** 1978. Metabolically deficient dwarf-colony mutants of *Escherichia coli*: deficiency and resistance to antibiotics of strains isolated from urine culture. *J Clin Microbiol* **8**:629–634.

71. **von Eiff C, Heilmann C, Proctor RA, Woltz C, Peters G, Götz F.** 1997. A site-directed *Staphylococcus aureus* hemB mutant is a small-colony variant which persists intracellularly. *J Bacteriol* **179**:4706–4712.

72. **Barth AL, Pitt TL.** 1995. Auxotrophic variants of *Pseudomonas aeruginosa* are selected from prototrophic wild-type strains in respiratory infections in patients with cystic fibrosis. *J Clin Microbiol* **33**:37–40.

73. **Giraud A, Matic I, Tenaillon O, Clara A, Radman M, Fons M, Taddei F.** 2001. Costs and benefits of high mutation rates: adaptive evolution of bacteria in the mouse gut. *Science* **291**:2606–2608.

74. **Denamur E, Bonacorsi S, Giraud A, Duriez P, Hilali F, Amorin C, Bingen E, Andremont A, Picard B, Taddei F, Matic I.** 2002. High frequency of mutator strains among human uropathogenic *Escherichia coli* isolates. *J Bacteriol* **184**:605–609.

75. **Baquero MR, Nilsson AI, Turrientes Mdel C, Sandvang D, Galán JC, Martínez JL, Frimodt-Møller N, Baquero F, Andersson DI.** 2004. Polymorphic mutation frequencies in *Escherichia coli*: emergence of weak mutators in clinical isolates. *J Bacteriol* **186**:5538–5542.

76. **Komp Lindgren P, Karlsson A, Hughes D.** 2003. Mutation rate and evolution of fluoroquinolone resistance in *Escherichia coli* isolates from patients with urinary tract infections. *Antimicrob Agents Chemother* **47**:3222–3232.

77. **Picard B, Duriez P, Gouriou S, Matic I, Denamur E, Taddei F.** 2001. Mutator natural *Escherichia coli* isolates have an unusual virulence phenotype. *Infect Immun* **69**:9–14.

78. **Labat F, Pradillon O, Garry L, Peuchmaur M, Fantin B, Denamur E.** 2005. Mutator phenotype confers advantage in *Escherichia coli* chronic urinary tract infection pathogenesis. *FEMS Immunol Med Microbiol* **44**:317–321.

79. **Cooper LA, Simmons LA, Mobley HL.** 2012. Involvement of mismatch repair in the reciprocal control of motility and adherence of

uropathogenic *Escherichia coli*. *Infect Immun* **80:**1969–1979.

80. **Galhardo RS, Hastings PJ, Rosenberg SM.** 2007. Mutation as a stress response and the regulation of evolvability. *Crit Rev Biochem Mol Biol* **42:**399–435.

81. **Kunin CM, Hua TH, Van Arsdale White L, Villarejo M.** 1992. Growth of *Escherichia coli* in human urine: role of salt tolerance and accumulation of glycine betaine. *J Infect Dis* **166:**1311–1315.

82. **Gawel D, Seed PC.** 2011. Urinary tract infection drives genome instability in uropathogenic *Escherichia coli* and necessitates translation synthesis DNA polymerase IV for virulence. *Virulence* **2:**222–232.

83. **Middendorf B, Hochhut B, Leipold K, Dobrindt U, Blum-Oehler G, Hacker J.** 2004. Instability of pathogenicity islands in uropathogenic *Escherichia coli* 536. *J Bacteriol* **186:**3086–3096.

84. **Hacker J, Bender L, Ott M, Wingender J, Lund B, Marre R, Goebel W.** 1990. Deletions of chromosomal regions coding for fimbriae and hemolysins occur *in vitro* and *in vivo* in various extraintestinal *Escherichia coli* isolates. *Microb Pathog* **8:**213–225.

85. **Anfora AT, Haugen BJ, Roesch P, Redford P, Welch RA.** 2007. Roles of serine accumulation and catabolism in the colonization of the murine urinary tract by *Escherichia coli* CFT073. *Infect Immun* **75:**5298–5304.

86. **Tree JJ, Ulett GC, Ong CL, Trott DJ, McEwan AG, Schembri MA.** 2008. Trade-off between iron uptake and protection against oxidative stress: deletion of cueO promotes uropathogenic *Escherichia coli* virulence in a mouse model of urinary tract infection. *J Bacteriol* **190:**6909–6912.

87. **Garcia EC, Brumbaugh AR, Mobley HL.** 2011. Redundancy and specificity of *Escherichia coli* iron acquisition systems during urinary tract infection. *Infect Immun* **79:**1225–1235.

88. **Zdziarski J, Brzuszkiewicz E, Wullt B, Liesegang H, Biran D, Voigt B, Grönberg-Hernandez J, Ragnarsdottir B, Hecker M, Ron EZ, Daniel R, Gottschalk G, Hacker J, Svanborg C, Dobrindt U.** 2010. Host imprints on bacterial genomes–rapid, divergent evolution in individual patients. *PLoS Pathog* **6:** e1001078. doi:10.1371/journal.ppat.1001078

89. **Bengtsson C, Bengtsson U, Bjorkelund C, Lincoln K, Sigurdsson JA.** 1998. Bacteriuria in a population sample of women: 24-year follow-up study. Results from the prospective population-based study of women in Gothenburg, Sweden. *Scand J Urol Nephrol* **32:**284–289.

90. **Hooton TM, Scholes D, Stapleton AE, Roberts PL, Winter C, Gupta K, Samadpour M, Stamm WE.** 2000. A prospective study of asymptomatic bacteriuria in sexually active young women. *N Engl J Med* **343:**992–997.

91. **Stapleton AE, Dziura J, Hooton TM, Cox ME, Yarova-Yarovaya Y, Chen S, Gupta K.** 2012. Recurrent urinary tract infection and urinary *Escherichia coli* in women ingesting cranberry juice daily: A randomized controlled trial. *Mayo Clin Proc* **87:**143–150.

92. **Trautner BW.** 2011. Asymptomatic bacteriuria: when the treatment is worse than the disease. *Nat Rev Urol* **9:**85–93.

93. **Cai T, Mazzoli S, Mondaini N, Meacci F, Nesi G, D'Elia C, Malossini G, Boddi V, Bartoletti R.** 2012. The role of asymptomatic bacteriuria in young women with recurrent urinary tract infections: to treat or not to treat? *Clin Infect Dis* **55:**771–777.

94. **Lane MC, Alteri CJ, Smith SN, Mobley HL.** 2007. Expression of flagella is coincident with uropathogenic *Escherichia coli* ascension to the upper urinary tract. *Proc Natl Acad Sci U S A* **104:**16669–16674.

95. **Chattopadhyay S, Weissman SJ, Minin VN, Russo TA, Dykhuizen DE, Sokurenko EV.** 2009. High frequency of hotspot mutations in core genes of *Escherichia coli* due to short-term positive selection. *Proc Natl Acad Sci U S A* **106:**12412–12417.

96. **Johnson JR, Delavari P, Kuskowski M, Stell AL.** 2001. Phylogenetic distribution of extraintestinal virulence-associated traits in *Escherichia coli*. *J Infect Dis* **183:**78–88.

97. **Connell I, Agace W, Klemm P, Shembri M, Mårild S, Svanborg C.** 1996. Type 1 fimbrial expression enhances *Escherichia coli* virulence for the urinary tract. *Proc Natl Acad Sci U S A* **93:**9827–9832.

98. **Sokurenko EV, Chesnokova V, Doyle RJ, Hasty DL.** 1997. Diversity of the *Escherichia coli* type 1 fimbriae lectin: differential binding to mannosides and uroepithelial cells. *J Biol Chem* **272:**17880–17886.

99. **Hasty DL, Wu XR, Sokurenko E.** 1998. Variants of the FimH adhesion confer distinct patterns of interaction of *E. coli* with urinary bladder. *Mol Biol Cell* **9**(11):501A.

100. **Sokurenko EV, Chesnokova V, Dykhuizen DE, Ofek I, Wu XR, Krogfelt KA, Struve C, Schembri MA, Hasty DL.** 1998. Pathogenic adaptation of *Escherichia coli* by natural variation of the FimH adhesin. *Proc Natl Acad Sci U S A* **95:**8922–8926.

101. **Weissman SJ, Beskhlebnaya V, Chesnokova V, Chattopadhyay S, Stamm WE, Hooton TM, Sokurenko EV.** 2007. Differential stability and trade-off effects of pathoadaptive

mutations in the *Escherichia coli* FimH adhesin. *Infect Immun* **75:**3548–3555.

102. **Sokurenko EV, Feldgarden M, Trintchina E, Weissman SJ, Avagyan S, Chattopadhyay S, Johnson JR, Dykhuizen DE.** 2004. Selection footprint in the FimH adhesin shows patho-adaptive niche differentiation in *Escherichia coli. Mol Biol Evol* **21:**1373–1383.

103. **Chattopadhyay S, Feldgarden M, Weissman SJ, Dykhuizen DE, van Belle G, Sokurenko EV.** 2007. Haplotype diversity in "source-sink" dynamics of *Escherichia coli* uroviru-lence. *J Mol Evol* **64:**204–214.

104. **Korotkova N, Chattopadhyay S, Tabata TA, Beskhlebnaya V, Vigdorovich V, Kaiser BK, Strong RK, Dykhuizen DE, Sokurenko EV, Moseley SL.** 2007. Selection for functional diversity drives accumulation of point muta-tions in Dr adhesins of *Escherichia coli. Mol Microbiol* **64:**180–194.

105. **Nowicki B, Hart A, Coyne KE, Lublin DM, Nowicki S.** 1993. Short consensus repeat-3 domain of recombinant decay-accelerating factor is recognized by *Escherichia coli* recom-binant Dr adhesin in a model of a cell-cell interaction. *J Exp Med* **178:**2115–2112.

106. **Medof ME, Walter EI, Rutgers JL, Knowles DM, Nussenzweig V.** 1987. Identification of the complement decay-accelerating factor (DAF) on epithelium and glandular cells and in body fluids. *J Exp Med* **165:**848–864.

107. **Westerlund B, Kuusela P, Risteli J, Risteli L, Vartio T, Rauvala H, Virkola R, Korhonen TK.** 1989. The O75X adhesin of uropathogenic *Escherichia coli* is a type IV collagen-binding protein. *Mol Microbiol* **3:**329–37.

108. **Guignot J, Breard J, Bernet-Camard MF, Peiffer I, Nowicki BJ, Servin AL, Blanc-Potard AB.** 2000. Pyelonephritogenic diffusely adher-ing *Escherichia coli* EC7372 harboring Dr-II ad-hesin carries classical uropathogenic virulence genes and promotes cell lysis and apoptosis in polarized epithelial caco-2/TC7 cells. *Infect Immun* **68:**7018–7027.

109. **Berger CN, Billker O, Meyer TF, Servin AL, Kansau I.** 2004. Differential recognition of members of the carcinoembryonic antigen family by Afa/Dr adhesins of diffusely adher-ing *Escherichia coli* (Afa/Dr DAEC). *Mol Microbiol* **52:**963–983.

110. **Korotkova N, Cota E, Lebedin Y, Monpouet S, Guignot J, Servin AL, Matthews S, Moseley SL.** 2006. A subfamily of Dr adhesins of *Escherichia coli* bind independently to decay-accelerating factor and the N-domain of carcinoembryonic antigen. *J Biol Chem* **281:**29120–29130.

111. **Servin AL.** 2005. Pathogenesis of Afa/Dr diffusely adhering *Escherichia coli. Clin Micro-biol Rev* **18:**264–292.

112. **Nowicki B, Labigne A, Moseley S, Hull R, Hull S, Moulds J.** 1990. The Dr hemagglutinin, afimbrial adhesins AFA-I and AFA-III, and F1845 fimbriae of uropathogenic and diarrhea-associated *Escherichia coli* belong to a family of hemagglutinins with Dr receptor recogni-tion. *Infect Immun* **58:**279–281.

113. **Le Bouguenec C, Garcia MI, Ouin V, Desperrier JM, Gounon P, Labigne A.** 1993. Characteriza-tion of plasmid-borne afa-3 gene clusters encoding afimbrial adhesins expressed by *Esch-erichia coli* strains associated with intestinal or urinary tract infections. *Infect Immun* **61:**5106–5114.

114. **Carnoy C, Moseley SL.** 1997. Mutational analysis of receptor binding mediated by the Dr family of *Escherichia coli* adhesins. *Mol Microbiol* **23:**365–379.

115. **Pettigrew D, Anderson KL, Billington J, Cota E, Simpson P, Urvil P, Rabuzin F, Roversi P, Nowicki B, du Merle L, Le Bouguénec C, Matthews S, Lea SM.** 2004. High resolution studies of the Afa/Dr adhesin DraE and its interaction with chloramphenicol. *J Biol Chem* **279:**46851–46857.

116. **Selvarangan R, Goluszko P, Singhal J, Carnoy C, Moseley S, Hudson B, Nowicki S, Nowicki B.** 2004. Interaction of Dr adhesin with collagen type IV is a critical step in *Escherichia coli* renal persistence. *Infect Immun* **72:**4827–4835.

117. **Hull RA, Hull SI, Falkow S.** 1984. Frequency of gene sequences necessary for pyelonephritis-associated pili expression among isolates of *Enterobacteriaceae* from human extraintestinal infections. *Infect Immun* **43:**1064–1067.

118. **Adlerberth I, Svanborg C, Carlsson B, Mellander L, Hanson LA, Jalil F, Khalil K, Wold AE.** 1998. P fimbriae and other adhesins enhance intestinal persistence of *Escherichia coli* in early infancy. *Epidemiol Infect* **121:**599–608.

119. **Connell H, Poulsen LK, Klemm P.** 2000. Expression of type 1 and P fimbriae *in situ* and localisation of a uropathogenic *Escherichia coli* strain in the murine bladder and kidney. *Int J Med Microbiol* **290:**587–597.

120. **Weissman SJ, Chattopadhyay S, Aprikian P, Obata-Yasuoka M, Yarova-Yarovaya Y, Stapleton A, Ba-Thein W, Dykhuizen D, Johnson JR, Sokurenko EV.** 2006. Clonal analysis reveals high rate of structural muta-tions in fimbrial adhesins of extraintestinal pathogenic *Escherichia coli. Mol Microbiol* **59:**975–988.

121. **Totsika M, Beatson SA, Holden N, Gally DL.** 2008. Regulatory interplay between pap

operons in uropathogenic *Escherichia coli*. *Mol Microbiol* **67**:996–1011.

122. **Chen SL, Hung CS, Xu J, Reigstad CS, Magrini V, Sabo A, Blasiar D, Bieri T, Meyer RR, Ozersky P, Armstrong JR, Fulton RS, Latreille JP, Spieth J, Hooton TM, Mardis ER, Hultgren SJ, Gordon JI.** 2006. Identification of genes subject to positive selection in uropathogenic strains of *Escherichia coli*: A comparative genomics approach. *Proc Natl Acad Sci U S A* **103**:5977–5982.

123. **Yang Z.** 2007. PAML 4: phylogenetic analysis by maximum likelihood. *Mol Biol Evol* **24**:1586–1591.

124. **Murray GL, Attridge SR, Morona R.** 2003. Regulation of *Salmonella typhimurium* lipopolysaccharide O antigen chain length is required for virulence; identification of FepE as a second Wzz. *Mol Microbiol* **47**:1395–1406.

125. **Nikaido H.** 1996. Outer Membrane, p 29–47. *In* Neidhardt FC, Curtiss R, Ingraham JL, Lin ECC, Low KB, Magasanik B, Reznikoff WS, Riley M, Schaecter M, Umbarger HE (ed), *Escherichia coli and Salmonella typhimurium: Cellular and Molecular Biology*, 2nd ed, ASM Press, Washington, DC.

126. **Snyder JA, Haugen BJ, Buckles EL, Lockatell CV, Johnson DE, Donnenberg MS, Welch RA, Mobley HL.** 2004. Transcriptome of uropathogenic *Escherichia coli* during urinary tract infection. *Infect Immun* **72**:6373–6381.

127. **Justice SS, Hung C, Theriot JA, Fletcher DA, Anderson GG, Footer MJ, Hultgren SJ.** 2004. Differentiation and developmental pathways of uropathogenic *Escherichia coli* in urinary tract pathogenesis. *Proc Natl Acad Sci U S A* **101**:1333–1338.

128. **Chattopadhyay S, Paul S, Kisiela DI, Linardopoulou EV, Sokurenko EV.** 2012. Convergent molecular evolution of genomic cores in *Salmonella enterica* and *Escherichia coli*. *J Bacteriol* **194**:5002–5011.

129. **Thomas WE, Trintchina E, Forero M, Vogel V, Sokurenko EV.** 2002. Bacterial adhesion to target cells enhanced by shear force. *Cell* **109**:913–923.

130. **Aprikian P, Tchesnokova V, Kidd B, Yakovenko O, Yarov-Yarovoy V, Trinchina E, Vogel V, Thomas W, Sokurenko E.** 2007. Interdomain interaction in the FimH adhesin of *Escherichia coli* regulates the affinity to mannose. *J Biol Chem* **282**:23437–23446.

131. **Tchesnokova V, Aprikian P, Yakovenko O, Larock C, Kidd B, Vogel V, Thomas W, Sokurenko E.** 2008. Integrin-like allosteric properties of the catch bond-forming FimH adhesin of *Escherichia coli*. *J Biol Chem* **283**:7823–7833.

132. **Le Trong I, Aprikian P, Kidd BA, Forero-Shelton M, Tchesnokova V, Rajagopal P, Rodriguez V, Interlandi G, Klevit R, Vogel V, Stenkamp RE, Sokurenko EV, Thomas WE.** 2010. Structural basis for mechanical force regulation of the adhesin FimH via finger trap-like beta sheet twisting. *Cell* **141**:645–655.

133. **Choudhury D, Thompson A, Stojanoff V, Langermann S, Pinkner J, Hultgren SJ, Knight SD.** 1999. X-ray structure of the FimC-FimH chaperone-adhesin complex from uropathogenic *Escherichia coli*. *Science* **285**:1061–1066.

134. **Tchesnokova V, Aprikian P, Kisiela D, Gowey S, Korotkova N, Thomas W, Sokurenko E.** 2011. Type 1 fimbrial adhesin FimH elicits an immune response that enhances cell adhesion of *Escherichia coli*. *Infect Immun* **79**:3895–3904.

135. **Langermann S, Palaszynski S, Barnhart M, Auguste G, Pinkner JS, Burlein J, Barren P, Koenig S, Leath S, Jones CH, Hultgren SJ.** 1997. Prevention of mucosal *Escherichia coli* infection by FimH-adhesin-based systemic vaccination. *Science* **276**:607–611.

136. **Sokurenko EV, Vogel V, Thomas WE.** 2008. Catch-bond mechanism of force-enhanced adhesion: counterintuitive, elusive, but ... widespread? *Cell Host Microbe* **4**:314–323.

137. **Essig M, Friedlander G.** 2003. Tubular shear stress and phenotype of renal proximal tubular cells. *J Am Soc Nephrol* **14**(Suppl 1):S33–S35.

138. **Brauner A, Katouli M, Tullus K, Jacobson SH.** 1990. Cell surface hydrophobicity, adherence to HeLa cell cultures and haemagglutination pattern of pyelonephritogenic *Escherichia coli* strains. *Epidemiol Infect* **105**:255–263.

139. **Johnson JR, Clabots C.** 2006. Sharing of virulent *Escherichia coli* clones among household members of a woman with acute cystitis. *Clin Infect Dis* **43**:e101–108.

140. **Owens RC Jr, Johnson JR, Stogsdill P, Yarmus L, Lolans K, Quinn J.** 2011. Community transmission in the United States of a CTX-M-15-producing sequence type ST131 *Escherichia coli* strain resulting in death. *J Clin Microbiol* **49**:3406–3408.

141. **Chattopadhyay S, Dykhuizen DE, Sokurenko EV.** 2007. ZPS: visualization of recent adaptive evolution of proteins. *BMC Bioinformatics* **8**:187.

142. **Sokurenko EV, Gomulkiewicz R, Dykhuizen DE.** 2006. Source-sink dynamics of virulence evolution. *Nat Rev Microbiol* **4**:548–555.

143. **Chattopadhyay S, Paul S, Dykhuizen DE, Sokurenko EV.** 2013. Tracking recent adaptive evolution in microbial species using TimeZone. *Nat Protoc* **8**:652–665.

Invasion of Host Cells and Tissues by Uropathogenic Bacteria

16

ADAM J. LEWIS,[1] AMANDA C. RICHARDS,[1] and MATTHEW A. MULVEY[1]

INTRODUCTION

The ability of bacterial pathogens to invade host cells can have profound effects on the establishment, persistence, and propagation of infections. By entering host cells and subsequently avoiding destruction within degradative lysosomes, bacteria can gain better access to scarce resources as well as protection from host defenses and antibiotics. Furthermore, host cell invasion can facilitate the dissemination of bacteria within and across tissue barriers. The actual benefits afforded to intracellular bacterial pathogens can be highly context-dependent and sometimes difficult to discern. Over the past three decades, a number of bacterial species that were conventionally thought to be strictly extracellular pathogens were found to have alternative intracellular lifestyles (1, 2). Among these facultative intracellular pathogens are strains of uropathogenic *Escherichia coli* (UPEC) and other bacteria that cause urinary tract infections (UTIs). These infections are very common, especially among females, and are prone to recur even after treatment with appropriate antibiotics (3, 4). Nearly one-third of women will have an acute UTI by the age of 24, and about 25% of these individuals will experience at least one recurrent

[1]Division of Microbiology and Immunology, Pathology Department, University of Utah School of Medicine, Salt Lake City, UT 84112.
Urinary Tract Infections: Molecular Pathogenesis and Clinical Management, 2nd Edition
Edited by Matthew A. Mulvey, David J. Klumpp, and Ann E. Stapleton
© 2017 American Society for Microbiology, Washington, DC
doi:10.1128/microbiolspec.UTI-0026-2016

UTI within 6 months of the initial infection. Many individuals endure painful bouts of recurrent and chronic UTIs throughout their lives (5). The capacity of some uropathogens to persist and even multiply within host cells may help explain why some UTIs repeatedly recur while also opening the door for new treatment options.

TAKING NOTICE OF INTRACELLULAR UROPATHOGENS

One of the first indications that uropathogenic bacteria could occupy intracellular niches within the urinary tract came from observations dating to the late 1970s. Using an experimental rat UTI model and transmission electron microscopy, researchers observed that UPEC could enter the large, terminally differentiated epithelial umbrella cells that line the luminal surface of the bladder urothelium (6). Several years later another group working with a mouse UTI model reported similar results (7). In each of these rodent infection models, the intracellular bacteria were observed both within membrane-bound vacuoles and free within the host cell cytosol. At the time, it was supposed that the bladder umbrella cells were killing the internalized bacteria as part of an innate host defense. This conclusion was in line with earlier work suggesting that uroepithelial cells have the capacity to act like phagocytes (8). In this 1974 study, it was noted that epithelial cells within the urothelium could engulf and destroy erythrocytes that were released due to hemorrhage of submucosal capillaries following the treatment of rats with bladder cytotoxins or carcinogens. The idea that UPEC strains could actually benefit from entry into host bladder cells did not gain a strong foothold until the late 1990s in the wake of observations made by researchers who were imaging the interactions between UPEC and the bladder mucosa in a mouse UTI model (9).

UPEC typically enters the urinary tract via an ascending route, transiting through the urethral opening and up the urethra before colonizing the bladder. Within the bladder, UPEC can utilize peritrichous filamentous adhesive organelles known as type 1 pili (or fimbriae) to engage the bladder umbrella cells (Fig. 1A). Each type 1 pilus is comprised of a 7-nm-wide rod made up of repeating FimA subunits linked via adapter subunits to a distal 3-nm-wide tip fibrillum containing the FimH adhesin (Fig. 1B) (10, 11). FimH can bind a variety of mannose-containing glycoprotein receptors, including the tetraspanin membrane protein uroplakin 1a (UP1a) (12). UP1a is one of four major uroplakin proteins that are embedded as two-dimensional quasi-crystalline arrays of 16-nm-wide hexameric complexes within the apical membranes of the terminally differentiated bladder umbrella cells (Fig. 1C) (13). The uroplakin complexes and specialized tight junctions that link the umbrella cells, as well as underlying layers of less differentiated epithelial cells, create an exceptionally strong permeability barrier (14–16). In 1998, high-resolution imaging of mouse bladders that were infected via transurethral catheterization showed UPEC tethered to the uroplakin-studded surfaces of bladder umbrella cells via numerous type 1 pili (Fig. 1A) (9). This study also revealed the host plasma membrane zippering around some of the adherent bacteria via contacts with the distal tips of the type 1 pili (Fig. 1D-E). Follow-up work indicated that these bacteria were internalized but not killed by the bladder cells (9, 17). In addition, the internalized bacteria were found to be markedly better at persisting within the bladder than their extracellular counterparts (9, 18–24).

THE FATES OF INTRACELLULAR UPEC

Following internalization by bladder epithelial cells in mice and in culture, UPEC is initially trafficked into membrane-bound compartments that are similar to late endosomes (25–27). Within these compartments, UPEC

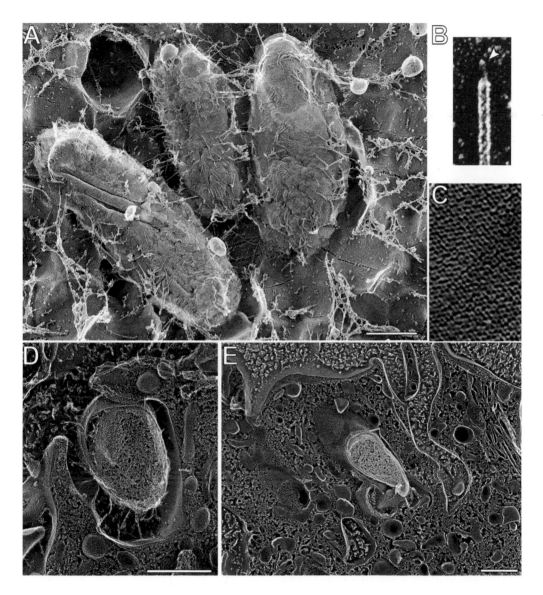

FIGURE 1 Type 1 pili mediate UPEC entry into bladder epithelial cells. (A) High-resolution deep-etch electron microscopy image showing UPEC (yellow) bound to a mouse bladder umbrella cell (blue) via multiple type 1 pili. (B) Close-up view of a type 1 pilus, showing the 3-nm-wide FimH-containing tip fibrillum structure (arrowhead). (C) Close-up view of the 16-nm-wide hexagonal uroplakin complexes that are embedded within the umbrella cell asymmetric unit membrane (AUM). (D, E) High-resolution freeze-fracture/deep-etch electron microscopy images showing the AUM enveloping bound UPEC. Scale bars = 0.5 μm. Images are reprinted from *Proc Natl Acad Sci USA* (18) and *Science* (9) with permission of the publishers.

growth is restricted and the pathogens appear to enter an almost quiescent state. However, UPEC can also occasionally break out into the host cytosol and subsequently undergo rampant multiplication, forming large biofilm-like intracellular bacterial communities (IBCs) that can contain more than 10,000 bacteria (Fig. 2A-B) (17, 21, 23, 26, 28).

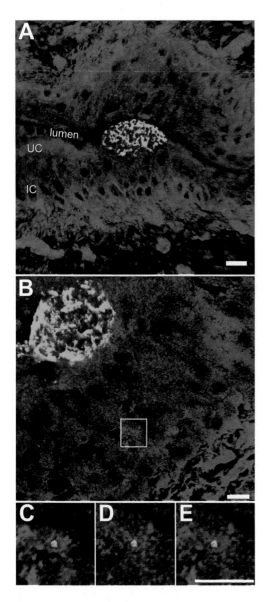

FIGURE 2 Localization of UPEC within the bladder
urothelium. (A, B) Confocal images of tissue sections from infected mouse bladders show IBCs
(green) within umbrella cells (UC). F-actin (red) is
sparse within these host cells but dense within
the underlying immature cells (IC). A single bacterium, localized within a LAMP-1-positive compartment (blue) and surrounded by F-actin, is
visible within one of the immature cells (box).
(C–E) Images show magnified views of the area
that is boxed in (B). Figures are reprinted from
Cellular Microbiology (26) with permission of the
publisher.

IBCs are clonal; thus, each is usually derived
from a single pathogen that manages to enter
the host cytosol (21). The development of
IBCs occurs primarily within the umbrella
cells, even though UPEC can also invade
the underlying, less differentiated epithelial
cells that comprise the bladder urothelium
(Fig. 2B-E) (17, 26, 27, 29). In mice, anywhere
from 3 to 700 IBCs can be detected in a single
bladder within just a few hours after inoculation with UPEC (21). IBCs begin as loose
assemblies of rod-shaped bacteria that can
multiply with doubling times of less than 30
minutes (23). This is not much slower than
the doubling times of *E. coli* strains in rich
broth culture, indicating that the host cytosol
likely has abundant nutrients that UPEC can
utilize. As bacteria within the IBCs multiply,
many produce daughter cells that are smaller
and more coccoid in shape. This morphological change may enable higher numbers of
bacteria to be packed within the host cell
while also providing added protection against
phagocytes (23, 30).

Eventually, the integrity of the infected
umbrella cells is compromised, and bacteria
begin to spill out into the surrounding environment (Fig. 3A-B) (17, 23). The emergent
bacteria are often highly motile and can go
on to infect neighboring cells or are flushed
from the urinary tract with the flow of urine.
At this stage, UPEC also temporarily transition into partially septated filamentous
forms that can attain lengths of greater than
100 μm. These remarkably long bacteria,
which are resistant to killing by phagocytes,
can worm their way through tight open-
ings in the host plasma membrane and
can extend relatively large distances within
and between host cells (Fig. 3C-D) (17, 23,
31). The formation of filamentous bacteria
during the final stages of IBC development
is important for the dissemination and persistence of UPEC within the urinary tract (31,
32).

IBCs provide UPEC with a means to
rapidly multiply within a protected niche, isolated from the shear flow of urine, infiltrating

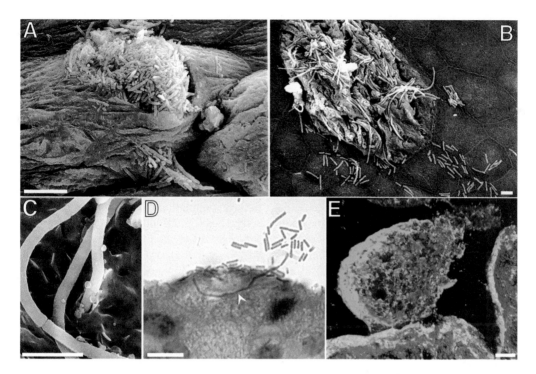

FIGURE 3 **The efflux and filamentation of UPEC coincident with the exfoliation of IBC-containing umbrella cells. (A–C) Scanning electron microscopy images show filamentous forms of UPEC, as well as their normal-sized counterparts, emerging from within IBCs. (D) Image from a hematoxylin- and eosin-stained bladder section highlights the ability of filamentous UPEC forms to extend long distances through umbrella cells. (E) Confocal image shows an IBC (blue) in close association with cytokeratin intermediate filaments (green) within an umbrella cell that is undergoing exfoliation. LAMP-1-positive compartments are red. Scale bars = 5 μm (A–C); 10 μm (D, E). Images are from mouse bladders recovered 6 hours after transurethral inoculation with UPEC. The figures are modified from *Cellular Microbiology* (26) or reprinted from *Infection and Immunity* (17) with permission of the publishers.**

phagocytes, and many other host defenses. Rounds of IBC growth and subsequent dispersal are observed in mouse models during the acute phase of a UTI, but the numbers of detectable IBCs eventually fade as the infection progresses (21, 23, 26, 33). This is likely due to multiple factors, including the upregulation of host defenses and the loss of susceptible umbrella cells, many of which are shed in response to a UTI (Fig. 3E) (9, 17, 23, 26). Though IBCs do not serve as stable long-term repositories of UPEC within the urinary tract, they can nonetheless have sizable effects on the progression and persistence of a UTI. In particular, the more IBCs that form in mice during

the first 24 hours of a bladder infection, the greater the chances are for the development of chronic UTIs (21). This correlation may reflect an ability of some of the bacteria that are released from IBCs to establish long-lived, mostly quiescent reservoirs within the urothelium (17, 21, 26, 27, 29). These intracellular bacterial reservoirs, bound within endosomal membranes, can persist in the bladder for many days and weeks, even when the host is treated with antibiotics that completely sterilize the urine (18–20, 22, 24, 29). The resurgence of UPEC from intracellular reservoirs is proposed to be an important source for relapsing and ostensibly recurrent UTIs.

INTRACELLULAR UROPATHOGENS IN THE HUMAN HOST

An ability to invade bladder epithelial cells and form both IBCs and quiescent reservoirs is widespread among UPEC isolates and has been documented in a range of genetically distinct mice and in various cell culture-based assays (34, 35). However, the relevance of host cell invasion by uropathogens to UTIs in the human population is the subject of some debate (for example, see references 36, 37). Nonetheless, evidence that intracellular bacteria contribute to disease within the human urinary tract is gaining traction.

One of the earliest examples of intracellular bacteria found within the human urinary tract comes from a 1985 study of patients with lower urinary tract symptoms (LUTS) (38). Individuals with LUTS can present with a variety of problems, including urinary urgency, frequency, dysuria, and bladder pain. Microscopic examination of bladder biopsies revealed the presence of intracellular bacteria in 8 out of 16 patients who had LUTS in the absence of bacteriuria (38). More recent studies have confirmed and extended these findings, showing that intracellular bacteria are especially common in LUTS patients with idiopathic overactive bladder (39–42). The apparent involvement of intracellular uropathogens in the etiology of LUTS for a large subset of patients is intriguing and suggests that the optimized use of host membrane-permeable antibiotics or other therapeutics that target invasive bacteria may be valuable treatment alternatives for these individuals.

The examination of LUTS patients also demonstrated that results from bacteriological analysis of urine samples do not necessarily reflect levels of bacterial colonization within the urinary tract (38). Specifically, culture-based diagnostic approaches do not take into account microbes that are associated with the bladder mucosa, and they can severely underestimate bacterial titers if IBCs or filamentous pathogens are present (41).

These issues may lead to the underdiagnosis of bacteriuria and UTI in a variety of patient populations. For example, in a recent study of 23 renal transplant recipients who were being screened for UTI, intracellular bacteria were observed in shed uroepithelial cells from 44% of the patients, but only one patient tested positive for bacteriuria by routine urine culture assays (43).

Intracellular bacteria, including IBCs, have also been documented in other diverse patient populations. In one study, microscopic examination of clean-catch urine specimens indicated the presence of IBCs within shed uroepithelial cells that were collected from young women who had acute uncomplicated UTI or a history of UTI (44). Of 65 women with UPEC-associated infections, 22% showed signs of IBCs and 45% had filamentous bacteria, which are often associated with IBC development. These were not observed in any samples recovered from 20 asymptomatic women. Interestingly, UTI symptoms were prolonged in women in whom IBC-containing uroepithelial cells were detected. Other work employing confocal microscopy revealed intracellular bacteria in 49 out of 133 (~37%) urine samples collected from children with UPEC-associated UTI (45, 46). In these individuals, the presence of intracellular *E. coli* was associated with recurrent UTIs, and this link was stronger in children who lacked any functional or morphological urinary abnormalities.

Cumulatively, these findings support the notion that uroepithelial cells can serve as staging grounds for rapid intracellular bacterial growth and as shelters for persistent bacterial reservoirs within the human urinary tract. Of note, recurrent UTIs are often caused by the same strain that was responsible for the initial infection, even when separated in time by many weeks or years (47–50). One interpretation of these data is that some recurrent UTIs develop due to the recrudescence of intracellular bacterial reservoirs that are not effectively cleared from the urinary tract following an initial infection. Though recurrent

UTIs also certainly arise by reinoculation of the urinary tract with bacteria coming from outside niches such as the vagina and gut, the potential for UPEC growth and long-term persistence within uroepithelial cells warrants attention when considering the nature of recalcitrant UTIs and the development of more efficacious treatment strategies.

MECHANISMS OF BLADDER CELL INVASION BY UPEC

The type 1 pilus-associated adhesin FimH is the major facilitator of UPEC entry into host cells. UPEC mutants that lack FimH are unable to effectively invade bladder epithelial cells either in cell culture model systems or in mice (9, 25). Furthermore, latex beads that are coated with FimH are readily internalized by bladder cells in culture, in contrast to beads that are coated with control proteins. The FimH-mediated uptake of bacteria or beads requires actin cytoskeletal rearrangements that drive the host plasma membrane to zipper around and eventually engulf the adherent particles (25, 51). This zippering mechanism resembles the host cell invasion processes used by a number of other bacterial pathogens, including *Listeria monocytogenes* and *Yersinia* spp. These invasive pathogens cause the directed reorganization of the host actin cytoskeleton by stimulating specific host receptors and downstream signaling cascades (52).

As the major receptor available to FimH on the bladder surface, UP1a is presumed to be an important mediator of UPEC entry into umbrella cells. UP1a has a single N-linked oligosaccharide side chain that is recognized by the mannose-binding domain of FimH (12, 53). During maturation and transport to the apical plasma membrane of umbrella cells, UP1a forms 16-nm-wide hexameric complexes with the integral membrane uroplakin proteins UP1b, UPII, and UPIIIa (13). These complexes are further organized into plaques that are about 0.5 μm in diameter

(see Fig. 1C). In umbrella cells, pairs of maturing plaques are assembled within discoidal vesicles, separated by interplaque hinge regions that lack the uroplakin complexes. As the bladder fills with urine, the discoidal vesicles are mobilized from the cytosol to add membrane and uroplakins to the luminal surface of the bladder (54). When the bladder is emptied, excess uroplakin plaques are internalized and degraded through a process referred to as compensatory endocytosis (55).

The bulk of each uroplakin protein on the umbrella cell surface is extracellular, with only the type-1 membrane protein UPIIIa having a sizable cytoplasmic tail (13). This gives the apical plasma membrane an asymmetric appearance when viewed from the side by transmission electron microscopy and earned it the label of asymmetric unit membrane (AUM) (56). Each hexagonal uroplakin complex within the AUM has a 3.7-nm wide central crevice that is lined in part by UP1a (13, 57). The localization of FimH at the distal tip of each type 1 pilus likely gives the adhesin better access to UP1a-associated mannose residues buried within the hexameric uroplakin complexes (9). Once formed, FimH interactions with UP1a may allow UPEC to enter umbrella cells by the same pathway used to take in uroplakin plaques during compensatory endocytosis. This rather passive entry mechanism could be expedited by the ability of FimH binding to elicit conformational changes within uroplakin transmembrane domains (58, 59). Work in bladder cell lines suggests that these changes can result in the phosphorylation of a threonine residue within the long cytoplasmic tail of UPIII by the host enzyme casein kinase II (CK2) (60). Subsequent downstream signaling events, including calcium fluxes, may in turn stimulate local cytoskeletal rearrangements, leading to bacterial internalization. The highly flexible (hypercompliant) nature of the AUM could further aid the internalization process by allowing multiple uroplakin plaques to deform and envelop UPEC (see Fig. 1D-E) (59).

In addition to umbrella cells, FimH can mediate UPEC entry into immature uroepithelial cells and many other host cell types that lack uroplakins or uroplakin plaques. Over the years, researchers have found that FimH can bind a wide range of mannose-containing glycoproteins as well as a few nonglycosylated components of the extracellular matrix. The known receptors for FimH are numerous and include α3 and β1 integrin subunits, the leukocyte adhesion molecules CD11b and CD18, the glycophosphatidylinositol-anchored protein CD48, the pattern recognition receptor Toll-like receptor 4 (TLR4), carcinoembryonic antigen-related cell adhesion family members, nonspecific cross-reacting antigen-50, glycoprotein 2, type I and type IV collagens, fibronectin, and laminin (61–72). Among these diverse receptors, α3 and β1 integrins have emerged as important mediators of host cell invasion by type 1-piliated bacteria (71).

Heterodimers of α and β integrin subunits serve as major surface-localized signaling conduits into and out of host cells, with especially important roles in connecting extracellular matrix (ECM) components with the actin cytoskeleton (73). These integrin subunits are expressed by umbrella cells and other epithelial cells found throughout the urinary tract (74, 75). Many bacterial pathogens, including *Yersinia* spp. and group A *Streptococcus*, can invade host cells by engaging integrins either directly or indirectly via association with ECM proteins (76, 77). The heterodimerization of α3 and β1 integrin subunits creates a ligand-binding pocket that can recognize a number of ECM factors, including collagen, laminin, and fibronectin (78). FimH interactions with α3β1 integrins occur independently of the canonical ECM-binding site (71). Rather, FimH binds α3 and β1 integrin subunits individually via N-linked high-mannose-type glycan structures present in their extracellular domains. Subsets of the glycans that are associated with α3β1 integrins expressed by bladder epithelial cells are structurally similar to those that decorate UP1a (71, 79–81).

In bladder cell infection models, α3 and β1 integrin subunits cluster around adherent and invading type 1-piliated bacteria, coincident with the accumulation of actin filaments (F-actin) (71). FimH-mediated invasion of host cells is inhibited by α3 and β1 integrin-specific blocking antibodies and by disruption of the β1 integrin gene. The use of conditional knockout mice shows that α3β1 integrins also promote UPEC entry into host cells within the bladder urothelium *in vivo* (M.A. Mulvey, unpublished observations). Signaling cascades downstream of integrin receptors are modulated by the phosphorylation of conserved serine, threonine, and tyrosine residues within the cytoplasmic tail of β integrin subunits (73). The mutation of these residues within β1 integrin can have variable effects, either stimulating or hindering UPEC entry depending on the nature of the altered residue(s) (71). Not unexpectedly, many of the host factors that are known to modulate actin dynamics in association with integrin signaling have been implicated as mediators of host cell invasion by UPEC (Fig. 4). These factors include a spectrum of signaling scaffolds and adaptor proteins (e.g., paxillin, AP-2, clathrin heavy chain, NUMB, Dab2, and ARH), kinases (e.g., FAK, MAP kinases, PI-3 kinase), Rho GTPases (e.g., Rac1, Cdc42, RhoA), and actin-binding proteins and nucleators (e.g., Arp2/3, WAVE2, α-actinin, vinculin) (25, 71, 82–84). Microtubules and microtubule-associated factors (e.g., kinesin-1, HDAC6) may also stimulate the entry process, acting indirectly to modulate actin dynamics (85, 86).

In addition to actin rearrangements, the internalization of UPEC by bladder cells likely requires the addition of host membrane to the cell surface to accommodate the zippering process (87, 88). This membrane can be derived from various sources, including endosomal compartments, lysosomes, or, in the case of bladder umbrella cells, discoid vesicles. The delivery of membrane to sites of UPEC entry involves small GTP-binding proteins such as Rab27b, which can also influ-

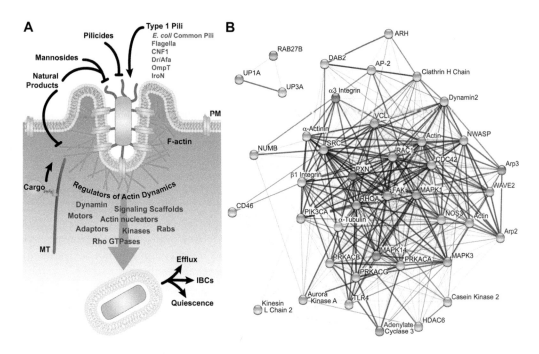

FIGURE 4 UPEC invasion of bladder epithelial cells. (A) Model depicts host and bacterial factors that have been identified as regulators of bladder cell invasion by UPEC. Potential therapeutics are also indicated. (B) The host factors that can modulate the FimH-dependent entry of UPEC into bladder cells are interconnected. The image in (B) was created using the STRING database (version 10.0) of known and predicted protein-protein interactions (198). Line thickness indicates the strength of the supporting data.

ence the intracellular trafficking and efflux of uroplakin plaques and UPEC alike (54, 89–91). The silencing of Rab27b expression inhibits UPEC entry into host cells, and internalized UPEC initially localizes within Rab27b-positive compartments (82, 89).

One of the final steps of the internalization process leading to the formation of nascent, UPEC-containing Rab27b-positive vesicles within the host cytosol is catalyzed by the large GTPase dynamin2 (82, 92, 93). The activity of dynamin2 is enhanced by S-nitrosylation of a single cysteine residue via reaction with nitric oxide (NO) that is generated by endothelial NO synthase (NOS3) (92, 93). Within the bladder, the levels of NO and other reactive nitrogen species rapidly increase in response to infection (94). Reactive nitrogen species are produced by the host as an antimicrobial defense, but UPEC

isolates are often highly resistant to the damaging effects these radicals (95–98). Consequently, UPEC can sidestep the dangers of eliciting NO production while simultaneously taking advantage of the benefits afforded to invasive pathogens by NO-enhanced dynamin2 activity.

To date, well over 40 host cell factors have been implicated as regulators of bladder cell invasion by UPEC downstream of FimH binding to either integrin subunits or UP1a. These host factors are interconnected within a web of signaling pathways (Fig. 4B). Rather than acting autonomously, different FimH receptors and associated signaling pathways may promote UPEC entry in a synergistic fashion. For example, the uptake of uroplakin plaques by compensatory endocytosis requires the activation of β1 integrin-associated signaling pathways (55). Consequently, it is feasible that FimH interactions with β1 integrins can promote the internalization of UP1a-bound UPEC into umbrella cells in part by stimulating compensatory endocytosis. The complexity of UPEC entry into host cells increases further when considering possible involvement of coreceptors such as the host complement receptor CD46 and contributions made by other bacterial factors that can also facilitate bacterial internalization (99, 100). The latter include Afa/Dr adhesins, *E. coli* common pili, flagella, outer membrane protein T, the salmochelin siderophore receptor IroN, and the Rho GTPase-activating toxin cytotoxic necrotizing factor 1 (see Fig. 4A) (101–110). The specific sets of bacterial and host factors that are engaged by UPEC to gain access to intracellular niches likely vary as the pathogens encounter changing environments and host cell types during the course of a UTI.

REGULATION OF INTRACELLULAR BACTERIAL GROWTH AND PERSISTENCE

Within bladder epithelial cells, UPEC is trafficked into membrane-bound compartments that are positive for the late endosomal markers Lysosome-Associated Membrane Protein-1 (LAMP-1), lysobisphosphatidic acid, and CD63 (Fig. 5) (27, 82). These compartments lack the lysosomal protease cathepsin D and may or may not be acidified. The endosomal trafficking of UPEC within bladder cells can also impinge on autophagic pathways and multivesicular bodies (1, 111, 112). UPEC avoids destruction within degra-

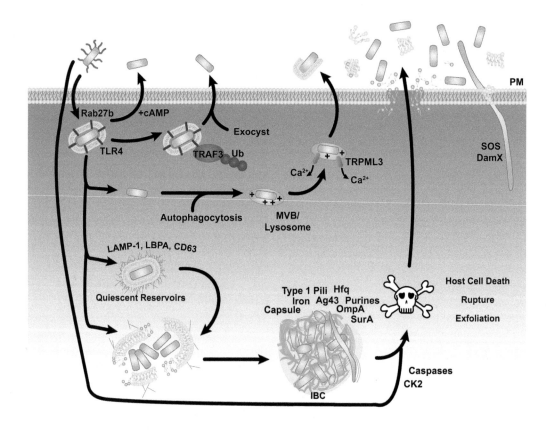

FIGURE 5 The fates of UPEC following entry into bladder epithelial cells. See text for details.

dative lysosomes in part by causing the up-regulation and recruitment of the host protein Rab35 (113). This small GTPase has a key role in the endosomal recycling of transferrin receptor, and its colocalization with UPEC-containing vacuoles aids iron acquisition by the pathogens and prevents fusion with degradative lysosomes.

Although UPEC can survive for long periods bound within host endosomes, its intravacuolar growth is restricted. This is partially attributable to host F-actin, which often surrounds UPEC-containing vacuoles within immature bladder epithelial cells (see Fig. 2B-E) (26). F-actin may limit bacterial growth by gating the trafficking of nutrients and/or other factors into and out of UPEC-containing vacuoles and by physically corralling the pathogens. The translocation of UPEC into the host cytosol where rapid growth and IBC development occurs is facilitated by the disruption of the actin cytoskeleton and endosomal membranes (26, 114). Within terminally differentiated umbrella cells, actin filaments are distributed primarily along basolateral surfaces and are sparse elsewhere (see Fig. 2) (115). This situation likely enables UPEC to more easily escape into the host cytosol and subsequently form IBCs. In contrast to umbrella cells, the much smaller immature transitional cells of the urothelium have a more dense arrangement of actin filaments. Bacteria that manage to invade these immature cells may become, in effect, locked within actin-bound endosomes. Although they are unable to effectively multiply, the near quiescent status of these intravacuolar bacteria renders them insensitive to many host defenses and antibiotics (17, 24). This is because current antibiotics are often only effective against growing bacteria, and many are unable to cross host membranes. The latter effect is amplified when considering bacteria that are buried within the urothelium, which itself functions as an especially strong permeability barrier (14). Intracellular UPEC reservoirs have a quantifiable survival advantage within antibiotic-treated hosts (19, 20, 22, 24). The resurgence of UPEC from these reservoirs may be triggered by the relocalization of F-actin due to terminal differentiation of umbrella cells or other processes and may contribute to the development of chronic and relapsing UTIs (29).

The development of quiescent intracellular UPEC reservoirs may also be facilitated by the activation of bacterial toxin-antitoxin (TA) systems. These systems consist of relatively stable toxins that are held in check by more labile antitoxins. Stressful conditions can result in degradation of the antitoxins and subsequent toxin activation. This, in turn, can cause bacteria to form quiescent, antibiotic-tolerant persister cells (116). Of the dozens of toxin-antitoxin systems that have been identified, only a subset is encoded by strains of UPEC (117, 118). A few of these (YefM-YoeB, YbaJ-Hha, and PasTI) have been shown to promote UPEC persistence within the urinary tract and may help regulate the establishment of quiescent reservoirs and IBC development (117).

The formation of IBCs within the cytosol of terminally differentiated bladder umbrella cells occurs in close association with host cytokeratin intermediate filaments, which can serve as scaffolding for biofilm development (see Fig. 3E) (26, 119). IBC maturation is also facilitated by multiple bacterial factors, including many that have been implicated in the production of extracellular biofilms. These include type 1 pili, the adhesin Ag43, capsule, OmpA, purine biosynthesis enzymes, and various regulators such as integration host factor, the QseC sensor kinase, the RNA chaperone Hfq, and the periplasmic prolyl isomerase and chaperone SurA (Fig. 5) (28, 120–128). Bacteria growing within IBCs appear to utilize non-glucose carbon sources such as galactoside and express stress-resistance genes such as *yeaR* that enable UPEC to better deal with oxidative and nitrosative stresses (97, 129). Not unexpectedly, bacterial iron acquisition systems are especially important to the intracellular survival and growth of UPEC (114, 129, 130).

ANTIBACTERIAL DEFENSES AND LIABILITIES

The host can deploy a wide array of defenses to interfere with the ability of UPEC and other uropathogens to colonize uroepithelial cells. Among these is the flow of urine that can wash away nonadherent microbes, the secretion of antiadherence factors such as Tamm-Horsfall protein and surfactant protein D, and the production of antibacterial proteins such as secretory IgA, ribonuclease 7, and defensins (131–136). When overwhelmed by bacteria, bladder epithelial cells can also initiate programmed cell death pathways that lead to their exfoliation and eventual clearance from the urinary tract with the flow of urine. This process entails the activation caspases, the exocytosis of lysosomes, and the disassembly of host tight junctions and other intercellular connections (9, 18, 137–141). The specific bacterial factors that elicit bladder cell exfoliation during a UTI are not yet well defined. However, exfoliation and host cell death may be enhanced by activation of the host kinase CK2 downstream of FimH-mediated interactions with uroplakin complexes, as well as by the secreted bacterial toxins cytotoxic necrotizing factor 1 and α-hemolysin (60, 142–146). The exfoliation of IBC-containing umbrella cells can rid the host of huge numbers of bacteria (see Fig. 3E). However, this defense may also facilitate the dissemination of UPEC to other hosts and can promote bacterial dispersal within the urinary tract by compromising the barrier function of the urothelium (17). This problem is countered in part by the remarkable ability of the otherwise extremely stable urothelium to rapidly regenerate when damaged (6, 9, 147–150).

Like exfoliation, other host defense mechanisms can also have downsides. For example, the production of antibacterial peptides and the induction of inflammatory responses are generally considered to be beneficial for the host (132, 151, 152). However, recent research in mice indicates that some antibacterial peptides, such as cathelicidin, can also potentiate UPEC colonization of the bladder while the excess stimulation of inflammatory responses can drive the development of chronic UTIs (33, 153). This effect could be in part attributable to the influx of neutrophils, which are important for the clearance of uropathogens but can also disrupt interepithelial junctions and thereby facilitate UPEC dissemination. The difference between a host defense and a liability within the urinary tract is therefore not always easily discerned. This is further exemplified by considering the intracellular trafficking of UPEC.

Recent reports indicate that bladder epithelial cells can redirect invading bacteria, forcing their expulsion via TLR4- and cAMP-dependent trafficking pathways before they can establish either IBCs or intracellular reservoirs (Fig. 5) (89, 90, 112, 154). One of these expulsion pathways involves the assembly of exocyst complexes around UPEC-containing vacuoles downstream of the TLR4-dependent ubiquitination of the immune regulator TRAF3 (154). Another pathway expels bacteria via a more circuitous route in which UPEC first enters the host cytosol before being trafficked through autophagosomes and multivesicular bodies to lysosomes that are jettisoned through a process that is regulated by the cation channel TRPML3 (112). Although these nonlytic expulsion pathways can hinder bacterial colonization of individual bladder cells, they may also enable uropathogens to better disseminate by moving in and out of host cells. Specifically, expulsion pathways may promote bacterial transmission through host cell layers and could be especially valuable for a microbe endeavoring to ascend through the urinary tract against the bulk flow of urine.

The efflux of UPEC from IBCs within dying umbrella cells may similarly facilitate bacterial dissemination within the urinary tract (see Fig. 3) (17, 31). The emergence of filamentous bacterial forms, in particular, may be advantageous by allowing UPEC to span distances between host cells without losing contact with the urothelium. The filamentous bacteria themselves are a consequence of TLR4-dependent host defenses that cause activation of DNA damage SOS responses in UPEC and subsequent inhibition of bacterial septation (23, 31, 32). In an alternative parallel pathway, activation of the bacterial cell division protein DamX can also induce transient filamentation upon exposure to liquid shear forces as encountered at the urothelium-urine interface (155). The altered morphology and surface characteristics of filamentous UPEC forms render these bacteria resistant to phagocytosis by neutrophils (23, 156). However, even if phagocytosis cannot be avoided, some UPEC strains can survive for some time within the phagosomal compartments of neutrophils and macrophages (61, 157–159). Transcriptional profiling recently identified 22 bacterial genes that seem to promote UPEC survival within macrophages (160). Among these were phage-shock-protein-related genes, which enable bacteria to better deal with extracytoplasmic stresses and pH changes.

Possibly the first description of intracellular bacterial reservoirs within the urinary tract comes from the analysis of intramacrophage *E. coli* communities. In patients suffering from a rare inflammatory condition known as urinary malakoplakia, viable intracellular *E. coli* populations were detected within macrophages in granulomatous ulcers isolated from bladder or kidney tissues (161–163). These internalized bacteria were protected from antibiotic treatments that sterilize the urine, leading the authors to suggest that intracellular bacterial reservoirs could contribute to the chronic and recurrent UTIs that often plague patients with malakoplakia. Although this work focused on bacterial survival within defective macrophages, it is highly reminiscent of findings made many years later with uroepithelial cells (9, 17, 24). The impact of enhanced UPEC survival within functionally normal immune cells on the progression and persistence of UTIs is

unclear. However, a recent study indicates that tissue-resident macrophages within the bladder can internalize and sequester large numbers of UPEC organisms, which in turn impedes the development of adaptive responses to UTI (164). These observations raise the possibility that internalized UPEC might be able to alter the antigen-presenting activities of macrophages.

BACTERIAL INVASION OF KIDNEY CELLS

Though most studies of host cell invasion by uropathogens have focused on the bladder, it has been appreciated for many years that bacteria can also enter renal epithelial cells (165–167). UPEC does not appear to multiply or take up long-term residence within kidney cells, but its translocation through collecting duct epithelial cells can facilitate bacterial dissemination into the renal interstitium and then into the bloodstream (168). This in turn can result in the development of bacteremia and urosepsis. Within the collecting ducts of the kidneys, intercalated cells are likely primary portals for UPEC translocation into the renal interstitium (169). Several host and bacterial factors have been identified as facilitators of UPEC translocation through renal epithelial cells. These include sets of bacterial adhesins made up of P pili in combination with type 1 pili, Dr/Afa adhesins, or S pili (170). Type 1 pili can also synergize with the complement component C3 and the C3 receptor CD46 to stimulate UPEC entry into renal epithelial cells (99, 100, 171). Another set of complement factors, C5a and its receptor C5aR1, promote UPEC colonization of the kidneys, in part, by enhancing bacterial survival within macrophages (172). The pattern recognition receptors TLR4 and TLR5 have also been implicated in the translocation of UPEC through renal epithelial cells, with the latter working in concert with bacterial flagella (168, 173, 174). The molecular machinery that controls the trafficking of UPEC through renal epithelial cells has not been defined.

OTHER INVASIVE UROPATHOGENS

UPEC is not alone in its ability to invade uro-epithelial cells. The Gram-positive opportunistic uropathogens *Staphylococcus saprophyticus*, *Streptococcus agalactiae*, and *Enterococcus faecalis* can invade bladder epithelial cells, and the latter has been isolated within shed urothelial cells from LUTS patients (175–177). *Proteus mirabilis*, which is often associated with the formation of urinary stones, can transiently invade both kidney and bladder epithelial cells (178, 179). Host cell entry by this pathogen is facilitated by sets of bacterial trimeric autotransporter proteins, the sigma factor RpoE, the putrescine importer PlaP, flagella, and regulators of swarm cell formation (180–183). *Klebsiella pneumoniae*, which is a common cause of nosocomial UTIs, enters bladder uroepithelial cells via an actin- and microtubule-dependent pathway that is triggered by interactions between host glycoprotein receptors and a FimH orthologue (184–186). This entry mechanism is comparable to that used by type 1 piliated UPEC isolates. Similar to UPEC, *K. pneumonia* and *E. faecalis* can both form IBC-like inclusions within bladder umbrella cells (177, 187). Interestingly, the presence of these and other opportunistic uropathogens during polymicrobial UTIs can select for more invasive UPEC isolates (188). This observation indicates that an ability to invade uroepithelial cells provides uropathogens with a *bona fide* competitive advantage within the urinary tract.

TARGETING INTRACELLULAR UROPATHOGENS

The treatment of UTIs is complicated by the ability of UPEC and other uropathogenic bacteria to invade uroepithelial cells, where they are protected from the effects of most antibiotics. In the case of UPEC, the formation of long-living intracellular reservoirs may make complete eradication of the pathogen from the urinary tract especially difficult (17, 24).

The development of pilicides and mannosides that interfere with the functions of adhesive organelles such as type 1 pili may prove useful in hindering bacterial invasion of uroepithelial cells as well as disrupting IBCs (189–191). Natural products from cranberry and other sources may likewise impede UPEC entry into host cells, either by preventing bacterial attachment or by disrupting signaling through β1 integrin or other key host receptors (see Fig. 4) (84, 192–195).

To target intracellular bacterial reservoirs within the bladder after they are already established, it may be possible to use a "shock and kill" approach akin to therapies that are being developed to eradicate latent HIV reservoirs. In this case, a bladder cell exfoliant such as chitosan or imidazolium salts is instilled into the bladder lumen, triggering the rapid release of umbrella cells and the subsequent proliferation and differentiation of newly exposed immature uroepithelial cells (149, 196, 197). This process eradicates any reservoir populations that may be present within the superficial layer of bladder cells but also induces the resurgence of UPEC from less mature uroepithelial cells as they differentiate and realign their actin filaments (29). The coordinate administration of antibiotics, which are entirely ineffective against intracellular UPEC populations, can then be used to clear the emergent pathogens. This strategy has worked well in mouse models, but its safety and efficacy in humans have not been addressed. Nonetheless, such exploratory studies in animal models are promising and will hopefully lead basic researchers and clinicians alike to consider treatment strategies that take advantage of our growing knowledge of the mechanisms and consequences of host cell invasion by uropathogens.

ACKNOWLEDGMENTS

This work was funded by National Institute of Allergy and Infectious Diseases (NIAID) grant AI095647; and National Institute of Diabetes and Digestive and Kidney Diseases

(NIDDK) grant T32 DK7115. The funders had no role in study design, data collection, and interpretation or in the decision to submit the work for publication.

CITATION

Lewis AJ, Richards AC, Mulvey MA. 2016. Invasion of host cells and tissues by uropathogenic bacteria. Microbiol Spectrum 4(6): UTI-0026-2016.

REFERENCES

1. **Bower JM, Eto DS, Mulvey MA.** 2005. Covert operations of uropathogenic *Escherichia coli* within the urinary tract. *Traffic* **6:**18–31.
2. **Silva MT.** 2012. Classical labeling of bacterial pathogens according to their lifestyle in the host: inconsistencies and alternatives. *Front Microbiol* **3:**71.
3. **Barber AE, Norton JP, Spivak AM, Mulvey MA.** 2013. Urinary tract infections: current and emerging management strategies. *Clin Infect Dis* **57:**719–724.
4. **Dielubanza EJ, Schaeffer AJ.** 2011. Urinary tract infections in women. *Med Clin North Am* **95:**27–41.
5. **Foxman B.** 2014. Urinary tract infection syndromes: occurrence, recurrence, bacteriology, risk factors, and disease burden. *Infect Dis Clin North Am* **28:**1–13.
6. **Fukushi Y, Orikasa S, Kagayama M.** 1979. An electron microscopic study of the interaction between vesical epitherlium and *E. coli*. *Invest Urol* **17:**61–68.
7. **McTaggart LA, Rigby RC, Elliott TS.** 1990. The pathogenesis of urinary tract infections associated with *Escherichia coli, Staphylococcus saprophyticus* and *S. epidermidis*. *J Med Microbiol* **32:**135–141.
8. **Wakefield JS, Hicks RM.** 1974. Erythrophagocytosis by the epithelial cells of the bladder. *J Cell Sci* **15:**555–573.
9. **Mulvey MA, Lopez-Boado YS, Wilson CL, Roth R, Parks WC, Heuser J, Hultgren SJ.** 1998. Induction and evasion of host defenses by type 1-piliated uropathogenic *Escherichia coli*. *Science* **282:**1494–1497.
10. **Russell PW, Orndorff PE.** 1992. Lesions in two *Escherichia coli* type 1 pilus genes alter pilus number and length without affecting receptor binding. *J Bacteriol* **174:**5923–5935.
11. **Jones CH, Pinkner JS, Roth R, Heuser J, Nicholes AV, Abraham SN, Hultgren SJ.** 1995. FimH adhesin of type 1 pili is assembled into a fibrillar tip structure in the *Enterobacteriaceae*. *Proc Natl Acad Sci USA* **92:**2081–2085.
12. **Zhou G, Mo WJ, Sebbel P, Min G, Neubert TA, Glockshuber R, Wu XR, Sun TT, Kong XP.** 2001. Uroplakin Ia is the urothelial receptor for uropathogenic *Escherichia coli*: evidence from *in vitro* FimH binding. *J Cell Sci* **114:**4095–4103.
13. **Wu XR, Kong XP, Pellicer A, Kreibich G, Sun TT.** 2009. Uroplakins in urothelial biology, function, and disease. *Kidney Int* **75:**1153–1165.
14. **Apodaca G.** 2004. The uroepithelium: not just a passive barrier. *Traffic* **5:**117–128.
15. **Hu P, Meyers S, Liang FX, Deng FM, Kachar B, Zeidel ML, Sun TT.** 2002. Role of membrane proteins in permeability barrier function: uroplakin ablation elevates urothelial permeability. *Am J Physiol Renal Physiol* **283:**F1200–F1207.
16. **Acharya P, Beckel J, Ruiz WG, Wang E, Rojas R, Birder L, Apodaca G.** 2004. Distribution of the tight junction proteins ZO-1, occludin, and claudin-4, -8, and -12 in bladder epithelium. *Am J Physiol Renal Physiol* **287:**F305–F318.
17. **Mulvey MA, Schilling JD, Hultgren SJ.** 2001. Establishment of a persistent *Escherichia coli* reservoir during the acute phase of a bladder infection. *Infect Immun* **69:**4572–4579.
18. **Mulvey MA, Schilling JD, Martinez JJ, Hultgren SJ.** 2000. Bad bugs and beleaguered bladders: interplay between uropathogenic *Escherichia coli* and innate host defenses. *Proc Natl Acad Sci USA* **97:**8829–8835.
19. **Kerrn MB, Struve C, Blom J, Frimodt-Moller N, Krogfelt KA.** 2005. Intracellular persistence of *Escherichia coli* in urinary bladders from mecillinam-treated mice. *J Antimicrob Chemother* **55:**383–386.
20. **Schilling JD, Lorenz RG, Hultgren SJ.** 2002. Effect of trimethoprim-sulfamethoxazole on recurrent bacteriuria and bacterial persistence in mice infected with uropathogenic *Escherichia coli*. *Infect Immun* **70:**7042–7049.
21. **Schwartz DJ, Chen SL, Hultgren SJ, Seed PC.** 2011. Population dynamics and niche distribution of uropathogenic *Escherichia coli* during acute and chronic urinary tract infection. *Infect Immun* **79:**4250–4259.
22. **Hvidberg H, Struve C, Krogfelt KA, Christensen N, Rasmussen SN, Frimodt-Moller N.** 2000. Development of a long-term ascending urinary tract infection mouse model for antibiotic treatment studies. *Antimicrob Agents Chemother* **44:**156–163.
23. **Justice SS, Hung C, Theriot JA, Fletcher DA, Anderson GG, Footer MJ, Hultgren SJ.**

2004. Differentiation and developmental pathways of uropathogenic *Escherichia coli* in urinary tract pathogenesis. *Proc Natl Acad Sci USA* **101**:1333–1338.

24. **Blango MG, Mulvey MA.** 2010. Persistence of uropathogenic *Escherichia coli* in the face of multiple antibiotics. *Antimicrob Agents Chemother* **54**:1855–1863.

25. **Martinez JJ, Mulvey MA, Schilling JD, Pinkner JS, Hultgren SJ.** 2000. Type 1 pilus-mediated bacterial invasion of bladder epithelial cells. *EMBO J* **19**:2803–2812.

26. **Eto DS, Sundsbak JL, Mulvey MA.** 2006. Actin-gated intracellular growth and resurgence of uropathogenic *Escherichia coli*. *Cell Microbiol* **8**:704–717.

27. **Mysorekar IU, Hultgren SJ.** 2006. Mechanisms of uropathogenic *Escherichia coli* persistence and eradication from the urinary tract. *Proc Natl Acad Sci USA* **103**:14170–14175.

28. **Anderson GG, Palermo JJ, Schilling JD, Roth R, Heuser J, Hultgren SJ.** 2003. Intracellular bacterial biofilm-like pods in urinary tract infections. *Science* **301**:105–107.

29. **Blango MG, Ott EM, Erman A, Veranic P, Mulvey MA.** 2014. Forced resurgence and targeting of intracellular uropathogenic *Escherichia coli* reservoirs. *PLoS One* **9**:e93327. doi:10.1371/journal.pone.0093327.

30. **Hunstad DA, Justice SS.** 2010. Intracellular lifestyles and immune evasion strategies of uropathogenic *Escherichia coli*. *Annu Rev Microbiol* **64**:203–221.

31. **Justice SS, Hunstad DA, Seed PC, Hultgren SJ.** 2006. Filamentation by *Escherichia coli* subverts innate defenses during urinary tract infection. *Proc Natl Acad Sci USA* **103**:19884–19889.

32. **Li B, Smith P, Horvath DJ Jr, Romesberg FE, Justice SS.** 2010. SOS regulatory elements are essential for UPEC pathogenesis. *Microbes Infect* **12**:662–668.

33. **Hannan TJ, Mysorekar IU, Hung CS, Isaacson-Schmid ML, Hultgren SJ.** 2010. Early severe inflammatory responses to uropathogenic *E. coli* predispose to chronic and recurrent urinary tract infection. *PLoS Pathog* **6**:e1001042. doi:10.1371/journal.ppat.1001042.

34. **Garofalo CK, Hooton TM, Martin SM, Stamm WE, Palermo JJ, Gordon JI, Hultgren SJ.** 2007. *Escherichia coli* from urine of female patients with urinary tract infections is competent for intracellular bacterial community formation. *Infect Immun* **75**:52–60.

35. **Barber AE, Norton JP, Wiles TJ, Mulvey MA.** 2016. Strengths and limitations of model systems for the study of urinary tract infections

and related pathologies. *Microbiol Mol Biol Rev* **80**:351–367.

36. **Barber AE, Mulvey MA.** 2014. Reply to Kaye and Sobel. *Clin Infect Dis* **58**:444–445.

37. **Kaye D, Sobel JD.** 2014. Persistence of intracellular bacteria in the urinary bladder. *Clin Infect Dis* **58**:444.

38. **Elliott TS, Reed L, Slack RC, Bishop MC.** 1985. Bacteriology and ultrastructure of the bladder in patients with urinary tract infections. *J Infect* **11**:191–199.

39. **Lakeman MM, Roovers JP.** 2016. Urinary tract infections in women with urogynaecological symptoms. *Curr Opin Infect Dis* **29**:92–97.

40. **Khasriya R, Sathiananthamoorthy S, Ismail S, Kelsey M, Wilson M, Rohn JL, Malone-Lee J.** 2013. Spectrum of bacterial colonization associated with urothelial cells from patients with chronic lower urinary tract symptoms. *J Clin Microbiol* **51**:2054–2062.

41. **Scott VC, Haake DA, Churchill BM, Justice SS, Kim JH.** 2015. Intracellular bacterial communities: a potential etiology for chronic lower urinary tract symptoms. *Urology* **86**:425–431.

42. **Cheng Y, Chen Z, Gawthorne JA, Mukerjee C, Varettas K, Mansfield KJ, Schembri MA, Moore KH.** 2016. Detection of intracellular bacteria in exfoliated urothelial cells from women with urge incontinence. *Pathog Dis* **74**.

43. **Kelley SP, Courtneidge HR, Birch RE, Contreras-Sanz A, Kelly MC, Durodie J, Peppiatt-Wildman CM, Farmer CK, Delaney MP, Malone-Lee J, Harber MA, Wildman SS.** 2014. Urinary ATP and visualization of intracellular bacteria: a superior diagnostic marker for recurrent UTI in renal transplant recipients? *Springerplus* **3**:200. doi:10.1186/2193-1801-3-200.

44. **Rosen DA, Hooton TM, Stamm WE, Humphrey PA, Hultgren SJ.** 2007. Detection of intracellular bacterial communities in human urinary tract infection. *PLoS Med* **4**:e329. doi:10.1371/journal.pmed.0040329.

45. **Robino L, Scavone P, Araujo L, Algorta G, Zunino P, Pirez MC, Vignoli R.** 2014. Intracellular bacteria in the pathogenesis of *Escherichia coli* urinary tract infection in children. *Clin Infect Dis* **59**:e158–e164.

46. **Robino L, Scavone P, Araujo L, Algorta G, Zunino P, Vignoli R.** 2013. Detection of intracellular bacterial communities in a child with *Escherichia coli* recurrent urinary tract infections. *Pathog Dis* **68**:78–81.

47. **Ikaheimo R, Siitonen A, Heiskanen T, Karkkainen U, Kuosmanen P, Lipponen P, Makela PH.** 1996. Recurrence of urinary tract infection in a primary care setting: analysis of

a 1-year follow-up of 179 women. *Clin Infect Dis* **22**:91–99.

48. **Russo TA, Stapleton A, Wenderoth S, Hooton TM, Stamm WE.** 1995. Chromosomal restriction fragment length polymorphism analysis of *Escherichia coli* strains causing recurrent urinary tract infections in young women. *J Infect Dis* **172**:440–445.

49. **Brauner A, Jacobson SH, Kuhn I.** 1992. Urinary *Escherichia coli* causing recurrent infections: a prospective follow-up of biochemical phenotypes. *Clin Nephrol* **38**:318–323.

50. **Jacobson SH, Kuhn I, Brauner A.** 1992. Biochemical fingerprinting of urinary *Escherichia coli* causing recurrent infections in women with pyelonephritic renal scarring. *Scand J Urol Nephrol* **26**:373–377.

51. **Wang H, Liang FX, Kong XP.** 2008. Characteristics of the phagocytic cup induced by uropathogenic *Escherichia coli*. *J Histochem Cytochem* **56**:597–604.

52. **Alonso A, Garcia-del Portillo F.** 2004. Hijacking of eukaryotic functions by intracellular bacterial pathogens. *Int Microbiol* **7**:181–191.

53. **Wu X-R, Sun T-T, Medina JJ.** 1996. In vitro binding of type 1-fimbriated *Escherichia coli* to uroplakins Ia and Ib: relation to urinary tract infections. *Proc Natl Acad Sci USA* **93**:9630–9635.

54. **Wankel B, Ouyang J, Guo X, Hadjiolova K, Miller J, Liao Y, Tham DK, Romih R, Andrade LR, Gumper I, Simon JP, Sachdeva R, Tolmachova T, Seabra MC, Fukuda M, Schaeren-Wiemers N, Hong WJ, Sabatini DD, Wu XR, Kong X, Kreibich G, Rindler MJ, Sun TT.** 2016. Sequential and compartmentalized action of Rabs, SNAREs, and MAL in the apical delivery of fusiform vesicles in urothelial umbrella cells. *Mol Biol Cell* **27**:1621–1634.

55. **Khandelwal P, Ruiz WG, Apodaca G.** 2010. Compensatory endocytosis in bladder umbrella cells occurs through an integrin-regulated and RhoA- and dynamin-dependent pathway. *EMBO J* **29**:1961–1975.

56. **Hicks RM.** 1975. The mammalian bladder: an accommodating organ. *Biol Rev Camb Philos Soc* **50**:1123.

57. **Min G, Stolz M, Zhou G, Liang F, Sebbel P, Stoffler D, Glockshuber R, Sun TT, Aebi U, Kong XP.** 2002. Localization of uroplakin Ia, the urothelial receptor for bacterial adhesin FimH, on the six inner domains of the 16 nm urothelial plaque particle. *J Mol Biol* **317**:697–706.

58. **Wang H, Min G, Glockshuber R, Sun T-T, Kong X-P.** 2009. Uropathogenic *E. coli* adhesin-induced host cell receptor conformation changes: implications in transmembrane signaling transduction. *J Mol Biol* **392**:352–361.

59. **Mathai JC, Zhou EH, Yu W, Kim JH, Zhou G, Liao Y, Sun TT, Fredberg JJ, Zeidel ML.** 2014. Hypercompliant apical membranes of bladder umbrella cells. *Biophys J* **107**:1273–1279.

60. **Thumbikat P, Berry RE, Zhou G, Billips BK, Yaggie RE, Zaichuk T, Sun T-T, Schaeffer AJ, Klumpp DJ.** 2009. Bacteria-induced uroplakin signaling mediates bladder response to infection. *PLoS Pathog* **5**:e1000415. doi:10.1371/journal.ppat.1000415.

61. **Baorto DM, Gao Z, Malaviya R, Dustin ML, van der Merwe A, Lublin DM, Abraham SN.** 1997. Survival of FimH-expressing enterobacteria in macrophages relies on glycolipid traffic. *Nature* **389**:636–639.

62. **Leusch HG, Drzeniek Z, Markos-Pusztai Z, Wagener C.** 1991. Binding of *Escherichia coli* and *Salmonella* strains to members of the carcinoembryonic antigen family: differential binding inhibition by aromatic alpha-glycosides of mannose. *Infect Immun* **59**:2051–2057.

63. **Sauter SL, Rutherfurd SM, Wagener C, Shiveley JE, Hefta SA.** 1991. Binding of nonspecific crossreacting antigen, a granulocyte membrane glycoprotein, to *Escherichia coli* expressing type 1 fimbriae. *Infection and Immunity* **59**:2485–2493.

64. **Gbarah A, Gahmberg CG, Ofek I, Jacobi U, Sharon N.** 1991. Identification of the leukocyte adhesion molecules CD11 and CD18 as receptors for type 1-fimbriated (mannose-specific) *Escherichia coli*. *Infect Immun* **59**:4524–4530.

65. **Pouttu R, Puustinen T, Virkola R, Hacker J, Klemm P, Korhonen TK.** 1999. Amino acid residue Ala-62 in the FimH fimbrial adhesin is critical for the adhesiveness of meningitis-associated *Escherichia coli* to collagens. *Mol Microbiol* **31**:1747–1757.

66. **Kukkonen M, Raunio T, Virkola R, Lahteenmaki K, Makela PH, Klemm P, Clegg S, Korhonen TK.** 1993. Basement membrane carbohydrate as a target for bacterial adhesion: binding of type I fimbriae of *Salmonella enterica* and *Escherichia coli* to laminin. *Mol Microbiol* **7**:229–237.

67. **Sokurenko EV, Courtney HS, Abraham SN, Klemm P, Hasty DL.** 1992. Functional heterogeneity of type 1 fimbriae of *Escherichia coli*. *InfectImmun* **60**:4709–4719.

68. **Mossman KL, Mian MF, Lauzon NM, Gyles CL, Lichty B, Mackenzie R, Gill N, Ashkar AA.** 2008. Cutting edge: FimH adhesin of type 1 fimbriae is a novel TLR4 ligand. *J Immunol* **181**:6702–6706.

69. **Hase K, Kawano K, Nochi T, Pontes GS, Fukuda S, Ebisawa M, Kadokura K, Tobe T, Fujimura Y, Kawano S, Yabashi A, Waguri S, Nakato G, Kimura S, Murakami T, Iimura M,**

Hamura K, Fukuoka S, Lowe AW, Itoh K, Kiyono H, Ohno H. 2009. Uptake through glycoprotein 2 of FimH(+) bacteria by M cells initiates mucosal immune response. *Nature* **462:**226–230.

70. Yu S, Lowe AW. 2009. The pancreatic zymogen granule membrane protein, GP2, binds *Escherichia coli* type 1 fimbriae. *BMC Gastroenterol* **9:**58. doi:10.1186/1471-230X-9-58.

71. Eto DS, Jones TA, Sundsbak JL, Mulvey MA. 2007. Integrin-mediated host cell invasion by type 1–piliated uropathogenic *Escherichia coli. PLoS Pathog* **3:**e100. doi:10.1371/journal.ppat.0030100.

72. Ielasi FS, Alioscha-Perez M, Donohue D, Claes S, Sahli H, Schols D, Willaert RG. 2016. Lectin-glycan interaction network-based identification of host receptors of microbial pathogenic adhesins. *MBio* **7:**e00584-16. doi:10.1128/mBio.00584-16.

73. Arnaout MA, Mahalingam B, Xiong JP. 2005. Integrin structure, allostery, and bidirectional signaling. *Annu Rev Cell Dev Biol* **21:**381–410.

74. Southgate J, Kennedy W, Hutton KA, Trejdosiewicz LK. 1995. Expression and *in vitro* regulation of integrins by normal human urothelial cells. *Cell Adhes Commun* **3:**231–242.

75. Kanasaki K, Yu W, von Bodungen M, Larigakis JD, Kanasaki M, Ayala de la Pena F, Kalluri R, Hill WG. 2013. Loss of beta1-integrin from urothelium results in overactive bladder and incontinence in mice: a mechanosensory rather than structural phenotype. *FASEB J* **27:**1950–1961.

76. Scibelli A, Roperto S, Manna L, Pavone LM, Tafuri S, Della Morte R, Staiano N. 2007. Engagement of integrins as a cellular route of invasion by bacterial pathogens. *Vet J* **173:**482–491.

77. Hauck CR, Borisova M, Muenzner P. 2012. Exploitation of integrin function by pathogenic microbes. *Curr Opin Cell Biol* **24:**637–644.

78. Elices MJ, Urry LA, Hemler ME. 1991. Receptor functions for the integrin VLA-3: fibronectin, collagen, and laminin binding are differentially influenced by Arg-Gly-Asp peptide and by divalent cations. *J Cell Biol* **112:**169–181.

79. Xie B, Zhou G, Chan SY, Shapiro E, Kong XP, Wu XR, Sun TT, Costello CE. 2006. Distinct glycan structures of uroplakins Ia and Ib: structural basis for the selective binding of FimH adhesin to uroplakin Ia. *J Biol Chem* **281:**14644–14653.

80. Litynska A, Przybylo M, Ksiazek D, Laidler P. 2000. Differences of alpha3beta1 integrin glycans from different human bladder cell lines. *Acta Biochim Pol* **47:**427–434.

81. Litynska A, Pochec E, Hoja-Lukowicz D, Kremser E, Laidler P, Amoresano A, Monti C. 2002. The structure of the oligosaccharides of alpha3beta1 integrin from human ureter epithelium (HCV29) cell line. *Acta Biochim Pol* **49:**491–500.

82. Eto DS, Gordon HB, Dhakal BK, Jones TA, Mulvey MA. 2008. Clathrin, AP-2, and the NPXY-binding subset of alternate endocytic adaptors facilitate FimH-mediated bacterial invasion of host cells. *Cell Microbiol* **10:**2553–2567.

83. Martinez JJ, Hultgren SJ. 2002. Requirement of Rho-family GTPases in the invasion of type 1-piliated uropathogenic *Escherichia coli. Cell Microbiol* **4:**19–28.

84. Shen XF, Teng Y, Sha KH, Wang XY, Yang XL, Guo XJ, Ren LB, Wang XY, Li J, Huang N. 2016. Dietary flavonoid luteolin attenuates uropathogenic *Escherichia coli* invasion of the urinary bladder. *Biofactors* [Epub ahead of print.] doi:10.1002/biof.1314.

85. Lewis AJ, Dhakal BK, Liu T, Mulvey MA. 2016. Histone deacetylase 6 regulates bladder architecture and host susceptibility to uropathogenic *Escherichia coli. Pathogens* **5:**E20. doi:10.3390/pathogens5010020.

86. Dhakal BK, Mulvey MA. 2009. Uropathogenic *Escherichia coli* invades host cells via an HDAC6-modulated microtubule-dependent pathway. *J Biol Chem* **284:**446–454.

87. Braun V, Niedergang F. 2006. Linking exocytosis and endocytosis during phagocytosis. *Biol Cell* **98:**195–201.

88. Swanson JA. 2008. Shaping cups into phagosomes and macropinosomes. *Nat Rev Mol Cell Biol* **9:**639–649.

89. Bishop BL, Duncan MJ, Song J, Li G, Zaas D, Abraham SN. 2007. Cyclic AMP-regulated exocytosis of *Escherichia coli* from infected bladder epithelial cells. *Nat Med* **13:**625–630.

90. Song J, Bishop BL, Li G, Grady R, Stapleton A, Abraham SN. 2009. TLR4-mediated expulsion of bacteria from infected bladder epithelial cells. *Proc Natl Acad Sci USA* **106:**14966–14971.

91. Guo X, Tu L, Gumper I, Plesken H, Novak EK, Chintala S, Swank RT, Pastores G, Torres P, Izumi T, Sun TT, Sabatini DD, Kreibich G. 2009. Involvement of vps33a in the fusion of uroplakin-degrading multivesicular bodies with lysosomes. *Traffic* **10:**1350–1361.

92. Wang Z, Humphrey C, Frilot N, Wang G, Nie Z, Moniri NH, Daaka Y. 2011. Dynamin2- and endothelial nitric oxide synthase-regulated invasion of bladder epithelial cells by uropathogenic *Escherichia coli. J Cell Biol* **192:**101–110.

93. Wang G, Moniri NH, Ozawa K, Stamler JS, Daaka Y. 2006. Nitric oxide regulates endocytosis by S-nitrosylation of dynamin. *Proc Natl Acad Sci USA* **103:**1295–1300.

94. **Lundberg JO, Ehren I, Jansson O, Adolfsson J, Lundberg JM, Weitzberg E, Alving K, Wiklund NP.** 1996. Elevated nitric oxide in the urinary bladder in infectious and noninfectious cystitis. *Urology* 48:700–702.

95. **Svensson L, Marklund BI, Poljakovic M, Persson K.** 2006. Uropathogenic *Escherichia coli* and tolerance to nitric oxide: the role of flavohemoglobin. *J Urol* 175:749–753.

96. **Poljakovic M, Svensson ML, Svanborg C, Johansson K, Larsson B, Persson K.** 2001. *Escherichia coli*-induced inducible nitric oxide synthase and cyclooxygenase expression in the mouse bladder and kidney. *Kidney Int* 59:893–904.

97. **Bower JM, Gordon-Raagas HB, Mulvey MA.** 2009. Conditioning of uropathogenic *Escherichia coli* for enhanced colonization of host. *Infect Immun* 77:2104–2112.

98. **Bower JM, Mulvey MA.** 2006. Polyamine-mediated resistance of uropathogenic *Escherichia coli* to nitrosative stress. *J Bacteriol* 188:928–933.

99. **Li K, Feito MJ, Sacks SH, Sheerin NS.** 2006. CD46 (membrane cofactor protein) acts as a human epithelial cell receptor for internalization of opsonized uropathogenic *Escherichia coli*. *J Immunol* 177:2543–2551.

100. **Li K, Zhou W, Hong Y, Sacks SH, Sheerin NS.** 2009. Synergy between type 1 fimbriae expression and C3 opsonisation increases internalisation of *E. coli* by human tubular epithelial cells. *BMC Microbiol* 9:64. doi:10.1186/1471-2180-9-64.

101. **He XL, Wang Q, Peng L, Qu YR, Puthiyakunnon S, Liu XL, Hui CY, Boddu S, Cao H, Huang SH.** 2015. Role of uropathogenic *Escherichia coli* outer membrane protein T in pathogenesis of urinary tract infection. *Pathog Dis* 73:ftv006. doi:10.1093/femspd/ftv006.

102. **Kakkanat A, Totsika M, Schaale K, Duell BL, Lo AW, Phan MD, Moriel DG, Beatson SA, Sweet MJ, Ulett GC, Schembri MA.** 2015. The role of H4 flagella in *Escherichia coli* ST131 virulence. *Sci Rep* 5:16149. doi:10.1038/srep16149.

103. **Saldana Z, De la Cruz MA, Carrillo-Casas EM, Duran L, Zhang Y, Hernandez-Castro R, Puente JL, Daaka Y, Giron JA.** 2014. Production of the *Escherichia coli* common pilus by uropathogenic *E. coli* is associated with adherence to HeLa and HTB-4 cells and invasion of mouse bladder urothelium. *PLoS One* 9:e101200. doi:10.1371/journal.pone.0101200.

104. **Visvikis O, Boyer L, Torrino S, Doye A, Lemonnier M, Lores P, Rolando M, Flatau G, Mettouchi A, Bouvard D, Veiga E, Gacon G, Cossart P, Lemichez E.** 2011. *Escherichia coli* producing CNF1 toxin hijacks Tollip to trigger Rac1-dependent cell invasion. *Traffic* 12:579–590.

105. **Doye A, Mettouchi A, Bossis G, Clement R, Buisson-Touati C, Flatau G, Gagnoux L, Piechaczyk M, Boquet P, Lemichez E.** 2002. CNF1 exploits the ubiquitin-proteasome machinery to restrict Rho GTPase activation for bacterial host cell invasion. *Cell* 111:553–564.

106. **Feldmann F, Sorsa LJ, Hildinger K, Schubert S.** 2007. The salmochelin siderophore receptor IroN contributes to invasion of urothelial cells by extraintestinal pathogenic *Escherichia coli in vitro*. *Infect Immun* 75:3183–3187.

107. **Rana T, Hasan RJ, Nowicki S, Venkatarajan MS, Singh R, Urvil PT, Popov V, Braun WA, Popik W, Goodwin JS, Nowicki BJ.** 2014. Complement protective epitopes and CD55-microtubule complexes facilitate the invasion and intracellular persistence of uropathogenic *Escherichia coli*. *J Infect Dis* 209:1066–1076.

108. **Goluszko P, Popov V, Selvarangan R, Nowicki S, Pham T, Nowicki BJ.** 1997. Dr fimbriae operon of uropathogenic *Escherichia coli* mediate microtubule-dependent invasion to the HeLa epithelial cell line. *J Infect Dis* 176:158–167.

109. **Das M, Hart-Van Tassell A, Urvil PT, Lea S, Pettigrew D, Anderson KL, Samet A, Kur J, Matthews S, Nowicki S, Popov V, Goluszko P, Nowicki BJ.** 2005. Hydrophilic domain II of *Escherichia coli* Dr fimbriae facilitates cell invasion. *Infect Immun* 73:6119–6126.

110. **Servin AL.** 2014. Pathogenesis of human diffusely adhering *Escherichia coli* expressing Afa/Dr adhesins (Afa/Dr DAEC): current insights and future challenges. *Clin Microbiol Rev* 27:823–869.

111. **Wang C, Mendonsa GR, Symington JW, Zhang Q, Cadwell K, Virgin HW, Mysorekar IU.** 2012. Atg16L1 deficiency confers protection from uropathogenic *Escherichia coli* infection *in vivo*. *Proc Natl Acad Sci USA* 109:11008–11013.

112. **Miao Y, Li G, Zhang X, Xu H, Abraham SN.** 2015. A TRP channel senses lysosome neutralization by pathogens to trigger their expulsion. *Cell* 161:1306–1319.

113. **Dikshit N, Bist P, Fenlon SN, Pulloor NK, Chua CE, Scidmore MA, Carlyon JA, Tang BL, Chen SL, Sukumaran B.** 2015. Intracellular uropathogenic *E. coli* exploits host Rab35 for iron acquisition and survival within urinary bladder cells. *PLoS Pathog* 11:e1005083. doi:10.1371/journal.ppat.1005083.

114. **Berry RE, Klumpp DJ, Schaeffer AJ.** 2009. Urothelial cultures support intracellular bacterial community formation by uropathogenic *Escherichia coli*. *Infect Immun* 77:2762–2772.

115. **Romih R, Veranic P, Jezernik K.** 1999. Actin filaments during terminal differentiation of urothelial cells in the rat urinary bladder. *Histochem Cell Biol* **112:**375–380.

116. **Fleming BA, Mulvey MA.** 2016. Toxin-antitoxin systems as regulators of bacterial fitness and virulence, p 437–445. *In* de Bruijn FJ (ed), *Stress and Environmental Regulation of Gene Expression and Adaptation in Bacteria.* John Wiley & Sons, Hoboken, NJ.

117. **Norton JP, Mulvey MA.** 2012. Toxin-antitoxin systems are important for niche-specific colonization and stress resistance of uropathogenic *Escherichia coli. PLoS Pathog* **8:**e1002954. doi:10.1371/journal.ppat.1002954.

118. **Fiedoruk K, Daniluk T, Swiecicka I, Sciepuk M, Leszczynska K.** 2015. Type II toxin-antitoxin systems are unevenly distributed among *Escherichia coli* phylogroups. *Microbiology* **161:**158–167.

119. **Chole RA, Faddis BT.** 2002. Evidence for microbial biofilms in cholesteatomas. *Arch Otolaryngol Head Neck Surg* **128:**1129–1133.

120. **Justice SS, Lauer SR, Hultgren SJ, Hunstad DA.** 2006. Maturation of intracellular *Escherichia coli* communities requires SurA. *Infect Immun* **74:**4793–4800.

121. **Anderson GG, Goller CC, Justice S, Hultgren SJ, Seed PC.** 2010. Polysaccharide capsule and sialic acid-mediated regulation promote biofilm-like intracellular bacterial communities during cystitis. *Infect Immun* **78:**963–975.

122. **Wright KJ, Seed PC, Hultgren SJ.** 2007. Development of intracellular bacterial communities of uropathogenic *Escherichia coli* depends on type 1 pili. *Cell Microbiol* **9:**2230–2241.

123. **Goller CC, Seed PC.** 2010. Revisiting the *Escherichia coli* polysaccharide capsule as a virulence factor during urinary tract infection: contribution to intracellular biofilm development. *Virulence* **1:**333–337.

124. **Nicholson TF, Watts KM, Hunstad DA.** 2009. OmpA of uropathogenic *Escherichia coli* promotes postinvasion pathogenesis of cystitis. *Infect Immun* **77:**5245–5251.

125. **Justice SS, Li B, Downey JS, Dabdoub SM, Brockson ME, Probst GD, Ray WC, Goodman SD.** 2012. Aberrant community architecture and attenuated persistence of uropathogenic *Escherichia coli* in the absence of individual IHF subunits. *PLoS One* **7:**e48349.

126. **Hadjifrangiskou M, Kostakioti M, Chen SL, Henderson JP, Greene SE, Hultgren SJ.** 2011. A central metabolic circuit controlled by QseC in pathogenic *Escherichia coli. Mol Microbiol* **80:**1516–1529.

127. **Kulesus RR, Diaz-Perez K, Slechta ES, Eto DS, Mulvey MA.** 2008. Impact of the RNA chaperone Hfq on the fitness and virulence potential of uropathogenic *Escherichia coli. Infect Immun* **76:**3019–3026.

128. **Shaffer CL, Zhang EW, Dudley AG, Dixon BR, Guckes KR, Breland EJ, Floyd KA, Casella DP, Algood HM, Clayton DB, Hadjifrangiskou M.** 2016. Purine biosynthesis metabolically constrains intracellular survival of uropathogenic *E. coli. Infect Immun.* [Epub ahead of print.] doi:10.1128/IAI.00471-16.

129. **Conover MS, Hadjifrangiskou M, Palermo JJ, Hibbing ME, Dodson KW, Hultgren SJ.** 2016. Metabolic requirements of *Escherichia coli* in intracellular bacterial communities during urinary tract infection pathogenesis. *MBio* **7:**e00104-16. doi:10.1128/mBio.00104-16.

130. **Reigstad CS, Hultgren SJ, Gordon JI.** 2007. Functional genomic studies of uropathogenic *Escherichia coli* and host urothelial cells when intracellular bacterial communities are assembled. *J Biol Chem* **282:**21259–21267.

131. **Kurimura Y, Nishitani C, Ariki S, Saito A, Hasegawa Y, Takahashi M, Hashimoto J, Takahashi S, Tsukamoto T, Kuroki Y.** 2012. Surfactant protein D inhibits adherence of uropathogenic *Escherichia coli* to the bladder epithelial cells and the bacterium-induced cytotoxicity: a possible function in urinary tract. *J Biol Chem* **287:**39578–39588.

132. **Zasloff M.** 2007. Antimicrobial peptides, innate immunity, and the normally sterile urinary tract. *J Am Soc Nephrol* **18:**2810–2816.

133. **Corthesy B.** 2010. Role of secretory immunoglobulin A and secretory component in the protection of mucosal surfaces. *Future Microbiol* **5:**817–829.

134. **Spencer JD, Schwaderer AL, Wang H, Bartz J, Kline J, Eichler T, DeSouza KR, Sims-Lucas S, Baker P, Hains DS.** 2013. Ribonuclease 7, an antimicrobial peptide upregulated during infection, contributes to microbial defense of the human urinary tract. *Kidney Int* **83:**615–625.

135. **Bates JM, Raffi HM, Prasadan K, Mascarenhas R, Laszik Z, Maeda N, Hultgren SJ, Kumar S.** 2004. Tamm-Horsfall protein knockout mice are more prone to urinary tract infection: rapid communication. *Kidney Int* **65:**791–797.

136. **Mo L, Zhu XH, Huang HY, Shapiro E, Hasty DL, Wu XR.** 2004. Ablation of the Tamm-Horsfall protein gene increases susceptibility of mice to bladder colonization by type 1-fimbriated *Escherichia coli. Am J Physiol Renal Physiol* **286:**F795–F802.

137. **Wood MW, Breitschwerdt EB, Nordone SK, Linder KE, Gookin JL.** 2012. Uropathogenic

E. coli promote a paracellular urothelial barrier defect characterized by altered tight junction integrity, epithelial cell sloughing and cytokine release. *J Comp Pathol* **147**:11–19.

138. **Klumpp DJ, Rycyk MT, Chen MC, Thumbikat P, Sengupta S, Schaeffer AJ.** 2006. Uropathogenic *Escherichia coli* induces extrinsic and intrinsic cascades to initiate urothelial apoptosis. *Infect Immun* **74**:5106–5113.

139. **Veranic P, Jezernik K.** 2001. Succession of events in desquamation of superficial urothelial cells as a response to stress induced by prolonged constant illumination. *Tissue Cell* **33**:280–285.

140. **Jezernik K, Sterle M, Batista U.** 1997. The distinct steps of cell detachment during development of mouse uroepithelial cells in the bladder. *Cell Biol Int* **21**:1–6.

141. **Aronson M, Medalia O, Amichay D, Nativ O.** 1988. Endotoxin-induced shedding of viable uroepithelial cells is an antimicrobial defense mechanism. *Infect Immun* **56**:1615–1617.

142. **Wiles TJ, Dhakal BK, Eto DS, Mulvey MA.** 2008. Inactivation of host Akt/protein kinase B signaling by bacterial pore-forming toxins. *Mol Biol Cell* **19**:1427–1438.

143. **Smith YC, Grande KK, Rasmussen SB, O'Brien AD.** 2006. Novel three-dimensional organoid model for evaluation of the interaction of uropathogenic *Escherichia coli* with terminally differentiated human urothelial cells. *Infect Immun* **74**:750–757.

144. **Smith YC, Rasmussen SB, Grande KK, Conran RM, O'Brien AD.** 2008. Hemolysin of uropathogenic *Escherichia coli* evokes extensive shedding of the uroepithelium and hemorrhage in bladder tissue within the first 24 hours after intraurethral inoculation of mice. *Infect Immun* **76**:2978–2990.

145. **Dhakal BK, Mulvey MA.** 2012. The UPEC pore-forming toxin alpha-hemolysin triggers proteolysis of host proteins to disrupt cell adhesion, inflammatory, and survival pathways. *Cell Host Microbe* **11**:58–69.

146. **Mills M, Meysick KC, O'Brien AD.** 2000. Cytotoxic necrotizing factor type 1 of uropathogenic *Escherichia coli* kills cultured human uroepithelial 5637 cells by an apoptotic mechanism. *Infect Immun* **68**:5869–5880.

147. **Mysorekar IU, Isaacson-Schmid M, Walker JN, Mills JC, Hultgren SJ.** 2009. Bone morphogenetic protein 4 signaling regulates epithelial renewal in the urinary tract in response to uropathogenic infection. *Cell Host Microbe* **5**:463–475.

148. **Mysorekar IU, Mulvey MA, Hultgren SJ, Gordon JI.** 2002. Molecular regulation of urothelial renewal and host defenses during

infection with uropathogenic *Escherichia coli*. *J Biol Chem* **277**:7412–7419.

149. **Veranic P, Erman A, Kerec-Kos M, Bogataj M, Mrhar A, Jezernik K.** 2009. Rapid differentiation of superficial urothelial cells after chitosan-induced desquamation. *Histochem Cell Biol* **131**:129–139.

150. **Shin K, Lee J, Guo N, Kim J, Lim A, Qu L, Mysorekar IU, Beachy PA.** 2011. Hedgehog/Wnt feedback supports regenerative proliferation of epithelial stem cells in bladder. *Nature* **472**:110–114.

151. **Lin AE, Beasley FC, Olson J, Keller N, Shalwitz RA, Hannan TJ, Hultgren SJ, Nizet V.** 2015. Role of hypoxia inducible factor-1α (HIF-1α) in innate defense against uropathogenic *Escherichia coli* infection. *PLoS Pathog* **11**:e1004818. doi:10.1371/journal.ppat.1004818.

152. **Chromek M, Slamova Z, Bergman P, Kovacs L, Podracka L, Ehren I, Hokfelt T, Gudmundsson GH, Gallo RL, Agerberth B, Brauner A.** 2006. The antimicrobial peptide cathelicidin protects the urinary tract against invasive bacterial infection. *Nat Med* **12**:636–641.

153. **Danka ES, Hunstad DA.** 2015. Cathelicidin augments epithelial receptivity and pathogenesis in experimental *Escherichia coli* cystitis. *J Infect Dis* **211**:1164–1173.

154. **Miao Y, Wu J, Abraham SN.** 2016. Ubiquitination of innate immune regulator TRAF3 orchestrates expulsion of intracellular bacteria by exocyst complex. *Immunity* **45**:94–105.

155. **Khandige S, Asferg CA, Rasmussen KJ, Larsen MJ, Overgaard M, Andersen TE, Moller-Jensen J.** 2016. DamX controls reversible cell morphology switching in uropathogenic *Escherichia coli*. *MBio* **7**:e00642-16. doi:10.1128/mBio.00642-16.

156. **Horvath DJ Jr, Li B, Casper T, Partida-Sanchez S, Hunstad DA, Hultgren SJ, Justice SS.** 2011. Morphological plasticity promotes resistance to phagocyte killing of uropathogenic *Escherichia coli*. *Microbes Infect* **13**:426–437.

157. **Nazareth H, Genagon SA, Russo TA.** 2007. Extraintestinal pathogenic *Escherichia coli* survives within neutrophils. *Infect Immun* **75**:2776–2785.

158. **Bokil NJ, Totsika M, Carey AJ, Stacey KJ, Hancock V, Saunders BM, Ravasi T, Ulett GC, Schembri MA, Sweet MJ.** 2011. Intramacrophage survival of uropathogenic *Escherichia coli*: differences between diverse clinical isolates and between mouse and human macrophages. *Immunobiology* **216**:1164–1171.

159. **Shepherd M, Achard ME, Idris A, Totsika M, Phan MD, Peters KM, Sarkar S, Ribeiro CA, Holyoake LV, Ladakis D, Ulett GC, Sweet MJ, Poole RK, McEwan AG, Schembri MA.**

2016. The cytochrome bd-I respiratory oxidase augments survival of multidrug-resistant *Escherichia coli* during infection. *Sci Rep* 6:35285.

160. **Mavromatis CH, Bokil NJ, Totsika M, Kakkanat A, Schaale K, Cannistraci CV, Ryu T, Beatson SA, Ulett GC, Schembri MA, Sweet MJ, Ravasi T.** 2015. The co-transcriptome of uropathogenic *Escherichia coli*-infected mouse macrophages reveals new insights into host-pathogen interactions. *Cell Microbiol* 17:730–746.

161. **Qualman SJ, Gupta PK, Mendelsohn G.** 1984. Intracellular *Escherichia coli* in urinary malakoplakia: a reservoir of infection and its therapeutic implications. *Am J Clin Pathol* 81: 35–42.

162. **Maderazo EG, Berlin BB, Morhardt C.** 1979. Treatment of malakoplakia with trimethoprim-sulfamethoxazole. *Urology* 13:70–73.

163. **Stanton MJ, Maxted W.** 1981. Malacoplakia: a study of the literature and current concepts of pathogenesis, diagnosis and treatment. *J Urol* 125:139–146.

164. **Mora-Bau G, Platt AM, van Rooijen N, Randolph GJ, Albert ML, Ingersoll MA.** 2015. Macrophages subvert adaptive immunity to urinary tract infection. *PLoS Pathog* 11: e1005044. doi:10.1371/journal.ppat.1005044.

165. **Palmer LM, Reilly TJ, Utsalo SJ, Donnenberg MS.** 1997. Internalization of *Escherichia coli* by human renal epithelial cells is associated with tyrosine phosphorylation of specific host cell proteins. *Infect Immun* 65:2570–2575.

166. **Warren JW, Mobley HL, Trifillis AL.** 1988. Internalization of *Escherichia coli* into human renal tubular epithelial cells. *J Infect Dis* 158: 221–223.

167. **Donnenberg MS, Newman B, Utsalo SJ, Trifillis AL, Hebel JR, Warren JW.** 1994. Internalization of *Escherichia coli* into human kidney epithelial cells: comparison of fecal and pyelonephritis-associated strains. *J Infect Dis* 169:831–838.

168. **Chassin C, Vimont S, Cluzeaud F, Bens M, Goujon JM, Fernandez B, Hertig A, Rondeau E, Arlet G, Hornef MW, Vandewalle A.** 2008. TLR4 facilitates translocation of bacteria across renal collecting duct cells. *J Am Soc Nephrol* 19: 2364–2374.

169. **Chassin C, Tourneur E, Bens M, Vandewalle A.** 2011. A role for collecting duct epithelial cells in renal antibacterial defences. *Cell Microbiol* 13:1107–1113.

170. **Szemiako K, Krawczyk B, Samet A, Sledzinska A, Nowicki B, Nowicki S, Kur J.** 2013. A subset of two adherence systems, acute pro-inflammatory pap genes and invasion coding dra, fim, or sfa, increases the risk of *Escherichia coli* translocation to the bloodstream. *Eur J Clin Microbiol Infect Dis* 32:1579–1582.

171. **Springall T, Sheerin NS, Abe K, Holers VM, Wan H, Sacks SH.** 2001. Epithelial secretion of C3 promotes colonization of the upper urinary tract by *Escherichia coli*. *Nat Med* 7:801–806.

172. **Choudhry N, Li K, Zhang T, Wu KY, Song Y, Farrar CA, Wang N, Liu CF, Peng Q, Wu W, Sacks SH, Zhou W.** 2016. The complement factor 5a receptor 1 has a pathogenic role in chronic inflammation and renal fibrosis in a murine model of chronic pyelonephritis. *Kidney Int* 90:540–554.

173. **Pichon C, Hechard C, du Merle L, Chaudray C, Bonne I, Guadagnini S, Vandewalle A, Le Bouguenec C.** 2009. Uropathogenic *Escherichia coli* AL511 requires flagellum to enter renal collecting duct cells. *Cell Microbiol* 11:616–628.

174. **Bens M, Vimont S, Ben Mkaddem S, Chassin C, Goujon JM, Balloy V, Chignard M, Werts C, Vandewalle A.** 2014. Flagellin/TLR5 signalling activates renal collecting duct cells and facilitates invasion and cellular translocation of uropathogenic *Escherichia coli*. *Cell Microbiol* 16:1503–1517.

175. **Szabados F, Kleine B, Anders A, Kaase M, Sakinc T, Schmitz I, Gatermann S.** 2008. *Staphylococcus saprophyticus* ATCC 15305 is internalized into human urinary bladder carcinoma cell line 5637. *FEMS Microbiol Lett* 285:163–169.

176. **Leclercq SY, Sullivan MJ, Ipe DS, Smith JP, Cripps AW, Ulett GC.** 2016. Pathogenesis of *Streptococcus* urinary tract infection depends on bacterial strain and beta-hemolysin/cytolysin that mediates cytotoxicity, cytokine synthesis, inflammation and virulence. *Sci Rep* 6:29000. doi:10.1038/srep29000.

177. **Horsley H, Malone-Lee J, Holland D, Tuz M, Hibbert A, Kelsey M, Kupelian A, Rohn JL.** 2013. Enterococcus faecalis subverts and invades the host urothelium in patients with chronic urinary tract infection. *PLoS One* 8: e83637. doi:10.1371/journal.pone.0083637.

178. **Chippendale GR, Warren JW, Trifillis AL, Mobley HL.** 1994. Internalization of *Proteus mirabilis* by human renal epithelial cells. *Infect Immun* 62:3115–3121.

179. **Schaffer JN, Norsworthy AN, Sun TT, Pearson MM.** 2016. Proteus mirabilis fimbriae-and urease-dependent clusters assemble in an extracellular niche to initiate bladder stone formation. *Proc Natl Acad Sci USA* 113:4494–4499.

180. **Alamuri P, Lower M, Hiss JA, Himpsl SD, Schneider G, Mobley HL.** 2010. Adhesion,

invasion, and agglutination mediated by two trimeric autotransporters in the human uropathogen *Proteus mirabilis*. *Infect Immun* **78**:4882–4894.

181. Liu MC, Kuo KT, Chien HF, Tsai YL, Liaw SJ. 2015. New aspects of RpoE in uropathogenic *Proteus mirabilis*. *Infect Immun* **83**:966–977.

182. Kurihara S, Sakai Y, Suzuki H, Muth A, Phanstiel Ot, Rather PN. 2013. Putrescine importer PlaP contributes to swarming motility and urothelial cell invasion in *Proteus mirabilis*. *J Biol Chem* **288**:15668–15676.

183. Allison C, Coleman N, Jones PL, Hughes C. 1992. Ability of *Proteus mirabilis* to invade human urothelial cells is coupled to motility and swarming differentiation. *Infect Immun* **60**:4740–4746.

184. Oelschlaeger TA, Tall BD. 1997. Invasion of cultured human epithelial cells by *Klebsiella pneumoniae* isolated from the urinary tract. *Infect Immun* **65**:2950–2958.

185. Rosen DA, Pinkner JS, Walker JN, Elam JS, Jones JM, Hultgren SJ. 2008. Molecular variations in *Klebsiella pneumoniae* and *Escherichia coli* FimH affect function and pathogenesis in the urinary tract. *Infect Immun* **76**:3346–3356.

186. Fumagalli O, Tall BD, Schipper C, Oelschlaeger TA. 1997. N-glycosylated proteins are involved in efficient internalization of *Klebsiella pneumoniae* by cultured human epithelial cells. *Infect Immun* **65**:4445–4451.

187. Rosen DA, Pinkner JS, Jones JM, Walker JN, Clegg S, Hultgren SJ. 2008. Utilization of an intracellular bacterial community pathway in *Klebsiella pneumoniae* urinary tract infection and the effects of FimK on type 1 pilus expression. *Infect Immun* **76**:3337–3345.

188. Croxall G, Weston V, Joseph S, Manning G, Cheetham P, McNally A. 2011. Increased human pathogenic potential of *Escherichia coli* from polymicrobial urinary tract infections in comparison to isolates from monomicrobial culture samples. *J Med Microbiol* **60**:102–109.

189. Mydock-McGrane L, Cusumano Z, Han Z, Binkley J, Kostakioti M, Hannan T, Pinkner JS, Klein RD, Kalas V, Crowley J, Rath NP, Hultgren SJ, Janetka JW. 2016. Anti-virulence C-mannosides as antibiotic-sparing, oral therapeutics for urinary tract infections. *J Med Chem* **59**:9390–9408.

190. Greene SE, Pinkner JS, Chorell E, Dodson KW, Shaffer CL, Conover MS, Livny J, Hadjifrangiskou M, Almqvist F, Hultgren SJ. 2014. Pilicide ec240 disrupts virulence circuits in uropathogenic *Escherichia coli*. *MBio* **5**:e02038. doi:10.1128/mBio.02038-14.

191. Chahales P, Hoffman PS, Thanassi DG. 2016. Nitazoxanide inhibits pilus biogenesis by interfering with folding of the usher protein in the outer membrane. *Antimicrob Agents Chemother* **60**:2028–2038.

192. Maki KC, Kaspar KL, Khoo C, Derrig LH, Schild AL, Gupta K. 2016. Consumption of a cranberry juice beverage lowered the number of clinical urinary tract infection episodes in women with a recent history of urinary tract infection. *Am J Clin Nutr* **103**:1434–1442.

193. Rafsanjany N, Senker J, Brandt S, Dobrindt U, Hensel A. 2015. *In vivo* consumption of cranberry exerts *ex vivo* antiadhesive activity against fimh-dominated uropathogenic *Escherichia coli*: a combined *in vivo*, *ex vivo*, and *in vitro* study of an extract from vaccinium macrocarpon. *J Agric Food Chem* **63**:8804–8818.

194. Hotchkiss AT Jr, Nunez A, Strahan GD, Chau HK, White AK, Marais JP, Hom K, Vakkalanka MS, Di R, Yam KL, Khoo C. 2015. Cranberry xyloglucan structure and inhibition of Escherichia coli adhesion to epithelial cells. *J Agric Food Chem* **63**:5622–5633.

195. Vollmerhausen TL, Ramos NL, Dzung DT, Brauner A. 2013. Decoctions from *Citrus reticulata* blanco seeds protect the uroepithelium against *Escherichia coli* invasion. *J Ethnopharmacol* **150**:770–774.

196. Erman A, Kerec Kos M, Zakelj S, Resnik N, Romih R, Veranic P. 2013. Correlative study of functional and structural regeneration of urothelium after chitosan-induced injury. *Histochem Cell Biol* **140**:521–531.

197. Wagers PO, Tiemann KM, Shelton KL, Kofron WG, Panzner MJ, Wooley KL, Youngs WJ, Hunstad DA. 2015. Imidazolium salts as small-molecule urinary bladder exfoliants in a murine model. *Antimicrob Agents Chemother* **59**:5494–5502.

198. Szklarczyk D, Franceschini A, Wyder S, Forslund K, Heller D, Huerta-Cepas J, Simonovic M, Roth A, Santos A, Tsafou KP, Kuhn M, Bork P, Jensen LJ, von Mering C. 2015. STRING v10: protein-protein interaction networks, integrated over the tree of life. *Nucleic Acids Res* **43**:D447–452.

Proteus mirabilis and Urinary Tract Infections

17

JESSICA N. SCHAFFER[1] and MELANIE M. PEARSON[1]

INTRODUCTION

Proteus mirabilis is well known in clinical laboratories and microbiology survey courses as the species that swarms across agar surfaces, overtaking any other species present in the process. Urease production and robust swarming motility are the two hallmarks of this organism. This species can be identified as a Gram-negative rod that is motile, urease-positive, lactose-negative, indole-negative, and produces hydrogen sulfide (1). It is a member of the same bacterial family (*Enterobacteriaceae*) as *E. coli*.

DISEASE

P. mirabilis is capable of causing symptomatic infections of the urinary tract including cystitis and pyelonephritis and is present in cases of asymptomatic bacteriuria, particularly in the elderly and patients with type 2 diabetes (2, 3). These infections can also cause bacteremia and progress to potentially life-threatening urosepsis. Additionally, *P. mirabilis* infections can cause the formation of urinary stones (urolithiasis).

[1]Department of Microbiology, New York University Langone Medical Center, New York, NY 10016
Urinary Tract Infections: Molecular Pathogenesis and Clinical Management, 2nd Edition
Edited by Matthew A. Mulvey, David J. Klumpp, and Ann E. Stapleton
© 2017 American Society for Microbiology, Washington, DC
doi:10.1128/microbiolspec.UTI-0017-2013

P. mirabilis is often isolated from the gastrointestinal tract, although whether it is a commensal, a pathogen, or a transient organism, is somewhat controversial (4). It is thought that the majority of *P. mirabilis* urinary tract infections (UTI) result from ascension of bacteria from the gastrointestinal tract while others are due to person-to-person transmission, particularly in healthcare settings (1). This is supported by evidence that some patients with *P. mirabilis* UTI have the same strain of *P. mirabilis* in their stool, while others have no *P. mirabilis* in their stools (5). In addition to UTI, this species can also cause infection in the respiratory tract, eye, ear, nose, skin, throat, burns, and wounds and has been implicated in neonatal meningoencephalitis, empyema, and osteomyelitis (1, 6). Several studies have linked *P. mirabilis* to rheumatoid arthritis, although others have failed to find an association (7, 8). It is thought that antibodies against hemolysin and urease enzymes are subsequently able to recognize self-antigens targeted in rheumatoid arthritis patients (8).

INCIDENCE

P. mirabilis causes between 1% and 10% of all UTIs, varying with the geographic location of the study, the types of samples collected, and the characteristics of the patients examined. In a recent large North American study, this species caused 4% of almost 3,000 UTI cases (9). In 2006, UTIs in the United States were the cause of 11 million physician visits and cost $3.5 billion dollars (10). This organism is more common in complicated UTIs (including patients with spinal cord injury or anatomical abnormality) and especially contributes to catheter-associated UTI (CAUTI), causing from 10% to 44% of long-term CAUTIs at a cost of $43 million to $256 million in the United States annually (6, 11, 12). The wide range of *P. mirabilis* CAUTI likely reflects differences in the population surveyed and the types of samples collected.

The highest incidence of *P. mirabilis* CAUTI occurs in elderly patients during long-term catheterization. *P. mirabilis* is also a common agent of Gram-negative bacteremia, particularly in patients with concurrent UTI; in recent studies, this species was found in between 5% and 20% of these cases and as high as a 50% mortality rate in geriatric patients (13–16).

VIRULENCE FACTORS

P. mirabilis virulence has primarily been tested using mouse or rat models of infection. Two models of ascending UTI are employed. Independent-challenge and co-challenge experiments insert bacteria directly into the bladder using a urethral catheter. In an independent challenge, each strain is tested for the ability to cause infection in the absence of other bacteria. In a co-challenge experiment, two different strains of bacteria are mixed prior to catheterization and must compete to colonize the urinary tract. A third model investigates a hematogenous route of infection, in which bacteria are injected intravenously and the ability of the bacteria to colonize the kidneys is examined.

Throughout this section, genes will be referenced by their PMI gene designations in the sequenced and annotated *P. mirabilis* genome, strain HI4320 (17), although other wild-type isolates have been studied.

Urease

The cytoplasmic nickel metalloenzyme urease acts by hydrolyzing urea into ammonia and carbon dioxide. The resulting ammonia is the preferred nitrogen source for many species of bacteria, and may be assimilated into biomolecules via glutamine synthetase (GlnA) or glutamate dehydrogenase (GdhA). Mobley provides further general information on urease (18) and *P. mirabilis* urease (19). A direct result of urease activity and ammonia generation is an increase in local pH. In the urinary tract alkaline pH leads to precipitation of calcium and magnesium ions and the formation of urinary stones composed of magnesium ammonium phosphate (struvite) and calcium phosphate (apatite) (20). In *P. mirabilis*, urease is encoded by the *ureDABCEFG* operon (PMI3682-88). A regulator, UreR (PMI3681), is encoded on the reverse strand adjacent to this operon (21). UreR is an AraC-type DNA binding protein which positively induces urease expression in the presence of urea; a *ureR* mutant lacks measurable urease activity (22). UreR also positively regulates its own expression when bound to the intergenic region between *ureR* and *ureD*, and *ureR* transcription is repressed by the global regulator H-NS (23). Urease expression is induced in the urinary tract of experimentally-infected mice (24, 25).

Virulence

Urease mediates virulence via the production of urinary stones. These stones can block urinary flow and cause tissue damage; they can also become quite large (>1 cm^2) (26, 27). The precipitated minerals may mix with bacteria adherent to a urinary catheter, forming a crystalline biofilm and eventually blocking urine flow through the catheter (28, 29). Similar intracellular crystals have been visualized in cultured urinary epithelial cells that have been experimentally infected with invasive *P. mirabilis* (30). Although other urease-positive bacterial species are associated with catheter-associated UTI, only *P. mirabilis* has a positive association with catheter obstruction (31). Bacteria can become embedded in these stones, which may protect pathogens from antibiotics or the immune system (32) (Fig. 1). Furthermore, urinary stones can act as a focal point for other species of bacteria to establish UTI (19). A *ureC* urease mutant is incapable of forming stones, and this has a direct impact on the ability of *P. mirabilis* to cause UTI. When this urease mutant was tested in independent challenge using a mouse model of ascending UTI, there was a highly-significant decrease in bacterial numbers compared to the wild-type parent in the bladder, kidneys, and urine (33). The effect was especially pronounced in the kidneys, where no mutant bacteria were detectable in most mice 48 hours post-infection. The murine 50% infective dose for the urease mutant (2.7×10^9 CFU) is 1000-fold higher than the wild type (2.2×10^6 CFU) (34). From 2 days to 2 weeks post inoculation, kidneys from mice infected with the urease mutant displayed less pathology (acute inflammation, epithelial necrosis) compared to wild-type infection. Likewise, a *ureR* mutant tested in co-challenge with the parent strain was unrecoverable in most mice (22). Urease activity during UTI may be influenced by polymicrobial infection; experimental co-infection of mice with *P. mirabilis* and urease-positive *Providencia stuartii* resulted in increased urolithiasis and bacteremia despite similar bacterial loads compared to monospecies infection (35). Indeed, *in vitro* co-culture of *P. mirabilis* and *P. stuartii* in human urine resulted in enhanced total urease activity compared to either species alone (35).

Due to the prominent role of urease in *P. mirabilis* virulence, this enzyme is an active target of investigation to identify clinically useful inhibitors (36). Since the ability of *P. mirabilis* to generate urinary stones and crystalline biofilms is dependent upon alkaline pH, another approach to prevent catheter blockage is to acidify the urine. Similarly, mineral nucleation can be inhibited by reducing mineral concentration in the urine, *i.e.*, by increasing fluid intake (37). These efforts are aimed at increasing the urinary nucleation pH (the pH at which minerals will precipitate from the urine); a lower nucleation pH is associated with increased stone formation. Preliminary results with patients consuming lemon juice are promising, with the result of increased nucleation pH (38). However, the effect of this treatment on catheter blockage has not yet been reported.

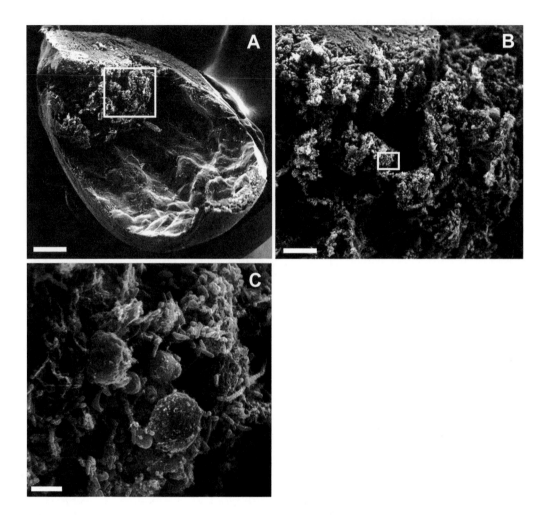

FIGURE 1 *P. mirabilis* **in urease-induced bladder stone. A, One-quarter bladder of experimentally infected mouse (bar, 500 μm). B, Higher magnification of the area indicated in panel A (bar, 100 μm). C, Higher magnification of the area indicated in panel B with individual bacteria visible (bar, 5 μm). Modified from** *Infect Immun* **(32), with permission.**

Flagella

Like many bacteria, *P. mirabilis* uses flagella to swim through liquids and toward chemical gradients (39). In liquid culture, *P. mirabilis* has a short rod shape and typically possesses a few peritrichous flagella. However, on rich solid media, *P. mirabilis* differentiates into very long [typically 20 micrometer (μm) to 80 μm, although cells longer than 100 μm occur], nonseptate polyploid cells with hundreds to thousands of flagella. These swarmer cells move as a population across surfaces, and will be discussed later in this chapter. Although the flagella produced by this organism are generally similar to flagella produced by other bacteria, there are two unusual characteristics of *P. mirabilis* flagella. First, all genes encoding flagellar components, including the class I flagellar master regulatory genes *flhDC* (PMI1671-72), are found within a single 54 kb locus in the chromosome (PMI1617-72). This is in contrast to most other flagella-producing bacteria, which have flagellar operons in disparate loci. Second, *P. mirabilis* encodes two flagellins, FlaA (PMI1620) and FlaB (PMI1619) (also known as FliC1 and FliC2, respectively) (40), which comprise the whip structure of the flagellum.

Antigenic variation

FlaA is the major flagellin. Despite the proximity of *flaA* and *flaB*, *flaA* is transcribed as a monocistronic message and *flaB* message is generally undetectable (41). Recombination between *flaA* and *flaB* can occur. This phenomenon was discovered when *flaA* mutants were often found to revert to a motile phenotype; these revertants produced antigenically distinct flagella that were the product of recombination resulting in a hybrid *flaAB* gene (41, 42). Later studies revealed that wild-type populations of *P. mirabilis* are heterogeneous, with a portion of the cells possessing hybrid *flaAB* genes (42, 43). RNA experiments suggested that between 1.0% and 1.5% of the total flagellin message in wild-type populations is *flaAB* (43). Recombination occurs between homologous regions of *flaA* and *flaB*, and leads to deletion of the intervening sequence. Hybrid flagellins have been detected in bacteria excreted in the urine of mice experimentally infected with *P. mirabilis*. There may be a selective advantage to particular recombination events, as the types of rearrangements typically found depend on the bacterial environment (broth, swarm agar, or murine urinary tract) (42). The recombinant flagella are functional and may serve as a method of immune evasion during UTI or to provide motility under adverse conditions. Indeed, bacteria possessing specific hybrid FlaAB flagella are more motile under conditions of high salinity (255–425 mM NaCl) or extreme pH (5.2 or 8.2) compared to bacteria with wild-type FlaA flagella; conversely, the wild-type flagellum confers a motility advantage in low salinity (85 mM NaCl) (44). Additionally, immune serum from mice experimentally infected with *P. mirabilis* reacts with both FlaA and FlaAB flagellins (45). A possible third flagel-

lin, designated FlaC or FliC3, was identified by DNA-DNA hybridization (40); however, *fliC3* was not annotated in the *P. mirabilis* HI4320 genome sequence and is not readily identifiable by nucleotide BLAST using queries that correspond to the probes used in the initial discovery of *fliC3*.

Regulation of flagella

As has been characterized for other bacteria, *P. mirabilis* flagellar genes are transcribed in a three-tier hierarchy (46). Regulation of flagella is mediated through the class 1 flagellar master regulator genes *flhDC*. The regulator functions as the heteromer $FlhD_2C_2$ or $FlhD_4C_2$ (47, 48). There are multiple inputs to regulation of *flhDC* as well as post-translational modification of $FlhD_4C_2$ and downstream regulation of the class II and class III flagellar genes. Perturbations in flagellar expression may lead to different outcomes with regard to swimming versus swarming, and will be discussed in detail below.

Expression of flagellar genes is regulated during experimental UTI. Within 24 hours post infection, flagella are repressed compared to an *in vitro* mid-logarithmic phase broth culture (25). However, by 7 days post infection, this repression is relieved (25) (Fig. 2). It is very likely that flagella are produced at some point during *P. mirabilis*-mediated UTI, as experimentally-infected mice produce antibodies that recognize flagella (45, 49). Flagella are also repressed during uropathogenic *E. coli* (UPEC)-mediated UTI (50), but are transiently expressed around 4 to 8 hours post infection (51). This time coincides with bacterial ascension from the bladder to the kidneys. It is possible that *P. mirabilis* flagella undergo similar transient expression during early stages of ascending UTI.

Role of flagella in virulence

There are differing conclusions in the literature about the role of motility in *P. mirabilis* virulence. In one study, the flagellum was found to contribute to ascending UTI (52). In that report, an *hpmA* (hemolysin, PMI2057)

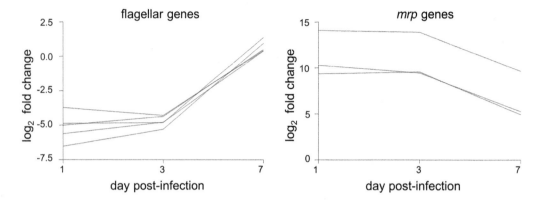

FIGURE 2 Adherence and motility genes are inversely regulated during UTI. Each line represents fold-change of a specific flagellar (left panel) or fimbrial (right panel) gene *in vivo* relative to mid-logarithmic phase culture *in vitro*. Genes in the *mrp* operon are highly induced early during infection, but expression falls by 7 days post infection. Flagellar genes are initially repressed, but expression increases late in infection. Modified from *Infect Immun* (25), with permission.

mutant was compared in independent challenge of CBA/J mice to the wild-type parent and an isogenic *hpmA flaD* (PMI1621) double mutant. FlaD is the capping protein of the flagellum; this mutant produces unassembled flagellin and is nonmotile. At 1 week post infection, the *hpmA flaD* mutant was recovered from the urine, bladder, or kidneys in numbers 100-fold lower than either the wild type or the *hpmA* single mutant. Functional flagella contribute to bacterial spread during UTI; immobilizing antibodies prevented the ability of *P. mirabilis* to traverse from one kidney to the other via the ureters using a rat model (53). However, another group has reported that nonmotile *P. mirabilis* strains also cause UTI (54, 55). In the first of these studies, both ascending UTI and hematogenous routes of infection were used to assess the ability of motile and nonmotile *P. mirabilis* isolates to colonize the bladders or kidneys of outbred mice at 7 days post infection (54). The nonmotile strain was as infective as the motile strains. In the second study, an isogenic mutant missing the 3′ portion of *flaA* and the 5′ portion of *flaB* was compared to the wild-type parent in the ascending UTI outbred mouse model (55). Again, at 7 days post infection, there was not a significant

difference between wild type and mutant in colonization of bladder or kidneys. Taken together, it is likely that flagella contribute to *P. mirabilis* UTI, but the effect may be modest and dependent on the strains of bacteria and animal models used. The specific contribution of swarming to virulence will be considered later in this chapter.

Signature-tagged mutagenesis (STM) studies have also identified mutants with diminished or absent production of flagella that were less competitive in the mouse UTI model (56, 57). In addition, STM identified a chemotaxis mutant (*cheW*), suggesting that not just motility but also the ability to move toward one or more unidentified signals contributes to bacterial fitness during UTI (56). Although eight likely methyl-accepting chemotaxis proteins (MCP) were identified in the HI4320 genome [two in the flagellar locus and six encoded elsewhere (17)], it is worth noting that the molecules sensed by these MCP have not been elucidated for *P. mirabilis*.

Swarming

When *P. mirabilis* is added to an agar surface, the bacteria grow in place for a time (which varies by medium, humidity, and temperature), differentiate into swarmer cells, and move forward as a population. The ability of *P. mirabilis* to swarm as an organized group across solid surfaces was first noted by Hauser in 1885 (58). DNA replication without septation occurs during swarm cell formation, which results in very long, polyploid cells. At defined intervals, the bacteria stop moving and revert to a shorter morphotype (consolidate). After a period, the bacteria redifferentiate into swarmer cells. These bacteria move across many media surfaces in a repeated process of swarming and consolidation, resulting in a characteristic bull's-eye pattern (Fig. 3). In fact, these bacteria are named for the Greek god Proteus, who was able to change shape at will to avoid questioning (58). Swarmer cells are phenotypically distinct from swimmer or vegetative cells, and are characterized by great length (typically 20 μm to 80 μm) and hyperflagellation (Fig. 4). The shift to the swarmer form is accompanied by changes in lipopolysaccharide (LPS), peptidoglycan, and membrane fatty acid composition (59, 60). Swarm cells move together as a group, forming rafts of parallel cells (61). A capsular polysaccharide

termed colony migration factor and an uncharacterized slime are associated with swarming cells and may be used to aid motility across surfaces (62–64). Numerous genes are differentially regulated during swarming, including genes that are not required for swarming (65, 66). Swarming motility has also been observed for other bacterial species, both Gram negative and Gram positive (67). Specifically, *P. mirabilis* shares common features with swarming by *E. coli* (68) and *Salmonella* (69, 70); however, *P. mirabilis* swarming is famously robust compared to these species and will occur on most laboratory media unless inhibitors are used. Swarming does not typically occur on chemically-defined minimal media (71), and may also be controlled in the laboratory by reducing the concentration of salt to ≤0.5 grams/Liter (g/L) or by adding inhibitors (*e.g.*, glycerol, *p*-nitrophenyl glycerin, or 4% agar) (72–76), although the success of these techniques may depend on incubation time, temperature, humidity, and strain of *P. mirabilis*. Mathematical models have been used to represent *P. mirabilis* swarming (77–80); these models recapitulate the terracing that occurs during swarm-consolidation cycles and are beginning to address issues such as water channeling.

Contribution of flagella to swarming

Transcription of flagella is greatly increased during the initial transition from swimming to swarming, and remains high (though cyclical) throughout the swarm cycle. When transcription of swarm and consolidate cells was measured by microarray, *flaA* (flagellin) was the third and sixth most highly-expressed transcript, respectively (65). Flagella are not only required for swarming motility, but they are also linked to swarmer cell differentiation. For example, a *flaD* mutant, which produces flagellin but does not assemble flagella, does not swarm and fails to elongate or undergo polyploidy (40). Interaction with a surface is important to swarmer cell development. Elongation of cells can be triggered by culture

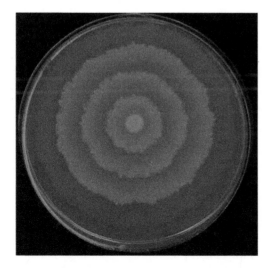

FIGURE 3 **Swarming colony of *P. mirabilis*.**

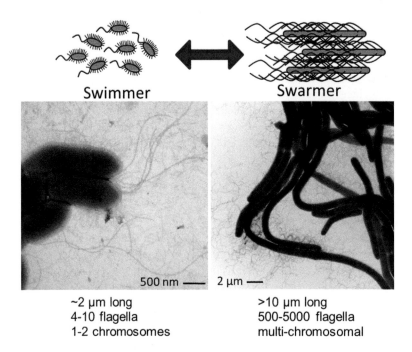

Swimmer ⟷ Swarmer

500 nm —   2 µm —

~2 µm long
4-10 flagella
1-2 chromosomes

>10 µm long
500-5000 flagella
multi-chromosomal

FIGURE 4 *P. mirabilis* **switches between swimming and swarming forms. On the left is a transmission electron micrograph (TEM) of broth-cultured, vegetative cells displaying peritrichous flagella. On the right is a TEM of differentiated swarm cells. Bundles of flagella are visible.**

in viscous fluids, such as by addition of poly-vinylpyrollidone (71) or by impeding flagellar rotation through addition of anti-flagellin antibodies (81). It is believed that solid or viscous surfaces are sensed by restricted rotation of flagella, which transmit the signal to transcribe genes associated with swarming (71, 82, 83).

Transposon mutagenesis has been used to identify genes that contribute to swarming, with many variable phenotypes noted (swarming null or crippled swarming, failure to elongate or constitutive elongation, positive or negative for swimming motility or chemotaxis) (73, 84–86) (Table 1). As a caveat, not all of the transposon mutants described in this chapter were complemented, and there could be polar effects. Most mutations of flagellar genes lead to a motility, swarm, and elongation null phenotype (81, 83), which is in keeping with the hypothesis that swarming differentiation requires surface sensing by flagella. However, mutants that were defec-tive in flagellar production or assembly were not universally non-motile, and had distinct phenotypes (81, 87). A *flaA* mutant was mo-tility and swarming null and failed to elon-gate under swarming conditions; however, this mutant occasionally reverted to being motility and swarming positive (87). This was later shown to be due to recombination with the *flaA* and *flaB* flagellin genes (42). Another swarming-null, motility-positive transposon insertion in *flgN* (flagella filament assembly, PMI1657) was identified by Gygi et al. (85). This mutation resulted in the secretion of unassembled flagellin. The few mature fla-gella assembled by the *flgN* mutant were apparently sufficient to mediate swimming motility but did not allow the hyperflagella-tion required for swarming motility to occur (85). Negative feedback occurs on *flhDC* when flagellar assembly is blocked in an *flhA* mutant; restoration of elongation but not flagellation occurs when *flhDC* is sup-plied *in trans* (88).

Constitutive elongation flagellar mutants

In contrast to the previously described flagellar mutants, *fliL* (PMI1636, hook basal body) and *fliG* (PMI1631, flagellar motor switch) mutants exhibited a constitutively elongated (pseudoswarmer) phenotype (81, 87). The *fliL* phenotype is especially interesting because mutations in other genes in its operon (*fliLMNOPQR*) lead to a failure to elongate (81). A nonpolar *fliL* mutant has relatively few flagella and low levels of *flaA* transcription despite its elongated phenotype (89). However, transcription of the class I master regulator *flhDC* and the class II flagellar cascade sigma factor *fliA* genes is increased in this strain compared to its wild-type parent. Complementation with *fliL* rescues the elongation phenotype but does not restore flagellin expression. Although FliM levels are not disrupted in the *fliL* mutant, complementation with both *fliL* and *fliM* is necessary to restore swarming motility, suggesting that there is an element of *fliM* DNA that contributes to *fliL* function (89). The *fliL* mutant also has increased expression by vegetative cells of virulence factors (*zapA* and *hpmA*) which are induced during wild-type swarming (81). RNA-seq analysis of *fliL* mutant pseudoswarmer cells found that the *umoA* regulator of swarm cell differentiation was induced (90), similar to the induction that occurs in wild-type cells during swarming (91). Although the mechanism of FliL-mediated swarming differentiation has not been elucidated, this protein has been proposed to sense the torque generated on the basal body when flagella encounter viscosity, or to count the rate of proton flow through the motor (81).

Regulation of *flhDC*

Cyclic regulation of flagellin contributes to the swarm-consolidation pattern of *P. mirabilis*, and perturbation of this regulation results in aberrant swarming. Artificial overexpression of the flagellar master regulator genes *flhDC* (PMI1671–2) results in earlier and faster swarming (88). A transposon screen by Clemmer and Rather identified two similar mutants with insertions in the *flhDC* promoter, which had constitutively high levels of expression of these genes during swarming (92). These mutants initiate swarming sooner and swarm at a higher velocity, but lack the consolidation phase and thus do not form a bull's-eye pattern when swarming. Likewise, factors that regulate *flhDC* expression, transcript stability, or posttranslational modification also influence swarming behavior. As might be expected, the FlhD and FlhC proteins have high turnover and are rapidly degraded during swarming, with a half-life of approximately 2 minutes (93). Three regulators that are known to repress *flhDC* in *E. coli* have little or no effect on swarming when mutated in *P. mirabilis*: *ompR*, *lrhA* and *hdfR* (65, 92). Known regulators of FlhD$_4$C$_2$ in *P. mirabilis* are summarized in Table 1 and discussed in detail below.

Lon protease

A mutation in the gene encoding Lon protease (PMI0117) results in cells that initiate swarming normally but migrate faster than the wild type; complementation with *lon* restored the wild-type phenotype (94). The *lon* mutant bacteria also had a tendency to differentiate into swarm cells (elongation, increased flagellation) under non-permissive conditions such as during broth culture. This regulation is likely due to Lon-dependent degradation of FlhD (94). Expression of *lon* is regulated during swarming (65, 94).

Lrp

The global regulator Lrp (PMI0696) is required to initiate cell elongation and swarming. The *lrp* mutant swarming phenotype can be complemented by artificial overexpression of *flhDC*, although this strain is not complemented for expression of hemolysin during swarming (95). Notably, Lrp accumulates at a higher rate in *P. mirabilis* compared to *E. coli* or *Vibrio cholerae* during culture in rich media (a condition that is permissive

TABLE 1 Genes that contribute to swarming in *P. mirabilis*

Name	Open reading frame	Function	Motility[e]			References
			Swim	Swarm	Elongate	
Flagella						
flhDC	PMI1671-2	flagellar transcriptional activator (cl. I)	–	–	–	81, 88
fliF	PMI1630	flagellar MS-ring protein (class II)	–	–	const[a]	57, 81
fliG	PMI1631	flagellar motor switch protein (cl. II)	–	–	const	81
fliL	PMI1636	flagellar basal body-associated protein (class II)	less	–	const	81, 87
fliM	PMI1637	flagellar motor switch protein (cl. II)	slight	–	–	81
flip	PMI1640	flagellar biosynthetic protein (cl. II)	+	–	–	81
fliQ	PMI1641	flagellar biosynthetic protein (cl. II)	–	–	–	81
flgH	PMI1648	flagellar L-ring (class II)	+	cr	const	87
flgE	PMI1651	flagellar hook protein (class II)	NR	–	NR	57
flgB	PMI1654	flagellar basal body rod protein (cl. II)	–	–	–	81
flgN	PMI1657	flagella filament assembly (cl. II)	+	–	NR	85
flhA	PMI1659	flagellar assembly (class II)	–	–	–	81, 83
flaA	PMI1620	flagellin (class III)	–[b]	–[b]	–	40, 41
flaD	PMI1621	flagellar capping protein (cl. III)	–	–	–	40
flgK	PMI1645	flagellar hook-associated protein 1 (III)	–	–	–	81
flgL	PMI1644	flagellar hook-associated protein 3 (III)	–	–	–	81
cheW	PMI1667	chemotaxis; required for CheA function (class III)	NR	–	NR	56
Flagellar regulation						
Lon	PMI0117	ATP-dependent Lon protease	+	++	const	94
Lrp	PMI0696	leucine-responsive regulator	less	–	–	95
rsbA	PMI1729	phosphotransfer intermediate protein	+	++	+	105, 106
rcsBC	PMI1730-1	capsular synthesis two-component system	+	++	+	92, 106
umoA	PMI3115	upregulator of flagellar master operon	less	cr	+	91
umoB	PMI3018	upregulator of flagellar master operon	less	cr	–	91
umoC	PMI1939	upregulator of flagellar master operon	less	cr	+	91
umoD	PMI0876	upregulator of flagellar master operon	less	cr	–	91
rppA	PMI1697	two-component system sensor kinase	NR	++	NR	108
disA	PMI1209	amino acid decarboxylase	++	++	+	97
LPS						
rfaD	PMI3176	ADP-L-glycero-D-manno-heptose-6-epimerase	+	cr	–	87
galU	PMI1490	UTP–glucose-1-phosphate uridylyltransferase	+	cr	–	87, 111
ugd	PMI3189	UDP-glucose 6-dehydrogenase	NR	–	–	111
waaL	PMI3163	O-antigen ligase	+	–	–	113
wzz/cld	PMI2183	O-antigen chain length determinant	+	–	–	113
Metabolism						
cyaA	PMI3333	adenylate cyclase	NR	cr	NR	57
aceE	PMI2046	pyruvate dehydrogenase E1 component	NR	cr	NR	57
sdhC	PMI0565	succinate dehydrogenase cytochrome b_{556} subunit	NR	cr	NR	57
fumC	PMI1296	fumarate hydratase, class II	NR	cr	NR	134
sdhB	PMI0568	succinate dehydrogenase iron-sulfur protein	NR	cr	NR	134
Gnd	PMI0655	6-phosphogluconate dehydrogenase, decarboxylating	NR	cr	NR	134
talB	PMI0006	transaldolase B	NR	cr	NR	134
pfkA	PMI3203	6-phosphofructokinase	NR	cr	NR	134
tpiA	PMI3205	triosephosphate isomerase	NR	cr	NR	134

(Continued on next page)

TABLE 1 Genes that contribute to swarming in *P. mirabilis* (Continued)

Name	Open reading frame	Function	Motility[e]			
			Swim	Swarm	Elongate	References
Amino acids						
serC	PMI0711	phosphoserine aminotransferase	NR	cr	NR	56
cysJ	PMI2250	sulfite reductase [NADPH] flavoprotein alpha component	NR	cr[c]	+	65
dppA	PMI2847	dipeptide ABC transporter, substrate-binding protein	NR	cr[c]	+	65
oppB	PMI1474	oligopeptide ABC transporter, permease protein	NR	cr[c]	+	65
glnA	PMI2882	glutamine synthetase	less	–	NR	120
hisG	PMI0665	ATP phosphoribosyltransferase	+	cr	+	120
Other						
gidA	PMI3055	tRNA uridine 5-carboxymethylaminomethyl modification enzyme	+	cr	–	87
pepQ	PMI3551	proline peptidase	+	cr	–	87
dapE	PMI1556	N-succinyl-diaminopimelate deacylase	+	cr	–	87
cmfA/ cpsF	PMI3190	colony migration factor (capsular polysaccharide)	+	cr	+	57, 63
ppaA	PMI3600	P-type ATPase zinc transporter	+	cr	slight	116, 117
znuC	PMI1151	high-affinity zinc uptake system ATP-binding protein	slight	cr	slight	118
nrpG	PMI2605	4'-phosphopantetheinyl transferase	NR	cr	NR	56, 119
hemY	PMI3329	porphyrin biosynthesis protein	NR	cr	NR	56
speB	PMI2093	Agmatinase	+	cr	delay	120, 121
ccmA	PMI1961	putative membrane protein, curved cell morphology	+	–	+	115
dsbA	PMI2828	thiol:disulfide interchange protein	NR	cr	NR	56
surA	PMI2332	peptidyl-prolyl cis-trans isomerase	+	–	NR	56, 61
mrcA	PMI3021	penicillin-binding protein 1A	NR	cr	NR	56
parE	PMIP32	plasmid stabilization (toxin-antitoxin)	NR	cr	NR	56
sufI	PMI2342	putative multicopper oxidase (suppressor of *ftsI*)	NR	–	NR	57
hexA	PMI1764	LysR-family transcriptional regulator	NR	crd	+	65
lrhA	PMI0629	LysR-family transcriptional regulator	NR	crd	+	65
	PMI1874	two-component system sensor kinase	NR	cr	NR	56
	PMI1046	putative polysaccharide deacetylase	NR	cr	NR	56
	PMI3692	putative lipoprotein	NR	cr	NR	56
Phenotypes observed by overexpression						
rsmA/csrA	PMI0377	carbon storage regulator	NR	–	–	100
mrpJ[d]	PMI0271	fimbrial operon regulator	–	–	aberr	162, 203
wosA	PMI0608	regulator of swarming motility	++	++	const	99

[a]Elongates on nonswarm agar but not in broth.
[b]Reverts to wild-type.
[c]Phenotype detected on MinA-T agar.
[d]12 of 14 additional *mrpJ* homologs also repress motility when overexpressed.
[e]const, constitutive; cr, crippled; ++, hyperswarm/hypermotility; aberr, aberrant; NR, not reported.

for swarming) and is subject to only weak auto-repression, unlike *E. coli* Lrp (96).

DisA

The DisA decarboxylase (PMI1209, decarboxylase inhibitor of swarming) is induced during swarming (97). Mutation of *disA* leads to increased *flhDC* transcript and correspondingly enhanced swarming and swimming motility. Overexpression of *disA* blocks flagellar class 2 (*fliA*) and class 3 (*flaA*) gene expression, but does not significantly alter *flhDC* transcription. DisA is predicted to be an amino acid decarboxylase. Addition of the decarboxylated amino acid phenethylamine reduced both swarming and *disA* expression (97, 98). Phenethylamine also inhibited transcription of class 2 and 3 flagellar genes yet had a minimal effect on *flhDC*. FlhC levels are not affected by *disA* overexpression, suggesting that DisA activity does not destabilize this protein. It has been proposed that the DisA decarboxylation product interferes with $FlhD_4C_2$ assembly or DNA binding (97).

WosA

The *wosA* gene (PMI0608, wild-type onset with superswarming) encodes a predicted membrane protein that induces *flhDC* expression and hyperswarming when overexpressed (99). It also causes constitutive swarm cell differentiation in liquid media. Inhibition of flagellar rotation, either in a *fliL* mutant or in viscous broth, increases *wosA* expression; this regulation is only partially dependent on the presence of flagella. Transcription is also increased over time during both broth and agar culture. However, a *wosA* mutant only has a modest decrease in swarming compared to wild type, suggesting that *wosA* is one of several inputs to *flhDC* regulation (99).

RsmA/CsrA

Overexpression of another gene, *rsmA* (PMI0377, repressor of secondary metabolites, also called *csrA*) results in repression of swarming motility and differentiation (100). The predicted protein is 96% identical to CsrA, a positive regulator of *flhDC* in *E. coli* (101). RsmA and CsrA appear to have opposite effects on *flhDC* regulation, and this warrants further investigation. However, attempts to mutate *rsmA* in *P. mirabilis* were unsuccessful. Expression of *P. mirabilis rsmA* complemented an *E. coli csrA* mutant with regard to glycogen storage, although the effect of *rsmA* on *flhDC* expression in *E. coli* was not reported (100). Overexpression of *rsmA* decreased the half-life of hemolysin

mRNA, indicating that RsmA, like CsrA, regulates by affecting mRNA stability.

Umo proteins

A screen to identify genetic regions that restore swarming motility to the motile but non-swarming *flgN* mutant (described above) revealed four additional genes that positively regulate *flhDC* (91). These four genes were designated *umoA*, *umoB*, *umoC*, and *umoD* for upregulator of the master operon (PMI3115, PMI3018, PMI1939, and PMI0876, respectively). Expression of *umoA* and *umoD* is induced during swarming and these genes are subject to negative feedback when flagellar assembly is blocked and positive feedback by *flhDC* (91). Despite the plethora of sequenced genomes available since the discovery of the *umo* genes, *umoA* remains unique to species within *Proteus* and *Morganella*. The UmoB homolog IgaA has been recognized as a regulator of the Rcs phosphorelay (102, 103), and UmoB homologs are widespread in the *Enterobacteriaceae*. The Rcs phosphorelay (104) has been implicated in swarming by *P. mirabilis* and is discussed below. In *P. mirabilis*, UmoB is a negative regulator of swarming inhibitor *disA* (discussed above) (98).

Two-component systems that regulate *flhDC*

RsbA-RcsBC

Generation of hyperswarming or precocious (early) swarming mutants has been used to identify repressors that contribute to swarming. RsbA (regulator of swarming behavior, PMI1729; also called *rcsD*) has twice been identified as a repressor of swarming (105, 106). The *rsbA* gene encodes a phosphotransfer intermediate that is part of the RcsBCD phosphorelay system. An *rsbA* mutant has a similar hyperswarming phenotype to that seen during flagellin overexpression (105, 106). Interestingly, overexpression of *rsbA* also results in precocious swarming (106). Liaw and colleagues examined the ability of transposon mutants to swarm in the presence of the swarming inhibitor *p*-nitrophenylglycerol (PNPG) (105) and identified a mutant with an insertion in *rsbA*. During swarming, this mutant expresses higher levels of flagellin, as well as other swarming co-regulated virulence factors including hemolysin, protease, and urease (105). An RsbA-mediated pathway may involve sensing of saturated fatty acids to determine a tendency toward swarming or biofilm formation (107). That is, in the presence of specific fatty acids (myristic acid, lauric acid, palmitic acid), swarming behavior is inhibited while biofilm formation and extracellular polysaccharide production is enhanced. The *rsbA* mutant is unresponsive to these fatty acids (*i.e.*, hyperswarms) and is deficient in biofilm formation under permissive conditions. To further investigate the contribution of the RcsBCD phosphorelay to swarming, Clemmer and Rather constructed an *rcsB* (PMI1730, response regulator) mutant; this strain also had a hyperswarming phenotype (92). Thus, the RcsBCD system is likely a repressor of swarming behavior.

RppAB

The *rppAB* genes (PMI1696-7) form a two-component system that regulates multiple cellular functions in *P. mirabilis* including LPS biosynthesis and repression of flagella. An *rppA* mutant was identified in a transposon screen for hyperswarming mutants (108). The *rppA* mutant also had increased expression of virulence factors associated with swarming (HpmA) and an altered LPS profile which conferred increased susceptibility to polymyxin B. RppA regulates *pmrI* (PMI1045, also called *arnA*), which is predicted to be a bacterial UDP-glucuronic acid decarboxylase and contributes to LPS modification (109). Expression of *rppA* is induced in the presence of polymyxin B and repressed when 10 millimolar (mM) Mg^{2+} is added (108). This profile is similar to what is observed for the PhoPQ two-component system of *Salmonella enterica* (110). Although RppAB

has some homology to PhoPQ, another two-component system encoded by *P. mirabilis* has greater similarity (*phoPQ*, PMI0884-5) (17). The function of this second locus has not yet been reported.

Non-flagellar loci that contribute to swarming

LPS

Several elongation-negative mutants with defects in LPS synthesis have been identified through transposon mutagenesis screens (PMI3176 *rfaD*, PMI1490 *galU*, PMI3189 *ugd*, PMI3163 *waaL/rfaL*, an O-acetyltransferase, and a probable O-antigen chain-length determinant mutant) (87, 111–113). Swarming is connected to LPS O-antigen but not to the related enterobacterial common antigen (ECA), as a *wzyE* (PMI3326) mutant swarms (113). Notably, mutation of *rcsB* or *rcsC*, and to a lesser extent *rcsF*, suppresses the *waaL* swarming deficient mutant (103, 113), showing that although O-antigen itself is not necessary for swarming, there is a regulatory link between LPS biosynthesis and swarming differentiation. UmoB and UmoD have been proposed to activate the Rcs system in an O-antigen dependent manner that is distinct from the canonical Rcs surface-sensed activation by RcsF (91, 103). Mutations in LPS also confer sensitivity to the cationic antimicrobial peptides (CAP) such as polymyxin B; *P. mirabilis* mutants defective in CAP resistance are either swarming negative or swarm poorly (111, 112). Mutations in at least two of these genes, *ugd* and *galU*, activate the alternative sigma factor RpoE, which leads to *flhDC* repression (111).

Capsule and cell morphology

A mutation in *cmfA* (colony migration factor, PMI3190, also called *cpsF*) results in a capsular polysaccharide defect. This mutant is able to elongate and become hyperflagellated, but exhibits reduced swarm velocity (63). The structure of this polysaccharide has been determined to be a tetrasaccharide repeat for one strain of *P. mirabilis* (64), and this polysaccharide contributes to virulence (114). Another transposon mutant with a motility and elongation-positive but swarming-deficient phenotype was localized to *ccmA* (curved cell morphology, PMI1961) (115). Although *ccmA* mutant cells become hyperflagellated and elongate under permissive conditions, the cells are curved and uneven in width. A second *ccmA* mutant in which 80% of the gene was deleted resulted in a less severe but still distinctive curved cell phenotype. Overexpression of *ccmA* results in cells with an ellipsoidal or spherical shape. CcmA is predicted to be an integral membrane protein, and its expression increases during swarming differentiation (65, 115). CcmA has been proposed to help maintain linearity during swarm cell elongation, perhaps by organizing peptidoglycan assembly (115). Immunoblotting with CcmA antibodies indicated that there are two forms of CcmA produced in wild-type cells; the larger size matched the predicted full-length protein (CcmA1), and the shorter protein corresponded to an alternative methionine at position 59 of the full-length protein (CcmA2) (115).

Zinc and iron acquisition

A study by Lai et al. described an aberrant swarm mutant with a disruption in a gene encoding a zinc-transporting membrane P-type ATPase (PMI3600 *ppaA*) (86, 116, 117). This mutant swarms at a lower velocity and does not fully elongate during swarming differentiation, yet swarms for longer intervals and has aberrant consolidation terracing. Despite producing lower levels of flagellin transcript and protein as well as repressing the motility regulator *lrp*, it has normal swimming motility. In wild-type *P. mirabilis, ppaA* expression is induced during swarming (116). Another zinc uptake mutant, *znuC* (PMI1151) was subsequently found to display aberrant swarming; furthermore, wild-type *P. mirabilis* has a similarly altered swarming pattern in the presence of the zinc chelator TPEN (118). Mutation of a nonribosomal peptide system involved in iron acquisition also leads to aberrant swarming (56, 119).

Multiple genes that contribute to swarming were identified in two STM screens (56, 57); however, the roles of these genes in swarming have not been elucidated. Two other elongation-negative transposon mutants had defects in cellular division (PMI3055 *gidA*) and a proline peptidase (PMI3551 *pepQ*) (87), although their roles in swarming differentiation have not been described further. These genes are listed in Table 1.

Extracellular contributors to swarming

Glutamine

Glutamine is required for swarmer cell differentiation (71, 120). When Allison et al. added a variety of components to a defined minimal growth medium that is normally nonpermissive for swarming, only glutamine triggered swarming behavior (71). Addition of the other 19 amino acids mixed together did not allow swarming within 24 hours. However, swarming occurred more rapidly and the characteristic bull's-eye pattern only developed when all 20 amino acids were present. Swarmer cell differentiation was not blocked in a glutamine transport mutant, but was inhibited by the addition of glutamine analog γ-glutamyl hydroxamate (71). Furthermore, glutamine acts as a chemoattractant for swarmer cells but not swimmer cells. In contrast, glycine, histidine, glutamate, alanine, aspartate, aspargine, tyrosine, and valine were chemoattractants solely for swimmer cells; only methionine and cysteine were chemoattractants for both cell types (71). A glutamine synthetase mutant (*glnA* PMI2882) is completely unable to swarm, even on rich media (120). Swarming is restored when exogenous L-glutamine, but not D-glutamine, is supplied.

Putrescine

Putrescine has been implicated as a trigger for differentiation in *P. mirabilis* (121). This molecule is continually produced and accumulates in the media during growth and is a component of the outer membrane in some Gram-negative bacteria, including *P. mirabilis* (122). In an effort to investigate cell-cell signaling by *P. mirabilis*, a *lacZ* fusion transposon screen was used to identify genes responsive to signals in spent culture supernatants (123). One mutant, with an insertion in *speA* (PMI2094, arginine decarboxylase), was repressed in the presence of spent wild-type supernatant (121). However, spent mutant supernatant did not repress *speA*. SpeA converts L-arginine to agmatine; the gene is next to *speB* (PMI2093), which converts agmatine to putrescine. Sturgill and Rather found that addition of putrescine (down to 25 μM), but not agmatine, repressed *speA::lacZ* expression. The *speA* mutant also displayed an aberrant swarm pattern comprised of very small, irregular swarm rings. An independent *speB* mutation had the same phenotype; taken together with the putrescine complementation of *speA::lacZ* repression, this suggested that the *speA::lacZ* phenotype was due to a polar effect on *speB*. Normal swarming was restored when extracellular putrescine was added (121), but not when added to a *plaP* (putrescine importer PMI0843) *speA* double mutant (124). Likewise, the *speB* mutants swarmed when inoculated onto agar in the vicinity of an undefined nonswarming *P. mirabilis* mutant that still produced extracellular signal (121). Swarming by a *speB* mutant can also be restored by the addition of ornithine, which can be converted to putrescine by the alternate SpeF pathway (120). Thus putrescine, which accumulates with increasing cell density, may be a signal that initiates swarming. Indeed, uptake of extracellular putrescine appears to be necessary for normal swarming, as the *plaP* putrescine importer mutant does not swarm as robustly as its wild-type parent (124). The effect could be mediated by putrescine forming a complex with LPS or capsular polysaccharide (125).

Arginine, histidine, malate, and ornithine

Previous approaches to defining the signals to initiate swarming involved adding substances to minimal chemically-defined media, or screening mutated bacteria for the loss of swarming ability. A new study was able to expand the list of factors that contribute to swarming by using a rich medium (LB) with low (10mM) NaCl (120). On this medium, *P. mirabilis* HI4320 does not swarm when incubated at 37°C. Under these conditions, addition of 20 mM L-glutamine, L-arginine, DL-histidine, malate, or DL-ornithine promoted swarming. Fumarate and agmatine promoted swarming to a lesser extent. None of these substances enhanced swimming motility, nor did they cause aberrant cell elongation in broth culture. A panel of clinical isolates responded to these swarming cues, although there was some variation in the response to each specific substance. Two of the stronger cues (ornithine and arginine) and the weaker cue agmatine are part of the putrescine biosynthetic pathway. However, as putrescine itself was not sufficient as a swarming cue on low salt LB, and ornithine and arginine stimulated swarming in different ways and induced unique responses when pH or media were changed, these substances most likely stimulate swarming through distinct pathways (120).

Specific mutants were used to further assess the roles of the five swarming cues (L-glutamine, L-arginine, DL-histidine, malate, or DL-ornithine) (120). All five of these cues are present in human urine. Wild-type *P. mirabilis* cannot swarm well on urine solidified with agar; this is likely due to increased pH and crystal formation, both due to urease activity. However, a *ureC* urease mutant swarms on urine agar, and swarming is further promoted by all five swarming cues. Mutation of glutamine (*glnA*) or histidine (*hisG*) biosynthetic pathways led to abolished or reduced swarming, respectively; swarming was restored with exogenous supply of the corresponding amino acid. Conversely, mutation of *gdhA* (glutamate dehydrogenase, PMI3008) had no effect on swarming behavior (120). This finding is intriguing when compared with transcription during UTI, where *gdhA* is induced and *glnA* repressed (this is discussed later) (25).

Quorum sensing

There is no strong evidence that *P. mirabilis* engages in quorum sensing, despite the coordinated behavior during swarming and the regularity with which *P. mirabilis* is found as part of multi-species communities. This species produces cyclic dipeptides [cyclo(ΔAla-l-Val) and cyclo(l-Pro-l-Tyr)] that were initially thought to be agonists of the *Vibrio fischeri* acylhomoserine lactone (AHL)-dependent LuxR quorum-sensing system (126). However, more recent work indicates that these cyclic dipeptides do not affect quorum sensing (127). AHL autoinducers regulate swarming via a LuxI/LuxR-type system in *Serratia marcescens* (128). Although the *P. mirabilis* genome sequence did not reveal any putative LuxI or LuxM-type synthases (17), various exogenously-supplied AHLs affect swarming and proteolytic activity (129). *P. mirabilis* does encode *luxS* and produces the LuxS-dependent quorum-sensing molecule AI-2 during swarming; however, a *luxS* mutant has neither a swarming nor a virulence defect (130). Despite these observations, the highly ordered *P. mirabilis* swarm cycle suggests a mechanism for multicellular coordination exists (131).

Transcription and metabolism during swarming

Despite the vigorous motility displayed during swarming, differentiated *P. mirabilis* bacteria are less metabolically active than consolidating cells (132). When 27 random DNA probes were hybridized with RNA isolated from different stages of the swarm process, most of the sequences were less transcribed during swarming (66). This finding was later confirmed by microarray analysis of the transcriptome of swarming *P. mirabilis*, in which there was a general repression of transcription of swarm cells compared to consolidate (541 genes repressed during swarming; 9 induced relative to consolidated cells) (65). Flagellin (*flaA*) is among the most highly expressed genes in both swarm and consolidate despite the lack of motion observed in consolidating cells. Consolidating cells are distinct from broth-cultured cells used to inoculate a swarm plate; expression by swarm and consolidate share more in common with each other than with broth-cultured bacteria. During consolidation, bacteria are metabolically active with increased expression of genes involved with central metabolism, respiration, nutrient uptake, and cell wall remodeling. The alternative sigma factor *rpoS*, associated with stationary phase and stress response, is among the genes induced during consolidation (65). Once initiated, protein synthesis is not required to maintain the swarm; swarming continues even in the presence of chloramphenicol (65, 133). Genes co-regulated with swarming are not necessarily required or involved with the swarming cycle, including virulence genes *hpmA* and *zapA* (these virulence genes will be discussed in another section). Investigation of genes regulated during swarming has led to the identification of virulence processes. Mutation of genes involved in peptide uptake (*oppB* PMI1474 and *dppA* PMI2847) and amino acid synthesis (*cysJ* PMI2250) that were regulated during swarming led to minor attenuation of swarming. However, the *dppA* and *cysJ* mutants were less fit in animal co-challenge. A similar result was found for the transcriptional regulator *hexA* (PMI1764) (65).

Swarming occurs under both aerobic and anaerobic conditions (133). STM studies for genes involved in virulence also led to the identification of central metabolism genes (*aceE* and *sdhC*) that cause aberrant swarming when mutated, suggesting that a complete (aerobic) TCA cycle contributes to swarming (57). To further investigate how central metabolism affects swarming, selected metabolic genes were mutated and were found to affect swarming in four distinct patterns (134). Two of these classes were characterized by altered distances between rings of swarming and consolidation; TCA cycle mutations (*fumC* PMI1296 and *sdhB* PMI0568) had decreased distances, and pentose phosphate pathway mutants (*gnd* PMI0655 and *talB* PMI0006) had increased distances. Mutations in glycolysis (*pfkA* PMI3203 and *tpiA* PMI3205) resulted in reduced swarming diameter. These mutations were rescued by complementation with the corresponding wild-type alleles or by addition of the missing biochemical intermediate to the growth medium. Specifically, the *fumC* mutant was rescued by addition of succinate or malate but not fumarate; this distinction indicates *fumC* is acting as part of the oxidative TCA cycle during swarming and not the reduced branched TCA cycle. Mutations in gluconeogenesis (*pckA* PMI3015) or the Entner-Doudoroff pathway (*edd* PMI2760) had no effect on swarming patterns, nor did a mutation in the fumarate reductase subunit gene *frdA* (PMI3588) which is involved in anaerobic respiration using the branched TCA pathway (134).

Swarming also occurs in the presence of the aerobic respiration poison sodium azide supplied at growth inhibitory concentrations (133, 134). The *frdA* (branched TCA) mutant also swarms on azide, but the *fumC* mutant, used in both aerobic and anaerobic TCA cycles, is unable to swarm (134). Thus, an alternative anaerobic electron acceptor has been proposed to provide energy during swarming. A transposon screen for mutants that are able to swarm on LB swarm agar but not in the presence of azide yielded 18 mutants. Not all of the mutated genes were expected to be involved in swarming *per se*, as mutations leading to increased permeability or susceptibility to azide would also be found. Two genes of particular interest that were identified are *hybB* (PMI0033 hydrogenase-2), which encodes an anaerobic cytochrome, and PMI2646, which encodes a putative quinine hydroxylase. To further

address whether fermentation occurs during swarming, bacteria were inoculated onto swarm agar with the pH indicator phenol red. During aerobic conditions, *P. mirabilis* produces alkaline conditions; fermentation would have resulted in secretion of acidic byproducts. Swarming under anaerobic conditions, however, results in acidity. Taken together, the authors concluded that anaerobic respiration with a complete oxidative TCA cycle generates the proton motive force required for flagellar rotation (134).

Cellular invasion

P. mirabilis uses its flagella to invade cultured cells derived from the urinary tract, including Vero (green monkey kidney parenchyma) (135, 136), EJ/28 and 5637 (transformed human transitional cell carcinoma of the urinary bladder) (66, 114, 135), NTUB1 (human urothelium) (105), and primary human renal proximal tubular epithelial cells (HRPTEC) (52, 137). In this experimental system, *P. mirabilis* is highly invasive, in numbers comparable to *Salmonella* Typhimurium (52) or *Salmonella typhi* (138). Allison and colleagues found the greatest invasive capability for *P. mirabilis* coincided with the swarmer cell phase (135). Likewise, a hyperswarming *rsbA* mutant is more invasive than its wild-type parent (105). Mobley et al. found that invasion was greatly impaired when flagella of swarmer cells were immobilized by the addition of antiserum (52). Furthermore, an isogenic flagella mutant (*hpmA flaD*) was less than 1% as invasive as the parent (*hpmA*). Centrifugation of the nonmotile bacteria onto the cultured cell monolayer partially restored invasiveness to about 10% of the parental level. Invasion is aided by non-flagellar components including the HpmA hemolysin, autotransporter protein AipA (139) and LPS modification protein PmrI (109). Using a variety of intestinal and urinary epithelial cell lines, Oelschlaeger and Tall were able to inhibit *P. mirabilis* invasion by blocking bacterial protein, RNA, or DNA synthesis, but not by blocking host cell (eukaryotic) protein synthesis (138). *P. mirabilis* colocalizes with mucins MUC2 and MUC5AC, and mucin expression has been found to correlate with invasion (140). Cellular invasion could be involved in the transit of *P. mirabilis* from the kidneys to the bloodstream. Alternately, *P. mirabilis* could have an intracellular population during bladder or kidney infections. Of course, cultured cells differ in fundamental ways from intact tissues (*e.g.* types and polarization of proteins, tight junctions). At this time, invasion of urinary cells by *P. mirabilis* during ascending UTI has not been investigated in much detail.

Role of swarming in virulence

Virulence factors, including urease, ZapA protease, and hemolysin are induced during swarming compared to broth culture or older bacteria in the interior of a swarm colony (65, 66, 141). Production of both ZapA and HpmA is increased during overexpression of the flagellar regulator *umoB*, and *hpmA* transcription is responsive to Lrp (141). The swarming ability of *P. mirabilis* is especially relevant in catheterized patients, as this organism is able to swarm across catheters made of silicone or latex (61, 142) (Fig. 5).

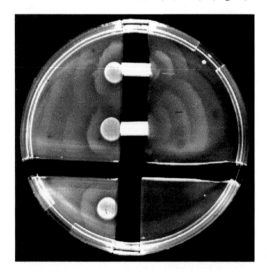

FIGURE 5 *P. mirabilis* **swarms across sections of latex catheter. Reproduced from** *Infect Immun* **(61), with permission.**

Since expression of several virulence genes is increased during swarming, it is possible that *P. mirabilis* swarming up catheters is primed to infect the urinary tract. However, the role of swarming during UTI is debated. In one study, outbred mice intravenously infected with *P. mirabilis* resulted in extensive kidney infection, and long-form, swarmer bacteria were observed in the kidney parenchymal tissue 15 days post infection (114). Transposon mutants that were motile but nonswarming caused lower rates of lethality and kidney abscesses in this model. Several of these mutants were also less virulent compared to wild type in an ascending UTI model using suckling mice (114). In contrast, swarmer cells were very rarely found in the urinary tracts of CBA/J mice infected via bladder catheterization with GFP-expressing *P. mirabilis* when examined 2 or 4 days post infection (143). In that study, 7 of 5087 (0.14%) bacteria counted in the bladders and kidneys had an elongated swarm form (>10 μm); no swarmer cells were observed in the ureters. Combined with the previously described reports in which nonmotile *P. mirabilis* was fully virulent, this suggests that swarming may not be an important contributor to UTI virulence. However, swarming might only be apparent when a catheter is in place, or during possible invasion of renal cells at later stages of disease progression.

It is noteworthy that the putrescine biosynthetic pathway also results in the formation of urea. *P. mirabilis* encodes genes that may catalyze the ultimate generation of putrescine and urea from ammonia and ATP (17). Because putrescine biosynthesis is required to initiate swarming, and excess urea could potentially drive the putrescine biosynthetic pathway in the opposite direction, it is possible that the abundant urea in the urinary tract represses swarming behavior.

Dienes lines and T6SS

P. mirabilis strains are self-recognizing during swarming; that is, any given pair of isolates, when inoculated on opposite ends of a swarm agar plate, will likely swarm up to but not into each other. When this occurs, a thin clear line of demarcation remains between the two strains, called a Dienes line (144). In contrast, swarms formed by identical strains will merge. Dienes line formation has been used as a method for typing clinical strains of *P. mirabilis* (145–147). When opposing swarms from two different strains meet, within 1 to 2 hours, one of the strains will form large, rounded cells (148, 149). Over a period of hours, the rounded cells die and lyse, while the other strain dedifferentiates into vegetative cells. This rounding is not observed when a swarming strain meets nonswarming cells. Dienes line formation is dependent upon cell–cell contact or at least very close cell proximity (<60 μm) between opposing strains (148). When a membrane permeable to most macromolecules but not bacteria is placed between swarming strains, no Dienes line or cell rounding occurs.

The Dienes phenomenon is likely a mechanism of territoriality. When two strains (one marked with red fluorescent protein and one marked with green fluorescent protein) are mixed and placed on swarm agar, one strain dominates the resulting swarm colony, while the other strain forms rounded cells when detected at all. However, in broth culture, neither strain is dominant, and when other strains were tested for biofilm formation, dominance of a given strain did not correlate with Dienes dominance (148).

Recent studies suggest that type VI-mediated secretion of toxic effector proteins is the main mechanism of Dienes line formation (149, 150). Type VI secretion systems (T6SS) comprise a method of protein export that generally requires cell-cell contact, whether between two bacteria or a bacterium and a eukaryotic host cell. They are widespread in Gram-negative bacteria and may be involved in virulence, commensalism, or bacterial competition (151–153). A core set of structural genes are essential for function and are thought to be involved in the production or function of the secretion apparatus. This includes the *hcp/vgrG* genes, which respectively encode proteins that form the tube through which export occurs and a needle-like structure used to penetrate the outer membrane of the target cell.

Type VI secretion has been studied in two *P. mirabilis* isolates, BB2000 and HI4320. Both strains encode the core structural genes essential for T6SS function, as well as multiple putative effector operons associated with the *hcp/vgrG* genes (17, 154–156). In T6SS characterized in other bacterial species, the multiple *hcp/vgrG* homologs are thought to act as adapters between the structural components and the effectors, allowing different effector proteins to be secreted (153). In both BB2000 and HI4320, the T6SS is essential for Dienes line formation and identification of self. However, the number of effector operons present and the significance of individual operons in self-recognition varies between the two strains (155, 156).

In BB2000, a transposon screen for mutants that form Dienes lines when in competition with the parent strain (that is, the strains no longer recognize the parent strain as "self") led to the *ids* operon (identification of self, *idsABCDEF* PMI2990–95) (149). The first two genes of this locus, *idsA* and *idsB*, have homology to *hcp* and *vgrG*, respectively. The *idsD* and *idsE* genes display the most variation between *P. mirabilis* strains, and are believed to be the determinants of self-recognition (149). Expression of the *ids* operon increases during late logarithmic and early stationary phase. During swarming, expression is highest in the center of the swarm colony where the highest cell density is observed, but expression also occurs within a subset of cells at the leading edge of approaching swarm fronts (150).

A second transposon mutagenesis screen identified two "no-merge" mutants which form Dienes lines with both the wild-type and the *idsA-F* mutant (155). These mutants mapped to a previously uncharacterized five gene operon, named *idr* for *id*entity recognition. As in the *ids* locus, the first two genes in the *idr* locus (*idrA* and *idrB*) have homology to *hcp* and *vgrG*, while the remaining three genes encode proteins of unknown function. IdsA, IdsB, IdsD, IdrA, and IdrB were detected using mass spectrometry in culture supernatants of wild-type BB2000 but not in a T6SS mutant, providing further evidence that these proteins are exported from the cell by the T6SS. It is apparent that competition and killing via T6SS effectors is a complex phenomenon in *P. mirabilis*. For example, wild-type BB2000 can dominate HI4320 in a swarm competition. However, BB2000 Δ*ids* mostly dominates HI4320 while HI4320 can dominate BB2000 *idrB*::tn5, suggesting that the *idr* operon is more important than the *ids* operon in BB2000 killing of HI4320 (155).

Unlike in BB2000, mutations in *idsB* and *idsD* in HI4320 do not result in a loss of self-recognition (156). However a transposon screen for HI4320 mutants which no longer recognize the wild type as self, and thus form a Dienes line with HI4320, identified hypothetical protein PMI0756 as a putative component of the T6SS. A secondary screen to identify additional mutants that no longer form a Dienes line with the PMI0756 mutant identified two groups of mutants. The first group of mutants clustered in the major T6SS structural operon (PMI0733-0749), while the second group of mutants clustered in the same operon as PMI0756 (PMI0750-0758). The first two genes in this operon are *hcp* and *vgrG* homologs, respectively; the other genes in the operon were named *pef* for *primary effector* operon. Additional targeted mutagenesis revealed that a mutation in any gene in the *pef* operon results in loss of self-recognition. Complementation of the *pefE* mutant with *pefE* is sufficient to restore immunity from wild-type killing, but is not sufficient to restore killing of non-immune strains. To restore both immunity to wild type and killing of the *pefE* mutant, complementation with *pefEFG* is needed (156). It is notable that although the *pef* and *idr* operons are both adjacent to the structural T6SS

apparatus operon in HI4320 and BB2000, respectively, the organization and sequences of these operons are completely different (aside from the *hcp* and *vgrG* homologs at the front of both operons).

Analysis of the HI4320 genome revealed three additional putative *hcp-vgrG* effector operons in addition to the *ids* and *pef* operons. The proteins encoded by the *hcp* genes are highly similar (with the exception of PMI1332, which appears to be truncated), while the genes predicted to encode VgrG homologs are similar at the N-terminus with decreasing similarity at the C-terminus. All five of the promoters for the *hcp-vgrG* operons are capable of driving luciferase expression during swarming (156), though they result in different patterns of expression.

Although significant gains have been made in the understanding of Dienes line formation and T6SS competition in *P. mirabilis*, a number of questions remain. The function of most of the genes in the effector operons remains unclear, and it is not understood how cell death and immunity are mediated. Nor is it clear why *P. mirabilis* possesses several effector operons, or how they are regulated. Perhaps most importantly, it is unknown when Type VI secretion-mediated killing occurs when *P. mirabilis* is in its natural environments or if this phenomenon has clinical relevance.

Fimbriae

Although the fimbriae of *P. mirabilis* are essential virulence factors in UTIs (Table 2), much remains unknown about their expression, physical characteristics, and biological functions. The first *P. mirabilis* fimbriae were identified by their ability to agglutinate red blood cells and bind uroepithelial cells. Using these methods, the mannose-resistant *Proteus*-like (MR/P), mannose-resistant *Klebsiella*-like (MR/K), and urothelial cell adhesin (UCA) fimbriae were described (157, 158). However, as *P. mirabilis* can simultaneously express multiple types of fimbriae,

TABLE 2 The fimbriae of *P. mirabilis*. The name, genomic location, Greek classification, determination of virulence, and MrpJ homolog of each fimbrial operon in *P. mirabilis*

Fimbria	Genes	Class[a]	Implicated in virulence?[b]	MrpJ homolog
MR/P'	PMI0254-PMI0261	π	ND	PMI0261
MR/P	PMI0262-PMI0271	π	yes, *IC, CO*	PMI0271
Fimbria 3	PMI0296-PMI0304	π	ND	PMI0296
UCA	PMI0532-PMI0536	γ₁	yes	PMI0532
Fimbria 5	PMI1060-PMI1067	π	ND	PMI1060
Fimbria 6	PMI1185-PMI1190	γ₁	ND	None
Fimbria 7	PMI1193-PMI1197	γ₁	ND	None
Fimbria 8	PMI1464-PMI1470	γ₁	yes	PMI1470
PMF	PMI1877-PMI1881	π	yes, *IC, CO*	None
Fimbria 10	PMI2207-PMI2214	γ₁	ND	PMI2209, PMI2207
PMP	PMI2216-PMI2224	π	yes	PMI2224
Fimbria 12	PMI2533-PMI2539	γ₂	ND	None
ATF	PMI2728-PMI2733	γ₁	no, *IC, CO*	PMI2733
Fimbria 14	PMI2997-PMI3003	ND	yes, *STM*	PMI3003
Fimbria 15	PMI3086-PMI3093	π	ND	None
Fimbria 16	PMI3348-PMI3352	γ₁	ND	None
Fimbria 17	PMI3435-PMI3440	γ₁	ND	None

[a]Greek classification was determined as in references 161 and 185. Fimbria 14 does not have an identified chaperone, and thus cannot be classified in the Greek system.
[b]*IC*, The mutant was tested in an independent challenge experiment; *CO*, The mutant was tested in a co-challenge experiment; *STM*, The mutant was tested in a signature-tagged mutagenesis experiment; ND, not determined.

identification of genes encoding specific fimbriae is often complicated (157). For example, the attempts to identify genes encoding UCA fimbriae led to the discovery of the *P. mirabilis* fimbriae (PMF) (159), and the genes associated with MR/K hemagglutination remain unidentified.

Prior to sequencing the *P. mirabilis* genome in 2008, five *P. mirabilis* fimbriae had been discovered. Sequencing revealed an additional 12 chaperone-usher fimbriae encoded in the *P. mirabilis* genome (17). In comparison, UPEC typically encode nine to 12 fimbrial operons (160). As the name suggests, chaperone-usher fimbriae are defined by the presence of a chaperone and an usher in the fimbrial operon. The chaperone protects the fimbrial subunits during transport from the cytoplasm to the cell surface which allows the fimbrial subunits to fold properly in the periplasm, and the usher, an integral membrane protein, releases the fimbrial subunit from the chaperone and guides it through assembly (161). Ten of the *P. mirabilis* chaperone-usher operons encode a homolog of MrpJ, a regulatory protein at the

beginning or end of the operon, which allows the coordination of adhesion and motility, and may allow fimbrial regulation of other fimbriae (162). The following sections describe the fimbriae of *P. mirabilis* in order of their discovery, as well as the role of MrpJ homologs in coordinating regulation of fimbriae and flagella.

Mannose-resistant *Proteus*-like (MR/P)

Genetic organization

MR/P is the best characterized fimbria in *P. mirabilis*. The *mrp* operon (*mrpABCDEFGHJ*) encodes the major structural subunit (*mrpA*), four minor subunits (*mrpBEFG*), a chaperone (*mrpD*), an usher (*mrpC*), a tip adhesin (*mrpH*), and a transcriptional regulator (*mrpJ*) (163, 164).

The operon is preceded by a σ^{70} promoter inside an invertible element whose orientation is controlled by MrpI, a recombinase transcribed divergently from the structural operon (164, 165). When the σ^{70} promoter faces the structural operon, MR/P is ON, while when the σ^{70} promoter faces *mrpI*,

MR/P is OFF (165). Transcription of *mrpI* is unaffected by the orientation of the invertable element, and *mrpI* is likely transcribed by a promoter outside of the invertable element (164). The MrpI C-terminal domain is homologous to the catalytic domains of the *E. coli* fimbrial recombinases FimB and FimE, while the N-terminal domain binds DNA (165). Like FimB, MrpI switches the invertible element both from ON to OFF and OFF to ON (165). However, when *mrpI* transcription decreases, *mrpA* transcription increases, suggesting MrpI favors switching from ON to OFF (166). In the absence of MrpI, the promoter orientation is fixed, suggesting that MrpI is the sole recombinase for the *mrp* promoter (165). To facilitate analysis of the role of MR/P fimbriae, two MR/P mutants were engineered. Both of these mutants have an insertion that inactivates *mrpI*; in the "MR/P L-ON" mutant, the *mrp* promoter is in the on position, while in the "MR/P L-OFF" mutant, the *mrp* promoter is in the off position.

Expression

Like many fimbriae, *in vitro* MR/P expression is encouraged through static culture. The percentage of bacteria with ON or OFF *mrp* promoters can be determined by a PCR assay, where the percentage of ON *mrp* promoters correlates to the percentage of cells expressing MR/P (Fig. 6) (165, 166). When a lower oxygen level is maintained in LB under shaking conditions, for example by culturing *P. mirabilis* in 5% O_2, the *mrp* promoter is up to 70% ON compared to 2% ON in aerated culture. The increased expression of MR/P under static conditions is due to decreased availability of oxygen gas, not increased levels of CO_2 (166).

Oxygen-limiting conditions enrich for expression of MR/P through two mechanisms. First, the expression of MR/P lends a competitive advantage during growth in low oxygen conditions. Although their growth during independent culture is similar, a *P. mirabilis* L-ON mutant will out-compete

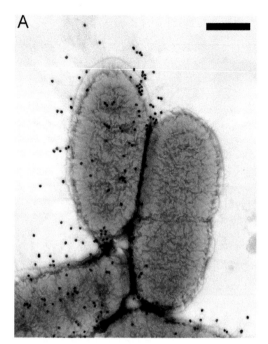

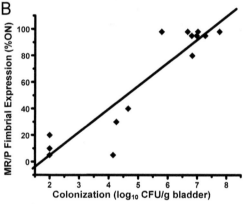

FIGURE 6 Expression of MR/P fimbriae is phase-variable and induced during UTI. A, Immunogold electron microscopy of wild-type *P. mirabilis* HI4320 labeled with gold particles targeting the MrpH tip adhesin. The cell on the left is expressing MR/P fimbriae, and the cell on the right is not (bar, 500 nm). **B,** The amount of MR/P fimbriae present positively correlates with murine bladder colonization. Data were obtained seven days post-inoculation. Modified from *J Bacteriol* (166), with permission.

a *P. mirabilis* L-OFF mutant in liquid culture (166, 167). It is hypothesized that this is due to fimbrial expression and flagellar repression driving electron transport and maintaining the proton gradient (166). Secondly, oxygen limitation increases the expression of MR/P. Under limited oxygen conditions, almost twice as much *mrpA* is transcribed as under atmospheric oxygen, possibly because less MrpI is expressed, leading to less switching of the *mrpA* promoter from ON to OFF (166).

Assembly

The role of each subunit in the assembly and structure of the MR/P fimbria is partially understood. MrpA has been identified as the major structural subunit by its strong expression and location in the *mrp* operon (164, 168).

MrpB is homologous to *E. coli* PapH, which terminates fimbrial growth while anchoring fimbriae to the cell wall and modulating fimbria length (169, 170). *P. mirabilis* Δ*mrpB* displays fewer but longer fimbriae on the surface of the cell. In *P. mirabilis* Δ*mrpB*, 5% of cells were fimbriated after three 48-hour passages in static culture, while 48% of wild-type cells were fimbriated. MR/P fimbriae from an *mrpB* mutant are 62-fold longer than wild-type fimbriae (18 μm compared to 0.29 μm) (171). These data suggest MrpB is needed to either initiate and/or terminate assembly of MR/P (171). The *mrpEFG* genes are predicted to encode minor structural subunits. When MrpG is found in *P. mirabilis* fimbriae, it is located on short fimbriae, possibly newly synthesized, and where fimbriae meet between aggregating bacteria (172).

MrpH is a two-domain adhesin, where the predicted MrpH C-terminus is similar to other *mrp* structural subunits and the larger N-terminal domain likely mediates receptor binding (Fig. 6). This two-domain structure and the presence of a proline at the last residue identify MrpH as an adhesin (163, 173). Less than 1% of *mrpH* mutant cells express MR/P fimbriae compared to 50% of the wild-type, suggesting that MrpH is essential for stable production of MR/P. Oddly, the *mrpH*

mutant still mediates weak hemagglutination, possibly due to the expression of a different fimbria in the absence of *mrpH* (163).

The N-terminus of MrpH has four conserved cysteine residues, which in other tip adhesins form disulfide bonds that are essential for the proper folding of the adhesin (174, 175). When the MrpH cysteine residues are mutated to serine, the MR/P fimbrial structure is produced and MrpA is found in fimbrial preparations, but there is less pellicle formation and hemagglutination, suggesting that the tip adhesin is inactivated (163). In *P. mirabilis* and other organisms, periplasmic disulfide bond formation is mediated by DsbA (PMI2828) (176). A *dsbA* mutant is deficient in MR/P production and in establishing infection in a mouse model of UTI, which is consistent with disulfide bond formation being an essential component of the MR/P structure (56, 166).

Biofilm formation

MR/P contributes to biofilm formation, presumably by mediating attachment to surfaces and autoaggregation (177, 178). *P. mirabilis* L-ON forms significantly more biofilm during culture in urine than a *P. mirabilis* L-OFF mutant or the wild type, which form similar levels of biofilms (177). Scanning electron microscopy showed that after two days, wild-type bacteria had adhered as small colonies to a cover slip, while *P. mirabilis* L-ON formed a full three-dimensional biofilm. By seven days however, the wild-type biofilm had become much thicker than the L-ON or the L-OFF biofilm, suggesting that the ability to turn MR/P on and off may be essential for robust biofilm maintenance (Fig. 7) (177).

Role in infection

MR/P fimbriae are expressed in the urinary tract and contribute to virulence. After transurethral infection with wild-type *P. mirabilis*, mice with high IgG antibody responses produced serum antibodies to MR/P (179). During infection, the *mrp* operon genes (including *mrpJ*), are upregulated 500- to

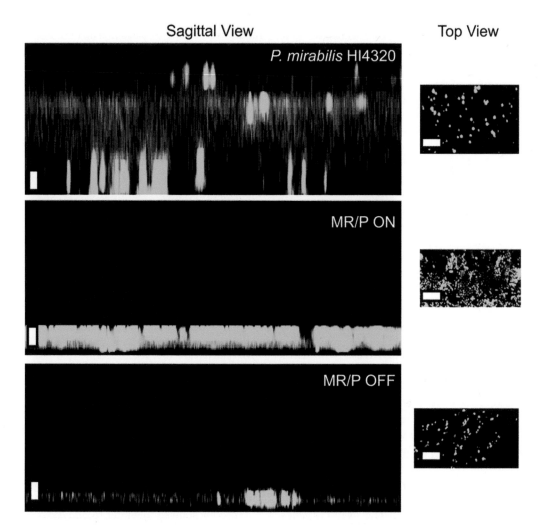

FIGURE 7 *P. mirabilis* biofilm formation is MR/P-dependent. *P. mirabilis* bacteria expressing GFP were grown on a cover glass in urine for 7 days. The resulting biofilm was imaged with confocal microscopy, and the 30 resulting z-stacks were stitched together to form the sagittal view. Wild-type *P. mirabilis* forms thick, robust biofilms. *P. mirabilis* MR/P L-ON forms dense, but thin, biofilms while *P. mirabilis* MR/P L-OFF forms weak biofilms. Reprinted from *Infect Immun* (177), with permission.

1300-fold compared to culture in LB, and are the nine most induced genes *in vivo* (25). The greatest upregulation *in vivo* occurs early in infection (1-3 days) and then falls later during infection (7 days), although these genes remain induced compared to *in vitro* culture (Fig. 2) (25). Greater than 90% of the *mrp* promoters are in the ON orientation in the bladder and urine during murine infection (165, 167). Direct observation of *P. mirabilis* in

the bladder, urine, and kidneys of mice revealed MR/P fimbriation in all parts of the urinary tract (177). However, in the kidneys up to 85% of bacteria do not express MR/P (167). Phase variation of the *mrp* promoter orientation may contribute to evasion of host defenses.

Multiple mutations in the *mrp* operon have been tested in mouse models of infection. All mutants that do not express MR/P are defective in infection. However, different mutations result in different colonization defects. The *P. mirabilis mrpG* mutant is inefficient at colonizing the mouse urinary tract in an independent challenge. Fewer *mrpG* mutant bacteria than wild type colonize the urine, bladder (>10,000-fold), and kidney (>250-fold) (172). This large of a defect has not been observed for any other *mrp* mutant.

An *mrpA* mutant compared to wild-type *P. mirabilis* showed decreased colonization of the bladder (28 fold), kidneys (18 fold), and urine (6 fold), as well as reduced pathology in the kidneys of mice in an independent challenge (180). In another study, independent challenge with wild-type *P. mirabilis* and an *mrpH* mutant revealed no difference in colonization of the mouse urinary tract. However, in a co-challenge infection, the *mrpH* mutant was outcompeted by the wild type in the bladder, kidneys, and urine (181).

Similarly, independent challenges between *P. mirabilis* L-ON, L-OFF, and wild-type strains revealed no differences in the ability of each strain to colonize the urinary tract (181). After a 24-hour co-challenge experiment, there was no difference in the ability of a wild-type, L-ON, or L-OFF *P. mirabilis* strain to colonize the bladder (177). However, there was a significant difference in the localization of adherent bacteria. The wild-type bacteria largely colonized the surface of uroepithelial umbrella cells and were occasionally attached to exfoliated cells. When examined by electron microscopy, the wild-type bacteria had a fuzzy appearance, indicating fimbriation (177). A *P. mirabilis* L-ON mutant was similarly able to colonize the intact bladder epithelium. However, the L-ON mutant had a distinct morphology from the wild-type. They were elongated (6-10 μm) and heavily fimbriated (177). The L-OFF mutant colonized a distinct location in the bladder. Instead of colonizing the intact epithelium, the *P. mirabilis* L-OFF mutant colonized exfoliated cells and areas of the bladder where the uroepithelium was disrupted (177). Exfoliation of umbrella cells exposed the lamina propia, and L-OFF bacteria were able to colonize the underlying basement membrane.

After a 7 day co-challenge experiment, wild-type *P. mirabilis* out-competed the L-OFF mutant in the bladder, kidneys, and urine. However, in a competition between the L-ON mutant and wild-type *P. mirabilis*, the L-ON mutant dominated the wild type in the bladder, but was similar to the wild type in the urine and the kidneys (167). These data support that MR/P fimbriae are essential for infection of the bladder and contribute to colonization of the kidneys.

Mannose-resistant *Klebsiella*-like (MR/K) hemagglutinin

Early characterization of the hemagglutination patterns of *P. mirabilis* revealed a MR/K hemagglutinin. MR/K hemagglutination is characterized by the ability of bacteria to agglutinate tannic acid-treated red blood cells but not untreated red blood cells (157). Expression of fimbriae 4 nanometers (nm) to 5 nm in diameter was found in approximately 60% of strains that showed MR/K hemagglutination patterns, suggesting that there may be more than one fimbria in *P. mirabilis* that is capable of mediating MR/K hemagglutination (157). The genes that encode the MR/K fimbria remain unidentified.

Uroepithelial cell adhesin (UCA)

Characterization
UCA, also known as the non-agglutinating fimbria (NAF) was identified by its ability to bind the surface of uroepithelial cells. *P. mirabilis* strains that express UCA bind shed uroepithelial cells from human urine (158). When the *uca* operon from an adherent *P. mirabilis* strain was expressed in *E. coli* K-12, which is not natively adherent to uroepithelial cells, the *E. coli* became adherent (182). A purified protein preparation of UcaA, the major subunit, also bound uroepithelial cells (158). As the major subunit does not

typically mediate adhesion, it is possible that the preparation contained the predicted tip adhesin. Electron microscopy revealed fimbriae 4 nm to 6 nm in diameter and 0.09 μm to 0.83 μm in length (158, 183). However, a recent study found that a *P. mirabilis* Δ*ucaA* mutant is impaired in colonization of the kidneys, but colonizes the bladder as well as wild-type *P. mirabilis*, suggesting that the UCA fimbriae may be important for kidney rather than bladder colonization (184). The major structural subunit of UCA fimbriae, UcaA, is variable (185), and it is possible that variant UCA fimbriae mediate adherence to different targets. Despite the evidence for a role for UCA in infection, the UCA fimbriae are not differentially regulated *in vivo* compared to culture in LB (25), and there is considerable variation in the amino acid sequence of UcaA in the seven currently sequenced *P. mirabilis* genomes (185). The structural proteins in the *uca* operon are homologous to the F17 family of fimbriae in uropathogenic *E. coli* (182). F17 fimbriae in enteropathogenic *E. coli* attach to intestinal epithelium via a terminal *N*-acetyl-D-glucosamine (186, 187). *P. mirabilis* is commonly found in the gut, but whether UCA fimbriae mediate the same role as F17 fimbriae in the gut is unknown (188).

Binding

UCA is the only *P. mirabilis* fimbria with a characterized binding target. Purified UcaA, presumably with the tip adhesin co-purified, binds asialo-GM1, asialo-GM2, and lactosyl-ceramide in *in vitro* biochemical assays (189). Asialo-GM1 is a Galβ1-3GalNAcβ1-4Galβ1-4Glcβ1-1′ceramide glycolipid commonly found on neuronal and a variety of immune cells. It has been found on the surface of polymorphonuclear leukocytes (PMN) (190), which respond during UTI, but at this time, there are no reports of *P. mirabilis* binding PMN. Asialo-GM1 has also been identified as a potential receptor for *Pseudomonas aeruginosa* fimbriae in the lung epithelium (191). *P. mirabilis* expressing UCA can be partially prevented *in vitro* from binding cultured uroepithelial cells either by use of an anti-UCA antibody (75% blocked) or an anti-asialo-GM1 antibody (50% blocked) (189). The presence of asialo-GM1 on the surface of the bladder has not been shown, and it remains unclear if asialo-GM1 is a valid target of UCA in the urinary tract (189).

Proteus mirabilis fimbria (PMF)

PMF was discovered due to cross-reactivity with both a *ucaA* degenerate DNA probe and a UcaA antibody (159). PMF is essential for full infection in a urethral challenge murine model. Infection with wild-type *P. mirabilis* leads to the production of serum antibodies against PMF (179). A *P. mirabilis* Δ*pmfA* mutant colonized the bladder less efficiently than the wild-type parent strain in both independent (83-fold) and co-challenge experiments (700-fold) (192, 193). The *pmfA* mutant also colonized the kidneys less effectively than the wild type, but only in the more sensitive co-challenge experiment (2630-fold) (192). In a hematogenous challenge, which tests *P. mirabilis*' ability to colonize the urinary tract from the bloodstream via the kidneys, fewer mice were colonized by the Δ*pmfA* mutant than the wild type. Specifically, the wild type colonized 94% of mice, while the Δ*pmfA* mutant colonized 45% of mice (192). In spite of the critical contribution of PMF toward infectious fitness in animal models, the *pmf* operon is downregulated *in vivo* compared to *in vitro* (25). In *E. coli*, flagella are downregulated *in vivo* compared to *in vitro*, but are essential to infection. This apparent paradox was explained by an *in vivo* imaging experiment, which revealed that flagella are temporarily upregulated *in vivo* to allow *E. coli* to ascend the ureters (51). PMF may have a similar short-lived but essential role in causing *P. mirabilis* UTIs.

A double *P. mirabilis* Δ*mrpA-D* Δ*pmfA* mutant is attenuated at colonizing the urinary tract compared to either the single *mrp* or *pmf* mutants. In independent and co-challenge experiments, *P. mirabilis* ΔmrpA-D ΔpmfA colonized the kidneys less effectively than either a *P. mirabilis* ΔmrpA-D or *P. mirabilis* ΔpmfA single mutant. However, *P. mirabilis* ΔmrpA-D ΔpmfA was less effective at colonizing the bladder than *P. mirabilis* ΔmrpA-D, but colonized the bladder similarly to *P. mirabilis* ΔpmfA (194).

Ambient temperature fimbria (ATF)

When the fimbrial profile of a *P. mirabilis* *mrpA* mutant was examined, an additional fimbria was discovered. The fimbria was named the <u>a</u>mbient <u>t</u>emperature <u>f</u>imbria (ATF) because it is maximally expressed at 25°C (195). The *atf* operon was initially thought to consist of only a major subunit (*atfA*), a chaperone (*atfB*), and an usher (*atfC*); however, subsequent sequencing of the *P. mirabilis* genome identified a minor pilin and a tip adhesin encoded in the *atf* operon (17, 196). A mutant with the *atfB* gene and parts of the *atfA* and *atfC* genes deleted is capable of infecting mice at the same level as wild type, which suggests that ATF may be important for environmental survival, but not for infection of the urinary tract. (197). However, in the absence of MR/P, ATF is upregulated both *in vivo* and *in vitro* (167, 177). Bacteria without MR/P are still able to bind the uroepithelium *in vivo*, but it is unclear if ATF is the only fimbria expressed under the circumstances or if it contributes to adherence (177).

P. mirabilis P-like fimbria (PMP)

The <u>P</u>. <u>m</u>irabilis <u>P</u>-like fimbria (PMP) is a prime example of the difficulties encountered when *P. mirabilis* expresses multiple fimbriae at the same time. PMP was first named in 1995, for a fimbria expressed by *P. mirabilis* isolated from canine urine (198). However, the authors noted that the PmpA protein sequence had been previously identified in a 1991 paper, which determined the N-terminal sequence of MrpA (49, 198). In 1993, the genes which encode MrpA were identified, but the nucleotide sequence encodes a different protein sequence than was initially identified (168). Thus, the first paper describing the isolation of MrpA was likely actually describing the isolation of PmpA.

Although the importance of PMP in human infection is unknown, genome sequencing has revealed the presence of *pmpA* in all the currently sequenced *P. mirabilis* genomes, and *pmpA* was detected in 57 of 58 clinical isolates using a PCR screen (185). Additionally, after infection with wild-type *P. mirabilis*, mice with the highest bacterial loads over a 6-week infection produced serum antibodies to PMP (49).

Fimbria 14

The presence of the fimbria 14 operon was first discovered upon sequencing of the *P. mirabilis* genome (17). Concurrent with the sequencing of the genome, a random transposon insertion in the *fim14* operon was identified which is attenuated in the mouse model compared to the wild-type parent (57). During infection, the *fim14* operon is upregulated compared to culture in LB, but it is not highly expressed (25). It is also induced when cultured *in vitro* at pH 8.0 compared to neutral pH, suggesting *fim14* may be responsive to the alkaline pH that is a result of urease activity in urine (185). Oddly, the *fim14* operon does not encode an obvious chaperone. Despite the apparent lack of an assembly system for fimbria 14, major structural subunit Fim14A protein has been detected in sheared surface protein preparations from strain HI4320 (185). It is possible that the structural components of Fimbria 14 are transported and assembled on the bacterial surface by a heterologous chaperone-usher pair.

Conservation and expression of the 17 fimbriae

Overall, the chaperone-usher fimbriae of *P. mirabilis* are highly conserved across isolates collected decades apart, from diverse geographical locations and different anatomical sites from either humans or mice. In a comparison between HI4320 and six additional sequenced genomes, 14 of the 17 fimbrial major subunits have ≥95% amino acid identity. UcaA, Fim6A, and Fim3A have greater diversity in amino acid identity. Only Fim3A, which is missing from two isolates, was not found in all sequenced isolates. In a PCR-based screen for the major fimbrial subunits of 58 clinical isolates, only *fim3A* was not detected in a majority of isolates (185). This high degree of conservation stands in contrast to the patterns detected in *E. coli* and *Salmonella enterica*, in which the majority of fimbrial genes are variable, and specific fimbrial operons are associated with isolation from a particular body site (199–201). The lifestyle of *P. mirabilis* outside of the urinary tract is poorly characterized; however, it is likely that the high degree of fimbrial conservation indicates that possessing a wide array of fimbriae is important for the survival of this species in the environment.

Transcript from all of the 17 major structural genes can be detected in logarithmic-phase aerated cultures. Fimbrial genes were generally induced during stationary-phase and static culture conditions. Mass spectrometry identified proteins from six different fimbriae (MR/P, PMF, UCA, ATF, Fimbria 8, and Fimbria 14) in sheared surface protein preparations of stationary-phase, aerated cultures (185). Again, this is in contrast to UPEC, where a single fimbria dominates under specific culture conditions (202).

MrpJ regulates the transition between swimming and swarming

To survive in both the urinary tract and the environment, *P. mirabilis* must coordinate adhesion to surfaces with swimming and swarming. Fimbriae are essential for adhesion while flagella are essential for swimming and swarming. *P. mirabilis* uses the product of a gene in the fimbrial operon to directly repress flagellar synthesis (162). Sequencing of the *mrp* operon revealed *mrpJ*, which

encodes a helix-turn-helix xenobiotic response element (HTH-XRE) transcriptional regulator downstream of the *mrpH* gene (181). When *mrpJ* is expressed *in vitro* at a level similar to the amount during experimental UTI, swarming, swimming, and FlaA levels are repressed (162). MrpJ binds to the *flhDC* promoter and represses transcription of *flhDC*, which results in repression of the downstream target of FlhD$_4$C$_2$, flagellin (*flaA*) (162, 203). During experimental murine infection, *mrpJ* transcription is highly induced within 1 day post infection and flagellar genes are repressed (Fig. 2) (25). At 7 days post infection, *mrp* gene expression falls (though it is still induced compared to broth culture) and flagellar gene expression rises. MrpJ is required for *P. mirabilis* virulence; during a co-challenge experiment, an *mrpJ* mutant was outcompeted in both the bladder (>100,000-fold) and kidneys (>10-fold) by the wild-type parent strain (203). It is unclear whether this reduction in virulence is due to changes in levels of FlaA and MrpA, the loss of coordination of motility and attachment, or loss of regulation of other virulence factors.

The *P. mirabilis* genome sequence revealed a surprising 14 additional MrpJ paralogs (17). Ten paralogs are located within fimbrial operons (Table 2), while four are orphans. Overexpression of 10 of the 14 MrpJ paralogs repressed swimming and swarming motility (162). Interestingly, one orphan paralog (PMI3508) repressed swimming but not swarming (162). MrpJ and its paralogs use multiple mechanisms and result in distinct phenotypes. For example, overexpression of *mrpJ* results in a radial swarm pattern with both swimming and swarming cells at the edge of the swarm front, while *ucaJ* overexpression completely represses FlaA and swarming, but does not affect differentiation into elongated swarmer cells. On the other hand, overexpressing *pmpJ* does not affect FlaA levels, but results in disorganized swarming, with both swimming and swarming cells at the edge of the swarm front (162)

(Fig. 8). Overexpression of *ucaJ* results in the fimbrial structures on the surface of vegetative cells changing from short, thick fimbriae to long, thin, fimbriae, suggesting that MrpJ and its paralogs might also coordinate expression of fimbriae (162).

Alignment of the 15 MrpJ sequences revealed a 10 to 26 amino acid N-terminal region followed by a predicted HTH motif and a 10 to 30 amino acid C-terminus. The HTH has a core region of nine conserved amino acids (SQQQFSRYE) (162). Single mutation of eight of the nine residues to alanine results in increased swimming motility when *mrpJ* is overexpressed. Insertion of 15 nucleotide linkers into the HTH domain also resulted in increased swimming motility, suggesting that this region is essential for function. Deletion of the C-terminal 27 amino acids, 25% of the protein, had no effect on MrpJ function (162). To completely abrogate MrpJ function, the C-terminal 45 amino acids had to be deleted, suggesting the C-terminus may not be important for MrpJ regulation of flagella (162).

Interestingly, although there is no sequence homology, MrpJ can be functionally complemented by PapX, a transcriptional regulator from the *E. coli pap* fimbrial operon (203). In *E. coli*, PapX directly binds the *flhDC* promoter to repress flagellar gene expression when P fimbriae are expressed (204, 205). Repression of flagella by components of fimbrial operons is emerging as a common regulatory theme. Searches of the recently annotated urinary pathogen *Providencia stuartii* 25827 genome, available in GenBank, revealed that *mrpJ* paralogs are also frequently associated with fimbrial operons in that species. Likewise, thirteen *mrpJ* paralogs are encoded by *Morganella morganii*, although only two are associated with fimbrial operons (206). Thus far, *mrpJ* paralogs have been identified by BLAST homology searches (207) in at least 20 other bacterial species. MrpJ functional homologs have also been discovered in enterohemorrhagic *E. coli*, and sequence homologs have been reported in *Xenorhabdus nematophila* and *Photorhabdus temperata* (208–210).

Autotransporters

The *P. mirabilis* genome encodes six putative autotransporters, also known as type V secretion proteins (211). Unlike other bacterial secretion systems, autotransporters do not encode specific proteins for their transport to the cell surface, but rather use conserved bacterial systems. Autotrans-

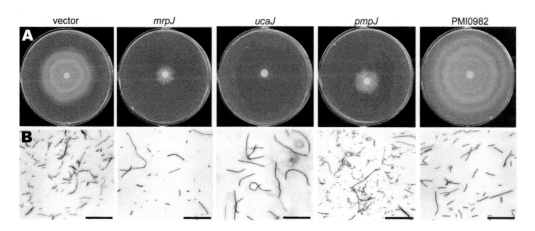

FIGURE 8 Overexpression of *mrpJ* and its paralogs results in distinct phenotypes. A, Swarming assays of *P. mirabilis* with an empty vector or expressing *mrpJ* or an *mrpJ* paralog. B, Gram-stained bacteria from the edge of the swarm front (bar, 50 μm). Modified from *Mol Microbiol* (162), with permission.

porters contain an N-terminal Sec-pathway-dependent signal sequence, an effector (α) domain, and a transmembrane (β) domain (211). Classical autotransporters encode a full β-barrel domain and can act as monomers, while trimeric autotransporters encode a shortened β-domain, which requires a trimer to form a β-barrel (212). *P. mirabilis* encodes three classical autotransporters and three trimeric autotransporters (17).

Classical autotransporters

Proteus toxic agglutinin (PMI2341, Pta) is the only characterized classical autotransporter from *P. mirabilis*. As its name implies, Pta acts as both a bacterial autoagglutinin and as a toxin to host cells. The Pta α-domain is a serine protease, which, despite remaining attached to the bacterial cell, can mediate cytotoxicity to host cells (213). Although the exact mechanism of Pta cytotoxicity is unknown, intoxication of host cells with Pta results in membrane damage, actin depolymerization, and eventual lysis (213). Pta expression is maximized during culture at 37°C and alkaline pH, conditions that occur during infection (213). During co-infection with wild-type *P. mirabilis*, a *pta* mutant is severely defective in colonizing the kidneys and spleen of mice, and has a minor defect in colonizing the bladder (213). The other two classical autotransporters (PMI0844 and PMI2126) are predicted to be proteases (17).

Trimeric Autotransporters

P. mirabilis encodes three putative trimeric autotransporters, of which AipA (adhesion and invasion mediated by the *Proteus* autotransporter, PMI2122) and TaaP (trimeric autoagglutination autotransporter of *Proteus*, PMI2575) have been characterized. Both proteins mediate binding to collagen and laminin *in vitro* (139). AipA mediates binding to HEK293 (kidney), UMUC-3 (bladder), and Vero (monkey kidney) cells *in vitro*. Co-challenge experiments revealed that a *P. mirabilis aipA* mutant colonizes the kidneys and the spleen less than the wild type

(139). TaaP does not bind any tested cell line, but mediates autoaggregation of *P. mirabilis*. A *P. mirabilis taaP* mutant is less successful at colonizing the bladder than the wild type in a co-challenge infection model (139). Also of note, TaaP is encoded within the horizontally-transmissible genetic element ICE*Pm1* (214).

Other adhesins

In addition to the chaperone-usher fimbriae, the *P. mirabilis* genome sequence revealed genes encoding one or two potential type IV pili. A putative type IV pilus on the surface of *P. mirabilis* was reported, although the size and expression pattern was consistent with the MR/P fimbria (215). However, *P. mirabilis* is capable of twitching motility, a flagellar-independent movement, which is associated with type IV pili (216, 217). More work is needed to determine if type IV pili are expressed and functional in *P. mirabilis*. Additionally, *P. mirabilis* encodes two genes that are homologous to the attachment and invasion locus genes of *Yersinia enterocolitica* (17). These genes are not upregulated in the murine urinary tract and have not been further characterized (25).

Hemolysin

In addition to the autotransporter toxin Pta, *P. mirabilis* encodes a *Serratia*-type calcium-independent hemolysin, which can lyse nucleated cells in addition to red blood cells (218, 219). The hemolysin is encoded by *hpmA*, while a second gene, *hpmB*, encodes a membrane transporter; both genes are highly conserved across *P. mirabilis* isolates (220). The levels of hemolysin in *P. mirabilis* correlate with its ability to invade cultured kidney cells, and an isogenic *P. mirabilis* Δ*hpmA* mutant is minimally invasive in cultured cells (136, 137). However, the contribution of HpmA to UTI remains unclear. In an independent challenge, *P. mirabilis hpmA* mutant bacteria colonize as well as the wild type, and *hpmA* is expressed at background levels in a murine infection model (25, 52, 219). In contrast, when wild-type or *hpmA* mutant *P. mirabilis* were tested in an IV injection mouse model, the *P. mirabilis hpmA* mutant had a 6-fold higher LD$_{50}$ than wild-type *P. mirabilis*, suggesting that hemolysin might be important during septic infection (219). A truncated hemolysin crystal structure has been solved, and a CXXC disulfide bond motif contributes to activity and stability of the protein (221). Co-challenge experiments with *P. mirabilis hpmA* mutant and wild-type *P. mirabilis* coupled with examination of tissue pathology are needed to further clarify the role of hemolysin in *P. mirabilis* infection.

Proteases

ZapA

For *P. mirabilis* to survive in the urinary tract, it must overcome host defenses, including antibodies and antimicrobial peptides. To accomplish this, *P. mirabilis* encodes an array of proteases (17). ZapA (mirabilysin) is a metalloprotease capable of mediating the degradation of numerous host proteins *in vitro* (222, 223). It was originally thought that *zapABCD* (PMI0276–0279) formed a single operon, with *zapBCD* encoding an ABC transporter that could export ZapA (222). However, Northern blot analysis detected only a single *zapA* transcript corresponding to the size of the *zapA* gene, and a putative promoter and ribosome binding site have been identified upstream of *zapBCD* (224). Transposon mutations in *zapBCD* affect production of ZapA activity, suggesting that ZapBCD is involved in the export of ZapA (222).

Early *in vitro* assays indicated that ZapA degrades mouse immunoglobulins, but recent work suggests that ZapA-mediated degradation of native immunoglobulins is less effective than degradation of the human antimicrobial proteins β-defensin-1 and LL-37 (also known as cathelicidin/CRAMP) (222,

223). Both β-defensin-1 and LL-37 are present in the urinary tract, with increased levels during UTI (225, 226). In particular, mice deficient in LL-37 are more susceptible to UPEC colonization of the kidney (225). ZapA mediated degradation of β-defensin-1 and LL-37 decreases their antimicrobial activity (223). Although *zapA* is expressed at low levels *in vivo*, a *P. mirabilis zapA* mutant is less efficient than wild type at colonizing the urine, bladder, and kidneys in independent challenge murine infections (25, 224). ZapA also contributes to acute prostatitis in a rat model of infection, and a *zapA* mutant is severely impaired in establishing chronic prostatitis in this model (227). In support of a role for ZapA in human infections, ZapA-like protease activity has been detected in the urine of individuals with *P. mirabilis* UTIs (228). ZapA is induced during swarming differentiation (66, 224), and expression remains high at the edge of an expanding swarm colony (65). ZapA is a possible therapeutic target, and toward this goal, chemical inhibitors have been identified (229).

Other proteases

Sequencing revealed four additional putative metalloproteases in the five open reading frames upstream of ZapA (17). The gene immediately upstream of *zapA* was named *zapE*, and the three putative metalloproteases appear to be *zapE* copies. One of the *zapE* copies (PMI0283) was identified as a virulence determinant using STM, although the mutagenesis employed a transposon, which might have affected the production of multiple metalloproteases (57). Additionally, a U32-family peptidase (PMI3442) was identified as important for infection in a separate STM study (230).

Metal Acquisition Systems

Iron

To acquire free iron in the host, pathogenic bacteria encode a number of different iron-acquisition systems, including siderophore-based systems, ferrous iron transport, and metal-type ABC transporters (231, 232). Siderophores are extracellular high affinity ferric iron (Fe^{3+}) chelators, which bind and solubilize iron. The siderophore-iron complex is then taken up by a specific outer membrane receptor, a process mediated by the TonB-ExbB-ExbD system (232). Uptake of ferrous iron (Fe^{2+}) is largely mediated by the FeoAB system, which is induced under anaerobic, iron-limiting conditions. Ferrous iron is then converted to ferric iron, although the mechanism remains unknown (232). Finally, metal-type ABC transporters are specific for iron, but may not require outer membrane receptors or siderophores. Additionally, bacteria may utilize bound iron in the forms of heme, transferrin, and lactoferrin (232). *P. mirabilis* is predicted to encode all four types of uptake systems, and can grow on hemoglobin, hemin, and ferric citrate as the sole iron source, but not on transferrin and lactoferrin (233) (Table 3). Additionally, *P. mirabilis* may use α-keto acids as iron chelators.

Prior to sequencing of the *P. mirabilis* genome, α-keto acids were the only known iron scavenging system in *P. mirabilis*. Initial studies show that α-keto acids, produced by an amino acid deanimase (Aad, PMI2834), could chelate iron from solutions (234, 235). Alpha-keto acids are also produced by *Providencia*, *Morganella*, and *Salmonella* species (234, 236). In *Salmonella typhimurium*, α-keto acids chelate iron in a TonB-dependent manner, but are less effective than siderophores (236, 237). In both *Salmonella* and *P. mirabilis*, α-keto acid production is not directly regulated by iron limitation (235, 236). Overall, α-keto acids appear to participate in iron chelation, but are probably not essential for growth during iron limitation.

P. mirabilis encodes two siderophore systems, the novel non-ribosomal peptide synthetase-independent siderophore system (proteobactin), and the non-ribosomal peptide siderophore system (Nrp). Proteobactin is predicted to be a hydroxycarboxylate

siderophore (238). The proteobactin locus involves three transcripts: one encodes an ABC transporter (PMI0229-030); the second transcript includes a single gene *pbtI* (PMI0231), which encodes an enzyme thought to catalyze synthesis of a proteobactin precursor; and the third transcript encompasses *pbtABCDEFGH* (PMI0232–0239), which encodes the remaining components necessary for siderophore synthesis and utilization (238).

The *P. mirabilis* Nrp system was identified in a swarming-defective *P. mirabilis* mutant (119). It was predicted to encode a siderophore based on its homology to the genes encoding the *Yersinia pestis* siderophore yersiniabactin (17). However, the *nrp* locus (PMI2596 and operon *nrpXYRSUTABG* PMI2597–2605) is missing genes essential for yersiniabactin production and encodes genes not found in the yersiniabactin operon (238). Since *P. mirabilis* is unable to produce or utilize yersiniabactin, the *nrp* operon is likely involved in the production and utilization of a siderophore distinct from yersiniabactin.

The Nrp and proteobactin systems are both involved in iron chelation, but only Nrp has been implicated in UTI (238). Wild-type *P. mirabilis* can chelate iron in the chrome azul S (CAS) assay (238). Single *P. mirabilis* *nrp* or *pbt* mutants have no defect in iron chelation, while a double mutant is incapable of iron chelation in the CAS assay (Fig. 9) (238). In an independent challenge experiment, neither a single *P. mirabilis* *pbt*, *nrp*, nor a double *nrp* *pbt* mutant had a defect in colonization compared to wild type. However, during a co-challenge experiment, the *P. mirabilis* *nrp* mutant was impaired in colonization of the bladder and kidneys (238). Interestingly, the *P. mirabilis* *nrp* *pbt* mutant was only impaired in colonization of the kidneys (238). These results suggest that the Nrp system is more important for colonizing the urinary tract than proteobactin.

P. mirabilis can grow on hemin, and a putative heme uptake system is upregulated

TABLE 3 Iron-related genes in *P. mirabilis*. Iron-related genes from *P. mirabilis* were identified by homology to other iron-related genes. Genes identified as iron-related by homology but not identified using one of the four conditions shown were excluded. A checkmark indicates that one or more of the genes in the row were identified using the condition specified

Gene Designation(s)	Proposed function	PMI number(s)	Upregulated in iron limitation	Upregulated *in vivo*	Implicated *in vivo*	Antigenic *in vivo*	References
Heme uptake							
	TonB-dependent receptor	PMI0409	✓			✓	45, 238
	hemin uptake protein	PMI1424	✓	✓			25, 238
hmuR1R2STUV	hemin uptake system	PMI1425-1430	✓	✓	✓	✓	25, 45, 238, 239
Ferrous iron uptake							
sitDCBA	Iron ABC transporter	PMI1024-1027	✓	✓			25, 238
feoAB	ferrus iron transport	PMI2920-2921		✓			25
Ferric citrate transport							
	exported protease	PMI3704	✓				238
	iron-related ABC transporter	PMI3705	✓		✓		57, 238
	iron receptor	PMI3706-3707	✓				238
	extracytoplasmic function family σ factor	PMI3708	✓				238
	TonB-like protein	PMI3709	✓				238
Siderophore production							
	TonB-dependent receptor	PMI2596	✓		✓	✓	45, 238
nrpXYRSUTABG	Nrp siderophore	PMI2597-2605	✓	✓	✓		25, 56, 238
pbtlABCDEFGH	proteobactin	PMI0231-0239	✓	✓			57, 238
Other TonB-dependent receptors							
irgA	TonB-dependent enterobactin receptor	PMI0842	✓	✓	✓	✓	25, 45, 238
ireA	ferric siderophore receptor	PMI1945	✓	✓		✓	25, 45, 238
	TonB-dependent receptor	PMI0363	✓	✓			25, 238
	TonB-dependent receptor	PMI1548-1551	✓	✓			25, 238
hasR	TonB-dependent receptor	PMI3120-3121			✓		56
ABC-transport system							
	iron-related ABC transporter	PMI0331	✓	✓			25, 238
	iron-related ABC transporter	PMI2957-2960	✓	✓	✓		25, 56, 238

(Continued on next page)

TABLE 3 Iron-related genes in *P. mirabilis*. Iron-related genes from *P. mirabilis* were identified by homology to other iron-related genes. Genes identified as iron-related by homology but not identified using one of the four conditions shown were excluded. A checkmark indicates that one or more of the genes in the row were identified using the condition specified *(Continued)*

Gene Designation(s)	Proposed function	PMI number(s)	Upregulated in iron limitation	Upregulated *in vivo*	Implicated *in vivo*	Antigenic *in vivo*	References
	Iron-related ABC transporter	PMI0229-0230	✓				238
Iron metabolism							
	iron utilization protein	PMI1437	✓	✓			25, 238
Iron sulfur cluster formation/uptake							
	iron sulfur cluster	PMI1411-1416	✓	✓			25, 238
	iron sulfur cluster	PMI3253	✓	✓			25, 238
	iron sulfur cluster	PMI0176-0172	✓	✓			25, 238

during iron limitation and *in vivo*. PMI1426 (*hmuR2*), which encodes a putative outer membrane receptor in the heme uptake system, binds heme, is implicated in infection, and has been identified as an antigen *in vivo* (25, 233, 238, 239).

Additionally, *P. mirabilis* encodes genes predicted to be involved in uptake of ferrous iron, ferric iron, and ferric citrate; iron sulfur cluster formation and uptake; as well as a number of predicted TonB-dependent receptors and iron-related ABC-transport systems (17, 238). Several of these genes have been identified as being iron regulated *in vitro*, or contribute to UTI by being positively regulated *in vivo*, expressed on the bacterial cell surface *in vivo*, or important for full colonization of the urinary tract (Table 2) (25, 45, 56, 57, 238). However, the functions of these systems have not been directly tested.

Zinc

P. mirabilis encodes a functional zinc uptake system, *znuACB* (118). Zinc levels are thought to be limited in the urinary tract and zinc uptake provides a competitive advantage to other pathogenic bacteria, including uropathogenic *E. coli* (240). In *P. mirabilis*, *znuCB* is upregulated *in vivo* and a *P. mirabilis znuC* mutant is out-competed by wild-type *P. mirabilis* during co-challenge infection (25, 118). Zinc is predicted to be needed for other

virulence factors, including flagella and *zapA*. The master flagellar regulator, $FlhD_4C_2$, has a putative zinc-binding site and a *P. mirabilis znuC* mutant expresses less flagellin, suggesting that zinc uptake is essential for flagellar synthesis (48, 119). As previously discussed, the metalloprotease ZapA provides a competitive advantage during infection, and its activity is thought to be zinc-dependent (223). Without a functional zinc-uptake system, *P. mirabilis* may improperly regulate expression of flagellin, ZapA, and possibly other factors, leading to weakened colonization of the urinary tract.

Phosphate Transport

P. mirabilis encodes a high-affinity inorganic phosphate (P_i) transporter very similar to the *E. coli* Pst operon (241). The *P. mirabilis pst* operon includes genes (PMI2893–PMI2896) encoding four proteins associated with phosphate transport (PstSCAB) and a negative transcriptional regulator (PMI2897) of the phosphate regulon, PhoU. When the *pst* operon is upregulated, so are other members of the phosphate regulon, including *phoA* (PMI2500), the gene encoding alkaline phosphatase. Thus, alkaline phosphatase activity can be measured as an indirect output of the *pst* operon and of phosphate limitation (241). When alkaline phosphatase activity is

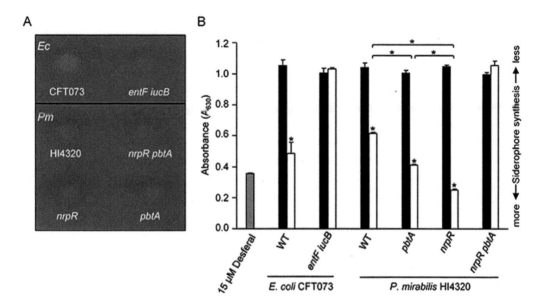

FIGURE 9 *P. mirabilis* **iron chelation is Nrp and proteobactin dependent. A, agar; and B, solution chrome azurol S (CAS) assays of uropathogenic** *E. coli* **CFT073 and** *P. mirabilis* **HI4320; a color change from blue to orange indicates iron chelation. In B),** *P. mirabilis* **supernatants from log-phase cultures grown in MOPS defined media either with 0.1 mM FeCl$_3$·6H$_2$O (black bars) or without supplementation (white bars) were concentrated 50-fold before being used in a liquid CAS assay (** *E. coli* **supernatants were not concentrated). Single** *P. mirabilis* ***nrpR* and *pbtA* mutants are not impaired in iron chelation, but the double** *P. mirabilis* ***nrpR pbtA* mutant is. Reprinted from** *Mol Microbiol* **(238), with permission.**

measured in wild-type *P. mirabilis*, there is very little activity in LB media and very high activity in phosphate-limited media. In human urine, there is intermediate induction of alkaline phosphatase, suggesting that urine is somewhat phosphate limited (241).

Two mutations in the *pst* operon were identified using STM as being essential for full infection of the murine urinary tract (56). Additional experiments showed that individual transposon insertions in *pstA* and *pstS* were outcompeted in the urine, bladder, and kidneys during co-challenge infection with wild type, which could be complemented by plasmid-encoded expression of the *pst* operon (56, 241). A general growth defect was not responsible for the phenotype observed in mice, as the *pst* mutants out-competed wild type during co-culture in human urine (241). However, both *pst* mutants were impaired in biofilm formation. While wild-type *P. mirabilis* formed biofilms with a dense

3D structure reminiscent of towering mushrooms with liquid channels throughout, the *pst* mutant biofilms lacked these distinctive structures (242). This may suggest a role for biofilms during UTI independent of catheter encrustation, or the *pst* operon may contribute to UTI through a yet to be characterized mechanism.

Virulence Factors Identified by STM

Three studies used an unbiased signature-tagged mutagenesis screen to identify genes important for infection in the murine urinary tract. These screens identified genes important for many of the virulence factors described above, including urease, motility, iron acquisition, fimbriae, phosphate transport, ZapE, and TaaP (56, 57, 230). Approximately 20% of the genes identified are involved in metabolism, including genes involved with glycolysis (*aceE* PMI2046), the TCA cycle (sdhC PMI0565), and the Entner–Doudoroff pathway (edd PMI2760). Genes associated with amino-acid metabolism and transport were identified, including D-methionine transporter metN (PMI2259), dihydrouridine synthase dusB (PMI3623), carbamoyl phosphate synthetase carA (PMI0020), and L-serine deaminase sdaA (PMI1607) (56, 57). Genes involved in GTP synthesis (guaB PMI1546), adenylate cyclase (cyaA PMI3333) and a predicted oxidoreductase with similarities to formate dehydrogenase (cbbBc PMI2378) were also identified (57). A gene recognized to regulate flhDC in E. coli (hdfR, PMI3295) was identified, although this mutation resulted in no apparent motility defect in P. mirabilis (56). A second transcriptional regulator (nhaR PMI0012) was also identified (57). Two genes likely involved in plasmid maintenance and that localized to the P. mirabilis plasmid pHI4320 were identified in these screens (parE (PMIP32) and pilX (PMIP08-09)), suggesting that this plasmid plays a role in virulence (56). The other genes identified largely encode hypothetical proteins, and the mechanism of their contribution to virulence has yet to be deduced.

Genome Organization

Following the first completely sequenced P. mirabilis genome, UTI isolate HI4320 (17), three new P. mirabilis genome sequences have been published: laboratory strain BB2000 (154), diarrhea isolate C05028 (243), and blood isolate PR03 (244). There are three more strains that have been partially sequenced and assembled (type strain ATCC 29906, urine isolate WGLW4, and mouse stool isolate WGLW6). The genome size of these seven strains ranges from 3.82 Mb to 4.06 Mb, and they have GC content ranging from 38.5% to 39.0%. Strain HI3420 also possesses a plasmid, pHI4320. STM studies indicate the plasmid contributes to virulence during UTI (56). The BB2000 genome does not contain any plasmids; the plasmid status of the other partially assembled

genomes has not been specified. Most of the notable features of the HI3420 genome have already been discussed in this chapter, including the 17 chaperone-usher fimbrial operons, contiguous flagellar locus, and other virulence factors. A type III secretion system is encoded by all currently-sequenced P. mirabilis genomes, and although there is evidence for transcription of the genes within this locus, it does not appear to contribute to UTI (245). Comparison of ten uropathogenic E. coli genomes led to the identification of 131 genes that are specific to UPEC compared to three fecal E. coli isolates (246); P. mirabilis encodes homologs of 25 of these genes (17). This species also encodes a pathogenicity island, ICEPm1, which is commonly found in P. mirabilis, Providencia stuartii, and Morganella morganii (214). Virulence genes, including the taaP autotransporter and the nrp siderophore system, are encoded in ICEPm1. This element can be transmitted to other P. mirabilis clinical isolates (247).

Gene Expression During UTI

Microarray analysis has been used to examine the transcriptome of P. mirabilis isolated from the urine of experimentally-infected mice (25). Compared to mid-logarithmic phase broth culture, 471 genes were induced and 82 were repressed in vivo. Many of the upregulated genes encode virulence factors that have been already discussed (urease, MR/P fimbriae, and iron uptake). These findings were generally similar to the transcriptome of uropathogenic E. coli in vivo (induction of fimbriae, iron and peptide uptake systems, pyruvate catabolism, osmoprotection; repressed flagella) (50).

There were several differences in metabolic gene expression between UPEC and P. mirabilis, most notably with respect to nitrogen assimilation: in P. mirabilis, gdhA (glutamate dehydrogenase, PMI3008) is induced and glnA (glutamine synthetase, PMI2882) is repressed (25); the opposite is true for UPEC. Since the glutamine synthetase system primarily operates during nitrogen starvation, it appears that the carbon-nitrogen balance in *P. mirabilis* is altered compared to *E. coli*. Genes in the TCA cycle corresponding to the steps from acetyl coenzyme A to α-ketoglutarate synthesis are also induced *in vivo* by *P. mirabilis*. The depletion of α-ketoglutarate pools from the TCA cycle due to GDH-mediated ammonia uptake was proposed to be responsible for the induction of this portion of the TCA cycle. Although urease seemed the likely candidate responsible for this pattern, culture of *P. mirabilis* in the presence of urease or exogenous ammonium did not recapitulate *in vivo* gene expression. Instead, the pattern was restored using citrate as a sole carbon source. A *gdhA* mutant is also not as fit during experimental murine co-challenge (25).

Biofilm Formation

CAUTI are generally initiated by biofilm formation on the urinary catheter. After catheter placement, host-derived factors from the urine and the urogenital tract are deposited on the catheter, creating a coating that facilitates bacterial attachment. A number of studies have reported on the contribution of specific factors to *P. mirabilis* biofilm formation *in vitro*, some of which have been described in detail above and elsewhere (urease, MR/P fimbriae, Pst, and RsbA) (248). Capsular polysaccharides purified from *P. mirabilis* have been reported to increase crystalline biofilm formation in artificial urine (249). More recently, *P. mirabilis* transposon mutants were screened within an artificial catheterized bladder model to find factors that contribute to catheter biofilms (250). In this assay, *nirB* nitrate reductase (PMI1479), *bcr* putative multidrug efflux pump (PMI0829), and two genes of unknown function (PMI1608 and PMI3402) were identified. Mutations in *nirB* and *bcr* resulted in decreased biomass and calcium associated with the catheter compared to wild type, but similar levels of un-attached biomass and of adherence to silicone surfaces (250).

CLINICAL ASPECTS OF *P. MIRABILIS* UTI

Prevention

The main risk factors for *P. mirabilis* infection are an anatomical or functional abnormality in the urinary tract or the presence of a catheter. The first cannot be easily prevented. The presence of a catheter is associated with a 3% to 10% risk of bacterial infection per day (11). Catheterization should be avoided whenever possible, and when it cannot be avoided, duration of catheterization should be limited. The preferred technique when a catheter is needed is clean, intermittent catheterization (11). It is recommended that catheters are changed regularly when they cannot be removed, but there is no strong clinical evidence to support this recommendation (11). However, as catheterization is indeed necessary in some patients, development of bacterial growth-resistant catheters and of a *P. mirabilis* vaccine is ongoing (see "Progress Towards a *P. mirabilis* Vaccine" section).

Treatment

The currently recommended treatment for acute uncomplicated cystitis is a 3-day course of double-strength sulfamethoxazole and trimethoprim (SMZ-TMP) if the local SMZ-TMP resistance rate is less than 20% (9, 251). However, this organism has a wide spectrum of local SMZ-TMP resistance rates ranging from 16% to 83% (9, 252) (see "Antibiotic Resistance" section below). If the SMZ-TMP resistance rate is above 10% to 20%, the recommended alternative antibiotic therapy for uncomplicated cystitis may include fluroquinolones, nitrofurantoin, or fosfomycin, all of which have some reported resistance (9, 251, 253). There are not specific recommendations for complicated- or catheter-associated UTIs. Instead, antibiotic treatment is based on the specific situation and the clinician's discretion.

When *P. mirabilis* infections are found in connection with a urinary stone, extra precautions need to be taken because bacteria are found inside the stone. The stone needs to be completely eradicated with shockwave lithotripsy or otherwise removed, and urine should be checked monthly for 3 months after antibiotic treatment is completed (254). Even in the absence of an obvious stone, follow-up urine cultures are recommended in *P. mirabilis* UTI to ensure that the infection is cleared from the urinary tract and to prevent stones from occurring.

Antibiotic Resistance

Worldwide, *P. mirabilis* has developed resistance to several classes of antibiotics, complicating treatment. In addition to the described resistance to SMZ-TMP, resistance to β-lactams (both penicillins and cephalosporins), fluroquinolones, nitrofurantoin, fosfomycin, aminoglycosides, tetracyclines, and sulfonamides has been reported (252, 253, 255). In particular, most isolates are resistant to tetracycline (1). The tendency of this organism to become encased in urinary stones or within crystalline biofilms on urinary catheters can shield bacteria and thus lead to treatment failures (see the "Urease" section above). *P. mirabilis* is also highly resistant to antimicrobial peptides including polymyxin B, protegrin, LL-37, and defensin (112, 223). This resistance relies on LPS modifications and extracellular proteases such as ZapA (112, 223, 256). The issues of increasing antibiotic resistance by this organism are otherwise largely similar to those for UPEC, which is covered in detail elsewhere in this book.

Progress Toward a *P. mirabilis* Vaccine

The difficulty in treating *P. mirabilis* and the increase in antibiotic-resistant infection speaks to the importance of developing a *P. mirabilis* vaccine. Vaccinating the general population is unlikely to be necessary as *P. mirabilis* only causes 3% to 5% of all UTI. This vaccine would be beneficial, however, for populations more prone to *P. mirabilis* UTI, including individuals with functional or structural urinary tract abnormalities, individuals starting long-term catheterization, and patients suffering from recurrent UTI (257). Vaccination of these patients could drastically decrease the incidence of *P. mirabilis* UTI.

An initial *P. mirabilis* infection only modestly protects against subsequent homologous and heterologous challenge despite the fact that infection generates a specific adaptive immune response (45, 257). Although this is also true for uropathogenic *E. coli*, vaccines are available that partially protect humans from recurrent *E. coli* UTI (258). This suggests that a vaccine is a viable option for *P. mirabilis*. A successful vaccine will activate a strong adaptive immune response, specifically, the production of mucosal IgA in the urinary tract against *P. mirabilis* antigens exposed to the immune system.

A number of *P. mirabilis* vaccines have been tested in a murine UTI model. The vaccines tested have varied the route of inoculation, the antigen, and the method of antigen display.

Although the protocol varies between studies, an initial vaccination is generally followed by 3 to 5 booster doses, given 7 to 10 days apart. The post-vaccination challenge occurs seven days after the last booster was administered (257, 259).

Initial *P. mirabilis* studies largely vaccinated via the transurethral and subcutaneous routes (260–262). However, in order to generate a stronger mucosal IgA response, oral and intranasal vaccines were introduced. Intranasal vaccination provides the best protection and generates a strong mucosal antibody response in both the urine and the vagina (257).

Complex antigens, including formalin-killed whole *P. mirabilis* and *P. mirabilis* outer membranes, provide some protection against homologous and heterologous challenge and generate a specific immune response (257, 261). However, as they are difficult to manipulate and failed to initiate an acceptable immune response, other vaccine types were tried.

Several *P. mirabilis* proteins have been tested in subunit vaccines, where a purified protein, sometimes conjugated to an adjuvant, is used to elicit an immune response. Fimbrial proteins (MrpA, UcaA, PmfA, and MrpH) have been the primary target of these vaccine efforts, though the Pta toxin has also been tested. Although it is difficult to compare results from different vaccine trials, MrpH seems to elicit the strongest response. A truncated MrpH-maltose binding protein fusion conjugated to cholera toxin prevented infection in 75% of mice (257). All these antigens provide some protection, with protection in the kidneys generally stronger than in the bladder (257, 263, 264). Additionally, all tested antigens induce specific antibody responses in either the blood or the urine of the host (257, 263, 264). However, none of the studies to date can correlate the induction of a specific antibody response with the level of protection, as has been seen with an *E. coli* FimH vaccine (257, 263–265).

Recent studies have also tried recombinant-vector vaccines, where the *P. mirabilis* antigen is expressed by different, non-pathogenic bacteria. An advantage of a recombinant-vector vaccine is that it can multiply in the host, increasing the exposure to the immune system, but there is no possibility of reversion as in an attenuated vaccine. Two recombinant-vector vaccines have been tested as a delivery vectors for MrpA: *Salmonella typhimurium* and *Lactococcus lactis*. Although the *S. typhimurium* vaccine protected better than the *L. lactis* vaccine, it is unclear whether the live vaccine approach provided greater protection than the recombinant MrpA vaccine (266, 267).

A defined multi-antigen vaccine has yet to be tested. In particular, because the *P. mirabilis* fimbriae may be redundant in their ability to adhere to the surfaces of the urinary tract (and at least one fimbria, MR/P, is phase-variable), a vaccine with proteins from multiple fimbriae should be tested. Recent work identifying antigenic proteins in the outer membrane of *P. mirabilis* during infection identified several other potential vaccine targets, including iron acquisition genes essential for UTI (45). Perhaps a combination of antigens including fimbrial and non-fimbrial targets will prove most effective.

Prevention of Urinary Catheter Biofilm Formation

Many efforts have been made to design catheters that are resistant to bacterial colonization, with the goals of preventing catheter blockage and reducing UTI in catheterized patients. Catheters made from latex, silicone, polyurethane, and composite biomaterials have been tested, but no single catheter material is sufficient to prevent bacterial colonization (6). Several groups have attempted to prevent biofilm formation on catheters by applying or embedding an antimicrobial solution on the catheter surface or in the catheter material. Antibiotic coatings, such as nitrofurazone or a combination of broad-spectrum antibiotics, have significantly reduced the number of CAUTIs in small clinical trials (6, 268). However, given the ubiquity of catheterization, this approach will likely result in antibiotic resistance, a difficulty that has not been resolved. A similar approach involving coating catheters with antiseptics, specifically various silver compounds, has been studied extensively. However, the combination of poorly designed clinical trials and conflicting results have made it difficult to determine the efficacy of these treatments (6, 268).

Several promising techniques have been tested *in vitro* but have not yet been tested in clinical trials. These include the use of catheters impregnated with combinations of chlorhexidine, silver sulfadiazine, and triclosan; the use of urease inhibitors to prevent crystalline biofilm formation; and the treatment of catheters with a cocktail of bacteriophages (6, 268). Most of the techniques tested have been chemical, not mechanical, which makes the promising *in vitro* results using low energy acoustic waves to prevent bacterial attachment and the development of a urinary catheter that uses inflation-generated catheter strain to clear *P. mirabilis* crystalline biofilms intriguing alternative approaches (268, 269).

Although complete prevention of biofilms on catheters is ideal, the most serious complications of CAUTI generally occur when the catheter becomes blocked by the presence of crystalline biofilms that are the result of urease activity. A sensor which detects the initial stages of crystalline biofilm formation early enough to allow the catheter to be changed prior to catheter blockage would help mitigate the effects of biofilm formation on catheters. An early prototype of this sensor detected crystalline biofilm formation around 12 days prior to blockage in a clinical trial (270). A more recent sensor, which is amenable to mass production, detected *P. mirabilis* crystalline biofilm formation 17 to 24 hours in advance (271). This technology could significantly decrease the morbidity and costs associated with CAUTI.

CONCLUSIONS

Significant recent advances have been made in our understanding of the lifestyle of this pathogen both with regard to swarming and to pathogenesis during UTI. However, there are still substantial gaps in our knowledge. What triggers the cyclic switch between swarming and consolidation? How important are motility and swarming to virulence, and at which points during virulence is motility involved? Why does this species encode such an extensive array of adherence factors, which of these contribute to UTI, and what are their host targets? Is invasion of either bladder or renal cells a component of path-

ogenesis? Very little investigation of the immune response during *P. mirabilis* UTI has occurred; is the response similar to the response for uropathogenic *E. coli*, or are there differences? Studies in this area could aid vaccine design. CAUTI are often polymicrobial. How is UTI affected by the interaction of *P. mirabilis* with other species? Finally, numerous genes have been identified by transposon mutagenesis, STM, and microarray analysis as being important for or regulated during swarming or UTI. However, the mechanism of most of these contributions is yet to be determined.

ACKNOWLEDGMENTS

This work was supported by grant AI083743 from the National Institutes of Health (M.M.P.).

Conflicts of interest: We declare no conflicts.

CITATION

Schaffer JN, Pearson MM. 2015. *Proteus mirabilis* and urinary tract infections. Microbiol Spectrum 3(5):UTI-0017-2013.

REFERENCES

1. **O'Hara CM, Brenner FW, Miller JM.** 2000. Classification, identification, and clinical significance of *Proteus, Providencia,* and *Morganella. Clin Microbiol Rev* **13:**534–546.
2. **Matthews SJ, Lancaster JW.** 2011. Urinary tract infections in the elderly population. *Am J Geriatr Pharmacother* **9:**286–309.
3. **Papazafiropoulou A, Daniil I, Sotiropoulos A, Balampani E, Kokolaki A, Bousboulas S, Konstantopoulou S, Skliros E, Petropoulou D, Pappas S.** 2010. Prevalence of asymptomatic bacteriuria in type 2 diabetic subjects with and without microalbuminuria. *BMC Res Notes* **3:**169.
4. **Janda JMA, Abbott SL.** 2006. *The Enterobacteria,* 2 ed. ASM Press, Washington, D.C.
5. **Mathur S, Sabbuba NA, Suller MT, Stickler DJ, Feneley RC.** 2005. Genotyping of urinary and fecal *Proteus mirabilis* isolates from individuals with long-term urinary

catheters. *Eur J Clin Microbiol Infect Dis* **24:** 643–644.

6. **Jacobsen SM, Stickler DJ, Mobley HLT, Shirtliff ME.** 2008. Complicated catheter-associated urinary tract infections due to *Escherichia coli* and *Proteus mirabilis*. *Clin Microbiol Rev* **21:**26–59.

7. **Gaston H.** 1995. *Proteus*—is it a likely aetiological factor in chronic polyarthritis? *Ann Rheum Dis* **54:**157–158.

8. **Rashid T, Ebringer A.** 2007. Rheumatoid arthritis is linked to *Proteus*—the evidence. *Clin Rheumatol* **26:**1036–1043.

9. **Karlowsky JA, Lagacé-Wiens PR, Simner PJ, DeCorby MR, Adam HJ, Walkty A, Hoban DJ, Zhanel GG.** 2011. Antimicrobial resistance in urinary tract pathogens in Canada from 2007 to 2009: CANWARD surveillance study. *Antimicrob Agents Chemother* **55:**3169–3175.

10. **Nielubowicz GR, Mobley HLT.** 2010. Host-pathogen interactions in urinary tract infection. *Nat Rev Urol* **7:**430–441.

11. **Nicolle LE.** 2005. Catheter-related urinary tract infection. *Drugs Aging* **22:**627–639.

12. **Hung EW, Darouiche RO, Trautner BW.** 2007. *Proteus* bacteriuria is associated with significant morbidity in spinal cord injury. *Spinal Cord* **45:**616–620.

13. **Adams-Sapper S, Sergeevna-Selezneva J, Tartof S, Raphael E, Diep BA, Perdreau-Remington F, Riley LW.** 2012. Globally dispersed mobile drug-resistance genes in gram-negative bacterial isolates from patients with bloodstream infections in a US urban general hospital. *J Med Microbiol* **61:**968–974.

14. **Mylotte JM.** 2005. Nursing home-acquired bloodstream infection. *Infect Control Hosp Epidemiol* **26:**833–837.

15. **Sader HS, Flamm RK, Jones RN.** 2014. Frequency of occurrence and antimicrobial susceptibility of Gram-negative bacteremia isolates in patients with urinary tract infection: results from United States and European hospitals (2009–2011). *J Chemother* **26:**133–138.

16. **Lubart E, Segal R, Haimov E, Dan M, Baumoehl Y, Leibovitz A.** 2011. Bacteremia in a multilevel geriatric hospital. *J Am Med Dir Assoc* **12:**204–207.

17. **Pearson MM, Sebaihia M, Churcher C, Quail MA, Seshasayee AS, Luscombe NM, Abdellah Z, Arrosmith C, Atkin B, Chillingworth T, Hauser H, Jagels K, Moule S, Mungall K, Norbertczak H, Rabbinowitsch E, Walker D, Whithead S, Thomson NR, Rather PN, Parkhill J, Mobley HLT.** 2008. Complete genome sequence of uropathogenic *Proteus*

mirabilis, a master of both adherence and motility. *J Bacteriol* **190:**4027–4037.

18. **Mobley HLT.** 2001. Urease. *In* Mobley HLT, Mendz GL, Hazell SL (ed), *Helicobacter pylori: physiology and genetics* doi:NBK2417. ASM Press, Washington, DC.

19. **Mobley HLT.** 1996. Virulence of *Proteus mirabilis*, p 245–269. *In* Mobley HL, Warren JW (ed), *Urinary Tract Infections: Molecular Pathogenesis and Clinical Management*. ASM Press, Washington, D.C.

20. **Griffith DP, Musher DM, Itin C.** 1976. Urease. The primary cause of infection-induced urinary stones. *Invest Urol* **13:**346–350.

21. **Nicholson EB, Concaugh EA, Foxall PA, Island MD, Mobley HLT.** 1993. *Proteus mirabilis* urease: transcriptional regulation by UreR. *J Bacteriol* **175:**465–473.

22. **Dattelbaum JD, Lockatell CV, Johnson DE, Mobley HLT.** 2003. UreR, the transcriptional activator of the *Proteus mirabilis* urease gene cluster, is required for urease activity and virulence in experimental urinary tract infections. *Infect Immun* **71:**1026.

23. **Poore CA, Mobley HLT.** 2003. Differential regulation of the *Proteus mirabilis* urease gene cluster by UreR and H-NS. *Microbiology* **149:** 3383–3394.

24. **Zhao H, Thompson RB, Lockatell V, Johnson DE, Mobley HLT.** 1998. Use of green fluorescent protein to assess urease gene expression by uropathogenic *Proteus mirabilis* during experimental ascending urinary tract infection. *Infect Immun* **66:**330–335.

25. **Pearson MM, Yep A, Smith SN, Mobley HLT.** 2011. Transcriptome of *Proteus mirabilis* in the murine urinary tract: virulence and nitrogen assimilation gene expression. *Infect Immun* **79:**2619–2631.

26. **Munns J, Amawi F.** 2010. A large urinary bladder stone: an unusual cause of rectal prolapse. *Arch Dis Child* **95:**1026.

27. **Chew R, Thomas S, Mantha ML, Killen JP, Cho Y, Baer RA.** 2012. Large urate cystolith associated with *Proteus* urinary tract infection. *Kidney Int* **81:**802.

28. **Jones BV, Mahenthiralingam E, Sabbuba NA, Stickler DJ.** 2005. Role of swarming in the formation of crystalline *Proteus mirabilis* biofilms on urinary catheters. *J Med Microbiol* **54:**807–813.

29. **Stickler DJ.** 2008. Bacterial biofilms in patients with indwelling urinary catheters. *Nat Clin Pract Urol* **5:**598–608.

30. **Torzewska A, Budzyńska A, Białczak-Kokot M, Różalski A.** 2014. *In vitro* studies of epithelium-associated crystallization caused

by uropathogens during urinary calculi development. *Microb Pathog* **71–72C:**25–31.

31. **Mobley HLT, Warren JW.** 1987. Urease-positive bacteriuria and obstruction of long-term urinary catheters. *J Clin Microbiol* **25:**2216–2217.

32. **Li X, Zhao H, Lockatell CV, Drachenberg CB, Johnson DE, Mobley HLT.** 2002. Visualization of *Proteus mirabilis* within the matrix of urease-induced bladder stones during experimental urinary tract infection. *Infect Immun* **70:**389–394.

33. **Jones BD, Lockatell CV, Johnson DE, Warren JW, Mobley HLT.** 1990. Construction of a urease-negative mutant of *Proteus mirabilis*: analysis of virulence in a mouse model of ascending urinary tract infection. *Infect Immun* **58:**1120–1123.

34. **Johnson DE, Russell RG, Lockatell CV, Zulty JC, Warren JW, Mobley HLT.** 1993. Contribution of *Proteus mirabilis* urease to persistence, urolithiasis, and acute pyelonephritis in a mouse model of ascending urinary tract infection. *Infect Immun* **61:**2748–2754.

35. **Armbruster CE, Smith SN, Yep A, Mobley HLT.** 2014. Increased incidence of urolithiasis and bacteremia during *Proteus mirabilis* and *Providencia stuartii* coinfection due to synergistic induction of urease activity. *J Infect Dis* **209:**1524–1532.

36. **Follmer C.** 2010. Ureases as a target for the treatment of gastric and urinary infections. *J Clin Pathol* **63:**424–430.

37. **Suller MT, Anthony VJ, Mathur S, Feneley RC, Greenman J, Stickler DJ.** 2005. Factors modulating the pH at which calcium and magnesium phosphates precipitate from human urine. *Urol Res* **33:**254–260.

38. **Khan A, Housami F, Melotti R, Timoney A, Stickler D.** 2010. Strategy to control catheter encrustation with citrated drinks: a randomized crossover study. *J Urol* **183:**1390–1394.

39. **Macnab RM.** 2003. How bacteria assemble flagella. *Annu Rev Microbiol* **57:**77–100.

40. **Belas R, Flaherty D.** 1994. Sequence and genetic analysis of multiple flagellin-encoding genes from *Proteus mirabilis*. *Gene* **148:**33–41.

41. **Belas R.** 1994. Expression of multiple flagellin-encoding genes of *Proteus mirabilis*. *J Bacteriol* **176:**7169–7181.

42. **Murphy CA, Belas R.** 1999. Genomic rearrangements in the flagellin genes of *Proteus mirabilis*. *Mol Microbiol* **31:**679–690.

43. **Manos J, Belas R.** 2004. Transcription of *Proteus mirabilis* *flaAB*. *Microbiology* **150:**2857–2863.

44. **Manos J, Artimovich E, Belas R.** 2004. Enhanced motility of a *Proteus mirabilis* strain expressing hybrid FlaAB flagella. *Microbiology* **150:**1291–1299.

45. **Nielubowicz GR, Smith SN, Mobley HLT.** 2008. Outer membrane antigens of the uropathogen *Proteus mirabilis* recognized by the humoral response during experimental murine urinary tract infection. *Infect Immun* **76:**4222–4231.

46. **Chevance FF, Hughes KT.** 2008. Coordinating assembly of a bacterial macromolecular machine. *Nat Rev Microbiol* **6:**455–465.

47. **Claret L, Hughes C.** 2000. Functions of the subunits in the FlhD(2)C(2) transcriptional master regulator of bacterial flagellum biogenesis and swarming. *J Mol Biol* **303:**467–478.

48. **Wang S, Fleming RT, Westbrook EM, Matsumura P, McKay DB.** 2006. Structure of the *Escherichia coli* FlhDC complex, a prokaryotic heteromeric regulator of transcription. *J Mol Biol* **355:**798–808.

49. **Bahrani FK, Johnson DE, Robbins D, Mobley HLT.** 1991. *Proteus mirabilis* flagella and MR/P fimbriae: isolation, purification, N-terminal analysis, and serum antibody response following experimental urinary tract infection. *Infect Immun* **59:**3574–3580.

50. **Snyder JA, Haugen BJ, Buckles EL, Lockatell CV, Johnson DE, Donnenberg MS, Welch RA, Mobley HLT.** 2004. Transcriptome of uropathogenic *Escherichia coli* during urinary tract infection. *Infect Immun* **72:**6373.

51. **Lane MC, Alteri CJ, Smith SN, Mobley HLT.** 2007. Expression of flagella is coincident with uropathogenic *Escherichia coli* ascension to the upper urinary tract. *Proc Natl Acad Sci USA* **104:**16669–16674.

52. **Mobley HLT, Belas R, Lockatell V, Chippendale G, Trifillis AL, Johnson DE, Warren JW.** 1996. Construction of a flagellum-negative mutant of *Proteus mirabilis*: effect on internalization by human renal epithelial cells and virulence in a mouse model of ascending urinary tract infection. *Infect Immun* **64:**5332–5340.

53. **Pazin GJ, Braude AI.** 1974. Immobilizing antibodies in urine. II. Prevention of ascending spread of *Proteus mirabilis*. *Invest Urol* **12:**129–133.

54. **Zunino P, Piccini C, Legnani-Fajardo C.** 1994. Flagellate and non-flagellate *Proteus mirabilis* in the development of experimental urinary tract infection. *Microb Pathog* **16:**379–385.

55. **Legnani-Fajardo C, Zunino P, Piccini C, Allen A, Maskell D.** 1996. Defined mutants of *Proteus mirabilis* lacking flagella cause

ascending urinary tract infection in mice. *Microb Pathog* **21**:395–405.

56. **Burall LS, Harro JM, Li X, Lockatell CV, Himpsl SD, Hebel JR, Johnson DE, Mobley HLT.** 2004. *Proteus mirabilis* genes that contribute to pathogenesis of urinary tract infection: identification of 25 signature-tagged mutants attenuated at least 100-fold. *Infect Immun* **72**:2922–2938.

57. **Himpsl SD, Lockatell CV, Hebel JR, Johnson DE, Mobley HLT.** 2008. Identification of virulence determinants in uropathogenic *Proteus mirabilis* using signature-tagged mutagenesis. *J Med Microbiol* **57**:1068–1078.

58. **Williams FD, Schwarzhoff RH.** 1978. Nature of the swarming phenomenon in *Proteus*. *Annu Rev Microbiol* **32**:101–122.

59. **Gué M, Dupont V, Dufour A, Sire O.** 2001. Bacterial swarming: a biochemical time-resolved FTIR-ATR study of *Proteus mirabilis* swarm-cell differentiation. *Biochemistry* **40**: 11938–11945.

60. **Strating H, Vandenende C, Clarke AJ.** 2012. Changes in peptidoglycan structure and metabolism during differentiation of *Proteus mirabilis* into swarmer cells. *Can J Microbiol* **58**:1183–1194.

61. **Jones BV, Young R, Mahenthiralingam E, Stickler DJ.** 2004. Ultrastructure of *Proteus mirabilis* swarmer cell rafts and role of swarming in catheter-associated urinary tract infection. *Infect Immun* **72**:3941–3950.

62. **Stahl SJ, Stewart KR, Williams FD.** 1983. Extracellular slime associated with *Proteus mirabilis* during swarming. *J Bacteriol* **154**: 930–937.

63. **Gygi D, Rahman MM, Lai HC, Carlson R, Guard-Petter J, Hughes C.** 1995. A cell-surface polysaccharide that facilitates rapid population migration by differentiated swarm cells of *Proteus mirabilis*. *Mol Microbiol* **17**: 1167–1175.

64. **Rahman MM, Guard-Petter J, Asokan K, Hughes C, Carlson RW.** 1999. The structure of the colony migration factor from pathogenic *Proteus mirabilis*: a capsular polysaccharide that facilitates swarming. *J Biol Chem* **274**: 22993–22998.

65. **Pearson MM, Rasko DA, Smith SN, Mobley HLT.** 2010. Transcriptome of swarming *Proteus mirabilis*. *Infect Immun* **78**:2834–2845.

66. **Allison C, Lai HC, Hughes C.** 1992. Co-ordinate expression of virulence genes during swarm-cell differentiation and population migration of *Proteus mirabilis*. *Mol Microbiol* **6**:1583–1591.

67. **Harshey RM.** 2003. Bacterial motility on a surface: many ways to a common goal. *Annu Rev Microbiol* **57**:249–273.

68. **Inoue T, Shingaki R, Hirose S, Waki K, Mori H, Fukui K.** 2007. Genome-wide screening of genes required for swarming motility in *Escherichia coli* K-12. *J Bacteriol* **189**:950–957.

69. **Wang Q, Frye JG, McClelland M, Harshey RM.** 2004. Gene expression patterns during swarming in *Salmonella typhimurium*: genes specific to surface growth and putative new motility and pathogenicity genes. *Mol Microbiol* **52**:169–187.

70. **Kim W, Surette MG.** 2004. Metabolic differentiation in actively swarming *Salmonella*. *Mol Microbiol* **54**:702–714.

71. **Allison C, Lai HC, Gygi D, Hughes C.** 1993. Cell differentiation of *Proteus mirabilis* is initiated by glutamine, a specific chemoattractant for swarming cells. *Mol Microbiol* **8**:53–60.

72. **Senior BW.** 1978. *p*-nitrophenylglycerol—a superior antiswarming agent for isolating and identifying pathogens from clinical material. *J Med Microbiol* **11**:59–61.

73. **Belas R, Erskine D, Flaherty D.** 1991. Transposon mutagenesis in *Proteus mirabilis*. *J Bacteriol* **173**:6289–6293.

74. **Liu MC, Lin SB, Chien HF, Wang WB, Yuan YH, Hsueh PR, Liaw SJ.** 2012. 10′(Z),13′(E)-heptadecadienylhydroquinone inhibits swarming and virulence factors and increases polymyxin B susceptibility in *Proteus mirabilis*. *PLoS One* **7**:e45563.

75. **Wang WB, Lai HC, Hsueh PR, Chiou RY, Lin SB, Liaw SJ.** 2006. Inhibition of swarming and virulence factor expression in *Proteus mirabilis* by resveratrol. *J Med Microbiol* **55**:1313–1321.

76. **Hernandez E, Ramisse F, Cavalho JD.** 1999. Abolition of swarming of *Proteus*. *J Clin Microbiol* **37**:3435.

77. **Ayati BP.** 2006. A structured-population model of *Proteus mirabilis* swarm-colony development. *J Math Biol* **52**:93–114.

78. **Esipov SE, Shapiro JA.** 1998. Kinetic model of *Proteus mirabilis* swarm colony development. *Journal of Mathematical Biology* **36**:249–268.

79. **Frénod E, Sire O.** 2009. An explanatory model to validate the way water activity rules periodic terrace generation in *Proteus mirabilis* swarm. *J Math Biol* **59**:439–466.

80. **Xue C, Budrene EO, Othmer HG.** 2011. Radial and spiral stream formation in *Proteus mirabilis* colonies. *PLoS Comput Biol* **7**:e1002332.

81. **Belas R, Suvanasuthi R.** 2005. The ability of *Proteus mirabilis* to sense surfaces and

regulate virulence gene expression involves FliL, a flagellar basal body protein. *J Bacteriol* **187**:6789–6803.

82. **Belas R.** 1996. *Proteus mirabilis* swarmer cell differentiation and urinary tract infection, p 271–298. *In* Mobley HL, Warren JW (ed), *Urinary Tract Infections: Molecular Pathogenesis and Clinical Management.* ASM Press, Washington, D.C.

83. **Gygi D, Bailey MJ, Allison C, Hughes C.** 1995. Requirement for FlhA in flagella assembly and swarm-cell differentiation by *Proteus mirabilis. Mol Microbiol* **15**:761–769.

84. **Belas R, Erskine D, Flaherty D.** 1991. *Proteus mirabilis* mutants defective in swarmer cell differentiation and multicellular behavior. *J Bacteriol* **173**:6279–6288.

85. **Gygi D, Fraser G, Dufour A, Hughes C.** 1997. A motile but non-swarming mutant of *Proteus mirabilis* lacks FlgN, a facilitator of flagella filament assembly. *Mol Microbiol* **25**:597–604.

86. **Allison C, Hughes C.** 1991. Closely linked genetic loci required for swarm cell differentiation and multicellular migration by *Proteus mirabilis. Mol Microbiol* **5**:1975–1982.

87. **Belas R, Goldman M, Ashliman K.** 1995. Genetic analysis of *Proteus mirabilis* mutants defective in swarmer cell elongation. *J Bacteriol* **177**:823–828.

88. **Furness RB, Fraser GM, Hay NA, Hughes C.** 1997. Negative feedback from a *Proteus* class II flagellum export defect to the *flhDC* master operon controlling cell division and flagellum assembly. *J Bacteriol* **179**:5585–5588.

89. **Lee YY, Patellis J, Belas R.** 2013. Activity of *Proteus mirabilis* FliL is viscosity dependent and requires extragenic DNA. *J Bacteriol* **195**: 823–832.

90. **Cusick K, Lee YY, Youchak B, Belas R.** 2012. Perturbation of FliL interferes with *Proteus mirabilis* swarmer cell gene expression and differentiation. *J Bacteriol* **194**:437–447.

91. **Dufour A, Furness RB, Hughes C.** 1998. Novel genes that upregulate the *Proteus mirabilis flhDC* master operon controlling flagellar biogenesis and swarming. *Mol Microbiol* **29**:741–751.

92. **Clemmer KM, Rather PN.** 2007. Regulation of *flhDC* expression in *Proteus mirabilis. Res Microbiol* **158**:295–302.

93. **Claret L, Hughes C.** 2000. Rapid turnover of FlhD and FlhC, the flagellar regulon transcriptional activator proteins, during Proteus swarming. *J Bacteriol* **182**:833–836.

94. **Clemmer KM, Rather PN.** 2008. The Lon protease regulates swarming motility and virulence gene expression in *Proteus mirabilis. J Med Microbiol* **57**:931–937.

95. **Hay NA, Tipper DJ, Gygi D, Hughes C.** 1997. A nonswarming mutant of *Proteus mirabilis* lacks the Lrp global transcriptional regulator. *J Bacteriol* **179**:4741–4746.

96. **Lintner RE, Mishra PK, Srivastava P, Martinez-Vaz BM, Khodursky AB, Blumenthal RM.** 2008. Limited functional conservation of a global regulator among related bacterial genera: Lrp in *Escherichia, Proteus* and *Vibrio. BMC Microbiol* **8**:60.

97. **Stevenson LG, Rather PN.** 2006. A novel gene involved in regulating the flagellar gene cascade in *Proteus mirabilis. J Bacteriol* **188**:7830–7839.

98. **Szostek BA, Rather PN.** 2013. Regulation of the swarming inhibitor *disA* in *Proteus mirabilis. J Bacteriol* **195**:3237–3243.

99. **Hatt JK, Rather PN.** 2008. Characterization of a novel gene, *wosA*, regulating FlhDC expression in *Proteus mirabilis. J Bacteriol* **190**: 1946–1955.

100. **Liaw SJ, Lai HC, Ho SW, Luh KT, Wang WB.** 2003. Role of RsmA in the regulation of swarming motility and virulence factor expression in *Proteus mirabilis. J Med Microbiol* **52**:19–28.

101. **Wei BL, Brun-Zinkernagel AM, Simecka JW, Pruss BM, Babitzke P, Romeo T.** 2001. Positive regulation of motility and *flhDC* expression by the RNA-binding protein CsrA of *Escherichia coli. Mol Microbiol* **40**:245–256.

102. **Cano DA, Dominguez-Bernal G, Tierrez A, Garcia-Del Portillo F, Casadesus J.** 2002. Regulation of capsule synthesis and cell motility in *Salmonella enterica* by the essential gene *igaA. Genetics* **162**:1513–1523.

103. **Morgenstein RM, Rather PN.** 2012. Role of the Umo proteins and the Rcs phosphorelay in the swarming motility of the wild type and an O-antigen (*waaL*) mutant of *Proteus mirabilis. J Bacteriol* **194**:669–676.

104. **Huang YH, Ferrieres L, Clarke DJ.** 2006. The role of the Rcs phosphorelay in *Enterobacteriaceae. Res Microbiol* **157**:206–212.

105. **Liaw SJ, Lai HC, Ho SW, Luh KT, Wang WB.** 2001. Characterisation of p-nitrophenylglycerol-resistant *Proteus mirabilis* super-swarming mutants. *J Med Microbiol* **50**:1039–1048.

106. **Belas R, Schneider R, Melch M.** 1998. Characterization of *Proteus mirabilis* precocious swarming mutants: identification of *rsbA*, encoding a regulator of swarming behavior. *J Bacteriol* **180**:6126–6139.

107. **Liaw SJ, Lai HC, Wang WB.** 2004. Modulation of swarming and virulence by fatty acids through the RsbA protein in *Proteus mirabilis. Infect Immun* **72**:6836–6845.

108. **Wang WB, Chen IC, Jiang SS, Chen HR, Hsu CY, Hsueh PR, Hsu WB, Liaw SJ.** 2008. Role of RppA in the regulation of polymyxin b susceptibility, swarming, and virulence factor expression in *Proteus mirabilis. Infect Immun* **76:**2051–2062.

109. **Jiang SS, Liu MC, Teng LJ, Wang WB, Hsueh PR, Liaw SJ.** 2010. *Proteus mirabilis pmrI*, an RppA-regulated gene necessary for polymyxin B resistance, biofilm formation, and urothelial cell invasion. *Antimicrob Agents Chemother* **54:**1564–1571.

110. **Kato A, Groisman EA.** 2008. The PhoQ/PhoP regulatory network of *Salmonella enterica. Adv Exp Med Biol* **631:**7–21.

111. **Jiang SS, Lin TY, Wang WB, Liu MC, Hsueh PR, Liaw SJ.** 2010. Characterization of UDP-glucose dehydrogenase and UDP-glucose pyrophosphorylase mutants of *Proteus mirabilis*: defectiveness in polymyxin B resistance, swarming, and virulence. *Antimicrob Agents Chemother* **54:**2000–2009.

112. **McCoy AJ, Liu H, Falla TJ, Gunn JS.** 2001. Identification of *Proteus mirabilis* mutants with increased sensitivity to antimicrobial peptides. *Antimicrob Agents Chemother* **45:** 2030–2037.

113. **Morgenstein RM, Clemmer KM, Rather PN.** 2010. Loss of the *waaL* O-antigen ligase prevents surface activation of the flagellar gene cascade in *Proteus mirabilis. J Bacteriol* **192:**3213–3221.

114. **Allison C, Emody L, Coleman N, Hughes C.** 1994. The role of swarm cell differentiation and multicellular migration in the uropathogenicity of *Proteus mirabilis. J Infect Dis* **169:** 1155–1158.

115. **Hay NA, Tipper DJ, Gygi D, Hughes C.** 1999. A novel membrane protein influencing cell shape and multicellular swarming of *Proteus mirabilis. J Bacteriol* **181:**2008–2016.

116. **Lai HC, Gygi D, Fraser GM, Hughes C.** 1998. A swarming-defective mutant of *Proteus mirabilis* lacking a putative cation-transporting membrane P-type ATPase. *Microbiology* **144** (Pt 7):1957–1961.

117. **Rensing C, Mitra B, Rosen BP.** 1998. A Zn(II)-translocating P-type ATPase from *Proteus mirabilis. Biochem Cell Biol* **76:**787–790.

118. **Nielubowicz GR, Smith SN, Mobley HLT.** 2010. Zinc uptake contributes to motility and provides a competitive advantage to *Proteus mirabilis* during experimental urinary tract infection. *Infect Immun* **78:**2823–2833.

119. **Gaisser S, Hughes C.** 1997. A locus coding for putative non-ribosomal peptide/polyketide synthase functions is mutated in a swarming-defective *Proteus mirabilis* strain. *Mol Gen Genet* **253:**415–427.

120. **Armbruster CE, Hodges SA, Mobley HLT.** 2013. Initiation of swarming motility by *Proteus mirabilis* occurs in response to specific cues present in urine and requires excess L-glutamine. *J Bacteriol* **195.**1305–1319.

121. **Sturgill G, Rather PN.** 2004. Evidence that putrescine acts as an extracellular signal required for swarming in *Proteus mirabilis. Mol Microbiol* **51:**437–446.

122. **Vinogradov E, Perry MB.** 2000. Structural analysis of the core region of lipopolysaccharides from *Proteus mirabilis* serotypes O6, O48 and O57. *Eur J Biochem* **267:**2439–2446.

123. **Sturgill GM, Siddiqui S, Ding X, Pecora ND, Rather PN.** 2002. Isolation of *lacZ* fusions to *Proteus mirabilis* genes regulated by intercellular signaling: potential role for the sugar phosphotransferase (Pts) system in regulation. *FEMS Microbiol Lett* **217:**43–50.

124. **Kurihara S, Sakai Y, Suzuki H, Muth A, Phanstiel Ot, Rather PN.** 2013. Putrescine importer PlaP contributes to swarming motility and urothelial cell invasion in *Proteus mirabilis. J Biol Chem* **288:**15668–15676.

125. **Rather PN.** 2005. Swarmer cell differentiation in *Proteus mirabilis. Environ Microbiol* **7:**1065–1073.

126. **Holden MT, Ram Chhabra S, de Nys R, Stead P, Bainton NJ, Hill PJ, Manefield M, Kumar N, Labatte M, England D, Rice S, Givskov M, Salmond GP, Stewart GS, Bycroft BW, Kjelleberg S, Williams P.** 1999. Quorum-sensing cross talk: isolation and chemical characterization of cyclic dipeptides from *Pseudomonas aeruginosa* and other gram-negative bacteria. *Mol Microbiol* **33:**1254–1266.

127. **Campbell J, Lin Q, Geske GD, Blackwell HE.** 2009. New and unexpected insights into the modulation of LuxR-type quorum sensing by cyclic dipeptides. *ACS Chem Biol* **4:**1051–1059.

128. **Eberl L, Winson MK, Sternberg C, Stewart GS, Christiansen G, Chhabra SR, Bycroft B, Williams P, Molin S, Givskov M.** 1996. Involvement of *N*-acyl-L-hormoserine lactone autoinducers in controlling the multicellular behaviour of *Serratia liquefaciens. Mol Microbiol* **20:**127–136.

129. **Stankowska D, Kwinkowski M, Kaca W.** 2008. Quantification of *Proteus mirabilis* virulence factors and modulation by acylated homoserine lactones. *J Microbiol Immunol Infect* **41:**243–253.

130. **Schneider R, Lockatell CV, Johnson D, Belas R.** 2002. Detection and mutation of a *luxS-*

encoded autoinducer in *Proteus mirabilis*. *Microbiology* **148:**773–782.

131. **Rauprich O, Matsushita M, Weijer CJ, Siegert F, Esipov SE, Shapiro JA.** 1996. Periodic phenomena in *Proteus mirabilis* swarm colony development. *J Bacteriol* **178:**6525–6538.

132. **Armitage JP.** 1981. Changes in metabolic activity of *Proteus mirabilis* during swarming. *J Gen Microbiol* **125:**445–450.

133. **Falkinham JO 3rd, Hoffman PS.** 1984. Unique developmental characteristics of the swarm and short cells of *Proteus vulgaris* and *Proteus mirabilis*. *J Bacteriol* **158:**1037–1040.

134. **Alteri CJ, Himpsl SD, Engstrom MD, Mobley HLT.** 2012. Anaerobic respiration using a complete oxidative TCA cycle drives multicellular swarming in *Proteus mirabilis*. *mBio* **3**.

135. **Allison C, Coleman N, Jones PL, Hughes C.** 1992. Ability of *Proteus mirabilis* to invade human urothelial cells is coupled to motility and swarming differentiation. *Infect Immun* **60:**4740–4746.

136. **Peerbooms PG, Verweij AM, MacLaren DM.** 1984. Vero cell invasiveness of *Proteus mirabilis*. *Infect Immun* **43:**1068–1071.

137. **Chippendale GR, Warren JW, Trifillis AL, Mobley HLT.** 1994. Internalization of *Proteus mirabilis* by human renal epithelial cells. *Infect Immun* **62:**3115–3121.

138. **Oelschlaeger TA, Tall BD.** 1996. Uptake pathways of clinical isolates of *Proteus mirabilis* into human epithelial cell lines. *Microb Pathog* **21:**1–16.

139. **Alamuri P, Lower M, Hiss JA, Himpsl SD, Schneider G, Mobley HLT.** 2010. Adhesion, invasion, and agglutination mediated by two trimeric autotransporters in the human uropathogen *Proteus mirabilis*. *Infect Immun* **78:**4882–4894.

140. **Mathoera RB, Kok DJ, Verduin CM, Nijman RJ.** 2002. Pathological and therapeutic significance of cellular invasion by *Proteus mirabilis* in an enterocystoplasty infection stone model. *Infect Immun* **70:**7022–7032.

141. **Fraser GM, Claret L, Furness R, Gupta S, Hughes C.** 2002. Swarming-coupled expression of the *Proteus mirabilis* hpmBA haemolysin operon. *Microbiology* **148:**2191–2201.

142. **Sabbuba N, Hughes G, Stickler DJ.** 2002. The migration of *Proteus mirabilis* and other urinary tract pathogens over Foley catheters. *BJU Int* **89:**55–60.

143. **Jansen AM, Lockatell CV, Johnson DE, Mobley HLT.** 2003. Visualization of *Proteus mirabilis* morphotypes in the urinary tract: the elongated swarmer cell is rarely observed in ascending urinary tract infection. *Infect Immun* **71:**3607–3613.

144. **Dienes L.** 1946. Reproductive processes in *Proteus* cultures. *Proc Soc Exp Biol Med* **63:**265–270.

145. **De Louvois J.** 1969. Serotyping and the Dienes reaction on *Proteus mirabilis* from hospital infections. *J Clin Pathol* **22:**263–268.

146. **Pfaller MA, Mujeeb I, Hollis RJ, Jones RN, Doern GV.** 2000. Evaluation of the discriminatory powers of the Dienes test and ribotyping as typing methods for *Proteus mirabilis*. *J Clin Microbiol* **38:**1077–1080.

147. **Senior BW, Larsson P.** 1983. A highly discriminatory multi-typing scheme for *Proteus mirabilis* and *Proteus vulgaris*. *J Med Microbiol* **16:**193–202.

148. **Budding AE, Ingham CJ, Bitter W, Vandenbroucke-Grauls CM, Schneeberger PM.** 2009. The Dienes phenomenon: competition and territoriality in swarming *Proteus mirabilis*. *J Bacteriol* **191:**3892–3900.

149. **Gibbs KA, Urbanowski ML, Greenberg EP.** 2008. Genetic determinants of self identity and social recognition in bacteria. *Science* **321:**256–259.

150. **Gibbs KA, Wenren LM, Greenberg EP.** 2011. Identity gene expression in *Proteus mirabilis*. *J Bacteriol* **193:**3286–3292.

151. **Silverman JM, Brunet YR, Cascales E, Mougous JD.** 2012. Structure and regulation of the type VI secretion system. *Annu Rev Microbiol* **66:**453–472.

152. **Jani AJ, Cotter PA.** 2010. Type VI secretion: not just for pathogenesis anymore. *Cell Host Microbe* **8:**2–6.

153. **Russell AB, Peterson SB, Mougous JD.** 2014. Type VI secretion system effectors: poisons with a purpose. *Nat Rev Microbiol* **12:**137–148.

154. **Sullivan NL, Septer AN, Fields AT, Wenren LM, Gibbs KA.** 2013. The complete genome sequence of *Proteus mirabilis* strain BB2000 reveals differences from the *P. mirabilis* reference strain. *Genome Announc* **1:**e00024-13.

155. **Wenren LM, Sullivan NL, Cardarelli L, Septer AN, Gibbs KA.** 2013. Two independent pathways for self-recognition in *Proteus mirabilis* are linked by type VI-dependent export. *mBio* **4**.

156. **Alteri CJ, Himpsl SD, Pickens SR, Lindner JR, Zora JS, Miller JE, Arno PD, Straight SW, Mobley HLT.** 2013. Multicellular bacteria deploy the type VI secretion system to preemptively strike neighboring cells. *PLoS Pathog* **9:**e1003608.

157. **Old DC, Adegbola RA.** 1982. Haemagglutinins and fimbriae of *Morganella*, *Proteus* and *Providencia*. *J Medl Microbiol* **15:**551.

158. **Wray SK, Hull SI, Cook RG, Barrish J, Hull RA.** 1986. Identification and characterization of a uroepithelial cell adhesin from a uropathogenic isolate of *Proteus mirabilis. Infect Immun* **54:**43–49.

159. **Bahrani FK, Cook S, Hull RA, Massad G, Mobley HLT.** 1993. *Proteus mirabilis* fimbriae: N-terminal amino acid sequence of a major fimbrial subunit and nucleotide sequences of the genes from two strains. *Infect Immun* **61:**884–891.

160. **Welch RA, Burland V, Plunkett G 3rd, Redford P, Roesch P, Rasko D, Buckles EL, Liou SR, Boutin A, Hackett J, Stroud D, Mayhew GF, Rose DJ, Zhou S, Schwartz DC, Perna NT, Mobley HLT, Donnenberg MS, Blattner FR.** 2002. Extensive mosaic structure revealed by the complete genome sequence of uropathogenic *Escherichia coli. Proc Natl Acad Sci USA* **99:**17020–17024.

161. **Nuccio SP, Baumler AJ.** 2007. Evolution of the chaperone/usher assembly pathway: fimbrial classification goes Greek. *Microbiol Mol Biol Rev* **71:**551–575.

162. **Pearson MM, Mobley HLT.** 2008. Repression of motility during fimbrial expression: identification of 14 *mrpJ* gene paralogues in *Proteus mirabilis. Mol Microbiol* **69:**548–558.

163. **Li X, Johnson DE, Mobley HLT.** 1999. Requirement of MrpH for mannose-resistant *Proteus*-like fimbria-mediated hemagglutination by *Proteus mirabilis. Infect Immun* **67:**2822–2833.

164. **Bahrani FK, Mobley HLT.** 1994. *Proteus mirabilis* MR/P fimbrial operon: genetic organization, nucleotide sequence, and conditions for expression. *J Bacteriol* **176:**3412–3419.

165. **Zhao H, Li X, Johnson DE, Blomfield I, Mobley HLT.** 1997. In vivo phase variation of MR/P fimbrial gene expression in *Proteus mirabilis* infecting the urinary tract. *Molec Microbiol* **23:**1009–1019.

166. **Lane MC, Li X, Pearson MM, Simms AN, Mobley HLT.** 2009. Oxygen-limiting conditions enrich for fimbriate cells of uropathogenic *Proteus mirabilis* and *Escherichia coli. J Bacteriol* **191:**1382–1392.

167. **Li X, Lockatell CV, Johnson DE, Mobley HLT.** 2002. Identification of MrpI as the sole recombinase that regulates the phase variation of MR/P fimbria, a bladder colonization factor of uropathogenic *Proteus mirabilis. Mol Microbiol* **45:**865–874.

168. **Bahrani FK, Mobley HLT.** 1993. *Proteus mirabilis* MR/P fimbriae: molecular cloning, expression, and nucleotide sequence of the major fimbrial subunit gene. *J Bacteriol* **175:**457–464.

169. **Båga M, Norgren M, Normark S.** 1987. Biogenesis of *E. coli* Pap pili: PapH, a minor pilin subunit involved in cell anchoring and length modulation. *Cell* **49:**241–251.

170. **Verger D, Miller E, Remaut H, Waksman G, Hultgren S.** 2006. Molecular mechanism of P pilus termination in uropathogenic *Escherichia coli. EMBO Rep* **7:**1228–1232.

171. **Li X, Mobley HLT.** 1998. MrpB functions as the terminator for assembly of *Proteus mirabilis* mannose-resistant *Proteus*-like fimbriae. *Infect Immun* **66:**1759–1763.

172. **Li X, Zhao H, Geymonat L, Bahrani F, Johnson DE, Mobley HLT.** 1997. *Proteus mirabilis* mannose-resistant, *Proteus*-like fimbriae: MrpG is located at the fimbrial tip and is required for fimbrial assembly. *Infect Immun* **65:**1327–1334.

173. **Kline KA, Falker S, Dahlberg S, Normark S, Henriques-Normark B.** 2009. Bacterial adhesins in host-microbe interactions. *Cell Host Microbe* **5:**580–592.

174. **Kuehn MJ, Normark S, Hultgren SJ.** 1991. Immunoglobulin-like PapD chaperone caps and uncaps interactive surfaces of nascently translocated pilus subunits. *Proc Natl Acad Sci U S A* **88:**10586–10590.

175. **Carnoy C, Moseley SL.** 1997. Mutational analysis of receptor binding mediated by the Dr family of *Escherichia coli* adhesins. *Mol Microbiol* **23:**365–379.

176. **Heras B, Shouldice SR, Totsika M, Scanlon MJ, Schembri MA, Martin JL.** 2009. DSB proteins and bacterial pathogenicity. *Nat Rev Microbiol* **7:**215–225.

177. **Jansen AM, Lockatell V, Johnson DE, Mobley HLT.** 2004. Mannose-resistant *Proteus*-like fimbriae are produced by most *Proteus mirabilis* strains infecting the urinary tract, dictate the in vivo localization of bacteria, and contribute to biofilm formation. *Infect Immun* **72:**7294–7305.

178. **Rocha SP, Elias WP, Cianciarullo AM, Menezes MA, Nara JM, Piazza RM, Silva MR, Moreira CG, Pelayo JS.** 2007. Aggregative adherence of uropathogenic *Proteus mirabilis* to cultured epithelial cells. *FEMS Immunol Med Microbiol* **51:**319–326.

179. **Johnson DE, Bahrani FK, Lockatell CV, Drachenberg CB, Hebel JR, Belas R, Warren JW, Mobley HLT.** 1999. Serum immunoglobulin response and protection from homologous challenge by *Proteus mirabilis* in a mouse model of ascending urinary tract infection. *Infect Immun* **67:**6683–6687.

180. **Bahrani FK, Massad G, Lockatell CV, Johnson DE, Russell RG, Warren JW, Mobley HLT.**

1994. Construction of an MR/P fimbrial mutant of *Proteus mirabilis*: role in virulence in a mouse model of ascending urinary tract infection. *Infect Immun* **62**:3363–3371.

181. **Li X, Johnson DE, Mobley HLT.** 1999. Requirement of MrpH for Mannose-Resistant Proteus-Like Fimbria-Mediated Hemagglutination by Proteus mirabilis. *Infect Immun* **67**:2822–2833.

182. **Cook SW, Mody N, Valle J, Hull R.** 1995. Molecular cloning of *Proteus mirabilis* uroepithelial cell adherence (*uca*) genes. *Infect Immun* **63**:2082–2086.

183. **Tolson DL, Barrigar DL, McLean RJ, Altman E.** 1995. Expression of a nonagglutinating fimbria by *Proteus mirabilis*. *Infect Immun* **63**:1127–1129.

184. **Pellegrino R, Scavone P, Umpiérrez A, Maskell DJ, Zunino P.** 2013. *Proteus mirabilis* uroepithelial cell adhesin (UCA) fimbria plays a role in the colonization of the urinary tract. *Pathog Dis* **67**:104–107.

185. **Kuan L, Schaffer JN, Zouzias CD, Pearson MM.** 2014. Characterization of 17 chaperone-usher fimbriae encoded by *Proteus mirabilis* reveals strong conservation. *J Med Microbiol* doi:10.1099/jmm.0.069971–0.

186. **Väisänen-Rhen V, Korhonen TK, Finne J.** 1983. Novel cell-binding activity specific for N-acetyl-D-glucosamine in an *Escherichia coli* strain. *FEBS Lett* **159**:233–236.

187. **Saarela S, Westerlund-Wikström B, Rhen M, Korhonen TK.** 1996. The GafD protein of the G (F17) fimbrial complex confers adhesiveness of *Escherichia coli* to laminin. *Infect Immun* **64**:2857–2860.

188. **Dorofeyev AE, Vasilenko IV, Rassokhina OA.** 2009. Joint extraintestinal manifestations in ulcerative colitis. *Dig Dis* **27**:502–510.

189. **Lee KK, Harrison BA, Latta R, Altman E.** 2000. The binding of *Proteus mirabilis* nonagglutinating fimbriae to ganglio-series asialoglycolipids and lactosyl ceramide. *Can J Microbiol* **46**:961–966.

190. **Ortaldo JR, Sharrow SO, Timonen T, Herberman RB.** 1981. Determination of surface antigens on highly purified human NK cells by flow cytometry with monoclonal antibodies. *J Immunol* **127**:2401–2409.

191. **Saiman L, Prince A.** 1993. *Pseudomonas aeruginosa* pili bind to asialoGM1 which is increased on the surface of cystic fibrosis epithelial cells. *J Clin Invest* **92**:1875–1880.

192. **Zunino P, Sosa V, Allen AG, Preston A, Schlapp G, Maskell DJ.** 2003. *Proteus mirabilis* fimbriae (PMF) are important for both bladder and kidney colonization in mice. *Microbiology* **149**:3231–3237.

193. **Massad G, Lockatell CV, Johnson DE, Mobley HLT.** 1994. *Proteus mirabilis* fimbriae: construction of an isogenic *pmfA* mutant and analysis of virulence in a CBA mouse model of ascending urinary tract infection. *Infect Immun* **62**:536–542.

194. **Zunino P, Sosa V, Schlapp G, Allen AG, Preston A, Maskell DJ.** 2007. Mannose-resistant *Proteus*-like and *P. mirabilis* fimbriae have specific and additive roles in *P. mirabilis* urinary tract infections. *FEMS Immunol Med Microbiol* **51**:125–133.

195. **Massad G, Bahrani FK, Mobley HLT.** 1994. *Proteus mirabilis* fimbriae: identification, isolation, and characterization of a new ambient-temperature fimbria. *Infect Immun* **62**:1989–1994.

196. **Massad G, Fulkerson JF Jr, Watson DC, Mobley HLT.** 1996. *Proteus mirabilis* ambient-temperature fimbriae: cloning and nucleotide sequence of the *atf* gene cluster. *Infect Immun* **64**:4390–4395.

197. **Zunino P, Geymonat L, Allen AG, Legnani-Fajardo C, Maskell DJ.** 2000. Virulence of a *Proteus mirabilis* ATF isogenic mutant is not impaired in a mouse model of ascending urinary tract infection. *FEMS Immunol Med Microbiol* **29**:137–143.

198. **Bijlsma IG, van Dijk L, Kusters JG, Gaastra W.** 1995. Nucleotide sequences of two fimbrial major subunit genes, *pmpA* and *ucaA*, from canine-uropathogenic *Proteus mirabilis* strains. *Microbiology* **141**(Pt 6):1349–1357.

199. **Spurbeck RR, Stapleton AE, Johnson JR, Walk ST, Hooton TM, Mobley HLT.** 2011. Fimbrial profiles predict virulence of uropathogenic *Escherichia coli* strains: contribution of Ygi and Yad fimbriae. *Infect Immun* **79**:4753–4763.

200. **Townsend SM, Kramer NE, Edwards R, Baker S, Hamlin N, Simmonds M, Stevens K, Maloy S, Parkhill J, Dougan G, Bäumler AJ.** 2001. *Salmonella enterica* serovar Typhi possesses a unique repertoire of fimbrial gene sequences. *Infect Immun* **69**:2894–2901.

201. **Wurpel DJ, Beatson SA, Totsika M, Petty NK, Schembri MA.** 2013. Chaperone-usher fimbriae of *Escherichia coli*. *PLoS One* **8**:e52835.

202. **Snyder JA, Haugen BJ, Lockatell CV, Maroncle N, Hagan EC, Johnson DE, Welch RA, Mobley HLT.** 2005. Coordinate expression of fimbriae in uropathogenic *Escherichia coli*. *Infect Immun* **73**:7588–7596.

203. **Li X, Rasko DA, Lockatell CV, Johnson DE, Mobley HLT.** 2001. Repression of bacterial motility by a novel fimbrial gene product. *EMBO J* **20**:4854–4862.

204. **Simms AN, Mobley HLT.** 2008. PapX, a P fimbrial operon-encoded inhibitor of motility in uropathogenic *Escherichia coli*. *Infect Immun* **76:**4833–4841.

205. **Reiss DJ, Mobley HLT.** 2011. Determination of target sequence bound by PapX, repressor of bacterial motility, in *flhD* promoter using systematic evolution of ligands by exponential enrichment (SELEX) and high throughput sequencing. *J Biol Chem* **286:**44726–44738.

206. **Chen YT, Peng HL, Shia WC, Hsu FR, Ken CF, Tsao YM, Chen CH, Liu CE, Hsieh MF, Chen HC, Tang CY, Ku TH.** 2012. Whole-genome sequencing and identification of *Morganella morganii* KT pathogenicity-related genes. *BMC Genomics* **13**(Suppl 7):S4.

207. **Altschul SF, Gish W, Miller W, Myers EW, Lipman DJ.** 1990. Basic local alignment search tool. *J Molec Biol* **215:**403.

208. **Meslet-Cladiere LM, Pimenta A, Duchaud E, Holland IB, Blight MA.** 2004. In vivo expression of the mannose-resistant fimbriae of *Photorhabdus temperata* K122 during insect infection. *J Bacteriol* **186:**611–622.

209. **Allison SE, Silphaduang U, Mascarenhas M, Konczy P, Quan Q, Karmali M, Coombes BK.** 2012. Novel repressor of *Escherichia coli* O157:H7 motility encoded in the putative fimbrial cluster OI-1. *J Bacteriol* **194:**5343–5352.

210. **He H, Snyder HA, Forst S.** 2004. Unique organization and regulation of the *mrx* fimbrial operon in *Xenorhabdus nematophila*. *Microbiology* **150:**1439–1446.

211. **Leyton DL, Rossiter AE, Henderson IR.** 2012. From self sufficiency to dependence: mechanisms and factors important for autotransporter biogenesis. *Nat Rev Microbiol* **10:**213–225.

212. **Cotter SE, Surana NK, St Geme JW 3rd.** 2005. Trimeric autotransporters: a distinct subfamily of autotransporter proteins. *Trends Microbiol* **13:**199–205.

213. **Alamuri P, Mobley HLT.** 2008. A novel autotransporter of uropathogenic *Proteus mirabilis* is both a cytotoxin and an agglutinin. *Mol Microbiol* **68:**997–1017.

214. **Flannery EL, Mody L, Mobley HLT.** 2009. Identification of a modular pathogenicity island that is widespread among urease-producing uropathogens and shares features with a diverse group of mobile elements. *Infect Immun* **77:**4887–4894.

215. **Silverblatt FJ.** 1974. Host-parasite interaction in the rat renal pelvis: a possible role for pili in the pathogenesis of pyelonephritis. *J Exp Med* **140:**1696–1711.

216. **Hola V, Peroutkova T, Ruzicka F.** 2012. Virulence factors in *Proteus* bacteria from biofilm communities of catheter-associated urinary tract infections. *FEMS Immunol Med Microbiol* **65:**343–349.

217. **Mattick JS.** 2002. Type IV pili and twitching motility. *Annu Rev Microbiol* **56:**289–314.

218. **Uphoff TS, Welch RA.** 1990. Nucleotide sequencing of the *Proteus mirabilis* calcium-independent hemolysin genes (*hpmA* and *hpmB*) reveals sequence similarity with the *Serratia marcescens* hemolysin genes (*shlA* and *shlB*). *J Bacteriol* **172:**1206.

219. **Swihart KG, Welch RA.** 1990. Cytotoxic activity of the *Proteus* hemolysin HpmA. *Infect Immun* **58:**1861–1869.

220. **Cestari SE, Ludovico MS, Martins FH, da Rocha SP, Elias WP, Pelayo JS.** 2013. Molecular detection of HpmA and HlyA hemolysin of uropathogenic *Proteus mirabilis*. *Curr Microbiol* **67:**703–707.

221. **Weaver TM, Hocking JM, Bailey LJ, Wawrzyn GT, Howard DR, Sikkink LA, Ramirez-Alvarado M, Thompson JR.** 2009. Structural and functional studies of truncated hemolysin A from *Proteus mirabilis*. *J Biol Chem* **284:**22297–22309.

222. **Wassif C, Cheek D, Belas R.** 1995. Molecular analysis of a metalloprotease from *Proteus mirabilis*. *J Bacteriol* **177:**5790.

223. **Belas R, Manos J, Suvanasuthi R.** 2004. *Proteus mirabilis* ZapA metalloprotease degrades a broad spectrum of substrates, including antimicrobial peptides. *Infect Immun* **72:**5159–5167.

224. **Walker KE, Moghaddame-Jafari S, Lockatell CV, Johnson D, Belas R.** 1999. ZapA, the IgA-degrading metalloprotease of *Proteus mirabilis*, is a virulence factor expressed specifically in swarmer cells. *Mol Microbiol* **32:**825–836.

225. **Chromek M, Slamova Z, Bergman P, Kovacs L, Podracka L, Ehren I, Hokfelt T, Gudmundsson GH, Gallo RL, Agerberth B, Brauner A.** 2006. The antimicrobial peptide cathelicidin protects the urinary tract against invasive bacterial infection. *Nat Med* **12:**636–641.

226. **Ganz T.** 2001. Defensins in the urinary tract and other tissues. *J Infect Dis* **183**(Suppl 1):S41–42.

227. **Phan V, Belas R, Gilmore BF, Ceri H.** 2008. ZapA, a virulence factor in a rat model of *Proteus mirabilis*-induced acute and chronic prostatitis. *Infect Immun* **76:**4859–4864.

228. **Senior BW, Loomes LM, Kerr MA.** 1991. The production and activity *in vivo* of *Proteus mirabilis* IgA protease in infections of the urinary tract. *J Med Microbiol* **35:**203–207.

229. **Carson L, Cathcart GR, Scott CJ, Hollenberg MD, Walker B, Ceri H, Gilmore BF.** 2011.

Comprehensive inhibitor profiling of the *Proteus mirabilis* metalloprotease virulence factor ZapA (mirabilysin). *Biochimie* **93**:1824–1827.

230. **Zhao H, Li X, Johnson DE, Mobley HLT.** 1999. Identification of protease and *rpoN*-associated genes of uropathogenic *Proteus mirabilis* by negative selection in a mouse model of ascending urinary tract infection. *Microbiology* **145**(Pt 1):185–195.

231. **Hood MI, Skaar EP.** 2012. Nutritional immunity: transition metals at the pathogen-host interface. *Nat Rev Microbiol* **10**:525–537.

232. **Andrews SC, Robinson AK, Rodriguez-Quiñones F.** 2003. Bacterial iron homeostasis. *FEMS Microbiol Rev* **27**:215–237.

233. **Piccini CD, Barbe FM, Legnani-Fajardo CL.** 1998. Identification of iron-regulated outer membrane proteins in uropathogenic *Proteus mirabilis* and its relationship with heme uptake. *FEMS Microbiol Lett* **166**:243–248.

234. **Drechsel H, Thieken A, Reissbrodt R, Jung G, Winkelmann G.** 1993. Alpha-keto acids are novel siderophores in the genera *Proteus*, *Providencia*, and *Morganella* and are produced by amino acid deaminases. *J Bacteriol* **175**:2727–2733.

235. **Massad G, Zhao H, Mobley HLT.** 1995. *Proteus mirabilis* amino acid deaminase: cloning, nucleotide sequence, and characterization of *aad*. *J Bacteriol* **177**:5878–5883.

236. **Reissbrodt R, Kingsley R, Rabsch W, Beer W, Roberts M, Williams PH.** 1997. Iron-regulated excretion of alpha-keto acids by *Salmonella typhimurium*. *J Bacteriol* **179**:4538–4544.

237. **Kingsley R, Rabsch W, Roberts M, Reissbrodt R, Williams PH.** 1996. TonB-dependent iron supply in *Salmonella* by alpha-ketoacids and alpha-hydroxyacids. *FEMS Microbiol Lett* **140**:65–70.

238. **Himpsl SD, Pearson MM, Arewang CJ, Nusca TD, Sherman DH, Mobley HLT.** 2010. Proteobactin and a yersiniabactin-related siderophore mediate iron acquisition in *Proteus mirabilis*. *Mol Microbiol* **78**:138–157.

239. **Lima A, Zunino P, D'Alessandro B, Piccini C.** 2007. An iron-regulated outer-membrane protein of *Proteus mirabilis* is a haem receptor that plays an important role in urinary tract infection and in *in vivo* growth. *J Med Microbiol* **56**:1600–1607.

240. **Sabri M, Houle S, Dozois CM.** 2009. Roles of the extraintestinal pathogenic *Escherichia coli* ZnuACB and ZupT zinc transporters during urinary tract infection. *Infect Immun* **77**:1155–1164.

241. **Jacobsen SM, Lane MC, Harro JM, Shirtliff ME, Mobley HLT.** 2008. The high-affinity phosphate transporter Pst is a virulence factor for *Proteus mirabilis* during complicated urinary tract infection. *FEMS Immunol Med Microbiol* **52**:180–193.

242. **O'May GA, Jacobsen SM, Longwell M, Stoodley P, Mobley HL, Shirtliff ME.** 2009. The high-affinity phosphate transporter Pst in *Proteus mirabilis* HI4320 and its importance in biofilm formation. *Microbiology* **155**:1523–1535.

243. **Shi X, Zhu Y, Li Y, Jiang M, Lin Y, Qiu Y, Chen Q, Yuan Y, Ni P, Hu Q, Huang S.** 2014. Genome sequence of *Proteus mirabilis* clinical isolate C05028. *Genome Announc* **2**.

244. **Khalid MI, Teh LK, Lee LS, Zakaria ZA, Salleh MZ.** 2013. Genome sequence of *Proteus mirabilis* strain PR03, isolated from a local hospital in Malaysia. *Genome Announc* **1**.

245. **Pearson MM, Mobley HLT.** 2007. The type III secretion system of *Proteus mirabilis* HI4320 does not contribute to virulence in the mouse model of ascending urinary tract infection. *J Med Microbiol* **56**:1277–1283.

246. **Lloyd AL, Rasko DA, Mobley HLT.** 2007. Defining genomic islands and uropathogen-specific genes in uropathogenic *Escherichia coli*. *J Bacteriol* **189**:3532–3546.

247. **Flannery EL, Antczak SM, Mobley HL.** 2011. Self-transmissibility of the integrative and conjugative element ICE*Pm*1 between clinical isolates requires a functional integrase, relaxase, and type IV secretion system. *J Bacteriol* **193**:4104–4112.

248. **Jacobsen SM, Shirtliff ME.** 2011. *Proteus mirabilis* biofilms and catheter-associated urinary tract infections. *Virulence* **2**:460–465.

249. **Dumanski AJ, Hedelin H, Edin-Liljegren A, Beauchemin D, McLean RJ.** 1994. Unique ability of the *Proteus mirabilis* capsule to enhance mineral growth in infectious urinary calculi. *Infect Immun* **62**:2998–3003.

250. **Holling N, Lednor D, Tsang S, Bissell A, Campbell L, Nzakizwanayo J, Dedi C, Hawthorne JA, Hanlon G, Ogilvie LA, Salvage JP, Patel BA, Barnes LM, Jones BV.** 2014. Elucidating the genetic basis of crystalline biofilm formation in *Proteus mirabilis*. *Infect Immun* **82**:1616–1626.

251. **Gupta K, Hooton TM, Naber KG, Wullt B, Colgan R, Miller LG, Moran GJ, Nicolle LE, Raz R, Schaeffer AJ, Soper DE.** 2011. International clinical practice guidelines for the treatment of acute uncomplicated cystitis and pyelonephritis in women: A 2010 update by the Infectious Diseases Society of America and the European Society for Microbiology and Infectious Diseases. *Clin Infect Dis* **52**:e103–120.

252. **Ma KL, Wang CX.** 2013. Analysis of the spectrum and antibiotic resistance of uropathogens in vitro: Results based on a retrospective study from a tertiary hospital. *Am J Infect Control* **41:**610-601–606.

253. **Schito GC, Naber KG, Botto H, Palou J, Mazzei T, Gualco L, Marchese A.** 2009. The ARESC study: an international survey on the antimicrobial resistance of pathogens involved in uncomplicated urinary tract infections. *Int J Antimicrob Agents* **34:**407–413.

254. **Bichler KH, Eipper E, Naber K, Braun V, Zimmermann R, Lahme S.** 2002. Urinary infection stones. *Int J Antimicrob Agents* **19:** 488–498.

255. **Adamus-Bialek W, Zajac E, Parniewski P, Kaca W.** 2013. Comparison of antibiotic resistance patterns in collections of *Escherichia coli* and *Proteus mirabilis* uropathogenic strains. *Mol Biol Rep* **40:**3426–3435.

256. **Kaca W, Radziejewska-Lebrecht J, Bhat UR.** 1990. Effect of polymyxins on the lipopolysaccharide-defective mutants of *Proteus mirabilis*. *Microbios* **61:**23–32.

257. **Li X, Lockatell CV, Johnson DE, Lane MC, Warren JW, Mobley HLT.** 2004. Development of an intranasal vaccine to prevent urinary tract infection by *Proteus mirabilis*. *Infect Immun* **72:**66–75.

258. **Brumbaugh AR, Mobley HLT.** 2012. Preventing urinary tract infection: progress toward an effective *Escherichia coli* vaccine. *Expert Rev Vaccines* **11:**663–676.

259. **Scavone P, Rial A, Umpierrez A, Chabalgoity A, Zunino P.** 2009. Effects of the administration of cholera toxin as a mucosal adjuvant on the immune and protective response induced by *Proteus mirabilis* MrpA fimbrial protein in the urinary tract. *Microbiol Immunol* **53:**233–240.

260. **Jones RJ.** 1976. Oral vaccination against *Proteus mirabilis*. *Br J Exp Pathol* **57:**395–399.

261. **Moayeri N, Collins CM, O'Hanley P.** 1991. Efficacy of a *Proteus mirabilis* outer membrane protein vaccine in preventing experimental *Proteus* pyelonephritis in a BALB/c mouse model. *Infect Immun* **59:**3778–3786.

262. **Pellegrino R, Galvalisi U, Scavone P, Sosa V, Zunino P.** 2003. Evaluation of *Proteus mirabilis* structural fimbrial proteins as antigens against urinary tract infections. *FEMS Immunol Med Microbiol* **36:**103–110.

263. **Alamuri P, Eaton KA, Himpsl SD, Smith SN, Mobley HLT.** 2009. Vaccination with *Proteus* toxic agglutinin, a hemolysin-independent cytotoxin in vivo, protects against *Proteus mirabilis* urinary tract infection. *Infect Immun* **77:**632–641.

264. **Scavone P, Sosa V, Pellegrino R, Galvalisi U, Zunino P.** 2004. Mucosal vaccination of mice with recombinant *Proteus mirabilis* structural fimbrial proteins. *Microbes Infect* **6:**853–860.

265. **Alteri CJ, Hagan EC, Sivick KE, Smith SN, Mobley HLT.** 2009. Mucosal immunization with iron receptor antigens protects against urinary tract infection. *PLoS Pathog* **5:**e1000586.

266. **Scavone P, Miyoshi A, Rial A, Chabalgoity A, Langella P, Azevedo V, Zunino P.** 2007. Intranasal immunisation with recombinant *Lactococcus lactis* displaying either anchored or secreted forms of *Proteus mirabilis* MrpA fimbrial protein confers specific immune response and induces a significant reduction of kidney bacterial colonisation in mice. *Microbes Infect* **9:**821–828.

267. **Scavone P, Umpierrez A, Maskell DJ, Zunino P.** 2011. Nasal immunization with attenuated *Salmonella* Typhimurium expressing an MrpA-TetC fusion protein significantly reduces *Proteus mirabilis* colonization in the mouse urinary tract. *J Med Microbiol* **60:**899–904.

268. **Siddiq DM, Darouiche RO.** 2012. New strategies to prevent catheter-associated urinary tract infections. *Nat Rev Urol* **9:**305–314.

269. **Levering V, Wang Q, Shivapooja P, Zhao X, López GP.** 2014. Soft Robotic Concepts in Catheter Design: An On-Demand Fouling-Release Urinary Catheter. *Adv Healthc Mater* **3:**1588–1596.

270. **Stickler DJ, Jones SM, Adusei GO, Waters MG, Cloete J, Mathur S, Feneley RC.** 2006. A clinical assessment of the performance of a sensor to detect crystalline biofilm formation on indwelling bladder catheters. *BJU Int* **98:** 1244–1249.

271. **Malic S, Waters MG, Basil L, Stickler DJ, Williams DW.** 2012. Development of an "early warning" sensor for encrustation of urinary catheters following *Proteus* infection. *J Biomed Mater Res B Appl Biomater* **100:**133–137.

Epidemiology and Virulence of Klebsiella pneumoniae

18

STEVEN CLEGG[1] and CAITLIN N. MURPHY[1]

INTRODUCTION

Awareness of the role of *Klebsiella pneumoniae* as an important opportunistic pathogen of the urinary tract in compromised individuals and hospitalized patients has increased over the last decades. The emergence of these bacteria exhibiting multiple antibiotic resistance phenotypes has made the treatment and management of *K. pneumoniae* urinary-tract infections (UTIs) difficult (1–4). Although not frequently encountered as a cause of community-associated UTIs, *K. pneumoniae* is a leading cause of enterobacterial nosocomially acquired UTIs and is frequently prevalent as an infectious agent of patients with indwelling urinary catheters (5). These catheter-associated urinary-tract infections (CAUTIs) can be source of invading organisms into the blood-stream among these compromised individuals, leading to serious infections with high morbidity and mortality rates. In addition, the frequency of UTIs in individuals residing in long-term-care facilities is increasing (6, 7). Therefore, the epidemiology of *Klebsiella* UTIs and CAUTIs indicates that factors that result in a decrease in the efficiency of the host immune system, for example, the insertion of an indwelling device such as a catheter, represents a significant increase in susceptibility to infection by these bacteria (Table 1).

[1]Department of Microbiology, University of Iowa College of Medicine, Iowa City, IA 52242.
Urinary Tract Infections: Molecular Pathogenesis and Clinical Management, 2nd Edition
Edited by Matthew A. Mulvey, David J. Klumpp, and Ann E. Stapleton
© 2017 American Society for Microbiology, Washington, DC
doi:10.1128/microbiolspec.UTI-0005-2012

TABLE 1 Incidence of UTIs caused by _Klebsiella pneumoniae_

UTI type	Occurrence	Reference
ICU-acquired UTIs	4%	157
Community-acquired UTI w/ catheter	8%	158
Nosocomial catheter-associated UTI	10% to 11%	159
Community-acquired UTI	8.7% to 16.9%	52, 55, 160, 161

The presence of indwelling urinary devices leads to an accumulation _in situ_ of host-derived material on the catheter surfaces (8, 9). This environment presents an excellent niche for the development of bacterial biofilms, particularly those of opportunistic pathogens such as _Klebsiella pneumoniae_. Biofilm formation plays a key role in many bacterial infections and contributes to the ability of bacteria to overcome host-defense mechanisms (8, 10, 11). CAUTIs are frequently associated with the development of _K. pneumoniae_ biofilms and factors that contribute to this process are likely to play an important role in pathogenesis (8, 12, 13). The development of techniques to measure and quantitate bacterial biofilm formation has facilitated the analysis of _K. pneumoniae_ biofilm formation on solid surfaces and this, in addition to historically older investigations into _Klebsiella_ virulence, will hopefully lead to better management of UTIs by these bacteria. Clearly, the isolation of the so-called "superbugs" that exhibit resistance to many recently developed antibiotics necessitates consideration of non-antibiotic-associated treatments of both UTI- and non-UTI-associated _K. pneumoniae_ infections. The major virulence factors and properties of _K. pneumoniae_ are discussed below as well as biofilm formation and antibiotic resistance.

As for any opportunist, the outcome of infection in the urinary tract by _K. pneumoniae_ is a function of the bacterial ability to resist clearance and/or killing by host-defense mechanisms. The bacteria causing these types of infection commonly do not produce single virulence factors that can be identified as a primary attribute but rely on a battery of gene products to facilitate successful colonization and growth in a host where the immune system may not be fully functional and damaged or impaired (Table 2). Colonization of the urinary tract may be asymptomatic or may progress to clinical disease. In contrast, relatively recent reports of _K. pneumoniae_ isolates causing acute pyogenic and disseminated infections are associated with the production of a distinct hypermucoviscous phenotype and are most commonly associated with one capsular type of bacterium (14–16). The virulence factors of these strains will not be the focus of this chapter since they do not represent the typical _K. pneumoniae_ isolates associated with UTIs. However, a comparison between the pyogenic strains and those causing UTIs will be made when the two groups share a common property. A summary of the virulence properties of _K. pneumoniae_ associated with UTIs is shown in Table 3.

CAPSULES AND LIPOPOLYSACCHARIDE

Capsule and Lipopolysaccharide Production

Strains of _K. pneumoniae_, particularly clinical isolates, frequently produce a viscous polysaccharide capsule (Table 3). The biochemical complexity of these capsules gives rise to the production of strain-specific antigenic types of capsular material. Currently, there are over 77 distinct antigenic types of capsule produced by _Klebsiella_ strains and these antigens have been used to discriminate between strains during clinical infections. Although infections in humans cannot

TABLE 2 Risk factors associated with _Klebsiella pneumoniae_ UTI

Diabetes mellitus
Urinary-tract obstruction
Chronic renal insufficiency
Immunosuppression
Catheterization

TABLE 3 Summary of virulence factors involved in *Klebsiella pneumoniae* pathogenesis

Virulence factor	Role in pathogenesis
Capsule	Inhibit and evade phagocytosis by host cells, induces dendritic cell maturation, neutralizes antibacterial activity of host defense
LPS	O antigen provides serum resistance
Siderophore	Scavenge essential iron for survival, hypermucoviscous phenotypes have been linked to increased iron-binding activity
Urease	Limited role in precipitation of inorganic salts leading to catheter encrustation
Type 1 fimbriae	Involved in the formation of intracellular bacterial communities
Type 3 fimbriae	Important for biofilm formation on biotic and abiotic surfaces, role in biofilm formation on urinary catheters *in vivo* remains to be elucidated
Biofilm formation	Formation promotes resistance to host killing and antimicrobials, experimentally shown to be facilitated in part by fimbriae and capsule
Antibiotic resistance	Carbapenem-resistance prevents many treatment options

be attributed solely to one distinct capsular serotype, most clinical isolates of *K. pneumoniae* belong to a relatively small number of serotypes. The recently described infections due to *K. pneumoniae* that result in disseminated pyogenic infections are commonly caused by K1 serotypes. The K2 serotype was most frequently associated with UTIs, but represented only approximately 13% of the isolates examined, with a broader range of serotypes being implicated in these types of infections (17). The relatively broad distribution of capsular types associated with *K. pneumoniae* UTIs was also reported in a study of 32 strains (18). Therefore, unlike the stains causing disseminated pyogenic infections and liver abscesses, the strains of *K. pneumoniae* implicated in UTIs represent a more diverse range of bacteria.

Biochemical analyses of *K. pneumoniae* capsules indicate that they are composed of complex polysaccharides consisting of repeating subunits. For example, the K21a, K36, and K50 serotypes possess di-mannose/rhamnose residues, whereas K2, K8, and K55 serotypes do not (19). The K53 capsule is comprised of D-glucuronic acid, D-galactose, D-mannose, and L-rhamnose and is similar to that found in the K74 serotypes that lack rhamnose (20, 21). The K60 capsule has been reported to be unusual in possessing three glucopyranosyl side chains, compared to one or two on most K-antigens, within a heptasaccharide-repeating subunit (22). The

K1 capsule from isolates responsible for disseminated infections possess fucose and this is not found in non-K1 isolates (23). The association of the capsule with the lipopolysaccharide (LPS) has been investigated in a K2 serotype in which the capsule is bound by an ionic interaction to the LPS through a negative charge of carboxyl groups on GalA (24). Indeed, it has been reported that an enzyme playing an important role in LPS biosynthesis influences the amount of cell-bound capsule present on *K. pneumoniae* (25). These results indicate the interrelationship at the biochemical and synthetic level between these two structures in *Klebsiella* strains. It is clear that the *K. pneumoniae* LPS and capsule are not produced completely independently of each other and the formation of one of these structures can impact the amount and presence of the other.

Compared to *E. coli* and *Salmonella enterica*, the analysis of *K. pneumoniae* O-antigens associated with the LPS has been limited. Nine O-antigenic types have been described, with O1 being the most common serogroup associated with human infections (19, 26). The antigenic specificity may be a function of the modification of a single repeat subunit. For example, the O-antigenic side chain of serotype O2a is a polymer of the disaccharide D-galactan subunit. In other serotypes of O-antigens this subunit may be O-acetylated or capped by different forms of the repeat subunit.

Genetics of Capsule Production

The genes encoding the production of capsules by *K. pneumoniae* were originally designated the *cps* gene cluster and were mapped close to *his* on the bacterial chromosome (27). Cloning of the *cps* cluster indicated that the genes occupied approximately 15kb of DNA and possessed all the determinants necessary to impart the serotype of a strain (26, 28). Expression of the *cps* genes was originally described to be under the control of RcsB, a protein involved in colonic acid biosynthesis (29), and RcsB enhanced transcription of the cloned *cps* genes in an *E. coli* background. Subsequently, three alleles of a different gene encoding RmpA/A2 have been described that also positively activate *cps*-gene expression (30–32). The distribution of these alleles varies between serotypes and the geographic region of the isolates; two of the alleles are plasmid encoded and the third is a chromosomal gene. The presence of a DNA-binding domain in the C-terminal region of RmpA has led to the suggestion that it activates transcription of *cps* genes by binding to the promoter region of these genes. Excess RmpA production leads to a hypermucous phenotype in *K. pneumoniae* and transcriptional fusions of *cps*-promoter regions are increased in expression in the presence of RmpA (32).

K. pneumoniae grown under iron-replete conditions exhibit a decreased production of capsular polysaccharide. The Fur protein has been shown to inhibit transcription of *rmpA* as well as other genes associated with iron-acquisition systems in *K. pneumoniae* (33). Therefore, as in other enterobacterial systems, the concentration of available iron in the environment of the *Klebsiella* strains can play an important role in virulence-gene expression, including the amount of capsule produced by clinical isolates. The role of iron in affecting *K. pneumoniae* virulence is discussed below.

As indicated above, K1 serotypes of *K. pneumoniae* are associated with causing severe disseminated infections. The mucoviscosity-associated gene (*magA*) was identified to be part of the *cps*-gene cluster in these isolates and encodes a polymerase necessary for K1 biosynthesis (34). This gene has been used to identify, by polymerase chain reaction (PCR), K1-postive strains, but recent evidence indicates that a similar allele is present in other K-serotypes. The allele specific for the K1 serotype has recently been renamed *wyz_K1* and the indication is that *wyz* alleles may be serotype-specific but functionally identical.

Role of Capsules in Virulence

The ability of specific *K. pneumoniae* capsules to inhibit or impair phagocytosis by host cells was reported over 20 years ago (35–39). Subsequently, it was demonstrated that capsules possessing the Man-α-2-Man sequence exhibited a greater binding and susceptibility to phagocytic cells, a phenomenon termed lectinophagocytosis (40). The K2 serotype does not possess the Man-α-2-Man disaccharide and is more frequently associated with human infections compared to serotypes that do possess this structure (e.g., K21a), indicating that, in part, the resistance to phagocytosis by some serotypes may be a function of decreased binding to these host cells. Simply exchanging the K21a *cps* genes for those from a K2 isolate, however, was not sufficient to restore full virulence to a strain and demonstrated that additional virulence factors must play a role during infection (26). Not only may the capsule influence susceptibility to phagocytosis by direct binding to these cells, but it has also been demonstrated that the composition of the capsule influences opsonophagocytosis. Serotypes or genetically engineered strains producing capsules possessing the di-Man/ Rha epitope exhibited a decreased ability to stimulate polymorphonuclear leucocytes (PMNs). These epitopes were not recognized by many components of the innate immune system, such as lung-surfactant proteins, mannose-binding lectins, and alternative complement activation constituents, in addition to the mannose-receptor of phagocytic cells (19). This has led to the suggestion that virulent serotypes such as K2 escape eradication by producing a surface capsule comprised of glycoepitopes not recognized by these factors. Overall, strains of *K. pneumoniae* can be effectively killed by host phagocytic cells once internalized into these cells. Therefore, mechanisms by which the capsule can effectively inhibit interaction of the bacteria with these cells will facilitate survival.

In addition to the role of capsules in the interaction between *K. pneumoniae* and macrophages or PMNs, recent evidence indicates that the capsules affect interaction with dendritic cells (41). In these studies, capsular material was shown to induce dendritic-cell maturation with increased expression of the markers CD83, CD86, and TLR4 but decreased production of CD14. The results indicate that *K. pneumoniae* capsular material induces a defective immunological response characterized by dendritic-cell maturation with increased pro-Th1 cytokine production. In these studies it could also be demonstrated that although LPS played no role in dendritic cell maturation, the LPS played a role in dendritic-cell activation. Very little is known about the interaction of *K. pneumoniae* with these important antigen-presenting cells and further studies may shed light on possible mechanisms that enable the bacteria to avoid recognition and processing by dendritic cells.

The role of *K. pneumoniae* capsule in experimental UTIs indicates that this structure plays a significant role in this type of infection (42). Using competitive-infection studies and non-capsulate variants in addition to the parental strains, it was possible to demonstrate that the capsulate strains outgrew the variants *in vivo*. However, no such advantage was observed when colonization of the murine gastrointestinal tract was monitored. In contrast, the non-capsulate variants were able to adhere *in vitro* to epithelial-cell lines better than capsulate strains, but this increase in adherence did not extrapolate to increased infectivity *in vivo*. Increased adherence *in vitro* by non-capsulate strains may be due to unmasking of surface-associated adherence factors that are covered by capsule in wild-type strains during growth under these conditions.

Antimicrobial peptides are predicted to play an important role against infection, particularly against mucosal pathogens, such as *K. pneumoniae*. Most of these peptides are cationic and the role of anionic capsules in neutralizing their affect has been investigated (43). One host defensin has been shown to stimulate the release of K2-capsular material from bacterial cells. The release of this anionic capsule from the bacterial cell *in vivo* has been postulated to neutralize the anti-bacterial activity of a host defensin by preventing the peptide from localizing to the bacterial membrane. Considering that *K. pneumoniae* is an opportunist that can colonize mucosal surfaces of a variety of organ systems, the ability of these bacteria to evade innate defense mechanisms represents one of the primary and early mechanisms in the establishment of infection.

Overall, the most virulent strains of *K. pneumoniae* produce a capsule composed of saccharides that do not facilitate binding to phagocytic cells and therefore are more resistant to phagocytosis. In addition, these capsules also present as poor opsonins and evade phagocytosis by complement activation. Since there are a greater variety of capsular serotypes implicated in UTIs compared to disseminating *K. pneumoniae* infections, the role of capsules in mucosal infections is less clear and a few studies of UTI-causing strains have indicated that serotypes from a wide variety of capsular types are found. The role of these capsules in protecting against innate defense mechanisms in the urinary tract has not been investigated in detail, but these structures could play a role in neutralizing host factors in this environment.

SIDEROPHORES

Siderophore Production

K. pneumoniae must compete with host cells *in vivo* for essential iron that is a necessary cofactor for bacterial metabolism. In a study of more than thirty strains representing 23 different serotypes of *K. pneumoniae* implicated in urinary tract infections, the siderophore enterochelin was the predominant iron-salvaging-system compound produced (18). This is in contrast to bacteremic strains, primarily K1 and K2 serotypes, in which aerobactin was correlated with virulence in an experimental murine model of infection (44). Enterochelin is a catechol-type of molecule comprised of three phenolic rings that are involved in the uptake of ferric ions (45). The ferrienterochelin complex is bound to an 81-kDa outer-membrane receptor that is found in most strains of both *K. pneumoniae* and *E. coli*. Aerobactin is a hydroxamate compound comprised of a citrate molecule with N6-hydroxyacetyl lysine and is derived by the oxidation of lysine.

Transport of ferric iron into the bacterial cells is dependent upon a family of proteins that includes TonB, Exb, and ATP-binding cassette (ABC) transporters (46). The iron-siderophore complex is recognized by a TonB-dependent outer-membrane receptor that mediates transport into the periplasm. Subsequently, the siderophore binds a periplasmic protein that facilitates transport to an ABC-transporter that mediates passage of iron through the inner membrane prior to reduction from ferric to ferrous iron. Although the role of iron-chelating systems in uropathogenic *K. pneumoniae* has not been experimentally examined in detail, both airway-infection models and bacteremic studies have suggested that mutants lacking the ability to take up iron are attenuated.

In addition to the two described siderophores aerobactin and enterochelin, it is possible that highly invasive strains of *K. pneumoniae* possess multiple iron-acquisition systems. Using experimental infection models it has been demonstrated that invasive strains express genes that encode proteins related to iron-binding molecules from other bacterial species and the relationship between these systems has yet to be examined (47, 48). It appears that the more invasive that some strains of *K. pneumoniae* have evolved to become, the greater the necessity to efficiently compete with the host for iron. This may have led to the evolution of a complexity of iron acquisition systems by these bacteria.

Genetics of Iron-Acquisition Systems

The ferric-uptake regulator (Fur) is a principal regulatory protein controlling gene expression in response to iron concentrations. In the presence of excess iron, i.e., iron-replete conditions, this protein can bind to specific DNA sequences (the Fur box) in promoter regions of iron-regulated genes. This binding results in repression of target-gene transcription and derepression under iron-depleted conditions. Fur is, therefore, a global regulator of iron-regulated genes in enterobacteria. For *K. pneumoniae* Fur has been shown to affect the expression of genes encoding capsule biosynthesis and iron-acquisition systems. For example the regulatory genes, *rpmA2* and *rcsA*, are themselves regulated by Fur in addition to at least six of eight genes involved in iron uptake (49). Fur-deletion mutants of *K. pneumoniae* exhibit reduced siderophore production as well as changes in adhesin production (50). Consequently, Fur can have multiple effects on gene expression in *K. pneumoniae* and many of these genes encode products involved in colonization and growth in the host.

Genome analysis of available *K. pneumoniae* sequences indicates the presence of 10 putative iron-uptake systems in some strains. Many (at least seven) of these systems exhibit properties associated with a dependency on the TonB machinery for their complete function. Although the precise function of each of these systems has yet to be elucidated, the presence of multiple systems in *K. pneumoniae* indicates that the uptake of iron from a limiting environment is an important and essential function to maintain the viability of the bacteria.

Role of Siderophore and Iron-Uptake Systems in Virulence

Growth of *K. pneumoniae in vivo* requires the capture and utilization of iron for essential metabolic processes. Since iron is also required by host cells, in order to survive the bacteria must compete with the host for available iron that is frequently in limiting supplies. Host iron-binding compounds, such as transferrin and lactoferrin, are strong chelators of elemental iron and *K. pneumoniae* has evolved to produce its own iron-binding systems to scavenge the necessary iron required for its survival. These iron-acquisition systems are described above. The importance of iron availability and *K. pneumoniae* pathogenesis was suggested by the early studies of Khimji and Miles (51) who demonstrated that excess iron in hosts' tissues increased bacterial pathogenicity.

The *K. pneumoniae* TonB system has been shown to play an important contribution in bacterial virulence. Using murine experimental models of infection, it has been demonstrated that a TonB mutant is significantly attenuated (47). In addition, this mutant could be used as a vaccine strain to prevent experimental infections by the parental virulent strain due to stimulation of a protective immune response in vaccinated animals. In contrast, individual mutations in genes involved in the iron-transport systems and siderophore production did not appear to have any effect on *K. pneumoniae* pathogenesis. However, a triple mutant, deleted in the *irp2*, *iuc*, and *iroA* genes exhibited decreased virulence (47). No studies have yet examined the role of these systems in UTIs caused by *K. pneumoniae*.

Recently, the isolation of hypervirulent strains of *K. pneumoniae* exhibiting a hypermucoviscous phenotype has been described (53). Investigations of virulence factors of these strains indicated that a small molecule (approximately 3kD) was produced that enabled the strains to grow and survive in human ascites. In addition, this molecule increased the growth of *K. pneumoniae in vivo* following subcutaneous inoculation of mice. Preliminary evidence indicates that this molecule is involved in iron uptake and is a siderophore. It is speculated that the enhanced virulence of these strains may, in part, be a function of increased siderophore production as well as an increased iron-binding activity compared to less virulent *K. pneumoniae* isolates (53).

UREASE

Genetics and Structure

Many strains of *Klebsiella* produce the extracellular enzyme urease that is a nickel-containing enzyme responsible for hydrolyzing urea to ammonia and carbamate (54). The molecular genetics of urease production and the structure of the enzyme have been described (54, 56). The enzyme is assembled by the interaction of four accessory proteins (UreD, UreE, UreF, and UreG) involved in delivering nickel to the urease apoprotein and one of these components (UreG) serves as a GTPase for activation. The molecular biology of urease production and activity have been described in a series of studies by Hausinger and colleagues (54, 56–60).

Role of Urease in Virulence

The significance of urease as a virulence factor has been suggested for bacteria that grow in the urinary tract and contribute to the formation of infection stones. Urea hydrolysis is proposed to lead to an increase in the localized pH resulting in precipitation of inorganic salts that are insoluble at a relatively high pH (pH 9.0). This precipitation leads to encrustation, particularly on indwelling urinary catheters that may influence the course of infection. First, urine flow is impaired under these conditions, leading to a deficiency in bacterial clearance from the site of infection. Second, the encrustations can facilitate the formation of biofilms on biotic or abiotic surfaces that themselves may inhibit effective treatment by antibiotics.

The role of urease production by *K. pneumoniae* in virulence and mediating the pathogenesis of the organisms, however, has yet to be demonstrated. Compared to species of *Proteus*, *K. pneumoniae* produces significantly less urease *in vitro*. In addition, it has been demonstrated that bacterially induced precipitation of salts onto abiotic surfaces *in vitro* by *K. pneumoniae* is less than that observed by urease-producing strains of *Proteus* (61, 62). More recently, these results were confirmed by Broomfield and co-workers (63) who demonstrated that stains of *Klebsiella* were less efficient at forming encrustations and subsequently blocking catheters in a laboratory model of infection, compared to strains of *Proteus* and *Providencia*. Direct experiments to investigate the role of urease production by *K. pneumoniae* *in vivo* using an animal model of infection have not yet been performed. However, the

evidence to date indicates that this enzyme may only have a limited role in virulence by these bacteria.

COLONIZATION AND ADHERENCE

As for many of members of the *Enterobacteriaceae*, a review of the available on-line genome sequences for strains of *K. pneumoniae* indicates the presence of multiple gene clusters that potentially encode fimbrial- and non-fimbrial-adherence antigens. The presence of these multiple-gene clusters appears to be the norm for most enteric bacteria, although investigations into the function of most of these genes and their products have been restricted to a small number of factors. For example, there are up to eleven putative fimbrial-gene clusters located on the *K. pneumoniae* genome, but only a few of these have been associated with specific adherence phenotypes (Fig. 1). Many of the gene clusters are not readily expressed under laboratory conditions of growth. However, it is likely that all of these systems encode functional adherence factors since their presence and expression represents a significant metabolic load to the bacterial cell. Most of the investigations into the role of *K. pneumoniae* adhesins that play a role in colonization of the urinary tract have focused upon two fimbrial types, type 1 and type 3 fimbriae,

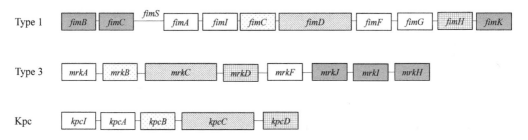

FIGURE 1 Organization of the *fim*, *mrk*, and *kpc* fimbrial gene clusters in *K. pneumoniae*. The direction of transcription is indicated by the arrows and functions of the gene products are as follows: ▯, Structural subunit; ▮, regulatory proteins that directly or indirectly affect the expression of the gene encoding the major subunit; ▯, adhesin; ▯, periplasmic chaperone; ▧, outer membrane-scaffolding (usher) protein.

which are produced by many urinary isolates of *K. pneumoniae* (Fig. 1).

Structure and Genetics of Fimbriae

All of the fimbriae that are produced by strains of *K. pneumoniae* are assembled by the chaperone/usher-assembly pathway and no adhesins similar to type IV pili of other bacteria have been described in *Klebsiella* (64–69). The fimbriae assembled by the chaperone/usher pathway are characterized by the presence of at least two genes that encode a periplasmic chaperone and a scaffolding protein necessary for ordered fimbrial assembly. Some gene clusters possess more than one gene encoding a periplasmic chaperone, but all the clusters are characterized by having a single gene encoding the scaffolding or usher protein. These proteins facilitate assembly of a functional fimbrial structure by the processes of donor-strand complementation and donor-strand exchange (64, 66). The structural subunits possess an incomplete immunoglobulin-like groove that is completed by a strand of the chaperone protein (donor-strand complementation). At the usher protein in the outer membrane, that strand is exchanged by a region from the most recently incorporated subunit (donor-strand exchange). The fimbrial appendages are composed of a major fimbrial subunit, adaptor molecules, and a specific adhesin, and the number of adaptor molecules associated with any particular fimbrial type is variable. In *K. pneumoniae*, each of the gene clusters encoding these components of the fimbrial-adherence factors are unlinked, but each cluster exists as a contiguous set of genes. There is no evidence to indicate that the individual genes, and their products, are interchangeable, and so the chaperone and usher molecules of one system are not functional in assembling the appendage of a heterologous fimbrial system. Similarly, the adhesin protein of one type of fimbria cannot be assembled onto the shaft of a different fimbrial type. Given these

observations and also the presence of multiple fimbrial-gene clusters in *K. pneumoniae*, the control of fimbrial production, localization, and assembly must represent a complex regulatory pathway. It has been suggested that the ability of enteric pathogens to colonize different ecological niches within the urinary tract, for example, may be a function of temporal expression of gene clusters encoding adherence factors facilitating attachment to different types of cells encountered throughout the organ system (70–73).

The three *K. pneumoniae* fimbrial adhesins that have been described in any detail are types 1 and 3 fimbriae and the Kpc fimbriae (65, 74) (Fig. 1). Type 1 fimbriae, or mannose-sensitive fimbriae, so-called because of their ability to bind soluble mannose as a competitive inhibitor to binding, are encoded by the *fim* gene cluster. This fimbrial type is produced by many members of the *Enterobacteriaceae* and yet, although characterized by the presence of functionally related proteins (e.g., periplasmic chaperone, scaffolding protein, adhesin, major structural protein, etc.), there is little amino acid-sequence relatedness between species (75–79). For example, the genes encoding the type 1 fimbriae of *Salmonella* cannot complement those encoding the same fimbrial type from *K. pneumoniae* (79). The regulation of the *K. pneumoniae* *fim* gene expression is controlled using an invertible DNA element (*fimS*) that facilitates transcription of the *fim* operon by a mechanism similar to that described in *E. coli* (65, 80, 81). In one orientation (phase "ON") the promoter enabling *fim*-gene expression allows transcription of the operon, whereas in the opposite orientation (phase "OFF"), transcription of the *fim* genes cannot occur. However, in addition the *K. pneumoniae* *fim*-gene cluster possesses a gene, *fimK*, not present in the *E. coli* cluster (82). This gene encodes a putative phosphodiesterase that may modulate the intracellular levels of cyclic-dimeric-guanosine monophosphate (c-di-GMP) that is an important bacterial second messenger (83–85).

As indicated below, the intracellular concentrations of c-di-GMP has been shown to effect many bacterial attributes including virulence-factor production.

Type 3 fimbriae are encoded by the *mrk*-gene cluster and may be present in *K. pneumoniae* as a plasmid-borne and/or a chromosomally borne gene cluster (86–88). MrkA is the major structural component of the fimbrial shaft and MrkD facilitates binding to extracellular-matrix proteins and functions as the fimbrial adhesin. Structural stability of the fimbriae is associated with β strands in the C-terminal region of MrkA (89). Regulation of *mrk*-gene expression is achieved using a combination of a phosphodiesterase (MrkJ) and a c-di-GMP-binding protein (MrkH) in conjunction with a DNA-binding protein (MrkI). These gene products are suggested to influence and respond to the intracellular concentration of c-di-GMP that modulates the level of *mrk* gene expression (90–93). The level of c-di-GMP within bacterial cells has been demonstrated to influence the expression of numerous genes and plays an important role in virulence-factor production (84, 85, 94).

The Kpc fimbriae are most frequently associated with K1-posiive strains of *K. pneumoniae* causing disseminated pyogenic infections (95). The production of these fimbriae is mediated by the *kpcABCD* gene cluster, although the conditions for optimal expression of these genes has yet to be elucidated. Preliminary investigations indicate that *kpc*-gene expression is mediated, in part, by the orientation of an invertible DNA element within the gene cluster. Like the *fimS* switch of *E. coli*, it is possible that this mechanism is responsible for a phase-variable phenotype of Kpc fimbriae in *K. pneumoniae*.

Role of Fimbriae in Colonization and Virulence

The role of type 1 and type 3 fimbriae in colonization and pathogenesis of urinary isolates of *K. pneumoniae* has been investigated. Since early studies had indicated that *E. coli* type1 fimbriae facilitated colonization and intracellular survival within the urinary tract of experimentally infected animals, the role of *K. pneumoniae* type 1 fimbria in uropathogenesis was examined (82, 96–98). The ability of type 1 fimbriae to mediate *in vivo* the formation of *K. pneumoniae* intracellular-bacterial communities (IBCs) within bladder-epithelial cells was compared to that of *E. coli* fimbriae. *E. coli* IBC formation is proposed to be a strategy whereby uropathogenic *E. coli* evade host innate immune mechanisms and subsequently migrate through the urinary tract (99, 100). Bacteria growing as IBCs exhibit properties similar to those growing as a biofilm community. Although *K. pneumoniae* did form IBCs, their frequency compared to those produced by *E. coli* was considerably less and this correlated with a decreased production of fimbriae *in vivo* (82). The reason for a decreased production of type 1 fimbriae in the murine model of infection by *K. pneumoniae* is unclear. *E. coli* type 1 fimbriae are less frequently produced in the kidney compared to the bladder and it has been suggested that *Klebsiella* has evolved to more efficiently colonize different regions of the urinary tract. However, the results investigating *K. pneumoniae* pathogenesis do indicate that IBC formation may be a general strategy among enteric pathogens implicated in UTIs.

Type 3 fimbriae have been shown to play a major role in the formation of *K. pneumoniae* biofilms *in vitro* on both abiotic and biotic surfaces (101–104). The MrkD adhesin facilitates binding to extracellular-matrix proteins, such as collagen, and also mediates binding to cells derived for the urinary tract (105, 106). Biofilm formation on abiotic surfaces such as plastic and silicon is mediated by the major structural fimbrial protein MrkA and can occur independently of MrkD (102). However, indwelling devices, such as urinary catheters, are coated *in situ* over time with host-derived factors and thus are more likely to represent a substrate for bacterial-biofilm formation that is mediated by MrkD binding (107). Examination of multiple clinical isolates of *K. pneumoniae* have demonstrated that the surface production of type 3 fimbriae was a significant indicator of a strain's ability to adhere to and form a biofilm on solid surfaces (108). Type 3 fimbriae are produced by several different genera of enterobacteria (109). Given the role of type 3 fimbriae in adherence to host-derived matrices, it has been speculated that type 3 fimbria-producing bacteria may be selected during growth on indwelling urinary catheters (110).

Evaluation of the contribution of both type 1 and type 3 fimbriae in biofilm formation has suggested differing results. In some studies, it has been demonstrated that strains unable to produce type 1 fimbriae are as proficient at forming biofilm as bacteria that can produce this type of fimbria, indicating that type 1 fimbriae play a minimal role. Also, it has been suggested that type 1-fimbrial expression is decreased during biofilm formation. In addition, several investigations have shown that type 3 fimbriae are essential for typical biofilm formation (91, 103). However, more recent investigations have demonstrated that both fimbrial types play a role in biofilm formation (104). One probable reason for the observed differences is the method of assessing biofilm formation *in vitro* among the investigations and also the nature of the substrate being used to examine biofilm formation. *K. pneumoniae* biofilm formation *in vivo* on urinary catheters is most likely to be an important stage in the pathogenesis of these organisms. However, there is currently no available animal infection-model system to mimic these conditions. Currently, we are developing a murine-infection model with silicone tubing implanted in the bladder to investigate the role of fimbrial adhesins *in vivo*.

As indicated above, urinary isolates of *K. pneumoniae* have the genetic capability of producing an array of adherence factors. The role of these factors in colonization and virulence has not been investigated and the difficulty of expressing these genes *in vitro* has made assessment of their function difficult. It is possible that many of these genes clusters are expressed only under conditions that are found *in vivo*. Therefore, the development of *in vivo* infection models along with increased assays to measure gene transcription under these conditions may provide mechanisms to examine the kinetics of fimbrial-gene expression *in vivo*. Examination of the transcriptome of *E. coli* during infection *in vivo* has recently been described and the application of these techniques to growth of *K. pneumoniae* in experimentally infected animals should be informative, and may address the question of multiple fimbrial-gene expression (111, 112).

BIOFILM FORMATION

K. pneumoniae-biofilm formation has been postulated to be an important stage in the pathogenesis of these bacteria, particularly in the case of CAUTIs. Investigations into bacterial-biofilm formation on solid surfaces has been a major focus of research into bacteria pathogens over the last decade (11, 12, 113–115). Biofilm formation is a complex process that has been divided into a number of stages involving attachment, microcolony production, mature-biofilm formation, and release of free-living planktonic bacteria from the biofilm. The formation of biofilms is believed to be a mechanism of promoting persistence since bacteria within a biofilm are proposed to be less susceptible to killing by host-defense mechanisms. Also, there is evidence that the biofilm community is more resistant to the action of many antibiotics and exhibits an increased resistance to antimicrobials. The mechanisms by which growth as a biofilm enables increased resistance can be attributed to several factors. The biofilm matrix that frequently consists of a dense matrix of proteins, polysaccharides, and DNA prevents the efficient diffusion of antibiotics resulting in significantly decreased exposure of bacteria. However, for some antibiotics, the rate of diffusion through a biofilm may not be impaired but the breakdown of these antimicrobials may be enhanced within the biofilm (116). Bacteria growing deep within biofilms grow at a much slower rate than planktonic bacteria or those close to the surface of the biofilm, again making the biofilm bacteria less susceptible to killing by antibiotics that act upon actively dividing bacteria. Also, for these slow-growing bacteria, specific antimicrobial-binding proteins are poorly expressed. Changes in gene expression by bacteria growing as a biofilm compared to free-living organisms may also lead to resistance to killing. Also, transfer of genetic material between bacterial cells can be enhanced in a biofilm, thus facilitating transfer of material conferring antibiotic resistance (10, 117–119). For all these reasons, the benefit of bacterial-biofilm formation provides an important attribute for increasing the persistence and establishment of chronic infections in the urinary tract.

The role of *K. pneumoniae* fimbrial adhesins in the initial stages of biofilm formation has been discussed in the preceding section and will not be considered here. However, as mentioned above, biofilm formation involves a complex series of stages and developmental processes that require differential gene expression and the production of specific gene products at defined stages. For this reason, many gene products are likely to be involved in biofilm formation and their production may also be transitory during the process. For *K. pneumoniae*, the important contributing factors to biofilm formation have only recently been explored and relatively little is known about the array of gene products that are required for this process. Most of this early work has focused upon constructing libraries of defined insertion mutants of *K. pneumoniae* and identifying from these libraries specific genes that play a role in facilitating biofilm formation. Alternatively, recombinant-plasmid libraries comprised of *K. pneumoniae* DNA segments

have been used to transform *E. coli* strains to detect genes that increase biofilm formation in specific transformants.

Signature-tagged mutagenesis (STM) has been used to identify genes that influence biofilm formation *in vitro* on extracellular-matrix material using a continuous-flow biofilm chamber (120). Using this procedure, mutations in three different groups of genes that resulted in decreased biofilm formation were identified. These groups were genes encoding transcriptional regulators (e.g., LuxR-, LysR-, and CRP-related genes), sugar phosphotransferases, and genes involved in the synthesis of extracellular structures such as fimbriae and capsule. LysR-related transcriptional regulators affect a diverse group of genes and functions and have been reported to influence biofilm formation in other pathogenic bacteria. More recently, OxyR, a LysR-related regulator, has been shown to influence *K. pneumoniae* colonization of both mucosal and abiotic surfaces (121). Often the genes encoding LysR regulators are located adjacent to and inversely transcribed from the target genes that they regulate. However, further analyses of the genes surrounding this group of mutants were not performed.

LuxR-like transcriptional regulators are frequently involved in affecting the genes involved in quorum sensing. They have been implicated in regulation of biofilm formation for a variety of organisms and members of this family usually respond to cell density by binding to autoinducers (122). The LuxR-like homolog identified in the STM studies did possess a putative DNA-binding domain, but lacked a conserved autoinducer-binding region. Therefore, its role in quorum sensing and biofilm formation has not been established. The role of quorum sensing in *K. pneumoniae* was further investigated by Balestrino and coworkers (123) who demonstrated that autoinducer-1 (AI-1), an acylhomoserine lactone, was not produced by *K. pneumoniae*, but autoinducers of the class AI-2 were detected in culture supernatants. The gene, *luxS*, encoding the synthesis of this molecule was able to complement a LuxS-negative strain of *E. coli*. Comparison of a LuxS *K. pneumoniae* mutant with its wild-type parent indicated that the mutant was impaired in its initial stages of biofilm formation in a flow chamber. In a second series of studies, it could be shown that LuxS was involved in altering the expression of two genes involved in LPS biosynthesis by *K. pneumoniae* and this phenotype may be responsible for the observed changes in biofilm formation (124). However, the mutant did not demonstrate any significant differences in its ability to colonize the murine intestine following oral inoculation and the role of quorum sensing in colonization and growth in the urinary tract has not been examined.

As indicated above, the third type of mutation identified using the STM screen was an insertion into a *K. pneumoniae* open-reading frame that contains a CRP-activation domain. CRP mediates catabolite repression and the role of this specific *K. pneumoniae* gene in affecting biofilm formation has not been determined. However, the cAMP-CRP system has been shown to influence biofilm formation in other enteric bacteria. In *E. coli*, cAMP-CRP activates genes that have been shown to mediate biofilm formation (125). In *Serratia marcescens*, mutations in *crp* result in a large increase in biofilm formation and this was, in part, due to an increased expression of *fimA* and production of type 1 fimbriae in this strain (126). The influence of sugars on biofilm formation has also been demonstrated in *Vibrio cholerae* (127). Therefore, it is likely that the carbohydrate concentration of the environment that *K. pneumoniae* is growing in will have a significant effect on its ability to form biofilms and persist in the urinary tract. Increase in *K. pneumoniae* biofilm formation as a response to carbohydrate presence is not necessarily associated with virulence in acute infections. Mutations in *K. pneumoniae* genes that convert mannose to fucose have been reported to result in increased biofilm

formation but decreased virulence in a murine model of sepsis (23). The observation has been made that increased gene expression resulting in efficient production of biofilms by pathogens is often associated with the establishment of chronic infections, whereas it has been reported that these genes are down regulated in acute infections (128).

More recently, a *K. pneumoniae*-derived clone library was constructed in *E. coli* to identify genes that played a role in biofilm formation on abiotic surfaces and might be implicated in the colonization of devices such as urinary catheters (129). These results confirmed that type 3 fimbriae production was an important property in mediating biofilm formation by *K. pneumoniae*. Three clones possessed genes implicated in arabinose or amino-acid biosynthesis and one additional gene encoding a large surface protein that had been implicated in collagen binding was also identified. However, only the genes encoding type 3 fimbriae were subsequently shown to increase biofilm formation on catheters.

With the exception of types 1 and 3 fimbriae, the only other well-defined phenotype that has been examined in relationship to *K. pneumoniae* biofilm formation is capsule production. For example, mutations within the genes *wza* and *wzc* that are involved in capsule biosynthesis resulted in a decreased ability of *K. pneumoniae* strains to form biofilms. In addition, mutations in the genes *treC* and *sugE* that alter both capsule production and affect the mucoviscosity of the bacteria, altered biofilm-forming properties of *K. pneumoniae*. Although it is unclear if capsule production influences biofilm formation in the urinary tract, these studies clearly indicated that *K. pneumoniae* biofilm formation is an important bacterial virulence factor in colonization of the gastrointestinal tract (130). In an interesting study, Dzul and colleagues (131) examined the role of *K. pneumoniae*-capsule production in bacterial spacing and interactions during biofilm formation. Their results indicate that the capsular material enables the bacteria to maintain shorter spacing between cells and thus tighter packing within the biofilm. The reason for this phenotype is unclear, but the authors indicate that greater capsule production could inhibit the separation of daughter cells during bacterial division or that the capsule could interfere with other factors, such as fimbriae, that might affect space-filling between cells. Indeed, the interaction of the capsule expression and fimbrial production has been examined in *K. pneumoniae* (132, 133). In these studies it was noted that the production of capsules interfered with adhesin function.

Another factor that influences *K. pneumoniae*-biofilm formation is cellobiose metabolism (134). Using a *K. pneumoniae* isolate causing pyogenic liver abscesses, it was demonstrated that a mutation in *celB* resulting in cellobiose deficiency was significantly reduced in biofilm formation and this phenotype could be restored by complementation using the cloned *celB* gene. In addition, the *celB* mutant was reduced in its ability to colonize the gastrointestinal tract of experimentally infected mice. These experiments provided another indication that a reduced-biofilm phenotype of *K. pneumoniae* correlates with decreased colonization activity *in vivo*.

It is now becoming apparent that colonization of and subsequent biofilm formation on indwelling urinary catheters by *K. pneumoniae* are important stages in the pathogenesis of the bacteria during CAUTIs. Also, the complexity of biofilm formation and the multiple genes involved in this process makes investigations into the molecular pathogenesis of this process very intriguing. The absence of a good *in vivo* model system to mimic CAUTI has limited the extrapolations that can be made from studying biofilms *in vitro*. However, recent developments have led to investigations of bacterial biofilm formation on silicone tubing that has been inserted into the bladders of mice (135). This technique is now being adapted to investigate *K. pneumoniae*-biofilm formation

in the urinary tract and may provide validation of many of the results observed using *in vitro* techniques.

ANTIBIOTIC RESISTANCE

The emergence of multidrug-resistant (MDR) *K. pneumoniae* isolates has presented a significant problem in the management and treatment of infection by these bacteria. *K. pneumoniae* infections of the urinary tract are a leading cause of Gram-negative CAUTIs in the hospital environment and long-term healthcare facilities. The carriage of resistance determinants on promiscuous plasmids has enabled the spread of resistance between strains. As an example of this rapid spread, it can be noted that *K. pneumoniae* resistance to the carbapenem family of antibiotics was initially described in the United States in 1997 and periodically thereafter for the next three years (136–138). However, over the last decade carbapenem-resistant strains of *K. pneumoniae* have emerged globally and this resistance has now been reported to be present in more than 33% of the isolates from medical centers in one major population area (139). Horizontal transfer of the plasmid carrying the resistance determinant between *E. coli* and *K. pneumoniae* residing in the gastrointestinal tract has been suggested (140). This transfer represents a major concern since the GI tract of hospitalized patients is frequently colonized with *K. pneumoniae* and represents an important reservoir of infection in this environment. Establishment of this plasmid in the normal flora will facilitate its transfer to strains causing extraintestinal infections.

With the establishment of rapid whole-genome-sequencing technologies it is now possible to compare the genomes of individual bacterial isolates. For *K. pneumoniae*, there are currently available six genome sequences that are accessible on-line, representing at least three different clinical isolates with differing antibiotic susceptibilities

(141). A comparison of the genome sequences in these strains indicates that approximately 70% of the proteins that could be encoded in these strains are conserved. This indicates a large degree of plasticity in the *K. pneumoniae* genome, a condition that would be conducive to allowing the spread of antibiotic resistance. Indeed, a comparison of these strains suggested that the acquisition of an MDR phenotype was not clonal but due to the horizontal gene transfer between strains with quite different genetic backgrounds (141). In fact, the presence of plasmids can only partially account for observed differences in antibiotic sensitivities. In *K. pneumoniae*, there are many examples of chromosomally borne two-component regulatory systems that appear to control the production and efficiency of efflux pumps (142–144). The activity of these pumps plays a vital role in determining resistance to a wide range of antimicrobials.

Bioinformatic analysis has also been used to identify possible regulators that control the expression of genes involved in antibiotic resistance. This approach has been used to characterize an AraC-type regulator, termed RarA, that controls the expression of an operon (*oqxAB*) encoding an efflux pump in *K. pneumoniae* (145). Overexpression of RarA results in a MDR phenotype that requires a functional efflux pump. RarA is also predicted to be encoded by the genomes of *Enterobacter* and *Serratia* species. This regulator is another protein that can be added to the family of transcriptional regulators that have been shown to control AcrAB efflux pump activity and include SoxS, RamA, MarA, and Rob.

DISCUSSION AND FUTURE PERSPECTIVES

K. pneumoniae has become of increasing concern in the clinical environment over the last two decades. The isolation of MDR strains causing nosocomially acquired infections and the relatively rapid spread of this resistance phenotype among strains has made the treatment and management of these types of infections difficult. The most common type of *K. pneumoniae* infection associated with hospitalized patients is CAUTIs. In most cases, removal of the infected device can lead to elimination of the infection, but while the device remains indwelling, there is a risk of more severe and disseminated infections. Many of these patients are immunologically compromised and are, therefore, at an increased risk of morbidity and mortality. If the bacteria migrate from the bladder to deeper tissues and the bloodstream it is possible that antibiotic therapy will not be successful in eradicating the organisms. For this reason, the development of alternative, non-antibiotic-based strategies to prevent urinary infections have been sought. For example, the use of probiotics to prevent *E. coli* UTIs has been investigated (146, 147). In addition, small molecule inhibitors and competitors of essential bacterial metabolic processes are beginning to be evaluated as methods of therapy (148–155). However, to date, no studies have been reported for the use of these molecules to prevent *K. pneumoniae* infections.

In order to fully understand the uropathogenesis of *K. pneumoniae* in CAUTIs it will be necessary to define those bacterial attributes that mediate adherence, colonization, and growth on both uroepithelial cells and catheters residing in the urinary tract. Murine models of experimental enterobacterial UTI have been developed and are widely used to investigate this type of infection, but do not involve the use of implanted devices. To date, there is no reported *in vivo* model to mimic urinary catheterization. Most models have used *in vitro* techniques employing silicone catheters and urine. The development of an animal model with silicone tubing implanted into the bladder has been developed to investigate growth on the tubing by Gram-positive bacteria (135). The adaptation of this model to investigate growth of *K. pneumoniae* on foreign bodies in the urinary tract *in vivo* is currently being examined by our group.

Bacterial growth on urinary catheters has been associated with biofilm formation in this environment. Research into this form of growth by bacterial pathogens has significantly increased over the last decade, and results have indicated that specific gene products can be produced only under these conditions. This is also likely to be the case for growth of *K. pneumoniae* in the urinary tract. Consequently, a more complete understanding of *K. pneumoniae* pathogenesis in CAUTIs will require investigations into this phase of growth. In addition, the insertion of medical devices into humans and animals rapidly results in the coating of these devices by host-derived material. Therefore, the biofilm growth on both abiotic and biotic surfaces should be examined. This is important since it is clear that that bacterial-gene expression during biofilm formation can be influenced by the substrate on which the bacteria are growing (156).

Another factor regarding growth of *K. pneumoniae* in the urinary tract is the ability of the bacteria to internalize within epithelial cells and avoid host defense mechanisms. Originally observed in *E. coli* and referred to as intracellular bacterial communities (IBCs), the bacteria exhibiting this phenotype were observed to demonstrate properties similar to organisms growing as biofilms (99). Uropathogenic *K. pneumoniae* isolates were subsequently shown to form IBCs in experimentally infected animals (82). Consequently, efficient elimination of bacteria from the urinary tract may have to employ strategies that take into consideration this phenomenon.

One troubling aspect of *K. pneumoniae* infections is the emergence of strains causing disseminated pyogenic infections. These isolates were originally described in patients from Asia and associated with diabetic individuals. However, there are increasing reports of such isolates from non-diabetic individuals and a more wide-spread geographic occurrence. Although these strains are not generally associated with UTIs,

they are clearly genetically related bacteria and present the potential to transfer, either directly or indirectly, genetic information into urinary isolates. The evolution of *K. pneumoniae* strains producing factors that mediate rapid dissemination from the urinary tract and subsequent tissue injury and destruction may be a possibility.

In conclusion, *K. pneumoniae* has become an important pathogen of immunocompromised individuals and presents significant potential problems with respect to treatment and elimination for the host. This is certainly the case for UTIs caused by these bacteria. Challenges that face the future management of these infections include the development of non-antibiotic based therapies since the ability of *K. pneumoniae* to rapidly evolve to antibiotic-resistant strains is alarming. The increased quality of healthcare has resulted in a greater population of susceptible hosts for *K. pneumoniae* infection. The prevention of infection and management of patients with infections will provide enormous challenges in the future.

ACKNOWLEDGMENTS

Conflicts of interest: We declare no conflicts.

DEDICATION

For Carleen Collins.

CITATION

Clegg S, Murphy CN. 2016. Epidemiology and virulence of *Klebsiella pneumoniae*. Microbiol Spectrum 4(1):UTI-0005-2012.

REFERENCES

1. **Ciobotaro P, Oved M, Nadir E, Bardenstein R, Zimhony O.** 2011. An effective intervention to limit the spread of an epidemic carbapenem-resistant *Klebsiella pneumoniae* strain in an acute care setting: from theory to practice. *Am J Infect Control* **39:** 671–677.

2. **da Silva RM, Traebert J, Galato D.** 2012. Klebsiella pneumoniae carbapenemase (KPC)-producing *Klebsiella pneumoniae*: A review of epidemiological and clinical aspects. *Expert Opin Biol Ther* **12:**663–671.

3. **Gijón D, Curiao T, Baquero F, Coque TM, Cantón R.** 2012. Fecal carriage of carbapenemase-producing Enterobacteriaceae: A hidden reservoir in hospitalized and nonhospitalized patients. *J Clin Microbiol* **50:**1558–1563.

4. **Won SY, Munoz-Price LS, Lolans K, Hota B, Weinstein RA, Hayden MK, Centers for Disease Control and Prevention Epicenter Program.** 2011. Emergence and rapid regional spread of *Klebsiella pneumoniae* carbapenemase-producing Enterobacteriaceae. *Clin Infect Dis* **53:**532–540.

5. **Stamm WE.** 1991. Catheter-associated urinary tract infections: epidemiology, pathogenesis, and prevention. *Am J Med* **91:**65S–71S.

6. **Borer A, Saidel-Odes L, Riesenberg K, Eskira S, Peled N, Nativ R, Schlaeffer F, Sherf M.** 2009. Attributable mortality rate for carbapenem-resistant *Klebsiella pneumoniae* bacteremia. *Infect Control Hosp Epidemiol* **30:**972–976.

7. **Centers for Disease Control and Prevention (CDC).** 2011. Carbapenem-resistant *Klebsiella pneumoniae* associated with a long-term-care facility --- West Virginia, 2009–2011. *MMWR Morb Mortal Wkly Rep* **60:**1418–1420.

8. **Djeribi R, Bouchloukh W, Jouenne T, Menaa B.** 2012. Characterization of bacterial biofilms formed on urinary catheters. *Am J Infect Control* **40:**854–859.

9. **Warren JW, Tenney JH, Hoopes JM, Muncie HL, Anthony WC.** 1982. A prospective microbiologic study of bacteriuria in patients with chronic indwelling urethral catheters. *J Infect Dis* **146:**719–723.

10. **Lazăr V, Chifiriuc MC.** 2010. Medical significance and new therapeutical strategies for biofilm associated infections. *Roum Arch Microbiol Immunol* **69:**125–138.

11. **Monds RD, O'Toole GA.** 2009. The developmental model of microbial biofilms: ten years of a paradigm up for review. *Trends Microbiol* **17:**73–87.

12. **Macleod SM, Stickler DJ.** 2007. Species interactions in mixed-community crystalline biofilms on urinary catheters. *J Med Microbiol* **56:**1549–1557.

13. **Trautner BW, Darouiche RO.** 2004. Role of biofilm in catheter-associated urinary tract infection. *Am J Infect Control* **32:**177–183.

14. **Fang CT, Chuang YP, Shun CT, Chang SC, Wang JT.** 2004. A novel virulence gene in *Klebsiella pneumoniae* strains causing primary liver abscess and septic metastatic complications. *J Exp Med* **199:**697–705.

15. **Fierer J, Walls L, Chu P.** 2011. Recurring *Klebsiella pneumoniae* pyogenic liver abscesses in a resident of San Diego, California, due to a K1 strain carrying the virulence plasmid. *J Clin Microbiol* **49:**4371–4373.

16. **Fung CP, Chang FY, Lee SC, Hu BS, Kuo BI, Liu CY, Ho M, Siu LK.** 2002. A global emerging disease of *Klebsiella pneumoniae* liver abscess: is serotype K1 an important factor for complicated endophthalmitis? *Gut* **50:**420–424.

17. **Podschun R, Sievers D, Fischer A, Ullmann U.** 1993. Serotypes, hemagglutinins, siderophore synthesis, and serum resistance of *Klebsiella* isolates causing human urinary tract infections. *J Infect Dis* **168:**1415–1421.

18. **Tarkkanen AM, Allen BL, Williams PH, Kauppi M, Haahtela K, Siitonen A, Orskov I, Orskov F, Clegg S, Korhonen TK.** 1992. Fimbriation, capsulation, and iron-scavenging systems of *Klebsiella* strains associated with human urinary tract infection. *Infect Immun* **60:**1187–1192.

19. **Sahly H, Keisari Y, Ofek I.** 2009. Manno (rhamno)biose-containing capsular polysaccharides of *Klebsiella pneumoniae* enhance opsono-stimulation of human polymorphonuclear leukocytes. *J Innate Immun* **1:**136–144.

20. **Dutton GG, Paulin M.** 1980. Structure of the capsular polysaccharide of *Klebsiella* serotype K53. *Carbohydr Res* **87:**107–117.

21. **Dutton GS, Paulin M.** 1980. Structure of the capsular polysaccharide of *Klebsiella* serotype K74. *Carbohydr Res* **87:**119–127.

22. **Dutton GG, Di Fabio J.** 1980. The capsular polysaccharide of *Klebsiella* serotype K60; a novel, structural pattern. *Carbohydr Res* **87:**129–139.

23. **Pan PC, Chen HW, Wu PK, Wu YY, Lin CH, Wu JH.** 2011. Mutation in fucose synthesis gene of *Klebsiella pneumoniae* affects capsule composition and virulence in mice. *Exp Biol Med (Maywood)* **236:**219–226.

24. **Fresno S, Jiménez N, Izquierdo L, Merino S, Corsaro MM, De Castro C, Parrilli M, Naldi T, Regué M, Tomás JM.** 2006. The ionic interaction of *Klebsiella pneumoniae* K2 capsule and core lipopolysaccharide. *Microbiology* **152:**1807–1818.

25. **Fresno S, Jiménez N, Canals R, Merino S, Corsaro MM, Lanzetta R, Parrilli M, Pieretti G, Regué M, Tomás JM.** 2007. A second galacturonic acid transferase is required for core lipopolysaccharide biosynthesis and complete capsule association with the cell surface

in *Klebsiella pneumoniae*. *J Bacteriol* **189**:1128–1137.

26. **Ofek I, Kabha K, Athamna A, Frankel G, Wozniak DJ, Hasty DL, Ohman DE.** 1993. Genetic exchange of determinants for capsular polysaccharide biosynthesis between *Klebsiella pneumoniae* strains expressing serotypes K2 and K21a. *Infect Immun* **61**:4208–4216.

27. **Laakso DH, Homonylo MK, Wilmot SJ, Whitfield C.** 1988. Transfer and expression of the genetic determinants for O and K antigen synthesis in *Escherichia coli* O9:K(A) 30 and *Klebsiella* sp. O1:K20, in *Escherichia coli* K12. *Can J Microbiol* **34**:987–992.

28. **Arakawa Y, Ohta M, Wacharotayankun R, Mori M, Kido N, Ito H, Komatsu T, Sugiyama T, Kato N.** 1991. Biosynthesis of *Klebsiella* K2 capsular polysaccharide in *Escherichia coli* HB101 requires the functions of rmpA and the chromosomal cps gene cluster of the virulent strain *Klebsiella pneumoniae* Chedid (O1:K2). *Infect Immun* **59**:2043–2050.

29. **Wacharotayankun R, Arakawa Y, Ohta M, Hasegawa T, Mori M, Horii T, Kato N.** 1992. Involvement of rcsB in *Klebsiella* K2 capsule synthesis in *Escherichia coli* K-12. *J Bacteriol* **174**:1063–1067.

30. **Cheng HY, Chen YS, Wu CY, Chang HY, Lai YC, Peng HL.** 2010. RmpA regulation of capsular polysaccharide biosynthesis in *Klebsiella pneumoniae* CG43. *J Bacteriol* **192**:3144–3158.

31. **Hsu CR, Lin TL, Chen YC, Chou HC, Wang JT.** 2011. The role of *Klebsiella pneumoniae* rmpA in capsular polysaccharide synthesis and virulence revisited. *Microbiology* **157**:3446–3457.

32. **Lai YC, Peng HL, Chang HY.** 2003. RmpA2, an activator of capsule biosynthesis in *Klebsiella pneumoniae* CG43, regulates K2 cps gene expression at the transcriptional level. *J Bacteriol* **185**:788–800.

33. **Lin CT, Wu CC, Chen YS, Lai YC, Chi C, Lin JC, Chen Y, Peng HL.** 2011. Fur regulation of the capsular polysaccharide biosynthesis and iron-acquisition systems in *Klebsiella pneumoniae* CG43. *Microbiology* **157**:419–429.

34. **Fang CT, Lai SY, Yi WC, Hsueh PR, Liu KL.** 2010. The function of wzy_K1 (magA), the serotype K1 polymerase gene in *Klebsiella pneumoniae* cps gene cluster. *J Infect Dis* **201**:1268–1269.

35. **Cryz SJ Jr, Fürer E, Germanier R.** 1984. Protection against fatal *Klebsiella pneumoniae* burn wound sepsis by passive transfer of anticapsular polysaccharide. *Infect Immun* **45**:139–142.

36. **Cryz SJ Jr, Fürer F, Germanier R.** 1984. Experimental *Klebsiella pneumoniae* burn wound sepsis: role of capsular polysaccharide. *Infect Immun* **43**:440–441.

37. **Domenico P, Johanson WG Jr, Straus DC.** 1982. Lobar pneumonia in rats produced by clinical isolates of *Klebsiella pneumoniae*. *Infect Immun* **37**:327–335.

38. **Simoons-Smit AM, Verweij-van Vught AM, MacLaren DM.** 1986. The role of K antigens as virulence factors in *Klebsiella*. *J Med Microbiol* **21**:133–137.

39. **Williams P, Lambert PA, Haigh CG, Brown MR.** 1986. The influence of the O and K antigens of *Klebsiella aerogenes* on surface hydrophobicity and susceptibility to phagocytosis and antimicrobial agents. *J Med Microbiol* **21**:125–132.

40. **Athamna A, Ofek I, Keisari Y, Markowitz S, Dutton GG, Sharon N.** 1991. Lectinophagocytosis of encapsulated *Klebsiella pneumoniae* mediated by surface lectins of guinea pig alveolar macrophages and human monocyte-derived macrophages. *Infect Immun* **59**:1673–1682.

41. **Evrard B, Balestrino D, Dosgilbert A, Bouya-Gachancard JL, Charbonnel N, Forestier C, Tridon A.** 2010. Roles of capsule and lipopolysaccharide O antigen in interactions of human monocyte-derived dendritic cells and *Klebsiella pneumoniae*. *Infect Immun* **78**:210–219.

42. **Struve C, Krogfelt KA.** 2003. Role of capsule in *Klebsiella pneumoniae* virulence: lack of correlation between *in vitro* and *in vivo* studies. *FEMS Microbiol Lett* **218**:149–154.

43. **Llobet E, Tomás JM, Bengoechea JA.** 2008. Capsule polysaccharide is a bacterial decoy for antimicrobial peptides. *Microbiology* **154**:3877–3886.

44. **Nassif X, Sansonetti PJ.** 1986. Correlation of the virulence of *Klebsiella pneumoniae* K1 and K2 with the presence of a plasmid encoding aerobactin. *Infect Immun* **54**:603–608.

45. **Neilands JB.** 1995. Siderophores: structure and function of microbial iron transport compounds. *J Biol Chem* **270**:26723–26726.

46. **Noinaj N, Guillier M, Barnard TJ, Buchanan SK.** 2010. TonB-dependent transporters: regulation, structure, and function. *Annu Rev Microbiol* **64**:43–60.

47. **Hsieh PF, Lin TL, Lee CZ, Tsai SF, Wang JT.** 2008. Serum-induced iron-acquisition systems and TonB contribute to virulence in *Klebsiella pneumoniae* causing primary pyogenic liver abscess. *J Infect Dis* **197**:1717–1727.

48. **Lawlor MS, O'connor C, Miller VL.** 2007. Yersiniabactin is a virulence factor for *Klebsiella*

pneumoniae during pulmonary infection. *Infect Immun* **75:**1463–1472.

49. **Lin CT, Wu CC, Chen YS, Lai YC, Chi C, Lin JC, Chen Y, Peng HL.** 2011. Fur regulation of the capsular polysaccharide biosynthesis and iron-acquisition systems in *Klebsiella pneumoniae* CG43. *Microbiology* **157:**419–429.

50. **Wu CC, Lin CT, Cheng WY, Huang CJ, Wang ZC, Peng HL.** 2012. Fur-dependent MrkHI regulation of type 3 fimbriae in *Klebsiella pneumoniae* CG43. *Microbiology* **158:**1045–1056.

51. **Khimji PL, Miles AA.** 1978. Microbial iron-chelators and their action on *Klebsiella* infections in the skin of guinea-pigs. *Br J Exp Pathol* **59:**137–147.

52. **Lee SJ, Lee DS, Choe HS, Shim BS, Kim CS, Kim ME, Cho YH.** 2011. Antimicrobial resistance in community-acquired urinary tract infections: results from the Korean Antimicrobial Resistance Monitoring System. *J Infect Chemother* **17:**440–446.

53. **Russo TA, Shon AS, Beanan JM, Olson R, MacDonald U, Pomakov AO, Visitacion MP.** 2011. Hypervirulent *K. pneumoniae* secretes more and more active iron-acquisition molecules than "classical" *K. pneumoniae* thereby enhancing its virulence. *PLoS One* **6:**e26734. doi:10.1371/journal.pone.0026734

54. **Boer JL, Hausinger RP.** 2012. *Klebsiella aerogenes* UreF: identification of the UreG binding site and role in enhancing the fidelity of urease activation. *Biochemistry* **51:**2298–2308.

55. **Miragliotta G, Di Pierro MN, Miragliotta L, Mosca A.** 2008. Antimicrobial resistance among uropathogens responsible for community-acquired urinary tract infections in an Italian community. *J Chemother* **20:**721–727.

56. **Carter EL, Boer JL, Farrugia MA, Flugga N, Towns CL, Hausinger RP.** 2011. Function of UreB in *Klebsiella aerogenes* urease. *Biochemistry* **50:**9296–9308.

57. **Boer JL, Quiroz-Valenzuela S, Anderson KL, Hausinger RP.** 2010. Mutagenesis of *Klebsiella aerogenes* UreG to probe nickel binding and interactions with other urease-related proteins. *Biochemistry* **49:**5859–5869.

58. **Carter EL, Flugga N, Boer JL, Mulrooney SB, Hausinger RP.** 2009. Interplay of metal ions and urease. *Metallomics* **1:**207–221.

59. **Carter EL, Hausinger RP.** 2010. Characterization of the *Klebsiella aerogenes* urease accessory protein UreD in fusion with the maltose binding protein. *J Bacteriol* **192:**2294–2304.

60. **Carter EL, Tronrud DE, Taber SR, Karplus PA, Hausinger RP.** 2011. Iron-containing urease in a pathogenic bacterium. *Proc Natl Acad Sci U S A* **108:**13095–13099.

61. **Hedelin H, Bratt CG, Eckerdal G, Lincoln K.** 1991. Relationship between urease-producing bacteria, urinary pH and encrustation on indwelling urinary catheters. *Br J Urol* **67:**527–531.

62. **McLean RJ, Downey J, Clapham L, Wilson JW, Nickel JC.** 1991. Pyrophosphate inhibition of *Proteus mirabilis*-induced struvite crystallization *in vitro*. *Clin Chim Acta* **200:**107–117.

63. **Broomfield RJ, Morgan SD, Khan A, Stickler DJ.** 2009. Crystalline bacterial biofilm formation on urinary catheters by urease-producing urinary tract pathogens: a simple method of control. *J Med Microbiol* **58:**1367–1375.

64. **Allen WJ, Phan G, Waksman G.** 2012. Pilus biogenesis at the outer membrane of Gram-negative bacterial pathogens. *Curr Opin Struct Biol* **22:**500–506.

65. **Clegg S, Wilson J, Johnson J.** 2011. More than one way to control hair growth: regulatory mechanisms in enterobacteria that affect fimbriae assembled by the chaperone/usher pathway. *J Bacteriol* **193:**2081–2088.

66. **Morrissey B, Leney AC, Rêgo AT, Phan G, Allen WJ, Verger D, Waksman G, Ashcroft AE, Radford SE.** 2012. The role of chaperone-subunit usher domain interactions in the mechanism of bacterial pilus biogenesis revealed by ESI-MS. *Mol Cell Proteomics* **11:**M111.015289. doi:10.1074/mcp.M111.015289

67. **Poole ST, McVeigh AL, Anantha RP, Lee LH, Akay YM, Pontzer EA, Scott DA, Bullitt E, Savarino SJ.** 2007. Donor strand complementation governs intersubunit interaction of fimbriae of the alternate chaperone pathway. *Mol Microbiol* **63:**1372–1384.

68. **Peabody CR, Chung YJ, Yen MR, Vidal-Ingigliardi D, Pugsley AP, Saier MH Jr.** 2003. Type II protein secretion and its relationship to bacterial type IV pili and archaeal flagella. *Microbiology* **149:**3051–3072.

69. **Rodgers K, Arvidson CG, Melville S.** 2011. Expression of a *Clostridium perfringens* type IV pilin by *Neisseria gonorrhoeae* mediates adherence to muscle cells. *Infect Immun* **79:**3096–3105.

70. **Dwyer BE, Newton KL, Kisiela D, Sokurenko EV, Clegg S.** 2011. Single nucleotide polymorphisms of fimH associated with adherence and biofilm formation by serovars of *Salmonella enterica*. *Microbiology* **157:**3162–3171.

71. **Stahlhut SG, Chattopadhyay S, Struve C, Weissman SJ, Aprikian P, Libby SJ, Fang FC, Krogfelt KA, Sokurenko EV.** 2009. Population variability of the FimH type 1 fimbrial adhesin in *Klebsiella pneumoniae*. *J Bacteriol* **191:**1941–1950.

72. **Stahlhut SG, Tchesnokova V, Struve C, Weissman SJ, Chattopadhyay S, Yakovenko O, Aprikian P, Sokurenko EV, Krogfelt KA.** 2009. Comparative structure-function analysis of mannose-specific FimH adhesins from *Klebsiella pneumoniae* and *Escherichia coli*. *J Bacteriol* **191:**6592–6601.

73. **Weissman SJ, Moseley SL, Dykhuizen DE, Sokurenko EV.** 2003. Enterobacterial adhesins and the case for studying SNPs in bacteria. *Trends Microbiol* **11:**115–117.

74. **Wu CC, Huang YJ, Fung CP, Peng HL.** 2010. Regulation of the *Klebsiella pneumoniae* Kpc fimbriae by the site-specific recombinase KpcI. *Microbiology* **156:**1983–1992.

75. **Adegbola RA, Old DC.** 1987. Antigenic relationships among type-1 fimbriae of Enterobacteriaceae revealed by immuno-electronmicroscopy. *J Med Microbiol* **24:**21–28.

76. **Clegg S, Hull S, Hull R, Pruckler J.** 1985. Construction and comparison of recombinant plasmids encoding type 1 fimbriae of members of the family Enterobacteriaceae. *Infect Immun* **48:**275–279.

77. **Clegg S, Purcell BK, Pruckler J.** 1987. Characterization of genes encoding type 1 fimbriae of *Klebsiella pneumoniae*, *Salmonella typhimurium*, and *Serratia marcescens*. *Infect Immun* **55:**281–287.

78. **Gerlach GF, Clegg S.** 1988. Characterization of two genes encoding antigenically distinct type-1 fimbriae of *Klebsiella pneumoniae*. *Gene* **64:**231–240.

79. **Gerlach GF, Clegg S, Ness NJ, Swenson DL, Allen BL, Nichols WA.** 1989. Expression of type 1 fimbriae and mannose-sensitive hemagglutinin by recombinant plasmids. *Infect Immun* **57:**764–770.

80. **Corcoran CP, Dorman CJ.** 2009. DNA relaxation-dependent phase biasing of the fim genetic switch in *Escherichia coli* depends on the interplay of H-NS, IHF and LRP. *Mol Microbiol* **74:**1071–1082.

81. **McVicker G, Sun L, Sohanpal BK, Gashi K, Williamson RA, Plumbridge J, Blomfield IC.** 2011. SlyA protein activates fimB gene expression and type 1 fimbriation in *Escherichia coli* K-12. *J Biol Chem* **286:**32026–32035.

82. **Rosen DA, Pinkner JS, Jones JM, Walker JN, Clegg S, Hultgren SJ.** 2008. Utilization of an intracellular bacterial community pathway in *Klebsiella pneumoniae* urinary tract infection and the effects of FimK on type 1 pilus expression. *Infect Immun* **76:**3337–3345.

83. **McDonough KA, Rodriguez A.** 2011. The myriad roles of cyclic AMP in microbial pathogens: from signal to sword. *Nat Rev Microbiol* **10:**27–38.

84. **Sondermann H, Shikuma NJ, Yildiz FH.** 2012. You've come a long way: c-di-GMP signaling. *Curr Opin Microbiol* **15:**140–146.

85. **Tamayo R, Pratt JT, Camilli A.** 2007. Roles of cyclic diguanylate in the regulation of bacterial pathogenesis. *Annu Rev Microbiol* **61:**131–148.

86. **Gerlach GF, Allen BL, Clegg S.** 1988. Molecular characterization of the type 3 (MR/K) fimbriae of *Klebsiella pneumoniae*. *J Bacteriol* **170:**3547–3553.

87. **Gerlach GF, Clegg S, Allen BL.** 1989. Identification and characterization of the genes encoding the type 3 and type 1 fimbrial adhesins of *Klebsiella pneumoniae*. *J Bacteriol* **171:**1262–1270.

88. **Sebghati TA, Korhonen TK, Hornick DB, Clegg S.** 1998. Characterization of the type 3 fimbrial adhesins of *Klebsiella* strains. *Infect Immun* **66:**2887–2894.

89. **Chan CH, Chen FJ, Huang YJ, Chen SY, Liu KL, Wang ZC, Peng HL, Yew TR, Liu CH, Liou GG, Hsu KY, Chang HY, Hsu L.** 2012. Identification of protein domains on major pilin MrkA that affects the mechanical properties of *Klebsiella pneumoniae* type 3 fimbriae. *Langmuir* **28:**7428–7435.

90. **Johnson JG, Clegg S.** 2010. Role of MrkJ, a phosphodiesterase, in type 3 fimbrial expression and biofilm formation in *Klebsiella pneumoniae*. *J Bacteriol* **192:**3944–3950.

91. **Johnson JG, Murphy CN, Sippy J, Johnson TJ, Clegg S.** 2011. Type 3 fimbriae and biofilm formation are regulated by the transcriptional regulators MrkHI in *Klebsiella pneumoniae*. *J Bacteriol* **193:**3453–3460.

92. **Wilksch JJ, Yang J, Clements A, Gabbe JL, Short KR, Cao H, Cavaliere R, James CE, Whitchurch CB, Schembri MA, Chuah ML, Liang ZX, Wijburg OL, Jenney AW, Lithgow T, Strugnell RA.** 2011. MrkH, a novel c-di-GMP-dependent transcriptional activator, controls *Klebsiella pneumoniae* biofilm formation by regulating type 3 fimbriae expression. *PLoS Pathog* **7:**e1002204. doi:10.1371/journal.ppat.1002204

93. **Wu CC, Lin CT, Cheng WY, Huang CJ, Wang ZC, Peng HL.** 2012. Fur dependent MrkHI regulation of type 3 fimbriae in *Klebsiella pneumoniae* CG43. *Microbiology* **158:**1045–1056.

94. **Römling U, Simm R.** 2009. Prevailing concepts of c-di-GMP signaling. *Contrib Microbiol* **16:**161–181.

95. **Wu CC, Huang YJ, Fung CP, Peng HL.** 2010. Regulation of the *Klebsiella pneumoniae* Kpc fimbriae by the site-specific recombinase KpcI. *Microbiology* **156:**1983–1992.

96. **Rosen DA, Pinkner JS, Walker JN, Elam JS, Jones JM, Hultgren SJ.** 2008. Molecular variations in *Klebsiella pneumoniae* and *Escherichia coli* FimH affect function and pathogenesis in the urinary tract. *Infect Immun* **76:** 3346–3356.

97. **Struve C, Bojer M, Krogfelt KA.** 2008. Characterization of *Klebsiella pneumoniae* type 1 fimbriae by detection of phase variation during colonization and infection and impact on virulence. *Infect Immun* **76:**4055–4065.

98. **Struve C, Bojer M, Krogfelt KA.** 2009. Identification of a conserved chromosomal region encoding *Klebsiella pneumoniae* type 1 and type 3 fimbriae and assessment of the role of fimbriae in pathogenicity. *Infect Immun* **77:**5016–5024.

99. **Rosen DA, Hooton TM, Stamm WE, Humphrey PA, Hultgren SJ.** 2007. Detection of intracellular bacterial communities in human urinary tract infection. *PLoS Med* **4:**e329. doi:10.1371/journal .pmed.0040329

100. **Schwartz DJ, Chen SL, Hultgren SJ, Seed PC.** 2011. Population dynamics and niche distribution of uropathogenic *Escherichia coli* during acute and chronic urinary tract infection. *Infect Immun* **79:**4250–4259.

101. **Jagnow J, Clegg S.** 2003. *Klebsiella pneumoniae* MrkD-mediated biofilm formation on extracellular matrix- and collagen-coated surfaces. *Microbiology* **149:**2397–2405.

102. **Langstraat J, Bohse M, Clegg S.** 2001. Type 3 fimbrial shaft (MrkA) of *Klebsiella pneumoniae*, but not the fimbrial adhesin (MrkD), facilitates biofilm formation. *Infect Immun* **69:**5805–5812.

103. **Ong CL, Ulett GC, Mabbett AN, Beatson SA, Webb RI, Monaghan W, Nimmo GR, Looke DF, McEwan AG, Schembri MA.** 2008. Identification of type 3 fimbriae in uropathogenic *Escherichia coli* reveals a role in biofilm formation. *J Bacteriol* **190:**1054–1063.

104. **Schroll C, Barken KB, Krogfelt KA, Struve C.** 2010. Role of type 1 and type 3 fimbriae in Klebsiella pneumoniae biofilm formation. *BMC Microbiol* **10:**179. doi:10.1186/1471-2180-10-179

105. **Tarkkanen AM, Allen BL, Westerlund B, Holthöfer H, Kuusela P, Risteli L, Clegg S, Korhonen TK.** 1990. Type V collagen as the target for type-3 fimbriae, enterobacterial adherence organelles. *Mol Microbiol* **4:**1353–1361.

106. **Tarkkanen AM, Virkola R, Clegg S, Korhonen TK.** 1997. Binding of the type 3 fimbriae of *Klebsiella pneumoniae* to human endothelial and urinary bladder cells. *Infect Immun* **65:**1546–1549.

107. **François P, Vaudaux P, Lew PD.** 1998. Role of plasma and extracellular matrix proteins

in the physiopathology of foreign body infections. *Ann Vasc Surg* **12:**34–40.

108. **Di Martino P, Cafferini N, Joly B, Darfeuille-Michaud A.** 2003. *Klebsiella pneumoniae* type 3 pili facilitate adherence and biofilm formation on abiotic surfaces. *Res Microbiol* **154:**9–16.

109. **Old DC, Adegbola RA.** 1985. Antigenic relationships among type-3 fimbriae of Enterobacteriaceae revealed by immuno-electronmicroscopy. *J Med Microbiol* **20:**113–121.

110. **Ong CL, Beatson SA, Totsika M, Forestier C, McEwan AG, Schembri MA.** 2010. Molecular analysis of type 3 fimbrial genes from *Escherichia coli*, *Klebsiella* and *Citrobacter* species. *BMC Microbiol* **10:**183. doi:10.1186/1471-2180 -10-183

111. **Hagan EC, Lloyd AL, Rasko DA, Faerber GJ, Mobley HL.** 2010. Escherichia coli global gene expression in urine from women with urinary tract infection. *PLoS Pathog* **6:**e1001187. doi:10.1371/journal.ppat.1001187

112. **Pearson MM, Yep A, Smith SN, Mobley HL.** 2011. Transcriptome of *Proteus mirabilis* in the murine urinary tract: virulence and nitrogen assimilation gene expression. *Infect Immun* **79:**2619–2631.

113. **Crawford RW, Reeve KE, Gunn JS.** 2010. Flagellated but not hyperfimbriated *Salmonella enterica* serovar Typhimurium attaches to and forms biofilms on cholesterol-coated surfaces. *J Bacteriol* **192:**2981–2990.

114. **Jackson G, Beyenal H, Rees WM, Lewandowski Z.** 2001. Growing reproducible biofilms with respect to structure and viable cell counts. *J Microbiol Methods* **47:**1–10.

115. **Kania RE, Lamers GE, van de Laar N, Dijkhuizen M, Lagendijk E, Huy PT, Herman P, Hiemstra P, Grote JJ, Frijns J, Bloemberg GV.** 2010. Biofilms on tracheoesophageal voice prostheses: A confocal laser scanning microscopy demonstration of mixed bacterial and yeast biofilms. *Biofouling* **26:**519–526.

116. **Anderl JN, Zahller J, Roe F, Stewart PS.** 2003. Role of nutrient limitation and stationary-phase existence in *Klebsiella pneumoniae* biofilm resistance to ampicillin and ciprofloxacin. *Antimicrob Agents Chemother* **47:**1251–1256.

117. **Hennequin C, Aumeran C, Robin F, Traore O, Forestier C.** 2012. Antibiotic resistance and plasmid transfer capacity in biofilm formed with a CTX-M-15-producing *Klebsiella pneumoniae* isolate. *J Antimicrob Chemother* **67:**2123–2130.

118. **Lewis K.** 2008. Multidrug tolerance of biofilms and persister cells. *Curr Top Microbiol Immunol* **322:**107–131.

119. **Long DY, Hu SP, Chen XC, Liu L, Chen YX.** 2010. Persisters and their effects on microbial

biofilm tolerance: A review. *Ying Yong Sheng Tai Xue Bao* **21:**2707–2714.

120. **Boddicker JD, Anderson RA, Jagnow J, Clegg S.** 2006. Signature-tagged mutagenesis of *Klebsiella pneumoniae* to identify genes that influence biofilm formation on extracellular matrix material. *Infect Immun* **74:**4590–4597.

121. **Hennequin C, Forestier C.** 2009. oxyR, a LysR-type regulator involved in *Klebsiella pneumoniae* mucosal and abiotic colonization. *Infect Immun* **77:**5449–5457.

122. **Fuqua C, Winans SC, Greenberg EP.** 1996. Census and consensus in bacterial ecosystems: The LuxR-LuxI family of quorum-sensing transcriptional regulators. *Annu Rev Microbiol* **50:**727–751.

123. **Balestrino D, Haagensen JA, Rich C, Forestier C.** 2005. Characterization of type 2 quorum sensing in *Klebsiella pneumoniae* and relationship with biofilm formation. *J Bacteriol* **187:**2870–2880.

124. **De Araujo C, Balestrino D, Roth L, Charbonnel N, Forestier C.** 2010. Quorum sensing affects biofilm formation through lipopolysaccharide synthesis in *Klebsiella pneumoniae*. *Res Microbiol* **161:**595–603.

125. **Jackson DW, Simecka JW, Romeo T.** 2002. Catabolite repression of *Escherichia coli* biofilm formation. *J Bacteriol* **184:**3406–3410.

126. **Kalivoda EJ, Stella NA, O'Dee DM, Nau GJ, Shanks RM.** 2008. The cyclic AMP-dependent catabolite repression system of *Serratia marcescens* mediates biofilm formation through regulation of type 1 fimbriae. *Appl Environ Microbiol* **74:**3461–3470.

127. **Houot L, Watnick PI.** 2008. A novel role for enzyme I of the *Vibrio cholerae* phosphoenolpyruvate phosphotransferase system in regulation of growth in a biofilm. *J Bacteriol* **190:**311–320.

128. **Bianconi I, Milani A, Cigana C, Paroni M, Levesque RC, Bertoni G, Bragonzi A.** 2011. Positive signature-tagged mutagenesis in *Pseudomonas aeruginosa*: tracking patho-adaptive mutations promoting airways chronic infection. *PLoS Pathog* **7:**e1001270. doi:10.1371/journal.ppat.1001270

129. **Stahlhut SG, Schroll C, Harmsen M, Struve C, Krogfelt KA.** 2010. Screening for genes involved in *Klebsiella pneumoniae* biofilm formation using a fosmid library. *FEMS Immunol Med Microbiol* **59:**521–524.

130. **Wu MC, Lin TL, Hsieh PF, Yang HC, Wang JT.** 2011. Isolation of genes involved in biofilm formation of a *Klebsiella pneumoniae* strain causing pyogenic liver abscess. *PLoS One* **6:**e23500. doi:10.1371/journal.pone.0023500

131. **Dzul SP, Thornton MM, Hohne DN, Stewart EJ, Shah AA, Bortz DM, Solomon MJ, Younger JG.** 2011. Contribution of the *Klebsiella pneumoniae* capsule to bacterial aggregate and biofilm microstructures. *Appl Environ Microbiol* **77:**1777–1782.

132. **Favre-Bonte S, Joly B, Forestier C.** 1999. Consequences of reduction of *Klebsiella pneumoniae* capsule expression on interactions of this bacterium with epithelial cells. *Infect Immun* **67:**554–561.

133. **Schembri MA, Blom J, Krogfelt KA, Klemm P.** 2005. Capsule and fimbria interaction in *Klebsiella pneumoniae*. *Infect Immun* **73:**4626–4633.

134. **Wu MC, Chen YC, Lin TL, Hsieh PF, Wang JT.** 2012. Cellobiose-specific phosphotransferase system of *Klebsiella pneumoniae* and its importance in biofilm formation and virulence. *Infect Immun* **80:**464–472.

135. **Guiton PS, Hung CS, Hancock LE, Caparon MG, Hultgren SJ.** 2010. Enterococcal biofilm formation and virulence in an optimized murine model of foreign body-associated urinary tract infections. *Infect Immun* **78:**4166–4175.

136. **Bradford PA, Bratu S, Urban C, Visalli M, Mariano N, Landman D, Rahal JJ, Brooks S, Cebular S, Quale J.** 2004. Emergence of carbapenem-resistant *Klebsiella* species possessing the class A carbapenem-hydrolyzing KPC-2 and inhibitor-resistant TEM-30 beta-lactamases in New York City. *Clin Infect Dis* **39:**55–60.

137. **Woodford N, Tierno PM Jr, Young K, Tysall L, Palepou MF, Ward E, Painter RE, Suber DF, Shungu D, Silver LL, Inglima K, Kornblum J, Livermore DM.** 2004. Outbreak of *Klebsiella pneumoniae* producing a new carbapenem-hydrolyzing class A beta-lactamase, KPC-3, in a New York Medical Center. *Antimicrob Agents Chemother* **48:**4793–4799.

138. **Yigit H, Queenan AM, Anderson GJ, Domenech-Sanchez A, Biddle JW, Steward CD, Alberti S, Bush K, Tenover FC.** 2001. Novel carbapenem-hydrolyzing beta-lactamase, KPC-1, from a carbapenem-resistant strain of *Klebsiella pneumoniae*. *Antimicrob Agents Chemother* **45:**1151–1161.

139. **Landman D, Bratu S, Kochar S, Panwar M, Trehan M, Doymaz M, Quale J.** 2007. Evolution of antimicrobial resistance among *Pseudomonas aeruginosa*, *Acinetobacter baumannii* and *Klebsiella pneumoniae* in Brooklyn, NY. *J Antimicrob Chemother* **60:**78–82.

140. **Goren MG, Carmeli Y, Schwaber MJ, Chmelnitsky I, Schechner V, Navon-Venezia S.** 2010. Transfer of carbapenem-resistant plasmid from *Klebsiella pneumoniae* ST258 to

Escherichia coli in patient. *Emerg Infect Dis* **16:**1014–1017.

141. **Kumar V, Sun P, Vamathevan J, Li Y, Ingraham K, Palmer L, Huang J, Brown JR.** 2011. Comparative genomics of *Klebsiella pneumoniae* strains with different antibiotic resistance profiles. *Antimicrob Agents Chemother* **55:**4267–4276.

142. **Coudeyras S, Nakusi L, Charbonnel N, Forestier C.** 2008. A tripartite efflux pump involved in gastrointestinal colonization by *Klebsiella pneumoniae* confers a tolerance response to inorganic acid. *Infect Immun* **76:**4633–4641.

143. **Piddock LJ.** 2006. Multidrug-resistance efflux pumps - not just for resistance. *Nat Rev Microbiol* **4:**629–636.

144. **Srinivasan VB, Vaidyanathan V, Mondal A, Rajamohan G.** 2012. Role of the two component signal transduction system CpxAR in conferring cefepime and chloramphenicol resistance in *Klebsiella pneumoniae* NTUH-K2044. *PLoS One* **7:**e33777. doi:10.1371/journal.pone.0033777

145. **Veleba M, Higgins PG, Gonzalez G, Seifert H, Schneiders T.** 2012. Characterization of RarA, a novel AraC-family multidrug resistance regulator in *Klebsiella pneumoniae. Antimicrob Agents Chemother* **56:**4450–4458.

146. **Amdekar S, Singh V, Singh DD.** 2011. Probiotic therapy: immunomodulating approach toward urinary tract infection. *Curr Microbiol* **63:**484–490.

147. **Hancock V, Vejborg RM, Klemm P.** 2010. Functional genomics of probiotic *Escherichia coli* Nissle 1917 and 83972, and UPEC strain CFT073: comparison of transcriptomes, growth and biofilm formation. *Mol Genet Genomics* **284:**437–454.

148. **Bunders CA, Richards JJ, Melander C.** 2010. Identification of aryl 2-aminoimidazoles as biofilm inhibitors in Gram-negative bacteria. *Bioorg Med Chem Lett* **20:**3797–3800.

149. **Chai SC, Wang WL, Ding DR, Ye QZ.** 2011. Growth inhibition of *Escherichia coli* and methicillin-resistant *Staphylococcus aureus* by targeting cellular methionine aminopeptidase. *Eur J Med Chem* **46:**3537–3540.

150. **Goller CC, Seed PC.** 2010. High-throughput identification of chemical inhibitors of *E. coli* Group 2 capsule biogenesis as anti-virulence agents. *PLoS One* **5:**e11642. doi:10.1371/journal.pone.0011642

151. **Han Z, Pinkner JS, Ford B, Obermann R, Nolan W, Wildman SA, Hobbs D, Ellenberger T, Cusumano CK, Hultgren SJ, Janetka JW.** 2010. Structure-based drug design and optimization of mannoside bacterial FimH antagonists. *J Med Chem* **53:**4779–4792.

152. **Kimura K, Iwatsuki M, Nagai T, Matsumoto A, Takahashi Y, Shiomi K, Omura S, Abe A.** 2011. A small-molecule inhibitor of the bacterial type III secretion system protects against *in vivo* infection with *Citrobacter rodentium. J Antibiot (Tokyo)* **64:**197–203.

153. **Segers K, Anné J.** 2011. Traffic jam at the bacterial sec translocase: targeting the SecA nanomotor by small-molecule inhibitors. *Chem Biol* **18:**685–698.

154. **Tang M, Odejinmi SI, Allette YM, Vankayalapati H, Lai K.** 2011. Identification of novel small molecule inhibitors of 4-diphosphocytidyl-2-C-methyl-D-erythritol (CDP-ME) kinase of Gram-negative bacteria. *Bioorg Med Chem* **19:**5886–5895.

155. **Wang Z, Humphrey C, Frilot N, Wang G, Nie Z, Moniri NH, Daaka Y.** 2011. Dynamin2- and endothelial nitric oxide synthase-regulated invasion of bladder epithelial cells by uropathogenic *Escherichia coli. J Cell Biol* **192:**101–110.

156. **Karatan E, Watnick P.** 2009. Signals, regulatory networks, and materials that build and break bacterial biofilms. *Microbiol Mol Biol Rev* **73:**310–347.

157. **Laupland KB, Zygun DA, Davies HD, Church DL, Louie TJ, Doig CJ.** 2002. Incidence and risk factors for acquiring nosocomial urinary tract infection in the critically ill. *J Crit Care* **17:**50–57.

158. **Milan PB, Ivan IM.** 2009. Catheter-associated and nosocomial urinary tract infections: antibiotic resistance and influence on commonly used antimicrobial therapy. *Int Urol Nephrol* **41:**461–464.

159. **Bouza E, San Juan R, Muñoz P, Voss A, Kluytmans J, Co-operative Group of the European Study Group on Nosocomial Infections.** 2001. A European perspective on nosocomial urinary tract infections II. Report on incidence, clinical characteristics and outcome (ESGNI-004 study). European Study Group on Nosocomial Infections. *Clin Microbiol Infect* **7:**532–542.

160. **Arpin C, Dubois V, Maugein J, Jullin J, Dutilh B, Brochet JP, Larribet G, Fischer I, Quentin C.** 2005. Clinical and molecular analysis of extended-spectrum {beta}-lactamase-producing enterobacteria in the community setting. *J Clin Microbiol* **43:**5048–5054.

161. **Bahadin J, Teo SS, Mathew S.** 2011. Aetiology of community-acquired urinary tract infection and antimicrobial susceptibility patterns of uropathogens isolated. *Singapore Med J* **52:**415–420.

Gram-Positive Uropathogens, Polymicrobial Urinary Tract Infection, and the Emerging Microbiota of the Urinary Tract

19

KIMBERLY A. KLINE[1] and AMANDA L. LEWIS[2]

GRAM-POSITIVE AND POLYMICROBIAL UTI EPIDEMIOLOGY

Uncomplicated UTI

Uncomplicated urinary-tract infection (UTI) is most common in young, sexually active, nonpregnant, premenopausal women. Gram-negative bacteria are isolated from 75% to 95% of these infections (1). The remaining proportions of uncomplicated UTI are associated with a variety of organisms, including the Gram-positive bacteria *Staphylococcus saprophyticus*, *Enterococcus faecalis*, *Streptococcus agalactiae* (group B *Streptococcus*, GBS), and other less frequently isolated organisms. In demographic groups such as pregnant women and the elderly, Gram-positive bacteria are found more often as etiologic agents of UTI. Symptoms associated with uncomplicated UTI caused by Gram-positive uropathogens are similar to those caused by Gram-negative organisms and usually include dysuria, urinary frequency, urinary urgency, and/or suprapubic pain. Fever, chills, costovertebral-angle tenderness, flank pain, and/or nausea are suggestive of upper urinary tract (kidney) involvement.

[1]Singapore Centre on Environmental Life Sciences Engineering, School of Biological Sciences, Nanyang Technological University, Singapore 637551; [2]Department of Molecular Microbiology, Washington University School of Medicine, St. Louis, MO 63110.
Urinary Tract Infections: Molecular Pathogenesis and Clinical Management, 2nd Edition
Edited by Matthew A. Mulvey, David J. Klumpp, and Ann E. Stapleton
© 2017 American Society for Microbiology, Washington, DC
doi:10.1128/microbiolspec.UTI-0012-2012

Point of Care Diagnosis of UTI

While the gold standard for UTI diagnosis is bacterial culture of the urine, dipstick urinalysis is commonly used in point-of-care diagnosis. In some clinical settings, such as with infants, leukocyte esterase (LE) and pyuria (by dipstick analysis) have a very high sensitivity and specificity for UTI (>90% as defined by the culture of an uropathogen from urine with >100,000 colony-forming units (CFU) per ml) (2, 3). However, in contexts such as pregnancy, dipstick analysis using LE, pyuria, or presence of nitrites is less reliable as an indication of UTI per the microbiological definition of 10^5 CFU/ml cutoff (4, 5). While dipstick urinalysis that is positive for LE and/or nitrites in a clean-catch urine sample is consistent with a UTI diagnosis, these tests can miss UTIs that meet the gold standard of bacteriuria diagnosis in relation to adverse outcomes in pregnancy (e.g., the microbiological 10^5 CFU/ml definition). One likely explanation is that nitrite tests are likely to be negative if the infecting organism does not reduce nitrate, as is the case for most Gram-positive uropathogens, including *S. saprophyticus,* enterococci, and group B *Streptococcus* (6, 7). Given the higher prevalence of Gram-positive bacteria as causes of UTI in certain populations such as the elderly, it is perhaps not surprising that some studies conclude that LE and nitrite are inadequate for UTI screening in this setting (8). In short, while dipstick urinalyses can help to quickly identify UTI caused by Gram-negative bacteria, they are less useful for infections involving Gram-positive uropathogens and perform poorly in ruling out these infections with certainty.

Complicated UTI

Complicated UTI is defined as cystitis or pyelonephritis that occurs in individuals with predisposing anatomic, metabolic, or functional risk factors that make UTI more difficult to treat. Complicated UTIs often occur in nosocomial and/or institutional settings, particularly in individuals with structural or functional alterations of the urinary tract (often associated with urinary catheterization), or other underlying renal, metabolic, or immunological disorders (9); these populations are at greater risk of Gram-positive and polymicrobial UTI (10, 11). Another less frequently recognized anatomic risk factor for UTI is female-genital cutting (FGM). A recent meta-analysis of five comparative studies showed that women who had experienced FGM were at 3-times higher risk of UTI compared to those who were uncut (12).

Catheter-associated UTI

Catheter-associated UTI (CAUTI) account for 40% of all nosocomial infections (13) and are the most common complication of indwelling urinary catheters (14, 15). Catheter-associated bacteria are thought to be derived largely from the patient's own gut microbiota (16). Bacteriuria occurs in 3% to 10% of patients following urinary catheterization (13, 17). Catheter-associated bacteriuria is often asymptomatic (15) and there is no good way to distinguish between pathogenic CAUTI and asymptomatic bacteriuria (ASB); even the presence of neutrophils in the urine (pyuria), which are a strong identifier of uncomplicated UTI, is not a good diagnostic indicator of CAUTI (13). Catheter-associated bacteria are largely in a biofilm state and are thus recalcitrant to antibiotic treatment (13, 18, 19). However, if left untreated, these infections can lead to severe complications, such as acute pyelonephritis, bacteremia, urosepsis, and death (13, 16). While the well-adapted uropathogenic *E. coli* (UPEC) cause the majority of noncatheter-associated UTI in the community, the diversity of species associated with CAUTI is greater. For example, enterococci are rarely associated with community-acquired UTI but play a prominent role in the pathogenesis of CAUTI and are among the predominant pathogens isolated from polymicrobial communities on the surface of indwelling urinary catheters and biliary stents (20–22).

LABORATORY MODELS TO STUDY GRAM-POSITIVE UTI

Model systems to recapitulate and study infection by Gram-positive uropathogens have been adapted from those used to study UTI caused by Gram-negative bacteria. Transurethral inoculation with 1.5 ml of *S. saprophyticus* at 1×10^9 CFU/ml into the bladders of female albino WISTAR rats showed that at 7 days post infection, both bladders and kidneys were colonized at similar levels, leukocytes were present in the urine, and bladder inflammation and epithelial damage were noted (23). Subsequent studies in which 50×1 of *S. saprophyticus* at 2×10^8 CFU/ml was transurethrally inoculated into 7–8 week old C3H/HeN female mice showed significantly higher CFU in the kidney compared to the bladder at 6 hours post infection and 2, 7, and 14 days post infection (24). Bacterial persistence in the kidneys was observed in C3H/HeN mice but not in C57BL/6 mice, indicating that host factors contribute to the ability of *S. saprophyticus* to cause UTI. Under the same infection conditions, GBS showed similar kidney tropism at 1, 7, and 14 days post infection (25). Similarly, *E. faecalis* preferentially infects the kidneys of C57Bl/6, outbred Harlan Sprague Dawley, and BALB/c female mice; however, these models require a 200 × l inoculum volume to consistently establish infection (26–28). Since *E. faecalis* is more commonly associated with CAUTI than ascending UTI, foreign body-associated UTI models have been developed in mice and rats to mimic the conditions of patients with indwelling urinary catheters (29, 30). In the murine model, catheter material is inserted transurethrally into the murine bladder prior to bacterial inoculation, where it remains throughout the course of infection. *E. faecalis* establishes a robust infection in the catheter-containing bladder, in the kidneys, and on the catheter material itself where it forms a biofilm that facilitates persistent infection in the face of robust catheter-driven inflammation (30, 31). The CAUTI-murine model has recently been used to test the efficacy of novel UTI therapeutics and will continue to be useful in the search for antimicrobial agents aimed at preventing or dispersing Gram-positive biofilms that arise in catheterized individuals (32).

Epidemiology and Animal Models for Polymicrobial UTI

The presence of multiple recognized uropathogens in midstream urine at titers >100,000 CFU/ml is consistent with a polymicrobial etiology of UTI. Polymicrobial infections occur most often among the elderly, immune compromised, and those with indwelling catheters, HIV, malignancy, and diabetes. Polymicrobial UTI is less common among young sexually active women. Since the highly polymicrobial microbiota of the gastrointestinal (GI) and reproductive tracts are thought to be a major inoculation source leading to UTI, and since truly dual-species or polymicrobial UTI do arise, several investigators have sought to examine the consequence of mixed microbial inoculation into the urinary tracts of model organisms. In a rat model, transurethral inoculation of *Staphylococcus saprophyticus* or *Proteus mirabilis* resulted in ascending pyelonephritis significantly more often when the two organisms were inoculated together compared to single-species infection, suggesting a synergistic virulence between the two species (33). *P. mirabilis* also synergizes with UPEC in the murine urinary tract, such that co-infection gave rise to greater CFU for both *P. mirabilis* and UPEC, compared to either single-species infection. The use of complementary, rather than competing, central-metabolism pathways in the urinary tract by UPEC and *P. mirabilis* may limit competition and thus promote synergy between these two organisms (34). Co-infection with the urease-positive Gram-negative organisms *P. mirabilis* and *Providencia stuartii* give rise to an increased incidence of urinary stones (urolithiasis) and bacteremia in a murine model of ascending UTI compared to mono-microbial infection (35), which may help explain why these organisms commonly co-occur in the urine of individuals with in-dwelling urinary catheters (20, 36, 37). Sim-ilar studies in mice showed that *Pseudomonas aeruginosa* and *E. faecalis* co-infection re-sulted in a more rapid development of pyelo-nephritis than observed when each species was inoculated alone (38). Moreover, co-infection studies with group B *Streptococcus* and UPEC in a murine UTI model have demonstrated that the presence of GBS can modulate host immunity and alter host sus-ceptibility to persistent high-titer infection of the bladder and kidneys by UPEC (39). Aged multiparous animals were particularly prone to UPEC infection in the context of GBS, demonstrating ~1,000-fold higher-titer UPEC infection in the presence of GBS com-pared to age-matched nulliparous controls (40). Microscopic and microbial culture ex-aminations, as well as culture-independent DNA-sequence analysis of bacterial biofilms found on urinary catheters, show that CAUTI biofilms are often polymicrobial in nature (41–45). By analogy to mixed-species effects in ascending UTI, the nature of the mixed-microbial community in CAUTI may also in-fluence the spectrum or severity of sequelae.

In the next section we will examine the epidemiology, virulence mechanisms, and host response to the most frequently isolated Gram-positive uropathogens: *Staphylococcus saprophyticus*, *Enterococcus faecalis*, and *Streptococcus agalactiae*.

URINARY TRACT INFECTION CAUSED BY STAPHYLOCOCCI

Epidemiology of *S. saprophyticus* UTI

Staphylococcus saprophyticus is a Gram-positive, coagulase-negative, nonhemolytic coccus. Colonies of *S. saprophyticus* are often yellow-pigmented (46). *S. saprophyticus* causes 5% to 20% of community-acquired

UTIs (47) and up to 42% of UTI among 16 to 25-year-old women (48). *S. saprophyticus* is second only to UPEC as the most common cause of uncomplicated UTI in this popula-tion (49, 50). Similar to UPEC infection, re-cent sexual intercourse is also a risk factor for *S. saprophyticus* UTI (51, 52). Over 40% of young, sexually active women are colonized with *S. saprophyticus* in the rectum, urethra, or cervix at any given time (52). It is thought that a major source of urethral inoculation is the GI microbiota. However, one study found no correlation between GI coloniza-tion and subsequent *S. saprophyticus* UTI (52). *S. saprophyticus* colonization and UTI display an interesting seasonal variation, with the greatest prevalence in the late summer and early fall (48, 52–54). The cause for this seasonality is not well understood. In the absence of complicating conditions, *S. sapro-phyticus* infection rarely causes UTI in males, but has been associated with urethritis and up to 17% of prostatitis (53, 55, 56). One study of older men presenting with UTI in a veteran's hospital found that the presence of *S. saprophyticus* was very rare (3 of 9,314 urine samples were *S. saprophyticus* positive) (57). When it is observed, male UTI caused by *S. saprophyticus* is found predominantly in elderly or institutionalized individuals (58). UTI symptoms caused by *S. saprophyticus* are similar in spectrum to those caused by *E. coli,* but can be more severe than in patients with *E. coli* UTI (59, 60); approximately 40% of patients with *S. saprophyticus* UTI present with acute pyelonephritis (58, 61).

Novobiocin resistance is a laboratory hall-mark for identification of *S. saprophyticus*. However, antibiotic resistance in *S. sapro-phyticus* is uncommon (62). As is the case for many coagulase-negative *Staphylococcus* spe-cies (CoNS) (63), methicillin-resistance can occur in *S. saprophyticus* and is found in ~1% to 8% of urine isolates (62, 64) via the ac-quisition of a penicillin-binding protein (PBP) with low β-lactam affinity encoded by the *mecA* gene (65, 66). *mecA* is carried on the staphylococcal-cassette chromosome

(SCC) *mec* (SCC*mec*) mobile-genetic element (MGE) (50, 67, 68).

S. saprophyticus UTI Virulence Factors

As the most frequent Gram-positive causative agent of UTI, *S. saprophyticus* is also the best studied with respect to virulence determinants necessary to cause infection. Scanning-electron micrographs of murine bladders infected with wild-type *S. saprophyticus* show the organism adhering over the entire surface of the bladder, but with apparent selective adherence to the tight junctions between individual epithelial cells (69). Several adhesins have been linked to *S. saprophyticus* colonization of the urinary tract (Fig. 1A). Conserved among all *S. saprophyticus* strains, Aas is a hemagglutinin that has autolytic and adhesive properties, binds to fibronectin and human ureters *in vitro*, and has been implicated in colonization of rat kidneys *in vivo* (70–74). Also widely conserved, the surface-associated lipase, Ssp, is found in >90% of *S. saprophyticus* strains, forms fimbria-like surface appendages, and is important for acute UTI as well as persistent kidney infection in a murine model (24, 74–76). However, the mechanism by which Ssp contributes to *in vivo* infection is not well-understood.

Cell wall-attached surface proteins also mediate adherence in *S. saprophyticus* (Fig. 1B). Sortase enzymes in Gram-positive bacteria covalently attach a subset of secreted proteins bearing sortase-recognition motifs, or sorting signals, to the exterior cell wall (77). Four sorting signal-containing proteins predicted to be cell wall-anchored via sortase enzymes have been described in *S. saprophyticus*: UafA, UafB, SdrI, and SssF. *S. saprophyticus* strains possess very few putative cell wall-associated sortase substrates compared to other staphylococci. The first sequenced *S. saprophyticus* strain ATCC 15305 encodes only one sortase substrate on its chromosome, UafA (78). *S. saprophyticus* MS1146 encodes UafA, as well as UafB and SssF, on

two different plasmids (79, 80). SdrI was characterized on the nonsequenced *S. saprophyticus* 7108 (81). Uro-adherence factor A (UafA) is a hemagglutinin that mediates adhesion to bladder-epithelial cells *in vitro* and is found in all *S. saprophyticus* strains examined to date (78). Plasmid-encoded UafB, found in ~5% of strains examined, is a serine-rich glycoprotein that binds fibronectin, fibrinogen, and human bladder-epithelial cells but does not promote bladder colonization in a murine UTI model (79). Like the related serine-rich platelet-binding proteins SraP in *Staphylococcus aureus* and GspB in *Streptococcus gordonii*, which rely on an accessory SecA2-secretion system (82–84), UafB is encoded in a genetic locus that also contains a gene for a putative accessory-secretion apparatus (85). In Gram-positive pathogens, these accessory Sec systems are associated with virulence (86). SdrI is a cell wall-associated serine-aspartate-rich protein, found in a minority of *S. saprophyticus* strains, that binds collagen, is associated with bacterial-surface hydrophobicity, and plays a role in acute UTI and persistent kidney infections (24, 79, 81, 87). SdrI shares sequence and structural homology with the adhesive Sdr proteins, including ClfA and ClfB, of *S. aureus* and *Staphylococcus epidermidis* (81, 88, 89). Plasmid-encoded SssF is highly conserved among *S. saprophyticus* strains and is involved in resistance to linoleic acid, but does not play a role in uropathogenesis in a murine model. Instead, SssF has been postulated to be important prior to urethral exposure, in biological niches such as the perineum or periurethral area where polyunsaturated fatty acids such as linoleic acid are particularly abundant (80).

In addition to cell wall-associated proteins, *S. saprophyticus* encodes urease that is important for efficient colonization of the bladder and kidneys, for inflammation in the bladder, and for dissemination to the spleen in a rat model of UTI (23). The presence of urease-producing *S. saprophyticus* has been associated with the formation of urinary

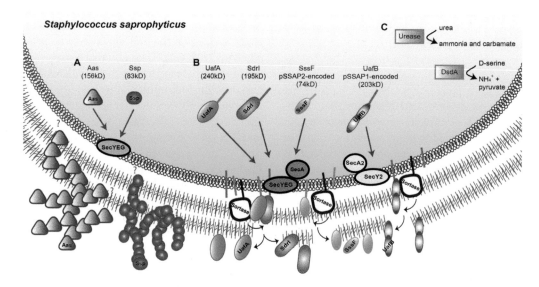

FIGURE 1 *S. saprophyticus* virulence factors. **(A) Secreted surface proteins: Aas, possesses N-terminal signal sequence, but no motifs such as a transmembrane domain, LPXTG sortase-recognition motif, or proline/glycine-rich cell wall-spanning domain to indicate the mode of attachment to the cell surface after translocation across the membrane (indicated by the purple question marks) (72). Immuno-electron microscopy shows Aas as part of a fuzzy surface layer that is absent when Aas is not expressed (75). Ssp has a YSIRK-containing signal sequence but no sortase-recognition motif so its mode of attachment to the cell surface is uncertain (indicated by the red question mark); it is easily sheared from the cell surface. Electron microscopy and immuno-electron micrographs also show Ssp to exist as part of fuzzy surface layer, apparently consisting of 50–75nm fibrillar structures; the nature of these fibers in not known (75, 76). (B) UafA, SdrI, SssF, and UafB contain an LPXTG motif and are predicted to be covalently attached to the cell wall (78, 79, 81). The small arrows near the membrane-anchored sortase enzymes indicate the two-step transpeptidation reaction whereby sortase substrates are first cleaved within the LPXTG motif to create a sortase-substrate intermediate (and releasing the membrane domain and positively charged cytoplasmic tail, indicated by the straight line in the membrane) that is then resolved, resulting in covalent linkage of the substrate to the cell wall (328). UafB is genetically linked to accessory secretion genes secA2 and secY2 that are predicted to encode a dedicated accessory secretion system for UafB (79). (C) Cytoplasmic enzymes that promote S saprophyticus survival in urine (329, 330).**

stones (90). The ability of *S. saprophyticus* to tolerate high concentrations of D-serine that occur in the urine is conferred by D-serine deaminase, found in *S. saprophyticus* but in no other staphylococci (91) (Fig. 1C). *S. saprophyticus* upregulates the virulence determinant Ssp in the presence of D-serine, and D-serine-deaminase mutants are outcompeted by the isogenic wild-type strain in the kidneys in a murine model of UTI (92). *S. saprophyticus* express a capsule, whose encoding genetic locus in the sequenced strain is found on an SCC genetic element and whose genetic arrangement is similar to

the *S. aureus cap5* (*cap8*) locus (78). The capsule of *S. saprophyticus* mediates resistance to complement-mediated opsonophagocytic killing by human neutrophils (93), but also prevents binding to urothelial cells, perhaps via masking and preventing interactions of adhesin(s) with the cells (78). Given the contributions of capsules of other uropathogens, such as UPEC (94, 95), it will be of interest to determine at which stages of UTI this polysaccharide is acting.

Phenol-soluble modulins (PSM) are staphylococcal pro-inflammatory cytolytic toxins characterized in *S. aureus* and *S. epidermidis*, are usually encoded within the core genomes, and are crucial in immune evasion as they are able to recruit, activate, and then lyse human neutrophils (96–99). Many PSMs also have antibacterial activities alone or in cooperation with host antimicrobial peptides that can kill other microbes within a niche (100–102). PSMs also mediate biofilm disassembly in their peptide form, whereas PSMs that assemble into amyloid-like extracellular fibrils under certain growth conditions promote biofilm stability (103–105). One exceptional noncore-genomic PSM has been found that is instead associated with the SCC*mec* MGE of *S. aureus*, termed PSM*mec*, genetically linking methicillin resistance and virulence (106). PSM*mec* have been identified in other CoNS, including multiple human *S. saprophyticus* isolates (50, 106, 107). A peptide Basic Local Alignment Search Tool (BLASTP) search of the sequenced genome of methicillin-sensitive *S. saprophyticus* ATCC 15305 (78) for peptides homologous to PSMβ1 or PSMβ2 of *S. aureus* (96) revealed the presence of 3 putative PSM peptides with 55% to 80% identity to PSMβ1/2 that do not appear to be associated with MGEs (KA Kline, unpublished). Whether *S. saprophyticus* PSMs are expressed and functional in UTI remains to be experimentally determined.

Non-saprophyticus Staphylococcal UTI

In contrast to *S. saprophyticus,* which is a predominant cause of community-acquired UTI, *S. aureus* UTI more often occurs in urinary-catheterized and pregnant individuals (108–110). The majority of *S. aureus* UTI isolates are methicillin-resistant and *S. aureus* bacteriuria is associated with subsequent development of invasive infection (108). Like *S. saprophyticus, S. aureus* also encodes an active urease enzyme. Two nickel ABC-transporters (Opp2 and Opp5a) have been identified as necessary for urease activity *in vitro*. These, along with a third ABC-transporter that imports nickel and cobalt when zinc is depleted, are both involved in

UTI colonization and virulence in a mouse model (111, 112). To our knowledge, no other *S. aureus* virulence factors have been examined during UTI.

Coagulase-negative *S. epidermidis* is a member of the human skin microbiota and is an important opportunistic pathogen, especially in biofilm-associated infections associated with indwelling medical devices. Thus, the biofilm-forming properties of *S. epidermidis* are an area of active investigation and are summarized in a recent review (113). CoNS, including *S. epidermidis,* are a leading cause of hospital-acquired infections where they are often methicillin-resistant (114) and are associated with 2.5% of CAUTI (115). In murine-UTI studies in the absence of catheter, *S. epidermidis* is able to colonize the bladder at a similar frequency, but with significantly delayed kinetics, compared to *E. coli* or *S. saprophyticus* (69).

Host Response to Staphylococcal UTI

The immune response to *S. saprophyticus* UTI in humans is not markedly different from Gram-negative UTI and is characterized by symptoms such as lower urinary-tract inflammation, pyuria, hematuria, and flank pain (49, 116). In the mouse-transurethral model of *S. saprophyticus* pathogenesis, inoculation of 10^7 bacteria into the murine bladder resulted in the recovery of 100-fold more CFU from the kidney than the bladder as early as 6 hours post infection and for as long as 2 weeks post infection (24). The kidney tropism of *S. saprophyticus* in a murine model may reflect the propensity of this organism to cause pyelonephritis in humans (49, 61). The innate inflammatory immune response to *S. saprophyticus* UTI reflected the higher kidney titers observed, with ~100-fold greater induction of numerous proinflammatory cytokines and significantly more neutrophils infiltrating in the kidney compared to the bladder, all of which peaked at 48 hours post infection. Enhanced macrophage recruitment to the bladder was also observed during acute stages of *S. sapro-phyticus* UTI (24). Recently, a role for the mitochondrial-respiratory chain has been implicated in the innate immune response to *S. saprophyticus* infection. A mouse strain carrying a heterogeneous knock-out of a sub-unit protein of the mitochondrial complex I (GRIM-19) is prone to spontaneous urinary-tract infection by *S. saprophyticus* (117). In response to infection, macrophages from GRIM-19+/- mice produce lower amounts of pro-inflammatory cytokines and are de-fective for intracellular killing of *S. sapro-phyticus*. (117). This report reflects the first mechanistic study of how the host handles *S. saprophyticus* infection.

URINARY TRACT INFECTION CAUSED BY ENTEROCOCCI

Epidemiology of Enterococcal UTI

Enterococci are a genus of Gram-positive lactic-acid bacteria that typically occur as diplococci or in short chains. These facul-tative anaerobes are γ-hemolytic and can tolerate a diversity of environmental con-ditions including temperature ranges from 10–45°C, pH ranges from 4.6–9.9, sodium-chloride concentrations up to 6.5%, bile salts up to 40%, and desiccation, despite the fact that they do not form spores (118). *Entero-coccus* species *E. faecalis* and *E. faecium* are responsible for a minority of community-acquired UTI, but together cause 15% to 30% of catheter-associated UTIs and are the third leading cause of hospital-acquired UTIs (13, 115, 119). Diabetic individuals are also at in-creased risk for UTI (120, 121), which serves as a nidus for the higher incidence of bac-teremia in this population (122, 123). While some studies indicate no increase in the frequency of enterococcal UTI in diabetic women compared to nondiabetic women (124, 125), others report that *Enterococcus* spp. are more often associated with UTI among diabetics and cause 13% of ASB in

diabetics compared to 4.9% in nondiabetics (121, 126). These differential findings in hu-mans have been reflected in animal studies. In a chemically induced (streptozocin) model of murine diabetes, diabetic mice were more susceptible to *E. faecalis* ascending UTI compared to nondiabetic mice (127), whereas pyelonephritis after intravenous injection of *E. faecalis* was similar in chemically in-duced (alloxan) diabetic rats and nondiabetic rats (128). Diabetes is also a risk factor for prostatitis, for which enterococci are respon-sible for up to 10% of cases (129, 130). The incidence of UTIs due to *E. faecalis* has risen steadily over the years and *E. faecalis* UTI now outnumbers *E. faecium* UTI 5:1 (131). Infection due to multiple-drug-resistant en-terococcal strains presents a significant med-ical problem, with vancomycin resistance increasingly prevalent among *E. faecium* isolates (131). *E. faecalis* readily adheres to and develops biofilms on abiotic surfaces such as urinary catheters. Many enterococcal-virulence factors involved in UTI described to date are also biofilm determinants. How-ever, it is unclear whether these virulence factors function in a similar manner during biofilm formation and infection in the absence of abiotic devices.

E. faecalis UTI Virulence Factors

The *E. faecalis* surface protein Esp is a large ~200 kilodalton (kD) surface protein that is enriched among enterococcal bloodstream and endocarditis isolates compared to fecal isolates (132). Esp is composed of multiple repetitive domains, sharing sequence simi-larity with Rib and C alpha-virulence de-terminants in group B streptococci, which confer protective immunity and mediate immune evasion (132–135). Esp promotes bladder colonization in a mouse model, as well as biofilm formation *in vitro* (136, 137).

Additional *E. faecalis* surface proteins that contribute to urovirulence in animal models include the collagen adhesin Ace and the enterococcal fibronectin-binding protein EfbA (28, 138, 139). Ace is regulated by the GrvRS two-component regulatory system and, accordingly, *gvrR* mutants are attenuated for biofilm formation *in vitro* and for virulence in a murine model of ascending UTI (140). The ArgR family transcription factor AhrC is important for early biofilm formation *in vitro*, and *ahrC* mutants are significantly attenuated for biofilm growth on catheters, as well as for bladder and kidney colonization during CAUTI (141). Nonproteinaceous *E. faecalis* factors monoglucosyl-diacylglycerol (MGlcDAG), diglucosyl-diacylglycerol (DGlcDAG), and D-alanylated lipoteichoic acid (LTA) appear to limit UTI virulence because mutations in synthetic genes for each of these products increase colonization of urothelial cells *in vitro* and *in vivo*, suggesting that their expression interferes with the host-pathogen interaction in the urinary tract (142, 143).

Similar to other sortase-assembled pili in Gram-positive bacteria, biogenesis of the *E. faecalis* endocarditis and biofilm-associated pilus (Ebp) relies on a pilus-associated sortase (Sortase C in *E. faecalis,* SrtC) for pilus polymerization and the housekeeping Sortase A (SrtA) for cell wall anchoring (77, 144–147). Temporal analysis of factors involved in biofilm formation *in vitro* showed that SrtA is important for the early attachment stage of biofilm formation, as is SrtC and the Ebp (144, 148). Furthermore, SrtA and the Ebp are important for biofilm formation *in vivo* during CAUTI (30, 149). However, since SrtA attaches multiple proteins to the cell wall, including aggregation substance (150) and Ebp, the contribution of SrtA may be attributable to multiple cell-wall proteins in addition to the pilus. In contrast, *E. faecalis* SrtC and the Ebp, but not SrtA, are important for ascending UTI in the absence of catheter in an outbred-mouse model (27, 151, 152), whereas Ebp is not required for ascending UTI in inbred mice (149). An *E. faecium ebp* mutant is attenuated in the ascending-UTI model (153). Thus, it appears clear that *E. faecalis* Ebp is required for biofilm forma-

tion during infection, but its contribution to infection in the absence of foreign devices may be species dependent.

Ebp pili can bind extracellular matrix (ECM) proteins and human platelets *in vitro* (138, 154). Recently, a predicted metal ion-dependent adhesion-site (MIDAS) motif in the von Willebrand factor A (VWA) domain within the N-terminus of EbpA, the likely tip adhesin of the Ebp, was found to be essential for Ebp-dependent CAUTI, biofilm formation, and associated bladder colonization *in vivo* (155). The VWA domain- and MIDAS motif-containing PilA of GBS, as well as the tip pilin RrgA of *Streptococcus pneumoniae,* bind extracellular matrix proteins (156, 157). The VWA domain of the PilA tip pilin of GBS is also important for bacterial adhesion to human alveolar and intestinal epithelial cells *in vitro*. The MIDAS motif of *E. faecalis* EbpA mediates binding to host fibrinogen, which is released in the bladder in response to catheter implantation and subsequently coats the catheter surface, providing a possible explanation of why *E. faecalis* is frequently associated with CAUTI (158).

Immune Responses to Enterococcal UTI

In a murine model, *E. faecalis* can colonize and persist in the kidneys, but is rapidly cleared from the bladder in the absence of apparent inflammation (26), suggesting lack of adherence by *E. faecalis* in the bladder. Nevertheless, urothelial cells containing intracellular *E. faecalis* have been observed in the urine of patients displaying lower urinary-tract symptoms (LUTS), such as pain and issues associated with urine storage and voiding, and the same *E. faecalis* strains can invade human urothelial cells *in vitro* (159). Kidney tropism is a common theme among Gram-positive bacteria in the murine-infection model (24, 25). Mild inflammation observed in the infected kidneys at 2 days post infection consisted primarily of monocytic cells and neutrophils, and was insufficient to clear the kidney infection.

Aggregation substance promotes *E. faecalis* survival within human neutrophils (160). *E. faecalis* can also survive longer within macrophages compared to nonpathogenic organisms, such as *E. coli* DH5α and *Lactococcus lactis* (161), and it has been proposed that glycosaminoglycans on the macrophage surface are receptors for *E. faecalis* (162). A number of enterococcal factors have been implicated in survival within macrophages, including extracellular polysaccharide, Ace, and oxidative-stress responses (139, 162–166). In addition, methionine-sulfoxide reductases A and B, predicted to reverse protein oxidation, not only promote survival within activated murine-peritoneal macrophages but also during ascending UTI (167). Together these studies present a picture in which macrophages are a key responder to *E. faecalis* UTI and where survival within macrophages may be important for the ability of enterococci to persist within this niche.

The host-signaling cascades leading to *E. faecalis*-mediated immune infiltration into the urinary tract is not yet well-delineated. In mice, toll-like receptor 2 (TLR2), which recognizes lipopeptides to initiate innate immune responses to Gram-positive bacteria, is not involved in the host response to *E. faecalis* either in TLR2-transiently transfected 293 cells *in vitro* or in the murine urinary tract (26, 168). However, *E. faecium*-induced signaling in murine-peritoneal macrophages requires both TLR2 and the intracellular adaptor protein MyD88 involved in TLR-mediated signaling, and these are thought to be important for signaling neutrophil recruitment during experimental peritonitis (169). After *E. faecalis* uptake into macrophages, proinflammatory-signaling cascades can also be induced via interaction between the intracellular macrophage nucleotide-binding oligomerization domain 2 (Nod2) protein and *E. faecalis* peptidoglycan fragments (85).

In contrast to ascending UTI, CAUTI is rarely symptomatic (15). Moreover, urinary catheterization alone, in the absence of bacterial infection, can give rise to dysuria, urinary urgency, urothelial damage, and bladder edema (15, 170–172). Despite being commonly asymptomatic, CAUTI is accompanied by pyuria. However, Gram-negative-associated CAUTI is more strongly associated with pyuria than those caused by Gram-positive organisms. CAUTI associated with Gram-positive organisms, such as enterococci or CoNS, are less inflammatory, as measured by leukocytes in the urine, compared to CAUTI associated with Gram-negative bacilli (173). These features of enterococcal CAUTI are also reflected in the murine CAUTI model where implantation of a foreign body (catheter) alone causes major histopathology including edema, urothelium damage, and proinflammatory-cytokine expression. *E. faecalis* can overcome the robust catheter-mediated inflammatory response and replicate to high numbers as a biofilm on the catheter, as well as in the bladder and kidney (31). Despite high *E. faecalis* CFU during CAUTI, markers of inflammation are not greatly enhanced in mice receiving the catheter implant together with *E. faecalis* compared to animals receiving catheter alone (30). Infection of catheterized mice with *E. faecalis* results in a modest augmentation of inflammation in the bladder, characterized by a more than 2-fold increase of interleukin 1β (IL-1β) and macrophage-inflammatory protein 1α (MIP-1α), despite bacterial loads in the bladder exceeding 10^6 CFU (30). These, differential inflammatory responses to enterococcal CAUTI versus ascending UTI may be due, in part, to differing responses of macrophage to *E. faecalis* in a biofilm state compared to planktonic cells *in vitro* (174). In addition, there is a significant decrease in the number of infiltrating-activated macrophages during *E. faecalis* CAUTI compared to catheterization in the absence of bacteria, suggesting an immunosuppressive capacity of *E. faecalis* (31). The gene encoding a Toll/interleukin-1 receptor (TIR) domain-containing protein (*tcpF*) is enriched among *E. faecalis* UTI isolates and can downregulate the host-inflammatory response, presumably by interfering via molecular mimicry with the TIR-TIR interactions between TLRs required for TLR dimerization and subsequent signaling (175, 176). Consistent with this finding, several studies of *E. faecalis* strains isolated from the gastrointestinal tract of healthy human-infant guts have shown that a subset of strains were capable of suppressing inflammatory-cytokine expression in human intestinal-epithelial cells *in vitro*, as well as suppressing cytokine responses in a dextran-sulfate-sodium-salt (DSS) model of inflammatory intestinal colitis *in vivo*, although the presence of *tcpF* was not examined in these strains (178, 179). Together these studies indicate that high-titer *E. faecalis* CAUTI is not highly inflammatory and can be immunosuppressive.

URINARY-TRACT INFECTION CAUSED BY GROUP B *STREPTOCOCCUS*

Epidemiology of Group B *Streptococcus* UTI

Streptococcus agalactiae, otherwise known as group B *Streptococcus* (GBS), is a Gram-positive β-hemolytic chain-forming coccus that is a common asymptomatic inhabitant of the lower gastrointestinal and female-reproductive tracts. GBS is estimated to cause approximately 1% to 2% of all monomicrobial UTIs (180). Other studies of elderly populations with UTI show an involvement of GBS in as many as 39% of nursing-home residents over 70 years of age (181). ASB and UTI caused by GBS are common not only among the elderly, but also in pregnant, diabetic, and immunocompromised individuals, as well as those with pre-existing urologic abnormalities; these are groups with higher risk of ascending pyelonephritis that can progress to bacteremia and/or urosepsis (10, 182–184). While GBS may represent only a small fraction of total UTIs, the burden of GBS UTI is a major public-health concern (10), with approximately 160,000 cases annually in the United States.

Among nonpregnant adults, the incidence of systemic GBS infections is estimated at approximately 4.4 cases per 100,000 individuals; 14% of these are cases of urosepsis (185). Common underlying conditions of individuals with GBS urosepsis include diabetes mellitus, malignancy, chronic kidney disease, recurrent urinary-tract infections, obstructive neuropathy, and neurogenic bladder (185, 186). In about 30% of cases, systemic infections caused by GBS in nonpregnant adults do not have an apparent focal origin such as cellulitis, pneumonia, or UTI. In contrast to the proportion of UTIs associated with GBS in young nonpregnant populations, estimated at 1% to 2% (180), it has been estimated that up to 7% of pregnant women have significant titers of GBS in urine (187, 188).

GBS Bacteriuria in Pregnancy

GBS is the leading cause of sepsis and meningitis in newborns and can be acquired by the newborn *in utero* or during passage through the colonized birth canal. In addition to colonizing the reproductive tract, GBS is often found to colonize the urinary tracts of pregnant women. Although GBS colonization of the urinary tract in pregnancy is often asymptomatic, GBS bacteriuria is an independent risk factor for maternal pyelonephritis and chorioamnionitis as well as neonatal GBS sepsis (187, 189–192). The Centers for Disease Control and Prevention (CDC) recommend universal screening of GBS vaginal-rectal colonization at 35–37 weeks of gestation and antibiotic prophylaxis for culture-positive women during labor and delivery.

Screening and treatment for ASB is also recommended in pregnancy, since women with ASB are at higher risk of preterm delivery, have a 20 to 30-fold increased risk of pyelonephritis, and antibiotic treatment of ASB has been demonstrated to reduce these risks significantly (193–198). The 2002 recommendations from the CDC stated that clinical microbiology labs should report *any* concentration of GBS detected in urine. The 2010 revised guidelines notes that this practice represents an

> "increased workload for clinical microbiology laboratories, which do not generally report bacterial growth in urine of other pathogens at concentrations <10^4 cfu/ml and rarely know whether urine samples are from pregnant women; as a result, some laboratories search for any GBS colonies in urine cultures from all women of reproductive age (199)."

Surprisingly, there is little published clinical data investigating whether low GBS titers (<10^4 CFU/ml) in urine is associated with adverse maternal or neonatal outcomes. One study found that babies of women with low-titer GBS bacteriuria were at higher risk of GBS disease compared to women without detectable GBS in urine (200). Other studies have not been performed to confirm or refute this finding, but it has been argued that this study *may* have been biased because only a subset of the women underwent urine culture (199). Clearly, additional studies are required to estimate the threat of low-level GBS in urine for neonatal GBS disease and determine the best course of action for screening and prophylactic measures. One study estimated that 4% of women who test negative for GBS rectal or vaginal colonization test positive for GBS in urine culture (201), suggesting that the urinary tract could be a distinct and independent niche for the bacterium that could be missed in routine third-trimester screening procedures. Currently, the CDC recommends that women with GBS bacteriuria at any time in pregnancy should be given prophylactic antibiotics at the time of labor and delivery.

Invasive GBS Disease in the Elderly

UTI caused by GBS is approximately 10 times more frequent than GBS neonatal infections and is common among the elderly. Case-fatality rates for invasive GBS infection are also higher among the elderly (~15%) com-pared to young infants with invasive infection (4% to 6%) (10). However, in contrast to screening and prophylactic measures during pregnancy and post-partum periods, similar screening and prevention strategies are lacking in settings where elderly patients are at risk of invasive-GBS disease. Attempts to develop an effective GBS vaccine have been successful in early trials in animals and humans (202–209), raising the possibility of future reductions in the incidence and severity of GBS disease in at-risk groups.

GBS Virulence Factors and Induction of Host Immune Responses

GBS can cause infections at a variety of body sites (skin, soft tissue, lung, peritoneum, urinary tract, etc.) and utilizes a wide range of virulence factors to injure and invade host tissues, resulting in disseminated infection (e.g., bacteremia, osteomyelitis, meningitis) (210). Although much is known about the molecular epidemiology of GBS, the organism has not been studied extensively in animal models of UTI; thus, relatively little is known about which virulence factors may play important roles in this context.

Epidemiological analyses have shown that a variety of GBS serotypes are associated with UTI. However, serotype III GBS was found to cause a disproportionate number of acute symptomatic disease compared to other serotypes, which were more likely to be associated with asymptomatic bacteriuria (183). These clinical studies have formed the basis for experimental studies of GBS urinary-tract infections using serotype III GBS strains in mouse models of transurethral infection (25, 211, 212). In the murine model, the presence of GBS robustly induces IL-1α (212), MIP-1α, MIP-1β, IL-9, and IL-10 (25). However, experimental studies have observed a striking lack of overall histological inflammation in the bladder during GBS cystitis (25) and marked differences in transcriptional responses compared to *E. coli* UTI (211). A number of GBS-virulence factors have been characterized in other models, but only a few studies have examined GBS-virulence factors by inoculating wild-type and mutant strains of bacteria into the mouse urinary tract (25, 39, 213). These studies have demonstrated that sialic-acid residues of the GBS capsular polysaccharide are necessary for optimal establishment of GBS in the urinary tract (25), whereas the β-hemolysin/cytolysin did not appear to have a significant effect on bacterial survival following transurethral infection (25, 213). Additional studies are needed to identify GBS-virulence factors of importance in the urinary tract and to understand the cellular, molecular, and biochemical details of host-microbe interactions in populations at-risk for GBS disease.

In the remainder of this chapter, we will 1) discuss several emerging, rare, and/or underreported Gram-positive pathogens of the urinary tract, 2) introduce several recent studies that demonstrate some unexpected inhabitants of the urinary tract using culture-independent approaches for bacterial detection in urine, and 3) examine evidence that host urogenital-colonization states may influence the risk of UTI.

VAGINAL MICROBIOTA, BACTERIAL VAGINOSIS (BV), AND UTI

Mounting clinical evidence argues that the composition of a woman's vaginal microbiota influences her risk of UTI. Women with a dominant population of vaginal lactobacilli are at a lower risk of UTI compared to women with more diverse microbiota, consisting of Gram-negative anaerobes, *Actinobacteria*, and other *Firmicutes* (214–216). This vaginal condition is referred to by some as bacterial vaginosis (BV) and is often labeled as an 'imbalance' or dysbiosis of the vaginal microbiota because it has been associated with a wide variety of adverse health outcomes for women and their babies (217, 218). Clinical features of BV (i.e., Amsel criteria) include vaginal pH >4.5, 'thin'

grayish, homogenous vaginal fluid, fishy odor upon potassium hydroxide (KOH) treatment of vaginal fluid, and the presence of epithelial cells studded with bacteria in wet mount (i.e., 'clue-cells') (219–221). Another method for BV diagnosis that has been used more extensively in recent years is the Nugent method (220), which is based on a morphotype-scoring system using Gram-stained slides where a higher score indicates BV. Only recently have studies in experimental models demonstrated that a single bacterium, *Gardnerella vaginalis*, one of the common bacteria to overgrow in BV, is sufficient to yield clinical features and biochemical phenotypes of BV in a murine vaginal-infection model (222, 223). Studies examining the association between BV and UTI have found that women with BV have anywhere from a 2.2- to 13.7-fold increased risk of UTI, depending on the population studied (214–216). Further supporting these findings, clinical studies now provide compelling preliminary data that vaginal probiotic administration may be beneficial for women who are prone to recurrent UTI (224). These studies are consistent with results of anaerobic culture studies presented above, showing that women with recurrent UTI were more likely to contain higher titers and a more diverse repertoire of mixed anaerobes in their urine than healthy (sexually active or inactive) women (225). Taken together, these results suggest a possible linkage between dysbiosis of the vaginal microbiota and dysbiosis of the bladder microbiota.

Unfortunately, our understanding of *why* women with bacterial vaginosis are more prone to urinary-tract infection is limited. It is thought that lactic acid, hydrogen peroxide, and other 'defensive' molecules produced by lactobacilli may create a hostile environment for potential pathogens, including uropathogens, in the vagina (226–228). However, experimental studies are needed to more fully understand the causal relationships linking specific vaginal bacteria with

UTI susceptibility in experimental models. While it remains a distinct possibility that lactobacilli act through specific mechanisms to discourage UTI (229), another possibility (though not a mutually exclusive one) is that one or more taxa of BV bacteria act through unknown mechanisms to enhance UTI susceptibility. As described below and illustrated in Table 1 and Fig. 2 (with emphasis on Gram-positive bacteria) many of the bacterial genera that have been described in the BV-associated vaginal environment have also been observed in both the female and male urinary tracts using culture-independent approaches. These findings raise the possibility that fastidious organisms associated with BV could play a more direct role in the etiology of host-uropathogen interactions within the urinary tract.

RARE, EMERGING, AND UNDER-REPORTED GRAM-POSITIVE AND POLYMICROBIAL ETIOLOGIES OF UTI

In the next section of this chapter, we present several specific examples of Gram-positive bacteria that are rare, emerging, or underreported, including species of *Aerococcus*, *Corynebacterium*, *Actinobaculum*, and the potential uropathogen *Gardnerella vaginalis*. These organisms may be missed as causes of UTI due to 1) misclassification due to lack of distinguishing phenotypic criteria, 2) dismissal of significant growth as 'microbiota contamination', or 3) lack of detection by standard approaches.

Aerococcus as a Cause of UTI

Aerococcus is a genus of microaerophilic, facultatively anaerobic, α-hemolytic, Gram-positive cocci that are catalase- and oxidase-negative and leucine-aminopeptidase positive. One unique characteristic of aerococci is that they divide on 2 planes at right angles, resulting in tetrads and irregular clusters. Species of *Aerococcus* are commonly isolated from air,

dust, and vegetation, and are also common isolates from the human vagina and urinary tract. Several species of *Aerococcus* can cause urinary-tract infections and urosepsis, including *A. urinae*, *A. viridans*, and *A. sanguinicola* (230–236). One study reported that 0.8% of all urine specimens cultured during a 4-month period in a Denmark hospital yielded growth of "*Aerococcus*-like" organisms (237). Patients with UTI caused by *Aerococcus* are most often elderly and many have urological abnormalities or other risk factors for UTI (230, 233, 237). Many of the cases described are invasive systemic infections in which *Aerococcus* is isolated from blood along with significant *Aerococcus* titers in urine.

Aerococci have a number of biochemical and physiological similarities with lactococci, pediococci, enterococci, and streptococci (238). In particular, phenotypic similarities between *Aerococcus* and viridans-group streptococci have made it difficult for clinical labs to distinguish between them using routine phenotypic tests. For example, one study examined the ability of three commonly used bacterial identification systems (API 20 STREP, ID 32 STREP, and VITEK 2 ID-GPC card, bioMérieux) to correctly identify 30 urinary-tract isolates representing different species of *Aerococcus*. This study revealed that *Aerococcus* isolates (with the exception of *A. viridans*) were commonly misidentified or identified with low discrimination (239). Molecular tools such as amplification and sequencing of 16S rRNA, which are not commonly employed in clinical microbiology labs, are needed for accurate identification of *Aerococcus* species. The result is that aerococci are often misclassified and/or discarded as likely contaminants.

Prompt and accurate identification of *Aerococcus* is necessary to avoid life-threatening systemic infection by this potential pathogen in susceptible individuals that present with uncomplicated UTI. Most isolates of *Aerococcus urinae* have been characterized as resistant to sulfonamides (240, 241). Case reports of *A. urinae* UTI demonstrate that when patients are treated with antibiotics effective against the organism, bacteriuria cleared and the patient recovered fully; however, in other situations when *A. urinae* was not recognized or effective antibiotic treatment was not promptly provided, patients often progressed from simple UTI to invasive systemic infections (234, 235, 240, 242, 243).

Corynebacterium urealyticum

Corynebacteria are Gram-positive, nonmotile, nonspore forming, facultatively anaerobic *Actinobacteria* that are common components of the skin microbiota and increasingly recognized as opportunistic pathogens (244, 245). *Corynebacterium urealyticum* has been associated with asymptomatic bacteriuria and, rarely, with acute and chronic infections of the urinary tract. *C. urealyticum* (previously known as *Corynebacterium* group D2) is the most common cause of alkaline-encrusting cystitis and pyelitis, a chronic inflammatory condition of the urinary tract in which the bacterium causes painful localized ulcerations with deposits of ammonium magnesium phosphate (a.k.a. struvite) that can be visualized on plain radiography (246–248). Alkaline urine and the presence of struvite crystals in urine sediments are characteristic features of *C. urealyticum*-encrusting cystitis and occur due to bacterial expression of urease activity, which catalyzes the conversion of urea to ammonia and carbon dioxide (CO_2). The presence of a previous urological procedure and/or mucosal lesion appears to be necessary for urea-splitting bacteria such as *C. urealyticum* to cause encrusting UTI. Immune-compromised patients with underlying urologic disease, those that have undergone long-term hospitalization, and/or previous treatment with broad-spectrum antibiotics are also at greater risk for developing alkaline-encrusted UTI. While this type of UTI is considered rare, one prospective study showed that 9.8% of renal-transplant recipients (n=163) had

C. urealyticum in urine, with nearly half of positive patients experiencing no symptoms until at least 1 month after first detection of the organism (249).

C. urealyticum infection can often be missed in routine urine culture because the organism is slow-growing and requires enriched media. Moreover, similar to other *Actinobacteria*, *C. urealyticum* requires at least 48 hours of growth on blood-agar plates to produce pinpoint colonies, which may be dismissed as "contaminating microbiota," since it is generally thought that species of *Corynebacteria* other than *C. diphtheriae* are nonpathogenic. The presence of alkaline urine or struvite crystals in urine sediment should prompt cultures of longer duration. A number of cases of *C. urealyticum* UTI also report that Gram-staining of urine specimens reveals Gram-positive bacilli and polymorphonuclear cells, despite routine urine cultures that come back negative. Molecular-detection techniques can also be employed when tests are negative, but there is strong suspicion of encrusting cystitis or pyelitis with *C. urealyticum*. Encrusting cystitis caused by *C. urealyticum* is treated by endoscopic removal of encrustations, acidification of the urinary tract, and treatment with appropriate antibiotics. A number of reports document resistance of *C. urealyticum* isolates to ampicillin, cephalothin, gentamicin, imipenem, tetracycline, ciprofloxacin, and ofloxacin, and sometimes to tetracycline, erythromycin, and rifampicin (250, 251); however, most isolates were susceptible to synercid and linezolid (250).

Actinobaculum schaalii

The genus *Actinobaculum* and the species *Actinobaculum schaalii* were first described in 1997 following the isolation of organisms described as *Actinomyces*-like or *Corynebacterium*-like from human blood and urine that displayed >6% sequence divergence from the nearest relative, *Actinobaculum suis* (252), a swine uropathogen also previously known as *Actinomyces*

TABLE 1 Genera of Gram-positive human urinary tract inhabitants and pathogens detected by culture-dependent and -independent techniques[a]

Phylum	Genus	Culture-dependent clinical reports			Culture-independent detection in urinary tract							
		Bladder/kidney infection	Bacteremia/urosepsis	Urethritis	Female, culture unknown	Female, culture positive	Female culture-negative, nitrite and leukocyte esterase negative or asymptomatic		Male asymptomatic		Urethral swab	Male symptomatic
					TUC	CC, TUC	CC	TUC	SPA	Urine	Urine	Urine
Firmicutes	Staphylococcus	5% to 20% of CAUTI 47	331	53, 55, 56	292, 332	n/r	318, 319	317, 332	317	319, 322–324	323	333
	Enterococcus	15% to 30% of CAUTI 13, 115, 119	331	334, 335	318	316, 336	319	n/r	n/r	322, 323	323	333
	Streptococcus	1% to 2% of UTI 180	180	337	292, 318, 332	n/r	316, 318, 319	317, 332	317	322–324	323	333
	Aerococcus	230, 232, 237, 239	233–236, 278	n/r	332, 338	317	318, 319	317	317	319, 322, 323	323	n/r
	Anaerococcus	n/r	n/r	n/r	292, 332, 338	336	316, 318, 319, 336	n/r	317	319, 322–324	323	n/r
	Peptostreptococcus	321, 339*	340	n/r	338	336	318, 319	n/r	n/r	319, 322, 324	n/r	n/r
	Peptoniphilus	n/r	341	n/r	338	336	317–319	n/r	n/r	319, 322	n/r	n/r
	Lactobacillus	321, 342, 343 *	n/r	n/r	338	336	316–319, 336	317, 332	317	319, 322–324	323	333
	Finegoldia	n/r	n/r	n/r	338	317, 318, 336	319	n/r	317	319, 322–324	323	333
	Veillonella	344, 345	340, 344	n/r	338	317	n/r	292, 332	n/r	322, 323	324	n/r
	Gemella	346	n/r	n/r	n/r	n/r	n/r	n/r	n/r	319, 322, 323	323	n/r

Actinobacteria	*Corynebacterium*	246–249, 347–352	353	354–356	292, 332, 338	n/r	316, 318, 319	317, 332	317	319, 322, 323	323	n/r
	Actinobaculum	155, 261, 262, 266, 267, 269, 272, 274, 276, 277, 357	270, 275, 278, 358	n/r	338	317, 336	318, 319	292, 317, 332	317	322	n/r	n/r
	Gardnerella	285, 300*	293, 300, 303, 359	307–309, 360–362	338	336	316, 318, 319, 336	317, 322–324, 332, 363	317	319, 322	323	n/r
	Atopobium	n/r	n/r	n/r	338	336	317, 319, 336	292, 317, 332	317	319, 322, 323, 364	323	364
	Actinomyces	365–369	370	n/r	338	n/r	319	292	n/r	322	n/r	n/r
	Bifidobacterium	n/r	n/r	n/r	338	336	n/r	292, 332	n/r	n/r	n/r	333
	Mycoplasma	371–374	n/r	177, 361, 375–381	n/r	n/r	n/r	n/r	n/r	319, 322, 323	323	n/r

[a]"culture negative" means specimens that did not reach the cutoff value, usually 10^5 CFU/ml urine in a clean-catch specimen or 10^4 CFU/ml for collections by catheter or suprapubic aspiration. n/r = not reported, CAUTI = community-acquired UTI, * = questionable clinical significance, CC = clean catch, TUC = transurethral catheter, SPA = suprapubic aspiration.

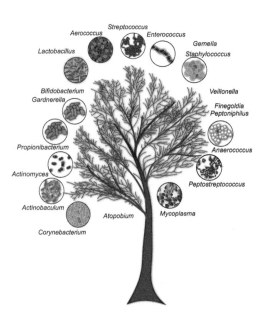

FIGURE 2 Gram-positive inhabitants and pathogens of the human urinary tract. Approximate phylogenetic relationships between Gram-positive bacteria are illustrated in this schematic representation. Please refer to the text and Table 1 for additional information and references describing these genera as uropathogens or inhabitants of the human urinary tract. On the left, Bifidobacterium and Gardnerella belong to the order Bifidobacteriales, which together with the orders Propionibacteriales, Actinomycetales, and Corynebacteriales belong to the class Actinobacteria (a.k.a. "high-GC Gram-positive bacteria"). Atopobium belongs to the order Coriobacteriales and the class Coriobacteriia. The classes Coriobacteriia and Actinobacteria both belong to the phylum Actinobacteria. The remaining genera, with the exception of Mycoplasma, belong to the phylum Firmicutes. Whereas Peptostreptococcus, Anaerococcus, Finegoldia, and Peptoniphilus belong to the order Clostridiales and the class Clostridia, Staphylococcus and Gemella belong to the order Bacillales and the class Bacilli. Members of the order Lactobacillales (Lactobacillus, Aerococcus, Enterococcus, and Streptococcus), are also classified as Bacilli. The genus Veillonella is also a member of the Firmicutes, but belongs to the class and order Negativicutes and Selenomonadales, respectively. On the other hand, Mycoplasma belongs to the phylum Tenericutes, the class Mollicutes, and the order Mycoplasmatales.

suis, Corynebacterium suis, or *Eubacterium suis* (253–255). Species of *Actinobaculum* are members of the family *Actinomycetacea,* which also includes the genera *Actinomyces, Arcanobacterium,* and *Mobiluncus* (256). These bacteria are facultative or obligate anaerobes and their detection in clinical specimens often requires a full 48 hours of growth on blood-agar plates in the presence of a 5% CO_2 atmosphere. Phylogenetically, these bacteria belong to the high guanine-cytosine (GC)-content Gram-positives, but phenotypically, they can appear 'Gram-variable' since their cell walls are relatively thin compared to other Gram-positive bacteria and thus more easily decolorized during the Gram-staining procedure. Currently, the genus *Actinobaculum* contains four species: *A. suis, A. schaalii, A. massiliense,* and *A. urinale.* To date, *A. urinale* and *A. massiliense* have only been described in a few cases of human pathology, including urinary-tract infections. In contrast, *A. suis* is a well-documented uropathogen of female pigs (254, 257–259) that appears to colonize the prepuce in a large proportion of wild and domesticated adult males (255, 260). Here we focus on *A. schaalii,* which appears to be an emerging human uropathogen, particularly among the elderly and those with underlying urologic abnormalities (261).

A. schaalii is a small rod-shaped (straight or slightly curved and sometimes branching), nonmotile, nonsporulating, weakly β-hemolytic, facultative anaerobe requiring CO_2 for growth and testing negative for catalase, oxidase, urease, and nitrite production. *A. schaalii* is resistant to trimethoprim and ciprofloxacin, antibiotics used as first-line treatments for urinary-tract infection. Although the natural history surrounding this species is still unclear, culture-dependent and -independent studies published to date suggest that this bacterium may be a commensal inhabitant of the human urinary tract. To date, there have been approximately 130 reported cases of human infection with *A. schaalii,* mostly as small retrospective studies and case reports (252, 261–279). Approximately 85% of *A. schaalii* infections reported so far occur in the urinary tract (i.e., single-organism cystitis, pyelonephritis, or urosepsis) or have disseminated (e.g., bacteremia, endocarditis) from an unknown primary source. These infections most often occur in elderly individuals and those with underlying urologic conditions such as chronic renal failure or urologic obstruction (261). In clinical cases of *A. schaalii* UTI, Gram-positive bacteria are often evident by direct microscopy, and yet, typical aerobic urine culture often produces a negative result (266, 268). Cases of apparent urosepsis where *A. schaalii* was isolated using anaerobic blood-culture techniques were documented only after culture conditions were altered, eventually leading to the identification of *A. schaalii* in urine.

A. schaalii has been considered an extremely rare uropathogen. However, recent literature suggests that *A. schaalii* may be underreported due to inadequate culture and identification techniques. Due to fastidious growth conditions required for *A. schaalii* culture, and the similar appearance of the colonies to resident microbiota of the skin and genitourinary tract, it is likely this organism has been overlooked or dismissed as "contamination" when it may be clinically significant in certain settings (252). Traditional phenotypic tests are still inadequate for identification of *A. schaalii*. Instead, studies have successfully used the API Coryne and Rapid ID 32A strip test systems (bioMérieux) together with molecular methodologies (sequencing of 16S ribosomal DNA and/or species-specific primers for quantitative polymerase chain reaction (PCR)) for identification and/or enumeration of *A. schaalii*.

A recent study examined 252 randomly selected urines from individuals in three hospitals in Viborg County, Denmark using a validated quantitative PCR that specifically analyzed the presence and titers of *A. schaalii*. The authors of this study found that 22% of persons >60 years of age harbored at least 10^4 CFU/ml of urine (268). Additional studies are needed to estimate the prevalence of *A. schaalii* in the urinary tract in other populations and to define the possible clinical significance of this organism.

Gardnerella vaginalis

Gardnerella vaginalis belongs to the class *Actinobacteria* (also frequently called 'high-GC Gram-positives') and the order *Bifidobacteriales*. *G. vaginalis* is best known for its connection to BV, a condition marked by *G. vaginalis* overgrowth in the vagina. BV has been linked to higher risks of a wide variety of adverse women's-health outcomes, including UTI (214–216, 280, 281). In addition to the ever more-apparent role of *G. vaginalis* in BV, a growing body of evidence implicates the organism as a potential uropathogen. *G. vaginalis* (see Fig. 3 for images of this organism) is a fastidious bacterium that cannot be recovered under the same conditions as typical uropathogens. It requires incubation in the presence of 5% CO_2 or, better yet, under anaerobic conditions. Although it can be recovered on sheep-blood agar under these conditions, the organism cannot survive in acidic urine or in urine stored at room temperature or at 37°C (rather than refrigerated) prior to recovery. *G. vaginalis* prevalence is significantly underestimated if plates are incubated for only 24 hours, but instead requires extended incubation (48–72 hours) (282).

In two large studies encompassing culture analysis of a total of ~33,000 urine specimens, *G. vaginalis* was reported in 0.6% to 2.3% of all samples at titers >10^4 CFU/ml of urine. *G. vaginalis* was often identified in pure culture and in a context of patient-reported symptoms and/or pyuria (283, 284). One of these studies (reporting *G. vaginalis* in 0.6% of urines) only examined plates at the 24-hour time point (284), suggesting that 0.6% may be an underestimation. The larger of these two studies examined hospital inpatients and reported that among the patients with *G. vaginalis* bacteriuria, 58% had evident pyuria and 10% had pyelonephritis

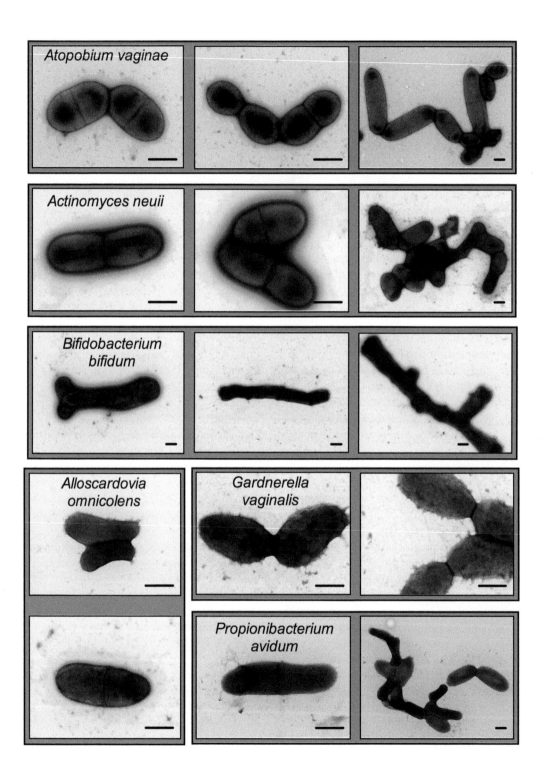

(285). These patients were also more likely to have a history of recurrent UTI. Other studies provide further support that *G. vaginalis* presence in urine is not simply the result of periurethral or vaginal contamination and further suggest a possible role of *G. vaginalis* in renal disease. Several studies in the 1960s to 1980s identified *G. vaginalis* identified in urine isolated by suprapubic aspiration, thus bypassing possible urogenital contaminants (286–290). These studies suggest that pregnancy renders women more likely to harbor *G. vaginalis* in their bladders and that recovery of the organism was especially common in women with underlying renal disease. Another study described *Gardnerella* as a common isolate from aspirates of individuals with reflux scarring and so-called "sterile pyelonephritis" (291). In addition to these culture-based studies, more recent studies using culture-independent approaches have echoed these early studies. In fact, one study in particular concludes that *G. vaginalis* is found at high levels in urines collected by suprapubic (needle) aspiration from many older adult women (292).

In addition to its apparent ability to infect the bladder and kidneys, *G. vaginalis* has also been described in bloodstream infections. One study reported thirty cases of bacteremia caused by *G. vaginalis* among obstetric patients over a 4-year period and suggests that bacteremia caused by this organism may be significantly underreported (293). Although cases of *G. vaginalis* bacteremia are enriched in the gynecologic setting, occurring after birth (294–296) or following procedures such as endometrial ablation (297) or vaginal myomectomy (298), *G. vaginalis* bloodstream infection is not limited to women (299–304). Indeed, the evidence suggests that female-to-male transmission of *G. vaginalis* can occur during sexual encounters and that regular condom use reduces the likelihood that *G. vaginalis* can be isolated from male urine (305, 306). In one example of bloodstream infection in a male, a previously healthy individual with flank pain was found to have urolithiasis (kidney stones) with fever, elevated peripheral neutrophils, and elevated creatinine (indicating damage to the kidneys) (301). Blood cultures revealed the man had urosepsis caused by a coccobacilli that produced pinpoint gray colonies on chocolate agar. In this case the hospital microbiology lab was not equipped to identify the organism. Instead, a national microbiology laboratory performed an extensive workup, identifying the organism as *G. vaginalis*. Another example involved an uncircumcised man with a previous history of diabetes mellitus and an ongoing sexual partner with recurrent BV. The man was shown to have bloodstream *G. vaginalis* and infective endocarditis, along with apparent septic emboli on one kidney and in the brain (300). In this case, routine culture of the urine was also unable to identify *G. vaginalis* as the culprit.

G. vaginalis has also been described in polymicrobial infections in individuals with underlying urologic abnormalities or following urologic procedures. One case series describes *G. vaginalis* as commonly co-occurring with members of the genus *Bacteroides* in abscesses following urological procedures (285). Sturm concluded this case series by recommending that microbiology labs should perform Gram-staining of urine specimens

FIGURE 3 **Transmission-electron micrographs of several urogenital isolates from the phylum *Actinobacteria*. Strains grown for 24–48 hours underwent negative staining with uranyl acetate and were examined by TEM. These strains were isolated from the urine or vaginas of pregnant or nonpregnant women and are available through BEI resources. Strain names are as follows: *Actinomyces neuii*, MJR8396A; *Alloscardovia omnicolens*, CMW7705A; *Atopobium vaginae*, CMW7778A; *Bifidobacterium bifidum*, MJR8628B; *Gardnerella vaginalis*, PSS7772B; *Propionibacterium avidum*, MJR7694. Scale bars are 500 nm. Shaded backgrounds contain images of the same strain.**

and should follow up results of pyuria in the context of visible coccobacilli by performing culture under conditions capable of *G. vaginalis* recovery. Given the apparently high risk of *G. vaginalis* infection following urological instrumentation, Sturm suggests that routine culture for this organism may be warranted during preoperative testing.

The role of *G. vaginalis* as a genitourinary pathogen is still controversial. Recent work in an experimental model of vaginal infection with *G. vaginalis* demonstrated that strain JCP8151B, isolated from a woman with BV, was sufficient to induce features of BV in mice (110). This is the first animal model in which any BV bacterium has been shown to induce the formation of "clue" cells – exfoliated vaginal-epithelial cells coated in bacteria – one of the hallmark microscopic features of BV. Interestingly, multiple studies of urinary-tract infections involving *G. vaginalis* report the presence of clue-like cells in urine, both from women (291) and from men (307–309). Interestingly, a wide variety of genera from the phylum *Actinobacteria* have been implicated in UTI or identified as part of the urinary microbiome, including not only *Gardnerella*, but also *Actinobaculum, Corynebacterium, Actinomyces, Atopobium, Alloscardovia,* and *Bifidobacterium* (see Fig. 3 for images of these bacteria). Few of these genera are accepted as true uropathogens. Thus, experimental studies of infection in animal models are needed to help settle the debate about their potential roles as uropathogens, either in the classical sense or by predisposing the urinary tract to infection or other urologic diseases.

Urine "Contamination" with Normal Microbiota or Polymicrobial UTI?

The potential biological and clinical significance of polymicrobial growth in urine depends on many factors. On one hand, the finding of multiple colony types after urine culture is a valid and justifiable concern to suspect contamination of the urine specimen with periurethral and/or vaginal microbiota. For this reason, most clinical microbiology labs will not evaluate plates with polymicrobial growth, but rather dismiss them as "contamination" and request another specimen. For instance, in suspected cases of uncomplicated cystitis, *E. faecalis* and GBS are often assumed to represent contamination of the urine specimen originating from the periurethral area during collection (1). In fact, one recent study evaluated titers of these Gram-positives in midstream urine and compared them to titers obtained when the same women were subjected to catheterization for urine collection. This study concludes that, "enterococci (in 10% of cultures) and group B streptococci (in 12% of cultures) were not predictive of bladder bacteriuria at any colony count (Spearman's $r=0.322$ for enterococci and 0.272 for group B streptococci)." However, only a few patients in this study had *Enterococcus* or GBS at levels considered significant for a UTI diagnosis ($>100,000$ CFU/ml). Further studies are needed to define 1) the relationship between Gram-positive bacteria in midstream and catheter-collected urine when patients meet the threshold for UTI (10^5 CFU/ml), 2) the likelihood of cystitis when these organisms are detected in pure culture versus in the context of multiple other colony types, and 3) whether these relationships are similar across all patient groups. We caution against dismissing Gram-positive pathogens as unimportant. For example, while *S. saprophyticus* is now established as the predominant Gram-positive uropathogen, it was originally considered to be a urinary contaminant (310, 311).

There are multiple cited cases of polymicrobial-bloodstream infection with identical organisms present in both urine and blood cultures [for examples see (312–314)]. In these cases, polymicrobial-bloodstream infection supports the interpretation of polymicrobial UTI, especially when multiple of the organisms identified in the blood are also identified in urine. Moreover, since cases of disseminated polymicrobial infection occur most often in compromised persons with underlying risk factors for UTI, the organisms involved are more likely to be members of the 'normal' microbiota that may be overlooked in urine specimens as non-pathogens. In another related case, a woman received a suprapubic catheter following urogynecological surgery and later developed an abscess near the location of bladder-catheter insertion. Although the hospital microbiology lab returned the result of this urine culture as "contaminated," identification of bacteria from the anaerobically cultured surgically drained pus demonstrated a polymicrobial infection with *E. coli, G. vaginalis,* and *Peptostreptococcus productus* (283).

It has been estimated that up to 20% of women presenting with classic symptoms of UTI have culture-negative urine (315). In the cited example, "culture-negative" included samples without a single dominant species (i.e., polymicrobial growth), samples with "secondary pathogens" at titers <10^4/ml urine, or "doubtful pathogens" at titers <10^5 CFU/ml urine. Unfortunately, the term sterile is often applied incorrectly to these "culture-negative" contexts. For example, Domann et al. found that 9.2% of urine specimens collected from renal-transplant recipients did not produce significant growth under aerobic conditions, but had evidence of intact bacterial rods and/or cocci that were identified as fastidious anaerobes using culture-independent molecular approaches (316). Alternatively, what might appear to be a monomicrobial infection when observed by aerobic culture may actually be a polymicrobial infection when characterized by culture-independent techniques. For example, one recent study detailed a case study where a woman had what appeared to be a typical culture-positive UTI with >10^5 CFU/ml of monomicrobial *E. coli* according to aerobic clinical lab results. The culture-independent approach performed in parallel revealed that urine specimens collected by catheterization and suprapubic aspiration

from this patient contained *Actinobaculum* and *Aerococcus* at levels that far exceeded *E. coli* (317). Other studies that have examined the incidence and significance of *Actinobaculum* and *Aerococcus* in urine specimens have shown that up to 90% and 69% of the time, respectively, significant titers of these organisms were found alongside significant titers of typical uropathogens such as *E. coli* (230, 268).

Another recent study performed anaerobic culture for fastidious anaerobes in urine of women who suffer from recurrent chronic episodes of cystitis and compared these results to similarly aged groups of young women without UTI (225). The study reports that women with recurrent cystitis were more likely to contain higher titers and a more diverse repertoire of mixed anaerobes in their urine than healthy women without UTI. In fact, this study had two healthy-control groups: one with women who reported regular sexual activity and another with women who reported no history of sexual intercourse. Results of the study strongly suggest that sexual activity does not provide a simple explanation for why women with recurrent chronic cystitis have fastidious bacteria in their urine. Taken together, these findings suggest that fastidious organisms and species assumed to represent contamination of urine specimens in urine cultures are routinely overlooked. We argue that additional studies specifically evaluating contested potential uropathogens in specific patient settings, such as those who suffer from recurrent UTI, those at risk for complicated UTI, and those with other urologic problems of uncertain etiology (e.g., interstitial cystitis, urinary-urgency incontinence (UUI), preeclampsia, etc.) are still needed.

"UNCULTIVATED" BACTERIAL INHABITANTS OF THE URINARY TRACT

Clinical microbiology labs employ various combinations of culture-dependent approaches in the analysis of urine specimens. These approaches include the use of selective and nonselective agar media for enumeration and identification of bacteria under *aerobic* conditions. Incubation of agar plates is typically performed at 35–37°C for 18–24 hours. Some labs use additional approaches to cast a slightly wider net to avoid false-negative results, often depending on clinical findings or other factors, extending aerobic incubation to 2–3 days, incubating plates in the presence of 5% CO_2, or performing Giemsa- or Gram-staining of urine specimens. In the last few years, culture-independent approaches have evaluated whether the urinary tract contains microbes that are not cultivated under typical aerobic conditions used for urine culture. Table 1 summarizes the results of the urinary-tract microbiome studies that have been performed to date, with an emphasis on Gram-positive organisms observed using culture-independent approaches (see right side of Table, 'culture-independent' detection).

In one study, Domann et al. compared the use of traditional urine culture to the use of a culture-independent denaturing high-performance liquid chromatography (DHPLC) to separate and identify PCR-amplified 16S rRNA sequences from genomic DNA purified from midstream-urine specimens (316) from renal-transplant recipients, a group at high risk for UTI. Interestingly, while the authors report 100% correspondence between culture-based method and detection of the corresponding *aerobic* bacteria in DHPLC, they also demonstrate that 9.2% of all patients examined were 'culture-negative' using standard approaches, yet appeared to contain significant numbers of fastidious bacteria detected using the DHPLC approach. Most of the culture-negative, DHPLC-positive specimens were polymicrobial, and many of the genera detected in these urines have been previously associated with BV (see above). Specifically, *Gardnerella vaginalis*, *Anaerococcus lactolyticus*, and *Lactobacillus iners* were among the Gram-positive bacteria detected in urines of renal-transplant recipients and *Prevotella, Bacteroides, Dialister, Lepto-*

trichia, and *Fusobacteria* were among the Gram-negative bacteria detected. Many of these urines were positive for leukocyte-esterase activity, strongly suggesting that the organisms present were the cause of symptomatic UTI as opposed to the possibility of urine contamination by vaginal bacteria. The authors state that standard curves of bacteria in pure culture were used to estimate CFU in urine using the DHPLC method, and that in many cases bacteria were in clinically relevant ranges of $>5 \times 10^4$ CFU/ml. Similar studies using midstream urine to assess bacterial diversity using 16S RNA sequencing have also implied that a variety of vaginal bacteria may inhabit the female urinary tract (318, 319). One limitation of these studies is that the collection method (midstream "clean-catch") does not adequately address the possibility of urine contamination with vaginal or peri-urethral microbes.

One line of evidence that microbes detected by culture-independent approaches are not simply a product of urine contamination by vaginal bacteria comes from a study in which bacteria recovered from catheters were identified by culture-dependent and -independent approaches (320). As with previous studies, the authors demonstrate the presence of many bacteria that are currently regarded as vaginal bacteria (see Table 1). As with previous studies, examples of patients were provided in whom culture-independent approaches detected high levels of bacteria such as *Actinobaculum* that were not amenable to culture using routine methods. The authors concluded that catheterized patients who had a UTI were more likely to have lower diversity of organisms recovered from catheters compared to patients who did not have an UTI. However, it is not clear from the methods at what time point the 16S data contributing to this interpretation were collected (e.g., during UTI or prior to UTI). Thus, it is not clear if higher diversity of the urinary-tract microbiota may protect against UTI or if the presence of an UTI drowns out the signals of other bacteria.

In an attempt to demonstrate beyond reasonable doubt that 'vaginal bacteria' in the urinary tract were in fact located in the bladder, a recent study by Wolfe et al. compared the bacterial composition of urine from asymptomatic adult women collected by various methods including midstream 'clean-catch,' transurethral catheterization, and direct suprapubic aspiration from the bladder (317). Suprapubic aspiration eliminates concerns of vaginal or periurethral contamination. All urines used in the study were culture-negative using the clinical microbiology laboratory-recommended cutoff of 10^4 CFU/ml. The authors showed that a subpopulation of urines collected by transurethral catheterization did not yield significant growth under typical clinical microbiology laboratory conditions (for urine), but contained Gram-positive and Gram-negative bacteria that were evident by direct microscopy and 16S rRNA amplification and sequencing. Surprisingly, bacterial genomic DNA could be amplified from 21/23 urine samples collected by direct-needle aspiration from the bladder. Organisms present in these specimens could often be observed in corresponding catheter and midstream specimens collected at the same time, but rarely from 'mock needle-stick' samples. Bacteria detected in urine specimens obtained by catheterization or aspiration included the anaerobic genera *Actinobaculum*, *Anaerococcus*, *Atopobium*, *Gardnerella*, *Prevotella*, *Sneathia*, and *Veillonella*, among others (see Table 1). These organisms are all fastidious anaerobes that would not be detected using typical aerobic urine culture and were not detected in mock needle-sticks that did not enter the bladder. In addition to these somewhat unexpected organisms, more typical potential uropathogens were also revealed by this study, including species of *Streptococcus* and *Staphylococcus*.

These findings are supported by a few much earlier studies. For example, in a 1979 study, 185 pregnant women (admitted for pregnancy complications) underwent suprapubic-bladder aspiration and their urine was subjected to anaerobic culture. A total of 6.4% of these women had anaerobes in their urine, including members of the genera *Peptostreptococcus*, *Veillonella*, *Clostridium*, and *Bifidobacterium* (321). These findings provide clear evidence that urine taken directly from the bladder, and thus not exposed to potential contaminants of the vagina, periurethral area, or distal urethra, often contains organisms that are undetectable using routine clinical-microbiology procedures.

Independent investigations have also evaluated the microbial content of *male* urine or urethral swabs using culture-independent approaches. These studies reveal strikingly similar patterns of fastidious and anaerobic bacteria seen in female-urogenital specimens, including more typical Gram-positive uropathogens (e.g., *Staphylococcus*, *Enterococcus*, *Streptococcus*), as well as other bacteria previously considered to be specific to the female-urogenital tract (e.g., *Gardnerella*, *Atopobium*, *Actinobaculum*, *Anaerococcus*, *Lactobacillus*, etc., see Table 1) (322, 323). Another recent study investigated the microbiota of urine and coronal sulcus specimens from adolescent males, showing that bacteria associated with BV are present in young men that report having *no history* of partnered sexual activity (324). These data provide further evidence that specific bacterial species not cultivated using routine approaches are in fact present in the genitourinary tracts of both men and women and are not necessarily acquired through sexual contact. Note that some of these species have also been reported in case studies of urinary-tract infection, most often in populations with underlying risk factors for UTI (see Table 1 references).

One of the major limitations of culture-independent urinary-microbiome studies is that it is not possible to know whether the bacteria are alive or dead. A skeptic might ask, "is the presence of microbial DNA in urine, even if taken by needle aspiration from the bladder, indicative of a live microbial community?" Teams led by Schreckenberger, Wolfe, and Brubaker recently used an expanded-quantitative urine culture (EQUC) method to address this important question (292, 325). These authors demonstrated previously that urine collected from women by transurethral catheterization and by suprapubic aspiration had very similar microbiome profiles. Thus, subsequent studies used catheter-based collection of urine because this was the least-invasive method that still provided confidence that the urine was not contaminated with vaginal bacteria. The authors performed 16S rDNA-based community profiling in parallel with EQUC and demonstrated that many of the fastidious bacteria detected in catheterized urine by culture-independent approaches can in fact be recovered in laboratory culture (292, 325). Whether these bacteria represent stable communities living in the bladder still remains to be fully characterized.

This same group most recently performed community profiling and EQUC on catheter-collected urine samples to test the hypothesis that the urinary microbiome may contribute to UUI (292). Overall, this paper sets an impressive standard for the field, providing a truly multidisciplinary approach to provide evidence that challenges previous assumptions regarding a condition (UUI) of uncertain etiology that has nevertheless been defined as noninfectious. The authors conclude that the urinary microbiomes of women with UUI were in fact different from women in the control group, having lesser proportions of *Lactobacilli* and greater proportions of *Gardnerella*, *Actinobaculum*, *Actinomyces*, *Aerococcus*, *Arthrobacter*, *Corynebacterium*, *Oligella*, *Staphylococcus*, and *Streptococcus*. This paper also reports that women in the UUI group were more likely to harbor *Lactobacillus gasseri* in their bladders compared to women in the control group, which were more likely to harbor *Lactobacillus crispatus*. In addition to the culture-independent component of this study, the authors also performed EQUC to demonstrate that many of the bacteria detected using molecular approaches could also be cultured in the laboratory. One potential limitation of this study is that the UUI group was found to be ~2-fold less likely to be using estrogen, despite being on average 14 years older than the control group (63 [±12] versus 49 [±14]; P <0.05). Differences in age and hormonal status have a wide variety of physiological implications and have been correlated with specific phenotypic features of the vaginal microbiota (e.g., presence or amounts of various bacteria) (326, 327). These factors could therefore be important confounders that may contribute to the differences observed between the cases and controls.

Another study from the same groups used the same population of women suffering from UUI who were randomized to two treatment arms of a large clinical trial, but instead employed quantitative PCR (qPCR) targeting the 16S ribosomal DNA (rDNA) to assign women simply as positive or negative for bacteria presence. Only women who were negative for UTI by the clinical (culture-based) definition at the time of urine-sample retrieval, prior to randomization, were included in the qPCR study. The authors report no statistically significant differences in median qPCR levels depending on whether women went on to have a post procedure UTI (UTI 2.58×10^5 vs no UTI 1.35×10^5 copies/ml) or whether treatment for UUI was successful. Rather, using a categorical variable (presence/absence), the authors demonstrate that women positive for bacterial DNA in catheterized urine suffered a higher number of UUI episodes (5.71 [±2.60] vs 4.72 [±2.86]). However, qPCR-positive women were less likely to suffer from post-procedure UTI compared to qPCR-negative women (10% vs 24%, respectively).

One of the limitations of the urinary-tract microbiome studies performed to date is that the abundance and composition of microbes in the urinary tract is relative and based on the total number of sequence reads, which can vary widely from individual to individual. Going forward, one of the challenges will be to measure absolute abundances of specific uncultivated species in larger epidemiological studies, and to evaluate the potential risks posed by urinary-tract colonization with different bacterial species or subsets. Experimental models in animals should also be employed to help to directly examine the hypothesis that some of these Gram-positive anaerobes are pathogenic in the urinary tract. Such models will also be required to test specific hypotheses related to the potential roles of urogenital bacteria in determining host susceptibility to UTI caused by common uropathogens.

Many of the fastidious anaerobes described in urine-microbiome studies (see Table 1) are taxa that have also been associated with bacterial vaginosis. Clinical studies demonstrating a higher risk of UTI in women with BV further suggest that some of these bacteria may simultaneously cause dysbiosis in the vagina and the bladder. However, contamination of urine with vaginal bacteria is not a viable explanation for these findings given that the bacteria were also detected in urines collected by catheterization and suprapubic aspiration (see Table 1). As described above, the presence of clue cells (i.e., squamous-epithelial cells covered in *G. vaginalis* and possibly other bacteria) is characteristic of women with BV (222). However, consistent with the idea of urinary-tract dysbiosis, women with "sterile pyelonephritis" who were identified as having a high infectious bladder burden of *G. vaginalis* also had clue cells in urine directly aspirated from the bladder (291). In fact, clue cells have also been reported in urethral and semen specimens from men who were identified as having *G. vaginalis*-associated infections (307–309). These findings support the interpretation that *G. vaginalis* may attach itself to epithelial cells of the urethra and bladder and induce exfoliation, thus resulting in urinary clue cells that much resemble the vaginal clue cells seen in women with BV.

Experimental studies are needed to determine whether these 'BV bacteria' have a direct causal role in determining host susceptibility to acute or recurrent UTI. Findings that women suffering from recurrent UTI have fewer recurrences following vaginal probiotic *L. crispatus* administration suggest that vaginal lactobacilli may have a protective role in UTI, or that they may displace other organisms that have a detrimental role. This interpretation is also indirectly supported by observations that *Lactobacillus* is the most common dominant organism in the urinary tract (among older women) and that having qPCR-positive urine was associated with a reduced likelihood of UTI (see above). Finally, the fact that these bacteria can also be found in the male urinary tract, even among adolescent males with no sexual history, strongly suggests that these 'BV bacteria' are not simply vaginal bacteria transiently exposed to the urinary tract during sexual activities, but rather that these organisms may have a wider tissue tropism that includes the urinary tract.

SUMMARY

In summary, we have 1) reviewed the epidemiology and virulence characteristics of the most-common Gram-positive uropathogens, including *Staphylococcus saprophyticus*, *Enterococcus faecalis*, and group B *Streptococcus*; 2) reviewed the natural history of several less-common uropathogens and the reasons why these bacteria are systematically overlooked; 3) summarized recent literature suggesting that a wide variety of Gram-positive bacteria are common inhabitants of the human urinary tract; and 4) presented evidence that polymicrobial infections involving Gram-positive bacteria may be more prevalent than previously appreciated, and also may impact pathologic outcomes in the urinary tract.

Fifty years ago it was thought that the major uropathogenic organisms did not include Gram-positive bacteria. Ten years ago it was assumed that the uninfected urinary tract was sterile. Currently, it is widely held that Gram-positive bacteria either alone or alongside Gram-negative uropathogens in the urine are likely to be contaminants of no consequence. The first two assumptions have been disproved as technology has advanced. The literature summarized in this chapter calls into question the last assumption. We propose that future studies will illuminate a previously unappreciated role for members of the polymicrobial microbiota in the urinary tract, vagina, and gut in UTI susceptibility and disease progression.

ACKNOWLEDGMENTS

K.A.K. is supported by the National Research Foundation (NRF) and Ministry of Education Singapore (MOE) under its Research Centre of Excellence programme, by the NRF under its Singapore NRF Fellowship programme (NRF2010NRF-NRFF001-226), and by the Singapore MOE under its Tier 2 funding scheme (MOE2014-T2-1-129). A.L.L. is supported by NIH grants R01 AI114635, R21 DK092586, and P50 DK064540.

CITATION

Kline KA, Lewis AL. 2016. Gram-positive uropathogens, polymicrobial urinary tract infection, and the emerging microbiota of the urinary tract. Microbiol Spectrum 4(2): UTI-0012-2012.

REFERENCES

1. **Hooton TM.** 2012. Clinical practice. Uncomplicated urinary tract infection. *N Engl J Med* **366:**1028–1037.
2. **Schroeder AR, Chang PW, Shen MW, Biondi EA, Greenhow TL.** 2015. Diagnostic accuracy of the urinalysis for urinary tract infection in infants <3 months of age. *Pediatrics* **135:**965–971.
3. **Clyne M.** 2014. Paediatrics: dipstick adequate for febrile UTI test. *Nat Rev Urol* **11:**304.
4. **Demilie T, Beyene G, Melaku S, Tsegaye W.** 2014. Diagnostic accuracy of rapid urine dipstick test to predict urinary tract infection among pregnant women in Felege Hiwot Referral Hospital, Bahir Dar, North West Ethiopia. *BMC Res Notes* **7:**481.
5. **Jido TA.** 2014. Urinary tract infections in pregnancy: evaluation of diagnostic –framework. *Saudi J Kidney Dis Transpl* **25:**85–90.
6. **Mehnert-Kay SA.** 2005. Diagnosis and management of uncomplicated urinary tract infections. *Am Fam Physician* **72:**451–456.
7. **Al Majid F, Buba F.** 2010. The predictive and discriminant values of urine nitrites in urinary tract infection. *Biomed Res* **21:**297–299.
8. **Arinzon Z, Peisakh A, Shuval I, Shabat S, Berner YN.** 2009. Detection of urinary tract infection (UTI) in long-term care setting: is the multireagent strip an adequate diagnostic tool? *Arch Gerontol Geriatr* **48:**227–231.
9. **Wagenlehner FM, Naber KG.** 2006. Current challenges in the treatment of complicated urinary tract infections and prostatitis. *Clin Microbiol Infect* **12**(Suppl 3):67–80.
10. **Edwards MS, Baker CJ.** 2005. Group B streptococcal infections in elderly adults. *Clin Infect Dis* **41:**839–847.
11. **Matthews SJ, Lancaster JW.** 2011. Urinary tract infections in the elderly population. *Am J Geriatr Pharmacother* **9:**286–309.
12. **Berg RC, Underland V, Odgaard-Jensen J, Fretheim A, Vist GE.** 2014. Effects of female genital cutting on physical health outcomes: A systematic review and meta-analysis. *BMJ Open* **4:**e006316. doi:10.1136/bmjopen-2014-006316
13. **Maki DG, Tambyah PA.** 2001. Engineering out the risk for infection with urinary catheters. *Emerg Infect Dis* **7:**342–347.
14. **Jain P, Parada JP, David A, Smith LG.** 1995. Overuse of the indwelling urinary tract catheter in hospitalized medical patients. *Arch Intern Med* **155:**1425–1429.
15. **Tambyah PA, Maki DG.** 2000. Catheter-associated urinary tract infection is rarely symptomatic: A prospective study of 1,497 catheterized patients. *Arch Intern Med* **160:**678–682.
16. **Warren JW.** 1997. Catheter-associated urinary tract infections. *Infect Dis Clin North Am* **11:**609–622.
17. **Stensballe J, Tvede M, Looms D, Lippert FK, Dahl B, Tonnesen E, Rasmussen LS.** 2007. Infection risk with nitrofurazone-impregnated urinary catheters in trauma patients: A randomized trial. *Ann Intern Med* **147:**285–293.
18. **Donlan RM, Costerton JW.** 2002. Biofilms: survival mechanisms of clinically relevant microorganisms. *Clin Microbiol Rev* **15:**167–193.

19. **Lewis K.** 2008. Multidrug tolerance of biofilms and persister cells. *Curr Top Microbiol Immunol* **322:**107–131.

20. **Dedeić-Ljubović A, Hukić M.** 2009. Catheter-related urinary tract infection in patients suffering from spinal cord injuries. *Bosn J Basic Med Sci* **9:**2–9.

21. **Desai PJ, Pandit D, Mathur M, Gogate A.** 2001. Prevalence, identification and distribution of various species of enterococci isolated from clinical specimens with special reference to urinary tract infection in catheterized patients. *Indian J Med Microbiol* **19:**132–137.

22. **Johnson AP, Warner M, Speller DC.** 1997. *In-vitro* activity of quinupristin/dalfopristin (Synercid) against isolates of *Streptococcus pneumoniae, Staphylococcus aureus* and *Enterococcus* spp. *J Antimicrob Chemother* **40:**604–605.

23. **Gatermann S, John J, Marre R.** 1989. *Staphylococcus saprophyticus* urease: characterization and contribution to uropathogenicity in unobstructed urinary tract infection of rats. *Infect Immun* **57:**110–116.

24. **Kline KA, Ingersoll MA, Nielsen HV, Sakinc T, Henriques-Normark B, Gatermann S, Caparon MG, Hultgren SJ.** 2010. Characterization of a novel murine model of *Staphylococcus saprophyticus* urinary tract infection reveals roles for Ssp and SdrI in virulence. *Infect Immun* **78:**1943–1951.

25. **Kline KA, Schwartz DJ, Lewis WG, Hultgren SJ, Lewis AL.** 2011. Immune activation and suppression by group B streptococcus in a murine model of urinary tract infection. *Infect Immun* **79:**3588–3595.

26. **Kau AL, Martin SM, Lyon W, Hayes E, Caparon MG, Hultgren SJ.** 2005. *Enterococcus faecalis* tropism for the kidneys in the urinary tract of C57BL/6J mice. *Infect Immun* **73:**2461–2468.

27. **Singh KV, Nallapareddy SR, Murray BE.** 2007. Importance of the ebp (endocarditis- and biofilm-associated pilus) locus in the pathogenesis of *Enterococcus faecalis* ascending urinary tract infection. *J Infect Dis* **195:**1671–1677.

28. **Torelli R, Serror P, Bugli F, Paroni Sterbini F, Florio AR, Stringaro A, Colone M, De Carolis E, Martini C, Giard JC, Sanguinetti M, Posteraro B.** 2012. The PavA-like fibronectin-binding protein of *Enterococcus faecalis*, EfbA, is important for virulence in a mouse model of ascending urinary tract infection. *J Infect Dis* **206:**952–960.

29. **Kim HY, Choe HS, Lee DS, Yoo JM, Lee SJ.** 2015. A novel rat model of catheter-associated urinary tract infection. *Int Urol Nephrol* **47:**1259–1263.

30. **Guiton PS, Hung CS, Hancock LE, Caparon MG, Hultgren SJ.** 2010. Enterococcal biofilm formation and virulence in an optimized murine model of foreign body-associated urinary tract infections. *Infect Immun* **78:**4166–4175.

31. **Guiton PS, Hannan TJ, Ford B, Caparon MG, Hultgren SJ.** 2013. *Enterococcus faecalis* overcomes foreign body-mediated inflammation to establish urinary tract infections. *Infect Immun* **81:**329–339.

32. **Guiton PS, Cusumano CK, Kline KA, Dodson KW, Han Z, Janetka JW, Henderson JP, Caparon MG, Hultgren SJ.** 2012. Combinatorial small-molecule therapy prevents uropathogenic *Escherichia coli* catheter-associated urinary tract infections in mice. *Antimicrob Agents Chemother* **56:**4738–4745.

33. **Hjelm E, Lundell-Etherden I, Mårdh PA.** 1987. Ascending urinary tract infections in rats induced by *Staphylococcus saprophyticus* and *Proteus mirabilis. Acta Pathol Microbiol Immunol Scand B* **95:**347–350.

34. **Alteri CJ, Himpsl SD, Mobley HL.** 2015. Preferential use of central metabolism *in vivo* reveals a nutritional basis for polymicrobial infection. *PLoS Pathog* **11:**e1004601. doi:10.1371/journal.ppat.1004601

35. **Armbruster CE, Smith SN, Yep A, Mobley HL.** 2014. Increased incidence of urolithiasis and bacteremia during *Proteus mirabilis* and *Providencia stuartii* coinfection due to synergistic induction of urease activity. *J Infect Dis* **209:**1524–1532.

36. **Kunin CM.** 1989. Blockage of urinary catheters: role of microorganisms and constituents of the urine on formation of encrustations. *J Clin Epidemiol* **42:**835–842.

37. **Mobley HL, Warren JW.** 1987. Urease-positive bacteriuria and obstruction of long-term urinary catheters. *J Clin Microbiol* **25:**2216–2217.

38. **Tsuchimori N, Hayashi R, Shino A, Yamazaki T, Okonogi K.** 1994. *Enterococcus faecalis* aggravates pyelonephritis caused by *Pseudomonas aeruginosa* in experimental ascending mixed urinary tract infection in mice. *Infect Immun* **62:**4534–4541.

39. **Kline KA, Schwartz DJ, Gilbert NM, Hultgren SJ, Lewis AL.** 2012. Immune modulation by group B *Streptococcus* influences host susceptibility to urinary tract infection by uropathogenic *Escherichia coli. Infect Immun* **80:**4186–4194.

40. **Kline KA, Schwartz DJ, Gilbert NM, Lewis AL.** 2014. Impact of host age and parity on susceptibility to severe urinary tract infection

in a murine model. *PLoS ONE* **9**:e97798. doi:10.1371/journal.pone.0097798

41. **Barford JM, Anson K, Hu Y, Coates AR.** 2008. A model of catheter-associated urinary tract infection initiated by bacterial contamination of the catheter tip. *BJU Int* **102**:67–74.

42. **Matsukawa M, Kunishima Y, Takahashi S, Takeyama K, Tsukamoto T.** 2005. Bacterial colonization on intraluminal surface of urethral catheter. *Urology* **65**:440–444.

43. **Nickel JC, Downey JA, Costerton JW.** 1989. Ultrastructural study of microbiologic colonization of urinary catheters. *Urology* **34**:284–291.

44. **Macleod SM, Stickler DJ.** 2007. Species interactions in mixed-community crystalline biofilms on urinary catheters. *J Med Microbiol* **56**:1549–1557.

45. **Frank DN, Wilson SS, St Amand AL, Pace NR.** 2009. Culture-independent microbiological analysis of foley urinary catheter biofilms. *PLoS ONE* **4**:e7811. doi:10.1371/journal.pone.0007811

46. **Nicolle LE, Hoban SA, Harding GK.** 1983. Characterization of coagulase-negative staphylococci from urinary tract specimens. *J Clin Microbiol* **17**:267–271.

47. **Hooton TM, Stamm WE.** 1997. Diagnosis and treatment of uncomplicated urinary tract infection. *Infect Dis Clin North Am* **11**:551–581.

48. **Wallmark G, Arremark I, Telander B.** 1978. *Staphylococcus saprophyticus*: A frequent cause of acute urinary tract infection among female outpatients. *J Infect Dis* **138**:791–797.

49. **Hovelius B, Mårdh PA.** 1984. *Staphylococcus saprophyticus* as a common cause of urinary tract infections. *Rev Infect Dis* **6**:328–337.

50. **Zong Z, Peng C, Lu X.** 2011. Diversity of SCCmec elements in methicillin-resistant coagulase-negative staphylococci clinical isolates. *PLoS ONE* **6**:e20191. doi:10.1371/journal.pone.0020191

51. **Gillespie WA, Sellin MA, Gill P, Stephens M, Tuckwell LA, Hilton AL.** 1978. Urinary tract infection in young women, with special reference to *Staphylococcus saprophyticus*. *J Clin Pathol* **31**:348–350.

52. **Rupp ME, Soper DE, Archer GL.** 1992. Colonization of the female genital tract with *Staphylococcus saprophyticus*. *J Clin Microbiol* **30**:2975–2979.

53. **Colodner R, Ken-Dror S, Kavenshtock B, Chazan B, Raz R.** 2006. Epidemiology and clinical characteristics of patients with *Staphylococcus saprophyticus* bacteriuria in Israel. *Infection* **34**:278–281.

54. **Ferry S, Burman LG, Mattsson B.** 1987. Urinary tract infection in primary health care in northern Sweden. I. Epidemiology. *Scand J Prim Health Care* **5**:123–128.

55. **Hovelius B, Thelin I, Mårdh PA.** 1979. *Staphylococcus saprophyticus* in the aetiology of nongonococcal urethritis. *Br J Vener Dis* **55**:369–374.

56. **Carson CC, McGraw VD, Zwadyk P.** 1982. Bacterial prostatitis caused by *Staphylococcus saprophyticus*. *Urology* **19**:576–578.

57. **Kauffman CA, Hertz CS, Sheagren JN.** 1983. *Staphylococcus saprophyticus*: role in urinary tract infections in men. *J Urol* **130**:493–494.

58. **Hovelius B, Colleen S, Mardh PA.** 1984. Urinary tract infections in men caused by *Staphylococcus saprophyticus*. *Scand J Infect Dis* **16**:37–41.

59. **Jellheden B, Norrby RS, Sandberg T.** 1996. Symptomatic urinary tract infection in women in primary health care. Bacteriological, clinical and diagnostic aspects in relation to host response to infection. *Scand J Prim Health Care* **14**:122–128.

60. **Rupp ME, Archer GL.** 1994. Coagulase-negative staphylococci: pathogens associated with medical progress. *Clin Infect Dis* **19**:231–243; quiz 244–245.

61. **Latham RH, Running K, Stamm WE.** 1983. Urinary tract infections in young adult women caused by *Staphylococcus saprophyticus*. *JAMA* **250**:3063–3066.

62. **Kahlmeter G, ECO.SENS.** 2003. An international survey of the antimicrobial susceptibility of pathogens from uncomplicated urinary tract infections: The ECO.SENS Project. *J Antimicrob Chemother* **51**:69–76.

63. **Diekema DJ, Pfaller MA, Schmitz FJ, Smayevsky J, Bell J, Jones RN, Beach M; SENTRY Participants Group.** 2001. Survey of infections due to *Staphylococcus* species: frequency of occurrence and antimicrobial susceptibility of isolates collected in the United States, Canada, Latin America, Europe, and the Western Pacific region for the SENTRY Antimicrobial Surveillance Program, 1997–1999. *Clin Infect Dis* **32**(Suppl 2):S114–132.

64. **Higashide M, Kuroda M, Ohkawa S, Ohta T.** 2006. Evaluation of a cefoxitin disk diffusion test for the detection of mecA-positive methicillin-resistant *Staphylococcus saprophyticus*. *Int J Antimicrob Agents* **27**:500–504.

65. **Hartman BJ, Tomasz A.** 1984. Low-affinity penicillin-binding protein associated with beta-lactam resistance in *Staphylococcus aureus*. *J Bacteriol* **158**:513–516.

66. **Matsuhashi M, Song MD, Ishino F, Wachi M, Doi M, Inoue M, Ubukata K, Yamashita N, Konno M.** 1986. Molecular cloning of the gene

of a penicillin-binding protein supposed to cause high resistance to beta-lactam antibiotics in *Staphylococcus aureus*. *J Bacteriol* **167**:975–980.

67. **Katayama Y, Ito T, Hiramatsu K.** 2000. A new class of genetic element, staphylococcus cassette chromosome mec, encodes methicillin resistance in *Staphylococcus aureus*. *Antimicrob Agents Chemother* **44**:1549–1555.

68. **Higashide M, Kuroda M, Omura CT, Kumano M, Ohkawa S, Ichimura S, Ohta T.** 2008. Methicillin-resistant *Staphylococcus saprophyticus* isolates carrying staphylococcal cassette chromosome mec have emerged in urogenital tract infections. *Antimicrob Agents Chemother* **52**:2061–2068.

69. **McTaggart LA, Rigby RC, Elliott TS.** 1990. The pathogenesis of urinary tract infections associated with *Escherichia coli, Staphylococcus saprophyticus* and *S. epidermidis*. *J Med Microbiol* **32**:135–141.

70. **Meyer HG, Wengler-Becker U, Gatermann SG.** 1996. The hemagglutinin of *Staphylococcus saprophyticus* is a major adhesin for uroepithelial cells. *Infect Immun* **64**:3893–3896.

71. **Gatermann S, Meyer HG.** 1994. *Staphylococcus saprophyticus* hemagglutinin binds fibronectin. *Infect Immun* **62**:4556–4563.

72. **Hell W, Meyer HG, Gatermann SG.** 1998. Cloning of aas, a gene encoding a *Staphylococcus saprophyticus* surface protein with adhesive and autolytic properties. *Mol Microbiol* **29**:871–881.

73. **Gatermann S, Marre R, Heesemann J, Henkel W.** 1988. Hemagglutinating and adherence properties of *Staphylococcus saprophyticus*: epidemiology and virulence in experimental urinary tract infection of rats. *FEMS Microbiol Immunol* **1**:179–185.

74. **Kleine B, Gatermann S, Sakinc T.** 2010. Genotypic and phenotypic variation among *Staphylococcus saprophyticus* from human and animal isolates. *BMC Res Notes* **3**:163.

75. **Gatermann S, Kreft B, Marre R, Wanner G.** 1992. Identification and characterization of a surface-associated protein (Ssp) of *Staphylococcus saprophyticus*. *Infect Immun* **60**:1055–1060.

76. **Sakinç T, Woznowski M, Ebsen M, Gatermann SG.** 2005. The surface-associated protein of *Staphylococcus saprophyticus* is a lipase. *Infect Immun* **73**:6419–6428.

77. **Schneewind O, Missiakas DM.** 2012. Protein secretion and surface display in Gram-positive bacteria. *Philos Trans R Soc Lond B Biol Sci* **367**:1123–1139.

78. **Kuroda M, Yamashita A, Hirakawa H, Kumano M, Morikawa K, Higashide M, Maruyama A, Inose Y, Matoba K, Toh H, Kuhara S, Hattori M, Ohta T.** 2005. Whole genome sequence of *Staphylococcus saprophyticus* reveals the pathogenesis of uncomplicated urinary tract infection. *Proc Natl Acad Sci U S A* **102**:13272–13277.

79. **King NP, Beatson SA, Totsika M, Ulett GC, Alm RA, Manning PA, Schembri MA.** 2011. UafB is a serine-rich repeat adhesin of *Staphylococcus saprophyticus* that mediates binding to fibronectin, fibrinogen and human uroepithelial cells. *Microbiology* **157**:1161–1175.

80. **King NP, Sakinç T, Ben Zakour NL, Totsika M, Heras B, Simerska P, Shepherd M, Gatermann SG, Beatson SA, Schembri MA.** 2012. Characterisation of a cell wall-anchored protein of *Staphylococcus saprophyticus* associated with linoleic acid resistance. *BMC Microbiol* **12**:8.

81. **Sakinç T, Kleine B, Gatermann SG.** 2006. SdrI, a serine-aspartate repeat protein identified in *Staphylococcus saprophyticus* strain 7108, is a collagen-binding protein. *Infect Immun* **74**:4615–4623.

82. **Bensing BA, Gibson BW, Sullam PM.** 2004. The *Streptococcus gordonii* platelet binding protein GspB undergoes glycosylation independently of export. *J Bacteriol* **186**:638–645.

83. **Bensing BA, López JA, Sullam PM.** 2004. The *Streptococcus gordonii* surface proteins GspB and Hsa mediate binding to sialylated carbohydrate epitopes on the platelet membrane glycoprotein Ibalpha. *Infect Immun* **72**:6528–6537.

84. **Siboo IR, Chambers HF, Sullam PM.** 2005. Role of SraP, a serine-rich surface protein of *Staphylococcus aureus*, in binding to human platelets. *Infect Immun* **73**:2273–2280.

85. **Kim YG, Shaw MH, Warner N, Park JH, Chen F, Ogura Y, Núñez G.** 2011. Cutting edge: Crohn's disease-associated Nod2 mutation limits production of proinflammatory cytokines to protect the host from *Enterococcus faecalis*-induced lethality. *J Immunol* **187**:2849–2852.

86. **Rigel NW, Braunstein M.** 2008. A new twist on an old pathway--accessory Sec [corrected] systems. *Mol Microbiol* **69**:291–302.

87. **Kleine B, Ali L, Wobser D, Sakinç T.** 2015. The N-terminal repeat and the ligand binding domain A of SdrI protein is involved in hydrophobicity of *S. saprophyticus*. *Microbiol Res* **172**:88–94.

88. **von Eiff C, Peters G, Heilmann C.** 2002. Pathogenesis of infections due to coagulase-negative staphylococci. *Lancet Infect Dis* **2**:677–685.

89. **McCrea KW, Hartford O, Davis S, Eidhin DN, Lina G, Speziale P, Foster TJ, Höök M.** 2000. The serine-aspartate repeat (Sdr) protein family in *Staphylococcus epidermidi*s. *Microbiology* **146:**1535–1546.

90. **Fowler JE Jr.** 1985. *Staphylococcus saprophyticus* as the cause of infected urinary calculus. *Ann Intern Med* **102:**342–343.

91. **Sakinç T, Michalski N, Kleine B, Gatermann SG.** 2009. The uropathogenic species *Staphylococcus saprophyticus* tolerates a high concentration of D-serine. *FEMS Microbiol Lett* **299:**60–64.

92. **Korte-Berwanger M, Sakinç T, Kline K, Nielsen HV, Hultgren S, Gatermann SG.** 2013. Significance of the D-serine-deaminase and D-serine metabolism of *Staphylococcus saprophyticus* for virulence. *Infect Immun* **81:**4525–4533.

93. **Park S, Kelley KA, Vinogradov E, Solinga R, Weidenmaier C, Misawa Y, Lee JC.** 2010. Characterization of the structure and biological functions of a capsular polysaccharide produced by *Staphylococcus saprophyticus*. *J Bacteriol* **192:**4618–4626.

94. **Bahrani-Mougeot FK, Buckles EL, Lockatell CV, Hebel JR, Johnson DE, Tang CM, Donnenberg MS.** 2002. Type 1 fimbriae and extracellular polysaccharides are preeminent uropathogenic *Escherichia coli* virulence determinants in the murine urinary tract. *Mol Microbiol* **45:**1079–1093.

95. **Anderson GG, Goller CC, Justice S, Hultgren SJ, Seed PC.** 2010. Polysaccharide capsule and sialic acid-mediated regulation promote biofilm-like intracellular bacterial communities during cystitis. *Infect Immun* **78:**963–975.

96. **Wang R, Braughton KR, Kretschmer D, Bach TH, Queck SY, Li M, Kennedy AD, Dorward DW, Klebanoff SJ, Peschel A, DeLeo FR, Otto M.** 2007. Identification of novel cytolytic peptides as key virulence determinants for community-associated MRSA. *Nat Med* **13:**1510–1514.

97. **Kaito C, Saito Y, Nagano G, Ikuo M, Omae Y, Hanada Y, Han X, Kuwahara-Arai K, Hishinuma T, Baba T, Ito T, Hiramatsu K, Sekimizu K.** 2011. Transcription and translation products of the cytolysin gene psm-mec on the mobile genetic element SCCmec regulate *Staphylococcus aureus* virulence. *PLoS Pathog* **7:**e1001267. doi:10.1371/journal.ppat.1001267

98. **Diep BA, Otto M.** 2008. The role of virulence determinants in community-associated MRSA pathogenesis. *Trends Microbiol* **16:**361–369.

99. **Mehlin C, Headley CM, Klebanoff SJ.** 1999. An inflammatory polypeptide complex from *Staphylococcus epidermidis*: isolation and characterization. *J Exp Med* **189:**907–918.

100. **Cogen AL, Yamasaki K, Muto J, Sanchez KM, Crotty Alexander L, Tanios J, Lai Y, Kim JE, Nizet V, Gallo RL.** 2010. *Staphylococcus epidermidis* antimicrobial delta-toxin (phenol-soluble modulin-gamma) cooperates with host antimicrobial peptides to kill group A *Streptococcus*. *PLoS ONE* **5:**e8557. doi:10.1371/journal.pone.0008557

101. **Cogen AL, Yamasaki K, Sanchez KM, Dorschner RA, Lai Y, MacLeod DT, Torpey JW, Otto M, Nizet V, Kim JE, Gallo RL.** 2010. Selective antimicrobial action is provided by phenol-soluble modulins derived from *Staphylococcus epidermidis*, a normal resident of the skin. *J Invest Dermatol* **130:**192–200.

102. **Marchand A, Verdon J, Lacombe C, Crapart S, Héchard Y, Berjeaud JM.** 2011. Anti-*Legionella* activity of staphylococcal hemolytic peptides. *Peptides* **32:**845–851.

103. **Tsompanidou E, Sibbald MJ, Chlebowicz MA, Dreisbach A, Back JW, van Dijl JM, Buist G, Denham EL.** 2011. Requirement of the agr locus for colony spreading of *Staphylococcus aureus*. *J Bacteriol* **193:**1267–1272.

104. **Periasamy S, Joo HS, Duong AC, Bach TH, Tan VY, Chatterjee SS, Cheung GY, Otto M.** 2012. How *Staphylococcus aureus* biofilms develop their characteristic structure. *Proc Natl Acad Sci U S A* **109:**1281–1286.

105. **Schwartz K, Syed AK, Stephenson RE, Rickard AH, Boles BR.** 2012. Functional amyloids composed of phenol soluble modulins stabilize *Staphylococcus aureus* biofilms. *PLoS Pathog* **8:**e1002744. doi:10.1371/journal.ppat.1002744

106. **Queck SY, Khan BA, Wang R, Bach TH, Kretschmer D, Chen L, Kreiswirth BN, Peschel A, Deleo FR, Otto M.** 2009. Mobile genetic element-encoded cytolysin connects virulence to methicillin resistance in MRSA. *PLoS Pathog* **5:**e1000533. doi:10.1371/journal.ppat.1000533

107. **Monecke S, Engelmann I, Archambault M, Coleman DC, Coombs GW, Cortez de Jäckel S, Pelletier-Jacques G, Schwarz S, Shore AC, Slickers P, Ehricht R.** 2012. Distribution of SCCmec-associated phenol-soluble modulin in staphylococci. *Mol Cell Probes* **26:**99–103.

108. **Muder RR, Brennen C, Rihs JD, Wagener MM, Obman A, Stout JE, Yu VL.** 2006. Isolation of *Staphylococcus aureus* from the urinary tract: association of isolation with symptomatic urinary tract infection and subsequent staphylococcal bacteremia. *Clin Infect Dis* **42:**46–50.

109. **Baraboutis IG, Tsagalou EP, Lepinski JL, Papakonstantinou I, Papastamopoulos V,**

Skoutelis AT, Johnson S. 2010. Primary *Staphylococcus aureus* urinary tract infection: The role of undetected hematogenous seeding of the urinary tract. *Eur J Clin Microbiol Infect Dis* **29**:1095–1101.

110. Gilbert NM, O'Brien VP, Hultgren S, Macones G, Lewis WG, Lewis AL. 2013. Urinary tract infection as a preventable cause of pregnancy complications: opportunities, challenges, and a global call to action. *Glob Adv Health Med* **2**:59–69.

111. Hiron A, Posteraro B, Carrière M, Remy L, Delporte C, La Sorda M, Sanguinetti M, Juillard V, Borezée-Durant E. 2010. A nickel ABC-transporter of *Staphylococcus aureus* is involved in urinary tract infection. *Mol Microbiol* **77**:1246–1260.

112. Remy L, Carrière M, Derré-Bobillot A, Martini C, Sanguinetti M, Borezée-Durant E. 2013. The *Staphylococcus aureus* Opp1 ABC transporter imports nickel and cobalt in zinc-depleted conditions and contributes to virulence. *Mol Microbiol* **87**:730–743.

113. Otto M. 2012. Molecular basis of *Staphylococcus epidermidis* infections. *Semin Immunopathol* **34**:201–214.

114. Widerström M, Wiström J, Sjöstedt A, Monsen T. 2012. Coagulase-negative staphylococci: update on the molecular epidemiology and clinical presentation, with a focus on *Staphylococcus epidermidis* and *Staphylococcus saprophyticus*. *Eur J Clin Microbiol Infect Dis* **31**:7–20.

115. Hidron AI, Edwards JR, Patel J, Horan TC, Sievert DM, Pollock DA, Fridkin SK; National Healthcare Safety Network Team; Participating National Healthcare Safety Network Facilities. 2008. NHSN annual update: antimicrobial-resistant pathogens associated with healthcare-associated infections: annual summary of data reported to the National Healthcare Safety Network at the Centers for Disease Control and Prevention, 2006–2007. *Infect Control Hosp Epidemiol* **29**:996–1011.

116. Hedman P, Ringertz O. 1991. Urinary tract infections caused by *Staphylococcus saprophyticus*. A matched case control study. *J Infect* **23**:145–153.

117. Chen Y, Lu H, Liu Q, Huang G, Lim CP, Zhang L, Hao A, Cao X. 2012. Function of GRIM-19, a mitochondrial respiratory chain complex I protein, in innate immunity. *J Biol Chem* **287**:27227–27235.

118. Fisher K, Phillips C. 2009. The ecology, epidemiology and virulence of *Enterococcus*. *Microbiology* **155**:1749–1757.

119. Richards MJ, Edwards JR, Culver DH, Gaynes RP. 1999. Nosocomial infections in medical intensive care units in the United States. National Nosocomial Infections Surveillance System. *Crit Care Med* **27**:887–892.

120. Nicolle LE, Friesen D, Harding GK, Roos LL. 1996. Hospitalization for acute pyelonephritis in Manitoba, Canada, during the period from 1989 to 1992; impact of diabetes, pregnancy, and aboriginal origin. *Clin Infect Dis* **22**:1051–1056.

121. Boyko EJ, Fihn SD, Scholes D, Abraham L, Monsey B. 2005. Risk of urinary tract infection and asymptomatic bacteriuria among diabetic and nondiabetic postmenopausal women. *Am J Epidemiol* **161**:557–564.

122. Carton JA, Maradona JA, Nuño FJ, Fernandez-Alvarez R, Pérez-Gonzalez F, Asensi V. 1992. Diabetes mellitus and bacteraemia: A comparative study between diabetic and non-diabetic patients. *Eur J Med* **1**:281–287.

123. MacFarlane IA, Brown RM, Smyth RW, Burdon DW, FitzGerald MG. 1986. Bacteraemia in diabetics. *J Infect* **12**:213–219.

124. Bonadio M, Meini M, Gigli C, Longo B, Vigna A. 1999. Urinary tract infection in diabetic patients. *Urol Int* **63**:215–219.

125. Bonadio M, Costarelli S, Morelli G, Tartaglia T. 2006. The influence of diabetes mellitus on the spectrum of uropathogens and the antimicrobial resistance in elderly adult patients with urinary tract infection. *BMC Infect Dis* **6**:54.

126. Ronald A. 2002. The etiology of urinary tract infection: traditional and emerging pathogens. *Am J Med* **113**(Suppl 1A):14S–19S.

127. Rosen DA, Hung CS, Kline KA, Hultgren SJ. 2008. Streptozocin-induced diabetic mouse model of urinary tract infection. *Infect Immun* **76**:4290–4298.

128. Raffel L, Pitsakis P, Levison SP, Levison ME. 1981. Experimental *Candida albicans, Staphylococcus aureus,* and *Streptococcus faecalis* pyelonephritis in diabetic rats. *Infect Immun* **34**:773–779.

129. Millán-Rodríguez F, Palou J, Bujons-Tur A, Musquera-Felip M, Sevilla-Cecilia C, Serrallach-Orejas M, Baez-Angles C, Villavicencio-Mavrich H. 2006. Acute bacterial prostatitis: two different sub-categories according to a previous manipulation of the lower urinary tract. *World J Urol* **24**:45–50.

130. Brede CM, Shoskes DA. 2011. The etiology and management of acute prostatitis. *Nat Rev Urol* **8**:207–212.

131. Huycke MM, Sahm DF, Gilmore MS. 1998. Multiple-drug resistant enterococci: The nature of the problem and an agenda for the future. *Emerg Infect Dis* **4**:239–249.

132. Shankar V, Baghdayan AS, Huycke MM, Lindahl G, Gilmore MS. 1999. Infection-derived *Enterococcus faecalis* strains are enriched in esp, a gene encoding a novel surface protein. *Infect Immun* **67**:193–200.

133. Li J, Kasper DL, Ausubel FM, Rosner B, Michel JL. 1997. Inactivation of the alpha C protein antigen gene, bca, by a novel shuttle/suicide vector results in attenuation of virulence and immunity in group B *Streptococcus*. *Proc Natl Acad Sci U S A* **94**:13251–13256.

134. Larsson C, Stålhammar-Carlemalm M, Lindahl G. 1996. Experimental vaccination against group B streptococcus, an encapsulated bacterium, with highly purified preparations of cell surface proteins Rib and alpha. *Infect Immun* **64**:3518–3523.

135. Madoff LC, Michel JL, Gong EW, Kling DE, Kasper DL. 1996. Group B streptococci escape host immunity by deletion of tandem repeat elements of the alpha C protein. *Proc Natl Acad Sci U S A* **93**:4131–4136.

136. Shankar N, Lockatell CV, Baghdayan AS, Drachenberg C, Gilmore MS, Johnson DE. 2001. Role of *Enterococcus faecalis* surface protein Esp in the pathogenesis of ascending urinary tract infection. *Infect Immun* **69**:4366–4372.

137. Tendolkar PM, Baghdayan AS, Gilmore MS, Shankar N. 2004. Enterococcal surface protein, Esp, enhances biofilm formation by *Enterococcus faecalis*. *Infect Immun* **72**:6032–6039.

138. Nallapareddy SR, Singh KV, Sillanpää J, Zhao M, Murray BE. 2011. Relative contributions of Ebp Pili and the collagen adhesin ace to host extracellular matrix protein adherence and experimental urinary tract infection by *Enterococcus faecalis* OG1RF. *Infect Immun* **79**:2901–2910.

139. Lebreton F, Riboulet-Bisson E, Serror P, Sanguinetti M, Posteraro B, Torelli R, Hartke A, Auffray Y, Giard JC. 2009. ace, Which encodes an adhesin in *Enterococcus faecalis*, is regulated by Ers and is involved in virulence. *Infect Immun* **77**:2832–2839.

140. Roh JH, Singh KV, La Rosa SL, Cohen AL, Murray BE. 2015. The two-component system GrvRS (EtaRS) regulates ace expression in *Enterococcus faecalis* OG1RF. *Infect Immun* **83**:389–395.

141. Frank KL, Guiton PS, Barnes AM, Manias DA, Chuang-Smith ON, Kohler PL, Spaulding AR, Hultgren SJ, Schlievert PM, Dunny GM. 2013. AhrC and Eep are biofilm infection-associated virulence factors in *Enterococcus faecalis*. *Infect Immun* **81**:1696–1708.

142. Diederich AK, Wobser D, Spiess M, Sava IG, Huebner J, Sakınç T. 2014. Role of glycolipids in the pathogenesis of *Enterococcus faecalis* urinary tract infection. *PLoS ONE* **9**:e96295. doi:10.1371/journal.pone.0096295

143. Wobser D, Ali L, Grohmann E, Huebner J, Sakınç T. 2014. A novel role for D-alanylation of lipoteichoic acid of enterococcus faecalis in urinary tract infection. *PLoS ONE* **9**:e107827. doi:10.1371/journal.pone.0107827

144. Nallapareddy SR, Singh KV, Sillanpää J, Garsin DA, Höök M, Erlandsen SL, Murray BE. 2006. Endocarditis and biofilm-associated pili of *Enterococcus faecalis*. *J Clin Invest* **116**:2799–2807.

145. Schlüter S, Franz CM, Gesellchen F, Bertinetti O, Herberg FW, Schmidt FR. 2009. The high biofilm-encoding Bee locus: A second pilus gene cluster in *Enterococcus faecalis*? *Curr Microbiol* **59**:206–211.

146. Nielsen HV, Flores-Mireles AL, Kau AL, Kline KA, Pinkner JS, Neiers F, Normark S, Henriques-Normark B, Caparon MG, Hultgren SJ. 2013. Pilin and sortase residues critical for endocarditis- and biofilm-associated pilus biogenesis in *Enterococcus faecalis*. *J Bacteriol* **195**:4484–4495.

147. Kline KA, Kau AL, Chen SL, Lim A, Pinkner JS, Rosch J, Nallapareddy SR, Murray BE, Henriques-Normark B, Beatty W, Caparon MG, Hultgren SJ. 2009. Mechanism for sortase localization and the role of sortase localization in efficient pilus assembly in *Enterococcus faecalis*. *J Bacteriol* **191**:3237–3247.

148. Guiton PS, Hung CS, Kline KA, Roth R, Kau AL, Hayes E, Heuser J, Dodson KW, Caparon MG, Hultgren SJ. 2009. Contribution of autolysin and Sortase a during *Enterococcus faecalis* DNA-dependent biofilm development. *Infect Immun* **77**:3626–3638.

149. Nielsen HV, Guiton PS, Kline KA, Port GC, Pinkner JS, Neiers F, Normark S, Henriques-Normark B, Caparon MG, Hultgren SJ. 2012. The metal ion-dependent adhesion site motif of the *Enterococcus faecalis* EbpA pilin mediates pilus function in catheter-associated urinary tract infection. *MBio* **3**:e00177-12. doi:10.1128/mBio.00177-12

150. Kristich CJ, Manias DA, Dunny GM. 2005. Development of a method for markerless genetic exchange in *Enterococcus faecalis* and its use in construction of a srtA mutant. *Appl Environ Microbiol* **71**:5837–5849.

151. Kemp KD, Singh KV, Nallapareddy SR, Murray BE. 2007. Relative contributions of *Enterococcus faecalis* OG1RF sortase-encoding genes, srtA and bps (srtC), to biofilm formation and a murine model of urinary tract infection. *Infect Immun* **75**:5399–5404.

152. Sillanpää J, Chang C, Singh KV, Montealegre MC, Nallapareddy SR, Harvey BR, Ton-That H, Murray BE. 2013. Contribution of individual Ebp Pilus subunits of *Enterococcus faecalis* OG1RF to pilus biogenesis, biofilm formation and urinary tract infection. *PLoS ONE* 8: e68813. doi:10.1371/journal.pone.0068813

153. Sillanpää J, Nallapareddy SR, Singh KV, Prakash VP, Fothergill T, Ton-That H, Murray BE. 2010. Characterization of the ebp(fm) pilus-encoding operon of *Enterococcus faecium* and its role in biofilm formation and virulence in a murine model of urinary tract infection. *Virulence* 1:236–246.

154. Nallapareddy SR, Sillanpää J, Mitchell J, Singh KV, Chowdhury SA, Weinstock GM, Sullam PM, Murray BE. 2011. Conservation of Ebp-type pilus genes among Enterococci and demonstration of their role in adherence of *Enterococcus faecalis* to human platelets. *Infect Immun* 79:2911–2920.

155. Branci S, Ewertsen C, Thybo S, Nielsen HV, Jensen F, Wettergren A, Larsen PN, Bygbjerg IC. 2012. Cystic echinococcosis of the liver: experience from a Danish tertiary reference center (2002–2010). *J Travel Med* 19:28–34.

156. Banerjee A, Kim BJ, Carmona EM, Cutting AS, Gurney MA, Carlos C, Feuer R, Prasadarao NV, Doran KS. 2011. Bacterial pili exploit integrin machinery to promote immune activation and efficient blood-brain barrier penetration. *Nat Commun* 2:462.

157. Hilleringmann M, Giusti F, Baudner BC, Masignani V, Covacci A, Rappuoli R, Barocchi MA, Ferlenghi I. 2008. Pneumococcal pili are composed of protofilaments exposing adhesive clusters of Rrg A. *PLoS Pathog* 4:e1000026. doi:10.1371/journal.ppat.1000026

158. Flores-Mireles AL, Pinkner JS, Caparon MG, Hultgren SJ. 2014. EbpA vaccine antibodies block binding of *Enterococcus faecalis* to fibrinogen to prevent catheter-associated bladder infection in mice. *Sci Transl Med* 6:254ra127.

159. Horsley H, Malone-Lee J, Holland D, Tuz M, Hibbert A, Kelsey M, Kupelian A, Rohn JL. 2013. *Enterococcus faecalis* subverts and invades the host urothelium in patients with chronic urinary tract infection. *PLoS ONE* 8: e83637. doi:10.1371/journal.pone.0083637

160. Rakita RM, Vanek NN, Jacques-Palaz K, Mee M, Mariscalco MM, Dunny GM, Snuggs M, Van Winkle WB, Simon SI. 1999. *Enterococcus faecalis* bearing aggregation substance is resistant to killing by human neutrophils despite phagocytosis and neutrophil activation. *Infect Immun* 67:6067–6075.

161. Gentry-Weeks CR, Karkhoff-Schweizer R, Pikis A, Estay M, Keith JM. 1999. Survival of *Enterococcus faecalis* in mouse peritoneal macrophages. *Infect Immun* 67:2160–2165.

162. Baldassarri L, Bertuccini L, Creti R, Filippini P, Ammendolia MG, Koch S, Huebner J, Orefici G. 2005. Glycosaminoglycans mediate invasion and survival of *Enterococcus faecalis* into macrophages. *J Infect Dis* 191:1253–1262.

163. La Carbona S, Sauvageot N, Giard JC, Benachour A, Posteraro B, Auffray Y, Sanguinetti M, Hartke A. 2007. Comparative study of the physiological roles of three peroxidases (NADH peroxidase, alkyl hydroperoxide reductase and thiol peroxidase) in oxidative stress response, survival inside macrophages and virulence of *Enterococcus faecalis*. *Mol Microbiol* 66:1148–1163.

164. Lebreton F, van Schaik W, Sanguinetti M, Posteraro B, Torelli R, Le Bras F, Verneuil N, Zhang X, Giard JC, Dhalluin A, Willems RJ, Leclercq R, Cattoir V. 2012. AsrR is an oxidative stress sensing regulator modulating *Enterococcus faecium* opportunistic traits, antimicrobial resistance, and pathogenicity. *PLoS Pathog* 8:e1002834. doi:10.1371/journal. ppat.1002834

165. Giard JC, Riboulet E, Verneuil N, Sanguinetti M, Auffray Y, Hartke A. 2006. Characterization of Ers, a PrfA-like regulator of *Enterococcus faecalis*. *FEMS Immunol Med Microbiol* 46:410–418.

166. Verneuil N, Rincé A, Sanguinetti M, Auffray Y, Hartke A, Giard JC. 2005. Implication of hypR in the virulence and oxidative stress response of *Enterococcus faecalis*. *FEMS Microbiol Lett* 252:137–141.

167. Zhao C, Hartke A, La Sorda M, Posteraro B, Laplace JM, Auffray Y, Sanguinetti M. 2010. Role of methionine sulfoxide reductases A and B of *Enterococcus faecalis* in oxidative stress and virulence. *Infect Immun* 78:3889–3897.

168. Zhang D, Zhang G, Hayden MS, Greenblatt MB, Bussey C, Flavell RA, Ghosh S. 2004. A toll-like receptor that prevents infection by uropathogenic bacteria. *Science* 303:1522–1526.

169. Leendertse M, Willems RJ, Giebelen IA, van den Pangaart PS, Wiersinga WJ, de Vos AF, Florquin S, Bonten MJ, van der Poll T. 2008. TLR2-dependent MyD88 signaling contributes to early host defense in murine *Enterococcus faecium* peritonitis. *J Immunol* 180:4865–4874.

170. Goble NM, Clarke T, Hammonds JC. 1989. Histological changes in the urinary bladder secondary to urethral catheterisation. *Br J Urol* 63:354–357.

171. **Peychl L, Zalud R.** 2008. [Changes in the urinary bladder caused by short-term permanent catheter insertion]. *Cas Lek Cesk* **147:**325–329.

172. **Delnay KM, Stonehill WH, Goldman H, Jukkola AF, Dmochowski RR.** 1999. Bladder histological changes associated with chronic indwelling urinary catheter. *J Urol* **161:**1106–1108; discussion 1108–1109.

173. **Tambyah PA, Maki DG.** 2000. The relationship between pyuria and infection in patients with indwelling urinary catheters: A prospective study of 761 patients. *Arch Intern Med* **160:**673–677.

174. **Daw K, Baghdayan AS, Awasthi S, Shankar N.** 2012. Biofilm and planktonic *Enterococcus faecalis* elicit different responses from host phagocytes *in vitro*. *FEMS Immunol Med Microbiol* **65:**270–282.

175. **Kraemer TD, Quintanar Haro OD, Domann E, Chakraborty T, Tchatalbachev S.** 2014. The TIR domain containing locus of *Enterococcus faecalis* is predominant among urinary tract infection isolates and downregulates host inflammatory response. *Int J Microbiol* **2014:**918143.

176. **Chan SL, Low LY, Hsu S, Li S, Liu T, Santelli E, Le Negrate G, Reed JC, Woods VL Jr, Pascual J.** 2009. Molecular mimicry in innate immunity: crystal structure of a bacterial TIR domain. *J Biol Chem* **284:**21386–21392.

177. **Wong ES, Hooton TM, Hill CC, McKevitt M, Stamm WE.** 1988. Clinical and microbiological features of persistent or recurrent nongonococcal urethritis in men. *J Infect Dis* **158:**1098–1101.

178. **Wang S, Ng LH, Chow WL, Lee YK.** 2008. Infant intestinal *Enterococcus faecalis* downregulates inflammatory responses in human intestinal cell lines. *World J Gastroenterol* **14:**1067–1076.

179. **Wang S, Hibberd ML, Pettersson S, Lee YK.** 2014. *Enterococcus faecalis* from healthy infants modulates inflammation through MAPK signaling pathways. *PLoS ONE* **9:** e97523. doi:10.1371/journal.pone.0097523

180. **Foxman B.** 2003. Epidemiology of urinary tract infections: incidence, morbidity, and economic costs. *Dis Mon* **49:**53–70.

181. **Beyer I, Mergam A, Benoit F, Theunissen C, Pepersack T.** 2001. Management of urinary tract infections in the elderly. *Z Gerontol Geriatr* **34:**153–157.

182. **Haft RF, Kasper DL.** 1991. Group B streptococcus infection in mother and child. *Hosp Pract (Off Ed)* **26:**111–122, 125–128, 133–134.

183. **Ulett KB, Benjamin WH Jr, Zhuo F, Xiao M, Kong F, Gilbert GL, Schembri MA, Ulett GC.**

184. **Trivalle C, Martin E, Martel P, Jacque B, Menard JF, Lemeland JF.** 1998. Group B streptococcal bacteraemia in the elderly. *J Med Microbiol* **47:**649–652.

2009. Diversity of group B streptococcus serotypes causing urinary tract infection in adults. *J Clin Microbiol* **47:**2055–2060.

185. **Farley MM, Harvey RC, Stull T, Smith JD, Schuchat A, Wenger JD, Stephens DS.** 1993. A population-based assessment of invasive disease due to group B *Streptococcus* in nonpregnant adults. *N Engl J Med* **328:**1807–1811.

186. **Chaiwarith R, Jullaket W, Bunchoo M, Nuntachit N, Sirisanthana T, Supparatpinyo K.** 2011. *Streptococcus agalactiae* in adults at Chiang Mai University Hospital: A retrospective study. *BMC Infect Dis* **11:**149.

187. **Muller AE, Oostvogel PM, Steegers EA, Dörr PJ.** 2006. Morbidity related to maternal group B streptococcal infections. *Acta Obstet Gynecol Scand* **85:**1027–1037.

188. **Persson K, Bjerre B, Elfström L, Polberger S, Forsgren A.** 1986. Group B streptococci at delivery: high count in urine increases risk for neonatal colonization. *Scand J Infect Dis* **18:**525–531.

189. **Nicolle LE.** 2008. Uncomplicated urinary tract infection in adults including uncomplicated pyelonephritis. *Urol Clin North Am* **35:**1–12, v.

190. **Nicolle LE.** 2007. Complicated pyelonephritis: unresolved issues. *Curr Infect Dis Rep* **9:**501–507.

191. **Anderson BL, Simhan HN, Simons KM, Wiesenfeld HC.** 2007. Untreated asymptomatic group B streptococcal bacteriuria early in pregnancy and chorioamnionitis at delivery. *Am J Obstet Gynecol* **196:**524e1–5.

192. **Kessous R, Weintraub AY, Sergienko R, Lazer T, Press F, Wiznitzer A, Sheiner E.** 2012. Bacteriuria with group-B streptococcus: is it a risk factor for adverse pregnancy outcomes? *J Matern Fetal Neonatal Med* **25:**1983–1986.

193. **Colgan R, Nicolle LE, McGlone A, Hooton TM.** 2006. Asymptomatic bacteriuria in adults. *Am Fam Physician* **74:**985–990.

194. **Kincaid-Smith P, Bullen M.** 1965. Bacteriuria in pregnancy. *Lancet* **1:**395–399.

195. **Smaill F.** 2001. Antibiotics for asymptomatic bacteriuria in pregnancy. *Cochrane Database Syst Rev* **2:**CD000490.

196. **Smaill F.** 2000. Antibiotics for asymptomatic bacteriuria in pregnancy. *Cochrane Database Syst Rev* **8:**CD000490.

197. **Mittendorf R, Williams MA, Kass EH.** 1992. Prevention of preterm delivery and low birth weight associated with asymptomatic bacteriuria. *Clin Infect Dis* **14:**927–932.

198. Romero R, Oyarzun E, Mazor M, Sirtori M, Hobbins JC, Bracken M. 1989. Meta-analysis of the relationship between asymptomatic bacteriuria and preterm delivery/low birth weight. *Obstet Gynecol* **73:**576–582.

199. Verani JR, McGee L, Schrag SJ; Division of Bacterial Diseases, National Center for Immunization and Respiratory Diseases, Centers for Disease Control and Prevention (CDC). 2010. Prevention of perinatal group B streptococcal disease–revised guidelines from CDC, 2010. *MMWR Recomm Rep* **59:**1–36.

200. Weng CK, K, Sheng X, Byington C. 2010. Pregnancy outcomes in women with Group B Streptococcal bacteriuria, abstr Annual Meeting of the Pediatric Academic Societies, Vancouver, Canada, May 1–4, 2010.

201. Hammoud MS, Al-Shemmari M, Thalib L, Al-Sweih N, Rashwan N, Devarajan LV, Elsori H. 2003. Comparison between different types of surveillance samples for the detection of GBS colonization in both parturient mothers and their infants. *Gynecol Obstet Invest* **56:**225–230.

202. Margarit I, Rinaudo CD, Galeotti CL, Maione D, Ghezzo C, Buttazzoni E, Rosini R, Runci Y, Mora M, Buccato S, Pagani M, Tresoldi E, Berardi A, Creti R, Baker CJ, Telford JL, Grandi G. 2009. Preventing bacterial infections with pilus-based vaccines: The group B streptococcus paradigm. *J Infect Dis* **199:**108–115.

203. Baker CJ, Rench MA, Paoletti LC, Edwards MS. 2007. Dose-response to type V group B streptococcal polysaccharide-tetanus toxoid conjugate vaccine in healthy adults. *Vaccine* **25:**55–63.

204. Palazzi DL, Rench MA, Edwards MS, Baker CJ. 2004. Use of type V group B streptococcal conjugate vaccine in adults 65–85 years old. *J Infect Dis* **190:**558–564.

205. Baker CJ, Paoletti LC, Rench MA, Guttormsen HK, Edwards MS, Kasper DL. 2004. Immune response of healthy women to 2 different group B streptococcal type V capsular polysaccharide-protein conjugate vaccines. *J Infect Dis* **189:**1103–1112.

206. Baker CJ, Rench MA, Fernandez M, Paoletti LC, Kasper DL, Edwards MS. 2003. Safety and immunogenicity of a bivalent group B streptococcal conjugate vaccine for serotypes II and III. *J Infect Dis* **188:**66–73.

207. Baker CJ, Rench MA, McInnes P. 2003. Immunization of pregnant women with group B streptococcal type III capsular polysaccharide-tetanus toxoid conjugate vaccine. *Vaccine* **21:**3468–3472.

208. Shen X, Lagergård T, Yang Y, Lindblad M, Fredriksson M, Wallerstrom G, Holmgren J. 2001. Effect of pre-existing immunity for systemic and mucosal immune responses to intranasal immunization with group B Streptococcus type III capsular polysaccharide-cholera toxin B subunit conjugate. *Vaccine* **19:**3360–3368.

209. Xue G, Yu L, Li S, Shen X. 2010. Intranasal immunization with GBS surface protein Sip and ScpB induces specific mucosal and systemic immune responses in mice. *FEMS Immunol Med Microbiol* **58:**202–210.

210. Maisey HC, Doran KS, Nizet V. 2008. Recent advances in understanding the molecular basis of group B Streptococcus virulence. *Expert Rev Mol Med* **10:**e27.

211. Tan CK, Carey AJ, Cui X, Webb RI, Ipe D, Crowley M, Cripps AW, Benjamin WH Jr, Ulett KB, Schembri MA, Ulett GC. 2012. Genome-wide mapping of cystitis due to *Streptococcus agalactiae* and *Escherichia coli* in mice identifies a unique bladder transcriptome that signifies pathogen-specific antimicrobial defense against urinary tract infection. *Infect Immun* **80:**3145–3160.

212. Ulett GC, Webb RI, Ulett KB, Cui X, Benjamin WH, Crowley M, Schembri MA. 2010. Group B Streptococcus (GBS) urinary tract infection involves binding of GBS to bladder uroepithelium and potent but GBS-specific induction of interleukin 1alpha. *J Infect Dis* **201:**866–870.

213. Kulkarni R, Randis TM, Antala S, Wang A, Amaral FE, Ratner AJ. 2013. β-Hemolysin/cytolysin of Group B Streptococcus enhances host inflammation but is dispensable for establishment of urinary tract infection. *PLoS ONE* **8:**e59091. doi:10.1371/journal.pone.0059091

214. Sumati AH, Saritha NK. 2009. Association of urinary tract infection in women with bacterial vaginosis. *J Glob Infect Dis* **1:**151–152.

215. Sharami SH, Afrakhteh M, Shakiba M. 2007. Urinary tract infections in pregnant women with bacterial vaginosis. *J Obstet Gynaecol* **27:**252–254.

216. Hillebrand L, Harmanli OH, Whiteman V, Khandelwal M. 2002. Urinary tract infections in pregnant women with bacterial vaginosis. *Am J Obstet Gynecol* **186:**916–917.

217. Huang B, Fettweis JM, Brooks JP, Jefferson KK, Buck GA. 2014. The changing landscape of the vaginal microbiome. *Clin Lab Med* **34:**747–761.

218. Cone RA. 2014. Vaginal microbiota and sexually transmitted infections that may influence transmission of cell-associated HIV. *J Infect Dis* **210**(Suppl 3):S616–621.

219. **Nugent RP, Krohn MA, Hillier SL.** 1991. Reliability of diagnosing bacterial vaginosis is improved by a standardized method of gram stain interpretation. *J Clin Microbiol* **29:**297–301.

220. **Joesoef MR, Hillier SL, Josodiwondo S, Linnan M.** 1991. Reproducibility of a scoring system for gram stain diagnosis of bacterial vaginosis. *J Clin Microbiol* **29:**1730–1731.

221. **Schwebke JR, Hillier SL, Sobel JD, McGregor JA, Sweet RL.** 1996. Validity of the vaginal gram stain for the diagnosis of bacterial vaginosis. *Obstet Gynecol* **88:**573–576.

222. **Gilbert NM, Lewis WG, Lewis AL.** 2013. Clinical features of bacterial vaginosis in a murine model of vaginal infection with *Gardnerella vaginalis. PLoS ONE* **8:**e59539. doi:10.1371/journal.pone.0059539

223. **Lewis WG, Robinson LS, Gilbert NM, Perry JC, Lewis AL.** 2013. Degradation, foraging, and depletion of mucus sialoglycans by the vagina-adapted *Actinobacterium Gardnerella vaginalis. J Biol Chem* **288:**12067–12079.

224. **Stapleton AE, Au-Yeung M, Hooton TM, Fredricks DN, Roberts PL, Czaja CA, Yarova-Yarovaya Y, Fiedler T, Cox M, Stamm WE.** 2011. Randomized, placebo-controlled phase 2 trial of a *Lactobacillus crispatus* probiotic given intravaginally for prevention of recurrent urinary tract infection. *Clin Infect Dis* **52:**1212–1217.

225. **Naboka IuL, Kogan MI, Vasil'eva LI, Gudima IA, Miroshnichenko EA, Ibishev KhS.** 2011. [Bacterial mixed infection in women with chronic recurrent cystitis]. *Zh Mikrobiol Epidemiol Immunobiol* **1:**8–12.

226. **Eschenbach DA, Davick PR, Williams BL, Klebanoff SJ, Young-Smith K, Critchlow CM, Holmes KK.** 1989. Prevalence of hydrogen peroxide-producing *Lactobacillus* species in normal women and women with bacterial vaginosis. *J Clin Microbiol* **27:**251–256.

227. **Klebanoff SJ, Coombs RW.** 1991. Viricidal effect of *Lactobacillus acidophilus* on human immunodeficiency virus type 1: possible role in heterosexual transmission. *J Exp Med* **174:**289–292.

228. **Mirmonsef P, Gilbert D, Zariffard MR, Hamaker BR, Kaur A, Landay AL, Spear GT.** 2011. The effects of commensal bacteria on innate immune responses in the female genital tract. *Am J Reprod Immunol* **65:**190–195.

229. **Reid G, Burton J.** 2002. Use of *Lactobacillus* to prevent infection by pathogenic bacteria. *Microbes Infect* **4:**319–324.

230. **Sierra-Hoffman M, Watkins K, Jinadatha C, Fader R, Carpenter JL.** 2005. Clinical significance of *Aerococcus urinae*: A retrospective review. *Diagn Microbiol Infect Dis* **53:**289–292.

231. **Senneby E, Petersson AC, Rasmussen M.** 2012. Clinical and microbiological features of bacteraemia with *Aerococcus urinae. Clin Microbiol Infect* **18:**546–550.

232. **Murray TS, Muldrew KL, Finkelstein R, Hampton L, Edberg SC, Cappello M.** 2008. Acute pyelonephritis caused by *Aerococcus urinae* in a 12-year-old boy. *Pediatr Infect Dis J* **27:**760–762.

233. **Ibler K, Truberg Jensen K, Ostergaard C, Sönksen UW, Bruun B, Schønheyder HC, Kemp M, Dargis R, Andresen K, Christensen JJ.** 2008. Six cases of *Aerococcus sanguinicola* infection: clinical relevance and bacterial identification. *Scand J Infect Dis* **40:**761–765.

234. **de Jong MF, Soetekouw R, ten Kate RW, Veenendaal D.** 2010. *Aerococcus urinae*: severe and fatal bloodstream infections and endocarditis. *J Clin Microbiol* **48:**3445–3447.

235. **Christensen JJ, Jensen IP, Faerk J, Kristensen B, Skov R, Korner B.** 1995. Bacteremia/septicemia due to *Aerococcus*-like organisms: report of seventeen cases. Danish ALO Study Group. *Clin Infect Dis* **21:**943–947.

236. **Christensen JJ, Gutschik E, Friis-Møller A, Korner B.** 1991. Urosepticemia and fatal endocarditis caused by *Aerococcus*-like organisms. *Scand J Infect Dis* **23:**717–721.

237. **Christensen JJ, Korner B, Kjaergaard H.** 1989. *Aerococcus*-like organism–an unnoticed urinary tract pathogen. *APMIS* **97:**539–546.

238. **Facklam R, Hollis D, Collins MD.** 1989. Identification of gram-positive coccal and coccobacillary vancomycin-resistant bacteria. *J Clin Microbiol* **27:**724–730.

239. **Cattoir V, Kobal A, Legrand P.** 2010. *Aerococcus urinae* and *Aerococcus sanguinicola*, two frequently misidentified uropathogens. *Scand J Infect Dis* **42:**775–780.

240. **Zhang Q, Kwoh C, Attorri S, Clarridge JE III.** 2000. *Aerococcus urinae* in urinary tract infections. *J Clin Microbiol* **38:**1703–1705.

241. **Schuur PM, Kasteren ME, Sabbe L, Vos MC, Janssens MM, Buiting AG.** 1997. Urinary tract infections with *Aerococcus urinae* in the south of The Netherlands. *Eur J Clin Microbiol Infect Dis* **16:**871–875.

242. **Heilesen AM.** 1994. Septicaemia due to *Aerococcus urinae. Scand J Infect Dis* **26:**759–760.

243. **Gritsch W, Nagl M, Hausdorfer J, Gschwendtner A, Pechlaner C, Wiedermann CJ.** 1999. Septicaemia and endomyocarditis caused by *Aerococcus urinae. Wien Klin Wochenschr* **111:**446–447.

244. **Soriano F, Tauch A.** 2008. Microbiological and clinical features of *Corynebacterium urealyticum*: urinary tract stones and genomics

as the Rosetta Stone. *Clin Microbiol Infect* **14:** 632–643.

245. **Cogen AL, Nizet V, Gallo RL.** 2008. Skin microbiota: A source of disease or defence? *Br J Dermatol* **158:**442–455.

246. **Chung SY, Davies BJ, O'Donnell WF.** 2003. Mortality from grossly encrusted bilateral pyelitis, ureteritis, and cystitis by *Corynebacterium* group D2. *Urology* **61:**463.

247. **Meria P, Margaryan M, Haddad E, Dore B, Lottmann HB.** 2004. Encrusted cystitis and pyelitis in children: An unusual condition with potentially severe consequences. *Urology* **64:**569–573.

248. **Johnson MH, Strope SA.** 2012. Encrusted cystitis. *Urology* **79:**e31–32.

249. **López-Medrano F, García-Bravo M, Morales JM, Andrés A, San Juan R, Lizasoain M, Aguado JM.** 2008. Urinary tract infection due to Corynebacterium urealyticum in kidney transplant recipients: An underdiagnosed etiology for obstructive uropathy and graft dysfunction-results of a prospective cohort study. *Clin Infect Dis* **46:**825–830.

250. **Sánchez Hernández J, Mora Peris B, Yagüe Guirao G, Gutiérrez Zufiaurre N, Muñoz Bellido JL, Segovia Hernández M, García Rodríguez JA.** 2003. *In vitro* activity of newer antibiotics against *Corynebacterium jeikeium, Corynebacterium amycolatum* and *Corynebacterium urealyticum. Int J Antimicrob Agents* **22:**492–496.

251. **Fernández-Natal I, Guerra J, Alcoba M, Cachón F, Soriano F.** 2001. Bacteremia caused by multiply resistant *Corynebacterium urealyticum*: six case reports and review. *Eur J Clin Microbiol Infect Dis* **20:**514–517.

252. **Lawson PA, Falsen E, Akervall E, Vandamme P, Collins MD.** 1997. Characterization of some *Actinomyces*-like isolates from human clinical specimens: reclassification of *Actinomyces suis* (Soltys and Spratling) as *Actinobaculum suis* comb. nov. and description of *Actinobaculum schaalii* sp. nov. *Int J Syst Bacteriol* **47:**899–903.

253. **Wegienek J, Reddy CA.** 1982. Nutritional and metabolic features of *Eubacterium suis. J Clin Microbiol* **15:**895–901.

254. **Wendt M, Liebhold M, Kaup F, Amtsberg G, Bollwahn W.** 1990. [*Corynebacterium suis* infection in swine. 1. Clinical diagnosis with special consideration of urine studies and cystoscopy]. *Tierarztl Prax* **18:**353–357.

255. **Jones JE, Dagnall GJ.** 1984. The carriage of *Corynebacterium suis* in male pigs. *J Hyg (Lond)* **93:**381–388.

256. **Schaal KP, Yassin AF, Stackebrandt E.** 2006. The Family Actinomycetaceae: The Genera *Actinomyces, Actinobaculum, Arcanobacterium,* *Varibaculum,* and *Mobiluncus,* p. 430–537. *In* Dworkin M, Falkow S, Rosenberg E, Schleifer K-H, Stackebrandt E (ed), *The Prokaryotes,* 3rd ed. Springer, New York.

257. **Waldmann KH.** 1987. [Pyelocystitis in breeding sows]. *Tierarztl Prax* **15:**263–267.

258. **Percy DH, Ruhnke HL, Soltys MA.** 1966. A case of infectious cystitis and pyelonephritis of swine caused by *Corynebacterium suis. Can Vet J* **7:**291–292.

259. **Kaup FJ, Liebhold M, Wendt M, Drommer W.** 1990. [*Corynebacterium suis* infections in swine. 2. Morphological findings in the urinary tract with special reference to the bladder]. *Tierarztl Prax* **18:**595–599.

260. **Pleschakowa V, Leibold W, Amtsberg G, Konine D, Wendt M.** 2004. [The prevalence of *Actinobaculum suis* in boars of breeding herds in the Omsk region (Russian Federation) by indirect immunofluorescence technique]. *Dtsch Tierarztl Wochenschr* **111:**67–69.

261. **Cattoir V.** 2012. *Actinobaculum schaalii:* review of an emerging uropathogen. *J Infect* **64:**260–267.

262. **Tschudin-Sutter S, Frei R, Weisser M, Goldenberger D, Widmer AF.** 2011. *Actinobaculum schaalii* - invasive pathogen or innocent bystander? A retrospective observational study. *BMC Infect Dis* **11:**289.

263. **Vanden Bempt I, Van Trappen S, Cleenwerck I, De Vos P, Camps K, Celens A, Van De Vyvere M.** 2011. *Actinobaculum schaalii* causing Fournier's gangrene. *J Clin Microbiol* **49:**2369–2371.

264. **Bank S, Hansen TM, Soby KM, Lund L, Prag J.** 2011. *Actinobaculum schaalii* in urological patients, screened with real-time polymerase chain reaction. *Scand J Urol Nephrol* **45:**406–410.

265. **Alvarez-Paredes L, López-García P, Ruiz-García M, Royo-García G.** 2012. [*Actinobaculum schaalii* infection]. *Enferm Infecc Microbiol Clin* **30:**505–506.

266. **Andersen LB, Bank S, Hertz B, Søby KM, Prag J.** 2012. *Actinobaculum schaalii,* a cause of urinary tract infections in children? *Acta Paediatr* **101:**e232–234.

267. **Bank S, Jensen A, Hansen TM, Søby KM, Prag J.** 2010. *Actinobaculum schaalii,* a common uropathogen in elderly patients, Denmark. *Emerg Infect Dis* **16:**76–80.

268. **Nielsen HL, Søby KM, Christensen JJ, Prag J.** 2010. *Actinobaculum schaalii:* A common cause of urinary tract infection in the elderly population. Bacteriological and clinical characteristics. *Scand J Infect Dis* **42:**43–47.

269. **Tavassoli P, Paterson R, Grant J.** 2012. *Actinobaculum schaalii:* An emerging uropathogen? *Case Rep Urol* **2012:**468516.

270. Beguelin C, Genne D, Varca A, Tritten ML, Siegrist HH, Jaton K, Lienhard R. 2011. *Actinobaculum schaalii*: clinical observation of 20 cases. *Clin Microbiol Infect* **17**:1027–1031.

271. Hoenigl M, Leitner E, Valentin T, Zarfel G, Salzer HJ, Krause R, Grisold AJ. 2010. Endocarditis caused by *Actinobaculum schaalii*, Austria. *Emerg Infect Dis* **16**:1171–1173.

272. Larios OE, Bernard KA, Manickam K, Ng B, Alfa M, Ronald A. 2010. First report of *Actinobaculum schaalii* urinary tract infection in North America. *Diagn Microbiol Infect Dis* **67**:282–285.

273. Fendukly F, Osterman B. 2005. Isolation of *Actinobaculum schaalii* and *Actinobaculum urinale* from a patient with chronic renal failure. *J Clin Microbiol* **43**:3567–3569.

274. Pajkrt D, Simoons-Smit AM, Savelkoul PH, van den Hoek J, Hack WW, van Furth AM. 2003. Pyelonephritis caused by *Actinobaculum schaalii* in a child with pyeloureteral junction obstruction. *Eur J Clin Microbiol Infect Dis* **22**:438–440.

275. Johnk ML, Olsen AB, Prag JB, Søby KM. 2012. [Severe phimosis as cause of urosepsis with *Actinobaculum schaalii*]. *Ugeskr Laeger* **174**:1539–1540.

276. Reinhard M, Prag J, Kemp M, Andresen K, Klemmensen B, Hojlyng N, Sorensen SH, Christensen JJ. 2005. Ten cases of *Actinobaculum schaalii* infection: clinical relevance, bacterial identification, and antibiotic susceptibility. *J Clin Microbiol* **43**:5305–5308.

277. García-Bravo M, González-Fernández MB, García-Castro MA, Jaime-Muniesa ML. 2011. [Urinary tract infection caused by *Actinobaculum schaalii* in an elderly patient]. *Rev Esp Quimioter* **24**:52–53.

278. Sturm PD, Van Eijk J, Veltman S, Meuleman E, Schulin T. 2006. Urosepsis with *Actinobaculum schaalii* and *Aerococcus urinae*. *J Clin Microbiol* **44**:652–654.

279. Haller P, Bruderer T, Schaeren S, Laifer G, Frei R, Battegay M, Flükiger U, Bassetti S. 2007. Vertebral osteomyelitis caused by *Actinobaculum schaalii*: A difficult-to-diagnose and potentially invasive uropathogen. *Eur J Clin Microbiol Infect Dis* **26**:667–670.

280. Amatya R, Bhattarai S, Mandal PK, Tuladhar H, Karki BM. 2013. Urinary tract infection in vaginitis: A condition often overlooked. *Nepal Med Coll J* **15**:65–67.

281. Harmanli OH, Cheng GY, Nyirjesy P, Chatwani A, Gaughan JP. 2000. Urinary tract infections in women with bacterial vaginosis. *Obstet Gynecol* **95**:710–712.

282. Lam MH, Birch DF. 1991. Survival of *Gardnerella vaginalis* in human urine. *Am J Clin Pathol* **95**:234–239.

283. Josephson S, Thomason J, Sturino K, Zabransky R, Williams J. 1988. *Gardnerella vaginalis* in the urinary tract: incidence and significance in a hospital population. *Obstet Gynecol* **71**:245–250.

284. Clarke RW, Collins LE, Maskell R. 1989. *Gardnerella vaginalis* as a urinary pathogen. *J Infect* **19**:191–193.

285. Sturm AW. 1989. *Gardnerella vaginalis* in infections of the urinary tract. *J Infect* **18**:45–49.

286. Lam MH, Birch DF, Fairley KF. 1988. Prevalence of *Gardnerella vaginalis* in the urinary tract. *J Clin Microbiol* **26**:1130–1133.

287. Gilbert GL, Garland SM, Fairley KF, McDowall DM. 1986. Bacteriuria due to ureaplasmas and other fastidious organisms during pregnancy: prevalence and significance. *Pediatr Infect Dis* **5**(Suppl 6):S239–243.

288. Savige JA, Gilbert GL, Fairley KF, McDowall DR. 1983. Bacteriuria due to *Ureaplasma urealyticum* and *Gardnerella vaginalis* in women with preeclampsia. *J Infect Dis* **148**:605.

289. McDowall DR, Buchanan JD, Fairley KF, Gilbert GL. 1981. Anaerobic and other fastidious microorganisms in asymptomatic bacteriuria in pregnant women. *J Infect Dis* **144**:114–122.

290. McFadyen IR, Eykyn SJ. 1968. Suprapubic aspiration of urine in pregnancy. *Lancet* **1**:1112–1114.

291. Fairley KF, Birch DF. 1983. Unconventional bacteria in urinary tract disease: *Gardnerella vaginalis*. *Kidney Int* **23**:862–865.

292. Pearce MM, Hilt EE, Rosenfeld AB, Zilliox MJ, Thomas-White K, Fok C, Kliethermes S, Schreckenberger PC, Brubaker L, Gai X, Wolfe AJ. 2014. The female urinary microbiome: A comparison of women with and without urgency urinary incontinence. *MBio* **5**:e01283-14. doi:10.1128/mBio.01283-14

293. Reimer LG, Reller LB. 1984. *Gardnerella vaginalis* bacteremia: A review of thirty cases. *Obstet Gynecol* **64**:170–172.

294. Datcu R, Charib K, Kjaeldgaard P. 2009. [Septic shock caused by *Gardnerellavaginalis* and *Peptostreptococcus* species after Cesarean section]. *Ugeskr Laeger* **171**:1012.

295. Amaya RA, Al-Dossary F, Demmler GJ. 2002. *Gardnerella vaginalis* bacteremia in a premature neonate. *J Perinatol* **22**:585–587.

296. La Scolea LJ Jr, Dryja DM, Dillon WP. 1984. Recovery of *Gardnerella vaginalis* from blood by the quantitative direct plating method. *J Clin Microbiol* **20**:568–569.

297. **McCool RA, DeDonato DM.** 2012. Bacteremia of *Gardnerella vaginalis* after endometrial ablation. *Arch Gynecol Obstet* **286:**1337–1338.

298. **Agostini A, Beerli M, Franchi F, Bretelle F, Blanc B.** 2003. *Gardnerella vaginalis* bacteremia after vaginal myomectomy. *Eur J Obstet Gynecol Reprod Biol* **108:**229.

299. **Alidjinou EK, Bonnet I, Canis F, Dewulf G, Mazars E, Cattoen C.** 2013. [*Gardnerella vaginalis* bacteremia in a male patient]. *Med Mal Infect* **43:**434–435.

300. **Yoon HJ, Chun J, Kim JH, Kang SS, Na DJ.** 2010. *Gardnerella vaginalis* septicaemia with pyelonephritis, infective endocarditis and septic emboli in the kidney and brain of an adult male. *Int J STD AIDS* **21:**653–657.

301. **Lagacé-Wiens PR, Ng B, Reimer A, Burdz T, Wiebe D, Bernard K.** 2008. *Gardnerella vaginalis* bacteremia in a previously healthy man: case report and characterization of the isolate. *J Clin Microbiol* **46:**804–806.

302. **Bastida Vilá MT, López Onrubia P, Rovira Lledos J, Martinez Martinez JA, Expositó Aguilera M.** 1997. *Garderella vaginalis* bacteremia in an adult male. *Eur J Clin Microbiol Infect Dis* **16:**400–401.

303. **Denoyel GA, Drouet EB, De Montclos HP, Schanen A, Michel S.** 1990. *Gardnerella vaginalis* bacteremia in a man with prostatic adenoma. *J Infect Dis* **161:**367–368.

304. **Legrand JC, Alewaeters A, Leenaerts L, Gilbert P, Labbe M, Glupczynski Y.** 1989. *Gardnerella vaginalis* bacteremia from pulmonary abscess in a male alcohol abuser. *J Clin Microbiol* **27:**1132–1134.

305. **Swidsinski A, Doerffel Y, Loening-Baucke V, Swidsinski S, Verstraelen H, Vaneechoutte M, Lemm V, Schilling J, Mendling W.** 2010. *Gardnerella* biofilm involves females and males and is transmitted sexually. *Gynecol Obstet Invest* **70:**256–263.

306. **Verstraelen H, Verhelst R, Vaneechoutte M, Temmerman M.** 2010. The epidemiology of bacterial vaginosis in relation to sexual behaviour. *BMC infectious diseases* **10:**81.

307. **Chowdhury MN.** 1986. *Gardnerella vaginalis* carriage in male patients. *Trop Geogr Med* **38:**137–140.

308. **Burdge DR, Bowie WR, Chow AW.** 1986. *Gardnerella vaginalis*-associated balanoposthitis. *Sex Transm Dis* **13:**159–162.

309. **Lefevre JC, Lepargneur JP, Bauriaud R, Bertrand MA, Blanc C.** 1991. Clinical and microbiologic features of urethritis in men in Toulouse, France. *Sex Transm Dis* **18:**76–79.

310. **Raz R, Colodner R, Kunin CM.** 2005. Who are you–*Staphylococcus saprophyticus*? *Clin Infect Dis* **40:**896–898.

311. **Torres Pereira A.** 1962. Coagulase-negative strains of *Staphylococcus* possessing antigen 51 as agents of urinary infection. *J Clin Pathol* **15:**252–253.

312. **Siegman-Igra Y, Kulka T, Schwartz D, Konforti N.** 1993. The significance of polymicrobial growth in urine: contamination or true infection. *Scand J Infect Dis* **25:**85–91.

313. **Bishara J, Leibovici L, Huminer D, Drucker M, Samra Z, Konisberger H, Pitlik S.** 1997. Five-year prospective study of bacteraemic urinary tract infection in a single institution. *Eur J Clin Microbiol Infect Dis* **16:**563–567.

314. **Woods TD, Watanakunakorn C.** 1996. Bacteremia due to *Providencia stuartii*: review of 49 episodes. *South Med J* **89:**221–224.

315. **Ferry SA, Holm SE, Stenlund H, Lundholm R, Monsen TJ.** 2007. Clinical and bacteriological outcome of different doses and duration of pivmecillinam compared with placebo therapy of uncomplicated lower urinary tract infection in women: The LUTIW project. *Scand J Prim Health Care* **25:**49–57.

316. **Domann E, Hong G, Imirzalioglu C, Turschner S, Kuhle J, Watzel C, Hain T, Hossain H, Chakraborty T.** 2003. Culture-independent identification of pathogenic bacteria and polymicrobial infections in the genitourinary tract of renal transplant recipients. *J Clin Microbiol* **41:**5500–5510.

317. **Wolfe AJ, Toh E, Shibata N, Rong R, Kenton K, Fitzgerald M, Mueller ER, Schreckenberger P, Dong Q, Nelson DE, Brubaker L.** 2012. Evidence of uncultivated bacteria in the adult female bladder. *J Clin Microbiol* **50:**1376–1383.

318. **Siddiqui H, Nederbragt AJ, Lagesen K, Jeansson SL, Jakobsen KS.** 2011. Assessing diversity of the female urine microbiota by high throughput sequencing of 16S rDNA amplicons. *BMC Microbiol* **11:**244.

319. **Lewis DA, Brown R, Williams J, White P, Jacobson SK, Marchesi JR, Drake MJ.** 2013. The human urinary microbiome; bacterial DNA in voided urine of asymptomatic adults. *Front Cell Infect Microbiol* **3:**41. doi:10.3389/fcimb.2013.00041

320. **Horwitz D, McCue T, Mapes AC, Ajami NJ, Petrosino JF, Ramig RF, Trautner BW.** 2015. Decreased microbiota diversity associated with urinary tract infection in a trial of bacterial interference. *J Infect* **71:**358–367.

321. **Dankert J, Mensink WF, Aarnoudse JG, Meijer-Severs GJ, Huisjes HJ.** 1979. The prevalence of anaerobic bacteria in suprapubic bladder aspirates obtained from pregnant women. *Zentralbl Bakteriol Orig A* **244:**260–267.

322. Nelson DE, Van Der Pol B, Dong Q, Revanna KV, Fan B, Easwaran S, Sodergren E, Weinstock GM, Diao L, Fortenberry JD. 2010. Characteristic male urine microbiomes associate with asymptomatic sexually transmitted infection. *PLoS One* 5:e14116. doi:10.1371/journal.pone.0014116

323. Dong Q, Nelson DE, Toh E, Diao L, Gao X, Fortenberry JD, Van der Pol B. 2011. The microbial communities in male first catch urine are highly similar to those in paired urethral swab specimens. *PLoS One* 6:e19709. doi:10.1371/journal.pone.0019709

324. Nelson DE, Dong Q, Van der Pol B, Toh E, Fan B, Katz BP, Mi D, Rong R, Weinstock GM, Sodergren E, Fortenberry JD. 2012. Bacterial communities of the coronal sulcus and distal urethra of adolescent males. *PLoS One* 7:e36298. doi:10.1371/journal.pone.0036298

325. Hilt EE, McKinley K, Pearce MM, Rosenfeld AB, Zilliox MJ, Mueller ER, Brubaker L, Gai X, Wolfe AJ, Schreckenberger PC. 2014. Urine is not sterile: use of enhanced urine culture techniques to detect resident bacterial flora in the adult female bladder. *J Clin Microbiol* 52:871–876.

326. Borgdorff H, Verwijs MC, Wit FW, Tsivtsivadze E, Ndayisaba GF, Verhelst R, Schuren FH, van de Wijgert JH. 2015. The impact of hormonal contraception and pregnancy on sexually transmitted infections and on cervicovaginal microbiota in african sex workers. *Sex Transm Dis* 42:143–152.

327. Lüthje P, Hirschberg AL, Brauner A. 2014. Estrogenic action on innate defense mechanisms in the urinary tract. *Maturitas* 77:32–36.

328. Marraffini LA, Dedent AC, Schneewind O. 2006. Sortases and the art of anchoring proteins to the envelopes of gram-positive bacteria. *Microbiol Mol Biol Rev* 70:192–221.

329. Mobley HLT. 2001. Urease, p 179–191. *In* Mobley HLT, Mendz GL, Hazell SL (ed), *Helicobacter pylori: Physiology and Genetics*, ASM Press, Washington, DC.

330. Roesch PL, Redford P, Batchelet S, Moritz RL, Pellett S, Haugen BJ, Blattner FR, Welch RA. 2003. Uropathogenic *Escherichia coli* use d-serine deaminase to modulate infection of the murine urinary tract. *Mol Microbiol* 49:55–67.

331. Ackermann RJ, Monroe PW. 1996. Bacteremic urinary tract infection in older people. *J Am Geriatr Soc* 44:927–933.

332. Pearce MM, Zilliox MJ, Rosenfeld AB, Thomas-White KJ, Richter HE, Nager CW, Visco AG, Nygaard IE, Barber MD, Schaffer J, Moalli P, Sung VW, Smith AL, Rogers R, Nolen TL, Wallace D, Meikle SF, Gai X, Wolfe AJ, Brubaker L; Pelvic Floor Disorders Network. 2015. The female urinary microbiome in urgency urinary incontinence. *Am J Obstet Gynecol* 213:347e1–11.

333. Nickel JC, Stephens A, Landis JR, Chen J, Mullins C, van Bokhoven A, Lucia MS, Melton-Kreft R, Ehrlich GD; MAPP Research Network. 2015. Search for microorganisms in men with urologic chronic pelvic pain syndrome: A culture-independent analysis in the MAPP Research Network. *J Urol* 194:127–135.

334. Marconi M, Pilatz A, Wagenlehner F, Diemer T, Weidner W. 2009. Impact of infection on the secretory capacity of the male accessory glands. *Int Braz J Urol* 35:299–308; discussion 308–309.

335. Ivanov YB. 2007. Microbiological features of persistent nonspecific urethritis in men. *J Microbiol Immunol Infect* 40:157–161.

336. Imirzalioglu C, Hain T, Chakraborty T, Domann E. 2008. Hidden pathogens uncovered: metagenomic analysis of urinary tract infections. *Andrologia* 40:66–71.

337. Pillai A, Deodhar L, Gogate A. 1990. Microbiological study of urethritis in men attending a STD clinic. *Indian J Med Res* 91:443–447.

338. Siddiqui H, Lagesen K, Nederbragt AJ, Jeansson SL, Jakobsen KS. 2012. Alterations of microbiota in urine from women with interstitial cystitis. *BMC Microbiol* 12:205.

339. Casullo VA, Bottone E, Herold BC. 2001. *Peptostreptococcus asaccharolyticus* renal abscess: A rare cause of fever of unknown origin. *Pediatrics* 107:E11.

340. Teo KP, Jacob SC, Lim SH. 1997. Post-caesarean septicaemia in Kandang Kerbau Hospital, Singapore, 1993–1995. *Med J Malaysia* 52:325–330.

341. Brown K, Church D, Lynch T, Gregson D. 2014. Bloodstream infections due to *Peptoniphilus* spp.: report of 15 cases. *Clin Microbiol Infect* 20: O857–860.

342. Bernier M, Njomnang Soh P, Lochet A, Prots L, Félice R, Senescau A, Fabre R, Philippon A. 2012. [*Lactobacillus delbrueckii*: probable agent of urinary tract infections in very old women]. *Pathol Biol (Paris)* 60:140–142.

343. Darbro BW, Petroelje BK, Doern GV. 2009. *Lactobacillus delbrueckii* as the cause of urinary tract infection. *J Clin Microbiol* 47:275–277.

344. Yagihashi Y, Arakaki Y. 2012. Acute pyelonephritis and secondary bacteraemia caused by *Veillonella* during pregnancy. *BMJ Case Rep* 2012. doi:10.1136/bcr-2012-007364

345. Meijer-Severs GJ, Aarnoudse JG, Mensink WF, Dankert J. 1979. The presence of antibody-coated anaerobic bacteria in asymptomatic bacteriuria during pregnancy. *J Infect Dis* 140:653–658.

346. García-Lechuz JM, Cuevas-Lobato O, Hernángomez S, Hermida A, Guinea J, Marín M, Peláez T, Bouza E. 2002. Extra-abdominal infections due to *Gemella* species. *Int J Infect Dis* 6:78–82.

347. Ahmad NM, Ahmad KM. 2005. *Corynebacterium minutissimum* pyelonephritis with associated bacteraemia: A case report and review of literature. *J Infect* 51:e299–303.

348. Audard V, Garrouste-Orgeas M, Misset B, Ali AB, Gattolliat O, Meria P, Carlet J. 2003. Fatal septic shock caused by *Corynebacterium* D2. *Intensive Care Med* 29:1376–1379.

349. Beteta López A, Gil Ruiz MT, Vega Prado L, Fajardo Olivares M. 2009. [Cystitis and haematuria due to *Corynebacterium striatum*. A case report and review]. *Actas Urol Esp* 33:909–912.

350. Craig J, Grigor W, Doyle B, Arnold D. 1994. Pyelonephritis caused by *Corynebacterium minutissimum*. *Pediatr Infect Dis J* 13:1151–1152.

351. Fontana I, Bertocchi M, Rossi AM, Gasloli G, Santori G, Ferro C, Patti V, Rossini A, Valente U. 2010. *Corynebacterium urealyticum* infection in a pediatric kidney transplant recipient: case report. *Transplant Proc* 42:1367–1368.

352. Pagnoux C, Bérezné A, Damade R, Paillot J, Aouizerate J, Le Guern V, Salmon D, Guillevin L. 2011. Encrusting cystitis due to *Corynebacterium urealyticum* in a patient with ANCA-associated vasculitis: case report and review of the literature. *Semin Arthritis Rheum* 41:297–300.

353. Matsunami M, Otsuka Y, Ohkusu K, Sogi M, Kitazono H, Hosokawa N. 2011. Urosepsis caused by *Globicatella sanguinis* and *Corynebacterium riegelii* in an adult: case report and literature review. *J Infect Chemother* 18:552–554.

354. Galan-Sanchez F, Aznar-Marin P, Marin-Casanova P, Garcia-Martos P, Rodriguez-Iglesias M. 2011. Urethritis due to *Corynebacterium glucuronolyticum*. *J Infect Chemother* 17:720–721.

355. Abdolrasouli A, Roushan A. 2013. *Corynebacterium propinquum* associated with acute, nongonococcal urethritis. *Sex Transm Dis* 40:829–831.

356. Swartz SL, Kraus SJ, Herrmann KL, Stargel MD, Brown WJ, Allen SD. 1978. Diagnosis and etiology of nongonococcal urethritis. *J Infect Dis* 138:445–454.

357. Greub G, Raoult D. 2002. "*Actinobaculum massiliae*," a new species causing chronic urinary tract infection. *J Clin Microbiol* 40:3938–3941.

358. Gomez E, Gustafson DR, Rosenblatt JE, Patel R. 2011. *Actinobaculum* bacteremia: A report of 12 cases. *J Clin Microbiol* 49:4311–4313.

359. Lagacé-Wiens PR, Ng B, Reimer A, Burdz T, Wiebe D, Bernard K. 2008. *Gardnerella vaginalis* bacteremia in a previously healthy man: case report and characterization of the isolate. *J Clin Microbiol* 46:804–806.

360. Aydin MD, Agaçfidan A, Güvener Z, Kadioglu A, Ang O. 1998. Bacterial pathogens in male patients with urethritis in Istanbul. *Sex Transm Dis* 25:448–449.

361. Iser P, Read TH, Tabrizi S, Bradshaw C, Lee D, Horvarth L, Garland S, Denham I, Fairley CK. 2005. Symptoms of non-gonococcal urethritis in heterosexual men: A case control study. *Sex Transm Infect* 81:163–165.

362. Kumar B, Sharma M. 1994. Carriage of *Gardnerella vaginalis* in the urethra of Indian men. *Indian J Med Res* 99:252–254.

363. McKechnie ML, Hillman R, Couldwell D, Kong F, Freedman E, Wang H, Gilbert GL. 2009. Simultaneous identification of 14 genital microorganisms in urine by use of a multiplex PCR-based reverse line blot assay. *J Clin Microbiol* 47:1871–1877.

364. Manhart LE, Khosropour CM, Liu C, Gillespie CW, Depner K, Fiedler T, Marrazzo JM, Fredricks DN. 2013. Bacterial vaginosis-associated bacteria in men: association of *Leptotrichia/Sneathia* spp. with nongonococcal urethritis. *Sex Transm Dis* 40:944–949.

365. Agrawal P, Vaiphei K. 2014. Renal actinomycosis. *BMJ Case Rep* 2014. doi:10.1136/bcr-2014-205892

366. Juhász J, Galambos J, Surján L Jr. 1980. Renal actinomycosis associated with bilateral necrosing renal papillitis. *Int Urol Nephrol* 12:199–203.

367. Herbland A, Leloup M, Levrat Q, Guillaume F, Verrier V, Bouillard P, Landois T, Ouaki CF, Lesieur O. 2014. Fulminant course of unilateral emphysematous pyelonephritis revealing a renal actinomycosis caused by *Actinomyces meyeri*, an unknown cause of septic shock. *J Intensive Care* 2:42.

368. Jang SM, Na W, Jun YJ, Paik SS. 2010. Primary vesical actinomycosis diagnosed by routine urine cytology. *Acta Cytol* 54:658–659.

369. Ieven M, Verhoeven J, Gentens P, Goossens H. 1996. Severe infection due to *Actinomyces bernardiae*: case report. *Clin Infect Dis* 22:157–158.

370. Horliana AC, Chambrone L, Foz AM, Artese HP, Rabelo Mde S, Pannuti CM, Romito GA. 2014. Dissemination of periodontal pathogens in the bloodstream after periodontal procedures: A systematic review. *PLoS ONE* 9: e98271. doi:10.1371/journal.pone.0098271

371. **Angelova I, Talakova Ch, Belovezhdov N.** 1988. [*Mycoplasma* infections of the kidneys]. *Vutr Boles* **27**:110–113.

372. **Küchle C, Abele-Horn M, Menninger M, Held E, Heesemann J.** 1997. [*Mycoplasmahominis*. A rare causative agent of acute pyelonephritis]. *Dtsch Med Wochenschr* **122**:542–544.

373. **Thomsen AC.** 1983. Occurrence and pathogenicity of *Mycoplasma hominis* in the upper urinary tract: A review. *Sex Transm Dis* **10** (Suppl 4):323–326.

374. **Wong SS, Yuen KY.** 1995. Acute pyelonephritis caused by *Mycoplasma hominis*. *Pathology* **27**:61–63.

375. **Anagrius C, Loré B, Jensen JS.** 2005. *Mycoplasma genitalium*: prevalence, clinical significance, and transmission. *Sex Transm Infect* **81**:458–462.

376. **Elsner P, Hartmann AA, Wecker I.** 1988. *Gardnerella vaginalis* is associated with other sexually transmittable microorganisms in the male urethra. *Zentralbl Bakteriol Mikrobiol Hyg A* **269**:56–63.

377. **Falk L, Fredlund H, Jensen JS.** 2005. Signs and symptoms of urethritis and cervicitis among women with or without *Mycoplasma genitalium* or *Chlamydia trachomatis* infection. *Sex Transm Infect* **81**:73–78.

378. **Högdahl M, Kihlström E.** 2007. Leucocyte esterase testing of first-voided urine and urethral and cervical smears to identify *Mycoplasma genitalium*-infected men and women. *Int J STD AIDS* **18**:835–838.

379. **Ishihara S, Yasuda M, Ito S, Maeda S, Deguchi T.** 2004. *Mycoplasma genitalium* urethritis in men. *Int J Antimicrob Agents* **24** (Suppl 1):S23–27.

380. **Moi H, Reinton N, Moghaddam A.** 2009. *Mycoplasma genitalium* is associated with symptomatic and asymptomatic non-gonococcal urethritis in men. *Sex Transm Infect* **85**:15–18.

381. **Shigehara K, Kawaguchi S, Sasagawa T, Furubayashi K, Shimamura M, Maeda Y, Konaka H, Mizokami A, Koh E, Namiki M.** 2011. Prevalence of genital *Mycoplasma*, *Ureaplasma*, *Gardnerella*, and human papillomavirus in Japanese men with urethritis, and risk factors for detection of urethral human papillomavirus infection. *J Infect Chemother* **17**:487–492.

Integrated Pathophysiology of Pyelonephritis

20

FERDINAND X. CHOONG,[1] HARIS ANTYPAS,[1] and
AGNETA RICHTER-DAHLFORS[1]

INTRODUCTION

Urinary tract infections (UTIs) occur when bacteria, often originating from the fecal flora, migrate via the urethra to the bladder causing symptomatic cystitis or asymptomatic bacteriuria (1). Further ascension via the ureters leads to infection of the normally sterile kidneys, termed pyelonephritis, which is commonly defined as a tubulointerstitial disorder whose gross pathology includes abscess formation in the renal parenchyma and edema (1, 2). These conditions often lead to irreversible scar formation, and may contribute towards the development of renal insufficiency (3). The acute form of pyelonephritis is clinically defined as a syndrome of bacteriuria with accompanied uni- and bi-lateral flank pain and tenderness, chills, and sudden increase in temperature. It may encompass, or progress to, urosepsis, septic shock, and death (4). Chronic pyelonephritis, on the other hand, is a radiological diagnosis defined by histological changes to the renal tissue resulting from infection. Such changes may include renal scarring, fibrosis, tissue destruction, and interstitial inflammation (3).

Normally, the kidney is considered relatively resistant to infection owing to a number of challenges presented to any bacteria entering the urinary tract.

[1]Swedish Medical Nanoscience Center, Department of Neuroscience, Karolinska Institutet, SE-171 77, Stockholm, Sweden.

Urinary Tract Infections: Molecular Pathogenesis and Clinical Management, 2nd Edition
Edited by Matthew A. Mulvey, David J. Klumpp, and Ann E. Stapleton
© 2017 American Society for Microbiology, Washington, DC
doi:10.1128/microbiolspec.UTI-0014-2012

To colonize in spite of the dynamically changing environment in these organs demands the adaptive flexibility of the invading pathogen. Moreover, the uroepithelium is far from uniform. Whereas the bladder is lined by a transitional stratified epithelium, the individual structure of the nephron contains different types of epithelia (5) (Fig. 1). The glomerular capillary tufts in the Bowman's capsule are lined with thin squamous epithelial cells, whereas the tubular systems consist of single-layer epithelium with segment-specific expression of structure and function (5). The physiological process of urine production and discharge exposes microbes to mechanical stress originating from the continuous cycles of urine production, storage, and voiding. The chemical composition of the glomerular filtrate, which eventually forms the urine, is also changing throughout the different segments of the urinary tract due to reabsorption of filtered water, ions, and other solutes (6–8). Abnormalities in structure and function of the urinary tract can therefore increase susceptibility to infection (9). Voiding dysfunction and vesicoureteral reflux have been ascribed as risk factors in children, whereas behavioral factors are more relevant in adults (10). Also, genetic susceptibilities have been linked to pyelonephritis. Deficient CXCR1 expression in both adults and children has been linked to the typical symptoms of acute pyelonephritis and renal scarring because of suppressed innate defense mechanisms in these individuals (11).

Bacteria need to possess high metabolic versatility as well as a good degree of resilience to fluctuating physical conditions in order to enter the organs of the urinary tract. It is thus unsurprising that in the majority of diagnosed community-acquired UTIs, uropathogenic *Escherichia coli* (UPEC) is implicated in up to 80% of cases as the causative agent (12). Other Gram-negative bacterial species able to colonize the urinary tract include *Klebsiella pneumoniae* (6.2%), *Enterococcus* (5.3%), *Proteus mirabilis* (2%), *Klebsiella oxytoca* (0.9%), and *Pseudomonas aeruginosa* (0.8%), whereas Gram-positive species include *Streptococcus agalactiae* (2.8%), *Staphylococcus saprophyticus* (1.4%), and the viridans streptococci group (0.9%) (13). With UPEC being the unanimously major causative microbe for UTI, as well as the primary pathogen applied in numerous molecular studies of the UTI process, this chapter focuses on the current knowledge gained over the years from UPEC-induced UTIs. Specifically, this section will cover a selection of relevant facets of pyelonephritis, which have gained attention in intravital and animal models. For further details regarding specific bacterial and host factors, the reader is referred to relevant chapters in this book.

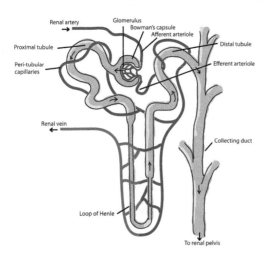

FIGURE 1 The nephron. The nephron is the basic filtration unit of the kidney, composed of tubular and vascular elements. Arrows denote the direction of fluid flow.

PATHOPHYSIOLOGY OF PYELONEPHRITIS

The Cinematic View of Pyelonephritis

The highly vascularized kidney consists of approximately 1 million nephrons that continuously filter blood from waste products. As filtrate flows through the nephron, tubulo-epithelial absorption and secretion turn the filtrate into urine. In pyelonephritis, invading bacteria cause deviation from this native-tissue physiology and histology, resulting in dramatic changes of the microenvironment at the infection site. In contrast to the widely used sacrificial animal model of pyelonephritis, which is limited to studies at defined time points or the end stage of infection, studies applying intravital imaging based on 2-photon microscopy allow for dynamic studies of organ physiology and disease pathophysiology at the cellular level in the live animal. Recently, this live-animal imaging technique was applied to perform a high-resolution study of a live-renal UPEC infection in real-time in the presence of all interplaying host factors, such as the immune, vascular, lymphatic, and nervous systems (Movie 1<http://asmscience.org/content/journal/microbiolspec/10.1128/microbiolspec.UTI-0014-2012>) (14–16). The slow infusion of green fluorescent protein (GFP)$^+$-expressing UPEC bacteria directly into the lumen of a superficial renal tubule allows for spatial and temporal control of the infection. Monitoring of this site revealed that very few bacteria initially adhered to the tubule epithelium in the face of the passing glomerular filtrate (Fig. 2A). These few bacteria, however, rapidly adapted to the microenvironment, and began colonizing the tubule (Figure 2B) (15).

As infection progresses, major alteration of the infected organ's physiology occur. Early

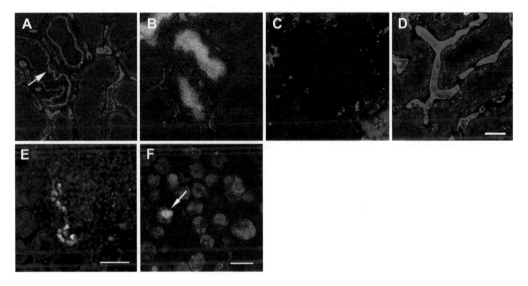

FIGURE 2 **Real-time, 2-photon microscopy of UPEC strain CFT073: GFP$^+$ (LT004) infecting the proximal tubule. Epithelium of the infected tubule (blue) and blood flow (red) are outlined by fluorophore-labeled dextran. Micro-infused bacteria are visualized by genetically encoded GFP$^+$ (green). A: Foci of infection 1 h post-infusion of LT004. Arrow points at bacteria adhering to the epithelium. B: Foci of infection 5 h post infusion. Massive green fluorescence indicates bacterial multiplication. C: Foci of infection 22 h post infusion. Lack of green fluorescence indicates clearance of bacteria. D: Non-infected renal tissue adjacent to the infection site shown in C. Scale bar = 30 μm. E: *Ex vivo* histological analysis of the foci of infection shown in C by confocal microscopy. Nuclear stain Hoechst 33342 (blue) and leukocyte marker α-CD18-Cy3 (red) have been added. Fluorophore-labeled dextran outlining the infected tubule is pseudocolored (yellow). Scale bar = 50 μm. F: Magnification of image E. The arrow highlights a neutrophil phagocytosing bacteria. Scale bar = 10 μm (From Månsson LE, et al. 2007. Real-time studies of the progression of bacterial infections and immediate tissue responses in live animals. Reprinted from *Cell Microbiol* (16), with permission.)**

tissue changes include vascular coagulation, epithelial breakdown, vascular leakage, immune cell recruitment, and general tissue destruction (15, 17). Coagulation in local peritubular capillaries, and subsequent vascular shut-down, occur within 5–6 h of infection, and these events are accompanied by a dramatic loss of local tissue oxygen (17). At this very early stage on infection, the host response is highly focused to the infection site. One significant finding is related to the rapidity of renal responses to a local infection, leading to bacterial clearance within 22 h (Figure 2C) with no effect on neighboring tubules (Figure 2D). Nevertheless, clearance appears to have come at the cost of local tissue destruction and vasculature shutdown. The resulting edema contains vast numbers of polymorphonuclear cells (PMNs) (Figure 2E, F), and this necrotic site showed great resemblance to abscesses seen on the superficial cortex in both the intravital model and the end-stage versions of the retrograde models of infection (16, 18). Whether the localized clearance of the infection is accompanied by bacterial colonization of distant nephrons is currently unknown.

Bacterial Colonization in the Dynamic Renal Tubular Environment

Whereas the reported shear stress for the renal tubule is low (19), bacteria must adhere to the tubulo-epithelium to withstand the associated hydrodynamic pressures. Shear stress in proximal tubules is estimated to be 0.17 dynes/cm^2; however, fluctuations in tubular reabsorption and secretion as the body regulates renal function alter the viscosity of the filtrate, which imply a degree of variability (19). Bacterial adhesion is thus an essential feature for successful colonization of the kidney. The wide and often redundant repertoire of UPEC attachment organelles includes P, type 1, F1C, S fimbriae, and Afa/Dr adhesins (20).

The P fimbriae, expressed in approximately 80% of the UPEC strains (21), are traditionally associated with pyelonephritis. A study undertaken comparing *E. coli* strains isolated from patients with acute pyelonephritis and those with asymptomatic bacteriuria showed that strains from the former group adhered in greater numbers to uroepithelium than the latter (22). Though renal colonization may also occur independently of PapG (18, 23–27), there is a strong indication that the presence of the P fimbriae increases the severity of infection, since PapG-positive clinical isolates induce more extensive renal damage as compared to the negative counterparts (21). When analyzing the role of P fimbriae for renal colonization using dynamic, intravital imaging, strains lacking PapG-mediated attachment showed compromised colonization kinetics with only a few bacteria visible before 8 h. This is in sharp contrast to the isogenic wild-type strain, which immediately established itself in the tubule. The important role of P fimbriae for early bacterial colonization *in vivo* was further demonstrated by the finding that successful, yet delayed, colonization only occurred in one-third of animals infected with the *papG* mutant strain compared to all animals infected with the P fimbriae-expressing counterpart. It is interesting to notice, however, that regardless of P-fimbriae-associated effects on the early colonization kinetics, the outcome 22 h post-infection was the same, with edema formation containing numerous neutrophils that had cleared the absolute majority of bacteria (14).

Synergistic Effects of P and Type 1 Fimbriae *In Vivo*

Traditionally, type 1 fimbriae have primarily been associated with bladder infections (18, 23–27). Gene expression studies on UPEC isolated from the kidney in the intravital model of infection showed, however, that bacteria also express type 1 during pyelonephritis. This suggests a novel role for type 1 fimbriae also in the upper compartments of the urinary tract. The ability to monitor bacterial behavior in real-time while exposed to physiological challenges of the nephron was critical to reveal a synergistic action between type 1 and P fimbriae. Intravital imaging showed that the combined action of the two fimbriae enabled UPEC as a population to resist the hydrodynamic pressure in the lumen of the proximal convoluted tubules. Whereas P fimbriae aid in the early stage of colonization by mediating binding to the epithelium, the type 1 fimbriae is essential for bacteria to colonize across the luminal center of the tubule while being exposed to the shear stress from the filtrate. Since bacteria are at risk to be washed away from the central part of the tubule lumen where no epithelium is available to hold on to, the population is forced to engage in inter-bacterial adhesion (Movie 2<http://asmscience.org/content/journal/microbiolspec/10.1128/microbiolspec.UTI-0014-2012>) (14). A role of type 1 fimbriae in inter-bacterial binding is corroborated by previous findings, showing that the FimH tip adhesin is essential for biofilm formation *in vitro* (28, 29). Inter-bacterial binding, as well as direct binding to biotic and abiotic surfaces, are key interactions for biofilm formation, making biofilms extraordinarily resistant to hydrodynamic-flow shear forces and allow bacteria to colonize perfused environments.

The binding capacity of the FimH tip adhesin is enhanced when exposed to shear stress of approximately the same magnitude as that which a bacterium experiences in the urinary tract (30, 31). This occurs through a force-enhanced allosteric catch-bond mechanism, operating via a finger-trap-like β sheet-twisting mechanism (30). The initial weak interaction of FimH shear-dependent binding is strengthened as the shear forces increase from 0.02 to 0.8 dynes/cm^2 (31). With the relaxation of shear strength, UPEC strains were observed to positively select for FimH variants that maintained attachment in comparison to fecal or vaginal *E. coli* isolates (32). The positive regulation of bacterial adhesion by hydrodynamic forces is particularly meaningful in the urinary tract where flow is pertinent albeit with fluctuations, allowing bacteria to bind and remain in the microenvironment rather than being washed out of the system.

Expression of P and type 1 fimbriae are controlled by 'phase variation', thereby allowing rapid adaptation and fine-tuning of the gene expression of the bacteria's adhesive nature in response to the microenvironment (33). At the same time, this allows for development of heterogeneous bacteria populations (34). The genetic *fim* switch controlling the expression of type 1 or P fimbriae is an example of phase variation. This invertible genetic element carries the main promoter for the expression of fimbrial-structural subunits (34). PapB represses the FimB-promoted off-to-on inversion of the *fim* switch (35). Similarly, the PapX protein in UPEC strain CFT073 binds to the *flhD* promoter and represses the transcription of the master regulator of flagella FlhD2C2, thereby negatively regulating motility (36). It can thus be envisaged that UPEC is able to tightly regulate the antagonistic forces of adhesion and motility in order to colonize the urinary tract.

Bacterial Toxins Affect the Kinetics of Host Responses

UPEC express several proteinaceous toxins. Though commonly referred to as virulence factors, their precise role in disease is still unclear. The virulence factor concept is further complicated by the notion that toxins are not ubiquitously expressed among all UPEC isolates. The lipoprotein α-hemolysin (Hly), considered to be an important UPEC virulence factor, is only expressed by circa 50% of *E. coli* isolates implicated in pyelonephritis. The traditional view of Hly as a pore-forming toxin had to be revised when it was found that Hly exerts biphasic, concentration-dependent effects on host systems (37). Whereas higher concentrations of Hly are cytolytic for a variety of cells, including erythrocytes, epithelial cells, polymorphonuclear leukocytes, monocytes, mast cells, basophils, and lymphocytes, sub-lytic concentrations were shown to induce pro-inflammatory responses in renal epithelial cells via a mechanism involving Ca^{2+} signaling (38–40). As Hly interacts with the plasma membrane, voltage-operated Ca^{2+} channels allows for the influx of extracellular Ca^{2+} while at the same time the IP_3 receptor aids in establishing an oscillating intracellular Ca^{2+} response. This signaling exerts frequency-dependent activation of nuclear factor kappa-light-chain-enhancer of activated B cells (NF-κB), with concomitant increased production of the pro-inflammatory cytokines interleukin (IL)-6 and IL-8 (40).

While colonizing the kidney, UPEC have indeed been shown to express Hly (16). To analyze a role for Hly *in vivo*, intravital imaging using either a wild-type or an isogenic *hlyA* mutant strain showed that bacterial colonization of the proximal tubule was equally efficient, whether or not Hly was expressed (40). A striking difference was, however, observed when studying the host response. In the absence of Hly, delayed kinetics of the tissue responses, which involved the onset of ischemia, obstruction, and immune cell recruitment, were observed (16). Indirectly, these findings correspond to reduced pro-inflammatory signaling in the absence of Hly; however, the precise role for Hly-induced Ca^{2+} signaling *in vivo* remains to be analyzed. Important to note is, however, that the same end point of infection at 24 h is achieved regardless of Hly expression.

Innate Immune Responses

Toll-like receptors (TLR) play an important function to alert the host innate immune system to the presence of pathogens through detection of pathogen-associated molecular patterns. Binding of these leads to the dimerization of TLRs and the activation of TLR signaling by co-receptor recruitment or engagement, typically inducing a pro-inflammatory responses (41). Among the 10 and 12 members of TLRs that have been characterized in humans and mice, respectively, TLR4, TLR5, and TLR11 have been associated with the urinary tract.

Uroepithelial cells express TLR4, which confers sensitivity to Gram-negative bacteria via lipopolysaccharide (LPS) detection. While the system is uniquely dependent on soluble CD14 in body fluids, the level of LPS detection of uroepithelial cells is as sensitive as that of macrophages (42–45). In uninfected kidneys, TLR4 is mainly found on the apical surface of the distal tubules (46). However, when the host is experiencing septicemia, TLR4 expression becomes ubiquitous across all segments of the kidney suggesting that TLR4 is up-regulated during inflammation (46, 47). This variability may explain the discrepancies of inconsistent reports on TLR4 presence on renal cells (42, 45). It remains, however, that TLR4 plays a role in protecting the host from UTI. Studies of the genetic relationship of TLR4 with UTI have shown an association with different types of UTI. For example, children with reduced TLR4 expression on neutrophils have a higher prevalence of developing asymptomatic bacteriuria (48, 49). Individuals with the TLR4 A(896)G allele are also more susceptible to recurrent UTI (50).

In mice models of UTI, TLR5, which binds bacterial flagellin (51), as well as TLR11 (ligand unknown) (52) are important pattern-recognition receptors (PRRs) involved in the innate immune system. TLR signaling induces the production of cytokines, which in turn coordinate the immune response. A well-known subset of this response is the extravasation of inflammatory cells such as (PMNs), or neutrophils, along a chemotactic gradient of cytokines to the site of infection in the tissue. Transepithelial PMN migration is promoted by the main human chemokine IL-8 and the corresponding receptor CXCR1 (53). The importance of this particular signaling pathway is highlighted by an increased susceptibility to pyelonephritis in individuals with polymorphisms and mutations in the *CXCR1* gene (54).

The neutrophil is the primary cell typed involved in clearing a bacterial infection. Intravital imaging suggested, however, the involvement of other cell types in both the early inflammatory response as well as in the clearance of bacteria. PMN recruitment is observed as early as 4 h post-infection; however, other cell types of hitherto unidentified origin also appear at the site. At 8 h post-infection, PMNs constituted only 20% to 40 % of recruited nucleated cells and several hours ensued before neutrophils became the predominant cell type at the site (16). PMNs play a significant role in bacterial clearance via phagocytosis, a process that may occur independently of prior opsonization.

While neutrophils are necessary for the clearance of infection, their recruitment is also linked to severe tissue damage. Neutrophil granules released in response to the infection contain anti-microbial peptides, proteins, and proteolytic enzymes, which may exert acute and/or permanent detrimental effects on the extracellular matrix, cell structures, or functions (53). Evidently, PMNs isolated from acute pyelonephritic exudates cause lysis of a number of cell types *in vitro* within 24–48 h (55). Suppression of suppuration was found to reduce tubular epithelial cell damage and renal scarring albeit a higher bacterial burden (56, 57). In contrast to the traditional bacteriocentric view on tissue damage, the integrative view describes tissue damage as the result of several facets of an infection originating from both pathogen and host (14). This includes the inflammatory response as well as physiological injuries, such as ischemia and obstruction, during an infection.

Antimicrobial proteins and peptides produced by different parts of the nephrons' urothelium constitute a major host defense mechanism against invading microbes. In contrast to TLRs, antimicrobial peptides can have a direct antimicrobial effect on pathogens, or indirectly function by modulating innate and adaptive immune responses. Antimicrobial peptides and proteins of importance in the kidney include the Tamm-Horsfall protein (THP), defensins, cathelicidins, lactoferrin, and lipocalin. THP is an evolutionarily conserved glycoprotein highly abundant in human urine. This protein is produced specifically by epithelial cells at the ascending loop of Henle (58, 59). THP stimulates both parts of the host's innate and adaptive immune responses (60, 61). It is involved in cytokine production, cell-specific stimulation of granulocytes towards IL-8 production (60, 62), up-regulation of co-stimulatory molecules, major histocompatibility complex (MHC) expression on dendritic cells (DC), and DC maturation via TLR4 signaling. Correspondingly, the wide-reaching and strong effect of THP on immune responses can lead to interstitial nephritis when excessively stimulated (60, 61).

Defensins constitute another group of antimicrobial peptides of renal importance. Whereas β-defensins are produced by the local renal epithelium (61, 62), α-defensins are secreted from infiltrating neutrophils. These peptides exert dual effect, either showing direct antimicrobial activity on invading bacteria, or causing an indirect enhancement of the innate and acquired immune response. Via induction of secondary signaling from cells and tissues, defensins are indirectly affecting immune cell recruitment, the regulation of acute inflammation, angiogenesis, and wound healing (61).

Cathelicidins are produced and subsequently released into the tubular lumen of the urinary tract (63). Some peptides of this family, such as LL-37, are also produced in a wide variety of locations in the host, including epithelial cells of the skin, the gastrointestinal tract, the epididymis, lungs, neutrophils, and myeloid bone marrow cells (64, 65). Cathelicidin-based antimicrobial response to invading pathogens constitutes a two-stage process. Prior to leukocyte infiltration early during infection, cathelicidins are mainly produced by the epithelium. As the infection progress, cathelicidin production shifts to the recruited neutrophils (63).

Intriguingly, the effect of cathelicidins is not impartial. The impact of the peptide on uropathogenic bacteria appears to be stronger than that with urogenital commensal bacteria (66). On the flipside, UPEC strains that cause severe UTI are likely to have a higher resistance to this peptide (66).

Infection-Associated Disruption of Mucosal Integrity

The intravital model of acute pyelonephritis reveals that local tissue destruction at infection site results from extremely rapid host responses (16). A cascade of signaling events is triggered, whose outcomes are observable as early as within 3 to 4 h from the entry of bacteria into the tissue. This includes massive rearrangement of the tubuloepithelial actin cytoskeleton, as well as sloughing of epithelial cells from the basement membrane (16, 17). The tight link between destruction of actin cytoskeleton and interference of the microfilament system, integrins, immunoglobulins, and cell-adhesion molecules are known to contribute to the disruption of epithelial-barrier function (67, 68). Additionally, intravital studies of non-infected kidneys have demonstrated that hypoxia-induced rearrangement of renal cortical-actin cytoskeleton interferes with cell–cell adhesion as well as cell–extracellular matrix adhesion, thereby disrupting tubular integrity (69–71).

A similar situation prevails in the UPEC-infected nephron, where epithelial signaling rapidly causes localized ischemia (see next section) (16). Dynamic imaging shows ischemic injuries in the infected nephron, including the typical signs of local vascular leakage and loss of epithelial membrane-barrier functions, as soon as 4 h after onset of infection (Fig. 3 and Movie 3<http://asmscience.org/content/journal/microbiolspec/10.1128/microbiolspec.UTI-0014-2012>). A dynamic view on these events can be obtained using

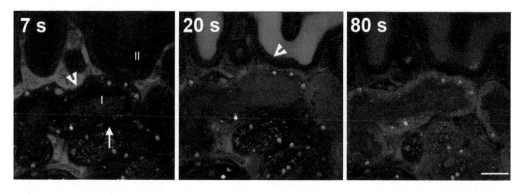

FIGURE 3 Epithelial-barrier disruption and impaired renal filtration due to UPEC infection. Real-time 2-photon imaging of the foci of infection 4 h post-infusion of UPEC strain LT004. Images obtained 7, 20, and 80 s after intravenous bolus infusion of fluorophore-labeled 10 kilodalton (kDa) dextran are shown (red). In the non-infected nephron (upper part of figure), efficient filtration is observed as the tubular appearance of the bright-red fluorescence arising from the labeled dextran (20 s) is followed by an obvious drop in intensity (80 s). This indicates renal clearance. Renal obstruction of the infected nephron (lower part of figure) is observed as only limited fluorescent dextran enters the tubule. Epithelial-barrier function is destroyed in the infected tubule, as dextran is observed to enter the epithelial layer (arrowhead, 7 s), suggestive of epithelial and endothelial dysfunction. In contrast, the healthy epithelia in the non-infected nephron exclude the dextran (arrowhead, 20 s). Vascular clotting can also be observed in the vasculature next to the infected nephron (arrow, 20 s). (From Melican K, et al. 2011. Uropathogenic *E. coli* P and type 1 fimbriae act in synergy in a living host to facilitate renal colonization leading to nephron obstruction. Reprinted from *PLoS Pathog* (14), with permission.)

2-photon microscopy to visualize the fate of systemically injected small molecular weight red-fluorescent dextran. Immediately upon injection, the red dextran is observed within the peritubular capillaries (Fig. 3, which represents selected images from Movie 3). In a non-infected nephron (upper part of Fig. 3 and Movie 3), dextran is filtered by the glomerulus and passes swiftly through the tubule lumens. The infected nephron showed a dramatically different situation (lower part of Fig. 3 and Movie 3). Loss of epithelial membrane-barrier function is obvious, as red dextran leaks into the epithelial cells. Careful inspection of the data reveals that leakage actually is initiated from the basolateral side, indicating a major contribution of vascular leakage in the interruption of epithelial membrane integrity. Inspection of the epithelial linings of the neighboring nephrons reveals that membrane disruption is strictly confined to the infected nephron.

A common assumption is that breakdown of epithelial integrity would open for further dissemination of bacteria, including systemic spread. It appears, however, that the tubular basement membrane confers an additional barrier function, attributed to the intact layer of connective tissue (17). Though bacteria can reach the basolateral side of the epithelium via paracellular migration, bacteria are confined within the tubule, lining up against the collagen IV-rich basement membrane. It can be hypothesized that the concerted actions of epithelial and basement-membrane barriers are important for the appropriate timing of the host response, keeping bacteria on-site to give sufficient time for an adequate, but not excessive, antimicrobial response to be mounted. Data shows that when UPEC eventually passes these barriers, neutrophils as well as mononucleated cells have arrived to the site to aid in solving the situation (16). Loss of tubular integrity may therefore permit unbiased paracellular movements allowing both local pathogen migration as well as the extravasation of host neutrophils at the site of infection.

From a clinical perspective, infection-associated breakdown of the epithelial integrity combined with cellular detachment and sloughing from the basement membrane strongly argue for a minor, if any, role for the tubule as a persistence reservoir for UPEC in the upper urinary tract. This is in contrast to the lower urinary tract where recurrent cystitis is considered to depend on UPEC persisting intracellularly in the underlying squamous epithelial layer of the bladder (72).

Clotting, Ischemia, and Hypoxia Prevents Dissemination

At the infected nephron, local hypoxia has been linked to the characteristic signs of ischemic injuries. Probing the local tissue oxygen tension using Clark-type microelectrodes (73) reveals that the infection site is completely devoid of oxygen within 4 h (oxygen tension = 0 mmHg) (17). *In vitro* recording of oxygen consumption demonstrated that increased metabolic activity of UPEC-infected primary proximal tubule cells contributed to this effect. The high metabolic activity correlated to up-regulation of the pro-inflammatory cytokines tumor necrosis factor (TNF)-α, IL-1β, and IL-6 *in vivo* and *in vitro*, demonstrating the important role of the proximal tubule epithelium as an early responder to infection.

Onset of ischemia occurs rapidly, prior to neutrophil infiltration, suggesting the involvement of rapid infection-associated cell signaling. Intravital imaging shows that black silhouettes of the size of platelets are present in the peritubular capillaries, and eventually, clots are observed in the peritubular capillaries (Fig. 4). Analysis of the local mRNA expression profile confirms that the vasculature shutdown is due to activation of the clotting cascade. Clot formation is usually associated with blood-borne infections (74); however, as local ischemia develops in the infected nephron, UPEC remain in the lumen of the proximal tubule with no direct contact to the

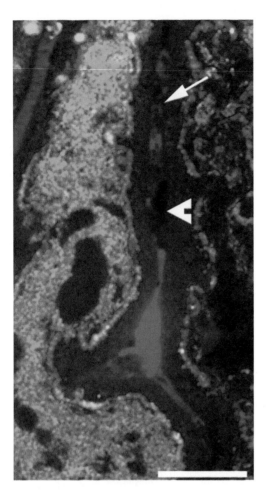

FIGURE 4 Clot formation in mucosal infections. Real-time 2-photon imaging of a UPEC strain LT004 (green)-infected nephron (blue) shows platelets (arrow) in the form of black silhouettes surrounded by blood (red) within the peri-tubular vasculature, 2.5 h post-infection. Black masses adhered to the vessel wall (arrow head) suggest the presence of platelet aggregates. The high-intensity-red fluorescence indicates stagnant blood flow and a lack of red blood cell movement. Scale bar = 30 μm. (From Melican K, et al. 2008. Bacterial infection-mediated mucosal signaling induces local renal ischemia as a defence against sepsis. Reprinted from *Cell Microbiol* (17), with permission.)

endothelium. Clot formation thus occurs as a consequence of infection-induced molecular epithelial–endothelial crosstalk of hitherto unknown nature.

Clot formation is shown to provide a novel innate immune-defense mechanism, as it protects the host from bacterial dissemination and systemic spread. By inhibiting clot formation via anticoagulant therapy, bacteria disseminate rapidly into the systemic circulation, causing death by septic shock within a few hours. Post-mortem examination of the succumbed animals shows bacterial spread to the blood, heart, liver, and spleen (17). Hence, clot formation adds to the list of host responses to mucosal infection as cessation of vascular flow helps to contain bacteria locally at the infection site, while neutrophil infiltration and other defense mechanisms become activated.

Infection-Associated Nephron Obstruction

Glomerular filtration is tightly associated with renal UPEC infection. Featured both as tubuloepithelially attached and free aggregates, the increasing bacterial population has severe impact on renal filtration, leading to obstruction of the nephron (14). Sloughing of the tissue, cast formation, and cellular blebbing following ischemic injury further adds to the tubular obstruction (75). This is a severe condition, since physical obstruction of the nephron can rapidly lead to acute renal failure (6).

Obstruction of the filtrate flow leads to major changes in hydrostatic pressure and dilation of the tubules (75). The increased intra-renal pressure can also effect renal filtration, leading to critical reduction of the glomerular-filtration rates (GFR) (75). In severe conditions, associated injuries may extend to arteriole vasoconstriction and a drop in renal blood flow (75, 76). Actually, minute injuries, such as single-nephron obstruction, are severely affecting the pathophysiology of renal injuries, as local induction of an inflammatory response, tubular-cell injury, changes in glomerular-capillary pressure and eventual disuse atrophy have been demonstrated (76–80). A strong link between infection and renal hydrodynamics is further illustrated as LPS is shown to impair ion-transport functions of renal tubules in a TRL4-dependent manner (81, 82).

Feedback between the tubules and renal vasculature, so-called tubuloglomerular feedback (TGF), is regulated by the juxtaglomerular apparatus (JGA) (83). The JGA is composed of specialized distal tubule epithelial cells called the macula densa, afferent arteriole, and juxtaglomerular cells. Found between the vascular pole of the renal corpuscle and the returning distal convoluted tubule, changes in glomerular filtrate volume and osmolarity is sensed by macula densa and juxtaglomerular cells, which in turn modulate the renal system through renin and vasopressin release, respectively (84). Whether the renal injuries associated with UPEC infection induces a JGA-directed response is currently unknown.

Inter-Organ Communication During Infection

Renal injury, such as ischemia, can initiate molecular crosstalk of inflammatory nature between the kidney and distant organs (85). Studies of such complex events call for the use of an all-inclusive intravital model of infection. A transcriptomic analysis of biopsies obtained from the spatio-temporally well-defined intravital model of pyelonephritis showed that the host response involves active crosstalk at the site of infection. Microarray data from biopsies isolated within the first 8 h of infection revealed circa 60 genes to be significantly differentially expressed as compared to non-infected controls (86). A Gene Ontology (GO) analysis showed these genes to cluster into functionally defined categories, many of which are associated to inflammation. The tissue response is, however, very versatile, and it embraces other GO categories, such as wound healing, response to hypoxia, cell death, and apoptosis. Collectively, this confirms the multifaceted events, such as ischemia and tissue

disruption, observed in the real-time model of infection.

A comparative tissue transcriptomics approach applied to study early (8 h post-infection) tissue responses in pyelonephritis revealed a common core of 80 significantly upregulated genes in Gram-negative infection and inflammation (86). Interestingly, 25% of identified genes are interferon (IFN)-γ regulated. A marked increase of IFN-γ serum levels is accompanied by upregulation of splenic *ifng* gene expression involving the IL-17/IL-23 pathway (86). This suggests a rapid engagement of the local infection site in communicating with distant sites. The mechanisms governing this communication remain to be identified.

ACUTE KIDNEY INJURY

Acute kidney injury (AKI) of septic or non-septic origin is a common clinical complication that can occur from a range of kidney pathologies, to which renal failure is the end result. The pathophysiology includes renal hypoperfusion and ischemic injury, cardiogenic or distributive shock, followed by tubular necrosis (87, 88). More recently, septic AKI has been linked to maintained global renal (medullar and cortical) blood flow (89), highlighting that symptoms and treatments of infection-associated kidney injuries differ amongst septic cases. In the U.S.A., AKI shows a varying incidence of 1% (community-acquired) up to 7.1% (hospital-acquired) of all hospital admissions (90, 91).

A number of mechanisms are proposed by which UPEC infections induce renal injury. Damaging toxins expressed by infecting bacteria are considered major factors for tissue destruction (55, 57, 92, 93). This is acting in parallel to the damaging effects from the strong inflammatory response. Upon immune suppression, renal scaring is reduced despite a consequential increased bacterial load (94, 95). Neutrophils isolated from acute pyelonephritic exudates are known to kill syngeneic renal cells *in vitro*, which is corroborated by *in vivo* studies, showing neutrophil-induced oxidative injury of renal cells (17). The intravital model of pyelonephritis identified local ischemia occurring only hours after the first bacterium entered the tubule, which together with bacterial multiplication in the tubule lumen led to severe obstruction (17). Both are complex syndromes to which severe episodes of either condition can result in renal scarring (96) and lead to end-stage renal failure.

Cell death associated to ischemic renal injury occurs either via necrosis or via induction of apoptotic-signaling pathways. Adenosine triphosphate (ATP) and guanosine-5′-triphosphate (GTP) act as small-molecule inducers of either one of the pathways (97). In the latter situation, hypoxia, resulting from a combination of reduced local blood flow and increased respiratory demand, depletes the pool of the small nucleotide GTP, thus altering the ratio of GTP/guanosine diphosphate (GDP) that is known to regulate the GTPase activity (97). It is postulated that GTP depletion induces apoptosis through Ras and Rho family GTPases (97).

Despite the original cause, AKI is not an isolated event. Inflammatory mechanisms causing increased cytokine production, oxidative stress, edema, and leukocyte trafficking may act in concert to instigate dysfunction of distal organs. For a comprehensive review of extrarenal organ dysfunction, the reader is referred to Yap et al. (87). Uncontrolled activation of inflammation in response to AKI leads to increased epithelial and endothelial apoptosis, increased membrane/vascular permeability, and effects on water channels (aquaporin) and sodium-potassium pumps (87, 98). In the lung, histological and physiological changes in response to septic AKI are associated with the activity IL-1, IL-6, TNF-α, and macrophages, as intervention therapies conferred attenuation of the pathways leading to dysfunction (99, 100). Within the gastrointestinal tract, AKI leads to increase in IL-17A, which brings about histological alterations through increased necrosis and apoptosis (101). Additionally, a major increase of pro-inflammatory cytokines such as IL-6 and TNF-α is observed (87). As cytokines are being drained into the liver through the portal circulation, this organ will also succumb to inflammation-related cell death and tissue damage (102–104). Pertaining to AKI and cardiac function, increased cytokines TNF-α, IL-6, and IL-1 and neutrophil infiltration are clinically associated with congestive heart failure (105).

TREATMENT

As the primary cause of pyelonephritis is a bacterial infection, the treatment of the condition is similar to that of UTI, which is mainly by the application of antibiotics. For in-depth reading into UTI treatment, the reader is referred to Wagenlehner et al. (106), which provides extensive statistics and efficacy analysis of clinically prescribed antibiotics. In Wagenlehner's analysis, the options for UTI treatment is based on the assignment of the patient into categories of uncomplicated and complicated UTI. (106, 107). However, treatment of both UTI categories follows the fundamental aim of providing fast and efficient intervention against recurrent infections and prevention of resistance generation, as well as exacerbation (106).

Wagenlehner et al. (106) defines the two categories as follows:

"Uncomplicated UTI denotes UTI without relevant structural and functional abnormalities arising from the urinary tract (uropathies), without relevant kidney diseases (nephropathies) and without relevant comorbidities. Conversely, complicated UTI is a complex condition of the following conditions: (1) Anatomical, structural or functional alterations of the urinary tract. (2) Impaired renal function by parenchymal and renal nephropathies. (3) Accompanying diseases or conditions that impair the patients' immune status."

Acute pyelonephritis is a complex syndrome, which may not consistently exhibit the same symptoms at the time of diagnosis. In addition, this stage of infection can open to a wide variety of complications. Treatment of acute pyelonephritis therefore follows under the category of complex UTI. Complex UTI follows a bidirectional approach in which treatment is directed to remove the disease-causing pathogen, as well as the complicating factors arising from the infection (107).

Calibration of the antibiotic treatment follows several criteria (106). An individual's risk when undergoing antibiotic treatment, the pathogen's sensitivity spectrum and concentration to antibiotics, clinical efficacy of an antibiotic, possible effects on the commensal microbiota, side effects, resistance development, and the possibility that the causative agent in complex UTI may be polymicrobial in nature with wide-spectrum antibiotic resistances (106).

Acute pyelonephritis may include or progress into permanent damage to the tissue, organ failure, and urosepsis. Urosepsis is treated by a combination of approaches, which include removal of the infection source, eradication of the existing infection, application of countermeasures against complications, and life-supportive care (106).

Folk remedies have also been used widely as a treatment for UTI infection. However, application is likely to be more suited to the category of uncomplicated UTI. Among the wide range in the market, one common folk remedy is cranberry juice. Also available in tablet form, cranberries contain fructose and proanthocyanidins, which inhibits bacterial adherence via type 1 and P fimbriae, respectively (108–110). However, clinical trials have yet to show definitive evidence supporting its efficacy as UTI treatment.

FUTURE PROJECTIONS

The pathophysiology of pyelonephritis is evidently a complex scenario, which changes dynamically over time. It involves a combi-

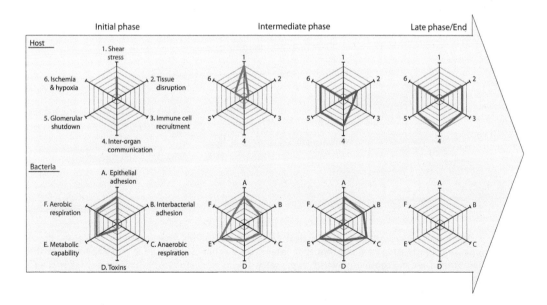

FIGURE 5 Pathophysiogram of pyelonephritis illustrates how the microenvironment of infected tissue changes dynamically during infection (upper panel), and the associated changes relevant for bacterial growth (lower panel). Axis of each plot represents the degree of intensity and/or involvement (arbitrary units) of each defined trait during the time course of the infection.

nation of factors originating from the infection as well as from physiological injuries, such as ischemia and obstruction. The latter are both well-studied physiological injuries, known to cause inflammation and tissue destruction in their own right (67–70, 77, 78, 111–114). Adding the tissue alterations associated with bacterial colonization, such as coagulation, epithelial breakdown, vascular leakage, immune cell recruitment, and tissue destruction, further complicates the picture. The homeostatic imbalance at the local site of infection, as well as involvement of distal signaling and engagement of multiple cell types, acts in concert over time to manage the infection. The pathophysiological changes occurring during infection can be schematically summarized in a 'pathophysiogram of pyelonephritis.' As Fig. 5 illustrates, the multi-faceted host response will dramatically alter the local microenvironment, forcing bacteria to adapt physiologically in order to maintain themselves. The integrated pathophysiology of pyelonephritis thus constitutes a number of integrated events in which the host and the microbes mutually influence each other over time. Full understanding of these events is required to coherently define relevant interactions and identify new potential angles for disease intervention.

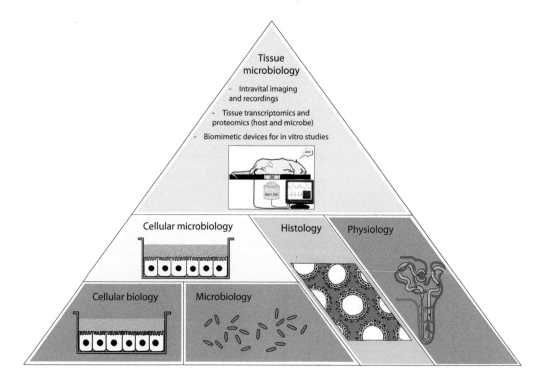

FIGURE 6 Schematic representation of classical research disciplines that coalesce to form 'tissue microbiology', a recently proposed concept that integrates a range of disciplines and expertise. At the base of the pyramid are microbiology, which represents the *in vitro* study of microbial pathogens, and cellular biology, which focuses on the study of host cell types. Cellular microbiology was a discipline formed where microbiology and cellular biology overlapped. Coined in 1996, the approach was aimed at studying host–pathogen interactions using another's perspectives, tools, and competences. Histology and physiology remained individual and required separate analysis. As knowledge and technological advancements make for unprecedented tools for intravital studies, tissue microbiology can now be established, advancing infection biology with an all-inclusive approach to generate an integrated view of the pathophysiology of infection.

In future microbial pathogenesis research, it can be predicted that focus will be placed on monitoring and mimicking host-pathogen interaction within the dynamic micro-ecology significant for infectious niches in the live host. This all-inclusive approach, which integrates all elements of 'cellular microbiology', 'histology', and 'physiology' when studying infections in real-time, is termed 'tissue microbiology' (Fig. 6) (17, 115, 116). During the time-course of infection, the assayed organ acts as the test tube in which the experimental parameters are set by the tissue's own response to infection. In this changing environment, the roles certain bacterial virulence factors play *in vivo* can be addressed, either as individual factors or as multiple factors acting in synergy. This strategy will allow for identification of any subtle effects virulence or fitness factors may exert on bacterial-colonization kinetics, or effects on the kinetics of the host response. Furthermore, this will allow for more precise definitions of the infectious niche. Instead of an organ-based definition, i.e., bladder versus kidney, these niches will be precisely defined as sub-compartments within the tissue. In pyelonephritis, this is exemplified by the center or periphery of a single tubule lumen (see previous section 'Synergistic effect of fimbriae to cope with urine shear stress).

Successful implementation of the novel area of tissue microbiology will, however, require the cooperation between different research disciplines, ranging from engineering sciences, pre-clinical research, and clinical research. Nanomedicine is an expanding field that is likely to produce novel tools with great promise to propel intravital studies into a new era. Also, we are likely to see the development of advanced biomimetic *in vitro* systems that closely resemble the *in vivo* microenvironment. Using these integrated *in vivo* and *in vitro* approaches, our understanding of host-pathogen interactions will advance, and so will clinical diagnostics, treatment, and patient care.

ACKNOWLEDGMENTS

We thank the members and partners of the Swedish Medical Nanoscience Center at Karolinska Institutet and at The Indiana Center for Biological Microscopy for helping to establish a milieu that promotes broad-minded thinking and cross-disciplinary science.

Conflicts of interest: We declare no conflicts.

CITATION

Choong FX, Antypas H, Richter-Dahlfors A. 2015. Integrated pathophysiology of pyelonephritis. Microbiol Spectrum 3(5):UTI-0014-2012.

REFERENCES

1. **Bostwick DG (ed).** 1999. *Uropathology.* Urologic Clinics of North America, **vol 26** (4):677–828. WB Saunders Co, Philadelphia, PA.
2. **Vercellone A, Stratta P.** 1990. Tubulo-interstitial nephropathies, p 197–205. *In* Amerio A, Coratelli P, Massry S (ed), *Proceedings of the 4th Bari Seminar in Nephrology.* Kluwer Academic Publishers, Boston, MA.
3. **Ronald AR, Nicolle LE..** 2002. Infections of the upper urinary tract, p 845–869. *In* Schrier RW (ed). *Diseases of the Kidney and Urinary Tract,* 7th ed. Lippincott Williams & Wilkins, Philadelphia, PA.
4. **Ifergan J, Pommier R, Brion MC, Glas L, Rocher L, Bellin MF.** 2012. Imaging in upper urinary tract infections. *Diagn Interv Imaging* **93:**509–519.
5. **Eaton DC, Pooler JD, Vander AJ.** 2009. *Vander's Renal Physiology,* 7th ed. McGraw-Hill Medical, New York, NY.
6. **Lote CJ.** 2000. Glomerular filtration, p 34–36. *In Principles of Renal Physiology.* Kluwer Academic Publishers, Boston, MA.
7. **Koeppen BM, Stanton BA.** 2007. Structure and function of the kidneys, p 228. *In Renal Physiology,* 7th ed. Mosby Elsevier, Philadelphia, PA.
8. **Koeppen BM.** 1987. Electrophysiology of ion transport in renal tubule epithelia. *Semin Nephrol* **7:**37–47.
9. **Jackson G, Grieble HG.** 1957. Pathogenesis of renal infection. *AMA Arch Intern Med* **100:** 692–700.

10. **Finer G, Landau D.** 2004. Pathogenesis of urinary tract infections with normal female anatomy. *Lancet Infect Dis* **4:**631–635.

11. **Frendéus B, Godaly G, Hang L, Karpman D, Lundstedt AC, Svanborg C.** 2000. Interleukin 8 receptor deficiency confers susceptibility to acute experimental pyelonephritis and may have a human counterpart. *J Exp Med* **192:**881–890.

12. **Ronald AR, Nicolle LE.** 2007. Infections of the upper urinary tract. *In* Schrier RW (ed), *Diseases of the Kidney and Urinary Tract*, 8th ed. Lippincott Williams & Wilkins, Philadelphia, PA.

13. **Foxman B.** 2010. The epidemiology of urinary tract infection. *Nat Rev Urol* **7:**653–660.

14. **Melican K, Sandoval RM, Kader A, Josefsson L, Tanner GA, Molitoris BA, Richter-Dahlfors A.** 2011. Uropathogenic *Escherichia coli* P and type 1 fimbriae act in synergy in a living host to facilitate renal colonization leading to nephron obstruction. *PLoS Pathog* **7:**e1001298. doi:10.1371/journal.ppat.1001298

15. **Melican K, Boekel J, Ryden-Aulin M, Richter-Dahlfors A.** 2010. Novel innate immune functions revealed by dynamic, real-time live imaging of bacterial infections. *Crit Rev Immunol* **30:**107–117.

16. **Månsson LE, Melican K, Boekel J, Sandoval RM, Hautefort I, Tanner GA, Molitoris BA, Richter-Dahlfors A.** 2007. Real-time studies of the progression of bacterial infections and immediate tissue responses in live animals. *Cell Microbiol* **9:**413–424.

17. **Melican K, Boekel J, Mnsson L, Sandoval R, Tanner G, Kllskog P, Molitoris B, Richter-Dahlfors A.** 2008. Bacterial infection-mediated mucosal signalling induces local renal ischaemia as a defence against sepsis. *Cell Microbiol* **10:**1987–1987.

18. **Mobley HL5, Jarvis KG, Elwood JP, Whittle DI, Lockatell CV, Russell RG, Johnson DE, Donnenberg MS, Warren JW.** 1993. Isogenic P-fimbrial deletion mutants of pyelonephritogenic *Escherichia coli*: The role of alpha Gal(1-4) beta Gal binding in virulence of a wild-type strain. *Mol Microbiol* **10:**143–155.

19. **Essig M, Friedlander G.** 2003. Tubular shear stress and phenotype of renal proximal tubular cells. *J Am Soc Nephrol* **14**(Suppl 1)**:**S33–S35.

20. **Wright KJ, Hultgren SJ.** 2006. Sticky fibers and uropathogenesis: bacterial adhesins in the urinary tract. *Future Microbiol* **1:**75–87.

21. **Donnenberg MS, Welch RA.** 1996. Virulence determinants of uropathogenic *Escherichia coli*. *In* Mobley HLT, Warren JW (ed), *Urinary Tract Infections: Molecular Pathogenesis and Clinical Management*. ASM Press, Washington, DC.

22. **Edén CS, Hanson LA, Jodal U, Lindberg U, Akerlund AS.** 1976. Variable adherence to normal human urinary-tract epithelial cells of *Escherichia coli* strains associated with various forms of urinary-tract infection. *Lancet* **1:**490–492.

23. **Johnson JR.** 1991. Virulence factors in *Escherichia coli* urinary tract infection. *Clin Microbiol Rev* **4:**80–128.

24. **Lane MC, Mobley HL.** 2007. Role of P-fimbrial-mediated adherence in pyelonephritis and persistence of uropathogenic *Escherichia coli* (UPEC) in the mammalian kidney. *Kidney Int* **72:**19–25.

25. **Plos K, Carter T, Hull S, Hull R, Svanborg Edén C.** 1990. Frequency and organization of pap homologous DNA in relation to clinical origin of uropathogenic *Escherichia coli*. *J Infect Dis* **161:**518–524.

26. **Hagberg L, Hull R, Hull S, Falkow S, Freter R, Svanborg Edén C.** 1983. Contribution of adhesion to bacterial persistence in the mouse urinary tract. *Infect Immun* **40:**265–272.

27. **Roberts JA, Marklund BI, Ilver D, Haslam D, Kaack MB, Baskin G, Louis M, Möllby R, Winberg J, Normark S.** 1994. The Gal(alpha 1-4)Gal-specific tip adhesin of *Escherichia coli* P-fimbriae is needed for pyelonephritis to occur in the normal urinary tract. *Proc Natl Acad Sci U S A* **91:**11889–11893.

28. **Pratt LA, Kolter R.** 1998. Genetic analysis of *Escherichia coli* biofilm formation: roles of flagella, motility, chemotaxis and type I pili. *Mol Microbiol* **30:**285–293.

29. **Schembri MA, Sokurenko EV, Klemm P.** 2000. Functional flexibility of the FimH adhesin: insights from a random mutant library. *Infect Immun* **68:**2638–2646.

30. **Le Trong I, Aprikian P, Kidd BA, Forero-Shelton M, Tchesnokova V, Rajagopal P, Rodriguez V, Interlandi G, Klevit R, Vogel V, Stenkamp RE, Sokurenko EV, Thomas WE.** 2010. Structural basis for mechanical force regulation of the adhesin FimH via finger trap-like beta sheet twisting. *Cell* **141:**645–655.

31. **Thomas WE, Trintchina E, Forero M, Vogel V, Sokurenko EV.** 2002. Bacterial adhesion to target cells enhanced by shear force. *Cell* **109:**913–923.

32. **Ronald LS, Yakovenko O, Yazvenko N, Chattopadhyay S, Aprikian P, Thomas WE, Sokurenko EV.** 2008. Adaptive mutations in the signal peptide of the type 1 fimbrial adhesin of uropathogenic *Escherichia coli*. *Proc Natl Acad Sci U S A* **105:**10937–10942.

33. **Snyder JA, Haugen BJ, Lockatell CV, Maroncle N, Hagan EC, Johnson DE, Welch RA, Mobley HL.** 2005. Coordinate expression of fimbriae in uropathogenic *Escherichia coli. Infect Immun* **73:** 7588–7596.

34. **Holden NJ, Gally DL.** 2004. Switches, crosstalk and memory in *Escherichia coli* adherence. *J Med Microbiol* **53:**585–593.

35. **Holden NJ, Uhlin BE, Gally DL.** 2001. PapB paralogues and their effect on the phase variation of type 1 fimbriae in *Escherichia coli. Mol Microbiol* **42:**319–330.

36. **Simms AN, Mobley HL.** 2008. PapX, a P fimbrial operon-encoded inhibitor of motility in uropathogenic *Escherichia coli. Infect Immun* **76:**4833–4841.

37. **Laestadius A, Richter-Dahlfors A, Aperia A.** 2002. Dual effects of *Escherichia coli* alpha-hemolysin on rat renal proximal tubule cells. *Kidney Int* **62:**2035–2042.

38. **Trifillis AL, Donnenberg MS, Cui X, Russell RG, Utsalo SJ, Mobley HL, Warren JW.** 1994. Binding to and killing of human renal epithelial cells by hemolytic P-fimbriated *E. coli. Kidney Int* **46:**1083–1091.

39. **Cavalieri SJ, Snyder IS.** 1982. Effect of *Escherichia coli* alpha-hemolysin on human peripheral leukocyte viability *in vitro. Infect Immun* **36:**455–461.

40. **Uhlén P, Laestadius A, Jahnukainen T, Söderblom T, Bäckhed F, Celsi G, Brismar H, Normark S, Aperia A, Richter-Dahlfors A.** 2000. Alpha-haemolysin of uropathogenic *E. coli* induces Ca2+ oscillations in renal epithelial cells. *Nature* **405:**694–697.

41. **Uematsu S, Akira S.** 2006. Toll-like receptors and innate immunity. *J Mol Med (Berl)* **84:**712–725.

42. **Bäckhed F, Söderhäll M, Ekman P, Normark S, Richter-Dahlfors A.** 2001. Induction of innate immune responses by *Escherichia coli* and purified lipopolysaccharide correlate with organ- and cell-specific expression of Toll-like receptors within the human urinary tract. *Cell Microbiol* **3:**153–158.

43. **Bäckhed F, Normack S, Schweda EK, Oscarson S, Richtor-Dahlfors A.** 2003. Structural requirements for TLR4-mediated LPS signalling: A biological role for LPS modifications. *Microbes Infect* **5:**1057–1063.

44. **Poltorak A, He X, Smirnova I, Liu MY, Van Huffel C, Du X, Birdwell D, Alejos E, Silva M, Galanos C, Freudenberg M, Ricciardi-Castagnoli P, Layton B, Beutler B.** 1998. Defective LPS signaling in C3H/HeJ and C57BL/10ScCr mice: mutations in Tlr4 gene. *Science* **282:** 2085–2088.

45. **Samuelsson P, Hang L, Wullt B, Irjala H, Svanborg C.** 2004. Toll-like receptor 4 expression and cytokine responses in the human urinary tract mucosa. *Infect Immun* **72:**3179–3186.

46. **El-Achkar TM, Huang X, Plotkin Z, Sandoval RM, Rhodes GJ, Dagher PC.** 2006. Sepsis induces changes in the expression and distribution of Toll-like receptor 4 in the rat kidney. *Am J Physiol Renal Physiol* **290:**F1034–F1043.

47. **Dagher PC, Basile DP.** 2008. An expanding role of Toll-like receptors in sepsis-induced acute kidney injury. *Am J Physiol Renal Physiol* **294:**F1048–F1049.

48. **Ragnarsdóttir B, Fischer H, Godaly G, Grönberg-Hernandez J, Gustafsson M, Karpman D, Lundstedt AC, Lutay N, Rämisch S, Svensson ML, Wullt B, Yadav M, Svanborg C.** 2008. TLR- and CXCR1-dependent innate immunity: insights into the genetics of urinary tract infections. *Eur J Clin Invest* **38**(Suppl 2):12–20.

49. **Ragnarsdóttir B, Samuelsson M, Gustafsson MC, Leijonhufvud I, Karpman D, Svanborg C.** 2007. Reduced toll-like receptor 4 expression in children with asymptomatic bacteriuria. *J Infect Dis* **196:**475–484.

50. **Karoly E, Fekete A, Banki NF, Szebeni B, Vannay A, Szabo AJ, Tulassay T, Reusz GS.** 2007. Heat shock protein 72 (HSPA1B) gene polymorphism and Toll-like receptor (TLR) 4 mutation are associated with increased risk of urinary tract infection in children. *Pediatr Res* **61:**371–374.

51. **Andersen-Nissen E, Hawn TR, Smith KD, Nachman A, Lampano AE, Uematsu S, Akira S, Aderem A.** 2007. Cutting edge: Tlr5-/- mice are more susceptible to *Escherichia coli* urinary tract infection. *J Immunol* **178:**4717–4720.

52. **Zhang D, Zhang G, Hayden MS, Greenblatt MB, Bussey C, Flavell RA, Ghosh S.** 2004. A toll-like receptor that prevents infection by uropathogenic bacteria. *Science* **303:**1522–1526.

53. **Heinzelmann M, Mercer-Jones MA, Passmore JC.** 1999. Neutrophils and renal failure. *Am J Kidney Dis* **34:**384–399.

54. **Lundstedt AC, McCarthy S, Gustafsson MC, Godaly G, Jodal U, Karpman D, Leijonhufvud I, Lindén C, Martinell J, Ragnarsdottir B, Samuelsson M, Truedsson L, Andersson B, Svanborg C.** 2007. A genetic basis of susceptibility to acute pyelonephritis. *PloS One* **2:**e825. doi:10.1371/journal.pone.0000825

55. **Williams TW, Lyons JM, Braude AI.** 1977. *In vitro* lysis of target cells by rat polymorphonuclear leukocytes isolated from acute pyelonephritic exudates. *J Immunol* **119:**671–674.

56. **Glauser MP, Lyons JM, Braude AI.** 1978. Prevention of chronic experimental pyelonephritis by suppression of acute suppuration. *J Clin Invest* **61**:403–407.

57. **Sullivan M, Harvey R, Shimamura T.** 1977. The effects of cobra venom factor, an inhibitor of the complement system, on the sequence of morphological events in the rat kidney in experimental pyelonephritis. *Yale J Biol Med* **50**:267–273.

58. **Orskov I, Ferencz A, Orskov F.** 1980. Tamm-Horsfall protein or uromucoid is the normal urinary slime that traps type 1 fimbriated *Escherichia coli. Lancet* **1**:887.

59. **Dulawa J, Jann K, Thomsen M, Rambausek M, Ritz E.** 1988. Tamm Horsfall glycoprotein interferes with bacterial adherence to human kidney cells. *Eur J Clin Invest* **18**:87–91.

60. **Zasloff M.** 2007. Antimicrobial peptides, innate immunity, and the normally sterile urinary tract. *J Am Soc Nephrol* **18**:2810–2816.

61. **Selsted ME, Ouellette AJ.** 2005. Mammalian defensins in the antimicrobial immune response. *Nat Immunol* **6**:551–557.

62. **Lehrer RI.** 2007. Multispecific myeloid defensins. *Curr Opin Hematol* **14**:16–21.

63. **Weichhart T, Haidinger M, Hörl WH, Säemann MD.** 2008. Current concepts of molecular defence mechanisms operative during urinary tract infection. *Eur J Clin Invest* **38**(Suppl 2):29–38.

64. **Chromek M, Slamová Z, Bergman P, Kovács L, Podracká L, Ehrén I, Hökfelt T, Gudmundsson GH, Gallo RL, Agerberth B, Brauner A.** 2006. The antimicrobial peptide cathelicidin protects the urinary tract against invasive bacterial infection. *Nat Med* **12**:636–641.

65. **Turner J, Cho Y, Dinh NN, Waring AJ, Lehrer RI.** 1998. Activities of LL-37, a cathelin-associated antimicrobial peptide of human neutrophils. *Antimicrob Agents Chemother* **42**:2206–2214.

66. **Gudmundsson GH, Agerberth B, Odeberg J, Bergman T, Olsson B, Salcedo R.** 1996. The human gene FALL39 and processing of the cathelin precursor to the antibacterial peptide LL-37 in granulocytes. *Eur J Biochem* **238**:325–332.

67. **Schwartz N, Hosford M, Sandoval RM, Wagner MC, Atkinson SJ, Bamburg J, Molitoris BA.** 1999. Ischemia activates actin depolymerizing factor: role in proximal tubule microvillar actin alterations. *Am J Physiol* **276**:F544–F551.

68. **Ashworth SL, Sandoval RM, Hosford M, Bamburg JR, Molitoris BA.** 2001. Ischemic injury induces ADF relocalization to the apical domain of rat proximal tubule cells. *Am J Physiol Renal Physiol* **280**:F886–F894.

69. **Molitoris BA.** 1991. Ischemia-induced loss of epithelial polarity: potential role of the actin cytoskeleton. *Am J Physiol* **260**:F769–F778.

70. **Molitoris BA, Marrs J.** 1999. The role of cell adhesion molecules in ischemic acute renal failure. *Am J Med* **106**:583–592.

71. **Goligorsky MS, Lieberthal W, Racusen L, Simon EE.** 1993. Integrin receptors in renal tubular epithelium: new insights into pathophysiology of acute renal failure. *Am J Physiol* **264**:F1–F8.

72. **Anderson GG, Palermo JJ, Schilling JD, Roth R, Heuser J, Hultgren SJ.** 2003. Intracellular bacterial biofilm-like pods in urinary tract infections. *Science* **301**:105–107.

73. **Palm F, Cederberg J, Hansell P, Liss P, Carlsson PO.** 2003. Reactive oxygen species cause diabetes-induced decrease in renal oxygen tension. *Diabetologia* **46**:1153–1160.

74. **Dixon B.** 2004. The role of microvascular thrombosis in sepsis. *Anaesth Intensive Care* **32**:619–629.

75. **Rose BD (ed).** 1987. Tubulointestinal diseases. In *Pathophysiology of Renal Disease*, 2nd ed. McGraw-Hill, New York, NY.

76. **Tanner GA.** 1982. Nephron obstruction and tubuloglomerular feedback. *Kidney Int Suppl* **12**:S213–S218.

77. **Tanner GA.** 1979. Effects of kidney tubule obstruction on glomerular function in rats. *Am J Physiol* **237**:F379–F385.

78. **Tanner GA, Knopp LC.** 1986. Glomerular blood flow after single nephron obstruction in the rat kidney. *Am J Physiol* **250**:F77–F85.

79. **Tanner GA.** 1985. Tubuloglomerular feedback after nephron or ureteral obstruction. *Am J Physiol* **248**:F688–F697.

80. **Tanner GA, Evan AP.** 1989. Glomerular and proximal tubular morphology after single nephron obstruction. *Kidney Int* **36**:1050–1060.

81. **Ortiz PA.** 2009. Toll-like receptor 4 (TLR-4) regulates renal ion transport. *Am J Physiol Renal Physiol* **297**:F864–F865.

82. **Good DW, George T, Watts BA III.** 2009. Lipopolysaccharide directly alters renal tubule transport through distinct TLR4-dependent pathways in basolateral and apical membranes. *Am J Physiol Renal Physiol* **297**:F866–F874.

83. **Ren Y, Garvin JL, Liu R, Carretero OA.** 2009. Cross-talk between arterioles and tubules in the kidney. *Pediatr Nephrol* **24**:31–35.

84. **Briggs JP, Schnermann J.** 1987. The tubuloglomerular feedback mechanism: functional and biochemical aspects. *Annu Rev Physiol* **49**:251–273.

85. **Li X, Hassoun HT, Santora R, Rabb H.** 2009. Organ crosstalk: The role of the kidney. *Curr Opin Crit Care* **15**:481–487.

86. Boekel J, Källskog O, Rydén-Aulin M, Rhen M, Richter-Dahlfors A. 2011. Comparative tissue transcriptomics reveal prompt inter-organ communication in response to local bacterial kidney infection. *BMC Genomics* 12:123. doi:10.1186/1471-2164-12-123

87. Yap SC, Lee HT. 2012. Acute kidney injury and extrarenal organ dysfunction: new concepts and experimental evidence. *Anesthesiology* 116:1139–1148.

88. El-Achkar TM, Hosein M, Dagher PC. 2008. Pathways of renal injury in systemic gram-negative sepsis. *Eur J Clin Invest* 38(Suppl 2): 39–44.

89. Jacobs R, Honore PM, Joannes-Boyau O, Boer W, De Regt J, De Waele E, Collin V, Spapen HD. 2011. Septic acute kidney injury: The culprit is inflammatory apoptosis rather than ischemic necrosis. *Blood Purif* 32:262–265.

90. Nash K, Hafeez A, Hou S. 2002. Hospital-acquired renal insufficiency. *Am J Kidney Dis* 39:930–936.

91. Kaufman J, Dhakal M, Patel B, Hamburger R. 1991. Community-acquired acute renal failure. *Am J Kidney Dis* 17:191–198.

92. Mobley HL, Green DM, Trifillis AL, Johnson DE, Chippendale GR, Lockatell CV, Jones BD, Warren JW. 1990. Pyelonephritogenic *Escherichia coli* and killing of cultured human renal proximal tubular epithelial cells: role of hemolysin in some strains. *Infect Immun* 58:1281–1289.

93. Roberts JA. 1983. Pathogenesis of pyelonephritis. *J Urol* 129:1102–1106.

94. Roberts JA, Roth JK Jr, Domingue G, Lewis RW, Kaack B, Baskin G. 1982. Immunology of pyelonephritis in the primate model. V. Effect of superoxide dismutase. *J Urol* 128:1394–1400.

95. Meylan PR, Markert M, Bille J, Glauser MP. 1989. Relationship between neutrophil-mediated oxidative injury during acute experimental pyelonephritis and chronic renal scarring. *Infect Immun* 57:2196–2202.

96. Kaack MB, Dowling KJ, Patterson GM, Roberts JA. 1986. Immunology of pyelonephritis. VIII. *E. coli* causes granulocytic aggregation and renal ischemia. *J Urol* 136:1117–1122.

97. Dagher PC. 2004. Apoptosis in ischemic renal injury: roles of GTP depletion and p53. *Kidney Int* 66:506–509.

98. Klein CL, Hoke TS, Fang WF, Altmann CJ, Douglas IS, Faubel S. 2008. Interleukin-6 mediates lung injury following ischemic acute kidney injury or bilateral nephrectomy. *Kidney Int* 74:901–909.

99. Donnahoo KK, Meng X, Ayala A, Cain MP, Harken AH, Meldrum DR. 1999. Early kidney TNF-alpha expression mediates neutrophil infiltration and injury after renal ischemia-reperfusion. *Am J Physiol* 277: R922–R929.

100. Kielar ML, John R, Bennett M, Richardson JA, Shelton JM, Chen L, Jeyarajah DR, Zhou XJ, Zhou H, Chiquett B, Nagami GT, Lu CY. 2005. Maladaptive role of IL-6 in ischemic acute renal failure. *J Am Soc Nephrol* 16:3315–3325.

101. Park SW, Chen SWC, Kim M, Brown KM, Kolls JK, D'Agati VD, Lee HT. 2011. Cytokines induce small intestine and liver injury after renal ischemia or nephrectomy. *Lab Invest* 91:63–84.

102. Cario E. 2005. Bacterial interactions with cells of the intestinal mucosa: Toll-like receptors and NOD2. *Gut* 54:1182–1193.

103. Pabst R. 1987. The anatomical basis for the immune function of the gut. *Anat Embryol (Berl)* 176:135–144.

104. Park SW, Kim M, Kim JY, Ham A, Brown KM, Mori-Akiyama Y, Ouellette AJ, D'Agati VD, Lee HT. 2012. Paneth cell-mediated multiorgan dysfunction after acute kidney injury. *J Immunol* 189:5421–5433.

105. Blake P, Hasegawa Y, Khosla MC, Fouad-Tarazi F, Sakura N, Paganini EP. 1996. Isolation of "myocardial depressant factor(s)" from the ultrafiltrate of heart failure patients with acute renal failure. *ASAIO J* 42:M911–M915.

106. Wagenlehner FME, Weidner W, Perletti G, and Naber KG. 2010. Emerging drugs for bacterial urinary tract infections. *Expert Opin Emerg Drugs* 15:375–397.

107. Wagenlehner FM, Naber KG. 2006. Treatment of bacterial urinary tract infections: presence and future. *Eur Urol* 49:235–244.

108. Foo LY, Lu Y, Howell AB, Vorsa N. 2000. A-Type proanthocyanidin trimers from cranberry that inhibit adherence of uropathogenic P-fimbriated *Escherichia coli. J Nat Prod* 63:1225–1228.

109. Foo LY, Lu Y, Howell AB, Vorsa N. 2000. The structure of cranberry proanthocyanidins which inhibit adherence of uropathogenic P-fimbriated *Escherichia coli* in vitro. *Phytochemistry* 54:173–181.

110. Zafriri D, Ofek I, Adar R, Pocino M, Sharon N. 1989. Inhibitory activity of cranberry juice on adherence of type 1 and type P fimbriated *Escherichia coli* to eucaryotic cells. *Antimicrob Agents Chemother* 33:92–98.

111. Bonventre JV, Zuk A. 2004. Ischemic acute renal failure: An inflammatory disease? *Kidney Int* 66:480–485.

112. **Sutton TA, Mang HE, Campos SB, Sandoval RM, Yoder MC, Molitoris BA.** 2003. Injury of the renal microvascular endothelium alters barrier function after ischemia. *Am J Physiol Renal Physiol* **285:**F191–F198.

113. **Evan AP, Tanner GA.** 1986. Proximal tubule morphology after single nephron obstruction in the rat kidney. *Kidney Int* **30:**818–827.

114. **Misseri R, Rink RC, Meldrum DR, Meldrum KK.** 2004. Inflammatory mediators and growth factors in obstructive renal injury. *J Surg Res* **119:**149–159.

115. **Richter-Dahlfors A, Rhen M, Udekwu K.** 2012. Tissue microbiology provides a coherent picture of infection. *Curr Opin Microbiol* **15:**15–22.

116. **Choong FX, Regberg J, Udekwu KI, Richter-Dahlfors A.** 2012. Intravital models of infection lay the foundation for tissue microbiology. *Future Microbiol* **7:**519–533.

HOST RESPONSES TO URINARY TRACT INFECTIONS AND EMERGING THERAPEUTICS

Susceptibility to Urinary Tract Infection: Benefits and Hazards of the Antibacterial Host Response

21

INES AMBITE,[1] KAROLY NAGY,[2] GABRIELA GODALY,[1] MANOJ PUTHIA,[1] BJÖRN WULLT,[1] and CATHARINA SVANBORG[1]

SUSCEPTIBILITY TO URINARY TRACT INFECTION

Definitions

Defining UTI susceptibility and appropriate therapeutic interventions is essential, in view of the prevalence of these infections and their significance to health and society (for relevant recent reviews see (1–5). The clinical appearance of infection reflects the host-response profile, the receptors engaged by specific virulence factors, and the activation of downstream-signaling pathways. Depending on the type of receptor engaged, different signaling pathways are activated, and cascades of innate immune-response effectors activate innate and specific immunity. Specific attachment alerts the host to the presence of virulent bacteria on the epithelial cell surface (6, 7) and perturbations of host-cell receptors by bacterial P fimbriae initiate the innate immune response (8, 9) (Fig. 1A) through specific transcription factors such as interferon regulatory factors (IRFs) and activator protein 1 (AP-1), which activate numerous, intricate antibacterial-effector functions (Fig. 2). Additional virulence factors disrupt the mucosal barrier and facilitate invasive

[1]Department of Microbiology, Immunology and Glycobiology, Institute of Laboratory Medicine, Lund University, Lund, S-223 62, Sweden; [2]Department of Urology, South-Pest Hospital, Budapest 1204, Hungary.

Urinary Tract Infections: Molecular Pathogenesis and Clinical Management, 2nd Edition
Edited by Matthew A. Mulvey, David J. Klumpp, and Ann E. Stapleton
© 2017 American Society for Microbiology, Washington, DC
doi:10.1128/microbiolspec.UTI-0019-2014

A

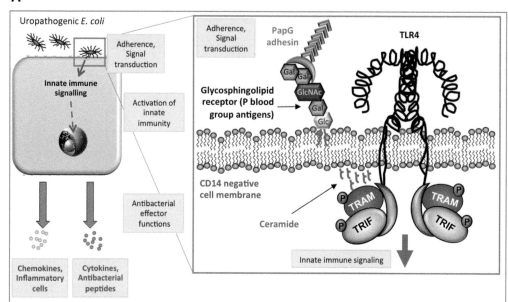

Genetic variants affecting the early innate immune response

Genetic variants in patients with APN or ABU			Murine phenotypes	
A₁P₁ blood group	Expression of receptors for P-fimbriae in the urothelium	APN	*Tlr4⁻/⁻*	ABU: high bacterial counts, low PMN influx, no symptoms of disease
TLR4 promoter SNPs	Low promoter activity and TLR4 expression	ABU	*Trif⁻/⁻*	ABU: high bacterial counts, low PMN influx
TRIF	No detected variants with clinical correlates	-	*Myd88⁻/⁻*	ABU: high bacterial counts, luminal PMN aggregates
MYD88*	No detected variants with clinical correlates	-	*Tram⁻/⁻*	ABU: high bacterial counts, low PMN influx

*Not included in the figure

(figure continues)

disease (2). Their interactions with host cells add to the complexity of the innate immune response through the activation of signaling pathways defined by their cellular receptors (Fig. 1B, see below). These include fimbriae (type 1, Dr, curli and Afa), flagella, toxins that permeabilize and kill host cells, like hemolysin and cytotoxic-necrotizing factor, and capsule formation, which confers resistance to bactericidal host molecules (such as defensins or antibodies). Interestingly, certain uropathogens actively inhibit innate immune signaling, for example, the Toll/interleukin-1 receptor (TIR) homologous protein TcpC secreted by the bacteria impairs Toll-like receptor (TLR) and myeloid differentiation primary response gene 88 (MYD88)-dependent innate immune responses and promote bacterial survival in the urinary tract (10). Differences in individual susceptibility are usually deduced from the frequency of infection, from the acute-disease severity and from the sequels. For the purpose of this review, we would like to emphasize that susceptibility parameters such as symptoms and tissue damage are a direct reflection of the host response to the infecting strain, usually its lack of efficiency. Acute pyelonephritis (APN) is a severe and potentially life-threatening infection of the kidneys with

systemic involvement and mortality, especially in patients who develop urosepsis. A rapid innate immune response and resulting inflammation in the infected kidney causes symptoms such as flank pain, and the spread of inflammatory mediators, such as interleukin (IL)-6, triggers the systemic host response (fever >38.5°C and general malaise) (11). The extent of renal tissue involvement varies between APN patients, with a subset developing recurrent infections and tissue damage. Acute infection foci may be detected by dimercaptosuccinic acid (DMSA) scans and more permanent damage by repeated scans or other imaging technology. Recurrent APN in childhood may cause end-stage renal disease, especially in societies with limited access to health care and antibiotic therapy. During the last decades, molecular determinants of APN susceptibility have been identified. These include exaggerated bacterial-virulence profiles and host genetic variants that reduce the efficiency of the effector phase of the antibacterial response, resulting in exaggerated and dysregulated inflammation in APN-susceptible patients (9, 12–17).

In patients with acute cystitis, bacteria trigger a local inflammatory response in the bladder mucosa. Classical symptoms include urgency, frequency, dysuria, and suprapubic

---

FIGURE 1 (A) Initiation of the innate immune response by UPEC. P fimbriae-mediated adherence and TLR4 activation. Bacterial adherence to epithelial surface receptors activates TLR4 and initiates innate immune signaling. Pathogen-specific recognition by the PapG adhesin of Galα1-4-Galβ-receptor motifs in the globoseries of glycosphingolipids. Release of ceramide activates TLR4 signaling, mainly through the TRIF/TRAM adaptors. The MYD88/TIRAP/NF-κB-dependent arm of TLR4 signaling, in contrast, is activated by type 1-fimbriated strains and, to some extent, also by ABU strains (not shown (36)). Genetic variants that affect the expression of receptors also influence the susceptibility to APN. Patients who express high levels of receptors are more susceptible to APN (blood group A$_1$ P$_1$), illustrating the relevance of this mechanism (145). In ABU patients, TLR4 expression is low and *TLR4*-promoter polymorphisms that reduce TLR4 expression are predominately found in this patient group. Clinical genetic screens have not detected ABU-associated polymorphisms in *MYD88* or *TRIF*. In the murine UTI model, *Tlr4* deletions abrogate the innate immune response, as do adaptor-gene deletions, to some extent. As a result, these mice develop ABU rather than symptomatic disease. Abbreviations: galactose (Gal), glucose (Glc), N-acetyl glucosamine (GlcNAc), Toll-like receptor (TLR), Toll/interleukin-1 receptor (TIR) domain-containing adapter-inducing interferon-β (TRIF), TRIF-related adaptor molecule (TRAM), phosphate group (P), acute pyelonephritis (APN), asymptomatic bacteriuria (ABU), single-nucleotide polymorphism (SNP), myeloid-differentiation primary-response protein 88 (MYD88), polymorphonuclear cells (PMN). Adapted from Ragnarsdóttir et al. (3), with permission.

(continued)

B

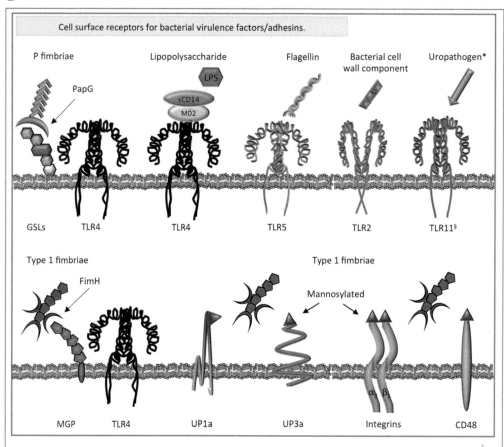

| Cell surface receptors for bacterial virulence factors/adhesins. |

Genetic variants affecting cell membrane receptors				
Genetic modifications in patients			**Murine phenotypes**	
TLR5	NA	APN	*Tlr5⁻/⁻*	Prolonged infection and inflammation at later time points
TLR2 coding region SNP	NA	ABU rUTI	*Tlr11⁻/⁻ §*	High bacterial counts, influx of leukocytes

*The bacterial ligand is not known, §Not expressed in human

pain (1). Basic host-susceptibility mechanisms remain to be defined, but these symptoms indicate a strong neurological component of the host response and symptomatology. Recurrent episodes of acute cystitis often occur in clusters, suggesting infection-induced attenuation of host resistance. Acute cystitis patients do not have fever or elevated systemic inflammatory parameters, but cystitis-like symptoms may precede APN. Genetic determinants of cystitis susceptibility are starting to appear, especially in the murine UTI model.

Asymptomatic bacteriuria (ABU) is detected in about 1% of girls, 2–11% of pregnant women, and about 20% of otherwise healthy men and women >70 years of age. The patients carry $\geq 10^5$ colony-forming units (cfu) per ml of urine and, despite the high bacterial load, they remain asymptomatic for long periods of time (1). Recent studies show that the asymptomatic state results from mutual adaptation between bacteria and the host. At least 50% of ABU strains appear to have evolved from virulent strains by reductive evolution. The loss of virulence has been explained by the accumulation of point mutations and gene deletions, resulting in a reduction in genome size (18–20) and attenuation of virulence, accompanied by a change in metabolic genes and pleiotropic regulators of bacterial gene expression. ABU strains evolve further in individual hosts (21), illus-

trating that the host environment directs the evolution of bacterial genomes in real time. Recently, active suppression by ABU strains of the very pathways that cause the disease response to virulent strains was discovered (22). Identified molecular characteristics of ABU susceptibility in the host include polymorphisms or mutations that reduce the magnitude of the innate immune response, creating a host susceptibility profile favoring asymptomatic carriage rather than symptomatic infection (15, 23).

UTI SUSCEPTIBILITY AND INNATE IMMUNE ACTIVATION

The mucosal immune system distinguishes pathogens from commensals and the pathogen-specific innate immune response eliminates the pathogens while commensals persist. To solve this apparent enigma, it is essential that molecular-recognition mechanisms allow the host to respond selectively to pathogenic strain by mechanisms other than "pattern recognition." If conserved microbial motifs were recognized, uropathogenic and ABU strains would both trigger the disease associated host response; something they obviously do not do (8, 24). Pattern-recognition receptors such as mCD14 are not expressed by uroepithelial cells, allowing the host to avoid immune activation by the

---

FIGURE 1 (continued) **(B) Uroepithelial receptors for bacterial ligands. Uroepithelial cells express a number of specific receptors for microbial ligands including TLRs and receptors for adhesins, toxins, and flagella, among others. Importantly, uroepithelial cells do not express CD14 and the initial recognition of Gram-negative bacteria does not involve LPS, unless soluble CD14 is present. TLR5 interacts with bacterial flagellin; TLR11 recognizes uropathogens through yet undetermined bacterial ligands and TLR2 is activated by bacterial cell wall components. Type 1 fimbriae activate TLR4 signaling through the FimH adhesin, which binds to a variety of mannosylated glycoproteins. Binding to uroplakin particles (UP1a, 1b, 2, and 3a) promotes bacterial internalization. FimH also binds to β1 and α3 integrins, which modulate F-actin dynamics in the mammalian cell. TNFα responses to type 1-fimbriated bacteria are triggered by the glycosyl-phosphatidyl-inositol-anchored CD48 receptor on mast cells and macrophages. The receptor epitopes of UPs, CD48, and integrins, are N-linked high-mannose oligosaccharides. Abbreviations: lipopolysaccharide (LPS), soluble CD14 (sCD14), glycosphingolipids (GSLs), Toll-like receptor (TLR), mannosylated cell-surface glycoprotein (MGP), uroplakin (UP), not applicable (NA), acute pyelonephritis (APN), single-nucleotide polymorphism (SNP), asymptomatic bacteriuria (ABU), recurrent urinary tract infection (rUTI). Adapted from Ragnarsdóttir et al. (3), with permission.**

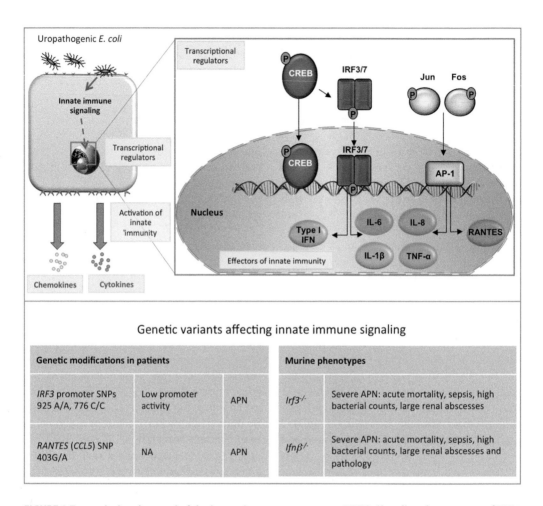

FIGURE 2 Transcriptional control of the innate immune response to UPEC. Signaling downstream of TLR4 activates the transcription of innate immune-effector molecules, such as chemokines, cytokines, and antibacterial peptides (3, 87). Transcription factors are activated by phosphorylation and nuclear translocation, including IRF3 and IRF7, as well as AP-1, a heterodimer of FOS and JUN. In addition, NF-κB is critically involved, (not shown). In clinical studies, promoter polymorphisms that reduce the expression of IRF3 have been associated with susceptibility to acute pyelonephritis. In the murine UTI model, mice lacking *Irf3* develop severe acute infection with mortality, followed by renal damage in surviving mice. Downstream mediators have also been shown to play an essential role for UTI susceptibility, including type 1 IFNs. Relevance of IFNβ has been demonstrated in the murine UTI model, where mutant mice develop severe acute infection with tissue damage. Clinical studies associating IFNβ with UTI susceptibility have not been reported. Abbreviations: phosphate group (P), cyclic AMP-response element-binding (CREB), interferon-regulatory factor (IRF), activator-protein 1 (AP-1), interferon (IFN), interleukin (IL), single-nucleotide polymorphism (SNP), acute pyelonephritis (APN), CC-chemokine ligand 5 (CCL5), not applicable (NA). Adapted from Ragnarsdóttir et al. (3), with permission.

normal microflora and providing a window for pathogen-specific TLR4 activation (25).

The importance of innate immunity for the host defense against UTI was first discovered in the early 1980s in the murine UTI model. In a screen of mice with different genetic defects (26), C3H/HeJ mice, then known as lipopolysaccharide (LPS) nonresponder mice, showed increased susceptibility to UTI with delayed bacterial clearance, as well as an impaired neutrophil response. These observations suggested that the antibacterial defense of the urinary tract was controlled by innate immunity (27). C3H/HeJ mice remained chronically infected, but without evidence of tissue damage, linking the lack of inflammation to a lack of tissue damage. The "Lps" gene defect in C3H/HeJ mice was subsequently identified as a point mutation in the intracellular TIR domain of TLR4 (28) and the importance of Tlr4 for host resistance to UTI was confirmed in $Tlr4^{-/-}$ mice of several background (9, 27, 29–33). Single-gene defects were subsequently shown to also control the severity of acute and chronic symptomatic UTI. The first example was $mCxcr2^{-/-}$ mice with deficient neutrophil function, which were shown to develop severe APN with urosepsis followed by abscess formation and renal scarring (12, 34, 35). This "single gene" effect of Tlr4 as a protector and $mCxcr2^{-/-}$ as a disease enhancer was unexpected, as UTI susceptibility was expected to be too complex for single-gene control. It is now well established that single-gene deletions affecting innate immunity may cause symptoms and pathology (3, 12) and numerous additional genetic determinants of UTI susceptibility have now been identified (see below).

TLR4 AND ADAPTOR-PROTEIN SIGNALING

Since the early studies, it has become widely accepted that Tlr4 controls many aspects of the innate immune response to UTI (36). TLR4 is a transmembrane glycoprotein composed of three major domains: a leucin-rich repeat (LRR) ectodomain, a transmembrane domain, and the intracellular TIR domain. The LRR domain is essential for microbial ligand recognition (37, 38) and forms a horse shoe-like structure as shown by crystallization (39). A stretch of about 22 uncharged amino acids forms a membrane-spanning α-helix (40) and the cytoplasmic domain

linking the transmembrane domain to the first secondary-structure element of the TIR domain varies from 20 amino acids in TLR4 to 30 in TLR5 (40).

The TLRs share the TIR domain with the IL-1 receptor (IL1-R) family and five adaptors (41). The five intracellular adaptors include the TIR domain–containing adapter-inducing interferon-β (TRIF), TRIF-related adaptor molecule (TRAM), MYD88, and TIR domain containing adaptor protein (TIRAP) (42, 43). The TIR domains are characterized by three main conserved sequences called box 1–3. Box 1 (FDAFISY) is the signature sequence of all TLRs and the most conserved. The proline-to-histidine mutation that renders C3H/HeJ mice hyporesponsive to Gram-negative stimuli is in box 2 (GYKLC-RD-PG) (44). Boxes 1 and 2 are involved in the binding of downstream-signaling proteins (45). After dimerization (46), TLR4 recruits IL-1R-associated kinase 4 (IRAK4) via the adaptors MYD88 and TIRAP. In addition, MYD88-independent TLR4 activation through TRIF/TRAM regulates other aspects of the innate response (47–49). TIR domain-containing proteins have also been identified in bacteria and viruses (10, 50).

MYD88 is common to all the TLRs as well as IL-1R and consists of a C-terminal TIR and an N-terminal death domain (DD), linked by an intermediate domain (51, 52). After binding of LPS to the TLR4 complex, TIRAP recruits MYD88 to a phosphatidylinositol 4,5-bisphosphate (PIP2)-binding domain in its N-terminus, which recruits TIRAP (53, 54) to PIP2-rich areas in the plasma membrane, thereby enabling the delivery of MYD88 to an activated TLR4 (55). TIRAP phosphorylation by Bruton's tyrosine kinase (BTK) is crucial for LPS-induced NF-kB activation (56). Signaling is initiated by the TLR4/MYD88 interaction through TIR–TIR association, resulting in the recruitment of IRAK4 to MYD88 through death-domain association. The intermediate domain is also crucial since an alternative splice variant of MYD88 lacking this region, MYD88s, fails to recruit IRAK4 and therefore acts as a negative regulator of TLR signaling (57).

TLR4 signaling is initiated by virulence ligands, followed by TIR-domain interactions with the adaptor proteins. In mice lacking functional Tlr4, the early innate immune response to UTI is low or absent, including mucosal cytokine production, expression of chemokine receptors, neutrophil recruitment, and tissue inflammation (25, 27, 29). As a result, bacteria persist in apparent symbiosis with the host in the lumen or as intracellular bacterial communities (58–60). Trif/Tram or Myd88/Tirap control the two main arms of immune-signaling downstream of Tlr4. Using isogenic *E. coli* mutants expressing either P fimbriae or type 1 fimbriae, P-fimbriated *E. coli* were shown to mainly activate Trif-dependent signaling and type 1 fimbriae Myd88-dependent signaling pathways (36). In mice lacking *Myd88* or *Trif*, infection resulted in an ABU-like phenotype with prolonged bacterial carriage, analogous to the Tlr4-deficient mice (61). Cytokine and neutrophil responses were virtually undetectable in *Myd88* mutant mice after infection with *E. coli* CFT073 that expresses type 1 fimbriae and were low also in *Trif* mutant mice after infection with P-fimbriated *E. coli* (36, 61).

Innate Immune Activation by P-Fimbriated *E. coli*

P fimbriae are major virulence factors of uropathogenic *E. coli* (UPEC), being expressed by up to 100% of strains causing APN and urosepsis in otherwise healthy individuals, but rarely by ABU strains (62, 63)˙ Virulence is explained by direct activation of host TLR4 signaling and inflammation, initiated by attachment to glycosphingolipid receptors (7) through the PapG adhesin located at the tip of the fimbrial rod (64, 65). Early studies identified P fimbriae as tissue-attacking virulence factors that trigger host cells to produce inflammatory mediators *in vitro* and inflammation in the murine urinary tract (66).

Subsequently, human inoculation studies have confirmed that P-fimbriae expression is sufficient to trigger an innate mucosal immune response in the human urinary tract (67). In the human inoculation model, P fimbriae act as independent virulence factors, fulfilling the molecular Koch postulates (67, 68). PapG-mediated adherence enables the host to sense the attacking uro-pathogens and to activate the innate mucosal response (Fig. 1a).

Innate immune recognition of P-fimbriated *E. coli* involves receptor motifs present in the extracellular oligosaccharide domains of the globoseries of glycosphingolipids (Galα1-4Galβ determinant, Fig. 1b) (64). Binding induces the hydrolysis of the receptor-anchoring domain (ceramide) and released ceramide activates TLR4 signaling (8, 69–71). P-fimbriated *E. coli* trigger a TRIF/TRAM-dependent signaling cascade. In addition to the TIR domain, TRIF has several N-terminal TNF receptor-associated factor 6 (TRAF6)-binding regions and a C-terminal receptor-interacting protein homotypic-interaction motif (RHIM) (49, 72, 73). The TRIF-dependent pathway requires TRAM (49, 72) bound to the plasma membrane through a myristoylated N-terminus. TRAM differs from the other adaptors by being TLR4-specific, acting as a bridge between TLR4 and TRIF, and protein kinase Cε (PKCε) phosphorylation of serine residues is essential for activation (74, 75). The TRIF pathway activates nuclear factor kappa B (NF-κB) and IRF3/7 dependent transcription (49). Following TRAM-dependent activation of TLR4 signaling, TRIF forms a complex with TRAF3, IRAK1, and an taIKK)-like kinase named the TRAF-family member-associated NF-κB activator (TANK)-binding kinase 1 (TBK1) and the IKK homolog IKKε, leading to the phosphorylation of IRF3 at its C terminus (49, 76), binding to the interferon-stimulated-response element (ISRE) on target genes, and transcription of interferon (IFN)-α/β (73). Similar to IRF3, IRF7 is activated via TBK1 and IKKε phosphorylation and induces the expression of target genes (77).

In response to P-fimbriated *E. coli*, the TRAM and TRIF are phosphorylated, mitogen-activated protein (MAP) kinases, phospholipase Cγ (PLCγ), p38, c-Jun N-terminal kinases (JNK), cyclic AMP response element-binding protein (CREB) and FOS–JUN, activates IRF3- and AP-1-dependent transcription of innate immune-response genes, including cytokines like IL-8, IL-6, and tumor necrosis factor (TNF) (78). Type-1 IFNs (IFNα and IFNβ) are produced in response to extracellular or intracellular bacterial pathogens and their expression is controlled by the IRF-transcription factors, which are activated by P-fimbriated *E. coli* (Fig. 2). The innate host response is further enhanced by signaling through the IFNα/β pathway and modified by IL-1β signaling and by the infiltration and activation of numerous immune-cell types (9, 79).

Innate Immune-Response Activation by Type 1 Fimbriae

Type 1 fimbriae are expressed by most *E. coli* strains and, as a consequence, epidemiological studies do not clearly resolve their association to human disease. The FimH adhesin binds to several mannosylated host cell glycoconjugates as receptors, such as secretory immunoglobulin (Ig) A (80); uroplakins (UPs) in a particle formed from UP1A, UP1b, UP2, and UP3a (81); CD48 (a membrane-anchored protein on mucosal mast cells) (82); N-oligosaccharides on integrins β₁ and α₃ (83); and to the Tamm–Horsfall protein (THP, also known as uromodulin) on the cell membrane (84). Type 1 fimbriae facilitate internalization into bladder-epithelial cells through FimH binding to the CD48 receptor, which resides in microdomains enriched in cholesterol and glycolipids in the cell membrane (85–87), and directs bacterial uptake via a route distinct from the classical opsonin-mediated phagosome–endosome route. Furthermore, binding of type 1-fimbriated *E. coli* to N-oligosaccharides on integrins β₁ and α₃ and phosphorylation of the integrin tail facilitates *E. coli* entry into host cells

regulated by the RAC1 and CDC42 GTPases, as well as PTK2 (also known as FAK) and SRC, which control actin rearrangement during FimH-dependent bacterial invasion (83, 88). UP1a supports bacterial uptake through RAC1, together with caveolin-1 and the formation of intracellular bacterial communities has been proposed to enhance bacterial persistence in mice (58, 60, 89, 90). Paradoxically, type 1-fimbrial attachment to UP-coated murine bladder-epithelial cells also triggers exfoliation of uroepithelial cells through rapid apoptosis (91), with cytochrome C release, activation of caspase 2, caspase 8, and BH3-interacting-domain death agonist (BID), as well as rescue by Bcl2l1 (Bcl2 antagonist of cell death also known as Bcl-xL) (92, 93). In addition, cyclic AMP regulates the incorporation of *E. coli* into fusiform vesicles and exocytosis of *E. coli* from bladder-epithelial cells. Most of the vesicles with incorporated *E. coli* were Rab27b–CD63-positive and knockdown of Rab27b markedly reduced bacterial invasion. Forskolin, which elevates intracellular levels of cyclic AMP, eliminated >99% of intracellular *E. coli* by exocytosis without affecting bacterial viability (94).

The role of type 1 fimbriae as innate immune-response activators is less clear, however. In cellular models, type 1 fimbriae trigger TLR4 signaling (32) through the MYD88 adapters (36), resulting in NF-κB activation (24, 25, 32, 36, 95–97). TLR4 has been suggested to control the response to FimH through release of calcium ions, elevation of cyclic AMP levels, activation of PKA (97, 98) and the phosphorylation of the transcription factor CREB. While MYD88 is a broad and essential innate immune-response regulator, mechanisms of MYD88 activation by type 1 fimbriae and downstream-signaling pathways are not fully understood. It is also not clear which receptor interactions give rise to an innate immune response and if signaling from different receptors converge on the MYD88-dependent pathway with NF-κB dependent transcription.

The human relevance of type 1 fimbriae and of these intricate cellular responses also remains unclear. Human-inoculation studies have failed to detect an effect of type 1-fimbrial expression alone on acute inflammation or bacterial persistence in the human bladder (99). Furthermore, UPEC rapidly stop expressing type 1 fimbriae in the urinary tract and ABU strains accumulate *fim* deletions, suggesting that inactivation of type 1-fimbrial expression is needed for long-term persistence in the human urinary tract (19). A positive effect of type 1 fimbriae in APN was suggested in children infected with *E. coli* O1K1H7, however. A variant expressing type 1 fimbriae was associated with increased disease severity in patients and in the murine UTI model (100), suggesting an effect on human disease in the background of a fully virulent genotype.

INDUCTION OF INNATE IMMUNITY BY OTHER VIRULENCE FACTORS

UPEC strains possess additional virulence factors that contribute to their ability to cause disease such as flagella (101), toxins, capsules, autotransporter proteins, and iron-acquisition systems (2). The autotransporter (AT) proteins, such as UpaH and Ag43, contribute to UPEC adhesion in the urinary tract, but AT proteins are also involved in UPEC-mediated aggregation, invasion, and biofilm formation (102). Interestingly, fimbriae were recently suggested to crosstalk with flagella, capsule, and the AT protein 43 in order to promote colonization (103). Bacterial growth and persistence in the urinary tract is further promoted by four different iron-acquisition-siderophore systems (enterobactin, the glucosylated enterobactin-derivative salmochelin, yersiniabactin, and aerobactin) (104). The yersiniabactin is also sequestering host-derived copper, thus protecting the UPEC against copper stress (105). In addition to fimbriae, curli may be involved (106), as well as the repeat-in-toxin TosA (107). Finally, for UPEC to survive in the nutritionally poor urinary tract, short peptides and amino acids are required as a carbon source during infection (108), as well as the presence of type II toxin-antitoxin systems (TA). In a murine study, three toxin-antitoxin systems were recently reported to contribute in extra-intestinal pathogenic *E. coli* (ExPEC) survival and persistence within the host urinary tract in the presence of antibiotics (109).

The contribution of these virulence factors as innate immune activators is less clear, however. Flagella are recognized by TLR5, resulting in signal transduction and the production of pro-inflammatory cytokines (110, 111). Curli are recognized by TLR1 and TLR2 (112), and cytotoxic-necrotizing factor (CNF)-1 activates GTP-binding proteins in the Rho family, by inducing glutamine deamination (113, 114). AT proteins, belonging to the type IV-secretion system, have cytopathic rather than immune-activating effects (114).

ANTIBACTERIAL-EFFECTOR FUNCTIONS IN THE URINARY TRACT

Cytokine Response Repertoire

The epithelial cytokine response to infection was first detected in the UTI model and the role of epithelial cells as active participants in the antibacterial host defense has been extensively studied in the urinary tract (78, 115, 116). UPEC provoke a rapid cytokine response in the urinary tract mucosa and the local and systemic effects of these inflammatory mediators set the stage for defensive and damaging consequences of infection. IL-6 is a pyrogen and acute-phase response activator and systemic levels of IL-6 have been shown to reflect disease severity and to predict the outcome of infection. Chemokine (C-X-C motif) ligand (CXCL)-8 (also known as IL-8) is essential to amplify inflammation past the early phase by recruitment of neutrophils to the site of infection, ultimately, eliminating bacteria from the tissues (117).

Two major chemokine super-family sub-groups can be distinguished. In (C-C motif) chemokine ligands (CCLs), two adjacent cysteines are located near the N terminus (e.g., MCP-1/CCL2, MIP-1α/CCL3, and RANTES/CCL5), but in the CXC chemokine family the cysteines are separated by one amino acid (e.g., GRO-α/CXCL1, IL-8/CXCL8, MIG/CXCL9, and IP-10/CXCL10). UPEC strains activate epithelial CXCL8 expression (118–120) and febrile UTIs trigger a broad repertoire of chemokines (121, 122). In addition to CXCL8, P-fimbriated *E. coli* stimulate CCL2, also known as monocyte-chemotactic protein-1 (MCP-1) (123). MCP-1 recruits monocytes, memory T cells, and dendritic cells to sites of inflammation and is released in response to ceramide 1-phosphate (8, 124). CCL3 is involved in acute inflammation through the recruitment and activation of polymorphonuclear leukocytes and is secreted by epithelial cells in response to P-fimbriated *E. coli*. CCL5 functions as a chemoattractant for blood monocytes, memory T-helper cells, and eosinophils through release of histamine from basophils and activation of eosinophils (5, 125, 126).

Type 1-fimbriated strains activate a different cytokine repertoire than P-fimbriated strains *in vitro* and studies in deliberately colonized patients demonstrated that type 1-fimbriated strains are poor host response inducers in the human urinary tract (11, 119, 127, 128).

Neutrophil Recruitment

UPEC trigger a rapid chemokine response and CXCL8 levels increase in blood and urine. Expression of the CXCL8-receptor CXCR1 is also stimulated by infection, and epithelial and neutrophil expression of this receptor is important for CXCL8-dependent neutrophil recruitment to sites of infection (Fig. 3). The need for CXCL8 and its receptors in neutrophil recruitment was first defined *in vitro* in a transwell model of the uroepithelium with bacterial stimulation

of confluent uroepithelial-cell layers (129). Numerous studies have since confirmed the importance of CXCL8 for the neutrophil-mediated antibacterial defense in UTI (12, 117, 122, 129). In humans, a rapid CXCL8 response to UTI has been observed and fimbriae-dependent CXCL8 and neutrophil responses in the human urinary tract were detected after therapeutic *E. coli* inoculation (67, 78).

Chemokine Receptors and UTI Susceptibility

CXCL8 is a powerful neutrophil chemoattractant, essential for transepithelial migration of neutrophils into the infected urinary tract. Human CXCL8 binds to 7-transmembrane G protein-coupled receptors, CXCR1 and CXCR2, expressed both on neutrophils and uro-epithelial cells (130, 131). In the murine UTI model, mCxcr2 is activated by several CXC chemokines containing the "ELR" (Glu-Leu-Arg) motif (34), including murine mCxcl1 (keratinocyte-derived chemokine, KC) (132), mCxcl2/3 (macrophage-inflammatory protein-2, MIP-2) (133), as well as human GRO proteins (hCXCL1/3) and hCXCL8 (132–135), which are functional homologues of human CXCL8. The neutrophil response to experimental UTI is delayed in $Cxcr2^{-/-}$ mice, and transepithelial-neutrophil migration is abrogated, resulting in massive neutrophil accumulation, first under the epithelium and later throughout the tissues. Acute neutrophil accumulation creates a renal edema and abscess formation after one week, followed after 35 days by significant pathology with subepithelial and perivascular fibrosis and loss of functional tissue (11, 35, 119).

Bone marrow transplantation was used to further investigate the role of mCxcr2 expression on hematopoietic cells (neutrophils) or non-hematopoietic cells (uro-epithelial cells). Irradiated $mCxcr2^{-/-}$ mice were reconstituted with mCxcr2-positive bone marrow from control mice and $mCxcr2^{+/+}$ mice with mCxcr2-negative bone marrow. The results

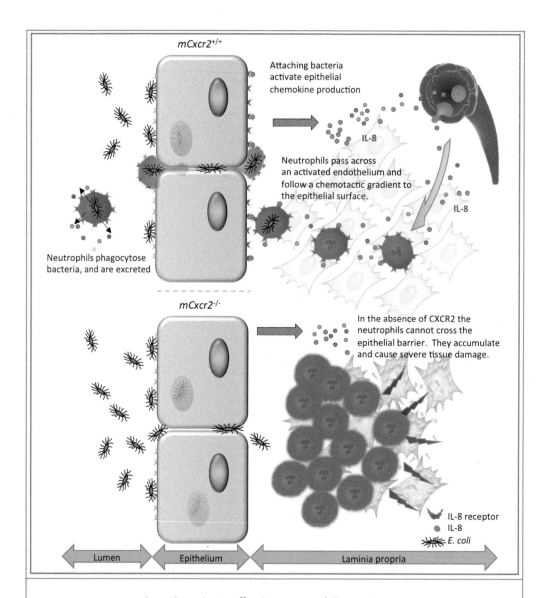

mCxcr2⁺/⁺

Attaching bacteria activate epithelial chemokine production

IL-8

Neutrophils pass across an activated endothelium and follow a chemotactic gradient to the epithelial surface.

IL-8

Neutrophils phagocytose bacteria, and are excreted

mCxcr2⁻/⁻

In the absence of CXCR2 the neutrophils cannot cross the epithelial barrier. They accumulate and cause severe tissue damage.

⋎ IL-8 receptor
● IL-8
✳ E. coli

Lumen ← → Epithelium ← → Laminia propria

Genetic variants affecting neutrophil recruitment

Genetic modifications in patients			Murine phenotypes	
CXCR1 intronic SNP	Low promoter activity, reduced CXCR1 expression	APN	*mCxcr2⁻/⁻*	Severe APN: acute mortality, sepsis, high bacterial counts, large renal abscesses and pathology
CXCR1 SNP in 3'UTR region	Increased 3'-mRNA processing	APN	mCxcr2: murine Cxcl8 (Il-8) receptor phenotype CXCR1: human CXCL8 (IL-8) receptor phenotype	
CXCL8 (IL-8) promoter and 3'UTR SNPs	NA	APN		

identified neutrophil mCxcr2 as a key for E. coli-induced neutrophil recruitment into the urinary tract and for renal scarring. Endothelial and epithelial mCxcr2 did not affect neutrophil recruitment, *per se*, but their neutrophils assembled under the epithelium and the mice failed to clear the infection (35). The results demonstrate that regulation of chemokine ligands and receptors by infection is a key to homeostasis or pathology in UTI.

Irf3 or Ifnβ Defects Increase APN Susceptibility

In a transcriptomic and proteomic screen to identify critical components in pathogen-induced host cell responses, we found that IRF3 controls pathogen-specific transcription in infected human uroepithelial cells. In the murine model, knockdown of the *Irf3* gene was found to dysregulate the innate immune response *in vivo*, resulting in severe, acute and chronic disease with urosepsis and massive abscess formation. Neutrophils were recruited to the site of infection, but in the absence of functional Irf3 signaling their antibacterial-effector functions were impaired, resulting in high bacterial numbers in renal tissues, bladders, and urine. Abscess formation and tissue damage were observed within 1 week of infection (9).

The IRF3–IRF7 complex controls the transcription of genes involved in the early innate response, including the production of CXCL2, IL-6, TNF, and IFNβ (136, 137). Transcriptomic analysis also revealed that IFNβ is activated by infection with UPEC and that signaling through the IFN receptor is enhanced by infection in human uroepithelial cells. *Ifnβ*$^{-/-}$ mice showed a severe-disease phenotype similar to that seen in *Irf3*$^{-/-}$ mice. Symptoms and tissue pathology were severe, suggesting that Ifnβ controls essential antibacterial-effector functions downstream of the Tlr4–Irf3–Irf7-signaling cascade (9). These findings reveal the importance for the host of regulating neutrophil recruitment and activation. They also illustrate how molecular defects affecting these mechanisms increase host susceptibility to APN, translating into acute and chronic disease.

Tlr5, Tlr11, Thp, and Cox2 in UTI Susceptibility

Mutant mice lacking Tlr5, Tlr11, Thp, and cytochrome c oxidase subunit II (Cox2) are also more susceptible to UTI than mice with the wild-type genotype. Tlr5 is expressed in both bladders and kidneys and is mobilized to the cell surface in response to infection. The difference between wild-type and *Tlr5*$^{-/-}$ mice is not apparent immediately after in-

---

FIGURE 3 Neutrophil-dependent clearance of infection. Chronic infection and renal scarring in *CXCR1*-deficient patients and *mCxcr2*-deficient mice. In UTI, neutrophils migrate to the mucosal epithelial barrier, which they cross into the lumen. As a result, infection causes pyuria, which often is used diagnostically, as the neutrophils first phagocytose and kill bacteria and then leave the tissue via this mechanism; tissue damage is prevented. Migration is directed by chemokines, first released by infected epithelial cells and subsequently amplified by neutrophils and other cells, such as mast cells, at the site of infection. In patients prone to APN, CXCR1 expression is reduced compared to age-matched controls and intronic and 3′UTR polymorphisms are more abundant than in controls without UTI. *CXCL8* polymorphisms have also been associated with APN susceptibility. In *mCxcr2*$^{-/-}$ mice lacking the chemokine receptor, neutrophil exit is prevented, however, and a backlog of neutrophils builds up in the tissues. The massive neutrophil infiltrate does not remove the bacteria, as neutrophils from mice lacking *mCxcr2* have an activation deficiency. Persisting bacteria continue to stimulate chemokine production and neutrophils continue to be recruited, resulting in chronic infection and renal scarring (35). Abbreviations: CXC-chemokine receptor (CXCR), interleukin (IL), single-nucleotide polymorphism (SNP), not applicable (NA), acute pyelonephritis (APN), untranslated region (UTR).

fection with *E. coli* CFT073, but in knockout mice (111), bacterial counts increase gradually in both bladders and kidneys. Bladders showed prominent submucosal edema with leukocyte infiltration, as well as focal micro-abscesses and mucosal accumulation of leukocyte-rich exudates. Bacteria were predominantly observed on the surface of the $Tlr5^{-/-}$ urothelium, but not in intracellular bacterial communities. Flagellin is the ligand of Tlr5 but there is no direct evidence of flagellin involvement in the phenotype of $Tlr5^{-/-}$ mice.

At physiological concentrations, THP abolishes the binding of type 1-fimbriated *E. coli* to UP1a and UP1b (84, 138) and in $Thp^{-/-}$ mice, type 1-fimbriated *E. coli* reached higher numbers in the bladders and persisted longer than in wild-type mice (139), but no effect on kidney infection was observed (139). $Cox2^{-/-}$ mice have severely reduced *Thp* mRNA expression, and increased susceptibility to type 1-fimbriated *E. coli* (140) and $Tlr4^{-/-}$ or $Myd88^{-/-}$ had impaired cytokine production in response to Thp (141). This may be relevant for human disease as human myeloid-dendritic cells need THP activation of TLR4-specific MYD88 signaling to acquire a fully mature dendritic-cell phenotype (141).

Antibacterial Peptides

Antimicrobial peptides (such as defensins and cathelicidins) are important for the first line of defense against microbes in the skin and at mucosal surfaces (142). Mice have only one urinary tract cathelicidin, cathelicidin-related, antimicrobial peptide (CRAMP) (143). CRAMP-deficient ($Camp^{-/-}$) mice are more susceptible to UTI, as shown by elevated bacterial numbers after UPEC infection (144), increased mortality with septicemia, and larger kidney size compared to wild-type mice. In an epidemiological study of sensitivity to the human cathelicidin LL-37, APN strains were found to be more resistant than lower urinary tract isolates (144).

HUMAN UTI SUSCEPTIBILITY; MOLECULAR DETERMINANTS AND GENETICS

The difference in susceptibility between the large group of patients with ABU and the small group of highly susceptible individuals, who develop recurrent, sometimes life-threatening, septic infections has proven extremely useful for identifying positive and negative regulators of UTI susceptibility. As described above, genes relevant for human disease have been identified in cellular models and mutant mice. Importantly, the genetic control of UTI susceptibility follows logically from the mechanisms of host response induction. The genetic determinants will be discussed here in the order that they are engaged during pathogenesis.

I. Receptors for Bacterial Adhesins, P Blood Group, and UTI Susceptibility

Host resistance to infection is influenced by the repertoire of host-cell receptors for bacterial adhesins (145) and by soluble, anti-adhesive molecules. Women with recurrent UTI have increased vaginal carriage of uropathogenic bacteria and their uroepithelial cells bind UPEC more efficiently than cells from non-UTI-prone controls (145–149). This difference reflects, in part, a P blood group-dependent variation in receptor expression (145, 146, 149). P-fimbriated *E. coli* bind to the globoseries of glycolipids, which are antigens in the P blood-group system. Individuals of blood-group P1 run about a 17-fold-increased risk of APN and have increased carriage of P-fimbriated *E. coli* in the intestinal flora (145, 146, 149). In contrast, individuals of blood group P lack the Gb_3 glucosyl-transferase and lack functional receptors for P fimbriae (7). Epithelial-receptor expression is also influenced by the ABO blood group and individuals expressing globo-A are preferentially infected by P-fimbriated *E. coli* expressing the prsG adhesin, which uses group A as a receptor (148).

By predicting the mucosal-receptor repertoire, the P blood group is therefore useful as a marker in epidemiologic studies.

The importance of receptor expression for UTI susceptibility has been confirmed *in vivo*, in the murine model using a pharmacological inhibitor of epithelial-glycolipid expression. The glucose analogue N-butyldeoxynojirimycin, which blocks the ceramide-specific glycosyl transferase involved in receptor biosynthesis (150), protected mice against infection and inflammation, confirming the importance of receptor expression *in vivo*. Bacterial adhesion is also modified by soluble receptor oligosaccharides or glycoconjugates. The THP, which is a normal urine constituent, carries terminal mannose residues that bind type 1-fimbriated *E. coli* (151), and sialic acid residues that bind S fimbriae (152, 153). There are no known soluble inhibitors in urine for P fimbriae, which show the strongest disease association.

II. Genetic Variation Affecting *TLR4* Signaling and Innate Immune Response Activation

The structure and function of TLRs is tightly regulated at the genetic level, as expected from their role as sentinels of the innate immune response. Still, there are differences in TLR4 expression between UTI-prone patients and controls, and genetic variants of *TLR4* have been identified in UTI-prone patients, especially promoter variants that reduce the efficiency of TLR4 expression.

Low TLR4 expression in patients with ABU

TLR4 controls the innate immune response to *E. coli* UTI and $Tlr4^{-/-}$ mice develop ABU rather than severe infection, suggesting that reduced mucosal TLR4 function may protect the host against symptomatic infection. To address if TLR4-expression differences characterize patients with UTI, we examined children with primary ABU (no previous symptomatic UTI) or repeated episodes of APN. The ABU-prone group had lower TLR4 expression than controls and children with APN (23). The results suggested low TLR4 expression might be advantageous, as the patients may enjoy the benefits of asymptomatic carriage.

TLR4-promoter variation and TLR4 expression

To explain the low TLR4 expression at the genetic level, we sequenced *TLR4* in patients with UTI. *TLR4* is ca. 19 kb and located on chromosome 9q33.1, and the protein comprises 839 amino acids. Polymorphic sites other than the known Asp299Gly were not observed and this site was polymorphic in one control only. The low expression levels were thus not explained by structural-gene variation.

We subsequently identified a new mechanism for human TLR variation, based on *TLR4*-promoter polymorphisms that influence expression dynamics *in vitro* and innate immune response dynamics in patients with ABU (15, 23). Our studies added a number of polymorphic sites in the *TLR4* promoter and emphasized sequence variation at the promoter level as a mechanism to vary the expression of critical innate immune-response genes like *TLR4*. The results suggest that genetic variation that reduces the efficiency of the *TLR4* promoter is an essential, largely overlooked mechanism to influence TLR4 expression and UTI susceptibility.

Adapter gene polymorphisms

As deletions of adapter protein (such as Trif and Tirap) resulted in reduced bacterial clearance and pathology in mice, *TRIF* sequences were obtained from ABU-prone children and healthy controls. Several polymorphic sites were identified but were not UTI associated (23). The ABU-prone children also had increased TRIF-protein levels, but TRAM and MYD88 expression was not changed (23). No association between *TIRAP* polymorphisms, 539C>T and 558C>T, and UTI susceptibility was found in a cross-sectional analysis of 1,261 women (aged 18–49 years) with ABU (154).

III. Genetic Variants Affecting the Effector Phase of the Innate Immune Response

IRF3 expression and promoter polymorphisms

The UTI susceptibility of $Irf3^{-/-}$ mice indicated that human UTI susceptibility might be influenced by genetic variants affecting IRF3 function (Fig. 2). To address this question, *IRF3* promoter-sequence variation was studied in UTI-prone children selected after >5 years of follow-up to establish their infection patterns. The APN-prone group had multiple episodes of APN but no intervening periods of ABU (9). Results were confirmed in a second population-based study of adults who were re-evaluated about 30 years after their first febrile UTI episode.

IRF3 promoter-genotype analysis revealed substantial differences between the APN-prone and ABU-prone groups. The homozygous A/A–C/C genotype at nucleotide position −925 and −776 was more common in the APN (79%) than in the ABU group, while the co-segregating single-nucleotide polymorphisms (SNPs) were more common in patients with ABU (69%). The transcriptional activity from the APN-promoter genotype was about 50% lower than the ABU-promoter genotype (P <0.001) (9). The results identify a strong association between reduced *IRF3*-promoter genotype and the risk of APN.

CXCR1 expression and genetic polymorphisms

Based on the susceptibility of $mCxcr2^{-/-}$ mutant mice, a clinical study of CXCR1 expression was performed in APN-prone children (12). CXCR1 expression was reduced in APN-prone children, suggesting that variant CXCR1-receptor expression might influence human disease susceptibility (12). Five SNPs were found in the *CXCR1* gene and the cumulative frequency was >50% in APN-prone children without vesicoureteral reflux (VUR). Four genetic variants were predicted to affect transcription-factor usage and one

was shown to reduce the efficiency of RUNX1-dependent transcription.

CXCR1 and related genes have also been examined in other patient groups with different diagnostic criteria (14, 155–157). In premenopausal women with recurrent UTI, CXCR2 levels were reduced compared with controls. No *CXCR1* SNPs were detected but reduced CXCR1 expression was observed in a subgroup of patients with onset of APN before the age of 15 years and before their first sexual intercourse. A difference in *CXCL8* SNPs was observed between patients with DMSA-confirmed APN or negative DMSA scan. No difference was observed between general UTI cases and controls but in patients with history of recurrent UTIs, no difference in CXCR1-expression levels but decreased TLR4 expression was observed compared to controls (157). Three *CXCR1* polymorphisms were more common in women with ABU caused by Gram-positive organisms (154) and one *CXCR1* polymorphism (827G>C) was associated with increased CXCL8 levels (154).

These studies identified reduced CXCR1 expression as a risk factor for APN and added the *CXCR2* receptor and the *CXCL8* chemokine to the repertoire of susceptibility genes. These genes influence neutrophil migration and function, which is essential for the innate defense against UTI. The studies also highlight the need to unify the diagnostic criteria and the polymorphisms so that studies addressing gene associations in UTI can be compared.

Other promoter polymorphisms

In addition to *TLR4* and *IRF3*, APN-prone children carry a *RANTES* promoter variant (−403G/A) (17). A *TNF* variant (−238G/A) is increased in UTI-prone patients with early rheumatoid arthritis, as are two transforming-growth factor (*TGF*)-$\beta 1$ variants; and two *VEGF*-promoter variants (16). Specifically, both the *TGF*-$\beta 1$ −509 T allele and *VEGF* −406CC genotype were associated with a 3.6-fold and 2.8-fold risk of renal scarring, but the SNPs were not linked to VUR. *TGF-β1* −509C/T polymorphism affects transcription because the T allele leads to increased *TGF-β1* gene expression (158) and *TGF-β1*-promoter variants −800G>A and −509TT genotype, as well as the Leu10Pro CT genotypes were associated with renal scarring after APN in children (159). Furthermore, reduced *TGF-β1* expression accompanying the −800GA genotype was proposed to protect against the development of fibrosis in an Irish cohort (160). Hussein et al. confirmed these results, showing that both *TGF-β1*- and *VEGF*-promoter variants were associated with post-UTI renal scar formation in children (16). However, these findings were not confirmed in Asian patients. The *TGF-β1* −509 C allele was more common in Korean patients (161), but the *VEGF* −460 CC promoter variant was increased in patients with VUR with or without UTI, indicating variation in risk of scarring between ethnic groups.

BACTERIAL INHIBITORS OF INNATE IMMUNITY

Co-Evolution of Bacteria and Host

The urinary tract provides unique opportunities to identify molecular mechanisms that promote symbiotic coexistence. Infection is usually caused by a single bacterial strain and changes in bacterial genotype and phenotype resulting from the mucosal host response can readily be monitored, especially in patients with ABU, who carry the same bacterial strain for extended periods of time (162, 163). The genome-wide evolution of a single ABU strain was recently characterized in different human hosts after deliberate inoculation of the urinary tract with this strain (21, 162, 163). Surprisingly, bacterial genome alterations were rapid and distinct for each host and involved functionally important genes potentially influencing the bacterial adaptation to the host environment, such as

regulators of gene expression, metabolism, and virulence. The results show for the first time that human hosts personalize their microflora on a continuous basis.

TIR-Domain Proteins in Bacteria Suppress the Innate Immune Response to Virulent Strains

The virulent bacteria are usually cleared by the host defense, making them less successful colonizers than ABU strains. On the other hand, virulent strains actively enhance their fitness for the urinary tract and have evolved mechanisms to specifically interfere with TLR-mediated immune responses. TIR-domain homologues have been detected in bacteria and viruses (50, 164–166) and TIR-containing proteins (Tcps) represent a new class of virulence factors, which suppress innate immunity (10). A TIR-homologous gene is present in the genome of the UPEC strain CFT073 and the bacteria secrete inhibitory homologues of the TIR domains that interfere with TLR-signaling through the MYD88 adaptor, due to direct binding of Tcps to MYD88. TcpC is prevalent in virulent uropathogenic *E. coli* strains and promote bacterial survival and kidney pathology *in vivo*.

Bacterial Control of Host Gene Expression Through RNA Polymerase II

A new mechanism of bacterial adaptation was recently identified in ABU strains, involving rapid suppression of RNA polymerase II (Pol II)-dependent host gene expression (22). After human inoculation with the prototype ABU strain *E. coli* 83972, over 60% of all genes showed reduced expression after 24 hours. This inhibition was verified by infection of human uroepithelial cells and activators and suppressors of Pol II were affected, as were pathogen-specific signaling pathways and innate immunity. The frequency of strains inhibiting Pol II was high in ABU or fecal isolates, suggesting that certain nonvirulent bacteria modify the host environment actively by controlling the quality and quantity of host gene expression.

NEW THERAPEUTIC APPROACHES THAT MODIFY INNATE OR SPECIFIC IMMUNITY

New approaches are needed for the treatment and prevention of UTI, especially in light of the rapid emergence of antibiotic resistance. A few promising examples are discussed below (for further comments about this area, please see references 167–170.)

1. Deliberate Establishment of ABU is Protective

Based on the protective effects of ABU, we have chosen to establish ABU-like, protective states in UTI-prone patients who do not spontaneously develop ABU. Initially, inoculation with *E. coli* isolates from the patient's own fecal flora was tested and the bacteria persisted for up to four months (171). To facilitate a broader clinical use of this methodology, we identified the ABU strain *E. coli* 83972 as a prototype strain to be used in different patient groups. The strain has a non-adhesive, P and type 1 fimbriae-negative phenotype, lacks expressed virulence factors, and lacks large conjugative plasmids. The genome sequence of *E. coli* 83972 identified multiple deletions and point mutations resulting in smaller overall genome size and inactivated virulence gene expression. *E. coli* 83972 was first used by us in 1991 (172) with the purpose of establishing prophylactic *E. coli* 83972 ABU in patients with recurrent UTI. Stable bacteriuria was established and the patients remained asymptomatic (172–174).

Subsequent studies have shown that *E. coli* 83972 establishes long-term bacteriuria following intravesical inoculation, and that asymptomatic carriers are protected against super-infections with more virulent strains (162, 171–177). In a placebo-controlled study, therapeutic efficacy was demonstrated in patients with dysfunctional voiding (163) and intravesical inoculation with the prototype ABU strain *E. coli* 83972 is recommended in the 2014 European guidelines for use in patients with dysfunctional bladder voiding. Inoculation is safe and protects the host against symptomatic infections, as shown in a placebo-controlled study (163).

2. Receptor Analogues Prevent Bacterial Attachment

As discussed above, oligosaccharide receptors or glycoproteins expressing defined oligosaccharide epitopes are secreted into the urine and act as competitive inhibitors of bacterial attachment (178). Original observations on mannose-mediated inhibition of infection were made by Aronson et al. Inhibition of type 1-fimbrial adhesion by THP has beneficial effects and the use of low molecular weight mannosides to inhibit FimH-mediated bacterial attachment has been readdressed in mice and has shown promise for treatment and prevention of recurrent UTI (179). Soluble Galα1-4Galβ-oligosaccharides inhibit P fimbriae-mediated attachment to human uroepithelial cells and therapeutic efficacy has also been demonstrated *in vivo* (180–183). The availability of Galα1-4Galβ-oligosaccharides in sufficient amounts for therapeutic clinical studies has been limited so far, however.

3. Host Immune Response-Effector Molecules as New "Antibiotics"

Antimicrobial-peptide activity is associated with the severity of recurrent UTI and their therapeutic use has considerable potential (143, 184). A gene therapy approach was used to induce β-defensin-2 expression in the rat urothelium (185), resulting in a decrease in bladder-associated bacteria and inflammatory scores during the first 24 and 48 hours of infection (168).

4. Specific Immunity and UTI Susceptibility: Vaccination Studies

Early studies in the murine UTI model showed that T-cell-deficient (nu/nu) and X-linked-immunodeficient (xid) mice are resistant to UTI (27, 186), suggesting that adaptive immunity is not essential to maintain sterility of the urinary tract. These findings were later confirmed in T-cell receptor (TCR)-$\alpha\beta$ and TCR$\gamma\delta$ knock-out and RAG-1 mice with defects in T-lymphocyte, Ig, or total lymphocyte function (186, 187). Other studies implicated adaptive-immunity involvement in UTI, where recruitment of activated T-cells and the development of specific IgG antibody in the serum and urine were suggested to prevent relapse (188). Furthermore, immunization against the FimH adhesin of type 1-fimbriated *E. coli* generates protective immunity in the murine model and non-human primates (189, 190) and siderophores have been identified as useful and efficient targets (191).

The specific immune response to UTI remains a target of investigation for many groups, however. One aim has been to compensate for innate immune defects by generating vaccine-induced protective immunity against the mucosal as well as the systemic phases of disease. In experimental models, several bacterial-surface antigens have been shown to trigger a protective immune response (192–195). Later, anti-fimbrial antibodies were shown to prevent infection or reduce the severity of disease (194, 196). Detailed studies by Uehling et al. assessed the efficiency of vaginal mucosal immunization with a multivalent bacterial vaccine in patients with recurrent UTI (197–201). A significant improvement was reported in women who received vaccine with boosters, were sexually active, <52 years old, and had not undergone hysterectomy (202). It therefore remains possible and interesting that vaccination might be especially useful to prevent infection or reduce its destructive effects.

MOLECULAR TOOLS FOR RISK ASSESSMENT AND PREDICTION OF UTI SUSCEPTIBILITY

Defining UTI susceptibility and risk for severe acute and chronic infection is essential as is the design of therapeutic interventions that fit the individual host-susceptibility profile. The new molecular markers of host susceptibility described in this study are well suited to improve the definition of susceptibility as a basis for therapeutic decisions, especially when antibiotics become less efficient. Appropriate technologies include genetics defining UTI-associated polymorphisms, gene-expression profiles defined at the RNA level by microarray technology, and proteomics for the detection of complex cytokine-response profiles.

Some examples of situations when these molecular tools might be of use are given below:

1. **Risk assessment:**

a. Prediction of recurrence risk, need for prophylaxis, and radiological examinations in young children with their first febrile UTI episode.

b. Preoperative assessment of the risk for serious UTI/urosepsis in connection with surgery.

2. **Differential diagnosis:**

a. ABU with fever of other origin. Urine cytokine measurements have been shown to distinguish febrile UTI from ABU with fever of other origin. Examples include IL-8, which is elevated in urine during febrile UTI but not ABU (203) and was significantly higher in children who later developed renal scarring than in those without scars (78, 115, 204). IL-6 is a marker of systemic involvement and IL-6 levels distinguish upper from lower urinary tract infection in children. In this clinical setting, a value >15 picograms (pg) per ml is a strong indicator of APN (205).

b. Distinguishing ABU from UTI in elderly patients presenting with diffuse systemic symptoms without apparent focal urinary signs.

c. Defining the origin of the fever in patients with indwelling urinary catheters and urological disorders.

d. Defining disease-severity patients with atypical symptoms or laboratory findings. Genetic variants that reduce the efficiency of the innate immune response also alter the level of inflammatory mediators, fever, and tissue inflammation, making symptoms and diagnostic criteria like neutrophil counts in urine and blood, fever, and CRP difficult to trust.

3. **Choice of therapy.** With a focus on limiting the use of antibiotics, it becomes even more urgent to assess the severity of infection by including molecular tools in the diagnostic arsenal. As antibiotics fail, it will become critical to identify exactly who would benefit from inoculation with "safe strains", vaccination, alternative therapies, etc.

It is essential to caution that these new techniques will not perfectly reproduce the diagnostic groups identified by current diagnostic tools or exactly the same UTI-prone patients as those who are currently being identified in the clinic. The prediction is that the number of subgroups among the UTI-prone population will likely increase to include new, interesting, and clinically relevant patient categories. The challenge will be to combine the new techniques with traditional diagnostic and risk-assessment tools in properly controlled epidemiological studies.

CONCLUDING REMARKS

In this review, we summarize the current understanding of innate immunity and UTI susceptibility, based on mechanisms of disease pathogenesis and innate immune responses in the urinary tract. We exemplify how genetic defects affecting innate immunity may be protective or lead to acute mortality, chronic infection, or tissue damage and how, by dissecting the increasingly well-defined molecular interactions between bacteria and host and the molecular features of excessive bacterial virulence or host-response malfunction, it is becoming possible to isolate the defensive from the damaging aspects of the host response. There is a great clinical need to identify UTI-susceptible patients and to distinguish different categories. We discuss molecular markers needed to assess host susceptibility and to identify patients who would benefit from therapies that restore their antibacterial-effector functions. Further clinical studies are needed to define how the detection of specific polymorphisms or protein levels can be used to predict risk and distinguish susceptible from resistant individuals. Distinguishing "good" from "bad" inflammation has been a long-term quest of biomedical science and, in UTI, patients need the "good" inflammatory response to resist infection while avoiding the "bad" aspects that cause chronicity and tissue damage.

ACKNOWLEDGMENTS

Conflicts of interest: We declare no conflicts.

CITATION

Ambite I, Nagy K, Godaly G, Puthia M, Wullt B, Svanborg C. 2016. Susceptibility to urinary tract infection: benefits and hazards of the antibacterial host response. Microbiol Spectrum 4(3):UTI-0019-2014.

REFERENCES

1. **Kunin CM.** 1997. *Urinary Tract Infections. Detection, Prevention and Management*, 5th ed. Williams & Wilkins, Baltimore, MD.
2. **Nielubowicz GR, Mobley HL.** 2010. Host-pathogen interactions in urinary tract infection. *Nat Rev Urol* 7:430–441.

3. Ragnarsdóttir B, Lutay N, Grönberg-Hernandez J, Köves B, Svanborg C. 2011. Genetics of innate immunity and UTI susceptibility. *Nat Rev Urol* 8:449–468.

4. Hannan TJ, Totsika M, Mansfield KJ, Moore KH, Schembri MA, Hultgren SJ. 2012. Host-pathogen checkpoints and population bottlenecks in persistent and intracellular uropathogenic *Escherichia coli* bladder infection. *FEMS Microbiol Rev* 36:616–648.

5. Abraham SN, St John AL. 2010. Mast cell-orchestrated immunity to pathogens. *Nat Rev Immunol* 10:440–452.

6. Svanborg C, Edén CS, Hanson LA, Jodal U, Lindberg U, Akerlund AS. 1976. Variable adherence to normal human urinary-tract epithelial cells of *Escherichia coli* strains associated with various forms of urinary-tract infection. *Lancet* 1:490–492.

7. Leffler H, Edén C. 1980. Chemical identification of a glycosphingolipid receptor for *Escherichia coli* attaching to human urinary tract epithelial cells and agglutinating human erythrocytes. *FEMS Microbiol Lett* 8:127–134.

8. Hedlund M, Svensson M, Nilsson A, Duan RD, Svanborg C. 1996. Role of the ceramide-signaling pathway in cytokine responses to P-fimbriated *Escherichia coli*. *J Exp Med* 183:1037–1044.

9. Fischer H, Lutay N, Ragnarsdóttir B, Yadav M, Jönsson K, Urbano A, Al Hadad A, Rämisch S, Storm P, Dobrindt U, Salvador E, Karpman D, Jodal U, Svanborg C. 2010. Pathogen specific, IRF3-dependent signaling and innate resistance to human kidney infection. *PLoS Pathog* 6:e1001109. doi:10.1371/journal.ppat.1001109

10. Cirl C, Wieser A, Yadav M, Duerr S, Schubert S, Fischer H, Stappert D, Wantia N, Rodriguez N, Wagner H, Svanborg C, Miethke T. 2008. Subversion of Toll-like receptor signaling by a unique family of bacterial Toll/interleukin-1 receptor domain-containing proteins. *Nat Med* 14:399–406.

11. Agace WW, Hedges SR, Ceska M, Svanborg C. 1993. Interleukin-8 and the neutrophil response to mucosal gram-negative infection. *J Clin Invest* 92:780–785.

12. Frendéus B, Godaly G, Hang L, Karpman D, Lundstedt AC, Svanborg C. 2000. Interleukin 8 receptor deficiency confers susceptibility to acute experimental pyelonephritis and may have a human counterpart. *J Exp Med* 192:881–890.

13. Lundstedt AC, McCarthy S, Gustafsson MC, Godaly G, Jodal U, Karpman D, Leijonhufvud I, Lindén C, Martinell J, Ragnarsdottir B, Samuelsson M, Truedsson L, Andersson B, Svanborg C. 2007. A genetic basis of susceptibility to acute pyelonephritis. *PLoS One* 2:e825. doi:10.1371/journal.pone.0000825

14. Artifoni L, Negrisolo S, Montini G, Zucchetta P, Molinari PP, Cassar W, Destro R, Anglani F, Rigamonti W, Zacchello G, Murer L. 2007. Interleukin-8 and CXCR1 receptor functional polymorphisms and susceptibility to acute pyelonephritis. *J Urol* 177:1102–1106.

15. Ragnarsdóttir B, Jonsson K, Urbano A, Grönberg-Hernandez J, Lutay N, Tammi M, Gustafsson M, Lundstedt AC, Leijonhufvud I, Karpman D, Wullt B, Truedsson L, Jodal U, Andersson B, Svanborg C. 2010. Toll-like receptor 4 promoter polymorphisms: common TLR4 variants may protect against severe urinary tract infection. *PLoS One* 5:e10734. doi:10.1371/journal.pone.0010734

16. Hussein A, Askar E, Elsaeid M, Schaefer F. 2010. Functional polymorphisms in transforming growth factor-beta-1 (TGFbeta-1) and vascular endothelial growth factor (VEGF) genes modify risk of renal parenchymal scarring following childhood urinary tract infection. *Nephrol Dial Transplant* 25:779–785.

17. Centi S, Negrisolo S, Stefanic A, Benetti E, Cassar W, Da Dalt L, Rigamonti W, Zucchetta P, Montini G, Murer L, Artifoni L. 2010. Upper urinary tract infections are associated with RANTES promoter polymorphism. *J Pediatr* 157:1038–1040e1.

18. Sundén F, Håkansson L, Ljunggren E, Wullt B. 2006. Bacterial interference—is deliberate colonization with *Escherichia coli* 83972 an alternative treatment for patients with recurrent urinary tract infection? *Int J Antimicrob Agents* 28:S26–29.

19. Zdziarski J, Svanborg C, Wullt B, Hacker J, Dobrindt U. 2008. Molecular basis of commensalism in the urinary tract: low virulence or virulence attenuation? *Infect Immun* 76:695–703.

20. Klemm P, Roos V, Ulett GC, Svanborg C, Schembri MA. 2006. Molecular characterization of the *Escherichia coli* asymptomatic bacteriuria strain 83972: the taming of a pathogen. *Infect Immun* 74:781–785.

21. Zdziarski J, Brzuszkiewicz E, Wullt B, Liesegang H, Biran D, Voigt B, Grönberg-Hernandez J, Ragnarsdottir B, Hecker M, Ron EZ, Daniel R, Gottschalk G, Hacker J, Svanborg C, Dobrindt U. 2010. Host imprints on bacterial genomes—rapid, divergent evolution in individual patients. *PLoS Pathog* 6:e1001078. doi:10.1371/journal.ppat.1001078

22. Lutay N, Ambite I, Grönberg-Hernandez JG, Rydström G, Ragnarsdóttir B, Puthia M, Nadeem A, Zhang J, Storm P, Dobrindt U, Wullt B, Svanborg C. 2013. Bacterial control of host gene expression through RNA polymerase II. *J Clin Invest* **123**:2366–2379.

23. Ragnarsdóttir B, Samuelsson M, Gustafsson MC, Leijonhufvud I, Karpman D, Svanborg C. 2007. Reduced toll-like receptor 4 expression in children with asymptomatic bacteriuria. *J Infect Dis* **196**:475–484.

24. Backhed F, Meijer L, Normark S, Richter-Dahlfors A. 2002. TLR4-dependent recognition of lipopolysaccharide by epithelial cells requires sCD14. *Cell Microbiol* **4**:493–501.

25. Samuelsson P, Hang L, Wullt B, Irjala H, Svanborg C. 2004. Toll-like receptor 4 expression and cytokine responses in the human urinary tract mucosa. *Infect Immun* **72**:3179–3186.

26. Hagberg L, Hull R, Hull S, McGhee JR, Michalek SM, Svanborg Eden C. 1984. Difference in susceptibility to gram-negative urinary tract infection between C3H/HeJ and C3H/HeN mice. *Infect Immun* **46**:839–844.

27. Hagberg L, Briles DE, Svanborg-Edén CS. 1985. Evidence for separate genetic defects in C3H/HeJ and C3HeB/FeJ mice, that affect susceptibility to gram-negative infections. *J Immunol* **134**:4118–4122.

28. Poltorak A, He X, Smirnova I, Liu MY, Van Huffel C, Du X, Birdwell D, Alejos E, Silva M, Galanos C, Freudenberg M, Ricciardi-Castagnoli P, Layton B, Beutler B. 1998. Defective LPS signaling in C3H/HeJ and C57BL/10ScCr mice: mutations in Tlr4 gene. *Science* **282**:2085–2088.

29. Shahin RD, Engberg I, Hagberg L, Svanborg Edén C. 1987. Neutrophil recruitment and bacterial clearance correlated with LPS responsiveness in local gram-negative infection. *J Immunol* **138**:3475–3480.

30. Hopkins WJ, Gendron-Fitzpatrick A, Balish E, Uehling DT. 1998. Time course and host responses to *Escherichia coli* urinary tract infection in genetically distinct mouse strains. *Infect Immun* **66**:2798–2802.

31. Frendéus B, Wachtler C, Hedlund M, Fischer H, Samuelsson P, Svensson M, Svanborg C. 2001. *Escherichia coli* P fimbriae utilize the Toll-like receptor 4 pathway for cell activation. *Mol Microbiol* **40**:37–51.

32. Hedlund M, Frendéus B, Wachtler C, Hang L, Fischer H, Svanborg C. 2001. Type 1 fimbriae deliver an LPS- and TLR4-dependent activation signal to CD14-negative cells. *Mol Microbiol* **39**:542–552.

33. Schilling JD, Martin SM, Hung CS, Lorenz RG, Hultgren SJ. 2003. Toll-like receptor 4 on stromal and hematopoietic cells mediates innate resistance to uropathogenic *Escherichia coli*. *Proc Natl Acad Sci U S A* **100**: 4203–4208.

34. Hang L, Frendeus B, Godaly G, Svanborg C. 2000. Interleukin-8 receptor knockout mice have subepithelial neutrophil entrapment and renal scarring following acute pyelonephritis. *J Infect Dis* **182**:1738–1748.

35. Svensson M, Irjala H, Alm P, Holmqvist B, Lundstedt AC, Svanborg C. 2005. Natural history of renal scarring in susceptible mIL-8Rh-/- mice. *Kidney Int* **67**:103–110.

36. Fischer H, Yamamoto M, Akira S, Beutler B, Svanborg C. 2006. Mechanism of pathogen-specific TLR4 activation in the mucosa: fimbriae, recognition receptors and adaptor protein selection. *Eur J Immunol* **36**:267–277.

37. Matsushima N, Tanaka T, Enkhbayar P, Mikami T, Taga M, Yamada K, Kuroki Y. 2007. Comparative sequence analysis of leucine-rich repeats (LRRs) within vertebrate toll-like receptors. *BMC Genomics* **8**:124.

38. Kawai T, Akira S. 2009. The roles of TLRs, RLRs and NLRs in pathogen recognition. *Int Immunol* **21**:317–337.

39. Kim HM, Park BS, Kim JI, Kim SE, Lee J, Oh SC, Enkhbayar P, Matsushima N, Lee H, Yoo OJ, Lee JO. 2007. Crystal structure of the TLR4-MD-2 complex with bound endotoxin antagonist Eritoran. *Cell* **130**:906–917.

40. Gay NJ, Gangloff M. 2008. Structure of toll-like receptors. *Handb Exp Pharmacol* **183**:181–200.

41. Anderson KV, Jürgens G, Nusslein-Volhard C. 1985. Establishment of dorsal-ventral polarity in the Drosophila embryo: genetic studies on the role of the Toll gene product. *Cell* **42**:779–789.

42. Dunne A, Ejdeback M, Ludidi PL, O'Neill LA, Gay NJ. 2003. Structural complementarity of Toll/interleukin-1 receptor domains in Toll-like receptors and the adaptors Mal and MyD88. *J Biol Chem* **278**:41443–41451.

43. Sheedy FJ, O'Neill LA. 2007. The Troll in Toll: Mal and Tram as bridges for TLR2 and TLR4 signaling. *J Leukoc Biol* **82**:196–203.

44. Poltorak A, Smirnova I, He X, Liu MY, Van Huffel C, McNally O, Birdwell D, Alejos E, Silva M, Du X, Thompson P, Chan EK, Ledesma J, Roe B, Clifton S, Vogel SN, Beutler B. 1998. Genetic and physical mapping of the Lps locus: identification of the toll-4 receptor as a candidate gene in the critical region. *Blood Cells Mol Dis* **24**:340–355.

45. **Slack JL, Schooley K, Bonnert TP, Mitcham JL, Qwarnstrom EE, Sims JE, Dower SK.** 2000. Identification of two major sites in the type I interleukin-1 receptor cytoplasmic region responsible for coupling to pro-inflammatory signaling pathways. *J Biol Chem* **275:**4670–4678.

46. **Ozinsky A, Smith KD, Hume D, Underhill DM.** 2000. Co-operative induction of pro-inflammatory signaling by Toll-like receptors. *J Endotoxin Res* **6:**393–396.

47. **Yamamoto M, Sato S, Mori K, Hoshino K, Takeuchi O, Takeda K, Akira S.** 2002. Cutting edge: A novel Toll/IL-1 receptor domain-containing adapter that preferentially activates the IFN-beta promoter in the Toll-like receptor signaling. *J Immunol* **169:**6668–6672.

48. **Hoebe K, Du X, Georgel P, Janssen E, Tabeta K, Kim SO, Goode J, Lin P, Mann N, Mudd S, Crozat K, Sovath S, Han J, Beutler B.** 2003. Identification of Lps2 as a key transducer of MyD88-independent TIR signalling. *Nature* **424:**743–748.

49. **Fitzgerald KA, Rowe DC, Barnes BJ, Caffrey DR, Visintin A, Latz E, Monks B, Pitha PM, Golenbock DT.** 2003. LPS-TLR4 signaling to IRF-3/7 and NF-kappaB involves the toll adapters TRAM and TRIF. *J Exp Med* **198:**1043–1055.

50. **Bowie A, Kiss-Toth E, Symons JA, Smith GL, Dower SK, O'Neill LA.** 2000. A46R and A52R from vaccinia virus are antagonists of host IL-1 and toll-like receptor signaling. *Proc Natl Acad Sci U S A* **97:**10162–10167.

51. **Burns K, Martinon F, Esslinger C, Pahl H, Schneider P, Bodmer JL, Di Marco F, French L, Tschopp J.** 1998. MyD88, an adapter protein involved in interleukin-1 signaling. *J Biol Chem* **273:**12203–12209.

52. **Kawai T, Adachi O, Ogawa T, Takeda K, Akira S.** 1999. Unresponsiveness of MyD88-deficient mice to endotoxin. *Immunity* **11:**115–122.

53. **Fitzgerald KA, Palsson-McDermott EM, Bowie AG, Jefferies CA, Mansell AS, Brady G, Brint E, Dunne A, Gray P, Harte MT, McMurray D, Smith DE, Sims JE, Bird TA, O'Neill LA.** 2001. Mal (MyD88-adapter-like) is required for Toll-like receptor-4 signal transduction. *Nature* **413:**78–83.

54. **Horng T, Barton GM, Medzhitov R.** 2001. TIRAP: an adapter molecule in the Toll signaling pathway. *Nat Immunol* **2:**835–841.

55. **Kagan JC, Medzhitov R.** 2006. Phosphoinositide-mediated adaptor recruitment controls Toll-like receptor signaling. *Cell* **125:**943–955.

56. **Gray P, Dunne A, Brikos C, Jefferies CA, Doyle SL, O'Neill LA.** 2006. MyD88 adapter-like (Mal) is phosphorylated by Bruton's tyrosine kinase during TLR2 and TLR4 signal transduction. *J Biol Chem* **281:**10489–10495.

57. **Janssens S, Burns K, Vercammen E, Tschopp J, Beyaert R.** 2003. MyD88S, a splice variant of MyD88, differentially modulates NF-kappaB- and AP-1-dependent gene expression. *FEBS Lett* **548:**103–107.

58. **Anderson GG, Palermo JJ, Schilling JD, Roth R, Heuser J, Hultgren SJ.** 2003. Intracellular bacterial biofilm-like pods in urinary tract infections. *Science* **301:**105–107.

59. **Justice SS, Hung C, Theriot JA, Fletcher DA, Anderson GG, Footer MJ, Hultgren SJ.** 2004. Differentiation and developmental pathways of uropathogenic Escherichia coli in urinary tract pathogenesis. *Proc Natl Acad Sci U S A* **101:**1333–1338.

60. **Wright KJ, Seed PC, Hultgren SJ.** 2007. Development of intracellular bacterial communities of uropathogenic *Escherichia coli* depends on type 1 pili. *Cell Microbiol* **9:**2230–2241.

61. **Yadav M, Zhang J, Fischer H, Huang W, Lutay N, Cirl C, Lum J, Miethke T, Svanborg C.** 2010. Inhibition of TIR domain signaling by TcpC: MyD88-dependent and independent effects on *Escherichia coli* virulence. *PLoS Pathog* **6:**e1001120. doi:10.1371/journal.ppat.1001120

62. **Leffler H, Svanborg-Edén C.** 1981. Glycolipid receptors for uropathogenic *Escherichia coli* on human erythrocytes and uroepithelial cells. *Infect Immun* **34:**920–929.

63. **Plos K, Connell H, Jodal U, Marklund B, Mårild S, Wettergren B, Svanborg C.** 1995. Intestinal carriage of P fimbriated *Escherichia coli* and the susceptibility to urinary tract infection in young children. *J Infect Dis* **171:**625–631.

64. **Roberts JA, Marklund BI, Ilver D, Haslam D, Kaack MB, Baskin G, Louis M, Möllby R, Winberg J, Normark S.** 1994. The Gal(alpha 1-4) Gal-specific tip adhesin of *Escherichia coli* P-fimbriae is needed for pyelonephritis to occur in the normal urinary tract. *Proc Natl Acad Sci U S A* **91:**11889–11893.

65. **Lindberg F, Lund B, Johansson L, Normark S.** 1987. Localization of the receptor-binding protein adhesin at the tip of the bacterial pilus. *Nature* **328:**84–87.

66. **Linder H, Engberg I, Hoschültzky H, Mattsby-Baltzer I, Svanborg C.** 1991. Adhesion-dependent activation of mucosal interleukin-6 production. *Infect Immun* **59:**4357–4362.

67. **Bergsten G, Samuelsson M, Wullt B, Leijonhufvud I, Fischer H, Svanborg C.**

2004. PapG-dependent adherence breaks mucosal inertia and triggers the innate host response. *J Infect Dis* **189:**1734–1742.

68. **Ambite I, Lutay N, Godaly G, Svanborg C.** 2015. Urinary tract infections and the mucosal immune system, p 2039–2058. *In* Mestecky J, Strober W, Russell MW, Cheroutre H, Lambrecht BN, Kelsall BL (ed), *Mucosal Immunology,* 4th ed. Academic Press, Boston.

69. **Hannun YA, Obeid LM.** 2008. Principles of bioactive lipid signalling: lessons from sphingolipids. *Nat Rev Mol Cell Biol* **9:**139–150.

70. **Hedlund M, Duan RD, Nilsson Å, Svanborg C.** 1998. Sphingomyelin, glycosphingolipids and ceramide signalling in cells exposed to P fimbriated *Escherichia coli. Mol Microbiol* **29:**1297–1306.

71. **Fischer H, Ellström P, Ekström K, Gustafsson L, Gustafsson M, Svanborg C.** 2007. Ceramide as a TLR4 agonist; a putative signalling intermediate between sphingolipid receptors for microbial ligands and TLR4. *Cell Microbiol* **9:**1239–1251.

72. **Yamamoto M, Sato S, Hemmi H, Uematsu S, Hoshino K, Kaisho T, Takeuchi O, Takeda K, Akira S.** 2003. TRAM is specifically involved in the Toll-like receptor 4-mediated MyD88-independent signaling pathway. *Nat Immunol* **4:**1144–1150.

73. **Sato S, Sugiyama M, Yamamoto M, Watanabe Y, Kawai T, Takeda K, Akira S.** 2003. Toll/IL-1 receptor domain-containing adaptor inducing IFN-beta (TRIF) associates with TNF receptor-associated factor 6 and TANK-binding kinase 1, and activates two distinct transcription factors, NF-kappa B and IFN-regulatory factor-3, in the Toll-like receptor signaling. *J Immunol* **171:**4304–4310.

74. **Rowe DC, McGettrick AF, Latz E, Monks BG, Gay NJ, Yamamoto M, Akira S, O'Neill LA, Fitzgerald KA, Golenbock DT.** 2006. The myristoylation of TRIF-related adaptor molecule is essential for Toll-like receptor 4 signal transduction. *Proc Natl Acad Sci U S A* **103:** 6299–6304.

75. **McGettrick AF, Brint EK, Palsson-McDermott EM, Rowe DC, Golenbock DT, Gay NJ, Fitzgerald KA, O'Neill LA.** 2006. Trif-related adapter molecule is phosphorylated by PKC {epsilon} during Toll-like receptor 4 signaling. *Proc Natl Acad Sci U S A* **103:**9196–9201.

76. **Oganesyan G, Saha SK, Guo B, He JQ, Shahangian A, Zarnegar B, Perry A, Cheng G.** 2006. Critical role of TRAF3 in the Toll-like receptor-dependent and -independent antiviral response. *Nature* **439:**208–211.

77. **Kawai T, Sato S, Ishii KJ, Coban C, Hemmi H, Yamamoto M, Terai K, Matsuda M, Inoue J, Uematsu S, Takeuchi O, Akira S.** 2004. Interferon-alpha induction through Toll-like receptors involves a direct interaction of IRF7 with MyD88 and TRAF6. *Nat Immunol* **5:**1061–1068.

78. **Agace W, Hedges S, Andersson U, Andersson J, Ceska M, Svanborg C.** 1993. Selective cytokine production by epithelial cells following exposure to *Escherichia coli. Infect Immun* **61:**602–609.

79. **Svensson M, Irjala H, Svanborg C, Godaly G.** 2008. Effects of epithelial and neutrophil CXCR2 on innate immunity and resistance to kidney infection. *Kidney Int* **74:**81–90.

80. **Wold AE, Mestecky J, Tomana M, Kobata A, Ohbayashi H, Endo T, Edén CS.** 1990. Secretory immunoglobulin-A carries oligosaccharide receptors for *Escherichia coli* type 1 fimbrial lectin. *Infect Immun* **58:**3073–3077.

81. **Xie B, Zhou G, Chan SY, Shapiro E, Kong XP, Wu XR, Sun TT, Costello CE.** 2006. Distinct glycan structures of uroplakins Ia and Ib: structural basis for the selective binding of FimH adhesin to uroplakin Ia. *J Biol Chem* **281:**14644–14653.

82. **Malaviya R, Gao Z, Thankavel K, van der Merwe PA, Abraham SN.** 1999. The mast cell tumor necrosis factor alpha response to FimH-expressing *Escherichia coli* is mediated by the glycosylphosphatidylinositol-anchored molecule CD48. *Proc Natl Acad Sci U S A* **96:**8110–8115.

83. **Eto DS, Jones TA, Sundsbak JL, Mulvey MA.** 2007. Integrin-mediated host cell invasion by type 1-piliated uropathogenic *Escherichia coli. PLoS Pathog* **3:**e100. doi:10.1371/journal. ppat.0030100

84. **Pak J, Pu Y, Zhang ZT, Hasty DL, Wu XR.** 2001. Tamm-Horsfall protein binds to type 1 fimbriated *Escherichia coli* and prevents *E. coli* from binding to uroplakin Ia and Ib receptors. *J Biol Chem* **276:**9924–9930.

85. **Baorto DM, Gao Z, Malaviya R, Dustin ML, van der Merwe A, Lublin DM, Abraham SN.** 1997. Survival of FimH-expressing enterobacteria in macrophages relies on glycolipid traffic. *Nature* **389:**636–639.

86. **Shin JS, Gao Z, Abraham SN.** 1999. Bacteria-host cell interaction mediated by cellular cholesterol/glycolipid-enriched microdomains. *Biosci Rep* **19:**421–432.

87. **Shin JS, Gao Z, Abraham SN.** 2000. Involvement of cellular caveolae in bacterial entry into mast cells. *Science* **289:**785–788.

88. **McLean GW, Carragher NO, Avizienyte E, Evans J, Brunton VG, Frame MC.** 2005. The role of focal-adhesion kinase in cancer - a new therapeutic opportunity. *Nat Rev Cancer* **5:**505–515.

89. **Mulvey MA, Schilling JD, Hultgren SJ.** 2001. Establishment of a persistent *Escherichia coli* reservoir during the acute phase of a bladder infection. *Infect Immun* **69:**4572–4579.

90. **Rosen DA, Hooton TM, Stamm WE, Humphrey PA, Hultgren SJ.** 2007. Detection of intracellular bacterial communities in human urinary tract infection. *PLoS Mede* **4:**e329. doi:10.1371/journal.pmed.0040329

91. **Mulvey MA, Lopez-Boado YS, Wilson CL, Roth R, Parks WC, Heuser J, Hultgren SJ.** 1998. Induction and evasion of host defenses by type 1-piliated uropathogenic *Escherichia coli*. *Science* **282:**1494–1497.

92. **Klumpp DJ, Rycyk MT, Chen MC, Thumbikat P, Sengupta S, Schaeffer AJ.** 2006. Uropathogenic *Escherichia coli* induces extrinsic and intrinsic cascades to initiate urothelial apoptosis. *Infect Immun* **74:**5106–5113.

93. **Thumbikat P, Berry RE, Zhou G, Billips BK, Yaggie RE, Zaichuk T, Sun TT, Schaeffer AJ, Klumpp DJ.** 2009. Bacteria-induced uroplakin signaling mediates bladder response to infection. *PLoS Pathog* **5:**e1000415. doi:10.1371/journal.ppat.10000415

94. **Bishop BL, Duncan MJ, Song J, Li G, Zaas D, Abraham SN.** 2007. Cyclic AMP-regulated exocytosis of *Escherichia coli* from infected bladder epithelial cells. *Nat Med* **13:**625–630.

95. **Thankavel K, Madison B, Ikeda T, Malaviya R, Shah AH, Arumugam PM, Abraham SN.** 1997. Localization of a domain in the FimH adhesin of *Escherichia coli* type 1 fimbriae capable of receptor recognition and use of a domain-specific antibody to confer protection against experimental urinary tract infection. *J Clin Invest* **100:**1123–1136.

96. **Schilling JD, Mulvey MA, Vincent CD, Lorenz RG, Hultgren SJ.** 2001. Bacterial invasion augments epithelial cytokine responses to *Escherichia coli* through a lipopolysaccharide-dependent mechanism. *J Immunol* **166:**1148–1155.

97. **Song J, Bishop BL, Li G, Duncan MJ, Abraham SN.** 2007. TLR4-initiated and cAMP-mediated abrogation of bacterial invasion of the bladder. *Cell Host Microbe* **1:**287–298.

98. **Song J, Duncan MJ, Li G, Chan C, Grady R, Stapleton A, Abraham SN.** 2007. A novel TLR4-mediated signaling pathway leading to IL-6 responses in human bladder epithelial cells. *PLoS Pathog* **3:**e60. doi:10.1371/journal.ppat.0030060

99. **Bergsten G, Wullt B, Schembri MA, Leijonhufvud I, Svanborg C.** 2007. Do type 1 fimbriae promote inflammation in the human urinary tract? *Cell Microbiol* **9:**1766–1781.

100. **Connell I, Agace W, Klemm P, Schembri M, Mårild S, Svanborg C.** 1996. Type 1 fimbrial adhesion enhances *Escherichia coli* virulence for the urinary tract. *Proc Natl Acad Sci U S A* **93:**9827–9832.

101. **Lane MC, Alteri CJ, Smith SN, Mobley HL.** 2007. Expression of flagella is coincident with uropathogenic *Escherichia coli* ascension to the upper urinary tract. *Proc Natl Acad Sci U S A* **104:**16669–16674.

102. **Allsopp LP, Beloin C, Moriel DG, Totsika M, Ghigo JM, Schembri MA.** 2012. Functional heterogeneity of the UpaH autotransporter protein from uropathogenic *Escherichia coli*. *J Bacteriol* **194:**5769–5782.

103. **Holden N, Totsika M, Dixon L, Catherwood K, Gally DL.** 2007. Regulation of P-fimbrial phase variation frequencies in *Escherichia coli* CFT073. *Infect Immun* **75:**3325–3334.

104. **Watts RE, Totsika M, Challinor VL, Mabbett AN, Ulett GC, De Voss JJ, Schembri MA.** 2012. Contribution of siderophore systems to growth and urinary tract colonization of asymptomatic bacteriuria *Escherichia coli*. *Infect Immun* **80:**333–344.

105. **Chaturvedi KS, Hung CS, Crowley JR, Stapleton AE, Henderson JP.** 2012. The siderophore yersiniabactin binds copper to protect pathogens during infection. *Nat Chem Biol* **8:**731–736.

106. **Kai-Larsen Y, Lüthje P, Chromek M, Peters V, Wang X, Holm A, Kádas L, Hedlund KO, Johansson J, Chapman MR, Jacobson SH, Römling U, Agerberth B, Brauner A.** 2010. Uropathogenic Escherichia coli modulates immune responses and its curli fimbriae interact with the antimicrobial peptide LL-37. *PLoS Pathog* **6:**e1001010. doi:10.1371/journal.ppat.1001010

107. **Vigil PD, Wiles TJ, Engstrom MD, Prasov L, Mulvey MA, Mobley HL.** 2012. The repeat-in-toxin family member TosA mediates adherence of uropathogenic *Escherichia coli* and survival during bacteremia. *Infect Immun* **80:**493–505.

108. **Alteri CJ, Smith SN, Mobley HL.** 2009. Fitness of *Escherichia coli* during urinary tract infection requires gluconeogenesis and the TCA cycle. *PLoS Pathog* **5:**e1000448. doi:10.1371/journal.ppat.1000448

109. **Norton JP, Mulvey MA.** 2012. Toxin-antitoxin systems are important for niche-specific colonization and stress resistance of uropathogenic *Escherichia coli*. *PLoS Pathog* **8:**e1002954. doi: 10.1371/journal.ppat.1002954

110. Hayashi F, Smith KD, Ozinsky A, Hawn TR, Yi EC, Goodlett DR, Eng JK, Akira S, Underhill DM, Aderem A. 2001. The innate immune response to bacterial flagellin is mediated by Toll-like receptor 5. *Nature* **410**:1099–1103.

111. Andersen-Nissen E, Hawn TR, Smith KD, Nachman A, Lampano AE, Uematsu S, Akira S, Aderem A. 2007. Cutting edge: Tlr5-/- mice are more susceptible to *Escherichia coli* urinary tract infection. *J Immunol* **178**:4717–4720.

112. Tükel C, Nishimori JH, Wilson RP, Winter MG, Keestra AM, van Putten JP, Bäumler AJ. 2010. Toll-like receptors 1 and 2 cooperatively mediate immune responses to curli, a common amyloid from enterobacterial biofilms. *Cell Microbiol* **12**:1495–1505.

113. Flatau G, Lemichez E, Gauthier M, Chardin P, Paris S, Fiorentini C, Boquet P. 1997. Toxin-induced activation of the G protein p21 Rho by deamidation of glutamine. *Nature* **387**:729–733.

114. Guyer DM, Radulovic S, Jones FE, Mobley HL. 2002. Sat, the secreted autotransporter toxin of uropathogenic *Escherichia coli*, is a vacuolating cytotoxin for bladder and kidney epithelial cells. *Infect Immun* **70**:4539–4546.

115. de Man P, van Kooten C, Aarden L, Engberg I, Linder H, Svanborg Edén C. 1989. Interleukin-6 induced at mucosal surfaces by gram-negative bacterial infection. *Infect Immun* **57**:3383–3388.

116. Hedges S, Svensson M, Svanborg C. 1992. Interleukin-6 response of epithelial cell lines to bacterial stimulation in vitro. *Infect Immun* **60**:1295–1301.

117. Godaly G, Bergsten G, Hang L, Fischer H, Frendéus B, Lundstedt AC, Samuelsson M, Samuelsson P, Svanborg C. 2001. Neutrophil recruitment, chemokine receptors, and resistance to mucosal infection. *J Leukoc Biol* **69**:899–906.

118. Baggiolini M, Walz A, Kunkel SL. 1989. Neutrophil-activating peptide-1/interleukin 8, a novel cytokine that activates neutrophils. *J Clin Invest* **84**:1045–1049.

119. Godaly G, Proudfoot AE, Offord RE, Svanborg C, Agace WW. 1997. Role of epithelial interleukin-8 (IL-8) and neutrophil IL-8 receptor A in *Escherichia coli*-induced transuroepithelial neutrophil migration. *Infect Immun* **65**:3451–3456.

120. Godaly G, Hang L, Frendéus B, Svanborg C. 2000. Transepithelial neutrophil migration is CXCR1 dependent in vitro and is defective in IL-8 receptor knockout mice. *J Immunol* **165**:5287–5294.

121. Olszyna DP, Prins JM, Dekkers PE, De Jonge E, Speelman P, Van Deventer SJ, Van Der Poll T. 1999. Sequential measurements of chemokines in urosepsis and experimental endotoxemia. *J Clin Immunol* **19**:399–405.

122. Otto G, Burdick M, Strieter R, Godaly G. 2005. Chemokine response to febrile urinary tract infection. *Kidney Int* **68**:62–70.

123. Godaly G, Svanborg C. 2007. Urinary tract infections revisited. *Kidney Int* **71**:721–723.

124. Arana L, Ordoñez M, Ouro A, Rivera IG, Gangoiti P, Trueba M, Gomez-Muñoz A. 2013. Ceramide 1-phosphate induces macrophage chemoattractant protein-1 release: involvement in ceramide 1-phosphate-stimulated cell migration. *Am J Physiol Endocrinol Metab* **304**:E1213–1226.

125. Chowdhury P, Sacks SH, Sheerin NS. 2004. Minireview: functions of the renal tract epithelium in coordinating the innate immune response to infection. *Kidney Int* **66**:1334–1344.

126. Weichhart T, Haidinger M, Hörl WH, Säemann MD. 2008. Current concepts of molecular defence mechanisms operative during urinary tract infection. *Eur J Clin Invest* **38**(Suppl 2):29–38.

127. Hedges S, Linder H, de Man P, Svanborg Edén C. 1990. Ciclosporin-dependent, nu-independent, mucosal interleukin 6 response to gram-negative bacteria. *Scand J Immunol* **31**:335–343.

128. Wullt B, Bergsten G, Connell H, Röllano P, Gebratsedik N, Hang L, Svanborg C. 2001. P-fimbriae trigger mucosal responses to *Escherichia coli* in the human urinary tract. *Cell Microbiol* **3**:255–264.

129. Godaly G, Frendéus B, Proudfoot A, Svensson M, Klemm P, Svanborg C. 1998. Role of fimbriae-mediated adherence for neutrophil migration across *Escherichia coli*-infected epithelial cell layers. *Mol Microb* **30**:725–735.

130. Murphy PM. 1997. Neutrophil receptors for interleukin-8 and related CXC chemokines. *Semin Hematol* **34**:311–318.

131. Ragnarsdóttir B, Fischer H, Godaly G, Grönberg-Hernandez J, Gustafsson M, Karpman D, Lundstedt AC, Lutay N, Rämisch S, Svensson ML, Wullt B, Yadav M, Svanborg C. 2008. TLR- and CXCR1-dependent innate immunity: insights into the genetics of urinary tract infections. *Eur J Clin Invest* **38**(Suppl 2):12–20.

132. Bozic CR, Kolakowski LF Jr, Gerard NP, Garcia-Rodriguez C, von Uexkull-Guldenband C, Conklyn MJ, Breslow R, Showell HJ, Gerard C. 1995. Expression and biologic characterization of the murine chemokine KC. *J Immunol* **154**:6048–6057.

133. Tekamp-Olson P, Gallegos C, Bauer D, McClain J, Sherry B, Fabre M, van Deventer S, Cerami A. 1990. Cloning and characterization of cDNAs for murine macrophage inflammatory protein 2 and its human homologues. *J Exp Med* **172**:911–919.

134. Bozic CR, Gerard NP, von Uexkull-Guldenhand C, Kolakowski LF Jr, Conklyn MJ, Breslow R, Showell HJ, Gerard C. 1994. The murine interleukin 8 type B receptor homologue and its ligands. Expression and biological characterization. *J Biol Chem* **269**:29355–29358.

135. Lee J, Cacalano G, Camerato T, Toy K, Moore MW, Wood WI. 1995. Chemokine binding and activities mediated by the mouse IL-8 receptor. *J Immunol* **155**:2158–2164.

136. Taniguchi T, Ogasawara K, Takaoka A, Tanaka N. 2001. IRF family of transcription factors as regulators of host defense. *Annu Rev Immunol* **19**:623–655.

137. Honda K, Taniguchi T. 2006. IRFs: master regulators of signalling by Toll-like receptors and cytosolic pattern-recognition receptors. *Nat Rev Immunol* **6**:644–658.

138. Schmid M, Prajczer S, Gruber LN, Bertocchi C, Gandini R, Pfaller W, Jennings P, Joannidis M. 2010. Uromodulin facilitates neutrophil migration across renal epithelial monolayers. *Cell Physiol Biochem* **26**:311–318.

139. Bates JM, Raffi HM, Prasadan K, Mascarenhas R, Laszik Z, Maeda N, Hultgren SJ, Kumar S. 2004. Tamm-Horsfall protein knockout mice are more prone to urinary tract infection: rapid communication. *Kidney Int* **65**:791–797.

140. Dou W, Thompson-Jaeger S, Lauderkind SJ, Becker JW, Montgomery J, Ruiz-Bustos E, Hasty DL, Ballou LR, Eastman PS, Srichai B, Breyer MD, Raghow R. 2005. Defective expression of Tamm-Horsfall protein/uromodulin in COX-2-deficient mice increases their susceptibility to urinary tract infections. *Amer J Physiol Renal Physiol* **289**:F49–60.

141. Saemann MD, Weichhart T, Zeyda M, Staffler G, Schunn M, Stuhlmeier KM, Sobanov Y, Stulnig TM, Akira S, von Gabain A, von Ahsen U, Hörl WH, Zlabinger GJ. 2005. Tamm-Horsfall glycoprotein links innate immune cell activation with adaptive immunity via a Toll-like receptor-4-dependent mechanism. *J Clin Invest* **115**:468–475.

142. Boman HG. 1991. Antibacterial peptides: key components needed in immunity. *Cell* **65**:205–207.

143. Zasloff M. 2013. The antibacterial shield of the human urinary tract. *Kidney Int* **83**:548–550.

144. Chromek M, Slamová Z, Bergman P, Kovács L, Podracká L, Ehrén I, Hökfelt T, Gudmundsson GH, Gallo RL, Agerberth B, Brauner A. 2006. The antimicrobial peptide cathelicidin protects the urinary tract against invasive bacterial infection. *Nat Med* **12**:636–641.

145. Lomberg H, Hanson LA, Jacobsson B, Jodal U, Leffler H, Edén CS. 1983. Correlation of P blood group phenotype, vesicoureteral reflux and bacterial attachment in patients with recurrent pyelonephritis. *N Engl J Med* **308**:1189–1192.

146. Lomberg H, Jodal U, Edén C, Leffler H, Samuelsson B. 1981. P1 blood group and urinary tract infection. *Lancet* **1**:551–552.

147. Fowler JE, Stamey TA. 1977. Studies of introital colonization in women with recurrent urinary infections. VII. The role of bacterial adherence. *J Urol* **117**:472–476.

148. Lindstedt R, Larson G, Falk P, Jodal U, Leffler H, Svanborg C. 1991. The receptor repertoire defines the host range for attaching *Escherichia coli* strains that recognize globo-A. *Infect Immun* **59**:1086–1092.

149. Stapleton A, Nudelman E, Clausen H, Hakomori S, Stamm WE. 1992. Binding of uropathogenic *Escherichia coli* R45 to glycolipids extracted from vaginal epithelial cells is dependent on histo-blood group secretor status. *J Clin Invest* **90**:965–972.

150. Svensson M, Platt F, Frendeus B, Butters T, Dwek R, Svanborg C. 2001. Carbohydrate receptor depletion as an antimicrobial strategy for prevention of urinary tract infection. *J Infect Dis* **183**(Suppl 1):S70–73.

151. Orskov I, Ferencz A, Orskov F. 1980. Tamm-Horsfall protein or uromucoid is the normal urinary slime that traps type 1 fimbriated *Escherichia coli*. *Lancet* **1**:887.

152. Korhonen TK, Väisänen-Rhen V, Rhen M, Pere A, Parkkinen J, Finne J. 1984. *Escherichia coli* fimbriae recognizing sialyl galactosides. *J Bacteriol* **159**:762–766.

153. Korhonen TK, Parkkinen J, Hacker J, Finne J, Pere A, Rhen M, Holthöfer H. 1986. Binding of *Escherichia coli* S fimbriae to human kidney epithelium. *Infect Immun* **54**:322–327.

154. Hawn TR, Scholes D, Wang H, Li SS, Stapleton AE, Janer M, Aderem A, Stamm WE, Zhao LP, Hooton TM. 2009. Genetic variation of the human urinary tract innate immune response and asymptomatic bacteriuria in women. *PLoS One* **4**:e8300. doi:10.1371/journal.pone.0008300

155. Smithson A, Sarrias MR, Barcelo J, Suarez B, Horcajada JP, Soto SM, Soriano A, Vila J, Martinez JA, Vives J, Mensa J, Lozano F. 2005. Expression of interleukin-8 receptors (CXCR1 and CXCR2) in premenopausal women with recurrent urinary tract infections. *Clin Diagn Lab Immunol* **12**:1358–1363.

156. Hawn TR, Scholes D, Li SS, Wang H, Yang Y, Roberts PL, Stapleton AE, Janer M, Aderem A, Stamm WE, Zhao LP, Hooton TM. 2009. Toll-like receptor polymorphisms and susceptibility to urinary tract infections in adult women. *PLoS One* **4**:e5990. doi:10.1371/journal.pone.0005990

157. Yin X, Hou T, Liu Y, Chen J, Yao Z, Ma C, Yang L, Wei L. 2010. Association of Toll-like receptor 4 gene polymorphism and expression with urinary tract infection types in adults. *PLoS One* **5**:e14223. doi:10.1371/journal.pone.0014223

158. Grainger DJ, Heathcote K, Chiano M, Snieder H, Kemp PR, Metcalfe JC, Carter ND, Spector TD. 1999. Genetic control of the circulating concentration of transforming growth factor type beta1. *Hum Mol Genet* **8**:93–97.

159. Cotton SA, Gbadegesin RA, Williams S, Brenchley PE, Webb NJ. 2002. Role of TGF-beta1 in renal parenchymal scarring following childhood urinary tract infection. *Kidney Intl* **61**:61–67.

160. Solari V, Owen D, Puri P. 2005. Association of transforming growth factor-beta1 gene polymorphism with reflux nephropathy. *J Urol* **174**:1609–1611; discussion 1611.

161. Yim HE, Bae IS, Yoo KH, Hong YS, Lee JW. 2007. Genetic control of VEGF and TGF-beta1 gene polymorphisms in childhood urinary tract infection and vesicoureteral reflux. *Pediatr Res* **62**:183–187.

162. Wullt B, Bergsten G, Connell H, Röllano P, Gebretsadik N, Hull R, Svanborg C. 2000. P fimbriae enhance the early establishment of *Escherichia coli* in the human urinary tract. *Mol Microbiol* **38**:456–464.

163. Sundén F, Håkansson L, Ljunggren E, Wullt B. 2010. *Escherichia coli* 83972 bacteriuria protects against recurrent lower urinary tract infections in patients with incomplete bladder emptying. *J Urol* **184**:179–185.

164. Harte MT, Haga IR, Maloney G, Gray P, Reading PC, Bartlett NW, Smith GL, Bowie A, O'Neill LA. 2003. The poxvirus protein A52R targets Toll-like receptor signaling complexes to suppress host defense. *J Exp Med* **197**:343–351.

165. Stack J, Haga IR, Schröder M, Bartlett NW, Maloney G, Reading PC, Fitzgerald KA, Smith GL, Bowie AG. 2005. Vaccinia virus protein A46R targets multiple Toll-like-interleukin-1 receptor adaptors and contributes to virulence. *J Exp Med* **201**:1007–1018.

166. Newman RM, Salunkhe P, Godzik A, Reed JC. 2006. Identification and characterization of a novel bacterial virulence factor that shares homology with mammalian Toll/interleukin-1 receptor family proteins. *Infect Immun* **74**:594–601.

167. Foxman B. 2010. The epidemiology of urinary tract infection. *Nat Rev Urol* **7**:653–660.

168. Brumbaugh AR, Mobley HL. 2012. Preventing urinary tract infection: progress toward an effective *Escherichia coli* vaccine. *Expert Rev Vaccines* **11**:663–676.

169. Williams GJ, Craig JC, Carapetis JR. 2013. Preventing urinary tract infections in early childhood. *Adv Exp Med Biol* **764**:211–218.

170. Wagenlehner FM, Vahlensieck W, Bauer HW, Weidner W, Piechota HJ, Naber KG. 2013. Prevention of recurrent urinary tract infections. *Minerva Urol Nefrol* **65**:9–20.

171. Hagberg L, Bruce A, Reid G, Svanborg C, Lincoln K, Lidin-Janson G. 1989. Colonization of the urinary tract with live bacteria from the normal fecal and urethral flora in patients with recurrent symptomatic urinary tract infections, p 194–197. *In* Kass EH, Svanborg C (ed), *Host-Parasite Interactions in Urinary Tract Infections.* University of Chicago Press, Chicago, IL.

172. Andersson P, Engberg I, Lidin-Janson G, Lincoln K, Hull R, Hull S, Svanborg C. 1991. Persistence of *Escherichia coli* bacteriuria is not determined by bacterial adherence. *Infect Immun* **59**:2915–2921.

173. Wullt B, Connell H, Rollano P, Månsson W, Colleen S, Svanborg C. 1998. Urodynamic factors influence the duration of *Escherichia coli* bacteriuria in deliberately colonized cases. *J Urol* **159**:2057–2062.

174. Hull R, Rudy D, Donovan W, Svanborg C, Wieser I, Stewart C, Darouiche R. 2000. Urinary tract infection prophylaxis using *Escherichia coli* 83972 in spinal cord injured patients. *J Urol* **163**:872–877.

175. Lindberg U, Claesson I, Hanson LA, Jodal U. 1978. Asymptomatic bacteriuria in schoolgirls. VIII. Clinical course during a 3-year follow-up. *J Pediatr* **92**:194–199.

176. Hagberg L, Leffler H, Svanborg-Edén C. 1984. Non-antibiotic prevention of urinary tract infection. *Infection* **12**:132–137.

177. Darouiche RO, Donovan WH, Del Terzo M, Thornby JI, Rudy DC, Hull RA. 2001. Pilot trial of bacterial interference for preventing urinary tract infection. *Urology* **58**:339–344.

178. Sharon N, Eshdat Y, Silverblatt FJ, Ofek I. 1981. Bacterial adherence to cell surface sugars. *Ciba Found Symp* **80**:119–141.

179. Cusumano CK, Pinkner JS, Han Z, Greene SE, Ford BA, Crowley JR, Henderson JP,

Janetka JW, Hultgren SJ. 2011. Treatment and prevention of urinary tract infection with orally active FimH inhibitors. *Sci Transl Med* 3:109–115.

180. Edén CS, Freter R, Hagberg L, Hull R, Hull S, Leffler H, Schoolnik G. 1982. Inhibition of experimental ascending urinary tract infection by an epithelial cell-surface receptor analogue. *Nature* 298:560–562.

181. Svanborg Edén C, Andersson B, Hagberg L, Hanson LA, Leffler H, Magnusson G, Noori G, Dahmén J, Söderström T. 1983. Receptor analogues and anti-pili antibodies as inhibitors of bacterial attachment *in vivo* and *in vitro*. *Ann N Y Acad Sci* 409:580–592.

182. Kihlberg J, Hultgren SJ, Normark S, Magnusson G. 1989. Probing of the combining site of the PapG adhesin of uropathogenic *Escherichia coli* bacteria by synthetic analogs of galabiose. *J Am Chem Soc* 111:6364–6368.

183. Leach JL, Garber SA, Marcon AA, Prieto PA. 2005. *In vitro* and *in vivo* effects of soluble, monovalent globotriose on bacterial attachment and colonization. *Antimicrob Agents Chemother* 49:3842–3846.

184. Chromek M, Brauner A. 2008. Antimicrobial mechanisms of the urinary tract. *J Mol Med (Berl)* 86:37–47.

185. Zhao J, Wang Z, Chen X, Wang J, Li J. 2011. Effects of intravesical liposome-mediated human beta-defensin-2 gene transfection in a mouse urinary tract infection model. *Microbiol Immunol* 55:217–223.

186. Svanborg Edén C, Briles D, Hagberg L, McGhee J, Michalec S. 1985. Genetic factors in host resistance to urinary tract infection. *Infection* 13(Suppl 2):S171–176.

187. Frendéus B, Godaly G, Hang L, Karpman D, Svanborg C. 2001. Interleukin-8 receptor deficiency confers susceptibility to acute pyelonephritis. *J Infect Dis* 183(Suppl 1):S56–60.

188. Thumbikat P, Waltenbaugh C, Schaeffer AJ, Klumpp DJ. 2006. Antigen-specific responses accelerate bacterial clearance in the bladder. *J Immunol* 176:3080–3086.

189. Langermann S, Palaszynski S, Barnhart M, Auguste G, Pinkner JS, Burlein J, Barren P, Koenig S, Leath S, Jones CH, Hultgren SJ. 1997. Prevention of mucosal *Escherichia coli* infection by FimH-adhesin-based systemic vaccination. *Science* 276:607–611.

190. Langermann S, Möllby R, Burlein JE, Palaszynski SR, Auguste CG, DeFusco A, Strouse R, Schenerman MA, Hultgren SJ, Pinkner JS, Winberg J, Guldevall L, Söderhäll M, Ishikawa K, Normark S, Koenig S. 2000. Vaccination with FimH adhesin protects cynomolgus monkeys from colonization and infection by uropathogenic *Escherichia coli*. *J Infect Dis* 181:774–778.

191. Hagan EC, Lloyd AL, Rasko DA, Faerber GJ, Mobley HL. 2010. *Escherichia coli* global gene expression in urine from women with urinary tract infection. *PLoS Pathog* 6:e1001187. doi:10.1371/journal.ppat.1001187

192. Hanson LA, Ahlstedt S, Fasth A, Jodal U, Kaijser B, Larsson P, Lindberg U, Olling S, Sohl-Akerlund A, Svanborg-Edén C. 1977. Antigens of *Escherichia coli*, human immune response, and the pathogenesis of urinary tract infections. *J Infect Dis* 136(Suppl):S144–149.

193. Kaijser B, Hanson LA, Jodal U, Lidin-Janson G, Robbins JB. 1977. Frequency of *E. coli* K antigens in urinary-tract infections in children. *Lancet* 1:663–666.

194. Silverblatt FJ, Cohen LS. 1979. Antipili antibody affords protection against experimental ascending pyelonephritis. *J Clin Invest* 64:333–336.

195. Kaijser B, Larsson P, Olling S, Schneerson R. 1983. Protection against acute, ascending pyelonephritis caused by *Escherichia coli* in rats, using isolated capsular antigen conjugated to bovine serum albumin. *Infect Immun* 39:142–146.

196. Pecha B, Low D, O'Hanley P. 1989. Gal-Gal pili vaccines prevent pyelonephritis by piliated *Escherichia coli* in a murine model. Single-component Gal-Gal pili vaccines prevent pyelonephritis by homologous and heterologous piliated *E. coli* strains. *J Clin Invest* 83:2102–2108.

197. Uehling DT, Hopkins WJ, Dahmer LA, Balish E. 1994. Phase I clinical trial of vaginal mucosal immunization for recurrent urinary tract infection. *J Urol* 152:2308–2311.

198. Uehling DT, Hopkins WJ, Balish E, Xing Y, Heisey DM. 1997. Vaginal mucosal immunization for recurrent urinary tract infection: phase II clinical trial. *J Urol* 157:2049–2052.

199. Hopkins WJ, Uehling DT, Wargowski DS. 1999. Evaluation of a familial predisposition to recurrent urinary tract infections in women. *Am J Med Genet* 83:422–424.

200. Uehling DT, Hopkins WJ, Beierle LM, Kryger JV, Heisey DM. 2001. Vaginal mucosal immunization for recurrent urinary tract infection: extended phase II clinical trial. *J Infect Dis* 183(Suppl 1):S81–83.

201. Uehling DT, Hopkins WJ, Elkahwaji JE, Schmidt DM, Leverson GE. 2003. Phase 2 clinical trial of a vaginal mucosal vaccine for urinary tract infections. *J Urol* 170:867–869.

202. Hopkins WJ, Elkahwaji J, Beierle LM, Leverson GE, Uehling DT. 2007. Vaginal

mucosal vaccine for recurrent urinary tract infections in women: results of a phase 2 clinical trial. *J Urol* **177**:1349–1353;quiz1591.

203. **Benson M, Jodal U, Agace W, Hellström M, Mårild S, Rosberg S, Sjöström M, Wettergren B, Jönsson S, Svanborg C.** 1996. Interleukin (IL)-6 and IL-8 in children with febrile urinary tract infection and asymptomatic bacteriuria. *J Infect Dis* **174**:1080–1084.

204. **Renata Y, Jassar H, Katz R, Hochberg A, Nir RR, Klein-Kremer A.** 2013. Urinary concentration of cytokines in children with acute pyelonephritis. *Eur J Pediatr* **172**:769–774.

205. **Rodríguez LM, Robles B, Marugán JM, Suárez A, Santos F.** 2008. Urinary interleukin-6 is useful in distinguishing between upper and lower urinary tract infections. *Pediatr Nephrol* **23**:429–433.

Innate Immune Responses to Bladder Infection

22

BYRON W. HAYES[1] and SOMAN N. ABRAHAM[1,2,3,4]

INTRODUCTION

The urinary tract (UT) consists of the kidneys, ureters, bladder, and urethra, all of which with the exception of the lower urethra are presumed to be sterile. Because of its close proximity to the gut, the lower UT is constantly exposed to a barrage of gut bacteria. However, the bladder remains for the most part free of infection. The resistance of the bladder to active microbial colonization is due to both anatomical design as well as secreted antimicrobial compounds of the urothelium. The apical face of the urothelium is covered by uroplakin plaques and is coated by mucus, which discourages adherence and invasion of most microorganisms (1, 2). Because of its role in storing urine for extended periods of time, the urothelium of the bladder has an additional role in protecting the underlying tissue from urine and its many toxic constituents. Since the bladder and urinary system as a whole need to constantly maintain the integrity of the urothelium, immune responses in the UT are often tightly regulated to minimize the extent of damage by quelling inflammation in a timely manner.

A rapid and vigorous response mediated by the immune system is largely responsible for guarding against bacterial infections that bypass the natural defenses of the urinary tract. In spite of this, some pathogens such as

[1]Departments of Pathology; [2]Molecular Genetics and Microbiology; [3]Immunology, Duke University Medical Center, Durham, NC 27710; [4]Program in Emerging Infectious Diseases, Duke-National University of Singapore, Singapore.
Urinary Tract Infections: Molecular Pathogenesis and Clinical Management, 2nd Edition
Edited by Matthew A. Mulvey, David J. Klumpp, and Ann E. Stapleton
© 2017 American Society for Microbiology, Washington, DC
doi:10.1128/microbiolspec.UTI-0024-2016

uropathogenic *Escherichia coli* (UPEC) are still able to colonize and infect this tract. UPEC is responsible for more than 80% of urinary tract infections (UTIs); the pathogen is believed to originate from the intestines because these bacteria can be regular components of the microbiome of the gastrointestinal tract (3). UPEC is especially adapted to infect the UT because it has the potential to multiply rapidly in urine and to bypass or overcome many of the natural barriers to infection. Beyond this, it appears capable of persisting for extended periods of time within the bladder epithelium, often in a quiescent state, thereby contributing to chronic or recurrent UTI (4).

One in three women will have at least one UTI by age 24, and over half of all women will have at least one UTI in their lifetime (5). The cost of treatment in the United States alone is several billion dollars each year and growing (5). There is emerging concern about the involvement of multidrug-resistant *E. coli*, many strains of which contribute to increased frequency of recurrence (6). In view of these concerns, there is a strong impetus to discover novel and alternative strategies to treat or manage UTIs. Valuable clues can be derived from understanding the natural immune processes and mechanisms involved in the host response to uropathogens in the UT. Our goal in this chapter is to briefly review key innate immune defense activities observed in the UT. Focus is placed on the bladder because most infections of the upper UT originate from bacteria ascending from the bladder (7). The adaptive immune responses will not be discussed here because they appear to be limited following bladder infections (8).

CELLULAR COMPONENTS OF THE INNATE IMMUNE RESPONSE TO INFECTION

Epithelial Cells

The first host cells invading uropathogens encounter upon entering the bladder are the superficial bladder epithelial cells (BECs) that form a tight, highly impregnable barrier. These cells are replete with receptors for various bacteria or their products. Evidence of pattern recognition receptor activity can be seen soon after infection as a large burst in secretion of interleukin-6 (IL-6) and IL-1β is detectable in the urine (9). These cytokines are typically the products of Toll-like receptor 4 (TLR4) and Nod-like receptor/caspase 1 activity, respectively (9, 10). Along with other inflammatory mediators secreted by BECs, these cytokines result in a vigorous influx of immune cells to the epithelium to counter the bacterial challenge. BECs complement this cytokine response with secretion of several factors capable of directly killing bacteria. One such example is the cathelicidin LL-37 (11). mRNA expression for cathelicidin was observed to rise within 5 minutes of stimulation of epithelial cells with UPEC (11). In addition, knockout of the mouse LL-37 ortholog resulted in a sharp increase in UPEC load in the bladder of mice within 1 hour following infection, revealing the rapid and powerful antibacterial contribution of this antimicrobial peptide (11). β-defensin is another antimicrobial peptide found in urine, but this agent mainly originates from kidney epithelial cells rather than BECs (12). Both these antimicrobial peptides also contribute to cytokine production and neutrophil recruitment in the bladder (13). Indeed, cathelicidin-deficient mice had lower levels of 16 inflammatory cytokines in bladder homogenates after infection, including IL-1β and IL-6.

Ribonuclease 7 is another antimicrobial agent that has swift, broad-spectrum microbicidal activity against many common uropathogens at basal levels (14). The soluble pattern recognition molecule pentraxin-related protein 3 (PTX3) was recently found to be relevant in defending against microbial infection of the UT (15). PTX3 is produced by many cell types including urothelial cells, and its activity appears to be through stimulating complement-mediated phagocytosis of pathogens. PTX3 levels after infection increased in the UT, and knockout mice had a decreased ability to clear infection (15). Genetic mutations in the *PTX3* locus in humans also coincides with increased likelihood of UTIs (15).

In addition to forming a highly effective barrier to infection and urine and functioning as a source of soluble antimicrobial agents, BECs have a critical role in regulating the bladder volume to accommodate urine. BECs contain a large pool of RAB27b$^+$ fusiform vesicles (16). When urine enters the bladder, these fusiform vesicles spontaneously exocytose into the plasma membrane in a cyclic AMP-dependent manner to provide the extra membranes required for bladder expansion. After voiding when the bladder volume needs to contract, the intracellular fusiform vesicles once again form by internalization of the RAB27b$^+$ membrane (16, 17). UPEC has been found to coopt this unique property to gain entry into BECs. Upon binding to the apical surface of BECs, UPEC is able to trigger a local burst of cyclic AMP resulting in exocytosis of fusiform vesicles at the adherence site; when these RAB27b$^+$ vesicles are subsequently retracted from the cell surface, the bacteria are drawn into the cell, where they reside encased in RAB27b$^+$ vesicles (17). Remarkably, BECs have an autonomous defense system capable of sensing intruding bacteria and initiating mechanisms to eject bacteria from their intracellular refuge. This activity is initiated within minutes of bacterial entry and is triggered by TLR4 localized in the vesicles encasing bacteria (18). This immune sensory molecule mobilizes the cellular trafficking machinery typically employed to export hormones to instead export UPEC from infected BECs. TLR4 signaling was found to trigger K33-linked polyubiquitination of the signaling substrate, TRAF3, which was then detected by RalGDS, a guanine nucleotide exchange factor that precipitated the assembly of the exocyst complex, a powerful exporter of subcellular vesicles (19).

Interestingly, not all intracellular bacteria in BECs are expelled from RAB27b$^+$ vesicles in this manner. It was observed that an appreciable number of UPEC organisms are capable of escaping intracellular vesicles in BECs and entering the cytosol, at which point a second export mechanism appears to be activated. This pathway is initiated by the cell's autophagy system, which encapsulates the bacterium in an autophagosome and transports it to the lysosome via the multivesicular body. However, UPEC has the capacity to neutralize the lysosome to survive this environment (20). BECs seem to counter this pathogen response by triggering expulsion of UPEC from malfunctioning lysosomes employing transient receptor potential mucolipin 3, a cation channel on the lysosome which becomes activated when the pH within the lysosome increases. UPEC expelled from lysosomes was found to be encased in a host cell membrane (20). While the expulsion of UPEC from RAB27b$^+$ vesicles occurs soon after bacterial entry, exocytosis from lysosomes is only initiated 4 hours or more after infection. Together these export mechanisms are highly effective, resulting in the expulsion by 24 hours of over 80% of bacteria that had gained entry into BECs (17).

Another powerful property possessed by BECs to combat infections is undergoing cell death and shedding into the urine, thus eliminating all of the adherent and intracellular bacteria associated with these cells. Urine of heavily infected subjects is rich in exfoliated BECs, indicating that exfoliation of BECs is a frequently employed host defense mechanism (21, 22). This capacity of infected BECs to trigger cell death is a double-edged sword because it allows for elimination of large congregates of bacteria associated with BECs but also exposes underlying cells to both toxic urine as well as any bacteria still present in the urine (23). To counter this drastic action, the basal epithelium promptly shifts to a rapid proliferative state to quickly replace the shed cells within hours and restore the urothelial barrier (24, 25).

Neutrophil, Mast Cells, and Macrophages

Arguably the first recruited immune cells to target infecting bacteria in the bladder are neutrophils. These innate immune cells are the primary cells responsible for bacterial elimination once the inflammatory response is triggered. When their pattern recognition receptors are stimulated upon entry of bacteria, BECs produce IL-8, which leads to the recruitment of neutrophils to the superficial epithelium and the bladder lumen, where infecting bacteria are found (26). Recruited neutrophils exit the blood vessels and migrate through the lamina propria to penetrate the basement membrane. Although neutrophil migration to the epithelium is primarily in response to IL-8, its ability also to express metalloproteinase-9 appears to be key to its ability to penetrate the basement membrane (27). Once the neutrophils cross the basement membrane they readily circumvent multiple layers of intermediate epithelial cells before reaching the superficial epithelium, which is the final epithelial cell layer before the neutrophils enter the bladder lumen. Since large numbers of neutrophils are found in the urine, it is clear that these neutrophils are capable of breaking through the tight junctions between adjoining superficial epithelial cells. Neutrophil recruitment into the UT is typically swift and often proportional to bacterial load (28). Neutrophil counts in the epithelial region following bladder infection reveal that these phagocytic cells arrive within 2 hours of infection but appear to peak by 6 hours (28). Their numbers correspondingly decrease as bacteria are cleared. A powerful indicator of the importance of neutrophils to bacterial clearance comes from studies of mouse strains with neutrophil response defects. These mice exhibited significantly increased susceptibility to UTIs and decreased ability to clear bacterial infections (29).

Much of the antibacterial actions of neutrophils in the UT can be ascribed to the capacity of these cells to engulf and intracellularly degrade bacteria. Promoting this activity is the neutrophil's capacity to secrete prestored PTX3, which serves as an opsonin, to coat bacteria and facilitate their uptake. Neutrophils from *PTX3* knockout mice are known to display decreased phagocytic activity that can be restored by opsonizing bacteria with soluble PTX3 (15). Although neutrophils are critical for bacterial clearance, they are also largely responsible for much of the cytotoxicity associated with bacterial infections of the UT (30). Neutrophil-mediated tissue damage is ascribable to reactive oxygen species and other cytotoxic products that these cells release upon reaching the urinary epithelium (30). Recently, treatment with nonsteroidal anti-inflammatory drugs or cyclooxygenase-2 (COX-2) inhibitors was found to exhibit a remarkable therapeutic effect in experimental mouse models (30). Use of inhibitors of COX-2 in infected mice revealed that COX-2 inhibitors exhibit a dramatic ability to limit neutrophil migration into the UT. Limited neutrophil influx in the UT was closely associated with markedly less pyuria and mucosal damage. The COX-2 inhibitors did not impact the exfoliation of infected BECs and other natural clearance activities in the UT. Interestingly, inhibiting COX-2 in a mouse recurrent UTI model was found to reduce the likelihood of chronic cystitis (30) because the neutrophil-associated tissue damage was a major contributor to chronic inflammation. In addition, serum from women who developed a recurrent infection had elevated levels of cytokines associated with neutrophil development and chemotaxis (30). In view of the severe pathology evoked by recruited neutrophils during UTIs, it is conceivable that the large presence of neutrophils in the urine is a byproduct of host defenses aimed at eliminating these cytotoxic cells from the surrounding tissue. It is unclear how effective neutrophils are at killing bacteria in urine, but the presence of large numbers of neutrophils in the urine will ensure that they are immediately eliminated during voiding (29).

One of the most prominent resident innate immune cell types in the UT is mast cells. These cells are found in much higher numbers in the bladder compared to the kidneys, perhaps suggesting a more prominent role in the former organ. In the bladder, mast cells are primarily localized to the lamina propria and in the detrusor muscles, although they appear capable of relocating and increasing in number following infection. Mast cells are sentinel cells that initiate the early immune responses to microbial or other insults. Prestored in their granules is a large pool of proinflammatory mediators that can be immediately released upon mast cell activation. The speed and magnitude of the ensuing inflammatory response is largely dependent on the contents of these mast cell granules. Within the first hour of infection, histamine and other prestored proinflammatory mediators typically found in granules can be found in urine (31, 32). Prestored mast cell products such as tumor necrosis factor contribute to the early neutrophil response in the bladder, because mast cell-deficient mice evoke a limited neutrophil response to bacterial infection compared to mast cell-sufficient mice (32). Not surprisingly, bacterial clearance in the bladder and kidneys of these mast cell-deficient mice is also limited (33). Although mast cell membranes are replete with pattern recognition receptors, they also express membrane receptors such as CD48, a glycosylphosphatidylinositol anchored molecule that serves as a receptor for the bacterial adhesin FimH, expressed by type I fimbriated *E. coli* (34). Since mast cells lie underneath the epithelium, it is unlikely that they make contact with bacteria or their products early in the infection. It is more likely that mast cells are activated by "danger signals" secreted by stressed or damaged epithelial cells such as ATP, IL-33, and β-defensin, all of which are potent mast cell activators (35–37).

In addition to their role in initiating inflammation during infection, mast cells also appear to be important in establishing homeostasis and accelerating tissue recovery after the infection subsides. Recently, it was reported that several hours after bladder infection, local mast cells were abruptly found to express the immunosuppressive cytokine, IL-10 (33). The timing of IL-10 release appeared to coincide with epithelial shedding and when recovery of the epithelium needed to occur. Because the constituents of urine can be highly toxic to the underlying tissue, it is imperative that the bladder epithelium be reconstituted as quickly as possible. Since tissue regeneration cannot occur during active inflammation, inflammation in the bladder epithelium needs to be sharply curtailed—hence the requirement for mast cell-derived IL-10. Interestingly, an apparent side effect of this mast cell-derived IL-10 appears to be premature abrogation of the adaptive immune response. Studies of mice revealed that mast cell-derived IL-10 negatively impacted functional activity of both dendritic cells and T cells in the iliac lymph nodes that drain the bladder (33). Additionally, mast cell-derived IL-10 could potentially contribute to premature resolution of inflammation in the bladder without complete eradication of infecting bacteria in the bladder. A common observation in mouse models of bladder infection is that a significant population of residual bacteria remain in the bladder long after apparent resolution of the infection (33, 38, 39).

Resident tissue and recruited macrophages also appear to play a critical role in the highly orchestrated neutrophil response to bladder infections. Macrophages that are resident in the lamina propria of the bladder tend to be Ly6C⁻ macrophages, which upon infection recruit both neutrophils and Ly6C⁺ macrophages out of the circulation into the bladder lamina propria through the secretion of chemokines such as CXCL1 and CXCL2 (27). Interestingly, there appears to be a checkpoint in the basement membrane which prevents newly recruited neutrophils from reaching the epithelium. It now appears that specific communication between resident Ly6C⁻ macrophages and newly recruited Ly6C⁺ macrophage subtypes is necessary before neutrophils can break through the collagen IV barrier that constitutes the basement membrane. Newly recruited macrophages secrete tumor necrosis factor, which causes the resident macrophage subset to secrete CXCL2, a powerful chemokine which in turn induces the neutrophil pool in the lamina propria to produce matrix metalloproteinase-9, which is necessary for neutrophil breakthrough across the basement membrane barrier to reach the epithelium (27). Thus, it appears that bladder macrophages are critical to regulatory control of the functional activity of recruited neutrophils. Concomitant with their role in promoting neutrophil influx and migration into the bladder epithelium, macrophages are known to temper neutrophil responses by ingesting and removing apoptotic neutrophils to promote early resolution of inflammation (40). Like mast cells, resident macrophages of the bladder appear to impede development of adaptive immune responses following infection. These resident macrophages appear to successfully compete with dendritic cells in bacterial uptake, but unlike dendritic cells, they are not efficient at antigen presentation, resulting in a poor adaptive immune response (41).

Studies of mice have revealed that other immune cells in the bladder that contribute to the innate immune response to bacterial infection include natural killer and γδ T cells (42–44). How these bladder-resident immune cells directly combat bacteria and also collaborate with other immune cells is currently being investigated.

IMMUNE BREAKDOWN AND UTI RISK FACTORS

UTIs remain frequent and even appear to be on the rise, especially within the rapidly growing aging population. Although UTIs are strongly associated with UPEC expressing a collection of virulence factors including type 1 fimbriae, flagella, and hemolysin, which together enable the bacteria to successfully colonize the unique microenvironment of the bladder and kidney (7), many of these infections can still be traced back to specific defects in one or more components of the innate immune system.

Often, there is a genetic basis for the breakdown in innate immune defenses. For example, polymorphisms in TLR1, TLR4, and CXCR1, among others, appear to predispose patients to recurrence of UTIs (45). The importance of these genes in the early recognition of infecting bacteria and/or in recruiting appropriate immune cells to infection sites can be readily gleaned from experimental infections in mice deficient in these genes (45). While not strictly considered a breakdown of immune defenses, the expression of certain blood group antigens on cells lining the UT can predispose individuals to UTIs. These antigens serve as receptors for specific fimbriae typically expressed by UPEC (46). Consequently, their presence on uroepithelial cell surfaces can promote bacterial adherence and promote infection at these sites. In view of their impact, some have proposed that P blood group status may be used as a risk marker for pyelonephritis (45).

As people age, the efficiency of their immune system in combating microbial challenge decreases due to immune senescence (47). For example, the migration of neutrophils and their associated microbicidal activity are severely impaired in elderly individuals (48); consequently, these individuals exhibit a limited capacity to effectively clear bacteria once infected. In aging women, in addition to defective immune cells, decreasing levels of estrogen appear to be a contributing factor, because postmenopausal women are more susceptible to UTIs than their premenopausal counterparts (49, 50). Estrogen receptors are present on epithelial cells lining the lower UT. When mice subjected to ovariectomy, to simulate menopause, were examined, they were found to exhibit higher bacterial loads in the bladder and an overall delayed clearance of UPEC compared to controls (51). These mice also experienced more severe inflammation and decreased exfoliation of their uroepithelial cells, a major contributing factor to bacterial load in the UT. Although estrogen supplementation does not reduce the level of bacteria in the urine, in a mouse model, estrogen supplementation decreased the levels of proinflammatory cytokines and decreased the extent of damage in the bladder during the natural immune response (51).

While specific defects in various components of the immune system predispose individuals for the development of UTIs, some highly host-adapted bacterial strains appear capable of establishing infection by actively suppressing most of the local immune responses in the UT. In these cases, not only is there no accompanying inflammatory response, but there is very little infection-associated pathology in the UT. Consequently, the infecting bacteria appear to grow unrestricted in the urine contained in the bladder lumen of the infected individual. Because the patients experience no discomforting symptoms, the infection is referred to as asymptomatic bacteriuria. Recent studies employing the asymptomatic *E. coli* 83972 strain have revealed the underlying basis for bacterial suppression of pathology in the UT. One important trait possessed by asymptomatic bacteria is that they fail to express most of the virulence factors traditionally associated with UPEC (52). This property allows the bacteria to avoid activating immune surveillance molecules found in cells that line the UT. An even more potent trait found in asymptomatic bacteria is that of suppressing RNA polymerase II-dependent gene expression in host cells (53). This bacteria-induced effect on polymerase II-dependent transcription factors was observed in both bladder and renal cells *in vitro*. In patients infected with *E. coli* 83972, the expression of 60% of all regulated genes was markedly reduced after 24 hours (53). Many of these regulated genes are found in disease-associated innate immune response pathways, defined by TLR4, interferon regulatory factor (IRF) -3, and IRF-7, and include

the immune surveillance molecules nucleotide-binding oligomerization domain protein 1 (NOD1) and TLR5 (45). This gene suppression was long-lasting, because it was evident in patients a week after inoculation of asymptomatic UPEC (53).

Since asymptomatic *E. coli* such as the 83972 strain is better suited for long-term survival in the UT than its uropathogenic counterparts that express many of the traditional virulence factors, it has been proposed that these bacteria be used as probiotics to displace more harmful strains or even prevent their infection of the bladder (53, 54). Since the bladder is mostly sterile, asymptomatic strains would have very little competition establishing themselves in the bladder of subjects prone to recurrent UTIs, and theoretically these bacteria could discourage the establishment of more virulent strains. However, a recent clinical trial using *E. coli* 83972 for prevention of catheter-associated urinary tract infections did not corroborate this theory; patients in which the asymptomatic strain successfully colonized still had uropathogens present, and more than half of these patients developed a UTI (55). Intriguingly, the individuals that developed a UTI had a lower assortment of bacteria in their urine, suggesting that microbial diversity may play a protective role in the UT (55).

CONCLUSION

Recent studies revealed that a coordinated multicomponent immune defense program is activated by resident BECs, mast cells, and macrophages when bacteria reach the bladder. In addition to directly attacking the invading pathogen, their individual responses are largely aimed at rapidly recruiting neutrophils into the bladder to clear the pathogens. Large populations of neutrophils recruited into the bladder are highly efficacious in eliminating infecting bacteria from the UT. However, these recruited neutrophils are also extremely cytotoxic to the surrounding tissue.

Indeed, much of the tissue damage incurred during UTIs is attributable to recruited neutrophils. Not surprisingly, each of the bladder-resident cell types have distinct mechanisms to reduce neutrophil numbers or activity at inflamed sites. Epithelial cells reduce neutrophil numbers by recruiting them into the urine, where they are eliminated during voiding. Mast cells impede neutrophils through secretion of IL-10 when epithelial cells begin to shed and compromise their barrier function. Macrophages eliminate spent neutrophils by phagocytosis, and subsets of these cells also appear capable of restricting inappropriate neutrophil recruitment into the epithelium by regulating the permeability of the basement membrane. Studies of mice and genetic analysis of patients prone to recurrent UTIs have revealed that defects in the one or more of these immune responses predispose patients to the severity and persistence of UTIs.

Cumulatively, these observations represent only a glimpse of the complexity of the immune responses evoked in the bladder. More extensive studies in animal models coupled with immune monitoring of UTI-prone populations during acute and chronic infections are required before meaningful therapeutic strategies can be developed.

ACKNOWLEDGMENTS

This work was funded by U.S. National Institutes of Health grants: U01-AI082107, R01-AI096305, and R56-DK095198.

CITATION

Hayes BW, Abraham SN. 2016. Innate immune response to bladder infection. Microbiol Spectrum 4(6):UTI-0024-2016.

REFERENCES

1. **Wu XR, Kong XP, Pellicer A, Kreibich G, Sun TT.** 2009. Uroplakins in urothelial biology, function, and disease. *Kidney Int* **75:**1153–1165.

2. **Grist M, Chakraborty J.** 1994. Identification of a mucin layer in the urinary bladder. *Urology* **44:**26–33.

3. **Ronald A.** 2003. The etiology of urinary tract infection: traditional and emerging pathogens. *Dis Mon* **49:**71–82.

4. **Rosen DA, Hooton TM, Stamm WE, Humphrey PA, Hultgren SJ.** 2007. Detection of intracellular bacterial communities in human urinary tract infection. *PLoS Med* **4:**e329. doi:10.1371/journal.pmed.0040329.

5. **Foxman B.** 2003. Epidemiology of urinary tract infections: incidence, morbidity, and economic costs. *Dis Mon* **49:**53–70.

6. **Foxman B.** 2010. The epidemiology of urinary tract infection. *Nat Rev Urol* **7:**653–660.

7. **Nielubowicz GR, Mobley HL.** 2010. Host-pathogen interactions in urinary tract infection. *Nat Rev Urol* **7:**430–441.

8. **Ratner JJ, Thomas VL, Sanford BA, Forland M.** 1981. Bacteria-specific antibody in the urine of patients with acute pyelonephritis and cystitis. *J Infect Dis* **143:**404–412.

9. **Song J, Duncan MJ, Li G, Chan C, Grady R, Stapleton A, Abraham SN.** 2007. A novel TLR4-mediated signaling pathway leading to IL-6 responses in human bladder epithelial cells. *PLoS Pathog* **3:**e60. doi:10.1371/journal.ppat.0030060.

10. **Nagamatsu K, Hannan TJ, Guest RL, Kostakioti M, Hadjifrangiskou M, Binkley J, Dodson K, Raivio TL, Hultgren SJ.** 2015. Dysregulation of *Escherichia coli* α-hemolysin expression alters the course of acute and persistent urinary tract infection. *Proc Natl Acad Sci USA* **112:**E871–E880.

11. **Chromek M, Slamová Z, Bergman P, Kovács L, Podracká L, Ehrén I, Hökfelt T, Gudmundsson GH, Gallo RL, Agerberth B, Brauner A.** 2006. The antimicrobial peptide cathelicidin protects the urinary tract against invasive bacterial infection. *Nat Med* **12:**636–641.

12. **Valore EV, Park CH, Quayle AJ, Wiles KR, McCray PB Jr, Ganz T.** 1998. Human beta-defensin-1: an antimicrobial peptide of urogenital tissues. *J Clin Invest* **101:**1633–1642.

13. **Danka ES, Hunstad DA.** 2015. Cathelicidin augments epithelial receptivity and pathogenesis in experimental *Escherichia coli* cystitis. *J Infect Dis* **211:**1164–1173.

14. **Spencer JD, Schwaderer AL, Wang H, Bartz J, Kline J, Eichler T, DeSouza KR, Sims-Lucas S, Baker P, Hains DS.** 2013. Ribonuclease 7, an antimicrobial peptide upregulated during infection, contributes to microbial defense of the human urinary tract. *Kidney Int* **83:**615–625.

15. **Jaillon S, Moalli F, Ragnarsdottir B, Bonavita E, Puthia M, Riva F, Barbati E, Nebuloni M,**

Cvetko Krajinovic L, Markotic A, Valentino S, Doni A, Tartari S, Graziani G, Montanelli A, Delneste Y, Svanborg C, Garlanda C, Mantovani A. 2014. The humoral pattern recognition molecule PTX3 is a key component of innate immunity against urinary tract infection. *Immunity* **40:**621–632.

16. Chen Y, Guo X, Deng FM, Liang FX, Sun W, Ren M, Izumi T, Sabatini DD, Sun TT, Kreibich G. 2003. Rab27b is associated with fusiform vesicles and may be involved in targeting uroplakins to urothelial apical membranes. *Proc Natl Acad Sci USA* **100:**14012–14017.

17. Bishop BL, Duncan MJ, Song J, Li G, Zaas D, Abraham SN. 2007. Cyclic AMP-regulated exocytosis of *Escherichia coli* from infected bladder epithelial cells. *Nat Med* **13:**625–630.

18. Song J, Bishop BL, Li G, Grady R, Stapleton A, Abraham SN. 2009. TLR4-mediated expulsion of bacteria from infected bladder epithelial cells. *Proc Natl Acad Sci USA* **106:** 14966–14971.

19. Miao Y, Wu J, Abraham SN. 2016. Ubiquitination of innate immune regulator TRAF3 orchestrates expulsion of intracellular bacteria by exocyst complex. *Immunity* **45:**94–105.

20. Miao Y, Li G, Zhang X, Xu H, Abraham SN. 2015. A TRP channel senses lysosome neutralization by pathogens to trigger their expulsion. *Cell* **161:**1306–1319.

21. Elliott TSJ, Reed L, Slack RCB, Bishop MC. 1985. Bacteriology and ultrastructure of the bladder in patients with urinary tract infections. *J Infect* **11:**191–199.

22. Cheng Y, Chen Z, Gawthorne JA, Mukerjee C, Varettas K, Mansfield KJ, Schembri MA, Moore KH. 2016. Detection of intracellular bacteria in exfoliated urothelial cells from women with urge incontinence. *Pathog Dis* **74:**74.

23. Mulvey MA, Lopez-Boado YS, Wilson CL, Roth R, Parks WC, Heuser J, Hultgren SJ. 1998. Induction and evasion of host defenses by type 1-piliated uropathogenic *Escherichia coli*. *Science* **282:**1494–1497.

24. Mysorekar IU, Isaacson-Schmid M, Walker JN, Mills JC, Hultgren SJ. 2009. Bone morphogenetic protein 4 signaling regulates epithelial renewal in the urinary tract in response to uropathogenic infection. *Cell Host Microbe* **5:**463–475.

25. Shin K, Lee J, Guo N, Kim J, Lim A, Qu L, Mysorekar IU, Beachy PA. 2011. Hedgehog/Wnt feedback supports regenerative proliferation of epithelial stem cells in bladder. *Nature* **472:**110–114.

26. Godaly G, Bergsten G, Hang L, Fischer H, Frendéus B, Lundstedt AC, Samuelsson M, Samuelsson P, Svanborg C. 2001. Neutrophil recruitment, chemokine receptors, and resistance to mucosal infection. *J Leukoc Biol* **69:**899–906.

27. Schiwon M, Weisheit C, Franken L, Gutweiler S, Dixit A, Meyer-Schwesinger C, Pohl JM, Maurice NJ, Thiebes S, Lorenz K, Quast T, Fuhrmann M, Baumgarten G, Lohse MJ, Opdenakker G, Bernhagen J, Bucala R, Panzer U, Kolanus W, Gröne HJ, Garbi N, Kastenmüller W, Knolle PA, Kurts C, Engel DR. 2014. Crosstalk between sentinel and helper macrophages permits neutrophil migration into infected uroepithelium. *Cell* **156:**456–468.

28. Shahin RD, Engberg I, Hagberg L, Svanborg Edén C. 1987. Neutrophil recruitment and bacterial clearance correlated with LPS responsiveness in local Gram-negative infection. *J Immunol* **138:**3475–3480.

29. Haraoka M, Hang L, Frendéus B, Godaly G, Burdick M, Strieter R, Svanborg C. 1999. Neutrophil recruitment and resistance to urinary tract infection. *J Infect Dis* **180:**1220–1229.

30. Hannan TJ, Roberts PL, Riehl TE, van der Post S, Binkley JM, Schwartz DJ, Miyoshi H, Mack M, Schwendener RA, Hooton TM, Stappenbeck TS, Hansson GC, Stenson WF, Colonna M, Stapleton AE, Hultgren SJ. 2014. Inhibition of cyclooxygenase-2 prevents chronic and recurrent cystitis. *EBioMedicine* **1:**46–57. doi:10.1016/j.ebiom.2014.10.011.

31. Abraham SN, St John AL. 2010. Mast cell-orchestrated immunity to pathogens. *Nat Rev Immunol* **10:**440–452.

32. Abraham S, Shin J, Malaviya R. 2001. Type 1 fimbriated *Escherichia coli*-mast cell interactions in cystitis. *J Infect Dis* **183**(Suppl 1)**:**S51–S55.

33. Chan CY, St John AL, Abraham SN. 2013. Mast cell interleukin-10 drives localized tolerance in chronic bladder infection. *Immunity* **38:**349–359.

34. Malaviya R, Gao Z, Thankavel K, van der Merwe PA, Abraham SN. 1999. The mast cell tumor necrosis factor α response to FimH-expressing *Escherichia coli* is mediated by the glycosylphosphatidylinositol-anchored molecule CD48. *Proc Natl Acad Sci USA* **96:**8110–8115.

35. Soruri A, Grigat J, Forssmann U, Riggert J, Zwirner J. 2007. beta-Defensins chemoattract macrophages and mast cells but not lymphocytes and dendritic cells: CCR6 is not involved. *Eur J Immunol* **37:**2474–2486.

36. Säve S, Persson K. 2010. Extracellular ATP and P2Y receptor activation induce a proinflammatory host response in the human urinary tract. *Infect Immun* **78:**3609–3615.

37. **Jang TY, Kim YH.** 2015. Interleukin-33 and mast cells bridge innate and adaptive immunity: from the allergologist's perspective. *Int Neurourol J* **19:**142–150.

38. **Mysorekar IU, Hultgren SJ.** 2006. Mechanisms of uropathogenic *Escherichia coli* persistence and eradication from the urinary tract. *Proc Natl Acad Sci USA* **103:**14170–14175.

39. **Mulvey MA, Schilling JD, Hultgren SJ.** 2001. Establishment of a persistent *Escherichia coli* reservoir during the acute phase of a bladder infection. *Infect Immun* **69:**4572–4579.

40. **Michlewska S, Dransfield I, Megson IL, Rossi AG.** 2009. Macrophage phagocytosis of apoptotic neutrophils is critically regulated by the opposing actions of pro-inflammatory and anti-inflammatory agents: key role for TNF-alpha. *FASEB J* **23:**844–854.

41. **Mora-Bau G, Platt AM, van Rooijen N, Randolph GJ, Albert ML, Ingersoll MA.** 2015. Macrophages subvert adaptive immunity to urinary tract infection. *PLoS Pathog* **11:** e1005044. doi:10.1371/journal.ppat.1005044.

42. **Jones-Carson J, Balish E, Uehling DT.** 1999. Susceptibility of immunodeficient gene-knockout mice to urinary tract infection. *J Urol* **161:**338–341.

43. **Sivick KE, Schaller MA, Smith SN, Mobley HL.** 2010. The innate immune response to uropathogenic *Escherichia coli* involves IL-17A in a murine model of urinary tract infection. *J Immunol* **184:**2065–2075.

44. **Engel D, Dobrindt U, Tittel A, Peters P, Maurer J, Gütgemann I, Kaissling B, Kuziel W, Jung S, Kurts C.** 2006. Tumor necrosis factor alpha- and inducible nitric oxide synthase-producing dendritic cells are rapidly recruited to the bladder in urinary tract infection but are dispensable for bacterial clearance. *Infect Immun* **74:**6100–6107.

45. **Godaly G, Ambite I, Svanborg C.** 2015. Innate immunity and genetic determinants of urinary tract infection susceptibility. *Curr Opin Infect Dis* **28:**88–96.

46. **Lichodziejewska-Niemierko M, Topley N, Smith C, Verrier-Jones K, Williams JD.** 1995. P1 blood group phenotype, secretor status in patients with urinary tract infections. *Clin Nephrol* **44:**376–379.

47. **Ginaldi L, Loreto MF, Corsi MP, Modesti M, De Martinis M.** 2001. Immunosenescence and infectious diseases. *Microbes Infect* **3:**851–857.

48. **Ginaldi L, De Martinis M, D'Ostilio A, Marini L, Loreto MF, Quaglino D.** 1999. The immune system in the elderly. III. Innate immunity. *Immunol Res* **20:**117–126.

49. **Hextall A, Cardozo L.** 2001. The role of estrogen supplementation in lower urinary tract dysfunction. *Int Urogynecol J Pelvic Floor Dysfunct* **12:**258–261.

50. **Foxman B.** 1999. Urinary tract infection in postmenopausal women. *Curr Infect Dis Rep* **1:**367–370.

51. **Wang C, Symington JW, Ma E, Cao B, Mysorekar IU.** 2013. Estrogenic modulation of uropathogenic *Escherichia coli* infection pathogenesis in a murine menopause model. *Infect Immun* **81:**733–739.

52. **Klemm P, Roos V, Ulett GC, Svanborg C, Schembri MA.** 2006. Molecular characterization of the *Escherichia coli* asymptomatic bacteriuria strain 83972: the taming of a pathogen. *Infect Immun* **74:**781–785.

53. **Lutay N, Ambite I, Grönberg Hernandez J, Rydström G, Ragnarsdóttir B, Puthia M, Nadeem A, Zhang J, Storm P, Dobrindt U, Wullt B, Svanborg C.** 2013. Bacterial control of host gene expression through RNA polymerase II. *J Clin Invest* **123:**2366–2379.

54. **Ferrières L, Hancock V, Klemm P.** 2007. Biofilm exclusion of uropathogenic bacteria by selected asymptomatic bacteriuria *Escherichia coli* strains. *Microbiology* **153:**1711–1719.

55. **Horwitz D, McCue T, Mapes AC, Ajami NJ, Petrosino JF, Ramig RF, Trautner BW.** 2015. Decreased microbiota diversity associated with urinary tract infection in a trial of bacterial interference. *J Infect* **71:**358–367.

Host Responses to Urinary Tract Infections and Emerging Therapeutics: Sensation and Pain within the Urinary Tract

23

LORI A. BIRDER[1] and DAVID J. KLUMPP[2]

UROTHELIUM AND LAMINA PROPRIA

The bladder wall contains the mucosal layer, the muscularis propria, and the adventitia/serosa. The urinary-bladder mucosa is defined as the portion of the lamina propria (LP) closest to the muscularis propria and contains the urothelium, a basement membrane, and the LP (1).

The major part of the urinary tract is lined with a fully differentiated urothelium. The urothelium lines the renal pelvis, ureters, bladder, upper urethra, and glandular ducts of the prostate, and forms the interface between the urinary space and the underlying vasculature, connective, nervous, and muscular tissues (1, 2). The urothelium is considered a stratified epithelium and contains multiple layers with epithelial subtypes, including a superficial or apical layer composed of large hexagonal cells (diameters of 25 μm to 250 μm) known as umbrella (also termed superficial or facet) cells. Adjacent to the umbrella cells is an intermediate layer of cells (possibly connected by gap junctions), and beneath this layer are basal cells attached to a basement membrane that also forms contacts with the underlying capillary bed. The thickness of the urothelium can significantly vary depending upon the degree

[1]Departments of Medicine and Pharmacology and Chemical Biology, University of Pittsburgh School of Medicine, Pittsburgh, PA 15261; [2]Departments of Urology and Microbiology-Immunology, Feinberg School of Medicine, Northwestern University, Chicago, IL 60610.

Urinary Tract Infections: Molecular Pathogenesis and Clinical Management, 2nd Edition
Edited by Matthew A. Mulvey, David J. Klumpp, and Ann E. Stapleton
© 2017 American Society for Microbiology, Washington, DC
doi:10.1128/microbiolspec.UTI-0023-2016

of bladder distension and the number of epithelial layers seems to be species dependent, and also may differ in various pathological conditions.

The apical surface of the fully differentiated superficial or umbrella cells possesses a unique asymmetric-unit membrane (AUM), whose protein components (uroplakins, UPs) have been well studied (3, 4). Tight junctions, localized between the superficial umbrella cells, are composed of multiple proteins. such as the occludins and claudins (5, 6). These proteins, along with UPs, which are crystalline proteins that assemble into hexagonal plaques, contribute to the urothelial-barrier function. Studies in uroplakin-null mice reveal that loss of UPs correlate with increased urothelial permeability (4).

At the trigone region, the epithelium is thought to exhibit a dilated or stretched appearance, due in part to the basal membrane that may limit the degree of contraction in this region of the bladder. The urothelium transitions at the bladder neck to a stratified or columnar epithelium of the urethra. It has been hypothesized that regions of epithelial transitions (involving differences in the substratum) may play a role in cell differentiation. Though the urethral epithelium has been less systematically studied, this region is accompanied by a lack of urothelial-specific differentiation markers (7). These mucus-secreting cells are often exposed to the lumen and may share similarities to that of the airway epithelium and gastrointestinal enterochromaffin cells. It has been suggested that the secretion product can adhere to the epithelial surface of the cells. In addition, these epithelium are also covered with microvilli, which may be involved in increasing cell-surface area, affecting bacterial adherence and fluid transport, and may also serve a sensor role for chemical, as well as mechanical, signals.

Our understanding of urothelial function and identification/expression of a number of receptors and ion channels have been performed *in vivo* using anesthetized and awake animals and *in vitro* using isolated tissue and cell-culture preparations. For example, flat sheet or cross-sectional preparations have allowed the mapping of receptive fields and localized application of various stimuli directly to the urothelium (8, 9). To evaluate the involvement of urothelial-derived factors in bladder function, a common technique is to study the surgically removed "urothelium," which is likely to impact or include the underlying LP. Further, while the use of isolated or cultured urothelial cells has been essential to our understanding of urothelial signaling, the extreme variation in culture conditions are very likely to influence the expression and function of receptors and ion channels, thus contributing to variability in results. However, it has been shown in a number of studies that urothelial cells in culture do express terminal differentiation markers similar to that *in vivo*, demonstrating the utility of using cells in isolation to examine urothelial function.

There is interest in how the urothelium is influenced by the underlying LP, which lies between the basement membrane of the mucosa and the detrusor muscle. This region is composed of several types of cells, including fibroblasts, interstitial cells, and sensory-nerve endings (10). A series of lymphatic channels and a vascular network is located beneath the urothelium and is likely a source of mediators that can affect cell survival and signaling (10, 11).

A layer of spindle-shaped cells within the upper LP have been described in a number of species, including humans, and is likely to play an important role in signaling mechanisms in the bladder wall (12, 13). While the terminology for categorizing these cells is still under debate, they have been termed interstitial cells (ICs), interstitial cells of Cajal (ICC), interstitial Cajal-like cells (ICLC) cells, myofibroblasts, or telocytes. In addition, there is evidence that these cells (typically identified using cellular markers, such as c-Kit or vimentin) are likely to have close contacts with both afferent and efferent bladder nerves (14). In addition, there is evidence that these cells are able to form a 'network' or functional syncytium connected by gap junctions, as shown by immunohisto-chemistry and functional recordings (using gap-junction blockers). There is evidence that these cells may be able to respond to a number of mediators, including acetylcho-line, adenosine triphosphate (ATP), and prostaglandin E2 (PGE2, released from the nearby urothelium, as well as bladder nerves) (15, 16). There may be subpopulations of IC cells within the LP that perform different functions. For example, recent studies have demonstrated a subpopulation of ICs identi-fied using antibodies against platelet-derived growth factor-receptor-α (PDGFRα), and distinct from "conventional" ICs. Similar to conventional ICs, PDGFRα+ cells (identified in the guinea pig bladder) had a branched-stellate morphology and formed cellular net-works in the LP. While the functions of IC cells in bladder physiology and pathology has not yet been established, these cells likely form a functional link with nearby urothelial cells and bladder nerves within the bladder wall.

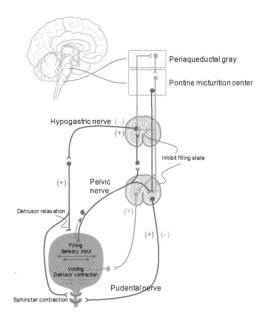

FIGURE 1 Opposing neural control during bladder filling and voiding. Circuits associated with filling (red) and voiding (green). During filling, receptor-mediated sensory signals from the bladder (blue arrow) are conveyed via the pelvic nerve, eliciting sphincter contraction and detrusor relaxation to retain urine. During voiding, sphincter relaxation and bladder contraction are achieved by inhibitory circuits that suppress the filling state and active circuits that promote detrusor contraction and thus urination (17).

AFFERENT NERVES

The bladder and lower urinary tract serves to store and evacuate urine and is controlled by a neural pathway organized at spinal and supraspinal levels (Fig. 1) (10, 17). These afferent pathways provide input to the reflex circuits that control bladder filling and emptying, and are the source of nonpainful sensations of fullness, as well as pain. The plexus of afferent nerves is most dense in the regions of the bladder neck and proximal urethra. The lower urinary tract is regulated by three sets of peripheral nerves that in-clude sacral parasympathetic (pelvic nerves), thoracolumbar sympathetic (hypogastric nerves and sympathetic chain), and somatic nerves (pudendal nerves). For example, affer-ents traveling in the pelvic nerve are thought to be involved in monitoring bladder vol-ume (during bladder storage) and bladder-contraction amplitude (during voiding). These nerves also contain the efferent para-sympathetic, sympathetic, and motor fibers supplying the bladder, urethra, and sphinc-ters. In the human lower urinary tract, affer-ents arise from dorsal-root ganglia (DRG) neurons at the S2–4 and T11–L2 spinal levels. Those axons within pelvic and pudendal nerves originate in the sacral DRG and those in the hypogastric nerves arise in rostral lumbar and caudal-thoracic DRG.

The two major subtypes of afferents are Aδ (myelinated) and C (unmyelinated) fibers (18, 19). Ultrastructure and immunohisto-chemical studies (used to identify the terminations of sensory afferents in the bladder wall) have shown that Aδ fibers are distributed mainly within the detrusor smooth muscle and are responsive mainly to detrusor stretch, which occurs during bladder filling. Often termed "in-series tension receptors," they are thought to be activated by bladder-wall tension caused by distension or contraction of the bladder. In contrast, C-type fibers (which are responsive to bladder filling) seem to be more widespread and are distributed in the detrusor muscle, within the LP, and in close proximity to the urothelium. Studies have demonstrated at least two subclasses of C-type bladder-afferent fibers (18–20). Peptidergic-containing afferents send their projections to lamina I within the spinal cord dorsal horn, in contrast to those that do not contain peptides (but bind isolectin B4), and project to the inner lamina II of the dorsal horn.

The majority of afferents are mechanosensitive and are able to respond to bladder filling with a range of thresholds from volumes that would be encountered under normal bladder filling to levels of distension that would be considered noxious and give rise to pain. In healthy volunteers, studies have shown that the first sensation of filling normally occurs when the bladder reaches 40% capacity, the first desire to void reported at 60% of capacity, and a strong desire to void at 90% capacity (21, 22). Those afferents with lower thresholds have small myelinated axons while unmyelinated fibers have generally higher thresholds for activation. High-threshold afferents are also likely to terminate in the deeper muscle layers or in the serosa, and respond to high levels of stretch that distort the bladder wall, but may also become sensitized in response to inflammation. In a study characterizing mechanosensitive mouse primary afferents (from the lumbar splanchnic and pelvic nerves), four afferent subclasses were distinguished based upon response to various mechanical stimuli (23). These afferent subclasses included serosal, muscle, muscle/urothelial, and urothelial-only. For example, some types of mechanoreceptor can be activated by stretch and by light stroking of the urothelium. Other urothelial endings respond to stroking, but not stretch, and some of these are stimulated by luminal chemicals such as capsaicin acid, and are temperature-sensitive.

This is in contrast to a subtype of afferents that are nonresponsive or silent under physiological conditions, but can be sensitized during inflammation. There are also recent studies that identified and characterized sacral afferents responding to 'flow' through the urethra (24). These are important observations whereby properties of these flow-responsive afferents seem to parallel that of cutaneous afferents. This could be important in terms of restoration of bladder emptying following spinal-cord injury.

Immunohistochemical studies have shown that afferent nerve fibers exhibit positive immunoreactivity for a number of neuropeptides including substance P (SP), calcitonin gene-related peptide (CGRP), corticotropin-releasing factor (CRF), and vasoactive-intestinal polypeptide (VIP) (10, 18). Peptidergic-axons seem to be localized throughout the bladder wall, but most species show a dense distribution within the LP next to the urothelium. In addition, afferents (in particular C-type fibers) express a wide range of receptors, such as TRPV1, TRPA1, tropomyosin-related kinase A (TrkA; responds to nerve growth factor), and muscarinic and purinergic subtypes. For example, P2X3 receptors are localized on afferent nerve fibers adjacent to the urothelium and in the suburothelial space in rodent and human bladders (25). Further, P2X3-null mice exhibit increases in bladder volume and decreased voiding frequency, suggesting that purinergic-signaling mechanisms play a role in bladder filling (26). These immunohistochemical studies show that LP nerve fibers label for receptors to a number of urothelially released mediators, suggesting a complexity of sensory-signaling mechanisms within the bladder wall.

There is considerable interest in mechanisms underlying sensitization of C-fiber afferents, as these nerves are thought to play a key role in symptoms of interstitial cystitis/bladder-pain syndrome (IC/BPS), as well as in patients with urgency sensations at lower-than-normal bladder volumes (27, 28). The pathophysiological mechanisms underlying visceral-pain hypersensitivity are not well known. For example, chemicals released from a variety of cells within the bladder wall, such as the urothelium, myofibroblasts, nerve endings, smooth muscle, and mast cells, may suppress or enhance afferent firing. In addition, mediators (such as prostaglandin, serotonin, ATP, histamine, bradykinin, and neurotrophic factors such as nerve growth factor (NGF)) are released during inflammation, injury, and ischemia, from a number of cell types, including blood vessels, muscle, and neurons. NGF, which is highly expressed within the urothelium, has attracted a great deal of interest as a key player in inflammation and nociceptive signaling. Increased levels of NGF (and other trophic factors) have been detected both in IC patient biopsies and in urine (29, 30). In addition, studies have shown that a number of agents, including NGF, can increase expression of the tetrodotoxin (TTX)-resistant sodium channel. This in turn appears to lower the threshold for afferent firing and affects the volume threshold for reflex voiding (resulting in symptoms such as urgency) (28). Thus, local mediators such as neurotrophins, amines, purines, prostanoids, proteases, and even cytokines, may act directly on sensory-nerve terminals, while others act indirectly, causing release of yet other agents from nearby cells.

Such changes can also occur at the spinal-cord level, whereby glial-cell activation (and release of a number of pro-inflammatory mediators) play a prominent role in a number of pain syndromes by modulating neuronal excitability (31, 32). Thus, glial cells placed at the interface of communication between the periphery and the central nervous system (CNS) may be important players in resetting and modulating lines of communication with deleterious effects. The net result can be sensitization of primary afferent nerves due to release of inflammatory mediators, reducing threshold of afferents and recruiting previously silent nociceptors, thus enhancing the gain of the transducer. In addition, studies have shown a regional variation exists in sensory innervation to the urothelium and the suburothelium.

UROTHELIAL BARRIER FUNCTION

The apical surface of the urothelium is covered with a sulfated polysaccharide glycosaminoglycan (GAG), or mucin layer, that is thought to act as a nonspecific antiadherence factor and as a defense mechanism against infection (1, 2, 33). In addition, during bladder filling, the umbrella cells become flat and squamous and this shape change as the cells elongate is accompanied by vesicular traffic (i.e., exocytosis/endocytosis), adding membrane to the apical surface, thereby increasing overall urinary-bladder surface area. This process of ongoing replacement of apical membrane by newly fused discoid vesicles also serves to maintain the urothelial barrier. There is evidence that this stretch-induced exocytosis is dependent on activation of epidermal growth-factor receptor (EGFR) (34). These processes allow the bladder to accommodate increasing volumes of urine during filling without compromising the barrier function. There is some evidence that superficial urothelial cells exhibit a lower level of endocytotic activity, which may be a protective mechanism against internalization of toxic substances excreted in the urine. Exocytosis/endocytosis (vesicular recycling) may also play an important role in modulating the release of a number of neurotransmitters/mediators, as well as that of uropathogenic *Escherichia coli* (UPEC), which can incorporate into urothelial-fusiform vesicles (35).

UROTHELIAL CELLS AND REPAIR

Epithelial integrity is maintained through a complex process of migration and proliferation (to restore cell numbers) and differentiation (to restore function) (36). Urothelial cells normally exhibit a low turnover rate (3–6 months), in fact the slowest turnover of any mammalian epithelial cells (37, 38). Some investigators have suggested a population of progenitor cells may reside in the intermediate cell layer (39). Urine-derived stem cells obtained from the upper urinary tract differentiate into urothelial cells with the goal toward using these cells for bladder repair and regeneration. However, a definitive identification of urothelial progenitor and stem cells remains elusive. It has been suggested that neither urine-derived factors nor cyclic mechanical changes contribute to urothelial proliferation and differentiation. However, differentiation of urothelial cells in culture can be stimulated by prostaglandin (which is abundant in the urine), and accelerated proliferation can occur in various bladder pathologies (7, 40). For example, using agents (protamine sulfate; cyclophosphamide) that damage the umbrella cell layer showed that the urothelium rapidly undergoes both functional and structural changes in order to restore the barrier in response to injury (41, 42). This led to different forms of treatment (such as removal of diseased urothelium by laser) in patients with the ulcerative form of bladder-pain syndrome. The rationale behind this approach is the ability of the urothelium to rapidly regenerate, and has been associated with nonrecurrence of pain months after treatment.

Following disruption of the barrier, in the early stages of regeneration, the superficial cells may appear smaller and often covered with microvilli (42). In some pathologies, a deficiency or defect in maturation or terminal differentiation of superficial umbrella cells have been reported, though the factors that may be involved are not yet known. The processes underlying urothelial repair is complex, involving several structural elements, signaling pathways, trophic factors, and the cellular environment. Furthermore, the interaction between these biochemical signals and mechanical forces in the bladder during the course of urothelial repair is not well understood. This is particularly important as the processes of epithelial repair occur within a cellular environment that is undergoing cyclic-mechanical deformation (cytoskeletal remodeling, adhesion/loss of adhesion, and generation of force/relaxation).

In addition, a number of these studies have focused on the response in cultured cells, and thus the role of various factors may be limited due to incomplete knowledge about the cellular source of signaling factors, as well as regulation of activity *in vivo*. There is evidence that the initiation of urothelial proliferation or differentiation of intermediate cells involves up-regulation of growth factors, such as fibroblast-growth factor and NGF (43, 44). In addition, members of the PPARγ and EGFR-signaling pathways may contribute to urothelial 're-epithelialization' in wound repair (45). There is also evidence that Hedgehog/Wnt signaling acting across the basal urothelial cell-stromal cell boundary, contributes to increase urothelial proliferation in response to injury (46).

Altered levels of circulating estrogens have been associated with changes to the urothelial structure, including epithelial shedding or mucosal atrophy (47). Other conditions, such as IC, senescence, or spinal-cord injury, are also associated with changes in the urothelial barrier (48, 49). Studies utilizing aged animals have demonstrated significant alterations to the bladder mucosa, including areas of mucosal denudation. In addition, there is evidence in many types of epithelium (including uroepithelium) that adhesion molecules, such as members of the cadherin family, play important roles in establishing and maintaining epithelial-cell contacts (50). Altered urothelial-cadherin expression has been reported in IC-patient bladder urothelium.

Both physiological and psychological stress can result in a failure of urothelial and sub-urothelial 'defensive' systems and thereby promote changes in both urothelial barrier and signaling function. Stress-mediated activation of the hypothalamic-pituitary-adrenal (HPA) axis can result in increased production of corticotrophin-releasing factor, which can regulate neuroendocrine and autonomic responses to stress. Psychological stress can exert deleterious effects on other epithelial structures, such as the skin, resulting in a down-regulation of antimicrobial peptides and increased severity of cutaneous infection in mice (51). Findings from this study suggest that psychological stress inhibits antimicrobial actions via activation of the HPA axis, in part by glucocorticoids. Thus, stress-related dysregulation of signaling between the brain and target structures (including the urinary bladder) can lead to a disruption of the epithelial barrier and increased prevalence of infection.

Though the urothelium maintains a tight barrier to ion and solute flux, a number of factors, such as tissue pH, mechanical or chemical trauma, hormonal changes, or bacterial infection, can modulate the barrier function of the urothelium. Bacteria adherence may involve alterations in urothelial proteins, including proteoglycans, and bacterial defense molecules may lead to distinctive changes in urothelial structure (52). In this regard, urinary tract infections produced by UPEC are initiated by bacterial adherence to uroplakin proteins on the apical surface (53). The UPEC express filamentous adhesive organelles (type 1 pili) that mediate bacterial attachment, invasion, and apoptosis of the urothelial cells. Urothelial differentiation (and increased uroplakin III expression) may play a pivotal role in sensitizing urothelial cells to UPEC-induced infection and possible cell death (54, 55). Even acute contact (within hours) of the mucosal surface by bacteria may result in altered urothelial-barrier function (56). For example, infection with UPEC alters bladder and ureteric contractility via host-urothelial interactions. UPEC can also internalize within umbrella cells, forming intracellular colonies (biofilm-like pods) of UPEC that has been implicated in the mechanism of chronic urinary tract infections. UPEC are able to commandeer the endocytic/exocytic machinery of urothelial cells, residing inside fusiform vesicles (35). This permits the bacteria to escape elimination during voiding and re-emerge into the urine during distension. When expelled into the urine during the storage phase, the urine may provide a nutrient-rich environment optimizing bacterial survival. It has also been shown that bacteria, taken up by the urothelium, can be presented to the immune system during successive cycles of filling and emptying.

Disruption of urothelial function can also be induced by more remote pathological conditions that influence neural or hormonal mechanisms. For example, spinal-cord transection in rats leads to a rapid alteration in the urothelial barrier, including ultrastructural changes and increased permeability (48). The changes are blocked by pretreatment with a ganglionic-blocking agent, suggesting an involvement of efferent autonomic pathways in the acute effects of spinal-cord injury on bladder urothelium. Other types of urothelial-neural interactions are also likely, based on the recent reports that various stimuli induce urothelial cells to release chemical mediators that can, in turn, modulate the activity of afferent nerves (57, 58). This has raised the possibility that the urothelium may have a role in sensory mechanisms in the urinary tract.

In summary, modification of the urothelium and/or loss of epithelial integrity in a number of pathological conditions can result in passage of toxic/irritating urinary constituents through the urothelium or release of neuroactive substances from the urothelium. This may lead to changes in the properties of sensory nerves and, in turn, sensory symptoms such as urinary frequency and urgency. Thus, chemical communication between the nervous system and the urothelial cells may play an important role in the generation of urinary-bladder dysfunction.

UROTHELIAL-CELL INTERACTIONS

While urothelial cells are often viewed as bystanders in the process of visceral sensation, recent evidence has supported the view that these cells function as primary transducers of some physical and chemical stimuli and are able to communicate with underlying cells, including bladder nerves, smooth muscle, and inflammatory cells. The urothelium is able to respond to a wide variety of mechanical stresses during bladder filling and emptying by activating a number of possible transducer proteins. Possibilities of mechanical signals include bladder pressure, tension in the urothelium or bladder wall, torsion, geometrical tension, movement of visceral organs, and even urine tonicity. Alterations in the composition of urine are a type of stress whose contents can vary in both their rate of delivery and the particular constituents.

Additional lines of evidence suggest that urothelial cells participate in the detection of both physical and chemical stimuli. Recent studies have shown that both afferent and autonomic efferent nerves are located in close proximity to the urothelium. Peptidergic, P2X- and TRPV1- immunoreactive nerve fibers presumed to arise from afferent neurons in the lumbosacral dorsal-root ganglia are distributed throughout the urinary-bladder musculature, as well as in a plexus beneath and extending into the urothelium (1). In humans with neurogenic detrusor overactivity, intravesical administration of resiniferatoxin, a C-fiber afferent neurotoxin, reduces the density of TRPV1 and P2X3 immunoreactive suburothelial nerves, indicating that these are sensory nerves (57). In addition, immunohistochemical studies have also revealed adrenergic (tyrosine hydroxylase)-positive, as well as cholinergic (choline acetyltransferase, ChAT)-positive, nerves in close proximity to the urothelium.

A network of cells with morphologic characteristics similar to those of myofibroblasts or interstitial cells is in the suburothelial space of the bladder in both humans and animals (13). These cells, which are extensively linked by gap junctions and have close contacts with nerves, can respond to neurotransmitters, such as ATP released from nerves or urothelial cells (58), suggesting that they could act as intermediaries in urothelial-nerve interactions. Thus, the anatomic substrates for bidirectional urothelial-neural communication exist within the urinary bladder. Further, any type of imbalance within this urothelial-suburothelial complex is likely to play a role in symptoms associated with bladder pathologies, including IC/BPS.

The involvement of urothelial function in sensory signaling is suggested by the finding that urothelial cells express various receptors that are linked to mechano- or nociceptive sensations. Examples of neuronal "sensor molecules" (receptors/ion channels) that have been identified in urothelium include receptors for purines (P2X (1-7) and P2Y (1,2,4)), adenosine (A_1, A_{2a}, A_{2b} and A_3), norepinephrine (α and β), acetylcholine (muscarinic and nicotinic), protease-activated receptors (PARs), amiloride- and mechanosensitive epithelial sodium channels (ENaC), bradykinin (B1 and B2), neurotrophins (p75, trkA, EGF family Erb1-3), corticotrophin-releasing factor (CRF1 and CRF2), estrogens (ERα and ERβ), endothelins, and various TRP channels (TRPV1, TRPV2, TRPV4, TRPM8, and TRPA1) (59–69). The expression of these various receptors enable the urothelium to respond to a number of "sensory inputs" from a variety of sources. These inputs include increased stretch during bladder filling, soluble factors (many found in the urine) such as epidermal-growth factor (EGF), or chemical mediators/peptides/transmitters such as substance P, calcitonin gene-related peptide (CGRP), corticotrophin-releasing factor (CRF), acetylcholine, and adenosine or norepinephrine released from nerves, inflammatory cells, and even blood vessels.

Various stimuli can lead to a number of outputs, including the secretion of numerous chemical substances, such as neurotrophins, peptides, ATP, acetylcholine, prostaglandins, prostacyclin, nitric oxide (NO), and cytokines. The net result can be complex, involving alteration of flow of ions and other substances across the urothelium, altering membrane turnover and modulating the activity of underlying smooth muscle and nearby sensory neurons. In this regard, this type of volume or signal transfer between the uroepithelium and underlying layers has been demonstrated by measuring the propagation of calcium and membrane-potential events. A number of stimuli are able to enhance this propagation, including mechanical stretch, and is enhanced in these preparations by pathological conditions.

For example, urothelial cells express the receptor proteins and mRNAs for all the muscarinic subtypes (M1-M5) and exhibit the machinery necessary for the synthesis and release of acetylcholine. While the significance of cholinergic signaling is still being investigated, acetylcholine is likely to act in a paracrine manner to stimulate underlying nerves and smooth muscle, as well as in an autocrine manner to stimulate urothelial (nicotinic; muscarinic) receptors. In this regard, studies have shown that muscarinic receptors are involved in urothelial-signaling by enhancing intrinsic detrusor contractions and afferent signaling (65). Other studies have suggested that cholinergic mechanisms may be involved in the release of inhibitory factors from the urothelium (62). The identity of these factors is not known, but it is likely that the mucosa is able to release agents, which depress muscle contractility.

Nerve-growth factor (NGF) is highly expressed within the uroepithelium in a number of species (30). Studies in patients with pelvic pain, including bladder-pain syndrome, show increased urothelial NGF expression (67). Intravesical NGF is able to sensitize bladder afferents to mechanical stimulation, and urothelial-specific overexpression of NGF results in increased bladder nerve 'sprouting' and increased voiding frequency. Taken together, these findings suggest that NGF may play a role in bladder pain and changes in bladder function. Investigators have also shown that the urothelium expresses both inducible nitric-oxide synthase (iNOS) and endothelial NOS (eNOS). Urothelial-derived NO can be released in response to mechanical, as well as chemical, stimulation and may either facilitate or inhibit the activity of bladder afferent nerves conveying bladder sensation. For example, reduced levels of NO (via experimental manipulation or pathology) results in bladder hyperactivity that is suggestive of an inhibitory role of NO in bladder function (68). In this regard, activation of urothelial receptors and release of inhibitory mediators may explain, in part, the mechanism of action for therapies (e.g., β3-adrenergic-receptor agonists) in treatment of bladder disorders, such as overactive bladder (OAB) (64, 69).

The mechanism underlying release of chemical mediators from the urothelium, including whether all sensory "inputs" stimulate membrane turnover (i.e., vesicular exocytosis) is not well understood. What little is known about the roles and dynamics of membrane-bound cytoplasmic vesicles in urothelial-cell physiology is derived from measurements of membrane capacitance and microscopy of fixed tissues and cells. For example, there is evidence that once released, ATP can act as an important autocrine mediator, which can induce membrane turnover as well as enhance both stretch-induced exocytosis and endocytosis. ATP was the first neurotransmitter shown to be released directly from the urothelium by several mechanisms, including transporters (members of the ATP-binding cassette, ABC, transporter superfamily), anion-selective channels, such as the maxi-anion channel, as well as vesicular exocytosis (60, 66). A number of channels, including the amiloride-sensitive apical sodium channel, ENaC, may be involved in mechanotransduction by controlling basolateral release of

ATP [65]. Urothelial release of ATP (and autocrine stimulation of urothelial-nucleotide receptors) is likely to contribute to release of a variety of mediators and the resulting signal depends upon a number of factors. These include the subtype of purinergic (or adenosine) receptor expressed, as well as expression of ATPases and other ectonucleotidases. These ecto-enzymes can be secreted or membrane-bound and act to degrade (ATP and UTP) to respective nucleotides, including conversion into adenosine, which activates its own class of P1 receptors also expressed within the urothelium. Studies have also shown that activation of P1 (adenosine) receptors within the urothelium may modulate umbrella-cell exocytosis (34). Thus, alterations in membrane turnover cannot only increase apical surface area (as described above), but also regulate the number and function of receptors and channels at the cell surface and participate in release of a variety of mediators.

There is evidence that epithelial cells in different organ systems may express similar receptor subtypes. Accordingly, epithelial cells could use multiple signaling pathways, whose intracellular mechanisms differ according to location and environmental stimuli. This would permit a greater flexibility for the cell to regulate function and respond to complex changes in their surrounding microenvironment. Whether urothelial-sensor molecules all feed into a diverse array of signaling pathways or share similarities with systems such as olfaction, whereby hundreds of receptors share identical transduction cascades, is yet to be uncovered.

Clinical Significance of Urothelial Signaling

Defects in urothelial-sensor molecules and urothelial-cell signaling are likely to contribute to the pathophysiology of bladder diseases. For example, a number of bladder conditions (IC/BPS, spinal-cord injury (SCI), chemically induced cystitis) are associated with augmented release of urothelial-derived ATP, which is likely to result in altered sensations or changes in bladder reflexes induced by excitation of purinergic receptors on nearby sensory fibers. ATP can also act in an autocrine manner that would act to facilitate its own release (and to contribute to basal levels of other mediators) from urothelial cells. Augmented expression/release of urothelial-derived chemical mediators is likely to reduce the threshold for activation of nearby bladder afferents. In addition, intercellular communication mediated by gap junctions in myofibroblasts could provide a mechanism for long-distance spread of signals from the urothelium to the detrusor muscle. Thus, the urothelium has the potential for amplifying signals, both within the urothelium and the bladder wall, and contributing to a gain of function in sensory processing. Factors that can impact this 'gain of function' include alterations in levels of trophic factors, as well as stress and steroid hormones. For example, altered levels of circulating estrogens may play a role in urinary-bladder dysfunction, including urgency and frequency. The resulting structural and functional abnormalities may lead to enhanced signaling between the urothelium and underlying cells. There is also a need for further understanding in epithelial repair mechanisms. For example, examining gene profiles in epithelial cells from various patient types may help to understand why some patients are more susceptible to infection, as well as the importance of chronic injury and repair in bladder disorders.

Changes in epithelial signaling/barrier function would not be unique to the urinary bladder. For example, airway epithelia in asthmatic patients and keratinocytes in certain types of skin diseases also exhibit a number of similar abnormalities and compromised repair processes (59, 63). This is particularly relevant given the high incidence of associated diseases that can include both visceral and somatic conditions, many of which exhibit a shared loss of epithelial-barrier function. Taken together, epithelial cells can respond to a number of challenges (including environmental pollutants and mediators released from nerves or nearby inflammatory cells), resulting in altered expression and/or sensitivity of various receptor/channels, as well as changes in release of mediators, all of which could impact function.

Urothelial Damage during UTI

The resulting loss of bladder-barrier function in conditions such as IC may have a direct role in pain and sensation. Elevated extracellular K^+ depolarizes many neurons and thereby activates voltage-gated membrane currents that trigger synaptic transmission (70). This effect has been exploited clinically to determine whether bladder pathology results in heightened responses in the "potassium-sensitivity test," where elevated K+ is instilled into the bladder (71). Indeed, intravesical potassium elicits a strong response in 75% of IC patients, suggesting that compromised urothelium plays a role in pain (72). These findings also support the hypothesis that pain in IC can be driven by urinary potassium.

UTI patients also respond to the potassium-sensitivity test (72), indicating increased urothelial permeability and suggesting a possible mechanism for dysuria and pelvic pain in UTI. Early studies in rodents revealed that experimental UTI resulted in urothelial damage that mimicked findings of UTI patient bladder biopsies (73–77). Building on these initial observations, Mulvey and colleagues showed via electron micrographs in a murine UTI model that UPEC strain NU14 interacted with urothelium via the FimH adhesin of type 1 pili (78). These pilus-urothelial interactions resulted in FimH-dependent sloughing of superficial urothelial cells that exposed underlying, less-differentiated urothelial cells. Given that tight junctions mediate interactions between adjacent superficial urothelial cells and these junctions are a major component of bladder-barrier function, the FimH-dependent loss of superficial urothelium during UTI is the likely basis of potassium sensitivity among UTI patients. Finally, using TUNEL staining to identify apoptotic nuclei in bladder sections, it was shown that urothelial cell sloughing was associated with apoptosis (78). TUNEL-positive urothelial nuclei were evident in bladders of mice infected with wild-type NU14, but were absent in mice infected with the FimH-deficient mutant NU14-1. Similarly, the K-12 strain AAEC185 induced apoptosis that was strictly dependent upon the presence of a plasmid encoding type 1 pili. Thus, FimH-dependent urothelial apoptosis is induced by both UPEC and nonpathogenic K-12 E. coli alike.

The potential role of FimH-dependent apoptosis as an initiator of bladder-barrier dysfunction spurred in vitro studies to define underlying mechanisms. Human urothelial-cell lines immortalized with human papillomavirus E6E7 proved particularly useful in these studies because, unlike carcinoma-derived lines, immortalized urothelial lines retain the capacity to enter into the urothelial-differentiation program and undergo FimH-dependent apoptosis with kinetics similar to the murine UTI model (79, 80). FimH-induced urothelial apoptosis was greater in urothelial cultures infected with UPEC strain NU14 than cultures infected with the K-12 strain HB101-expressing type 1 pili, and this finding correlated with a capacity of UPEC to suppress inflammatory responses at the level of NF-κB activation (79). Modulation of urothelial inflammatory responses by NU14 proved to be common among clinical isolates (81), and subsequent studies identified genes mediating inflammatory modulation by screening UPEC transposon-insertion libraries for mutants that induced elevated cytokine secretion in urothelial cultures (82, 83). In contrast to inflammatory modulation by secreted virulence factors that act on intracellular-host targets (e.g., YopJ of Y. pestis (84)), these genetic screens identified

UPEC genes involved in biosynthesis of LPS, peptidoglycan, and outer-membrane proteins. Taken together, these findings suggest that UPEC have evolved systems for triggering and modulating urothelial apoptosis by the combined effects of FimH and diverse bacterial-surface molecules.

Initial observations in murine UTI demonstrated that the pan-caspase inhibitor BAF blocked urothelial sloughing, suggesting that FimH triggers a caspase cascade in urothelial cells that mediates apoptosis (85). This finding was confirmed using immortalized urothelial cultures. NU14 induced caspase 2 and caspase 8 in urothelial cells, consistent with activation of both intrinsic and extrinsic apoptotic pathways, respectively (86). Bid translocation to mitochondria was induced by either NU14 or recombinant FimH. Collectively, these observations demonstrated that FimH functions as a tethered toxin capable of triggering urothelial apoptosis by distinct apoptotic pathways that converge on caspase 3 (86). Despite implicating caspases, however, the urothelial initiators of apoptosis remained unclear.

Electron micrographs from Mulvey and colleagues provided compelling evidence for direct interactions between type 1 complexes and 16-nm uroplakin complexes densely expressed on the apical surface of superficial umbrella cells (78). These data bolstered previous biochemical findings that FimH bound to blots of purified uroplakins Ia and Ib (87). These observations, therefore, established that uroplakins Ia and Ib serve as receptors for type 1 pili and suggested that uroplakin complexes mediated urothelial responses to UPEC, including apoptosis. Using a combination of brute-force biochemistry, biophysical studies, and elegant cell biology, Dr. Sun's group demonstrated that the 16-nm particle is a complex of four major uroplakins, UPIa, UPIb, UPII, and UPIII (4). Despite this structural understanding of the 16-nm particle and evidence for interactions with FimH, it was unknown whether uroplakins functioned as signal transducers.

Clues to FimH-induced signals came from considering the membrane topology of uroplakins because only UPIII is predicted to have an appreciable cytosolic domain that might transduce signals (88). Consistent with this possibility, *in vitro* differentiation of urothelial cultures enhanced FimH-dependent cell death that correlated with increased UPIII expression, and acid removal of superficial urothelial cells abrogated NU14-induced apoptosis in murine UTI without impacting bladder colonization (55, 89). A direct role for UPIII as a mediator of signals during UTI was confirmed in studies that detected FimH-induced increases of urothelial intracellular calcium (55) via phosphorylation of T_{244} on the cytosolic tail of UPIII by casein kinase II. UPEC-induced apoptosis was significantly reduced in urothelial cultures by knocking down UPIII expression, and apoptosis was reduced both in urothelial cultures and *in vivo* by the presence of a specific inhibitor of casein kinase II. Together, these findings make a case for UPEC-induced urothelial dysfunction as a mediator of bladder symptoms during UTI.

UTI as a Visceral-Pain Model

Despite our growing understanding of UTIs at the levels of microbial pathogenesis and inflammatory responses, until recently little was known of sensory mechanisms during UTI or how these mechanisms contribute to pain. Previous studies of inflammatory pain have demonstrated that components of the inflammatory milieu activate sensory nerves in the periphery (Fig. 2). A variety of factors released from sites of tissue insult or the resulting leukocytic influx can activate sensory neurons including ATP, nitric oxide, cytokines, and eicosanoids (90). These inflammatory factors induce pain either directly by triggering neuronal-action potentials or indirectly by reducing firing thresholds and thereby rendering sensory neurons more susceptible to firing in response to otherwise subthreshold stimuli. Since pyuria is a prominent feature of UTI (91, 92), it is possible that UTI-associated inflammation drives pain. And, in the case of the bladder, it is conceivable that inflammatory mediators that reduce thresholds for sensory-neuron firing could result in pain due to bladder filling.

Strictly speaking, pain is a perception and thus can only be reported by humans, so indirect correlates of pain must be utilized for animal models, including UTI. Since the bladder and other visceral organs share innervation with specific skin surfaces, increased tactile sensitivity is employed widely as a measure of visceral pain, and "allodynia" evoked by mechanical stimulation to the pelvic region was increased during chemical cystitis (93). Pelvic pain in murine UTI was evaluated by quantifying allodynia, and bladder instillation of UPEC strain NU14 induced pelvic allodynia, whereas the asymptomatic bacteriuria strain 83972 did not (94). Thus, pelvic allodynia during murine UTI recapitulated the clinical response to NU14 and 83972, indicating that murine UTI is an appropriate model for evaluating the role of FimH-induced urothelial dysfunction in

bladder pain. Surprisingly, pelvic allodynia induced by infection with wild-type NU14 was not diminished in mice infected with the mutant NU14-1 lacking functional type 1 pili (94). Conversely, wild-type 83972 do not express type 1 pili, but infection with 83972 harboring a plasmid encoding the type 1 pilus did not induce pelvic allodynia in response to bladder instillation. Since UTI-induced urothelial apoptosis is strictly dependent upon type 1 pili, these findings suggest that pelvic symptoms of UTI are distinct from the urothelial dysfunction induced by FimH (Fig. 3).

TLR4-Dependent UTI Pain is Independent of Inflammation

The finding that pelvic allodynia induced by UPEC was independent of FimH indicated that other factors mediate UTI pain (94). This possibility was consistent with previous studies demonstrating that UTI induced thermal sensitivity in the mouse hindpaw, and this sensitivity did not develop in mice lacking the LPS receptor, TLR4 (95). Since TLR4 is a major mediator of inflammatory

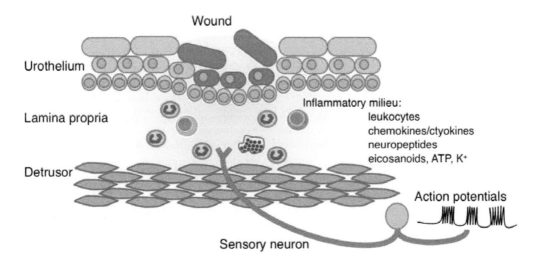

FIGURE 2 Inflammatory pain. Factors within the inflammatory microenvironment can trigger pain responses. These factors may be released by damaged cells at the site of injury or by leukocytes. Such inflammatory mediators can trigger action potentials in sensory neurons or reduce firing thresholds for other environmental or physiologic stimuli.

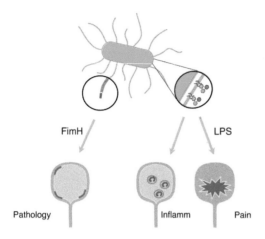

FimH   LPS

Pathology   Inflamm   Pain

FIGURE 3 **UPEC induces pain separable from other facets of UTI pathogenesis. FimH acts as a tethered toxin that mediates urothelial apoptosis and consequent bladder-barrier dysfunction. LPS plays dual roles through its interactions with TLR4. In addition to the well-characterized role as a trigger for inflammation, LPS mediates pelvic-pain responses. Reproduced from (102), with permission.**

responses induced by LPS, it suggested that inflammation mediated the pain response. This hypothesis was tested by comparing the UPEC strain NU14 with the asymptomatic bacteriuria isolate 83972. Although the pelvic allodynia of NU14 infection was dependent upon TLR4, both NU14 and 83972 induced comparable levels of neutrophil myeloperoxidase (MPO), indicating that inflammation was similarly induced by both strains, despite the disparate allodynia induced by these two strains (94). In this same study, purified LPS was instilled into the bladder to confirm a role for LPS as the mediator of UTI pain responses. Like whole bacteria, inflammation was similar at the level of MPO, but NU14 LPS evoked pelvic allodynia, whereas no such allodynia was associated with 83972 LPS (94). These findings indicate that FimH induces urothelial dysfunction that is separable from infection pain (Fig. 3). Moreover, while UPEC LPS induces inflammation through TLR4, these effects appear separable from LPS: TLR4 interactions that mediate UTI pain.

O-Antigen is a Rheostat for LPS-Induced Pain

The observation that purified LPS preparations could induce distinct pelvic-pain responses raised the question of which LPS structural element(s) mediated these differential responses. Initial clues came from the targeted *waaL* mutant of NU14 with a deletion of the gene encoding O-antigen ligase (82). Although infection with NU14 caused acute pelvic allodynia that decayed over several days, like clinical UTI, the *waaL* mutant elicited no allodynia during an initial infection (96). While this finding suggested that O-antigen modulates pain states, it was also possible that the absence of allodynia in response to *waaL* was due to the high degree of virulence attenuation, where wild-type NU14 outcompetes *waaL* nearly 1,000-fold (82). However, serial infections revealed another pain phenotype of the *waaL* mutant. Whereas serial NU14 infections at two-week intervals resulted in resolving, acute allodynia each time, serial *waaL* infection resulted in chronic allodynia that persisted long after bacterial clearance (96). This finding suggested that O-antigen status defined, at least in part, the pain phenotype of *E. coli*.

The possibility that O-antigen modulates infection pain was confirmed using two alternative systems (Fig. 3). The K-12 strain SΦ874 lacks the entire O-antigen gene cluster, *wz**, and this defect was complemented with the plasmid pWQ288 encoding *Klebsiella pneumoniae* O2a (97–99). Although infection with either SΦ874 or SΦ874/pWQ288 resulted in a similarly transient colonization of the bladder, the pain phenotypes were striking: SΦ874/pWQ288 evoked no pelvic allodynia, but SΦ874 lacking O-antigen caused a durable pelvic allodynia within 4 days that persisted for weeks (96). To cement the role of O-antigen as a modulator of infection pain states, a targeted deletion of the *wz** cluster was generated in NU14 and complemented with fosmids containing a control insert, NU14, or the asymptomatic bacteriuria (ASB) isolate 83972. While the wz* mutant containing a control fosmid exhibited a chronic-pain phenotype, complementation with wz*$_{NU14}$ yielded an acute-pain phenotype, and complementation with the ASB-derived wz*$_{83972}$ resulted in a null-pain phenotype (96). Therefore, a single bacterial strain may induce a range of pain states, depending upon the presence and type of O-antigen.

Like acute allodynia induced by NU14, chronic allodynia induced by strains lacking O-antigen was largely dependent upon TLR4, reduced approximately 70% in TLR4-deficient mice, relative to wild-type mice (96). While this raised the possibility that chronic pain was associated with chronic inflammation, that idea was not borne out. Pathologic assessment was not different, either acutely or after resolution of inflammation between mice infected with NU14, waaL, 83972 or SΦ874, despite the significant differences in allodynia or absence thereof (96). Consistent with this, urinary MPO was also not different, thus bolstering the prior findings from acute NU14 infection suggesting that UTI pain states are not correlated with inflammation (94, 96). Bone-marrow chimeras of wild-type and TLR4-deficient strain also demonstrated that wild-type recipients develop chronic allodynia in response to SΦ874, regardless of donor bone-marrow status, whereas TLR4-deficient recipients had significantly compromised allodynia, even when reconstituted with wild-type bone marrow (96). Recent studies confirmed that nonhematopoietic cells mediate UTI pain using the visceromotor reflex (VMR) to quantify abdominal-muscle contraction as a measure of evoked pain during bladder filling. Normal mice exhibit a VMR response characterized by a sigmoidal increase in abdominal-muscle contractions as a function of applied bladder distension, and cystitis due to infection with UPEC strain UTI89 shifted the sigmoidal response upward/leftward, indicating increased bladder pain during filling (100). Protamine sulfate instillation was previously shown to erode the superficial urothelium, and VMR testing revealed that protamine-treated mice failed to exhibit increased bladder pain following UTI89 infection. These findings suggest that urothelium is the inducer of UTI pain. Together, these studies indicate that O-antigen functions as a rheostat to modulate LPS pain signals via TLRs (Fig. 4), and these functions appear independent of gross inflammation and are not mediated by a hematopoietic lineage. Rather, the urothelium itself is the transducer of the UTI pain signal, consistent with its functions role as an extension of the bladder-sensory system.

Chronic Pain Resulting from Transient Infection

The chronic allodynia elicited by SΦ874 or ΔwaaL persists long after bacterial clearance from the bladder (96), suggesting that a transient bacterial infection can cause chronic pain. However, a hallmark of chronic pain is alterations within the CNS that lead to enhanced signaling through sensory pathways, so-called "central sensitization," such that subthreshold noxious stimuli become painful. To determine whether the chronic allodynia induced by SΦ874 or ΔwaaL is associated with altered central physiology, spinal responses were quantified in an *ex vivo* sacral spinal cord preparation to evaluate both baseline and evoked excitability (96). Acute UTI with NU14 resulted in increased spontaneous firing of action potentials that resolved by fourteen days after infection, consistent with allodynia. Similarly, spinal cords of mice with chronic allodynia from SΦ874 or ΔwaaL infection also exhibited increased spontaneous firing, so spontaneous activity revealed no difference between acute and chronic states. However, distinct differences were observed when examining evoked responses.

To the quantify the full spinal-reflex arc, a series of stimulus pulses was administered at increasing intensity to spinal inputs and then

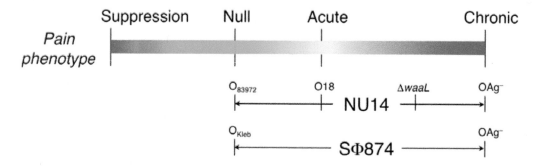

FIGURE 4 LPS O-antigen acts as a rheostat to modulate UTI pain. Mice exhibited allodynia that varied with O-antigen. The acute-pain phenotype of UPEC strain NU14 was rendered chronic by deleting *waaL* or the entire O-antigen gene cluster, but was suppressed by expressing 83972 O-antigen genes. K-12 strain SΦ874 induced chronic pain that was suppressed by expressing *Klebsiella* O-antigen O2a. Adapted from (96), with permission.

recorded as spinal outputs (96). During the stimulus series, control spinal cords responded less robustly with each successive stimulus, characteristic of a typical desensitization response to a repetitive stimulus mediated by inhibitory circuits (101). Spinal cords of acute NU14-infected mice displayed similar desensitization. In contrast, SΦ874 or ΔwaaL spinal cords failed to exhibit normal desensitization to successive stimulus pulses, and this desensitization defect manifested at multiple stimulus intensities. Thus, evoked potentials reveal that mice exhibiting chronic allodynia also exhibit increased spinal excitability and suggest that a loss of normal inhibitory mechanisms may underlie that increased excitability. But, no matter the precise mechanism, these findings support the idea that a transient bladder infection may initiate chronic pelvic pain, a finding that has significant implications for UTI and other infectious diseases.

Although the central effects and mechanisms mediating *E. coli*-induced chronic pain remain to be elucidated, peripheral mechanisms are coming into focus with recent studies. TRPV1 is required for the establishment of post-UTI chronic pain, but is dispensable for pain maintenance ((102) and J. M. Rosen et al., submitted). Conversely, CCR2 is required for pain maintenance. Together with TLR4, these findings establish a three-receptor peripheral cascade that mediates the development of post-UTI chronic pain and results in spinal central sensitization with features of neuropathic pain (Fig. 5).

Bacterial Analgesia in the Bladder

During acute murine UTI, initial experiments with purified 83972 LPS demonstrated a small-but-significant decrease in allodynia induced by NU14, raising the possibility that the null pain phenotype of some *E. coli* might actually represent analgesic activity (94). Exploring this more fully, a panel of ASB *E. coli* isolates was evaluated for effects on cystitis-induced pelvic allodynia (103). ASB *E. coli* exhibited a broad range of analgesic activity against NU14-induced from no analgesia to modest pain inhibition to analgesia that exceeded intravesical lidocaine. Furthermore, ASB *E. coli* exhibited analgesic activity against acute pain induced by acute infection with *Proteus mirabilis*, *Enterococcus faecalis*, and *Klebsiella pneumoniae*. Interestingly, the antimicrobial ciprofloxacin that is commonly prescribed as a UTI therapy had no effect on pain despite its microbiologic impact, suggesting that uropathogens initiate a pain response with kinetics defined by the bacterial-pain phenotype (103).

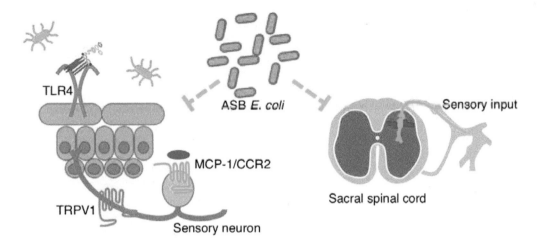

FIGURE 5 **Chronic-pain pathway and ASB *E. coli* analgesia. A peripheral three-receptor cascade, consisting of TLR4, TRPV1, and CCR2, mediates development of chronic pain that is associated with central sensitization of the sacral spinal cord receiving bladder sensory input due to altered sensory input (green) or inhibitory control (red). Analgesic activity of ASB *E. coli* disrupts pain in the periphery and/or spinal cord.**

CNS Circuits in Cystitis Pain

The role of the CNS in bladder control has been studied extensively (17). The normal bladder filling and voiding functions require that urinary sphincters and bladder smooth muscle exhibit opposing tonality, and this inverse regulation is provided by circuits in the spinal cord as well as higher-order CNS circuits, including the Barrington's nucleus (rodent correlate to the Pontine micturition center) and the periaqueductal region of the brain (104). Pseudorabies virus (PRV) induces a neurogenic cystitis that recapitulates many aspects of IC, including pelvic pain (105–108). PRV neurogenic cystitis is associated with increased expression and activation of calcium/calmodulin-dependent kinase 2 (CaMKII) in the sacral spinal-cord dorsal horn, and a CaMKII inhibitor decreased PRV-induced pelvic allodynia (109), thus demonstrating a role for specific spinal mediators of cystitis pain. In the brain, lesions in the Barrington's nucleus abrogate bladder inflammation in response to PRV, demonstrating that brain circuits can mediate bladder inflammation (110). This suggests that higher-order structures also modulate bladder pain during UTI, and new studies are beginning to support this possibility.

Cyclophosphamide (CYP)-induced cystitis has long been used as a bladder-inflammation model, and circuits mediating CYP effects on bladder and voiding function have been a focus of intense study (111). CYP cystitis was shown to induce enhanced VMR, and increased VMR was associated with increased activation of extracellular signal-regulated kinase 1/2 (Erk1/2) in the dorsal horn and deeper laminae of the lumbosacral spinal cord (112). Erk activity is modulated by upstream kinases MEK1/2, and a MEK inhibitor reduced the CYP VMR response, thus indicating that Erk mediates visceral hyperalgesia in a cystitis model. Extending these studies, UTI89 infection was shown to increase VMR, albeit modestly and only at high intravesical pressures (113). Metabotropic glutamate receptors are neurotransmitter receptors, and the metabotropic receptor mGluR5 is associated with nociceptive pathways in several models. mGluR5-deficient mice exhibited reduced VMR (113), demonstrating a role for mGluR5 in bladder-distension pain. An mGluR5 antagonist, fenobam, significantly reduced UTI89-associated VMR, suggesting that mGluR5 also mediates UTI pain, although fenobam similarly reduced VMR in untreated mice, and it was not reported whether mGluR5-deficient mice were resistant to UTI89-induced VMR. Nonetheless, these findings spurred additional studies to dissect the role of mGluR5 in bladder responses.

The central nucleus of the amygdala is a brain region previously implicated in visceral pain in colitis models and also sends outputs to the periaqueductal gray region (114). Since pharmacologic manipulation of mGluR5 in the amygdala modulated responses to rectal distension, amygdala mGluR5 might also modulate bladder responses ((115) and references therein). Indeed, delivering the mGluR1/5 agonist DHPG directly to the right amygdala via a cannula increased VMR during bladder filling, consistent with previous observations of right amygdala mGluR5 in somatic pain. In contrast, the mGluR5 antagonist MPEP reduced VMR. Bolstering the role of amygdala mGluR5 in bladder pain, mGluR5 expression was selectively ablated by delivering a lentivirus encoding Cre recombinase to the right amygdala in transgenic mice bearing a floxed mGluR5 gene. Deletion of mGluR in the amygdala reduced bladder-filling VMR and reduced the accumulation of activated Erk1/2 in the spinal cord in response to bladder filling. Finally, using recombinant herpesviruses encoding rhodopsin-2 as an optogenetic switch responsive to a blue laser, laser excitation of mice expressing the optogenetic switch in the amygdala elicited increased VMR during bladder filling. Together, these studies implicate multiple signaling pathways as spinal mediators of cystitis nociception and suggest the amygdala is a center for bladder responses mediated by mGluR5. Moreover, these studies indicate that the tools are now in place for precise characterization of UTI pain circuits.

EMERGING CONCEPT: BACTERIAL-PAIN PHENOTYPES

Studies reviewed here suggest bacteria exhibit distinct pain phenotypes in the bladder (102). Among *E. coli*, bladder colonization may be associated with an acute pain response, a chronic pain response, or a null response, modulated at least in part by O-antigen (Fig. 4). Moreover, *E. coli* strains obtained from patients with ASB exhibit analgesic activity. There may also be pain phenotypes specific to particular phases of UTI. For example, we observed no pelvic allodynia in mice infected with the pyelonephritis strain CFT073 (C. N. Rudick and D. J. Klumpp, unpublished observations), and pyelonephritis patients are presumably asymptomatic as UPEC ascend through the lower urinary tract. However, the potential to induce acute pain in the urinary tract is not unique to UPEC, but is shared with *P. mirabilis*, *E. faecalis*, and *K. pneumoniae*. Similarly, we recently evaluated *milleri* group *Streptococci* for pain responses in the bladder (C. N. Rudick and D. J. Klumpp, in preparation). Like *E. coli*, *milleri* strains of *S. anginosus*, *S. intermedius*, and *S. constellatus* exhibited chronic-, acute-, and null-pain phenotypes, respectively. And in another parallel with *E. coli*, *S. constellatus* exhibited analgesic activity against *S. intermedius*-induced pain. Given the diversity of species and genera exhibiting distinct pain phenotypes in the bladder, we speculate that pain phenotypes extend beyond the bladder to other niches. Indeed, it is conceivable that bacteria exhibit distinct phenotypes for modulation of diverse neurally regulated processes ranging from nonpainful bladder-voiding function to intestinal motility to airway function.

FUTURE DIRECTIONS AND IMPLICATIONS

The study of infection pain is in its infancy, although UTI studies lead this clinically relevant aspect of microbial pathogenesis.

Much of our current understanding of inflammatory pain (Fig. 2) is derived from reductionist pain models that lack the complexity of bacterial infections, yet infection pain is often assumed to be inflammatory. TLR4 appears to mediate UTI pain-independent hematopoietic lineages and independent of inflammation at the levels of gross pathology. However, since cytokines modulate peripheral sensitivity, it remains to be seen if bacteria with distinct pain phenotypes mediate these differential effects by subtle modulation of specific bladder cytokines, despite a lack of correlation between gross inflammatory markers and pain phenotypes. It will also be important to dissect how specific O-antigen structures modulate bacterial-pain phenotypes at the level of LPS-TLR4 interactions. Is such modulation unique to pain, or are other TLR4-dependent processes subject to such profound modulation by LPS structure?

TLR4 signaling is required for UTI pain, and urothelium is an apparent transducer of UTI pain. It remains unclear how urothelial signals are conveyed to bladder sensory afferents. Sacral sensory neurons are responsive to UPEC LPS, but TLR4 staining has proven challenging to detect on bladder-sensory fibers (R. E. Yaggie and D. J. Klumpp, unpublished observations). Thus, it remains unclear whether bladder-sensory fibers play a direct role in pain responses to bacteria that might complement the role of urothelium. It will be interesting to determine whether known urothelial signals play a role in pain, such as ATP that mediates responses to bladder filling through purinergic receptors, alone or in concert with modulation by urothelial cytokines.

A surprising aspect of these early studies in UTI pain is that a transient infection can result in chronic allodynia durable long after bacterial clearance. Mechanistically, it is unclear whether a transient infection causes rapid central sensitization that has been previously associated only with persistent noxious stimuli. Are microbial pain stimuli inherently different from reductionist models? Or do certain pathogens cause extended release of pain mediators locally? The finding of compromised inhibitory control in spinal cords of mice with chronic allodynia following a single, transient UTI suggests that bacteria can trigger central sensitization in the absence persistence. This possibility has important medical implications for urology and medicine generally, for it suggests that some chronic-pain conditions can be initiated by a transient microbial infection. It will be important to determine whether strains exist among clinical isolates that recapitulate the chronic-pain phenotypes of O-antigen mutants. Indeed, we have identified one such UPEC strain (C. N. Rudick and D. J. Klumpp, in preparation) and future studies will define the relative roles of pathogens and host susceptibility in pain responses to UTI and whether a clinical UTI can precipitate chronic pelvic pain. Finally, the observation that some *E. coli* are analgesic raises important mechanistic questions and has significant clinical implications (Fig. 5). For example, do analgesic bacteria mediate their effects against pain at the level of TLRs or other elements in the sensory system? In addition, ASB *E. coli* analgesic activity offers a novel and obvious therapeutic opportunity for the treatment of chronic pelvic pain. Together, these studies identify pain responses as a new area in UTI pathogenesis rich in mechanistic questions and clinical opportunities.

CITATION

Birder LA , Klumpp DJ. 2016. Host responses to urinary tract infections and emerging therapeutics: sensation and pain within the urinary tract. Microbiol Spectrum 4(5):UTI-0023-2016.

REFERENCES

1. **Birder L, Andersson KE.** 2013. Urothelial signaling. *Physiol Rev* **93:**653–680.

2. Khandelwal P, Abraham SN, Apodaca G. 2009. Cell biology and physiology of the uroepithelium. *Am J Physiol Renal Physiol* **297:** F1477–1501.

3. Sun TT, Liang FX, Wu XR. 1999. Uroplakins as markers of urothelial differentiation. *Adv Exp Med Biol* **462:**7–18.

4. Wu XR, Kong XP, Pellicer A, Kreibich G, Sun TT. 2009. Uroplakins in urothelial biology, function, and disease. *Kidney Int* **75:**1153–1165.

5. Acharya P, Beckel J, Ruiz WG, Wang E, Rojas R, Birder L, Apodaca G. 2004. Distribution of the tight junction proteins ZO-1, occludin, and claudin-4, -8, and -12 in bladder epithelium. *Am J Physiol Renal Physiol* **287:** F305–318.

6. Carattino MD, Prakasam HS, Ruiz WG, Clayton DR, McGuire M, Gallo LI, Apodaca G. 2013. Bladder filling and voiding affect umbrella cell tight junction organization and function. *Am J Physiol Renal Physiol* **305:** F1158–1168.

7. Sun TT. 2006. Altered phenotype of cultured urothelial and other stratified epithelial cells: implications for wound healing. *Am J Physiol Renal Physiol* **291:**F9–21.

8. Fry CH, Young JS, Jabr RI, McCarthy C, Ikeda Y, Kanai AJ. 2012. Modulation of spontaneous activity in the overactive bladder: the role of P2Y agonists. *Am J Physiol Renal Physiol* **302:**F1447–1454.

9. Truschel ST, Ruiz WG, Shulman T, Pilewski J, Sun TT, Zeidel ML, Apodaca G. 1999. Primary uroepithelial cultures. A model system to analyze umbrella cell barrier function. *J Biol Chem* **274:**15020–15029.

10. Andersson KE. 2002. Bladder activation: afferent mechanisms. *Urology* **59**(5 Suppl 1):43–50.

11. Heppner TJ, Layne JJ, Pearson JM, Sarkissian H, Nelson MT. 2011. Unique properties of muscularis mucosae smooth muscle in guinea pig urinary bladder. *Am J Physiol Regul Integr Comp Physiol* **301:**R351–362.

12. Koh BH, Roy R, Hollywood MA, Thornbury KD, McHale NG, Sergeant GP, Hatton WJ, Ward SM, Sanders KM, Koh SD. 2012. Platelet-derived growth factor receptor-α cells in mouse urinary bladder: a new class of interstitial cells. *J Cell Mol Med* **16:**691–700.

13. McCloskey KD. 2010. Interstitial cells in the urinary bladder–localization and function. *Neurourol Urodyn* **29:**82–87.

14. Johnston L, Woolsey S, Cunningham RM, O'Kane H, Duggan B, Keane P, McCloskey KD. 2010. Morphological expression of KIT positive interstitial cells of Cajal in human bladder. *J Urol* **184:**370–377.

15. Cheng S, Scigalla FP, Speroni di Fenizio P, Zhang ZG, Stolzenburg JU, Neuhaus J. 2011. ATP enhances spontaneous calcium activity in cultured suburothelial myofibroblasts of the human bladder. *PLoS One* **6:**e25769. doi:10.1371/journal.pone.0025769

16. Nile CJ, de Vente J, Gillespie JI. 2010. Stretch independent regulation of prostaglandin E(2) production within the isolated guinea-pig lamina propria. *BJU Int* **105:**540–548.

17. Fowler CJ, Griffiths D, de Groat WC. 2008. The neural control of micturition. *Nat Rev Neurosci* **9:**453–466.

18. Kanai A, Andersson KE. 2010. Bladder afferent signaling: recent findings. *J Urol* **183:**1288–1295.

19. Zagorodnyuk VP, Brookes SJ, Spencer NJ. 2010. Structure-function relationship of sensory endings in the gut and bladder. *Auton Neurosci* **153:**3–11.

20. Zagorodnyuk VP, Gibbins IL, Costa M, Brookes SJ, Gregory SJ. 2007. Properties of the major classes of mechanoreceptors in the guinea pig bladder. *J Physiol* **585:**147–163.

21. Birder L, Wyndaele JJ. 2013. From urothelial signalling to experiencing a sensation related to the urinary bladder. *Acta Physiol (Oxf)* **207:**34–39.

22. De Wachter S, Wyndaele JJ. 2008. How sudden is a compelling desire to void? An observational cystometric study on the suddenness of this sensation. *BJU Int* **101:**1000–1003.

23. Xu L, Gebhart GF. 2008. Characterization of mouse lumbar splanchnic and pelvic nerve urinary bladder mechanosensory afferents. *J Neurophysiol* **99:**244–253.

24. Snellings AE, Yoo PB, Grill WM. 2012. Urethral flow-responsive afferents in the cat sacral dorsal root ganglia. *Neurosci Lett* **516:** 34–38.

25. Cook SP, McCleskey EW. 2000. ATP, pain and a full bladder. *Nature* **407:**951–952.

26. Cockayne DA, Hamilton SG, Zhu QM, Dunn PM, Zhong Y, Novakovic S, Malmberg AB, Cain G, Berson A, Kassotakis L, Hedley L, Lachnit WG, Burnstock G, McMahon SB, Ford AP. 2000. Urinary bladder hyporeflexia and reduced pain-related behaviour in P2X3-deficient mice. *Nature* **407:**1011–1015.

27. Chai TC, Keay S. 2004. New theories in interstitial cystitis. *Nat Clin Pract Urol* **1:**85–89.

28. de Groat WC, Yoshimura N. 2009. Afferent nerve regulation of bladder function in health and disease. *Handb Exp Pharmacol* **4:**91–138.

29. Liu HT, Tyagi P, Chancellor MB, Kuo HC. 2010. Urinary nerve growth factor but not prostaglandin E2 increases in patients with

interstitial cystitis/bladder pain syndrome and detrusor overactivity. *BJU Int* **106**:1681–1685.

30. **Ochodnicky P, Cruz CD, Yoshimura N, Cruz F.** 2012. Neurotrophins as regulators of urinary bladder function. *Nat Rev Urol* **9**:628–637.

31. **Bradesi S.** 2010. Role of spinal cord glia in the central processing of peripheral pain perception. *Neurogastroenterol Motil* **22**:499–511.

32. **Gosselin RD, Suter MR, Ji RR, Decosterd I.** 2010. Glial cells and chronic pain. *Neuroscientist* **16**:519–531.

33. **Grist M, Chakraborty J.** 1994. Identification of a mucin layer in the urinary bladder. *Urology* **44**:26–33.

34. **Prakasam HS, Gallo LI, Li H, Ruiz WG, Hallows KR, Apodaca G.** 2014. A1 adenosine receptor-stimulated exocytosis in bladder umbrella cells requires phosphorylation of ADAM17 Ser-811 and EGF receptor transactivation. *Mol Biol Cell* **25**:3798–3812.

35. **Bishop BL, Duncan MJ, Song J, Li G, Zaas D, Abraham SN.** 2007. Cyclic AMP-regulated exocytosis of *Escherichia coli* from infected bladder epithelial cells. *Nat Med* **13**:625–630.

36. **Romih R, Korosec P, de Mello W Jr, Jezernik K.** 2005. Differentiation of epithelial cells in the urinary tract. *Cell Tissue Res* **320**:259–268.

37. **Hicks RM, Ketterer B, Warren RC.** 1974. The ultrastructure and chemistry of the luminal plasma membrane of the mammalian urinary bladder: a structure with low permeability to water and ions. *Philos Trans R Soc Lond B Biol Sci* **268**:23–38.

38. **Martin H.** 1972. [Comparative histological and clinical studies on the glomerula of plasmacytoma kidneys using the semi-thin section technic]. *Folia Haematol Int Mag Klin Morphol Blutforsch* **98**:195–206.

39. **Gandhi D, Molotkov A, Batourina E, Schneider K, Dan H, Reiley M, Laufer E, Metzger D, Liang F, Liao Y, Sun TT, Aronow B, Rosen R, Mauney J, Adam R, Rosselot C, Van Batavia J, McMahon A, McMahon J, Guo JJ, Mendelsohn C.** 2013. Retinoid signaling in progenitors controls specification and regeneration of the urothelium. *Dev Cell* **26**:469–482.

40. **Varley CL, Stahlschmidt J, Lee WC, Holder J, Diggle C, Selby PJ, Trejdosiewicz LK, Southgate J.** 2004. Role of PPARgamma and EGFR signalling in the urothelial terminal differentiation programme. *J Cell Sci* **117**:2029–2036.

41. **Kreft ME, Romih R, Kreft M, Jezernik K.** 2009. Endocytotic activity of bladder superficial urothelial cells is inversely related to their differentiation stage. *Differentiation* **77**:48–59.

42. **Kreft ME, Jezernik K, Kreft M, Romih R.** 2009. Apical plasma membrane traffic in

superficial cells of bladder urothelium. *Ann N Y Acad Sci* **1152**:18–29.

43. **de Boer WI, Vermeij M, Diez de Medina SG, Bindels E, Radvanyi F, van der Kwast T, Chopin D.** 1996. Functions of fibroblast and transforming growth factors in primary organoid-like cultures of normal human urothelium. *Lab Invest* **75**:147–156.

44. **Bassuk JA, Cockrane K, Mitchell ME.** 2003. Induction of urothelial cell proliferation by fibroblast growth factor-7 in RAG1-deficient mice. *Adv Exp Med Biol* **539**:623–633.

45. **Varley CL, Stahlschmidt J, Smith B, Stower M, Southgate J.** 2004. Activation of peroxisome proliferator-activated receptor-gamma reverses squamous metaplasia and induces transitional differentiation in normal human urothelial cells. *Am J Pathol* **164**:1789–1798.

46. **Shin K, Lee J, Guo N, Kim J, Lim A, Qu L, Mysorekar IU, Beachy PA.** 2011. Hedgehog/Wnt feedback supports regenerative proliferation of epithelial stem cells in bladder. *Nature* **472**:110–114.

47. **Robinson D, Cardozo L.** 2011. Estrogens and the lower urinary tract. *Neurourol Urodyn* **30**:754–757.

48. **Apodaca G, Kiss S, Ruiz W, Meyers S, Zeidel M, Birder L.** 2003. Disruption of bladder epithelium barrier function after spinal cord injury. *Am J Physiol Renal Physiol* **284**:F966–976.

49. **Parsons CL, Lilly JD, Stein P.** 1991. Epithelial dysfunction in nonbacterial cystitis (interstitial cystitis). *J Urol* **145**:732–735.

50. **Georgopoulos NT, Kirkwood LA, Walker DC, Southgate J.** 2010. Differential regulation of growth-promoting signalling pathways by E-cadherin. *PLoS One* **5**:e13621. doi:10.1371/journal.pone.0013621

51. **Aberg KM, Radek KA, Choi EH, Kim DK, Demerjian M, Hupe M, Kerbleski J, Gallo RL, Ganz T, Mauro T, Feingold KR, Elias PM.** 2007. Psychological stress downregulates epidermal antimicrobial peptide expression and increases severity of cutaneous infections in mice. *J Clin Invest* **117**:3339–3349.

52. **Rostand KS, Esko JD.** 1997. Microbial adherence to and invasion through proteoglycans. *Infect Immun* **65**:1–8.

53. **Schilling JD, Hultgren SJ, Lorenz RG.** 2002. Recent advances in the molecular basis of pathogen recognition and host responses in the urinary tract. *Int Rev Immunol* **21**:291–304.

54. **Thumbikat P, Berry RE, Schaeffer AJ, Klumpp DJ.** 2009. Differentiation-induced uroplakin III expression promotes urothelial cell death in response to uropathogenic *E. coli. Microbes Infect* **11**:57–65.

55. **Thumbikat P, Berry RE, Zhou G, Billips BK, Yaggie RE, Zaichuk T, Sun TT, Schaeffer AJ, Klumpp DJ.** 2009. Bacteria-induced uroplakin signaling mediates bladder response to infection. *PLoS Pathog* **5:**e1000415. doi:10.1371/journal.ppat.1000415

56. **Wood MW, Breitschwerdt EB, Nordone SK, Linder KE, Gookin JL.** 2012. Uropathogenic *E. coli* promote a paracellular urothelial barrier defect characterized by altered tight junction integrity, epithelial cell sloughing and cytokine release. *J Comp Pathol* **147:**11–19.

57. **Brady CM, Apostolidis A, Yiangou Y, Baecker PA, Ford AP, Freeman A, Jacques TS, Fowler CJ, Anand P.** 2004. P2X3-immunoreactive nerve fibres in neurogenic detrusor overactivity and the effect of intravesical resiniferatoxin. *Eur Urol* **46:**247–253.

58. **Ikeda Y, Fry C, Hayashi F, Stolz D, Griffiths D, Kanai A.** 2007. Role of gap junctions in spontaneous activity of the rat bladder. *Am J Physiol Renal Physiol* **293:**F1018–1025.

59. **Bossé Y, Paré PD, Seow CY.** 2008. Airway wall remodeling in asthma: from the epithelial layer to the adventitia. *Curr Allergy Asthma Rep* **8:**357–366.

60. **Burnstock G.** 2001. Purine-mediated signalling in pain and visceral perception. *Trends Pharmacol Sci* **22:**182–188.

61. **Du S, Araki I, Mikami Y, Zakoji H, Beppu M, Yoshiyama M, Takeda M.** 2007. Amiloride-sensitive ion channels in urinary bladder epithelium involved in mechanosensory transduction by modulating stretch-evoked adenosine triphosphate release. *Urology* **69:**590–595.

62. **Hawthorn MH, Chapple CR, Cock M, Chess-Williams R.** 2000. Urothelium-derived inhibitory factor(s) influences on detrusor muscle contractility *in vitro. Br J Pharmacol* **129:**416–419.

63. **Hendrix S.** 2008. Neuroimmune communication in skin: far from peripheral. *J Invest Dermatol* **128:**260–261.

64. **Igawa Y, Aizawa N, Homma Y.** 2010. Beta3-adrenoceptor agonists: possible role in the treatment of overactive bladder. *Korean J Urol* **51:**811–818.

65. **Ikeda Y, Kanai A.** 2008. Urotheliogenic modulation of intrinsic activity in spinal cord-transected rat bladders: role of mucosal muscarinic receptors. *Am J Physiol Renal Physiol* **295:**F454–461.

66. **Knight GE, Bodin P, De Groat WC, Burnstock G.** 2002. ATP is released from guinea pig ureter epithelium on distension. *Am J Physiol Renal Physiol* **282:**F281–288.

67. **Lowe EM, Anand P, Terenghi G, Williams-Chestnut RE, Sinicropi DV, Osborne JL.** 1997. Increased nerve growth factor levels in the urinary bladder of women with idiopathic sensory urgency and interstitial cystitis. *Br J Urol* **79:**572–577.

68. **Pandita RK, Mizusawa H, Andersson KE.** 2000. Intravesical oxyhemoglobin initiates bladder overactivity in conscious, normal rats. *J Urol* **164:**545–550.

69. **Yamaguchi O, Chapple CR.** 2007. Beta3-adrenoceptors in urinary bladder. *Neurourol Urodyn* **26**(6 Suppl)**:**752–756.

70. **Hille B.** 1992. *Ionic Channels of Excitable Membranes,* 2nd ed. Sinauer Associates, Sunderland, MA.

71. **Parsons CL, Stein PC, Bidair M, Lebow D.** 1994. Abnormal sensitivity to intravesical potassium in interstitial cystitis and radiation cystitis. *Neurourol Urodyn* **13:**515–520.

72. **Parsons CL, Greenberger M, Gabal L, Bidair M, Barme G.** 1998. The role of urinary potassium in the pathogenesis and diagnosis of interstitial cystitis. *J Urol* **159:**1862–1866; discussion 1866–1867.

73. **Elliott TS, Reed L, Slack RC, Bishop MC.** 1985. Bacteriology and ultrastructure of the bladder in patients with urinary tract infections. *J Infect* **11:**191–199.

74. **Elliott TS, Slack RC, Bishop MC.** 1985. Bladder changes associated with urinary tract infections. *Lancet* **1:**1509.

75. **Fukushi Y, Orikasa S, Kagayama M.** 1979. An electron microscopic study of the interaction between vesical epithelium and *E. coli. Invest Urol* **17:**61–68.

76. **McTaggart LA, Rigby RC, Elliott TS.** 1990. The pathogenesis of urinary tract infections associated with *Escherichia coli, Staphylococcus saprophyticus* and *S. epidermidis. J Med Microbiol* **32:**135–141.

77. **Orikasa S, Hinman F Jr.** 1977. Reaction of the vesical wall to bacterial penetration: resistance to attachment, desquamation, and leukocytic activity. *Invest Urol* **15:**185–193.

78. **Mulvey MA, Lopez-Boado YS, Wilson CL, Roth R, Parks WC, Heuser J, Hultgren SJ.** 1998. Induction and evasion of host defenses by type 1-piliated uropathogenic *Escherichia coli. Science* **282:**1494–1497.

79. **Klumpp DJ, Weiser AC, Sengupta S, Forrestal SG, Batler RA, Schaeffer AJ.** 2001. Uropathogenic *Escherichia coli* potentiates type 1 pilus-induced apoptosis by suppressing NF-kappaB. *Infect Immun* **69:**6689–6695.

80. **Mudge CS, Klumpp DJ.** 2005. Induction of the urothelial differentiation program in the absence of stromal cues. *J Urol* **174:**380–385.

81. **Billips BK, Forrestal SG, Rycyk MT, Johnson JR, Klumpp DJ, Schaeffer AJ.** 2007. Modu-

lation of host innate immune response in the bladder by uropathogenic *Escherichia coli*. *Infect Immun* **75**:5353–5360.

82. **Billips BK, Schaeffer AJ, Klumpp DJ.** 2008. Molecular basis of uropathogenic *Escherichia coli* evasion of the innate immune response in the bladder. *Infect Immun* **76**:3891–3900.

83. **Hunstad DA, Justice SS, Hung CS, Lauer SR, Hultgren SJ.** 2005. Suppression of bladder epithelial cytokine responses by uropathogenic *Escherichia coli*. *Infect Immun* **73**:3999–4006.

84. **Orth K, Palmer LE, Bao ZQ, Stewart S, Rudolph AE, Bliska JB, Dixon JE.** 1999. Inhibition of the mitogen-activated protein kinase kinase superfamily by a *Yersinia* effector. *Science* **285**:1920–1923.

85. **Mulvey MA, Hultgren SJ.** 2000. Cell biology. Bacterial spelunkers. *Science* **289**:732–733.

86. **Klumpp DJ, Rycyk MT, Chen MC, Thumbikat P, Sengupta S, Schaeffer AJ.** 2006. Uropathogenic *Escherichia coli* induces extrinsic and intrinsic cascades to initiate urothelial apoptosis. *Infect Immun* **74**:5106–5113.

87. **Wu XR, Sun TT, Medina JJ.** 1996. *In vitro* binding of type 1 fimbriated *Esherichia coli* to uroplakins Ia and Ib: relation to urinary tract infections. *Proc Natl Acad Sci U S A* **93**:9630–9635.

88. **Liang FX, Riedel I, Deng FM, Zhou G, Xu C, Wu XR, Kong XP, Moll R, Sun TT.** 2001. Organization of uroplakin subunits: transmembrane topology, pair formation and plaque composition. *Biochem J* **355**:13–18.

89. **Thumbikat P, Berry RE, Schaeffer AJ, Klumpp DJ.** 2008. Differentiation-induced uroplakin III expression promotes urothelial cell death in response to uropathogenic *E. coli*. *Microbes Infect* **11**:57–65.

90. **Scholz J, Woolf CJ.** 2007. The neuropathic pain triad: neurons, immune cells and glia. *Nat Neurosci* **10**:1361–1368.

91. **Agace WW, Hedges SR, Ceska M, Svanborg C.** 1993. Interleukin-8 and the neutrophil response to mucosal gram-negative infection. *J Clin Invest* **92**:780–785.

92. **Schaeffer AJ, Matulewicz RS, Klumpp DJ.** 2016. Infections of the urinary tract, p 237–303. *In* Wein AJ, Kavoussi LR, Partin AW, Peters CA (ed), *Campbell-Walsh Urology*, 11th ed, **vol 1**. Elsevier, Philadelphia, PA.

93. **Laird JM, Souslova V, Wood JN, Cervero F.** 2002. Deficits in visceral pain and referred hyperalgesia in Nav1.8 (SNS/PN3)-null mice. *J Neurosci* **22**:8352–8356.

94. **Rudick CN, Billips BK, Pavlov VI, Yaggie RE, Schaeffer AJ, Klumpp DJ.** 2010. Host-pathogen interactions mediating pain of urinary tract infection. *J Infect Dis* **201**:1240–1249.

95. **Bjorling DE, Wang ZY, Boldon K, Bushman W.** 2008. Bacterial cystitis is accompanied by increased peripheral thermal sensitivity in mice. *J Urol* **179**:759–763.

96. **Rudick CN, Jiang M, Yaggie RE, Pavlov VI, Done J, Heckman CJ, Whitfield C, Schaeffer AJ, Klumpp DJ.** 2012. O-antigen modulates infection-induced pain states. *PLoS One* **7**:e41273. doi:10.1371/journal.pone.0041273

97. **Batchelor RA, Haraguchi GE, Hull RA, Hull SI.** 1991. Regulation by a novel protein of the bimodal distribution of lipopolysaccharide in the outer membrane of *Escherichia coli*. *J Bacteriol* **173**:5699–5704.

98. **Kos V, Cuthbertson L, Whitfield C.** 2009. The *Klebsiella pneumoniae* O2a antigen defines a second mechanism for O antigen ATP-binding cassette transporters. *J Biol Chem* **284**:2947–2956.

99. **Neuhard J, Thomassen E.** 1976. Altered deoxyribonucleotide pools in P2 eductants of *Escherichia coli* K-12 due to deletion of the dcd gene. *J Bacteriol* **126**:999–1001.

100. **Stemler KM, Crock LW, Lai HH, Mills JC, Gereau RW IV, Mysorekar IU.** 2013. Protamine sulfate induced bladder injury protects from distention induced bladder pain. *J Urol* **189**:343–351.

101. **Zucker RS, Regehr WG.** 2002. Short-term synaptic plasticity. *Annu Rev Physiol* **64**:355–405.

102. **Rosen JM, Klumpp DJ.** 2014. Mechanisms of pain from urinary tract infection. *Int J Urol* **21** (Suppl 1):26–32.

103. **Rudick CN, Taylor AK, Yaggie RE, Schaeffer AJ, Klumpp DJ.** 2014. Asymptomatic bacteriuria *Escherichia coli* are live biotherapeutics for UTI. *PLoS One* **9**:e109321. doi:10.1371/journal.pone.0109321

104. **Sasaki M.** 2004. Feed-forward and feedback regulation of bladdercontractility by Barrington's nucleus in cats. *J Physiol* **557**:287–305.

105. **Lai H, Gereau RW IV, Luo Y, O'Donnell M, Rudick CN, Pontari M, Mullins C, Klumpp DJ.** 2015. Animal models of urologic chronic pelvic pain syndromes: findings from the Multidisciplinary Approach to the Study of Chronic Pelvic Pain Research Network. *Urology* **85**:1454–1465.

106. **Rudick CN, Bryce PJ, Guichelaar LA, Berry RE, Klumpp DJ.** 2008. Mast cell-derived histamine mediates cystitis pain. *PLoS One* **3**:e2096. doi:10.1371/journal.pone.0002096

107. **Rudick CN, Chen MC, Mongiu AK, Klumpp DJ.** 2007. Organ cross talk modulates pelvic pain. *Am J Physiol Regul Integr Comp Physiol* **293**:R1191–1198.

108. **Rudick CN, Pavlov VI, Chen MC, Klumpp DJ.** 2012. Gender specific pelvic pain severity in neurogenic cystitis. *J Urol* **187**:715–724.

109. **Yang W, Rudick CN, Hoxha E, Allsop SA, Dimitrakoff JD, Klumpp DJ.** 2012. Ca2+/calmodulin-dependent protein kinase II is associated with pelvic pain of neurogenic cystitis. *Am J Physiol Renal Physiol* **303:**F350–356.

110. **Jasmin L, Janni G, Manz HJ, Rabkin SD.** 1998. Activation of CNS circuits producing a neurogenic cystitis: evidence for centrally induced peripheral inflammation. *J Neurosci* **18:**10016–10029.

111. **Arms L, Vizzard MA.** 2011. Neuropeptides in lower urinary tract function. *Handb Exp Pharmacol* **202:**395–423.

112. **Lai HH, Qiu CS, Crock LW, Morales ME, Ness TJ, Gereau RW IV.** 2011. Activation of spinal extracellular signal-regulated kinases (ERK) 1/2 is associated with the development of visceral hyperalgesia of the bladder. *Pain* **152:**2117–2124.

113. **Crock LW, Stemler KM, Song DG, Abbosh P, Vogt SK, Qiu CS, Lai HH, Mysorekar IU, Gereau RW IV.** 2012. Metabotropic glutamate receptor 5 (mGluR5) regulates bladder nociception. *Mol Pain* **8:**20.

114. **Han JS, Neugebauer V.** 2004. Synaptic plasticity in the amygdala in a visceral pain model in rats. *Neurosci Lett* **361:**254–257.

115. **Crock LW, Kolber BJ, Morgan CD, Sadler KE, Vogt SK, Bruchas MR, Gereau RW IV.** 2012. Central amygdala metabotropic glutamate receptor 5 in the modulation of visceral pain. *J Neurosci* **32:**14217–14226.

Drug and Vaccine Development for the Treatment and Prevention of Urinary Tract Infections

24

VALERIE P. O'BRIEN,[1] THOMAS J. HANNAN,[2]
HAILYN V. NIELSEN,[1] and SCOTT J. HULTGREN[1]

THE URGENT NEED FOR NEW THERAPIES AND VACCINES

Urinary tract infections (UTI) are one of the most common bacterial infections, with roughly eleven-million cases reported in the U.S. each year that cost an estimated $5 billion annually (1, 2). More than one in every two women will experience at least one UTI in her lifetime, and nearly one in three women will have received antibiotic treatment for a UTI before age 24 (3, 4). The clinical manifestations of symptomatic UTI include infection-induced inflammation of the urethra (urethritis), urinary bladder (cystitis), and kidneys (pyelonephritis) and are diagnosed by the presence of high levels of bacteria in the urine (bacteriuria) with concomitant symptoms. Symptoms of cystitis include frequent urination, burning sensation and pain during urination (dysuria), suprapubic pain and/or lower abdominal discomfort, and cloudy and/or bloody, foul-smelling urine. Symptoms of pyelonephritis include the presence of bacteriuria and pyuria (white blood cells in the urine) that is accompanied by flank pain and fever, but may or may not include other symptoms of cystitis. The vast majority of UTI manifest as cystitis and

[1]Department of Molecular Microbiology, Center for Women's Infectious Disease Research; [2]Department of Pathology & Immunology, Washington University Medical School, St. Louis, MO 63110.
Urinary Tract Infections: Molecular Pathogenesis and Clinical Management, 2nd Edition
Edited by Matthew A. Mulvey, David J. Klumpp, and Ann E. Stapleton
© 2017 American Society for Microbiology, Washington, DC
doi:10.1128/microbiolspec.UTI-0013-2012

urethritis, affecting primarily the lower urinary tract, but this can potentially lead to bacterial ascension to the kidneys and pyelonephritis, particularly in pregnant women, diabetics, and children with vesicoureteral reflux (VUR) (5, 6). As a result, renal scarring and loss of function is a potentially serious complication of any UTI, particularly in infants, where diagnosis of UTI may be delayed.

UTI are not only common, but also highly recurrent. In particular, sexually active women, the elderly, and pre-pubertal children are highly susceptible to chronically recurrent UTI, resulting in increased use of antibiotics and negatively affecting quality of life (3). Approximately 20% to 30% of adult women with an initial UTI will experience a recurrence within 3–4 months (7). In children, about one in three experiencing a UTI before the age of one will experience a recurrence within three years, and 18% will have a recurrence within a few months (8). Uncomplicated UTI, which are infections that are not associated with urethral instrumentation or abnormal anatomy or physiology of the urinary tract, predominantly affect women, young children, and the elderly. Risk factors for uncomplicated cystitis in adult women include environmental factors such as frequent sexual activity, exposure to spermicides, menopause, and a history of childhood UTI, as well as genetic factors such as Toll-like receptor polymorphisms and a maternal history of UTI (9, 10). In contrast, patients at risk for what is termed "complicated" UTI include patients with spinal cord injuries, patients undergoing urethral catheterization, diabetics, and individuals with underlying urologic abnormalities such as vesicoureteral reflux (VUR) (3). Uropathogenic *Escherichia coli* (UPEC) cause 85% or more of uncomplicated UTI cases, while other Gram-negative rods and Gram-positive cocci, such as *Staphylococcus saprophyticus* and enterococci, are responsible for the remaining 5% to 15% of cases (11).

The epidemiology of UTI changes significantly in the health care environment. Urethral catheterization is strongly associated with UTI, and the risk of infection increases with the length of catheterization (12). Catheter-associated UTI (CAUTI) account for 30% to 40% of health care-associated infections in the United States, making them the most common nosocomial infection, with more than one-million cases occurring yearly in hospitals and nursing homes (13). Although enterococci contribute only minimally to the burden of uncomplicated UTI, data from a national surveillance network of 463 hospitals in the United States revealed that 15% of CAUTI are caused by enterococci, second among bacterial isolates only to *E. coli* (21%) (14). Furthermore, CAUTI affect both sexes, as long-term urinary catheterization of both men and women almost invariably leads to detection of bacteria in the urine (bacteriuria) and carries a daily risk of 3% to 7% for the development of symptomatic CAUTI (15). While CAUTI is most often asymptomatic, the high incidence in catheterized patients greatly increases their risk for relatively rare but serious sequelae such as bacteremia, urosepsis, and death (16). Furthermore, CAUTI serve as reservoirs for the dissemination of antimicrobial-resistant nosocomial pathogens in the health care environment (17).

Although antibiotic therapy has historically been very successful in combating both uncomplicated and complicated UTI, many individuals suffer from chronically recurrent cystitis, requiring long-term antibiotic prophylaxis (18). Furthermore, the widespread use of antibiotics has led to accelerating antibiotic resistance and the emergence and spread of multidrug-resistant (MDR) uropathogens (19). As early as 1957, Weyrauch and colleagues foresaw this problem in their discussion of the results of a UTI vaccine trial (20). In their study, they found that intramuscular injection with heat-killed *E. coli* was protective or partially protective against pyelonephritis in 12 of 16 rabbits. However, unvaccinated rabbits treated with prophylactic tetracycline were completely resistant to pyelonephritis, leading the authors to predict that prophylactic antibiotic treatment would remain the best strategy for preventing UTI in humans. Despite the authors' admonition that "every effort must be made to avoid indiscriminate use" of antibiotic prophylaxis in order to prevent resistance, drug-resistant UTI has exploded into a major public health concern. For instance, a recent five-year nested case-control study of drug resistance in uncomplicated febrile UTI in adults found that 12% of patients with UPEC UTI had fluoroquinolone-resistant urine cultures; fluoroquinolone use in the previous six months was a significant independent risk factor for being afflicted with a fluoroquinolone-resistant UTI (21). In another study, more than 9,000 patient urine samples were analyzed for drug resistance; of the samples containing uropathogens, 22.1% were multi-drug resistant (resistant to third-generation cephalosporins, ciprofloxacin, and aminoglycosides) (22). In the past decade, the *E. coli* clone O25:H4-ST131 (*E. coli* ST131) emerged globally as an important MDR UPEC strain (23). Unlike other antibiotic-resistant UPEC strains, ST131 is highly virulent in the urinary tract and is not only found in health care settings, but is also isolated from the community (19, 24).

In summary, urinary tract infection is a significant cause of morbidity in women throughout their lifespan, in infant boys, and in older men. Serious sequelae include frequent recurrences, pyelonephritis with sepsis, renal damage in young children, pre-term birth, and complications of frequent antimicrobial use including high-level antibiotic resistance and *Clostridium difficile* colitis (25, 26). High recurrence rates and increasing antimicrobial resistance among uropathogens threaten to greatly increase the economic burden of this common infection. It has become increasingly evident that prophylactic use of antibiotics to prevent UTI is not a sustainable solution. The high incidence and recurrence rate of UTI, along with the rapid rise of MDR uropathogens and CAUTI, necessitate new drugs and vaccine therapies

for the prevention of these infections. In this chapter, we will review UTI pathogenesis, focusing on UPEC as a model organism for uncomplicated UTI and *Enterococcus faecalis* as a model organism for complicated, CAUTI. We will then describe the development of anti-virulence therapies, including new classes of small-molecule inhibitors that target uropathogenic virulence factors. Finally, we will discuss vaccines for the prevention of recurrent UTI, including both whole cell and specific-antigen vaccines. It is our hope that this chapter will draw attention to recent advances in the field of UTI therapeutics while highlighting specific topics that require further study.

RECENT DISCOVERIES IN UTI PATHOGENESIS

Since 90% of symptomatic UTI present as simple cystitis/urethritis, an ideal drug or vaccine target would be one that is critical for both establishing and maintaining bladder colonization, thus preventing UTI altogether. Quantum-leap advances in molecular-biological and imaging technologies in the past 15 years, along with the maturation of genomic science, have led to an unprecedented expansion of our understanding of UTI pathogenesis, and to the identification of previously unknown virulence mechanisms. For example, we now know that UPEC invade bladder epithelial cells and have the capacity to rapidly replicate within the cytoplasm of superficial facet cells of the bladder urothelium, producing between 10,000 and 100,000 daughter cells from a single invasive bacterium within 12–16 hours (27–30). The establishment of this protected intracellular niche, known as the intracellular-bacterial community or IBC, helps UPEC gain a foothold in the lower urinary tract. Although discovered in mice, exfoliated bladder epithelial cells containing IBCs have been observed in urine sediments obtained from women and children with recurrent UTI, but not in healthy controls or in cases of UTI caused by Gram-positive pathogens (31–33), indicating that the murine model is a relevant and powerful tool for studying UTI pathogenesis. Indeed, mice are naturally susceptible to UPEC UTI and recapitulate many of the known characteristics of UTI in humans (34, 35).

While the translation of findings from animal models to the clinic is always fraught with difficulties, a major challenge in developing new therapeutics is not just the species differences between mice and humans in our animal models, but also the fact that these experimental infections are typically performed in naive animals. Experimental models of UTI in naive mice and primates have revealed similar pathogenic mechanisms of both cystitis and pyelonephritis (36–41). In contrast, recently developed models of recurrent UTI and post-menopausal UTI in mice have found urogenital mucosal immune responses very different from what is seen in naive mice (42–45). The epidemiology of UTI suggests that in a damaged or sensitized mucosal environment the requirements for bacterial virulence factors are diminished, potentially making any therapeutic intervention that targets the bacteria a tremendous challenge. In this section, we will briefly summarize our current knowledge of acute UTI pathogenesis in the latest animal models of uncomplicated UTI and CAUTI, highlighting the bacterial factors and host-pathogen processes that are promising drug and vaccine targets (Fig. 1 and Table 1).

Uncomplicated UTI: UPEC

UPEC adhesins

Whole-genome sequencing of several "prototypical" UPEC strains in the past 10 years has revealed the presence in each strain of multiple known and putative adhesins, a number of which have been demonstrated to contribute to UPEC's ability to colonize the urinary tract. These include adhesive fibers called pili (fimbriae). A molecular machine known as the chaperone-usher pathway (CUP) mediates the assembly of pili on the bacterial outer membrane of diverse genera of Gram-negative bacteria (46–51) (detailed in "Structure, Function, and Assembly of Adhesive Organelles by Uropathogenic Bacteria" by Thanassi et al.) and summarized in Fig. 2). Pili are long fibers that extend beyond the bacterial capsule. They contain adhesins at their tips that are thought to play an important role in host-pathogen interactions (52). Each sequenced UPEC strain encodes a multitude of CUP operons (53–55). For example, the cystitis strain UTI89 encodes 10 CUP operons, but of those that are broadly conserved among UPEC isolates, only two, type 1 and P pili, have so far been strongly implicated in UTI pathogenesis (Fig. 2). CUP adhesins are known to recognize specific receptors with stereochemical specificity. For example, FimH, the tip adhesin of the type 1 pilus, has been shown to bind mannosylated glycoproteins (56–58), as well as N-linked oligosaccharides on $\alpha 3$ and $\beta 1$ integrins (59), and the pattern-recognition receptor Toll-like receptor 4 (TLR4) (60), all of which are expressed on the luminal surface of human and murine bladders. In contrast, the P pilus adhesin, PapG, is known to bind to Gal-α-1,4-Gal in globosides in the human kidney (61).

Type 1 pili

Several lines of evidence point to the type 1 pilus as a critical virulence factor in the establishment of UTI by UPEC in humans. Type 1 pili have been shown to be expressed during human UTI. A number of studies investigating whether type 1 pili are expressed by UPEC during UTI have found that the frequency of positive type 1 pilus immunostaining in urine sediments from women with acute UTI ranged from 40% to 76% (62–64), comparable to what was found in acutely infected mice (65). These results can be explained by the recent finding that urine decreases UPEC expression of type 1 pili (66). Type 1 pili have long been known

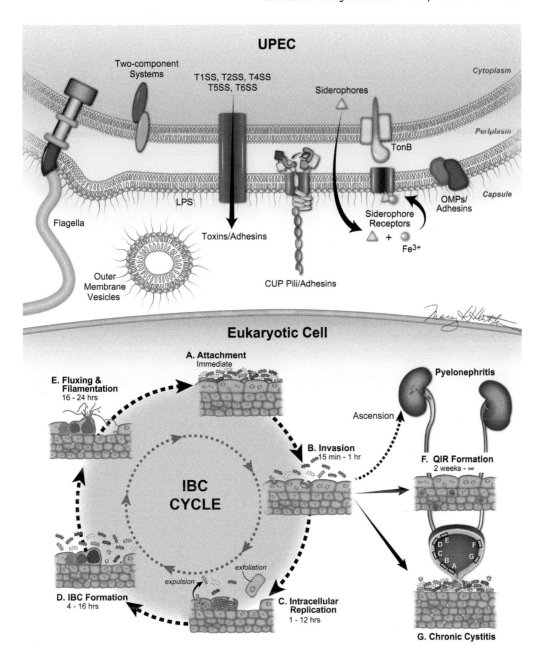

FIGURE 1 Targeting UPEC virulence factors that are critical for pathogenesis. Uropathogenic *E. coli* (UPEC) elaborate a variety of surface structures and two-component systems that play critical roles in UTI pathogenesis. The stages of pathogenesis, as determined from animal models and clinical data, include initial bladder colonization and the IBC cycle (*A-E*), the chronic bladder outcomes of quiescent intracellular-reservoir (QIR) formation (*F*) and chronic cystitis (*G*), and ureteral ascension and pyelitis/pyelonephritis with increased risk for bacteremia/septicemia. UPEC surface structures that play a role in UTI pathogenesis include lipopolysaccharide (LPS), polysaccharide capsule, flagella, outer-membrane vesicles, pili, non-pilus adhesins, outer-membrane proteins (OMPs), toxins, secretion systems, and TonB-dependent iron-uptake receptors, including siderophore receptors. These virulence components are attractive drug and vaccine candidates.

TABLE 1 Prevalence and sites of action of selected uropathogenic *E. coli* (UPEC) virulence factors and their use as candidate vaccine antigens[a]

Category	Virulence factor	Prevalence (%)	Animal model	Bladder	Kidneys	References
CUP Pili/Adhesins	Type 1	86–100	Mice	S[b],C[c],P[d]	S,C	36, 39, 65, 70, 300, 407
			Monkeys	P	NR[e]	38
	P	12–87	Rodents	N[f]	S,P	37, 41, 106, 357, 365, 368
			Monkeys	C	S,P	40, 102, 105, 298, 325, 366, 367
	Ygi	57–66	Mice	N	C(R)[g]	117
	Yad	36–53	Mice	N	N	117
	F1C	8–50				
	S	3–31	Rodents	NR	S(R),P	114, 370
	Dr	1–15	Mice	N	S,P	371, 408
Adhesins	FdeC	99	Mice	C	C,P	122
	TosA	25	Mice	S	S	118, 121
Toxins	α-Hemolysin	18–68	Rodents	L[h]	S,P,L	114, 193, 409
	CNF1	13–54	Mice	S,C(R),L	C	87, 88
Endotoxin	WaaL	Conserved	Mice	S,P	P	287, 339
Autotransporter (T5SS) serine proteasess	Sat	26–52	Mice	N	L	196
	PicU	19–31	Mice	N	N	197, 410
	Vat/Tsh	54–68		NR	NR	197
Autotransporter (T5SS) adhesins	UpaG	21	Mice	N	N	148
	UpaC	47	Mice	N	N	145
	UpaB	58	Mice	S,C	N	145
	UpaH	76	Mice	C	N	146
Two-component systems	QseBC	Conserved	Mice	S,C	S,C	157, 163
	PhoPQ	Conserved	Mice	S(R)	N	168
	BarA-UvrY	Conserved	Mice	S	S	169
	Cpx	Conserved	Mice, zebrafish embryos	S,C		167
Iron acquisition	FepA	Conserved		NR	NR	
	FyuA	71–96		NR	NR	
	IreA	20–26	Mice	N,P	NR	373, 411
	IroN	42–78	Mice	C	N,P	180, 412
	IutA	14–85	Mice	C,P	C,P	184, 373
	Iha	16–74	Mice	C	C	123
	Heme: ChuA	84–90	Mice	IBCi,C	C	135, 184
	Heme: HmaA	61–81	Mice	N,P	S,P	198, 373
	TonB	Conserved	Mice	C(R)	S(R),C(R)	184
Capsule	K1	19–56	Mice	S(R)	NR	142
	K2	9–13	Mice	C(R)	C(R)	143
Motility	Flagellin	Conserved	Mice	S,C	S,C	153, 192
Secretion systems	T2SS		Mice	N	N	199
	T4SS	NR	Mice	N	S(R)	199
	T6SS	NR	Mice	N	N	413
Immune modulatory	TcpC	21–40	Mice	N	S(R),L	200, 202, 414
	SisA/B	A:67–86; B:22–28	Mice	S,C	S,Dj(R)	415

(Continued on next page)

TABLE 1 Prevalence and sites of action of selected uropathogenic *E. coli* (UPEC) virulence factors and their use as candidate vaccine antigens[a] *(Continued)*

Category	Virulence factor	Prevalence (%)	Animal model	Bladder	Kidneys	References
OMPs	OmpA	Conserved	Mice	S(R),C	S	416
	OmpT	70–94	Mice	S	NR	417, 418
Filamentation	SulA	Conserved	Mice	S	NR	154

[a]Data from references 41, 117, 119, 176, 373, 399–406.
[b]S, mutant strain was attenuated in single-infection studies.
[c]C, mutant strain was attenuated in competition studies.
[d]P, vaccination with virulence factor was protective against challenge infection.
[e]NR, not reported.
[f]N, no phenotype.
[g](R), attenuation of mutant strain was rescued by complementation.
[h]L, mutant strain caused less organ damage in single-infection studies.
[i]IBC, mutant strain formed smaller intracellular-bacterial communities.
[j]D, mutant strain caused more organ damage in single-infection studies.

to play a critical and essential role in establishing cystitis in a murine model of experimental UTI, and vaccination of mice and cynomolgus monkeys with the type 1 pilus-tip adhesin, FimH, protects against experimental cystitis (36, 38, 39, 65, 67–71). Type 1 pili have also been shown to be required for UPEC adherence to human urothelial tissue culture cells, and expression of FimH was required for bacterial adherence to human bladder tissue *in situ* (39, 56, 72). Lastly, there is strong evidence that FimH has undergone pathoadaptive mutation in UPEC clinical isolates, with several amino-acid residues found to be under positive selection (73–77). Mutation of these residues resulted in reduced virulence in a murine model of cystitis, providing further support that FimH plays an important role *in vivo* during human UTI (73). Recently, specific pathoadaptive FimH alleles were found to affect the structural conformation and mannose-binding affinity of clinical UPEC isolates, as well as the virulence of the isolates in mouse models of acute and chronic cystitis (78).

FimH mediates UPEC adherence to and invasion of urothelial cells

During experimental bladder infection, the type 1 pilus-associated tip adhesin FimH mediates adherence and invasion of the superficial umbrella cells of the urothelium (27, 65). The specific receptor for type 1 pili appears to vary with the differentiation state of the urothelial cells. In mature superficial umbrella cells, the FimH receptor is the mannosylated-uroplakin protein UPIa (58). However, the immature urothelial cells generally used for *in vitro* studies, such as 5637 bladder transitional carcinoma cells, do not typically express uroplakins on the cell surface, and FimH was found to bind to mannosylated $\alpha 3$ and $\beta 1$ integrins *in vitro* (59). Klumpp and colleagues have demonstrated that binding of FimH to the uroplakin receptor complex via UPIa leads to the phosphorylation of UPIIIa, the only one of the four major uroplakins with a potential cytoplasmic-signaling domain, resulting in an increase in intracellular calcium and enhanced invasion (79, 80). However, in immature urothelial cells, bacterial invasion subsequent to FimH binding has been reported to involve components of clathrin-coated pits such as clathrin and the cargo-adaptor protein AP-2 (81), caveolae and lipid rafts (82), the action of microtubules (83), and actin rearrangement involving focal-adhesion kinase, phosphotidylinositol-3-kinase, and the Rho GTPases Rac1 and Cdc42 (72, 84, 85). The toxin cytotoxic-necrotizing factor 1 (CNF1) has been reported to enhance UPEC invasion of urothelial cells *in vitro* by constitutively activating Rho GTPases (86), but CNF1 has not been shown to play a clear role *in vivo* (87, 88).

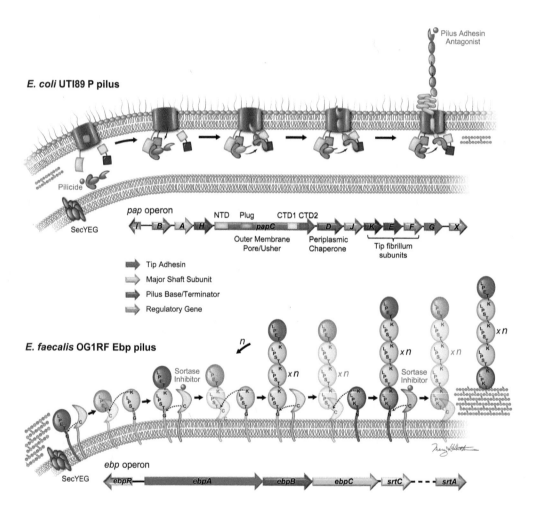

P pili

The role of P pili in UTI is complex and not fully understood. While P pilus-expressing UPEC are strongly associated with first-time pyelonephritis in children (89, 90), they are less well conserved in women with acute and recurrent UTI, being expressed in only 40% to 50% of isolates, regardless of upper urinary tract involvement (91). This is likely due to the chronic inflammatory changes that occur in the urinary tract of patients with a history of severe or recurrent UTI (35), which may lessen the requirement for P pili in colonizing the kidney. P pili mediate adhesion to Gal-α-1,4-Gal-containing globoseries glycosphingo-lipids elaborated on the surface of urinary tract epithelial cells (92–95). Patients with upper urinary tract symptoms during UTI mount humoral-antibody responses to P pili, indicating that they are expressed during infection (96, 97). Moreover, in humans who are non-secretors of ABO antigens, sialyl galactosyl globosides, which are P pilus receptors, are found more abundantly on the surface of epithelial cells in the kidneys and lower urogenital tract compared to "secretors," perhaps explaining why some studies have identified the non-secretor status as a significant risk factor for recurrent UTI (98–101). Concordant with these findings, P pili-expressing UPEC have a greater capacity to bind to vaginal epithelial cells from non-secretors than from secretors (98).

There are at least three alleles of the P pilus tip adhesin PapG, each of which differs in its binding specificity to globosides. Of these, $PapG_{II}$, which binds the human kidney receptor GbO4, is required for the establishment of pyelonephritis in cynomolgus monkeys (102, 103). Replacement of $PapG_{II}$ with $PapG_{III}$, which binds to Forssman antigen (GbO5) and is the predominant PapG allele

---

FIGURE 2 Models of pilus assembly in Gram-negative and Gram-positive pathogens. *Top panel,* **Model of P pilus formation by the chaperone-usher pathway in uropathogenic *E. coli*.** After secretion of pilus subunits into the periplasm via the general Sec machinery, periplasmic chaperones (*dark green*) serve as folding templates, providing a beta-sheet that enables proper folding of the pilin subunits into immunoglobulin-like domains, but in a non-conical orientation, in a mechanism called donor-strand complementation. Assembly and anchoring of the pilus occurs at an outer-membrane pore known as the usher (*orange*). The pilus-tip adhesin (*red*) is the first subunit to interact with the usher, via a preferential interaction between the tip adhesin/periplasmic-chaperone complex and the usher N-terminal-periplasmic domain (NTD, *light blue*), and this interaction initiates assembly by causing a conformational change in the usher that "unplugs" (Plug, *dark blue*) the pore and displaces the tip-adhesin subunit/chaperone complex to two C-terminal-usher domains, CTD1 (*yellow*) and CTD2 (*purple*) (50, 421, 422). The next pilin subunit/chaperone complex then binds to the NTD and if it has an N-terminal extension that is able to complete the immunoglobulin fold of the preceding subunit in a canonical fashion, this provides the free energy to displace the chaperone, in a process called donor-strand exchange, and drive assembly (47–49, 423). In P pili, this occurs repeatedly, incorporating anywhere from hundreds to thousands of PapA major-pilin subunits (green) in the pilus, until PapH (brown) is incorporated into the pilus. PapH is a terminator because it is unable to undergo donor-strand exchange (424). Small-molecule inhibitors (*pink*) that disrupt pilus assembly ("pilicides") or adhesin binding to its receptor ("pilus-adhesin antagonists") have been identified (223, 237). *Bottom panel,* Model of sortase-mediated assembly of the endocarditis- and biofilm-associated pilus (Ebp pilus) in *E. faecalis* (217). Unlike CUP pili in Gram-negative bacteria, sortase-assembled pilus subunits are covalently linked. Pilin subunits are first secreted to the outside of the cell via the general Sec machinery, and are retained in the membrane via a hydrophobic domain within their cell wall-sorting sequence. Sortase C (SrtC, *yellow*) cleaves the EbpA (*red*) LPETG sequence, resulting in an EbpA-SrtC thioacyl intermediate that is resolved by the EbpC (*green*) Lys186 nucleophile. Pilus polymerization occurs when SrtC processes the EbpC LPSTG sequence at the base of a growing, membrane-associated pilus forming a pilus-SrtC intermediate that is resolved by the Lys186 of an incoming EbpC subunit. EbpB (*brown*) incorporates at the base of a pilus fiber when its Lys179 nucleophile resolves a pilus-SrtC intermediate. Sortase A (SrtA, *blue*) processing of the EbpB LPKTN sequence leads to eventual incorporation of the mature pilus into the cell wall. Sortase inhibitors (*pink*) may be useful for disrupting the virulence potential of Gram-positive uropathogens.

found in cystitis strains (104), shortened the course of bacteriuria after bladder inoculation of primates and diminished both renal damage and the development of a serum titers against P pili. In a primate model of uncomplicated cystitis, neither PapG$_{II}$ nor PapG$_{III}$ were required for robust bladder infection (105). However, PapG$_{II}$ conferred a competitive advantage in the primate bladder when co-inoculated with an isogenic strain lacking P pili (PapG$_{III}$ was not tested in competitive infection). Deletion of P pilus operons, including one expressing PapG$_{II}$, from a virulent UPEC strain did not affect pathogenesis in a CBA murine model of infection (106). This may be because the UPEC strain used in the study contains alternative adhesins capable of colonizing the kidney epithelium, or perhaps the CBA-mouse model does not reflect the importance of PapG$_{II}$ – GbO4 interactions, since the GbO4 receptor is likely not as highly expressed in the murine kidney due to the presence of a functional Forssman synthetase in non-primate mammals (107). Furthermore, the C3H/HeJ strain, which is closely related to the CBA strain, has been reported to be genetically susceptible to vesicoureteric reflux (108), so the requirement for kidney adhesins may be diminished in this model. In contrast, a mutant UPEC strain lacking PapG$_{II}$ and other putative kidney-colonization factors was defective in colonizing the kidneys of Balb/c mice. Live multiphoton studies of pyelonephritis in rats suggest that P and type 1 pili may work in concert to colonize the renal tubules by facilitating bacterial attachment and biofilm production, respectively (109). Therefore, although P pili are only expressed in about half of all recurrent UTI isolates, drugs that target both type 1 and P pili would likely have broadly protective effects in both the bladder and kidneys.

Other pili and non-pilus adhesins

Compared to commensal strains, UPEC have been shown to contain numerous CUP-assembled pilus operons, in addition to those encoding type 1 and P pili. These pilus operons were identified by homology to the four minimum essential components of CUP-assembled fimbrial systems: the major pilin, the pilus adhesin, the outer-membrane usher, and the periplasmic chaperone. However, the functions and receptors of these additional UPEC-associated pili are poorly understood. S and F1C pili and Afa/Dr adhesins are enriched among UPEC, but are less well conserved than type 1 and P pili. They have been demonstrated to impart the capacity to bind to human kidney epithelia in frozen-tissue sections (110, 111) and may play distinct roles in various stages of UTI (112–116). Recently, the Yad pilus was shown to contribute to adherence of UPEC to bladder epithelial cells *in vitro*, but was not required for experimental urinary tract infection in mice, whereas the Ygi pilus conferred a modest competitive advantage in kidney colonization (117). While pili are likely involved in the initial attachment of UPEC to the urinary tract mucosa, the bacteria elaborate numerous other outer-membrane protein adhesins that may play an important role in disease pathogenesis. Recently, a novel adhesin, TosA, which is secreted by a cognate Type 1 secretion system, was described (118). TosA is found in about 30% of urinary-tract isolates and is expressed during UTI (118–120). However, the role of TosA in UTI is unclear. Although an isogenic UPEC mutant lacking this RTX protein is dramatically impaired in its ability to colonize the urinary tract, the authors of the study were unable to complement the mutant, and vaccination of mice with this protein had no impact on the course of UTI by a TosA-containing UPEC strain (121). Another recently identified adhesin, FdeC, is highly conserved among all *E. coli* pathotypes and intestinal commensals and is reportedly expressed only upon contact with host cells (122). The presence of FdeC conferred a competitive advantage in colonization of the bladder and kidneys of the mouse model, but vaccination of mice with FdeC antigen only protected against kidney infection, with no effect on bladder infection. Finally, the iron-regulated adhesin Iha has been shown to mediate adherence to bladder epithelial cells and confer a slight, but significant, competitive advantage to UPEC in the mouse model of UTI (123). Thus, despite the considerable progress in our understanding of UPEC pathogenesis over the past 15 years, type 1 and P pili remain the most promising candidate adhesins for drug and vaccine intervention.

Cyclic adenosine monophosphate (AMP) and UPEC expulsion

After internalization, UPEC have been found to reside within Rab27b/CD63/Caveolin-1-positive fusiform vesicles, which resemble secretory lysosomes and are normally involved in regulating the surface area of the apical-plasma membrane. However, UPEC can be expelled by a mechanism that requires Toll-like receptor 4 (TLR4), cyclic AMP, Rab27b, and caveolin-1 (85, 124–126). Treatment of mice prior to infection with the drug forskolin, which increases cytosolic cyclic AMP, reduces the intracellular bacterial burden in the bladder. Forskolin has also been shown to suppress UPEC-induced inflammation *in vitro* in primary kidney tubular epithelial cells, and forskolin administration at one hour post-infection reduced acute kidney bacterial burden and inflammation in UPEC-infected C57BL/6 mice (127). Furthermore, TLR4-signaling-incompetent C3H/HeJ mice have higher intracellular bacterial burdens than TLR4-signaling-competent C3H/HeN mice (128). Thus, TLR4-dependent antagonism of invasion and active expulsion of internalized bacteria by urothelial cells is an important early innate defense against acute infection of the bladder, and these expulsion pathways are attractive targets for drug prophylaxis.

UPEC escape the endocytic vesicle

Although UPEC may be expelled from host cells after invasion, it is clear that a fraction of invasive bacteria survive within the super-ficial umbrella cells, eluding expulsion and phagolysosomal death, and escape into the cytoplasm, where several groups have demonstrated that they can replicate rapidly to form IBCs (29, 30, 129–132). Although the mechanism of escape into the cytoplasm is not understood, IBC formation does not typically occur in undifferentiated urothelial cells unless they are treated with either membrane- or actin-destabilizing agents (133, 134). This suggests that the actin network, which is denser in undifferentiated urothelial cells compared to superficial umbrella cells, may restrict bacterial escape from the vesicle and/or proliferation of UPEC within the cytoplasm. It may also be that the difference in FimH receptors in undifferentiated (α3 and β1 integrins) and differentiated (UPIa) urothelial cells results in distinct and divergent UPEC-invasion pathways and intracellular trafficking. While pore-forming toxins have been shown to be required for bacterial escape into the cytoplasm in other infectious-disease model systems, α-hemolysin, a UPEC pore-forming toxin which is expressed in IBCs (135), is not required for vesicular escape, as a mutant strain of UTI89 lacking the α-hemolysin gene forms IBCs equally as well as the wild-type strain (136).

IBC formation

Upon escape into the urothelial cell cytoplasm, UPEC replicate quickly to form IBCs, with a doubling time of 30–35 minutes (30). A survey of published studies finds that the number of IBCs detected at 6 hours post-infection (hpi) in the bladders of individual mice ranges from 3 to 700 (median: ~40) after infection of 7–10-week-old C3H/HeN mice with 10^7 colony-forming units of the UPEC strain UTI89 (28, 71, 73, 137–139). Microscopy studies of mouse bladders after infection with a mixed inoculum of green fluorescent protein-expressing (GFP+) and non-expressing (GFP-) UPEC have demonstrated that IBCs are clonal, originating from a single invasive bacterium (28). As a result,

IBC formation appears to constitute a population bottleneck that initially limits bacterial diversity, followed by a rapid expansion of the clonal IBC population. Novel anti-infective drugs that prevent IBC formation may thus target a vital molecular bottleneck, which is thought to be the Achilles' heel of a pathogen during infection (140, 141). UPEC aggregation into IBCs resembles biofilm formation, as it requires continued type 1 pilus expression after invasion (71) and is accompanied by the production of structural components otherwise associated with UPEC biofilm, such as antigen 43 and a polysaccharide-rich matrix (29). Capsular-synthesis genes also play a role, as a K1 capsule-deficient mutant of the human cystitis isolate UTI89 is markedly deficient in its ability to aggregate and form IBCs (142), and a K2 capsule-deficient mutant of the human pyelonephritis isolate CFT073 is significantly outcompeted by wild type CFT073 in the bladder and kidneys (143). Secreted amyloid fibers (curli) and several other UPEC-autotransporter proteins, including UpaC, UpaG, and UpaH, have been implicated in biofilm growth; however, their role in IBC formation is unknown (144–148). The IBC pathway has been observed in all mouse strains tested, and 15 of 18 human clinical UPEC isolates formed IBCs in experimental infections of C3H/HeN mice, including some isolates without common putative UPEC-virulence factors such as α-hemolysin (149). Those strains unable to form IBCs were also unable to invade the mouse urothelium. UPEC within IBCs are protected from both phagocytosis by polymorphonuclear leukocytes (PMNs) (30) and many antibiotics, particularly the first-line drug trimethoprim-sulfamethoxazole, which has increased efficacy against UTI because it concentrates in the urine but is relatively cell-impermeant (131, 150). A recent study demonstrated that 16 antibiotics capable of killing the virulent cystitis isolate UTI89 *in vitro* are relatively ineffective in eliminating intracellular bacteria either from bladder epithelial cells *in vitro*

or from bladder tissue during *in vivo* infection, even though they achieved urine levels far exceeding the minimum-inhibitory concentrations for UTI89 (131). Thus, harboring bacteria that are protected from antibiotics within IBCs or a persistent intracellular niche (131, 133, 151) may provide a source of surviving pathogens within the bladder that can cause a relapse (treatment failure) or recurrent cystitis, respectively, once antibiotics are removed.

The IBC pathway occurs in humans and with other Gram-negative uropathogens that express type 1 pili

IBC development is not limited to experimental UPEC infection in mice. Translational studies found evidence of IBCs in 18% of urine sediments from women with recurrent cystitis with UPEC (a rate of detection similar to that seen in the urine of mice acutely infected with UPEC), but never in urine from healthy controls or when the causative agent of the UTI was a Gram-positive organism (31). Other studies have identified IBCs in the urine of children with recurrent UTI (32, 33). Furthermore, other Gram-negative uropathogens that express type 1 pili, such as *Klebsiella pneumonia*, *Enterobacter* spp., and *Citrobacter freundii*, also utilize the IBC pathway (Rosen and Hultgren, unpublished data; 139, 152). Together, these findings suggest that the IBC pathway is an important mechanism for the establishment of UTI in mammalian bladders by Gram-negative uropathogens that express type 1 pili and invade the urothelium. Therefore, the IBC pathway is an important and relevant target for therapeutic intervention.

UPEC dispersal and further IBC formation

Dispersal of UPEC from the IBC is also critical for bacterial persistence. IBC maturation involves a partially understood differentiation program during the first 12–16 hours of experimental infection of the mouse bladder. During this time, the rapidly replicating bacteria first take on a coccoid morphology, become more rod-shaped again as the IBC matures, and then begin to flux away from the IBC. UPEC then emerge out of the dying urothelial cells, often in filamentous form, and colonize and invade neighboring cells, thus initiating a second round of IBC formation (30). Flagellar motility does not appear to be required for UPEC dispersal or initiation of the second round of IBC formation (153). However, deletion of the cell division-inhibitor gene, *sulA*, disrupts the ability of UTI89 to filament, a property that has been associated with resistance to neutrophil attack. The *sulA* mutant is also defective in bladder colonization and IBC formation at 24 hpi, but not at 6 hpi, suggesting that UPEC filamentation is necessary for virulence after the first round of IBC formation in the immunocompetent host (154). IBCs are transient, cycling through formation and dispersal primarily during the first 2–3 days of experimental UPEC infection in the immunocompetent mouse (30). However, the immunodeficient-mouse strain C3H/HeJ, which lacks the ability to sense bacterial lipopolysaccharide (LPS), had microscopic evidence of bladder IBCs 4 weeks after experimental infection, indicating that host responses to LPS during infection alter the susceptibility of the bladder urothelium to IBC formation (42). Thus, the IBC pathway is important for the establishment of acute infection in the host, and resembles biofilm formation in the sense that both aggregation and dispersal of UPEC are critical for acute pathogenesis.

Central metabolism and two-component systems

Recently, central-metabolism pathways, such as the tricarboxylic acid cycle, have been shown to be important for acute UPEC virulence and IBC formation in the urinary tract, but not for planktonic growth in urine (155–159). The QseBC two-component system, which is found in many Gram-negative pathogens including UPEC, plays a critical role in regulating virulence-factor expression (160–162). Two-component systems typically consist of an inner membrane sensor kinase and a cytoplasmic response regulator. In response to a stimulus, the sensor kinase regulates by phosphorylation the activation state of the response regulator, thereby regulating gene-expression programs. Dysregulation of the QseBC system by deletion of the sensor kinase QseC causes pleiotropic effects in the bacterial cell, including reduced expression of virulence factors (such as type 1 pili) and reduced virulence and IBC formation *in vivo* (156, 157). A *qseC*-deletion mutant forced to express type 1 pili also had an acute virulence defect when in competition with wild-type UPEC, suggesting that the misregulation of additional factors beyond type 1 pili was responsible for the attenuation (163). Surprisingly, we found that the altered virulence-factor regulation in the Δ*qseC* mutant was due to defects in central metabolism, as two different mutants unable to complete the tricarboxylic acid (TCA) cycle phenocopied the Δ*qseC* mutant (156). The QseBC system was recently found to have robust and highly sensitive cross-regulation with another two-component system, PmrAB, which is activated by ferric iron and which mediates polymyxin resistance. Addition of ferric iron to UTI89 growth medium induced *qseBC* expression in a PmrB-dependent manner (164). Other two-component systems also contribute to UPEC virulence. Cpx is an envelope stress-response system known to regulate the expression of P pili (165, 166). It was recently shown that deletion of the Cpx system resulted in impaired UPEC colonization of the murine bladder and impaired virulence in zebrafish embryos (167). Finally, other two-component systems, including PhoP-PhoQ, BarA-UvrY, and KguS-KguR, have been found to contribute to UPEC virulence (168–171). Interestingly, constitutive activation of the Pho regulon in CFT073 by inactivation of the phosphate-specific transport system Pst was recently shown to result in a loss of expression of type 1 pili, resulting in significant attenuation in a mouse model of

UTI (171). Therefore, compounds that alter UPEC virulence gene expression or central-metabolism pathways, either directly or by misregulating two-component systems, are potential novel therapeutics.

Metal ions

Iron acquisition is another critical requirement for bacterial virulence (172–174). Iron acquisition-associated genes common to all *E. coli* strains are under strong positive selection in UPEC clinical isolates (54). UPEC typically have multiple, seemingly redundant iron acquisition systems, and these have been shown to be highly upregulated in the IBC (135). As many as four siderophores (small-molecule iron chelators) are commonly produced by UPEC strains, and scavenging ferric iron (Fe^{3+}) is thought to be their main function. Among the siderophores, enterobactin is broadly conserved among *E. coli* strains, while yersiniabactin, salmochelin, and aerobactin-synthesis genes are enriched in UPEC. To prevent microbial iron scavenging, urothelial cells in close proximity to the IBC upregulate genes for the transferrin receptor and for lipocalin 2, host factors that are involved in preventing bacterial acquisition of iron (135). However, metabolomic studies have found that UPEC clinical isolates preferentially synthesize yersiniabactin and salmochelin, each of which is associated with resistance to the antibacterial effects of lipocalin 2 (175–177). Furthermore, UPEC can scavenge iron from heme, and a deletion mutant lacking the heme transporter ChuA (which is highly expressed in the IBC) forms significantly smaller IBCs *in vivo* (135, 178). Siderophore production in *E. coli* is mediated in part by the small regulatory noncoding RNA RhyB; a recent study found that deleting *rhyB* in CFT073 reduced siderophore production *in vitro* and *in vivo* and reduced bladder and kidney colonization in CBA/J mice (178).

The salmochelin receptor IroN may have multiple functions, as it has been shown to enhance bacterial invasion of bladder epithelial cells *in vitro*, and a mutant UPEC strain lacking IroN was attenuated in a mouse model of cystitis (179, 180). Yersiniabactin also plays a role in sequestering the toxic effects of copper (II) ions, possibly enhancing resistance to phagocyte killing (181). Interestingly, the broadly conserved siderophore enterobactin actually contributes to copper sensitivity, suggesting that the apparent redundancy of siderophores may actually be a bacterial adaptation to inhabiting different host niches. In mice, the asymptomatic bacteriuria isolate UPEC 83972 outcompeted a mutant strain lacking salmochelin and enterobactin in the urine, bladder, and kidney (182). In the pyelonephritis isolate CFT073, which does not synthesize yersiniabactin, aerobactin appears to play an important role in bladder fitness, suggesting that these two siderophores may have overlapping functions (183, 184). Thus, bacterial iron acquisition by multiple systems has been selected for in UPEC, possibly in part due to their role in IBC formation. Their redundancy points to their importance, which may make targeting them with vaccines or therapeutics problematic. Including multiple siderophore-receptor antigens in a single vaccine might possibly overcome this challenge. However, all siderophore receptors require the TonB inner-membrane protein to transduce the energy needed for import. Deletion of TonB from a UPEC strain greatly reduced virulence in the kidney and, to a lesser extent, the bladder during experimental mouse infection (184). Therefore, targeting TonB with small-molecule inhibitors may be an effective anti-infective strategy.

Modeling the outcomes of acute cystitis

Experimental mouse models of infection have revealed that UPEC are capable of chronic colonization of the urinary bladder in several different ways. In immunocompetent mice, the outcome of cystitis is typically either resolution of acute infection with elimination of bacteriuria, or persistent bacteriuria and chronic cystitis (42). However, even with resolution of active infection, UPEC are capable of persisting latently within Lamp1⁺ vesicles inside urothelial cells (133, 151). These latent reservoirs have been termed the quiescent-intracellular reservoir (QIR) and have the capacity to seed recurrent infections (126, 151). Treatment of mice with urothelial exfoliations, including protamine sulfate and chitosan, has shown some promise in eliminating this bacterial reservoir from the bladder (151, 185).

In contrast to a latent reservoir, some mouse strains (including the C3H/HeN and CBA/J strains, which are commonly used for mouse models of UTI) are prone to developing high-titer persistent bacteriuria and chronic cystitis, which appears to last for the life of the animal, in response to UPEC infection in an infectious dose-dependent fashion (42, 186). C57BL/6 mice also develop chronic cystitis when "superinfected" with two UPEC exposures 24 hours apart (187). Inflammation is most severe during early acute infection in this model and plays a non-productive role, actually contributing to the development of chronic cystitis. This is potentially a very interesting model, as placebo-controlled studies have found that approximately 50% of women remain bacteriuric several weeks after a symptomatic UTI if not treated with antibiotics, despite overall improvement of symptoms (188, 189). Antibiotic therapy readily cures the infection in mice. However, if chronic cystitis is allowed to ensue for at least 7–14 days prior to antibiotic therapy, the mice become highly sensitized to severe, recurrent cystitis upon a second bacterial challenge administered 6 months or more after antibiotic therapy to clear the initial infection (O'Brien, Hannan, and Hultgren, unpublished data; 42). A recent proteomics investigation revealed bladder mucosal remodeling in these so-called "Sensitized" mice that renders the host more susceptible to neutrophil damage as a consequence of inflammation (45). In contrast, mice that are treated with antibiotics after

only 1 day of infection, or that spontaneously resolve bacteriuria during the first two weeks of infection, are more resistant to challenge than naive, age-matched mice. This model may be an invaluable tool not only for understanding host mechanisms of chronic and recurrent UTI, but also for developing therapies and vaccines that combat recurrent UTI (35). For instance, treatment with cyclooxygenase-2 (COX-2) inhibitors prior to challenge infection was found to prevent recurrent UTI in "Sensitized" mice by preventing bladder epithelial transmigration of neutrophils and subsequent mucosal wounding (45). This finding may explain the results of a small study that found that women who received ibuprofen, a non-specific inhibitor of cyclooxygenases 1 and 2, resolved UTI symptoms as quickly as women who received the antibiotic ciprofloxacin did (190).

Ascension to and colonization of the kidneys

The main complication of untreated cystitis in humans is ascension of bacteria in the ureters and colonization of the kidney parenchyma (medulla and cortex), which can lead to marked kidney inflammation with progressive loss of nephron function and even sepsis. Flagella are highly expressed by UPEC *in vivo* during ascension to the murine kidney (191). However, the contribution of flagella to kidney colonization is unclear, as different mutants impaired in flagellar motility or chemotaxis were not all defective in kidney colonization in competition-infection experiments (153, 192). This may be because the mouse strains used in these studies, C3H/HeN and CBA/J, are genetically susceptible to vesicoureteric reflux (108), and therefore flagellar motility may not be important for ascension in these models. The UPEC pore-forming toxin α-hemolysin is associated with renal damage and scarring (193). At small physiological doses it induces Ca²⁺ oscillations in renal tubular epithelial cells and thereby potentially enhances ascension and colonization of the ureters and kidney parenchyma by disrupting the normal flow of urine (194). Recently, α-hemolysin was found to induce proinflammatory Caspase-1/Caspase-4-dependent cell death in bladder epithelial cells, resulting in cell exfoliation. UPEC strains overexpressing α-hemolysin were attenuated in acute and chronic infection in mice, suggesting that acute bladder exfoliation is a host defense mechanism (195). Other UPEC toxins, such as Sat, PicU, and Tsh, are also not required for infection, but may contribute to renal pathology (196, 197). Ygi pili, type II- and IV-secretion systems, and multiple iron and heme acquisition components have all been shown to contribute to kidney colonization in animal models (117, 184, 198, 199). Recently, a Toll/interleukin-1 receptor (TIR) domain-containing protein that is secreted by an unknown mechanism, TcpC, was discovered in a subset of UPEC and found to associate directly with MyD88 and TLR4 (200, 201). Loss of function studies in mice indicate that TcpC enhances bacterial virulence by suppressing the early innate immune response to UPEC infection, resulting in higher bacterial burdens in the kidney and more severe kidney pathology over time (202).

Complicated UTI: *Enterococcus faecalis*

CAUTI
In contrast to the healthy-bladder environment affected by UPEC in uncomplicated UTI, the placement of a urinary catheter effects pathologic changes in the bladder that may contribute to the greater variety of competent uropathogens that infect the catheterized bladder. In addition to causing mechanical damage, urinary catheterization interferes with micturition (urination), a natural impediment to bacterial colonization of the bladder. Even in the absence of infection, bladder catheterization may lead to tissue edema and hyperplasia of the urothelium, in addition to hematuria and alteration of urine composition (203, 204). Probably the most significant factor contributing to infection development is the presence of an abiotic surface in the bladder, the catheter, which promotes bacterial-biofilm formation and is recognized as an important component of CAUTI. UPEC and enterococci are the two most common isolates from symptomatic CAUTI (14). Enterococci, commensal gut bacteria, have emerged as important human pathogens in the last 40 years, especially in the health care environment. Several aspects of modern medicine have contributed to this recent rise in infections caused by *E. faecium* and *E. faecalis*, the most commonly isolated species. Widespread use of antibiotics has likely selected for enterococci that display intrinsic or acquired resistance to many common classes of antimicrobials. Furthermore, medical devices (such as indwelling urinary catheters) and invasive surgical procedures compromise natural barriers to infection and are being used more frequently in the health care setting. Despite the increasing incidence of enterococcal infection, little is known about the molecular mechanisms these bacteria use to cause disease. However, it is clear that biofilm formation and the elaboration of secreted and surface proteins and organelles contribute significantly to enterococcal pathogenesis.

The role of secreted and surface structures in enterococcal-biofilm formation
Biofilm formation is a critical aspect of device-related infections, including enterococcal CAUTI. As extracellular pathogens, enterococci may also rely on growth in the biofilm state to infect host tissues. Thus, putative molecular determinants of enterococcal virulence, identified using a variety of methodologies, are typically examined *in vitro* in assays of biofilm formation. These determinants must be secreted and, in many cases, covalently linked to the cell wall. This latter function is carried out by a group of membrane proteins known as sortases, which recognize conserved-cell wall-sorting sequences (CWSS) on membrane-linked proteins and catalyze the attachment of their extracellular domains to the cell wall. Bacte-

rial proteins well-studied in biofilm assays include enterococcal surface protein (Esp), of unknown function, and gelatinase, a secreted zinc metalloprotease that hydrolyzes gelatin, collagen, and casein (205). Extracellular DNA, autolysin, the housekeeping sortase SrtA, the endocarditis and biofilm-associated (Ebp) pilus, and the Ebp pilus-associated sortase SrtC have also been shown to play a role in biofilm formation (206, 207).

The role of adherence and biofilm formation in CAUTI

Many of the same virulence factors involved in biofilm formation, including Ace, Esp, AhrC, Eep, SrtA, Ebp pili, and SrtC, have been shown to play a role in a ureteric-reflux model of ascending pyelonephritis (207–211). However, the contribution of these factors was not large, and robust bladder infection could not be achieved in rodent models of uncomplicated UTI (212). More recently, a more relevant model of foreign-body cystitis has been developed in mice for testing the role of *E. faecalis* virulence factors in CAUTI (213). In this model, the presence of urinary-catheter material in the urinary bladder allows for biofilm formation on the implant and robust, high-titer bladder infection that persists as long as the catheter remains. Transient immunosuppression of mice concurrent with catheter implantation exacerbates *E. faecalis* infection in C57BL/6J mice, suggesting that the presence of the catheter and not the inflammatory response is driving CAUTI (214). Ebp pili, and specifically the metal ion-dependent adhesion site (MIDAS) motif found within the predicted von Willebrand factor A domain of the fibrinogen-binding tip adhesin protein EbpA, are essential for both bladder and implant colonization in this model (215, 216). However, the *in vitro* biofilm determinants autolysin and gelatinase were not required (213). Thus, enterococci utilize Ebp pili to take advantage of the presence of foreign abiotic surfaces and damaged bladder-mucosal barriers to cause CAUTI.

Sortase-mediated pilus assembly

Ebp pilus assembly is directed by the action of two different sortases (Fig. 2) (217). These membrane-linked transpeptidase enzymes catalyze the covalent assembly of pilus subunits into a functional pilus that is covalently attached to the peptidoglycan cell wall. Sortases are nearly ubiquitous among Gram-positive organisms and have duplicated and diversified among and within species to perform specific functions (218). Therefore, the development of small molecules to inhibit these enzymes has the potential to yield a wide array of therapeutics that range from broadly anti-Gram positive to specifically targeting virulence processes of single species. A sortase inhibitor targeting Ebp pilus assembly could potentially be beneficial for preventing health care-associated infections, including CAUTI. The Ebp pilins also make attractive vaccine candidates for several reasons, particularly for those individuals in a long-term health care setting (e.g., a nursing home), which strongly increases the risk of acquiring CAUTI. The *ebp* locus is present in ~95% of *E. faecalis* isolates regardless of source and is highly conserved (219). These proteins are expressed and function on the bacterial cell surface and are important virulence factors. Indeed, immunization with pilins from Group B streptococci has been shown to be protective in relevant infection models (220, 221). The protective effects of immunization with *E. faecalis* Ebp pilus components in experimental-disease models are described later in the chapter.

TRANSLATING DISCOVERIES IN PATHOGENESIS: THE DEVELOPMENT OF ANTI-VIRULENCE THERAPIES

Biarylmannose-Derivative FimH Antagonists (Mannosides)

The mannose-binding pocket of FimH is invariant in all strains of uropathogenic *E. coli* (56), and mutations in these residues disrupt mannose binding and attenuate virulence (56, 73, 75). With information gained from the crystal structures of FimH bound to α-D-mannose and mannose derivatives called mannosides (56, 222–224), we and others have rationally designed biarylmannose-derivative FimH binding inhibitors (223, 225, 226). Using a reiterative process of structure-based design, combinatorial chemistry, and *in vitro* cell-based screening, lead compounds with excellent cellular potency, low molecular weight, and optimized oral pharmacokinetics were identified. Experimental and pre-clinical translational studies have demonstrated that these optimized mannoside compounds can be given orally to mice either to prevent cystitis or to successfully treat an established bladder infection (150, 225, 227, 228).

Since the mannose-binding pocket of FimH is invariant, mannosides have potent efficacy in preventing acute UTI caused by divergent strains, including the trimethoprim-sulfamethoxazole (TMP-SMZ)-resistant UPEC strain PBC-1 (150) and the multi-drug resistant UPEC strain ST131(228). Mannoside treatment prior to infection of C3H/HeN mice prevents UPEC invasion of the urothelium and IBC formation, a process that protects UPEC from the effects of many antibiotics (131). Thus, in mannoside-treated mice, UPEC are confined to the bladder-extracellular niche, where they are left exposed to high levels of antibiotics that are commonly used to treat UTI. As a result, although TMP-SMZ alone had no effect on bladder colonization by the resistant strain PBC-1, mannoside-potentiated killing by TMP-SMZ (which concentrates in the urine to levels well above the minimum-inhibitory concentration of PBC-1) to successfully prevent the establishment of UTI by this strain. In a similar way, mannoside also potentiated killing by TMP-SMZ to prevent CAUTI in a foreign-body model of experimental infection (227). Finally, a recent study found that mannoside is also efficacious against the multi-drug-resistant UPEC clone ST131

in acute and chronic experimental infection in C3H/HeN mice. The clinical isolate used in this study was EC958, which is an extended-spectrum β-lactamase strain that is resistant to eight classes of antibiotics, including fluoroquinolones. One prophylactic mannoside dose significantly decreased acute bacterial burdens in the bladder and treatment of chronically infected mice with a single dose of oral mannoside reduced bladder bacterial burdens greater than 1,000-fold (228). If translated to clinical practice, mannosides have tremendous and exciting potential to be an efficacious, safe, and cost-effective new therapy either used in combination with commonly used first-line antibiotics to successfully treat existing uncomplicated cystitis and CAUTI, or used alone as a daily prophylaxis against chronically recurrent cystitis. By reducing the use of antibiotics, and particularly the use of fluoroquinolone antibiotics, in the treatment and prevention of UTI, mannosides could have an immediate and long-term impact on the development of antibiotic resistance in UPEC clinical isolates, which is currently as high as 30% in some studies (21). Furthermore, the unique mechanism of mannoside action, i.e., inhibiting the function of the extracellular FimH pilus tip adhesin by blocking the invariant lectin pocket, likely circumvents the development of resistance due to mutation of the binding pocket, porin mutations, or efflux.

Galabiose PapG Antagonists

In 1982, the efficacy of glycolipids in preventing P pilus binding and *in vivo* infection was established (229). Determination of the crystal structure of PapG$_{II}$ bound to its receptor, GbO4, elucidated critical details of the adhesin-receptor interaction and allowed for further rational design of galabiose-derived receptor analogs (92, 230, 231). Further studies have identified high-affinity multivalent inhibitors that also inhibit galabiose binding by *Streptococcus suis* (232–

234), though they were not consistently effective in a mouse model of peritonitis. Therapy that combines bioavailable PapG antagonists with mannosides has tremendous potential to treat and prevent UTI.

Inhibitors of the Chaperone-Usher Pathway: Pilicides and Curlicides

Since Gram-negative pili are assembled by the chaperone-usher pathway (CUP), inhibitors of this pathway could be broadly effective against a number of pathogens that require pili for pathogenesis. In collaboration with Fredrik Almqvist, a medicinal chemist, we developed ring-fused 2-pyridone small-molecule inhibitors that target the CUP periplasmic chaperones (235, 236). We have called these compounds "pilicides." By screening for inhibitors that prevented type 1 pilus-mediated hemagglutination and *in vitro* biofilm formation, we identified pilicides with activity not only against type 1 pili, but also against P pili (237). NMR and crystallographic studies of the interaction of a pilicide with the P pilus cognate-periplasmic chaperone, PapD, found that pilicide compounds interacted with a highly conserved region (238) that interacts with the N-terminal domain of the outer-membrane usher (237) (Fig. 2). One highly potent inhibitor of type 1 piliation and biofilm formation *in vitro*, pilicide ec240, was used for an *in vitro* transcriptomic and proteomic investigation of pilicide effects on UPEC virulence (239). The ec240 pilicide was found to decrease motility and dysregulate CUP pili, including type 1, P, and S pili.

Curli are amyloid fibers produced by many Enterobacteriaceae that are assembled at the outer membrane by a nucleation pathway of fibrillization (240). By screening for inhibitors of curli-mediated biofilm, we identified 2-pyridone derivatives capable of inhibiting both curli and pili formation. One such "curlicide" that inhibits both type 1 pilus production and curli biogenesis rendered UPEC relatively avirulent in a mouse model

of experimental cystitis (144). Further optimization of these pilicide and curlicide compounds has increased their potency dramatically (241–243), and these lead compounds are promising candidates for future drug development.

Intravesical Therapy with ASB Strain 83972

Another strategy that is currently under investigation is the use of an avirulent asymptomatic bacteriuria (ASB) strain, 83972, which has adapted for long-term colonization of the human urinary tract without causing significant symptoms or pathology, as a therapy for recurrent UTI (244). Although ASB strain 83972 is discussed in more detail in "Asymptomatic Bacteriuria and Bacterial Interference" by Nicolle et al., we will briefly discuss its therapeutic potential here. Strain 83972 has lost the capacity to express type 1, P, and F1C pili, and can outcompete UPEC strains in human urine whether growing planktonically or in a biofilm (245–247). Therefore, it is hypothesized that colonization of the bladder by ASB strain 83972 prevents virulent UPEC strains from colonizing the urinary tract, thereby preventing recurrent symptomatic UTI. This therapy is currently undergoing clinical trials and has shown promise in "at-risk" populations, such as those with incomplete bladder emptying or neurogenic bladder from a spinal-cord injury (248–250). The development of urinary symptoms after 83972 inoculation is rare, and apparently not caused by bacterial reversion to virulence (251). Interestingly, intravesical administration of 83972 to mice with acute UPEC UTI was found to reduce visceral pain, suggesting that 83972 may be an effective treatment for UTI symptoms (252). A variation of this approach was recently employed by Schembri and colleagues. The PapG$_{II}$ receptor, GbO4, is also the receptor for Shiga toxin. Adapting existing technology, Schembri and colleagues engineered a strain of 83972 to synthesize a galabiose analog that is linked to LPS on the surface of the cell (253, 254). This strain is able to inhibit binding of PapG$_{II}$-expressing UPEC to kidney epithelial cells, and when co-inoculated into the mouse urinary tract with virulent UPEC, the galabiose-expressing strain significantly reduces the UPEC-bacterial load in the urine compared to the wild-type 83972 strain. Conversely, others have transformed strain 83972 with a plasmid expressing type 1 pili and have demonstrated that this new strain forms better biofilm on urinary catheters and as a result is more efficacious in preventing their colonization by enterococci (255, 256).

Nutraceuticals

So-called "nutraceuticals" are foods or food products that are thought to provide medical benefits and are often sold in a medicinal form. Since use of these products typically does not require regulatory authority approval (e.g., approval by the Food and Drug Administration in the United States), the efficacy of these compounds is often based merely on anecdotal reports. Two common nutraceuticals that have been investigated for preventing recurrent UTI are probiotic *Lactobacillus* preparations and cranberry products. These products may be advantageous because they are generally safe and readily available. However, there exists only limited evidence for their effectiveness.

Probiotics

Clinical evidence suggests that the vagina and periurethral area, which is normally colonized by *Lactobacillus* spp. in healthy women, can act as a UPEC reservoir that could potentially seed recurrent infections. Women with recurrent UTI are more likely to have vaginal or periurethral UPEC colonization than women without recurrent UTI (257), and periurethral-UPEC carriage dramatically increases in the days prior to a recurrent episode (258). Vaginal *Lactobacillus* suppositories might help clear this UPEC

reservoir, preventing recurrences. However, probiotic therapy with *Lactobacillus* spp. has had mixed results and a recent meta-analysis found no evidence for efficacy (259–261). Clinical trials for *Lactobacillus* prophylaxis for infectious diseases of the urogenital tract, including UTI, are ongoing both in the United States and abroad. A Phase I trial to test the safety of an *L. crispatus* vaginal suppository (LACTIN-V, Osel, Mountain View, California) found that no severe adverse events occurred, although seven women (out of 15) developed asymptomatic pyuria (262). In the randomized, placebo-controlled Phase II trial, women with a history of recurrent UTI received antibiotic therapy for acute UTI, and then either LACTIN-V or placebo. The probiotic was protective, with recurrent UTI occurring in 15% of women receiving LACTIN-V and 27% of women receiving placebo (263). The mechanism of *Lactobacillus*-mediated protection from UTI is not clear, and may involve hydrogen peroxide production. Some *Lactobacillus* strains produce surfactants and anti-adhesive molecules (260). *Lactobacillus acidophilus* surfactant was shown to inhibit initial deposition rates and adhesion numbers for several uropathogens, including *E. coli*, *E. faecalis*, and *Proteus mirabilis* (264). This raises the possibility that some *Lactobacillus* strains might be more effective than others at preventing UTI.

Cranberry products

Cranberry products are a common folk-remedy for preventing recurrent UTI, but the efficacy of cranberry products in UTI prophylaxis is largely unproven. Cranberries contain two compounds that have been shown to inhibit UPEC adherence to eukaryotic cells *in vitro*: fructose (found in all fruits), which weakly blocks type 1 pilus-mediated binding, and A-type proanthocyanidins, which have been shown to block P pilus-mediated binding (265–268). UPEC that was grown in human urine collected after consumption of cranberry juice had significantly reduced adherence to human red blood cells, resin beads coated with P-receptor oligosaccharides, and urothelial bladder cells compared to UPEC grown in normal urine (269, 270). Cranberry products are also less expensive and better tolerated than antibiotics, and thus are an intriguing candidate for UTI prophylaxis. However, the literature regarding the clinical efficacy of cranberry-prophylactic therapy remains inconclusive. For example, a recent meta-analysis of 10 randomized clinical trials found some benefit for women with recurrent UTI, but studies of protection in elderly or catheterized patients are lacking (271). A 2009 randomized controlled trial found that cranberry extract had similar efficacy to low-dose trimethoprim for preventing recurrent UTI in older women (272), while two randomized controlled clinical trials showed no significant effect of cranberry on UTI recurrence in adult pre-menopausal women (273, 274). Comparisons of these different studies may be confounded by differences in type of cranberry product consumed (e.g., juice vs. extract) and dosage regimens. Therefore, more studies are necessary to determine the effectiveness of cranberry products for preventing recurrent UTI.

Estrogen Therapy

Another therapeutic approach is the intravaginal application of estrogen. In a controlled trial of post-menopausal women with recurrent UTI, intravaginal application of a topical estriol cream significantly reduced the incidence of recurrence (275). The efficacy was attributed to a restoration of low vaginal pH and vaginal *Lactobacillus* colonization. In contrast, several studies have indicated that systemic estrogen-replacement therapy is not protective against recurrent UTI (276). A recent study found that vaginal estradiol therapy in post-menopausal women altered the expression of antimicrobial peptides and cell-junction proteins in epithelial cells isolated from voided urine, suggesting that estrogen therapy modulates the mucosal

barrier of the lower urinary tract (277). In support of these translational findings, several studies of experimental UTI in ovariectomized mice have demonstrated that altering estrogen levels has profound effects on UTI pathogenesis (277–279). Thus, vaginal estrogen therapy remains a safe and viable therapeutic option for post-menopausal women suffering from recurrent UTI.

Possible Applications to CAUTI

In contrast to uncomplicated UTI, CAUTI is dependent upon the presence of a foreign body, and removal of the catheter is often curative. Although it is not clear whether the biofilm forms first on the catheter, which in turn allows colonization of the damaged bladder mucosa, or vice versa, biofilm formation is clearly strongly associated with CAUTI, and the most common CAUTI pathogens, UPEC and *E. faecalis*, are good at making biofilm. The use of antibiotic- or silver-impregnated catheters has shown some efficacy in reducing the occurrence of bacteriuria in catheterized patients, but it is unclear whether they lower the rate of symptomatic CAUTI and associated complications (17). For those individuals who require long-term catheter placement, and particularly those with spinal-cord injuries or in nursing homes, the additional use of anti-infective drugs such as mannosides and sortase inhibitors in combination may help to prevent a large percentage of CAUTI, as well as infections of other implants. Furthermore, vaccines that target UPEC, *E. faecalis*, and *Proteus* spp. may also benefit these patients, reducing the incidence of CAUTI and potentially also the risk for bloodstream infections.

UTI VACCINES

Historical Perspective

Vaccines have been used against UTI for more than a century, though initially their intended purpose was therapeutic rather than prophylactic. In 1909, two case reports of pregnant women with pyelonephritis described significant clinical improvement after therapeutic systemic vaccination with *E. coli* (previously known as *Bacillus coli*) isolated from the urinary tract of the same patient (280, 281). Despite these anecdotal reports, by the 1920s therapeutic vaccination against UTI was largely seen as ineffective. A survey of a thousand American physicians reported little to no use of vaccine therapy for cystitis and pyelonephritis (282), and it was said that "the day of extravagant expectations from vaccine therapy [for UTI] is for the moment past (283)." As our understanding of vaccines advanced, UTI vaccines as a prophylactic therapy (i.e., "clinical" vaccines) reemerged as a topic of interest by the 1950s (20, 284), and have been the focus of much research, refinement, and testing in animals for the past 60 years. Efficacy in animals has been shown for UTI vaccines in every classical category: attenuated, inactivated, subunit, toxoid, and conjugate (Table 2). However, few modern vaccines have been tested in humans, and only one is currently commercially available.

Challenges in Developing UTI Vaccines

Clinical and technical challenges associated with developing a clinical UTI vaccine include our lack of understanding of the mechanisms that induce protective immunity in the urinary tract; the diverse patient subpopulations that would benefit from a vaccine; the heterogeneity of UPEC strains, which complicates the choice of the best target antigens; the route of administration (mucosal vs. systemic); and the choice of adjuvant, if needed. Furthermore, experimental conditions tightly control for genetics (often using inbred animals) and environment, whereas the human population is outbred and regularly encounters a broad diversity of environmental variables that could affect the mucosal immune system. When testing UTI vaccines in humans, care

must be taken to assess potential confounding variables. It is known that many common vaccines (e.g., influenza, pneumococcal, and zoster vaccines) do not induce optimal immune responses in a large portion of infants and the elderly (285). In addition, hysterectomy may decrease the effectiveness of vaginal-mucosal vaccines, as the cervix has been shown to be "the major inductive and effector site for cell-mediated immunity in the lower female genital tract (286)."

Recurrent UTI and protective immunity

The high frequency of recurrent UTI indicates that many patients are unable to mount an effective adaptive-immune response that prevents re-infection. The reasons for this are unknown. One possibility is that uropathogens may mask themselves from and/or directly suppress the immune system. For example, the attenuated UTI vaccine NU14 Δ*waaL* stimulates the immune system much more than wild-type NU14, which requires waaL O antigen ligase for host-immune suppression (287). Recent studies have found evidence that UPEC infection can promote a tolerogenic bladder environment, suggesting that UPEC exposure can both promote and prevent inflammation (288, 289). Alternatively, the body may downregulate its own immune response to uropathogens in order to maintain the integrity of the mucosal barrier (42, 290). Animal models of recurrent UTI have begun to shed light on the mechanisms of adaptive immunity in response to urinary tract infection. C57BL/6J mice that are repeatedly infected with UPEC become more resistant to experimental UTI (44). However, in C3H/HeN mice, which are genetically susceptible to chronic cystitis in an infectious dose-dependent manner, the outcome of the first infection in naive mice determines susceptibility to subsequent UTI (42, 45). Mice that spontaneously resolve the first episode of cystitis are resistant to bacterial challenge, but those that develop chronic infection upon the first infection are highly susceptible to severe, recurrent infection after antibiotic therapy, even when challenged by less virulent strains that do not cause severe infection in naive mice (O'Brien, Hannan, and Hultgren, unpublished data; 42). Understanding this puzzle will be critical in order to rationally design UTI vaccines with maximal therapeutic efficacy.

Who should receive an UTI vaccine?

Cystitis accounts for 90% of all UTI and recurs at high frequency, with 20% to 30% of women experiencing a recurrence within 3–4 months (7). These women are excellent candidates to receive a cystitis vaccine to lower the rate and severity of subsequent recurrences. The target population for a pyelonephritis vaccine is more limited. Children with vesicoureteral reflux (VUR), diabetics, and newly pregnant women or women of child-bearing age (pre-natal) might benefit from a pyelonephritis vaccine. Both children with VUR and diabetics are at higher risk for developing pyelonephritis (291, 292) and therefore might benefit from vaccination. Importantly, the use of antibiotic prophylaxis to reduce the frequency and severity of VUR-associated UTI in children is controversial among clinicians, and a systematic review of 20 randomized, controlled trials found no clear benefit for antibiotic prophylaxis (293, 294). Thus, new treatment strategies, such as vaccination, are needed for this patient population. Bacteriuria that progresses to pyelonephritis during pregnancy is associated with poor outcomes, including perinatal death, low birth weight, prematurity, and preterm low birth weight (295, 296). However, aggressive treatment of asymptomatic bacteriuria in pregnancy ensures that only a small percentage of pregnant women progress to pyelonephritis (297). Pregnant women with asymptomatic bacteriuria could potentially benefit from a pyelonephritis vaccine, if it were extremely effective and safe for both the mother and fetus. Systemic vaccination that induces immunoglobulin (Ig)G antibodies could also have the potential benefit of conferring passive immunity to the developing fetus, which could protect newborns during their first year or so of life. Indeed, a systemic P pilus-subunit vaccine administered to pregnant rhesus monkeys protected the newborns from pyelonephritis and induced a significant antigen-specific IgG response in the sera of both the mothers and newborns (298).

Choice of immunogen(s)

Effective UTI vaccines should target one or more surface-exposed bacterial structures that are either uniformly expressed by the uropathogen in the host or are expressed during critical stages of infection (Fig. 1). UTI vaccines in development can generally be categorized into two broad categories: "whole-agent" or "whole-cell" vaccines, which include whole bacteria (either live-attenuated or inactivated) and bacterial lysates, and "specific-antigen" vaccines, which include one or more antigens (subunit, toxoid, or conjugate vaccines). Although the majority of vaccines currently licensed in the United States are whole-agent vaccines, most of these target viral pathogens. Among the eight bacterial pathogens for which there are licensed vaccines (compared to 16 vaccines targeting viruses), only three are targeted with whole-cell vaccines. This is because, for bacterial pathogens, specific-antigen vaccines are generally much safer than whole bacterial cell vaccines, particularly when vaccinating systemically with Gram-negative bacteria, which can lead to endotoxemia. However, in the absence of whole organisms, isolated antigens typically do not elicit robust or long-lasting immune responses, and thus must be administered with adjuvants to increase the inflammatory response to the antigen, thereby enhancing immunogenicity (299).

Specific-antigen vaccines can only be as good as the antigen(s) selected. The process of selecting the best antigen(s) represents a critical and formidable challenge early in vaccine development. An ideal vaccine target antigen would be highly and broadly expressed and would be required either for

TABLE 2 Selected candidate vaccines targeting uropathogens

Vaccine/antigen	Animal model[a]	Human trials	Site of protection	Route[b]	Adjuvant[c]	Antibody responses	References
Multi-species vaccines							
Whole cell: inactivated	**SolcoUrovac:** 6 E. coli strains and 1 strain each of P. mirabilis, M. morganii, E. faecalis and K. pneumoniae	Trial in children and women Phase I & II clinical trials	Rodent bladder and kidney; Human bladder	IM V VS	None	Increased total and antigen-specific urinary IgG & IgA in mice Increased urinary sIgA Increased total IgG and IgA in vaginal wash and urine	326, 329–335, 419, 420
Whole cell: inactivated	**StroVac:** E. coli, P. mirabilis, M. morganii, K. pneumoniae and E. faecalis	Distributed by Strathmann GmbH (Hamburg) in parts of Europe, Latin America, and the Middle East	Bladder	IM	Aluminum phosphate	NRd	StroVac Product Insert
UPEC-targeted vaccines							
Whole cell: inactivated	Heat- or formalin-killed E. coli	Led to the development of SolcoUrovac/StroVac	Bladder, kidney	IV IP B IM V B	Freund's	Anti-E. coli IgG and IgA in urine	20, 318, 319, 321–324
Whole cell: attenuated	NU14 ΔwaaL	NR	Bladder	B	None	NR	287
Whole cell: attenuated	S. enterica serovar Typhimurium expressing S pili	NR	Kidney	Oral	None	Antigen-specific IgG and IgA in serum	370
Whole pili	P pili: purified recombinant synthetic peptides	NR	Kidney	ID IM IV SQ LN	Freund's	Anti-P IgG in serum and urine anti-P pilus IgM in serum	37, 298, 357, 365–368
Specific antigen: subunit	FimH adhesin: FimCH FimH truncate	Phase I and II clinical trials performed by MedImmune	Bladder	SQ IM IN	Systemic: Freund's MF59	Anti-FimH IgG in urine and serum Anti-FimH IgA in vaginal washes	38, 39, 300

Vaccine type	Antigen	NHP	License recently acquired by Sequoia Sciences		Route	Mucosal: CpG	Immune response	Ref
Specific antigen: subunit	PapDG adhesin	NHP	NR	Kidney	IP	Aluminum phosphate	Anti-PapDG IgG in serum	40
Specific antigen: subunit	IroN siderophore receptor	M	NR	Kidney	SQ	None	Anti-IroN IgG in serum	180
Specific antigen: subunit	FyuA siderophore receptor	M	NR	Kidney	IN	CT	Anti-FyuA IgG in serum (correlated with renal bacterial load) and IgA in urine	374
Specific antigen: subunit	FdeC adhesin	M	NR	Kidney	IN	CT	NR	122
Specific antigen: subunit	Dr adhesins	M	NR	Reduced mortality	NR	Freund's	Anti-Dr IgG in serum, but no significant effect on colonization	371
Specific antigen: subunit	S pili	R	NR	Kidney	SQ	None	Anti-S IgG in serum	370
Specific antigen: toxoid	α-hemolysin	M	NR	Kidney	IM	Freund's	Anti-α-hemolysin IgG in serum	193
Specific antigen: conjugated	O antigen	R, NHP	NR	Bladder, kidney	SQ SQ+IV LN+SQ	Freund's	Anti-O8 IgG in serum	341–343
Specific antigen: conjugated	K13 antigen	M, R	NR	Kidney	SQ	DT Conjugated to bovine serum albumin (BSA)	Anti-K13 IgG and IgM in serum	344, 345
Specific antigen (multi-epitope) subunit	IreA, Hma and IutA iron receptors	M	NR	Bladder, kidney	IN	CT	Antigen-specific IgG and IgM in serum and IgA in urine	373
Proteus-targeted vaccines								
Whole cell: inactivated	Heat-killed P. mirabilis	R	NR	Kidney	SQ	None	NR	380
Whole cell: attenuated	L. lactis MrpA	M	NR	Kidney	IN	None	Anti-MrpA IgG and IgA in serum	384
Specific antigen: subunit	MrpH adhesin	M	NR	Bladder, kidney	IN, V	CT	High levels of IgG in serum, but no correlation with protection	383

(Continued on next page)

TABLE 2 Selected candidate vaccines targeting uropathogens (Continued)

Vaccine/antigen	Animal model[a]	Human trials	Site of protection	Route[b]	Adjuvant[c]	Antibody responses	References
Specific antigen: MrpA pilus subunit	M	NR	Bladder, kidney	IN, SQ, V	Freund's CT	Anti-MrpA IgG in serum and urine, Anti-MrpA IgA in serum and urine, but no correlation with protection	385, 386, 388
Specific antigen: UcaA adhesin subunit	M	NR	Bladder, kidney	SQ	Freund's	Anti-UcaA IgG in serum, but no correlation with protection	385
Single antigen: Proteus toxin agglutinin (Pta) Toxoid	M	NR	Kidney, spleen	IN	CT	Anti-Pta IgG in the serum	381

[a]Animal models: M, mice; NHP, non-human primates; R, rats; Rb, rabbits.
[b]Routes of delivery: B, intravesical; ID, intradermal; IM, intramuscular; IN, intranasal; IP, intraperitoneal; IV, intravenous; LN, intra-lymph node; SQ, subcutaneous; V, vaginal instillation; VS, vaginal suppositories.
[c]Adjuvants: CT, cholera toxin; DT, diphtheria toxoid.
[d]NR, not reported.

the initiation and/or maintenance of infection or for disease symptoms. For example, toxoid (inactivated toxin) vaccines are highly successful against diseases in which the toxin itself is the main cause of disease, such as tetanus and diphtheria. The development of subunit (protein antigen), conjugated (carbohydrate antigen conjugated to antigenic protein), and DNA (protein expressed from a DNA vector inside the host) vaccines has been the focus of much research, but has been met with limited success. The challenge is that bacterial infections are often much more complex than viral infections, involving multiple host niches and more antigenic variation. Additionally, bacteria have evolved to have redundant virulence mechanisms, a fact highlighted by the multiple adhesins and iron-acquisition systems in many UPEC isolates (53–55). Furthermore, mechanistic studies of bacterial pathogenesis must be carried out in human cell lines or in animal models, and may not accurately reflect the requirement for vaccine targets to initiate infection and cause human disease. Finally, with the exception of the iron-acquisition factors and type 1 pili, the majority of putative UPEC-virulence factors that have been described are found in about 50% or fewer of all isolates (Table 1), and thus would only be useful in a multi-subunit vaccine.

Once a candidate antigen is chosen, it must be tested in animal-infection models, but the results of such efficacy studies may be difficult to parse. Several potential confounders that can vary among studies must be assessed. Among vaccines comprising whole bacteria, outer-membrane vesicle, and membrane-protein extract preparations, the method of bacterial preparation can have profound effects on the efficacy of the vaccine, as the antigens present will vary with the culture conditions. For example, UPEC grown statically at 37°C in LB liquid media will predominantly express type 1 pili, whereas UPEC cultured on tryptic soy agar plates at 37°C will express P pili. This bias can be overcome with live-attenuated vaccines, which can replicate, mimic the natural route of infection, and change their gene expression accordingly once introduced into the host. Other variables in testing UTI vaccine efficacy in animal models include the animal-infection model and choice of uropathogen used for bacterial challenge, the culture conditions of the challenge-bacterial inoculum, and, as we discuss below, the vaccine-inoculation regimen, including the route and frequency of immunization and choice of adjuvant.

Route of administration

Both mucosal (vaginal and intranasal) and systemic UTI vaccines have been effective in animal models. In general, mucosal vaccines elicit both IgA and IgG responses and systemic vaccines elicit only IgG responses. Since IgA is thought to be protective against intimate and invasive infection of the gut, and is found in high concentrations at mucosal sites, it has been traditionally assumed that IgA is the most effective means of inducing mucosal immunity. However, one group compared systemic and mucosal routes of vaccination with the FimH adhesin and found that although only the mucosal route induced elevated levels of vaginal wash and urine antigen-specific IgA in mice, both vaccine delivery routes resulted in similar serum and urine antigen-specific IgG responses and protection against experimental UTI (300). It is possible that this finding may be explained by the experimental model, in which UPEC were instilled directly into the bladder, thus avoiding the initial stages of periurethral colonization and urethral ascension, where IgA may be more important for protection. However, systemic vaccination and serum IgG responses have provided protection against other mucosal pathogens, such as rotavirus and human papillomavirus (301, 302). Therefore, antigen-specific IgG may be equally or more important than IgA for host defense at some mucosal surfaces. Also, data from experimental UPEC infection in mice suggest that the urogenital mucosa may become "sensi-

tized" to uropathogens subsequent to an initial chronic bladder infection (42, 45). In these individuals, a vaginal-mucosal route of vaccine delivery may exacerbate this sensitization.

Vaccine adjuvants

The choice of adjuvant can be critical for adequate stimulation of the immune system, but relatively little is known about how adjuvants work (303), and only a few are approved for use in humans. In order to be approved for use in clinical vaccines, adjuvants must have low toxicity. Adjuvants currently approved for use in humans include aluminum salts (e.g., alum), the squalene-based MF59, the LPS-derived monophosphoryl lipid A (MPL), and liposomes (304). The aluminum-based adjuvants aluminum phosphate and aluminum hydroxide are commonly used in systemic vaccines in humans. The specific functions of aluminum adjuvants continue to be debated, but in general, it is accepted that they form a depot at the injection site, allowing for efficient uptake of antigen by antigen-presenting cells (APCs). They also stimulate the immune system by inducing eosinophilia and activating complement and macrophages (305). Of note, alum is a poor stimulator of cellular (TH1) immune responses (306). MF59 is a squalene-based oil-in-water emulsion. After intramuscular injection, the squalene-emulsion droplets are internalized by dendritic cells and enhance antigen presentation (307). Gene-expression analysis of mouse muscle found that, compared to alum and the TLR9 agonist CpG, MF59 induces more changes in gene expression and is a stronger inducer of genes involved with cytokine responses, leukocyte migration, and antigen presentation (308). As a result, MF59 elicited a more rapid influx of myeloid cells to the site of injection. Monophosphoryl lipid A has been modified to reduce its toxicity, while retaining its ability to induce inflammation. It is an agonist for TLR4, although it is unclear whether this agonism is the main cause of its efficacy as an adjuvant. Due to its hydrophobicity, it has a strong propensity to aggregate into microparticulates that are potent activators of the NLRP3 inflammasome (309). Liposomes are thought to enhance immunogenicity, both by enhancing phagocytosis by APCs and by enabling direct cytoplasmic delivery of antigens by membrane fusion (310). Recently, combinations of the above adjuvants have been the subject of much research. In particular, MPL in liposomes has shown great promise, as liposomes have the dual benefit of masking the residual toxicity of MPL while enhancing its potency as an adjuvant (310).

Mast cell-derived adjuvants

Mast cells are important players in the bladder innate-immune response. Not only have they been implicated in early defense against UTI in mice, but mast cell-derived factors play an important role in directing the adaptive-immune response (311–313). Mast cell-derived adjuvants are an interesting recent development in vaccinology. Nasal instillation of vaccine antigens along with small-molecule mast-cell activators resulted in antigen-specific serum IgG and mucosal (nasal, vaginal, fecal) IgA responses that correlated with increased dendritic cell and lymphocyte recruitment to the lymph nodes (312). Recently, Abraham and colleagues described the synthesis of submicrometer particles that model mast-cell granules and showed their successful use as an adjuvant in a mouse model of influenza (314). An advantage of these particles is that they can be engineered to contain particular cytokines in order to skew the adaptive-immune response. To the best of our knowledge, mast cell-related adjuvants have not yet been tested in UTI vaccines. However, they are an intriguing candidate for further study.

Innate Immunity and the Rise of Systems Vaccinology

In the past, vaccine development was most often a hit-or-miss venture, with little understanding of why some vaccines are efficacious and others are not. To a certain extent this is still true today, but in recent years, vaccinologists and immunologists have begun to understand the role of the innate-immune system in vaccine efficacy. The innate-immune system relies on pattern-recognition receptors (PRRs) expressed by innate-immune cells in order to detect pathogen-associated molecular patterns (PAMPs). An important category of PRRs are the Toll-like receptors (TLRs), which can detect molecular patterns commonly found in bacteria, viruses, fungi, and parasites; C-type lectins and nucleotide oligomerization domain (NOD)-like receptors are also important innate-sensing receptors. Signaling by PRRs on innate-immune cells can trigger an adaptive-immune response that differs based on the PRR and the dendritic-cell subtype (315). Recently, systems biology approaches have been used to assess the effects of vaccination on the immune system, with a particular focus on the early innate response, which can predict vaccine immunogenicity (285). This "systems-vaccinology" approach was used by Pulendran and colleagues to investigate changes in the human immune system after vaccination with a live-attenuated yellow fever vaccine. By performing multiplex-cytokine assays and microarrays with blood collected at baseline and at different time-points post-vaccination, the authors were able to characterize a "molecular signature" involving the innate sensing of viruses that very accurately predicted the development of antiviral immunity (316). The same group subsequently used systems vaccinology to compare immune responses among vaccines for yellow fever, influenza, and meningococcus, revealing distinct transcriptional responses that correlated with antibody responses specific to each vaccine (317). Systems-vaccinology approaches may be useful for assessing the immunological profiles of UTI vaccines, predicting efficacy in vaccinated individuals, and determining the best adjuvant for a given vaccine. It is interesting to note that the mechanism of protection of the only commercially available UTI vaccine, StroVac, is currently unknown. Systems vaccinology may be the key to elucidating the efficacy of this and other UTI vaccines.

Whole-Cell Vaccines in Development

Whole-cell vaccines have been among the most successful vaccines developed to date. Indeed, the only UTI vaccine currently available for use in humans is the polyvalent-inactivated whole-cell vaccine StroVac. Vaccines comprising whole uropathogens, whether attenuated or inactivated, expose the host to a variety of virulence factors. These preparations may or may not include, depending upon how the preparation is grown and processed, pili and other adhesins, outer-membrane proteins, toxins such as hemolysin and CNF1, siderophore receptors, flagellin, and LPS (Fig. 1). Of all vaccine types, live-attenuated vaccines have the potential to most closely mimic natural infections, and thus elicit strong immune responses. However, they cannot be given to immunocompromised patients, and there may be a risk of reversion to virulence in healthy individuals. Inactivated vaccines are generally safer than live ones, but this can be accompanied by the tradeoff of eliciting a weaker immune response than live-attenuated vaccines.

Inactivated vaccines

Inactivated *E. coli* vaccines have been investigated since at least the 1950s and have been found to protect animals from UTI. An early UTI vaccine consisted of intravenously injected, heat-killed *E. coli*, and was protective against pyelonephritis in rabbits (20). In the 1970s, systemic vaccination with heat-killed or formalin-killed *E. coli* strains grown in trypticase soy broth (TSB, which induces P pili expression) protected rats from retrograde *E. coli* pyelonephritis (318) and ascending UTI (319), but did not protect rabbits

against hematogenous pyelonephritis (320). Rats that were vaccinated by intravesical instillation of formalin-killed *E. coli* were protected against ascending UTI and resolved UTI faster than non-vaccinated controls (319, 321). To the best of our knowledge, vaginal-mucosal immunization with an inactivated-UTI vaccine was first published in 1982, when vaginal instillation with formalin-killed *E. coli* protected rats from cystitis (322). In a later study, vaginally instilled, but not systemically injected, formalin-killed *E. coli* inhibited bacterial adhesion to rat bladder mucosae (323). In 1987, vaginal immunization with formalin-killed *E. coli* protected cynomolgus monkeys from cystitis (324), and in 1995, intramuscular injection of formalin-killed *E. coli* protected monkeys from pyelonephritis-associated renal scarring (325).

Urovac and StroVac

With several decades of research showing the efficacy of inactivated-UTI vaccines in animals, Solco Basel Co. developed SolcoUrovac for use in humans. SolcoUrovac was a polyvalent whole-cell vaccine consisting of 10 strains of heat-killed uropathogens: six from *E. coli* of different serotypes and one each from *Klebsiella pneumoniae*, *Proteus mirabilis*, *Morganella morganii*, and *Enterococcus faecalis*. It was initially administered by three intramuscular (intragluteal) injections at weekly intervals. The first results of clinical trials with SolcoUrovac, which were performed in Europe, showed that the vaccine was protective against recurrent UTI (326). Current information about SolcoUrovac is not available as Solco Basel appears to be defunct and SolcoUrovac is unavailable in Europe at this time. However, another intramuscular polyvalent inactivated UTI vaccine, called StroVac (Strathmann AG, Hamburg, Germany), is apparently approved for use in Europe. StroVac contains the same 10 strains in a different formulation [http://www.strathmann.de/index.php/en/component/content/article/112-pflichtangaben/367-strovac-pflichttext-, reference in German].

While these vaccines have shown promise (261, 327), to our knowledge they have never undergone large phase III studies to demonstrate efficacy. However, the European Association of Urology's "Guidelines on Urological Infections" mention SolcoUrovac and StroVac as options to consider in the non-antibiotic prophylaxis of recurrent UTI (http://uroweb.org/wp-content/uploads/18-Urological-Infections_LR.pdf).

Vaginal mucosal delivery of SolcoUrovac

In the initial human studies with intramuscular injections of SolcoUrovac, some women experienced adverse effects such as pain (5.4%), fever (3.5%) and swelling at the injection site (1.5%) (326). Thus, David Uehling, a pioneer in the field of UTI vaccines from the University of Wisconsin, tested the efficacy of vaginally administered SolcoUrovac, hypothesizing that mucosal administration in the vagina would reduce adverse effects. Vaginal instillation was effective in mice (328), cynomolgus monkeys (329), and women (330), paving the way for phase II clinical trials of vaginally instilled SolcoUrovac in the United States. SolcoUrovac's phase II clinical trials with the vaginal suppository form of the vaccine, which were published between 1996 and 2007, were only partially successful (331–334). The most effective treatment course was determined to be six vaginal suppositories given at weeks 0, 1, 2, 6, 10, and 14. With this vaccination regimen, the percentage of women having a recurrence during the 6-month trial declined from 83% to 89% in the placebo-treated groups to 45% to 54% in the vaccinated, boosted groups. However, these differences were not always significant. In one trial, the authors identified six patient sub-populations with significantly lower re-infection rates after vaccination: "women younger than 52 years, without a childhood history of recurrent UTIs, [having] 6 or more UTIs in the previous year, without a hysterectomy, using estrogen, [or] using birth-control pills (331)." Adverse events (e.g.,

low-grade fever, nausea, vaginal irritation) did occur, but no patient was unable to complete the treatment.

The adaptive response to SolcoUrovac

After intramuscular vaccination, mice had 10-fold more total IgG and 2-fold more total IgA in the urine; IgM was not found in the urine (335).Vaginal immunization of monkeys was protective but did not increase anti-*E. coli* serum, vaginal wash, or urine antibody levels (serum IgG, IgM, and IgA; vaginal wash and urine IgA and IgG) (329). Interestingly, although some vaccinated women did have increased antibody titers over time, there were no significant differences in any of the tested antibody levels among treatment groups in any phase II trial (331–334). As expected by the formulation, SolcoUrovac was most effective against UPEC uropathogens. In one study, the percentage of women who experienced a UPEC UTI after vaccination and boosting was 27.5%, compared to 70% of women in the placebo group. However, the percentage of infections caused by uropathogens other than UPEC increased dramatically in vaccinated women and overall the vaccine did not statistically significantly prevent recurrent UTI when all uropathogens, not just UPEC, were considered (331). Perhaps because of this shift in uropathogens, to the best of our knowledge vaginal SolcoUrovac has not progressed to Phase III trials or beyond.

CP923

In 2007, Johnson and colleagues described a candidate vaccine consisting of a formalin-killed derivative of the *E. coli* sepsis strain CP9 that is deficient in capsule and O-antigen, termed CP923 (336). Compared to formalin-killed CP9, intranasal vaccination with formalin-killed CP923 resulted in a significantly greater systemic-antibody response that was able to bind to a subset of heterologous UPEC and bacteremia strains. The mucosal immune response and protection against urinary-tract infection were not assessed, but this study shows the potential benefit of using genetically modified UPEC for inactivated vaccines.

Attenuated vaccines

Attenuated vaccines have the potential benefit of progressing through early steps in disease pathogenesis in the relevant host niche. Recently, the Klumpp group identified as a vaccine candidate a mutant of the UPEC strain NU14 that lacks the O-antigen ligase *waaL* (287). This gene was identified in a screen of transposon mutants that had lost the ability to suppress IL-8 production by bladder epithelial cells (337). In a murine-UTI model, NU14 Δ*waaL* was significantly more inflammatory and less virulent than wild-type NU14. Vaccination with NU14 Δ*waaL* protected mice from challenge infection with NU14, CFT073, and four UPEC isolates from the *E. coli* Reference Collection (287, 338). However, protection waned over time and was absent by 8 weeks. Interestingly, vaccination also significantly reduced the level of persistent bladder colonization (indicative of a QIR population) by NU14 14 days after challenge, even though the NU14 Δ*waaL* (vaccine) strain itself is unable to persist past acute infection. The authors hypothesized that the lack of O antigen on LPS in NU14 Δ*waaL* may allow for Toll-like receptor 4 recognition of LPS lipid A, or may increase the exposure of bacterial-surface antigens to antigen-presenting cells, thereby stimulating protective immunity (287). However, the recent finding that O antigen modulates infection-induced bladder pain, and that serial infections with NU14 Δ*waaL* result in chronic bladder pain, diminishes the promise of NU14 Δ*waaL* as a vaccine candidate (339).

Specific-Antigen Vaccines in Development

Specific-antigen vaccines (such as toxoid, conjugate, and subunit vaccines) have become more popular in recent years due to advances in protein purification and the development of recombinant-DNA technology. Single-antigen vaccines typically have lower rates of adverse events than whole-cell vaccines, and several antigens may be combined in a single multi-epitope vaccine to increase efficacy, e.g., in a recent extraintestinal pathogenic E. coli vaccine. (340). Candidate antigens may be revealed by UTI-virulence studies *in vitro* and in animal models; alternatively, reverse vaccinology allows researchers to predict effective antigens computationally.

Conjugate vaccines

Conjugate UTI vaccines against UPEC capsule and LPS components have shown protection in animal models after same-strain challenge, but have not been tested clinically in humans. In early studies, intraperitoneal injection of *E. coli* endotoxin protected rats from pyelonephritis (284), and the protection was later determined to be mediated by antibodies against O antigen (341). Subcutaneous and bladder injection of purified O antigen from *E. coli* O6 serotype protected rats from bladder infection with the same O6 strain (342). Decades later, O polysaccharide prepared from *E. coli* O8 LPS, conjugated to bovine serum albumin, reduced renal scarring and intratubular-neutrophil infiltration in rhesus monkeys that were challenged with an O8 UPEC strain (343). Purified *E coli* K13 polysaccharide conjugated to bovine serum albumin protected rats from pyelonephritis (344). A different group conjugated *E. coli* K13 to diphtheria toxoid and found that the vaccine decreased renal bacteria load and disease-severity scores in mice after challenge with a K13 UPEC strain (345). A considerable challenge in formulating a vaccine targeting capsule or O antigen, the most exposed component of LPS, is the fact that there is a great heterogeneity of serotypes among *E. coli* isolates. For example, 6 different O serotypes account for only 75% of UPEC isolates (346), making the formulation of a broadly protective conjugate vaccine impractical. Furthermore, some capsule serotypes, such as K1, are thought to evade the host-immune response by molecular mimicry, potentially making them poor vaccine candidates (347).

Toxoid vaccines and outer-membrane vesicles

UPEC toxins have not been demonstrated to play a required role in UTI pathogenesis, so they are not ideal vaccine candidates. For example, a purified α-hemolysin toxoid vaccine prevented renal injury, but not colonization, in mice after challenge with a hemolytic UPEC strain (193). Rather than being secreted as "naked" proteins, UPEC toxins such as α-hemolysin and CNF1 are associated with outer-membrane vesicles (OMVs), which bleb from the surface of Gram-negative bacteria during all stages of growth (348). OMVs also contain adhesins, enzymes, and nonprotein antigens like LPS (348), and recently, a high-throughput tandem mass spectrometry approach was used to define the UPEC outer membrane proteome (349). OMVs may be a mechanism for UPEC and other bacteria to protect their toxins while they are *en route* to host cells, and to deliver "concentrated bursts of effector molecules" to modulate host-cell processes (350). OMVs are intriguing vaccine candidates, and because they contain LPS and other pro-inflammatory virulence factors, they should not require adjuvants to stimulate the immune system. Several successful meningococcal-OMV vaccines have been developed, and other OMV vaccines have been effective in mice (351–355). UPEC OMVs are thus candidate vaccine antigens, though to our knowledge they have not been tested.

Pili as vaccine candidates

Pili are adhesive surface organelles that mediate the colonization of mucosal surfaces. Adhesins make an attractive antigen candidate, because antibodies raised against an adhesin should be able to block adhesin-host cell-receptor binding, thus disrupting bacterial colonization of the host (356). Several types of pili have been investigated as UTI vaccine candidates. Pilus vaccines have been tested since well before the effects of growth conditions on pilus production were fully understood, and as such, many early pilus vaccine papers do not describe the bacterial growth conditions for the challenge inocula. When possible, we will report the relevant information. To our knowledge, the first pilus-UTI vaccine was published in 1979. Rats that were vaccinated intradermally with pili purified from two clinical isolates of *E. coli* were protected from pyelonephritis; anti-pili antibodies raised in rabbits were also protective in rats (357).

FimH

Vaccines targeting the type 1 pilus adhesin FimH, which plays a critical role in UTI pathogenesis in the lower urinary tract in animal models, have been tremendously effective in animals that are challenged with bacteria grown in type 1 pilus-inducing conditions. Since the tip adhesin is functionally critical, but not highly abundant, purified adhesin was found to be better than whole pili at eliciting antibodies that blocked receptor binding (39). FimH can be purified bound to its periplasmic chaperone FimC, or as a naturally occurring, mannose-binding FimH truncate (FimHt) (358). Both antigens protected subcutaneously vaccinated mice from cystitis (39). The FimCH vaccine also protected intramuscularly injected cynomolgus monkeys from bacteriuria and pyuria though the number of animals tested was by necessity small (38). Of note, only one out of four FimCH-vaccinated monkeys developed bacteriuria and pyuria upon challenge infection with type 1 pili-expressing UPEC (compared to four out of four control monkeys), and this was also the only FimCH-vaccinated monkey without increased anti-FimH IgG in vaginal secretions. Another group compared the efficacy of a recombinant FimHt vaccine administered either intranasally or intramuscularly to mice. Both routes were protective against cystitis, but the intranasal vaccine induced greater anti-rFimHt IgA in vaginal washes (300). Yet another group demonstrated that subcutaneous administration of recombinant *fimH* fused to the flagellin subunit *fliC*, a TLR5 agonist and candidate adjuvant, protected mice against cystitis upon challenge with a type 1 pilus-expressing clinical isolate; vaccination with admixed FimH, FliC, and Montanide ISA 206 adjuvant was also protective (359, 360). These investigations of the cellular immune response to vaccination are unique among UTI vaccine studies, which generally test the humoral response only. Immunization resulted in T_H1 and T_H2 responses as assessed by cytokine responses in splenocyte-proliferation assays and ratio of IgG1 to IgG2a. Lastly, another group has recently developed a mammalian codon-optimized *fimH* plasmid construct for use in a DNA vaccine, whereby plasmid DNA injected into mice can induce a protective immune response (361, 362). Mice that received the DNA vaccine via footpad injection had significantly reduced bladder colonization 48 hours post-challenge and significantly increased urine IgA titers (363).

While it has been reported that monoclonal antibodies raised against FimH do not block, but rather enhance, adherence to bladder epithelial cells *in vitro* (364), the above studies demonstrate that polyclonal IgG raised against FimH block bacterial binding to bladder epithelium and is clearly protective *in vivo*. Possible explanations for this discrepancy include steric hindrance preventing the antibody-adhesin complex from binding to the uroplakin-receptor pocket and the effects of opsonization. The FimCH vaccine was originally licensed by MedImmune (Gaithersburg, Maryland, USA) and entered Phase I and II trials in the early 2000s. The vaccine was found to be safe in Phase I trials, but was dropped from development during Phase II trials. Sequoia Sciences (St. Louis, Missouri, USA) has since acquired the license and the vaccine is re-entering clinical trials

in women with recurrent UTI, using a new adjuvant. In two pre-clinical rabbit studies conducted by Sequoia, serum IgG anti-FimH titers were greater than 1:1,000,000, with no apparent adverse effects from the vaccination (personal communication, Gary Eldridge, Sequoia Sciences).

P pili

P pilus-subunit vaccines to protect against pyelonephritis became a hot topic in the 1980s and the initial studies showed promise. Vaccination with purified recombinant P pili blocked renal colonization in mice when the challenge bacteria were grown under P pilus-inducing conditions (37, 365). Vaccination with purified P pili protected monkeys (366, 367) and the unvaccinated infants of vaccinated monkeys (298) from pyelonephritis when the challenge bacteria were grown under P pilus-inducing conditions. Finally, synthetic P pilus peptides that were prepared by solid-phase Merrifield synthesis and conjugated to carrier proteins prevented urine and renal colonization in mice (368). However, studies utilizing whole-purified P pilus-UTI vaccines have not been published since the late 1980s. This is likely due to the high degree of antigenic variation among UPEC strains in the major P pilin subunit, PapA, which is the most abundant pilin protein in P pilus preparations. Indeed, natural P pilus-specific antibodies from patients with pyelonephritis do not seem to target the binding pocket of PapG as they are unable to prevent P pili-mediated hemagglutination (96). Consistent with this, an inactivated whole-cell vaccine consisting of formalin-killed P-fimbriated *E. coli* offered only limited protection against renal dysfunction and scarring in monkeys (325). Thus, whole-cell vaccines may not be an effective way of inducing anti-pilin antibodies, even if they are being expressed on the bacterial surface.

PapDG vaccine

By 1988, the composition of P pili had been determined, and the tip-adhesin protein

PapG was identified as a vaccine candidate (61, 369). Lund and colleagues suggested that PapG could be purified in a complex with its periplasmic-chaperone protein PapD, analogous to the FimCH vaccine. In 2004 it was shown that intraperitoneal administration of a purified PapDG vaccine protected cynomolgus monkeys from pyelonephritis (40). The efficacy was presumed to be the result of PapG-specific antibodies blocking the pilus adhesins and thereby preventing colonization, though the specific mechanism of protection is unknown. To the best of our knowledge, no further studies have been conducted with the PapDG-pyelonephritis vaccine.

Other pili

Among the adhesins expressed by some UPEC strains are S pili and Dr adhesins (and others), each of which has been used as a vaccine antigen. Rats vaccinated with purified recombinant S pili had reduced kidney colonization (370). In the same study, an avirulent strain of *Salmonella enterica* serovar Typhimurium was genetically transformed to produce S pili, and live bacteria were orally administered to rats, which had reduced kidney colonization compared to mock-vaccinated and purified S pilus-vaccinated mice (370). In addition, mice vaccinated with purified recombinant Dr adhesins had reduced UTI-associated mortality (371). However, these adhesins are even less broadly conserved among UPEC than are P pili antigens, and thus, these targets would only be useful in a multi-epitope vaccine. Pilus antigens from non-UPEC uropathogens have shown efficacy in mouse models of infection, as described below.

Subunit vaccines and reverse vaccinology

Recently, investigators have used information gathered through bioinformatic, genomic, and proteomic analyses to identify novel candidate antigens, in an approach termed "reverse vaccinology." The first web-based reverse-vaccinology program, Vaxign, was used to predict 22 UPEC outer-membrane proteins as potential vaccine targets (372). Some of these targets had been previously shown to be immunogenic and protective in animal models, while others remain to be tested. A large-scale reverse-vaccinology screen was used to identify vaccine-antigen candidates in *E. coli* CFT073, which is predicted to encode 5379 proteins. The criteria for candidate antigens were pathogen specificity, high *in vivo* expression, induction during growth in human urine, antigenicity, and surface exposure. Six candidates were identified, each an outer membrane-receptor protein involved in bacterial iron or heme acquisition. When purified and administered intranasally, the candidate antigens Hma, IreA, and IutA protected mice from challenge infection (373). Vaccination with Hma, a heme receptor, protected the kidneys; IreA, a putative siderophore, protected the bladder; and IutA, a siderophore receptor for aerobactin, protected both the bladder and the kidneys. The three antigens also induced antigen-specific IgA in the urine and class-switching from IgM to IgG in the serum (373). A subsequent study investigated additional UPEC outer membrane iron receptors as vaccine candidates. The yersiniabactin receptor FyuA, purified and administered intranasally, protected mice from developing pyelonephritis upon challenge with 536, a UPEC strain that expresses FyuA. Vaccination-elicited anti-FyuA IgA in urine and IgG in serum, and serum-antibody levels were correlated with kidney bacterial burden (374). A recent RNAseq analysis confirmed that the yersiniabactin system is highly expressed by UPEC during uncomplicated cystitis in women (375). Another group had previously found that systemic vaccination with the siderophore-receptor IroN protected against renal colonization in mice (180). This last study did not explicitly use a reverse-vaccinology approach, but IroN was chosen because of its prevalence among clinical UPEC isolates, its role in urovirulence, and its expression in bodily fluids.

Another reverse-vaccinology approach employed by a group at Novartis (Siena, Italy) involved comparing the genome of a neonatal meningitis-associated K1 strain of *E. coli* with known pathogenic and nonpathogenic *E. coli* strains. Potential antigens were chosen if they were predicted to be surface-associated or secreted, with no more than three transmembrane domains, and were absent from nonpathogenic strains. Two hundred and thirty candidates were identified in this manner and tested for protection in a murine-sepsis model; nine were protective (376). One protective antigen, named FdeC for factor-adherence *E. coli*, was found to be expressed by most UPEC, but only upon host cell contact, helping to mediate *E. coli* adhesion to mammalian cells. Intranasal vaccination with recombinant FdeC significantly reduced kidney colonization in mice that were challenged with UPEC strains 536 or CFT073 (122). Another protective antigen, SslE (secreted and surface-associated lipoprotein from *E. coli*; also known as YghJ), was found to be involved in the degradation of mucin substrates (377). Intranasal vaccination with recombinant SslE significantly reduced kidney and spleen colonization in mice that were challenged intravesically with the UPEC strain 536, which expresses a different variant of SslE.

Vaccines Against Non-UPEC UTI

Uropathogenic *E. coli* cause approximately 85% of uncomplicated UTI, and so it is not surprising that most tested UTI vaccines have used UPEC strains and antigens. However, vaccines targeting other uropathogens have been protective in animal models. The polyvalent-inactivated vaccine SolcoUrovac/StroVac (described in detail above) contains one strain each of *Klebsiella pneumoniae*, *Proteus mirabilis*, *Morganella morganii*, and *Enterococcus faecalis*. Vaginally instilled, formalin-killed *K. pneumoniae* inhibited bacterial adhesion to rat bladder mucosae (323). In addition, a recent immunoproteome analysis identified candidate antigens for a *K. pneumoniae* vaccine (378), but to the best of our knowledge, these antigens have not been tested in a UTI model. Most of the other non-UPEC vaccines have targeted *P. mirabilis*, which causes about 3% of uncomplicated and 13% of complicated UTI (379).

Proteus vaccines

Vaccines against *P. mirabilis* infection have been tested since at least the 1960s, when heat-killed *P. mirabilis* protected rats from pyelonephritis by promoting renal clearance of bacteria (380). A preparation of *P. mirabilis* outer-membrane protein promoted renal clearance in mice and protected mice from renal infection and death (379). Purified inactivated-*Proteus* toxic agglutinin (Pta), a cytotoxic surface-associated alkaline protease, was conjugated with cholera toxin and protected mice from kidney colonization (381). Finally, several *P. mirabilis* adhesins have been tested as vaccine antigens. *P. mirabilis* expresses MR/P (mannose-resistant, *Proteus*-like) fimbriae on the cell surface, and most of the bacterial population synthesizes MR/P fimbriae during UTI (382). Vaccination with purified MR/P fimbriae was protective against ascending *P. mirabilis* UTI in a murine model (383). An attenuated-mucosal vaccine consisting of *Lactococcus lactis* expressing the recombinant MrpA subunit of MR/P fimbriae significantly reduced renal colonization in mice after *P. mirabilis* challenge (384). Systemically injected, purified recombinant MrpA also protected mice from ascending *P. mirabilis* UTI (385). Intranasal vaccination with recombinant MrpA protected mice from ascending *P. mirabilis* UTI (386, 387); the addition of a cholera toxin adjuvant did not enhance protection (388). MrpH is the MR/P fimbrial-tip adhesin, similar to FimH and PapG (described above). Vaccination with recombinant MrpH was protective against ascending *P. mirabilis* UTI in a murine model (383). A fusion protein comprised of recombinant MrpH and UPEC FimH protected intranasally-vaccinated mice from challenge with either P. mirabilis or UPEC (389). Other *P. mirabilis* antigens have also been tested. Vaccination with the urothelial cell adhesin subunit UcaA protected mice from *P. mirabilis* infection in a hematogenous infection model (385). However, intranasal vaccination with flagellin was not protective in mice. Interestingly, flagellin co-administered with MprA negated the protective effect of MprA vaccination, suggesting an immunomodulatory effect (387).

Enterococcus vaccines

The endocarditis- and biofilm-associated pilus (Ebp pilus) (Fig. 2) is an attractive vaccine candidate. Recently, the minor pilus subunit EpbA was found to be an adhesin that binds to fibrinogen, a host protein that is released upon bladder catheterization (216). Systemic vaccination of C57BL/6 mice with *E. faecalis* EbpA prior to catheter implantation and *E. faecalis* infection significantly reduced bacterial colonization of the catheter and protected against CAUTI. Furthermore, serum from vaccinated mice was found to block EbpA binding to fibrinogen *in vitro*.

Immunotherapeutic Compounds

OM-89/UroVaxom
The immunotherapeutic formulation OM-89 (marketed in Europe by EurimPharm GmbH as UroVaxom) is a bacterial extract prepared from 18 strains of *E. coli*. For the purpose of preventing recurrent UTI, it is administered orally, typically as a daily dose for three months, and is recommended by the European Association of Urology for women with recurrent uncomplicated UTI. Several meta-analyses have assessed the effectiveness of OM-89 in preventing recurrent UTI in humans (261). A meta-analysis of five placebo-controlled double-blind studies found that OM-89 was superior to placebo with regards to reducing UTI frequency and dysuria,

bacteriuria, and leukocyturia (390). Another meta-analysis of five studies found that the mean number of UTI episodes and the use of antibiotics were reduced in patients treated with OM-89 (391). OM-89 is generally safe and well-tolerated; the most frequent adverse events are headache and gastrointestinal events. While early studies looked at UTI recurrence over just six months from the start of treatment, OM-89 is effective for up to 12 months when booster doses are administered for the first 10 days in months 7, 8, and 9 (392). Several other immunostimulatory compounds exist for the prevention of recurrent UTI. Uromune, a preparation of *E. coli*, *K. pneumoniae*, *P. vulgaris*, and *E. faecalis*, was recently evaluated in a multicenter, retrospective, observational study. Women with a history of recurrent UTI who received daily Uromune prophylaxis for three months had significantly fewer recurrences over a 15-month period than women who received daily trimethoprim-sulfamethoxazole for six months (393). Other similar formulations, such as Urostim and Urvakol, have been developed, but few controlled studies are available and they will not be discussed further here.

The mechanism by which immunostimulatory compounds protect against recurrent UTI remains unclear. In mice, OM-89 activated macrophages (394) and induced a T_H1-type immune response as determined by increased IgG2a in the serum and interferon (IFN)-γ in spleen-cell supernatant (395). OM-89 also increased interleukin (IL)-6 and IFN-γ levels and decreased inflammation in the mouse bladder (396). The antibody response to OM-89 varies in different studies. In mice, OM-89 mainly induced IgG in the serum, and increased IgM only weakly (397). Also in mice, strain-specific IgG and IgA were increased in immune sera, and total and strain-specific IgG and IgA were increased in the urogenital tract. In addition, the antisera could recognize other human uropathogens such as *E. faecalis*, *K. pneumoniae*, and *P. mirabilis*. However, cross-reactivity was stronger with intraperitoneal injection of OM-89, rather than oral administration (395, 398). Interestingly, in a meta-analysis of two studies, there was no significant difference in urine and vaginal fluid anti-*E. coli* IgG and IgA between OM-89 treated and placebo-treated patients (391), which would suggest that OM-89's efficacy may not be antibody-mediated. As OM-89 is a lysate of *E. coli* administered daily for several months, its efficacy could be the result of induced LPS tolerance. Similarly, a TLR4 polymorphism that is associated with reduced inflammatory signaling in response to LPS is also associated with protection from recurrent UTI (10).

CONCLUSIONS

The Challenge is Great

The progress in our understanding of UTI pathogenesis over the past 15 years has been truly remarkable and has begun to change how UTI is viewed and treated in the clinic. Sophisticated animal models of infection and translational studies have revealed that, rather than being a simple extracellular infection of the urinary tract mucosa, infection with UPEC and a number of other Gram-negative uropathogens (which together cause more than 80% of all UTI) proceeds through dynamic intracellular and extracellular-host niches during the course of acute and chronic infection. Whole-genome sequencing and gene-expression analysis of uropathogens have allowed an unprecedented look into the lifestyle of these versatile pathogens, thus enabling genetic and computational approaches to identify novel virulence mechanisms and vaccine candidates. The determination of structural details of uropathogenic adhesins has led to the rational design of anti-infectives and preventative strategies. However, with increased understanding comes knowledge of the imposing challenges facing scientists in the effort to develop broadly protective therapies. These challenges stem

from the fact that the development of symptomatic UTI is exceedingly complex, hinging upon two factors that are highly variable and yet define the host-pathogen interaction: the virulence of the infecting uropathogen and the character of the bladder mucosal-immune response. On the one hand, uropathogens are a very heterogeneous collection of bacterial isolates that have the capacity to inhabit diverse host niches, including the gut, the urogenital tract, the bloodstream, and the meninges. Therefore, even among isolates from a single species (such as UPEC), uropathogens differ widely in their genetic and epigenetic makeup, and thus are represented by a large and varied number of serotypes and virulence-factor profiles. On the other hand, the host genetic and environmental variables that determine the extent and character of the bladder mucosal-immune response to infection are considerable and very poorly understood. As a result, two individuals may be infected with the same strain but have very different responses to infection, ranging from an asymptomatic-carrier state and/or asymptomatic bacteriuria, to severe cystitis and pyelonephritis with renal scarring (35).

A Call to Arms: Investigators in Bladder Mucosal Immunity Needed!

Developing new and efficacious therapies should be the highest priority in UTI research, as such therapies have the real potential to positively affect the quality of life of millions of individuals and decrease the overall use of antibiotics. However, many challenges must be overcome. The significant contribution of the host to recurrent UTI makes it unlikely that any new vaccine or therapeutic alone will completely eliminate recurrent UTI in all patients, unless it is also able to alter the innate mucosal-immune response to uropathogens. As the clinical trials of the vaginal vaccine Urovac have demonstrated, even a whole-cell vaccine broadly targeting several UPEC serotypes, as well as four other uropathogens, was only able to significantly protect against recurrent UPEC UTI, and then only in a subset of women. Importantly, what the researchers saw in many of these women was a shift towards recurrent UTI caused by less common uropathogens, suggesting a defect or sensitization of the mucosal-immune response. In the absence of vaccination, UPEC were able to predominate in the urogenital niche of these subjects, but with vaccination, other uropathogens replaced UPEC (331). Thus, as new and more efficacious strategies are being developed to combat recurrent UTI by UPEC, such as pilus-adhesin antagonists (e.g., mannosides) and UPEC-subunit vaccines, we must anticipate the likelihood that a subset of patients will continue to suffer from recurrent UTI with less-common uropathogens. For these patients, therapeutic interventions targeting the bladder mucosal immune response may provide additional benefit and relief from symptoms. However, our understanding of bladder-mucosal immunity is currently insufficient to allow informed predictions, and there is a paucity of investigators in this field. Despite the pioneering efforts of Dr. Svanborg and others, the field is relatively small. For example, of the 10 investigator teams of the Mucosal Immunity Study Team National Institutes of Health (NIH)-National Institute of Allergy and Infectious Diseases (NIAID) U01 consortium (www.mucosal.org), only one group is focused on investigating the urogenital tract. Until our understanding of bladder mucosal immunity matures, which will require a critical mass of investigators in this field, novel approaches to the treatment and prevention of UTI may be slow in coming.

ACKNOWLEDGMENTS

We thank Ender Volkan for assistance with Fig. 2 and Karen Dodson and Thomas Hooton for critical reading of this manuscript and helpful discussion. This work was supported by the National Institutes of Health and Office of Research on Women's Health Specialized Center of Research (DK64540, DK51406, AI48689, AI29549, AI49950, and AI95542 to SJH, and a Mentored Clinical Scientist Research Career Development Award K08 AI083746 to TJH), and a National Science Foundation Graduate Research Fellowship (VPO).

Conflicts of interest: Scott Hultgren is a part owner of Fimbrion and may financially benefit if the company is successful in marketing the mannosides, pilicides, and curlicides that are related to this article. He may also receive royalty income based on the FimH vaccine technology that he developed, which was licensed by Washington University to Sequoia Sciences.

CITATION

O'Brien VP, Hannan TJ, Nielsen HV, Hultgren SJ. 2016. Drug and vaccine development for the treatment and prevention of urinary tract infections. Microbiol Spectrum 4(1):UTI-0013-2012.

REFERENCES

1. **Griebling TL.** 2005. Urologic diseases in America project: trends in resource use for urinary tract infections in women. *J Urol* **173:** 1281–1287.

2. **Foxman B.** 2014. Urinary tract infection syndromes: occurrence, recurrence, bacteriology, risk factors, and disease burden. *Infect Dis Clin North Am* **28:**1–13.

3. **Foxman B.** 2002. Epidemiology of urinary tract infections: incidence, morbidity, and economic costs. *Am J Med* **113**(Suppl 1A):5S–13S.

4. **Foxman B, Barlow R, D'Arcy H, Gillespie B, Sobel JD.** 2000. Urinary tract infection: self-reported incidence and associated costs. *Ann Epidemiol* **10:**509–515.

5. **Hooton TM, Stamm WE.** 1997. Diagnosis and treatment of uncomplicated urinary tract infection. *Infect Dis Clinics North Am* **11:**551–581.

6. **Celik O, Ipekci T, Aydogdu O, Yucel S.** 2014. Current medical diagnosis and management of vesicoureteral reflux in children. *Nephrourol Mon* **6:**e13534.

7. **Foxman B.** 1990. Recurring urinary tract infection: incidence and risk factors. *Am J Public Health* **80:**331–333.

8. **Nuutinen M, Uhari M.** 2001. Recurrence and follow-up after urinary tract infection under the age of 1 year. *Pediatr Nephrol* **16:**69–72.

9. **Scholes D, Hooton TM, Roberts PL, Stapleton AE, Gupta K, Stamm WE.** 2000. Risk factors for recurrent urinary tract infection in young women. *J Infect Dis* **182:**1177–1182.

10. **Hawn TR, Scholes D, Li SS, Wang H, Yang Y, Roberts PL, Stapleton AE, Janer M, Aderem A, Stamm WE, Zhao LP, Hooton TM.** 2009. Toll-like receptor polymorphisms and susceptibility to urinary tract infections in adult women. *PLoS One* **4:**e5990.

11. **Ronald A.** 2002. The etiology of urinary tract infection: traditional and emerging pathogens. *Am J Med* **113**(Suppl 1A):14S–19S.

12. **Sedor J, Mulholland SG.** 1999. Hospital-acquired urinary tract infections associated with the indwelling catheter. *Urol Clin North Am* **26:**821–828.

13. **Edwards JR, Peterson KD, Mu Y, Banerjee S, Allen-Bridson K, Morrell G, Dudeck MA, Pollock DA, Horan TC.** 2009. National Healthcare Safety Network (NHSN) report: data summary for 2006 through 2008, issued December 2009. *Am J Infect Control* **37:**783–805.

14. **Hidron AI, Edwards JR, Patel J, Horan TC, Sievert DM, Pollock DA, Fridkin SK.** 2008. NHSN annual update: antimicrobial-resistant pathogens associated with healthcare-associated infections: annual summary of data reported to the National Healthcare Safety Network at the Centers for Disease Control and Prevention, 2006-2007. *Infect Control Hosp Epidemiol* **29:**996–1011.

15. **Lo E, Nicolle L, Classen D, Arias KM, Podgorny K, Anderson DJ, Burstin H, Calfee DP, Coffin SE, Dubberke ER, Fraser V, Gerding DN, Griffin FA, Gross P, Kaye KS, Klompas M, Marschall J, Mermel LA, Pegues DA, Perl TM, Saint S, Salgado CD, Weinstein RA, Wise R, Yokoe DS.** 2008. Strategies to prevent catheter-associated urinary tract infections in acute care hospitals. *Infect Control Hosp Epidemiol* **29**(Suppl 1):S41–50.

16. **Warren JW, Damron D, Tenney JH, Hoopes JM, Deforge B, Muncie HL Jr.** 1987. Fever, bacteremia, and death as complications of bacteriuria in women with long-term urethral catheters. *J Infect Dis* **155:**1151–1158.

17. **Wagenlehner FM, Cek M, Naber KG, Kiyota H, Bjerklund-Johansen TE.** 2012. Epidemiology, treatment and prevention of healthcare-associated urinary tract infections. *World J Urol* **30:**59–67.

18. **Raz R, Gennesin Y, Wasser J, Stoler Z, Rosenfeld S, Rottensterich E, Stamm WE.** 2000. Recurrent urinary tract infections in postmenopausal women. *Clin Infect Dis* **30:**152–156.

19. **Rogers BA, Sidjabat HE, Paterson DL.** 2011. *Escherichia coli* O25b-ST131: a pandemic, multiresistant, community-associated strain. *J Antimicrob Chemother* **66:**1–14.

20. **Weyrauch HM, Rosenberg ML, Amar AD, Redor M.** 1957. Effects of antibiotics and vaccination on experimental pyelonephritis. *J Urol* **78:**532–539.

21. **van der Starre WE, van Nieuwkoop C, Paltansing S, van't Wout JW, Groeneveld GH, Becker MJ, Koster T, Wattel-Louis GH, Delfos NM, Ablij HC, Leyten EM, Blom JW, van Dissel JT.** 2010. Risk factors for fluoroquinolone-resistant *Escherichia coli* in adults with community-onset febrile urinary tract infection. *J Antimicrob Chemother* **66:**650–656.

22. **Taneja N, Rao P, Arora J, Dogra A.** 2008. Occurrence of ESBL & Amp-C beta-lactamases & susceptibility to newer antimicrobial agents in complicated UTI. *Indian J Med Res* **127:**85–88.

23. **Nicolas-Chanoine MH, Blanco J, Leflon-Guibout V, Demarty R, Alonso MP, Canica MM, Park YJ, Lavigne JP, Pitout J, Johnson JR.** 2008. Intercontinental emergence of *Escherichia coli* clone O25:H4-ST131 producing CTX-M-15. *J Antimicrob Chemother* **61:**273–281.

24. **Totsika M, Beatson SA, Sarkar S, Phan MD, Petty NK, Bachmann N, Szubert M, Sidjabat HE, Paterson DL, Upton M, Schembri MA.** 2011. Insights into a multidrug resistant *Escherichia coli* pathogen of the globally disseminated ST131 lineage: genome analysis and virulence mechanisms. *PLoS One* **6:**e26578.

25. **Foxman B.** 2010. The epidemiology of urinary tract infection. *Nat Rev Urol* **7:**653–660.

26. **Aldeyab MA, Kearney MP, Scott MG, Aldiab MA, Alahmadi YM, Darwish Elhajji FW, Magee FA, McElnay JC.** 2012. An evaluation of the impact of antibiotic stewardship on reducing the use of high-risk antibiotics and its effect on the incidence of Clostridium difficile infection in hospital settings. *J Antimicrob Chemother* doi:dks330 [pii] 10.1093/jac/dks330.

27. **Mulvey MA, Lopez-Boado YS, Wilson CL, Roth R, Parks WC, Heuser J, Hultgren SJ.** 1998. Induction and evasion of host defenses

by type 1-piliated uropathogenic *Escherichia coli*. *Science* **282**:1494–1497.

28. **Schwartz DJ, Chen SL, Hultgren SJ, Seed PC.** 2011. Population dynamics and niche distribution of uropathogenic *Escherichia coli* during acute and chronic urinary tract infection. *Infect Immun* **79**:4250–4259.

29. **Anderson GG, Palermo JJ, Schilling JD, Roth R, Heuser J, Hultgren SJ.** 2003. Intracellular bacterial biofilm-like pods in urinary tract infections. *Science* **301**:105–107.

30. **Justice SS, Hung C, Theriot JA, Fletcher DA, Anderson GG, Footer MJ, Hultgren SJ.** 2004. Differentiation and developmental pathways of uropathogenic *Escherichia coli* in urinary tract pathogenesis. *Proc Natl Acad Sci U S A* **101**:1333–1338.

31. **Rosen DA, Hooton TM, Stamm WE, Humphrey PA, Hultgren SJ.** 2007. Detection of intracellular bacterial communities in human urinary tract infection. *PLoS Med* **4**:e329.

32. **Robino L, Scavone P, Araujo L, Algorta G, Zunino P, Vignoli R.** 2013. Detection of intracellular bacterial communities in a child with *Escherichia coli* recurrent urinary tract infections. *Pathog Dis* **68**:78–81.

33. **Robino L, Scavone P, Araujo L, Algorta G, Zunino P, Pirez MC, Vignoli R.** 2014. Intracellular bacteria in the pathogenesis of *Escherichia coli* urinary tract infection in children. *Clin Infect Dis* **59**:e158–164.

34. **Hung CS, Dodson KW, Hultgren SJ.** 2009. A murine model of urinary tract infection. *Nat Protoc* **4**:1230–1243.

35. **O'Brien VP, Hannan TJ, Schaeffer AJ, Hultgren SJ.** 2015. Are you experienced? Understanding bladder innate immunity in the context of recurrent urinary tract infection. *Curr Opin Infect Dis* **28**:97–105.

36. **Connell I, Agace W, Klemm P, Schembri M, Marild S, Svanborg C.** 1996. Type 1 fimbrial expression enhances *Escherichia coli* virulence for the urinary tract. *Proc Natl Acad Sci U S A* **93**:9827–9832.

37. **O'Hanley P, Lark D, Falkow S, Schoolnik G.** 1985. Molecular basis of *Escherichia coli* colonization of the upper urinary tract in BALB/c mice. Gal-Gal pili immunization prevents *Escherichia coli* pyelonephritis in the BALB/c mouse model of human pyelonephritis. *J Clin Invest* **75**:347–360.

38. **Langermann S, Mollby R, Burlein JE, Palaszynski SR, Auguste CG, DeFusco A, Strouse R, Schenerman MA, Hultgren SJ, Pinkner JS, Winberg J, Guldevall L, Soderhall M, Ishikawa K, Normark S, Koenig S.** 2000. Vaccination with FimH adhesin protects cynomolgus monkeys from colonization and infection by uropathogenic *Escherichia coli*. *J Infect Dis* **181**:774–778.

39. **Langermann S, Palaszynski S, Barnhart M, Auguste G, Pinkner JS, Burlein J, Barren P, Koenig S, Leath S, Jones CH, Hultgren SJ.** 1997. Prevention of mucosal *Escherichia coli* infection by FimH-adhesin-based systemic vaccination. *Science* **276**:607–611.

40. **Roberts JA, Kaack MB, Baskin G, Chapman MR, Hunstad DA, Pinkner JS, Hultgren SJ.** 2004. Antibody responses and protection from pyelonephritis following vaccination with purified *Escherichia coli* PapDG protein. *J Urol* **171**:1682–1685.

41. **Tseng CC, Huang JJ, Wang MC, Wu AB, Ko WC, Chen WC, Wu JJ.** 2007. PapG II adhesin in the establishment and persistence of *Escherichia coli* infection in mouse kidneys. *Kidney Int* **71**:764–770.

42. **Hannan TJ, Mysorekar IU, Hung CS, Isaacson-Schmid ML, Hultgren SJ.** 2010. Early severe inflammatory responses to uropathogenic *E. coli* predispose to chronic and recurrent urinary tract infection. *PLoS Pathog* **6**:e1001042.

43. **Anand M, Wang C, French J, Isaacson-Schmid M, Wall LL, Mysorekar IU.** 2012. Estrogen affects the glycosaminoglycan layer of the murine bladder. *Female Pelvic Med Reconstr Surg* **18**:148–152.

44. **Thumbikat P, Waltenbaugh C, Schaeffer AJ, Klumpp DJ.** 2006. Antigen-specific responses accelerate bacterial clearance in the bladder. *J Immunol* **176**:3080–3086.

45. **Hannan TJ, Roberts PL, Riehl TE, van der Post S, Binkley JM, Schwartz DJ, Miyoshi H, Mack M, Schwendener RA, Hooton TM, Stappenbeck TS, Hansson GC, Stenson WF, Colonna M, Stapleton AE, Hultgren SJ.** 2014. Inhibition of Cyclooxygenase-2 Prevents Chronic and Recurrent Cystitis. *EBioMedicine* **1**:46–57.

46. **Kuehn MJ, Ogg DJ, Kihlberg J, Slonim LN, Flemmer K, Bergfors T, Hultgren SJ.** 1993. Structural basis of pilus subunit recognition by the PapD chaperone. *Science* **262**:1234–1241.

47. **Sauer FG, Futterer K, Pinkner JS, Dodson KW, Hultgren SJ, Waksman G.** 1999. Structural basis of chaperone function and pilus biogenesis. *Science* **285**:1058–1061.

48. **Sauer FG, Pinkner JS, Waksman G, Hultgren SJ.** 2002. Chaperone priming of pilus subunits facilitates a topological transition that drives fiber formation. *Cell* **111**:543–551.

49. **Remaut H, Tang C, Henderson NS, Pinkner JS, Wang T, Hultgren SJ, Thanassi DG,**

Waksman G, Li H. 2008. Fiber formation across the bacterial outer membrane by the chaperone/usher pathway. *Cell* 133:640–652.

50. **Phan G, Remaut H, Wang T, Allen WJ, Pirker KF, Lebedev A, Henderson NS, Geibel S, Volkan E, Yan J, Kunze MB, Pinkner JS, Ford B, Kay CW, Li H, Hultgren SJ, Thanassi DG, Waksman G.** 2011. Crystal structure of the FimD usher bound to its cognate FimC-FimH substrate. *Nature* 474:49–53.

51. **Waksman G, Hultgren SJ.** 2009. Structural biology of the chaperone-usher pathway of pilus biogenesis. *Nat Rev Microbiol* 7:765–774.

52. **Kuehn MJ, Heuser J, Normark S, Hultgren SJ.** 1992. P pili in uropathogenic *E. coli* are composite fibres with distinct fibrillar adhesive tips. *Nature* 356:252–255.

53. **Brzuszkiewicz E, Bruggemann H, Liesegang H, Emmerth M, Olschlager T, Nagy G, Albermann K, Wagner C, Buchrieser C, Emody L, Gottschalk G, Hacker J, Dobrindt U.** 2006. How to become a uropathogen: comparative genomic analysis of extraintestinal pathogenic *Escherichia coli* strains. *Proc Natl Acad Sci U S A* 103:12879–12884.

54. **Chen SL, Hung CS, Xu J, Reigstad CS, Magrini V, Sabo A, Blasiar D, Bieri T, Meyer RR, Ozersky P, Armstrong JR, Fulton RS, Latreille JP, Spieth J, Hooton TM, Mardis ER, Hultgren SJ, Gordon JI.** 2006. Identification of genes subject to positive selection in uropathogenic strains of *Escherichia coli*: a comparative genomics approach. *Proc Natl Acad Sci U S A* 103:5977–5982.

55. **Welch RA, Burland V, Plunkett G 3rd, Redford P, Roesch P, Rasko D, Buckles EL, Liou SR, Boutin A, Hackett J, Stroud D, Mayhew GF, Rose DJ, Zhou S, Schwartz DC, Perna NT, Mobley HL, Donnenberg MS, Blattner FR.** 2002. Extensive mosaic structure revealed by the complete genome sequence of uropathogenic *Escherichia coli*. *Proc Natl Acad Sci U S A* 99:17020–17024.

56. **Hung CS, Bouckaert J, Hung D, Pinkner J, Widberg C, DeFusco A, Auguste CG, Strouse R, Langermann S, Waksman G, Hultgren SJ.** 2002. Structural basis of tropism of *Escherichia coli* to the bladder during urinary tract infection. *Molec Microbiol* 44:903–915.

57. **Choudhury D, Thompson A, Stojanoff V, Langermann S, Pinkner J, Hultgren SJ, Knight SD.** 1999. X-ray structure of the FimC-FimH chaperone-adhesin complex from uropathogenic *Escherichia coli*. *Science* 285:1061–1066.

58. **Zhou G, Mo W-J, Sebbel P, Min G, Neubert TA, Glockshuber R, Wu X-R, Sun T-T, Kong X-P.** 2001. Uroplakin Ia is the urothelial

receptor for uropathogenic *Escherichia coli*: evidence from in vitro FimH binding. *Journal of Cell Science* 114:4095–4103.

59. **Eto DS, Jones TA, Sundsbak JL, Mulvey MA.** 2007. Integrin-mediated host cell invasion by type 1-piliated uropathogenic *Escherichia coli*. *PLoS Pathog* 3:e100.

60. **Mossman KL, Mian MF, Lauzon NM, Gyles CL, Lichty B, Mackenzie R, Gill N, Ashkar AA.** 2008. Cutting edge: FimH adhesin of type 1 fimbriae is a novel TLR4 ligand. *J Immunol* 181:6702–6706.

61. **Lund B, Lindberg F, Marklund BI, Normark S.** 1987. The PapG protein is the alpha-D-galactopyranosyl-(1–4)-beta-D-galactopyranose-binding adhesin of uropathogenic *Escherichia coli*. *Proc Natl Acad Sci U S A* 84:5898–5902.

62. **Lichodziejewska M, Topley N, Steadman R, Mackenzie RK, Jones KV, Williams JD.** 1989. Variable expression of P fimbriae in *Escherichia coli* urinary tract infection. *Lancet* 1:1414–1418.

63. **Pere A, Nowicki B, Saxen H, Siitonen A, Korhonen TK.** 1987. Expression of P, type-1, and type-1C fimbriae of *Escherichia coli* in the urine of patients with acute urinary tract infection. *J Infect Dis* 156:567–574.

64. **Kisielius PV, Schwan WR, Amundsen SK, Duncan JL, Schaeffer AJ.** 1989. In vivo expression and variation of *Escherichia coli* type 1 and P pili in the urine of adults with acute urinary tract infections. *Infect Immun* 57:1656–1662.

65. **Hultgren SJ, Porter TN, Schaeffer AJ, Duncan JL.** 1985. Role of type 1 pili and effects of phase variation on lower urinary tract infections produced by *Escherichia coli*. *Infect Immun* 50:370–377.

66. **Greene SE, Hibbing ME, Janetka J, Chen SL, Hultgren SJ.** 2015. Human urine decreases function and expression of type 1 pili in uropathogenic *Escherichia coli*. *MBio* 6:e00820.

67. **Hagberg L, Hull R, Hull S, Falkow S, Freter R, Svanborg Eden C.** 1983. Contribution of adhesion to bacterial persistence in the mouse urinary tract. *Infect Immun* 40:265–272.

68. **Abraham JM, Freitag CS, Clements JR, Eisenstein BI.** 1985. An invertible element of DNA controls phase variation of type 1 fimbriae of *Escherichia coli*. *Proc Natl Acad Sci U S A* 82:5724–5727.

69. **Alkan ML, Wong L, Silverblatt FJ.** 1986. Change in degree of type 1 piliation of *Escherichia coli* during experimental peritonitis in the mouse. *Infect Immun* 54:549–554.

70. **Bahrani-Mougeot FK, Buckles EL, Lockatell CV, Hebel JR, Johnson DE, Tang CM,**

Donnenberg MS. 2002. Type 1 fimbriae and extracellular polysaccharides are preeminent uropathogenic *Escherichia coli* virulence determinants in the murine urinary tract. *Mol Microbiol* **45**:1079–1093.

71. Wright KJ, Seed PC, Hultgren SJ. 2007. Development of intracellular bacterial communities of uropathogenic *Escherichia coli* depends on type 1 pili. *Cell Microbiol* **9**:2230–2241.

72. Martinez JJ, Mulvey MA, Schilling JD, Pinkner JS, Hultgren SJ. 2000. Type 1 pilus-mediated bacterial invasion of bladder epithelial cells. *EMBO J* **19**:2803–2812.

73. Chen SL, Hung CS, Pinkner JS, Walker JN, Cusumano CK, Li Z, Bouckaert J, Gordon JI, Hultgren SJ. 2009. Positive selection identifies an in vivo role for FimH during urinary tract infection in addition to mannose binding. *Proc Natl Acad Sci U S A* **106**:22439–22444.

74. Schembri MA, Sokurenko EV, Klemm P. 2000. Functional flexibility of the FimH adhesin: insights from a random mutant library. *Infect Immun* **68**:2638–2646.

75. Sokurenko EV, Chesnokova V, Dykhuizen DE, Ofek I, Wu XR, Krogfelt KA, Struve C, Schembri MA, Hasty DL. 1998. Pathogenic adaptation of *Escherichia coli* by natural variation of the FimH adhesin. *Proc Natl Acad Sci U S A* **95**:8922–8926.

76. Sokurenko EV, Courtney HS, Maslow J, Siitonen A, Hasty DL. 1995. Quantitative differences in adhesiveness of type 1 fimbriated *Escherichia coli* due to structural differences in fimH genes. *J Bacteriol* **177**:3680–3686.

77. Sokurenko EV, Courtney HS, Ohman DE, Klemm P, Hasty DL. 1994. FimH family of type 1 fimbrial adhesins: functional heterogeneity due to minor sequence variations among fimH genes. *J Bacteriol* **176**:748–755.

78. Schwartz DJ, Kalas V, Pinkner JS, Chen SL, Spaulding CN, Dodson KW, Hultgren SJ. 2013. Positively selected FimH residues enhance virulence during urinary tract infection by altering FimH conformation. *Proc Natl Acad Sci U S A* **110**:15530–15537.

79. Thumbikat P, Berry RE, Zhou G, Billips BK, Yaggie RE, Zaichuk T, Sun TT, Schaeffer AJ, Klumpp DJ. 2009. Bacteria-induced uroplakin signaling mediates bladder response to infection. *PLoS Pathog* **5**:e1000415.

80. Wang H, Min G, Glockshuber R, Sun TT, Kong XP. 2009. Uropathogenic *E. coli* adhesin-induced host cell receptor conformational changes: implications in transmembrane signaling transduction. *J Mol Biol* **392**:352–361.

81. Eto DS, Gordon HB, Dhakal BK, Jones TA, Mulvey MA. 2008. Clathrin, AP-2, and the NPXY-binding subset of alternate endocytic adaptors facilitate FimH-mediated bacterial invasion of host cells. *Cell Microbiol* **10**:2553–2567.

82. Duncan MJ, Li G, Shin JS, Carson JL, Abraham SN. 2004. Bacterial penetration of bladder epithelium through lipid rafts. *J Biol Chem* **279**:18944–18951.

83. Dhakal BK, Mulvey MA. 2009. Uropathogenic *Escherichia coli* invades host cells via an HDAC6-modulated microtubule-dependent pathway. *J Biol Chem* **284**:446–454.

84. Martinez JJ, Hultgren SJ. 2002. Requirement of Rho-family GTPases in the invasion of Type 1-piliated uropathogenic *Escherichia coli*. *Cell Microbiol* **4**:19–28.

85. Song J, Bishop BL, Li G, Duncan MJ, Abraham SN. 2007. TLR4-initiated and cAMP-mediated abrogation of bacterial invasion of the bladder. *Cell Host Microbe* **1**:287–298.

86. Doye A, Mettouchi A, Bossis G, Clement R, Buisson-Touati C, Flatau G, Gagnoux L, Piechaczyk M, Boquet P, Lemichez E. 2002. CNF1 exploits the ubiquitin-proteasome machinery to restrict Rho GTPase activation for bacterial host cell invasion. *Cell* **111**:553–564.

87. Rippere-Lampe KE, O'Brien AD, Conran R, Lockman HA. 2001. Mutation of the gene encoding cytotoxic necrotizing factor type 1 (cnf(1)) attenuates the virulence of uropathogenic *Escherichia coli*. *Infect Immun* **69**:3954–3964.

88. Johnson DE, Drachenberg C, Lockatell CV, Island MD, Warren JW, Donnenberg MS. 2000. The role of cytotoxic necrotizing factor-1 in colonization and tissue injury in a murine model of urinary tract infection. *FEMS Immunol Med Microbiol* **28**:37–41.

89. Kallenius G, Mollby R, Svenson SB, Helin I, Hultberg H, Cedergren B, Winberg J. 1981. Occurrence of P-fimbriated *Escherichia coli* in urinary tract infections. *Lancet* **2**:1369–1372.

90. Winberg J. 1984. P-fimbriae, bacterial adhesion, and pyelonephritis. *Arch Dis Child* **59**:180–184.

91. Norinder BS, Koves B, Yadav M, Brauner A, Svanborg C. 2011. Do *Escherichia coli* strains causing acute cystitis have a distinct virulence repertoire? *Microb Pathog* doi:S0882-4010(11)00157-4 [pii] 10.1016/j.micpath.2011.08.005.

92. Dodson KW, Pinkner JS, Rose T, Magnusson G, Hultgren SJ, Waksman G. 2001. Structural basis of the interaction of the pyelonephritic *E. coli* adhesin to its human kidney receptor. *Cell* **105**:733–743.

93. **Kallenius G, Svenson S, Mollby R, Cedergren B, Hultberg H, Winberg J.** 1981. Structure of carbohydrate part of receptor on human uroepithelial cells for pyelonephritogenic *Escherichia coli. Lancet* **2**:604–606.

94. **Leffler H, Lomberg H, Gotschlich E, Hagberg L, Jodal U, Korhonen T, Samuelsson BE, Schoolnik G, Svanborg-Eden C.** 1982. Chemical and clinical studies on the interaction of *Escherichia coli* with host glycolipid receptors in urinary tract infection. *Scand J Infect Dis Suppl* **33**:46–51.

95. **Stapleton AE, Stroud MR, Hakomori SI, Stamm WE.** 1998. The globoseries glycosphingolipid sialosyl galactosyl globoside is found in urinary tract tissues and is a preferred binding receptor In vitro for uropathogenic *Escherichia coli* expressing pap-encoded adhesins. *Infect Immun* **66**:3856–3861.

96. **de Ree JM, van den Bosch JF.** 1987. Serological response to the P fimbriae of uropathogenic *Escherichia coli* in pyelonephritis. *Infect Immun* **55**:2204–2207.

97. **Kantele A, Papunen R, Virtanen E, Mottonen T, Rasanen L, Ala-Kaila K, Makela PH, Arvilommi H.** 1994. Antibody-secreting cells in acute urinary tract infection as indicators of local immune response. *J Infect Dis* **169**:1023–1028.

98. **Lomberg H, Cedergren B, Leffler H, Nilsson B, Carlstrom AS, Svanborg-Eden C.** 1986. Influence of blood group on the availability of receptors for attachment of uropathogenic *Escherichia coli. Infect Immun* **51**:919–926.

99. **Stapleton A, Hooton TM, Fennell C, Roberts PL, Stamm WE.** 1995. Effect of secretor status on vaginal and rectal colonization with fimbriated *Escherichia coli* in women with and without recurrent urinary tract infection. *J Infect Dis* **171**:717–720.

100. **Sheinfeld J, Schaeffer AJ, Cordon-Cardo C, Rogatko A, Fair WR.** 1989. Association of the Lewis blood-group phenotype with recurrent urinary tract infections in women. *N Engl J Med* **320**:773–777.

101. **Kinane DF, Blackwell CC, Brettle RP, Weir DM, Winstanley FP, Elton RA.** 1982. ABO blood group, secretor state, and susceptibility to recurrent urinary tract infection in women. *Br Med J (Clin Res Ed)* **285**:7–9.

102. **Roberts JA, Marklund BI, Ilver D, Haslam D, Kaack MB, Baskin G, Louis M, Mollby R, Winberg J, Normark S.** 1994. The Gal(alpha 1-4)Gal-specific tip adhesin of *Escherichia coli* P-fimbriae is needed for pyelonephritis to occur in the normal urinary tract. *Proc Natl Acad Sci USA* **91**:11889–11893.

103. **Roberts JA, Kaack MB, Baskin G, Marklund BI, Normark S.** 1997. Epitopes of the P-fimbrial adhesin of *E. coli* cause different urinary tract infections. *J Urol* **158**:1610–1613.

104. **Johnson JR, Russo TA, Brown JJ, Stapleton A.** 1998. papG alleles of *Escherichia coli* strains causing first-episode or recurrent acute cystitis in adult women. *J Infect Dis* **177**:97–101.

105. **Winberg J, Mollby R, Bergstrom J, Karlsson KA, Leonardsson I, Milh MA, Teneberg S, Haslam D, Marklund BI, Normark S.** 1995. The PapG-adhesin at the tip of P-fimbriae provides *Escherichia coli* with a competitive edge in experimental bladder infections of cynomolgus monkeys. *J Exp Med* **182**:1695–1702.

106. **Mobley HL, Jarvis KG, Elwood JP, Whittle DI, Lockatell CV, Russell RG, Johnson DE, Donnenberg MS, Warren JW.** 1993. Isogenic P-fimbrial deletion mutants of pyelonephritogenic *Escherichia coli*: the role of alpha Gal(1-4) beta Gal binding in virulence of a wild-type strain. *Mol Microbiol* **10**:143–155.

107. **Xu H, Storch T, Yu M, Elliott SP, Haslam DB.** 1999. Characterization of the human Forssman synthetase gene. An evolving association between glycolipid synthesis and host-microbial interactions. *J Biol Chem* **274**:29390–29398.

108. **Murawski IJ, Maina RW, Malo D, Guay-Woodford LM, Gros P, Fujiwara M, Morgan K, Gupta IR.** 2010. The C3H/HeJ inbred mouse is a model of vesico-ureteric reflux with a susceptibility locus on chromosome 12. *Kidney Int* **78**:269–278.

109. **Melican K, Sandoval RM, Kader A, Josefsson L, Tanner GA, Molitoris BA, Richter-Dahlfors A.** 2011. Uropathogenic *Escherichia coli* P and Type 1 fimbriae act in synergy in a living host to facilitate renal colonization leading to nephron obstruction. *PLoS Pathog* **7**:e1001298.

110. **Korhonen TK, Parkkinen J, Hacker J, Finne J, Pere A, Rhen M, Holthofer H.** 1986. Binding of *Escherichia coli* S fimbriae to human kidney epithelium. *Infect Immun* **54**:322–327.

111. **Virkola R, Westerlund B, Holthofer H, Parkkinen J, Kekomaki M, Korhonen TK.** 1988. Binding characteristics of *Escherichia coli* adhesins in human urinary bladder. *Infect Immun* **56**:2615–2622.

112. **Backhed F, Alsen B, Roche N, Angstrom J, von Euler A, Breimer ME, Westerlund-Wikstrom B, Teneberg S, Richter-Dahlfors A.** 2002. Identification of target tissue glycosphingolipid receptors for uropathogenic, F1C-fimbriated *Escherichia coli* and its role in mucosal inflammation. *J Biol Chem* **277**:18198–18205.

113. Selvarangan R, Goluszko P, Singhal J, Carnoy C, Moseley S, Hudson B, Nowicki S, Nowicki B. 2004. Interaction of Dr adhesin with collagen type IV is a critical step in *Escherichia coli* renal persistence. *Infect Immun* **72**:4827–4835.

114. Marre R, Hacker J, Henkel W, Goebel W. 1986. Contribution of cloned virulence factors from uropathogenic *Escherichia coli* strains to nephropathogenicity in an experimental rat pyelonephritis model. *Infect Immun* **54**:761–767.

115. Blanco M, Blanco JE, Alonso MP, Mora A, Balsalobre C, Munoa F, Juarez A, Blanco J. 1997. Detection of pap, sfa and afa adhesin-encoding operons in uropathogenic *Escherichia coli* strains: relationship with expression of adhesins and production of toxins. *Res Microbiol* **148**:745–755.

116. Qin X, Hu F, Wu S, Ye X, Zhu D, Zhang Y, Wang M. 2013. Comparison of adhesin genes and antimicrobial susceptibilities between uropathogenic and intestinal commensal *Escherichia coli* strains. *PLoS One* **8**:e61169.

117. Spurbeck RR, Stapleton AE, Johnson JR, Walk ST, Hooton TM, Mobley HL. 2011. Fimbrial profiles predict virulence of uropathogenic *Escherichia coli* strains: contribution of ygi and yad fimbriae. *Infect Immun* **79**:4753–4763.

118. Vigil PD, Alteri CJ, Mobley HL. 2011. Identification of in vivo-induced antigens including an RTX family exoprotein required for uropathogenic *Escherichia coli* virulence. *Infect Immun* **79**:2335–2344.

119. Vigil PD, Stapleton AE, Johnson JR, Hooton TM, Hodges AP, He Y, Mobley HL. 2011. Presence of putative repeat-in-toxin gene tosA in *Escherichia coli* predicts successful colonization of the urinary tract. *MBio* **2**:e00066-00011.

120. Engstrom MD, Alteri CJ, Mobley HL. 2014. A conserved PapB family member, TosR, regulates expression of the uropathogenic *Escherichia coli* RTX nonfimbrial adhesin TosA while conserved LuxR family members TosE and TosF suppress motility. *Infect Immun* **82**:3644–3656.

121. Vigil PD, Wiles TJ, Engstrom MD, Prasov L, Mulvey MA, Mobley HL. 2012. The repeat-in-toxin family member TosA mediates adherence of uropathogenic *Escherichia coli* and survival during bacteremia. *Infect Immun* **80**:493–505.

122. Nesta B, Spraggon G, Alteri C, Moriel DG, Rosini R, Veggi D, Smith S, Bertoldi I, Pastorello I, Ferlenghi I, Fontana MR, Frankel G, Mobley HL, Rappuoli R, Pizza M, Serino L, Soriani M. 2012. FdeC, a novel broadly conserved *Escherichia coli* adhesin eliciting protection against urinary tract infections. *MBio* **3**.

123. Johnson JR, Jelacic S, Schoening LM, Clabots C, Shaikh N, Mobley HL, Tarr PI. 2005. The IrgA homologue adhesin Iha is an *Escherichia coli* virulence factor in murine urinary tract infection. *Infect Immun* **73**:965–971.

124. Bishop BL, Duncan MJ, Song J, Li G, Zaas D, Abraham SN. 2007. Cyclic AMP-regulated exocytosis of *Escherichia coli* from infected bladder epithelial cells. *Nat Med* **13**:625–630.

125. Song J, Bishop BL, Li G, Grady R, Stapleton A, Abraham SN. 2009. TLR4-mediated expulsion of bacteria from infected bladder epithelial cells. *Proc Natl Acad Sci U S A* **106**:14966–14971.

126. Mulvey MA, Schilling JD, Hultgren SJ. 2001. Establishment of a persistent *Escherichia coli* reservoir during the acute phase of a bladder infection. *Infect Immun* **69**:4572–4579.

127. Wei Y, Li K, Wang N, Cai GD, Zhang T, Lin Y, Gui BS, Liu EQ, Li ZF, Zhou W. 2015. Activation of endogenous anti-inflammatory mediator cyclic AMP attenuates acute pyelonephritis in mice induced by uropathogenic *Escherichia coli*. *Am J Pathol* **185**:472–484.

128. Schilling JD, Mulvey MA, Vincent CD, Lorenz RG, Hultgren SJ. 2001. Bacterial invasion augments epithelial cytokine responses to *Escherichia coli* through a lipopolysaccharide-dependent mechanism. *J Immunol* **166**:1148–1155.

129. Wang C, Mendonsa GR, Symington JW, Zhang Q, Cadwell K, Virgin HW, Mysorekar IU. 2012. Atg16L1 deficiency confers protection from uropathogenic *Escherichia coli* infection in vivo. *Proc Natl Acad Sci U S A* **109**:11008–11013.

130. Li B, Smith P, Horvath DJ Jr, Romesberg FE, Justice SS. 2010. SOS regulatory elements are essential for UPEC pathogenesis. *Microbes Infect* **12**:662–668.

131. Blango MG, Mulvey MA. 2010. Persistence of uropathogenic *Escherichia coli* in the face of multiple antibiotics. *Antimicrob Agents Chemother* **54**:1855–1863.

132. Wieser A, Guggenberger C, Pritsch M, Heesemann J, Schubert S. 2011. A novel ex vivo set-up for dynamic long-term characterization of processes on mucosal interfaces by confocal imaging and simultaneous cytokine measurements. *Cell Microbiol* **13**:742–751.

133. Eto DS, Sundsbak JL, Mulvey MA. 2006. Actin-gated intracellular growth and resurgence of uropathogenic *Escherichia coli*. *Cell Microbiol* **8**:704–717.

134. **Berry RE, Klumpp DJ, Schaeffer AJ.** 2009. Urothelial cultures support intracellular bacterial community formation by uropathogenic *Escherichia coli*. *Infect Immun* **77:**2762–2772.

135. **Reigstad CS, Hultgren SJ, Gordon JI.** 2007. Functional genomic studies of uropathogenic *Escherichia coli* and host urothelial cells when intracellular bacterial communities are assembled. *J Biol Chem* **282:**21259–21267.

136. **Hannan TJ, Mysorekar IU, Chen SL, Walker JN, Jones JM, Pinkner JS, Hultgren SJ, Seed PC.** 2008. *LeuX* tRNA-dependent and -independent mechanisms of *Escherichia coli* pathogenesis in acute cystitis. *Mol Microbiol* **67:**116–128.

137. **Justice SS, Lauer SR, Hultgren SJ, Hunstad DA.** 2006. Maturation of intracellular *Escherichia coli* communities requires SurA. *Infect Immun* **74:**4793–4800.

138. **Cusumano CK, Hung CS, Chen SL, Hultgren SJ.** 2010. Virulence plasmid harbored by uropathogenic *Escherichia coli* functions in acute stages of pathogenesis. *Infect Immun* **78:**1457–1467.

139. **Rosen DA, Pinkner JS, Walker JN, Elam JS, Jones JM, Hultgren SJ.** 2008. Molecular variations in Klebsiella pneumoniae and *Escherichia coli* FimH affect function and pathogenesis in the urinary tract. *Infect Immun* **76:**3346–3356.

140. **Cegelski L, Marshall GR, Eldridge GR, Hultgren SJ.** 2008. The biology and future prospects of antivirulence therapies. *Natl Rev Microbiol* **6:**17–27.

141. **Monack DM, Mueller A, Falkow S.** 2004. Persistent bacterial infections: the interface of the pathogen and the host immune system. *Natl Rev Microbiol* **2:**747–765.

142. **Anderson GG, Goller CC, Justice S, Hultgren SJ, Seed PC.** 2010. Polysaccharide capsule and sialic acid-mediated regulation promote biofilm-like intracellular bacterial communities during cystitis. *Infect Immun* **78:**963–975.

143. **Buckles EL, Wang X, Lane MC, Lockatell CV, Johnson DE, Rasko DA, Mobley HL, Donnenberg MS.** 2009. Role of the K2 capsule in *Escherichia coli* urinary tract infection and serum resistance. *J Infect Dis* **199:**1689–1697.

144. **Cegelski L, Pinkner JS, Hammer ND, Cusumano CK, Hung CS, Chorell E, Aberg V, Walker JN, Seed PC, Almqvist F, Chapman MR, Hultgren SJ.** 2009. Small-molecule inhibitors target *Escherichia coli* amyloid biogenesis and biofilm formation. *Natl Chem Biol* **5:**913–919.

145. **Allsopp LP, Beloin C, Ulett GC, Valle J, Totsika M, Sherlock O, Ghigo JM, Schembri MA.** 2011. Molecular characterization of UpaB

and UpaC - two new autotransporter proteins of uropathogenic *Escherichia coli* CFT073. *Infect Immun* doi:IAI.05322-11 [pii] 10.1128/IAI.05322-11.

146. **Allsopp LP, Totsika M, Tree JJ, Ulett GC, Mabbett AN, Wells TJ, Kobe B, Beatson SA, Schembri MA.** 2010. UpaH is a newly identified autotransporter protein that contributes to biofilm formation and bladder colonization by uropathogenic *Escherichia coli* CFT073. *Infect Immun* **78:**1659–1669.

147. **Ulett GC, Valle J, Beloin C, Sherlock O, Ghigo JM, Schembri MA.** 2007. Functional analysis of antigen 43 in uropathogenic *Escherichia coli* reveals a role in long-term persistence in the urinary tract. *Infect Immun* **75:**3233–3244.

148. **Valle J, Mabbett AN, Ulett GC, Toledo-Arana A, Wecker K, Totsika M, Schembri MA, Ghigo JM, Beloin C.** 2008. UpaG, a new member of the trimeric autotransporter family of adhesins in uropathogenic *Escherichia coli*. *J Bacteriol* **190:**4147–4161.

149. **Garofalo CK, Hooton TM, Martin SM, Stamm WE, Palermo JJ, Gordon JI, Hultgren SJ.** 2007. *Escherichia coli* from urine of female patients with urinary tract infections is competent for intracellular bacterial community formation. *Infect Immun* **75:**52–60.

150. **Cusumano CK, Pinkner JS, Han Z, Greene SE, Ford BA, Crowley JR, Henderson JP, Janetka JW, Hultgren SJ.** 2011. Treatment and prevention of urinary tract infection with orally active FimH inhibitors. *Sci Transl Med* **3:**109ra115.

151. **Mysorekar IU, Hultgren SJ.** 2006. Mechanisms of uropathogenic *Escherichia coli* persistence and eradication from the urinary tract. *Proc Natl Acad Sci U S A* **103:**14170–14175.

152. **Rosen DA, Pinkner JS, Jones JM, Walker JN, Clegg S, Hultgren SJ.** 2008. Utilization of an intracellular bacterial community pathway in *Klebsiella pneumoniae* urinary tract infection and the effects of FimK on type 1 pilus expression. *Infect Immun* **76:**3337–3345.

153. **Wright KJ, Seed PC, Hultgren SJ.** 2005. Uropathogenic *Escherichia coli* Flagella Aid in Efficient Urinary Tract Colonization. *Infect Immun* **73:**7657–7668.

154. **Justice SS, Hunstad DA, Seed PC, Hultgren SJ.** 2006. Filamentation by *Escherichia coli* subverts innate defenses during urinary tract infection. *Proc Natl Acad Sci USA* **103:**19884–19889.

155. **Alteri CJ, Smith SN, Mobley HL.** 2009. Fitness of *Escherichia coli* during urinary tract infection requires gluconeogenesis and the TCA cycle. *PLoS Pathog* **5:**e1000448.

156. **Hadjifrangiskou M, Kostakioti M, Chen SL, Henderson JP, Greene SE, Hultgren SJ.** 2011. A central metabolic circuit controlled by QseC in pathogenic *Escherichia coli*. *Mol Microbiol* **80:**1516–1529.

157. **Kostakioti M, Hadjifrangiskou M, Pinkner JS, Hultgren SJ.** 2009. QseC-mediated dephosphorylation of QseB is required for expression of genes associated with virulence in uropathogenic *Escherichia coli*. *Mol Microbiol* **73:**1020–1031.

158. **Alteri CJ, Himpsl SD, Mobley HL.** 2015. Preferential use of central metabolism in vivo reveals a nutritional basis for polymicrobial infection. *PLoS Pathog* **11:**e1004601.

159. **Hryckowian AJ, Welch RA.** 2013. RpoS contributes to phagocyte oxidase-mediated stress resistance during urinary tract infection by *Escherichia coli* CFT073. *MBio* **4:**e00023-00013.

160. **Clarke MB, Hughes DT, Zhu C, Boedeker EC, Sperandio V.** 2006. The QseC sensor kinase: a bacterial adrenergic receptor. *Proc Natl Acad Sci U S A* **103:**10420–10425.

161. **Rasko DA, Moreira CG, Li de R, Reading NC, Ritchie JM, Waldor MK, Williams N, Taussig R, Wei S, Roth M, Hughes DT, Huntley JF, Fina MW, Falck JR, Sperandio V.** 2008. Targeting QseC signaling and virulence for antibiotic development. *Science* **321:**1078–1080.

162. **Sperandio V, Torres AG, Kaper JB.** 2002. Quorum sensing *Escherichia coli* regulators B and C (QseBC): a novel two-component regulatory system involved in the regulation of flagella and motility by quorum sensing in *E. coli*. *Mol Microbiol* **43:**809–821.

163. **Kostakioti M, Hadjifrangiskou M, Cusumano CK, Hannan TJ, Janetka JW, Hultgren SJ.** 2012. Distinguishing the contribution of type 1 pili from that of other QseB-misregulated factors when QseC is absent during urinary tract infection. *Infect Immun* **80:**2826–2834.

164. **Guckes KR, Kostakioti M, Breland EJ, Gu AP, Shaffer CL, Martinez CR 3rd, Hultgren SJ, Hadjifrangiskou M.** 2013. Strong cross-system interactions drive the activation of the QseB response regulator in the absence of its cognate sensor. *Proc Natl Acad Sci U S A* **110:**16592–16597.

165. **Hung DL, Raivio TL, Jones CH, Silhavy TJ, Hultgren SJ.** 2001. Cpx signaling pathway monitors biogenesis and affects assembly and expression of P pili. *EMBO J* **20:**1508–1518.

166. **Hernday AD, Braaten BA, Broitman-Maduro G, Engelberts P, Low DA.** 2004. Regulation of the pap epigenetic switch by CpxAR: phosphorylated CpxR inhibits transition to the phase ON state by competition with Lrp. *Mol Cell* **16:**537–547.

167. **Debnath I, Norton JP, Barber AE, Ott EM, Dhakal BK, Kulesus RR, Mulvey MA.** 2013. The Cpx stress response system potentiates the fitness and virulence of uropathogenic *Escherichia coli*. *Infect Immun* **81:**1450–1459.

168. **Alteri CJ, Lindner JR, Reiss DJ, Smith SN, Mobley HL.** 2011. The broadly conserved regulator PhoP links pathogen virulence and membrane potential in *Escherichia coli*. *Mol Microbiol* **82:**145–163.

169. **Palaniyandi S, Mitra A, Herren CD, Lockatell CV, Johnson DE, Zhu X, Mukhopadhyay S.** 2012. BarA-UvrY two-component system regulates virulence of uropathogenic *E. coli* CFT073. *PLoS One* **7:**e31348.

170. **Cai W, Wannemuehler Y, Dell'anna G, Nicholson B, Barbieri NL, Kariyawasam S, Feng Y, Logue CM, Nolan LK, Li G.** 2013. A novel two-component signaling system facilitates uropathogenic *Escherichia coli*'s ability to exploit abundant host metabolites. *PLoS Pathog* **9:**e1003428.

171. **Crepin S, Houle S, Charbonneau ME, Mourez M, Harel J, Dozois CM.** 2012. Decreased expression of type 1 fimbriae by a pst mutant of uropathogenic *Escherichia coli* reduces urinary tract infection. *Infect Immun* **80:**2802–2815.

172. **Skaar EP.** 2010. The battle for iron between bacterial pathogens and their vertebrate hosts. *PLoS Pathog* **6:**e1000949.

173. **Williams PH, Carbonetti NH.** 1986. Iron, siderophores, and the pursuit of virulence: independence of the aerobactin and enterochelin iron uptake systems in *Escherichia coli*. *Infect Immun* **51:**942–947.

174. **Valdebenito M, Bister B, Reissbrodt R, Hantke K, Winkelmann G.** 2005. The detection of salmochelin and yersiniabactin in uropathogenic *Escherichia coli* strains by a novel hydrolysis-fluorescence-detection (HFD) method. *Int J Med Microbiol* **295:**99–107.

175. **Bachman MA, Oyler JE, Burns SH, Caza M, Lepine F, Dozois CM, Weiser JN.** 2011. Klebsiella pneumoniae yersiniabactin promotes respiratory tract infection through evasion of lipocalin 2. *Infect Immun* **79:**3309–3316.

176. **Henderson JP, Crowley JR, Pinkner JS, Walker JN, Tsukayama P, Stamm WE, Hooton TM, Hultgren SJ.** 2009. Quantitative metabolomics reveals an epigenetic blueprint for iron acquisition in uropathogenic *Escherichia coli*. *PLoS Pathog* **5:**e1000305.

177. **Raffatellu M, George MD, Akiyama Y, Hornsby MJ, Nuccio SP, Paixao TA, Butler BP, Chu H, Santos RL, Berger T, Mak TW, Tsolis RM,**

Bevins CL, Solnick JV, Dandekar S, Baumler AJ. 2009. Lipocalin-2 resistance confers an advantage to Salmonella enterica serotype Typhimurium for growth and survival in the inflamed intestine. *Cell Host Microbe* **5:**476–486.

178. Porcheron G, Habib R, Houle S, Caza M, Lepine F, Daigle F, Masse E, Dozois CM. 2014. The small RNA RyhB contributes to siderophore production and virulence of uropathogenic *Escherichia coli. Infect Immun* **82:**5056–5068.

179. Feldmann F, Sorsa LJ, Hildinger K, Schubert S. 2007. The salmochelin siderophore receptor IroN contributes to invasion of urothelial cells by extraintestinal pathogenic *Escherichia coli* in vitro. *Infect Immun* **75:**3183–3187.

180. Russo TA, McFadden CD, Carlino-MacDonald UB, Beanan JM, Olson R, Wilding GE. 2003. The Siderophore receptor IroN of extraintestinal pathogenic *Escherichia coli* is a potential vaccine candidate. *Infect Immun* **71:**7164–7169.

181. Chaturvedi KS, Hung CS, Crowley JR, Stapleton AE, Henderson JP. 2012. The siderophore yersiniabactin binds copper to protect pathogens during infection. *Nat Chem Biol* doi:nchembio.1020 [pii] 10.1038/nchembio.1020.

182. Watts RE, Totsika M, Challinor VL, Mabbett AN, Ulett GC, De Voss JJ, Schembri MA. 2012. Contribution of siderophore systems to growth and urinary tract colonization of asymptomatic bacteriuria *Escherichia coli. Infect Immun* **80:**333–344.

183. Garcia EC, Brumbaugh AR, Mobley HL. 2011. Redundancy and specificity of *Escherichia coli* iron acquisition systems during urinary tract infection. *Infect Immun* **79:**1225–1235.

184. Torres AG, Redford P, Welch RA, Payne SM. 2001. TonB-dependent systems of uropathogenic *Escherichia coli*: aerobactin and heme transport and TonB are required for virulence in the mouse. *Infect Immun* **69:**6179–6185.

185. Blango MG, Ott EM, Erman A, Veranic P, Mulvey MA. 2014. Forced resurgence and targeting of intracellular uropathogenic *Escherichia coli* reservoirs. *PLoS One* **9:**e93327.

186. Hopkins WJ, Gendron-Fitzpatrick A, Balish E, Uehling DT. 1998. Time course and host responses to *Escherichia coli* urinary tract infection in genetically distinct mouse strains. *Infect Immun* **66:**2798–2802.

187. Schwartz DJ, Conover MS, Hannan TJ, Hultgren SJ. 2015. Uropathogenic *Escherichia coli* superinfection enhances the severity of mouse bladder infection. *PLoS Pathog* **11:** e1004599.

188. Ferry SA, Holm SE, Stenlund H, Lundholm R, Monsen TJ. 2004. The natural course of uncomplicated lower urinary tract infection in women illustrated by a randomized placebo controlled study. *Scand J Infect Dis* **36:**296–301.

189. Mabeck CE. 1972. Treatment of uncomplicated urinary tract infection in non-pregnant women. *Postgrad Med J* **48:**69–75.

190. Bleidorn J, Gagyor I, Kochen MM, Wegscheider K, Hummers-Pradier E. 2010. Symptomatic treatment (ibuprofen) or antibiotics (ciprofloxacin) for uncomplicated urinary tract infection?–results of a randomized controlled pilot trial. *BMC Med* **8:**30.

191. Lane MC, Alteri CJ, Smith SN, Mobley HL. 2007. Expression of flagella is coincident with uropathogenic *Escherichia coli* ascension to the upper urinary tract. *Proc Natl Acad Sci U S A* **104:**16669–16674.

192. Lane MC, Lockatell V, Monterosso G, Lamphier D, Weinert J, Hebel JR, Johnson DE, Mobley HL. 2005. Role of motility in the colonization of uropathogenic *Escherichia coli* in the urinary tract. *Infect Immun* **73:**7644–7656.

193. O'Hanley P, Lalonde G, Ji G. 1991. Alpha-hemolysin contributes to the pathogenicity of piliated digalactoside-binding *Escherichia coli* in the kidney: efficacy of an alpha-hemolysin vaccine in preventing renal injury in the BALB/c mouse model of pyelonephritis. *Infect Immun* **59:**1153–1161.

194. Uhlen P, Laestadius A, Jahnukainen T, Soderblom T, Backhed F, Celsi G, Brismar H, Normark S, Aperia A, Richter-Dahlfors A. 2000. Alpha-haemolysin of uropathogenic *E. coli* induces Ca2+ oscillations in renal epithelial cells. *Nature* **405:**694–697.

195. Nagamatsu K, Hannan TJ, Guest RL, Kostakioti M, Hadjifrangiskou M, Binkley J, Dodson K, Raivio TL, Hultgren SJ. 2015. Dysregulation of *Escherichia coli* alpha-hemolysin expression alters the course of acute and persistent urinary tract infection. *Proc Natl Acad Sci U S A* **112:**E871–880.

196. Guyer DM, Radulovic S, Jones FE, Mobley HL. 2002. Sat, the secreted autotransporter toxin of uropathogenic *Escherichia coli*, is a vacuolating cytotoxin for bladder and kidney epithelial cells. *Infect Immun* **70:**4539–4546.

197. Heimer SR, Rasko DA, Lockatell CV, Johnson DE, Mobley HL. 2004. Autotransporter genes pic and tsh are associated with *Escherichia coli* strains that cause acute pyelonephritis and are expressed during urinary tract infection. *Infect Immun* **72:**593–597.

198. Hagan EC, Mobley HL. 2009. Haem acquisition is facilitated by a novel receptor Hma and required by uropathogenic *Escherichia coli* for kidney infection. *Mol Microbiol* **71:**79–91.

199. **Kulkarni R, Dhakal BK, Slechta ES, Kurtz Z, Mulvey MA, Thanassi DG.** 2009. Roles of putative type II secretion and type IV pilus systems in the virulence of uropathogenic *Escherichia coli. PLoS One* 4:e4752.

200. **Cirl C, Wieser A, Yadav M, Duerr S, Schubert S, Fischer H, Stappert D, Wantia N, Rodriguez N, Wagner H, Svanborg C, Miethke T.** 2008. Subversion of Toll-like receptor signaling by a unique family of bacterial Toll/interleukin-1 receptor domain-containing proteins. *Nat Med* 14:399–406.

201. **Snyder GA, Cirl C, Jiang J, Chen K, Waldhuber A, Smith P, Rommler F, Snyder N, Fresquez T, Durr S, Tjandra N, Miethke T, Xiao TS.** 2013. Molecular mechanisms for the subversion of MyD88 signaling by TcpC from virulent uropathogenic *Escherichia coli. Proc Natl Acad Sci U S A* 110:6985–6990.

202. **Yadav M, Zhang J, Fischer H, Huang W, Lutay N, Cirl C, Lum J, Miethke T, Svanborg C.** 2010. Inhibition of TIR domain signaling by TcpC: MyD88-dependent and independent effects on *Escherichia coli* virulence. *PLoS Pathog* 6.

203. **Delnay KM, Stonehill WH, Goldman H, Jukkola AF, Dmochowski RR.** 1999. Bladder histological changes associated with chronic indwelling urinary catheter. *J Urol* 161:1106–1108; discussion 1108–1109.

204. **Goble NM, Clarke T, Hammonds JC.** 1989. Histological changes in the urinary bladder secondary to urethral catheterisation. *Br J Urol* 63:354–357.

205. **Mohamed JA, Huang DB.** 2007. Biofilm formation by enterococci. *J Med Microbiol* 56:1581–1588.

206. **Guiton PS, Hung CS, Kline KA, Roth R, Kau AL, Hayes E, Heuser J, Dodson KW, Caparon MG, Hultgren SJ.** 2009. Contribution of autolysin and Sortase a during *Enterococcus faecalis* DNA-dependent biofilm development. *Infect Immun* 77:3626–3638.

207. **Singh KV, Nallapareddy SR, Murray BE.** 2007. Importance of the ebp (endocarditis- and biofilm-associated pilus) locus in the pathogenesis of *Enterococcus faecalis* ascending urinary tract infection. *J Infect Dis* 195:1671–1677.

208. **Kemp KD, Singh KV, Nallapareddy SR, Murray BE.** 2007. Relative contributions of *Enterococcus faecalis* OG1RF sortase-encoding genes, srtA and bps (srtC), to biofilm formation and a murine model of urinary tract infection. *Infect Immun* 75:5399–5404.

209. **Nallapareddy SR, Singh KV, Sillanpaa J, Zhao M, Murray BE.** 2011. Relative contributions of Ebp Pili and the collagen adhesin ace to host extracellular matrix protein adherence and experimental urinary tract infection by *Enterococcus faecalis* OG1RF. *Infect Immun* 79:2901–2910.

210. **Sillanpaa J, Nallapareddy SR, Singh KV, Prakash VP, Fothergill T, Ton-That H, Murray BE.** 2010. Characterization of the ebp(fm) pilus-encoding operon of Enterococcus faecium and its role in biofilm formation and virulence in a murine model of urinary tract infection. *Virulence* 1:236–246.

211. **Frank KL, Guiton PS, Barnes AM, Manias DA, Chuang-Smith ON, Kohler PL, Spaulding AR, Hultgren SJ, Schlievert PM, Dunny GM.** 2013. AhrC and Eep are biofilm infection-associated virulence factors in *Enterococcus faecalis. Infect Immun* 81:1696–1708.

212. **Kau AL, Martin SM, Lyon W, Hayes E, Caparon MG, Hultgren SJ.** 2005. *Enterococcus faecalis* tropism for the kidneys in the urinary tract of C57BL/6J mice. *Infect Immun* 73:2461–2468.

213. **Guiton PS, Hung CS, Hancock LE, Caparon MG, Hultgren SJ.** 2010. Enterococcal biofilm formation and virulence in an optimized murine model of foreign body-associated urinary tract infections. *Infect Immun* 78:4166–4175.

214. **Guiton PS, Hannan TJ, Ford B, Caparon MG, Hultgren SJ.** 2013. *Enterococcus faecalis* overcomes foreign body-mediated inflammation to establish urinary tract infections. *Infect Immun* 81:329–339.

215. **Nielsen HV, Guiton PS, Kline KA, Port GC, Pinkner JS, Neiers F, Normark S, Henriques-Normark B, Caparon MG, Hultgren SJ.** 2012. The metal ion-dependent adhesion site motif of the *Enterococcus faecalis* EbpA pilin mediates pilus function in catheter-associated urinary tract infection. *MBio* 3.

216. **Flores-Mireles AL, Pinkner JS, Caparon MG, Hultgren SJ.** 2014. EbpA vaccine antibodies block binding of *Enterococcus faecalis* to fibrinogen to prevent catheter-associated bladder infection in mice. *Sci Transl Med* 6:254ra127.

217. **Nielsen HV, Flores-Mireles AL, Kau AL, Kline KA, Pinkner JS, Neiers F, Normark S, Henriques-Normark B, Caparon MG, Hultgren SJ.** 2013. Pilin and sortase residues critical for endocarditis- and biofilm-associated pilus biogenesis in *Enterococcus faecalis. J Bacteriol* 195:4484–4495.

218. **Clancy KW, Melvin JA, McCafferty DG.** 2010. Sortase transpeptidases: insights into mechanism, substrate specificity, and inhibition. *Biopolymers* 94:385–396.

219. **Nallapareddy SR, Sillanpaa J, Mitchell J, Singh KV, Chowdhury SA, Weinstock GM, Sullam PM, Murray BE.** 2011. Conservation of Ebp-type pilus genes among Enterococci and demonstration of their role in adherence of *Enterococcus faecalis* to human platelets. *Infect Immun* **79:**2911–2920.

220. **Maione D, Margarit I, Rinaudo CD, Masignani V, Mora M, Scarselli M, Tettelin H, Brettoni C, Iacobini ET, Rosini R, D'Agostino N, Miorin L, Buccato S, Mariani M, Galli G, Nogarotto R, Nardi Dei V, Vegni F, Fraser C, Mancuso G, Teti G, Madoff LC, Paoletti LC, Rappuoli R, Kasper DL, Telford JL, Grandi G.** 2005. Identification of a universal Group B streptococcus vaccine by multiple genome screen. *Science* **309:**148–150.

221. **Margarit I, Rinaudo CD, Galeotti CL, Maione D, Ghezzo C, Buttazzoni E, Rosini R, Runci Y, Mora M, Buccato S, Pagani M, Tresoldi E, Berardi A, Creti R, Baker CJ, Telford JL, Grandi G.** 2009. Preventing bacterial infections with pilus-based vaccines: the group B streptococcus paradigm. *J Infect Dis* **199:**108–115.

222. **Bouckaert J, Berglund J, Schembri M, De Genst E, Cools L, Wuhrer M, Hung CS, Pinkner J, Slattegard R, Zavialov A, Choudhury D, Langermann S, Hultgren SJ, Wyns L, Klemm P, Oscarson S, Knight SD, De Greve H.** 2005. Receptor binding studies disclose a novel class of high-affinity inhibitors of the *Escherichia coli* FimH adhesin. *Mol Microbiol* **55:**441–455.

223. **Han Z, Pinkner JS, Ford B, Obermann R, Nolan W, Wildman SA, Hobbs D, Ellenberger T, Cusumano CK, Hultgren SJ, Janetka JW.** 2010. Structure-based drug design and optimization of mannoside bacterial FimH antagonists. *J Med Chem* **53:**4779–4792.

224. **Wellens A, Garofalo C, Nguyen H, Van Gerven N, Slattegard R, Hernalsteens JP, Wyns L, Oscarson S, De Greve H, Hultgren S, Bouckaert J.** 2008. Intervening with urinary tract infections using anti-adhesives based on the crystal structure of the FimH-oligomannose-3 complex. *PLoS One* **3:**e2040.

225. **Klein T, Abgottspon D, Wittwer M, Rabbani S, Herold J, Jiang X, Kleeb S, Luthi C, Scharenberg M, Bezencon J, Gubler E, Pang L, Smiesko M, Cutting B, Schwardt O, Ernst B.** 2010. FimH antagonists for the oral treatment of urinary tract infections: from design and synthesis to in vitro and in vivo evaluation. *J Med Chem* **53:**8627–8641.

226. **Schwardt O, Rabbani S, Hartmann M, Abgottspon D, Wittwer M, Kleeb S, Zalewski A, Smiesko M, Cutting B, Ernst B.** 2011. Design, synthesis and biological evaluation of mannosyl triazoles as FimH antagonists. *Bioorg Med Chem* **19:**6454–6473.

227. **Guiton PS, Cusumano CK, Kline KA, Dodson KW, Han Z, Janetka JW, Henderson JP, Caparon MG, Hultgren SJ.** 2012. Combinatorial small-molecule therapy prevents uropathogenic *Escherichia coli* catheter-associated urinary tract infections in mice. *Antimicrob Agents Chemother* **56:**4738–4745.

228. **Totsika M, Kostakioti M, Hannan TJ, Upton M, Beatson SA, Janetka JW, Hultgren SJ, Schembri MA.** 2013. A FimH Inhibitor Prevents Acute Bladder Infection and Treats Chronic Cystitis Caused by Multidrug-Resistant Uropathogenic *Escherichia coli* ST131. *J Infect Dis* **208:**921–928.

229. **Eden CS, Freter R, Hagberg L, Hull R, Hull S, Leffler H, Schoolnik G.** 1982. Inhibition of experimental ascending urinary tract infection by an epithelial cell-surface receptor analogue. *Nature* **298:**560–562.

230. **Larsson A, Ohlsson J, Dodson KW, Hultgren SJ, Nilsson U, Kihlberg J.** 2003. Quantitative studies of the binding of the class II PapG adhesin from uropathogenic *Escherichia coli* to oligosaccharides. *Bioorg Med Chem* **11:**2255–2261.

231. **Ohlsson J, Larsson A, Haataja S, Alajaaski J, Stenlund P, Pinkner JS, Hultgren SJ, Finne J, Kihlberg J, Nilsson UJ.** 2005. Structure-activity relationships of galabioside derivatives as inhibitors of *E. coli* and S. suis adhesins: nanomolar inhibitors of S. suis adhesins. *Org Biomol Chem* **3:**886–900.

232. **Joosten JA, Loimaranta V, Appeldoorn CC, Haataja S, El Maate FA, Liskamp RM, Finne J, Pieters RJ.** 2004. Inhibition of Streptococcus suis adhesion by dendritic galabiose compounds at low nanomolar concentration. *J Med Chem* **47:**6499–6508.

233. **Salminen A, Loimaranta V, Joosten JA, Khan AS, Hacker J, Pieters RJ, Finne J.** 2007. Inhibition of P-fimbriated *Escherichia coli* adhesion by multivalent galabiose derivatives studied by a live-bacteria application of surface plasmon resonance. *J Antimicrob Chemother* **60:**495–501.

234. **Pieters RJ, Slotved HC, Mortensen HM, Arler L, Finne J, Haataja S, Joosten JA, Branderhorst HM, Krogfelt KA.** 2013. Use of tetravalent galabiose for inhibition of streptococcus suis serotype 2 infection in a mouse model. *Biology (Basel)* **2:**702–718.

235. **Emtenas H, Ahlin K, Pinkner JS, Hultgren SJ, Almqvist F.** 2002. Design and parallel solid-phase synthesis of ring-fused 2-pyridinones that target pilus biogenesis in pathogenic bacteria. *J Comb Chem* **4:**630–639.

236. Svensson A, Larsson A, Emtenas H, Hedenstrom M, Fex T, Hultgren SJ, Pinkner JS, Almqvist F, Kihlberg J. 2001. Design and evaluation of pilicides: potential novel antibacterial agents directed against uropathogenic *Escherichia coli*. *Chembiochem* **2:**915–918.

237. Pinkner JS, Remaut H, Buelens F, Miller E, Aberg V, Pemberton N, Hedenstrom M, Larsson A, Seed P, Waksman G, Hultgren SJ, Almqvist F. 2006. Rationally designed small compounds inhibit pilus biogenesis in uropathogenic bacteria. *Proc Natl Acad Sci USA* **103:**17897–17902.

238. Hung DL, Knight SD, Woods RM, Pinkner JS, Hultgren SJ. 1996. Molecular basis of two subfamilies of immunoglobulin-like chaperones. *EMBO J* **15:**3792–3805.

239. Greene SE, Pinkner JS, Chorell E, Dodson KW, Shaffer CL, Conover MS, Livny J, Hadjifrangiskou M, Almqvist F, Hultgren SJ. 2014. Pilicide ec240 disrupts virulence circuits in uropathogenic *Escherichia coli*. *MBio* **5:**e02038.

240. Chapman MR, Robinson LS, Pinkner JS, Roth R, Heuser J, Hammar M, Normark S, Hultgren SJ. 2002. Role of *Escherichia coli* curli operons in directing amyloid fiber formation. *Science* **295:**851–855.

241. Chorell E, Bengtsson C, Sainte-Luce Banchelin T, Das P, Uvell H, Sinha AK, Pinkner JS, Hultgren SJ, Almqvist F. 2011. Synthesis and application of a bromomethyl substituted scaffold to be used for efficient optimization of anti-virulence activity. *Eur J Med Chem* **46:**1103–1116.

242. Chorell E, Pinkner JS, Bengtsson C, Banchelin TS, Edvinsson S, Linusson A, Hultgren SJ, Almqvist F. 2012. Mapping pilicide anti-virulence effect in *Escherichia coli*, a comprehensive structure-activity study. *Bioorg Med Chem* **20:**3128–3142.

243. Chorell E, Pinkner JS, Phan G, Edvinsson S, Buelens F, Remaut H, Waksman G, Hultgren SJ, Almqvist F. 2010. Design and synthesis of C-2 substituted thiazolo and dihydrothiazolo ring-fused 2-pyridones: pilicides with increased antivirulence activity. *J Med Chem* **53:**5690–5695.

244. Sunden F, Hakansson L, Ljunggren E, Wullt B. 2006. Bacterial interference–is deliberate colonization with *Escherichia coli* 83972 an alternative treatment for patients with recurrent urinary tract infection? *Int J Antimicrob Agents* **28**(Suppl 1):S26–29.

245. Ferrieres L, Hancock V, Klemm P. 2007. Biofilm exclusion of uropathogenic bacteria by selected asymptomatic bacteriuria *Escherichia coli* strains. *Microbiology* **153:**1711–1719.

246. Roos V, Ulett GC, Schembri MA, Klemm P. 2006. The asymptomatic bacteriuria *Escherichia coli* strain 83972 outcompetes uropathogenic *E. coli* strains in human urine. *Infect Immun* **74:**615–624.

247. Klemm P, Roos V, Ulett GC, Svanborg C, Schembri MA. 2006. Molecular characterization of the *Escherichia coli* asymptomatic bacteriuria strain 83972: the taming of a pathogen. *Infect Immun* **74:**781–785.

248. Sunden F, Hakansson L, Ljunggren E, Wullt B. 2010. *Escherichia coli* 83972 bacteriuria protects against recurrent lower urinary tract infections in patients with incomplete bladder emptying. *J Urol* **184:**179–185.

249. Darouiche RO, Green BG, Donovan WH, Chen D, Schwartz M, Merritt J, Mendez M, Hull RA. 2011. Multicenter randomized controlled trial of bacterial interference for prevention of urinary tract infection in patients with neurogenic bladder. *Urology* **78:**341–346.

250. Darouiche RO, Thornby JI, Cerra-Stewart C, Donovan WH, Hull RA. 2005. Bacterial interference for prevention of urinary tract infection: a prospective, randomized, placebo-controlled, double-blind pilot trial. *Clin Infect Dis* **41:**1531–1534.

251. Koves B, Salvador E, Gronberg-Hernandez J, Zdziarski J, Wullt B, Svanborg C, Dobrindt U. 2014. Rare emergence of symptoms during long-term asymptomatic *Escherichia coli* 83972 carriage without an altered virulence factor repertoire. *J Urol* **191:**519–528.

252. Rudick CN, Taylor AK, Yaggie RE, Schaeffer AJ, Klumpp DJ. 2014. Asymptomatic bacteriuria *Escherichia coli* are live biotherapeutics for UTI. *PLoS One* **9:**e109321.

253. Paton AW, Morona R, Paton JC. 2000. A new biological agent for treatment of Shiga toxigenic *Escherichia coli* infections and dysentery in humans. *Nat Med* **6:**265–270.

254. Watts RE, Tan CK, Ulett GC, Carey AJ, Totsika M, Idris A, Paton AW, Morona R, Paton JC, Schembri MA. 2012. *Escherichia coli* 83972 expressing a P fimbriae oligosaccharide receptor mimic impairs adhesion of uropathogenic *E. coli*. *J Infect Dis* doi:jis493 [pii] 10.1093/infdis/jis493.

255. Trautner BW, Cevallos ME, Li H, Riosa S, Hull RA, Hull SI, Tweardy DJ, Darouiche RO. 2008. Increased expression of type-1 fimbriae by nonpathogenic *Escherichia coli* 83972 results in an increased capacity for catheter adherence and bacterial interference. *J Infect Dis* **198:**899–906.

256. Trautner BW, Darouiche RO, Hull RA, Hull S, Thornby JI. 2002. Pre-inoculation of

urinary catheters with *Escherichia coli* 83972 inhibits catheter colonization by *Enterococcus faecalis. J Urol* **167**:375–379.

257. **Navas-Nacher EL, Dardick F, Venegas MF, Anderson BE, Schaeffer AJ, Duncan JL.** 2001. Relatedness of *Escherichia coli* colonizing women longitudinally. *Mol Urol* **5**:31–36.

258. **Czaja CA, Stamm WE, Stapleton AE, Roberts PL, Hawn TR, Scholes D, Samadpour M, Hultgren SJ, Hooton TM.** 2009. Prospective cohort study of microbial and inflammatory events immediately preceding *Escherichia coli* recurrent urinary tract infection in women. *J Infect Dis* **200**:528–536.

259. **Baerheim A, Larsen E, Digranes A.** 1994. Vaginal application of lactobacilli in the prophylaxis of recurrent lower urinary tract infection in women. *Scand J Prim Health Care* **12**:239–243.

260. **Reid G.** 2001. Probiotic agents to protect the urogenital tract against infection. *Am J Clin Nutr* **73**:437S–443S.

261. **Beerepoot MA, Geerlings SE, van Haarst EP, van Charante NM, ter Riet G.** 2013. Non-antibiotic prophylaxis for recurrent urinary tract infections: a systematic review and meta-analysis of randomized controlled trials. *J Urol* **190**:1981–1989.

262. **Czaja CA, Stapleton AE, Yarova-Yarovaya Y, Stamm WE.** 2007. Phase I trial of a Lactobacillus crispatus vaginal suppository for prevention of recurrent urinary tract infection in women. *Infect Dis Obstet Gynecol* **2007**:35387.

263. **Stapleton AE, Au-Yeung M, Hooton TM, Fredricks DN, Roberts PL, Czaja CA, Yarova-Yarovaya Y, Fiedler T, Cox M, Stamm WE.** 2011. Randomized, placebo-controlled phase 2 trial of a Lactobacillus crispatus probiotic given intravaginally for prevention of recurrent urinary tract infection. *Clin Infect Dis* **52**:1212–1217.

264. **Velraeds MM, van de Belt-Gritter B, van der Mei HC, Reid G, Busscher HJ.** 1998. Interference in initial adhesion of uropathogenic bacteria and yeasts to silicone rubber by a Lactobacillus acidophilus biosurfactant. *J Med Microbiol* **47**:1081–1085.

265. **Gupta K, Chou MY, Howell A, Wobbe C, Grady R, Stapleton AE.** 2007. Cranberry products inhibit adherence of p-fimbriated *Escherichia coli* to primary cultured bladder and vaginal epithelial cells. *J Urol* **177**:2357–2360.

266. **Howell AB, Reed JD, Krueger CG, Winterbottom R, Cunningham DG, Leahy M.** 2005. A-type cranberry proanthocyanidins and uropathogenic bacterial anti-adhesion activity. *Phytochemistry* **66**:2281–2291.

267. **Sobota AE.** 1984. Inhibition of bacterial adherence by cranberry juice: potential use for the treatment of urinary tract infections. *J Urol* **131**:1013–1016.

268. **Zafriri D, Ofek I, Adar R, Pocino M, Sharon N.** 1989. Inhibitory activity of cranberry juice on adherence of type 1 and type P fimbriated *Escherichia coli* to eucaryotic cells. *Antimicrob Agents Chemother* **33**:92–98.

269. **Di Martino P, Agniel R, David K, Templer C, Gaillard JL, Denys P, Botto H.** 2006. Reduction of *Escherichia coli* adherence to uroepithelial bladder cells after consumption of cranberry juice: a double-blind randomized placebo-controlled cross-over trial. *World J Urol* **24**:21–27.

270. **Howell AB, Foxman B.** 2002. Cranberry juice and adhesion of antibiotic-resistant uropathogens. *JAMA* **287**:3082–3083.

271. **Jepson RG, Craig JC.** 2008. Cranberries for preventing urinary tract infections. *Cochrane Database Syst Rev* doi:10.1002/14651858. CD001321.pub4:CD001321.

272. **McMurdo ME, Argo I, Phillips G, Daly F, Davey P.** 2009. Cranberry or trimethoprim for the prevention of recurrent urinary tract infections? A randomized controlled trial in older women. *J Antimicrob Chemother* **63**:389–395.

273. **Stapleton AE, Dziura J, Hooton TM, Cox ME, Yarova-Yarovaya Y, Chen S, Gupta K.** 2012. Recurrent urinary tract infection and urinary *Escherichia coli* in women ingesting cranberry juice daily: a randomized controlled trial. *Mayo Clin Proc* **87**:143–150.

274. **Barbosa-Cesnik C, Brown MB, Buxton M, Zhang L, DeBusscher J, Foxman B.** 2011. Cranberry juice fails to prevent recurrent urinary tract infection: results from a randomized placebo-controlled trial. *Clin Infect Dis* **52**:23–30.

275. **Raz R, Stamm WE.** 1993. A controlled trial of intravaginal estriol in postmenopausal women with recurrent urinary tract infections. *N Engl J Med* **329**:753–756.

276. **Raz R.** 2011. Urinary tract infection in postmenopausal women. *Korean J Urol* **52**:801–808.

277. **Luthje P, Brauner H, Ramos NL, Ovregaard A, Glaser R, Hirschberg AL, Aspenstrom P, Brauner A.** 2013. Estrogen supports urothelial defense mechanisms. *Sci Transl Med* **5**:190ra180.

278. **Curran EM, Tassell AH, Judy BM, Nowicki B, Montgomery-Rice V, Estes DM, Nowicki S.** 2007. Estrogen increases menopausal host susceptibility to experimental ascending urinary-tract infection. *J Infect Dis* **195**:680–683.

279. **Wang C, Symington JW, Ma E, Cao B, Mysorekar IU.** 2013. Estrogenic modulation of uropathogenic *Escherichia coli* infection pathogenesis in a murine menopause model. *Infect Immun* **81:**733–739.

280. **Hicks HT.** 1909. Pyelitis of pregnancy treated with coli vaccine. *Br Med J* **1:**203–204.

281. **Routh CF.** 1910. Vaccine treatment of pyelonephritis in pregnancy. *Br Med J* **1:**191.

282. **Hektoen L IE.** 1929. Vaccine therapy: Result of a questionnaire to american physicians. *JAMA: J Am Med Assoc* **92:**864–869.

283. **Benians TH.** 1926. The role of vaccine therapy: in coliform infections of the urinary tract. *Postgrad Med J* **1:**94–96.

284. **Braude AI, Shapiro AP, Siemienski J.** 1955. Hematogenous pyelonephritis in rats. I. Its pathogenesis when produced by a simple new method. *J Clin Invest* **34:**1489–1497.

285. **Pulendran B, Li S, Nakaya HI.** 2010. Systems vaccinology. *Immunity* **33:**516–529.

286. **Pudney J, Quayle AJ, Anderson DJ.** 2005. Immunological microenvironments in the human vagina and cervix: mediators of cellular immunity are concentrated in the cervical transformation zone. *Biol Reprod* **73:**1253–1263.

287. **Billips BK, Yaggie RE, Cashy JP, Schaeffer AJ, Klumpp DJ.** 2009. A live-attenuated vaccine for the treatment of urinary tract infection by uropathogenic *Escherichia coli*. *J Infect Dis* **200:**263–272.

288. **Chan CY, St John AL, Abraham SN.** 2013. Mast cell interleukin-10 drives localized tolerance in chronic bladder infection. *Immunity* **38:**349–359.

289. **Lutay N, Ambite I, Gronberg Hernandez J, Rydstrom G, Ragnarsdottir B, Puthia M, Nadeem A, Zhang J, Storm P, Dobrindt U, Wullt B, Svanborg C.** 2013. Bacterial control of host gene expression through RNA polymerase II. *J Clin Invest* **123:**2366–2379.

290. **Duell BL, Carey AJ, Tan CK, Cui X, Webb RI, Totsika M, Schembri MA, Derrington P, Irving-Rodgers H, Brooks AJ, Cripps AW, Crowley M, Ulett GC.** 2012. Innate Transcriptional Networks Activated in Bladder in Response to Uropathogenic *Escherichia coli* Drive Diverse Biological Pathways and Rapid Synthesis of IL-10 for Defense against Bacterial Urinary Tract Infection. *J Immunol* **188:**781–792.

291. **Majd M, Rushton HG, Jantausch B, Wiedermann BL.** 1991. Relationship among vesicoureteral reflux, P-fimbriated *Escherichia coli*, and acute pyelonephritis in children with febrile urinary tract infection. *J Pediatr* **119:**578–585.

292. **Nicolle LE, Friesen D, Harding GK, Roos LL.** 1996. Hospitalization for acute pyelonephritis in Manitoba, Canada, during the period from 1989 to 1992; impact of diabetes, pregnancy, and aboriginal origin. *Clin Infect Dis* **22:**1051–1056.

293. **Pennesi M, Travan L, Peratoner L, Bordugo A, Cattaneo A, Ronfani L, Minisini S, Ventura A.** 2008. Is antibiotic prophylaxis in children with vesicoureteral reflux effective in preventing pyelonephritis and renal scars? A randomized, controlled trial. *Pediatrics* **121:**e1489–1494.

294. **Nagler EV, Williams G, Hodson EM, Craig JC.** 2011. Interventions for primary vesicoureteric reflux. *Cochrane Database Syst Rev* doi:10.1002/14651858.CD001532.pub4:CD001532.

295. **Schieve LA, Handler A, Hershow R, Persky V, Davis F.** 1994. Urinary tract infection during pregnancy: its association with maternal morbidity and perinatal outcome. *Am J Public Health* **84:**405–410.

296. **Gilbert NM, O'Brien VP, Hultgren S, Macones G, Lewis WG, Lewis AL.** 2013. Urinary tract infection as a preventable cause of pregnancy complications: opportunities, challenges, and a global call to action. *Glob Adv Health Med* **2:**59–69.

297. **Gratacos E, Torres PJ, Vila J, Alonso PL, Cararach V.** 1994. Screening and treatment of asymptomatic bacteriuria in pregnancy prevent pyelonephritis. *J Infect Dis* **169:**1390–1392.

298. **Kaack MB, Roberts JA, Baskin G, Patterson GM.** 1988. Maternal immunization with P fimbriae for the prevention of neonatal pyelonephritis. *Infect Immun* **56:**1–6.

299. **McKee AS, Munks MW, Marrack P.** 2007. How do adjuvants work? Important considerations for new generation adjuvants. *Immunity* **27:**687–690.

300. **Poggio TV, La Torre JL, Scodeller EA.** 2006. Intranasal immunization with a recombinant truncated FimH adhesin adjuvanted with CpG oligodeoxynucleotides protects mice against uropathogenic *Escherichia coli* challenge. *Can J Microbiol* **52:**1093–1102.

301. **Stanley M.** 2007. Prophylactic HPV vaccines. *J Clin Pathol* **60:**961–965.

302. **Westerman LE, McClure HM, Jiang B, Almond JW, Glass RI.** 2005. Serum IgG mediates mucosal immunity against rotavirus infection. *Proc Natl Acad Sci U S A* **102:**7268–7273.

303. **Levitz SM, Golenbock DT.** 2012. Beyond empiricism: informing vaccine development through innate immunity research. *Cell* **148:**1284–1292.

304. Wilson-Welder JH, Torres MP, Kipper MJ, Mallapragada SK, Wannemuehler MJ, Narasimhan B. 2009. Vaccine adjuvants: current challenges and future approaches. *J Pharm Sci* **98:**1278–1316.

305. Gupta RK. 1998. Aluminum compounds as vaccine adjuvants. *Adv Drug Deliv Rev* **32:**155–172.

306. Petrovsky N, Aguilar JC. 2004. Vaccine adjuvants: current state and future trends. *Immunol Cell Biol* **82:**488–496.

307. Dupuis M, Murphy TJ, Higgins D, Ugozzoli M, van Nest G, Ott G, McDonald DM. 1998. Dendritic cells internalize vaccine adjuvant after intramuscular injection. *Cell Immunol* **186:**18–27.

308. Mosca F, Tritto E, Muzzi A, Monaci E, Bagnoli F, Iavarone C, O'Hagan D, Rappuoli R, De Gregorio E. 2008. Molecular and cellular signatures of human vaccine adjuvants. *Proc Natl Acad Sci U S A* **105:**10501–10506.

309. Sharp FA, Ruane D, Claass B, Creagh E, Harris J, Malyala P, Singh M, O'Hagan DT, Petrilli V, Tschopp J, O'Neill LA, Lavelle EC. 2009. Uptake of particulate vaccine adjuvants by dendritic cells activates the NALP3 inflammasome. *Proc Natl Acad Sci U S A* **106:**870–875.

310. Alving CR, Rao M, Steers NJ, Matyas GR, Mayorov AV. 2012. Liposomes containing lipid A: an effective, safe, generic adjuvant system for synthetic vaccines. *Expert Rev Vaccines* **11:**733–744.

311. Malaviya R, Ikeda T, Abraham SN, Malaviya R. 2004. Contribution of mast cells to bacterial clearance and their proliferation during experimental cystitis induced by type 1 fimbriated *E. coli*. *Immunol Lett* **91:**103–111.

312. McLachlan JB, Shelburne CP, Hart JP, Pizzo SV, Goyal R, Brooking-Dixon R, Staats HF, Abraham SN. 2008. Mast cell activators: a new class of highly effective vaccine adjuvants. *Nat Med* **14:**536–541.

313. Shelburne CP, Nakano H, St John AL, Chan C, McLachlan JB, Gunn MD, Staats HF, Abraham SN. 2009. Mast cells augment adaptive immunity by orchestrating dendritic cell trafficking through infected tissues. *Cell Host Microbe* **6:**331–342.

314. St John AL, Chan CY, Staats HF, Leong KW, Abraham SN. Synthetic mast-cell granules as adjuvants to promote and polarize immunity in lymph nodes. *Natl Mater* **11:**250–257.

315. Pulendran B, Ahmed R. 2006. Translating innate immunity into immunological memory: implications for vaccine development. *Cell* **124:**849–863.

316. Querec TD, Akondy RS, Lee EK, Cao W, Nakaya HI, Teuwen D, Pirani A, Gernert K, Deng J, Marzolf B, Kennedy K, Wu H, Bennouna S, Oluoch H, Miller J, Vencio RZ, Mulligan M, Aderem A, Ahmed R, Pulendran B. 2009. Systems biology approach predicts immunogenicity of the yellow fever vaccine in humans. *Natl Immunol* **10:**116–125.

317. Li S, Rouphael N, Duraisingham S, Romero-Steiner S, Presnell S, Davis C, Schmidt DS, Johnson SE, Milton A, Rajam G, Kasturi S, Carlone GM, Quinn C, Chaussabel D, Palucka AK, Mulligan MJ, Ahmed R, Stephens DS, Nakaya HI, Pulendran B. 2014. Molecular signatures of antibody responses derived from a systems biology study of five human vaccines. *Natl Immunol* **15:**195–204.

318. Brooks SJ, Lyons JM, Braude AI. 1974. Immunization against retrograde pyelonephritis. II. Prevention of retrograde *Escherichia coli* pyelonephritis with vaccines. *Am J Pathol* **74:**359–364.

319. Kaijser B, Larsson P, Olling S. 1978. Protection against ascending *Escherichia coli* pyelonephritis in rats and significance of local immunity. *Infect Immun* **20:**78–81.

320. Zaruba K, Vejbora O, Chobola M. 1971. The effect of preliminary sensitization and immunization on bacteriuria in rabbits with experimental pyelonephritis. *Clin Exp Immunol* **9:**399–405.

321. Jensen J, Balish E, Mizutani K, Uehling DT. 1982. Resolution of induced urinary tract infection: an animal model to assess bladder immunization. *J Urol* **127:**1220–1222.

322. Uehling DT, Jensen J, Balish E. 1982. Vaginal immunization against urinary tract infection. *J Urol* **128:**1382–1384.

323. Uehling DT, Jensen J, Balish E. 1985. Immunization against urinary tract infections. *J Urol (Paris)* **91:**23–26.

324. Uehling DT, Hopkins WJ, Jensen J, Balish E. 1987. Vaginal immunization against induced cystitis in monkeys. *J Urol* **137:**327–329.

325. Roberts JA, Kaack MB, Baskin G, Svenson SB. 1995. Vaccination with a formalin-killed P-fimbriated *E. coli* whole-cell vaccine prevents renal scarring from pyelonephritis in the non-human primate. *Vaccine* **13:**11–16.

326. Grischke EM, Ruttgers H. 1987. Treatment of bacterial infections of the female urinary tract by immunization of the patients. *Urol Int* **42:**338–341.

327. Kochiashvili D, Khuskivadze A, Kochiashvili G, Koberidze G, Kvakhajelidze V. 2014. Role of the bacterial vaccine Solco-Urovac(R) in treatment and prevention of recurrent urinary tract infections of bacterial origin. *Georgian Med News:*11–16.

328. **Uehling DT, James LJ, Hopkins WJ, Balish E.** 1991. Immunization against urinary tract infection with a multi-valent vaginal vaccine. *J Urol* **146:**223–226.

329. **Uehling DT, Hopkins WJ, James LJ, Balish E.** 1994. Vaginal immunization of monkeys against urinary tract infection with a multi-strain vaccine. *J Urol* **151:**214–216.

330. **Uehling DT, Hopkins WJ, Dahmer LA, Balish E.** 1994. Phase I clinical trial of vaginal mucosal immunization for recurrent urinary tract infection. *J Urol* **152:**2308–2311.

331. **Hopkins WJ, Elkahwaji J, Beierle LM, Leverson GE, Uehling DT.** 2007. Vaginal mucosal vaccine for recurrent urinary tract infections in women: results of a phase 2 clinical trial. *J Urol* **177:**1349–1353; quiz 1591.

332. **Uehling DT, Hopkins WJ, Balish E, Xing Y, Heisey DM.** 1997. Vaginal mucosal immunization for recurrent urinary tract infection: phase II clinical trial. *J Urol* **157:**2049–2052.

333. **Uehling DT, Hopkins WJ, Beierle LM, Kryger JV, Heisey DM.** 2001. Vaginal mucosal immunization for recurrent urinary tract infection: extended phase II clinical trial. *J Infect Dis* **183**(Suppl 1):S81–83.

334. **Uehling DT, Hopkins WJ, Elkahwaji JE, Schmidt DM, Leverson GE.** 2003. Phase 2 clinical trial of a vaginal mucosal vaccine for urinary tract infections. *J Urol* **170:**867–869.

335. **Kruze D, Holzbecher K, Andrial M, Bossart W.** 1989. Urinary antibody response after immunisation with a vaccine against urinary tract infection. *Urol Res* **17:**361–366.

336. **Russo TA, Beanan JM, Olson R, Genagon SA, MacDonald U, Cope JJ, Davidson BA, Johnston B, Johnson JR.** 2007. A killed, genetically engineered derivative of a wild-type extraintestinal pathogenic *E. coli* strain is a vaccine candidate. *Vaccine* **25:**3859–3870.

337. **Billips BK, Forrestal SG, Rycyk MT, Johnson JR, Klumpp DJ, Schaeffer AJ.** 2007. Modulation of host innate immune response in the bladder by uropathogenic *Escherichia coli.* *Infect Immun* **75:**5353–5360.

338. **Ochman H, Selander RK.** 1984. Standard reference strains of *Escherichia coli* from natural populations. *J Bacteriol* **157:**690–693.

339. **Rudick CN, Jiang M, Yaggie RE, Pavlov VI, Done J, Heckman CJ, Whitfield C, Schaeffer AJ, Klumpp DJ.** 2012. O-antigen modulates infection-induced pain states. *PLoS One* **7:**e41273.

340. **Wieser A, Romann E, Magistro G, Hoffmann C, Norenberg D, Weinert K, Schubert S.** 2010. A multiepitope subunit vaccine conveys protection against extraintestinal pathogenic *Escherichia coli* in mice. *Infect Immun* **78:**3432–3442.

341. **Sanford JP, Hunter BW, Souda LL.** 1962. The role of immunity in the pathogenesis of experimental hematogenous pyelonephritis. *J Exp Med* **115:**383–410.

342. **Uehling DT, Wolf L.** 1969. Enhancement of the bladder defense mechanism by immunization. *Invest Urol* **6:**520–526.

343. **Roberts JA, Kaack MB, Baskin G, Svenson SB.** 1993. Prevention of renal scarring from pyelonephritis in nonhuman primates by vaccination with a synthetic *Escherichia coli* serotype O8 oligosaccharide-protein conjugate. *Infect Immun* **61:**5214–5218.

344. **Kaijser B, Larsson P, Olling S, Schneerson R.** 1983. Protection against acute, ascending pyelonephritis caused by *Escherichia coli* in rats, using isolated capsular antigen conjugated to bovine serum albumin. *Infect Immun* **39:**142–146.

345. **Kumar V, Ganguly N, Joshi K, Mittal R, Harjai K, Chhibber S, Sharma S.** 2005. Protective efficacy and immunogenicity of *Escherichia coli* K13 diphtheria toxoid conjugate against experimental ascending pyelonephritis. *Med Microbiol Immunol* **194:**211–217.

346. **Stenutz R, Weintraub A, Widmalm G.** 2006. The structures of *Escherichia coli* O-polysaccharide antigens. *FEMS Microbiol Rev* **30:**382–403.

347. **Johnson JR.** 1991. Virulence factors in *Escherichia coli* urinary tract infection. *Clin Microbiol Rev* **4:**80–128.

348. **Ellis TN, Kuehn MJ.** 2010. Virulence and immunomodulatory roles of bacterial outer membrane vesicles. *Microbiol Mol Biol Rev* **74:**81–94.

349. **Wurpel DJ, Moriel DG, Totsika M, Easton DM, Schembri MA.** 2015. Comparative analysis of the uropathogenic *Escherichia coli* surface proteome by tandem mass-spectrometry of artificially induced outer membrane vesicles. *J Proteomics* **115:**93–106.

350. **Wiles TJ, Kulesus RR, Mulvey MA.** 2008. Origins and virulence mechanisms of uropathogenic *Escherichia coli.* *Exp Mol Pathol* **85:**11–19.

351. **Oster P, Lennon D, O'Hallahan J, Mulholland K, Reid S, Martin D.** 2005. MeNZB: a safe and highly immunogenic tailor-made vaccine against the New Zealand Neisseria meningitidis serogroup B disease epidemic strain. *Vaccine* **23:**2191–2196.

352. **Rosenqvist E, Hoiby EA, Wedege E, Bryn K, Kolberg J, Klem A, Ronnild E, Bjune G, Nokleby H.** 1995. Human antibody responses to meningococcal outer membrane antigens after three doses of the Norwegian group B meningococcal vaccine. *Infect Immun* **63:**4642–4652.

353. **Sierra GV, Campa HC, Varcacel NM, Garcia IL, Izquierdo PL, Sotolongo PF, Casanueva GV, Rico CO, Rodriguez CR, Terry MH.** 1991. Vaccine against group B Neisseria meningitidis: protection trial and mass vaccination results in Cuba. *NIPH Ann* **14:**195–207; discussion 208–110.

354. **Camacho AI, de Souza J, Sanchez-Gomez S, Pardo-Ros M, Irache JM, Gamazo C.** 2011. Mucosal immunization with Shigella flexneri outer membrane vesicles induced protection in mice. *Vaccine* **29:**8222–8229.

355. **Roberts R, Moreno G, Bottero D, Gaillard ME, Fingermann M, Graieb A, Rumbo M, Hozbor D.** 2008. Outer membrane vesicles as acellular vaccine against pertussis. *Vaccine* **26:**4639–4646.

356. **Wizemann TM, Adamou JE, Langermann S.** 1999. Adhesins as targets for vaccine development. *Emerg Infect Dis* **5:**395–403.

357. **Silverblatt FJ, Cohen LS.** 1979. Antipili antibody affords protection against experimental ascending pyelonephritis. *J Clin Invest* **64:**333–336.

358. **Jones CH, Pinkner JS, Nicholes AV, Slonim LN, Abraham SN, Hultgren SJ.** 1993. FimC is a periplasmic PapD-like chaperone that directs assembly of type 1 pili in bacteria. *Proc Natl Acad Sci U S A* **90:**8397–8401.

359. **Asadi Karam MR, Oloomi M, Mahdavi M, Habibi M, Bouzari S.** 2013. Vaccination with recombinant FimH fused with flagellin enhances cellular and humoral immunity against urinary tract infection in mice. *Vaccine* **31:**1210–1216.

360. **Karam MR, Oloomi M, Mahdavi M, Habibi M, Bouzari S.** 2013. Assessment of immune responses of the flagellin (FliC) fused to FimH adhesin of Uropathogenic *Escherichia coli. Mol Immunol* **54:**32–39.

361. **Bagherpour G, Fooladi AA, Mehrabadi JF, Nourani MR, Einollahi B.** 2011. Evaluation of mammalian codon usage of fimH in DNA vaccine design. *Acta Microbiol Immunol Hung* **58:**259–271.

362. **Ferraro B, Morrow MP, Hutnick NA, Shin TH, Lucke CE, Weiner DB.** 2011. Clinical applications of DNA vaccines: current progress. *Clin Infect Dis* **53:**296–302.

363. **Imani Fooladi AA, Bagherpour G, Khoramabadi N, Fallah Mehrabadi J, Mahdavi M, Halabian R, Amin M, Izadi Mobarakeh J, Einollahi B.** 2014. Cellular immunity survey against urinary tract infection using pVAX/fimH cassette with mammalian and wild type codon usage as a DNA vaccine. *Clin Exp Vaccine Res* **3:**185–193.

364. **Tchesnokova V, Aprikian P, Kisiela D, Gowey S, Korotkova N, Thomas W, Sokurenko E.** 2011. Type 1 fimbrial adhesin FimH elicits an immune response that enhances cell adhesion of *Escherichia coli. Infect Immun* **79:**3895–3904.

365. **Pecha B, Low D, O'Hanley P.** 1989. Gal-Gal pili vaccines prevent pyelonephritis by piliated *Escherichia coli* in a murine model. Single-component Gal-Gal pili vaccines prevent pyelonephritis by homologous and heterologous piliated *E. coli* strains. *J Clin Invest* **83:**2102–2108.

366. **Roberts JA, Hardaway K, Kaack B, Fussell EN, Baskin G.** 1984. Prevention of pyelonephritis by immunization with P-fimbriae. *J Urol* **131:**602–607.

367. **Roberts JA, Kaack MB, Baskin G, Korhonen TK, Svenson SB, Winberg J.** 1989. P-fimbriae vaccines. II. Cross reactive protection against pyelonephritis. *Pediatr Nephrol* **3:**391–396.

368. **Schmidt MA, O'Hanley P, Lark D, Schoolnik GK.** 1988. Synthetic peptides corresponding to protective epitopes of *Escherichia coli* digalactoside-binding pilin prevent infection in a murine pyelonephritis model. *Proc Natl Acad Sci U S A* **85:**1247–1251.

369. **Lund B, Lindberg F, Marklund BI, Normark S.** 1988. Tip proteins of pili associated with pyelonephritis: new candidates for vaccine development. *Vaccine* **6:**110–112.

370. **Schmidt G, Hacker J, Wood G, Marre R.** 1989. Oral vaccination of rats with live avirulent Salmonella derivatives expressing adhesive fimbrial antigens of uropathogenic *Escherichia coli. FEMS Microbiol Immunol* **1:**229–235.

371. **Goluszko P, Goluszko E, Nowicki B, Nowicki S, Popov V, Wang HQ.** 2005. Vaccination with purified Dr Fimbriae reduces mortality associated with chronic urinary tract infection due to *Escherichia coli* bearing Dr adhesin. *Infect Immun* **73:**627–631.

372. **He Y, Xiang Z, Mobley HL.** 2010. Vaxign: the first web-based vaccine design program for reverse vaccinology and applications for vaccine development. *J Biomed Biotechnol* **2010:**297505.

373. **Alteri CJ, Hagan EC, Sivick KE, Smith SN, Mobley HL.** 2009. Mucosal immunization with iron receptor antigens protects against urinary tract infection. *PLoS Pathog* **5:**e1000586.

374. **Brumbaugh AR, Smith SN, Mobley HL.** 2013. Immunization with the yersiniabactin receptor, FyuA, protects against pyelonephritis in a murine model of urinary tract infection. *Infect Immun* **81:**3309–3316.

375. **Brumbaugh AR, Smith SN, Subashchandrabose S, Himpsl SD, Hazen TH, Rasko DA, Mobley HL.** 2015. Blocking yersiniabactin import

attenuates extraintestinal pathogenic *Escherichia coli* in cystitis and pyelonephritis and represents a novel target to prevent urinary tract infection. *Infect Immun* 83:1443–1450.

376. **Moriel DG, Bertoldi I, Spagnuolo A, Marchi S, Rosini R, Nesta B, Pastorello I, Corea VA, Torricelli G, Cartocci E, Savino S, Scarselli M, Dobrindt U, Hacker J, Tettelin H, Tallon LJ, Sullivan S, Wieler LH, Ewers C, Pickard D, Dougan G, Fontana MR, Rappuoli R, Pizza M, Serino L.** 2010. Identification of protective and broadly conserved vaccine antigens from the genome of extraintestinal pathogenic *Escherichia coli*. *Proc Natl Acad Sci U S A* 107:9072–9077.

377. **Nesta B, Valeri M, Spagnuolo A, Rosini R, Mora M, Donato P, Alteri CJ, Del Vecchio M, Buccato S, Pezzicoli A, Bertoldi I, Buzzigoli L, Tuscano G, Falduto M, Rippa V, Ashhab Y, Bensi G, Fontana MR, Seib KL, Mobley HL, Pizza M, Soriani M, Serino L.** 2014. SslE elicits functional antibodies that impair in vitro mucinase activity and in vivo colonization by both intestinal and extraintestinal *Escherichia coli* strains. *PLoS Pathog* 10:e1004124.

378. **Kurupati P, Teh BK, Kumarasinghe G, Poh CL.** 2006. Identification of vaccine candidate antigens of an ESBL producing Klebsiella pneumoniae clinical strain by immunoproteome analysis. *Proteomics* 6:836–844.

379. **Moayeri N, Collins CM, O'Hanley P.** 1991. Efficacy of a Proteus mirabilis outer membrane protein vaccine in preventing experimental Proteus pyelonephritis in a BALB/c mouse model. *Infect Immun* 59:3778–3786.

380. **Hunter BW, Akins LL, Sanford JP.** 1964. The Role of Immunity in the Pathogenesis of Experimental Retrograde Pyelonephritis. *J Exp Med* 119:869–879.

381. **Alamuri P, Eaton KA, Himpsl SD, Smith SN, Mobley HL.** 2009. Vaccination with proteus toxic agglutinin, a hemolysin-independent cytotoxin in vivo, protects against Proteus mirabilis urinary tract infection. *Infect Immun* 77:632–641.

382. **Li X, Zhao H, Geymonat L, Bahrani F, Johnson DE, Mobley HL.** 1997. Proteus mirabilis mannose-resistant, Proteus-like fimbriae: MrpG is located at the fimbrial tip and is required for fimbrial assembly. *Infect Immun* 65:1327–1334.

383. **Li X, Lockatell CV, Johnson DE, Lane MC, Warren JW, Mobley HL.** 2004. Development of an intranasal vaccine to prevent urinary tract infection by Proteus mirabilis. *Infect Immun* 72:66–75.

384. **Scavone P, Miyoshi A, Rial A, Chabalgoity A, Langella P, Azevedo V, Zunino P.** 2007. Intranasal immunisation with recombinant Lactococcus lactis displaying either anchored or secreted forms of Proteus mirabilis MrpA fimbrial protein confers specific immune response and induces a significant reduction of kidney bacterial colonisation in mice. *Microbes Infect* 9:821–828.

385. **Pellegrino R, Galvalisi U, Scavone P, Sosa V, Zunino P.** 2003. Evaluation of Proteus mirabilis structural fimbrial proteins as antigens against urinary tract infections. *FEMS Immunol Med Microbiol* 36:103–110.

386. **Scavone P, Sosa V, Pellegrino R, Galvalisi U, Zunino P.** 2004. Mucosal vaccination of mice with recombinant Proteus mirabilis structural fimbrial proteins. *Microbes Infect* 6:853–860.

387. **Scavone P, Umpierrez A, Rial A, Chabalgoity JA, Zunino P.** 2014. Native flagellin does not protect mice against an experimental Proteus mirabilis ascending urinary tract infection and neutralizes the protective effect of MrpA fimbrial protein. *Antonie Van Leeuwenhoek* 105:1139–1148.

388. **Scavone P, Rial A, Umpierrez A, Chabalgoity A, Zunino P.** 2009. Effects of the administration of cholera toxin as a mucosal adjuvant on the immune and protective response induced by Proteus mirabilis MrpA fimbrial protein in the urinary tract. *Microbiol Immunol* 53:233–240.

389. **Habibi M, Asadi Karam MR, Shokrgozar MA, Oloomi M, Jafari A, Bouzari S.** 2015. Intranasal immunization with fusion protein MrpH.FimH and MPL adjuvant confers protection against urinary tract infections caused by uropathogenic *Escherichia coli* and Proteus mirabilis. *Mol Immunol* 64:285–294.

390. **Bauer HW, Rahlfs VW, Lauener PA, Blessmann GS.** 2002. Prevention of recurrent urinary tract infections with immuno-active *E. coli* fractions: a meta-analysis of five placebo-controlled double-blind studies. *Int J Antimicrob Agents* 19:451–456.

391. **Naber KG, Cho YH, Matsumoto T, Schaeffer AJ.** 2009. Immunoactive prophylaxis of recurrent urinary tract infections: a meta-analysis. *Int J Antimicrob Agents* 33:111–119.

392. **Bauer HW, Alloussi S, Egger G, Blumlein HM, Cozma G, Schulman CC.** 2005. A long-term, multicenter, double-blind study of an *Escherichia coli* extract (OM-89) in female patients with recurrent urinary tract infections. *Eur Urol* 47:542–548; discussion 548.

393. **Lorenzo-Gomez MF, Padilla-Fernandez B, Garcia-Criado FJ, Miron-Canelo JA,**

Gil-Vicente A, Nieto-Huertos A, Silva-Abuin JM. 2013. Evaluation of a therapeutic vaccine for the prevention of recurrent urinary tract infections versus prophylactic treatment with antibiotics. *Int Urogynecol J* **24:**127–134.

394. **Huber M, Ayoub M, Pfannes SD, Mittenbuhler K, Weis K, Bessler WG, Baier W.** 2000. Immunostimulatory activity of the bacterial extract OM-8. *Eur J Med Res* **5:**101–109.

395. **Huber M, Krauter K, Winkelmann G, Bauer HW, Rahlfs VW, Lauener PA, Blessmann GS, Bessler WG.** 2000. Immunostimulation by bacterial components: II. Efficacy studies and meta-analysis of the bacterial extract OM-89. *Int J Immunopharmacol* **22:**1103–1111.

396. **Lee SJ, Kim SW, Cho YH, Yoon MS.** 2006. Anti-inflammatory effect of an *Escherichia coli* extract in a mouse model of lipopolysaccharide-induced cystitis. *World J Urol* **24:** 33–38.

397. **Sedelmeier EA, Bessler WG.** 1995. Biological activity of bacterial cell-wall components: immunogenicity of the bacterial extract OM-89. *Immunopharmacology* **29:**29–36.

398. **Huber M, Baier W, Serr A, Bessler WG.** 2000. Immunogenicity of an *E. coli* extract after oral or intraperitoneal administration: induction of antibodies against pathogenic bacterial strains. *Int J Immunopharmacol* **22:**57–68.

399. **Johnson JR, O'Bryan TT, Delavari P, Kuskowski M, Stapleton A, Carlino U, Russo TA.** 2001. Clonal relationships and extended virulence genotypes among *Escherichia coli* isolates from women with a first or recurrent episode of cystitis. *J Infect Dis* **183:**1508–1517.

400. **Kudinha T, Kong F, Johnson JR, Andrew SD, Anderson P, Gilbert GL.** 2012. Multiplex PCR-based reverse line blot assay for simultaneous detection of 22 virulence genes in uropathogenic *Escherichia coli*. *Appl Environ Microbiol* **78:**1198–1202.

401. **Kanamaru S, Kurazono H, Ishitoya S, Terai A, Habuchi T, Nakano M, Ogawa O, Yamamoto S.** 2003. Distribution and genetic association of putative uropathogenic virulence factors iroN, iha, kpsMT, ompT and usp in *Escherichia coli* isolated from urinary tract infections in Japan. *J Urol* **170:**2490–2493.

402. **Foxman B, Zhang L, Palin K, Tallman P, Marrs CF.** 1995. Bacterial virulence characteristics of *Escherichia coli* isolates from first-time urinary tract infection. *J Infect Dis* **171:**1514–1521.

403. **Usein CR, Damian M, Tatu-Chitoiu D, Capusa C, Fagaras R, Tudorache D, Nica M, Le Bouguenec C.** 2001. Prevalence of virulence genes in *Escherichia coli* strains isolated from Romanian adult urinary tract infection cases. *J Cell Mol Med* **5:**303–310.

404. **Ikaheimo R, Siitonen A, Karkkainen U, Makela PH.** 1993. Virulence characteristics of *Escherichia coli* in nosocomial urinary tract infection. *Clin Infect Dis* **16:**785–791.

405. **Marrs CF, Zhang L, Tallman P, Manning SD, Somsel P, Raz P, Colodner R, Jantunen ME, Siitonen A, Saxen H, Foxman B.** 2002. Variations in 10 putative uropathogen virulence genes among urinary, faecal and periurethral *Escherichia coli*. *J Med Microbiol* **51:** 138–142.

406. **Ruiz J, Simon K, Horcajada JP, Velasco M, Barranco M, Roig G, Moreno-Martinez A, Martinez JA, Jimenez de Anta T, Mensa J, Vila J.** 2002. Differences in virulence factors among clinical isolates of *Escherichia coli* causing cystitis and pyelonephritis in women and prostatitis in men. *J Clin Microbiol* **40:**4445–4449.

407. **Schaeffer AJ, Schwan WR, Hultgren SJ, Duncan JL.** 1987. Relationship of type 1 pilus expression in *Escherichia coli* to ascending urinary tract infections in mice. *Infect Immun* **55:**373–380.

408. **Goluszko P, Moseley SL, Truong LD, Kaul A, Williford JR, Selvarangan R, Nowicki S, Nowicki B.** 1997. Development of experimental model of chronic pyelonephritis with *Escherichia coli* O75:K5:H-bearing Dr fimbriae: mutation in the dra region prevented tubulointerstitial nephritis. *J Clin Invest* **99:** 1662–1672.

409. **Smith YC, Rasmussen SB, Grande KK, Conran RM, O'Brien AD.** 2008. Hemolysin of uropathogenic *Escherichia coli* evokes extensive shedding of the uroepithelium and hemorrhage in bladder tissue within the first 24 hours after intraurethral inoculation of mice. *Infect Immun* **76:**2978–2990.

410. **Parham NJ, Srinivasan U, Desvaux M, Foxman B, Marrs CF, Henderson IR.** 2004. PicU, a second serine protease autotransporter of uropathogenic *Escherichia coli*. *FEMS Microbiol Lett* **230:**73–83.

411. **Russo TA, Carlino UB, Johnson JR.** 2001. Identification of a new iron-regulated virulence gene, ireA, in an extraintestinal pathogenic isolate of *Escherichia coli*. *Infect Immun* **69:**6209–6216.

412. **Russo TA, McFadden CD, Carlino-MacDonald UB, Beanan JM, Barnard TJ, Johnson JR.** 2002. IroN functions as a siderophore receptor and is a urovirulence factor in an extraintestinal pathogenic isolate of *Escherichia coli*. *Infect Immun* **70:**7156–7160.

413. **Lloyd AL, Henderson TA, Vigil PD, Mobley HL.** 2009. Genomic islands of uropathogenic *Escherichia coli* contribute to virulence. *J Bacteriol* **191:**3469–3481.

414. **Starcic Erjavec M, Jesenko B, Petkovsek Z, Zgur-Bertok D.** 2010. Prevalence and associations of tcpC, a gene encoding a Toll/interleukin-1 receptor domain-containing protein, among *Escherichia coli* urinary tract infection, skin and soft tissue infection, and commensal isolates. *J Clin Microbiol* **48:**966–968.

415. **Lloyd AL, Smith SN, Eaton KA, Mobley HL.** 2009. Uropathogenic *Escherichia coli* Suppresses the host inflammatory response via pathogenicity island genes sisA and sisB. *Infect Immun* **77:**5322–5333.

416. **Nicholson TF, Watts KM, Hunstad DA.** 2009. OmpA of uropathogenic *Escherichia coli* promotes postinvasion pathogenesis of cystitis. *Infect Immun* **77:**5245–5251.

417. **Zhao T, Fang XX, Liu XL, Peng L, Long M, Zhang WB, Luo J, Cao H.** [Construction and functional studies of uropathogenic *E. coli* strains with ompT gene knockout]. *Nan Fang Yi Ke Da Xue Xue Bao* **32:**956–959.

418. **Hui CY, Guo Y, He QS, Peng L, Wu SC, Cao H, Huang SH.** 2010. *Escherichia coli* outer membrane protease OmpT confers resistance to urinary cationic peptides. *Microbiol Immunol* **54:**452–459.

419. **Kruze D, Biro K, Holzbecher K, Andrial M, Bossart W.** 1992. Protection by a polyvalent vaccine against challenge infection and pyelonephritis. *Urol Res* **20:**177–181.

420. **Nayir A, Emre S, Sirin A, Bulut A, Alpay H, Tanman F.** 1995. The effects of vaccination with inactivated uropathogenic bacteria in recurrent urinary tract infections of children. *Vaccine* **13:**987–990.

421. **Ng TW, Akman L, Osisami M, Thanassi DG.** 2004. The usher N terminus is the initial targeting site for chaperone-subunit complexes and participates in subsequent pilus biogenesis events. *J Bacteriol* **186:**5321–5331.

422. **Volkan E, Ford BA, Pinkner JS, Dodson KW, Henderson NS, Thanassi DG, Waksman G, Hultgren SJ.** 2012. Domain activities of PapC usher reveal the mechanism of action of an *Escherichia coli* molecular machine. *Proc Natl Acad Sci U S A* **109:**9563–9568.

423. **Barnhart MM, Pinkner JS, Soto GE, Sauer FG, Langermann S, Waksman G, Frieden C, Hultgren SJ.** 2000. PapD-like chaperones provide the missing information for folding of pilin proteins. *Proc Natl Acad Sci U S A* **97:** 7709–7714.

424. **Verger D, Miller E, Remaut H, Waksman G, Hultgren S.** 2006. Molecular mechanism of P pilus termination in uropathogenic *Escherichia coli*. *EMBO Rep* **7:**1228–1232.

Index

ABC transporters, 237, 241, 573
  Klebsiella pneumoniae, 440
  Proteus mirabilis, 414–416
  Staphylococcus aureus, 465
Abdominal stoma conduit, bacteriuria and, 93
Abscess
  perirenal, 150
  prostatic, 151
  renal, 150
Accidental pathogen, 332
Ace, 301–304, 306, 466–467
Acetylcholine, 572–573
Acinetobacter baumannii, 180
ACTH (adrenocorticotrophic hormone), 138–139
Actinobacteria (phylum), 473, 476, 478, 480
Actinobaculum, 473–477, 482–484
Actinomyces, 478, 484
Activator protein 1 (AP-1), 525
Acute kidney injury (AKI), 513–514
Adaptive immunity
  involvement in UTI, 543
  recurrent UTI and, 610–611
Adenovirus, acute hemorrhagic cystitis and, 70
Adherence
  interplay between adherence and motility, 245
  in renal tubule, 506–507
Adherence factors. *See* Adhesins
Adhesins, 277–312
  E. coli, 70–71
    prostatitis isolates, 125–131
    uropathogenic *E. coli* (UPEC), 592–599
  Enterococcus, 466
  Gram-negative uropathogens, 279–301
    autotransporters, 293–300
    Iha, 300
    pili, 279–293
    TosA, 300–301
  Gram-positive uropathogens, 281, 301–311
    EfbA, 307
    MSCRAMMs, 301–307
    pili, 307–311
  Klebsiella pneumoniae, 442–445
  overview, 277–279
  pathoadaptive mutation and, 341–343
  prostatitis isolates, 125–131

Staphylococcus saprophyticus, 463–464
  as virulence factors, 334–335
Adjuvants, vaccine, 615–616
Adrenocorticotrophic hormone (ACTH), 138–139
Aerobactin, 126, 129, 238–239, 440, 602
Aerococcus, 472–473, 484
Aeromonas hydrophilia aerolysin, 268–269
Afa/Dr family of pili, 281, 283–284
Afferent nerves, 567–569
Ag43, 298–299, 534
aggR gene, 125
Aging female
  anatomy and physiology, 13–14
  asymptomatic bacteriuria
    diagnosis, 89
    incidence, 93–94
    microbiology of, 99
    prevalence, 92, 94–95
  immune breakdown in, 560
  incidence of UTI, 79
  prevalence of UTI, 13–14
Aging male, asymptomatic bacteriuria in
  diagnosis, 89
  incidence, 93–94
  microbiology of, 99
  prevalence, 92, 95
Agmatine, *Proteus mirabilis* swarming and, 397
AIDA-I, 293, 295
AipA autotransporter, 412
AKI (acute kidney injury), 513–514
Akt, 268
Algorithm, for management of urosepsis, 148–150
Allelic variation under positive selection, 214–215
Allodynia, 577–583
Alloscardovia omnicolens, 478
α-hemolysin (Hly), 126, 128–129, 246, 253, 264–269
  host response to, 507–508
  mechanism, 267–269
  renal damage and, 603–604
  structure, 266–267
  toxoid vaccine, 620
  uropathogenic *E. coli* (UPEC), 264–269, 507–508
  as virulence factor, 264–266
α-keto acids, 414
Aluminum-based adjuvants, 615

Ambient temperature fimbria (ATF), 409
Amikacin
  origin of, 195
  resistance, 196
Aminoglycoside acetyltransferase, 191
Aminoglycosides
  for catheter-associated UTI, 34
  indications, 195
  mechanism of action, 196
  origin of, 195
  resistance, 144, 591
    mechanisms, 196
    prevalence, 196
  side effects, 195
  for uncomplicated pyelonephritis, 54
  for urosepsis, 141–143, 146
  for UTI in children, 74
Amoxicillin
  for asymptomatic bacteriuria, 106
  resistance, 58
Amoxicillin-clavulanic acid
  for asymptomatic bacteriuria, 106
  for ESBL-producing *E. coli*, 49
  for pyelonephritis during pregnancy, 31
  for uncomplicated cystitis, 51–52
  for UTI during pregnancy, 31
  for UTI in children, 74
Ampicillin resistance, 124, 145
Amplification of gene copies, 337
Amygdala, 582
Anaerococcus, 483
  A. lactolyticus, 482
Analgesia in the bladder, bacterial, 580, 582–583
Anatomy, 3–18
  abnormalities, 10–14
    in aging female, 13–14
    asymptomatic bacteriuria and, 100
    bladder-outlet obstruction, 12–13
    calyceal diverticula, 11
    in children, 71
    medullary sponge kidney, 10–11
    pelvic anatomy, 13
    ureteral obstruction, 11–12
    vesicoureteric reflux, 12
  bladder, 4–5
  microscopic, 6–8
  upper urinary-collecting system, 4
  ureterovesical junction, 4, 5
  urethra, 5–6
  vagina, 6
Animal models, for polymicrobial UTI, 461–462
Anterior vaginal-wall prolapse (cystocele), 14
Antibacterial peptides, UTI susceptibility and, 538
Antibiotic prophylaxis. *See* Antimicrobial prophylaxis
Antibiotics. *See* Antimicrobial(s)

Anticholinergic medications, for neurologic patients, 17
Antigen 43, 249
Antimicrobial(s)
  growth promoters, 170
  intracellular bacterial communities, effectiveness against, 600
  stewardship, 170
  use in food animals, 170
Antimicrobial peptides, 556
  Klebsiella pneumoniae interactions with, 439
  in neutrophil granules, 509
  produced by nephron urothelium, 509
Antimicrobial prophylaxis, 56–58, 591
  in catheterized patients, 33
  continuous prophylaxis, 57
  postcoital prophylaxis, 56–57
  self-diagnosis and self-treatment, 57–58
  special considerations about, 58
  with vesicoureteral reflux, 75
Antimicrobial resistance, 179–198
  avian pathogenic *E. coli* (APEC), 183–187, 197
  ExPEC, 162–167, 170
  Klebsiella pneumoniae, 448–449
  mechanisms of, 187–198
    aminoglycosides, 195–196
    β-lactams, 191–193
    fluoroquinolones, 190–191
    fosfomycin, 194
    nitrofurantoin, 193–194
    overview, 187–188
    silver-containing agents, 197–198
    tetracyclines, 194–195
    trimethoprim-sulfamethoxazole, 189–190
  origins of, 180–187
  overview, 48–49
  in UTI prophylaxis, 58
  vancomycin-resistant *S. aureus* (VRSA), 144–145
Antimicrobial therapy
  in children, 73–74
  in pregnant women, 31–32
  pyelonephritis, 514–515
  for urosepsis, 141–147
  vaginal microbiotic alterations with, 82
AP-1 (activator protein 1), 525
APEC. *See* Avian pathogenic *E. coli*
Apoptosis, FimH-dependent urothelial, 575–578
Arginine, *Proteus mirabilis* swarming and, 397
Arthrobacter, 484
Aspiratory samples, 89, 90
Asymmetric unit membrane (AUM), 365, 566
Asymptomatic bacteriuria, 87–111
  analgesic activity, 580, 582–583
  ASB strains as therapy, 607–608
  bacterial interference, 110–111
  in children, 71
  co-evolution of bacteria and host, 541

Corynebacterium urealyticum, 473
defined, 87
deliberate establishment of, 542
in diabetics, 34–35
diagnosis, 88–91
   aspiration samples, 90
   catheter specimens, 90
   inflammatory markers, 90–91
   in men, 89–90
   pyuria, 90–91
   quantitative urine culture, 88
   voided urine specimens, 88–90
   in women, 88–89
differential diagnosis, 543–544
epidemiology, 91–94
   incidence, 93–94
   prevalence, 36, 91–93, 529
group B *Streptococcus*, 469–470
innate immune system and, 561
microbiology of, 97–100
   catheterized patients, 99–100
   diabetic women, 98–99
   healthy women, 98
   infants and children, 98
   institutionalized elderly, 99
   older women and men, 99
   pregnant women, 98
as model for pathoadaptive mutation in UPEC,
   339–340
molecular characteristics of susceptibility, 529,
   532, 534, 539–541
morbidity and mortality, 100–104
   after invasive genitourinary procedures, 104
   catheterized patients, 103–104
   diabetic women, 101–102
   elderly patients, 102–103
   healthy women, 101
   infants and children, 100–101
   pregnant women, 102
mutation in strains, 345–346
natural history, 97–104
   microbiology, 97–100
   morbidity and mortality, 100–104
overview, 28–29, 87
pathogenesis, 94–97
   host factors, 94–95
   organism factors, 95–97
in pregnant women, 31, 469–470, 611
prevalence of, 36, 91–93, 529
Proteus mirabilis, 383
reductive evolution in, 529
screening for and treatment of, 104–110
   after invasive genitourinary procedures, 109
   children, 104–105
   diabetic patients, 107
   elderly institutionalized patients, 107
   indwelling urethral catheter patients, 107–108

   older women, 107
   pregnant women, 105–107
   renal transplant patients, 109–110
   spinal cord injury patients, 109
in vivo evolution of *E. coli* strains, 215
ATF (ambient temperature fimbria), 409
Atopobium, 478, 483
ATP, as autocrine mediator, 573–574
Atrophic vaginitis, 82
Attachment inhibitors, 60, 542
Attenuated vaccines, 619
Augmentation cystoplasty, bacteriuria and, 93
AUM (asymmetric unit membrane), 365
Autotransporter toxins, 247
Autotransporters, 248–249, 271, 293–300, 534
   functions in uropathogenic *E. coli* (UPEC),
      298–300
   Ag43, 298–299
   FdeC, 300
   UpaB, 299
   UpaG, 299
   UpaH, 299
   overview, 293–295
   Proteus mirabilis, 411–412
   secretion pathway, 296–298
   structure, 295–296
Avian pathogenic *E. coli* (APEC), 166, 168, 208
   antimicrobial resistance, 183–187, 197
   relationship to NMEC and UPEC strains,
      216
   sequencing of O1:K1:H7 strain, 212

Bacillus cereus pili, 308
Bacterial interference, 110–111
Bacterial prostatitis. *See* Prostatitis
Bacterial spectrum in urosepsis, 143–144
Bacterial vaginosis (BV), 82, 471–472, 477, 480,
   482–483, 485
Bacteriuria
   asymptomatic, 28–29, 45, 55
   catheter-associated, 32–33, 460
   Gardnerella vaginalis, 477
   interpretation of urine culture results, 44–45
   microbiology of, 97–100
   natural history of, 97–104
   patterns of response to therapy, 35
   in pregnancy, 15, 469–470
   in prostatitis, 122, 124
   rapid detection strategies, 45
   threshold for diagnosis, 27
Bacteroides, 479, 482
Bam complex, 297–298
BarA-UvrY two-component system, 601
Barrier function, urothelial, 569
Bedwetting, 72
Behavioral risk factors for asymptomatic
   bacteriuria, 94

Benign prostatic hyperplasia (BPH) and bladder-outlet obstruction, 13
β-defensin, 542, 556
β-hemolysin, group B *Streptococcus*, 471
Beta-lactam antibiotics
  mode of action of, 191
  recurrence of UTI associated with, 82
  resistance
    mechanism, 192
    prevalence, 192–193
  for uncomplicated cystitis, 50–52
  for uncomplicated pyelonephritis, 53–54
  for urosepsis, 141–142
  for UTI during pregnancy, 31
β-lactam inhibitor (BLI), for urosepsis, 146
β-lactamases, 180, 192–193
  extended-spectrum beta lactamases (ESBL), 28, 49, 74, 162–163, 165–166, 168, 192
  New Delhi metallo (NDM) β-lactamases, 163, 193, 214
Biarylmannose-derivative FimH antagonists, 605–606
Bifidobacterium, 478, 483
Biofilms, 90, 95, 97
  catheter-associated bacteria, 460–462
  CAUTI and, 605, 609
  curli, 247
  Enterococcus, 466–467, 604–605, 609
  Klebsiella pneumoniae, 436, 444–448
  matrix composition, 247
  prevention of urinary catheter biofilm formation, 421–422
  Proteus mirabilis, 385, 405–406, 417, 419, 421–422
  Staphylococcus epidermidis, 465
  uropathogenic *E. coli* (UPEC), 247–248
    gene expression within, 248
    inhibition of formation, 248
    regulation of formation, 248
  in urosepsis, 143
Bladder
  anatomy, 4–5
  diabetic cystopathy, 14
  filling and emptying, neural control of, 567–568
  micturition cycle, 9–10
  neurogenic, 42, 71, 218
  primary bladder-neck obstruction (PBNO), 16
  urothelium, 7
Bladder cells, bacterial invasion of, 360–371
Bladder epithelial cells, 556–557
Bordetella pertussis, 293, 296
Bruton's tyrosine kinase (BTK), 531
BV. *See* Bacterial vaginosis
"By-product of commensalism" hypothesis, 224

Cadherin, 570
Calcitonin gene-related peptide (CGRP), 568, 572
Calcium (Ca^{2+}) signaling, 508

Calyceal diverticula, 11
Calyces, 4
Candida spp., asymptomatic bacteriuria and, 97
Capsule
  E. coli prostatitis isolates, 125–128
  group B *Streptococcus*, 471
  group II type, 248, 249
  Klebsiella pneumoniae, 436–439
  K1-type, 249
  Proteus mirabilis swarming and, 395–396
  shift in antigenicity, 338
  Staphylococcus saprophyticus, 464
  uropathogenic *E. coli* (UPEC), 248, 249–250
Carbapenem
  resistance, 144, 448
  for urosepsis, 146
CARS (counter-regulatory anti-inflammatory response syndrome), 138
CAS (chrome azul S) assay, 414, 417
Caspases, 576
Cathelicidin, 370, 509–510, 538, 556
Catheter-acquired bacteriuria
  morbidity, 103–104
  treatment, 107–108
Catheter-associated urinary tract infections (CAUTIs)
  biofilms and, 460–462, 605, 609
  daily risk, 590
  Enterococcus faecalis, 306, 310–311, 461, 466–469, 604–605, 609
  Escherichia coli strains infecting, 218
  Gram-positive uropathogens, 460–462, 465
  Klebsiella pneumoniae, 435–436, 445, 448–450
  overview, 32–34
  percentage of health care-associated infections, 590
  polymicrobial, 462
  prevention using probiotic *E. coli*, 561
  Proteus mirabilis, 384, 385, 419–422
  role of adherence and biofilm formation in, 605
  Staphylococcus epidermidis, 465
  treatment, 33–34
Catheterization
  clean intermittent catheterization (CIC), for neurologic patients, 17
  risk increase for asymptomatic bacteriuria, 95
Catheters
  bacterial interferences and, 110–111
  biofilms on, 90, 95, 97, 447–448
  indwelling, 32–34, 93
    bacterial interference, 111
    bacteriuria, 95, 103, 107–108
  obstruction, 385
  prevention of urinary catheter biofilm formation, 421–422
  silver-coated, 197

urine samples for asymptomatic bacteriuria
  diagnosis, 90
Cats, as ExPEC reservoir, 165–166
Cavernitis, 151
CCLs, 535
ccmA, 396
CDC42, 533
Cefixime, for bacterial prostatitis, 124–125
Cefotaxime resistance, 144–145, 192
Ceftazidime resistance, 144
Ceftriaxone, for uncomplicated pyelonephritis, 54
Cell membrane receptors, genetic variants affecting,
  528
Cell-cell communication, 251
Cellobiose metabolism, by *Klebsiella pneumoniae*,
  448
Cellular invasion, by *Proteus mirabilis*, 399–400
Central nervous system circuits in cystitis pain,
  581–582
Cephalexin
  for asymptomatic bacteriuria, 106
  for prophylaxis, 57
Cephalosporin
  for pyelonephritis during pregnancy, 31
  resistance, 49, 144, 591
  for urosepsis, 146
  for UTI in children, 74
Cephalothin resistance, 124
Ceramide, 289, 532
CGRP (calcitonin gene-related peptide), 568, 572
Chaperone/usher (CU) assembled pili, 243, 251,
  279–289
  adhesins, 283–284, 592
  assembly at outer membrane, 285–286
  chaperone-subunit complex formation, 284–285
  fiber, 281–283
  functions of, 287–289
  gene clusters, 282
  inhibitors, 607
  Klebsiella pneumoniae, 443
  P pili, 289
  pilus usher, 286–287
  structure, 281–284
  type 1 pili, 287–288
Chemotaxis, 245
Children
  asymptomatic bacteriuria
    diagnosis, 88
    incidence, 93
    microbiology, 99
    morbidity, 100–101
    prevalence, 92
    screening, 104–105
    treatment, 104–105
  urinary tract infection in, 69–75
    clinical presentation, 72
    diagnosis, 72–73

epidemiology, 69–70
imaging, 74–75
laboratory work-up, 72–73
management and treatment, 73–74
overview, 30–31
pathogenesis, 70–71
prophylaxis, 75
treatment, 31
vesicoureteral reflux, 70–72
vesicoureteral reflux, 70–72, 100, 105
Chitosan, as bladder cell exfoliant, 372
Chrome azul S (CAS) assay, 414, 417
Chronic pelvic pain syndrome (CPPS)
  E. coli in, 129–132
  pathogenesis of, 131–132
  virulence factors in, 130
Chronic prostatitis/chronic pelvic pain syndrome
  (CP/CPPS), 36
Chronic urinary retention, 13
ChuA, 239, 240
Ciprofloxacin
  for asymptomatic bacteriuria, 108
  for prophylaxis, 57–58
  resistance, 144–145, 191, 476, 591
  for uncomplicated cystitis, 50–52
  for uncomplicated pyelonephritis, 53–54
  for urosepsis, 142–147
Circumcision, 71
Citrobacter
  asymptomatic bacteriuria, 97
  C. freundii
    adhesins, 280
    intracellular bacterial communities (IBCs), 600
Clathrin-coated pits, 595
Claudins, 566
Clean intermittent catheterization (CIC), for
  neurologic patients, 17
Clean-catch technique, 43–44
ClfA, 301–302, 304
Clinical diagnosis of urinary tract infection, 41–42
Clinical syndromes
  acute pyelonephritis, 30
  asymptomatic bacteriuria, 28–29
  complicated UTI, 29–30
  prostatitis, 30
  special patient groups
    catheterized patients, 32–34
    children, 30–31
    diabetes mellitus, 34–35
    men, 31
    pregnant women, 31–32
  uncomplicated cystitis, 29
  urosepsis, 30
Clostridium difficile
  antimicrobial resistance, 180
  colitis, 591
  diarrhea, UTI prophylaxis and, 58

Clot formation, in pyelonephritis, 511–512
Clue cells, 480, 485
CmfA, 395–396
Cna, 301–302, 304
CNF-1. *See* Cytotoxic necrotizing factor-1
Coagulase-negative staphylococci, asymptomatic
    bacteriuria and, 97
Colibactin, 126, 128–129
Colicin, 126
Collagen hug model of ligand binding, 302–304
Colony migration factor, 389, 395–396
Colony morphology, variation in *E. coli*, 337–338
Commensals, 529
    as opportunistic pathogens, 332
    virulence as by-product of commensalism, 224
Community-acquired UTI, 28
Companion animal reservoirs of ExPEC, 165–166
Comparative genomics, of UPEC strains, 211
Compensatory endocytosis, 365
Complement-mediated phagocytosis, 556–557
Complicated urinary tract infection (UTI)
    defined, 28, 460
    Enterococcus faecalis, 604–605
    Gram-positive uropathogens, 460
    overview of clinical syndrome, 29–30
    treatment, 29–30
Condom catheter, 89, 95
Conjugate vaccines, 619–620
Conjugative plasmids, 180, 182
Conjugative transposons, 182
Contamination, of voided urine sample, 43–44
Contraceptive method, effect on vaginal microbiota,
    82
Convergent evolution, 350
Corticotropin-releasing factor/hormone, 139, 568,
    571–572
Corynebacterium, 484
    C. diphtheriae pili, 278, 307, 308, 310
    C. renale pili, 278
    C. urealyticum, 473
Cost of urinary tract infections, 36–37, 79–80, 236,
    556
Counter-regulatory anti-inflammatory response
    syndrome (CARS), 138
Cox2, 537–538
COX-2 (cyclooxygenase-2) inhibitors, 558
CP923, 618–619
CP/CPPS (chronic prostatitis/chronic pelvic pain
    syndrome), 36
cps gene cluster, 438
Cpx, 601
Cranberry, 55, 75, 372, 515, 608–609
Crosstalk, molecular, 513
csg gene cluster, 290–292
CsrA, 394
CTX-M
    CTX-M-1, 168

CTX-M-9, 168
CTX-M-15, 162–163
CTX-M-32, 168
    Escherichia coli, 213–215
    extended-spectrum beta-lactamase (ESBL), 49,
        192
CueO, 339
Culture-negative urine, 481
Curli, 247, 278, 289–293
    assembly machinery, 292
    assembly on the bacterial surface, 291–292
    extracellular nucleation-precipitation, 292
    functions in UPEC, 292–293
    structure, 292–293
    as virulence factor, 534
Curlicide, 607
CXCL1, 559
CXCL2, 537, 559–560
CXCL8, 534–535
CXCR1, 504, 508, 535, 540, 560
CXCR2, 535
Cyclic adenosine monophosphate (cAMP) and
    UPEC expulsion, 599
Cyclooxygenase-2 (COX-2) inhibitors, 558, 603
Cyclophosphamide-induced cystitis, 581
Cystic fibrosis, 338
Cystitis
    acute cystitis, modeling the outcomes of, 602–603
    acute hemorrhagic cystitis, 70
    acute uncomplicated cystitis
        antimicrobial treatment, 29
        clinical diagnosis, 42
        overview, 29
        treatment, 49–53
    clinical diagnosis, 42
    clinical picture, 150
    CNS circuits in pain, 581–582
    Corynebacterium urealyticum encrusting cystitis,
        473
    cyclophosphamide-induced cystitis, 581
    dipstick diagnosis, 46–47
    incidence, 36
    interpretation of urine culture results, 44
    modeling outcomes of acute, 602–603
    polymicrobial, 481
    symptoms, 29, 527, 589
Cystocele, 14
Cytokines
    as markers of the septic response, 138
    pro-inflammatory, 138–139
    response to infection, 534–535
Cytolethal-distending toxin 1, 126
Cytotoxic necrotizing factor-1(CNF-1), 126, 128–129,
    246–247, 265, 269–270, 534, 595

DAF (decay-accelerating factor), 342–343
DamX, 371

Debilitated patients, *Escherichia coli* strains
infecting, 218
Decay-accelerating factor (DAF), 342–343
Defensins, 439, 509, 538
DegP periplasmic protease, 285
Denaturing high-performance liquid chromatography (DHPLC), 482
Dendritic-cell maturation, *Klebsiella pneumoniae*
induction of, 439
Detrusor-external sphincter dyssynergia (DESD), 16–17
Dexamethasone, 53
DHFR (dihydrofolate reductase), 182, 189
DHPLC (denaturing high-performance liquid
chromatography), 482
DHPPP (dihydro-6-hydroxymethylpterin-
pyrophosphate), 189
DHPS (dihydropteroate synthase), 189
Diabetes mellitus, 14–15
asymptomatic bacteriuria, 95
diagnosis, 89
microbiology of, 98–99
morbidity, 101–102
prevalence, 92
treatment, 107
Enterococcus UTI, 466
UTI in, 34–35
asymptomatic bacteriuria, 34–35
complications, 34
risk of recurrent, 34
treatment, 34
Diabetic cystopathy, 14
Diagnosis, 27, 41–48
asymptomatic bacteriuria, 88–91
aspiration samples, 90
catheter specimens, 90
inflammatory markers, 90–91
in men, 89–90
pyuria, 90–91
quantitative urine culture, 88
voided urine specimens, 88–90
in women, 88–89
clinical, 41–42
Gram-positive UTI, 460
laboratory, 43–48, 72–73, 460
dipsticks, 46–48, 460
interpretation of culture results, 44–45
LE (leukocyte esterase), 46–48
nitrite testing, 46–48
pyuria, 45
rapid detection for bacteriuria, 45
voided urine collection techniques, 43–44
voided urine contamination, 43–44
voided urine culture, 43
point of care for UTI, 460
self, 57–58
UTI in children, 72–73
Dialister, 482

Dibekacin, origin of, 195
Dienes line, 401–402
Differential diagnosis, molecular tools for, 543–544
Dihydrofolate reductase (DHFR), 182, 189
Dihydro-6-hydroxymethylpterin-pyrophosphate
(DHPPP), 189
Dihydropteroate synthase (DHPS), 189
Dimercaptosuccinic acid (DMSA) scans, 527
Dipsticks, 43, 46–48, 73, 460
DisA decarboxylase, 393–394
D-mannose, 60
DMSA (dimercaptosuccinic acid) scans, 527
DNA gyrase, 190, 336
DNA microarrays, 254
DNA topoisomerase IV, 190
DNA-mismatch repair genes, 338–339
Dock, lock, and latch model, 304
Dogs, as ExPEC reservoir, 165–166
Doppler ultrasonography, for renal scarring
detection, 75
Doripenem, 143
Dorsal-root ganglia, 567
Doxycycline, for bacterial prostatitis, 124–125
Dr adhesins
mutations in, 342–343
vaccine targeting, 622
DraE adhesin, 343
DsdA, 339
Dynamin2, 367
Dysfunctional voiding, 16
Dysuria
in cystitis, 29
in prostatitis, 122, 124

EAEC (enteroaggregative *Escherichia coli*)
prostatitis, 125
Ebp, 307–311, 467, 596, 605, 624
Eco-evo view of bacterial pathogens, 331–333
Ecto-enzymes, 574
Efflux pumps
in aminoglycosides, 196
β-lactam resistance, 192
fluoroquinolone resistance, 190–191
Klebsiella pneumoniae, 449
EHEC (enterohemorrhagic *E. coli*), quorum-sensing
system in, 251
EibD autotransporter, 296
Epidermal-growth factor (EGF), 572
Elderly individuals
immune breakdown in, 560
invasive group B *Streptococcus* disease in, 470
Encopresis, 30
Encrusting cystitis, *Corynebacterium urealyticum*
and, 473
Endocytosis
compensatory, 365
urothelial, 569

Endothelial NOS (eNOS), 573
Endotoxin, 138
  endotoxemia, 611
  vaccine, 619
Enteroaggregative *Escherichia coli* (EAEC)
      prostatitis, 125
Enterobacter
  antimicrobial resistance, 48
  E. cloacae
      antimicrobial resistance, 145
      pyelonephritis, 150
  intracellular bacterial communities (IBCs), 600
  urinary tract infection in children, 70
Enterobacteriaceae
  asymptomatic bacteriuria, 97
  fluoroquinolone resistance, 190
  incompatibility groups, 187
  plasmid replicons, 187
  UTI in children, 70
Enterobacterial-repetitive intergenic-consensus
      (ERIC) polymerase chain reaction (PCR),
      213, 217
Enterobactin, 238–239, 534, 602
Enterochelin, 440
Enterococcus, 459, 461–462, 466–469
  asymptomatic bacteriuria, 97
  catheter-associated UTI (CAUTI), 461,
      466–469
  E. faecalis
      Ace, 301–304, 306
      adhesins, 281, 301–302, 307–311
      antimicrobial resistance, 144–145
      biofilm, 604–605, 609
      catheter-associated UTI (CAUTI), 306,
          310–311, 461, 591, 604–605, 609
      Ebp pilus, 605
      inhibition biofilm formation, 248
      invasion of host cells, 372
      laboratory model for study of, 461
      nosocomial UTIs, 301, 306
      pili, 307–311, 467, 596
      sortases, 604–605
      urine contamination, 480
      UTI, 459, 461–462, 466–469
  E. faecium, 604
      adhesins, 281
      antimicrobial resistance, 144–145
      pili, 307–311
  epidemiology of UTI, 466
  frequency of urinary tract colonization, 504
  immune responses to, 467–469
  urosepsis, 143
  UTI in children, 70
  vaccines, 624
  virulence factors, 466–467
Enterohemorrhagic *E. coli* (EHEC), quorum-
      sensing system in, 251

Enteropathogenic *Escherichia coli* (EPEC) intimin
      protein, 300
Environmental reservoirs of ExPEC, 164–165
EPEC (enteropathogenic *Escherichia coli*) intimin
      protein, 300
Epidemiology, 36–37
  asymptomatic bacteriuria, 91–94
      incidence, 93–94
      prevalence, 91–93
  ExPEC strains, 215–219
  urinary tract infection, 590, 592
      in children, 69–70
      of enterococcal, 466
      group B *Streptococcus* UTI, 469
      polymicrobial, 461–462
      Staphylococcus saprophyticus, 462–463
  urosepsis, 136, 137
Epidermal growth-factor receptor (EGFR), 569–570
Epididymitis, clinical picture, 151
Epithelial cells, bladder, 556–557
Epithelial integrity, infection-associated breakdown
      of, 511
EQUC (expanded-quantitative urine culture), 484
ERIC (enterobacterial-repetitive intergenic-
      consensus) polymerase chain reaction
      (PCR), 213, 217
Erk, 581
ESBL (extended-spectrum beta lactamases), 28, 49,
      74, 162–163, 165–166, 168, 192
Escherichia coli. *See also* Extraintestinal pathogenic
      E. coli; Uropathogenic *Escherichia coli*
  acute hemorrhagic cystitis, 70
  adhesins, 280, 281, 293, 295, 334–335, 341–343
  AIDA-I, 293, 295
  α-hemolysin (Hly), 126, 128–129, 246, 253,
      264–269, 507–508
  analgesic activity, 580, 582–583
  antimicrobial resistance, 48–49, 124, 144–146,
      162–167, 170, 179–198
  asymptomatic bacteriuria, 28, 71, 88, 95–97,
      97–100, 561
  avian pathogenic *E. coli* (APEC), 166, 168, 208
      antimicrobial resistance, 183–187, 197
      relationship to NMEC and UPEC strains, 216
      sequencing of O1:K1:H7 strain, 212
  bacterial interference, 110–111
  biofilms, 143
  catheter-associated UTI (CAUTI), 33
  chromosome, 209
  colony type variations, 337–338
  commensal strains, 208
  curli, 289–293
  in diabetics, 14–15
  diseases caused by, 208
  eco-evo categories of, 332
  enteroaggregative *E. coli* (EAEC) prostatitis,
      125

enterohemorrhagic *E. coli* (EHEC), quorum-
sensing system in, 251
enteropathogenic *E. coli* (EPEC) intimin protein,
300
epidemiology, 36
extended-spectrum beta lactamases (ESBL), 28,
49, 74, 162–163, 165 166, 168
fimbriae
innate immune activation, 532–534
P fimbriae, 532–533
type 1 fimbriae, 533–534
genetic population structure, 208–209
horizontal gene transfer, 335
lactobacilli inhibition of, 81
multidrug-resistant strains, 213–214, 556
NDM-1 β-lactamase, 193
neonatal meningitis *Escherichia coli* (NMEC),
208, 216, 218–219
number of cells on Earth, 208
periurethral colonization, 70
phylogenetic groups, 128
pili, 280–281
polymicrobial extraintestinal infections, 221–223
population-genetics structure of *E. coli* species,
208–209
probiotic, 333–334, 561
prostatitis, 121–132, 151
pyelonephritis, 150
renal and perirenal abscess, 151
sexual transmission, 333
swarming by, 389
urosepsis, 143
UTI in children, 74
Variome database, 351
virulence factors, 96
in vivo evolution of strains, 215
Escherichia fergusonii, 212
Esp, 466
EstA autotransporter, 295–296
Estradiol-releasing vaginal ring (Estring), 82
Estriol cream, 609
Estrogen
decline of contributing factor to UTI, 560–561
effect of loss on vaginal microbiota, 81–82
lactobacilli effect on, 14
supplementation, 561, 609
therapy, 609
urothelial structure, effect on, 570
Estrogen cream, 36, 82, 609
Estrogen-replacement therapy, 609
Evolution
allelic variation under positive selection, 214–215
"by-product of commensalism" hypothesis, 224
co-evolution of bacteria and host, 541
ExPEC virulence, 224
host environment and, 529
molecular-convergent, 350

pathoadaptive mutations in UPEC, 331
reductive, 529
role of recombination in, 209
in vivo evolution of *E. coli* strains, 215
Exfoliation
of bladder epithelial cells, 557
as host defense, 370
Exocytosis, 573–574
stretch-induced, 569, 573
vesicular, 573
Expanded-quantitative urine culture (EQUC), 484
Extended-spectrum beta lactamases (ESBL), 28, 49,
74, 162–163, 165–166, 168, 192
External sphincterotomy, 17
External-urethral sphincter, 6, 9, 16–17
Extraintestinal pathogenic *E. coli* (ExPEC),
161–170. *See also* Uropathogenic
Escherichia coli
avian pathogenic *E. coli* (APEC), 166
companion animal reservoirs, 165–166
environmental reservoirs, 164–165
foodborne reservoirs, 167–169
beef and cattle sources, 169
pork and pig sources, 169
poultry sources, 167–169
human reservoir, 163–164
important lineages, 162–163
mutation rate, 214–215
population phylogenomics, 207–225
allelic variation under positive selection,
214–215
"by-product of commensalism" hypothesis,
224
CGA (clonal group A), 212–213, 217, 219, 221
clinical relevance of intrinsic virulence,
219–221
diversity within clones, 215
epidemiologic data based on proxy markers,
215–219
infants, UPEC and NMEC strains infecting,
218–219
origin of UPEC strains, 216–217
phylogenetics, 209–212
polymicrobial infections, 221–223
population-genetics structure, 208–209
relatedness of UPEC, NMEC, and APEC
strains, 216
strains infecting debilitated patients, 218
UPEC-specific genes, search for, 211–214
virulence factors, phylogeny and, 217–218
public health perspectives, 169–170
UpaG autotransporter, 299
virulence factors, 123, 125–128, 208

F-actin, 366, 369
F1C fimbriae, 96, 125–126, 128, 244
FdeC, 300, 623

Febrile urinary tract infections, 29–30, 543, 591
  in children, 69
  clinical presentation, 72
Female-genital cutting (FGM), 460
Fenobam, 582
Ferric citrate transport, in *Proteus mirabilis*,
      414–415
Ferric iron uptake, in *Klebsiella pneumoniae*, 440
Ferric-uptake regulator (Fur), 237–238, 440
Ferrous iron uptake, 240, 414–415
FGL chaperones, 285
Fibroblast-growth factor, 570
fim operon, 209, 213–214
fim switch, 507
FimB, 507
Fimbria 14 operon, *Proteus mirabilis*, 409
Fimbriae. *See also* Pili
  ambient temperature fimbria (ATF), 409
  biofilms and, 247–248
  F1C, 96, 125–126, 128, 244
  gene clusters, 244, 251
  innate immune activation, 532–534
  Klebsiella pneumoniae, 442–445
      gene clusters, 442–444
      role in virulence, 444
      structure and genetics, 443–444
  lack of gene expression *in vivo*, 253–254
  mannose-resistant (MR), 402–407
  motility, effect on, 245
  mutations, 341–343, 346–348
  origin of term, 278
  P fimbriae, 7, 243–244, 343, 506–507, 532–533
  phase variation, 507
  pilicides, 244
  Pix fimbria, 244
  Proteus mirabilis, 14, 402–411, 623–624
      ambient temperature fimbria (ATF), 409
      conservation and expression of, 409–410
      mannose-resistant *Klebsiella*-like (MR/K),
          402–403, 407
      mannose-resistant *Proteus*-like (MR/P),
          402–407
      Proteus mirabilis fimbria (PMF), 408–409
      Proteus mirabilis P-like fimbria (PMP), 409
      regulation of transition between swimming
          and swarming, 410–411
      uroepithelial cell adhesin (UCA), 402–403,
          407–408
      vaccine potential of, 421
  pyelonephritis and, 506–507
  S fimbriae, 244, 539
  synergistic effects of P and type 1 fimbriae *in vivo*,
      506
  type 1 fimbriae, 243, 245
      innate immune-response activation by,
          533–534
      Klebsiella pneumoniae, 442–445

mutations, 341–342, 346–348
      phase variation, 507
      Salmonella, 443
      synergy with P fimbriae, 506–507
      uropathogenic *E. coli* (UPEC), 243, 506–507
  uropathogenic *E. coli* (UPEC), 7, 242–244,
      506–507, 532–533
  Yad fimbria, 244
  Ygi fimbria, 244
fim-gene cluster, *Klebsiella pneumoniae*, 443
FimH, 60, 71, 125, 127, 284, 287, 360–361, 365–368,
      507, 533, 559
  antagonists, 605–606
  in *E. coli* prostatitis, 130
  immunization against, 543
  inhibition of attachment, 542
  mediation of UPEC adherence to and invasion of
      urothelial cells, 595
  mutations in, 341–342, 346–348
  P fimbriae, 96
  pathoadaptive mutation in UPEC clinical isolates,
      595
  phase variation, 7
  uroplakin interaction, 7
  urothelial dysfunction and, 575–578
  vaccines targeting, 620–621
FlaA, 387–389, 398
FlaAB, 387
FlaB, 387–388, 390
FlaC, 387
FlaD, 389, 399
Flagella
  ascension to kidneys and, 603
  loss of expression, 338
  phase variation, 7
  regulation of motility in uropathogenic *E. coli*
      (UPEC), 242
  as virulence factor, 534
Flagellin, 245, 624
flgN, 390
flhA, 390
flhDC, 387, 390–391, 394–395
fliG, 391
fliL, 391
flu gene, 298–299
Fluoroquinolones
  for catheter-associated UTI (CAUTI), 33
  mode of action, 190
  for prophylaxis, 57–58
  for prostatitis, 31
  resistance, 48–49, 162, 181, 338, 591
      mechanism, 190–191
      prevalence, 191
  side effects, 51, 58
  for uncomplicated cystitis, 50–52
  for uncomplicated pyelonephritis, 53–55
  for urosepsis, 141–146

Focal nephritis, 150
Folk remedies, 515
Foodborne reservoirs of human ExPEC, 167–169
Forskolin, 599
Fosfomycin
    for asymptomatic bacteriuria, 106
    for ESBL-producing *E. coli*, 49
    mode of action, 194
    for prophylaxis, 57
    for *Proteus mirabilis*, 419
    resistance
        mechanism, 194
        prevalence, 194
    for uncomplicated cystitis, 49, 51
Fournier's gangrene, 151–152
frdA, 399
Fumarate, *Proteus mirabilis* swarming and, 397
fumC, 399
Functional analysis, pathoadaptive mutations and, 346–348
Fur (ferric-uptake regulator), 237–238, 440
"Fur boxes," 237
Fusiform vesicles, 557, 569, 599
Fusobacteria, 482
FyuA, 240, 622

Galabiose PapG antagonists, 606–607
Gap junctions, 567, 572
Gardnerella vaginalis, 471, 477–480, 482, 485
GBS. *See* Group B *Streptococcus*
Gene activation, 336, 339–340
Gene amplification, 337
Gene cassettes, 182
Gene clusters
    fimbriae, 244, 251
    fimbrial in *Klebsiella pneumoniae*, 442
    pili, 282
Gene expression
    global regulators of, 250–251
    Proteus mirabilis during UTI, 418–419
    uropathogenic *E. coli* (UPEC) gene expression
        in vivo, 253–254
Gene therapy, 542
Genetic drift, 337, 350
Genetic population structure, *Escherichia coli*, 208–209
Genetic variation, pathoadaptive evolution and, 337
Genetics
    asymptomatic bacteriuria and genetic factors, 94
    immune breakdown, 560
    predisposition to UTIs, 17, 35
    renal scarring and, 72
    susceptibility, to pyelonephritis, 504
Genome evolution, role of recombination in, 209
Genome organization, *Proteus mirabilis*, 418
Genome sequences, for uropathogenic *E. coli*
    (UPEC) strains, 251–253

Genome stability, UPEC, 339
Genomic islands, *Escherichia coli*, 210–215, 252–253
Genotoxins, 129
Gentamicin
    for catheter-associated UTI (CAUTI), 33
    discovery, 195
    resistance, 144, 145, 196
Glial cells, 569
Glomerular filtration
    in pyelonephritis, 512–513
    rate, urosepsis and, 140
Glutamine, *Proteus mirabilis* swarming and, 396
Gram-negative uropathogens, adhesins expressed
    by, 279–301
    autotransporters, 293–300
    Iha, 300
    pili, 279–293
    TosA, 300–301
Gram-positive uropathogens, 459–486
    Actinobaculum schaalii, 473–477
    Aerococcus, 472–473
    catheter-associated UTI (CAUTI), 460
    complicated UTI, 460
    Corynebacterium urealyticum, 473
    Enterococcus, 459, 461–462, 466–469
        catheter-associated UTI (CAUTI), 461,
            466–469
        epidemiology of UTI, 466
        immune responses to, 467–469
        virulence factors, 466–467
    Gardnerella vaginalis, 471, 477–480
    group B *Streptococcus* (GBS), 459–460, 462,
        469–471
        bacteriuria in pregnancy, 469–470
        epidemiology, 469
        immune response to, 470–471
        invasive disease in elderly, 470
        virulence factors, 470–471
    laboratory models for study of UTI, 461
    point of care diagnosis of UTI, 460
    polymicrobial UTI, 461–462, 480–481
    Staphylococcus saprophyticus, 459–466
        epidemiology of UTI, 462–463
        host response to UTI, 465–466
        laboratory models of UTI, 461
        polymicrobial UTI, 461
        virulence factors, 463–465
    uncomplicated UTI, 459
    uncultivated bacterial inhabitants of the urinary
        tract, 481–485
    urine contamination with, 480–481
Group B *Streptococcus* (GBS), 459–460, 462,
    469–471
    bacteriuria in pregnancy, 469–470
    in children, 70
    epidemiology, 469
    immune response to, 470–471

Group B *Streptococcus* (GBS) *(continued)*
    invasive disease in elderly, 470
    treatment during pregnancy, 32
    urine contamination, 480
    UTI, 70, 459–460, 462, 469–471
    vaccine, 470
    virulence factors, 470–471
Growth in urine, 236–237
Growth promoter, 170
GspB, 463
GTPase, 514, 533
Guarding reflex, 9

Haemophilus
    capsule biogenesis system of, 249
    H. influenzae Hia adhesin, 295–296
Hammock theory, 14
HCN3 (hyperpolarization-activated cation-3), 9
Hedgehog/Wnt signaling, 570
Helicobacter pylori, VacA cytotoxin of, 293
Hematuria, in prostatitis, 124
Heme uptake
    Proteus mirabilis, 414–416
    receptor-mediated, 239
Hemolysin
    α-hemolysin (Hly), 126, 128–129, 246, 253,
        264–269
        host response to, 507–508
        mechanism, 267–269
        renal damage and, 603–604
        structure, 266–267
        toxoid vaccine, 620
        uropathogenic *E. coli* (UPEC), 264–269,
            507–508
        as virulence factor, 264–266
    β-hemolysin, group B *Streptococcus*, 471
    Proteus mirabilis, 412–413
Hia adhesin, 295–296
High-pathogenicity island (HPI), 214
Histamine, 559
Histidine, *Proteus mirabilis* swarming and, 397
Histone-like nucleoid structuring protein (H-Ns),
    250
HmA, 239, 240, 622
Homologous recombination, gene exchange by, 335
Horizontal gene transfer (HGT), 25, 35–336, 181
Hospital-acquired UTI, 28
Host defenses, 71, 370. *See also* Innate immune
        system
    bacteriuria and, 94
    exfoliation of bladder epithelial cells, 557
    innate immune response to bladder infection,
        555–562
    in pyelonephritis, 507–508
Host factor Q-beta (Hfq), 250–251
Host response to staphylococcal UTI, 465–466
HPI (high-pathogenicity island), 214

hpmA, 387–388, 391, 398–400, 412–413
Human reservoir of ExPEC, 163–164
hybB, 399
Hydrogen peroxide production by lactobacilli, 80–81
Hydronephrosis
    imaging guidelines, 74
    infected, 152
    perinatal, 71
    in pregnancy, 15
    vesicoureteral reflux with, 71
Hypercolonization phenotype, 339
Hyperglycosuria, 14
Hyperpolarization-activated cation-3 (HCN3), 9
Hypogastric nerves, 567
Hypothalamic-pituitary-adrenal axis
    stress activation of, 571
    system inflammation and, 139
Hypoxia, nephron, 511, 514

IBCs. *See* Intracellular bacterial communities
Ibuprofen, 52, 603
IDSA (Infectious Diseases Society of America), 48
IgA, 615–616, 618, 622, 624
IgA1 protease, 293
IgG, 615–616, 618, 622, 624
Iha, 300, 335, 599
IL-1R-associated kinase 4 (IRAK4), 531–532
Imaging children after their first urinary tract
        infection, 74
Imidazolium salts, as bladder cell exfoliant, 372
Immune response. *See also* Innate immune response
    enterococcal UTI, 467–469
    group B *Streptococcus*, 470–471
    in pyelonephritis, 508–510
    to *Staphylococcus saprophyticus*, 466–467
Immunocompromised individuals, *E. coli* strains
        infecting, 218
Immunotherapeutic compounds, 624–625
Incidence
    in aging females, 79
    of asymptomatic bacteriuria, 93–94
    of *Proteus mirabilis* UTI, 384
    of recurrent urinary tract infections, 590
    of urinary tract infection, 36, 236
    of uropathogenic *Escherichia coli* (UPEC) UTI,
        556
Incontinence, urinary-urgency, 481, 484
Inducible nitric-oxide synthase (iNOS), 573
Indwelling catheters, 32–34, 93
    bacterial interference, 111
    bacteriuria, 95, 103, 107–108
Infants
    asymptomatic bacteriuria
        microbiology, 99
        morbidity, 100–101
    UTI in, 69–75 (*See also* Children)
    vesicoureteral reflux, 100

Infectious Diseases Society of America (IDSA), 48
Inflammation
    acute kidney injury, 513–514
    in *Enterococcus* UTIs, 467–469
    mast cells and, 559, 562
    neutrophil-associated tissue damage, 558, 561–562
    UTI pain and, 577–578
Inflammatory markers, in asymptomatic
        bacteriuria, 90–91
Infundibulum, 4
Innate immune response
    bacteriuria and, 94
    to bladder infection, 555–562
        cellular components of, 556–560
        epithelial cells, 556–557
        immune breakdown, 560–561
    genetic variants affecting the early, 526
    in pyelonephritis, 508–510
    to *Staphylococcus saprophyticus*, 466–467
Innate immune system
    bacterial inhibitors of, 541–542
        RNA polymerase II expression, 541–542
        TIR-domain proteins, 541
    therapeutic approaches that modify, 542–543
        deliberate establishment of asymptomatic
                bacteriuria, 542
        gene therapy, 542
        receptor analogues, 542
        vaccination, 543
    UTI susceptibility and activation, 529–534
        autotransporters, 534
        curli, 534
        iron acquisition systems, 534
        P fimbriae, 532–533
        TLR4, 531–532
        toxins, 534
        type 1 fimbriae, 533–534
"In-series tension receptors," 568
Insertion sequences (IS elements), 181
Integral theory, 13–14
Integrase, 182–183
Integrins, 366–368, 372, 592
Integrons, 182–184
Interferon (IFN)-β, 537
Interferon (IFN)-γ, 513
Interferon regulatory factors (IRFs), 525, 533
Interleukins
    IL-1, 138, 511, 514
    IL-4, 139
    IL-6, 508, 511, 514, 527, 534, 556
        levels in pyelonephritis, 543
        urine, 90–91
    IL-8, 508, 509, 534, 543, 558
    IL-10, 139, 559
    IL-1β, 556
Intermittent catheterization
    bacterial interference, 111

bacteriuria and, 103–104
Internal-urethral sphincter, 5
Inter-organ communication during infection, 513
Interstitial cystitis
    interstitial cystitis/bladder-pain syndrome
        (IC/BPS), 569
    urothelial damage during, 57–576
Intimin, 300
Intracellular bacterial communities (IBCs)
    expulsion of UPEC, 599
    formation, 599–600
    Klebsiella pneumoniae, 444, 450
    UPEC dispersal and further IBC formation,
        600–601
    uropathogenic *E. coli* (UPEC), 7, 35, 236, 239,
        246, 249–250, 361–364, 369–370, 571,
        591, 599–601
Intrinsic virulence, ExPEC, 220–221
Invasin, 300
Invasion of host cells, 359–372. *See also* Intracellu-
        lar bacterial communities
    antibacterial defenses and liabilities, 370–371
    fates of intracellular UPEC, 360–363
    intracellular bacterial communities (IBCs),
        361–364, 369–370
    of kidney cells, 371
    mechanisms of bladder cell invasion, 365–368
    regulation of intracellular growth and persis-
        tence, 368–369
    relevance to UTIs, 364–365
    targeting intracellular pathogens, 372
Invasive genitourinary procedures, bacteriuria and,
        104, 109
IRAK4 (IL-1R-associated kinase 4), 531–532
IreA, 240, 622
IRF3, 537, 540
IroN, 602, 622
Iron acquisition
    Escherichia coli, 214, 238–239, 534, 602
    ferrous iron uptake, 240
    Klebsiella pneumoniae, 440–441
        genetics, 440–441
        role in virulence, 441
        siderophore production, 440
    pathogenesis, 602
    Proteus mirabilis, 396, 413–416
    receptor-mediated heme uptake, 239
    redundant systems, 239
    regulation of iron uptake, 237–238
    siderophores, 238–239, 534
        E. coli prostatitis, 125–129, 130
        Klebsiella pneumoniae, 440–441
        pathogenesis, 602
        Proteus mirabilis, 414–415
        uropathogenic *E. coli* (UPEC), 238–239, 534,
            602
    uropathogenic *E. coli* (UPEC), 238–239, 534, 602

Iron receptors as vaccine candidates, 240
IS elements (insertion sequences), 181
Ischemia, nephron, 510–511, 513–514
IutA, 240, 622

Juxtaglomerular apparatus (JGA), 513

Kanamycin resistance, 196
KguS-KguR two-component system, 601
Kidney. *See also* Pyelonephritis
  acute kidney injury, 513–514
  anatomy, 4
  medullary sponge, 10–11
  renal scarring, 72, 75, 100, 509, 513, 531, 540, 590,
    603
  UPEC ascension to and colonization of the
    kidneys, 603–604
Kidney cells, bacterial invasion of, 371
Klebsiella
  antimicrobial resistance, 48
  asymptomatic bacteriuria, 95
  UTI in children, 70
Klebsiella oxytoca
  antimicrobial resistance, 145
  frequency of urinary tract colonization, 504
Klebsiella pneumoniae, 435–450
  adhesins, 280
  antimicrobial resistance, 144, 145, 448–449
  asymptomatic bacteriuria, 97
  biofilms, 436, 444–448
  capsule, 436–439
    genetics of production, 438
    production of, 436–437
    role in virulence, 438–439
  catheter-associated UTI (CAUTI), 435–436, 445,
    448–450
  colonization and adherence, 442–445
  fimbriae, 442–445
    gene clusters, 442–444
    role in virulence, 444
    structure and genetics, 443–444
  frequency of urinary tract colonization, 504
  future perspectives on, 449–450
  genome plasticity, 449
  hypermucoviscous phenotype, 436, 438
  inhibition biofilm formation, 248
  intracellular bacterial communities (IBCs), 444,
    450, 600
  invasion of host cells, 372
  LPS, 437
  NDM-1 β-lactamase, 193
  as opportunistic pathogen, 436
  overview, 435–436
  pyelonephritis, 150
  renal and perirenal abscess, 151
  siderophores, 440–441
  urease, 441–442

  virulence factors, 436–449
Koch's postulates, molecular, 236, 264–265, 532
Kpc fimbriae, 444

Laboratory diagnosis, 43–48
  dipsticks, 46–48
  interpretation of culture results, 44–45
  LE (leukocyte esterase), 46–48
  nitrite testing, 46–48
  pyuria, 45
  rapid detection for bacteriuria, 45
  urinary tract infection in children, 72–73
  voided urine collection techniques, 43–44
  voided urine contamination, 43–44
  voided urine culture, 43
Lactic acid, lactobacillus production of, 81
Lactobacillus, 482–485
  decline in low-estrogen state, 14, 82
  as dominant vaginal microbe, 80–81
  exclusion of uropathogens by, 81
  hydrogen peroxide produced by, 80–81
  L. acidophilus, 608
  L. crispatus, 80–81, 83, 484–485, 608
  L. gasseri, 484
  L. iners, 482
  L. jensenii, 80–81
  nonoxynol-9 toxic effect on, 82
  probiotic, 83, 485, 608
  protective role, 14, 80–81, 471–472
Lactoferrin, 441, 509
Lamina propria, 565–568
Lateral gene transfer. *See* Horizontal gene transfer
Lectinophagocytosis, 438
Leptotrichia, 482
Leukocyte esterase (LE), 43, 46–48, 73, 460, 482
Levofloxacin
  for bacterial prostatitis, 124–125
  resistance, 191
  for uncomplicated cystitis, 50–51
  for uncomplicated pyelonephritis, 53–54
  urinary-bactericidal titer (UBT), 143
  for urosepsis, 143
Lipid A, 138
Lipocalin, 509
Lipopolysaccharides (LPS), 138
  hemolysis (HylA), 268
  Klebsiella pneumoniae, 437
  LPS-smooth/-rough phenotypic variation,
    337–338
  pain induction by, 578–580
  Proteus mirabilis swarming and, 395
Lipoprotein receptor-related proteins (Lrp),
    243–244, 391–393
Liposomes, 616
Listeria monocytogenes, host cell invasion, 365–366
LL-37, 509, 556
Lon protease, 391

Loss-of-function mutations, 337
Lower urinary tract dysfunction, 71
Lrp, 243–244, 391–393
LuxS, 447
Lysosomes, 368–369, 557
LysR regulators, 446

Macrophages
    Enterococcus faecalis survival within, 468
    response to bladder infection, 559–560, 562
    uropathogenic *E. coli* (UPEC) survival within,
        371
magA, 438
Malate, *Proteus mirabilis* swarming and, 397
Mannose-resistant *Klebsiella*-like (MR/K)
        hemagglutinin, 407
Mannose-resistant *Proteus*-like (MR/P) fimbriae
    assembly, 405
    biofilm formation, 405
    expression, 404–405
    genetic organization, 403–404
    role in infection, 405–407
Mannosides, 372, 605–606
Mast cell-derived adjuvants, 616
Mast cells, response to bladder infection, 559,
        562
MCP-1 (monocyte-chemotactic protein-1), 535
mecA gene, 462
Mechanoreceptor, 568
Mechanosensitive afferent nerves, 568
Medullary sponge kidney, 10–11
Men
    aging, asymptomatic bacteriuria in
        diagnosis, 89
        incidence, 93–94
        microbiology of, 99
        prevalence, 92, 95
    asymptomatic bacteriuria
        incidence, 93–94
        prevalence, 92
    clinical diagnosis in, 42
    overview of clinical syndromes, 31
    voided urine specimens for asymptomatic
        bacteriuria diagnosis, 89–90
Metabolism
    during *Proteus mirabilis* swarming, 398–399
    of uropathogenic *E. coli* (UPEC) during infection,
        241–242
Metal acquisition systems. *See also* Iron acquisition
    Proteus mirabilis, 413–416
    zinc acquisition
        Escherichia coli, 240–241
        Proteus mirabilis, 396, 416
Metallo-β-lactamase, 193
Metalloproteases
    metalloproteinase-9, 558, 560
    Proteus mirabilis, 413, 416

Metals
    competition for, 237
    iron acquisition, 237–240
    zinc acquisition, 240–241
Methicillin resistance, 462, 465
Methicillin-resistant *S. aureus* (MRSA), 70, 144–145,
        465
Methyl-accepting chemotaxis, 388
MF59, 615
MHC class I molecules, curli adherence and, 293
Microbiology of bacteriuria, 97–100
Micturition, normal, 9–10
Midstream clean catch technique, 43–44
Mismatch-repair system, 338–339
MLEE (multilocus enzyme electrophoresis), 208
MLST. *See* Multilocus-sequence typing
Mobilome, 181
Molecular Koch's postulates, 236, 264–265, 532
Molecular tools for risk assessment and prediction
        of UTI susceptibility, 543–544
Monocyte-chemotactic protein-1 (MCP-1), 535
Monophosphoryl lipid A (MPL), 615–616
Morganella morganii, 411
Motility, uropathogenic *Escherichia coli* (UPEC),
        244–246
    adherence and, 245
    chemotaxis, 245
    PapX as inhibitor of, 246
    at population level, 245
    regulation of flagellar, 242
MPL (monophosphoryl lipid A), 615–616
mrk-gene cluster, *Klebsiella pneumoniae*, 444
MrpA, 421, 623–624
MrpH, 421, 623–624
MrpJ, 410–411
MRSA (methicillin-resistant *S. aureus*), 70, 144–145,
        465
MSCRAMMs, 301–307
    Ace, 301–304, 306
    assembly, 305
    collagen hug model, 302–304
    dock, lock, and latch model, 304–305
    functions in uropathogenic bacteria, 305–307
    overview, 301–302
    SdrI, 307
    structure, 302, 303
    UafA, 304–305, 306
    UafB, 306–307
Mucosal integrity, infection-associated disruption
        of, 510–511
Mucoviscosity-associated gene (*magA*), 438
Multidrug-resistant (MDR) pathogens, 590–591
    emergence of, 180
    enterococcal strains, 466
    Escherichia coli, 556
    Klebsiella pneumoniae, 448–449
Multilocus enzyme electrophoresis (MLEE), 208

Multilocus-sequence typing (MLST)
  E. coli isolates in prostatitis, 124, 125
  ExPEC, 161–162
MutaFlor, 333
Mutation(s)
  pathoadaptive, 331–351
  synonymous and nonsynonymous changes,
      340–341, 345–346, 350
Mutation rates
  asymptomatic bacteriuria strains, 345–346
  commensals, 345
  with mutator phenotype, 338
  Shigella, 345
  uropathogenic *E. coli* (UPEC) strains, 214–215,
      345
Mutator phenotype, 338–339
Mycobacterium tuberculosis, 70, 180
Myeloid differentiation primary response gene 88
    (MYD88), 526–527, 531–533, 541, 604
Myeloperoxidase, 81

N-acetyl-neuraminic acid regulator (NanR),
    249–250
Nalidixic acid, 190, 191
NalP protease, 293, 296
Natural history of bacteriuria, 97–104
Natural killer cells, 560
NDM (New Delhi metallo) β-lactamases, 163, 193,
    214
Necrotizing fasciitis, 151–152
Negative (purifying) selection, 341
Neisseria
  capsule biogenesis system of, 249
  N. meningitidis NalP protease, 293, 296
Neomycin resistance, 196
Neonatal meningitis *Escherichia coli* (NMEC), 208
  relationship to UPEC and APEC strains, 216
  strains infecting infants, 218–219
Neonates, urinary tract infection prevalence in, 69
Nephritis, focal, 150
Nephron
  breakdown of epithelial integrity, 511
  clotting, ischemia, and hypoxia in, 511–512
  number of, 504
  obstruction, 512–513
  structure of, 504
Nerve growth factor, 569, 573
Nerves, afferent, 567–569
Netlimicin, 195
Neuro-endocrine axis, systemic inflammation and,
    139
Neurogenic bladder
  in children, 71
  diagnosis of UTI, 42
  E. coli strains infecting patients, 218
Neurologic patients, 16–17
Neutrophils

macrophage ingestion of apoptotic, 560, 562
neutrophil-associated tissue damage, 558,
    561–562
neutrophil-dependent clearance of infection, 536
neutrophil-induced oxidative injury of renal
    cells, 513–514
recruitment of, 508–509, 535–537
response to bladder infection, 558, 561–562
New Delhi metallo (NDM) β-lactamases, 163, 193,
    214
NF-κB (nuclear factor kappa-light-chain-enhancer
    of activated B cells), 138–139, 508
Nissle-1917 strain, 333–334
Nitric oxide, 573
Nitrites, 43, 46–48, 73, 460
Nitrofuran reductase, 193
Nitrofurantoin
  for asymptomatic bacteriuria, 105–106, 110
  creatinine clearance (CrCl) and, 50, 58
  for prophylaxis, 57–58
  for *Proteus mirabilis*, 419
  resistance
    mechanism, 193
    prevalence, 193–194
  side effect, 50–51, 58
  for uncomplicated cystitis, 49–51
  for UTI during pregnancy, 31
NMEC. *See* Neonatal meningitis *Escherichia coli*
NOD1, 556, 561
Nonoxynol-9, 82
Non-ribosomal peptide siderophore system (Nrp),
    414, 417
Norfloxacin
  for asymptomatic bacteriuria, 110
  for prophylaxis, 33
  resistance, 191
Nosocomial urinary tract infection, in children, 70
Novobiocin resistance, 462
Nrp (non-ribosomal peptide siderophore system),
    414, 417
Nuclear factor kappa-light-chain-enhancer of
    activated B cells (NF-κB), 138–139, 508
Nutraceuticals, 608–609
Nutrient requirements, uropathogenic *Escherichia
    coli* (UPEC), 236–241
  competition for metals, 237
  growth in urine, 236–237
  iron acquisition, 237–240
  zinc acquisition, 240–241
Nutritional competence, 223

O-antigen
  loss of, 337–338
  as rheostat for LPS-induced pain, 578–580
Obstruction
  nephron, 512–513
  primary bladder-neck, 16

ureteral, 11–12
ureteropelvic-junction, 15
urosepsis and, 139–141
Occludins, 566
Ofloxacin resistance, 191
Oligella, 484
OM-89, 624–625
Opportunistic pathogens
features of, 332
Klebsiella pneumoniae, 436
uropathogenic *E. coli* (UPEC) as, 332–333
Opsonophagocytosis, 438, 464
Orchitis, clinical picture, 151
Ornithine, *Proteus mirabilis* swarming and, 397
Orthoptic bladder substitution, bacteriuria and, 93
Outer membrane, assembly of the pilus fiber at, 285–286
Outer membrane proteins, uropathogenic *E. coli* (UPEC), 248–249
Outer membrane-associated adhesins of UPEC, 300
Outer-membrane iron receptors, 240
Outer-membrane vesicles (OMVs), 620
Overactive bladder, 573
OXA-derived β-lactamases, 192

P blood group
as pyelonephritis risk factor, 560
UTI susceptibility and, 538–539
P fimbriae, 7, 125, 128–129
innate immune activation, 532–533
mutation, 343
P blood group and UTI susceptibility, 538–539
uropathogenic *E. coli* (UPEC), 243–244
P pili
assembly, 596
functions, 289
pathogenesis, 597–598
vaccines targeting, 621–622
P1 (adenosine) receptors, 574
pAA virulence plasmid, 125
Pain
bacterial analgesia in the bladder, 580–581
bacterial-pain phenotypes, 582
central sensitization, 579–580
chronic from transient infection, 579–581
CNS circuits in cystitis pain, 581–582
future directions and implications, 582–583
inflammatory, 577–578
interstitial cystitis/bladder-pain syndrome, 569
LPS-induced, 578–580
pathophysiological mechanisms, 569
TLR4-dependent UTI pain, 577–578, 583
UTI as visceral-pain model, 576–577
PAIs (pathogenicity islands), 70, 129, 183–186, 212, 220, 224, 243, 252–253, 335
p-aminobenzoic acid (PABA), 189
PAML program, 214

PAMPs (pathogen-associated molecular patterns), 616
Pangenome, 209–210
pap operon, 213
PapA, 289
PapB, 244, 507
PapD chaperone, 285
PapDG vaccine, 621–622
PapE, 289
PapF, 289
PapG, 243, 289, 335, 506, 526, 532, 592, 597–598
antagonists, 606–607
mutation, 343
vaccines targeting, 621–622
PapI, 343
PapK, 289
PapX, 244, 246, 507
Pathoadaptive mutations
detection of, 348–350
functional analysis and, 346–348
genetic mechanisms for, 336–348
in uropathogenic *E. coli*, 331–351
Pathogen-associated molecular patterns (PAMPs), 616
Pathogenesis
asymptomatic bacteriuria, 94–97
host factors, 94–95
organism factors, 95–97
chronic pelvic pain, 131–132
complicated UTI, 604–605
Enterococcus faecalis, 604–605
recent discoveries in, 591–605
uncomplicated UTI, 592–604
urinary tract infection in children, 70–71
uropathogenic *E. coli* (UPEC), 592–604
ascension to and colonization of kidneys, 603–604
central metabolism and two-component systems, 601–602
cyclic AMP, 599
escape from endocytic vesicle, 599
FimH, 595
IBCs, 599–601
metal ions, 602
modeling outcomes of acute cystitis, 602–603
P pili, 596–598
type 1 pili, 592, 595
vagina role in, 80
Pathogenicity islands (PAIs), 70, 129, 183–186, 212, 220, 224, 243, 252–253, 335
Pathophysiogram of pyelonephritis, 515–516
Pathophysiology
pyelonephritis, 503–517
of urosepsis, 137–139
PathoSystems Resource Integration Center (PATRIC), 351
Pattern-recognition receptors (PRRs), 508, 529, 559, 616

Paxillin, 269
PBNO (primary bladder-neck obstruction), 16
PBP (penicillin-binding protein), 147, 191–192, 462
PCR testing, 45
PDGFRα (platelet-derived growth factor-
    receptor-α), 567
Pelvic anatomy, abnormal, 13
Pelvic nerves, 567
Penicillinase, 180
Penicillin-binding protein (PBP), 147, 191–192, 462
Pentraxin-related protein 3 (PTX3), 556–558
Peptostreptococcus, 483
Perirenal abscess, 150–151
Peristalsis, 9
Persistence of bacteria, 35
Pertactin, 293, 296
Phagocytosis
    complement-mediated, 556–557
    Klebsiella pneumoniae impairment of, 438
    by macrophages, 560, 562
    by neutrophils, 558
Pharmacokinetic/pharmacodynamic (PK/PD)
    properties, in treatment of severe UTI, 142
Phase variation, 7, 507
Phenazopyridine, 52
Phenol-soluble modulins (PSM), 464–465
Phenotypic variations, observations of in UPEC,
    337–338
PhoP, 250
PhoP-PhoQ two-component system, 601
Phosphate transport, in *Proteus mirabilis*, 416–417
Photorhabdus temperata, 411
Phylogenetics, of *Escherichia coli*, 209–212
Physiology
    abnormalities, 14–17
        diabetes mellitus, 14–15
        dysfunctional voiding, 16
        neurologic patients, 16–17
        pregnancy, 15
        primary bladder-neck obstruction, 16
        ureteropelvic-junction obstruction, 15
    microscopic, 6–8
    micturition, 9–10
    urine transport, 9–10
Pic (protease involved in colonization), 246–247
PicU (protein involved in intestinal colonization),
    265
Pili. *See also* Fimbriae
    Afa/Dr family, 281, 283–284
    Bacillus cereus, 308
    chaperone/usher (CU) assembled, 243, 251,
        279–289
        adhesin, 283–284
        assembly at outer membrane, 285–286
        chaperone-subunit complex formation,
            284–285
        fiber, 281–283

functions of, 287–289
    gene clusters, 282
    P pili, 289, 596–598
    pilus usher, 286–287
    structure, 281–284
    type 1 pili, 287–288, 592, 595
Corynebacterium, 278, 307, 308, 310
Enterococcus faecalis, 307–311, 467
    gene clusters, 282
    Gram-negative uropathogens, 279–301
    Gram-positive uropathogens, 281, 301–311
        assembly, 310
        function in UTIs, 310–311
        structure, 308–310
    mannose-resistant (MR), 278, 403–407
    mannose-sensitive (MS), 278
    models of pilus assembly in Gram-negative and
        Gram-positive pathogens, 596
    origin of term, 278
    pathogenesis, 592–599
    Proteus mirabilis, 281, 412
    S pili, vaccine targeting, 622
    type 1
        bladder cell invasion and, 360–361, 365–368,
            371
        functions, 287–288
        pathogenesis, 592, 595
        structure, 286–287
        vaccines targeting, 620–621
    type IV, 243, 412
    as vaccine candidates, 620–622
Pilicides, 60, 244, 372, 607
Pilus adhesin, 283–284
Piperacillin/tazobactam
    resistance, 144
    for urosepsis, 146
Pivmecillinam
    for asymptomatic bacteriuria, 106
    for uncomplicated cystitis, 49
Pix pilus, 244
Plasmid replicons, 187
Plasmids
    fluoroquinolone resistance, 190–191
    horizontal gene transfer, 335
    incompatibility groups, 187
    resistance, 183–187
Platelet-derived growth factor-receptor-α
    (PDGFRα), 567
Point of care diagnosis of UTI, 460
Polymicrobial UTI, 461, 479, 480–481
Polymorphonuclear cells (PMNs), recruitment of,
    508–509
Population phylogenomics of extraintestinal
    pathogenic *Escherichia coli*, 207–225
Population-genetics analyses, 350
Positive selection
    allelic variation under, 214–215

detection methods, 350–351
Postcoital prophylaxis, 56–57
Posterior fourchette-to-anus distance, 13
Postmenopausal women
    immune breakdown in, 560
    recurrent UTI in, 58–59
    vaginal microbiota alterations in, 81–82
Post-renal obstruction, urosepsis and, 139
Potassium sensitivity test, 575
Poultry, sources of ExPEC, 167–169
Pregnant women
    antimicrobial use in, 31–32
    asymptomatic bacteriuria, 31, 611
        diagnosis, 88–89
        incidence, 93
        microbiology of, 98
        morbidity, 102
        prevalence, 92
        screening for, 105–107
        treatment, 105–107
    group B *Streptococcus* (GBS) in, 32, 469–470
    physiology and anatomy changes, 15
    pyelonephritis, 31, 102, 105–107
Prevalence
    asymptomatic bacteriuria, 36, 91–93, 529
    prostatitis, 36
    uropathogenic *Escherichia coli* (UPEC)
        pathoadaptive mutations, 343–346
        virulence and fitness determinants, 594–595
    UTI in aging female, 13–14
    UTI in children, 69
Prevention of recurrent urinary tract infection, 36,
        55–60
    antimicrobial prophylaxis, 56–58
        continuous prophylaxis, 57
        postcoital prophylaxis, 56–57
        self-diagnosis and self-treatment, 57–58
        special considerations about, 58
    antimicrobial-sparing approaches, 55
    attachment inhibitors, 60
    in postmenopausal women, 58–59
    probiotics, 59
    vaccines, 59–60
Prevotella, 482, 483
Primary bladder-neck obstruction (PBNO), 16
Probiotics, 75, 83, 608
    E. coli, 561
    Lactobacillus, 83, 485, 608
    Nissle-1917 strain, 333–334
Professional pathogens, 332
Progesterone, 102
Promoter polymorphisms, 539–541
Prophage, horizontal gene transfer, 335
Prophylaxis
    antimicrobial, 56–58, 591
        in catheterized patients, 33
        continuous prophylaxis, 57

postcoital prophylaxis, 56–57
    self-diagnosis and self-treatment, 57–58
    special considerations about, 58
    with vesicoureteral reflux, 75
    children, urinary tract infection in, 75
    cranberry products, 608–609
    Lactobacillus, 608
    postcoital, 56–57
    probiotics as, 608
    UTI in children, 75
Propionibacterium avidum, 478
Prostate massage, 151
Prostatic abscess, 151
Prostatitis, 121–132
    categories of syndromes, 121
    chronic, 129–132
    clinical picture of acute, 151
    E. coli in acute bacterial prostatitis, 121–132
        characteristics of *E. coli* isolates, 125–127
        clinical presentation, 124
        overview, 122
        in previously healthy young men, 123–126
        studies on, 122–123
        treatment, 124–125
        virulence, 123, 125–129
    Enterococcus, 466
    overview, 30
    prevalence, 36
    symptoms, 30
    treatment, 31
Protease involved in colonization (Pic), 246–247
Proteases, *Proteus mirabilis*, 413
Protein involved in intestinal colonization (PicU),
        265
Protein-kinase C, 139
Proteobactin, 414–417
Proteus
    antimicrobial resistance, 48
    renal and perirenal abscess, 151
    UTI in children, 70
Proteus mirabilis, 383–422
    adhesins, 280, 281, 623–624
    antimicrobial resistance, 145
    asymptomatic bacteriuria, 95, 97
    autotransporters, 295
    biofilms, 385, 405–406, 419
    catheter obstruction, 97
    characteristics of, 383
    clinical aspect of UTI, 419–422
        antibiotic resistance, 420
        prevention, 419
        treatment, 419–420
        urinary catheter biofilm formation,
            preventing, 421–422
        vaccine, 420–421
    disease, 383–384
    fimbriae, 14, 402–411, 623–624

Proteus mirabilis (continued)
  ambient temperature fimbria (ATF), 409
  conservation and expression of, 409–410
  mannose-resistant *Klebsiella*-like (MR/K),
      402–403, 407
  mannose-resistant *Proteus*-like (MR/P),
      402–407
  Proteus mirabilis fimbria (PMF), 408–409
  Proteus mirabilis P-like fimbria (PMP), 409
  regulation of transition between swimming
      and swarming, 410–411
  uroepithelial cell adhesin (UCA), 402–403,
      407–408
  vaccine potential of, 421
flagella, 386–388
  antigenic variation, 387
  cellular invasion and, 399–400
  characteristics, 386–387
  constitutive elongation mutants, 391
  contribution to swarming, 389–395
  regulation, 387
  role in virulence, 387–388
frequency of urinary tract colonization, 504
Iha, 300
incidence of UTI, 384
invasion of host cells, 372
MrpJ motility inhibitor, 246
pili, 281, 412
polymicrobial UTI, 461
quorum sensing, 398
swarming, 388–402
  bull's-eye pattern on media, 389
  capsule and, 395–396
  cellular invasion and, 399–400
  Dienes line formation, 401–402
  DisA, 393–394
  DNA replication without septation during,
      389
  extracellular contributors to, 396–397
  flagella contribution to, 389–395
  flhDC regulation, 391
  genes contributing to, table of, 392–393
  glutamine, 396
  Lon protease, 391
  LPS and, 395
  Lrp, 391–393
  metal acquisition, 396
  non-flagellar loci contributing to, 395–396
  overview of, 388–389
  putrescine, 396–397
  role in virulence, 400
  RppAB, 395
  RsbA-RscBC, 394–395
  RsmA/csrA, 394
  switch between swimming and swarming
      forms, 390, 410–411
  transcription and metabolism during, 398–399

  Umo proteins, 394
  WosA, 394
vaccines against, 623–624
virulence factors, 384–419
  autotransporters, 411–412
  biofilm, 405–406, 417, 419, 421–422
  fimbriae, 402–411
  flagella, 386–388
  gene expression during UTI, 418–419
  genome organization, 418
  hemolysin, 412–413
  identified by signature-tagged mutagenesis,
      417–418
  metal acquisition systems, 413–416
  phosphate transport, 416–417
  proteases, 413
  swarming, 388–402
  type IV pili, 412
  urease, 384–386
Providencia stuartii, 385
  asymptomatic bacteriuria, 97
  polymicrobial UTI, 461
  urinary stones, 461
PRRs (pattern-recognition receptors), 508, 529, 559,
      616
Pseudomonas
  asymptomatic bacteriuria, 97
  UTI in children, 70
Pseudomonas aeruginosa
  antimicrobial resistance, 48
  autotrophs, 338
  biofilms, 143
  EstA autotransporter, 295–296
  frequency of urinary tract colonization, 504
  hospital-acquired UTIs, 28
  inhibition biofilm formation, 248
  LPS variation, 338
  mucoid phenotype, 336
  polymicrobial UTI, 462
  urosepsis, 143
Pseudorabies virus, 581
PSM (phenol-soluble modulins), 464–465
Pta autotransporter, 412
PTK2, 533
PTX3 (pentraxin-related protein 3), 556–558
Pudendal nerves, 567
Putrescine, *Proteus mirabilis* swarming and,
      396–397
P2X3, 568
Pyelitis, *Corynebacterium urealyticum*, 473
Pyelonephritis, 503–517
  Actinobaculum schaalii, 477
  acute, 503, 515, 527
    clinical diagnosis, 42
    complicated pyelonephritis, clinical diagnosis,
        42
    incidence, 36

overview of clinical syndrome, 30
susceptibility, 527, 529–531, 536, 539–541
symptoms, 30, 527
treatment, 53–55
uncomplicated pyelonephritis, 42, 53–55
acute kidney injury (AKI), 513–514
asymptomatic bacteriuria, 29
bacterial colonization in renal tubular
environment, 506
bacterial toxins and host responses, 506–508
chronic, 503
cinematic view of, 504–517
clinical diagnosis, 42
clinical picture, 150
defined, 503
differential diagnosis, 543
emphysematous, 150
future projections on, 515–517
Gardnerella vaginalis, 479
genetic risk factors, 35
group B *Streptococcus*, 469
innate immune responses, 508–510
inter-organ communication, 513
interpretation of urine culture results, 44
mucosal integrity disruption, 510–511
nephron-obstruction, 512–513
P blood group as risk factor for, 560
P fimbriae and type 1 fimbriae synergy, 506–507
pathophysiogram, 515
pathophysiology, 503–517
in pregnant women, 15, 31, 102, 105–107
prevention of dissemination, 511–512
Proteus mirabilis, 383
quantitative urine culture, 73
risk factors, 504
Staphylococcus saprophyticus, 462, 466
sterile, 479
symptoms, 503, 589
treatment, 514–515
acute uncomplicated pyelonephritis, 42,
53–55
in children, 74
urosepsis, 139
vaccine, 611, 617, 619–621
vesicoureteric reflux and, 71, 72
Pyonephrosis, 152
Pyuria, 460
asymptomatic bacteriuria, 90–91, 107
catheter-associated UTI (CAUTI), 468
in children, 73
Gardnerella vaginalis, 477
laboratory diagnosis, 45
prostatitis, 122, 124, 126

QepA1, 163
QseBC two-component system, 601
Quantitative urine culture, 44–45, 73, 88, 484

Quiescent intracellular reservoirs (QIRs), 8, 603
Quinolones
Proteus mirabilis, 419
resistance, 336
Quorum sensing
Klebsiella pneumoniae, 446
Proteus mirabilis, 398
quorum-sensing (Qse) system, 251

R (resistance) plasmids, 183–184, 195–196
Rab35, 369
Rab27b, 366–367, 533
RAB27b$^+$ vesicles, 557
RAC1, 533
Random-amplified polymorphic DNA (RAPD), 208,
217
RANTES promoter variant, 540
Rapid detection strategies for bacteriuria, 45
RarA, 449
rbs operon, 223
RcsBC, 394–395
Receptor analogues, to prevent bacterial attachment, 542
Receptor-mediated uptake of heme, 239
Recombination, role in genome evolution, 209
Recurrent urinary tract infections
asymptomatic bacteriuria strains as therapy,
607–608
in children, 69–70
defined, 35
genetic predisposition to, 35
incidence of, 590
overview, 35–36
prevention, 36, 55–60
antimicrobial prophylaxis, 56–58
antimicrobial-sparing approaches, 55
attachment inhibitors, 60
in postmenopausal women, 58–59
probiotics, 59
vaccines, 59–60
protective immunity and, 610–611
Reductive evolution, 529
Reflux. *See* Vesicoureteral reflux
Reinfection, 35
Relapse, 35
Renal abscess, 150–151
Renal function, impairment in urosepsis, 139–141
Renal papillae, 4
Renal pelvis, 4
Renal pelvis aspiration, 90
Renal scarring, 72, 75, 100, 509, 513, 531, 540, 590,
603
Renal transplant patients
asymptomatic bacteriuria, 29
bacteriuria, 104, 109–110
Renal-cortical scintigraphy, 74–75
Repair, urothelial, 570–572

Repeats-in-toxin (RTX) family, 25, 300–301
  E. coli hemolysin, 264–269
  UpxA, 269
Resiniferatoxin, 572
Resistance, urosepsis treatment and, 144
Reverse vaccinology, 622–623
RfaH, 248, 269
Rheumatoid arthritis, 384
Rho GTPases, 246–247
Ribonuclease 7, 556
Ribosomal DNA (rDNA) restriction fragment-
    length polymorphism analyses, 208
Risk assessment, molecular tools for, 543
Risk factors, 35, 590
Risk increase for asymptomatic bacteriuria, 95
RmpA, 438
RmtB, 163
RNA polymerase II, 541–542
RNA-seq, 254
RpoS, 223
RppAB, 395
RrgA, 467
RsbA, 394–395
RseA, 250
RsmA, 394
RTX family of toxins. *See* Repeats-in-toxin (RTX)
    family

S fimbriae, 244, 539
S pili, vaccine targeting, 622
SadA, 299
Salmochelin, 126, 128, 238–239
  in *E. coli* prostatitis, 130
  uropathogenic *E. coli* (UPEC), 534, 602
Salmonella
  curli, 289, 291
  SadA, 299
  serovar Typhimurium
    cellular invasion, 399
    iron chelation, 414
    Sit genes, 240
  swarming by, 389
  typhi, cellular invasion and, 399
Schistosomiasis haematobium, 70
SdrG, 301, 304
SdrI, 307, 463
Sec systems, 463
Secondary vesicoureteric reflux, 15
Secreted autotransporter toxin (Sat), 246–247, 265,
    271
Selfish DNA, 209
Self-preservation and nutritional competence
    (SPANC), 223
Sensation within the urinary tract, 565–583
Sepsis. *See also* Urosepsis
  classifications of, 135
  cytokines as markers, 138

diagnostic criteria for, 136, 137
epidemiology of, 136–137
Surviving Sepsis Campaign guidelines, 147–148
Septic shock
  clinical diagnostic criteria of, 137
  prevention, 152
Serine-protease autotransporter toxins of
    Enterobacteriaceae (SPATE), 247, 271
Serratia marcescens
  antimicrobial resistance, 145
  swarming, 398
Sewage, as ExPEC reservoir, 164
Sexual activity
  as asymptomatic bacteriuria risk factor, 94
  Gardnerella vaginalis transmission, 479
  postcoital prophylaxis, 56–57
  as UTI risk factor, 236
Shear stress for the renal tubule, 506–507
Shigella
  as professional pathogen, 332
  S. flexneri, infectious dose of, 334
SHV-1 β-lactamases, 192
Siderophores
  E. coli prostatitis, 125–129, 130
  Klebsiella pneumoniae, 440–441
  pathogenesis, 602
  Proteus mirabilis, 414–415
  uropathogenic *E. coli* (UPEC), 238–239, 534,
    602
Sigma E, 250
Signaling, 566, 570, 572–575
Signature-tagged mutagenesis (STM)
  Klebsiella pneumoniae biofilm gene identification,
    446–448
  Proteus mirabilis, 388, 417–418
Silver compounds, 197
Single-nucleotide polymorphisms (SNPs), 17, 337
SIRS (systemic inflammatory response syndrome),
    135, 138
Sit system, 240
16S rRNA methylation, 196
Small colony-forming variants, 338
Sneathia, 483
SNPs (single-nucleotide polymorphisms), 17, 337
SolcoUrovac, 617–618
Soluble tumor necrosis factor receptors (sTNFR-1),
    102
Sortases, 305, 310, 604–605
  Enterococcus faecalis, 467
  Staphylococcus saprophyticus, 463
SPA (suprapubic-bladder aspiration) in children,
    72–73
Space of Retzius, 5
SPANC (self-preservation and nutritional
    competence), 223
SPATE (serine-protease autotransporter toxins of
    Enterobacteriaceae), 247, 271

Spermicide use, as risk factor for asymptomatic bacteriuria, 94

Spinal cord injuries, asymptomatic bacteriuria and, 109
  incidence, 93–94
  prevalence, 93

SraP, 463

SslE, 623

Ssp, 463

SssF, 463

Staphylococcus, 483, 484
  S. aureus
    antimicrobial resistance, 144–145, 180
    autotrophs, 338
    ClfA, 301–302, 304
    Cna, 301–302, 304
    inhibition biofilm formation, 248
    methicillin-resistant, 70, 144–145, 465
    phenol-soluble modulin (PSM), 465
    renal and perirenal abscess, 151
    SraP, 463
    α-toxin, 268–269
    toxin A, 138
    urosepsis, 143
    UTI, 70, 465
    vancomycin-resistant *S. aureus* (VRSA), 144–145
  S. epidermidis, 465
    inhibition biofilm formation, 248
    SdrG, 301, 304
  S. saprophyticus, 459–466
    adhesins, 281, 306–307
    antibiotic resistance in, 462
    in children, 70
    community-acquired UTIs, 301
    epidemiology of UTI, 462–463
    frequency of urinary tract colonization, 504
    host response to UTI, 465–466
    invasion of host cells, 372
    laboratory models of UTI, 461
    polymicrobial UTI, 461
    SdrI, 307
    UafA, 304, 306
    UafB, 306–307
    virulence factors, 463–465

STM. *See* Signature-tagged mutagenesis

STNFR-1 (soluble tumor necrosis factor receptors), 102

Streptococcus, 483, 484
  group A, host cell invasion by, 366
  S. agalactiae, 459, 469 (*see also* group B *Streptococcus*)
    asymptomatic bacteriuria, 95
    frequency of urinary tract colonization, 504
    invasion of host cells, 372
  S. gordonii GspB, 463
  S. pneumoniae RrgA, 304, 467

S. pyogenes Spy0128 pilin, 304

S. suis, 606

viridans streptococci group, urinary tract colonization by, 504

Streptomycin
  discovery, 195
  resistance, 196

Stress, urothelial consequences of, 571

Stretch-induced exocytosis, 569–570

StroVac, 617–618

Struvite crystals, *Corynebacterium urealyticum*, 473

Substance P, 568, 572

Sulfamethoxazole and trimethoprim
  for asymptomatic bacteriuria, 106, 108, 109
  for bacterial prostatitis, 124–125
  mode of action, 189
  for prophylaxis, 56–58
  for prostatitis, 31
  for *Proteus mirabilis*, 419
  resistance, 48–49, 51, 58, 124, 213, 219, 419–420, 606
    mechanism, 189–190
    prevalence, 190
  for uncomplicated cystitis, 49–51
  for uncomplicated pyelonephritis, 53–54
  for UTI children, 74

Sulfamethoxazole and trimethoprim (SMZ-TMP) resistance, 419–420

Sulfamethoxazole-resistance genes, 182

Sulfonamide resistance, 124, 472

Suprapubic aspirate, 72–73, 89, 90

Surviving Sepsis Campaign, 147–148

Susceptibility to urinary tract infection
  antimicrobial peptides, 538
  asymptomatic bacteriuria, 529, 532, 534, 539–541
  chemokine receptors and, 535, 537
  cytokine response, 534–535
  definitions, 525–529
  innate immune activation and, 529–534
    autotransporters, 534
    curli, 534
    iron acquisition systems, 534
    P fimbriae, 532–533
    TLR4, 531–532
    toxins, 534
    type 1 fimbriae, 533–534
  Irf3 or IfnB defects, 537
  molecular determinants and genetics, 538–541
    adaptor gene polymorphisms, 539
    CXCRI expression, 540
    genetic variation, 539–541
    IRF3 expression, 540
    low TLR4 expression, 539
    promoter polymorphisms, 539–541
    receptors for bacterial adhesins, 538–539
  molecular tools for risk assessment and predictions, 543–544

Susceptibility to urinary tract infection (*continued*)
neutrophil recruitment, 535–536
pyelonephritis, 527, 529–531, 536, 539–541
therapeutic approaches to modify, 542–543
deliberate establishment of asymptomatic
bacteriuria, 542
gene therapy, 542
receptor analogue, 542
vaccination, 543
Tlr5, Tlr11, Thp, and COX2 defects, 537–538
Swarming motility of *Proteus mirabilis*, 383,
388–402
Systemic inflammatory response syndrome (SIRS),
135, 138
Systems vaccinology, 616–617

T cells, γβ, 560
TA (toxin-antitoxin) systems, 369
TaaP autotransporter, 412
Tamm-Horsfall protein (THP), 509, 533, 537–538
TCA (tricarboxylic acid) cycle, 601
TcpC, 527, 604
Temperature-sensitive hemagglutinin (Tsh), 247
Tendonitis, fluoroquinolones and, 58
Tension receptors, 568
Tetracyclines
mode of action, 194
resistance, 124, 182, 420
mechanism, 194–195
prevalence, 195
TGF (tubuloglomerular feedback), 513
TGF-β1 polymorphism, 541
Tigecycline, 194
Tight junctions, 7, 566, 575
TimeZone software package, 350
TIR domain proteins, 541
TIR domain-containing adaptor adaptor protein
(TIRAP), 531–532
Tissue microbiology, 516–517
TNF-α (tumor necrosis factor-α), 138–139, 511, 514,
560
Toll-like receptors (TLRs)
in *Enterococcus* UTIs, 466–467
as pattern-recognition receptors (PRRs), 616
role in pyelonephritis response, 508
TcpC impairment of, 527
TLR1, 534, 560
TLR2, 534
TLR4, 526, 530–533, 556–557, 560, 592, 599, 604
low expression in patients with ABU, 539
MPL as agonist, 615
promoter variation, 539
recognition of LPS lipid A, 619
TLR4-dependent host defenses, 370–371
TLR4-dependent UTI pain, 577–578, 583
TLR5, 534, 537–538, 561
TLR11, 537–538

TonB system, 602
Klebsiella pneumoniae, 440–441
TonB-dependent receptors, in *Proteus mirabilis*,
414–415
TonB-ExbB-ExbD complex, 237
Topical vaginal estrogen therapy, 82
Torsades de pointe, fluoroquinolones and, 58
TosA, 252, 269, 300–301
Toxic-shock syndrome toxin 1, 138
Toxin-antitoxin (TA) systems, 369
Toxins
autotransporter, 247
E. coli prostatitis isolates, 125–129
pyelonephritis and, 507–508
RTX family, 252
uropathogenic *E. coli* (UPEC), 246–247, 263–272,
604–605
actions and epidemiology, table of, 265
autotransporter family, 271
cytotoxic necrotizing factor type 1 (CNF1),
265, 269–270
hemolysin (HylA), 264–269, 507–508
PicU, 265
secreted autotransporter toxin (Sat), 265, 271
type V secretion family, 270–271
Vat, 265
Toxoid, 619–620
TraDIS (transposon-directed insertion-site
sequencing), 255
TRAM, 526, 531–533, 539
Transcription, during *Proteus mirabilis* swarming,
398–399
Transcriptional profiling, 253–254
Transcriptomics
in pyelonephritis, 513
uropathogenic *E. coli* (UPEC), 253
Transfer RNA (tRNA) genes, 252
Transferrin, 441
Transitional stratified epithelium, 504
Transposase, 184
Transposon mutagenesis, 254
Transposon-directed insertion-site sequencing
(TraDIS), 255
Transposons, 181–182, 189–190, 195–196
Transurethral-bladder catheterization, in children,
72–73
Treatment, 48–55
acute uncomplicated cystitis, 49–53
acute uncomplicated pyelonephritis, 53–55
antimicrobial resistance, 48–49, 179–198
anti-virulence therapies, 605–609
asymptomatic bacteriuria, 104–110
after invasive genitourinary procedures, 109
children, 104–105
diabetic patients, 107
elderly institutionalized patients, 107
indwelling urethral catheter patients, 107–108

older women, 107
  pregnant women, 105–107
  renal transplant patients, 109–110
  spinal cord injury patients, 109
bacterial prostatitis, 124–125
of catheter-associated UTI (CAUTI), 609
chaperone/usher pathway inhibitors, 607
in children, 73–74
cranberry products, 608–609
curlicides, 607
drug and vaccine development, 589–626
estrogen therapy, 609
FimH antagonists, 605–606
ibuprofen, 603
intravesicular therapy, 607–608
need for new therapies, 589–591
nutraceuticals, 608–609
PapG antagonists, 606–607
pilicides, 607
probiotics, 608
pyelonephritis, 514–515
self, 57–58
vaccines (*see* Vaccines)
Tricarboxylic acid (TCA) cycle, 601
TRIF-related adaptor molecule (TRAM), 526,
  531–533, 539
Trigone of the bladder, 5
Trimethoprim resistance, 476
Trimethoprim-sulfamethoxazole. *See*
  Sulfamethoxazole and trimethoprim
Tubular necrosis, 513
Tubuloglomerular feedback (TGF), 513
Tumor necrosis factor (TNF)-α, 138–139, 511, 514,
  560
Type V secretion system, 293
  autotransporters in *Proteus mirabilis*, 411–412
  toxins in, 270–271
Type IV secretion system, 534
Type IV pili, 243, 412
Type 1 fimbriae
  innate immune-response activation by, 533–534
  Klebsiella pneumoniae, 442–445
  mutations, 341–342, 346–348
  phase variation, 507
  Salmonella, 443
  synergy with P fimbriae, 506–507
  uropathogenic *E. coli* (UPEC), 243, 506–507
Type 1 pili
  bladder cell invasion and, 360–361, 365–368,
    371
  functions, 287–288
  pathogenesis, 592, 595
  structure, 286–287
  vaccines targeting, 620–621
Type VI secretion system, 401–402
Type 3 fimbriae, of *Klebsiella pneumoniae*,
  442–445

Type III secretion system, 251, 418
Type II toxin-antitoxin systems (TA), UPEC, 534

UafA, 301–304, 306, 463
UBT (urinary-bactericidal titer), 143
UCA (uroepithelial cell adhesin), 407–408
UcaA, 624
Umbrella cells, 7, 360–363, 365–366, 368–370, 372,
  565–566, 569–571, 595, 599
Umo proteins, 394
Uncomplicated urinary tract infection
  defined, 28
  demographics of, 459
  Gram-positive uropathogens, 459
  overview of clinical syndrome, 29
  risk factors, 590
  symptoms, 459
  uropathogenic *E. coli* (UPEC), 592–604
Uncultivated bacterial inhabitants of the urinary
    tract, 481–485
UpaB, 248–249, 299
UpaC, 248–249
UpaG, 299
UpaH, 299, 534
UPEC. *See* Uropathogenic *Escherichia coli*
UPEC-specific genes, search for, 211–214
UPJO (ureteropelvic-junction obstruction), 15
UPIa, 360, 365–368
Upper urinary-collecting system, 4
UpxA, 269
Urease
  Corynebacterium urealyticum, 473
  Klebsiella pneumoniae, 441–442
    genetics and structure, 441
    role in virulence, 441–442
  Proteus mirabilis, 383, 384–386
  Staphylococcus aureus, 465
  Staphylococcus saprophyticus, 463
  virulence mechanism, 385
Ureter, 4
  obstruction, 11–12, 139–140
  urine transport, 9
Ureteral stents, bacteriuria and, 93
Ureteropelvic junction, 4
Ureteropelvic-junction obstruction (UPJO), 15
Ureterovesical junction, 4, 5, 71
Urethra
  anatomy, 5–6
  length, 13, 27, 236
Urethral catheterization, interpretation of urine
    culture results from, 44–45
Urethra-to-anus distance, 13
Urinary continence, female
  in aging female, 13–14
  hammock theory, 14
  integral theory, 13–14
Urinary malakoplakia, 371

Urinary retention, chronic, 13
Urinary stones, 461
    Proteus mirabilis and, 385–386, 420
    Staphylococcus saprophyticus, 463–464
Urinary-bactericidal titer (UBT), 143
Urinary-urgency incontinence (UUI), 481, 484
Urine, uropathogenic *Escherichia coli* growth in,
    236–237
Urine cultures
    in children, 72–73
    expanded-quantitative urine culture (EQUC),
        484
    interpretation of results, 44–45
    quantitative, 44–45, 73, 88, 484
    voided urine, 43
Urine interleukin-8 (IL-8), 91
Urine sample
    collection in children, 72–73
    contamination, 480–481
Urine transport, 9–10
Uro-adherence factor A (UafA), 301–304, 306, 463
Uro-adherence factor B (UafB), 463
Uroepithelial cell adhesin (UCA), 407–408
Uromune, 624
Uropathogenic *Escherichia coli* (UPEC), 208. *See
    also Escherichia coli*
    adhesins, 70–71, 506–507, 592–599
    allelic variation under positive selection, 214–215
    antimicrobial resistance, 179–198
    asymptomatic bacteriuria, 96
    attachment, 60
    autotransporter adhesins, 298–300
    in chronic pelvic pain syndrome, 129–132
    curli, 290, 292–293
    cytokine response to infection, 534–535
    dispersal, 600
    ecological cycle, 333
    escape from endocytic vesicle, 599
    expulsion of internalized, 599
    fimbriae, 242–244, 506–507
        F1C fimbria, 244
        gene clusters, 244, 251
        lack of gene expression *in vivo*, 253–254
        P fimbriae, 7, 243–244, 506–507, 532–533
        phase variation, 507
        pilicides, 244
        Pix fimbria, 244
        S fimbria, 244
        type 1 fimbriae, 243, 245, 506–507
        Yad fimbria, 244
        Ygi fimbria, 244
    frequency of urinary tract colonization, 504
    genome sequences, 251–253
    heterogeneity of strains, 610
    incidence of UTI, 556, 590
    initiation of the innate immune response by,
        526

intracellular bacterial communities (IBCs), 7, 35,
        236, 239, 246, 249–250, 361–364, 369–
        370, 571, 591, 599–600
    invasion of host cells, 359–372
        antibacterial defenses and liabilities, 370–371
        fates of intracellular UPEC, 360–363
        intracellular bacterial communities (IBCs),
            361–364, 369–370
        of kidney cells, 371
        mechanisms of bladder cell invasion, 365–368
        regulation of intracellular growth and persis-
            tence, 368–369
        relevance to UTIs, 364–365
        targeting intracellular pathogens, 372
    iron acquisition, 238–239, 534, 602
    lysosome neutralization, 557
    as a model organism for uncomplicated UTI, 591
    motility, 244–246
        adherence and, 245
        chemotaxis, 245
        PapX as inhibitor of, 246
        at population level, 245
        regulation of flagellar, 242
    multi-drug resistance, 591
    mutation rate, 214
    nutrient requirements, 236–241
        competition for metals, 237
        growth in urine, 236–237
        iron acquisition, 237–240
        zinc acquisition, 240–241
    as opportunistic pathogen, 332–333
    origin of strains, 216–217
    pathoadaptive mutations, 331–351
        amplification of gene copies, 337
        detection of, 348–350
        evidence of occurrence, 337–338
        functional analysis of traits, 346–348
        gene inactivation, 336–337, 339–340
        genetic variation, 337, 340
        genome instability, 339
        genome-wide screens for prevalence of,
            343–346
        horizontal gene transfer compared, 336
        mutator phenotype, 338–339
        phenotypic variations, observations of,
            337–338
    pathogenesis, 592–604
        ascension to and colonization of kidneys,
            603–604
        central metabolism and two-component
            systems, 601–602
        cyclic AMP, 599
        escape from endocytic vesicle, 599
        FimH, 595
        IBCs, 599–601
        metal ions, 602
        modeling outcomes of acute cystitis, 602–603

P pili, 596–598
type 1 pili, 592, 595
phylogeny and virulence factors, 217–218
pili function, 287–289
P pili, 289
type 1 pili, 287–288
polymicrobial UTI, 461–462
prostatitis, 125–127, 130–131
pyelonephritis, 150, 505–513
QIR, 603
relationship to APEC and NMEC strains, 216
siderophores, 534, 602
strains both resistant and virulent, 219
strains infecting infants, 218–219
toxins, 263–272, 603–604, 620
actions and epidemiology, table of, 265
autotransporter family, 271
cytotoxic necrotizing factor type 1 (CNF1),
265, 269–270
hemolysin (HylA), 264–269, 507–508
PicU, 265
secreted autotransporter toxin (Sat), 265, 271
type V secretion family, 270–271
Vat, 265
transcriptional control of the innate immune
response to, 530
uroplakin interaction, 7
urothelial damage, 575–578
urothelium interactions, 571, 575–579
UTI in children, 70–71
vaccines, 60, 612–623
virulence and fitness determinants, 235–255
autotransporter, 534
biofilm, 247–248
capsule, 249–250
curli, 534
flagella, 534
gene expression *in vivo*, 253–254
genome sequences, 251–253
global regulators of gene expression, 250–251
innate immunity induction by, 534
metabolism, 241–242
motility, 244–246
nutrient requirements, 236–241
outer membrane proteins, 248–249
overview, 236
P fimbriae, 7, 506–507, 532–533
pathogenesis, 592–599
phylogeny and, 217–218
prevalence and sites of action, 594–595
siderophores, 534
surface structures, 242–244
toxins, 246–247, 534
Uroplakins, 6–8, 360, 365–368, 533, 555, 566, 571,
576, 595
Urosepsis, 135–153
Actinobaculum schaalii, 477

algorithm for management of, 148–150
antimicrobial therapy, 141–147
aminoglycosides, 142–143
bacterial spectrum of pathogens, 143–144
beta-lactams, 142
biofilm infection, 143
fluoroquinolones, 142–143
pharmacokinetic/pharmacodynamics and, 142
selection of antimicrobials for empiric
therapy, 144–147
clinical pictures of severe urogenital infections,
150–152
acute prostatitis, 151
cavernitis, 151
cystitis, 150
epididymitis/orchitis, 151
Fournier's gangrene, 151–152
prostatic abscess, 151
pyelonephritis, 150
renal and perirenal abscess, 150–151
cytokines as markers, 138
definition, 135–136
diagnostic criteria of sepsis and septic shock, 137
epidemiology, 136–137
Gardnerella vaginalis, 479
neuro-endocrine axis and, 139
overview, 30
pathophysiology, 137–139
prevention, 152
renal function alterations, 139–141
Surviving Sepsis Campaign (SSC), 147–148
treatment, 515
Uroseptic shock, 30
Urostim, 624
Urothelial plaques, 7
Urothelial-cell turnover, 7
Urothelium, 6
barrier function, 569
cell interactions, 572–580
cells and repair, 570–572
damage during UTI, 575–576
FimH mediation of UPEC adherence to, 595
functioning, 566
signaling, 566, 572–575
structure of, 565–566
turnover rate, 570
Urovac, 617–618, 625
UroVaxom, 624
Urvakol, 624
Ushers, 286–287, 443
UUI (urinary-urgency incontinence), 481, 484

VacA cytotoxin, 293
Vaccines
candidate vaccines, table of, 612–613
challenges in developing, 610–617
adjuvant choice, 615–616

Vaccines *(continued)*
  choice of immunogens, 611, 614–615
  choice of recipients, 611
  recurrent UTI and protective immunity,
    610–611
  route of administration, 615
  Enterococcus vaccines, 624
  ExPEC, 170
  group B *Streptococcus*, 470
  historical perspective, 609–610
  iron receptors as candidates for, 240
  OM-89/UroVaxom, 624–625
  P pili, 289, 621
  Proteus mirabilis, 420–421, 623–624
  pyelonephritis, 611
  reverse vaccinology, 622–623
  specific-antigen vaccines in development, 611,
    614, 619–623
    conjugate vaccines, 619–622
    FimH, 620–621
    P pili, 621
    PapDG vaccine, 621–622
    pili as candidates, 620–622
    reverse vaccinology, 622–623
    subunit vaccines, 614, 622–623
    toxoid vaccines, 620
  susceptibility to UTI and, 543
  systems vaccinology, 616–617
  uropathogenic *E. coli* (UPEC), 612–623
  whole-cell vaccines in development, 611, 617–619
    attenuated vaccines, 619
    CP923, 618–619
    inactivated vaccines, 617
    SolcoUrovac, 617–618
    StroVac, 617–618
Vacuolating autotransporter toxin (Vat), 265
Vagina, 6
  protective role of lactobacilli in, 80–81
  role in the pathogenesis of urinary tract infection,
    80
Vaginal estrogen treatment, 36, 609
Vaginal *Lactobacillus* suppositories, 608
Vaginal microbiota, 79–83
  alterations associated with antimicrobial therapy,
    82
  alterations associated with loss of estrogen, 81–82
  alterations associated with UTI, 81
  bacterial vaginosis, 471–472
  clinical implications, 82–83
  contraceptive method effect on, 82
  protective role of lactobacilli, 80–81
Vancomycin resistance, 145, 466
Vancomycin-resistant *S. aureus* (VRSA), 144–145
Variome Project, 351
Vasoactive-intestinal polypeptide (VIP), 568
Vat (vacuolating autotransporter toxin), 265
VEGF-promoter variants, 541

Veillonella, 483
Vesicoureteral junction, 100
Vesicoureteral reflux, 4, 5, 12, 590, 611
  antimicrobial prophylaxis, 75
  in children, 70, 71–72, 100, 105
  primary, 71–72
  as pyelonephritis risk factor, 504
  secondary, 72
Vesicular exocytosis, 573
Vesicular recycling, 569
Vesicular trafficking, 8
VIP (vasoactive-intestinal polypeptide), 568
Viridans streptococci group, urinary tract
    colonization by, 504
Virulence
  acute prostatitis *E. coli*, 123, 125–129
  chronic prostatitis *E. coli*, 130
  determinants of uropathogenic *E. coli* (UPEC),
    235–255
  evolution of
    genetic mechanisms, 333–336
    horizontal gene transfer, 335–336
    pathoadaptive mutations, 336–351
  extraintestinal, 219–221
    "by-product of commensalism" hypothesis,
      224
    clinical correlation, 220–221
    hierarchical organization of factors involved,
      221–222
    intrinsic virulence, 220–221
    measuring virulence, 219–220
  Proteus mirabilis, 384–419
Virulence factors. *See also specific virulence factors*
  acquisition of, 334
  adhesins, 334–335
  Enterococcus faecalis urinary tract infection, 466
  group B *Streptococcus*, 470–471
  horizontal gene transfer and, 335–336
  Klebsiella pneumoniae, 436–449
  phylogeny, correlation with, 217–218
  Staphylococcus saprophyticus UTI, 463–465
  uropathogenic *E. coli* (UPEC), 130, 592–599
Visceral pain, UTI as model of, 576–577
Visceral-pain hypersensitivity, 569
Voided urine specimens
  asymptomatic bacteriuria, 88–90
    men, 89–90
    women, 88–89
  collection techniques, 43–44
  contamination, 88–89
Voiding cystourethrography, 74
Voiding dysfunction, 16, 504
VRSA (vancomycin-resistant *S. aureus*), 144–145

Waterways, *E. coli* in, 165
Whole-cell vaccines in development, 611, 617–618
Whole-genome sequencing, 45

Wildlife, as *E. coli* reservoir, 165
Wolffian ducts, 5
Women
  aging female
    anatomy and physiology, 13–14
    asymptomatic bacteriuria, 89, 92–95, 99
    immune breakdown in, 560
    incidence of UTI, 79
    UTI prevalence in, 13–14
  asymptomatic bacteriuria in aging
    diagnosis, 89
    incidence, 93–94
    microbiology of, 99
    prevalence, 92, 94–95
  asymptomatic bacteriuria in healthy
    microbiology of, 98
    morbidity, 101
    treatment, 105
  lifetime risk of UTI, 79
  postmenopausal
    immune breakdown in, 560
    recurrent UTI in, 58–59
    vaginal microbiota alterations in, 81–82
  pregnant
    antimicrobial use in, 31–32
    asymptomatic bacteriuria, 31, 611
      diagnosis, 88–89
      incidence, 93

      microbiology of, 98
      morbidity, 102
      prevalence, 92
      screening for, 105–107
      treatment, 105–107
    group B *Streptococcus* (GBS) in, 32, 469–470
    physiology and anatomy changes, 15
    pyelonephritis, 31, 102, 105–107
  voided urine specimens for asymptomatic
    bacteriuria diagnosis, 88–89
WosA, 394

Xenorhabdus nematophila, 411

Yad fimbriae, 244
YadA adhesin, 295
Yersinia
  Y. pestis YadA adhesin, 295
  Y. pseudotuberculosis invasion protein, 300
Yersiniabactin, 126, 128–129, 238–239, 253, 534, 602
Ygi fimbriae, 244

ZapA, 391, 398, 413, 416, 420
Zinc acquisition
  Escherichia coli, 240–241
  Proteus mirabilis, 396, 416
ZnuACB system, 241–242
Zonal-Phylogeny analysis, 350